Understanding Physics for Advanced Level

Jim Breithaupt MSc, C Phys, M Inst P
Head of Physics,
Wigan College of Technology

Stanley Thornes (Publishers) Ltd

First edition published in 1987 by Century Hutchinson Ltd
Second edition published in 1990 by:
Stanley Thornes (Publishers) Ltd
Old Station Drive, Leckhampton, Cheltenham, GL53 0DN, England

First edition 1987
Reprinted 1988, 1989
Second edition 1990

British Library Cataloguing in Publication Data

Breithaupt, Jim
Understanding physics for advanced level.
1. Physics
I. Title
530

ISBN 0–7487–0510–4

Typeset by August Filmsetting, Haydock, St Helens in 10 on 11pt Lasercomp Melior

Printed and bound in Great Britain at Scotprint Limited, Musselburgh

Acknowledgements

I would like to express my gratitude to students and colleagues, past and present, who have contributed through endless discussions to many of the features in this book. In particular, I wish to thank Colin Cartwright of Dundee College of Technology who read the manuscript in detail and made many useful suggestions and improvements. Also, I wish to thank Jean Sinclair of Wigan College for her advice and encouragement, particularly in electronics, and I am grateful to Aijaz Baig, also of Wigan College, who read and made helpful comments on the Skills section. I am also grateful to Christopher Lowe, Senior Physics Technician at Wigan College, for his assistance in setting up and testing many of the experiments described in this book.

I am indebted to my wife for secretarial assistance and for her cheerful encouragement throughout the preparation of the book. The production of a textbook from a manuscript requires the expertise of many people, and I am grateful to them, in particular to Pat Rowlinson of Hutchinson Education who directed the publishing team and offered invaluable advice to me throughout. I am also grateful to Stephen Pevsner and Joan Miller for editing the book and ironing out any difficulties as they arose.

I am grateful to the following Examination Boards for permission to use questions from recent examination papers: Associated Examining Board, University of Cambridge Local Examinations Syndicate, Joint Matriculation Board, University of London School Examinations Board, Northern Ireland Schools Examination Council, Oxford and Cambridge Schools Examination Board (including Nuffield questions), Oxford Delegacy of Local Examinations, Scottish Examinations Board (Higher and Sixth Year Studies questions), Southern Universities Joint Board for School Examinations, and the Welsh Joint Education Committee.

Finally, I wish to thank those listed below for their kind permission to reproduce photographs for which they hold the copyright.

Dr A.A. Smith 7.29, 7.30, 7.31 7.32, 7.33; AERE Harwell Photographic Group, 28.39; Alan Thomas I4(a), (b), (c), 4.22(a), (b), 7.1(a), (b), 7.14, 9.16, 13.4(a), (b), 23.9, 28.17; Ann Ronan Picture Library I2, 2.4, 13.22, 17.7, 19.2, 27.26, 28.1; Associated Press Photo 16.25; Austin Rover E.1; Barnaby's Picture Library B.1; British Steel Corporation 29.22; Camera Press Ltd A.1, 5.1, 5.15, D.1, 18.9; Colin Davey/Camera Press London 18.8, 27.1; David Lee Photography Ltd 4.1; Dept of Medical Illustration, St Bartholomew's Hospital 18.7; Professor H M Flower, Imperial College of Science and Technology 7.36; Griffin and George 25.15, 26.29; Professor C Henderson, Aberdeen University 7.18; Kodansha 1.18, 1.22; London Regional Transport 4.19; Macmillan Publishers Ltd 7.19; Memtek 16.23; Paul Brierley 19.16; Pilkington Brothers Ltd 7.38(a), (b); PSSC Physics, 2nd edition, 1965; D C Heath and Co with Education Development Center 17.14, 17.16, 17.17, 17.21, 17.25; Royal Society 7.35(b), (c); Satchwells 25.24; Science Museum 12.5; Science Photo Library I1 (Ronald Rayer), 2.22 (Martin Bond), 5.17 (NASA), 7.37 (Sheila Terry), C.1 (NASA), 14.7 (Russ Kinne), 15.28 (Jerry Mason), 21.2 (Dr R Clark and M Goff), 21.23 (George Hayling), 25.1(a) (H Schneebeli), 25.1(b) (Jerry Mason), 28.20(a), (b), 28.29, 31.14 (Brookhaven National Laboratory); SPL 26.16, 29.13; R Staines 19.1, 19.21; United Kingdom Atomic Energy Authority F.1; UPI/Bettmann Newsphotos 30.15; Westland Helicopters Ltd 18.1; Yamaha-Kemble Music (UK) Ltd A1.22; George Herringshaw/ Associated Sports Photography A2.1; Philips Electrical Ltd A2.14; British Aerospace (Space Systems) Limited A2.19; STC Telecommunications, Cable Products Division A2.28, A2.34.

Jim Breithaupt

Contents

Preface

The step from GCSE to A-level physics can be daunting, even for the most motivated students. This textbook is designed to help students achieve that transition. Physics syllabuses have changed considerably in recent years, with the emphasis now placed on understanding and applying key concepts. All A-level physics syllabuses have a core of common topics, and this book is built around these core studies. The book also reflects the importance that is now placed on electronics as a major feature of all A-level physics courses. The book covers the latest syllabuses of each Examining Board, and it is organized into eight major sections to reflect these syllabuses. Special attention has been given to making the book highly visual and readable!

The introductory section, 'The world of physics', attempts to explain in general terms how ideas and techniques are developed in physics. Career opportunities are also discussed and some advice on study techniques is offered. The major subject sections (A to G) are divided into chapters which are subdivided into topics. Within each chapter, equations are highlighted and key points and principles are in **bold print** for emphasis. Plenty of experiments are described, and worked examples are given throughout. Each chapter ends with a summary of key points and equations, accompanied by a set of short questions. At the end of each major section, there is a set of multiple choice questions followed by a selection of longer questions from the different Examination Boards. Using these questions, all students ought to find ample scope to develop and test their grasp of the subject.

The last major section, 'Skills of physics' (Section G), deals with topics often overlooked in the classroom or laboratory. All Examination Boards assess skills such as laboratory techniques, mathematical methods, data handling and communication skills. Section G offers students the opportunities to sharpen their skills in these key areas.

Most A-level physics students do not carry on to study the subject to degree level. Of course, many post-A-level courses rely heavily on a thorough grasp of A-level physics, and this book ought to form a sound base for further study on such courses. At the same time, the needs of future physicists have not been neglected since coverage is given to recent developments in the subject. The book is also designed for use by students taking BTEC national courses as well as those studying for Scottish Higher Grade or SYS examinations. The book is also extremely well suited for students taking AS-level examinations in physics.

Two appendices have been added to cater for SYS students and optional studies in Telecommunications. Appendix 1, 'More about waves', deals with the equations describing wave motion in sufficient detail for SYS students. It should also be useful to A-level students intending to progress to a degree course in physics, physical science or electronics. Appendix 2, 'Telecommunications' option in the ULSEB and Cambridge A-level physics courses. Since telecommunications is a major growth area, I hope this section will prove useful for other A-level and BTEC National courses. Both appendices are accompanied by questions, with answers to numerical problems given.

Studying physics is an exciting experience for those prepared to probe the ideas in the subject. Discovering new ideas and applications can be very rewarding in many ways. Finding the right answers is only a part of physics though! Hitting on the right questions to ask is often the key to new discoveries. I hope this book provides the answers for your examination course, but more importantly, I hope it provokes many more questions from your curiosity and imagination!

Introduction: The world of physics

The Scientific Age

Ideas and imagination are products of the creative mind. Our present understanding of the world about us has been developed through the efforts of many creative minds. We live in the Scientific Age, and science has dominated society over the past few centuries. Of course, creativity has always been a part of human activity, ranging from the cave drawings of stone-age people to the musical works of the great classical composers. Before the Scientific Age, ideas about the natural world had to conform to the views of authority. For example, the astronomers of Ancient Greece considered that the Earth was at the centre of a huge invisible sphere to which they imagined the stars were attached. The Earth was at the centre because it was the most important body in the Universe. This model of the Universe was the established view until a few centuries ago – the beginning of the Scientific Age. Before the Age of Science anyone who disagreed with the established view was liable to be treated harshly. In the 16th century, Giordano Bruno was burned at the stake because he refused to accept the established view of the Universe which was part of the Church's authority. He insisted on teaching the Copernican model developed a century earlier by Nicolas Copernicus. This held that the Earth and the planets revolved round the Sun. At that time the Church would not accept the Copernican model because Heaven was beyond the Celestial Sphere – and if the Sphere didn't exist, then neither did Heaven; so the Sphere must exist, and Bruno was burned!

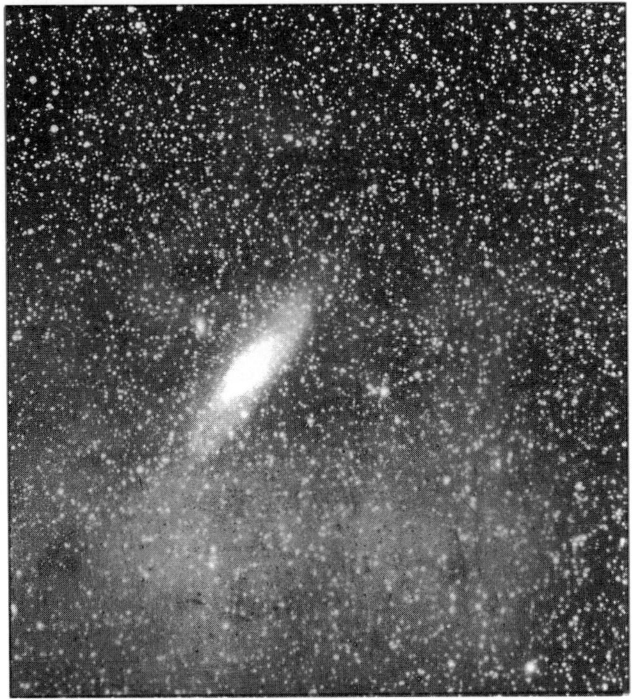

Figure I1

Science has demystified our understanding of the world in many ways. Predicting the future is no longer taken as seriously as it was by the astrologers of ancient courts. The Earth is the third out of nine planets of an ordinary star near the edge of a spiral galaxy (the Milky Way) which is one of many galaxies that make up the Universe. Yet scientific knowledge has caused new mysteries: how did life on Earth reach its present state? What is the fundamental nature of matter? Science seems to create more questions than it answers.

The role of experimental science is the key to the Scientific Age. Any scientific experiment involves observing or measuring physical variables under controlled conditions. The measurements or observations are used to form links between the physical variables. The next step is to explain those links by using or making a theory. Predictions from the theory are then tested by more experiments; if a prediction fails, the theory is wrong so the theory must be modified or replaced. If the predictions hold, the theory has withstood the test, although we cannot say the theory is absolutely correct because it may fail a future test. The more tests a theory withstands, the greater its status becomes. The **principle of conservation of energy** is a theory that has been subjected to many tests. Its success in all tests so far has elevated it to a key principle of physics because it is the starting point for many other theories in science. Remember, though, it is only a theory; in all the tests so far, energy is conserved. Nobody knows whether some future test will show that energy is not conserved in some particular situation. The study of β-decay made scientists think that energy conservation might not hold in the nucleus; measurements showed that β-particles from a single radioactive isotope had a range of energies up to a maximum, but each disintegrating nucleus released a fixed amount of energy equal to the maximum possible β-particle energy. When a β-particle is released with less than the maximum possible energy, what happens to the rest of the energy released by the nucleus? Rather than believe that energy can disappear, scientists preferred to believe that each disintegrating nucleus releases a 'mystery' particle as well as a β-particle. The mystery particle was even given a name – the neutrino. At first the neutrino could not be detected, so new experiments were devised and tested. Eventually, acceptable experimental evidence for the neutrino was obtained and the principle of conservation of energy was saved. But for a time there was a question mark over it.

Great scientists have signposted the route to our present understanding of the natural world. Galileo Galilei (1564–1642) was responsible more than any other scientist for starting the Scientific Age. Galileo invented the telescope in 1609 and used it to study the stars and planets. One of his discoveries was the four inner moons of Jupiter, which he named *Io*, *Callisto*, *Ganymede* and *Europa*. He observed the motion of the moons in their orbits about Jupiter. He carried out dynamics experiments to try to understand his observations and he established basic ideas about force and motion. Galileo was a skilled experimenter but he lacked the means to measure short intervals of time, so he devised a water clock that involved collection of water dripping from a tank at a constant rate. He used the clock to time a brass

Figure I2 *Galileo Galilei (1564–1642)*

Galileo spent almost 20 years furthering his studies on motion. His development of the telescope in 1609 set his studies in a new context as he discovered solar disturbances, Jupiter's moons, and much more. He came into conflict with the Church because he supported the Copernican view. In 1632 the Inquisition forced him to recognize the Church's authority; but he laid the foundation for Newton's great works 50 years later.

sphere rolling down an incline, and so showed that falling bodies increase their velocities at a steady rate. Try his water clock experiment!

Isaac Newton (1642–1727) built on Galileo's discoveries with achievements that laid the foundations of science and mathematics for the next two centuries. His work is immensely important and is described in more detail in Chapter 2. Newton developed Galileo's ideas into a mathematical theory of mechanics, the *Principia Mathematica*. The laws of motion and the theory of gravitation in the *Principia* were just two of the products of Newton's creative mind, and he used these ideas to explain a wide range of observations and experiments. Predictions made using Newton's laws, such as the return of comets, were confirmed with great accuracy. The theories Newton produced withstood all the tests during the following two centuries – until they were challenged by Albert Einstein. For his achievements, Newton is considered by many to be the greatest scientist of all time.

Einstein's ideas revolutionized the way that scientists think about space and time. Before Einstein, time was thought to be universal; identical clocks in different situations were thought to run at the same rate. Einstein thought not and he formulated the **special theory of relativity**. According to Einstein, clocks moving at different velocities do run at different rates. Imagine boarding a high speed train from London to Glasgow, departing from London at noon and arriving in Glasgow three hours later by your watch to find the station clock showing 5.00 pm. It doesn't happen and it won't, because Newton's theory only breaks down at speeds approaching the speed of light – and British Rail isn't that fast! Tests of Einstein's theory have been carried out and they confirm the relativity theory. One such piece of evidence is the measurement of the lifetime of short-lived particles called muons (or mu-mesons; see p. 445). Muons produced at the top of the atmosphere from collisions between cosmic ray particles and air molecules last long enough to reach the Earth's surface, but muons produced by accelerator experiments have much shorter lives. The cosmic muons last much longer because they are moving at speeds approaching the speed of light – much faster than muons from accelerator experiments. However, Newton's laws are still valid for speeds much lower than the speed of light.

The role of physics

Physics is one of the key disciplines of science. The boundaries between physics and the other key disciplines are not rigid or fixed. For example, medical physics obviously involves human biology as well as physics. Although each scientific discipline has its own distinctive features, the roots of all science are firmly based in experiment.

What are the distinctive features of physics? The subject can be described as the study of matter and energy. However, energy is an important aspect of any scientific discipline, so perhaps it is easier to consider what physicists do.

Figure I3 *What physicists do*

Understanding natural phenomena has always been a central aim of physics. Since Galileo, the study of Astronomy has led to more and more discoveries that have tested the laws of physics. The *Voyager* space journey past Jupiter, then Saturn and then Uranus answered some questions but provoked many more. Astronomers across the world have linked their observatories to identify stars with possible planetary systems. Distant galaxies called 'quasars' at the very edge of the Universe have been located, and the laws of physics have been used to describe the origin of the Universe – in the 'big bang' theory. Yet physics on such a grand scale also meets the subject on the smallest possible scale in the drive to discover the nature of matter.

Ever since John Dalton put forward the **atomic theory**, early in the 19th century, the quest to unlock the secrets of the atom has been a key aim of science. The atom was 'opened up' to reveal some of its secrets as a result of the experiments by J.J. Thomson and Lord Rutherford; their investigations proved that each atom contains a positively-charged nucleus surrounded by electrons. Then it was discovered from further experiments that each nucleus is composed of neutrons and protons. Now scientists think that neutrons and protons and other particles since discovered are made up from three fundamental particles called **quarks**. The story is a bit like opening a box to find another box inside, with another box inside that, and so on. Who knows if the quark model is the last box? The energy needed to rearrange quarks is enormous, and it is thought that the 'big bang' has been the only time when such conditions occurred naturally.

Between the extremes of quarks and quasars, the study of matter and energy has advanced our understanding of the natural world in many ways. Trying to understand why materials behave as they do is an important part of physics. Why does water turn to steam at a certain temperature? What is the cause of superconductivity – the absence of electrical resistance in certain materials at a low temperature? Physicists spend a lot of time and effort trying to answer questions such as these.

The use of physics in other branches of science or outside science is perhaps one of the reasons why physics is at the centre of the Scientific Age that we live in. The consequences of new discoveries can be far-reaching and can affect everyone. Often the discoverers do not realize the full significance of their work. When Michael Faraday discovered the principles of electromagnetic induction in 1829, few people at the time could have realized just how important electricity would become a century or more later. 'What use is electricity, Mr. Faraday?' he was asked. His reply is classic: 'What use is a new-born baby?'. Who knows where a new discovery can lead? Look at the microchip revolution. Just over 30 years ago, the transistor was invented. Now thousands of transistors can be made on a single chip called an **integrated circuit**. Use of integrated circuits has brought about a revolution in the way information is transmitted and used. Even day-today activities such as shopping or visiting the bank have altered as a result of the chip revolution.

We are all users of physics now. Even before a baby's birth, physics is at work with ultrasonic scanning machines to picture the unborn child in the womb. Issues such as nuclear power make a background knowledge of physics important for everyone. The individual without knowledge of the subject is at a disadvantage. In medieval times, the key to advancement was the ability to read and write – preferably in Latin. In the Scientific Age, a scientific background is essential and physics is a key element in that background.

Physics in perspective

What do you hope to do after your A-level course? Your present studies may not leave you much time to think about the future, but of course you must do so before the end of your present course. Ten years hence, you could be in a job that doesn't even exist now. So how can you plan ahead if you can't be sure of the future? Flexibility is the key, and a person with a science background is ideally placed to meet the demands of new technologies. Some students may have a specific career in mind. For example, a student aiming to become a doctor must study at a medical school to qualify. Even so, new techniques and knowledge require continual retraining and updating.

Flexibility means being able to adapt to changing circumstances. New technologies force the pace of change. If you don't adapt, you lose the benefits of new technology. For example, microcomputers are used throughout industry and commerce, so people need to develop keyboard skills to be in a position to use them effectively.

The first step towards flexibility is keeping an open mind on new techniques. If a new technique is rejected because the present approach works well anyway, then the possiblity of new ideas following from the new technique is blocked off. Studying physics is an excellent way to develop openmindedness. Problem solving in physics has a multitude of routes and you may discover unorthodox ways of solving a problem which shed new light on your approach.

Developing a wide background is important for flexibility. Most science students study three A-level subjects drawn from the main scientific disciplines (including mathematics). The chance to specialize in the sciences offers the opportunity to further your intellectual skills. But such skills are enhanced by a wide background of knowledge outside science. Also, applying scientific ideas and communicating them requires skills and knowledge from outside science. You should try to fit in a general studies programme or something similar alongside your mainstream studies; you may not have much choice but, nevertheless, you should find such a programme helpful. If you can't fit such a programme in with your main subjects, then keep in touch with current affairs through newspapers and TV programmes. A wide background knowledge is helpful when you are in a job and you need to retrain or update your skills. Your main subjects at A-level open the doors to a worthwhile career for you, but a wide background is important to help you to cope with inevitable changes during your career.

Keeping your options open is essential unless you have a definite career in mind and you (and your teachers) are sure that you can achieve your ambition. But remember that the way a job is done now is not likely to be the way it will be done in ten years' time. Can you think of any career unlikely to be affected by new technologies? Engineer, manager, doctor, tailor, soldier and so on – just ask someone in such a career if their method of working has changed in the last ten years. Is it likely to change much in the next ten years? Their replies may surprise you!

Consider the careers of some of the individuals who were in a typical A-level Physics class ten years ago. The names have been changed to protect the innocent...

Graham's story 'I took double maths and physics at A-level. I really enjoyed maths so I went on to read the subject at university. At that stage I had no real idea of what I would do after that but I knew that there would be plenty of opportunities. After university, I joined an aerospace company in the design department. We're now working on a joint project within a European consortium.'

Jane's story 'Chemistry was my favourite subject at A-level and I took physics and biology to support it. After my A-levels, I went off to university to study biochemistry. I stayed on to do research after my first degree then joined a firm developing biotechnology.'

David's story 'I studied maths, physics and economics at A-level but didn't get high enough grades for the Higher Education course I wanted. So I took a one year repeat A-level course which improved my understanding a lot. I went on to university to study computer science and then joined a software firm. I've recently set up my own company marketing and selling software to businesses large and small.'

Christine's story 'I enjoyed the electronics topics we covered at A-level so I applied for a sponsorship with an electronics company. I worked full-time with them for a year then I went to polytechnic to study electronics and business studies. I worked for the company during the vacations so gained useful experience. After polytechnic, I rejoined the company in their production team, and I have since progressed into management.'

What can you learn from these and other comments by past A-level students? Few people can mark out their careers precisely. In ten years' time, you may be in a very different career to the one you're perhaps planning now. The ideal situation to be in at your present stage is perhaps to be studying subjects you enjoy, including some background studies – and the combination of subjects you have chosen ought to give you a range of career options.

Without A-level mathematics an A-level physics student would not be able to consider a career in engineering, physics, mathematics, business studies, economics, to name just a few. The most popular combination of A-level sciences without mathematics is physics, chemistry and biology; those studying such a combination would probably hope for a career in a medical or related sciences or in the life sciences. Potential medical students must study A-level chemistry, so those taking mathematics, physics and chemistry are not precluded from a medical career. Some universities and polytechnics do offer a 'conversion' year to allow students without a science or mathematics background to study engineering subjects.

With A-level mathematics an A-level physics student is able to keep a wide range of career options open by choosing a third A-level subject such as biology or chemistry. Students taking double maths and physics would not be able to consider post A-level courses in medicine, biological science or chemistry etc. If you have a special interest (e.g. geology or electronics) then you may be able to take a third A-level in line with that interest. However, few universities or polytechnics that provide courses in such subjects insist on A-level qualifications in these subjects. For example, for entry to most electronics courses, mathematics and physics at A-level are essential but A-level electronics is not.

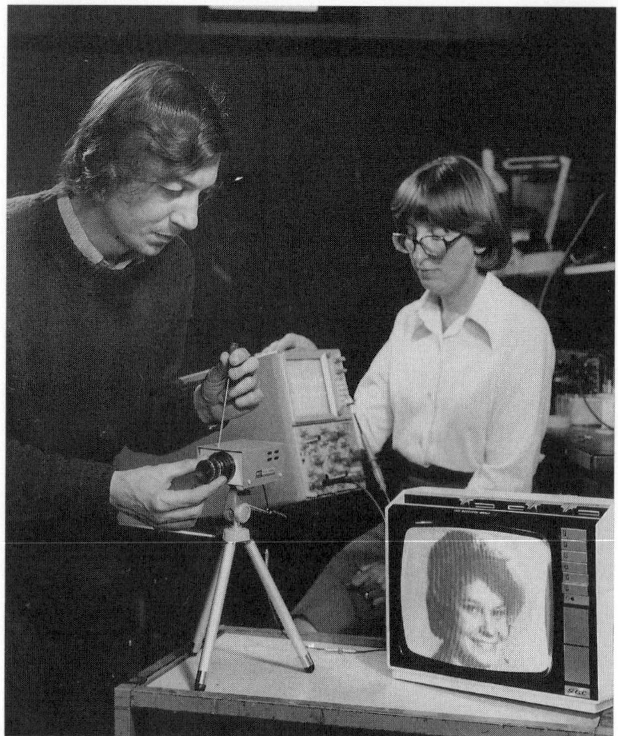

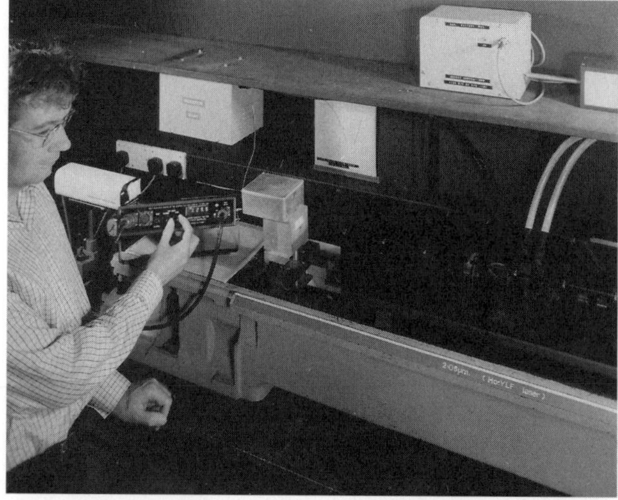

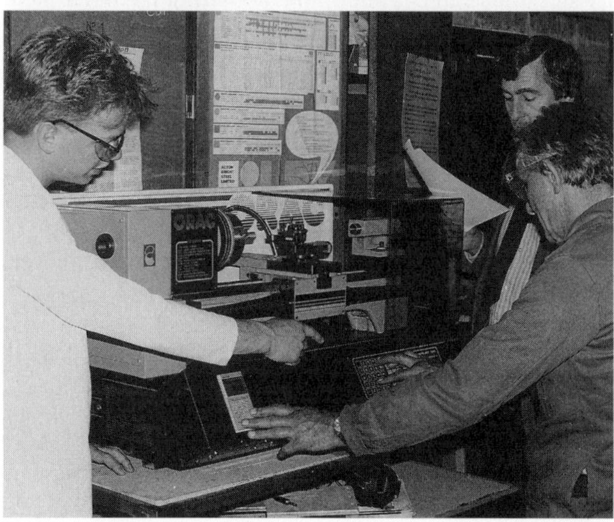

Figure I4 *Using physics*

Planning what to do after your A-level course ought to start well before your examinations. If you have a set aim in mind, then you need to make sure you can achieve the necessary grades; if you are unsure, then line up some other possibilities just in case. If you are following an A-level course with plenty of options afterwards, you need to think about which direction to go in. Don't dither! But first ask yourself some general questions so you don't end up as a 'square peg in a round hole'. Here are a few useful questions.

1 Is the 'outdoor life' appealing? If not, forget about civil engineering, environmental studies, geology, forestry and so on.
2 Do you like taking things to bits and seeing how they work? Do you try to find out why they work or to improve them? Do you like designing and making gadgets? There must be some branch of engineering for you if your answers are 'yes'.
3 Can you relate easily to people of all ages? The ability to work with people is important in any career, but especially so in health services such as physiotherapy, dentistry, medicine and so on. The patient needs to have confidence in the specialist and so the specialist has to encourage the patient to 'confide'. If you're keen on a career in the health services, do some voluntary work to develop your inter-personal skills.
4 Are you hooked on computers? If so, then watch out if you plan to become a computer professional. Teaching yourself to drive is not the best way to become a racing driver! Likewise, learning computing using a home microcomputer can cause problems when you start a computing course after A-levels. You might have to unlearn questionable techniques before you can start to learn the principles of programming.

Your answers to simple questions like the ones above should enable you to narrow the options if you haven't already done so. So now you need to think about how you want to spend the next few years. Perhaps you can talk these matters over with parents, older brothers or sisters. If you have a specific career in mind, you may not have any choice about how you spend the next few years. For example, if you want to become a doctor (medical), you must go to medical school. But you could delay your entry for a year to gain some work experience. Here are some more questions.

5 Do you want to move straight on to full-time study after your A-level course? Some university and polytechnic departments encourage a year of work experience before entry to their courses. Other departments might not allow a year off so would not be prepared to reserve a place a year or more in advance. Check with your careers teacher.
6 Sponsorships are available from many large companies but there is usually a good deal of competition for them – not surprising in the view of the grant levels for university and polytechnic students. The sponsor usually supplements the grant and in return requires the student to work for the company in vacations and for a year before entering university or polytechnic. Most sponsorships do not commit the student to the company after university or polytechnic – but if the company is satisfied with the student's progress they would offer attractive prospects to keep the student. If you are interested in sponsorships, ask your careers teacher and watch out for advertisements in the national press.
7 Going straight into a job is attractive to many A-level students when jobs are in short supply. A job with day-release for further training is one way to further your career plans, although working at a full-time job *and* studying for qualifications is very demanding. No wonder that those who succeed by this route often go to the top; but it can be much more demanding than full-time study, especially when the pressure is on at work. Without the opportunity for part-time study, you may find your job does not have many prospects and your hopes could become thwarted if you're competing for promotion with those who have improved their qualifications.

Most A-level students move onto higher education in universities or polytechnics. A-level courses are demanding and the examinations are a hurdle you must jump to move into higher education. Meeting the entry requirements of a particular course ensures you have the ability to benefit from that course. If you work hard, you ought to pass the course, so success at A-level is the key. Applications for entry to university or polytechnic must be made through the appropriate admissions system – UCCA or PCAS. Your careers teacher will advise you about different courses, deadlines and application forms. Take advantage of career visits to your school or college by university or polytechnic course tutors. Write off to universities or polytechnics that have courses that are of interest to you, and go on relevant careers visits if possible. By the time you start the second year of your A-level course, you ought to have decided on the type of course (i.e. subject area in general) you intend to apply for. Then you must choose particular courses and get your application forms in before the deadline date!

Career opportunities for A-level physics students with A-level mathematics will no doubt continue to be good. Employers look favourably on A-level physics because it is an indicator of ability to benefit from further training, and the subject provides a good background for new technology. A list of careers available to an A-level physics student would probably fill a book. Figure I5, on the next page, shows some of the possible career routes. No doubt your careers teacher can supply more information, but remember that all careers involve updating and retraining. If you enjoy physics, then consider a career as a physicist; at university or polytechnic you study the subject in depth, and there are plenty of opportunities for graduate physicists throughout industry, in research establishments, in education and in the health services. Physicists are valued not just for their intellectual capabilities and experimental skills; understanding fundamental principles enables the physicist to grasp and readily use new technology as well.

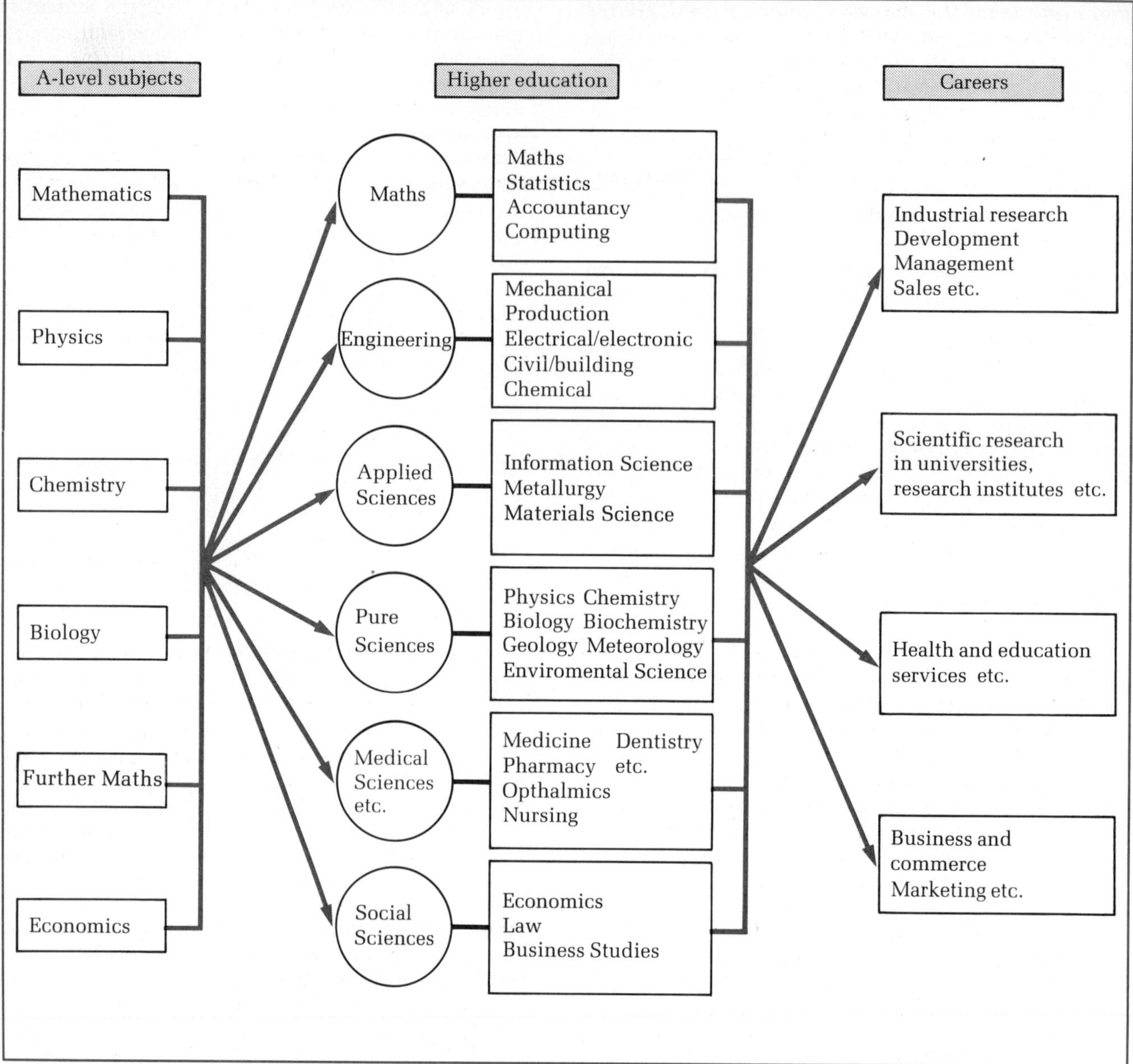

Figure I5 *Possible career routes with A-level physics*

The differences between GCSE and A-level physics Successful completion of a GCSE physics or integrated science course is an essential requirement for A-level physics. You can't afford to take things easy at the start of the A-level course; it's all too easy to fall into the trap of thinking that the examination is two years away and so there's plenty of time. There isn't! The two year A-level course is really a five term course. By the end of your first term you should have completed 20% of the coursework, so you can't afford to waste time at any stage, especially at the start. You need to keep up with your studies throughout the course. Fall behind and you soon find the work piles up. With three A-levels to study for, you've got your work cut out.

What are the essentials for success at A-level?

1 Mathematics is an important feature of A-level physics. You need to be competent at basic arithmetic, algebra and graph work. Chapter 32 should be helpful in this respect. If you are not studying A-level mathematics you will need to work that much harder than your 'mathematical' friends.
2 Understanding key ideas and being able to link them together is an import part of A-level physics. Your previous course in physics has lots of facts and formulae; A-level has even more, and you must try to understand the principles behind the formulae.
3 Initiative on your part is essential. There is a much greater emphasis on individuality than in your previous course. You should be prepared to question ideas that are new to you. Teaching methods differ at A-level and you have to organize your own notes. Great emphasis is placed on problem solving and on laboratory skills. Your teacher is not there just to transmit knowledge etc., but is your adviser and guide. Make use of the

resources available to you, not just in the laboratory or the library but also any video or computer programs.

4 Organizing your studies yourself around your timetable is necessary otherwise you will flounder. With three A-level subjects and possibly some minority studies on your timetable, you need to plan in advance to ensure work is handed in on time. The ability to organize your private study time is as important as the content of your course. At university or polytechnic, the ability is taken for granted. If you can't organize yourself at university or polytechnic, you won't survive the first year, so you must learn to organize for yourself at A-level. Draw up a timetable for your private study time and allocate equal time, say about six hours per week, to each subject. Of course, as you approach examinations and tests, you need to spend extra time on private study – your social life suffers at such times! However, if you plan and work to a schedule, you should be able to fit in all those roller discos, charity runs etc. With an organized schedule it is easier to cope with the inevitable emergencies such as missing classes through illness.

A-level physics syllabuses vary slightly between boards. All syllabuses at A-level have a core of topics that are emphasized in this book. Each syllabus has its own features built around the core topics so you should obtain a copy of your examination syllabus to guide your revision. All A-level syllabuses place great emphasis on understanding key concepts which you must be able to explain and to use. Most courses build on basic principles from previous studies, so a thorough understanding of these principles is essential. Check your previous course notes when you come to each topic at A-level and identify points that you found difficult; working at these points should help your progress in the topic at A-level. Try past questions from O-level, CSE and GCSE papers to build up your confidence. The emphasis on understanding ideas at A-level is reflected in the way most syllabuses are organized. This book attempts to follow the syllabuses in organization, but you ought not to expect to progress through your course simply from Section A to Section B to Section C and so on. You may find that you start your course with the study of wave motion then you move on to d.c. electricity and electronics and then to materials. Your teacher controls the sequence of the course, and this book is not organized for any particular sequence.

Section A: Mechanics This involves basic ideas of force, motion and energy. Your particular course might not feature mechanics as a distinct topic in which case you will find the mechanics topics are threaded in with the other branches of the course. Nevertheless, you should find the separate section on mechanics in this book is helpful. The topic of mechanics may be left until later in your course so you may need to consult topics in Section A before dealing with them distinctly. For example, if you study materials before mechanics, you might find it is useful to read up on force and energy during your study of materials.

Section B: Materials Properties such as mechanical and thermal behaviour and the behaviour of fluids are covered in this section. In this book, electrical properties are included in the section on electricity and optical properties in the section on wave motion. The emphasis is on understanding and explaining the properties of materials in molecular terms.

Section C: Fields The interactions between different objects are explained using the 'field' concept. A field is a 'thinking model' to describe and understand interactions. There are different types of field, which include gravitational, electric, magnetic and nuclear force fields. A sound understanding of the link between force and energy is vital to grasp field concepts. In this book, nuclear force fields are described in the section 'Inside the atom' rather than in Section C.

Section D: Waves Most branches of physics involve some aspect of wave motion. The study of oscillations involves basic ideas about force and motion. Mechanical oscillations are studied to understand how vibrations affect bodies. Where oscillating bodies are linked together, energy can be transferred through the system by means of wave motion. Sound waves and electromagnetic waves are essential for communications so the study of wave motion concentrates on these important waveforms too.

Section E: Electricity and electronics These branches of physics are at the core of new technology. Basic principles of d.c. electricity must be understood before studying electrical measurements and alternating current. The far-reaching consequences of electronics can only be appreciated in full if you have a sound grasp of d.c. electricity. Electronics is a rapidly-changing subject and the emphasis at A-level is on general principles developed in terms of present-day devices. Devices soon become obsolete in electronics but the general principles remain.

Section F: Inside the atom Our understanding of the nature of matter has been developed as a result of lots of investigations over the past century or so. Many of these investigations have led to important applications. For example, J.J. Thomson's experiments into electron beams led to the development of the TV tube. Energy from the nucleus is vital for present and future supplies of fuel. Ideas drawn from mechanics and fields are used in this section but they are developed in directions that may seem strange. For example, the idea of converting mass into energy is not easy to understand – but what *is* mass and what *is* energy? Inside the atom, our understanding of nature is taken to the limit – with yet more to be discovered.

How you use this text book will depend on you, your teacher and your syllabus. The book is designed to cover the core topics of all A-level physics syllabuses with the emphasis on explaining difficult ideas. You may or may not use the book a lot in class, depending on your teacher. You will certainly need to use the book to reinforce your studies; for example, if you are uncertain about a topic you have covered in class, you can use the book to read up at home on that topic. Occasionally you may need to refer to a library book, especially if you are investigating or researching a project. To help you to check your grasp of the topics in each section, there is a selection of short questions at the end of each chapter. Numerical answers are given at the back of the book. Suppose you have studied 'Uniform electric fields' in class and you are unsure of some basic points. How can the book help you?

1 Turn to the introductory page on fields (p. 153) and read through the introduction to check that you are familiar with the key ideas of force and energy necessary to understand fields.

2 Then turn to Chapter 13, 'Electric fields', and find the part on uniform electric fields. The topic headings at the start of the chapter are useful here. Read through this part on the topic then check your class notes against your reading. It's all too easy to make errors when writing notes at speed. If necessary, amend or add to your notes and highlight the key points in colour.

3 Try some questions on the topic. Use the short questions at the end of the chapter, and make use of any suitable long questions at the end of the 'Fields' section.

Making and using notes is a skill you need to develop. At first you may need to rewrite your notes to make them useful for revision. With practice, you ought to become capable of making notes in class that do not need rewriting afterwards – although you do need to check them. Add to your notes by reading your textbook. Many students keep their main file of notes at home and take a separate file with only the current topic to and from classes. This has two chief advantages; you carry less weight round in your bag, and your main notes are secure at home. If you lose your complete set of notes a month before the examination, then you will need to spend valuable time and energy catching up in a hurry. As each topic is completed, you can file it at home – after checking you understand the topic. Your notes are for you to use; keep them well-organized in your file, highlight key points and equations, use clear, numbered headings and make topic charts linking the key aspects of each topic. (See Figure 1.6.)

Now that you are organized, let's consider the essentials of how you can get to grips with the subject. Your studies in class ought to include plenty of practical work and you should see plenty of demonstrations in the 'chalk and talk' sessions. If you are not given notes, you need to make your own. You may be asked to work through a series of problems, written or practical. The idea is to develop your understanding of a concept or to further your skill in a particular area. Be prepared to ask if you do not understand. Physics is a 'linear' subject and you build on one topic to understand. If you don't understand some basic point, then you may find it holds up your progress in a later topic. That's why it is important to use your notes and textbook to ensure you have grasped each topic as it is taught.

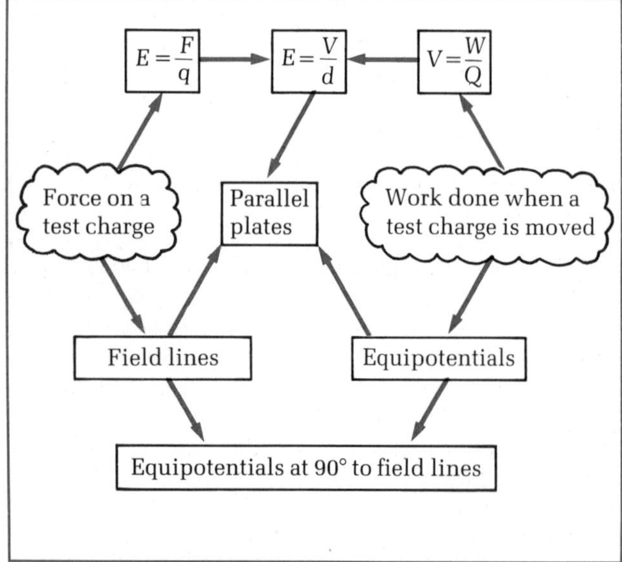

Figure I6 *Linking a topic*

Understanding the subject is achieved by different routes for different people. All routes require a body of facts to work on. Memorize basic facts and formulae as you meet them in your course so that you can think about how they fit together – without continually having to check the facts against your notes. For example, suppose you are attempting to grasp the essential ideas in the topic 'Uniform electric fields':

1 What is a line of force?

2 What is the definition of electric field strength?

3 What is the pattern of the lines of force in a uniform field?

4 Does the force on a test charge vary with position in a uniform field?

The answers to the questions are some of the key facts for uniform electric fields. You must be able to remember the key facts and link them together. A 'spider diagram' is a useful aid to see and remember how key facts and ideas are linked together.

Check your ability to recall key facts and ideas by testing yourself. Write down the basic facts from memory and check them against your notes. Check at regular intervals to ensure you can remember key facts and ideas and the links between them. Reread your notes on the topic regularly once you have grasped the topic. After a while, you ought to find that the ideas become second nature and can be recalled with little effort.

If you think you have mastered a topic, ask yourself these general questions.

1 Can you solve problems on the topic? Short questions can test your recall of facts and your understanding of the topic. Longer questions can test your understanding of links between topics as well as within a given topic.

2 Can you explain the topic to a classmate? Imagine you are writing an account of the topic for a classmate who has been absent. See what you can produce then check it against your notes.

3 Can you apply the topic to unusual situations? Try questions on the topic that involve unusual applications (i.e. ones you have not met before). Such questions test your understanding of basic principles by asking you to apply basic principles to an unusual situation.

4 Can you understand these basic principles as applied to the experiments and demonstrations that you have seen? Could you explain each experiment, principle etc., to that classmate who keeps missing classes?

Practical work is an integral part of A-level physics and you have to learn to use equipment such as oscilloscopes, multimeters, spectrometers, travelling microscopes, micrometers and so on. You ought to find Chapter 34 helpful for developing laboratory skills as you work through the experiments one by one. Microcomputers are used to stimulate experiments, and computer programs can be useful for revision or for situations beyond normal resources (e.g. nuclear reactor simulation). Microcomputers in the physics laboratory can be 'interfaced' to equipment for making measurements; such a system can usually be programmed to measure several different physical quantities over the same period, and the program can be written to include data analysis routines (see p. 402 for further details on interfacing.)

Microcomputer programs to illustrate difficult ideas and concepts are helpful. For example, you may have seen a program that simulates radioactive decay. When the program runs, a grid of boxes appears on the screen with each box filled by a solid circle, ●. Each solid circle represents a radioactive nucleus that disintegrates to form a stable nucleus. When this happens, the solid circle becomes an open circle, ○. The program is written so that the solid circles change at random to open circles. A graph on the screen shows how the number of radioactive atoms remaining changes with time.

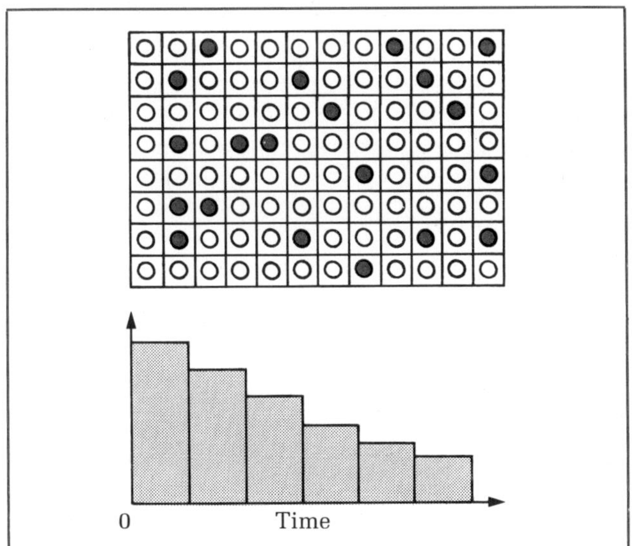

Figure I7 *Radioactivity program*

Revision and examination skills

Revision should be a continual process throughout your course. As examinations approach, revision practice of past papers becomes more and more important. Physics examinations test candidates in a variety of ways.

1 Factual recall
2 Understanding ideas
3 Laboratory skills
4 Comprehension and writing skills
5 Powers of organization
6 Mathematical skills
7 Data analysis skills

Each examination board with its particular syllabus tests your knowledge and skills with its own distinct range of papers.

Internal assessment of practical work or project work is becoming more and more popular. Such work is assessed during your course by your teacher, and your mark counts towards your final grade. Exam boards take great care to ensure that the standards of internal assessment are the same from one school (or college) to any other. Project reports have to be written up and handed in for assessment; the report itself is usually sent off to the examining board together with your teacher's assessment of it. For both project work and internal assessment experiments your organizing ability is important, as well as all the other skills listed earlier. (See pp. 525–9 for more information on projects.)

Objective (i.e. multiple choice) test papers require fast recall and understanding of basic ideas across a wide range of topics. Questions can be searching, but you must not spend too long on any one. Work through all the questions in order; if a question takes more than its fair share of time, abandon it and return to it later. When you have worked through the paper in this way, return to the questions you couldn't manage first time. If necessary, make an 'educated guess' from the alternatives you think may be correct.

Short questions are usually in compulsory sections, each section on a given paper drawn from a wide range of topics. Long, detailed proofs are unlikely to be required in short questions which are more likely to test factual recall and understanding of key ideas. Often the recall element is developed into a test of your understanding of the topic. An example is given below.

Question **a)** Write down an equation linking the pressure of an ideal gas to its density and the r.m.s. speed of its molecules.
b) Explain why the Earth is able to retain an atmosphere whereas the Moon cannot.

Clearly your answer to **b)** requires an understanding that molecules have a range of speeds and that the mean square speed is proportional to the (absolute) temperature. You need to link these ideas up with ideas about escape velocity to explain why gas molecules released on the lunar surface escape into space.

Longer questions test your in-depth knowledge and understanding of topics. Detailed proofs or questions about experiments or demonstrations are likely topics for long questions. Another source is your ability to link topics together and to use basic principles in situations you have not met before. Usually you have a choice of questions to answer from so answer your 'best' question first. Don't be tempted to save it for a grand finish because you may run out of time.

Some long questions require little planning if the question is structured in detail, part-by-part. At the other extreme there is the type of question that runs to a few lines only but expects a 30-minute answer. Clearly, planning is vital before you begin your written answer. (See p. 525).

Comprehension papers often present basic principles in situations that you would not otherwise meet on your syllabus. You are expected to recognize the basic principles and apply them to the situation. Obviously comprehension and writing skills are under test, but so too is your grasp of basic ideas and your factual recall. (Comprehension skills are discussed in more detail on pp. 529–30).

Data analysis questions are set by some examination boards in separate data anlaysis sections or papers. However, analysing data can be a feature of longer questions, comprehension papers and practical examinations. In addition, data analysis is a key part of experimental physics so is needed for practicals and projects. Examination questions that test data analysis skills directly supply experimental measurements, usually with brief details of how the measurements were obtained. You are then expected to analyse the data and establish links between the physical variables involved. In this situation, it may be helpful to recall similar techniques you have used during the course. For example, you may have used log-log graphs (see p. 253) to establish the link between the current and p.d. across a light bulb. If you are asked in an examination question to plot a log-log graph for an entirely different topic, your previous work gives you the general idea. Data analysis skills need to be developed throughout your course; often, such skills are not taught formally so you may find Chapter 33 helpful.

Practical examinations can either be in the form of a series of short experiments or a single much longer experiment. Short experiments test a particular technique or skill (using a vernier, for example), and usually require recall of basic knowledge. Some short experiments involve making then explaining observations to test your understanding of basic ideas; usually the instructions are set out in the question step-by-step, and you are expected to know how to use the instruments to carry out each step.

For the long experiment in a practical examination, planning and organization are important skills. Knowledge of the correct techniques for making accurate measurements is expected. (See Chapter 34.) You also need to know how to reduce and estimate errors, and the skills for analysing data are important. So too is the need to explain your methods clearly. (Advice on writing up practicals is given in Chapter 35.)

SECTION A

Mechanics

Dynamics · Force and energy · Energy resources · Statics · Rotational dynamics

Magnetic levitation producing frictionless motion. The photograph shows a prototype 'maglev' train undergoing trials. Note the absence of conventional wheels. A strong magnetic field generated by the train supports the train above the track.

Mechanics involves the study of basic ideas of force and energy, developed to understand all aspects of the physical world, from inside the atom to the edge of the Universe. Aspects of mechanics are used in many other branches of physics to develop models and theories. When necessary, scientists have re-examined the basic concepts of mechanics, and have discovered new links. For example, the link between energy and mass discovered by Einstein was not even suspected a century ago. Mechanics is an everyday branch of physics, used to explain, understand and predict; its language is essentially mathematical but however complicated the mathematics is made, the key principles such as conservation of energy remain valid.

The emphasis in the Mechanics section of this book is to establish and to develop the links between motion, force and energy. A sound grasp of key principles of mechanics is essential for topics in all other sections; even if your particular syllabus does not involve the study of mechanics as a distinct part, you will still meet the ideas of mechanics threaded in with other topics.

1

Dynamics

1.1 Vectors

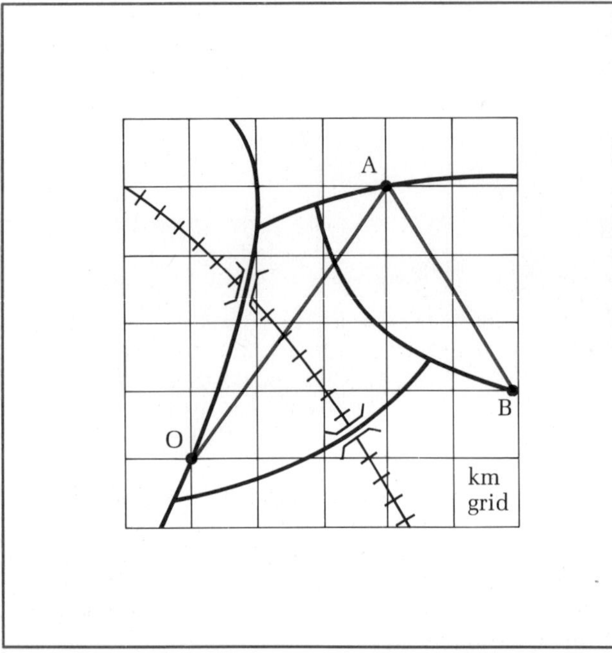

Figure 1.1 *Map of locality*

Imagine you are planning to visit a friend who lives several kilometres away from your own home. Whichever route you choose, your starting point and finishing point is unchanged by your route. Your **displacement** is the distance directly from start to finish. Even though you may take a roundabout route over a longer distance, your displacement from start to finish is the same as if you were to take the direct route. For example, suppose the map in Figure 1.1 shows your home area. Your home at O and your friend's home at A are both marked on the map. If A is 4 km north and 3 km east, then the displacement from O to A, **OA**, can be stated in one of two ways.

1 Using coordinates, which are similar to map references, we would state that **OA** = (3 km east, 4 km north). Even simpler, we could write **OA** = (+3 km, +4 km), where we use '+' for east or north, and '–' for west or south. But you also must remember that the first part of the bracket is for east or west.

2 Using Pythagoras' theorem to calculate the shortest distance from O to A, and using trigonometry to calculate the direction, so the distance is 5 km from O to A $(=\sqrt{3^2+4^2})$, and the direction is 37° east of due north.

A **vector** is any quantity with size and direction. Displacement is an example of a vector quantity. Other examples are velocity, acceleration, momentum, force, and field strength. Quantities without directional properties are called **scalars**. Examples include speed, mass, energy and distance.

Using vectors

Consider the map in Figure 1.1 again. Suppose you and your friend set out from A to visit a disco at B which is 3 km south and 2 km east of A. At B, your total displacement from home (i.e. O) is 1 km north and 5 km east. So B is 5.1 km directly from O (using Pythagoras' theorem, $OB=\sqrt{5^2+1^2}$).

Displacement from O to A = 3 km east, 4 km north = (+3 km, +4 km)
Displacement from A to B = 2 km east, 3 km south = (+2 km, −3 km)

Displacement from O to B = 5 km east, 1 km north = (+5 km, +1 km)

Hence the displacement from O to B, **OB**, is equal to the displacement from O to A, **OA**, + the displacement from A to B, **AB**.

OB = OA + AB

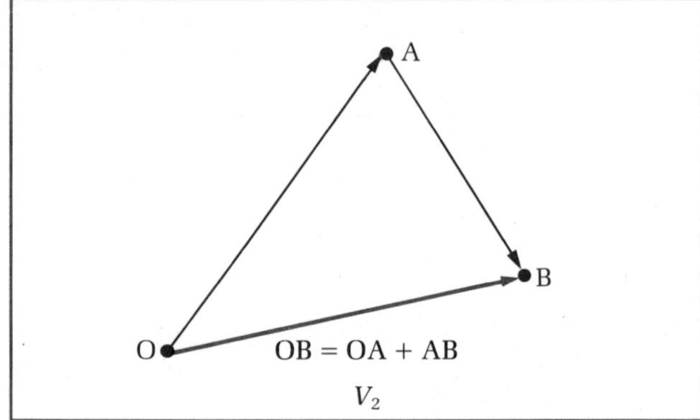

Figure 1.2 *Adding displacement vectors*

This example shows two important features often used with vectors.

1) Resolving a vector into components at right angles to one another

Displacement **OA** in the map example is made up of a component of 4 km due north and a component at right angles of 3 km due east. More generally, suppose we wish to resolve a vector into a component along a given line and another component at right angles to the line. Let θ be the angle between the vector and the given line, as in Figure 1.3.

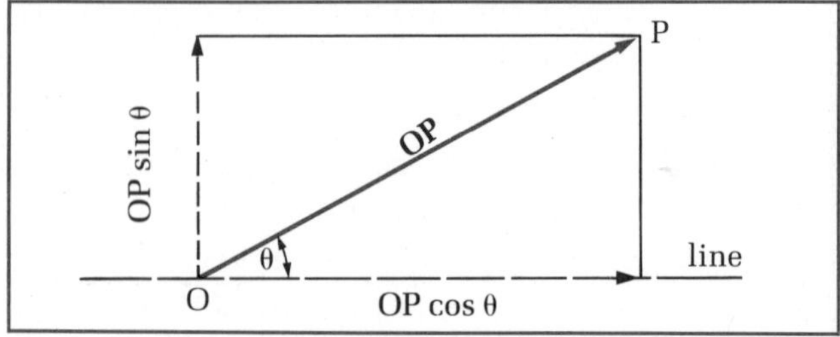

Figure 1.3 *Resolving a vector*

The vector **OP** has a component OP cos θ along the given line, and a component OP sin θ at right angles to the line. In other words, to go from O to P, we can go **along** the line by a length OP cos θ and then **at right angles** to the line by a length OP sin θ.

2) Adding vectors

The parallelogram rule The map example shows how displacement vectors can be added together. The displacement from O to B is equal to the displacement from O to A + the displacement from A to B; in other words, **OB** = **OA** + **AB**. As Figure 1.3 shows, the two vectors to be added form two sides of a triangle. The sum of the two vectors, their **resultant**, is the third side of the triangle. To use this method of adding two vectors, we need actually to draw the vectors in the correct directions and to choose a scale. This is often easier to do by making the two vectors to be added into adjacent sides of a parallelogram, as in Figure 1.4. The resultant is the diagonal from the origin of the two vectors. The *parallelogram of forces* rule for adding two force vectors is discussed further on p. 32.

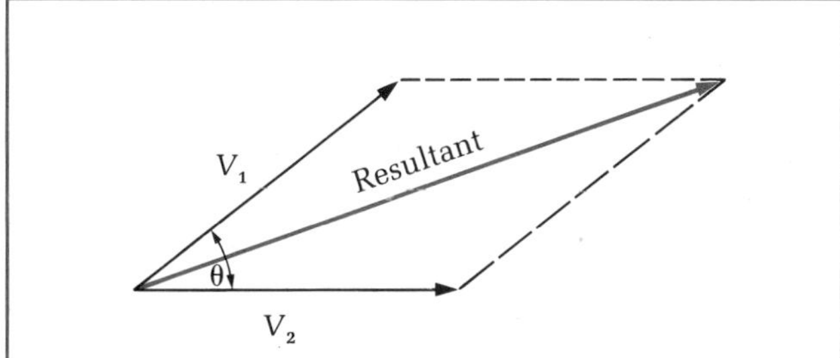

Figure 1.4 *The parallelogram rule*

The calculator method This involves resolving one of the vectors along and at right angles to the line of the other. This method is useful if a calculator is available because it saves having to make accurate diagrams, although a sketch is essential. Let θ be the angle between the two vectors which we shall call $\mathbf{V}_1$ and $\mathbf{V}_2$. Suppose we resolve $\mathbf{V}_1$ into components $V_1 \cos\theta$ and $V_1 \sin\theta$ along and at right angles to the line of $\mathbf{V}_2$ respectively, as in Figure 1.5. The resultant $(\mathbf{V}_1 + \mathbf{V}_2)$ therefore has a component $(V_2 + V_1 \cos\theta)$ along the line of $\mathbf{V}_2$, and a component $V_1 \sin\theta$ at right angles to the line of $\mathbf{V}_2$. By using Pythagoras' theorem, the size of the resultant, $\mathbf{V}_r$, is obtained by squaring and adding its components to give V_r^2. In other words

$$V_r^2 = (V_1 \sin\theta)^2 + (V_1 \cos\theta + V_2)^2.$$

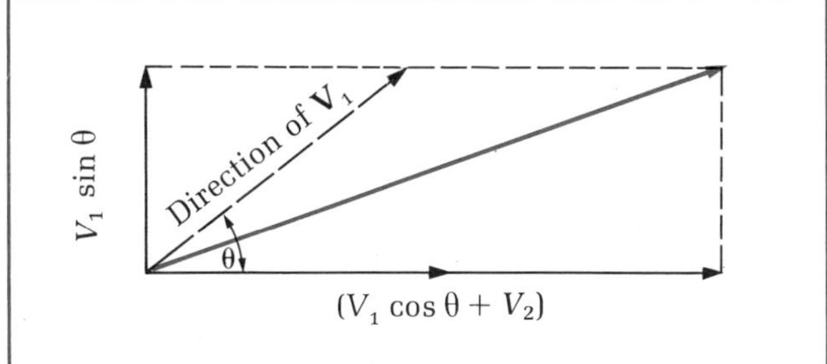

Figure 1.5 *Adding vector components*

Worked example A ship leaves port P and travels 30 km due north. Then it changes course and travels 20 km in a direction 30° east of due north (N30°E) to reach port R. Calculate the displacement from P to R.

Solution Let Q be the point where the ship changes course.

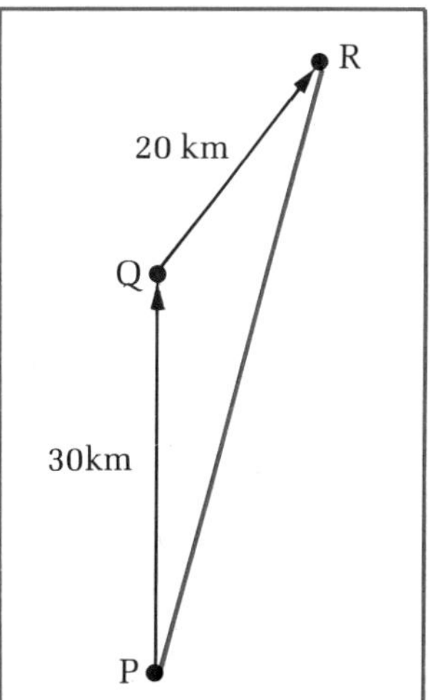

Figure 1.6

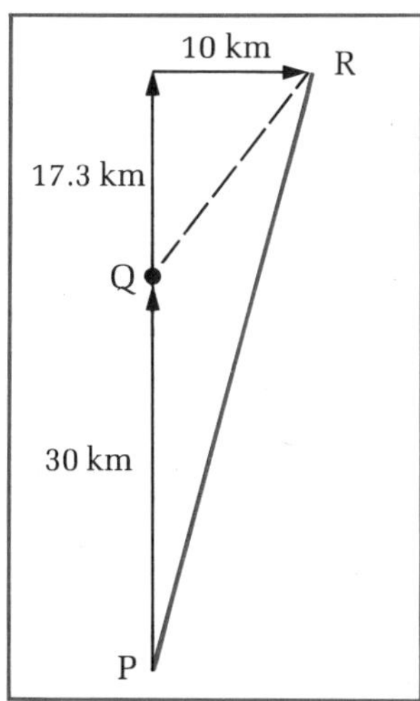

Figure 1.7

Displacement **PQ** = 30 km due north
Displacement **QR** = 20 km in a direction 30° east of due north (N 30°E)
hence **QR** = 20 cos 30° due north, 20 sin 30° due east (in km).

To determine the resultant **PR**, resolve **QR** parallel and perpendicular to **PQ**, as shown in Figure 1.7.

PR has a component, parallel to **PQ**, equal to 30 + (20 cos 30°) = 47.3 km

PR has a component, perpendicular to **PQ**, equal to 20 sin 30° = 10 km

Therefore the distance from P to R directly is $\sqrt{10^2 + 47.3^2} = 48.4$ km.

The direction from P to R is 12° east of due north since tan 12° = 10/47.3, as shown by Figure 1.7.

1.2 Speed and velocity

Speed is change of distance per unit time, often considered as per second. The unit of speed is m s^{-1}.

Velocity is change of displacement per unit time, generally per second. Its unit is also m s^{-1}.

Velocity is a vector; speed is a scalar. To see the difference, consider an object moving uniformly along a circular path. Its speed does not change. However, because its direction keeps changing, its velocity continuously changes. Just as displacement is distance in a given direction, so velocity is speed in a given direction.

Even for an object moving along a straight line, there are two possible directions in which the object can move. One direction is opposite to the other. We can use a simple code for direction; this involves making one direction the positive direction, and the other the negative direction. For example, for vertical motion, we usually make 'upwards positive' and 'downwards negative'. So an object thrown upwards would have an initial velocity which is positive, but its acceleration would be negative. Gravity acts downwards. When such an object is 30 m above its starting point, its displacement at that point is + 30 m. But if its displacement is written as − 30 m, then that means the object is 30 m below its starting point.

Displacement-time graphs are useful to represent motion. Figure 1.8 shows the graph of displacement against time for an object thrown directly upwards. As the object rises it slows down; its change of displacement per second becomes less and less, so the curve becomes less and less steep until it becomes flat. At this point, the object is at its greatest height; its speed at that instant is zero. Then it falls back towards its starting point, gathering speed as it falls. So after reaching maximum displacement, its displacement returns to zero. Because its speed increases as it falls, its change of displacement per second increases. So the curve becomes steeper and steeper after the peak.

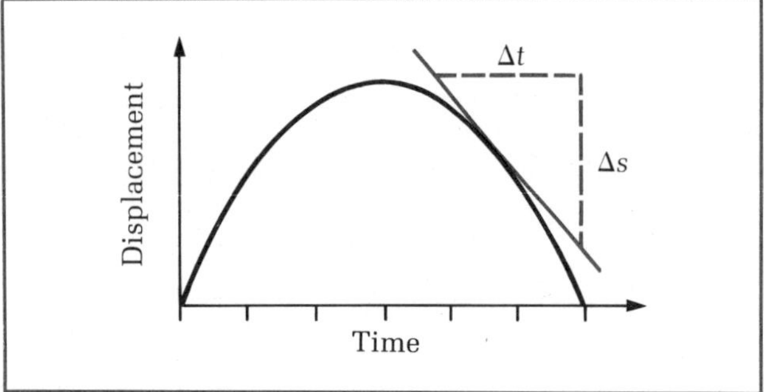

Figure 1.8 *A displacement-time graph*

The gradient of a displacement-time graph gives the velocity of the object. To determine the gradient at any point on the curve, we need to draw a tangent to the curve at that point. The tangent has the same gradient as the curve at that point. Using Figure 1.8, the gradient is given by $\frac{\Delta s}{\Delta t}$, where Δs = change of displacement and is called 'delta s', and Δt (delta t) is the time taken for the change. We often use the Δ symbol to show change. If Δs is an increase in s, its value is positive; a negative value indicates a decrease. On the way up, the gradient becomes less and less positive. This means that its velocity becomes less and less. At its maximum displacement, its velocity is zero just for an instant. Then its velocity becomes more and more but in the opposite direction. So after the peak, the graph has a negative gradient which becomes steeper and steeper.

Relative velocity

Sometimes it is necessary to consider the velocity of an object relative to a second moving object. This is called the **relative velocity**. Imagine a chase along a straight motorway where a sports car S moving due north at 120 km h^{-1} is being chased by a police car travelling at 150 km h^{-1}. Each minute, the sports car travels 2 km and the police car 2.5 km. So the police car catches up by 0.5 km per minute. In other words, it closes on the sports car at a relative speed of 30 km h^{-1}. In this example, the relative speed is the difference between the actual speeds.

In velocity terms, we need to make one direction along the motorway positive and the other direction negative to give a 'direction code' so we make due N positive and due S negative. The velocity of the sports car, v_s, is +120 km h^{-1}, and of the police car, v_p, is +150 km h^{-1}. As seen from the sports car, the police car has a relative velocity of +30 km h^{-1} which equals $v_p - v_s$. As seen from the police car, the sports car has a relative velocity of −30 km h^{-1} which equals $v_s - v_p$. From the viewpoint of the police car, the sports car is moving towards the police car; the direction of the sports car as seen from the police car is therefore negative! This simple example illustrates the general rule for relative velocity.

The relative velocity of object A as seen from object B $= v_A - v_B$ **where** v_A **is the velocity of A, and** v_B **is that of B.**

The rule can be used for two velocities at any direction to one another, not just along a straight line. Even in the case of straight line motion, care must be taken with the direction code. For example, consider Figure 1.9 again, where a helicopter flies above the motorway with a velocity of 180 km h^{-1} due south to intercept the sports car.

The helicopter's velocity is −180 km h^{-1}. Therefore, the relative velocity of the sports car as seen from the helicopter is (+120) − (−180) = +300 km h^{-1}. In other words, the sports car moves north at a relative speed of 300 km hr^{-1} to meet the helicopter.

1.3 Accelerated motion

Acceleration is defined as change of velocity per unit time (per second). Its unit is m s^{-2}. Velocity is speed in a given direction so acceleration can involve change of speed or change of direction or both. For example, a freely falling object accelerates because its speed increases. However, an object travelling on a circular path at steady speed continuously changes direction, so it too is being accelerated. In fact, its acceleration is towards the centre of the circle. Circular motion is discussed in more detail on p. 42.

Consider the example of a car which accelerates from rest to a speed of 10 m s^{-1} in 5 s without change of direction (Figure 1.10). Its speed increases by 2 m s^{-1} every second, so its acceleration is 2 m s^{-2}. Suppose the car then travels at steady speed for a further 25 s. Then at 30 s from the start, suppose the driver brakes to a halt in 10

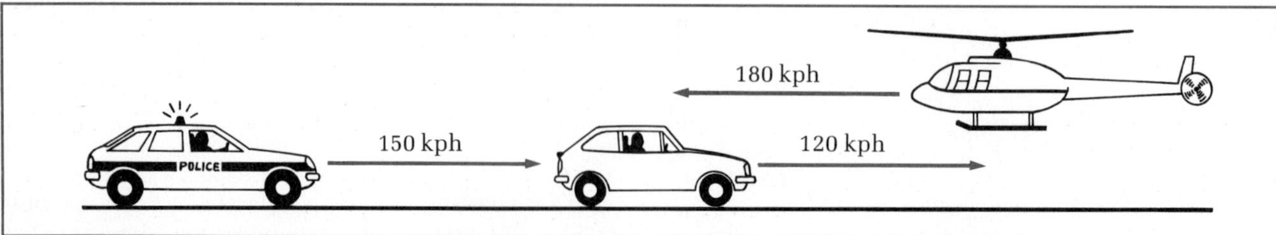

Figure 1.9 *Relative velocity*

Figure 1.10 *Acceleration*

seconds. During breaking, the speed falls by $1\,m\,s^{-1}$ every second so the 'acceleration' in this period has a value of $-1\,m\,s^{-2}$. The minus sign indicates the *fall* of speed.

Velocity-time graphs are useful to represent the motion of an object. For example, Figure 1.11 shows the velocity-time graph for the car in the above example. This graph is in three sections. Section OA represents the initial period of acceleration. Section AB represents the period of steady speed. Section BC represents the braking period. For period OA, the gradient of OA gives the acceleration. This is because the gradient triangle OAX has a height AX which represents the change of velocity, and it has a base OX which represents the time taken. Therefore the acceleration (= change of velocity/time taken) is represented by the gradient of OA (= height AX/base OX).

The gradient of a velocity-time graph gives the acceleration.

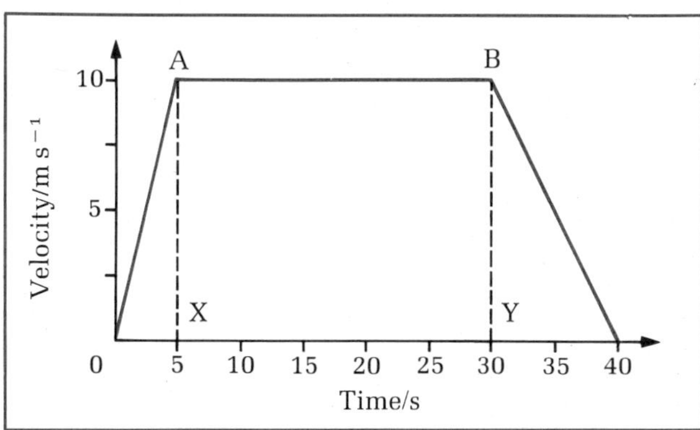

Figure 1.11 *A velocity-time graph*

The section AB of Figure 1.11 has zero gradient because the acceleration in that period is zero. Finally, section BC has a negative gradient which represents the deceleration in that period.

For an object with constant acceleration *a*

$$a = \frac{v - u}{t}$$

where u = initial velocity,
v = final velocity,
t = time taken.

The equation cannot be used where the acceleration changes. Figure 1.12 shows a velocity-time graph where the acceleration continuously increases. The gradient increases with time so the acceleration increases with time. To determine the acceleration at any point on the curve, a tangent to the curve at that point must be drawn. The tangent is then used to form a gradient triangle as in Figure 1.12. The acceleration is then determined from the gradient of this triangle.

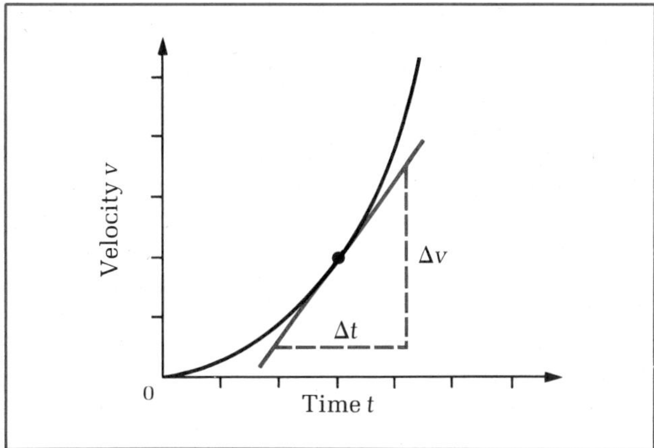

Figure 1.12 *Increasing acceleration*

The area under a velocity-time curve gives the displacement.

Again, consider Figure 1.11. Section AB represents steady speed which lasts for 25 s. Since the speed is $10\,m\,s^{-1}$ from A to B, then the distance travelled must be 250 m. The area of the rectangle ABXY represents the speed × time taken. Section OA shows a steady increase of speed from 0 to $10\,m\,s^{-1}$ so the average speed in this section is $5\,m\,s^{-1}\left(=\frac{0+10}{2}\right)$. Therefore the distance travelled in section OA must be 25 m. This is represented by the area of the triangle OAX because $\frac{\text{height} \times \text{base}}{2}$ represents average speed × time taken.

The final section has an area BYZ which represents displacement during braking. Thus the distance travelled during braking is 50 m $\left(=\frac{\text{height} \times \text{base}}{2}=\frac{10 \times 10}{2}\right)$. The total distance travelled for the whole journey from O to C is therefore 325 m (= 25 m + 250 m + 50 m), and this is completed in a total time of 40 s. Therefore, the average speed for the whole journey is 325/40 $m\,s^{-1}$ which equals $8.125\,m\,s^{-1}$.

In this example, the direction of motion is unchanged, so distance travelled is the same as displacement. Where direction does change, it is important to remember the difference between distance and displacement.

Making a tapechart

One way of investigating motion is to use a tickertimer. The timer prints dots on a tape at a steady rate of 50 per second. If the tape is pulled through the timer with the timer on, dots will be printed along the tape. The spacing of the dots depends on how fast the tape is pulled through the timer.

Using a sloping runway, the tape is attached to a trolley at the top of the runway. When the trolley is released, it accelerates down the runway pulling the tape through the timer as it goes. This apparatus is shown in Figure 1.13. The tape shows an increased spacing between the dots as in Figure 1.14.

To make a tapechart, the tape is marked and cut into ten-dot lengths from the start. With ten dots on each length, each length takes the same time to pass through the timer. So each length is a measure of the speed at that part of the tape. The tape lengths are fixed side-by-side to make a tapechart as in Figure 1.15. The profile of the tapechart shows how the speed alters. For the trolley on the runway, its tape chart shows a steady increase of speed.

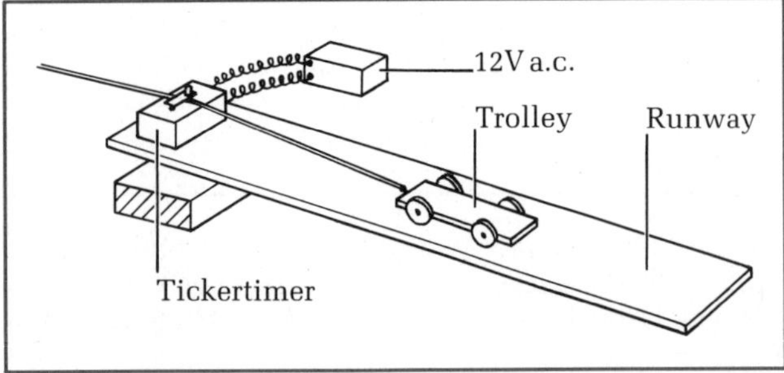

Figure 1.13 *Making a tape chart*

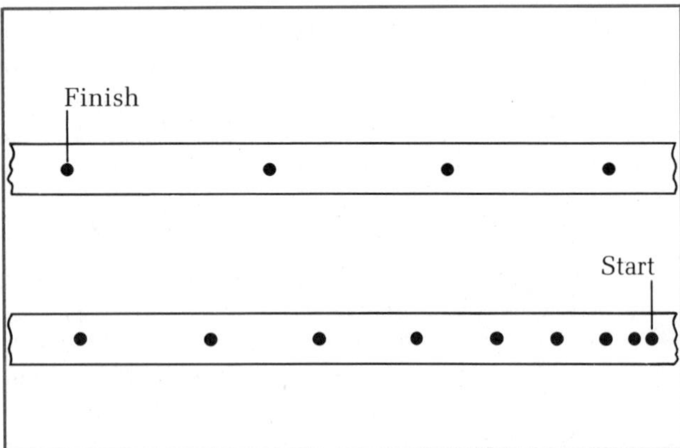

Figure 1.14 *Tape produced by an accelerating trolley*

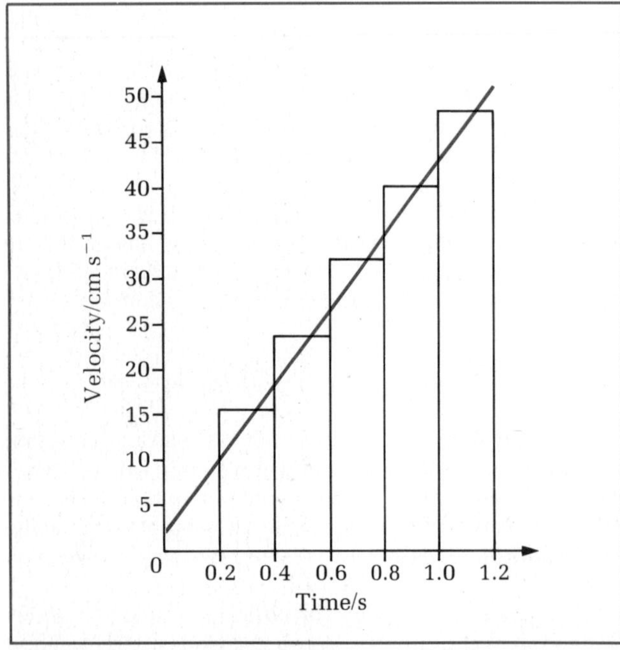

Figure 1.15 *Tape chart of accelerated motion*

To determine the acceleration, the tapechart axes must be calibrated (i.e. marked with values) for speed and time. The horizontal axis represents time; each tape width corresponds to an increase in time of 0.2 s (=time for 10 dots). The upright (vertical) axis represents speed; a 1 cm length with 10 dots on represents a speed of 5 cm s^{-1} (=1 cm/0.2 s). So each centimetre along the speed axis represents a speed of 5 cm s^{-1}. To determine the acceleration, the gradient of the line passing through the top of each length is measured, as shown on Figure 1.15.

1.4 Dynamics equations

For constant acceleration, sometimes called **uniform acceleration**, consider an object which accelerates from initial velocity u to final velocity v without change of direction. Let t be the time taken. Its velocity-time graph is shown in Figure 1.16.

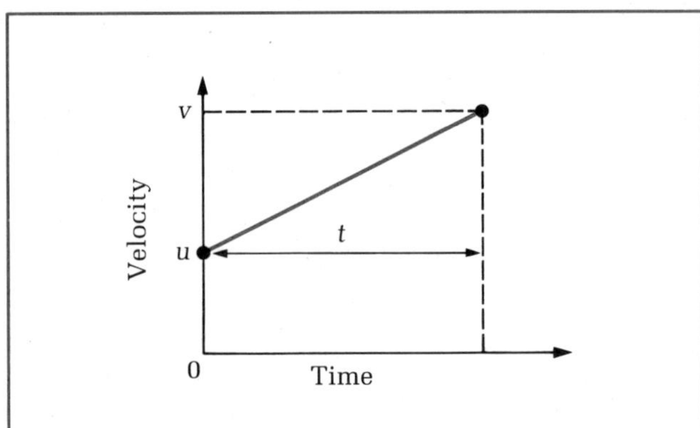

Figure 1.16 *Constant acceleration*

a) The acceleration (a) may be calculated from $a=\frac{v-u}{t}$

Rearranging this equation gives

$$at=v-u$$

and then making v the subject gives

$$v=u+at \qquad (1)$$

b) The displacement (s) may be calculated from

$$s=\frac{(u+v)}{2}t \qquad (2)$$

because $\frac{(u+v)}{2}$ is the average velocity.

c) By combining the two equations above to eliminate v, a further useful equation is produced. To do this, substitute $u+at$ in place of v in equation (2). This gives

$$s=\frac{(u+u+at)}{2}t$$

which simplifies to become

$$s=ut+\tfrac{1}{2}at^2 \qquad (3)$$

d) A fourth useful equation is obtained by combining (1) and (2) to eliminate t. This can most easily be done by multiplying $\frac{(v-u)}{t}$ by the right hand side of (2)

$$as = \frac{(v-u)}{t}\frac{(u+v)t}{2}$$

which simplifies to become

$$as = \frac{v^2-u^2}{2}$$

which gives

$$v^2 = u^2 + 2as \qquad (4)$$

These four equations are invaluable in any situation where the acceleration is constant. Which equation to use depends on the information given and the required quantity.

For changing acceleration, the four dynamics equations of the previous paragraph cannot be used if the acceleration changes. Acceleration is defined as change of velocity per second, and is represented by the slope of the velocity-time graph. If the acceleration changes, then the slope of the velocity-time graph changes, as in Figure 1.12.

To represent *change* of any quantity, the triangular symbol Δ is used. It is called big delta. If the velocity of an object changes by Δv in a time Δt, the acceleration is given by

$$a = \frac{\Delta v}{\Delta t}$$

Where the time interval Δt is very short, then $\frac{\Delta v}{\Delta t}$ is written as $\frac{dv}{dt}$ where $\frac{d}{dt}$ is the mathematical way of writing 'change per second'. So $\frac{dv}{dt}$ is the mathematical way of writing 'change of velocity per second', and equals the slope of the velocity-time graph.

$$a = \frac{dv}{dt}$$

The equation $v = u + at$ is a special case of $\frac{dv}{dt} = a$. This is because a, the acceleration, is constant for $v = u + at$. From $v = u + at$, in a time interval Δt, the velocity changes by $a\Delta t$. So $\Delta v = a\Delta t$. Rearranging, $a = \Delta v/\Delta t$ which gives $a = \frac{dv}{dt}$ for short intervals.

1.5 Motion due to gravity

Free fall

Everything on or near the Earth feels the pull of gravity. Where the only force on a given object is the pull of gravity, the object falls freely towards the Earth. Parachutists who plummet from great heights, opening their parachutes at the last moment for a safe descent, are sometimes said to fall freely. In fact, air resistance prevents them from reaching ever-greater speeds even before opening the parachute. So, even without the parachute open, they are not in free fall because air resistance acts on them. An example of a freely falling object is when a steel ball bearing is allowed to fall. Here the air resistance is so small that the pull of gravity is effectively the only force on the ball as it falls.

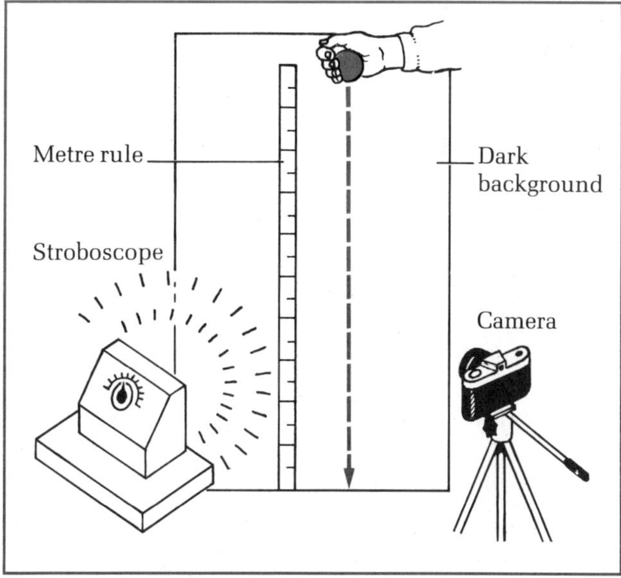

Figure 1.17 *Investigating free fall*

Investigating free fall

One way of investigating the free fall of a steel ball is to make a **multiflash photograph** of the ball's flight, as shown in Figure 1.18. The photograph must be taken in a dark room. The camera shutter is kept open while the ball falls. A **stroboscope**, essentially a regularly flashing light, is used. Each time the light flashes on, the ball is photographed on the camera film. Thus, the ball's position against the scale is recorded at regular intervals of time.

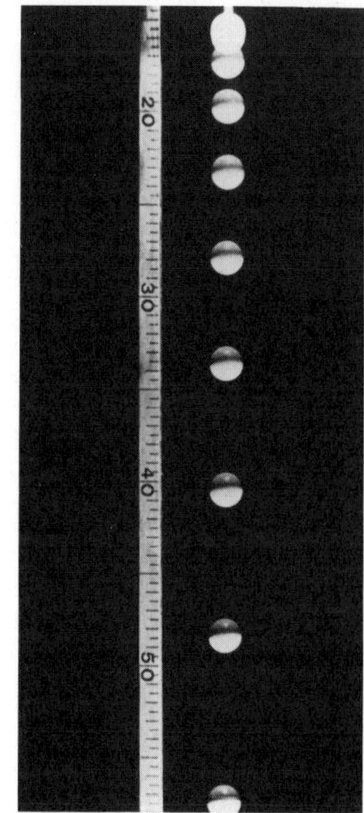

Figure 1.18 *Multiflash photo of free fall. The actual scale is shown by the cm rule in the photograph. The stroboscope was operated at 31 Hz. The first 3 images overlap. Measurements from the photographs are plotted on Figure 1.19*

The photograph clearly shows that the ball's speed increases as it falls. Measurements may be taken from the film to find out if the ball accelerated uniformly. The displacement s of the ball from its position of release is given by

$$s=\tfrac{1}{2}at^2$$

if the acceleration is constant and t is the time from release. The initial speed is zero, assuming it was released from rest. By measuring the rate of flashing from the stroboscope scale, the time interval between flashes can be calculated. Thus the time t from release of each 'image' on the film can be determined. The graph of s against t^2 ought to give a straight line *if* the acceleration is constant. In fact, Figure 1.19 shows the graph for the photograph of Figure 1.18. The graph shows a straight line through the plotted points. From $s=\tfrac{1}{2}at^2$, the gradient of the graph must equal $\tfrac{1}{2}a$. Thus the acceleration may be determined. The graph of Figure 1.19 has a gradient of 4.9 m s^{-2} so the acceleration value is 9.8 m s^{-2}.

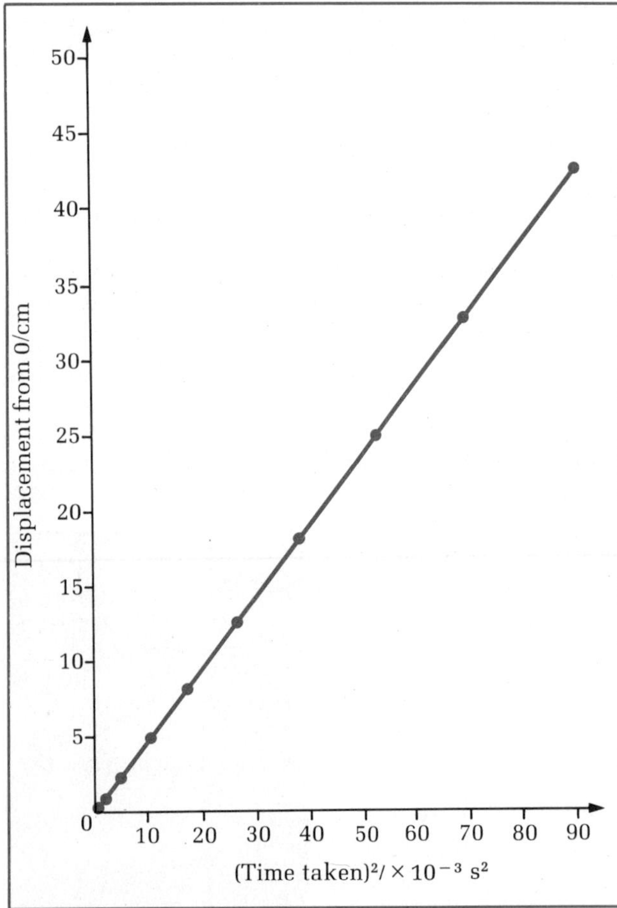

Figure 1.19 *Displacement v (time)² for free fall*

In fact, all freely falling objects fall at a constant acceleration of 9.8 m s^{-2}. This value is denoted by g. There are slight variations in the value of g over the Earth's surface. Its value at the North pole is 9.81 m s^{-2} which is slightly more than its value at the Equator (9.78 m s^{-2}).

Velocity-time graphs for freely falling objects must always have a constant gradient because the gradient represents g. For example, consider the motion of a ball

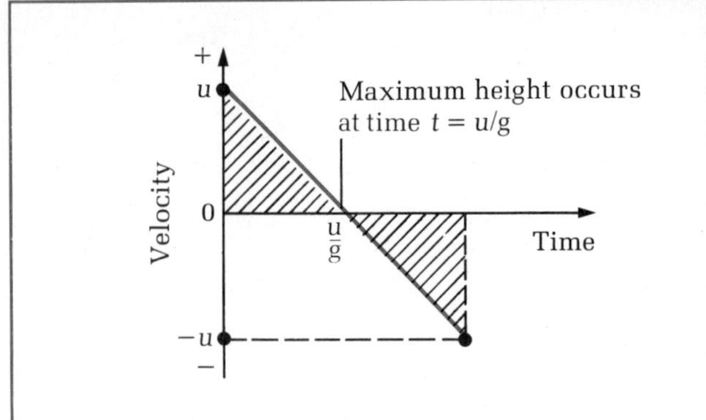

Figure 1.20 *Velocity-time graph for an object thrown straight upwards*

thrown straight up. As it gains height its speed falls, becoming zero at maximum height. Then, the ball falls with increasing speed until it returns to the thrower. For a velocity-time graph, a 'code' is needed for direction; define upwards as positive, downwards as negative. So the initial velocity is large and positive. At maximum height, the velocity is zero. As it falls, its velocity becomes increasingly more negative. Figure 1.20 shows how the velocity alters with time. The gradient is constant, and is negative, because it represents g, and g is constant and acts downwards. The area under the line represents the displacement. Because the area above the time axis (which counts as positive) is equal to the area under the axis (which counts as negative), the total displacement from launch to impact is zero. This is what you would expect because the ball falls back to the thrower.

Terminal speed

When an object falls, if the pull of gravity is opposed by air resistance the object will not accelerate at the value g. This is because air resistance increases with speed. As the speed increases, so does the air resistance until the air resistance equals the pull of gravity. Once this equality is reached, no further acceleration is possible. Thus the object continues its fall at constant speed, called its *terminal speed*. The same general ideas apply when an object falls through water or any other fluid. The object reaches terminal speed more quickly in water than in air because the 'water resistance' is much greater than the 'air resistance'.

Projectile motion

Throw a ball into the air at any non-zero angle to the vertical, and the ball will follow a curved path. The exact path is determined by the initial speed and angle at which the ball is thrown. Gravity does the rest, and takes the ball on its path. The motion of any object acted upon by gravity only is called **projectile motion**.

Consider the example of a ball rolled down a ramp on a table, then rolling off the table at the edge, as in Figure 1.21. The ball follows a flight path which is horizontal initially where it rolls off the edge. Then its path becomes steeper and steeper as the pull of gravity acts on it alone. Suppose someone just 'nudged' a second ball resting near the edge, causing the second ball to fall off the table at the same time as ball A. The second ball, ball B, falls at the

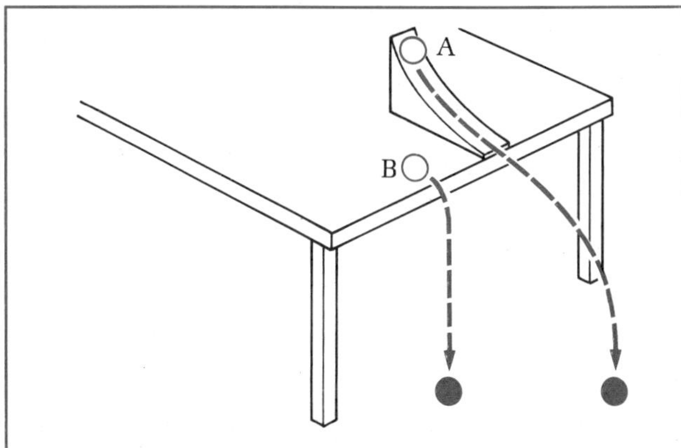

Figure 1.21 *All falling objects fall at the same rate*

same rate as ball A. Both A and B hit the floor at the same time – assuming the floor is level. Try it yourself.

A multiflash photograph of the above example is shown in Figure 1.22. This shows that A, the projected ball, falls at the same rate as B, the ball which drops straight down. Each flash of the stroboscope has pictured A at the same level as B. Thus they fall at the same rate. However, whereas B drops straight down, A moves across at the same time as it falls. In fact, the photograph shows that A moves across at a steady rate. In other words, A is pictured at regular horizontal spacing.

For any projectile, its vertical motion is at constant acceleration *g*, its horizontal motion is at constant speed.

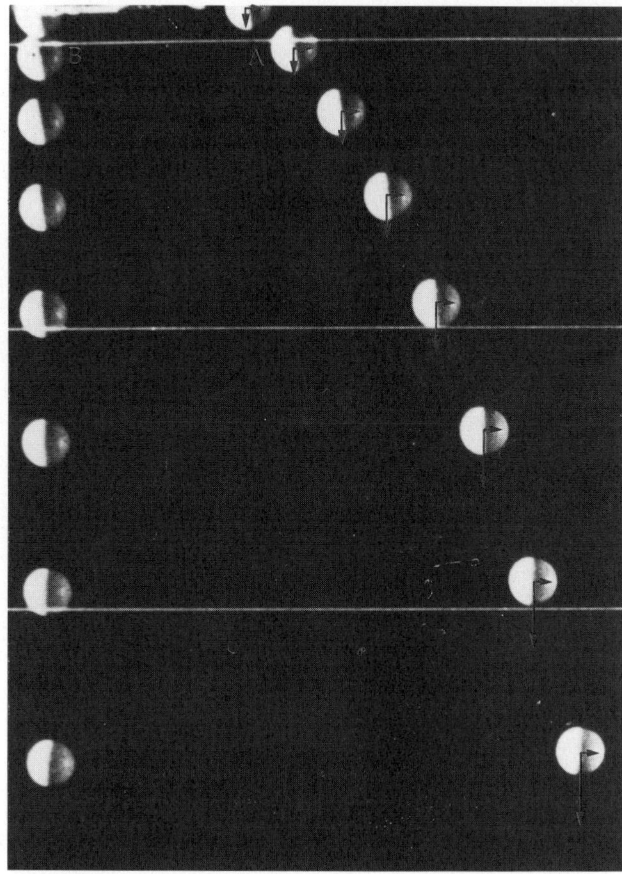

Figure 1.22 *Multiflash photograph of two falling objects*

This means that the horizontal component of its velocity stays constant but the vertical component changes at a rate of $9.8\ \mathrm{m\,s^{-2}}$. Consider the multiflash photograph of Figure 1.22. The position of ball A is shown at each flash interval. The velocity vector is shown at each position, using its horizontal and vertical components. The horizontal component is unchanged throughout. However, the vertical component increases by the *same* amount from one position to the next. The path therefore becomes steeper as the ball falls.

Why should the horizontal motion be unchanged? The reason is that the pull of gravity acts only downwards – and not sideways! So the pull of gravity cannot affect the sideways (i.e. horizontal) motion.

Because a projectile has a constant acceleration g we may use the dynamics equations of p. 6 to calculate the exact flight path. However, to do this, we must consider the horizontal components of the motion separately from the vertical components. Consider again the example of Figure 1.22. Suppose the ball's initial velocity off the table edge was $2.0\ \mathrm{m\,s^{-1}}$ in the horizontal direction. Assume $g = 10\ \mathrm{m\,s^{-2}}$ to simplify calculations. Also, suppose the table edge was 1.25 m above the floor. The time of flight t and the position of impact are to be calculated.

Initial velocity $= 2.0\ \mathrm{m\,s^{-1}}$ in a horizontal direction,
Initial position $= 1.25$ m above the floor.

Vertical motion

Initial vertical speed, $u_y = 0$ (because its initial velocity is horizontal)
Vertical acceleration, $g = -10\ \mathrm{m\,s^{-2}}$ (– for downwards)
Vertical displacement to impact, $s_y = -1.25$ m (again – for downwards)

To calculate the time of flight t use $s = ut + \frac{1}{2}at^2$ applied to the vertical motion.

$$-1.25 = (0 \times t) + (\tfrac{1}{2} \times -10 \times t^2)$$

Hence $5t^2 = 1.25$ so $t^2 = 0.25$
Thus $t = 0.5$ s

Horizontal motion

Initial horizontal speed $u_x = 2.0\ \mathrm{m\,s^{-1}}$
Horizontal acceleration $a_x = 0$ because g is vertical!

To calculate the horizontal displacement to impact s_x, use $s = ut + \frac{1}{2}at^2$ applied to the horizontal motion. Thus,

$$s_x = (2 \times 0.5) + (\tfrac{1}{2} \times 0 \times 0.5^2) = 1.0\ \text{m}$$

Measuring projectile flight paths

Figure 1.23 *Projectiles*

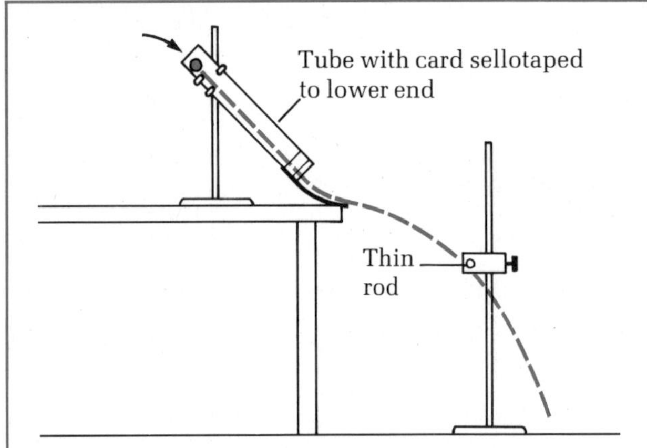

Figure 1.24 *Measuring projectile paths*

Use a cardboard tube as a ramp. If necessary, roll up a sheet of thick paper to form the tube. Clamp the tube at right angles to the edge of a flat table as in Figure 1.24, so that the lower end of the tube is near the table edge. A steel ball-bearing, released from rest from the top of the inside of the tube, ought to roll down the tube and roll off the table edge. The ball ought to follow exactly the same path each time it is released in this way.

A thin rod clamped to a stand is positioned along the flight path so the ball hits the rod. The horizontal and vertical distances from the rod to the table edge can be measured. A plumbline to give a vertical line and a set square to give a horizontal line (off the plumbline) are useful, so that the position of several points along the flight path may be determined and the flight path can be plotted.

Equations for projectile motion Consider a projectile P launched from point O with initial speed u at an angle θ above the horizontal. Let x, y be used to denote the horizontal and vertical components of displacement from O at time *t*, as in Figure 1.25. The initial velocity has vertical component $u \sin \theta$ and a horizontal component $u \cos \theta$. We use the usual direction code, '+ for up, − for down'.

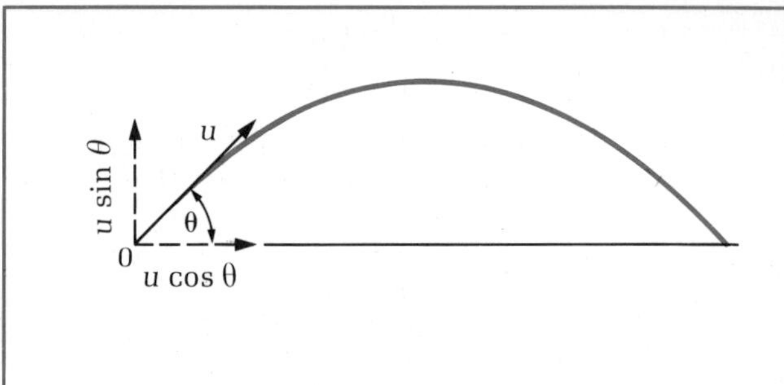

Figure 1.25 *Projectile motion*

	Vertical component	*Horizontal component*
Initial velocity	$u \sin \theta$	$u \cos \theta$
Velocity at P	v_y	v_x
Displacement at P	y	x
Acceleration	−g	0

Using $s = ut + \frac{1}{2}at^2$ applied separately to the vertical and horizontal motion,

$$y = (u \sin \theta)t - \tfrac{1}{2}gt^2$$

and $x = (u \cos \theta)t$

Using $v = u + at$ applied separately again,

$$v_y = (u \sin \theta) - gt$$
$$v_x = u \cos \theta$$

The speed at P may then be calculated from the Pythagoras' rule.

Speed at $P = \sqrt{(v_x^2 + v_y^2)}$

Worked example A projectile is launched from the top of a tower of height 30 m with an initial speed of 20 m s^{-1} at an angle of 30° above the horizontal. Calculate the time of flight to the ground, and the speed of impact.

Solution Break the motion down into components in a table

	Vertical component	*Horizontal component*
Initial velocity (in m s^{-1})	20 sin 30°	20 cos 30°
Acceleration	−g	0
Displacement to impact (in m)	−30	x (not given)

Using $s = ut + \frac{1}{2}at^2$ applied to the vertical motion only,

$$-30 = (20 \sin 30°)t - \tfrac{1}{2}gt^2$$

where t = time to impact.

Assuming $g = 10$ m s^{-2}, $-30 = (20 \sin 30°)t - 5t^2$

and since sin 30° = 0.5, $-30 = 10t - 5t^2$

hence $5t^2 - 10t - 30 = 0$

so $t^2 - 2t - 6 = 0$

The equation for *t* is a quadratic equation and its solution is therefore

$$t = \frac{2 \pm \sqrt{(2^2 - 4 \times 1 \times -6)}}{2} = \frac{2 \pm \sqrt{28}}{2}$$

which gives $t = 3.65$ s or -1.65 s.

The value of -1.65 s is not acceptable because t must be greater than 0,
thus $t = 3.65$ s.
To calculate the speed of impact, use $v = u + at$ applied to the vertical motion to determine the vertical component of the impact velocity v_y

$$v_y = (20 \sin 30°) - (g \times 3.65)$$
$$v_y = 10 - 36.5 = -26.5 \text{ m s}^{-1}$$

The horizontal component $v_x = 20 \cos 30° = 17.32 \text{ m s}^{-1}$
Therefore, the speed of impact v is given by

$$v = \sqrt{(26.5^2 + 17.32^2)} = 31.65 \text{ m s}^{-1}.$$

1.6 Summary

Definitions

1 *Displacement* is distance in a given direction.
2 *Velocity* is change of displacement per second.
3 *Acceleration* is change of velocity per second.

Equations

1 In general,

$$a = \frac{dv}{dt}$$
$$v = \frac{ds}{dt}$$

2 For constant acceleration,

$$v = u + at$$
$$s = (u + v)t/2$$
$$v^2 = u^2 + 2as$$
$$s = ut + \tfrac{1}{2}at^2$$

where a = acceleration,
v = velocity,
s = displacement,
$\frac{d}{dt}$ = change per second,
u = initial velocity,
t = time taken.

Short questions

Assume $g = 10 \text{ m s}^{-2}$

1.1 A car travelling at a steady speed of 30 m s^{-1} along a motorway overtakes a coach travelling in the same direction at 25 m s^{-1}. How far ahead of the coach will the car be
a) 1 minute after overtaking,
b) 30 minutes after overtaking?

1.2 Two ships, A and B, leave port P at the same time. Ship A travels due north at a steady speed of 15 km h^{-1} and ship B travels N 60°E at a steady speed of 10 km h^{-1}. What is the distance and direction from A to B after 1 hour? What is the velocity of B relative to A?

1.3 A plane with a top speed of 50 m s^{-1} leaves an airport on course for another airport 120 km due north. The plane's journey is speeded up by a steady wind blowing due north, and its journey takes 35 minutes.
a) Assuming it travelled at top speed, calculate the wind speed.
b) How long would it take to make the return journey if the wind velocity were unchanged?

1.4 A *Concorde* aircraft leaves New York at 9.40 am GMT and travels at a steady speed of 400 m s^{-1} on a direct course for London, 4800 km away. A similar aircraft travels on the opposite route, scheduled to arrive at the same time, on local time, as the first aircraft, on GMT. Assuming the second *Concorde* travels at the same speed as the first, what is its departure time from London? New York is five hours behind London.

1.5 A boat has a speed of 1.5 m s^{-1} in still water. It is used to cross a river 500 m wide along which there is a strong current of speed 0.9 m s^{-1}. The boat is directed towards the opposite bank but the water current carries it downstream. Calculate
a) the boat's speed relative to the bank,
b) the distance the boat is carried downstream from its point of departure to its point of arrival at the opposite bank,
c) the time taken to cross.

1.6 In **1.5**, what direction must the boat take, relative to the water, to cross directly? How long would it take for such a crossing?

1.7 A train pulls out of a station and accelerates steadily for 20 s until its speed reaches 8 m s^{-1}. It then travels at a constant speed for 100 s, then it decelerates steadily to rest in a further time of 30 s.
a) Sketch a speed-time graph for the journey.
b) Calculate the acceleration and the distance travelled in each part of the journey.
c) Calculate the average speed for the journey.

1.8 A ball is released from rest at the side of the top of a tall tower. The ball hits the ground 2.5 s later. A second ball is released from rest at the same point 1 s after the first is released. How far above the ground is the second ball when the first one makes its initial impact?

1.9 A tennis ball is dropped from a height of 2.0 m above a hard level floor and it rebounds to a height of 1.5 m. Sketch the velocity-time graph for the motion of the tennis ball from the instant it is released to the point where it regains maximum height. Calculate
a) the total time taken,
b) the speed just after impact.

1.10 To measure her reaction time, a student enlists the aid of a friend who holds a 300 mm rule vertically at its zero mark at the upper end. The student is prepared to catch the rule between her thumb and finger at the initial position of the lower end. The rule is released and caught at its 100 mm mark by the student. Calculate the student's reaction time.

1.11 A rocket accelerates from rest at ground level at 7 m s^{-2} for 30 s vertically. Calculate
a) the speed and height of the rocket at 30 s after launch,
b) the maximum height gained by the rocket if its fuel is used up after 30 s,
c) the speed of impact at the ground,
d) the time of flight.

1.12 A car driver, travelling in his car at a steady speed of 8 m s^{-1}, sees a dog walking across the road 30 m ahead. The driver's reaction time is 0.2 s, and the brakes are capable of producing a deceleration of 1.2 m s^{-2}. Calculate the distance from where the car stops to where the dog is crossing, assuming the driver reacts and brakes as quickly as possible.

1.13 In an experiment to measure the effect of 'drag' on a moving object, a steel ball bearing is released from rest at the top of a long tube filled with water. Pairs of light gates at intervals down the tube are used to measure the speed of the ball as it falls. The results are given as a speed-time graph as below. Sketch graphs to show the variation of
a) acceleration with time,
b) distance travelled with time.

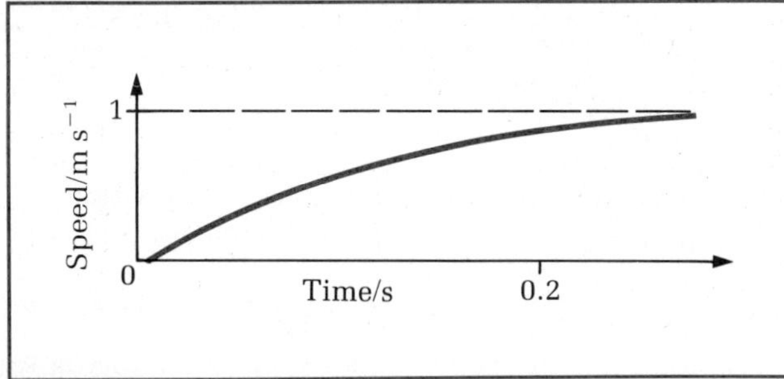

Figure 1.26

1.14 A stone thrown horizontally at a speed of 24 m s^{-1} from the top of a cliff takes 4.0 s to hit the sea. Calculate the height of the clifftop above the sea, and the distance from the base of the cliff to the point of impact.

1.15 A cricket ball is thrown at a speed of 30 m s^{-1} in a direction 30° above the horizontal. Calculate
a) the maximum height,
b) the distance from the thrower to where the ball returns to the same level,
c) the time taken to return to the same level.

1.16 A transport plane travelling at a steady speed of 50 m s^{-1} at an altitude of 300 m releases a parcel when directly above a point X on level ground. Calculate
a) the time of flight of the parcel,
b) the speed of impact of the parcel,
c) the distance from X to the point of impact.

1.17 A stuntman on a motorcycle plans to ride at top speed up and off the top of a ramp 2 m high inclined at 45° to the ground. The motorcycle is capable of a top speed of 30 m s^{-1}. Calculate the distance travelled from the ramp to the point of impact on the ground.

2

Force and energy

2.1 Force and motion

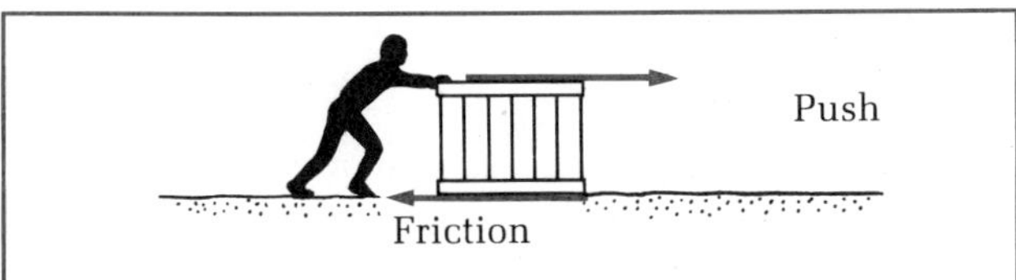

Figure 2.1 *Friction at work*

The link between force and motion is not easy to find in everyday life. Push a heavy box across a rough floor and you might mistakenly conclude that motion *needs* force. In fact the 'hidden' force of friction confuses the link. To move the heavy box, the push force must be large enough to overcome the friction between the floor and the box. If the floor is slippery, then the friction is low and so only a small push is needed to move the box.

If we can remove friction, the link between force and motion can be more easily studied. The principle of 'magnetic levitation' can be used to remove friction, as in the photograph on page 1. The cab is lifted off the track by magnetic repulsion, so contact friction is removed. Try walking on an icy pavement and you might well experience the link between force and motion in the absence of friction.

The linear air track, shown in Figure 2.2 is yet another way of removing friction. Tiny air jets support the 'rider' just above the track. Once set moving, the rider moves at steady velocity along the track. It moves by equal distances in equal times so its velocity is constant.

Whenever any object moves at constant velocity, *either* no forces at all act on it *or* the forces acting on it balance out. Where several forces act on an object, their combined effect is known as the **resultant** force. If the resultant force on an object is zero (i.e. if the forces balance one another out), then the velocity of the object stays constant. Therefore, when a heavy box is pushed across a rough floor at steady velocity, the friction force must be equal and opposite to the push force.

The general link between force and motion is *not* that force is needed to maintain motion. It is that force is required to *change* motion. In other words, to change the velocity of an object, the object must be acted on by a resultant force. If there is no resultant force on an object, its velocity must stay the same. If there is a resultant force on an object, its velocity must change.

2.2 Momentum and Newton's laws of motion

Mass

The effect of a resultant force on an object's velocity depends not just on the force but also on the *mass* of the object. Mass is a measure of the quantity of matter of an object. The unit of mass is the *kilogram* (kg). See p. 511 for the definition of the kilogram. The more mass an object has, the harder it is to change its velocity. The resistance to change of velocity is known as **inertia**. A trick using inertia is to pull a card from under a coin, the card and coin initially resting on a tumbler or cup. The inertia of the coin keeps it at rest while the card is removed. Then the coin drops into the tumbler. Try it!

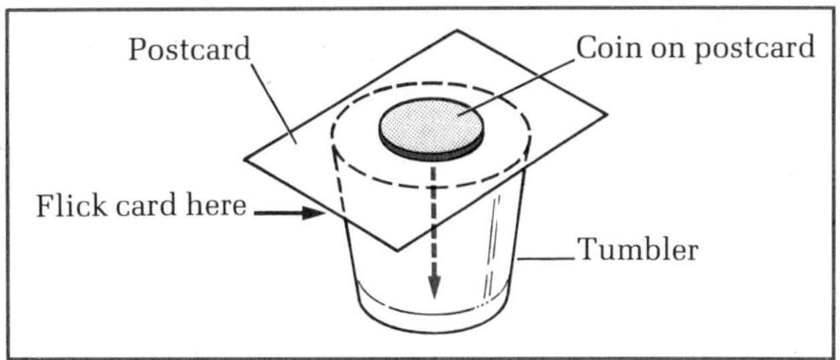

Figure 2.3 *An 'inertia' trick*

Momentum

The momentum of a moving object is defined as its **mass × velocity.**

The unit of momentum is kg m s^{-1}. Momentum is a vector quantity; its direction is the same as the velocity direction. The importance of momentum was first realized by the great scientist Sir Isaac Newton. Newton established the basic principles which link force and motion. He stated these basic principles in three laws, known as **Newton's laws of motion.** Then he applied the laws to explain many aspects of dynamics from the 'down-to-earth' (e.g. falling objects) to the motion of the planets themselves. Other scientists since that time have successfully used Newton's laws in countless situations. Not for more than two centuries did scientists need to question Newton's laws. In fact, it was not until the twentieth century when another great scientist, Albert Einstein, showed that space and time were linked. Einstein must rank alongside Newton for Einstein showed how to deal with objects moving at speeds approaching the speed of light – 300 000 km s^{-1}. Where speeds are low compared with the speed of light, Einstein showed that Newton's

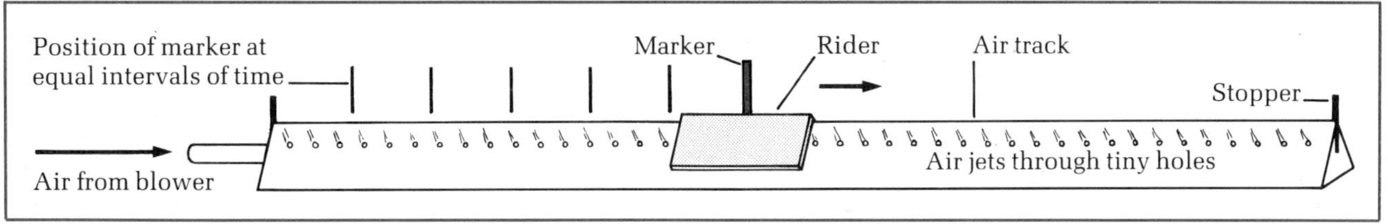

Figure 2.2 *The linear air track*

Isaac Newton was an intellectual giant whose record of achievement stands out like a lighthouse in the dark. He graduated from Cambridge University in 1661. Shortly afterwards, the Great Plague forced scholars away and Newton returned to his home in Lincolnshire. In the two years of isolation to follow, Newton revolutionized mathematics and physics. He invented the calculus, discovered the nature of white light and established his Theory of Gravitation.

He returned to Cambridge in 1667 and became Professor of Mathematics in 1669 – aged 27. His greatest work, the Principia, *was still to come. In this work, published in 1687, he established the mathematical principles of physics which were to be the dominant guidelines of physics for the next two centuries.*

Figure 2.4 *Issac Newton 1642–1727*

laws were quite satisfactory. More important, his **theories of relativity** provide equations which take the place of Newton's laws at high speeds. Such situations are not very common so we can use Newton's laws with confidence in most circumstances.

Newton's laws of motion

Newton's first law Every object continues at rest or with uniform velocity unless acted upon by a resultant force.

An object sliding freely on ice moves at constant velocity. This is because there is no friction between the object and the ice, so there is no resultant force on the object.

Newton's second law The rate of change of momentum of an object is proportional to the resultant force which acts on the object.

Consider an object of fixed mass m acted upon by a constant resultant force F. Suppose the object is accelerated from rest to speed v in time t.

Change of momentum
$= \text{final momentum} - \text{initial momentum}$
$= mv - 0$
so change of momentum per second
$= \text{change of momentum/time taken}$
$= \dfrac{mv}{t}$
but the object's acceleration
$a = \text{change of velocity/ time taken}$
$= \dfrac{v}{t}$
so change of momentum per second $= ma$.

From Newton's second law, the resultant force F is proportional to the change of momentum per second, so the resultant force is proportional to mass × acceleration.

F is proportional to ma, which becomes
$F = kma$

as a result of introducing a constant of proportionality k. The value of k is set at 1 by defining the unit of force, the **newton** (N), as the force which would give a 1 kg mass an acceleration of $1\ \text{m s}^{-2}$.

Thus, for fixed mass

$F = ma$ where F = resultant force in newtons,
m = mass in kg,
a = acceleration.

Worked example A pellet of mass 0.001 kg is fired from an air rifle at a speed of $110\ \text{m s}^{-1}$ into a wood block. The pellet penetrates the block to a depth of 0.050 m. Calculate the impact force.

Solution Initial speed $u = 110\ \text{m s}^{-1}$,
Final speed $v = 0$,
Distance $s = 0.050$ m
To calculate the acceleration a, use $v^2 = u^2 + 2as$
which gives $0 = 110^2 + (2a \times 0.050)$
so $a = \dfrac{-110 \times 110}{(2 \times 0.05)} = 1.21 \times 10^5\ \text{m s}^{-2}$
To calculate the force F, use $F = ma$
which gives $F = 0.001 \times 1.21 \times 10^5 = 121$ N.

Newton's third law When two objects interact, they exert equal and opposite forces on one another.

When you sit on a chair, your weight pushes on the chair. The chair pushes back on you with an equal and opposite force. If the chair suddenly gives way, the push back on you is removed – so the force of gravity pulls you to the floor with a bump!

Worked example A car pulling a trailer accelerates from rest to a speed of $8\ \text{m s}^{-1}$ in 40 s. If the masses of the car and trailer are respectively 1000 kg and 200 kg.

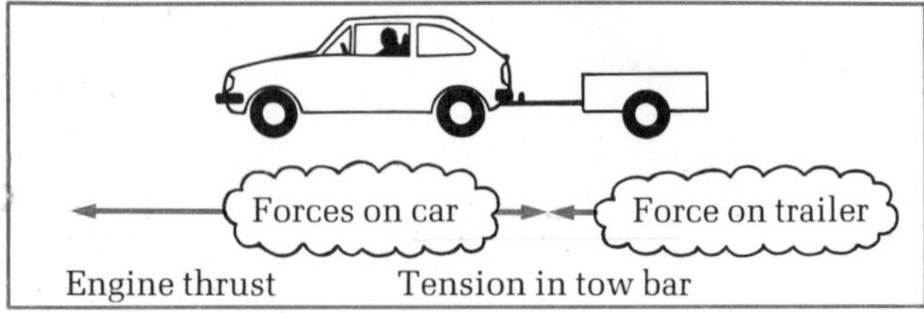

Figure 2.5 *Car and trailer*

Calculate

a) the least thrust force of the car necessary to give this acceleration,

b) the tension in the tow bar.

Solution

a) Initial speed $u=0$
Final speed $v=8\,\text{m s}^{-1}$
Time taken $t=40\,\text{s}$
To calculate the acceleration, a, use $v=u+at$
hence $8=0+(a\times 40)$
so $a=8/40=0.2\,\text{m s}^{-2}$.
To calculate the least thrust force of the car engine, assume frictional forces are negligible so the resultant force equals the least thrust force.
Hence least thrust force $=ma=1200\times 0.2=240\,\text{N}$.
The engine thrust must be at least 240 N, and would need to be greater if friction was significant. The total mass being accelerated is 1200 kg.

b) The tension in the tow bar accelerates the trailer. Considering the trailer alone, and neglecting friction, the resultant force on the trailer is equal to the tension.
Hence tension $=m_t a=200\times 0.2=40\,\text{N}$
($m_t=$ trailer mass).

Note that an equal and opposite force of 40 N is exerted by the trailer on the car. So the car itself is pushed forward by the engine thrust at 240 N against a backward pull of 40 N from the trailer. The car itself is acted upon by a resultant force of 200 N which is equal to the mass of the car × its acceleration. Check it!

Testing $F=ma$

The trolley is pulled along a flat runway by means of a constant force. The force is applied by stretching a length of elastic by a constant amount. A tape attached to the trolley passes through a tickertimer, as shown in Figure 2.6. The timer prints dots on the tape at a rate of 50 per second. In this way, the motion of the trolley is recorded on the tape. Then a tapechart may be made from the tape, as explained on p. 6. For the best results, the runway should be sloped just enough so that the trolley, if given a push, runs at steady speed unaided down the slope.

The procedure is then repeated for 'double-decker' or even 'triple-decker' trolleys. Also, the force may be doubled or trebled by using two or three elastic lengths. The lengths are used 'in parallel', as in Figure 2.6, and stretched by the same amount as before. In this way, several tapecharts may be made, each one representing a different force and mass. Since the gradient of each tapechart gives the acceleration, the gradient ought to be proportional to the (force/mass), i.e. to the number of elastic lengths/the number of trolleys. See Figure 2.7 which shows typical results for a single trolley. The results show that by doubling the force, the acceleration is doubled; by trebling the force, the acceleration is trebled.

Weight and weightlessness

The **weight** of an object is the force of gravity on it. An unsupported object just above the Earth's surface accelerates freely because of the pull of gravity. Since its acceleration is g, then from the equation $F=ma$, the force of gravity on it must be mg. In other words, its weight is equal to its mass × g.

$$\text{weight (in newtons)}=\text{mass (in kg)}\times g\ (\text{in m s}^{-2})$$
$$W=mg$$

The correct unit of weight is the newton, but sometimes the 'kilogram weight' or 'kilogram force' is used in everyday circumstances. The 'kilogram weight' means 'the weight of one kilogram', and it is not a scientific unit. The

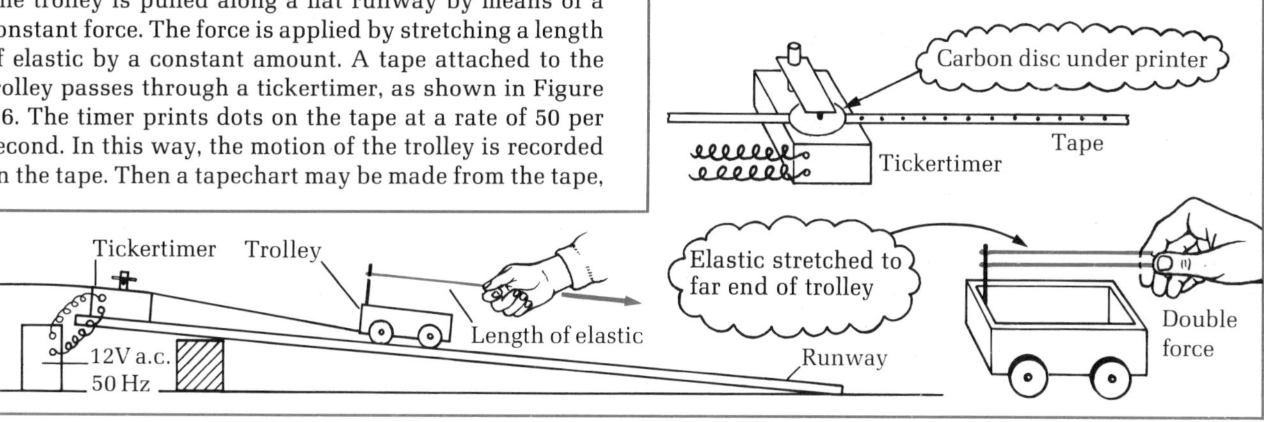

Figure 2.6 *Investigating* F = ma

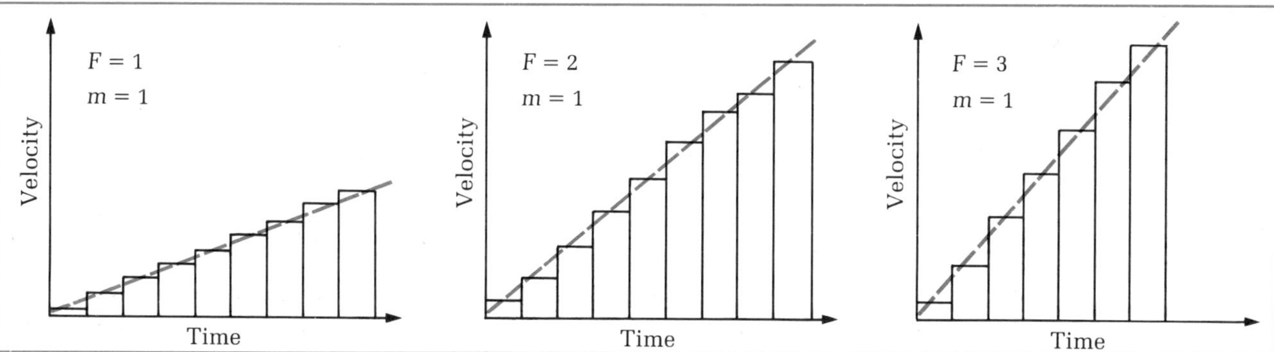

Figure 2.7 *Tapecharts for* F = ma

reason is the variation of g at the Earth's surface. At the poles of the Earth, a 1 kg mass has weight of 9.81 N because $g = 9.81\ \text{m s}^{-2}$ at the poles. However, at the Equator, the value of g is $9.78\ \text{m s}^{-2}$, so here a 1 kg mass has weight of 9.78 N. Thus the 'kilogram weight' changes according to position. On the Moon's surface, where $g = 1.6\ \text{m s}^{-2}$, a 'kilogram weight' actually weighs only 1.6 N. One certain way of losing weight is to go to the Moon – unfortunately though, this would not be due to loss of mass!

Weightlessness is the feeling we experience when unsupported. Imagine standing in a lift on a set of bathroom scales, as in Figure 2.8. When the lift is stationary or moving at steady speed, the scales register your correct weight. However, when the lift *accelerates* downwards, the reading on the scales decreases. You feel lighter. In fact, if the lift cable unfortunately snapped, both you and the lift would fall freely. During your fall, the reading on the scales would be zero because you are unsupported in free fall.

Suppose the lift accelerates downwards, and let its acceleration $= a$. From Figure 2.8, the resultant force on you is $(mg) - S$, where S is the support force on you from the scales. Hence $ma = (mg) - S$, which gives $S = (mg) - (ma)$. So the reading on the scales is reduced from the correct weight (mg) to $mg - ma$.

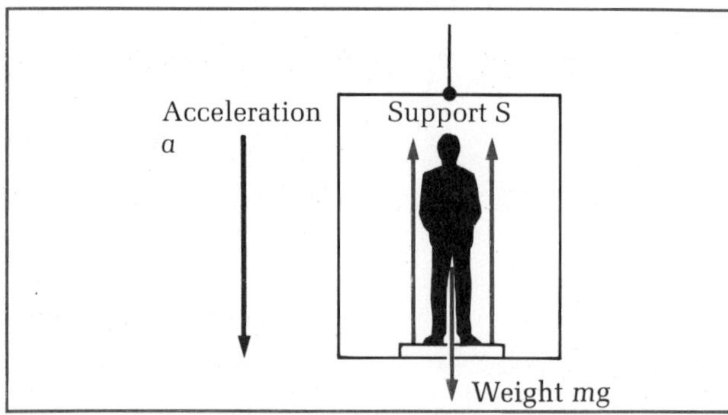

Figure 2.8 *Weightlessness*

Impulse

When a cricketer catches a fast-moving ball, the impact is greatly reduced if he pulls his hands back as the ball is caught. To stop the ball, its momentum must be reduced to zero. By pulling the hands back, the time taken to stop the ball is increased. Therefore, the change of momentum per second is reduced. Since the force equals the change of momentum per second, the force of impact is reduced by this technique.

Where forces act only briefly, as in the above example, an **impulse** is exerted by the force. The impulse of a force is defined as

Impulse = force × time

The unit of impulse is the **newton second** (N s). From Newton's second law which states that force = change of momentum/time taken we can write,

Force × time = change of momentum

Hence impulse = change of momentum

To show the use of impulse, consider the example of a constant force of 5 N acting on a 2 kg mass for a time of 10 s. The mass is initially at rest; to calculate the speed v after 10 s from rest, either

a) use $F = ma$, then use $v = u + at$
so acceleration $a = F/m = 5/2 = 2.5\ \text{m s}^{-2}$,
which gives $v = 0 + 2.5 \times 10 = 25\ \text{m s}^{-1}$.

b) use impulse (Ft) = change of momentum
so change of momentum $= 5 \times 10 = 50$ N s,
which gives $mv = 50$, so $v = 50/2 = 25\ \text{m s}^{-1}$.

Clearly the two approaches used are little different in terms of ease, but now suppose the above example is extended. After 10 s from rest with a constant force of 5 N, suppose a constant force of 10 N acts for a further 15 s. The variation of force with time is shown in Figure 2.9. The impulse approach used to calculate the final speed is now much easier than the other way.

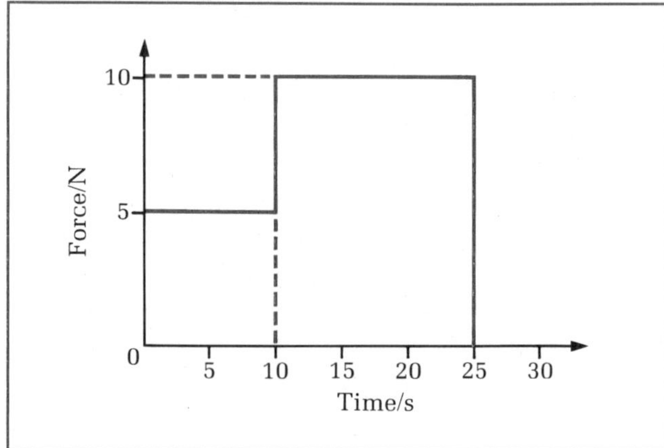

Figure 2.9 *Force-time graph*

a) Using $F = ma$, repeat (a) above to give the speed at 10 s of $25\ \text{m s}^{-1}$. Then consider the next period of 15 s. The initial speed is $25\ \text{m s}^{-1}$, and the force is now 10 N. Thus the acceleration now becomes $10/2 = 5\ \text{m s}^{-2}$. Hence the final speed $v = 25 + 5 \times 15 = 100\ \text{m s}^{-1}$. This approach involves considering each stage in turn.

b) Using impulse, the impulse in the first stage is 50 N s. In the second stage, the impulse is $10 \times 15 = 150$ N s. So the total impulse $= 50 + 150 = 200$ N s. Therefore the total change of momentum $= 200\ \text{kg m s}^{-1}$. Hence the final speed = 200/mass which gives final speed = $100\ \text{m s}^{-1}$.

Note that the area under the force-time graph is the impulse.

Newton's second law for changing mass

The equation $F = ma$ can only be used where the mass is constant. Where the mass changes, such as a rocket burning fuel causing its mass to decrease, the equation $F = ma$ cannot be used. It is then necessary to look again at how $F = ma$ was obtained from Newton's second law.

In more general terms, the change of momentum may be written as $\Delta(mv)$ and the time taken as Δt. Remember the symbol Δ is used to indicate *change*. So the force F is given by

$$F=\frac{\Delta(mv)}{\Delta t}$$

For short time intervals, the equation may be written as

$$F=\frac{d(mv)}{dt}$$

where $\frac{d}{dt}$ is the mathematical way of writing 'change per second'.

So $\frac{d(mv)}{dt}$ means 'change of momentum per second'.

For constant mass, $\frac{d(mv)}{dt}$ becomes $m\frac{dv}{dt}$ which equals ma. This is because acceleration (a) is the change of velocity per second $\frac{dv}{dt}$. So the general form of Newton's second law gives $F=ma$ for constant mass. However, the strictly correct equation for Newton's second law is

$$F=\frac{d(mv)}{dt}$$

Worked example Calculate the force required to hold a water hose of diameter 8.0 cm when it emits a jet of water at a speed of 20 m s^{-1}. The density of water is 1000 kg m^{-3}.

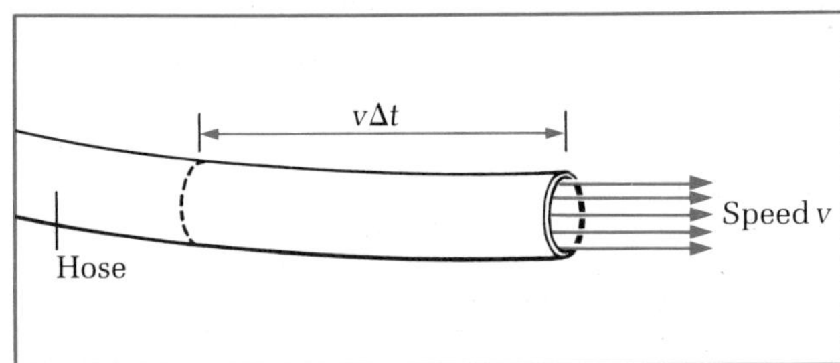

Figure 2.10 *Force of a water jet*

Solution In time Δt, the hose emits a tube of water of length $v\Delta t$ and area of cross-section A.
The mass of water in the tube is therefore $\rho\times$ volume which equals $\rho v\Delta tA$.
Hence the momentum of the tube is mass × velocity which equals $(\rho v\Delta tA)v$.
The hose loses this amount of momentum in time Δt. Thus the loss of momentum per second is $(\rho v\Delta tA)v/\Delta t$ which equals ρv^2A.
Since the force = change of momentum per second, the force $=\rho v^2A$.

$\rho=1000$ kg m^{-3}
$v=20$ m s^{-1}

$$A=\frac{\pi(\text{diam})^2}{4}=\frac{\pi(8\times10^{-2})^2}{4}\text{ m}^2.$$

$$\text{Hence force}=1000\times20^2\times\frac{\pi(8\times10^{-2})^2}{2}=2010\text{ N}.$$

2.3 Conservation of momentum

When an object is acted upon by a resultant force, its momentum changes. If the resultant force is zero, the momentum does not change. Now consider several objects which interact with one another. If no external resultant force acts on this system of objects, the total momentum does not change. However, within the system, the interactions between the objects causes transfer of momentum between them. But the total momentum stays constant.

The principle of conservation of momentum states that for a system of interacting objects, the total momentum remains constant, provided no external resultant force acts on the system.

When any two objects in the system interact, they exert equal and opposite forces on one another. As a result, one of the objects gains momentum from the other object. The total momentum stays the same because the gain of momentum by one object is due to an equal loss of momentum from the other.

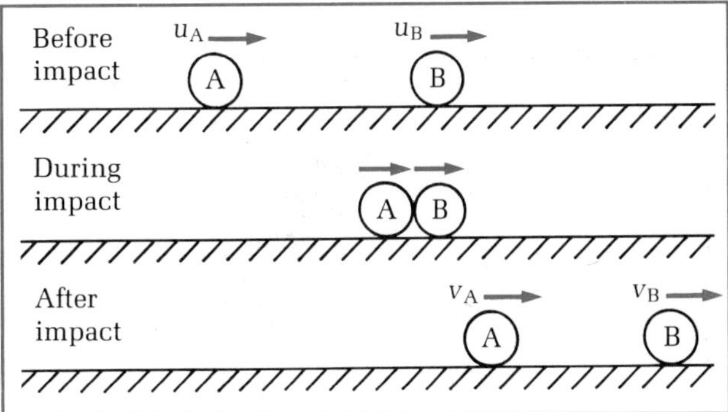

Figure 2.11 *Conservation of momentum*

Consider the example of two billiard balls in collision, as in Figure 2.11. The impact force F on ball A from B changes the velocity of A from u_A to v_A, so

$$F=\frac{m_Av_A-m_Au_A}{t}$$

where t is the time of contact between A and B, and m_A is the mass of A. An equal and opposite force F' on ball B from A changes the velocity of B from u_B to v_B, so

$$F'=\frac{m_Bv_B-m_Bu_B}{t}.$$

Because the two forces F and F' are equal and opposite to one another, then $F=-F'$ which gives

$$\frac{m_Av_A-m_Au_A}{t}=\frac{-(m_Bv_B-m_Bu_B)}{t}\text{ hence}$$

$$m_Av_A+m_Bv_B=m_Au_A+m_Bu_B$$

So **total final momentum = total initial momentum**

In other words, the total momentum is unchanged by the collision.

An **explosion** where objects fly apart is another example where momentum is conserved. No external forces act during an explosion, only internal forces which push the objects apart. Consider a bullet of mass m fired from a gun of mass M. After the explosion which drives the bullet out of the gun barrel, the gun barrel recoils. The momentum carried away by the bullet is mv where v is the speed of the bullet. The gun barrel recoils at speed V, so the momentum carried away by it is $-MV$ ($-$ since its direction is opposite to the bullet's momentum). Therefore the total final momentum $= mv - MV$. Since the bullet and gun were at rest before the explosion, the total initial momentum is zero. Hence, by the conservation of momentum principle, $mv - MV = 0$ which gives,

$$MV = mv.$$

A less explosive example is when you jump off a rowing boat. Your jump to the safety of the dry ground pushes the boat backwards so the boat recoils – sometimes with disastrous consequences. Take care if you ever try it! The momentum carried away by you is equal and opposite to that carried away by the boat. The same equation as above applies.

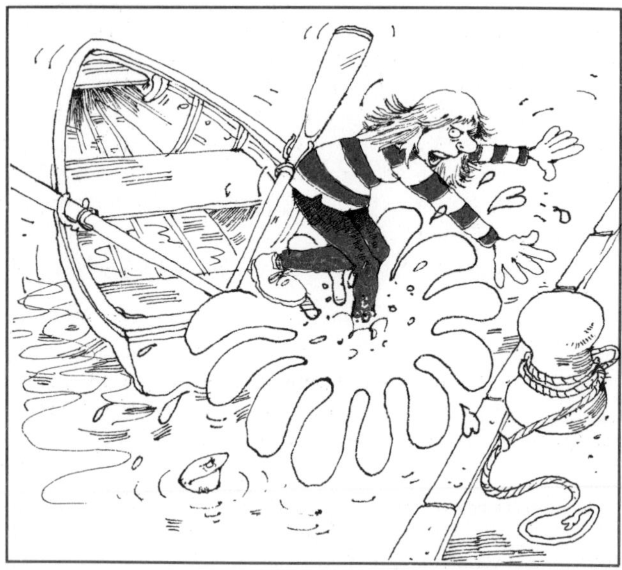

Figure 2.12 *A momentum problem*

Momentum in two dimensions

A good snooker player can 'pocket' a ball at an awkward angle. An impact between two balls at an angle, as in Figure 2.13 is an example of an *oblique* collision. In any type of collision, momentum is always conserved. For oblique collisions, the vector nature of momentum is important. In the example of Figure 2.13, the two final momentum vectors must add up to be equal to the initial momentum vector. A multiflash photograph of such a collision would show the position of each ball at regular intervals of time. The speed of each ball can then be gauged by the spacing from one position to the next. Given the mass of each ball, the sum of the final momentum vectors should equal the initial momentum vector. Use is made of this principle when collisions between sub-atomic particles are photographed and then analysed.

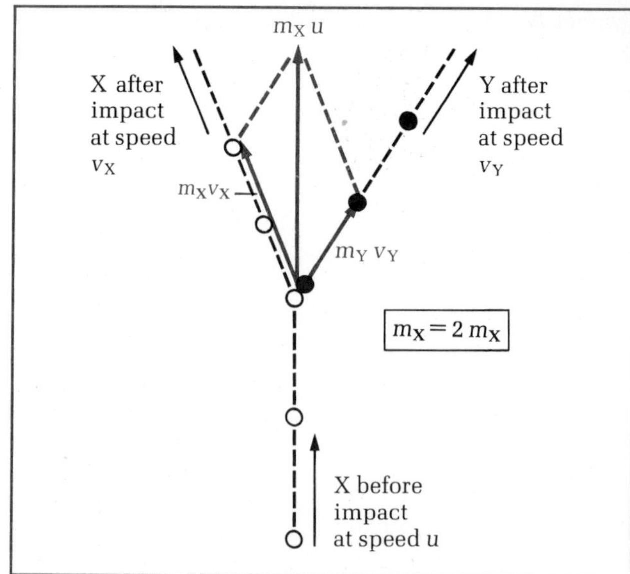

Figure 2.13 *Momentum in two dimensions*

2.4 Work, energy and power

Work

Whenever you lift a heavy weight off the ground you do work. The heavier the weight, or the higher it is lifted, the more work is done. Whenever you push a heavy crate across a rough floor you do work. The further the crate is pushed or the rougher the floor, the more work is done. Work is done whenever a force moves its point of application along its line of action. The weight when lifted is moved upwards; the crate when pushed across the floor moves in the direction of the push.

The **work done** by a force when it moves is defined as

Force × distance moved along its line of action.

The unit of work is the **joule**, defined as the work done when a force of 1 N moves its point of application by 1 m along the line of action of the force.

$W = Fs$ where W = work done in joules, F = force in newtons, s = distance moved in metres.

When a weight of 1 N is lifted vertically by a height of 1 m, 1 J of work is done by the lifting force. When an 8 N weight is raised vertically by 3 m, 24 J of work is done. If

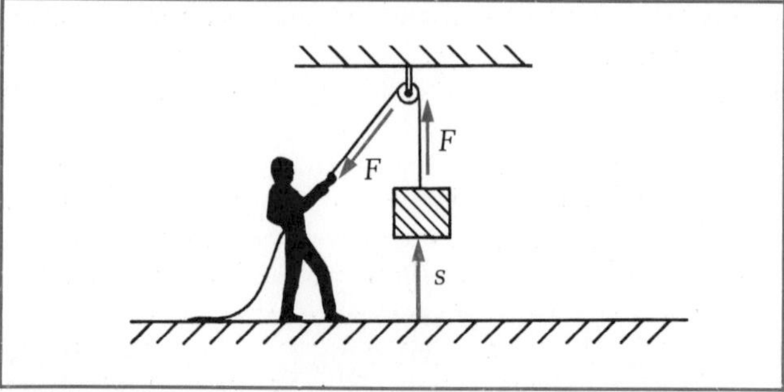

Figure 2.14 *Doing work*

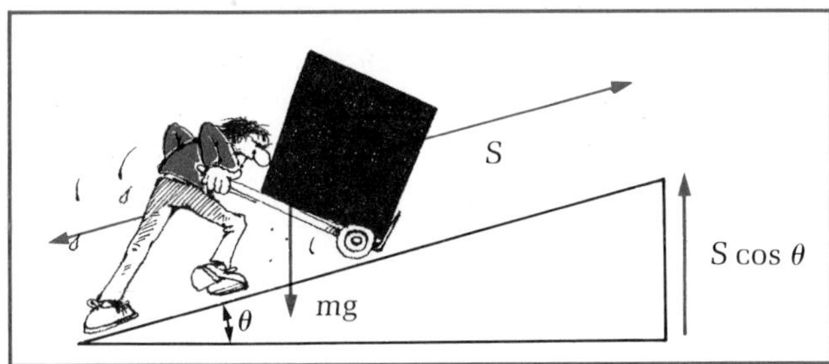

Figure 2.15 *An easy way of doing work*

however, the 8 N weight is moved by 3 m *horizontally*, no work is done. This is because the weight, and hence the support force, are vertical – but the movement is horizontal.

Where a force moves in a direction at an angle θ to the line of action of the force, as in Figure 2.15, the work done by the force is $Fs\cos\theta$. This is because the distance moved by the force along its line of action is $s\cos\theta$. For example, if a person of weight 600 N walks up a steady 1 in 100 incline for 200 m, the work done is $600 \times 200 \cos\theta$, where θ is the angle of the incline to the *vertical*. Since $\cos\theta = \frac{1}{100}$, the work done is therefore $600 \times 200 \times \frac{1}{100} = 1200$ J. The same amount of work is done if that person moves directly upwards by 2 m. However, using the incline is easier because the work can be done at a slower rate!

Energy

Energy is the capacity to do work A body with no energy can do no work. For example, a 'flat' car battery will not have the capacity to start a car. However, a fully charged car battery ought to contain enough energy to do the work involved in starting the car. When one body A does work on another body B, A transfers energy to B as a result. The energy transferred from A to B is equal to the work done by A on B. Energy is therefore measured in joules. If a car battery initially contains 2.0 MJ and does 10 kJ of work when the car is started, then the battery's energy falls to 1.99 MJ (= 2.0 MJ − 10 kJ) afterwards.

Many different forms of energy exist. The energy stored in a car battery is in the form of chemical energy. Whenever electricity is drawn from the battery, chemical reactions take place inside the battery, changing chemical energy into electrical energy. Other forms of energy include light energy, sound energy, heat energy, nuclear energy, kinetic energy and potential energy. **Kinetic energy** is energy due to motion. **Potential energy** is energy due to position. An object raised uphill has potential energy. The work done to raise the object is equal to the increase of potential energy of the object.

Energy can be changed from any form into other forms. A weightlifter changes energy from his muscles (chemical energy) into potential energy of the weight when the weight is lifted. If the weightlifter releases the weights, the potential energy is converted to kinetic energy as the weights fall. When the weights hit the floor, the kinetic energy gained is converted into heat energy and sound energy.

Figure 2.16 *P.E. changing to K.E.*

The principle of conservation of energy states that energy can neither be created nor destroyed.

When energy is changed from one form into other forms, the total amount of energy after the change must equal the initial amount of energy. Conservation of energy is a very important principle (VIP!) and every experiment designed and carried out to test it supports this key principle. To show how the principle may be used, consider again the weightlifter. Suppose he lifts a 500 N weight by a height of 1.5 m. Then the work done by the weightlifter on the weight is 750 J (= 500 × 1.5). Therefore, the potential energy gain of the weight is 750 J. When the weight is released, the potential energy is converted to kinetic energy. Just before impact, the potential energy loss is 750 J and the kinetic energy gain is 750 J. In other words, all the initial potential energy is converted to kinetic energy by the fall.

For a mass m moving at speed v, its kinetic energy is given by

$$\text{K.E.} = \tfrac{1}{2}mv^2$$

To prove this equation, suppose mass m is accelerated by a constant force F from rest to speed v in time t. Since the initial speed $u = 0$, then from the dynamic equations,

$$\text{acceleration, } a = \frac{v-u}{t} = \frac{v}{t} \quad \text{since } u = 0$$

$$\text{distance moved, } s = \frac{(u+v)}{2}t = \frac{vt}{2} \quad \text{since } u = 0.$$

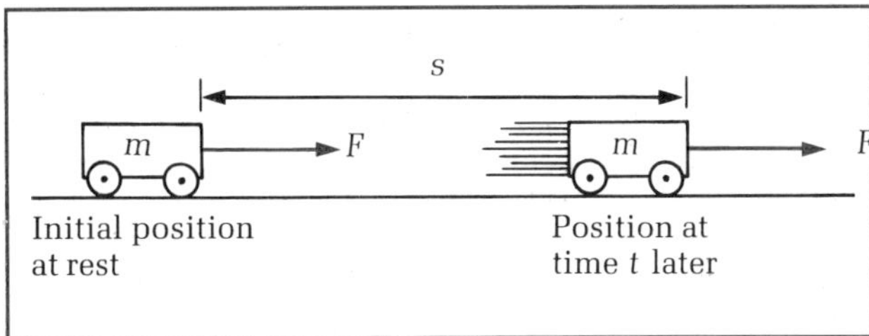

Figure 2.17 *Work done producing K.E.*

Because the work done = force × distance = Fs and for constant mass, $F = ma$

$$\text{then the work done} = mas = m\frac{v}{t}\frac{vt}{2} = \tfrac{1}{2}mv^2.$$

Finally, since the work done is converted to kinetic energy, thc K.E. $= \tfrac{1}{2}mv^2$.

For a mass m moved through a height gain h, its potential energy is given by

P.E. $= mgh$.

To prove this equation, the force necessary to raise the weight must be equal to mg. By raising the weight through a height h, the work done by the lifting force = force × distance so the work done = mgh. Since the work done is converted to potential energy, the gain of potential energy is therefore mgh.

Worked example A pendulum bob of mass 0.2 kg is suspended by means of a string of length 2.0 m from a fixed point. The bob is pulled to the side with the string taut until the string makes an angle of 30° with the vertical. Calculate

a) the gain of potential energy when the bob is moved aside as above,

b) the kinetic energy of the bob after being released (from the position described above) when the bob passes through the position where the string is vertical,

c) the speed as the bob passes through the position where the string is vertical.

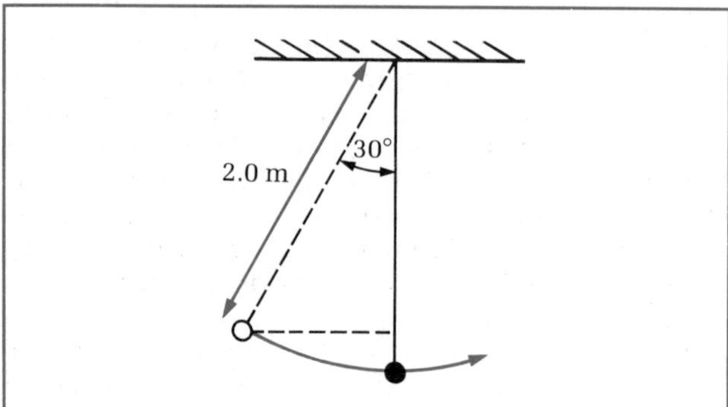

Figure 2.18 *P.E. changing to K.E.*

Solution

a) When the bob is directly beneath the point of suspension of the string, it is exactly 2.0 m beneath the fixed point of suspension. When the bob is pulled to the side as above, its distance beneath the fixed point is 2.0 cos 30°. Thus the height gain H of the bob is $2.0 - 2.0 \cos 30° = 0.268$ m
The gain of P.E. of the bob $= mgh = 0.2 \times g \times 0.268 = 0.536$ J (assuming $g = 10\ \text{m s}^{-2}$).

b) The gain of K.E. when the bob moves from the side to the centre = loss of P.E. thus its K.E. passing through the centre = 0.536 J (since the initial K.E. is zero).

c) Let its speed as it passes through the centre $= v$, thus $\frac{1}{2}mv^2 = 0.536$ thus $\frac{1}{2} \times 0.2 \times v^2 = 0.536$, which gives $v^2 = 5.36$ so $v = 2.32\ \text{m s}^{-1}$.

Power

Power is the rate of doing work In other words, power is the work done per second. The unit of power is the **watt** (W), defined as a rate of doing work equal to $1\ \text{J s}^{-1}$. The kilowatt (kW) which equals 1000 W, and the megawatt (MW), equal to 1 000 000 W, are often used. If our weightlifter described earlier uses 750 J to raise the weights in a time of 3.0 s, then his rate of doing work is 750/3 = 250 W. So his muscle power is 250 W. If a second weightlifter performs the same task in 2.5 s, then the muscle power of the second weightlifter is 300 W (= 750 J/2.5 s). Clearly, the second weightlifter is more powerful than the first.

Try to estimate your personal muscle power! For arm power, ask a friend to time how long it takes you to raise a known weight by a measured height. Be sure to use a weight which you can reasonably lift. Your arm power equals the P.E. gain (mgh)/time taken in seconds, where mg is the weight lifted, in newtons, and h is the height raised, in metres. Don't be too disheartened if your arm power is less than 100 W – even if a light bulb *is* more powerful than you, remember that even trained weightlifters are not much better!

2.5 Elastic and inelastic collisions

Drop a bouncy rubber ball from a measured height onto a hard floor, and the ball will bounce back to almost the same height. Try the same with a cricket ball and there will be very little bounce! **A perfectly elastic ball** would be one that bounces back to exactly the same height. Its K.E. just after impact must equal its K.E. just before impact. Otherwise, it cannot regain its initial height. Thus, there is no loss of K.E. in an perfectly elastic collision.

The cricket ball does not rebound much, so its K.E. just after the impact is less than its K.E. just before the impact. Some K.E. is converted into other forms of energy as a result of the impact. This type of impact is an example of a **partially inelastic collision** – one in which there is *some* loss of K.E. *and* the colliding objects move apart afterwards. In other words, there is some bounce in the impact. The K.E. lost is converted to heat energy and sound energy.

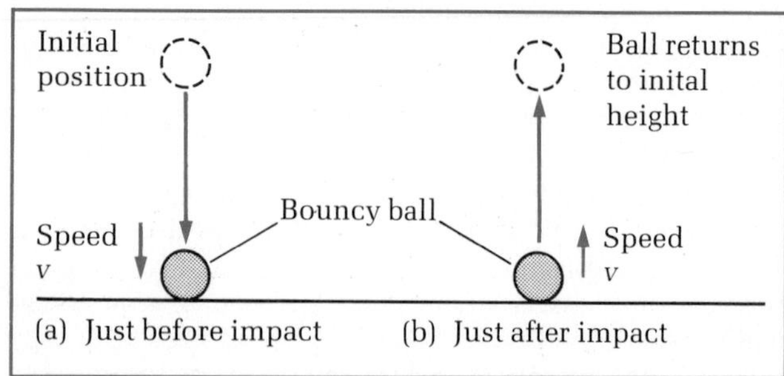

Figure 2.19 *An elastic impact*

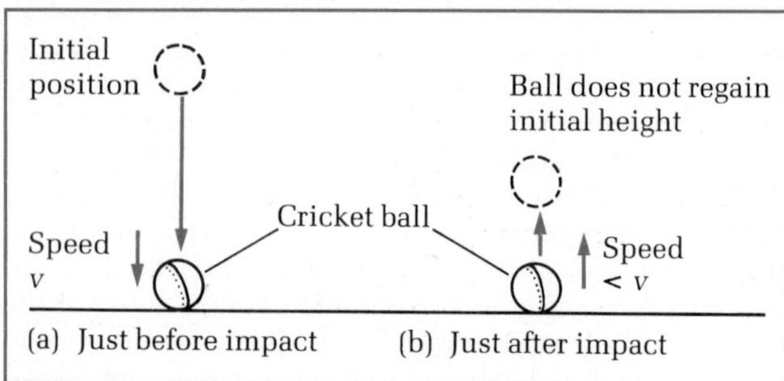

Figure 2.20 *A partly inelastic impact*

A totally inelastic collision happens when the colliding objects stick together on impact and do not move apart afterwards. They remain stuck together. An example is when a railway engine collides with and couples to a truck, as in Figure 2.21.

Figure 2.21 *A totally inelastic impact*

Worked example A railway engine of mass 20 000 kg moving at 3 m s^{-1} collides with and couples to an initially stationary train of wagons of total mass 10 000 kg. Calculate,

a) the speed of the engine and wagons just after impact,
b) the loss of K.E. as a result of the impact.

Solution

a) To calculate the speed v just after impact, use the principle of conservation of momentum which is true for any type of collision.
Total initial momentum $= 20\,000 \times 3$
$= 60\,000$ kg m s^{-1}.
Total final momentum = Total mass $\times v = 30\,000v$.
Since the total final momentum = the total initial momentum
$30\,000v = 60\,000$
hence $v = 2$ m s^{-1}.

b) The initial K.E. $= \frac{1}{2} \times 20\,000 \times 3^2 = 90\,000$ J
The final K.E. $= \frac{1}{2} \times 30\,000 \times 2^2 = 60\,000$ J
hence the loss of K.E. = 30 000 J.

A dramatic example of a totally inelastic collision was a demonstration by British Rail to test the safety of casks used to transport radioactive material by rail. A diesel locomotive, driven by remote control, was allowed to crash at high speed into a stationary cask on the track. The spectacular impact was observed and filmed, and the cask was inspected thoroughly after the impact. Even though the locomotive was destroyed, the cask itself was intact afterwards. (See Figure 2.22.)

Figure 2.22 *After a high-speed impact, the cask is intact*

Investigating collisions

Each type of collision mentioned above may be investigated using the arrangement of Figure 2.23. A trolley A is to be given a push towards a stationary target trolley B. A tape is attached to each trolley, and each tape is passed through a tickertimer which prints dots on the tape at a rate of 50 per second. When A is pushed towards B, the motion of each trolley before and after the impact is recorded on the tapes. The speed of A before and after the impact, and of B after the impact can therefore be calculated.

a) For a partially inelastic impact, A is allowed to make a hard impact with B so that they move apart after the impact. Figure 2.23 shows an example of the tapes produced.

b) For a totally inelastic impact, a cork is fitted to B and a pin to A, as shown in Figure 2.24. When A is pushed into B, the pin pierces the cork so A and B move away together. Figure 2.24 shows an example of the tapes produced.

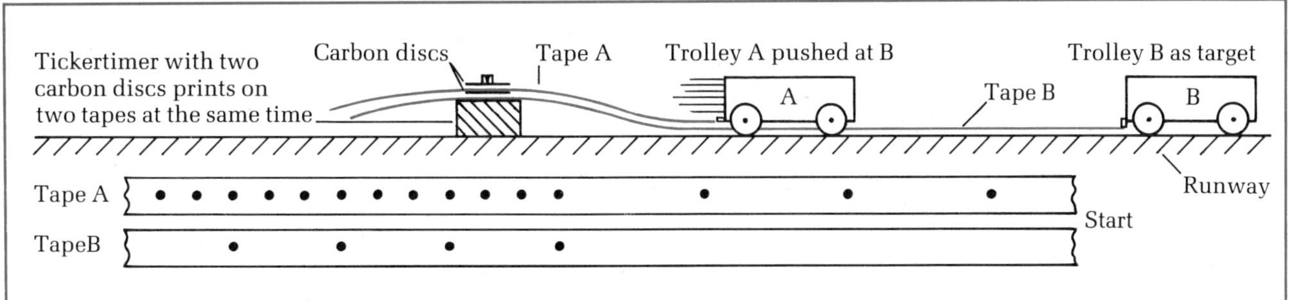

Figure 2.23 *Investigating collisions*

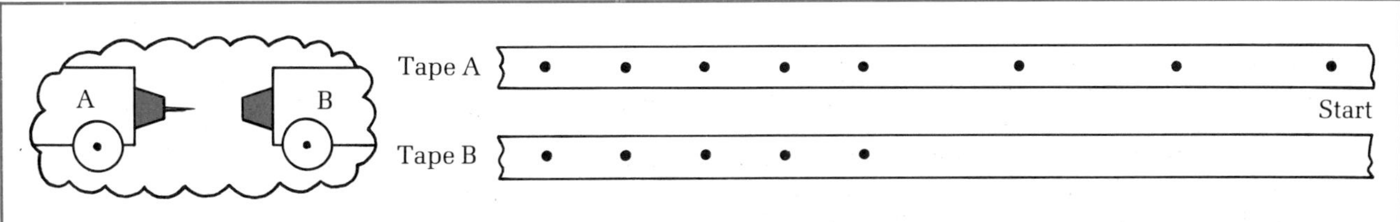

Figure 2.24 *A totally inelastic collision*

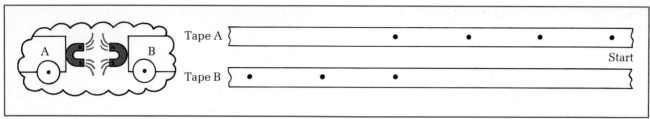

Figure 2.25 *An elastic collision*

c) For an elastic collision, horse shoe magnets are fitted to the two trolleys so that they repel without direct impact (i.e. without contact). Figure 2.25 shows an example of the tapes produced.

The mass of each trolley must be measured and recorded. The tapes above were produced by trolleys of 1 kg mass each.

1 In all cases, the total initial momentum = the total final momentum:

$$m_A u_A = m_A v_A + m_B v_B$$

where m_A and m_B are the masses of A and B respectively, and u_A and v_A represent initial and final speeds of A and v_B represents the final speed of B.

2 Only in (c) above is the total initial K.E. = the total final K.E.

$$\tfrac{1}{2}m_A u_A^2 = \tfrac{1}{2}m_A v_A^2 + \tfrac{1}{2}m_B v_B^2$$

Momentum is always conserved in any collision. Kinetic energy is only conserved in elastic collisions.

2.6 Energy and efficiency

Any machine which does work uses energy. It takes energy in one form and converts it to other forms. For example, an electric winch uses electrical energy when it does work to raise a weight. Not all the energy used by the machine is used as work, some will be wasted. The wasted energy usually results from frictional forces in the machine which produce heat energy. Sound energy may also be produced. The energy used by the machine is converted to useful energy, enabling work to be done, and waste energy.

The efficiency of a machine

$$= \frac{\text{useful work done by the machine}}{\text{energy used by the machine}} \times 100\%.$$

Values of efficiency can never be greater than 100%. If that were not the case, a machine with efficiency greater than 100% would produce more useful energy than it was supplied with. Useful energy could therefore be produced for free! Alas, this can never be, since conservation of energy tells us that the energy output (useful and wasted) must equal the energy input. One enterprising 19th century 'inventor' in America managed to convince visitors to an exhibition that his machine could produce power for free – until they discovered a man pedalling away furiously in a hidden room next to the machine!

If energy can neither be created or destroyed, why then is there so much concern about future energy supplies? Most machines produce some waste energy in the form of heat energy. The energy is wasted because the heat energy cannot be reconverted completely back to a useful form of energy such as electrical or chemical energy. By burning oil, chemical energy of oil can be converted entirely to heat energy. Unfortunately, heat energy cannot be converted entirely into electrical or other useful energy. There is always some waste energy produced which cannot be used. Energy tends to spread out, and it is impossible to recover it entirely in useful form. For this reason, our energy resources need to be used carefully and economically.

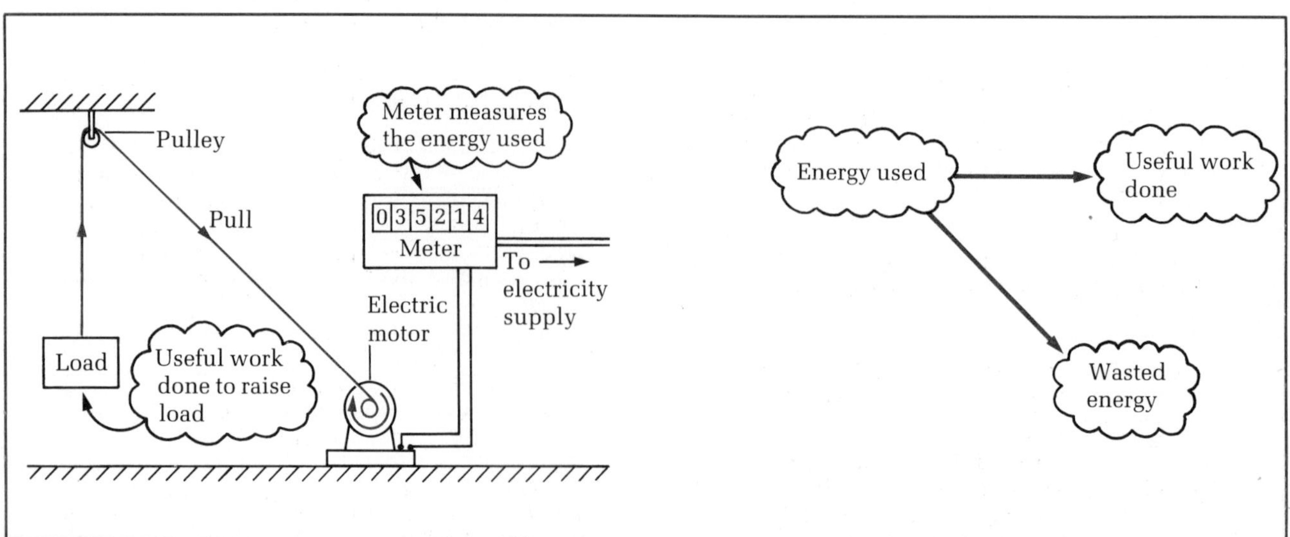

Figure 2.26 *An electric winch*

2.7 Summary

Definitions

1 *Momentum* is mass × velocity.
2 *Work done* is force × distance (moved in direction of force).
3 *Power* is the rate of doing work.
4 *Impulse* is force × time.

Equations

1 In general, force = change of momentum per second

$$F = \frac{d(mv)}{dt}$$

2 For constant mass, the above equation may be written

force = mass × acceleration
$F = ma$

3 K.E. $= \frac{1}{2}mv^2$ P.E. $= mgh$
4 Power required to move a steady force F at speed $v = Fv$.

Principles and laws

1 *Newton's 1st law* A body continues at rest or constant velocity unless acted on by a resultant force.
2 *Newton's 2nd law* The resultant force on a body is proportional to its rate of change of momentum.
3 *Newton's 3rd law* When two bodies interact, they exert equal and opposite forces on one another.
4 *Principle of conservation of momentum* The total momentum of a system is constant, provided no external resultant force acts on it.
5 *Principle of conservation of energy* Energy can neither be created or destroyed.

Short questions

Assume $g = 10\,\mathrm{m\,s^{-2}}$

2.1
a) A car of mass 600 kg accelerates from rest to $8\,\mathrm{m\,s^{-1}}$ in 40 s. Calculate the resultant force acting on the car.
b) If the car is fitted with a trailer of mass 100 kg, how long would it now take the car and trailer to accelerate to $8\,\mathrm{m\,s^{-1}}$? What would be the tension in the coupling during this time?

2.2 A packing case of mass 40 kg is pushed across a rough floor at a steady speed of $1.5\,\mathrm{m\,s^{-1}}$. When the push force is removed, the case slides a further distance of 1.2 m before stopping. Calculate
a) the friction force acting on the case when it slides,
b) the work done per second to push the case at a steady speed of $1.5\,\mathrm{m\,s^{-1}}$.

2.3 A mass of 1 kg is suspended from a 50 N spring balance, and then a further mass of 2 kg is suspended by a thread from the first mass. The thread is then burned through. What is the acceleration of each mass at the instant the thread snaps?

2.4 A ship is towed at steady speed into port by two tugs, each tug pulling the ship by a steel cable. If each cable is under a tension of 8000 N and the angle between the two cables is 60°, calculate
a) the combined force on the ship due to the two cables,
b) the 'drag' force on the ship,
c) the resultant force on the ship if one of the cables snaps.

2.5 A spring with a natural length of 300 mm hanging vertically extends by 40 mm when a 1.0 N weight is suspended from its lower end. However, in a descending lift, a smaller extension of 35 mm was measured. Calculate the acceleration of the lift.

2.6 An air rifle pellet of mass 0.001 kg travelling at a speed of $80\,\mathrm{m\,s^{-1}}$ is stopped by a wooden block. The pellet goes 25 mm into the block. Calculate
a) the time of impact,
b) the loss of momentum of the bullet,
c) the average force of the block on the bullet.

2.7 A point object of mass 5 kg, initially at rest, is acted on by a resultant force of 10 N for 4 s then it is brought to rest again by a 2 N resultant force. Calculate
a) its maximum momentum and speed,
b) the time for which it is in motion,
c) the distance travelled.

2.8 An object of mass 2.5 kg moving horizontally in a direction due north at an initial speed of $4\,\mathrm{m\,s^{-1}}$ is acted on by an applied force of 0.10 N for 20 s. The force is directed horizontally in the direction due east. Calculate
a) the velocity of the object after the applied force is removed,
b) the increase of K.E. of the object,
c) the distance moved by the object eastwards.

2.9 A jet of water from a fire hose is capable of reaching a height of 20 m. What is the minimum speed of water from the hose? Given the area of cross-section of the hose outlet is $4 \times 10^{-4}\,\mathrm{m^2}$, calculate
a) the mass of water leaving the hose each second,
b) the momentum leaving the hose each second,
c) the force on the hose due to the water jet.
The density of water is $1000\,\mathrm{kg\,m^{-3}}$.

2.10 A car of mass 500 kg accelerates steadily from rest down a steady incline at 20° to the horizontal then onto a level road. If the car takes 15 s to travel 100 m from rest to the bottom of the incline, calculate
a) the speed and K.E. at the bottom,
b) the loss of P.E.,
c) the work done against friction,
d) the average frictional force.

2.11 Two pendulum bobs, A and B, hang side by side in contact with one another at rest, each supported by a thread 2.0 m long. A has mass 0.200 kg and B has mass 0.080 kg. Bob A is then pulled to one side with its thread taut until it is 0.100 m higher. Then it is released so it collides with B. After the collision, B gains 0.150 m of height. Calculate
a) the speed of A before impact,
b) the speed of B just after impact,
c) the speed of A just after impact,
d) the height gained by A after impact,
e) the loss of K.E. as a result of the impact.

2.12 Estimate
a) the height gain of a pole vaulter running at a top speed of $10\,\mathrm{m\,s^{-1}}$ before 'launch'; how can the pole vaulter gain extra height?

b) the initial K.E. of a 1.5 kg discus, projected at 45° above the horizontal, which returns to the ground 60 m away,
c) the muscle power of a 60 kg sprinter who can reach a speed of 10 m s^{-1} in 2.5 s.

2.13 A pile driver supported by a crane is used to drive a vertical steel girder of mass 900 kg into the ground. The pile driver has an 80 kg steel 'hammer' which is raised 4.0 m above the top end of the girder then released to strike the end of the girder and drive it further into the ground. Each impact drives the end of the girder 0.4 m further into the ground. Calculate
a) the speed of the pile driver just before impact,
b) the speed just after impact,
c) the force of friction on the girder as it penetrates the ground.

2.14 A car of weight 7000 N travels at a steady speed of 8 m s^{-1} up a steady incline at 15° above the horizontal. The car's motion is opposed by a constant frictional force of 500 N. Calculate
a) the gain of P.E. per second,
b) the work done per second against friction,
c) the car's engine power.

2.15 If the car in **2.14** travelled *down* the incline with the same engine thrust and friction, what would its acceleration be?

2.16 A railway wagon of mass 1500 kg travelling at a speed of 2 m s^{-1} collides with three identical wagons initially at rest. The wagons couple together as a result of the impact. Calculate
a) the speed of the wagons after impact,
b) the loss of K.E. due to the impact. Account for the loss of K.E.

2.17 Repeat **2.16** if the three wagons were initially moving at 1 m s^{-1} in the opposite direction to the single wagon.

2.18 Coal is transported by a horizontal conveyor belt at a rate of 30 kg s^{-1} when the belt moves at a steady speed of 1.8 m s^{-1}. Calculate
a) the force needed to keep the belt moving,
b) the K.E. per second gained by coal falling onto the belt,
c) the power required to keep the belt moving at 1.8 m s^{-1}. Explain why (b) and (c) differ.

2.19 Two frictionless gliders, P and Q, on a linear air track are fitted with magnets so they repel one another. They are held near one another by a thread from one to the other, and they are positioned at rest. Then the thread is burned through so that they move away from one another. P moves away at a speed of 0.20 m s^{-1} and Q moves away at 0.12 m s^{-1}. Given the mass of P is 0.80 kg, calculate
a) the mass of Q,
b) the initial P.E. of the system,
c) the speed at which P would have moved away if Q had been held fixed when the thread was burned through.

2.20 An electric winch needs to use 200 W of electrical power to raise a 100 N load at a steady speed of 0.5 m s^{-1}. Calculate the efficiency of the winch.

3

Energy resources

3.1 Fuel supplies

Primary and secondary fuels

Modern living uses a lot of energy, about 5000 J per person every second. Work out for yourself what this amounts to in national terms for 55 million people. We get our energy from the Sun, most of it having been stored in *fuel* in the Earth over millions of years or more. The term 'fuel' is used for any material from which energy can be obtained. The Sun produces its energy by fusing light nuclei together, as explained on p. 474, and astronomers reckon its power output is about 4×10^{26} J every second. But only a tiny fraction of that huge amount falls on the Earth, insufficient to meet our demands unless vast areas of land were covered with 'solar energy' collectors. So we must use our fuel reserves such as coal and oil, built up using solar energy which arrived millions of years ago. Energy can never be created or destroyed, but we can convert it from any one form into other forms; however, fuel *is* destroyed by use. When coal is burned, the total quantity of energy afterwards (in whatever forms) is equal to the total quantity of energy initially available – but the coal has been used to release energy when its atoms combine with atoms of oxygen from the air. The products of combustion, ash and waste gases such as carbon monoxide, are of no further use as fuel.

Primary fuels are those such as coal or oil which occur naturally, and they may be used to produce **secondary fuels** such as electricity or petrol which do not occur naturally. When you boil water in an electric kettle for your cup of coffee, you need a certain amount of energy to heat and boil the water. Your fuel is electricity in this instance, and it supplies you with **useful energy** – the energy needed for the given task. But energy must be used to supply electricity, and the fuel used to produce electricity has to be obtained and transported. So your cup of coffee actually uses up a lot more energy than is required just to heat and boil the water. To supply a given amount of useful energy, the total quantity of energy needed is called the **primary energy**. The primary energy is therefore equal to the useful energy plus the energy used (and possibly wasted) to obtain and transport the fuel.

Most of the World's primary energy is obtained at present from four primary fuels which are oil, gas, coal and uranium. Figure 3.1 shows the percentage of each type of fuel used in 1982. A small percentage of our primary energy comes from **renewable resources** such as hydroelectricity. Renewable resources are continually replenished naturally without using fuel from the Earth, and make use of processes which are part of our natural environment. Renewable resources will undoubtedly become more important as oil and gas become more scarce. However, of equal importance is the efficient use of our fuel reserves.

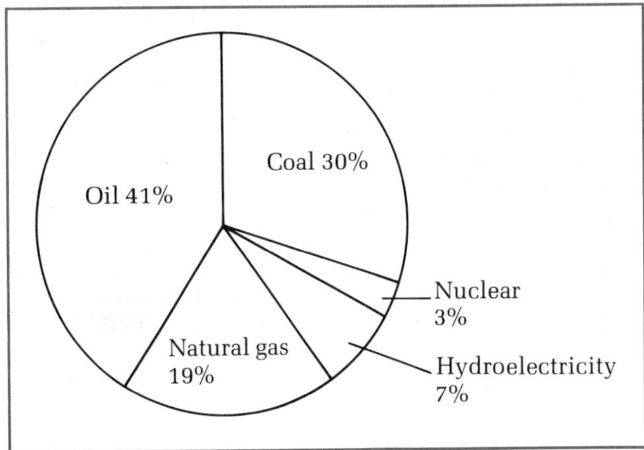

Figure 3.1 *World fuel use, 1982*

Measuring fuel

To compare fuel use and fuel reserves, a common unit is helpful if not essential. Often, each type of fuel is measured in units which have no scientific roots but are convenient for that type of fuel. So our electricity bills are in terms of *kilowatt hours*, our gas bills are in *therms*, and coal is by *weight*. To make a comparison, we need to know how many joules of energy each fuel unit would supply. For example, 1 kilowatt hour is equal to 3.6 MJ because if 1 kilowatt of electrical power is used for 1 hour, then the electrical energy supplied is 1000×3600 J. So if a 3 kW electric kettle is used for 5 minutes, the electrical energy used is 0.25 ($= 3 \times 5/60$) kilowatt hours which is equal to 0.9 MJ. Another fuel unit often used is the **tonne of coal equivalent** which is equal to 26.6 GJ ($= 26.6 \times 10^9$ J), and is defined in terms of the primary energy from 1 tonne of coal. For example, if in a certain year, the total amount of oil used in the UK was equivalent to 130 million tonnes of coal, then the total oil usage for that year is simply written as **130 Mtce** (which means 130 million tonnes of coal equivalent). The problem with fuel units such as therms or the Mtce is that each industry uses its own unit. So when you compare use of different types of fuels, be prepared for some conversion calculations. Life would be much simpler if they all used joules.

1 thousand therms = 4 tonnes of coal = 2.3 tonnes of oil = 30 000 kWh of electricity.

See if you can find recent electricity or gas bills to compare the cost of each type of energy. How many joules of electrical energy can you buy for £1? Compare this with gas.

3.2 Fuel reserves

The demand for fuel

World demand for fuel has increased by a factor of five over the past seventy years. Figure 3.2 shows that demand doubled from 1925 to about 1950, then doubled again from 1950 to about 1965.

People in the richer parts of the world use much more energy than people in the so-called 'Third World' where

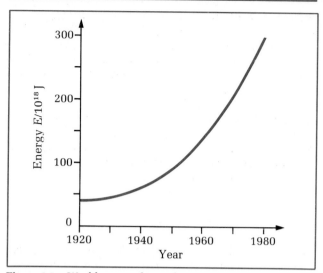

Figure 3.2 *World energy demand*

there are far fewer cars or modern homes. As countries become richer, the demand for fuel becomes greater because of the increasing use of labour-saving machines. In 1980, the world demand for energy was about 350×10^{18} J, and it could double by the year 2000. Of course predictions often turn out to be wrong but past trends are usually the best guide to the future – assuming no shortage. World demand might well be restricted by shortages in the future, but new reserves are continually being discovered with the aid of new technology.

The UK pattern of fuel use in 1983 is shown in Figure 3.3. About 40% of our energy needs was met by North Sea oil; nuclear power provided about 4%. About 35% of primary fuel used was exported or used for non-energy purposes such as making plastics or fertilizers. About 20% was used to convert primary fuels into useful energy. The remaining 45% was used either by industry (20%) or by transport and the domestic consumer.

Power stations

Most power stations in the UK are either nuclear or coal-fired or oil-fired. Nuclear power stations provided about 15 to 20% of our total electricity requirements in 1983. Electricity demand varies during the year and also during each day. Imagine how demand must leap when five million homes each switch on an electric kettle at the same time, which can and does happen at the end of popular TV programmes. Work out for yourself the amount of energy used by five million electric kettles, each using 3000 W of power for five minutes. Yet such huge surges in demand are met every day. Figure 3.4 shows how the demand for electricity changes during a typical day.

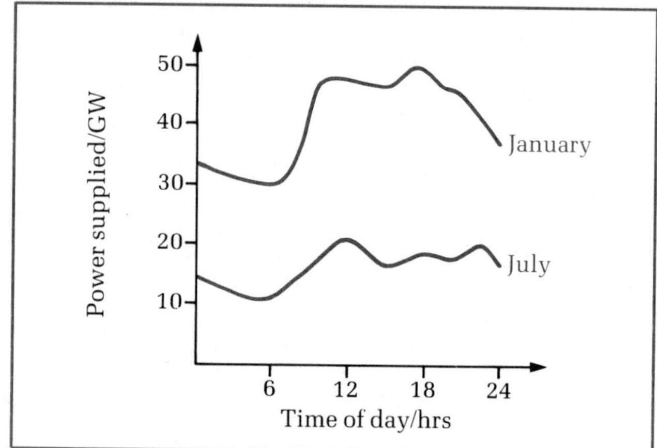

Figure 3.4 *Typical national power demand*

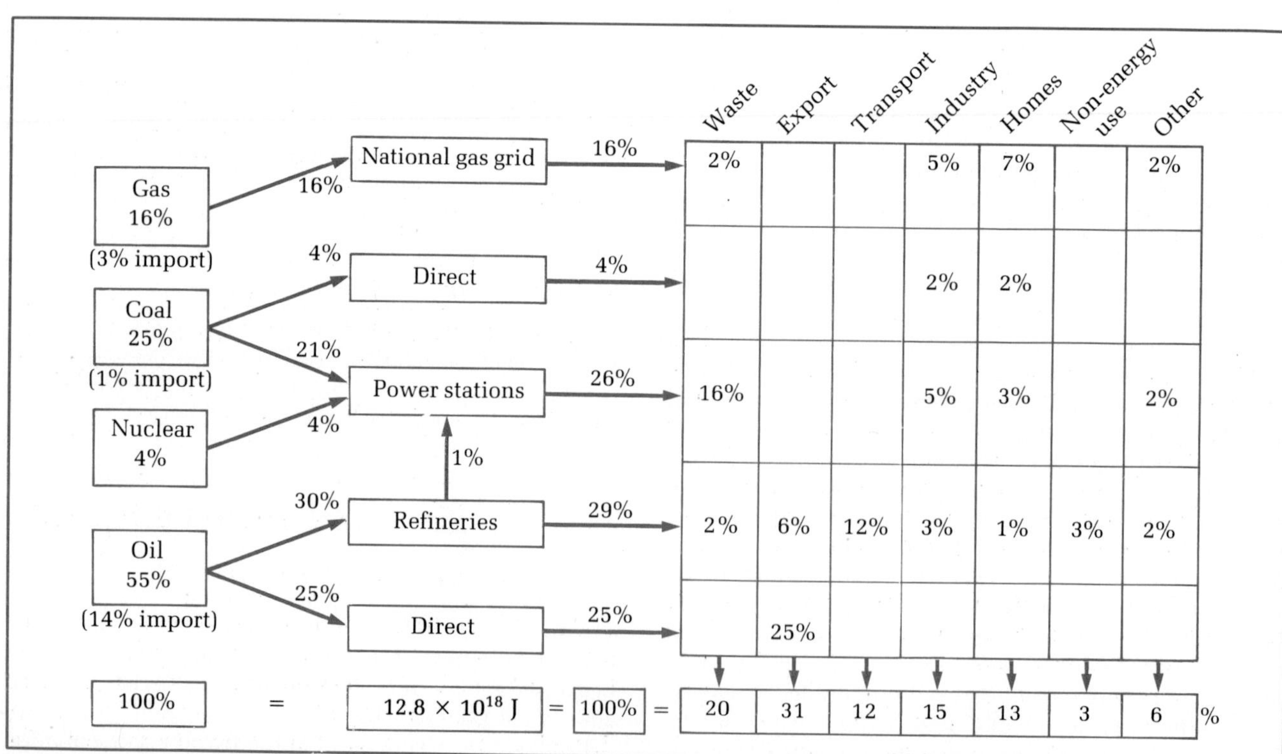

Figure 3.3 *UK energy flow, 1983*

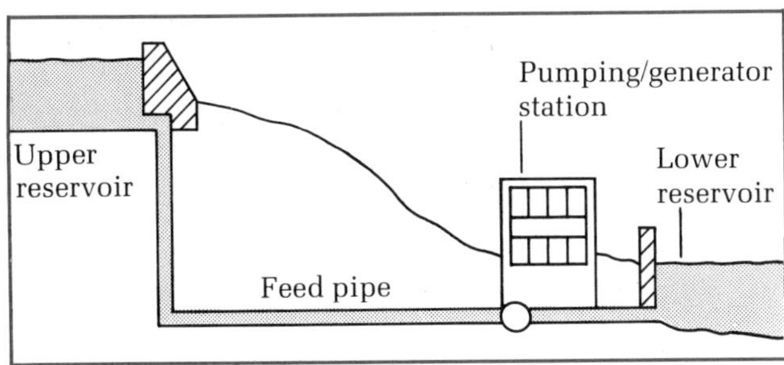

Figure 3.5 *Pumped storage*

Nuclear power stations produce energy at a steady rate, and cannot be switched off when demand is low. So when the energy which they produce is not wanted, it can be stored by pumping water uphill. Then when demand suddenly increases, the water flow is reversed and the stored energy is converted back to electrical energy. Such **pumped storage schemes** were first developed in North Wales with a 360 MW pumped storage station using four reversible turbines capable of pumping water uphill through a height gain of 300 m. The principle of this system is shown in Figure 3.5.

Worked example A pumped storage station has an efficiency of 25% for converting stored energy into electrical energy. The upper reservoir is 200 m above the lower reservoir. Calculate the volume of water which must pass through its turbines from the upper reservoir each second for a power output of 100 MW. Assume the density of water is 1000 kg m^{-3}.

Solution In one second, the electrical energy produced is 100 MJ so the potential energy loss of the water must be 400 MJ each second (since the efficiency is 25%). Now we need to calculate the weight of water which would release 400 MJ of P.E. as a result of falling through a drop of 200 m. Let the weight be W.

Therefore $W \times 200 = 400 \times 10^6$
$W = 2 \times 10^6$ N

so the mass of water must equal 2×10^5 kg, taking g as 10 m s^{-2}.

Thus the volume of water which must pass through in 1 second is equal to 200 m^3 (since volume = mass/density).

Pollution

Using fuel creates waste products which can affect us. For example, oil and coal-fired power stations produce waste gases such as sulphur dioxide which can dissolve in rain droplets to produce *acid rain*. So plant life far away from the power station can be harmed by acid rain unless the waste gases are cleaned before release. *Smog* produced by smoke in fog conditions used to be a big problem in cities before smoke control was enforced, but vehicle exhaust gases can create smog problems as well. Clearly some form of control by modifying the exhaust system is a desirable measure.

Nuclear power stations have always been strictly monitored to ensure that no radioactive waste is released into the surroundings. The used fuel from nuclear power stations is highly radioactive and very hot when removed from the reactor. So it is first stored in 'cooling ponds', and then kept in safe conditions until its radioactivity has decreased. After that, the fuel is reprocessed to separate unused uranium or plutonium from the other radioactive waste. The plutonium is actually created from uranium (in fact from the U-238 content) inside the reactor; because plutonium itself is a fuel (for the 'fast breeder' type reactor), then the plutonium waste from thermal reactors can be reused as fuel in fast breeder reactors. The other radioactive waste left after removal of uranium and plutonium must then be stored in secure conditions for many years until it is harmless.

Reserves

Known reserves of oil and gas will probably be used up by the middle of the next century. New reserves may well be discovered but coal will continue to be an important fuel long after that. Even uranium will be rapidly used if thermal reactors are not replaced by fast breeder reactors. The thermal type reactor uses only the U-235 content of uranium, and U-235 makes up less than 1% of uranium ore. But the other 99% of uranium ore is U-238 which is converted into plutonium fuel by the nuclear reactions inside the thermal reactor. Then the plutonium fuel can be used in the fast breeder reactor to produce useful energy. In this way, a much greater percentage of the uranium ore is used as fuel. Figure 3.6 shows the known reserves and use of each type of fuel. You can see for yourself just how long each type of fuel will be available, assuming no further reserves are discovered. But demand is increasing every year, so renewable resources such as hydroelectric power will become more and more important. So, too, will the more efficient use of our fuel reserves.

	World reserves	World use 1982	UK reserves	UK use 1982
Oil	147 000	4700	3–10 000	111
Gas	95 000	2200	2–4000	71
Coal	1 452 000	3300	66 000	111
Nuclear	54 000 (thermal) 2 670 000 (FBR)	300	Imported	16
Hydro	Renewable	750	Renewable	3
1 Mtce = 26.6 × 10^{15} J				

Figure 3.6 *Fuel reserves and use (in Mtce units)*

3.3 Efficiency and its limits

Energy flow diagrams

When a steam engine is used to generate electricity, more than 50% of the energy from the fuel is wasted. Figure 3.7 (on page 28) shows the energy transformations which take place. Figure 3.7 is an example of an energy flow diagram, sometimes called a **Sankey diagram.** The total energy input must equal the total energy output, but

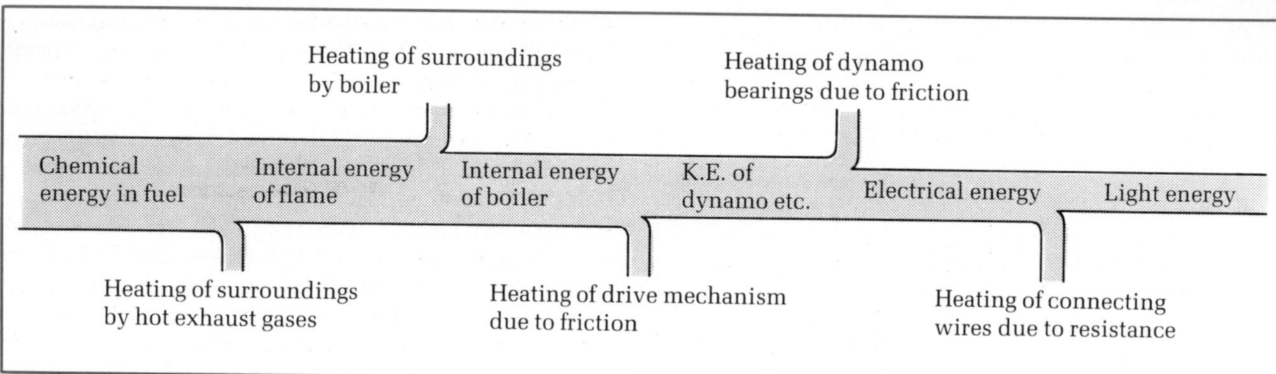

Figure 3.7 *Energy flow diagram for a steam engine used to generate electricity for lighting*

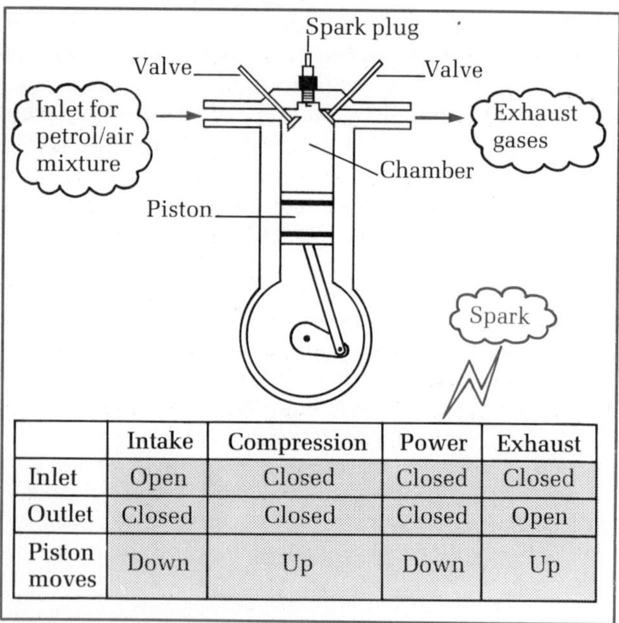

	Intake	Compression	Power	Exhaust
Inlet	Open	Closed	Closed	Closed
Outlet	Closed	Closed	Closed	Open
Piston moves	Down	Up	Down	Up

Figure 3.8 *The four-stroke petrol engine*

much of the input is converted to wasted heat energy. So a steam engine with an efficiency of 20% wastes 80% of its fuel energy. Car engines are little better with typical efficiency of 25%. Even if friction between the moving parts in the engine and transmission could be eliminated, a car engine could never be more than about 67% efficient. The reason is that the combustion of the petrol-air mixture produces heat energy as well as kinetic energy of the pistons. The heat energy is taken away by the exhaust gases and by the cooling system. Nevertheless, huge improvements in efficiencies of car engines have been made through improved lubricants and lighter car bodies.

The rules of thermodynamics

Energy can never be created or destroyed but it does seem to have a habit of spreading out and becoming less useful to us. When we burn fuel, we don't actually create energy; we simply change it from one form into other forms. The rules which govern energy flow are part of the subject of thermodynamics.

The first law of thermodynamics tells us that the internal energy of a system can be used to do work or to supply heat energy or both. We use the term 'internal energy' for the energy due to random motion of the particles of the system. The change of the internal energy must be equal to the sum of the work done and the heat supplied so our fuel can be used to do work and supply heat energy. Energy can never be created or destroyed so if the internal energy of a system falls, then the energy lost from the system must be accounted for by the work done and the heat supplied by the system.

The second law of thermodynamics is where we run into difficulty because it tells us that, no matter how hard we might try to stop it, energy always spreads out. In more exact terms, we can never convert heat energy completely back into other forms, even though other forms of energy can be converted completely to heat energy.

For example, let's suppose we have invented an ideal engine which wastes as little as possible of its fuel energy. Our engine does work by taking energy from the high temperature source created by burning the fuel. But to make energy flow from the high temperature source, we need a low temperature sink – because a temperature difference is essential for heat transfer. To keep the transfer process operating, some of the energy from the source must be allowed to flow to the sink. So only part of the energy from the source can be used by the engine to do work. The first person to realize that no engine could ever be 100% efficient was Sadi Carnot in 1824. He showed that the highest possible efficiency depended on the temperatures of the source and sink.

$$\text{Maximum efficiency} = \frac{T_1 - T_2}{T_1} \times 100\%$$

where T_1 = source temperature, in kelvins,
T_2 = sink temperature, in kelvins.

So a turbine driven by superheated steam at 700 K condensed at 300 K would have a maximum efficiency of 57% $\left(= \frac{700 - 300}{700} \times 100\%\right)$. See p. 139 for a further discussion of the second law of thermodynamics.

An example of more efficient use of fuel is provided by the development of **combined heat and power (CHP) stations**. These are electricity generating stations where waste heat energy is delivered to local buildings. In many power stations, waste heat energy is released into the atmosphere. Cooling water taken from nearby reservoirs is used to condense the steam to water at about 30°C

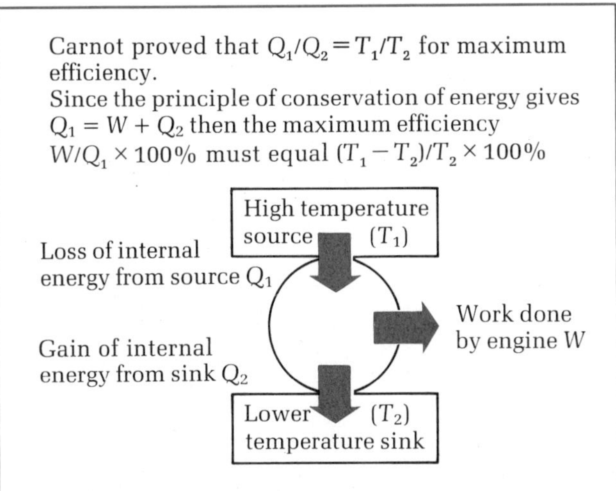

Carnot proved that $Q_1/Q_2 = T_1/T_2$ for maximum efficiency.
Since the principle of conservation of energy gives $Q_1 = W + Q_2$ then the maximum efficiency $W/Q_1 \times 100\%$ must equal $(T_1 - T_2)/T_2 \times 100\%$
High temperature source (T_1)
Loss of internal energy from source Q_1
Work done by engine W
Gain of internal energy from sink Q_2
Lower temperature sink (T_2)

Figure 3.9 *Carnot's ideal engine*

before being pumped back to the reservoir. In a CHP station, the condensation occurs at about 70°C so the cooling water is much hotter and can be piped to local buildings to provide heating. The efficiency of CHP turbines is reduced a little by condensing the steam at 70°C instead of 30°C – but the huge amount of otherwise wasted heat energy can be used.

3.4 Energy options

Oil and gas

Oil reserves are being used up at a fast rate, and even though new reserves are continually being discovered, shortages are predicted. Yet oil is essential to make such things as fertilizers and plastics – so is it sensible to use scarce oil resources for transport and heating purposes? Reserves of natural gas, too, are dwindling so a return to so-called 'town gas' produced from coal is a possibility.

Coal

Estimated coal reserves are based on 10% recovery, so improved mining methods would make our reserves even greater. Britain is literally built on coal and new coalfields such as those at Selby are being developed to meet future demands. The Selby coalfield covers an area of over 200 km² and is thought to have 600 million tonnes of recoverable coal which is to be extracted at a rate of 10 million tonnes per year. Coal will undoubtedly remain a major source of energy well into the 21st century.

Nuclear power

Uranium reserves are sufficient to meet future demand for 50 years or so if thermal reactors only are used. Thermal reactors use only uranium-235 as fuel, and natural uranium contains only 1% of U-235. The rest is U-238 which is converted into plutonium when fuel rods of natural uranium are placed inside thermal reactors. The U-235 content of the fuel rod is fissioned (see p. 471) to provide energy but the U-238 content is converted to plutonium which doesn't fission in a thermal reactor. Specially designed reactors called *fast breeders* can fission plutonium, so if fast breeder reactors are developed, the world reserves can be made to last much longer. Used in a fast breeder reactor, 1 tonne of natural uranium can provide the same amount of energy as 1 million tonnes of coal. **Fusion power** plants are still at the design stage, trying to recreate the conditions which exist in the Sun's core. Temperatures greater than 10 million kelvins are required, but the benefits will be enormous because the fuel is mostly hydrogen – available by the bucketful from the sea! And the waste products produced by fusing light atoms are non-radioactive. See p. 475 for further details.

Hydroelectric power

Lots of rainfall in the mountains is the essential requirement for hydroelectricity. Hydroelectric power stations in the UK supplied about 0.5% of our primary energy needs in 1983 (compared with nuclear power stations which supplied about 4%). How much potential energy do you think would be released by rainwater to a depth of 1 cm over an area of 1 km² running downhill, by a vertical distance of 500 m?
The volume of rainwater involved equals $1000 \times 1000 \times 0.01\ \text{m}^3$ (= area × depth).
So the mass of rainwater equals 1000×10^4 kg (= density of water × volume).
So the P.E. released must equal $10^7 \times 10 \times 500$ J ($= mgh$) giving 5×10^{10} J equivalent to about 2 tonnes of coal!

Now find an atlas which gives the annual rainfall over the mountains, and estimate how much P.E. is released each year.

Geothermal power

The Earth's interior is still very hot and gives an average heat flow of $0.1\ \text{W m}^{-2}$. But in some parts of the Earth, the heat flow has increased the temperature of underground rock basins to 200°C or more. By pumping water down to the hot rocks, steam can be drawn back up to generate electricity. A few sites in the UK have been pinpointed, and could give the equivalent of several million tonnes of coal.

Tidal power

The tide rises and falls twice each day. By trapping each high tide and allowing it to flow out through generators, tidal energy can be converted to electrical energy. The most promising site in Britain is the Severn Estuary because the incoming tide is reflected along its banks, causing its height to increase further.

Let's estimate the energy available by allowing water to fall 10 m, if the area behind the tidal barrier is 10 km by 10 km. The volume of water is $10\,000 \times 10\,000 \times 10\ \text{m}^3$ so the P.E. loss when its level falls by 10 m is 5×10^{13} J, assuming the density of water is $1000\ \text{kg m}^{-3}$ (it is actually a little greater for sea water). If the level drops by 10 m, the centre of gravity must fall by 5 m. The loss of P.E. happens once every 12 hours, so how much power would be available from such a scheme?

Solar power

Solar water panels are useful for home heating. Each panel has a blackened surface which absorbs energy directly from solar radiation. Water passes over the surface so the water is heated up. In summer, water temperatures can reach 70°C in solar panels. Each square metre can receive up to 1400 J every second from solar radiation, so the larger the panel area, the more power it can produce.

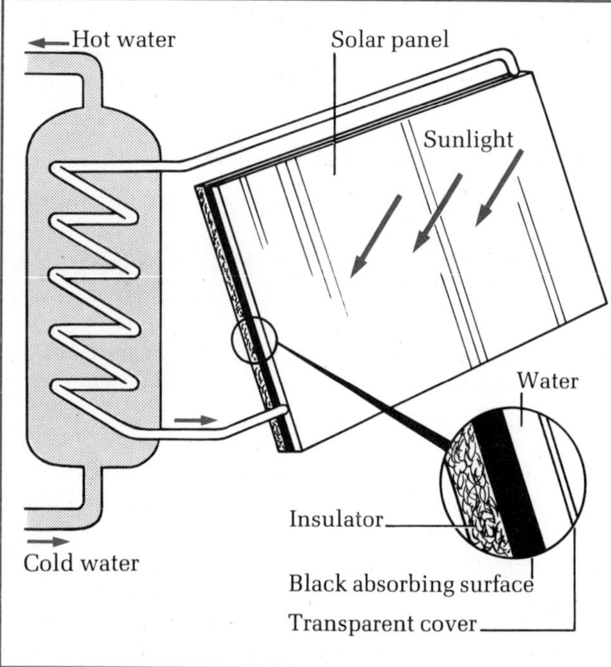

Figure 3.10 *A solar panel*

Worked example Estimate the maximum power available from 10 m² of solar panels and calculate the volume of water per second which must pass through if the inlet and outlet temperatures are 20°C and 70°C. Assume the water carries away energy at the same rate as the maximum power available. The specific heat capacity of water is 4200 J kg^{-1} K^{-1}.

Solution Each square metre can receive 1400 J s^{-1}, so the maximum power available from the solar panels must equal 14 kW (= 1400 × 10 J s^{-1}).

Let the mass of water passing through each second be m.

Using $Q = mc\Delta\theta$ (see p. 110 if necessary), then

$$14 \times 10^3 = m \times 4200 \times (70 - 20)$$

so $$m = \frac{14 \times 10^3}{4200 \times 50} = 6.67 \times 10^{-2}\ \text{kg}$$

Since the density of water is 1000 kg m^{-3}, then the volume per second is given by mass/density.

Hence volume per second of water = 6.67×10^{-5} m³.

Solar cells are thin wafers made from silicon, and they convert solar energy directly into electrical energy. Electrons in the silicon gain energy from light photons (see p. 382 to create a voltage. Solar cells are used to power satellites which have large panels of solar cells which are kept facing the Sun. Solar power reaching the Earth amounts to 1.4 kW m^{-2}; work out for yourself what area of cells would be needed for a 100 MW solar cell power station.

Wind power

Windmills have been used for centuries but their modern equivalent, the **aerogenerator**, looks rather different. Sited in windy coastal areas, single generators are capable of producing up to 1 MW or more. Let's consider how much kinetic energy we can get from wind.

Start with K.E. = $\frac{1}{2}mv^2$, so for air of density ρ moving at speed v, its K.E. per unit volume = $\frac{1}{2}\rho v^2$.

Consider an aerogenerator like the one pictured in Figure 3.11. The blades sweep out an area A in one turn, so in one second, the volume of air passing through the area is vA (since a 'cylinder' of air of length v and area A passes through in one second). So in one second, the K.E. passing through the area is given by

K.E. per second
= K.E. per unit volume × volume per second
$= \frac{1}{2}\rho v^2 vA = \frac{1}{2}\rho v^3 A$

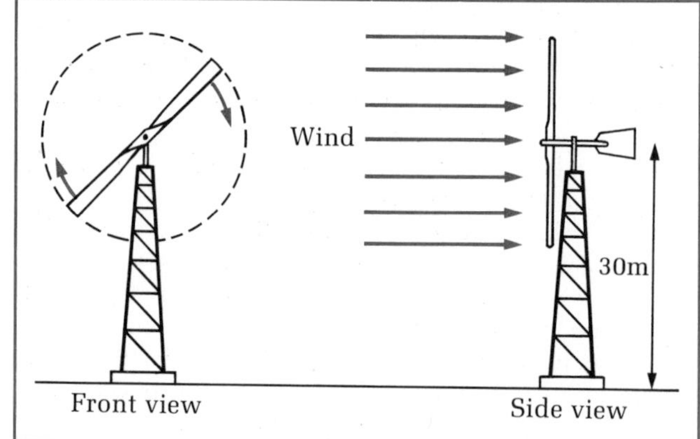

Figure 3.11 *An aerogenerator*

The idea of an aerogenerator is to convert the K.E. of the wind into electrical energy. So high wind speeds and long blades give high power. Estimate the maximum power available for a wind speed of 10 m s^{-1} and blade diameter of 50 m. Assume the density of air is 1 kg m^{-3}.

With up to 10 000 suitable sites around the coast of Britain, wind power could provide a significant contribution to our energy needs.

Wave power

Scientists reckon that up to 70 kW of power can be obtained from each metre length of wavefront along certain parts of the UK coastline. To convert wave energy into electrical energy much ingenuity is called for. One reasonably successful device is called the 'Salter duck', designed several metres high to float partly submerged. The wave energy makes one part move relative to the rest, and the relative motion is used to generate electricity. The chief problem is the unpredictable nature of the waves.

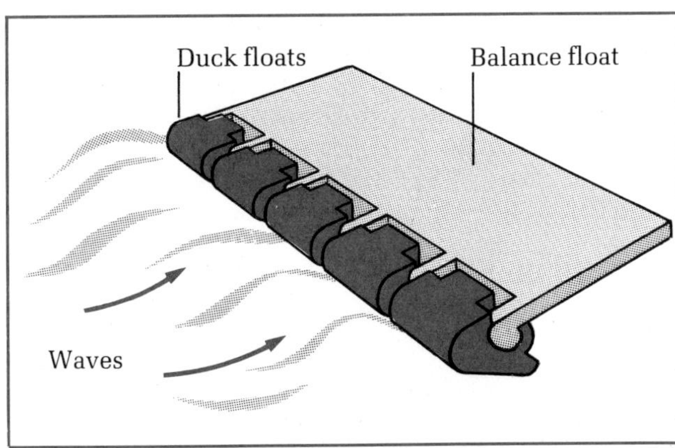

Figure 3.12 *Wave energy*

3.5 Summary

1 *Primary energy* is the total energy needed to obtain and supply useful energy.
2 *Fuel* is the material which energy can be obtained from. Fuel is a non-renewable resource because once used it cannot be reused.
3 *The second law of thermodynamics* tells us that the energy from fuel can be converted totally into heat energy but we can never convert heat energy totally into other forms of energy.
4 *Renewable resources* offer the possibility of energy supplies which are continually replenished naturally.
   a) Solar power is provided directly by solar radiation.
   b) Tidal power is produced by the gravitational pull of the Moon on the Earth's seas as the Moon orbits the Earth.
   c) Hydroelectric, wind and wave power are all produced by the Earth's atmosphere acting like a huge engine driven by the Sun's energy.

Short questions

Assume $g = 10\,\text{m}\,\text{s}^{-2}$

3.1 Draw a Sankey diagram representing the energy changes that take place when a battery-powered vehicle travels uphill at steady speed.

3.2 A therm of gas gives the same energy as 30 kilowatt hours of electricity. Look at a recent gas bill and compare it with a recent electricity bill in terms of cost for equal amounts of energy.

3.3 A coal-fired power station has an output of 100 MW. Given that its efficiency is 45%, how much coal must be supplied each day? Assume 1 tonne of coal gives 3×10^{10} J of energy.

3.4 Calculate the energy required to transport 1000 tonnes of oil along a 100 km pipeline, given that 0.05 kW hours of energy is used to shift each tonne of oil along each km of pipeline. Given that 1 tonne of oil releases 4.2×10^{10} J if burned, what percentage of the total energy available from 1000 tonnes of oil is used to shift the oil along the pipeline?

3.5 A hydroelectric power station has efficiency of 25%. The water driving the turbines falls through a height of 300 m before reaching the turbines. Calculate the volume of water that must pass through the turbines each second to give a power output of 2 MW. Assume the density of water is $1000\,\text{kg}\,\text{m}^{-3}$.

3.6 The solar energy flux near the Earth is $1.4\,\text{kW}\,\text{m}^{-2}$. A solar power station consists of concave mirrors that focus sunlight onto a steam boiler. What must be the minimum mirror area to give an output of 1 MW, assuming 100% efficiency? Why, in practice, should the mirror area be greater?

3.7 A solar panel attached to the roof of a house is used to heat water from 5°C to 40°C. If the water flows through the panel at a rate of $0.012\,\text{kg}\,\text{s}^{-1}$, calculate the heat gained per second by the water. Assume the specific heat capacity of water is $4200\,\text{J}\,\text{kg}^{-1}\,\text{K}^{-1}$.

3.8 An aerogenerator has a power output that is proportional to (wind speed)^2, and its efficiency varies with wind speed. On a day when there is a steady wind of speed $9\,\text{m}\,\text{s}^{-1}$, the power output is 40 kW operating at an efficiency of 20%. If the wind speed on the next day is $13.5\,\text{m}\,\text{s}^{-1}$ and the efficiency increases to 25%, what is the new power output?

3.9 Estimate the energy released from a tidal power station if 100 km^2 of water is raised to a height of 1.5 m by the tide behind a tidal barrier. What would be the mean power output of such a station if its efficiency is 25% and there are two tides per day?

3.10 An open boat of width 1.0 m has a total weight of 3000 N. Used near a beach, it bobs up and down through 0.5 m once every 5 s. Calculate the loss of P.E. every time it drops from a crest to a trough. Hence estimate the mean power available per metre of beach waves.

4

Statics

4.1 Equilibrium of forces

Imagine the huge forces which must act on each section of a suspension bridge, as in Figure 4.1. Yet the bridge is designed so that all the forces on each section exactly balance each other out. Each section is then in **equilibrium** because there is no resultant force on it. Any object at rest or moving at constant velocity is said to be in equilibrium because there is no resultant force. If at rest, the object is said to be in **static equilibrium**.

Figure 4.1 *The Humber suspension bridge*

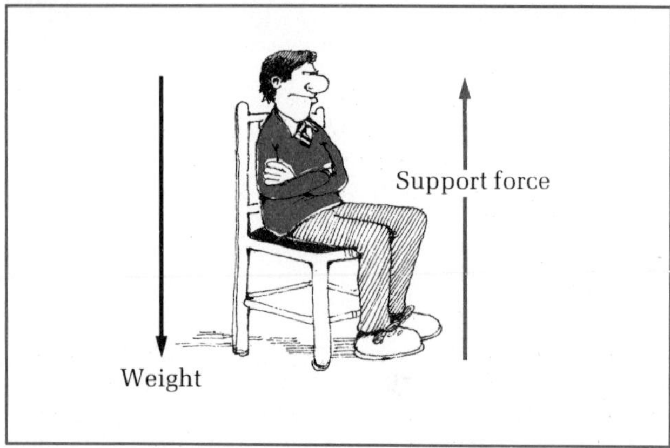

Figure 4.2 *Equal and opposite forces*

A simple form of equilibrium is where an object is acted upon by two forces only. For example, when sitting on a chair, you are in equilibrium because the support force from the chair is equal and opposite to your weight. Force is a vector quantity, so in this example the two force vectors are exactly equal and opposite to one another. Thus, the vector sum of the two forces, the resultant force, is zero.

Where more than two forces act on an object, the situation is more complicated. Imagine the object is small enough to be treated as a point object. All the forces must act at that point, and in equilibrium they balance one another out. In other words, the vector sum of all the forces must be zero, i.e. the resultant force must be zero. Figure 4.3 shows how three forces acting on a point O may be arranged to balance one another out. The reading of each spring balance gives the force pulling in that direction on point O.

Let the three forces be represented by force vectors, $\mathbf{F}_1$, $\mathbf{F}_2$ and $\mathbf{F}_3$ as in Figure 4.3. Because the force vectors add up to zero, then $\mathbf{F}_1+\mathbf{F}_2+\mathbf{F}_3=0$. This equation can be seen in a different way by using the **parallelogram rule** to add two vectors together, as in Figure 4.4. To add $\mathbf{F}_1$ and $\mathbf{F}_2$, the two vectors are used to form adjacent sides of a parallelogram. The parallelogram is then completed from the two vectors. The resultant, $\mathbf{F}_1+\mathbf{F}_2$, is given by the diagonal from the common origin of the two vectors. In fact, the resultant $\mathbf{F}_1+\mathbf{F}_2$ ought to be equal and opposite to the third force vector $\mathbf{F}_3$. This is because the three forces balance one another out. So $\mathbf{F}_1+\mathbf{F}_2=-\mathbf{F}_3$ which gives $\mathbf{F}_1+\mathbf{F}_2+\mathbf{F}_3=0$ as before.

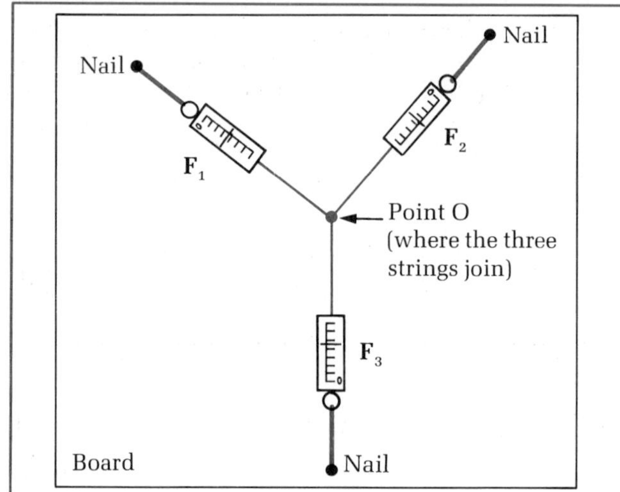

Figure 4.3 *Investigating equilibrium*

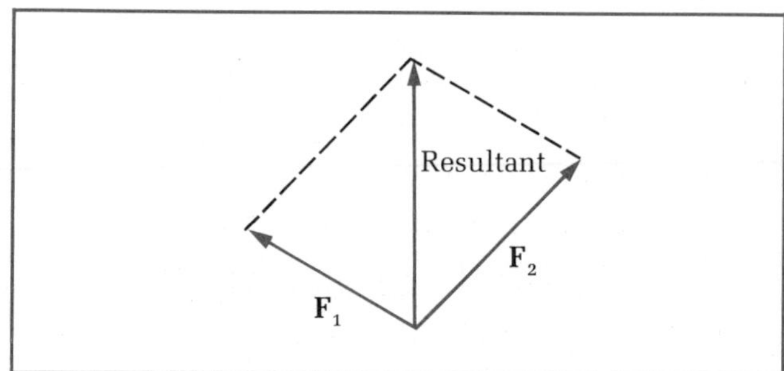

Figure 4.4 *The parallelogram rule*

Investigating equilibrium of point objects

Use an arrangement like the one shown in Figure 4.3. Try three spring balances pulling on a single point in different directions. On the sheet of paper under the strings, mark the point at which the strings are joined. The three spring balances pull on this point. Mark the direction of each string from the point. Note the reading of each spring balance. Remove the sheet of paper, and draw the three vectors representing the pull force of each spring balance. Then use the parallelogram rule to show that any two of the three force vectors give a resultant which is equal and opposite to the third force vector.

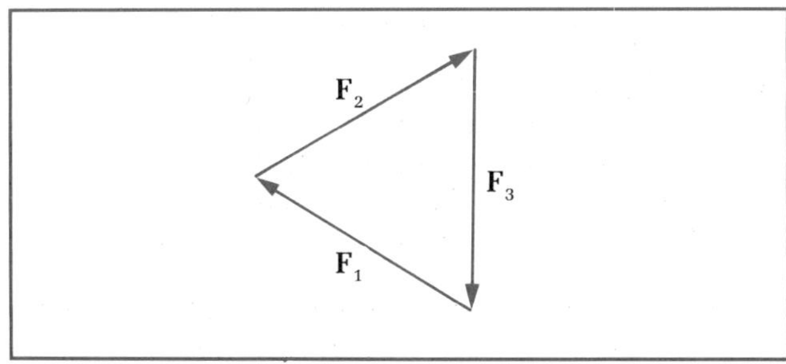

Figure 4.5 *Triangle of forces*

Alternatively, show that the three force vectors form a triangle when drawn as in Figure 4.5.

Now try more than three spring balances pulling on a single point. Use a fresh sheet of paper under the strings, and mark the force directions as before. Note the readings of the spring balances, and then remove the sheet to make a vector diagram. This time, show that all the force vectors make up a closed figure. In fact, by taking the force vectors in order as you move round the fixed point, the closed figure ought to make up a polygon. Figure 4.6 shows the vector polygon for five forces pulling on a fixed point.

For any point object acted upon by several forces, it is in equilibrium only if the force vectors form a closed polygon.

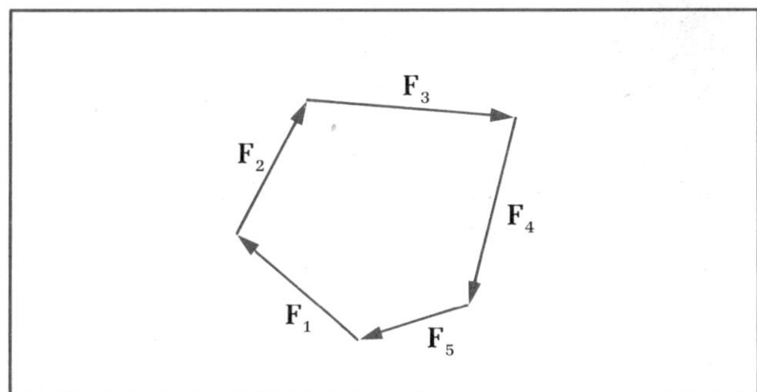

Figure 4.6 *The closed polygon*

Resolving forces is another way of checking if forces on a point object balance out. Any force may be resolved into two perpendicular components. Essentially, this means that the force vector may be replaced by two components at right angles to each other. Suppose one of the components is to be parallel to a line at angle θ to the line of action of the force, as in Figure 4.7. The other component is therefore at right angles to the first line. The parallel component is $F \cos\theta$, and the perpendicular component is $F \sin\theta$.

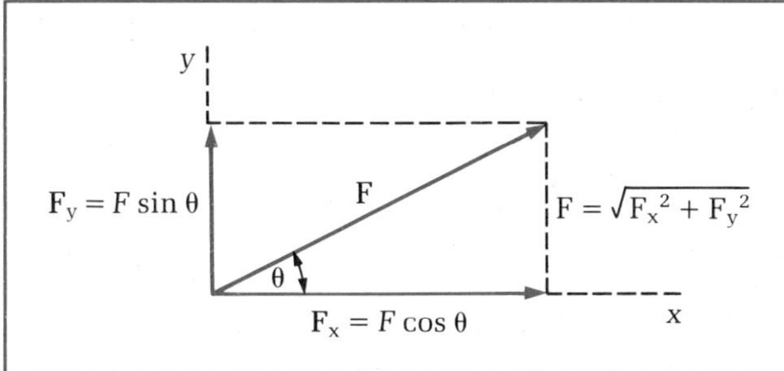

Figure 4.7 *Resolving a force*

Now consider a point object acted upon by several forces, as in Figure 4.5 where there are three forces involved. Resolve each force vector into two perpendicular components along and at right angles to a given line. This is shown in Figure 4.8 where the given line is the x-axis, so each force is resolved into an x-component and a y-component.

	x-component	*y*-component
Force F_1	$-F_1 \cos\theta_1$	$+F_1 \sin\theta_1$
Force F_2	$+F_2 \cos\theta_2$	$+F_2 \sin\theta_2$
Force F_3	0	$-F_3$

where θ_1 and θ_2 are as in Figure 4.8.

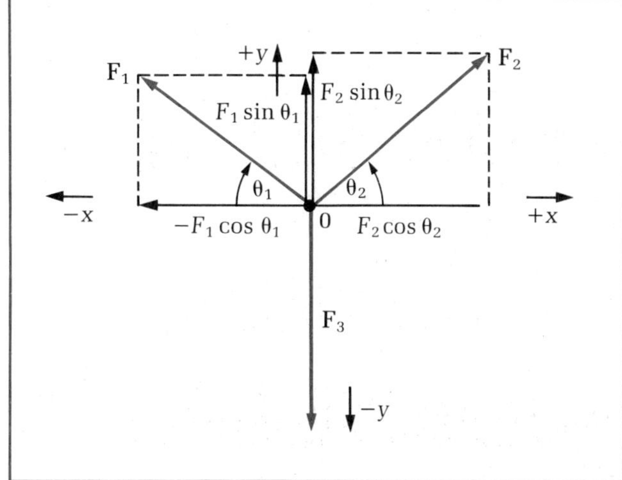

Figure 4.8 *Equilibrium*

If the forces balance out, then their x-components must balance out and their y-components must balance out separately. For the example above, this means that

a) the x-components balance out:
$-F_1 \cos\theta_1 + 0 + F_2 \cos\theta_2 = 0$
so $F_1 \cos\theta_1 = F_2 \cos\theta_2$;

b) the y-components balance out:
$F_1 \sin\theta_1 - F_3 + F_2 \sin\theta_2 = 0$
so $F_3 = F_1 \sin\theta_1 + F_2 \sin\theta_2$.

If the forces *do not* balance out, their resultant has an x-component given by adding the individual x-components, and it has a y-component given by adding the individual y-components. For the example above, the resultant has an x-component of $-F_1 \cos\theta_1 + F_2 \cos\theta_2$ and a y-component of $F_1 \sin\theta_1 - F_3 + F_2 \sin\theta_2$. The size of the resultant can then be determined by Pythagoras' rule, as shown in Figure 4.7.

Worked example A point object O is acted upon by a 6 N force and an 8 N force at 30° to one another. What single force will balance these two forces out?

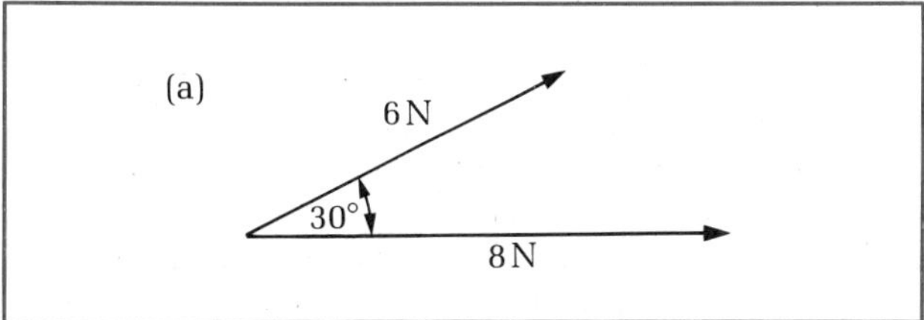

Figure 4.9a

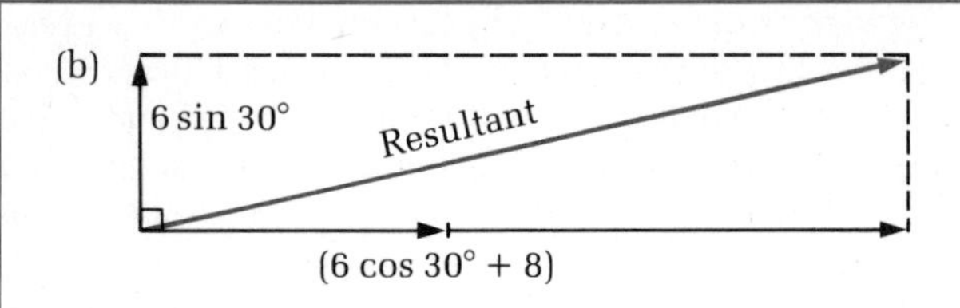

Figure 4.9b

Solution Resolve the 6 N force along and perpendicular to the 8 N force, as in Figure 4.9(b). Thus the resultant of the two forces, R, has a component $(8+6\cos 30°)$ along the line of the 8 N force, and a component of $6 \sin 30°$ at right angles to it. The size of the resultant is therefore given by

$$R=\sqrt{(6\sin 30°)^2+(8+6\cos 30°)^2}=13.5\text{ N}$$

The resultant R makes an angle θ to the line of the 8 N force, where θ is given by

$$\tan\theta=\frac{6\sin 30°}{8+6\cos 30°} \text{ which gives } \theta=12.8°=12°\,48'.$$

The third force must be exactly equal and opposite to the resultant of the 8 N and 6 N forces. Its size must therefore be 13.5 N, and its direction 180° from θ.

4.2 Moments and turning forces

Figure 4.10 *Turning effects*

When a small yacht is blown by a crosswind, it is essential for the crew to lean outwards to balance the turning effect. The force of the crosswind tries to turn the yacht over but its effect is counterbalanced by the crew.

Any body, other than a point object, acted upon by forces can be turned. If the body is to stay in equilibrium, the turning effects of the different forces must balance.

The turning effect of a single force depends not just on the size of the force and its direction; it also depends on the distance from the point of application to the pivot (i.e. to the point about which turning would take place). Try to lever a nail from a piece of wood, as in Figure 4.11, using a claw hammer. The longer the handle, the easier it is to remove the nail when you apply a force to the end of the handle.

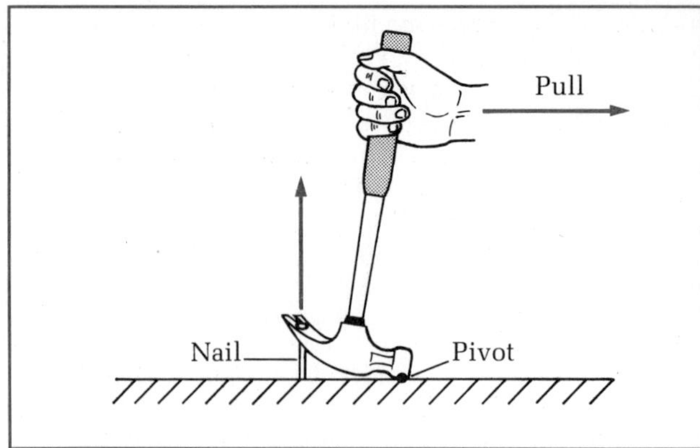

Figure 4.11 *Using moments*

The **moment** of a (turning) force about a given pivot is defined as follows.

Moment of a force = force × perpendicular distance from the pivot to the line of action of the force.

The unit of moment is the **newton metre** (N m).

Consider the example of a tower crane, as shown in Figure 4.12. You may well have watched a tower crane in operation. Steel lattice sections form the actual tower, which supports a horizontal jib. A hoist on the jib is used to raise the load. The hoist is on a trolley which can travel along the jib controlled from the driver's cabin. To counterbalance the jib and load, a concrete block is suitably positioned near the far end of the jib.

What happens when the hoist lifts a heavy load off the ground? The load would pull the jib off the tower if the jib were not firmly fixed to the tower. But the joints between the jib and tower must take the strain. The steel 'tie-cables' are important here. The cables help to take the strain by pulling downwards on the top of the tower. Thus downward thrust acts through the system where the jib is connected to the tower. So the turning effect of the load is reduced by the use of the cables.

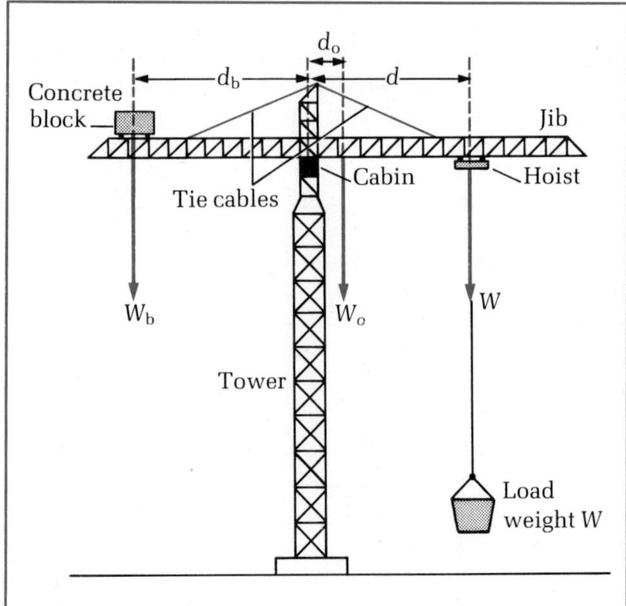

Figure 4.12 *A tower crane*

The position where the jib is fixed to the tower is like a pivot. As shown in Figure 4.12, the load weight tries to turn the jib clockwise about the pivot. So, too, does the jib's own weight since it is not balanced at its centre. The jib is stopped from turning clockwise by the concrete block and cables. The block provides a moment which tries to turn the jib anticlockwise. When the load changes, the tension in each cable alters to maintain the balance.

The principle of moments states that for any body in equilibrium, the sum of the clockwise moments about any pivot must equal the sum of the anticlockwise moments about that pivot.

For the tower crane of Figure 4.12, assuming that the cables have equal tension,

1 the load weight W has a clockwise moment $W \times d$ about the pivot,
2 the jib weight W_o has a clockwise moment $W_o \times d_o$ about the pivot (the distance d_o is the distance from the pivot to the centre of gravity of the jib; see p. 36 if necessary),
3 the concrete block weight W_b has an anticlockwise moment $W_b \times d_b$ about the pivot. Hence in balance, $W_b \times d_b = W \times d + W_o \times d_o$.

A simpler case than Figure 4.12 is shown in Figure 4.13, in which there are only two forces in balance, and $Wd = W_b x$.

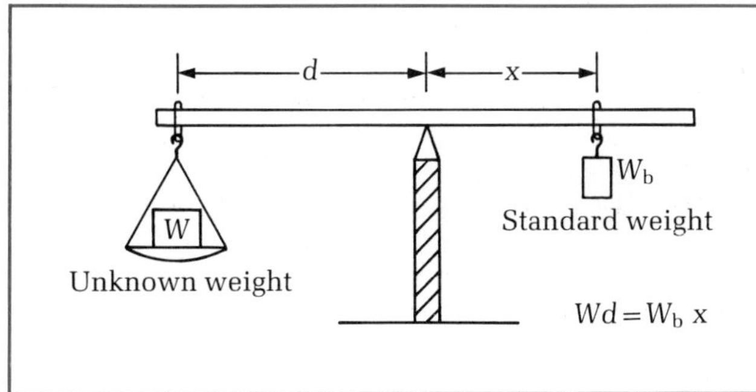

Figure 4.13 *The beam balance*

There are many applications of using moments around us, even though they may not involve calculations. Changing the wheel of a car for example may involve using moments at several stages, from jacking the car up to unscrewing the wheel's locking nuts, as in Figure 4.14.

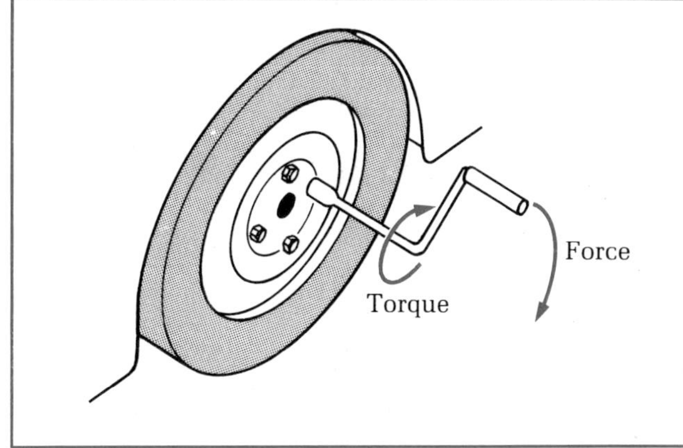

Figure 4.14 *Applying a torque*

Where a turning force is used to try to turn an object about an **axis**, the moment of the force is often described as a **torque**. Torque is therefore calculated in the same way as moment from 'force × distance'. An electric motor driving a machine at steady speed is said to exert a constant torque because its drive belt pulls on the machine with a constant turning force.

A couple is a pair of equal and opposite forces acting on a body along different parallel lines of action. Figure 4.15 shows a couple acting on the coil of a moving coil meter, caused by the force on each side when current passes round the coil.

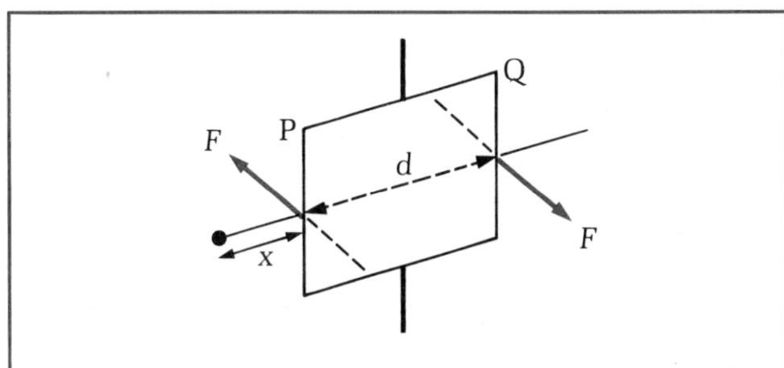

Figure 4.15 *A couple*

The moment of a couple about any point = force × shortest distance between their lines of action

It makes no difference which point you choose to take moments about; the moment of a couple is the same about any point. For example, in Figure 4.15, take moments about an arbitrary point X at distance x from one of the sides as shown. The moment about side P is Fx; but the moment about side Q is $F(d+x)$ in the opposite turning direction. So the total moment is $F(d+x) - Fx$ which equals Fd! The total moment of a couple, Fd, does not depend on the point you choose to take moments about.

Work is done by a turning force only when it moves. The turning forces acting on a body in equilibrium do no work because the body does not turn. If, however, the balance of moments changes so that the sum of the clockwise moments no longer equals the sum of the anticlockwise moments, then the body will turn. Where a body is turned by a single turning force F, as in Figure 4.16, the distance *moved* by the turning force depends on the angle $\Delta\theta$ which the body turns through. If d is the shortest distance from the axis to the line of action of the force, then the distance moved by the turning force is the arc length $d\Delta\theta$. Hence the work done by the turning force is $Fd\Delta\theta$ which is the same as the torque (Fd) × the angle turned ($\Delta\theta$).

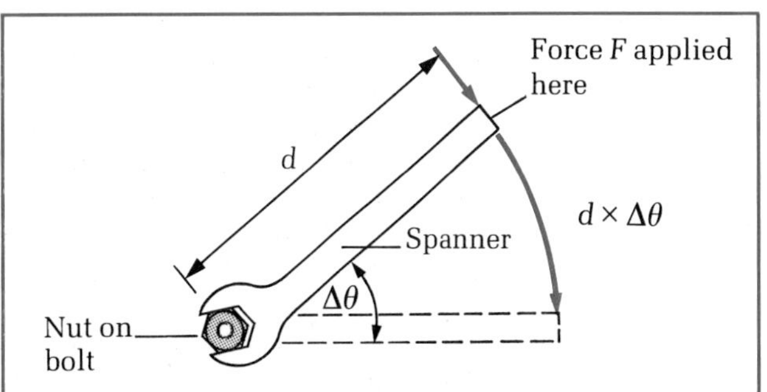

Figure 4.16 *Torque at work*

Work done by a turning force = torque × angle turned through

In equilibrium, the clockwise moments balance the anticlockwise moments, and no work is done. The unit of moment (and torque) is the newton metre. It would be misleading to use the joule as the unit of moment or torque because no work is done in equilibrium.

4.3 Centre of gravity

An object is said to be **top heavy** if it easily topples over. A vital requirement of the design of a double decker bus is to make sure it will not topple over, even when travelling around very sharp bends. The designers must ensure that the heavy parts of the bus, such as the engine, are as low as possible. In other words, the centre of gravity should be as low as possible for greatest stability.

The centre of gravity of a body is the point where its weight is considered to act.

A body in equilibrium is acted upon by two or more forces which balance out. The weight of the body is usually one of these forces – unless it is beyond the Earth's gravity. The idea of centre of gravity enables the weight to be treated as any other force would be, with a known line of action, and applied at a given point.

Consider the example of the shelf in Figure 4.17. The moment of the tension in the strut is balanced by the moment of the weight about the hinge. How can the weight's moment be determined? This can be done by considering the entire weight of the shelf acting at its centre of gravity. Because the shelf is of uniform width, its centre of gravity is midway between the hinge and the far edge. So the moment of the weight about the hinge is $\frac{Wd}{2}$ where d is the shelf width and W is its weight. Using the principle of moments therefore gives $Td\sin\theta = \frac{Wd}{2}$ for the tension T.

The centre of gravity of the shelf in the last example is the point at which the shelf, removed from the wall, can be balanced with a single force. Its centre of gravity can be located by finding the point on the shelf at which it can be supported by just one finger! Try the same with a flat board or a plate. Support the plate at different points with one finger, until you find the point at which it balances. The point of balance must be the centre of gravity since the weight has no turning effect about that point. When balanced at its centre of gravity, those parts of the plate on one side of the centre of gravity exactly balance the parts on the other side. The entire weight of the plate is supported by a single force applied at the centre of gravity.

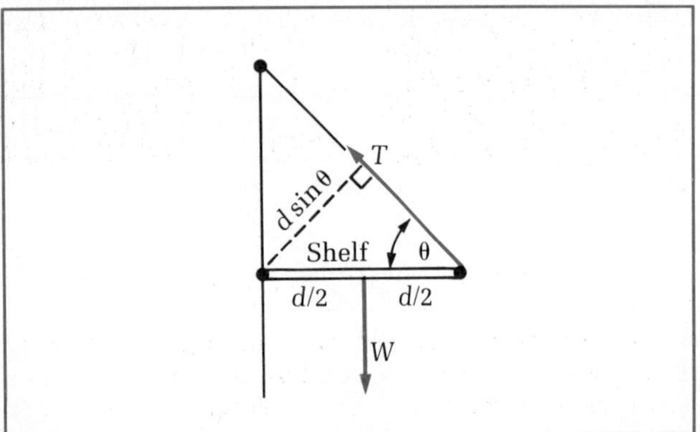

Figure 17 *Using the principle of moments*

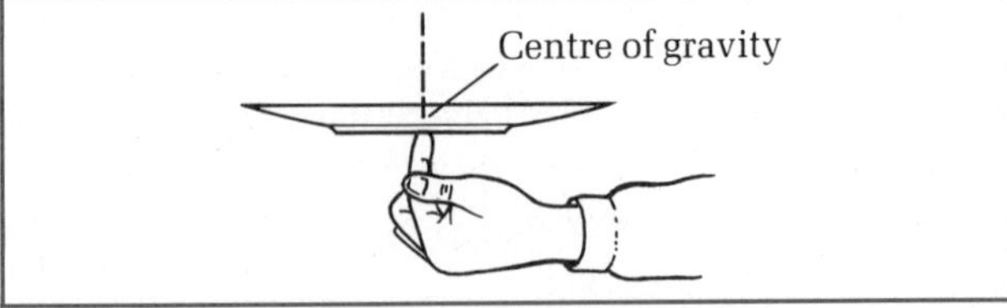

Figure 4.18 *Balancing*

The **stability** of a body depends on its centre of gravity. If you have ever tilted over whilst sitting on a high stool, you will be aware that if you tilt too far, you fall over! That is what happens when your centre of gravity goes beyond the pivot. This idea is used to test the stability of buses and other high-sided vehicles. The vehicle is put on an adjustable slope, as shown in Figure 4.19. The vehicle is tilted more and more until it is just on the point of overturning. At this point, the centre of gravity is directly above the point of turning. Up to this point, the line of action of the weight through the centre of gravity passes through the vehicle base, so the weight gives a moment which returns the vehicle to equilibrium. When turned too far, the weight gives a moment which topples the vehicle onto its side.

Figure 4.19 *Testing the stability of a bus*

Worked example A uniform metre rule is pivoted at a distance of 0.20 m from one end. It is then balanced horizontally by adjusting a 2.0 N weight between the pivot and the near end. At balance, the 2 N weight was positioned at 0.05 m from the end. Calculate the weight of the rule.

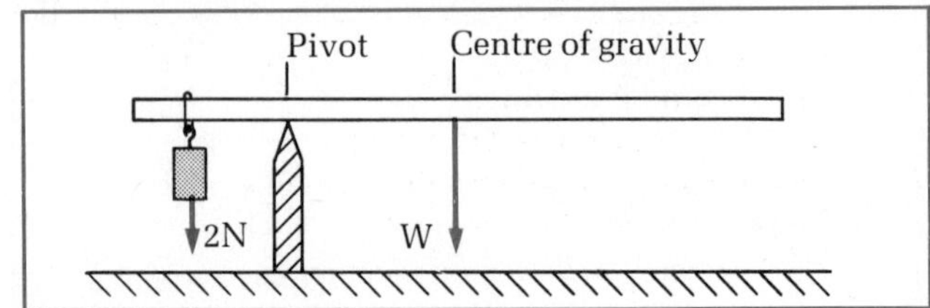

Figure 4.20 *Example*

Solution The rule is uniform in thickness so its centre of gravity is at the 50 cm mark.

The moment of the 2 N weight about the pivot = 2.0 × 0.15 N m = 0.30 N m.

The moment of the rule, of weight W, about the pivot = $W \times 0.30$.

Because the rule is balanced horizontally, the 2 N moment (anticlockwise turning) equals the moment of the weight (clockwise).

$W \times 0.30 = 0.30$ hence $W = 1.0$ N.

Worked example A bridge crane consists of a horizontal steel gantry of weight 20 kN across 2 steel pillars 10 m apart, as shown in Figure 4.21. The lifting gear (weight 5 kN) can be moved across the gantry on tracks. The crane is used to raise a 50 kN container when the lifting hook is 3 m from the nearest pillar. Calculate the force exerted by the gantry on each pillar.

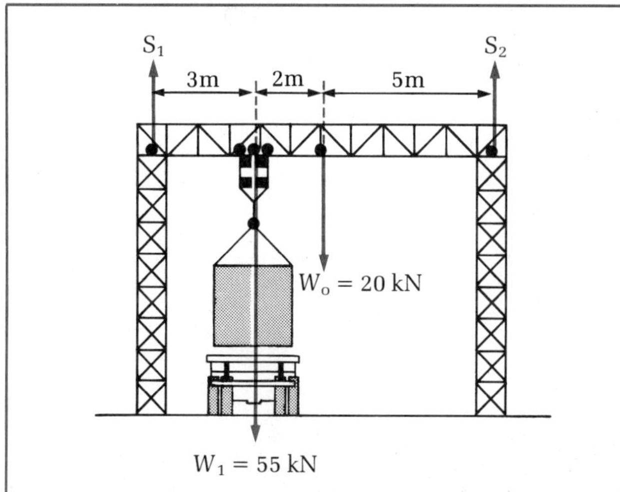

Figure 4.21 *A bridge crane*

Solution Let W_0 = the weight of the gantry which we assume acts at its midpoint. Let W_1 = the weight of the lifting gear and container, acting 3 m from the nearest pillar.

The total weight acting on the two pillars is $W_0 + W_1$ which equals 75 kN. The total weight of 75 kN acting on the two pillars is shared unequally between the two pillars. However, the support forces must add up to equal the total weight. Hence $S_1 + S_2 = 75$ kN.

By taking moments about any point, a second equation for S_1 and S_2 may be derived. Then, with the equation above, the values of S_1 and S_2 can be calculated. By choosing the 'pivot' (i.e. the point about which to take moments) at one of the two points of support, the second equation contains only one unknown force. Let's take moments about S_1's point of contact.

1 The total clockwise moment
= $W_0 \times 5$ (due to the gantry) + $W_1 \times 3$ (due to the lifting gear and container).
Hence the total clockwise moment
= $(20 \times 5) + (55 \times 3)$ = 265 kN m.

2 The total anticlockwise moment = $S_2 \times 10$.
Using the principle of moments,

The total anticlockwise moment = total clockwise moment

$S_2 \times 10 = 265$ kN m

Hence $S_2 = 26.5$ kN.

Finally, because $S_1 + S_2 = 75$ kN, then $S_1 = 48.5$ kN.

Therefore, the gantry exerts a force of 48.5 kN on the pillar where S_1 acts, and a force of 26.5 kN on the other pillar.

4.4 Friction

Rub two dry surfaces together vigorously and the surfaces will become warm. The force of friction between the surfaces when they move relative to one another is the cause. When a match is struck, the heat produced by friction causes the match to ignite. Friction acts whenever two surfaces move or try to move relative to one another. Sometimes the force applied is not large enough to overcome the friction so no movement takes place. However, with enough force applied, the force of friction can be overcome and then movement takes place. Try pushing a box across a rough floor to appreciate this point; the box does not move until the push becomes greater than a certain value.

In some circumstances, friction is unwanted because it wastes energy by producing heat energy. In other circumstances, friction is vital. For example, when the brakes of a car are applied, friction between the wheels and the road surface is essential if the car is to be slowed down. If the road is slippery, the wheels may lose their grip. Hence the importance of sufficient tread on all wheels of a car. Without the vital gaps between the tread on a car tyre, water will not be pushed away effectively from the area of contact. So there may be a film of water between the tyre and the road, with disastrous consequences if braking is required! The channels or gaps in the tread allow the water to be pushed away from the contact area, so a good contact is maintained between the road and the tyre.

Figure 4.22 *Tyre treads*

Measuring friction

To study friction between two solid surfaces, cover the underside of an open wooden box with one of the surfaces under study. It may be necessary to glue it. Then place the box on the other surface so that the two surfaces are in contact.

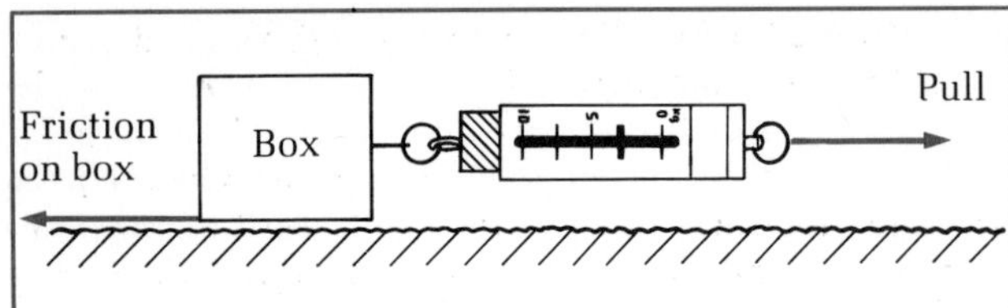

Figure 4.23 *Investigating friction*

With the lower surface horizontal, pull the box by a spring balance, as shown in Figure 4.23. At any stage, the friction force is equal to the reading on the spring balance, provided the box does not accelerate. Try each of the following tests.

1 Apply an increasingly stronger pull on the box, noting the spring balance reading and whether or not the box moves.
2 Repeat the above test for different amounts of weight in the box.
3 Smear the surfaces with oil and see what difference is made when you repeat the first test.
4 Try a different combination of surface materials.

When the pull is increased from zero, you ought to find that there is no movement until the pull reaches a certain value. At this value, the frictional force is as high as possible for those conditions. This is called the **limiting** frictional force. Once the pull becomes greater than the limiting frictional force, the surfaces slide over one another.

The second of the above tests ought to show that the limiting frictional force is proportional to the weight of the box. More generally, where surfaces are pushed together, the limiting frictional force is proportional to the force with which the surfaces are pushed together. This push force of one surface on the other is at right angles to the boundary, so it is called the *normal* force. The test box is said to experience a normal reaction force from the lower surface supporting it. Where the boundary is horizontal, the normal reaction force is equal (and opposite) to the weight.

For any pair of surfaces, the limiting frictional force is proportional to the normal reaction force between the surfaces. In other words, for any pair of surfaces, the ratio of $\dfrac{\text{the limiting friction force } (F)}{\text{the normal reaction force } (N)}$ is a constant. The value of this constant depends on the surfaces involved, and so it is known as the **coefficient of friction** (μ) for the two surfaces. Use your results from the test experiments to calculate the coefficient of friction for the different combinations used. Determining coefficients of friction is important if we are to be able to predict if surfaces will slip. For example, safety in the home is greatly improved by designing kitchen and bathroom floorcoverings with high friction. Then, even when damp, 'slipping up' is less likely.

From the definition of the coefficient of friction, the limiting frictional force F is equal to μN (where N is the normal reaction between the surfaces). The applied force tries to slide the two surfaces over one another. If the applied force is greater than the limiting frictional force ($=\mu N$), then the surfaces will slip.

For no slip, the frictional force between the two surfaces prevents movement, so must oppose the applied force exactly. If the applied force is increased, the frictional force increases by the same amount, until limiting friction is reached. The frictional force can never be greater than the limiting friction. The point of slipping is when the frictional force has reached its limit (i.e. the limiting friction value μN).

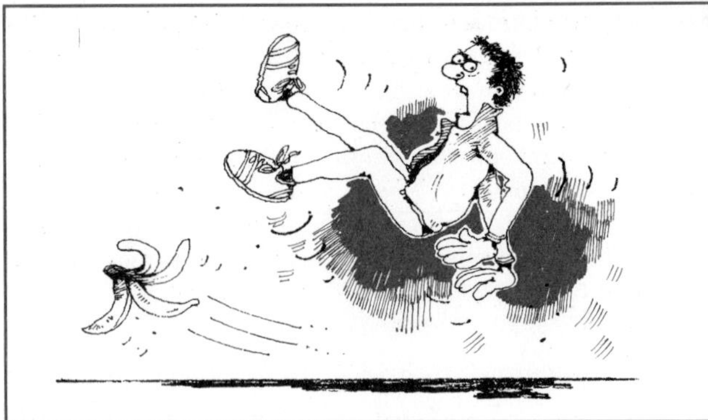

Figure 4.24 *Low friction*

Worked example A sloped conveyor is to be used to transport people from the ground floor of a building to the first floor. If it is to be sloped at 30° to the horizontal, what should be the coefficient of friction of the conveyor surface?

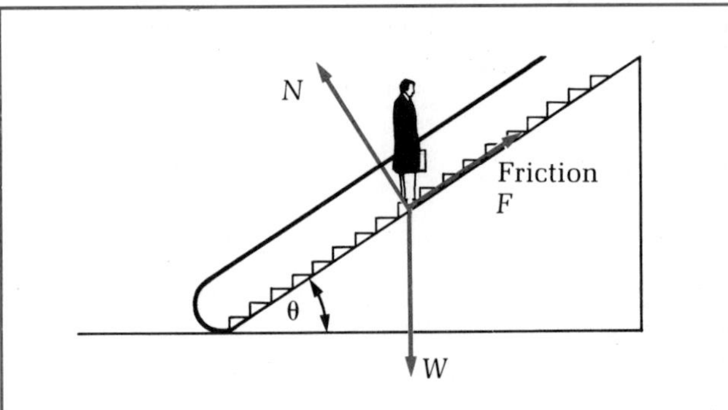

Figure 4.25 *Using friction*

Solution Surface friction ought to be sufficient to enable a passenger to be carried up the conveyor without slipping down the incline. Resolve the weight into two components parallel and perpendicular to the slope.
Let W be the weight of the passenger.
The perpendicular component $W\cos 30°$ equals the normal reaction N (i.e. the push from the surface). Hence the limiting friction force is $\mu W\cos 30°$ where μ is the coefficient of friction between the passenger's footwear and the conveyor.
The parallel component is $W\sin 30°$, and this acts down the slope. It is the force trying to slide the surfaces apart. For no slip, this force must not be greater than the limiting friction force,

i.e. $W\sin 30°$ must not be greater than $\mu W\cos 30°$

Therefore the lowest value of μ for no slip is when $\mu W\cos 30° = W\sin 30°$. Hence μ must be at least equal to $W\sin 30°/W\cos 30°$ which equals $\tan 30°$. The coefficient of friction must be at least 0.577 ($=\tan 30°$) or else passengers will slip. Compare this value with the values obtained from your tests.

Understanding friction is one of the aims of the science of **tribology.** Friction is thought to be due to the attraction between molecules at the surface. Even smooth surfaces are quite uneven at the microscopic level. When two surfaces are in contact, the surfaces actually touch at relatively few points. When pressed together, the surfaces come into contact at more points. So there is more attraction between their molecules. As a result, it is harder to slide one surface over the other when pressed together.

4.5 The conditions for equilibrium

For a point object in equilibrium, the forces acting on it must balance one another. There must be zero resultant force. For a body (i.e. a 'non-point' object or extended object) in equilibrium, the turning effects of the forces must also balance out. There must be zero resultant torque as well as zero resultant force for a body to be in equilibrium.

For example, consider the forces acting on a picture hanging against a wall by means of a string as shown in Figure 4.26.

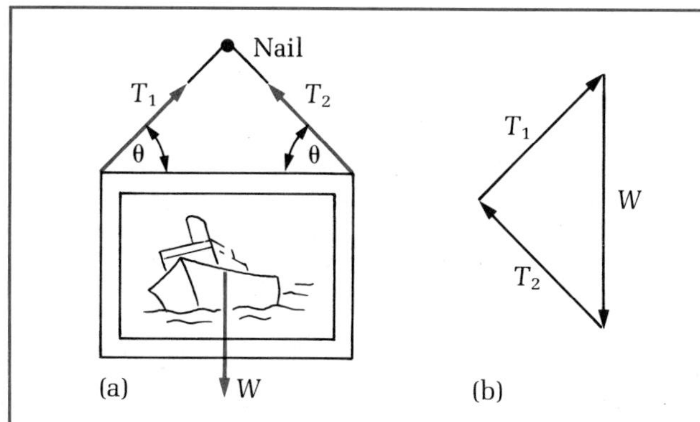

Figure 4.26 *Forces on a picture*

The string is supported by a nail in the wall, the middle of the string being at the nail. The picture is acted upon by three forces which are its weight W, and the tension in each string, T_1 and T_2.

1 The forces must balance out. The resultant of the three force vectors, $\mathbf{W}+\mathbf{T}_1+\mathbf{T}_2$ must be zero. Thus the three force vectors must form a triangle, as in Figure 4.26(b). Alternatively, we can state that

a) the vertical components must balance out

i.e. $W=T_1 \sin\theta+T_2 \sin\theta$

where θ is the angle of each string to the vertical.

b) the horizontal components must balance out

i.e. $T_1 \cos\theta=T_2 \cos\theta$
so $T_1=T_2$.

2 The turning effects of the forces must balance out. In fact, because the lines of action of the three forces pass through a single point (the nail), there can be no overall turning effect.

Condition 1 For a body to be in equilibrium, the force vectors must form a closed polygon. In other words, the resultant force must be zero.

Condition 2 For a body to be in equilibrium, the principle of moments applied about any point must be satisfied. In other words, the resultant torque about any point must be zero.

Worked example A winch fixed to a wall is used for lifting weights. A weight of 200 N is lifted by pulling on the rope with an equal force at 45° to the vertical. Calculate

a) the compression force in the support strut,

b) the force of the tie on the wall.

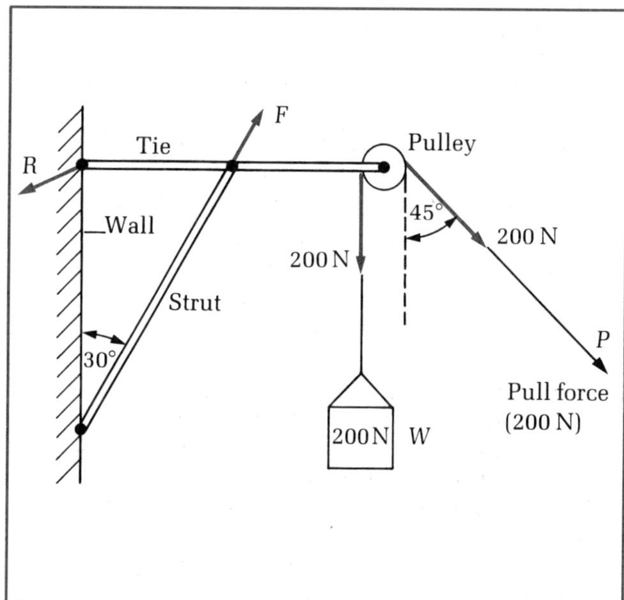

Figure 4.27 *Using a winch*

Solution

a) Take moments about the end of the tie fixed to the wall.
Clockwise moment due to weight $=200\times1=200$ N m.
Clockwise moment due to pull on rope $=200\times1\cos45°$ $=141$ N m.
The perpendicular distance from the line of the pull force to the pivot is $1\cos45°$, so the moment is $200\times1\cos45°$. The same answer is obtained by multiplying the component of the pull force perpendicular to the tie ($200\cos45°$) by the length of the tie (1 m).
Anticlockwise moment due to the force F in the strut $=F\times0.5\cos30°$; the perpendicular distance from the pivot to the strut is $0.5\cos30°$. So the moment is $F\times0.5\cos30°$.
Using the principle of moments,
$F\times0.5\cos30° = 200+141=341$ N m.
Hence $F=341/(0.5\cos30°)=788$ N, so the compression force in the strut is 788 N.

b) The tie is in equilibrium so the force vectors of the forces acting on it form a closed polygon. Since the tie's weight is negligible, there are just four forces acting on the tie. They are the pull P, the weight W, the strut force F and the force R holding the tie to the wall. Three of the four forces, P, W and F are known, so they can be drawn as three sides of a four-sided polygon as in

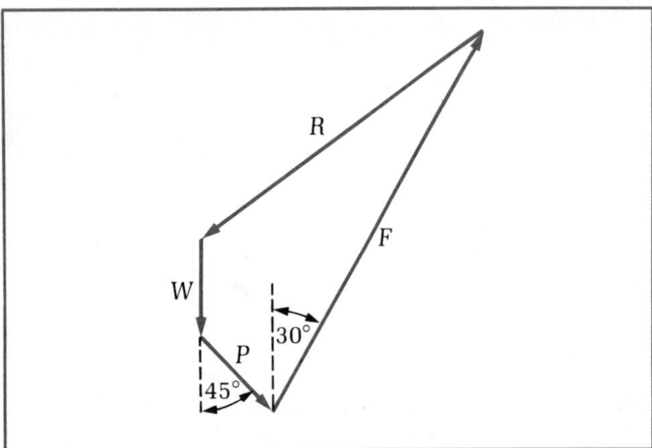

Figure 4.28 *Example*

Figure 4.28. Hence the fourth side is the unknown force R. From Figure 4.28, $R = 600$ N. The force of the tie on the wall is therefore 600 N, equal and opposite to R.

4.6 Summary

1 *Resolving* a force F along and at right angles to a given line:
parallel component $= F\cos\theta$
perpendicular component $= F\sin\theta$
where θ is the angle between the given line and the force.

2 *Adding force vectors*: use the parallelogram rule, or resolve one force along and at right angles to the other to give the components of the resultant.

3 *Moment of a force* = force × perpendicular distance from pivot to line of action.

4 *Coefficient of friction* $= \dfrac{\text{limiting friction force}}{\text{normal reaction}}$.

5 *Equilibrium conditions*:
a) the resultant force must be zero,
b) the resultant torque must be zero.

Short questions

Assume $g = 10\ \text{m s}^{-2}$

4.1 Calculate the resultant force produced by forces of 3 N and 4 N acting on a point object if the lines of action of the forces are
a) at right angles to one another,
b) at 120° to one another.

4.2 A point object acted on by forces of 4 N, 5 N and 6 N is in equilibrium. If the 6 N force is removed, what is resultant force now on the object?

4.3 In an experiment to measure the tension in the bowstring of an archer's bow, the bowstring is drawn back using a spring balance to pull on the middle of the string. The spring balance reads 110 N when the bowstring is pulled back to make a V at 120°. Calculate
a) the tension in the bow string at that angle,
b) the force on an arrow when the bowstring is drawn back as above at the instant of release.

4.4 A uniform metre rule is balanced horizontally on a pivot at its 24 cm mark by hanging a 1.50 N weight on a thread from the 4 cm mark of the rule. Calculate the weight of the rule.

4.5 A uniform metre rule is balanced horizontally on a pivot through its centre. A glass stopper is then suspended by a thread from the 10 cm mark of the rule. The rule is rebalanced by suspending a 0.50 N weight W, on a thread from the 74.0 cm mark on the rule. When the glass stopper, still hanging from the 10 cm mark, is suspended completely in water, the weight W must be moved to the 64.5 cm mark. Calculate
a) the weight of the glass stopper,
b) the density of the stopper.

4.6 A spanner of length 0.20 m is used to tighten a wheel nut on a car. If the torque is not to exceed 40 N m, calculate the force that must be applied to the end of the spanner to give the maximum allowed torque.

4.7 A rectangular concrete paving stone has dimensions 750 mm × 600 mm × 75 mm. Given the density of concrete is 2500 kg m^{-3}, calculate
a) the weight of the paving stone,
b) the minimum force needed to raise one side of the paving stone when it lies flat.

4.8 High-sided lorries are very vulnerable to crosswinds on motorways. Consider a lorry with a total weight of 1.5×10^5 N and of length 15 m, width 2 m and with sides of height 8 m. Assume the pressure exerted by wind at speed v hitting a vertical surface normally is ρv^2, where ρ is the density of air ($= 1.2$ kg m^{-3}).
a) Calculate the force exerted on one side by a crosswind of speed 15 m s^{-1}.
b) Assume that the wind force F exerts a turning moment about one side equal to $Fh/2$ where h is the height of the side. Is the wind force calculated in (a) strong enough to lift one side?
c) Even if the wind force did not lift one side, what might be the result of the wheel grip on the road differing on each side?

4.9 A wardrobe of total weight 400 N is shifted steadily across a room by pushing horizontally on one side with a force of 160 N.
a) Calculate the coefficient of friction between the floor and the wardrobe.
b) If the wardrobe had been emptied to reduce its weight to 150 N, what force would then have been necessary to shift it?

4.10 On a production line, a conveyor belt sloping at 20° to the horizontal takes boxes up for packing. What must be the minimum value of the coefficient of friction between the belt and the boxes if the boxes are not to slip down?

4.11 A uniform horizontal shelf of width 0.4 m is attached to a wall by a hinge, and is supported by a strut of length

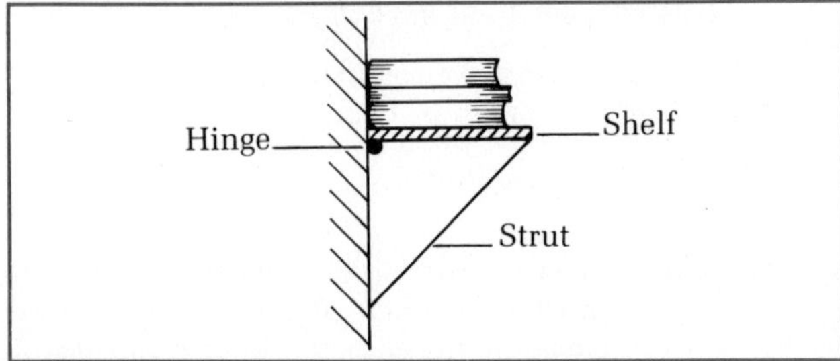

Figure 4.29

0.5 m as in the diagram. The shelf weight is 10 N and it is loaded with books of weight 50 N. The books are stacked on the shelf up against the wall to leave 0.1 m at the edge of the shelf clear.

a) Show that the turning moment of the books about the hinge is 7.5 N m.

b) Calculate the tension in the strut.

4.12 A tractor of weight 15 000 N crosses a single span bridge of weight 8000 N and of length 21 m. The bridge span is supported half a metre from either end. The tractor's front wheels take 1/3 of the total weight of the tractor, and the rear wheels are 3 m behind the front wheels. Calculate the force on the bridge supports when the rear wheels are at the middle of the bridge span.

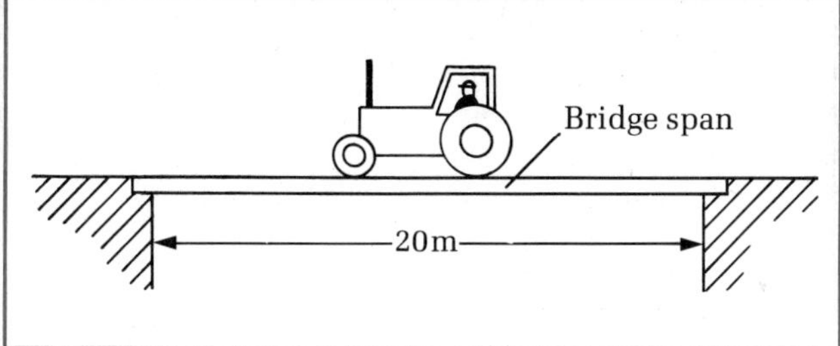

Figure 4.30

4.13 A pulley is attached to a wall using a strut and tie, as in the diagram below. The pulley is used to raise steadily a 200 N weight as shown. When the weight is being raised

a) what is the total force on the pulley mounting at X?

b) Is (i) XY, (ii) XZ in tension or compression?

c) Calculate the force in (i) XY, (ii) XZ.

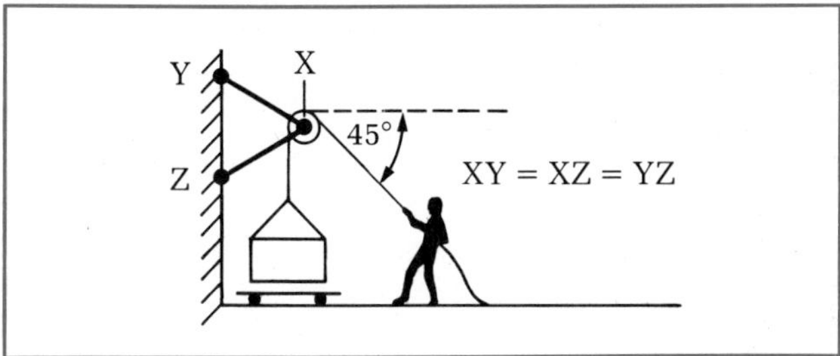

Figure 4.31

4.14 A uniform ladder of length 10 m and weight 500 N is carried horizontally by two people, one holding the ladder at one end and the other holding the ladder 2 m from the other end. What proportion of the ladder's weight does each person support?

4.15 The ladder of **4.14** is propped up against a smooth vertical wall, its lower end being on a rough horizontal floor. The coefficient of friction between the ladder and floor is 0.2. What is the maximum angle of the ladder to the wall, for it not to slip?

5

Rotational dynamics

5.1 Uniform circular motion

Figure 5.1 *A hammer thrower*

To whirl and hurl the hammer, a champion hammer thrower must rotate very fast and release the hammer when it is at the correct point along its circular path. To start with, the hammer is swung round above the thrower's head, and then as it speeds up, the thrower turns with it, releasing the hammer in a dramatic fling. When the hammer is being speeded up, the thrower accelerates it along its circular path. Rotating the hammer at steady speed and constant radius would make it move with **uniform circular motion**. Our champion hammer thrower needs to make it move non-uniformly if, from rest, it is to achieve a very high speed at the point of release.

Consider an object which does move with uniform circular motion. For example, a point on the tread of a car tyre moves with uniform circular motion if the car travels at steady speed. **The time period, *T*,** sometimes simply stated as the **period**, is the time taken for a rotating object to make one complete rotation. In this time, the rotating object turns through 360° which is equal to 2π radians. The **radian** is a unit for measuring angles, and it is based on a scale of 2π radians being equal to 360°. We use radians because then the formula for the circumference of a circle $2\pi R$ (where R is the radius) can be developed to give a useful formula for the length of a circular arc, as in Figure 5.2.

Length of a circular arc

$$s = R\theta$$

where θ is the angle which the arc subtends to the centre of the circle of radius R.

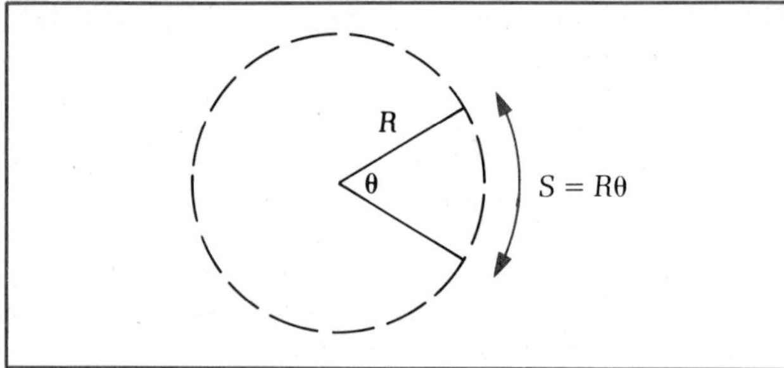

Figure 5.2 *Using radians*

The **frequency,** f, of rotation is the number of rotations per second made by the rotating object. The unit of frequency is the **hertz** (Hz), defined as 1 rotation per second. If the time period is measured, the frequency may be calculated from

$$f = \frac{1}{T}$$

For a point object moving on a circular path at uniform speed, its speed, v, may be calculated if the radius R and time period T are known. In one complete rotation, the object travels a distance of $2\pi R$ (i.e. the circumference) in time T. Therefore, its speed is given by,

$$v = \frac{2\pi R}{T}$$

The radial line from the centre of the circle to the point object sweeps round the centre, turning through 2π radians in time T. The **angular speed** of the point object, ω, is the angle per second swept out by the radial line. So ω can be calculated from

$$\omega = \frac{2\pi}{T} = 2\pi f$$

The unit of angular speed, ω, is rad s^{-1} (radian per second). The angular speed is linked to the speed, v, because $v = 2\pi R/T$.

Therefore, $v = R\left(\frac{2\pi}{T}\right) = R\omega$

$$v = R\omega$$

This equation can also be derived from the equation for arc length $s = R\theta$. If the object travels a distance s in time t along the circular path, then its speed $v = s/t$. Hence $v = R\theta/t$. Because θ/t is the angle per second which the radial line turns through, then θ/t equals the angular speed, ω. Therefore, $v = R\omega$.

Now let's return to our whirling hammer thrower! At what point on its circle of rotation should the hammer be released? You may think the best point of release is when the thrower's arms point along the measuring tape (i.e. the direction in which the hammer is *meant* to go). If the hammer thrower thinks the same, then the spectators at the side should dive for cover! The velocity of the hammer is, at any point on the circular path, along the **tangent** at that point. The thrower should release the hammer one quarter of a cycle before he faces the throwing direction. At this point the velocity of the hammer is in the intended direction. Releasing the grip at this point removes the force on it from the thrower, so the hammer is projected with its velocity in the correct direction.

An object moving with uniform circular motion has a constant speed. However, its velocity is continuously changing because its direction changes continuously. Because its velocity keeps changing, the object experiences acceleration towards the centre of rotation. To keep the object on a circular path, a force must act on it to prevent it flying off at a tangent. The hammer thrower has to pull with a considerable force on the hammer to make it go round. Whatever the cause of the force making an object move on a circular path, it is referred to as the **centripetal force** because it tries to pull the object towards the centre. The acceleration towards the centre is due to the centripetal force, so is referred to as the **centripetal acceleration**.

For an object in uniform circular motion,

a) its velocity is tangential

b) its acceleration is towards the centre (i.e. inwards along the radius).

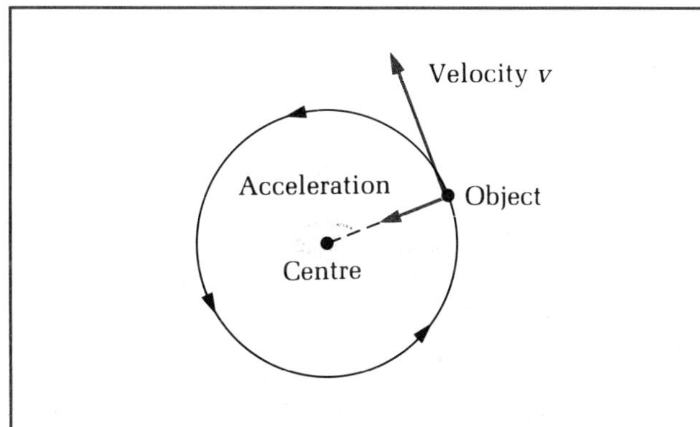

Figure 5.3 *Centripetal acceleration*

The centripetal force may be due to a single force or to a combination of several forces. If a cork is whirled round on a string in a horizontal circle, the centripetal force is provided by the tension in the string (provided the string is horizontal). For a racing car moving round a curve on a banked track, the centripetal force is supplied by a combination of friction and normal reaction from the track, see p. 45. Whatever the cause of the centripetal force, F, it is always linked to the centripetal acceleration, a, by the equation $F = ma$ where m is the mass of the object moving with uniform circular motion.

Equation for centripetal acceleration

For any point object moving with uniform circular motion at speed v, its centripetal acceleration, a, is given by

$$a = -\frac{v^2}{R}$$

where R is the circle radius.

The $-$ sign indicates that the acceleration is directed *inwards* (i.e. to the centre of rotation). Since $v = \omega R$, where ω is the angular speed, the above equation for centripetal acceleration may also be written as

$$a = -\frac{(\omega R)^2}{R} = -\omega^2 R$$

To prove the equation $a = -v^2/R$, consider a point object P at successive positions A and B along its path. Let the time taken to pass from A to B be a short time interval δt (see Figure 5.4).

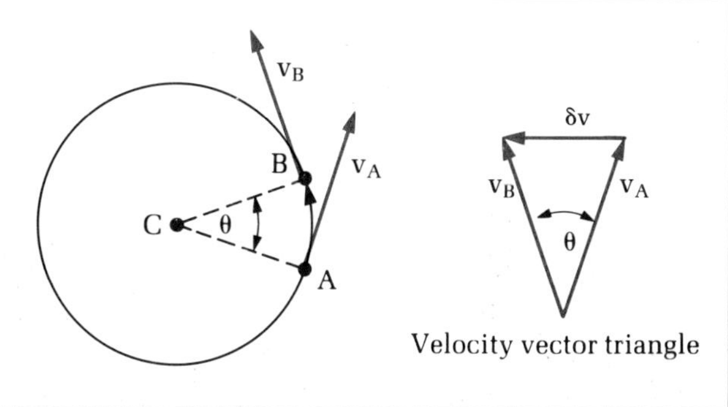

Figure 5.4 *Proving* $a = v^2/r$

In the time interval δt, P moves a distance $v\delta t$ from A to B along the circle. From A to B, the radial line from A to B turns through an angle θ. Since distance along a circular arc, s, is given by $s = R\theta$ so $v\delta t = R\theta$. Therefore, the angle θ is given by

$$\theta = \frac{v\delta t}{R}$$

The velocity vector also turns through an angle θ from A to B. The change of velocity $\delta\mathbf{v}$ is equal to the difference between the velocity at B and that at A.

$$\delta\mathbf{v} = \mathbf{v}_B - \mathbf{v}_A$$

This equation is represented by the velocity vector triangle of Figure 5.4. Because $\mathbf{v}_A$ and $\mathbf{v}_B$ are equal in magnitude, vector $\delta\mathbf{v}$ points towards the centre of the circle C.

The triangles ABC and the velocity vector triangle have the same shape. They are mathematically similar. Each has two sides of equal length, and at the same angle to one another. Therefore the ratio of the short side to the long side is the same for the two triangles. So

$$\frac{\delta v}{v} = \frac{AB}{AC}$$

but since δt is very short, angle θ is very small so $AB = v\delta t$. Also AC = radius R. Hence

$$\frac{\delta v}{v} = \frac{v\delta t}{R}$$

and because the acceleration a is given by change of velocity δv/time taken dt, then

$$a = \frac{\delta v}{\delta t} = \frac{v^2}{R}$$

Finally, we put the $-$ sign in to indicate that the acceleration is towards the circle centre (since $\delta\mathbf{v}$ is towards the circle centre), giving

$$a = -\frac{v^2}{R}$$

Investigating centripetal force

Figure 5.5 shows an experiment to investigate centripetal force. When the motor is switched on, it drives the turntable at steady speed. The speed may be adjusted using the variable resistor on the control box. The metal frame is firmly mounted on the turntable. When the turntable is driven, the slider on the slider bar is thrown to the end of the bar.

Before switching on the motor, a spirit level should be used to check that the slider bar is horizontal. If not, adjustments to it need to be made to set it horizontal. The bar should be well oiled to remove friction. The slider is pulled in to the centre by a thread over a pulley at the centre. A weight is supported by the other end of the thread. A stopper is screwed down on the slider bar, so that the slider is pulled in only as far as the stopper. (See Figure 5.5(a).)

The stopper is set to make the slider rest at a suitable distance from the centre. The distance (= radius of rotation of the slider) is then measured. A known weight is hung from the thread over the pulley. The motor is then switched on and its speed is gradually increased until the slider *just* moves outwards from the stopper. At this speed of rotation, the time period is determined by timing 20 complete turns. Typical results are shown in Figure 5.5(b).

Different values of radius and weight can then be tried, in each case measuring the time period as above. By keeping the radius constant and changing the weight in steps, the variation of centripetal force with rotation frequency may be investigated. Another possibility is to keep the weight constant and vary the radius in steps. For each radius, the time period can be determined to give a set of results of radius and rotation frequency.

Figure 5.5(b) shows typical results for constant radius. The slider *just* moves outwards when the weight can no longer hold it in. At the point of moving outwards, the weight is only just unable to supply sufficient centripetal force. Hence, at that point

Weight $W = m\omega^2 R$

Since the angular speed ω is equal to $2\pi/T$ and $T = 1/f$, then $\omega = 2\pi f$ where T is the time period and f is the rotation frequency.

Therefore

$$W = m(2\pi f)^2 R = (4\pi^2 mR)f^2$$

A graph of weight W against f^2 ought to give a straight line passing through the origin. Figure 5.5(c) shows a graph of W against f^2 for the results of Figure 5.5(b). Given the mass m of the slider and the radius of rotation R as in Figure 5.5(b) use the graph to check that its slope is equal to $4\pi^2 mR$.

Examples of circular motion

The conical pendulum The pendulum bob swings round on a circle at steady speed. The forces on it are its weight (mg) and the tension (T) in the string. Because the bob moves along a horizontal circle, the centripetal force (due to the combined effect of weight and tension) is horizontal. Resolve the weight and tension into horizontal and vertical components.

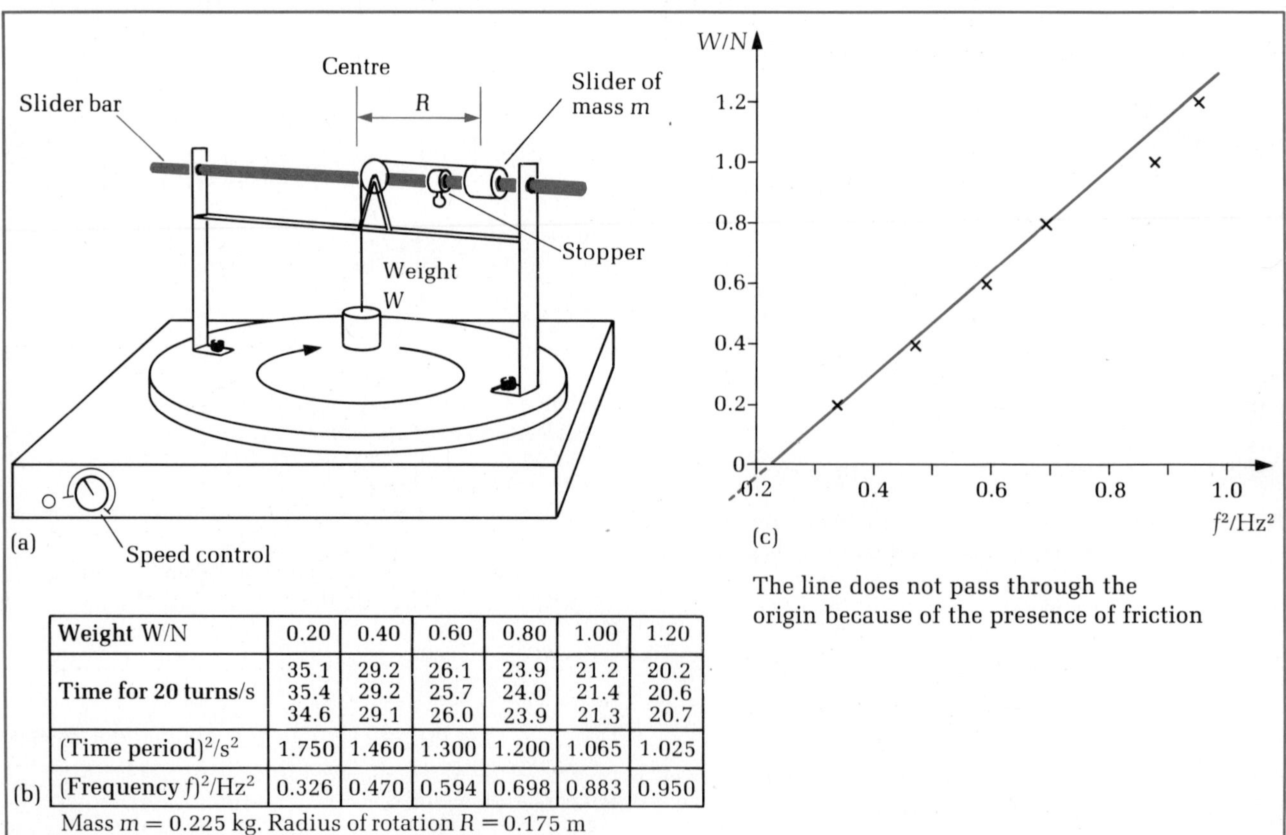

(b)

Weight W/N	0.20	0.40	0.60	0.80	1.00	1.20
Time for 20 turns/s	35.1 35.4 34.6	29.2 29.2 29.1	26.1 25.7 26.0	23.9 24.0 23.9	21.2 21.4 21.3	20.2 20.6 20.7
(Time period)^2/s^2	1.750	1.460	1.300	1.200	1.065	1.025
(Frequency f)^2/Hz^2	0.326	0.470	0.594	0.698	0.883	0.950

Mass m = 0.225 kg. Radius of rotation R = 0.175 m

Figure 5.5 *Testing* $F = mv^2/r$

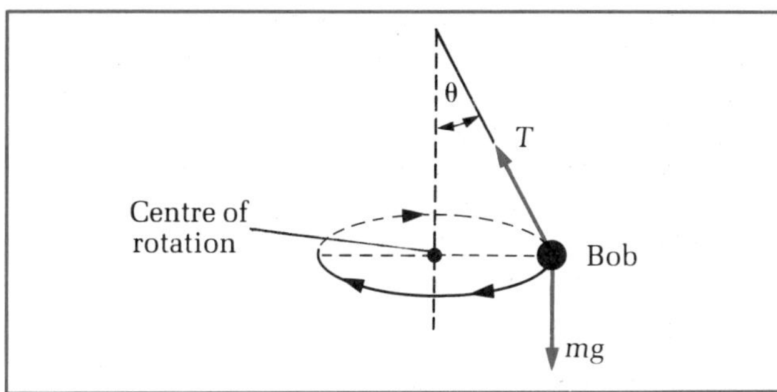

Figure 5.6 *The conical pendulum*

Vertically: $T\cos\theta = mg$
(because the vertical component of T balances the weight)

Horizontally: $T\sin\theta = \dfrac{mv^2}{R}$
(because the horizontal component of T supplies the centripetal force)

The two equations may be combined to give

$$\tan\theta = \frac{\sin\theta}{\cos\theta} = \frac{(mv^2/R)}{mg} = \frac{v^2}{gR}$$

Banked tracks When a racing car speeds round a curve, if the track is banked the car can travel faster than if it were flat. On a flat track, the centripetal force is provided by friction between the tyres and the track. If limiting friction is exceeded, then the car will crash. When the track is banked, the centripetal force can be provided wholly by the normal reaction $(N_1 + N_2)$ on the car, as in Figure 5.7. At a certain speed, there is no sideways friction on the tyres.

At this speed, v, resolving the forces acting gives

Vertically: $(N_1 + N_2)\cos\theta = mg$
(because the vertical component of the normal reaction forces balances the weight)

Horizontally: $(N_1 + N_2)\sin\theta = \dfrac{mv^2}{R}$

The two equations may be combined to give

$$\tan\theta = \frac{v^2}{gR}$$

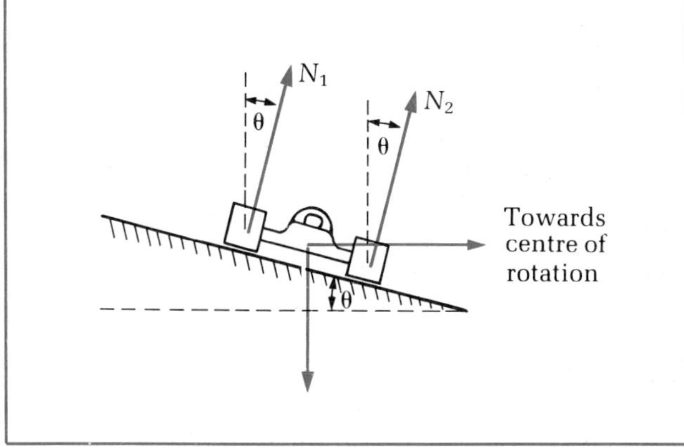

Figure 5.7 *A racing car taking a bend*

5.2 Angular acceleration

When a flywheel speeds up, every point of the flywheel moves at increasing speed. Consider a point on the flywheel rim, at distance R from the axis of rotation. If the flywheel speeds up from initial angular speed ω_i to angular speed ω_f in time t, then the speed of the point increases from speed $u = \omega_i R$ to speed $v = \omega_f R$ in time t. The speeding-up process accelerates the point along its circular path. So the acceleration of the point along its path may be calculated from $(v - u)/t$ which equals

$$\frac{(\omega_f - \omega_i)R}{t}$$

The **angular acceleration** of a rotating object is defined as the change of angular velocity per second. The unit of angular acceleration is rad s^{-2}.

Where the angular speed increases from initial value ω_i to final value ω_f in time t, the angular acceleration may be calculated from $(\omega_f - \omega_i)/t$. Angular acceleration α is given by

$$\alpha = \frac{(\omega_f - \omega_i)}{t}$$

For the flywheel example above, every point of the flywheel experiences the same angular acceleration. However, the acceleration of a point along its circular path (i.e. its linear acceleration) is equal to αR because

$$\alpha = \frac{(\omega_f - \omega_i)}{t}$$

Linear acceleration of a point along its circular path,

$$a = \alpha R$$

where R is the distance from the axis to the point.

Worked example A flywheel is speeded up from 5 revs per minute to 11 revs per minute in 100 s. The radius of the flywheel is 0.08 m. Calculate

a) the angular acceleration of the flywheel,

b) the acceleration of a point on the rim along its circular path.

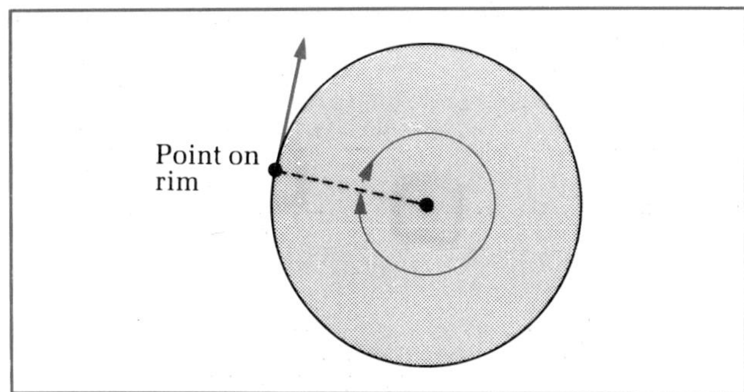

Figure 5.8 *Example*

Solution

a) The angular speed values given must be changed to rad s^{-1}. An angular speed of 1 rev per minute involves the flywheel turning through 2π radians in 60 s. Therefore, 1 rev per minute equals $2\pi/60$ rad s^{-1}.

The initial angular speed $\omega_i = 5 \times 2\pi/60$ rad s^{-1}.
The final angular speed $\omega_f = 11 \times 2\pi/60$ rad s^{-1}.
Hence the change of angular speed

$$= \frac{(11 \times 2\pi)}{60} - \frac{(5 \times 2\pi)}{60} = 0.628 \text{ rad s}^{-1}$$

So angular acceleration,

$$\alpha = \frac{(\omega_f - \omega_i)}{t} = 0.628/100 = 6.28 \times 10^{-3} \text{ rad s}^{-2}$$

b) The acceleration of the rim $= \alpha R = 6.28 \times 10^{-3} \times 0.08 = 5.0 \times 10^{-4}$ m s^{-2}

Of course the point on the rim experiences a centripetal acceleration as well, but here we are only concerned with its increasing speed, i.e. its acceleration along its path.

Dynamics equations for constant angular acceleration may be derived in much the same way as for straight line motion with constant acceleration.

a) From the definition of angular acceleration α, by rearrangement, we obtain

$$\omega = \omega_0 + \alpha t$$

where ω_0 = initial angular speed,
and ω = angular speed at time t.

b) The average value of the angular speed is obtained by averaging the initial and final values, giving $(\omega + \omega_0)/2$. The angle which the object turns through, the angular displacement θ, is equal to the average angular speed × time taken which gives

$$\theta = \frac{(\omega + \omega_0)t}{2}$$

c) The two equations above can be combined to eliminate ω, to give

$$\theta = \omega_0 t + \tfrac{1}{2}\alpha t^2$$

d) Alternatively, the first two equations may be combined to eliminate t, giving

$$\omega^2 = \omega_0^2 + 2\alpha\theta$$

The four equations are directly comparable with the four linear dynamics equations. In fact, the task of remembering them is made much easier by translating from linear to angular terms, as follows.

Linear dynamics	**Angular dynamics**
Displacement s	→ Angular displacement θ
Speed or velocity v	→ Angular speed or angular velocity ω
Acceleration a	→ Angular acceleration α

so, for example,

$$v = u + at \rightarrow \omega = \omega_0 + \alpha t$$
$$\text{and } s = ut + \tfrac{1}{2}at^2 \rightarrow \theta = \omega_0 t + \tfrac{1}{2}\alpha t^2$$

The other two equations may be translated in a similar way.

Worked example A spin drier tub accelerates from rest to an angular speed of 800 revs per minuts in 40 s. Calculate the angle which it turns through, and the number of turns made whilst it speeds up.

Solution The initial angular speed $w_0 = 0$.
The final angular speed $\omega = 800 \times 2\pi/60$ rad s^{-1}
(Since 1 rev per minute $= 2\pi/60$ rad s^{-1}.)
The time taken $t = 40$ s.

To calculate θ, use $\theta = \frac{(\omega + \omega_0)t}{2}$

which gives

$$\theta = \frac{[(800 \times 2\pi/60) + 0]}{2} \times 40 = 1.68 \times 10^3 \text{ rad.}$$

Therefore, the number of turns made $= \theta/2\pi = 267$ turns.

5.3 Torque and moment of inertia

To make a flywheel rotate, a turning force must be applied to it. The turning effect depends not just on the force but also on where it is applied. The **torque** of a turning force is the moment of the force about the axis. Therefore, torque is defined as follows.

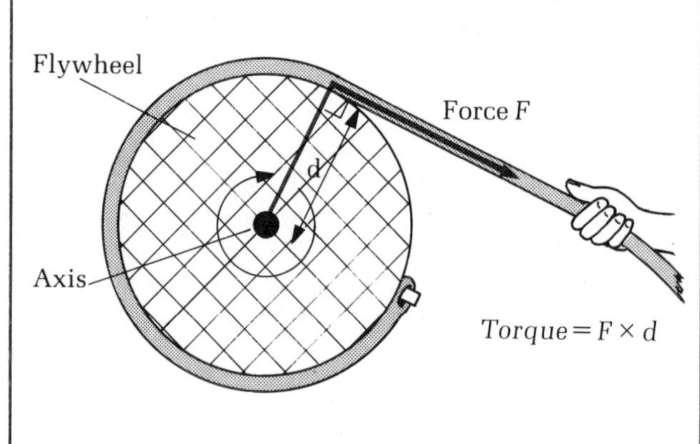

Figure 5.9 *Applying a torque*

Torque = force × perpendicular distance from the axis to the line of action of the force.

The unit of torque is the **newton metre** (N m).

If a large torque is required to start a flywheel turning, the flywheel must have considerable inertia. In other words, its resistance to change of its motion is large.

Every object has the property of inertia because every object has mass. However, the inertia of a rotating body depends on the **distribution** of mass as well as the amount of mass.

Consider a flat rigid body which can be rotated about an axis perpendicular to its plane, as shown in Figure 5.10. Suppose it is initially at rest, so a torque must be applied to it to make it rotate. Assuming there is no friction on its bearing, the applied torque will increase its angular speed. When the torque is removed, the angular speed stops increasing so it turns at constant frequency once the torque is removed.

The body in Figure 5.10 may be thought of as a network of point masses, m_1, m_2, m_3, etc., at distances r_1, r_2, r_3, etc. from the axis. Each point turns on a circular path about the axis. At angular speed ω, the speed of each point is given by $v = \omega r$ so the speed of m_1 is ωr_1, the speed of m_2 is ωr_2, etc.

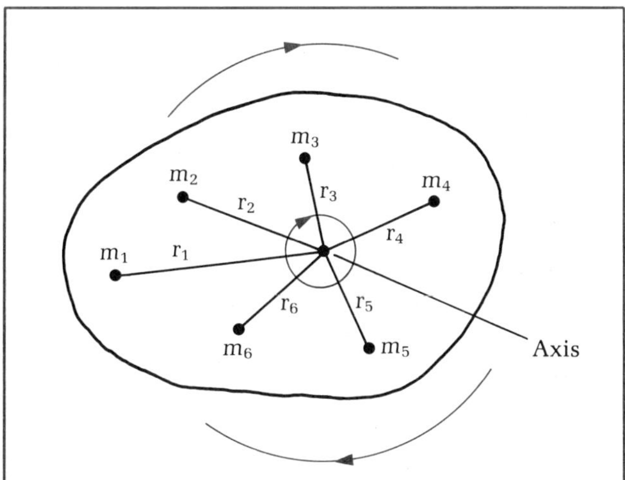

Figure 5.10 *A rigid body considered as a network of point masses*

When the body speeds up, every point in it accelerates. If the angular acceleration of the body is α, then the acceleration of each point mass is given by $a=\alpha r$. So the acceleration of m_1 is αr_1, the acceleration of m_2 is αr_2 etc. Using $F=ma$, the force needed to accelerate each point mass is therefore given by $m_1\alpha r_1$ for m_1, $m_2\alpha r_2$ for m_2 etc. So the moment needed for each point mass to be given angular acceleration α can be expressed since moment = force × distance. Thus the moment for m_1 is $(m_1\alpha r_1)r_1$, the moment for m_2 is $(m_2\alpha r_2)r_2$ etc.

To recap: Let α = the angular acceleration of the body.

Point mass	***Distance from axis***	***Acceler-ation***	***Force needed***	***Moment needed***
m_1	r_1	αr_1	$(m_1\alpha r_1)$	$(m_1\alpha r_1{}^2)$
m_2	r_2	αr_2	$(m_2\alpha r_2)$	$(m_2\alpha r_2{}^2)$
m_3	r_3	αr_3	$(m_3\alpha r_3)$	$(m_3\alpha r_3{}^2)$
m_4	r_4	αr_4	$(m_3\alpha r_4)$	$(m_4\alpha r_4{}^2)$
m_5	r_5	αr_5	$(m_5\alpha r_5)$	$(m_5\alpha r_5{}^2)$
⋮	⋮	⋮	⋮	⋮
m_N	r_N	αr_N	$(m_N\alpha r_N)$	$(m_N\alpha r_N{}^2)$

Total moment needed (i.e. torque T) is the sum of the individual moments needed for all the point masses.

Hence torque $T=(m_1r_1{}^2)\alpha+(m_2r_2{}^2)\alpha+(m_3r_3{}^2)\alpha+\ldots (m_Nr_N{}^2)\alpha$

so $T=(m_1r_1{}^2+m_2r_2{}^2+m_3r_3{}^2+\ldots+m_Nr_N{}^2)\alpha$

which may be rewritten as

$$T=(\sum_i{}_1 m_ir_i{}^2)\alpha$$

using $\sum_i{}_1 m_ir_i{}^2$ for $(mr_1{}^2+m_2r_2{}^2+\ \ldots+m_Nr_N{}^2)$.

The moment of inertia (I) of a body about a given axis is defined as $\sum_i m_ir_i^2$ for all the points in the body, where m_i and r_i are the mass and the distance of each point from the axis. The unit of I is kg m^2.

The equation for the torque T in terms of angular acceleration α may now be written as

$$T=I\alpha \text{ or } \alpha=T/I$$

When a constant torque is applied to a rotating body, the angular acceleration $(=T/I)$ therefore depends not just on the torque. It also depends on the moment of inertia of the body about the given axis, and this in turn depends on the distribution of mass about the axis.

$$T=I\alpha$$

where T = *resultant torque,*
α = angular acceleration,
I = moment of inertia about the axis.

The value of I about a given axis depends on the distribution of mass about that axis. Two bodies with equal mass distributed in different ways will have different values of I. For example, compare a hoop and a disc of the same mass, as in Figure 5.11. About the axis shown, the moment of inertia of the hoop is simply MR^2 where M is its mass and R is its radius. This is because *all* the mass of the hoop is at the same distance (= R) from the axis. $\sum mr^2$ is therefore just MR^2 for the hoop. However, the disc mass is distributed between the centre and the rim, so its moment of inertia about the given axis is less than MR^2. In fact, detailed theory shows that the value of I for the disc about the axis shown is $\frac{1}{2}MR^2$. In general, the further the mass is distributed from the axis, the greater is the moment of inertia about that axis.

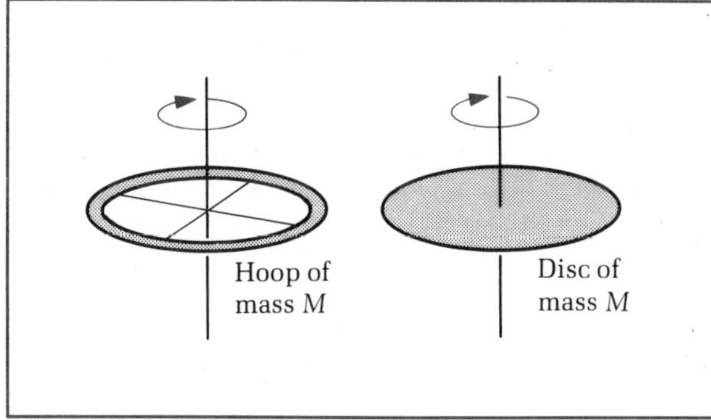

Figure 5.11 *Distribution of a mass*

For any simple shape, it is possible to calculate I using ferent values of I. A long thin cylinder is much easier to rotate about an axis along its length than about an axis at right angles to its length. When a value is given for moment of inertia of a body, it is important that the axis should be given as well as a numerical value and unit. To be told that the moment of inertia of a beam is 10 kg m^2 without being told which axis the value is about is of no use.

Moment of inertia formulae

For any simple shape, it is possible to calculate I using an appropriate mathematical formula for that shape and axis. In general terms, such formulae include geometrical factors as well as mass. For example, the hoop as shown in Figure 5.11 has a moment of inertia given by MR^2.

To show how such formulae are derived for other simple shapes, consider the example of a uniform beam with an axis at right angles to its midpoint, as in Figure 5.12 on page 48. Divide the beam into short sections, each of the same length and mass. Let the mass of each section be m, and length Δx. Let x = the distance of each section from the axis. Let L be the length of the beam and M total mass.

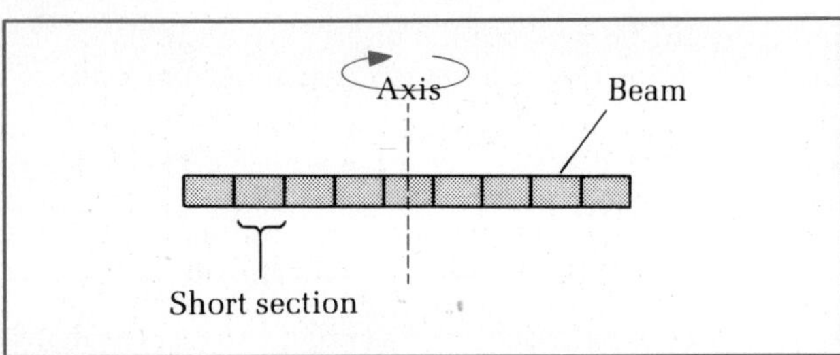

Figure 5.12 *Moment of inertia of a beam*

The moment of inertia of each section $= mx^2$. However, the mass m of each section is equal to $\rho\Delta x$ where ρ is the mass per unit length. Therefore, the moment of inertia of each section is $\rho x^2\Delta x$ about the given axis. So the total moment of inertia of all the sections is given by $I=\sum(\rho x^2\Delta x)$. The mathematical technique of integration can now be used to add up $\rho x^2\Delta x$ for all the sections. The result, for $x=-L$ to $x=L$, is

$$\frac{2\rho(L/2)^3}{3}$$

The total moment of inertia is therefore $\frac{ML^2}{12}$ where M is the total mass $(=\rho L)$. Using this formula, the moment of inertia of the beam (about the given axis) can be calculated if the mass M and length L are known.

Formulae for other simple shapes can be derived in the same sort of way. The emphasis in physics and engineering is on the *use* rather than the derivation of the formulae. By using the appropriate formula, the moment of inertia of a simple shape can be calculated from its dimensions and mass. Then, when subjected to torque, the effect on the motion can be determined.

Worked example A solid cylinder of radius 0.20 m and mass 30.0 kg is mounted on frictionless bearings so that the axis of rotation is along the axis of the cylinder. A rope with one end attached to the cylinder surface is wrapped round the stationary cylinder, and a constant force, W, of 20 N is applied to the other end of the rope. Calculate the angular speed of the cylinder and the speed of the rope just after 1.5 m of rope has unwrapped. Assume $g=10\ \text{m s}^{-2}$. The moment of inertia of a solid cylinder is $\frac{1}{2}MR^2$ (where M is its mass and R its radius) about its own axis.

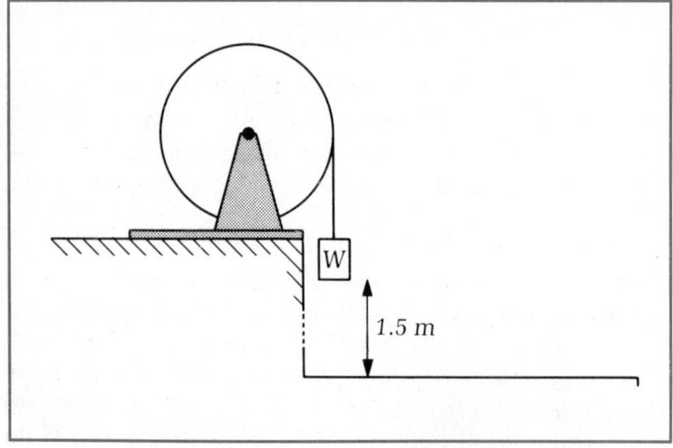

Figure 5.13 *Example*

Solution Firstly, calculate the moment of inertia of the cylinder about the given axis.

Thus $I=\frac{1}{2}MR^2=\frac{1}{2}\times 30\times 0.2^2=0.60\ \text{kg m}^2$.

Then calculate the torque and angular acceleration. The torque is provided by the force pulling on the rope on the cylinder surface.

Hence, the torque $=$ force $\times$ radius $=20\times 0.20=4.0\ \text{N m}$.

The angular acceleration $\alpha=\text{torque}/I=4.0/0.60=6.67\ \text{rad s}^{-2}$.

To calculate the angular speed, we need to know the angle which the cylinder turns through when the weight unwraps by 1.5 m (assuming no slipping).

In one turn, the rope unwraps by a distance $2\pi R=2\pi\times 0.2$ m. So unwrapping by distance $2\pi\times 0.2$ m makes the cylinder turn through 2π radians Hence, for a distance of 1.5 m, the cylinder turns through an angle of 1.5/0.2 radians.

$\theta=1.5/0.2$ radians

$\alpha=6.67\ \text{rad s}^{-2}$

$\omega_i=0$ (initial angular speed)

Use $\omega_f^2=\omega_i^2+2\alpha\theta$ to calculate the final angular speed ω_f.

$$\omega_f^2=0+2(6.67\times 1.5/0.2)=100\ \text{rad}^2\text{s}^{-2}$$

which gives $\omega_f=10\ \text{rad s}^{-1}$.

Finally, to calculate the speed of the rope, use $v=\omega_f R$ because the speed of the rope is the same as that of the rim.

Hence, the speed just after 1.5 m $=10\times 0.2=2.0\ \text{m s}^{-1}$.

5.4 Rotational energy

To make a body which is initially at rest start rotating about a fixed axis, it is necessary to apply a torque to the body. The torque does work on the body and, if there is no friction, the work done increases the kinetic energy of the body. The body rotates faster and faster.

The kinetic energy of a rotating body $=\frac{1}{2}I\omega^2$

where ω is its angular speed
I is its moment of inertia about the given axis.

K.E. of rotation $=\frac{1}{2}I\omega^2$

To prove this equation, consider the body as a network of point masses m_1, m_2, m_3, m_4, etc. When the body rotates at angular speed ω, the speed of each point mass is given by $v=\omega r$ (see Figure 5.10).

speed of $m_1=\omega r_1$

where r_1 is the distance of m_1 from the axis

speed of $m_2=\omega r_2$

where r_2 is the distance of m_2 from the axis, etc.

Since the K.E. of each point mass is given by $\frac{1}{2}mv^2$, then

K.E. of $m_1=\frac{1}{2}m_1\omega^2 r_1^2=\frac{1}{2}m_1 r_1^2\omega^2$

K.E. of $m_2=\frac{1}{2}m_2\omega^2 r_2^2=\frac{1}{2}m_2 r_2^2\omega^2$, etc.

Hence the total K.E. $=$ K.E. of m_1+ K.E. of m_2+ K.E. of $m_3+\ldots$ for all the point masses.

which gives total K.E. $=\frac{1}{2}m_1r_1^2\omega^2+\frac{1}{2}m_2r_2^2\omega^2+\ldots$

so total K.E. $=\frac{1}{2}I\omega^2$

since $I=\sum_i m_ir_i^2\ (=m_1r_1^2+m_2r_2^2+\ldots)$.

The **work done** by a constant torque T when the body is turned through angle θ is given by $T\theta$, as explained on p. 63. Assuming there is no frictional torque, $T = I\alpha$, so the work done by T equals $(I\alpha)\theta$. However, the dynamics equations for angular motion links $\alpha\theta$ to the final and initial angular speeds,

$$\omega^2 = \omega_0^2 + 2\alpha\theta.$$

Hence $\alpha\theta = \frac{1}{2}\omega^2 - \frac{1}{2}\omega_0^2$.
Since the work done by $T = I\alpha\theta$, substituting for θ gives

Work done by $T = \frac{1}{2}I\omega^2 - \frac{1}{2}I\omega_0^2$.

Therefore, in the absence of friction, the work done by the torque increases the kinetic energy of rotation.

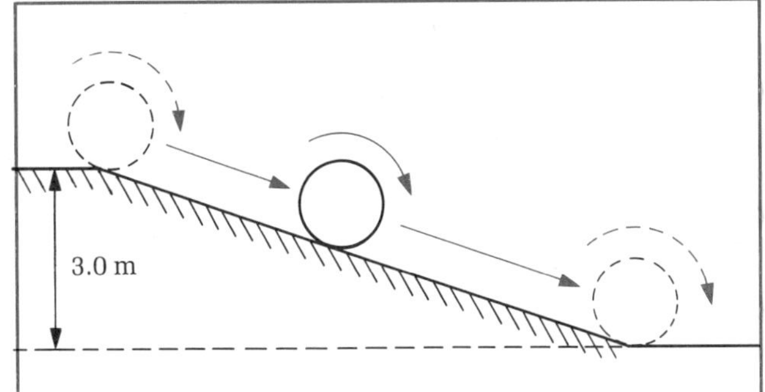

Figure 5.14 *Example*

Worked example A barrel of moment of inertia 0.4 kg m² about its own axis is to be rolled down a slope of height 3.0 m. The barrel has mass 20 kg and radius 0.2 m. Calculate its speed at the foot of the slope, after it has been released from rest at the top. Assume $g = 10\ \text{m s}^{-2}$.

Solution Consider the energy changes between the top and the bottom

P.E. loss =	gain of K.E. +	gain of K.E.
(mgh)	of linear motion	of rotation
	($\frac{1}{2}mV^2$)	($\frac{1}{2}I\omega^2$)

where h is the height fallen, m is the mass, I is the given moment of inertia, ω is the angular speed and V is the speed at the bottom of the slope.

Substituting values gives

$$20 \times 10 \times 3.0 = \tfrac{1}{2} \times 20 \times V^2 + \tfrac{1}{2} \times 0.4 \times \omega^2$$

so $600 = 10V^2 + 0.2\omega^2$

Next, we must use the link between angular speed ω and speed V which is the equation $V = \omega R$.

Hence $V = 0.2\omega$ or $\omega = V/0.2$.

Substituting in the energy equation gives

$$600 = 10V^2 + 0.2 \times (V/0.2)^2$$

$$\text{so } 600 = 10V^2 + 5V^2 = 15V^2$$

which gives $V^2 = 600/15 = 40\ \text{m}^2\ \text{s}^{-2}$.

Hence $V = 6.32\ \text{m s}^{-1}$.

5.5 *Angular momentum*

An ice skater spinning rapidly is a dramatic sight. The onlooker sees the skater turning slowly at first, then quite suddenly the skater goes into a rapid spin. This sudden

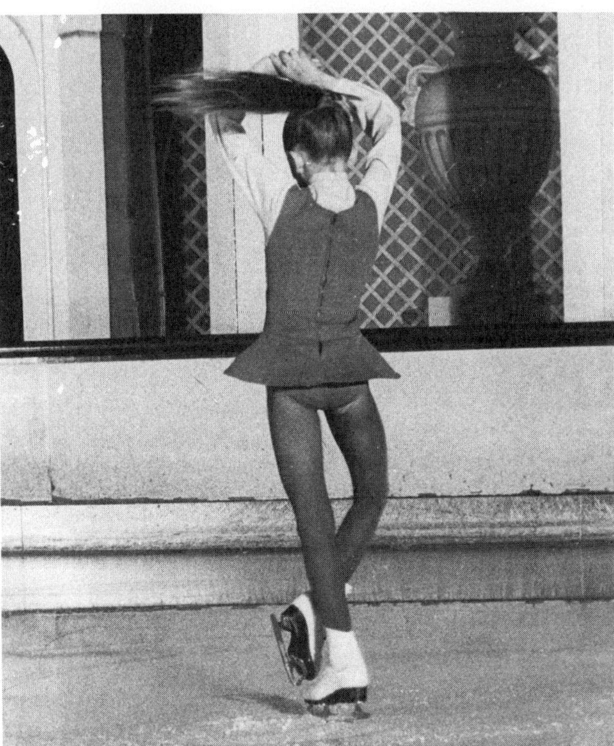

Figure 5.15 *An ice skater spinning*

change is brought about by the skater pulling both arms (and possibly a leg!) towards the axis of rotation. In this way the moment of inertia of the skater about the axis is reduced. As a result, the skater spins faster. To slow down, the skater only needs to throw out the arms and maybe a leg. In this way, the moment of inertia is increased. So the skater slows down.

To understanding such effects, consider a rotating body with no resultant torque on it. Provided its moment of inertia stays the same, then its angular speed ω does not change. This can be seen by using the equation $T = I\alpha$ where α is the angular acceleration. In more general terms, this equation ought to be written as

$$T = \frac{d}{dt}(I\omega)$$

where $\frac{d}{dt}$ is the mathematical way of writing 'change per unit time'.

In situations where I does *not* change, then $\frac{d}{dt}(I\omega)$ is equal to $I\frac{d\omega}{dt}$ because I is a constant here. Now, $\frac{d\omega}{dt}$ is the mathematical way of writing 'change of angular speed per unit time', which is the angular acceleration α. So the equation $T = \frac{d}{dt}(I\omega)$ becomes $T = I\alpha$ where I stays the same during the time in which the torque acts.

The angular momentum of a rigid rotating body is defined as $I\omega$, where ω is its angular velocity and I is its moment of inertia about a given axis. The unit of angular momentum is kg m² rad s⁻¹.

Angular momentum $= I\omega$

From the general equation $T=\frac{d}{dt}(I\omega)$, we see that when no resultant torque acts ($T=0$) the angular momentum of the body is constant. The spinning ice skater has constant angular momentum. By pulling in the arms, the skater's moment of inertia is made smaller. So the angular speed ω increases, because $I\omega$ (the angular momentum) must not change.

What have spinning ice skaters got to do with astronomy? In 1967, astronomers first discovered a type of star called a *pulsar*. Regular pulses of radio energy were detected from these stars, which some astronomers dubbed 'LGM' stars. It seemed as if 'little green men' were trying to contact us! That theory was soon abandoned when it was shown that pulsars are in fact rapidly rotating neutron stars which emit radio energy in a beam at an angle to the axis. Each time the beam sweeps round to point towards Earth, radio energy is directed towards us, rather like a light beam from a lighthouse. Neutron stars are the remnants of large stars. When a massive star runs out of fuel, a huge explosion takes place. The remnants of the explosion are pulled in together by their gravitational attraction, perhaps equivalent to a mass equal to the Sun shrinking to only 15 km or so in diameter. The moment of inertia is therefore made much smaller so the angular speed increases. The pulse frequency from a pulsar is of the order of 1 to 10 Hz, so the rate of rotation is of that order, much much greater than the Sun's rate which is about once every 25 days!

The angular momentum of a point mass is defined as its momentum × its distance from the axis of rotation. For a point mass m rotating at angular speed ω at distance r from the axis, its momentum is $m\omega r$ (speed $v=\omega r$) so its angular momentum is $m\omega r^2$.

Angular momentum of a point mass $= mr^2\omega$

For a network of point masses m_1, m_2, ... which make up a rigid body, the total angular momentum is therefore $(m_1r_1^2\omega)+(m_2r_2^2\omega)+\ldots$ which equals $I\omega$ because I is defined as $\sum_i m_ir_i^2 (=m_1r_1^2+m_2r_2^2+m_3r_3^2+\ldots)$.

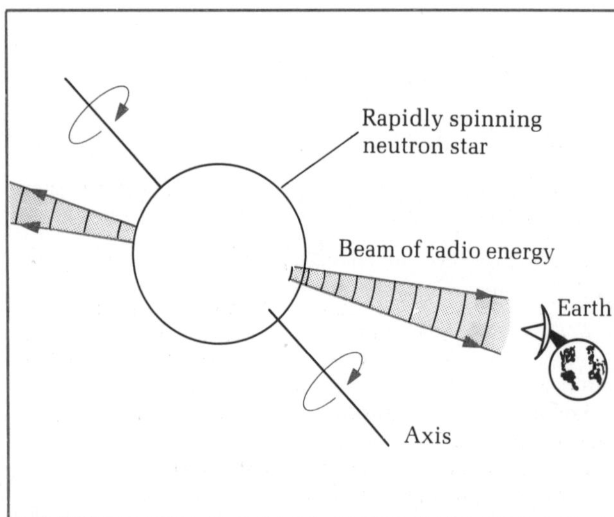

Figure 5.16 *A pulsar*

The principle of conservation of angular momentum states that any system of rotating bodies must have constant total angular momentum, provided there is no resultant torque on the system.

The skater is just one example of the conservation of angular momentum. No friction acts on the skater, so the total angular momentum is constant. The pulsar is another example where no external torque acts, so the total angular momentum is constant.

Where a system is made up of more than one spinning body, then when two of the bodies interact (e.g. collide), one might lose angular momentum to the other. But the total amount of angular momentum must stay the same. For example, if a spinning satellite is taken on board a space repair laboratory, the whole laboratory may be set spinning. The angular momentum of the satellite is transferred to the laboratory when the satellite is stopped inside. Unless rocket motors are used to prevent it from turning, then the whole laboratory would spin.

Worked example A frictionless turntable rotates at an angular speed of 12 revolutions per minute when a mass of 50 g is dropped from rest just above the disc. The mass falls onto the disc at a distance of 0.10 m from its axis. Because of the impact, the rate of rotation of the disc drops to 9 revolutions per second. Use this information to calculate the moment of inertia of the disc.

Solution Let $I=$ the moment of inertia of the turntable about its axis.

The moment of inertia of the mass about the axis is $mr^2=0.05\times0.1^2$.

The initial angular momentum $=I\times(12\times2\pi/60)$ since the initial angular speed $=12\times2\pi/60$ rad s^{-1}.

The final angular momentum $=(I+mr^2)\times9\times2\pi/60$ since the total moment of inertia is now $(I+mr^2)$.

Figure 5.17 *Spacelab taking a satellite on board*

Since the total angular momentum does not change when the mass is dropped onto the turntable, then

$$I \times (12 \times 2\pi/60) = (I + mr^2) \times 9 \times 2\pi/60$$

so cancelling by $2\pi/60$, $12I = 9(I + mr^2)$

which gives $3I = 9mr^2$

so $I = 3mr^2 = 3 \times 0.05 \times 0.1^2 = 1.5 \times 10^{-3}\ \text{kg m}^2$.

Therefore the moment of inertia I of the turntable about its axis is 1.5×10^{-3} kg m³.

5.6 Comparison of linear and rotational motion

Apply a **force** to an object initially at rest on a frictionless surface. The object will **accelerate.** Provided the force is constant, the acceleration is constant.

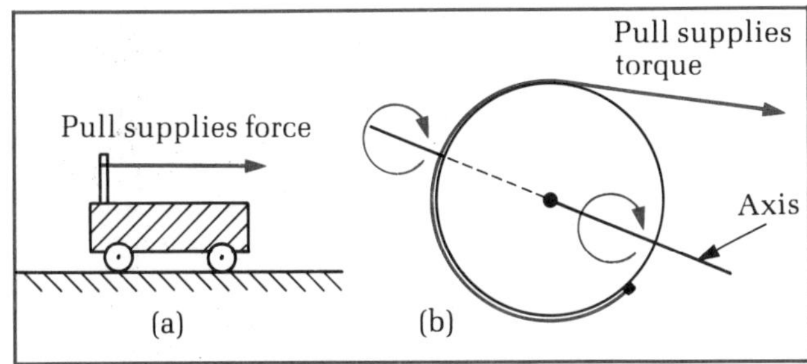

Figure 5.18 *Comparison of linear and rotational motion*

Apply a **torque** to an object initially at rest on frictionless bearings. The object will **rotate** faster and faster. Provided the torque is constant, the angular acceleration is constant.

Comparing the two situations described above, the force in linear motion plays the same role as the torque in rotational motion. We have already seen that the linear dynamics equations have their angular counterparts, as explained on p. 46. The comparison may be taken further, and becomes very useful when tackling rotational problems.

Linear motion	**Rotational motion**
Displacement s	Angular displacement θ
Speed and velocity v	Angular speed and angular velocity ω
Acceleration a	Angular acceleration α
Force F	Torque T
Mass m	Moment of inertia I
Momentum mv	Angular momentum $I\omega$
$F = \frac{d}{dt}(mv)$	$T = \frac{d}{dt}(I\omega)$
K.E. $= \frac{1}{2}mv^2$	K.E. $= \frac{1}{2}I\omega^2$
Work done $= Fs$	Work done $= T\theta$

Worked example To investigate the friction on the bearings of a car wheel, the car is jacked up to allow one of its wheels to be turned whilst off the ground. The wheel is set turning at a frequency of 5.0 Hz. When the applied torque is removed the wheel slows down, taking 150 s to stop. Given that the moment of inertia of the wheel about the given axis is 0.5 kg m², calculate **a)** the frictional torque which brings it to rest, **b)** the work done by such a torque in 1 minute when the wheel turns at 5 Hz.

Solution This is like the linear situation where a mass moving at constant speed is brought to rest by a constant force. We would use $F = ma$ to calculate the force, and work $= Fs$ to calculate the work done. So in this example, we can use $T = I\alpha$ to calculate the torque, and then work = torque × angle turned for the work done.

a) To calculate the angular acceleration α,
initial angular speed, $\omega_0 = 2\pi f = 2\pi \times 5 = 31.4\ \text{rad s}^{-1}$
final angular speed, $\omega = 0$
time taken = 150 s
Use $\omega = \omega_0 + \alpha t$ to give $\alpha = (0 - 31.4)/150\ \text{rad s}^{-2}$

b) To calculate the torque T, use $T = I\alpha$
which gives $T = 0.5 \times (-31.4/150) = 0.1047$ N m.
To calculate the work done, use work $= T\theta$ where θ is the angle turned through in 1 minute at an angular speed of 31.4 rad s⁻¹.
So $\theta = 31.4 \times 60$ rad, which gives work done $= 0.1047 \times 31.4 \times 60$ J.
Hence work done = 197 J in each minute.

5.7 Summary

For a particle moving along a circular path

1 *Angular speed* $\omega = 2\pi/T = 2\pi f$
2 *Speed along path* $= \omega R$
3 *Centripetal acceleration* $= -v^2/R$

For a body rotating about a fixed axis

1 *Moment of inertia I* is defined as $\sum_i m_i r_i^2$ for all points of the object, where m_i and r_i are the masses and distances of the points from the axis.
2 *Resultant torque* = rate of change of angular momentum.
3 If the resultant torque is zero, the angular momentum remains constant.
4 See the comparison of linear and rotational terms in the adjacent column.

Short questions

Assume $g = 10\ \text{m s}^{-2}$

5.1 A spin drier tub operates at 800 revolutions per minute. The tub has a diameter of 0.80 m. Calculate
a) the speed of a point on the tub wall,
b) the centripetal acceleration of the tub wall.

5.2 On a fairground big dipper, a train freewheels down a steep incline into a dip before rising out of the dip. The track at the dip has a radius of curvature of 20 m. Assuming the train is almost stationary at the top of the incline before descending, and given that the train drops 20 m, calculate
a) the maximum speed the train could have at the dip,
b) the centripetal acceleration of the train as it passes through the dip,
c) the extra 'g' force on the passengers.

5.3 A pendulum bob of mass 0.15 kg is suspended from a fixed point by a thread of length 2.0 m. The bob is given a push so that it moves along a circular path in a horizontal plane at a steady speed, taking 18.0 s to make 10 complete revolutions. Calculate
a) the speed of the bob,
b) the centripetal acceleration of the bob,
c) the tension in the thread.

5.4 A stone of mass 0.40 kg is tied firmly to one end of a rope, then the stone is whirled round in a vertical circle of radius 0.6 m at a steady rate of 8 turns per second. Calculate
a) the centripetal acceleration of the stone,
b) the maximum and minimum tension in the thread.

5.5 A bobsleigh shoots round a banked track on a horizontal curve of radius of curvature 150 m at a speed of 24 m s^{-1}. Calculate the angle of the track to the horizontal.

5.6 Calculate the difference between the value of g at the Earth's equator and the Earth's poles. Assume the Earth is a sphere of radius 6400 km.

5.7 A car of mass 600 kg travels at steady speed, without slipping, round a roundabout of diameter 100 m. The centripetal force is supplied by sideways friction between the car tyres and the road. If the sideways friction is not to exceed 0.2 × the car's weight, calculate the maximum speed that the car can travel at without slipping.

5.8 A solid aluminium cylinder and a hollow steel cylinder have the same mass and radius. The two cylinders are placed side by side at the top of a slope, and are then released at the same time. Which one reaches the bottom of the slope first? Explain your answer.

5.9 Write down the angular quantity or expression that is analagous to each of the following linear quantities.
a) force
b) momentum
c) speed
d) acceleration
e) K.E. $= \frac{1}{2}mv^2$
f) power = force × speed (for an applied force moving at steady speed)
For **a)** to **d)**, give the units of the angular quantities.

5.10 A flywheel of moment of inertia 1.2 kg m^2 about its mounting has a spindle of diameter 15 mm and is mounted on a horizontal axis on friction-free bearings. A rope attached to the spindle, as in the diagram, is wrapped several times round the spindle and is then pulled from rest with a steady force of 5.0 N for 10 s. Calculate
a) the angular acceleration of the flywheel,
b) the rate at which the wheel turns after 10 s from rest.

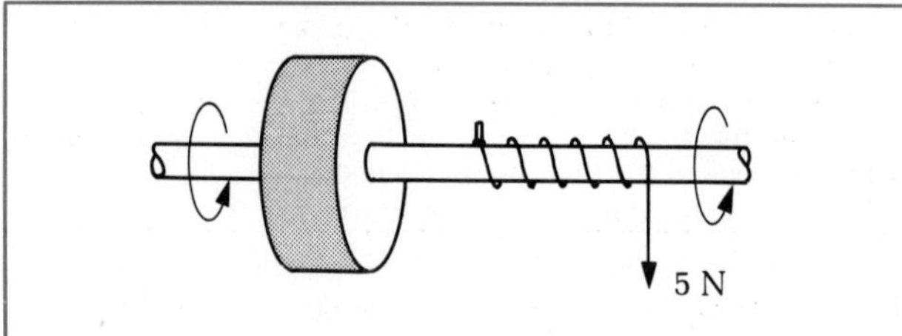

Figure 5.19

5.11 A garden roller consists of a solid steel cylinder of mass 40 kg and diameter 0.5 m. The roller, fitted with a light handle, is pulled to the top of a uniform slope of length 30 m. Given the angle of the slope to the horizontal is 10°, calculate
a) the fall of height and the loss of P.E. when the roller is released at rest and rolls to the lower end of the slope,
b) its speed at the bottom of the slope.
Assume that the moment of inertia of a solid cylinder about the cylinder axis is $\frac{1}{2}MR^2$ where M is the mass and R is the radius of the cylinder.

5.12 A prototype vehicle driven by a flywheel engine is designed to travel at a steady speed of 8 m s^{-1} for at least 1 hour, its engine giving a steady power output of 4 kW. Calculate
a) the total amount of energy the flywheel must supply for a one-hour journey,
b) the flywheel's initial rate of rotation at the start of the journey, given that its moment of inertia is 25 kg m^2.

5.13 A flywheel is speeded up from rest to a rate of rotation of 16 Hz in 40 s. Calculate
a) its angular acceleration,
b) the number of turns made in that time.

5.14 A cycle wheel of radius 0.40 m has moment of inertia of 0.32 kg m^2 about its spindle. In a test, the wheel is set turning at a frequency of 8 Hz and it takes 720 s to stop unaided. Calculate
a) the angular deceleration of the wheel,
b) the frictional torque which slows it down,
c) the power needed to keep it turning at a steady speed of 8 Hz.

5.15 A frictionless turntable is set rotating at a steady angular speed of 20 rad s^{-1}. A small 0.5 kg mass is dropped onto the disc from rest just above it, at a distance of 0.25 m from the centre of the disc. As a result, the angular speed of the turntable decreases to 17.5 rad s^{-1}. Calculate the moment of inertia of the turntable about an axis through the centre at right angles to its plane.

A

Multiple choice questions

Assume $g = 10\ m\ s^{-2}$

AM.1 Two objects, X and Y, were dropped from rest from a tall tower on a wind-free day. In the graph below are plotted their squared velocities as a function of their height above the ground.

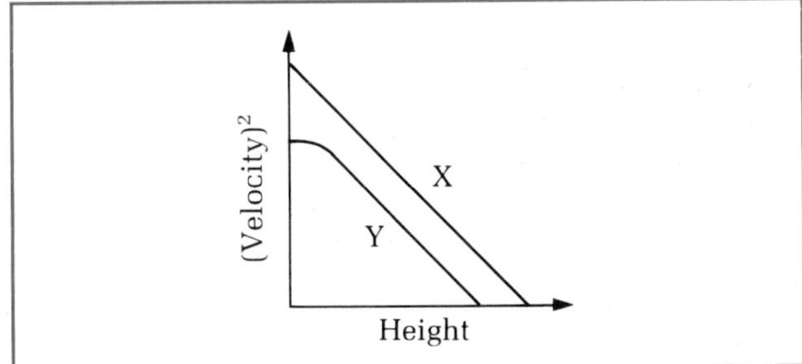

Figure AM.1

From the information given in the graph and knowledge of the properties of bodies falling under the influence of gravity it is possible to say that the two objects

A experienced unequal viscous drag
B had different densities
C had different masses
D hit the ground at the same time
E were not dropped simultaneously (*London*)

AM.2 A ball X is thrown directly upwards and takes 4.0 s to return to the thrower. A second ball Y is thrown directly upwards 1.0 s after X is thrown at the same initial speed as X. At the instant X returns to the thrower, which alternative gives the correct height and velocity of Y?

	A	B	C	D	E
Height/m	5	5	10	15	15
Velocity/m s^{-1}	10 up	10 down	10 up	10 down	10 up

AM.3 Two particles P and Q, at the same initial position, accelerate uniformly from rest along a straight line. After 1 s, P is 0.5 m ahead of Q. The separation of P and Q after 2 s from the start is.

A 0.5 m **B** 1.0 m **C** 1.5 m **D** 2.0 m **E** 2.5 m

AM.4 In which one of the following are all three quantities vectors?

A displacement, velocity, energy
B displacement, velocity, momentum
C velocity, acceleration, power
D Force, work, energy
E weight, work, power (*AEB June 85*)

AM.5 A ball-bearing X is projected horizontally from a certain point at the same time as a ball-bearing Y of the same size but twice the mass is released from rest and allowed to fall vertically from the same level. Which of the following will occur? (Air resistance may be neglected.)

A Y will hit the floor just before X.
B X will hit the floor just before Y.
C X and Y will hit the floor at the same time.
D X hits the floor while Y is halfway to the floor.
E Y hits the floor while X is halfway to the floor.

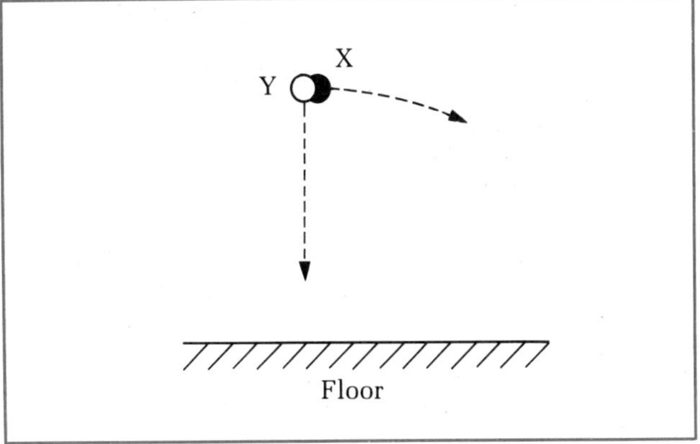

Figure AM.2

(*AEB June 83*)

AM.6 A helicopter of mass 800 kg rises from rest to a height of 60 m in 10 s. Assuming constant upwards acceleration and taking the acceleration of free fall as 9.8 m s^{-2}, the thrust in newtons on the helicopter blades on take-off is

A 480 **B** 960 **C** 7840 **D** 8320 **E** 8800 (*NI*)

AM.7 A block X of mass 3 kg is accelerated from rest by a force of 5 N for 2 s. An identical block Y is accelerated from rest by a force of 5 N for 4 s. The ratio of the kinetic energy of X to the kinetic energy of Y after acceleration is

A 5 : 3 **B** 3 : 4 **C** 2 : 1 **D** 1 : 2 **E** 1 : 4 (*Scottish H*)

AM.8 Two trolleys are travelling in the same direction, as shown below. The trolleys collide.

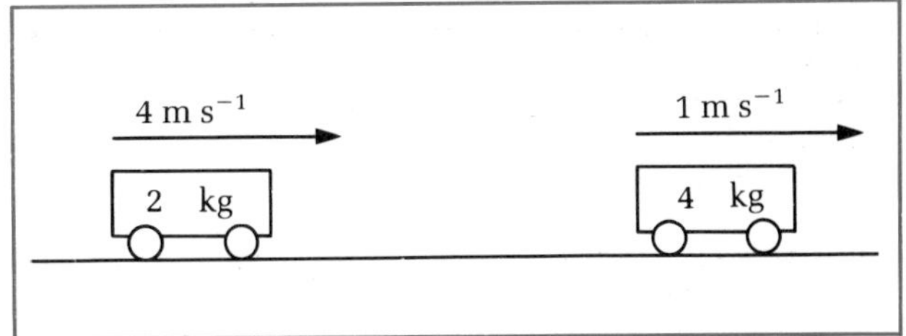

Figure AM.3

After impact they move off together. The amount of kinetic energy lost is

A 4 J **B** 6 J **C** 12 J **D** 14 J **E** 18 J (*Scottish H*)

AM.9 An explosive charge separates a space rocket, of mass *M*, from its launcher, of mass 4*M*. If the launcher moves away with a velocity of 10 m s^{-1} relative to its original motion, the velocity of the rocket, in m s^{-1}, also relative to the original motion, is

A −2.5 **B** −4.0 **C** −10 **D** −20 **E** −40 (*AEB June 82*)

AM.10

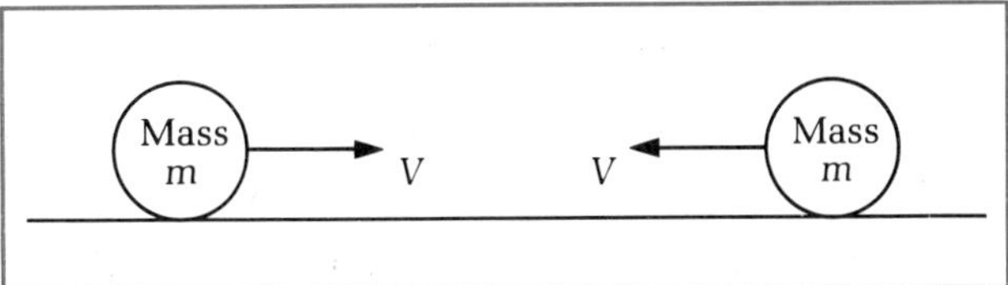

Figure AM.4

Two perfectly elastic spheres having the same mass travel with the same speed *V* and collide head-on, on a smooth horizontal surface as shown above. Which of the following statements is correct?

A The sum of the momenta before impact is 2 mV.
B The sum of the momenta after the impact is 2 mV.
C The sum of their kinetic energy after impact is zero.
D The sum of their kinetic energy after impact is MV^2.
E They coalesce on impact giving a resultant speed of zero.

(*AEB June 84*)

AM.11 A steel ball of mass 0.005 kg falls vertically through a liquid at terminal speed of $0.1\,\mathrm{m\,s^{-1}}$. The energy dissipated per second in the liquid by the ball's motion, in mJ is

A 0.5 **B** 1.0 **C** 5.0 **D** 10 **E** 50

AM.12 A 'block and tackle' used to raise a 300 N weight by 15 m requires an effort of 50 N over a distance of 120 m. The efficiency of the system is

A 6% **B** 8% **C** 25% **D** 75% **E** 100%

AM.13 A brick slides across a wooden floor and strikes a box. They move together to the right, gradually slowing down.

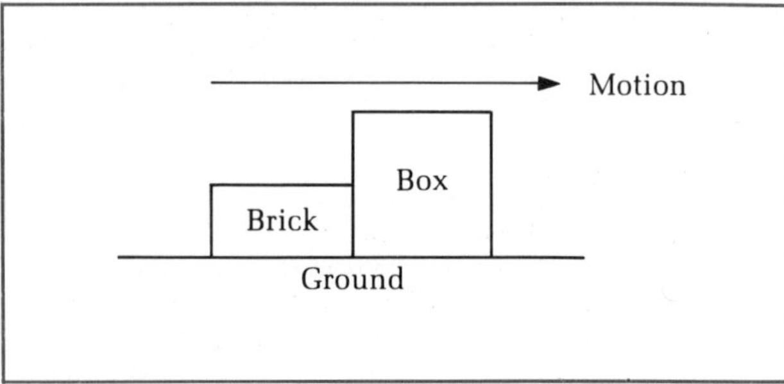

Figure AM.5

1 The push of the brick on the box could be greater than the push of the box on the brick.
2 The push of the brick on the box could be less than the push of the box on the brick.
3 The push of the brick on the box could be equal to the push of the box on the brick.

A 1, 2, 3, correct **B** 1, 2, correct **C** 2, 3, correct
D 1 only correct **E** 3 only correct

(London)

AM.14 The torque produced by the forces *F* about the point O is

A $(Fx+Fd)$ **B** $(Fd-Fx)$ **C** $(Fx-Fd)$
D $F(x+\frac{d}{2})$ **E** Fd

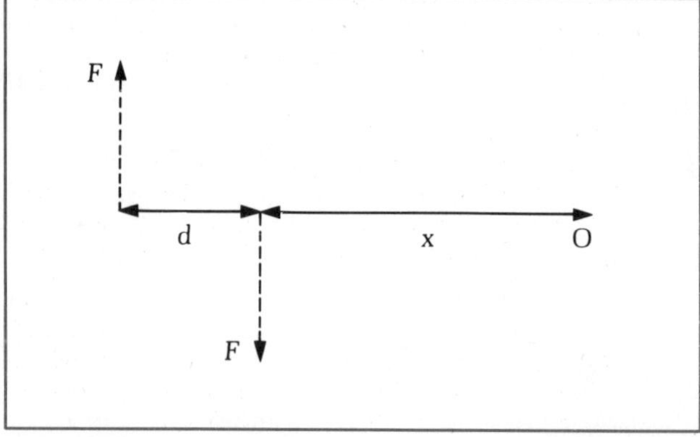

Figure AM.6 (*AEB June 84*)

AM.15 The resultant of two forces of magnitude *F* and 2*F* can be

A less than *F* **B** more than 3*F* **C** zero
D perpendicular to *F* **E** perpendicular to 2*F*

AM.16 A weight *W* hangs vertically from a string XYZ by a thread tied to the string at Y. The section XY of the string is horizontal and the section YZ is at 20° to the horizontal, as shown in the diagram. The tension in section XY is

A *W* sin 20° **B** *W*/sin 20° **C** *W* tan 20°
D *W*/tan 20° **E** *W*

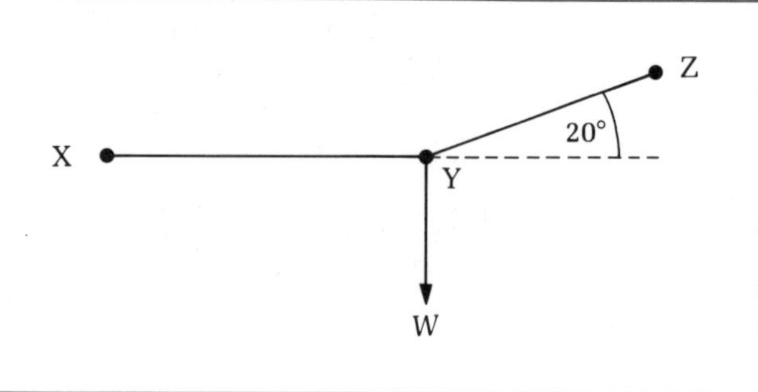

Figure AM.7

AM.17 A simple pendulum consists of a bob of mass *m* at the end of a thread of length *l*. The bob is held so the thread is horizontal with the thread taut. The bob is then released. The tension in the thread as the bob passes through equilibrium is

A *mg* **B** 1.5 *mg* **C** 2 *mg*
D 2.5 *mg* **E** 3 *mg*

AM.18 A body of mass 1 kg is supported from a fixed point by a light inextensible string of length 1 m. The body is made to rotate about the fixed point in a vertical circle so that its speeds at the highest and lowest points of the circular path are the same. If the gravitation field strength is $10\,\mathrm{N\,kg^{-1}}$ the difference in tension, in N, in the string at these two points will be

A 0 **B** 1 **C** 2 **D** 10 **E** 20

(*AEB June 85*)

AM.19 When an aeroplane turns on a horizontal path at steady speed V, the resultant force on it is due to

1 the lift force **2** its weight **3** the drag force

A 1 and 2 **B** 1 and 3 **C** 1, 2 and 3
D 1 only **E** 2 only

AM.20 A stone thrown from X follows a parabolic path, reaching its greatest height at Y and returning to the same level as X at Z.

1 The time from X to Y is greater than the time from Y to Z.
2 Its speed at X is the same as its speed at Z.
3 Its speed at Y is zero.

A 1 and 2 **B** 1 and 3 **C** 1, 2 and 3
D 1 only **E** 2 only

A
Long questions

AL.1

a) A car of mass 1000 kg is initially at rest. It moves along a straight road for 20 s and then comes to rest again. The speed-time graph for the movement is given below.

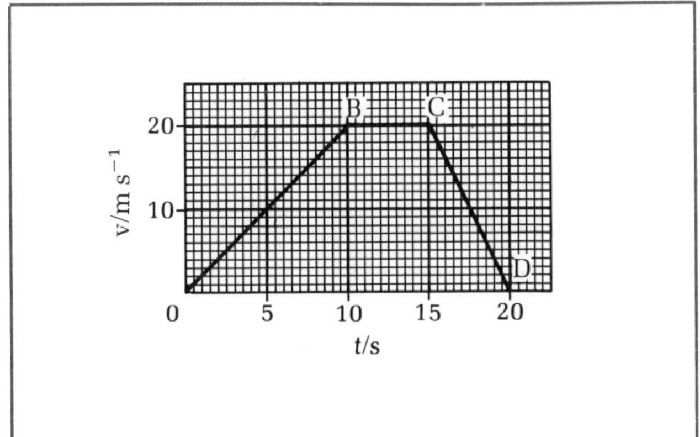

Figure AL.1

i) What is the total distance travelled?

ii) What resultant force acts on the car during the part of the motion represented by CD?

iii) What is the momentum of the car when it has reached its maximum speed? Use this momentum value to find the constant resultant accelerating force.

iv) During the part of the motion represented by 0B on the graph, the constant resultant force found in (iii) is acting on the moving car although it is moving through air. Sketch a graph to show how the driving force would have to vary with time to produce this constant acceleration. Explain the shape of your graph.

b) If, when travelling at this maximum speed, the 1000-kg car had struck and remained attached to a stationary vehicle of mass 1500 kg, with what speed would the interlocked vehicles have travelled immediately after collision?

Calculate the kinetic energy of the car just prior to this collison and the kinetic energy of the interlocked vehicles just afterwards. Comment upon the values obtained.

Explain how certain design features in a modern car help to protect the driver of a car in such a collison.

(London)

AL.2

a) An isolated body of mass m, initially at rest, is acted upon by a constant force F. Derive an expression for the distance s travelled from the rest position in time t.

Write down an expression for the work done on the body by the force F and show that this work is equal to $\frac{1}{2}mv^2$, where v is the final velocity.

b) Define linear momentum and state the law concerning its conservation. How is linear momentum related to force?

c) Explain why, when catching a fast moving ball, the hands are drawn back while the ball is being brought to rest. Discuss whether your explanation has any bearing on the use of crushable boxes for packing eggs.

d) A particle, travelling along a straight path with velocity v, explodes into two equal parts. The forces due to the explosion act along the original direction of travel. The explosion causes the kinetic energy of the system to be doubled. Determine the subsequent velocities of the two segments.

(AEB)

AL.3 State the principle of conservation of linear momentum.

Describe an experiment to demonstrate momentum conservation, and give an estimate of the accuracy attainable.

An aircraft jet engine takes the air at the rate of 50 kg s^{-1}, and burns fuel at the rate of 1.0 kg s^{-1}. The products of combustion are ejected with a velocity of 500 m s^{-1} *relative to the aircraft.* Calculate the thrust exerted by the engine when the aircraft is stationary on the runway. Calculate also the power used in providing the kinetic energy of the ejected gases. Why is this less than the total power produced by the combustion of the fuel?

The aircraft is now travelling through still air at a speed of 80 m s^{-1}, calculate the change of momentum per second of the air taken in by the engine and of the fuel consumed. Hence calculate the thrust exerted by the engine. Determine also the power used to drive the aircraft forward.

(SUJB)

AL.4 The ideas of

conservation of momentum
conservation of total energy
conservation of kinetic energy

are useful for analysing collisions as varied as traffic accidents and events between sub-atomic particles.

Choose **three** collisons. Your examples should include elastic and inelastic collisions and microscopic and macroscopic events. Discuss as fully as possible the application of the above ideas to each collision chosen, using estimates and calculations to illustrate the usefulness of the ideas.

(O and C Nuffield)

AL.5 This question is about the use of windmills as a source of energy.

It is proposed that a cluster of these built out to sea could make a substantial contribution to our needs. Answer the following questions which are about the physics and economics of this idea. The data given at the end of the question will be useful. Wherever appropriate, you should make quantitative estimates to support your arguments.

a) If a blade of area A intercepts a wind of speed v, it can generate power given by $\frac{1}{2}\rho Av^3=\frac{1}{2}\rho v^2(Av)$, where ρ is the density of air. What physical quantities do $\frac{1}{2}\rho v^2$ and Av each represent? Explain why they are multiplied to give the power.

b) The wind speed (v) actually varies a lot. As a result the mean power is given by $K(\frac{1}{2}\rho Av_m^3)$, where K is a constant with approximate value 2.4 and v_m is the mean speed over the whole year. Show, by calculation from the data given in the diagram at the top of the next page, that K is about 2.4.

c) The speed of the wind over the sea varies with height according to $v_H=CH^{0.14}$, where H is the height and C is a constant. Use this equation to show why there is some advantage in building a high structure for a windmill (say 50 m rather than 10 m high). What other

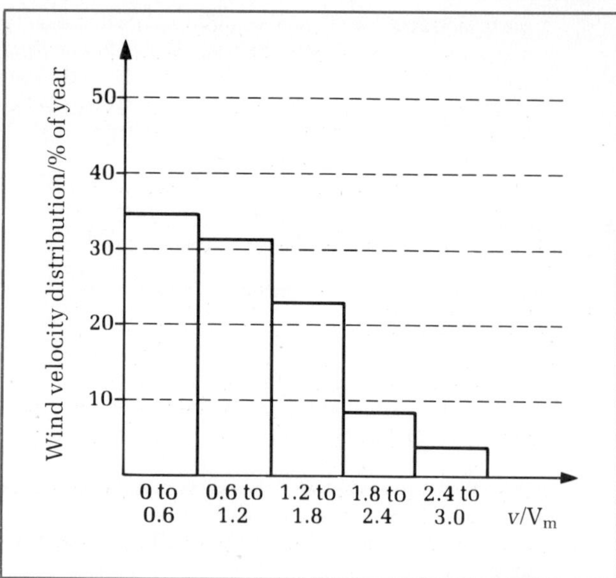

Figure AL.2

factors would you need to take into consideration in deciding the best height to build a windmill?

d) How big would the total blade area of the windmill need to be, given that:

v_m for the region was 9.0 m s^{-1} at 50 m height
the windmill is to be 50 m high
the windmill is to be designed to operate at about 1.5v_m
the power generated is to be 2.5 MW at the design speed?

Say, with reasons, whether you would expect your design to be practicable.

e) The windmills in a cluster cannot be too close behind one another; they must be set apart with a spacing of between 7 and 10 times the largest blade dimension. Suggest one reason for this.

f) Make a rough estimate of the cost of electrical energy generated by a cluster of 400 such windmills designed to produce a mean power of 100 MW. Assume that the capital cost is to be met by a loan on which interest has to be paid at 15% per year.

Data

Density of air $\approx$ 1 kg m^{-3};
area off east coast of England at suitable depth near the Wash is 4000 km^2;
cost of one off-shore windmill to generate 2.5 MW is about £600 per kW of which 40% is cost of the tower;
windmill designed for 1.5 v_m will operate for effectively 40% of the time;
cost of cables for transmission over 30 km out to sea is about £50 per kW;
number of seconds in a year $\approx 3 \times 10^7$.

(O and C Nuffield)

AL.6

a) A particle A of mass *m*, moving with velocity *u*, in the direction shown (Figure AL.3 (a)), makes a head-on, elastic collison with a particle B of mass *M* that is originally at rest. After the collison, A and B move off with velocities *v* and *V* respectively (Figure AL.3 (b)).

i) Write down the equations that summarise the application of the principles of conservation of energy and momentum to this collision.

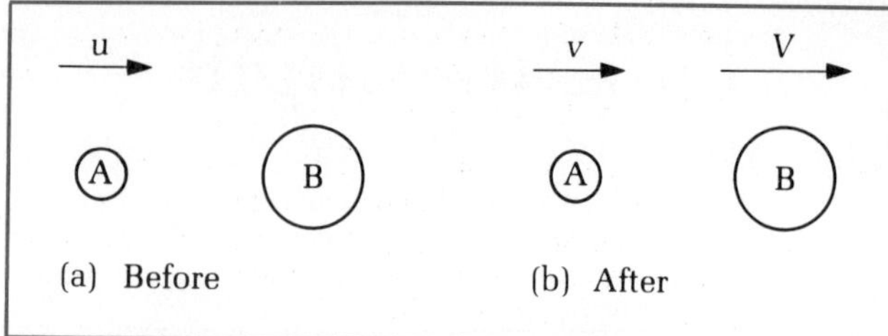

Figure AL.3

ii) It can be shown from these equations that $v = V - u$. Using this result, or otherwise, find an expression for the fractional loss of kinetic energy of A, in terms of *m* and *M* only.

iii) Evaluate this fractional loss for $M = 50\,m$.

b) An alpha-particle is projected directly towards a stationary gold nucleus, which is free to move, and collides with it.

i) Sketch a graph showing how the electrostatic force between the alpha-particle and the nucleus depends on their separation x.

ii) Discuss whether the existence of this force would make it incorrect to calculate the fractional loss of kinetic energy of the alpha-particle in its collision with the gold nucleus by the method you have followed above.

c) Evidence for the existence of the nucleus came from observations of the way in which alpha-particles were scattered by thin metal foils.

i) Give a labelled sketch of an apparatus suitable for studying this scattering.

ii) What aspects of the results of such experiments indicated that atoms contain small, massive, positively-charged nuclei?

(Cambridge)

AL.7

a) What do you understand by the *principle of conservation of energy*?

b) Explain how the principle applies to: (i) an object falling from rest in vacuo, (ii) a man sliding from rest down a vertical pole, if there is a constant resistive force opposing the motion. Sketch graphs, using one set of axes for (i) and another set for (ii), showing how each form of energy you consider varies with time, and point out the important features of the graphs.

c) A motor car of mass 600 kg moves with constant speed up an inclined straight road which rises 1.0 m for every 40 m travelled along the road. When the brakes are applied with the power cut off, there is a constant resistive force and the car comes to rest from a speed of 72 km h^{-1} in a distance of 60 m. By using the principle of conservation of energy, calculate the resistive force and the deceleration of the car.

(JMB)

AL.8

a) A particle of mass *m* moves with a constant speed *v* in a circular path of radius *r*. Show that a force is necessary to maintain the circular motion and derive an expression for its magnitude. State clearly the direction of the force on the particle.

b) A small mass hangs by a string from a fixed point and moves in a circular path at constant speed in a horizontal plane. Draw a diagram showing the forces acting upon the mass and derive an equation showing how the angle of inclination of the string to the vertical depends upon the speed of the mass and the radius of the circle in which it moves.

c) In the arrangement described in **b)** the mass is 0.50 kg, the radius of the circle is 1.00 m and the mass makes 30 revolutions per minute. The string suddenly breaks and the mass falls freely to the ground through a vertical distance of 1.00 m. Calculate
  i) the inclination of the string to the vertical immediately before the string breaks,
  ii) the horizontal distance travelled by the mass from its position when the string breaks to the point of impact,
  iii) the change in the kinetic energy of the mass during its free fall. *(JMB)*

AL.9

a) A body, X, of mass m_1 is moving with a velocity u towards a stationary body, Y, of mass m_2 along a line joining their centres of mass. X collides with Y and adheres to it, the two bodies moving with a velocity v after the collision along the same line. Write down an equation in terms of m_1, m_2, u and v relating the momentum of the system before the collision to the momentum of the system after the collision.

b) Describe, with the aid of a labelled diagram, an experimental arrangement that you could use in a school laboratory to test the validity of the equation you have written in (a). Describe the procedures you would adopt to make the relevant measurements and show how you would use them to draw your conclusions.

c) One trolley of mass 0.25 kg and another of mass 0.50 kg, each of length 0.20 m, are held in contact on a horizontal runway. A spring inside one of the trolleys is released and they are forced apart with a total kinetic energy of 0.75 J.
  i) By considering energy and momentum changes, calculate the velocity of each trolley immediately after separation.
  ii) Each trolley travels a distance of 1.0 m after separation, strikes a rigid vertical obstacle and on rebounding loses 19% of its kinetic energy. Ignoring both the effect of friction and the duration of the collisions with the obstacles, calculate the position on the runway at which the trolleys next collide. *(JMB)*

AL.10

a) The passenger cars on a roller-coaster at a fun-fair (in the diagram below) are pulled up to a point A and released.

A safety officer has to measure experimentally the speed of the cars as they pass the lowest point B on the track.
Describe how the safety officer could carry out the experiment. Your description should include:
  i) a sketch of the apparatus used,
  ii) the procedure carried out and the measurements taken,
  iii) an indication of how the final result would be obtained.

b) During one run, a car and passengers of mass 800 kg are released from rest at point A, a height of 20 m above the ground. The car travels a distance of 120 m along the track until it reaches point C, a height of 15 m above the ground. A constant frictional force of 250 N acts between the car and the track as the car moves from A to C.
  i) State the energy changes which occur as the car moves from A to C.
  ii) Calculate the work done against friction in moving the car from A to C.
  iii) Find the kinetic energy of the car on reaching point C.

c) At the fair, a boy and a girl are seen bouncing on separate trampolines as shown below.

Figure AL.5

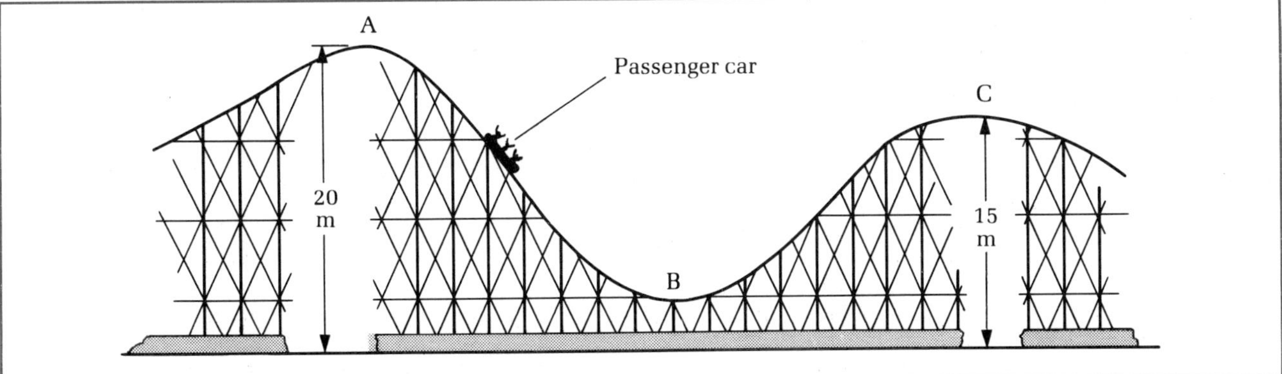

Figure AL.4

These graphs show how the forces exerted by the trampolines on the children vary with time. Graph A represents the force on the girl and graph B the force on the boy.

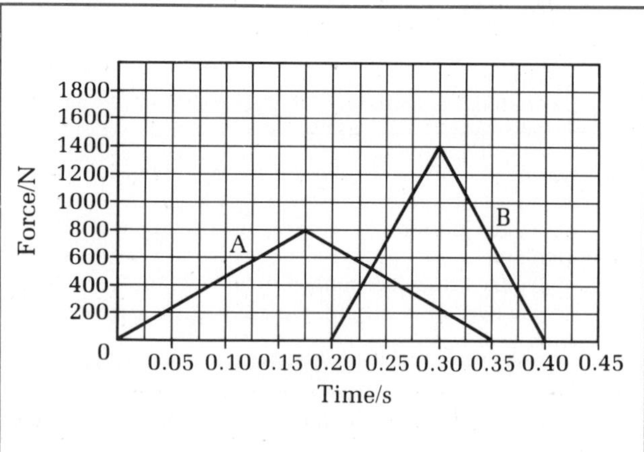

Figure AL.6

i) 'Each child undergoes the same change in momentum during one bounce.'
Justify this statement.

ii) The boy has a greater mass than the girl. Both children drop from the same height on to their trampolines.
Which child will rebound from the trampoline with the greater speed?
Explain your answer. *(Scottish H)*

AL.11 A shell of mass 0.2 kg is fired vertically upwards with a speed of $100\,\mathrm{m\,s^{-1}}$, and 15 s later arrives back at ground level with a speed of $80\,\mathrm{m\,s^{-1}}$. Draw sketches showing the directions of the forces acting on the shell (a) during the upward flight, and (b) during the downward flight.

Find the total work done against air resistance during the flight.

Find the total change in momentum and hence find the mean force acting on the shell during the flight.

Sketch the form of the speed-time graph for the whole flight and, with the aid of your graph, discuss whether the time for the upward journey is greater than, equal to, or less than, the time for the downward journey.
[Assume that g is constant.] *(WJEC)*

AL.12 Show, by means of sketches accompanied by brief comments, how the law of conservation of linear momentum can be deduced through a series of experiments on the linear air track or on any other convenient mechanical system.

Describe carefully how you would measure the essential quantities in your experiments.

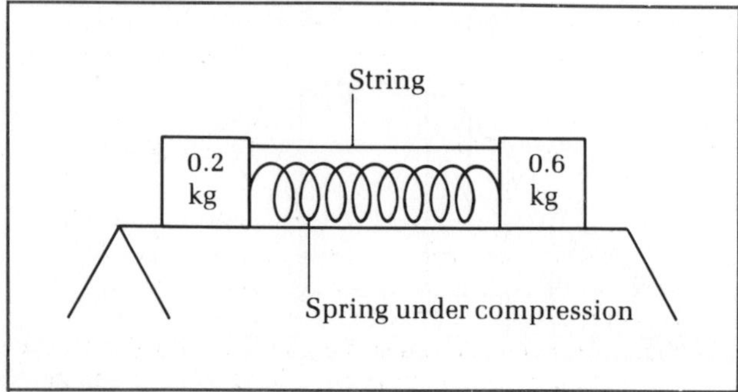

Figure AL.7

The diagram shows two riders on a linear track with a light spring under compression placed between them. Originally the arrangement is stationary with a light string holding the riders in place; the spring is not attached to either rider. The string is then cut, and it is found that the heavier rider acquires a velocity of $0.03\,\mathrm{m\,s^{-1}}$. Find the velocity acquired by the second rider and deduce the potential energy originally stored in the spring.

After rebounding elastically from buffers at the ends of the track, the two riders collide perfectly inelastically. Describe the state of motion after this collision.

(WJEC)

AL.13

a) What physical events occur at a molecular level during the non-elastic collision of two solid bodies?

b) Outline an experiment that could be carried out with laboratory equipment to determine the fraction of the available kinetic energy dissipated during the collision of two solid bodies. Describe the apparatus used, the procedure, and how the result would be obtained.

c) A car travelling at a steady speed of $30\,\mathrm{m\,s^{-1}}$ is acted on by air resistance which may be regarded as equivalent to that experienced by a flat surface of area $0.50\,\mathrm{m^2}$ held with its plane normal to the direction of motion. As a simplification, the incident air is taken as stationary before impact with the car and assumed to accelerate to the speed of the car on impact. All other aerodynamic effects are neglected.
(Take the density of air to be $1.2\,\mathrm{kg\,m^{-3}}$.)
Calculate:

i) the mass of air striking the surface per second,
ii) the force acting on the surface,
iii) the work done against this force during a 100 km journey made at a constant speed of $30\,\mathrm{m\,s^{-1}}$,
iv) the power exerted to sustain this constant speed.

In order to overcome this air-resistance, 15% of the mechanical output of the car engine is used. What is the total petrol used during the 100 km journey if the engine produces 40 MJ for each litre of petrol consumed?

How is the residue of an engine's mechanical output utilized after air-resistance has been accounted for?

(Oxford)

AL.14

a) A very bouncy ball is held at a height of 1.00 m and then released. It bounces to a height of 0.83 m before falling to the floor again.

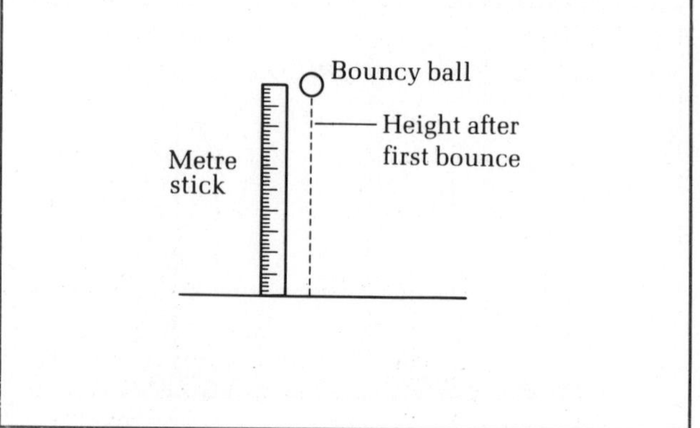

Figure AL.8

i) Calculate how long it took for the ball to drop to the floor from both of these heights.
ii) State how long it took to rise from the floor to its maximum height of 0.83 m after its first contact with the floor.
iii) Calculate the speed of the ball just before contact with the floor.
(A) for the first time;
(B) for the second time.
iv) State the speed with which it left the floor after the first contact.

b) i) Use the results of your calculations in part (a) to draw a velocity-time graph of the motion of the ball from release until it strikes the floor for the second time.
You may assume that the time of contact is very small compared with the other times involved.
ii) On the same axes, use a dotted line to sketch the graph you would expect to obtain if the experiment were repeated with a soft sponge ball.

c) A boy attempts to measure the value of g by dropping a piece of chalk on to the turntable of a record player, rotating at 45 revolutions per minute.
His only measuring instruments are a metre stick and a protractor.
He uses the relationship $s = \frac{1}{2}gt^2$.

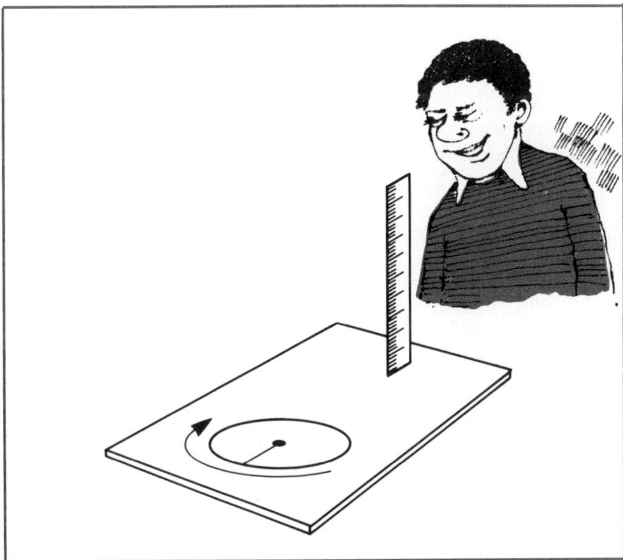

Figure AL.9

i) Describe how he could obtain the necessary values of distance and time from the experiment to allow him to find a value for g.
ii) State one hint you would give him to ensure that his value for g is as accurate as possible.

(Scottish H)

AL.15 A rubber ball, initially at rest, is dropped from a height of 2.5 m onto a horizontal floor. It rebounds several times to a height which you may assume is unchanged after each bounce.

a) Calculate how long the ball takes to reach the floor for the first time.

b) Sketch graphs showing how the following properties of the ball vary with time for the first three bounces: (i) height above the floor, (ii) velocity, (iii) acceleration. Where possible indicate numerical values on the axes.

c) Discuss the relationship between kinetic energy, potential energy and total energy of the ball during the motion. Include a consideration of the short time during which the ball is in contact with the floor.

d) State **three** ways in which the motion, although periodic, differs from simple harmonic motion.

A second ball, less elastic than the first, is now used. Each time the ball hits the floor it rebounds with 0.95 times the speed with which it hits the floor. Starting from the same initial conditions as the first ball, find how many bounces the ball makes before its rebound height is less than 1.5 m.

(Oxford and Cambridge)

AL.16

a) State the conditions under which (i) *linear momentum* and (ii) *kinetic energy* is conserved in a collision between two masses.

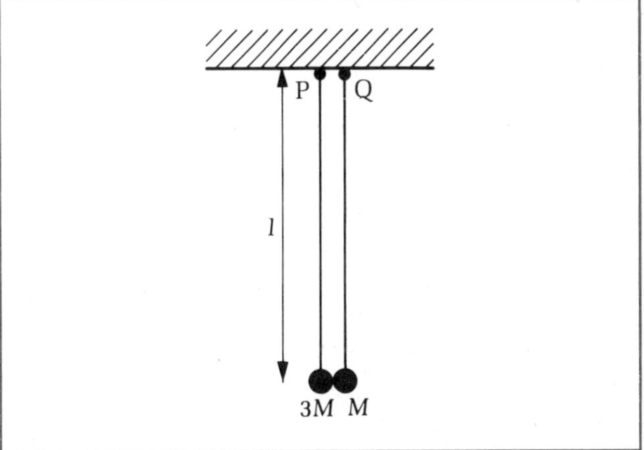

Figure AL.10

b) Two small spheres of mass $3M$ and M hang on strings from two fixed points P and Q at the same level so that their centres of mass are also at the same level and the spheres are in contact. It may be assumed that the radii of the spheres are very small indeed compared with the distance l of the centres of mass below P and Q. The mass $3M$ is drawn aside, keeping its string taut until it makes an angle θ with the vertical and then released from the rest.

Obtain expressions for
i) the speed with which the mass $3M$ collides with the stationary mass M, neglecting air resistance,
ii) the speeds which the masses have immediately after the collision, assuming it is perfectly elastic.

With the aid of diagrams, discuss the subsequent motion as fully as possible. Explain why it is periodic if θ is small and state the period.

(Oxford and Cambridge)

AL.17 This question is about the use of solar cells as sources of electric power.

In tests on a single solar cell, the cell delivers power to a load resistance. If the load resistance is changed, the potential difference across it and the current through it will also change. Figure AL.11, on the next page, shows the relation between p.d. and current for a single cell operating at a light intensity equal to that of bright sunlight at the Earth's surface (100 mW cm $^{-2}$).

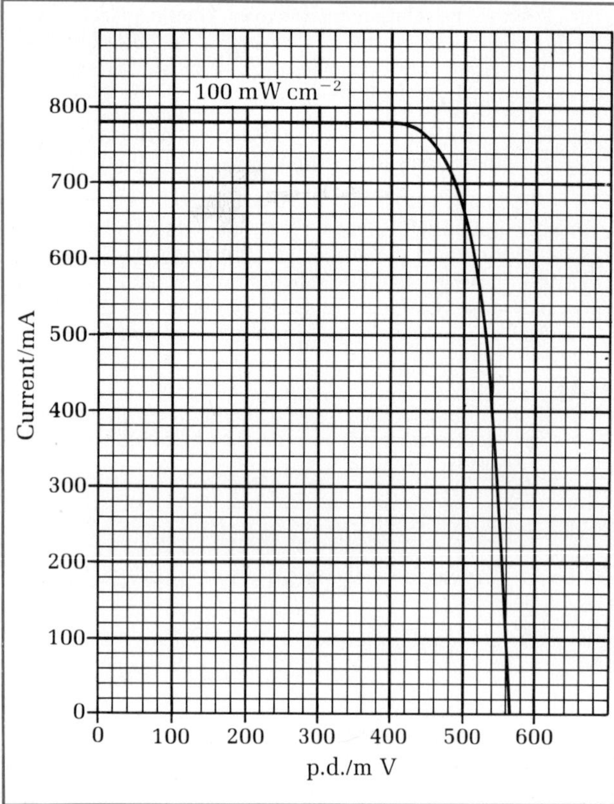

Figure AL.11

a) Draw the circuit, using a variable resistor and suitable meter(s), which you would use if you wanted to take electrical measurements on the cell to check the shape of the curve.
Specify suitable sensitivity ranges and resistances for the meter(s) and the range of values needed for the resistor.

b) This cell behaves very differently from a normal dry cell. Sketch the current–p.d. curve you would expect to obtain if a dry cell were connected to a variable load.

c) The solar cell delivers maximum power to an external load at about 450 mV.
Sketch a curve of power delivered as a function of p.d. across the cell terminals to show how this maximum occurs, and so estimate the value of the resistor and of the power delivered to it at the maximum.

d) Given that the surface area of the solar cell is 25 cm², what is the maximum efficiency of conversion of light energy to electrical energy?

e) The circuit below shows an arrangement of ten solar cells, each of the type described above, connected to supply current to a load but also to recharge a nickel-cadmium cell which acts as a reserve energy store.

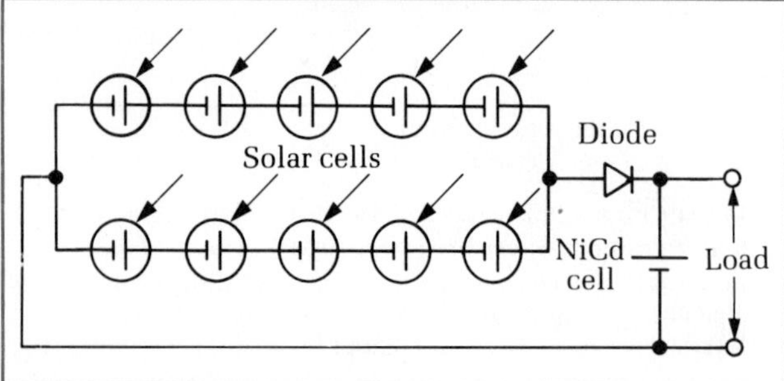

Figure AL.12

i) Why do you think five cells have been connected in series?
ii) Why do you think two such sets of five have been connected in parallel?
iii) What is the maximum power that this arrangement of solar cells can supply?
iv) What is the function of the diode?
v) If the sun shines only for an average of 3 hours per day, what is the average continuous power that the system can supply? (Assume that the nickel-cadmium cell can store and re-convert energy with negligible loss.)
vi) Estimate the area of solar cells required to provide the electric power needs of an average household, assuming there are three hours of sunshine per day.

(O and C Nuffield)

AL.18

a) Write an expression for the kinetic energy of a rigid body rotating about a fixed axis in terms of ω, its angular velocity, and I, the moment of inertia of the body about that axis.

A rotating stool is mounted on a frictionless vertical screw which is fastened to the ground. The moment of inertia of the seat about the axis of the screw is 3.0×10^{-2} kg m² and its mass is 1.2 kg. The seat is made to spin with an initial angular velocity of 2.0 revolutions per second so that it rises.

Calculate
i) the initial kinetic energy of the seat,
ii) the height through which it rises before coming to rest, and
iii) the initial angular momentum of the seat.

b) Explain what is meant by the torque, T, exerted on a body about an axis. Give an equation relating T to α, the angular acceleration of the body about that axis.

A Catherine wheel is a firework consisting of a uniform disc of diameter 0.12 m which can rotate freely about a horizontal axis through its centre, perpendicular to its plane, by ejecting combustion products at one point on its edge. The moment of inertia of the disc about its axis is 5.0×10^{-4} kg m², and the diameter and the moment of inertia may be assumed constant during the first 0.50 s after ignition. Initially the combustion products are expelled tangentially with a speed of 2.0 m s^{-1}, a mass of 5.0×10^{-3} kg being ejected in the first 0.50 s. Calculate the average acceleration during this period.

(JMB)

AL19 Explain what is meant by *angular acceleration* and *torque*, and give the SI unti for each quantity.

Write down the mathematical relationship between angular acceleration α, torque T and moment of inertia I.

Upon what factors does the moment of inertia of a body depend?

A particle A of mass 0.022 kg is attached to a light string which passes through a smooth hole O in a smooth horizontal table and is held at a point below the table (see Figure AL.13). The particle is set into uniform motion in a circle of radius 0.5 m about O, making one revolution every 0.80 s.

a) Draw a diagram showing the magnitudes and directions

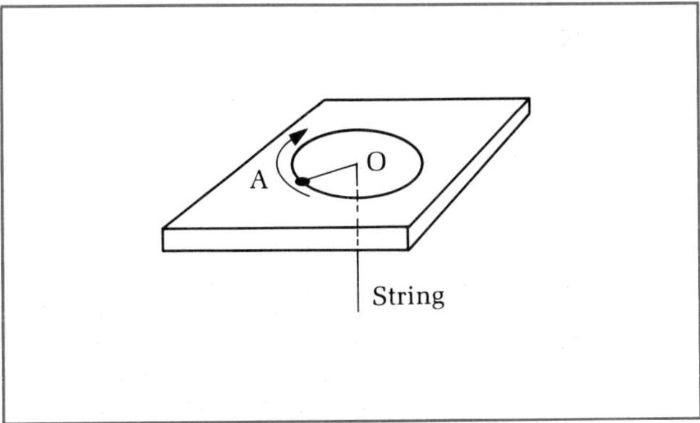

Figure AL.13

of the forces acting on the particle when it is rotating and is in the position shown.

The string is now pulled down slowly so that the radius of rotation is reduced to 0.25 m.

b) Explain why the angular momentum of the system remains constant during this operation, and hence find the new angular speed of the particle.

c) Calculate the work done in pulling the string down.

(Cambridge)

AL.20

a) Describe an experiment to determine the moment of inertia of a flywheel about its axis. The account should include details of the measurements taken, explain how the effects of friction are allowed for, and show how the result is calculated.

b) A flywheel is driven about a vertical axis at a constant angular speed of 200 rad s^{-1}. A stationary metal disc with a hole through its centre is dropped onto the rotating flywheel, as shown in Figure AL.14. The friction between the disc and the flywheel causes the disc to accelerate uniformly to an angular speed of 200 rad s^{-1} in a time of 0.5 s. The moment of inertia of the disc about an axis through its centre and perpendicular to its plane is 6×10^{-3} kg m^2.

Find:

i) the couple acting on the disc,

ii) the kinetic energy gained by the disc,

iii) the total energy transferred by the couple.

Explain why the answers to (ii) and (iii) are not equal.

(Oxford)

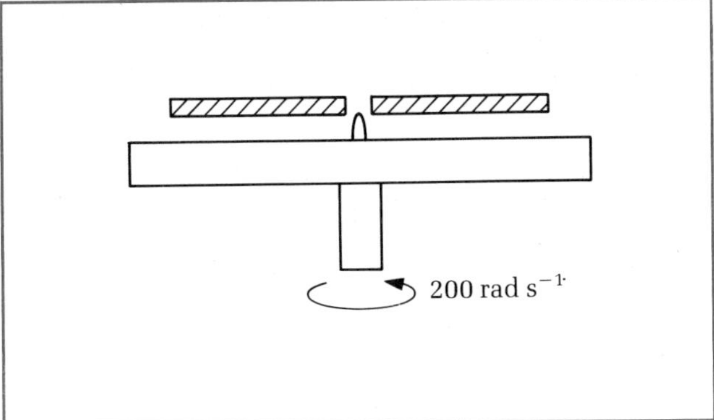

Figure AL.14

SECTION B

Materials

Molecules and materials · Strength of solids · Fluids · Thermal properties · Gases · Thermodynamics

Effective use of materials requires knowledge of their properties. Such knowledge is achieved by measuring and testing materials, then developing models and theories to explain and understand the measurements. Understanding materials in molecular terms is essential if the limits of their use are to be appreciated. When considering which materials to use in any given situation, the demands on them must be specified first. Then the user can choose materials capable of meeting those demands.

In this section, the emphasis is on structure and energy so the ideas of force and energy are used extensively. Energy at the molecular level determines the physical state of a material; the molecular structure of the material in that state determines its physical properties such as strength and stiffness if it is a solid, or its flow properties if it is a liquid.

Windsurfing; the photograph shows a windsurfer controlling the sailboard, so that the wind forces the board in the required direction. Sailboards use lightweight modern materials capable of withstanding the force of the wind and the waves.

6

Molecules and materials

6.1 Atoms and molecules

Elements and compounds

Look around you and count how many different materials you can see. Add to your list other materials you can think of, such as oil, bone, helium, alcohol. The list could, given time, be almost endless, yet all the materials known to us are composed of just 92 or so elements. Centuries ago it was thought that there were just four basic elements, **earth**, **water**, **air** and **fire.** Modern science, by which we mean discoveries made from about the 17th century, easily dismissed such primitive ideas by showing for example that air was composed of at least two elements (oxygen and nitrogen). However, many steps over many years were necessary before it was shown that there are just about 92 elements.

Most substances are in the form of **compounds**, which means that they can be broken down into other substances. An **element** is a substance which cannot be broken down into anything else. For example, water is a compound because it can be broken down into oxygen and hydrogen. But oxygen and hydrogen are both elements because they are impossible to break down. Many of the elements are metals, but not all metals are elements. For example, copper and tin are both elements but brass is a mixture, called an **alloy**, of copper and tin.

Suppose you could cut a metal element into smaller and smaller pieces, cutting each small piece into even smaller pieces. Eventually you would be down to the smallest possible pieces which could be identified with the metal used: single atoms. Cut the piece any smaller and the atom would be broken into electrons, protons and neutrons which could have been produced from any element.

Atoms are the smallest particles of an element which can be identified as being from that element.

Atoms join together to form molecules which are the 'building blocks' of all substances. For example, each molecule of water is composed of two hydrogen atoms and one oxygen atom. Another example is oxygen gas which has molecules each composed of two oxygen atoms. Imagine some of the possible combinations of just a few types of atom; with 92 or so different types of atoms known, the range of possible combinations is huge. Hence the enormous variety of materials.

Molecules are the smallest particles of an element or compound which can exist independently.

For example, oxygen atoms at room temperature do not exist independently. They join in pairs to form oxygen molecules. So the symbol O_2 represents a molecule of oxygen which is composed of two oxygen atoms. Each element has its own symbol; each compound has the symbols of the elements into which it can be broken down, usually written to show how many of each type of atom there are in each molecule. So the symbol for carbon dioxide is written as CO_2, indicating that each molecule has one carbon atom and two oxygen atoms joined together.

Avogadro's ideas

What evidence can be offered to show the existence of molecules? We could never really hope to cut a material down into molecules, as suggested above. Electron microscopes and X-ray devices enable scientists to investigate and even picture materials at molecular levels. Yet molecular theory has been a part of modern science for almost 200 years, well before such powerful machines were invented.

The idea of molecules was used by Amadeo Avogadro in 1811 to explain the **law of combination of gases**, the discovery that when gases combine, they do so in volumes in simple proportions. For example, when hydrogen and oxygen gases combine to form water, the volume of hydrogen used is twice that of the oxygen used. Avogadro put forward a **hypothesis** (or unproved theory) to explain this law. His theory was that **equal volumes of gases at the same conditions contain equal numbers of molecules.** So for example, because one volume of oxygen combines with exactly two volumes of hydrogen, then each oxygen molecule reacts with two hydrogen molecules. Avogadro established another important idea – that molecules are combinations of atoms – to help explain the combination of gases. However, he was unable to supply direct evidence for the existence of molecules, neither could he estimate how many molecules were in a given volume. But his ideas contributed enormously to scientific progress.

Brownian motion

Observation of individual smoke particles using a microscope gives nearly direct evidence for molecules. Figure 6.1 shows an arrangement for viewing smoke particles in air. A beam of light is directed at a cell containing the smoke. Some of the light is reflected by the particles into the microscope. An observer looking down the microscope sees the particles as pinpoints of light. The particles move about in an erratic, unpredictable way, continually changing direction at random. Their motion is

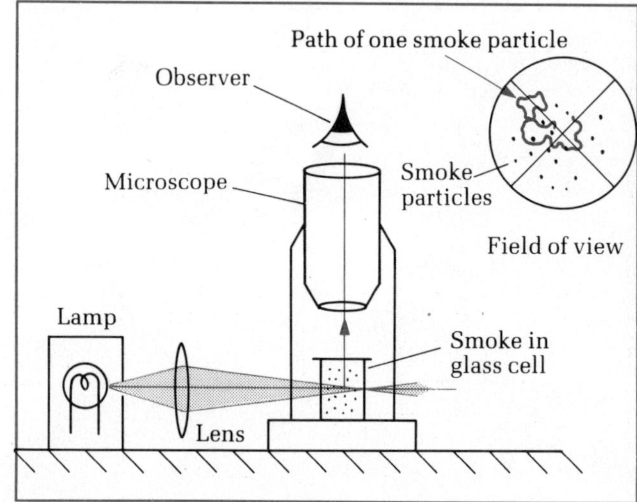

Figure 6.1 *Brownian motion*

called **Brownian motion**, after Robert Brown who first observed it, in 1827, with pollen grains in water.

The cause of the motion is the continual impacts on each smoke particle by air molecules which are too small to be seen. The smoke particles are much larger than the air molecules so each smoke particle is bombarded from different sides by air molecules; but the bombardments are random and uneven. The result is that each smoke particle is pushed around haphazardly. The random nature of the bombardments is because the air molecules move about at random; the uneven nature is because the smoke particles are small enough to make even bombardment unlikely.

How would the motion be affected if the air temperature were increased, or larger smoke particles were used instead? Higher temperature would make the air molecules move faster so the impacts would be more forceful. The motion of the smoke particles would then be even more agitated. Larger smoke particles would be more evenly bombarded which would reduce their agitation. A greater mass would make them more difficult to push around anyway.

Brownian motion provided the best evidence for molecules in the 19th century. Yet many scientists were unconvinced until Albert Einstein in 1906 worked out the complete theory of Brownian motion.

The Periodic Table of the elements

During the 19th century scientists were able to use Avogadro's ideas to establish the *relative* masses of different types of atoms. For example, accurate measurements of gas densities gave the result that oxygen was 16 times more dense than hydrogen at the same temperature and pressure. Using the idea that equal volumes of gases contain equal numbers of molecules, then a given number of oxygen molecules must have 16 times the mass of the same number of hydrogen molecules. So each oxygen molecule has a mass which is 16 times that of a hydrogen molecule. Using methods like this, scientists were able to determine the relative mass of each type of atom. They gave the results in terms of the lightest known atom, which is hydrogen. Atomic mass is now defined on the carbon-12 scale.

The chemical properties of each element were used to see if there was a pattern if the elements were listed in order of increasing atomic mass. The similar nature of elements eight places apart was noticed for the lighter elements; for example, the third element in the list, lithium, is a very reactive metal. So too is the eleventh (sodium) and the nineteenth (potassium). But the pattern did not continue through all the 63 elements known at that time. Dmitri Mendeleev, in 1869, was the first person to realize why; undiscovered elements made the pattern of known elements confusing. He arranged the list of all the known elements in the form of a table, now known as the **Periodic Table** shown in outline in Figure 6.2.

The elements were listed in order of increasing atomic mass, left to right along each row, then down to the next row, and so on. But the key was to arrange them so that each column contained elements with similar chemical properties, even though gaps had to be left in the rows in some places. Mendeleev boldly predicted that each gap indicated an undiscovered element, and sure enough in time his predictions were confirmed. The Periodic Table is now known to consist of 92 naturally occurring elements and some artificial elements. Artificial elements are all placed after uranium in the Periodic Table. They are created by nuclear reactions.

The nuclear model of the atom

The idea that atoms could not be divided or changed was a key starting point for scientists in the 19th century. They supposed that atoms were indestructible, capable only of being joined together to form molecules. In 1897 J. J. Thomson discovered that matter contains tiny negatively charged particles which we now call **electrons**. He showed that electrons from different elements were identical and so concluded that all atoms contained electrons (see p. 418). Experiments by Ernest Rutherford over the years 1908 to 1912, described on p. 442, showed that every atom contains a point-like positively charged nucleus where most of the mass of the atom is concentrated. We know now that the nucleus contains two types

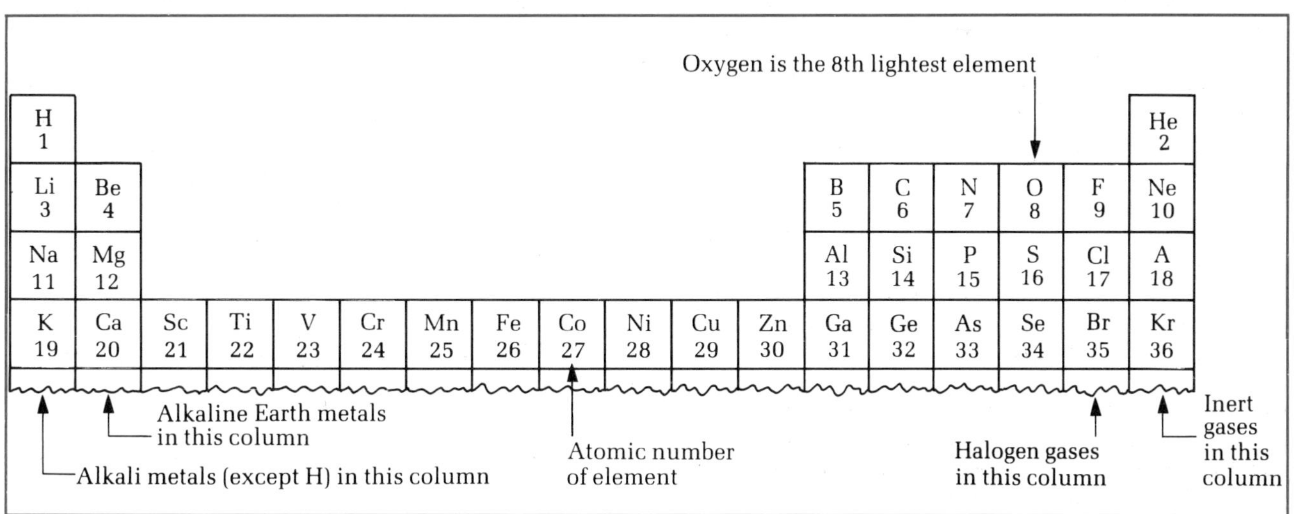

Figure 6.2 *The Periodic Table (top half only)*

	Mass	Charge
Proton	1.0073 u	+e
Neutron	1.0087 u	0
Electron	0.00055u	−e

$e = 1.6 \times 10^{-19}$ C $1u = 1.66 \times 10^{-27}$ kg

Figure 6.3 *Particles in the atom*

of particles, **protons** and **neutrons**, and that the electrons are outside the nucleus. The electrostatic force between each electron and the nucleus prevents each electron from leaving the atom.

The charge of the proton is equal and opposite to that of the electron. Neutrons are uncharged. So an uncharged atom has the same number of electrons around its nucleus as there are protons in its nucleus. Atoms become charged by adding or removing electrons usually. Charged atoms are called **ions**.

The mass of the proton is approximately the same as the mass of the neutron. The mass of an electron is much less at about 1/2000 of the proton mass. (See Figure 6.3.)

The Periodic Table lists the elements in order of increasing atomic mass, so it starts with hydrogen, then helium, then lithium and so on. The order number of each element is called its **atomic number** Z; so for hydrogen, Z=1. For helium Z=2, for lithium Z=3, etc. Rutherford proved that the atomic number is equal to the number of protons in the nucleus.

Atomic number *Z* is equal to the number of protons in the nucleus.

Each hydrogen atom has only one proton in its nucleus. Each helium atom has two protons in its nucleus. Yet the mass of a helium atom is about four times that of a hydrogen atom. Why? Electrons are too light to account for the difference. Neutrons are responsible. A helium atom with a mass about four times that of a hydrogen atom must have two neutrons in its nucleus as well as two protons.

Mass number *A* (sometimes called atomic mass) is the number of neutrons and protons in the nucleus.

We use the atomic number Z and the mass number *A* to identify each type of atom. So the symbol $^{A}_{Z}X$ identifies an atom of element X which has Z protons and $(A-Z)$ neutrons. An uncharged atom would have Z electrons round its nucleus. The electrons of each atom are arranged in **shells** around the nucleus, each shell able to hold up to a certain number of electrons. Each shell is an electron energy level, and the electrons normally occupy the innermost shells available since these are the lowest energy levels. The maximum number of electrons which each shell can hold is worked out from the Periodic Table. The innermost shell which is nearest the nucleus can take up to two electrons; the next shell out can take up to eight electrons, etc. So an uncharged sodium atom (Z=11) which has eleven electrons would have two electrons in the innermost shell, eight electrons in the next shell, then a single electron in the otherwise empty third shell. Sodium is very reactive because that single electron can easily be removed.

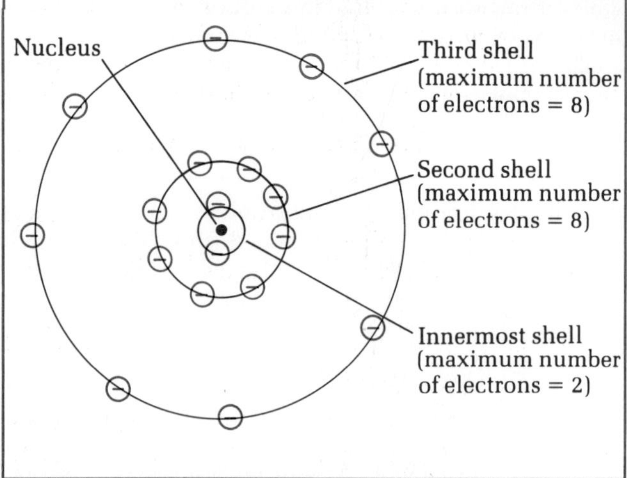

Figure 6.4 *Electron shells of the atom*

Quantum mechanics, a branch of modern physics, gives an explanation of the rules which govern the arrangement of electrons around the nucleus. See p. 455 for further discussion of the ideas about energy levels.

6.2 Measurement of molecules

Molecular mass

The hydrogen atom was the first choice for the standard of atomic mass because it is the lightest. However, it proved to be an unreliable standard because some hydrogen atoms contain extra neutrons (see **isotopes**, p. 445). The accepted standard of atomic mass is now the carbon-12 atom, $^{12}_{6}C$. Although carbon atoms with extra neutrons do exist, they can be removed to leave carbon-12 atoms only. So the unit of atomic mass is now based on the carbon-12 scale.

One unit of atomic mass (*u*) = $\frac{1}{12}$ of the mass of a carbon-12 atom.

So if we state that the atomic mass of nitrogen is 14 u, then we mean that each nitrogen atom has a mass which is 14/12 times the mass of a carbon-12 atom. To determine atomic mass in kilograms, we need to know the mass of a single carbon-12 atom. For this we must know the number of atoms in a given mass of carbon-12.

The *Avogadro constant L* is defined as the number of atoms in 0.012 kg of carbon-12.

X-ray methods have been used to determine the value of L accurately, and its accepted value is 6.023×10^{23}. Now we can work out the mass of a carbon-12 atom in kilograms because 0.012 kg of carbon contains 6.023×10^{23} atoms. So each carbon-12 atom has a mass of 1.99×10^{-26} kg ($=0.012/L$ kg). Therefore 1 unit of atomic mass is one-twelfth of 1.99×10^{-26} kg ($=0.001/L$ kg) $= 1.66 \times 10^{-27}$ kg.

$$1 \text{ unit of atomic mass} = \frac{0.001}{L} \text{ kg}$$

Given the atomic mass of an element we can now determine how many atoms are in a known mass of the element. For example, suppose we wish to know how many atoms are in 1 kg of sodium, given that the atomic mass of sodium is 23 u. Each sodium atom has a mass of $\frac{0.023}{L}$ kg. So 1 kg of sodium contains $\frac{L}{0.023}$ atoms, giving 2.62×10^{25} atoms. Counting in terms of L makes calculations easier.

One mole of any substance is the amount containing *L* particles of the substance.

Any type of particle can be counted in terms of L so the amount of any substance can be expressed in terms of moles, but we need to specify the type of particle involved. For example, one mole of electrons is very different in substance to 1 mole of uranium atoms. One mole of atoms of a given element has a mass in grams numerically equal to the atomic mass. For example, L sodium atoms have a mass of 0.023 kg, which equals 23 grams. So, 1 mole of sodium atoms has a mass numerically equal to the atomic mass of sodium. The same idea applies to molecules. For a compound with molecular mass M (on the carbon-12 scale), each molecule has a mass of M units of atomic mass. So the mass of one molecule in kilograms is $M \times \frac{0.0001}{L}$. Hence the mass of one mole of the compound (i.e. of L molecules) is $M \times 0.001$ kg which equals M grams.

The *molar mass* of a substance is defined as the mass of 1 mole of the substance.

Use of density to calculate molecular size

Given the molar mass of a solid or liquid of known density, we can calculate the volume each molecule occupies. As each molecule in a solid or liquid is in contact with neighbouring molecules, we can estimate the distance between neighbouring molecules. For example, the molar mass of water is 0.018 kg and its density is 1000 kg m^{-3}. So 1 m^3 of water has a mass of 1000 kg which is 1000/0.018 moles. Each mole contains L molecules so 1 m^3 of water contains $1000 \times L/0.018$ molecules. Therefore the volume occupied by one molecule is given by

$$\text{Average volume } V \text{ of 1 molecule} = \frac{0.018}{L \times 1000} \text{ m}^3$$

$$= 3.0 \times 10^{-29} \text{ m}^3$$

To estimate the approximate spacing between adjacent molecules in water, assume the centre of each molecule is at average distance d from its neighbour's centre. So the rough volume of each molecule is d^3, as shown in Figure 6.5.

$$d^3 = 3.0 \times 10^{-29} \text{ m}^3$$
$$d = 3.1 \times 10^{-10} \text{ m}$$

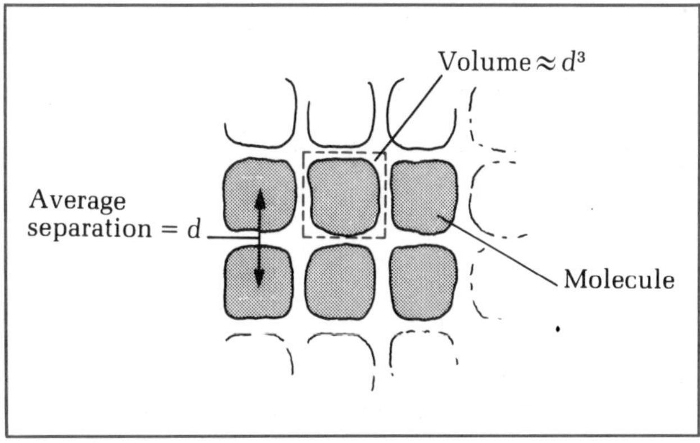

Figure 6.5 *Molecular size*

Worked example Estimate the spacing between adjacent atoms of copper, given that the molar mass of copper is 0.064 kg and its density is 8000 kg m^{-3}. Assume that the Avogadro constant L has a value of 6.023×10^{23}.

Solution One mole of copper atoms has a mass of 0.064 kg. The volume V of 0.064 kg of copper is given by mass/density. Hence $V = 0.064/8000 = 8.0 \times 10^{-6}$ m^3. So L atoms of copper have a volume of 8×10^{-6} m^3. Therefore 1 atom of copper occupies a volume v given by $8 \times 10^{-6}/L$. Hence $v = 1.33 \times 10^{-29}$ m^3. Assuming average spacing d between adjacent copper atoms gives volume per atom $= d^3 = 1.33 \times 10^{-29}$ m^3. Hence $d = 2.37 \times 10^{-10}$ m.

Usually molar mass is stated in units written as kg mol^{-1}. So the molar mass of copper should be given as 0.064 kg mol^{-1}. This simply means that each mole of copper has a mass of 0.064 kg. The symbol mol is used for moles just as the symbol kg is used for kilograms.

Estimating molecular size using the oil film experiment

A drop of oil on clean water spreads, if allowed, until it is a film on the water surface only one molecule thick. By measuring the thickness of the film we can obtain an estimate of the size of an oil molecule. Figure 6.6, on page 68, shows an arrangement for making such a measurement. First, the water surface must be cleaned by skimming it with clean metal bars. Then the surface is sprinkled with a light powder. When a tiny oil drop, of known volume, is placed at the centre of the surface it spreads out to form a large circular patch on the surface. The patch is easily seen because the oil pushes the powder away.

To estimate the patch thickness, the oil drop diameter must be measured before the drop is allowed to fall. The drop needs to be very small unless a huge water surface is available, so a magnifying glass is necessary to measure the drop diameter against a suitable scale. The patch diameter is easily measured using a metre rule marked in millimetres.

To calculate the patch thickness, let d be the diameter of the oil drop, and let D be the patch diameter.

The volume of the oil drop $= \frac{4}{3}\pi\left(\frac{d}{2}\right)^3$, assuming the drop is spherical.

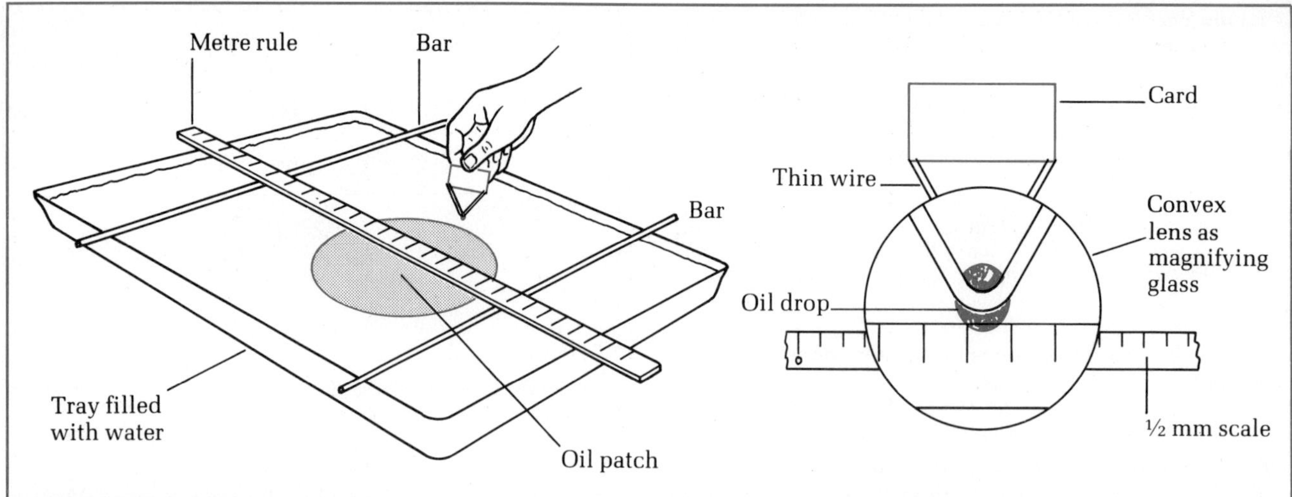

Figure 6.6 *The oil film experiment*

The volume of the oil patch = thickness(t) × area$\left(\frac{\pi D^2}{4}\right)$.

Since the patch volume equals the oil drop volume,

$$\frac{\pi D^2 t}{4} = \frac{4}{3}\pi\left(\frac{d}{2}\right)^3$$

which gives

$$t = \frac{2d^3}{3D^2}$$

The method gives an estimate for the size of an oil molecule calculated from the formula for thickness t. Accuracy cannot be justified because of the rough method for finding the oil drop diameter d, so giving the oil molecule size correct to several decimal places would be misleading since it implies a very accurate measurement. Nevertheless, the method does give a direct estimate of the order of magnitude of molecular size.

Worked example An oil drop of diameter approximately 0.5 mm spreads on clean water to form a circular patch of diameter 400 mm. Estimate the size of a molecule of the oil drop.

Solution Oil drop volume $= \frac{4}{3}\pi\left(\frac{5 \times 10^{-4}}{2}\right)^3$

Oil patch volume $= t \times \frac{\pi(0.4)^2}{4}$ where t is the patch thickness.

Since the oil patch volume = the oil drop volume, then

$$t = \frac{2 \times (5 \times 10^{-4})^3}{3 \times (0.4)^2} = 5 \times 10^{-10}\,\text{m}$$

6.3 States of matter

Ice is a solid. Its molecules are locked together in a rigid structure. When ice is heated sufficiently, it changes to water. The energy supplied by the heating process enables the molecules to break away from each other, so its rigid structure falls apart. When water is heated sufficiently, its temperature rises to boiling point when the water changes to steam. Once again the heating process supplies energy which enables the water molecules to break from each other. Ice, water and steam are examples of the three states of matter.

The solid state Solids have fixed volume and fixed shape.

The liquid state Liquids have fixed volume but take the shape of their container.

The gaseous state Gases have no fixed volume and no fixed shape.

To change a solid to a liquid or to change a liquid to a gas, energy must be supplied to break the bonds which hold the molecules together. When a gas changes to a liquid or a liquid changes to a solid, energy is released because bonds are formed to hold the molecules together. For example, consider what happens when salt is heated. To melt salt, it must be heated very strongly; at room temperature, the atoms in salt are locked together in a very rigid structure with strong forces holding the atoms in place. To enable the atoms to break the grip of these strong forces, the salt must be strongly heated; this makes the atoms vibrate so much that they break free.

The forces which hold atoms and molecules together are due to the charged particles in each atom. In other words, the forces are *electrostatic* in origin. Electrons are mostly responsible, but there are several ways in which they cause bonding forces.

Ionic bonding

Crystals of common salt (i.e. sodium chloride) are bonded in this way. An uncharged sodium atom has a single electron in its third shell, the inner shells being full. Each uncharged chlorine atom has seven electrons in its third shell; again the inner shells are full. Now the third shell is full with eight electrons, and since atoms prefer full shells, then a chlorine atom likes to gain an extra electron whereas each sodium atom likes to lose an electron. So when sodium and chlorine atoms form a

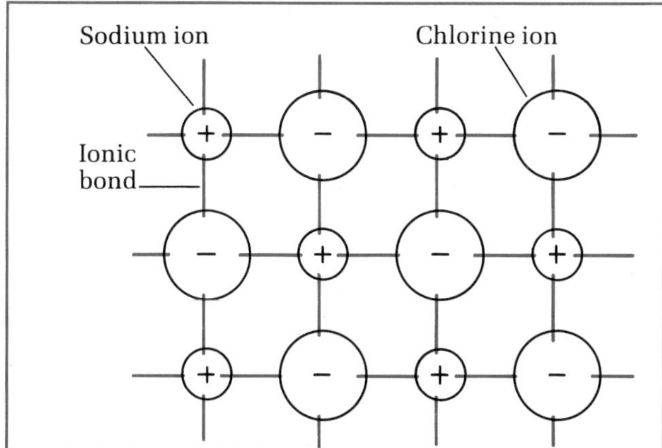

Figure 6.7 *Sodium chloride crystal*

sodium chloride crystal, each sodium atom gives up an electron to a chlorine atom. The sodium atoms become positive ions and the chlorine atoms become negative ions. The electrical forces between the ions causes them to become regularly arranged, as shown in Figure 6.7. The force between adjacent oppositely charged ions is called an **ionic bond**.

Put an ionic crystal in water and it will dissolve. The effect of the water is to weaken the electrical forces between the ions. The ions break off and the crystal dissolves to form a solution. Most inorganic crystals are ionic.

Covalent bonding

When atoms are unable to gain electrons to complete part-filled shells, they can share electrons. Shared electrons act as bonds between the atoms; this is referred to as a **covalent bond**. For example, oxygen molecules are each composed of two oxygen atoms joined by a covalent bond. Each uncharged oxygen atom has 16 electrons ($Z=16$), arranged with two in the innermost shell, eight in the next and then six in the third shell. If an oxygen atom shares two of its third shell electrons with another oxygen atom, which also contributes two electrons for sharing, each of the two oxygen atoms has a full third shell. Each covalent bond requires one electron from each atom. So the two oxygen atoms form two covalent bonds since each atom contributes two electrons for sharing, as shown in Figure 6.8.

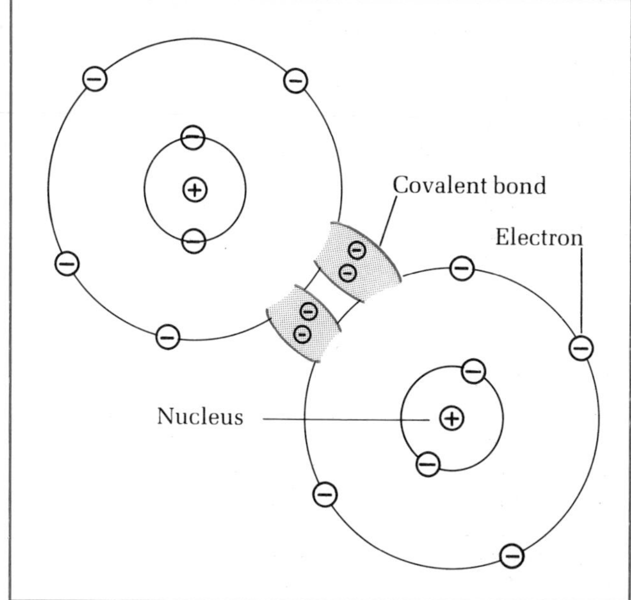

Figure 6.8 *An oxygen molecule*

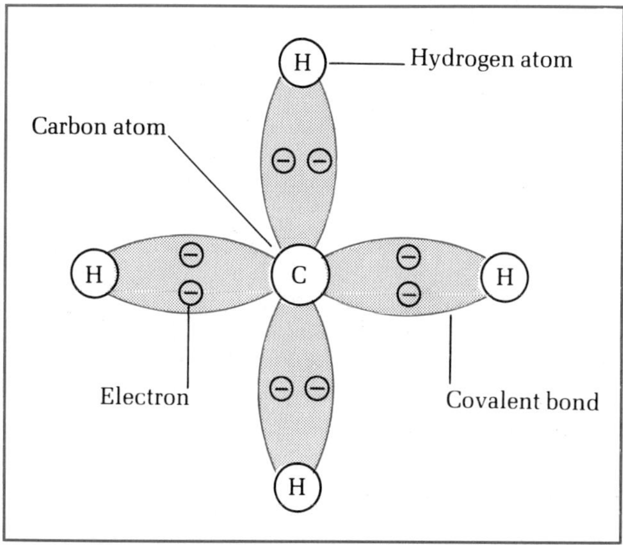

Figure 6.9 *A methane molecule*

Molecules of organic compounds are held together by covalent bonds between their atoms. Organic compounds contain carbon atoms, and an uncharged carbon atom ($Z=6$) has two electrons in its innermost shell and four in its second shell. The second shell can take up to eight electrons, so to fill it a carbon atom forms four covalent bonds with other atoms. For example, methane gas molecules each have a carbon atom joined to four hydrogen atoms; each hydrogen atom forms a covalent bond with the carbon atom to satisfy the full shell requirement of the hydrogen atom. So by forming four covalent bonds, the carbon atom fills its second shell. (See Figure 6.9.)

Metallic bonding

In a metal the atoms have lost their outermost electrons which move freely inside the metal. The metal atoms therefore become positive ions; they are held in place in regular order by the electrical forces between the ions and the free electrons. The atoms are arranged in an order which can differ from one metal to another. So when a

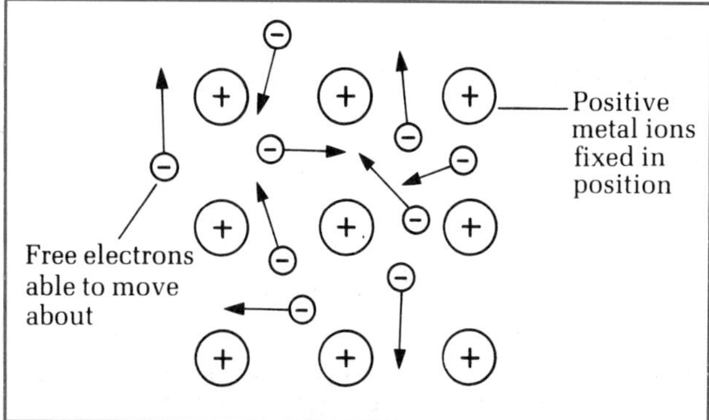

Figure 6.10 *Metallic bonding*

metal solidifies, lots of tiny crystals, called **grains**, are formed inside the metal (see p. 82).

All metals conduct electricity. The reason is that they all contain free electrons. When a potential difference is applied across a metal, the free electrons inside the metal move towards the positive terminal. So an electric current in a metal is due to the movement of free electrons.

Van der Waals bonding

Uncharged atoms can and do exert forces on one another. Because they are uncharged, each atom has the same number of electrons as protons. But the electrons of one atom can experience a force due to the protons of an adjacent atom. So two uncharged atoms can exert a weak attractive force on one another. Such a force is called a **Van der Waals bond.** Uncharged liquid molecules exert Van der Waals forces on one another; if a molecule tries to move away from its neighbours at the liquid surface, the molecule experiences a pull back into the liquid because of Van der Waals forces between it and its neighbours. Surface tension of liquids is explained in terms of these weak attractive forces (see p. 71).

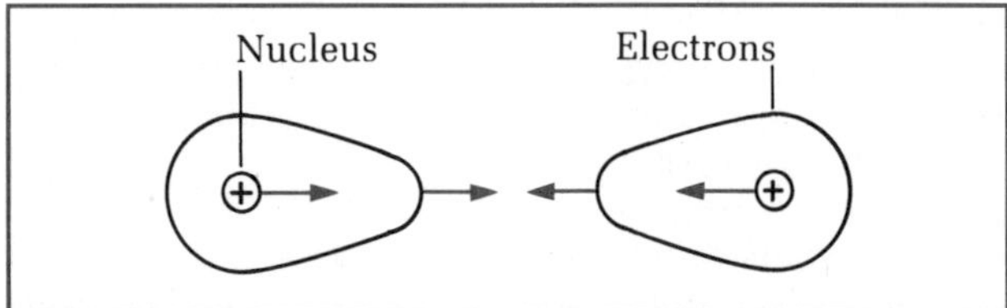

Figure 6.11 *Van der Waals force between two uncharged molecules*

6.4 Intermolecular forces

Solids and liquids, unlike gases, are very difficult to compress. Molecules in solids and liquids are in contact with each other. Very strong forces are needed to push molecules in a solid or liquid even closer; adjacent atoms are prevented from pushing into one another by their outer electrons. When an attempt is made to push two atoms in a solid closer together, their outer electrons act as repulsive barriers which prevent them from pushing into one another.

In general, two atoms exert forces on one another, even if uncharged. If the atoms are far apart, the forces are too weak to be significant. But brought closer, the atoms attract one another, as was explained in the previous section. If the atoms come very close, effectively in contact, then their electron shells try to repel them. So two atoms near each other will attract one another and move together, but only as close as the electron shells will allow. They are then in equilibrium because the overall force between them is zero. If they move closer than equilibrium then they repel; if they move away from equilibrium then they attract one another. Figure 6.12 shows how the force between two atoms or two molecules varies with the distance between them.

The force curve can be thought of as being made up from two parts; a weak attraction curve due to one of the types of bonds explained previously, and a strong, short-range repulsion curve due to electron-electron interaction when the atoms get very close (i.e. closer than equilibrium).

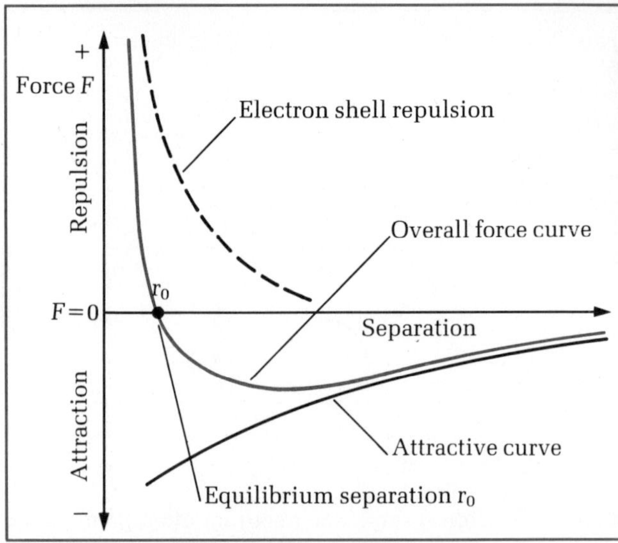

Figure 6.12 *Intermolecular force curve*

So the overall force curve shows repulsion at very short range and weak attraction at long range. The equilibrium position is where the overall force is zero; the two parts cancel one another at equilibrium. The distance apart at equilibrium therefore is the average spacing between the molecules of the material in bulk.

Intermolecular potential energy

Imagine pushing two magnets, arranged to repel, closer together. You must do work against the repulsion to push them together. The work done on the magnets gives them increased potential energy. If the two magnets, still arranged to repel, move apart, their potential energy decreases.

Now imagine we rearrange the two magnets so that they attract one another. If they are moved apart, their P.E. rises because work must be done against the attractive force between them. If they are moved together, their P.E. falls.

Let's now apply these general ideas to two molecules moved from very far apart to very close together. When very far apart their P.E. due to interaction is zero. When they are moved towards each other their P.E. falls because of the attraction between them. The P.E. continues to fall until they reach the position where the overall force changes from attraction to repulsion (as shown by Figure 6.12). If they are moved closer still, the P.E. rises because work must be done on the system to push them together against the repulsion forces. Figure 6.13 shows how the P.E. varies with separation.

The **minimum** of the P.E. curve occurs where the force changes from attraction to repulsion. In other words, it is where the overall force is zero. So minimum P.E. is at the equilibrium separation of the two molecules. To separate two molecules from equilibrium separation to infinity, work must be done to increase their P.E. from the minimum value, $-\varepsilon$, to zero; ε is sometimes called the **bond energy**.

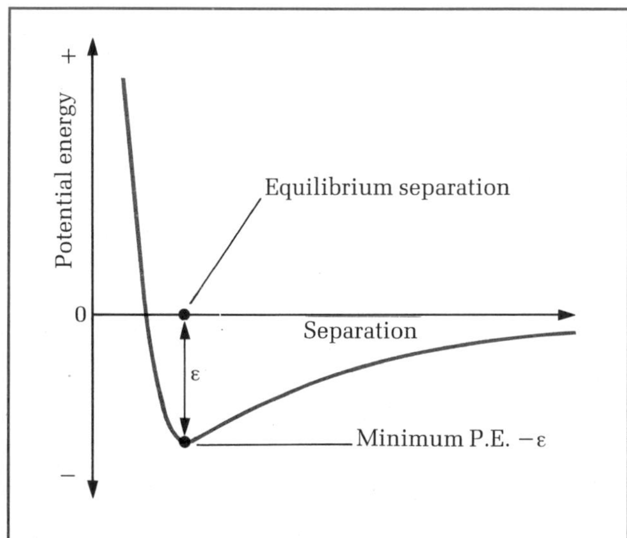

Figure 6.13 *P.E. curve*

The **gradient** of the P.E. curve at any point on the curve gives the force which must be applied to hold the molecules at that separation. Suppose they move away from each other slightly so their separation increases by Δr. Let the change of P.E. which results be ΔP.E. The force holding them and moving them slightly must therefore have done work equal to ΔP.E. Since work done = force × distance, then the force which moved them is given by $\frac{\Delta \text{P.E.}}{\Delta r}$ which is the gradient of the P.E. curve. The force F between the molecules due to their interaction is equal and opposite to the force applied to move them. So the force F between the molecules is given by $-\frac{\Delta \text{P.E.}}{\Delta r}$.

The force due to the interaction between the two molecules = − gradient of the P.E. curve.

a) At equilibrium separation r_0, $F = 0$ so the P.E. curve gradient = 0.

b) At separation $r > r_0$, $F < 0$ (i.e. attractive) so the P.E. curve gradient > 0.

c) At separation $r < r_0$, $F > 0$ (i.e. repulsive) so the P.E. curve gradient < 0.

Using the curves to explain some properties of materials

Surface tension Molecules inside a liquid are spaced on average at equilibrium separation. The molecules move about but the average force on a molecule inside the liquid is zero, because such molecules are at equilibrium separation. But molecules near the surface have a bit more space than those inside the liquid. So the average separation between surface molecules is a little greater than equilibrium spacing r_0. Therefore weak attractive forces link surface molecules due to their increased separation. So molecules at and near the surface are linked in a state of tension. Surface tension is evident when you fill a beaker with water until the water is just about to 'brim' over. Try it! Surface tension keeps the water on the brim until the water pressure becomes too high at the rim.

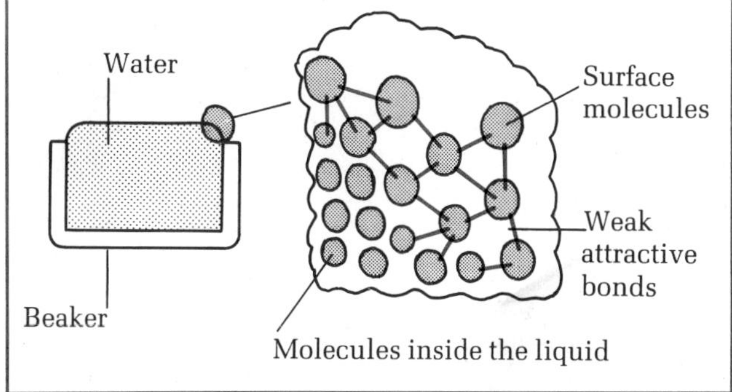

Figure 6.14 *Surface tension*

Hooke's law The tension in a spring is proportional to its extension from its natural length. The same rule applies to a stretched wire. The reason is that the intermolecular force curve is straight at and near equilibrium.

The straight line near equilibrium means that the force is proportional to the displacement from equilibrium. So the bonds between atoms in solids are like tiny Hooke's law springs, provided they are not extended too much. Therefore provided the whole structure is not stretched too much, Hooke's law is obeyed.

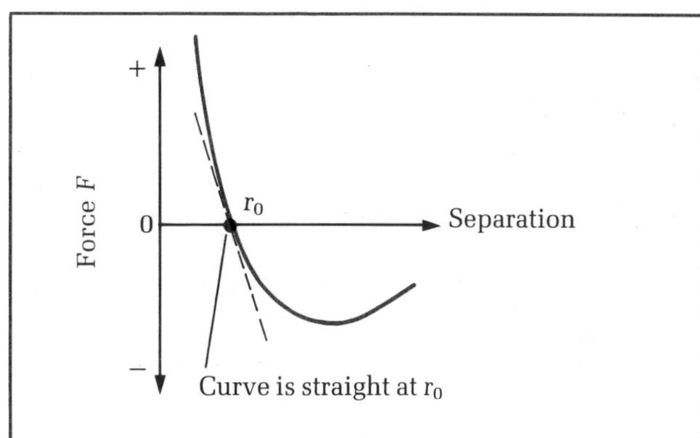

Figure 6.15 *Hooke's law*

Latent heat To change the state of a given material from solid to liquid or liquid to gas, the material must be supplied with energy to enable its atoms to break free from one another. Before change of state, the atoms inside the material are at equilibrium separation with minimum P.E. The bonds between the atoms must be broken to bring about the change of state. Since energy ε must be supplied to break each bond, latent heat energy must be supplied to the material.

When a gas changes to a liquid or a liquid changes to a solid, bonds are reformed and so energy is released as each bond is formed. So in this case latent heat energy is released by the material.

Thermal expansion Solids expand when heated. The atoms vibrate more and more as more and more energy is supplied. So the separation between any pair of adjacent atoms varies about the equilibrium value as they vibrate. The P.E. of the pair increases on both sides of equilibrium, but the increase is sharper at separation which is less than equilibrium. A given temperature rise increases the P.E. of the pair of atoms by a given amount on average; but because the P.E. curve is sharper on one side, the

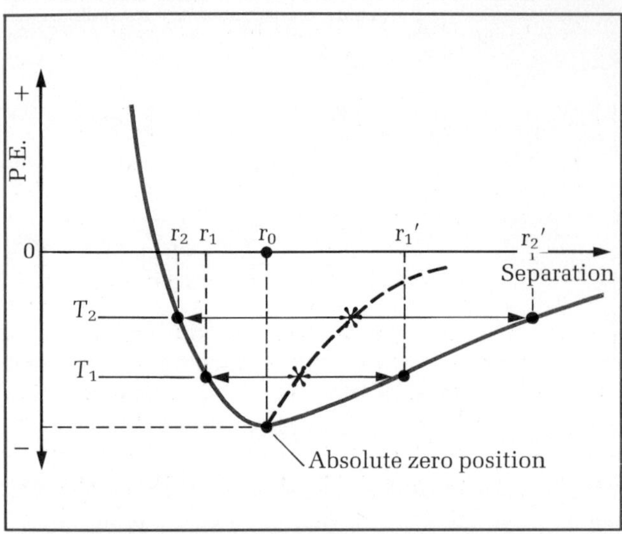

Figure 6.16 *Thermal expansion*

average separation of the pair increases. So for a whole network of atoms, an overall expansion takes place.

a) At absolute zero, the atoms do not vibrate so their separation is r_0.

b) At temperature T_1, the atoms vibrate so that their separation varies from r_1 to r_1'. Because the P.E. curve is sharper at separation less than r_0, r_1 is closer to r_0 than r_1'. So the average separation, midway between r_1 and r_1', is greater than r_0.

c) At increased temperature T_2, the vibration increases so their separation varies from r_2 to r_2'. The shape of the P.E. curve is such that the average separation, midway between r_2 and r_2' is even greater.

6.5 Summary

1 A *molecule* is the smallest part of an element or compound which can exist alone.

2 An *atom* is the smallest part of an element which can be identified with the element.

3 The *Avogadro constant* L is the number of atoms in 0.012 kg of carbon-12.

4 One *mole* of a substance is the amount containing L particles of the substance, where L is the Avogadro constant.

5 The *molar mass* of a substance is defined as the mass of one mole of the substance.

6 The intermolecular force curve:
 a) $F=0$ at equilibrium separation r_0.
 b) At separation $>r_0$, $F<0$ i.e. attraction.
 c) At separation $<r_0$, $F>0$ i.e. repulsion.

7 The intermolecular P.E. curve:
 a) Its gradient equals $-F$.
 b) At equilibrium the P.E. is a minimum so $F=0$ because the gradient is zero.
 c) At separation $>r_0$, the gradient >0 so $F<0$ i.e. attractive.
 d) At separation $<r_0$, the gradient <0 so $F>0$ i.e. repulsive.

Short questions

Assume $L = 6.02 \times 10^{23}\ \text{mol}^{-1}$.

6.1 In an experiment to estimate the size of a molecule, an oil drop of diameter 0.5 mm is placed at the centre of a clean water surface. The drop spreads out to form a circular patch of diameter 650 mm. Calculate the thickness of the oil patch, and hence estimate the size of an oil molecule. State any assumptions made in giving your estimate.

6.2 Water has a density of $1000\ \text{kg m}^{-3}$ and its molar mass is 0.018 kg. Air at room temperature and atmospheric pressure has a density of $1.2\ \text{kg m}^{-3}$, and its molar mass is 0.029 kg. Compare the spacing of water molecules with the spacing of air molecules.

6.3 Copper has a molar mass of 0.0635 kg and its density is $8920\ \text{kg m}^{-3}$. Calculate
a) the number of atoms in 1 kg of copper,
b) the average volume occupied by a copper atom,
c) the average spacing of copper atoms.

6.4
a) In what ways is the ionic bond different to the Van der Waals bond?
b) The atoms in a crystal of sodium chloride are bonded ionically. The atoms of a crystal of argon (melting point 84 K) are joined by Van der Waals bonds. Explain, in terms of bond strength, why the melting point of sodium chloride crystals is much higher than that of argon crystals.

6.5 For two uncharged molecules, sketch curves to show
a) the variation of intermolecular force with separation,
b) the variation of P.E. with separation of the molecules.
On each curve, indicate the equilibrium separation and the range of separation for (i) attraction, (ii) repulsion.

6.6 Use the curves drawn in **6.5** to explain the following.
a) When a solid is heated, it expands slightly.
b) When a vapour condenses, heat energy is released.

6.7 Aluminium has a molar mass of 0.027 kg and its density in the solid state is $2700\ \text{kg m}^{-3}$.
a) Calculate the number of aluminium atoms in $1\ \text{m}^3$ of aluminium.
b) Estimate the diameter of an atom of aluminium.
c) Estimate the number of atoms per m^2 at the surface of the solid.

6.8
a) Show how the force between two uncharged molecules is related to the P.E. of the molecules,
b) If two uncharged molecules at equilibrium separation are moved away from one another, their P.E. rises. In terms of intermolecular forces, explain why this happens.

6.9 In an experiment involving electrolysis of copper sulphate solution, both electrodes are copper plates. Measurements showed that 3000 coulombs of electric charge deposited 1 g of copper on the cathode from the solution. Given that each copper ion deposited releases two electrons, calculate a value for L. The molar mass of copper is 0.0635 kg, and the charge of the electron is 1.6×10^{-19} C.

7

Strength of solids

7.1 Behaviour of solids

Before choosing materials for any given purpose, we must first of all list the demands to be made of them. Does the material need to be strong? What about thermal properties? Must it be a good electrical insulator? Once we have established the demands we can then choose the materials with the most suitable properties. For example, consider the different materials used to make a bicycle. You might like to list the different materials used giving reasons for the choice of each. Tubular steel is a much better choice than wood for the frame since wood is unlikely to take the stresses – unless thick beams are used and then it would be far too heavy. Why tubular steel? Solid steel bars would be heavy and would add to the cost; tubular steel gives a much lighter frame, still strong enough to meet the stresses likely. The wheels are usually stainless steel or aluminium with air-filled rubber tyres. Solid rubber tyres could be used but would give an uncomfortable ride. You ought to be able to find about ten different materials quite easily on most bikes.

Mechanical properties

Collect together some samples of different solids such as rubber, perspex, steel, wood, etc. How would you describe each material? Strong, stiff, bouncy, hard, brittle and smooth would probably be amongst the words you

Figure 7.1 **a** *A racing bicycle*

b *An earlier bicycle*

choose to describe the mechanical properties of the materials. Let's consider more precisely what these words mean.

Strength is important if we wish to know how much force can be applied without breaking the material. Some materials, like concrete, can be strong in compression but not in tension (i.e. when stretched). So concrete is used for load-bearing walls in some buildings but not for roof beams – unless it is reinforced. The force required to break a given material is a measure of its strength. Since breaking force depends on the shape and size as well as the type of material, then it varies from one object to another even if they are made from the same material. For example, the breaking force of a steel cable is much greater than the breaking force of a steel piano wire. A more useful measure of strength is the **breaking stress** which is the breaking force per unit area of material, the force being at right angles to the area, as in Figure 7.2. So if we are given that the breaking stress of a certain type of steel is $10^9\,\mathrm{N\,m^{-2}}$, then we can work out the breaking force for any wire or cable of known diameter of that material. A wire of diameter 0.5 mm and hence area of cross-section $\pi(5\times10^{-4})^2/4$, of that material would have a breaking force given by

$$\text{breaking force} = \begin{matrix}\text{breaking}\\ \text{stress}\end{matrix} \times \begin{matrix}\text{area of}\\ \text{cross-section}\end{matrix}$$

$$= 10^9 \times \frac{\pi(5\times10^{-4})^2}{4}\,\mathrm{N} = 197\,\mathrm{N}.$$

The breaking stress of a wire is the breaking force/area of cross-section of the wire, with the force applied at right angles to the area.

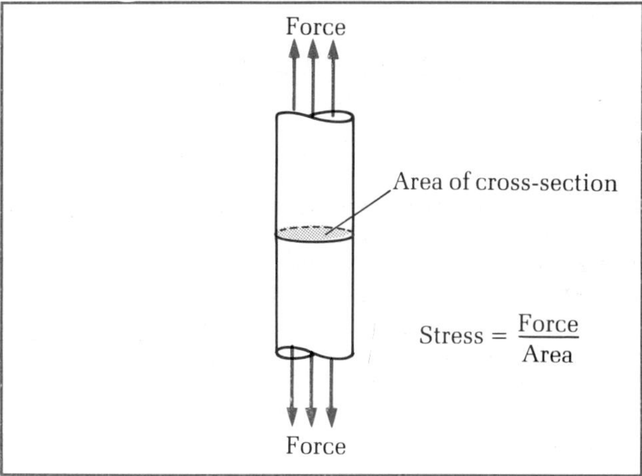

Figure 7.2 *Stress*

Stiffness is a measure of the difficulty of changing the shape of an object. A plastic guitar string is much easier to stretch than a steel string because steel is stiffer than plastic. For wires of the same diameter and length, made of different materials, the stiffest material is the one which extends least when each is put under the same tension. Figure 7.12 shows one possible arrangement for testing the stiffness of wires. Figure 7.3 shows another arrangement for testing the stiffness of 'softer' materials like polythene or rubber. If a copper wire is compared with a steel wire of the same diameter and length by

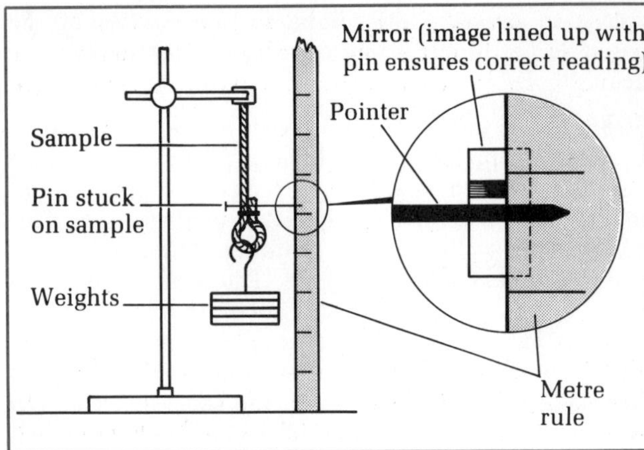

Figure 7.3 *Testing stiffness*

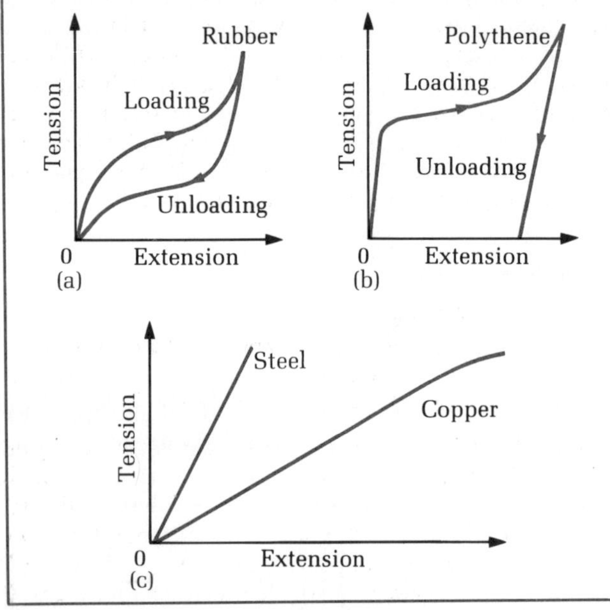

Figure 7.4 *Typical curves*

measuring the extension of each for different tensions, the results would be shown as in Figure 7.4(c). The stiffer wire, steel, has a larger value of tension per unit extension. We shall see later that the stiffness is measured by the Young modulus of the material.

Brittle materials are very stiff but not very strong. They snap if too much force is applied. A dry biscuit is a simple example! It is almost impossible to change its shape by forcing it, and if forced too much it just snaps.

Elasticity is the property which allows a material to regain its shape after being distorted. Some materials, like rubber bands, are much more elastic than others. The **elastic limit** of a material is the maximum amount by which it can be stretched and still regain its original shape after the distorting forces are removed. If a material is stretched beyond its elastic limit its shape is permanently changed. The arrangement of Figure 7.3 can be used to test for elasticity. Figure 7.4(b) and (c) show the properties of an elastic and an inelastic material when under tension and when released from tension.

Plastic behaviour occurs when a material is deformed beyond its elastic limit. If its elastic limit is low compared with its extension at breaking point, it stretches a lot after becoming plastic in behaviour. Can you think of any material like this? Try adding loads to plasticine.

Hardness is a measure of the difficulty of scratching a material. Rub a coin against a perspex ruler. The coin scratches the ruler, not the other way round. It is easier to dislodge atoms from the surface of the ruler than from the coin, so the coin is harder than the ruler. Diamonds are the hardest known material; the extreme hardness is due to the way the carbon atoms are bonded together. Yet graphite, which is also composed only of carbon atoms, is much softer. Carbon atoms in graphite are arranged in layers with relatively weak forces between the layers, so graphite can easily be rubbed off and can be used as a lubricant. Carbon atoms in diamond are joined together in a tetrahedral structure, as explained on p. 81. It is very difficult to remove carbon atoms from this structure.

Creep occurs when a material, acted on by constant forces, changes its shape even though the forces on it remain constant. A material which creeps will deform more and more with time when it is acted on by constant forces. Polythene carrier bags creep if loaded too much (see Figure 7.6).

Investigating the properties of different materials

Measuring tension–extension curves Use in turn a spring, a rubber band, strips of different materials cut to the same length, etc. For each one, measure and note its initial length when arranged unloaded as in Figure 7.3. Then add known weights in steps, at each step remeasuring the length to give its extension from the initial length. Use a metre rule marked in millimetres. The tension is equal to the total weight supported. If the maximum load used doesn't break the material, then repeat the measurements in steps by unloading it. If it breaks, start again but don't load it as much. Plot your results in the form of curves of tension against extension. Figure 7.4 shows some typical curves. Which material is the stiffest? Which has the least strength? Has any material been extended beyond its elastic limit?

Testing for hardness Drop a metal punch down a hollow tube onto the material, as shown in Figure 7.5(a). The punch makes a tiny indentation on the surface of the

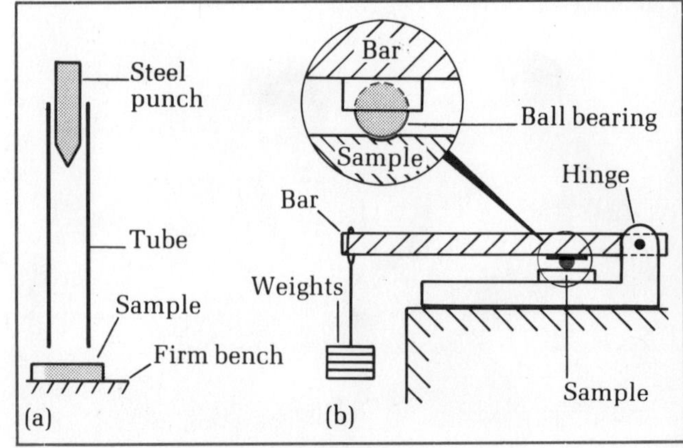

Figure 7.5 *Hardness testing*

material. The size of the indentation can be measured with a microscope, and gives a measure of the hardness of the material. The bigger the hole, the softer the material. The punch must be dropped from rest from the same height each time for a true comparison. Another method is to press a small ball bearing onto the surface using a lever to apply a constant force, as in Figure 7.5(b).

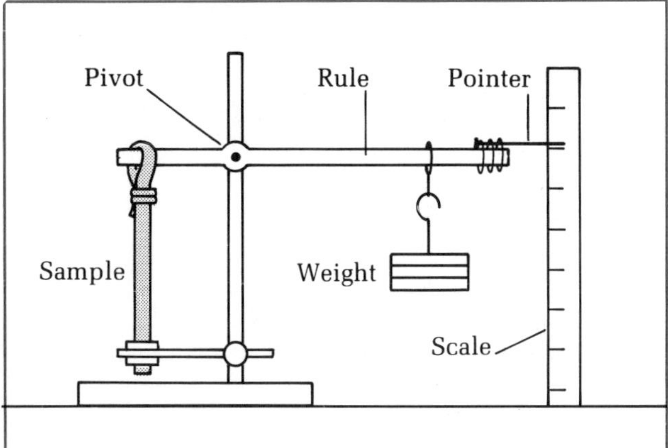

Figure 7.6 *Testing creep*

Measuring creep A strip of polythene sheet can be tested using the arrangement of Figure 7.6. A suitable weight hung from the pivoted rule causes a constant tension to be exerted on the strip of polythene on the other side of the pivot. A pin attached to the free end of the pivoted rule can be used as a pointer against a vertical millimetre scale to give a measure of the extension of the strip. So the pointer reading is noted at regular intervals. A graph of extension against time can then be plotted.

Since the readings may need to be taken over a period of a few days, a data recorder is useful here. A data recorder is a device which measures and records voltages at regular intervals chosen by the operator. The measurements are stored in its memory and can be read from its memory later. For example, suppose a given data recorder can take up to 1000 readings at one of three possible rates, 10 readings per second, 10 readings per minute, and 10 readings per hour. Which rate would you choose for measurements over a period of a few days?

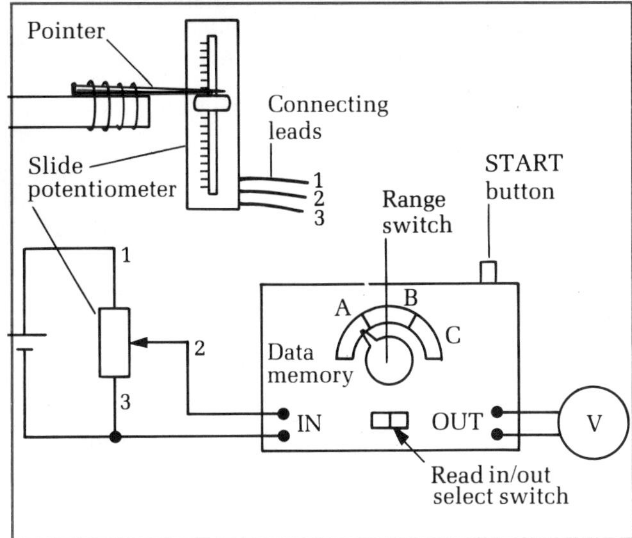

Figure 7.7 *Using a data recorder*

To use a data recorder here, we need to convert the movement of the pointer into a voltage which depends on pointer position. A slide potentiometer would be suitable, as shown in Figure 7.7.

7.2 Elasticity

Springs

Stretch a spring then release it and it returns to its initial unstretched length, provided it has not been overstretched. Springs are therefore elastic because they regain their unstretched length, provided they are not overstretched. When a spring is overstretched it pulls out from its coiled shape into a long wire, so it cannot coil back again when the cause of its tension is removed. Using an arrangement like Figure 7.3, it is not too difficult to show that the extension of a spring (i.e. its increase of length from its natural or unstretched length) is proportional to the tension in it. This is known as **Hooke's law** after its discoverer, Robert Hooke, a 17th century scientist. The link between tension and extension for a spring may be written as an equation.

$$T = ke = k(l - l_0)$$

where T = tension
e = extension from its natural length l_0,
l = stretched length,
k = the stiffness constant.

***The stiffness constant* k** differs from one spring to another. Its unit is N m^{-1}. It is sometimes confused with the spring modulus λ (used often in Applied Mathematics) which equals kl_0. Figure 7.8 shows a graph of how the tension varies with the extension. Since it is a graph with tension on the vertical axis, the gradient of the graph equals k.

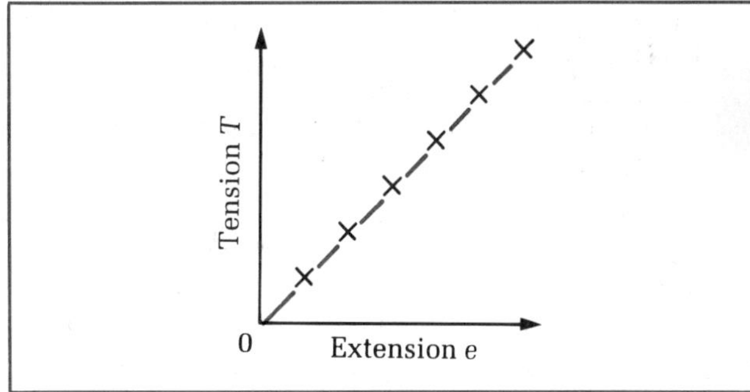

Figure 7.8 *Hooke's law*

Stress and strain

When a wire is put under tension, its extension depends on other factors besides the tension. The diameter and length of the wire are relevant, and so too is the material of the wire. By taking account of the diameter and length of the wire we can apply the measurements of tension and extension to any size of the same material. We can also compare different materials, given a wire of each material.

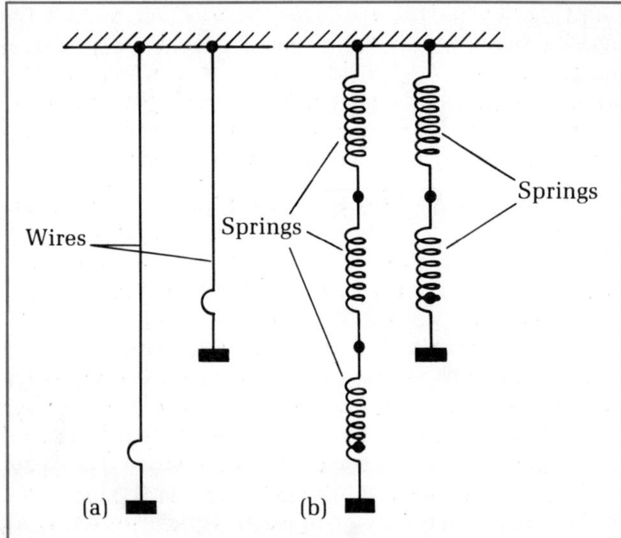

Figure 7.9 *Strain (wires and springs shown before loading)*

Suppose two wires of the same material and diameter with different lengths are loaded with identical weights, as in Figure 7.9(a). Which wire stretches more? A comparison with the two spring systems shown in Figure 7.9(b) is helpful here. Each spring is identical so the arrangement with more springs end to end stretches more. The load weights are equal so each spring stretches equally; so the arrangement with three springs extends $1\frac{1}{2}$ times as much as the arrangement with only two springs. The longer wire of Figure 7.9(a) stretches more than the other wire because it has more units of length. However the extension per unit length is the same.

Strain is defined as the extension per unit length.

No unit is required because strain is a ratio. If we state that the strain of a wire is 0.001, we mean that its extension is 0.001 × its initial length.

Now consider two wires of the same material, the same length but different diameters, as in Figure 7.10(a). When equally loaded, which wire stretches more? Again we can make a comparison with spring systems; this time the springs hang side by side as in Figure 7.10(b). The system with three parallel springs extends less than the other system with two parallel springs. In each system, each spring takes an equal share of the load weight, so each spring of the 3-spring system supports one-third of the load weight. Each spring of the 2-spring system supports half of the load weight. So the extension of the 2-spring system is $\frac{3}{2}\times$ the extension of the 3-spring system. In Figure 7.10(a), the thicker wire stretches less because it has more cross-sectional area over which to spread the tension due to the load weight.

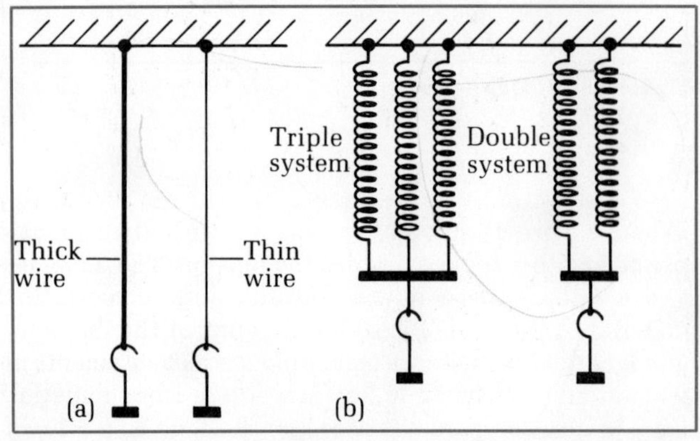

Figure 7.10 *Stress (wires and springs shown before loading)*

(Tensile) stress is defined as the tension per unit area applied normal to that area.

The unit of stress is the **pascal** (Pa), equal to 1 N m^{-2}.

Consider a uniform wire of diameter d and unstretched length l_0 hung vertically from a fixed point. Suppose a weight W is hung from its lower end, causing the wire to lengthen to a new length l, as in Figure 7.11.

$$\text{The strain} = \frac{\text{extension } e}{l_0} = \frac{l - l_0}{l_0}$$

$$\text{The stress} = \frac{\text{tension}}{\text{area of cross-section}} = \frac{W}{(\pi d^2/4)}$$

since the tension is equal to the weight W and the area of cross-section is equal to $\frac{\pi d^2}{4}$.

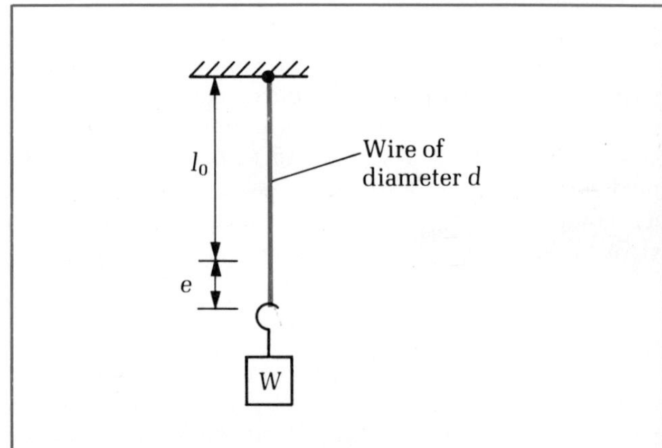

Figure 7.11 *A loaded wire*

The Young modulus

When stress is applied to a material, strain is produced in the material. The strain is proportional to the stress, provided the stress does not exceed a limit known simply as the **limit of proportionality.** Within this limit, the value of $\frac{\text{stress}}{\text{strain}}$ is a constant for that material, and is known as the Young modulus for the material.

$$\textbf{The Young modulus } E = \frac{\textbf{stress}}{\textbf{strain}}$$

provided the limit of proportionality is not exceeded.

The unit of E is the same as the unit of stress because strain has no unit. So E is given in pascals or newtons per metre squared.

For a uniform wire or bar, the Young modulus is therefore given in terms of its length, etc., by the following equation.

$$\text{The Young modulus } E = \frac{\text{stress}}{\text{strain}} = \frac{T/A}{e/l_0}$$

so $$E = \frac{T}{e} \times \frac{l_0}{A}$$

where T = tension,
l_0 = initial length,
e = extension,
A = area of cross-section.

Worked example A steel wire of diameter 0.40 mm and initial length 2.0 m is hung vertically from a fixed support. When an 80 N weight is suspended from the lower end of the wire so that it hangs freely, the wire extends by 6.4 mm. Calculate the Young modulus for the material of the wire.

Solution First calculate the strain from extension/initial length,

$$\text{Strain} = \frac{6.4 \times 10^{-3}}{2} = 3.2 \times 10^{-3}$$

Next calculate the stress from weight/area of cross section,

$$\text{Stress} = \frac{80}{(\pi/4)(4 \times 10^{-4})^2} = 6.37 \times 10^8\ \text{Pa}$$

Finally calculate the Young modulus from stress/strain,

$$E = \frac{6.37 \times 10^8}{3.2 \times 10^{-3}}\ \text{Pa} = 2.0 \times 10^{11}\ \text{Pa}$$

Note the enormous value of E when given in Pa. Sometimes E is given in units of kN mm^{-2} instead of Pa. Rework the calculation to show that the answer of 2×10^{11} Pa is the same as 200 kN mm^{-2}.

Investigating stress and strain for different materials

Method 1 Using very long wires The longer the wire, the more it extends for a given tension. Using an arrangement like Figure 7.12, the length of wire is limited only by the length of the bench. Use safety goggles in case the wire whips back if it snaps. When weights are hung from the end of the wire over the pulley, the wire stretches and the marker fixed to it therefore moves. Start with the smallest weight necessary to straighten the wire as shown, and read the marker position against the millimetre scale underneath. Measure the distance from the fixed end of the wire to the marker for the unstretched length. Increase the weight in steps, at each step note the marker reading and the corresponding weight. The extension is equal to the change of the marker reading. Finally use a micrometer to measure the wire diameter.

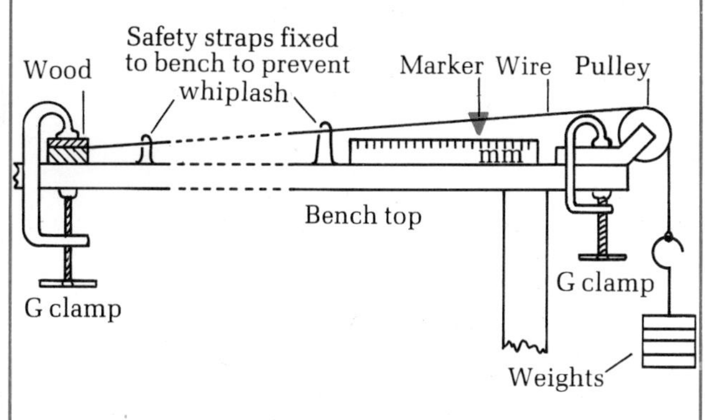

Figure 7.12 *A long wire*

Method 2 Using Searle's apparatus Two wires are suspended side by side from fixed points, as shown in Figure 7.13. One wire is the test wire and the other is a control wire. Any give in the fixed supports affects both wires equally. Use safety goggles. The control wire supports a micrometer which can be turned to lift one side of a spirit level up and down. The other side of the spirit level is hinged to a bar which is supported by the test wire. When the test wire is loaded, it extends so its side of the spirit level drops a little. The micrometer on the control side is then turned to keep the spirit level horizontal. The change of the micrometer reading gives the extension of the test wire.

Both wires are loaded just enough to straighten them out, then a micrometer is used to measure the wire diameter at several points. The initial length of the test wire is also measured using metre rules marked in millimetres. The control wire micrometer is then turned to make the spirit level horizontal, and its reading is then taken. Then the test wire is loaded in steps, at each step the spirit level is made horizontal and the micrometer reading noted. Readings can also be taken at steps during unloading. The extension at any point is the change of the micrometer reading from its initial value.

For both methods, the measurements can be used to plot a graph of load against extension as shown by Figure 7.16. To calculate the Young modulus, the gradient of the

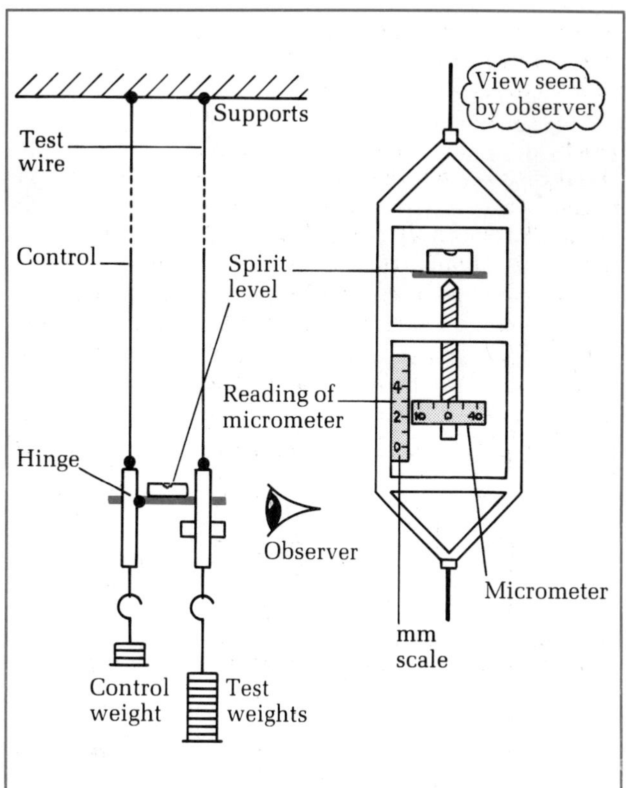

Figure 7.13 *Searle's apparatus*

initial section of the graph is calculated. Then E can be calculated as follows.

$$E = \frac{\text{tension } T}{\text{extension } e} \times \frac{\text{initial length } l_0}{\text{area of cross-section } A}$$

Hence $E = \text{gradient} \times \frac{l_0}{A}$

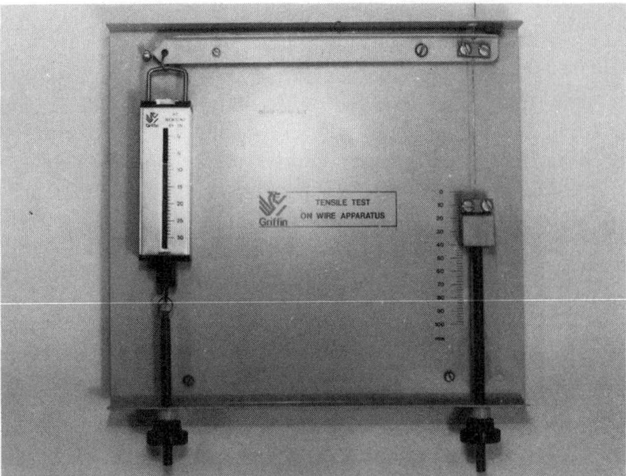

Figure 7.14 *A tensometer. The right hand screw is used to set the extension at the required value. The tension is then measured with the newton balance on the left*

Method 3 Using a tensometer This is used to extend a sample of material in steps, at each step the tension is measured. It can be used to test the material beyond its maximum strength to its breaking point. Figure 7.15 shows a stress-strain curve obtained using a tensometer.

Stress–strain curves

By plotting stress against strain we can compare different materials without sample dimensions causing confusion. Figure 7.15 shows a typical stress-strain curve for a metal. The important features are as follows.

1 Stress is proportional to strain up to the limit of proportionality, P. The gradient of the line OP is equal to the Young modulus.
2 The elastic limit, E, beyond which the sample becomes permanently stretched.

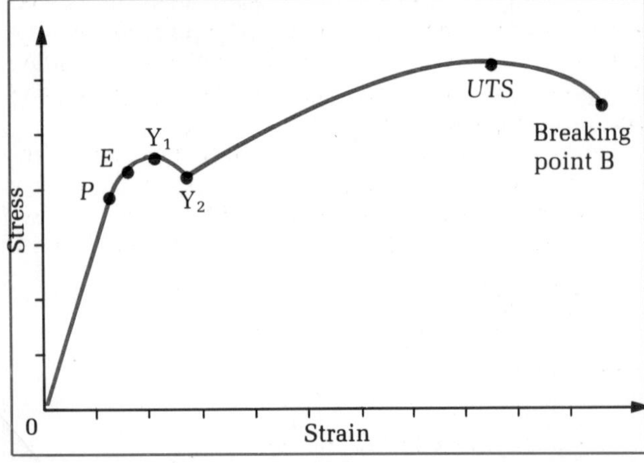

Figure 7.15 *Stress v strain for a metal*

3 The yield point, reached beyond the elastic limit, is where the sample suddenly gives a little. The stress increases the strain until the upper yield point Y_1 where the stress drops a little when the sample is stretched a little. Further stretching increases the stress from the lower yield point Y_2.
4 Plastic behaviour. Beyond the elastic limit, the wire loses its elasticity. Its behaviour is described as **plastic** because it does not regain its initial shape when the applied forces are removed.
5 The maximum stress, which is the point where the sample has its greatest strength, is sometimes called the ultimate tensile strength (UTS) of the material.
6 The breaking stress is the stress at the breaking point B of Figure 7.15.

Energy stored in a stretched wire

When a wire is stretched without exceeding its elastic limit, the work done on the wire is stored as elastic potential energy in the wire. The graph of tension against extension shown in Figure 7.16 is useful here because it is a force against distance graph. The area under a force v distance graph gives the work done by the force.

Consider a wire which is extended from zero extension to extension e_0 without exceeding the limit of proportionality. The graph is a straight line up to the limit of proportionality. So the tension T is proportional to the extension e at any point, as given by

$$T = \left(\frac{AE}{l_0}\right)e = ke$$

where A = area of cross-section,
l_0 = initial length,
E = the Young modulus,
k = a constant (AE/l_0).

The work done W is given by the area under the line which in this case is a triangle. The base of the triangle represents the final extension e_0; the height of the triangle represents the final tension T_0 which equals ke_0. Hence the work done, represented by the area ($=\frac{1}{2}$ height × base), is equal to $\frac{1}{2}T_0e_0$. So the energy stored in the wire can be calculated from $\frac{1}{2}T_0e_0$.

$$\text{Energy stored} = \tfrac{1}{2}T_0e_0 = \tfrac{1}{2}\left(\frac{AE}{l_0}\right)e_0^2$$

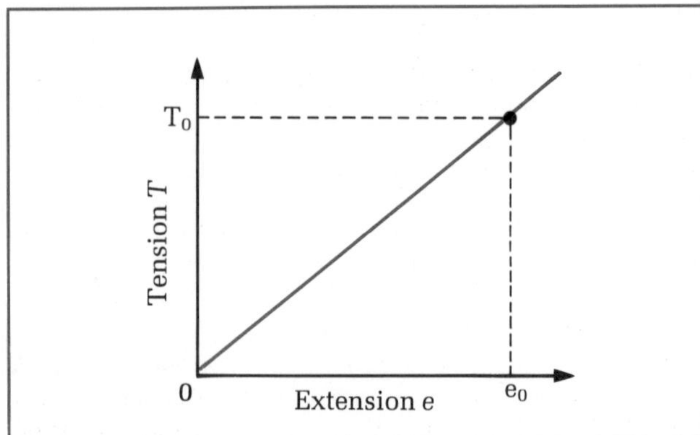

Figure 7.16 *Tension v extension for a wire*

The energy stored per unit volume of the wire is given by $(\frac{1}{2}T_0e_0)/(Al_0)$ since the volume of the wire is equal to Al_0. Rearranging this expression gives

$$\text{Energy stored per unit volume} = \tfrac{1}{2} \times \frac{T_0}{A} \times \frac{e_0}{l_0}$$
$$= \tfrac{1}{2}\ \text{stress} \times \text{strain}$$

since T_0/A is equal to the stress and e_0/l_0 is the strain.

Energy stored per unit volume $= \frac{1}{2}$ stress × strain

Worked example A uniform iron bar of diameter 8.0 mm and initial length 500 mm is heated uniformly until it expands by 0.4 mm. It is then clamped firmly at its ends and allowed to cool. Because it is unable to contract, tension builds up. Calculate the tension and energy stored in the bar when cooled. Assume the Young modulus for the bar is 1.8×10^{11} Pa.

Solution Extension $e_0 = 0.4$ mm,
initial length $l_0 = 500$ mm,

$$\text{Area of cross-section } A = \frac{\pi}{4}(8 \times 10^{-3})^2 = 5.02 \times 10^{-5}\ \text{m}^2.$$

When cooled, the tension is given by

$$T_0 = \frac{AE}{l_0}e_0$$

$$= \frac{5.02 \times 10^{-5} \times 1.8 \times 10^{11} \times 4 \times 10^{-4}}{0.500}\ \text{N}$$

which gives $T_0 = 7.24 \times 10^3$ N.

The energy stored is given by $\frac{1}{2}T_0e_0$, giving

Energy stored $= \frac{1}{2} \times 7.24 \times 10^3 \times 4 \times 10^{-4} = 1.45$ J.

Molecular interpretation of stress v strain curves

As explained on p. 71, the bonds between atoms in a solid can be pictured as tiny springs holding the atoms in place. However, the force curve for these 'atomic springs' is very different overall from the straight line given by Hooke's law, except where the force curve is near equilibrium (i.e. where it cuts the horizontal axis).

Near equilibrium, the extension of each tiny 'atomic spring' is proportional to the force in the spring. So the extension of a solid is proportional to the tension, provided the atoms are not pulled apart too far.

The limit of proportionality is reached when the separation of the atoms takes them beyond the straight section of the force curve, so the extension of the solid is no longer proportional to the tension.

Plastic behaviour occurs when a solid is extended beyond its elastic limit. Why does the solid not regain its initial shape when the applied forces are removed? The atoms of a solid are arranged in fixed positions. Change of shape must mean that atoms have been moved to new positions relative to one another. In a single crystal the atoms are arranged regularly in planes to form a lattice. When adjacent planes are pushed in opposite directions, the atoms in one plane can slip over the atoms in the other plane, if the forces are great enough. The force must be large enough to break the bonds linking the atoms of one plane to the atoms of the other plane. The atoms are moved and reform new bonds preventing movement back again. In a metal there are lots of tiny crystals packed together. The metal is said to be **polycrystalline.** The single crystal easily extends beyond its elastic limit. However the polycrystalline sample does not easily show plastic behaviour. The tendency of one crystal to slip is opposed by neighbouring crystals in the polycrystalline material because its tiny crystals are arranged in random directions.

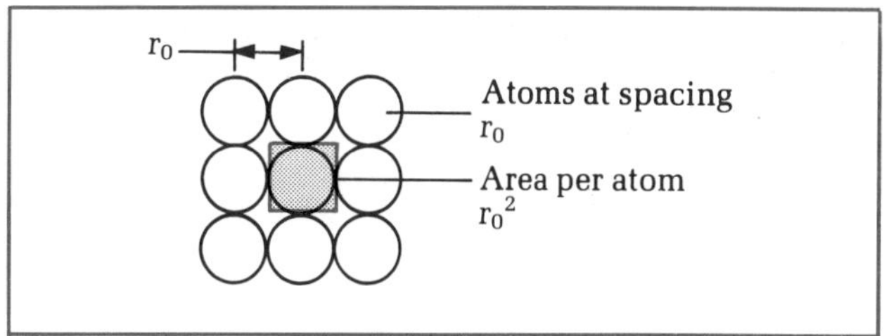

Figure 7.17 *Calculating* k

The stiffness constant k for an 'atomic spring' (i.e. the spring constant) can be calculated from the Young modulus E and the equilibrium spacing r_0 of the atoms. Assuming the limit of proportionality is not reached, consider what happens when the atomic spacing increases by Δr. The force F between two atoms is then given by $k\Delta r$. Each atom takes up an area r_0^2, so the force per unit area is equal to $\frac{k\Delta r}{r_0^2}$. This represents the stress.

Now the strain is given by (change of separation/ initial separation) which equals $\frac{\Delta r}{r_0}$.

Since the Young modulus $= \frac{\text{stress}}{\text{strain}}$, then

$$E = \frac{k\Delta r/r_0^2}{\Delta r/r_0} = \frac{k}{r_0}$$

so $k = Er_0$ gives an estimate of the stiffness.

Typical values for steel for E and r_0 are 10^{11} Pa and 10^{-9} m respectively, giving $k \approx 100\ \text{N m}^{-1}$.

The breaking stress can be estimated if the bond energy ε, the Young modulus and the atomic spacing are known. To break a bond between two atoms, energy of $\frac{1}{2}\varepsilon$ must be supplied to each atom of the bond. Each atom occupies a volume of approx. r_0^3. So the energy supplied per unit volume at breaking point is $\frac{1}{2}\varepsilon/r_0^3$.

Let's assume that the breaking point is not much further than the limit of proportionality. Then the energy stored per unit volume is also given by $\frac{1}{2}$ stress × strain which equals $\frac{1}{2}$ (stress)²/E, since strain = stress/E. Hence

$$\frac{(\text{stress})^2}{2E} = \frac{\varepsilon}{2r_0^3}$$

which gives

$$\text{Stress at breaking point} = \sqrt{\frac{\varepsilon E}{r_0^3}}$$

Using the value of ε (estimated from latent heat) of approximately 1.5×10^{-21} J and E and r_0 as before, we obtain an estimate of the breaking stress.

$$\text{Breaking stress} \approx \sqrt{\frac{1.5 \times 10^{-21} \times 10^{11}}{10^{-27}}} \approx 4 \times 10^{8}\,\text{Pa}$$

This value is no more than an 'order of magnitude' estimate. Other factors such as 'slip' play a part in determining the actual breaking stress.

7.3 X-ray diffraction

The oil film experiment described on p. 67 gives a rough estimate of the size of a molecule. The experiment was first performed by Lord Rayleigh in the 19th century. The discovery of X-ray diffraction by crystals gave a much more accurate way of measuring the size of atoms and finding out how they are arranged. Max von Laue in 1912 was the first to show that X-rays could be diffracted. He used crystals and photographic film to detect the pattern of diffracted X-rays. The pattern produced by X-rays diffracted by a gold crystal is shown in Figure 7.18.

The method was developed by William and Lawrence Bragg to work out the atomic arrangement for different materials. By 1960 techniques had been improved so much that scientists were able to unravel the structure of complicated molecules such as DNA and RNA which carry the genetic code in every living cell.

X-rays are electromagnetic waves with wavelengths less than about 1 nm. An optical diffraction grating does not diffract X-rays because its line spacing is much much greater than the wavelength range of X-rays. But a crystal with its regularly arranged atomic planes very close together *does* act like a grating. Any suitable regularly arranged object gives a diffraction pattern if the regular spacings are of the same order as the wavelength used. For example, try viewing a bright point of light through a piece of gauze. The regular mesh of the gauze diffracts the light into a pattern similar to that shown in Figure 7.19. Turn the gauze and the pattern seen also turns.

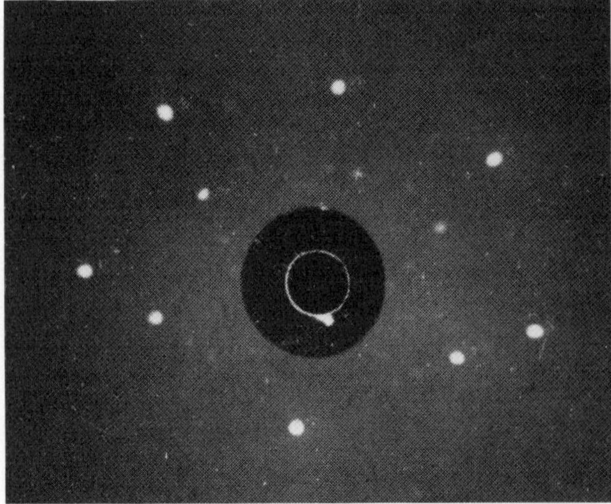

Figure 7.18 *Laue pattern for a single crystal*

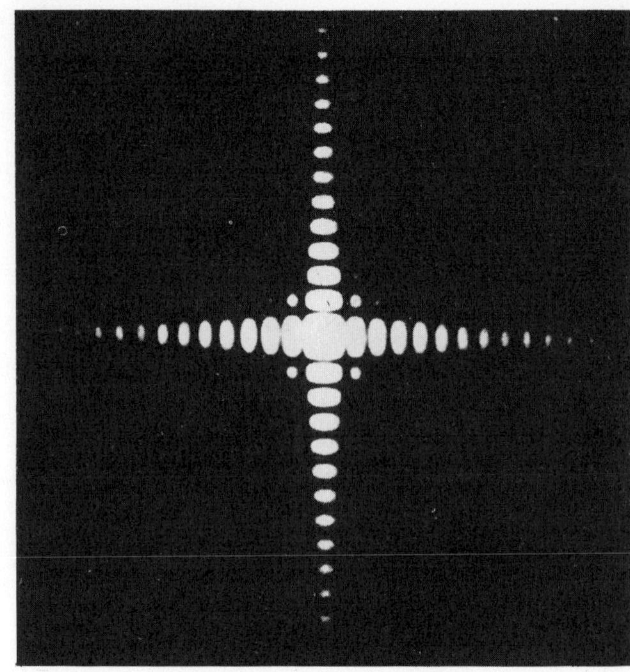

Figure 7.19 *Diffraction pattern due to a mesh*

When X-rays with a single precise wavelength are directed in a narrow beam at a single crystal, X-rays emerge from the crystal in certain directions only. A photographic film behind the crystal can record the directions of the emergent beams. By measuring the angle of diffraction of each beam, the arrangement and spacing of the atoms can be worked out. Different atomic arrangements are possible, each one giving a characteristic crystal shape and a characteristic X-ray diffraction pattern.

Crystal structures

When we think of crystals we usually think in terms of obvious ones like salt crystals or calcite crystals which can be seen unaided. Yet metals are composed of tiny crystals, called **grains**, packed together in random directions in the metal. Using X-rays it has been shown that the tiny crystals inside metals have the same sort of structure as big crystals.

FCC structure or face centred cubic, has atoms arranged as at the corners of a cube and at the centre of the faces, as represented by Figure 7.20. Aluminium has an FCC structure.

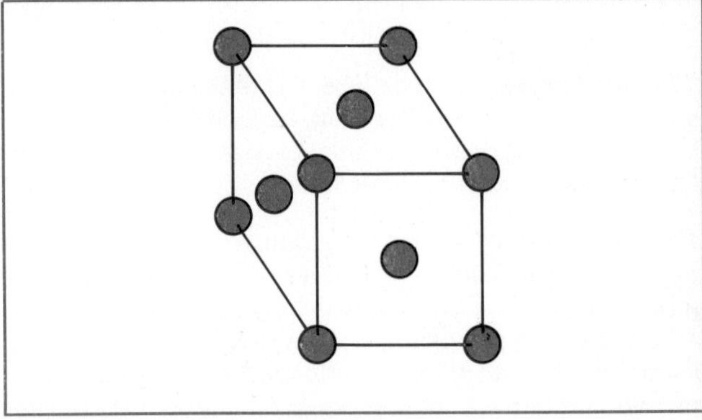

Figure 7.20 *FCC structure (atoms on front faces only are shown)*

BCC structure or body centred cubic, is shown in Figure 7.21. The atoms are arranged as at the corners of a cube and at the centre. Iron has a BCC structure up to 800°C. Above that temperature its structure changes to FCC. If iron wire is heated above 800°C then suddenly quenched in water, it becomes very brittle because the atoms have not been allowed enough time to revert to the BCC structure.

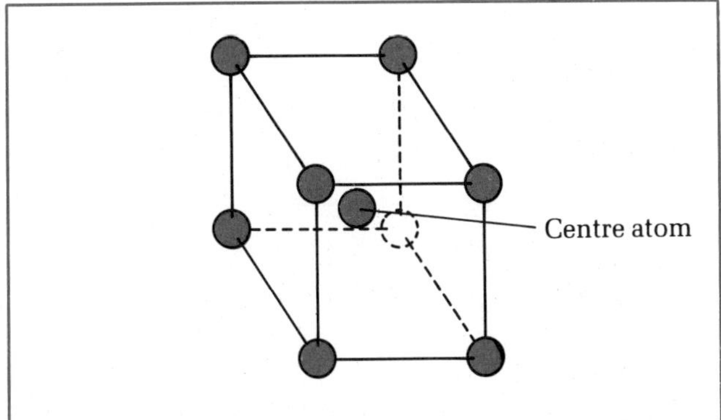

Figure 7.21 *BCC structure*

HCP structure or hexagonal close packed, is where the atoms fit together as closely as possible. Zinc and magnesium have this structure. The atoms are in layers, each atom in contact with six neighbouring atoms of the same layer. The layers are stacked so that the atoms of one layer fit into the hollows of the adjacent layers.

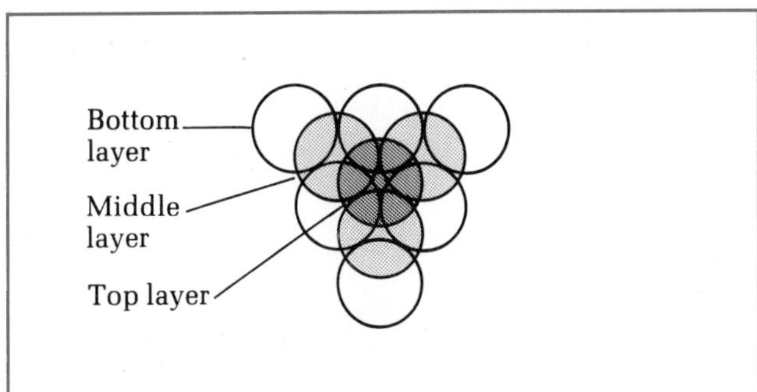

Figure 7.22 *HCP structure*

Tetrahedral structure is shown in Figure 7.23. Diamond is an example. Each atom is in contact with four neighbouring atoms to form a tetrahedral shape which is difficult to break up.

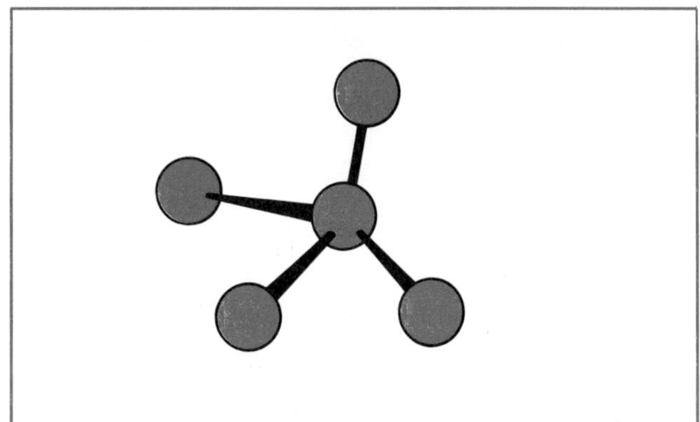

Figure 7.23 *Tetrahedral structure*

Microwave analogue of X-ray diffraction

We can make a scaled-up model of X-ray diffraction by a crystal using microwaves and polystyrene models. A transmitter is used to produce microwaves of wavelength 3 cm. A wax lens is used to direct a parallel beam of microwaves at a crystal model made from polystyrene. A second wax lens is used to focus the diffracted microwaves into a receiver. The crystal model is on a turntable. The receiver and its lens can be scanned across at different angles to the incident beam.

Crystal model made of polystyrene tiles as 'layers of atoms' The tiles are spaced regularly at 2–3 cm apart as in Figure 7.24. With the turntable rotating at a steady rate, the receiver is scanned across. Strong signals are detected only when the receiver is at certain positions. Why? Each tile reflects weakly so the arrangement produces a number of reflected beams in any given direction. However, the reflected beams are not in general in phase so they cancel one another out. Reinforcement only takes place in certain directions.

Consider layers at spacing d with the incident beam at angle θ to the layers. As shown in Figure 7.24, the extra distance travelled by reflections from one layer compared with the next is $2d\sin\theta$. Only if the extra distance $2d\sin\theta$ is a whole number of wavelengths can reinforcement occur. Therefore, for reinforcement,

$$2\,d\sin\theta = m\lambda$$

where d = layer spacing,
θ = angle between each layer and the incident beam,
λ = X-ray wavelength,
m = 1, 2, 3, etc.

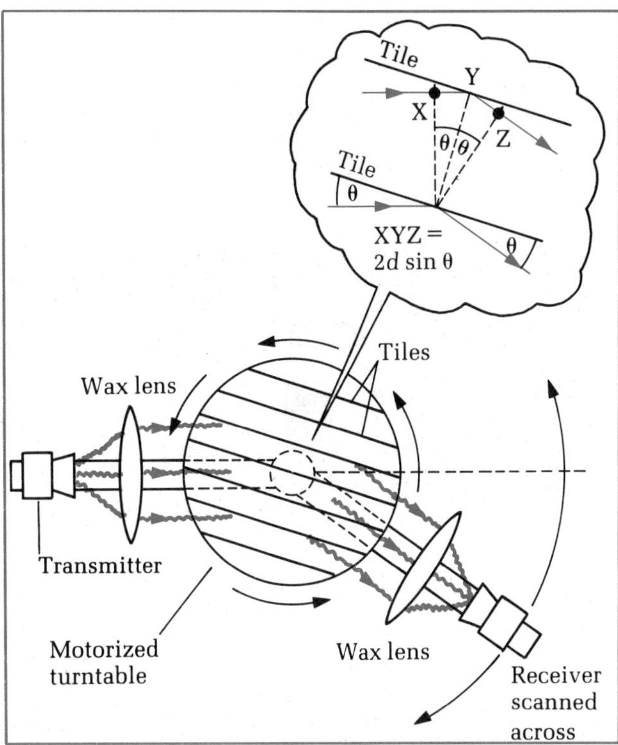

Figure 7.24 *Top view of a microwave analogue*

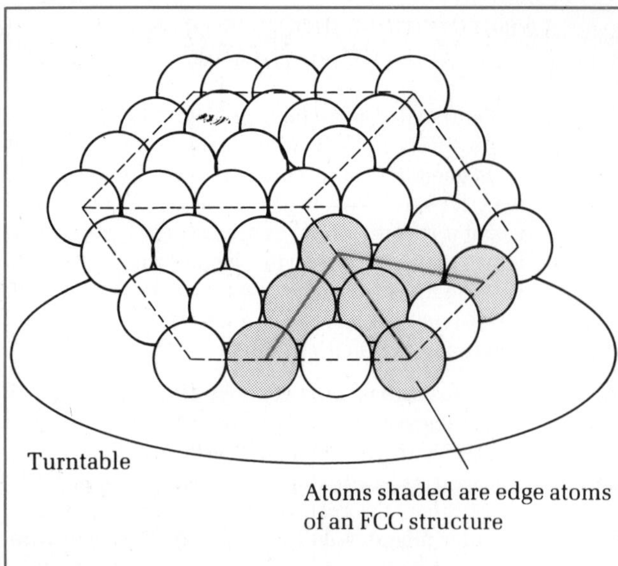

Figure 7.25 *FCC model*

With vertical tiles at a spacing of 2.0 cm for example, then the above equation gives $2 \times 2 \sin\theta = m \times 3$ (using $\lambda = 3$ cm). So $\sin\theta = 0.75m$, giving $\theta = 49°$ for $m = 1$ as the only diffracted order.

FCC crystal model This can be made using identical polystyrene balls glued together. Each layer is made of balls glued so each is in contact with six others. Then alternate layers are placed directly over one another, with the layer between fitting in the hollows. A structure like Figure 7.25 results. Each FCC unit in the structure has its faces at 45° to the vertical.

The model is set rotating on the turntable and the receiver is scanned across. Once again, strong signals are detected for certain positions of the receiver only. By measuring these positions, the spacing between layers of atoms causing these signals can be calculated. Like the tiles used earlier, the layers of atoms responsible must lie vertically.

X-ray diffraction by metals

Each tiny crystal in a given metal has the same arrangement of atoms, but each is in a different direction. Imagine lots of model crystals like those in Figure 7.25, each the same but at different angles to one another. The layers which are vertical in Figure 7.25 will occur in all directions. That layers at angle θ to the incident beam cause reinforcement. Thus the diffracted beams form a cone. So in the case of the X-ray beam directed at lots of tiny metal crystals, each diffracted order is in a cone, as shown in Figures 7.26 and 7.27. The angle of the cone is 2θ where θ is given by the equation above.

A photographic film behind the metal would show rings when developed, each ring being produced by a diffracted cone. In practice, a circular strip of film is placed in a horizontal ring around the sample of metal, as shown in Figure 7.27. The X-ray beam is directed at the sample and the diffracted orders produce lines on the film. From the line positions, the angle θ can be worked out to enable the layer spacing d to be calculated.

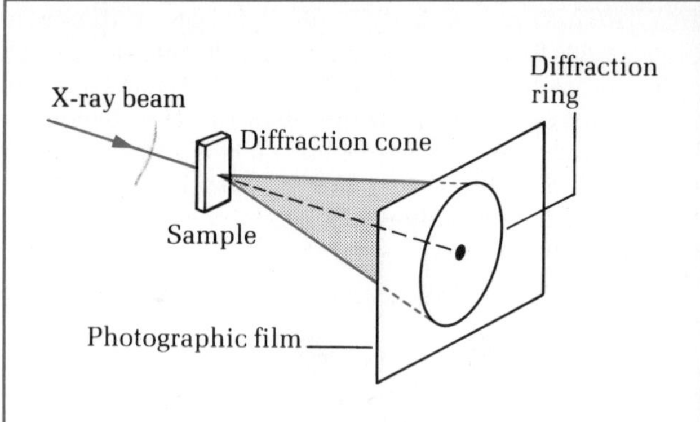

Figure 7.26 *A polycrystalline sample*

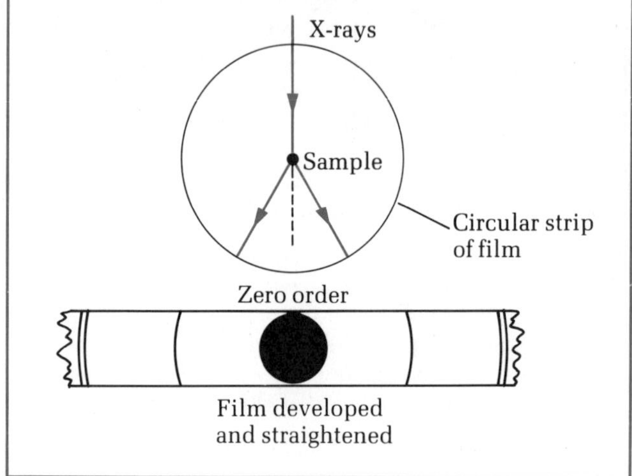

Figure 7.27 *Using a strip of film*

7.4 Structure of metals

Grains

The tiny crystals inside a metal are called **grains**, each with the same arrangement of atoms but in a different direction to adjacent grains. Grain sizes vary even in the same metal, and they can be as small as 0.01 mm. Microscopes are used to study grains and, as Figure 7.29 shows, there is no regularity to their shape.

Reflection microscopes are designed to observe the light reflected from specially prepared surfaces. The metal surface to be studied must be cleaned to remove surface impurities, then it needs to be polished using special techniques until it is very smooth. Finally the polished surface is etched to remove unwanted polishing matter. Each grain reflects light in a slightly different way from its neighbours so under a high power light microscope, the grains are easily visible. However, to observe defects inside the grains, even the most powerful light microscopes may not be good enough.

Electron microscopes can give magnification up to a million times or more. Even atomic planes can be resolved with electron microscopes. An electron microscope uses high velocity electrons instead of light. At high velocity electrons have a much shorter wavelength than light (see wave particle duality for electron wavelength) so they can

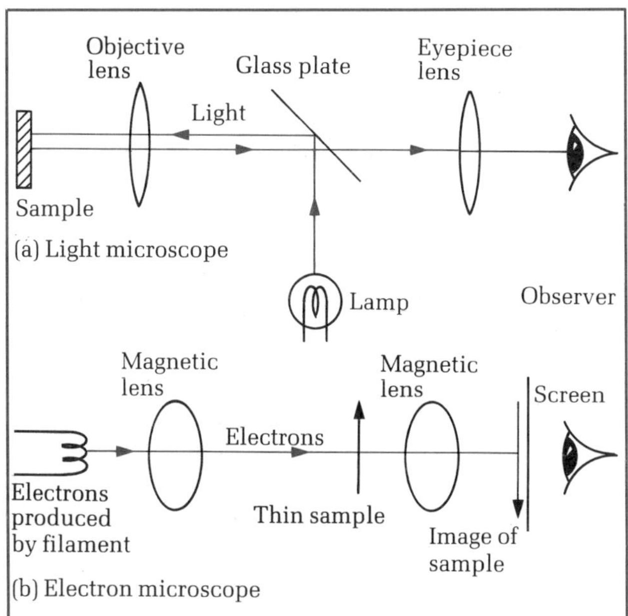

Figure 7.28 *Comparison of light and electron microscopes*

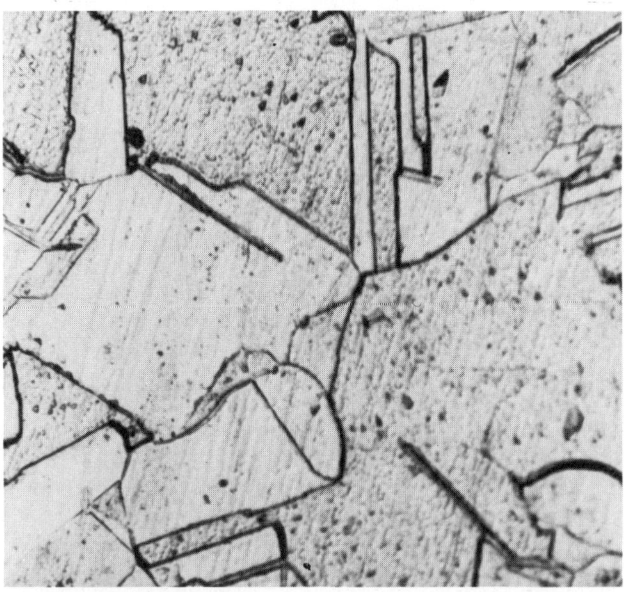

Figure 7.29 *Austenite*

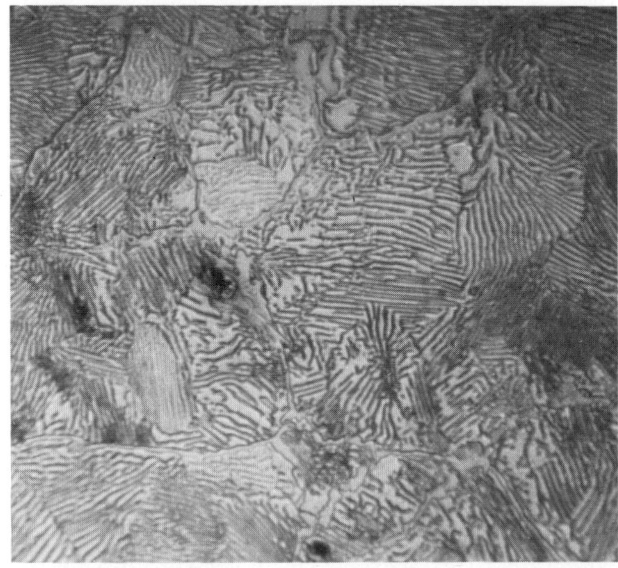

Figure 7.30 *Pearlite*

Figure 7.31 *Martensite*

be used to form images of structures not much larger than individual atoms. Magnetic fields are used to control and focus the electrons, as shown by Figure 7.28.

Strength and hardness

These are just two of the mechanical properties of a solid metal affected by grain size. The smaller the grains, the stronger the material is. With small grains fracture of the material is more difficult because there are more grain boundaries. When the atoms in one grain slide apart along a crystal plane, the movement is not easily passed on to the adjoining grains. The atomic planes of adjoining grains are in different directions so fractures tend to be halted at grain boundaries. So the more boundaries there are, the stronger the material is.

Steel is one of the most important materials that we use. It is composed of iron with a small percentage of carbon. Carbon atoms stuck between layers of iron atoms give added strength by obstructing layers when they try to slide over one another. The percentage of carbon in steel is fixed when the steel is manufactured, and it has an important effect on the grain structure and hence strength of the steel.

When steel is heated above 800°C but not above its melting point, its grains become relatively large. They form a characteristic pattern called **austenite**, shown in Figure 7.29. The iron atoms form an FCC structure with carbon atoms at the cube centres. If the steel is then allowed to cool slowly, the grains become much smaller, forming a pattern called **pearlite**, as in Figure 7.30. Here the iron atoms form a BCC structure; the transition temperature is about 800°C.

If the cooling process is rapid from the austenite phase, for example by **quenching** the steel in water, there is not enough time for the transition to occur. A grain pattern known as **martensite** forms, shown in Figure 7.31. How do you think its mechanical properties compare with pearlite? Some experiments are suggested later.

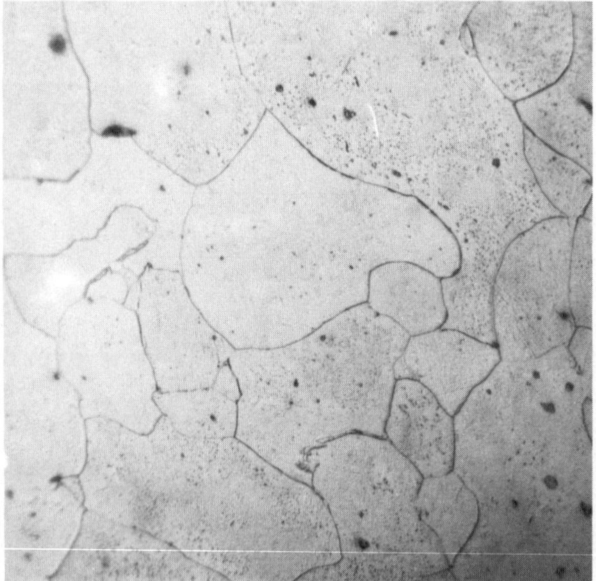

Figure 7.32 *Ferrite*

Yet another type of grain structure is possible by **annealing** steel. Here the cooling from the austenite phase is very slow. Larger grains than pearlite are formed, in a pattern known as **ferrite** as shown in Figure 7.32. Very slow cooling gives enough time for adjoining grains to line up to form larger grains than in pearlite. How do you think the properties of ferrite compare with the others? See below.

Investigating steel

Quenching Heat one end of a 20 cm length of steel wire in a bunsen flame until it becomes cherry red in colour (about 800°C). Then plunge the end of the hot wire quickly into cold water. When the rest of the wire has cooled, try to bend the quenched end. Quenching makes it hard and brittle so it ought to snap.

Annealing Heat the other end of the wire until it is red hot, and keep it at red heat for about 15 s. Withdraw it very slowly so that it cools gradually. When cool, it should easily bend because it has been annealed.

Tempering Reheat the annealed end then quench it. When it is cool, use emery paper to make it shiny, then reheat it in a small bunsen flame, until a blue tint is seen. Remove the wire from the flame and allow it to cool. It should now be tough and springy.

Dislocations

Slip lines can often be seen on micrographs (microscope photographs) of metals. Each slip line is along the boundary between two planes of atoms which have moved relative to each other. When a metal is stretched beyond its elastic limit, planes of atoms slip over one another by one or more 'spaces', as shown in Figure 7.33. So a slip line across a surface is where there is a step across the surface.

Twin lines or bands are another common feature. Here adjacent planes are displaced a little from one another from one plane to the next, to the next, etc., across a number of planes. So looking down on them is like looking down on terraces from high up.

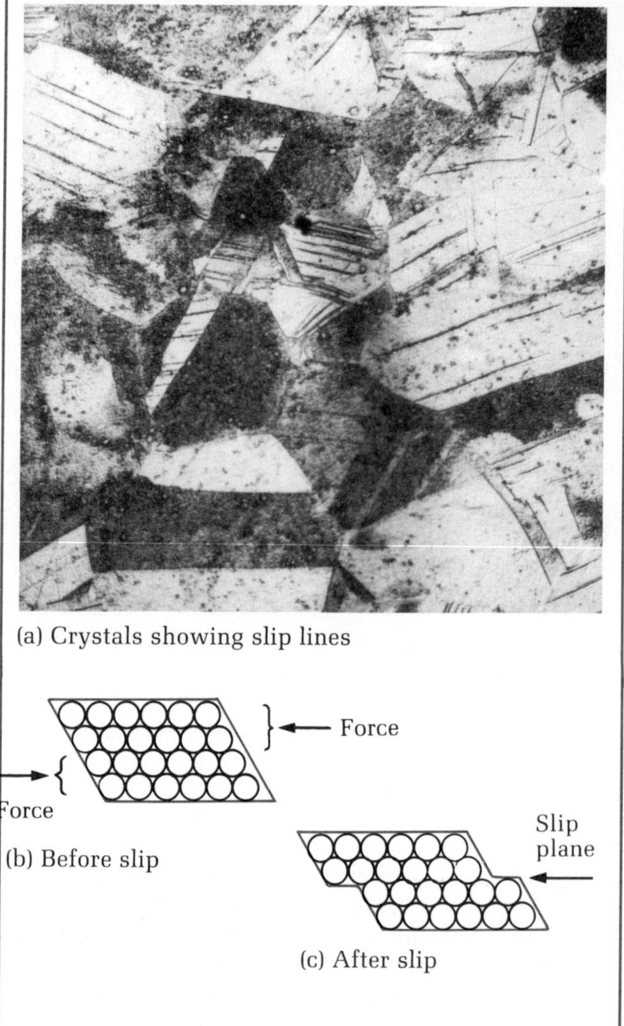

(a) Crystals showing slip lines

Force

Force

(b) Before slip

Slip plane

(c) After slip

Figure 7.33 *Slip*

To move one plane of atoms relative to its neighbouring plane to produce slip, the interatomic bonds between the planes must be broken. When slip takes place the bonds are broken in sequence, not all at once; so huge forces are not necessary for slip to be produced. It's a bit like trying to move a carpet: you can pull it all in one go, or you can use your feet to shuffle a 'ruck' across the floor. The equivalent of a ruck for planes of atoms is called a **dislocation**. A dislocation is where a plane of atoms is suddenly squeezed out by other planes, as in Figure 7.34. When under stress, a dislocation moves through the atomic planes without too much force – like the ruck in a carpet!

Consider the crystal of Figure 7.34 acted upon by forces as shown. The top half of plane A is broken off from its other half, and is pushed onto the initial dislocation, allowing the top half of B to join the lower half of A. The process continues across the crystal until the dislocation has reached the other side. So although no atom moves by more than one atomic space, the dislocation can travel right across the crystal. At the sides, slip lines are left when the forces are removed.

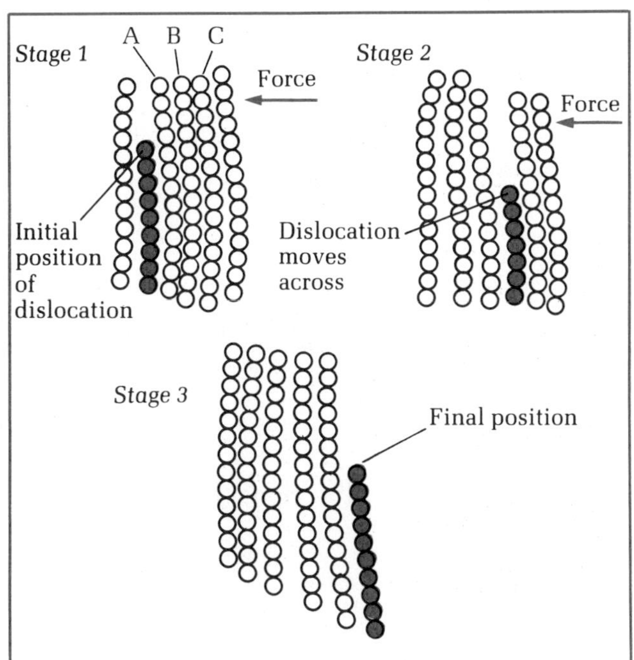

Figure 7.34 *Dislocations on the move*

A bubble raft on an overhead projector is a useful way of picturing dislocations. To make a bubble raft, a shallow glass dish is partly filled with soap solution initially clear of bubbles. Then an air jet through a syringe is used to create lots of small bubbles on the surface of the liquid, as shown in Figure 7.35. The bubbles pack themselves together in rows, regularly arranged, but here and there a row is squeezed out. Such dislocations may be caused by big bubbles or missing bubbles. Push the pattern a little with the end of a ruler to see the dislocations move. In a crystal, dislocations are due to **vacancies** (i.e. missing atoms) or **interstitial atoms** (i.e. atoms out of place). Figure 7.36 shows an electron micrograph of dislocations in a very deformed piece of metal. There are so many dislocations that they have become tangled up. The result is a stronger metal. So a few dislocations per unit volume make a metal weak but lots of dislocations increase its strength.

Work hardening is a process which makes a metal stronger. The metal is *worked* or deformed when cold to make it stronger and harder. The effect of working the metal is to increase the number of dislocations, so increasing its strength. Also, working the metal increases the number of atoms out of place in between slip planes, so slip becomes harder to achieve.

Metal fatigue occurs when a metal is repeatedly worked, causing it to fracture. Repeated working causes hardening and so **cracks** may develop. Once a small crack starts to develop, continued working on the metal makes the crack grow. Stress build-up at the crack makes it develop. The growth of the crack reduces the cross-sectional area of unaffected metal which makes the stress even bigger. Eventually the metal can no longer take the stress and it fractures. Detecting small cracks in metals is very important for safety reasons. For example, the joints of an aircraft's wings need to be checked for cracks at regular intervals.

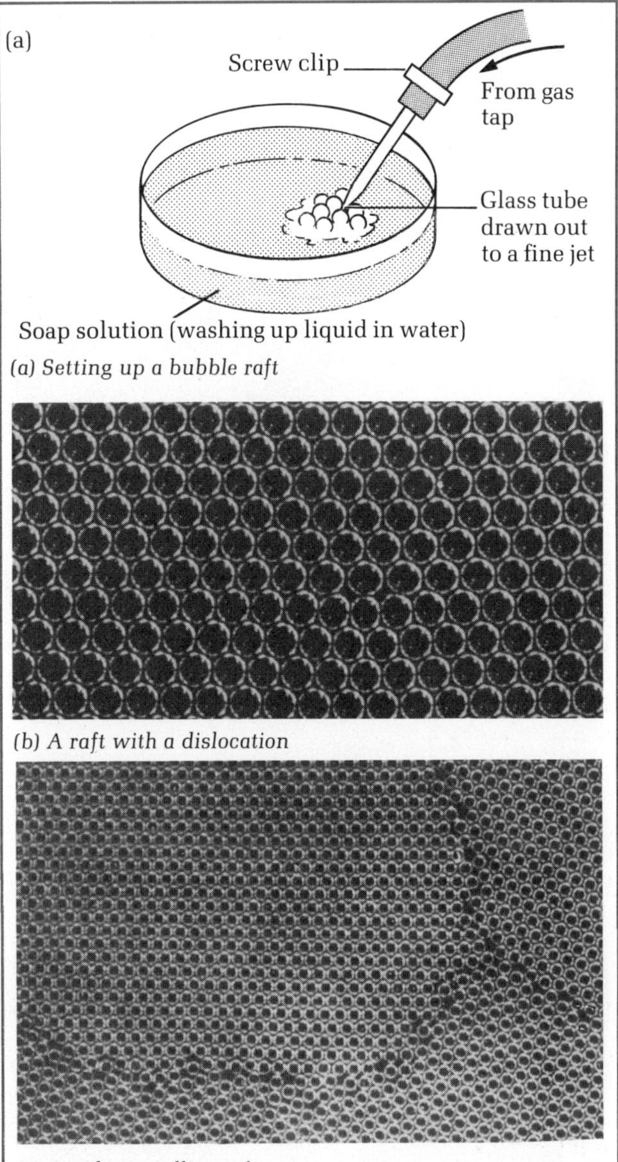

(a) Setting up a bubble raft

(b) A raft with a dislocation

(c) A polycrystalline raft

Figure 7.35 *A bubble raft*

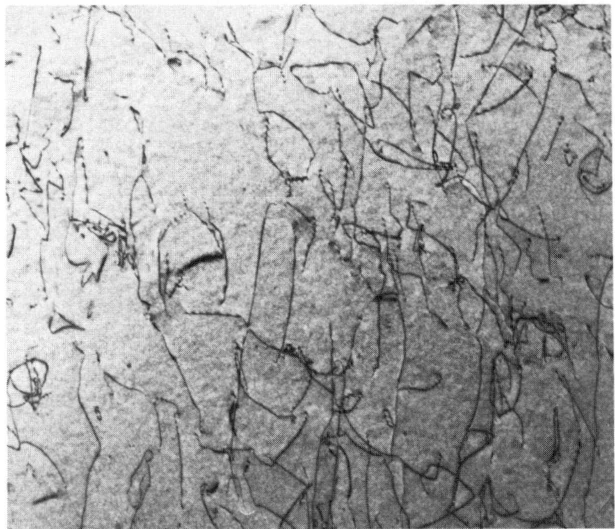

Figure 7.36 *Electron micrograph of dislocations*

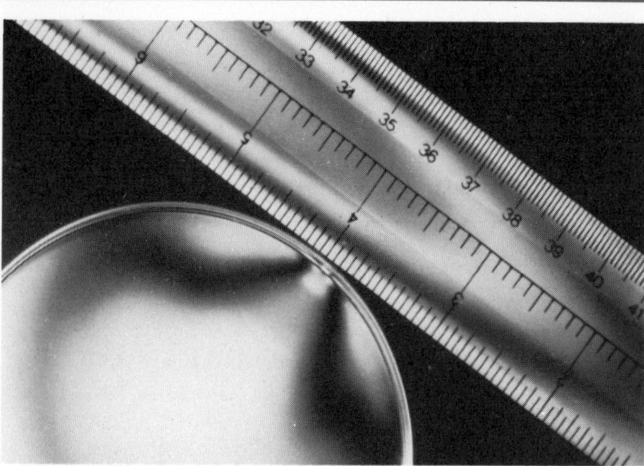

Figure 7.37 *Stress fringes in perspex*

Photoelasticity can be used to test for stress at the design stage. A model of the item under design is made from perspex. When placed between crossed polaroids, coloured fringes are seen where the model is under stress. The fringes are closest where the stress is greatest. Try it yourself using a perspex ruler placed between two crossed polaroids. A small cut in the ruler should show the stress build-up at the cut when the ruler is flexed.

7.5 Glass

Glass is made by fusing silica with certain oxides. X-ray diffraction by glass gives a pattern like that given by water. Glass molecules form a disordered structure. Glass is an example of an **amorphous** structure because there is no pattern for the arrangement of its atoms. It is sometimes called a **solid liquid** because its molecules are irregularly arranged as in a liquid but fixed in their relative positions.

(a) Water

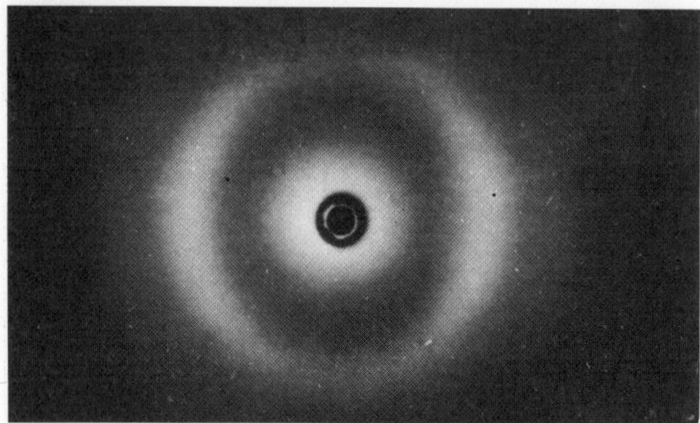

(b) Glass

Figure 7.38 *Comparison of X-ray diffraction pattern for glass and water*

Glass is very hard and brittle unless heated, when it softens and melts. When softened by heat, it can be shaped as required or drawn into thin fibres. Cracks on the surface of glass weaken it much more than on a metal. Cracks on a glass surface under stress develop across the glass quickly. Glass cutters use this to their advantage; to cut a glass pane, a scratch is deliberately drawn across the surface. When the pane is tapped, the glass fractures along the scratch line.

Investigating the strength of glass

Glass fibres Heat the end of a glass rod so that it softens enough to bend it into a hook shape to support weights. When cooled, heat the centre of the rod until soft, then remove the rod from the flame and quickly draw it out into a fibre. Take care not to flex the fibre or it will snap. Clamp its straight end so the fibre hangs vertically. Wearing safety goggles, load the fibre with weights, step by step, until it snaps. Note the final weight, and use a micrometer to measure the diameter of the broken fibre. Hence calculate the breaking stress of glass. Compare with the value for steel which is about 10^8 Pa.

Cracks Wear safety goggles again. Support a glass rod horizontally at either end as in Figure 7.40. Load its centre

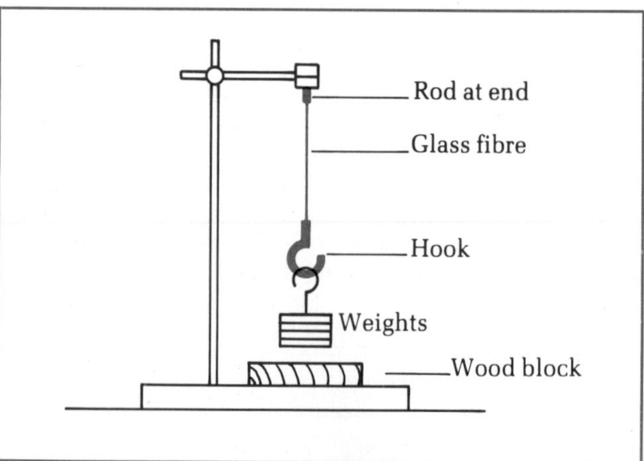

Figure 7.39 *Strength of a glass fibre*

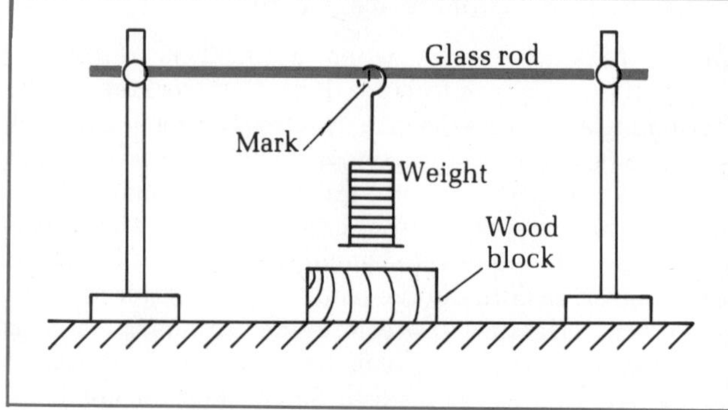

Figure 7.40 *Testing cracks*

with weights step by step until it snaps. Repeat the test with an identical rod marked with a glass cutter underneath at the middle. The cut rod should snap more easily because the crack weakens it.

7.6 Polymers

Polymers are long molecules with lots of atoms joined to one another like links in a chain. Each molecule is composed of a string of carbon atoms joined by covalent bonds, as in Figure 7.41. Other atoms join on to the carbon atoms, but each carbon atom is always linked to two other carbon atoms. Figure 7.41 shows the way in which a polyethylene molecule is made up. Different types of atoms joined on to the carbon string give different polymer molecules, each with its own properties. You only need to look around to see different polymers in use: nylon, polythene (which is made of polyethylene molecules), perspex, PVC, rubber, etc.

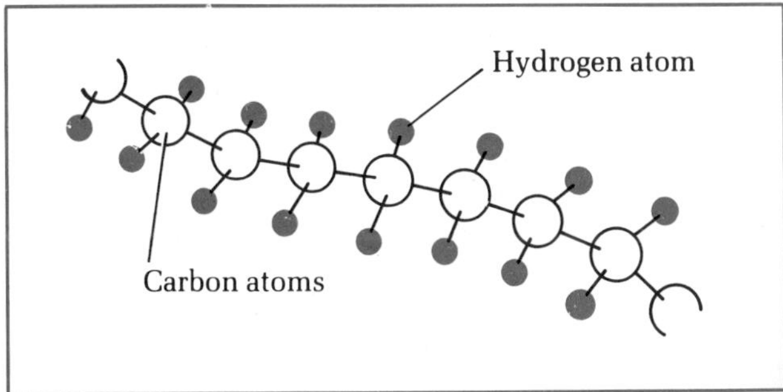

Figure 7.41 *A polyethylene molecule*

Rubber

This is an example of a naturally occurring polymer. It is formed as a gum resin by rubber trees. In a piece of rubber, the molecules are all tangled up with one another, until the rubber is stretched and then they straighten out. X-ray diffraction pictures of unstretched rubber are like the pictures for water or glass, showing that the molecules are in disorder. However, stretched rubber does show a 'crystal-like' diffraction pattern when X-rays are directed at it. So there is some degree of order amongst the molecules in stretched rubber; this is because they are parallel to one another when stretched. Stretched rubber can be termed **crystalline** because there is some order present. When a piece of stretched rubber is released, it regains its original shape because the molecules curl up and tangle together again. The stretchiness of rubber occurs because its molecules are tangled together in the unstretched state. Its elasticity occurs because the molecules curl up when released.

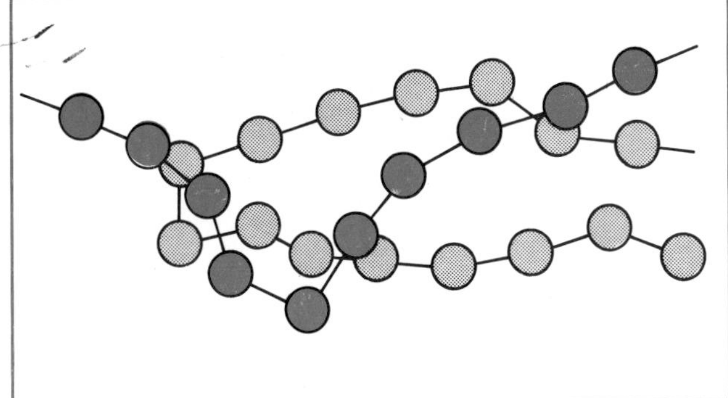

Figure 7.42 *Tangled up rubber molecules*

Vulcanized rubber is stiffer and stronger than raw rubber. It is made by mixing and heating raw rubber with sulphur. The sulphur atoms act as strong cross-links between molecules, so vulcanized rubber is harder to stretch because it is more difficult to untangle the molecules. However it still keeps its elasticity because the tendency of the molecules to curl up is unaffected.

Investigating rubber

Strength and stiffness Load against extension curves can be obtained using an arrangement like Figure 7.3. The unloading curve differs from the loading curve as in Figure 7.43 but it does return to zero extension when unloaded completely. Clearly Hooke's law is not obeyed since the material becomes stiffer the more it is stretched. Can you explain this in terms of the molecules? Once the molecules have been straightened they are very difficult to stretch any further.

Hysteresis and resilience Stretch a rubber elastic band quickly several times, then hold it to your lower lip. You ought to notice that it has become warm due to being stretched. Why? The clue is the load *v* extension curves for rubber. The area under a load *v* extension curve represents the work done. On loading, energy is supplied; on unloading, energy is released. Because the loading curve is higher than the unloading curve of Figure 7.43, then more energy is supplied than is released. The difference is accounted for by the increase of internal energy of the rubber molecules (i.e. K.E. of vibration, etc.), so the temperature of the rubber rises. The loop formed by the loading and unloading curves is called a **hysteresis loop** because the unloading curve lags behind the loading curve. The greater the area of the hysteresis loop, the greater is the increase of internal energy when loaded then unloaded. **Resilience** is the ability of the material to stand up to hysteresis. High resilience means that the material can be repeatedly stretched without losing its strength. Car tyres must be very resilient otherwise they would not be safe.

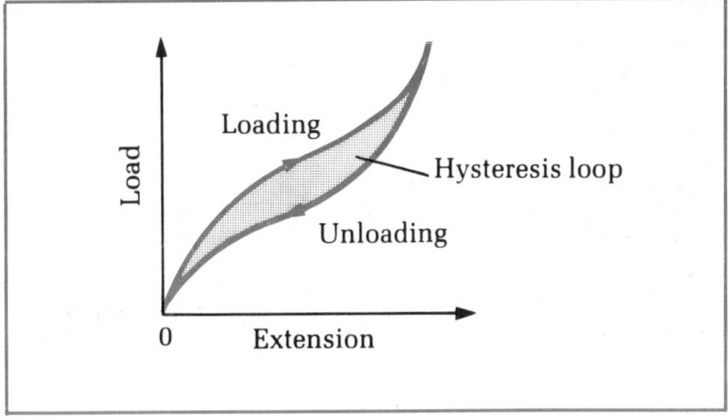

Figure 7.43 *Hysteresis of rubber*

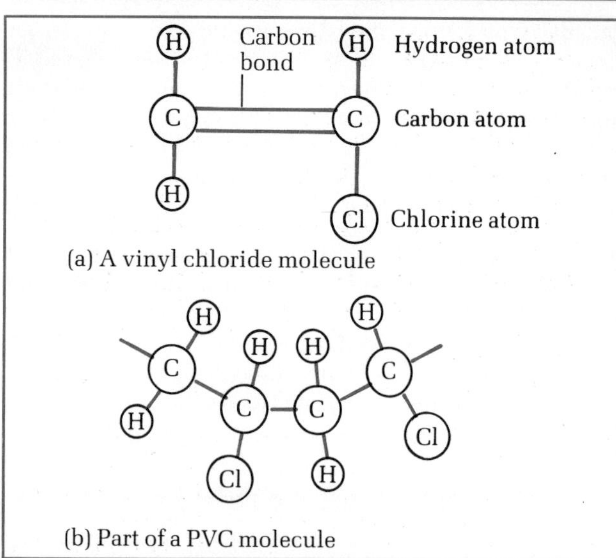

Figure 7.44 *Two plastics molecules*

Synthetic polymers

These are usually known as **plastics**, and they are made by joining smaller molecules end-to-end to produce long polymer molecules. For example, polyvinyl chloride (PVC) is composed of molecules formed by joining vinyl chloride molecules end-to-end as in Figure 7.44.

Some plastics can easily be deformed. Their molecules are all tangled up. When stress is applied then removed, they do not regain their initial shape. Other plastics like nylon can be made so that they are partly crystalline – their molecules are lined up. This makes them strong and stiff. Plastics can also be made stronger by cross-linking their molecules – the same idea as for rubber. Bakelite is an example of a polymer which is cross-linked.

Cross-linking is due to Van der Waals bonds between atoms of neighbouring molecules. Atoms within a molecule are held to one another by strong covalent bonds which are difficult to break. However, weak attractive forces between the atoms of one molecule and of adjacent molecules supply cross-links. A plastic material like polythene sheet is easily stretched at first then it becomes much stiffer. The initial stretching is easy because straightening out the molecules involves breaking weak cross-links only. But once straight, further stretching is difficult because the covalent bonds are being stretched. Once stretched, polythene sheet loses its shape, unlike rubber. The polyethylene molecules form new cross-links in the stretched state, and do not curl up like rubber molecules when released.

Thermoplastic polymers can be repeatedly heated and cooled without forming cross-links. When heated they soften and can be moulded to any required shape. Polycarbonate is a thermoplastic material used to make hard hats, instrument cases, etc. Polyethylene is also thermoplastic but cross-linking occurs if it is exposed to ultra-violet radiation. This is the reason why polyethylene exposed to daylight hardens.

Thermosetting materials form cross-links when initially heated; this makes the material very hard, if required. Cross-links are not broken by reheating the material so it does not soften when reheated. Formica, bakelite and rubber are thermosetting. So, too, is the material used for knobs and handles in the kitchen. Cross-links give stiffness and heat resistance, but too much heating just breaks the molecules up. A plastic pan handle made from urea formaldehyde plastic gives off a very unpleasant smell if it becomes too hot!

7.7 Summary

Definitions

1 *Grains* are the tiny crystals inside metals.
Dislocations occur where atom planes are squeezed out by other atom planes.
Quenching makes a metal hard and brittle.
Annealing makes a metal easy to bend.
Tempering makes a metal tough and springy.

2 *Elastic* behaviour is shown where a material regains its shape when the applied forces are removed.
Plastic behaviour is shown where the material permanently deforms.

3 *Thermoplastic* materials soften when heated so can be remoulded. No cross-links are formed.
Thermosetting materials become hard due to initial heating when cross-links are formed. Reheating does not cause them to soften.

Equations

1 *Hooke's law* for springs states

$T = ke$ where T = tension,
e = extension,
k = stiffness constant of the spring.

2 *Tensile stress* is defined as tension per unit area of cross-section, applied at right angles to the area.

3 *Strain* is defined as extension per unit length.

4 *The Young modulus E* for a material is defined as stress/strain, provided the limit of proportionality is not exceeded.

5 *The Young modulus* for a wire is given by

$$E = \frac{Tl}{Ae}$$

where T = tension in the wire,
l = unloaded length of wire,
e = extension,
A = area of cross-section.

6 *Energy stored* in a spring or wire $= \frac{1}{2}Te$.

7 Energy stored per unit volume in a stretched wire $= \frac{1}{2}$ stress × strain.

8 X-ray diffracted orders occur at angles given by

$$2d \sin\theta = m\lambda$$

where d = layer spacing,
λ = X-ray wavelength,
θ = angle between incident beam and the layers,
$m = 1, 2, 3$, etc.

Short questions

7.1 In an experiment to test the strength of a nylon wire, the wire was hung vertically and then loaded by hanging weights from its lower end. Before loading, the wire diameter was measured at several points along its length, giving values of 0.81 mm, 0.80 mm, 0.81 mm, 0.79 mm and 0.80 mm. Then the load was increased in 1 N steps until the wire snapped at 25 N. Calculate the breaking stress of nylon. How accurate is your value?

7.2 Two identical springs each have a natural length of 300 mm and a stiffness constant 400 N m^{-1}. Calculate the length of each spring when hung vertically, as below,
a) with one spring supporting a 10 N weight joined end-on to the other spring,
b) side-by-side from the same level to support a horizontal 10 N bar.

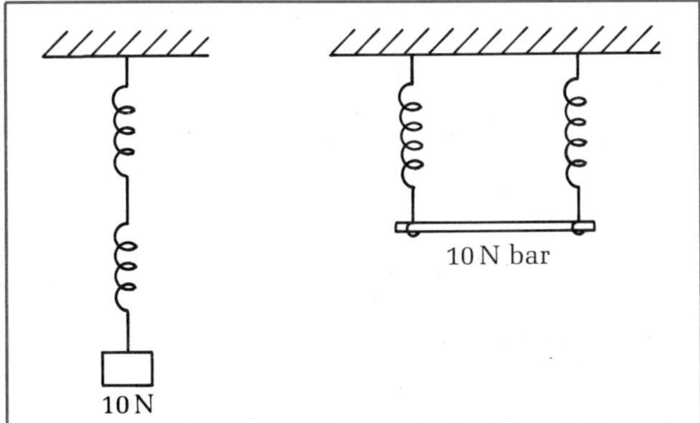

Figure 7.45

7.3 A steel wire of diameter 0.50 mm and length 1.84 m is suspended vertically from a fixed point and used to support a 60 N weight hung from the lower end of the wire. The Young modulus for steel is 2.0×10^{11} N m^{-2}. Calculate
a) the extension of the wire,
b) the elastic energy stored in the wire, given the elastic limit for steel is at a strain of 0.0025.

7.4 A crane used to lift iron girders on a building site has a lifting cable, made of steel, of diameter 40 mm. When the crane is used to lift an iron girder of weight 10 kN, the cable length with the girder just off the ground is 75.0 m. Calculate
a) the extension of the cable for this length and load,
b) the elastic energy stored.
Use the Young modulus value given in question **7.3**.

7.5 A nylon string of diameter 0.5 mm is fitted to a guitar. The string is fixed to a point near one end of the guitar and then attached to a tension key near the other end of the guitar. The string length from the fixed point to the tension key is 850 mm. If each turn of the tension key extends the string by 4.2 mm, calculate the tension in the string after 12 turns. (The Young modulus for nylon is 3×10^{9} N m^{-2}.)

7.6 The end of a copper wire of diameter 0.30 mm and length 1.50 m is fused to one end of a steel wire of the same diameter and 1.20 m in length. The wire is then hung vertically by attaching the free copper end to a fixed point and attaching a small weight to the lower end of the steel wire. The attachments reduce the length of each wire by 20 mm. A load weight of 30 N is then hung from the lower end of the steel wire. Calculate
a) the stress in each wire,
b) the strain in each wire,
given that the Young modulus for steel is 2.0×10^{11} N m^{-2} and for copper is 1.3×10^{11} N m^{-2},
c) the total extension of the two wires.

7.7 A block of solid copper of size 10 mm × 10 mm × 20 mm is placed in a vice with its longest side parallel to the jaws of the vice. Each turn of the vice moves the jaws 0.5 mm closer together. The vice is tightened up by half a turn. Calculate
a) the compressive force in the copper,
b) the elastic energy stored in the block assuming its elastic limit is not exceeded.
Use the Young modulus value for copper as in question **7.6**.

7.8 An iron bar with a circular cross-section of diameter 240 mm is heated to a temperature of 600 K. The bar is then clamped firmly at either end at points 0.40 m apart using a rigid steel frame. The temperature of the bar is then reduced to 300 K. Calculate the tension in the bar after it is cooled to 300 K. Assume the Young modulus for iron is 2.0×10^{11} N m^{-2} and its thermal expansion is 0.012 mm per metre length for each 1 K rise of temperature.

7.9 The load v extension curve for a strip of rubber is shown below. The graph also shows the unloading curve.

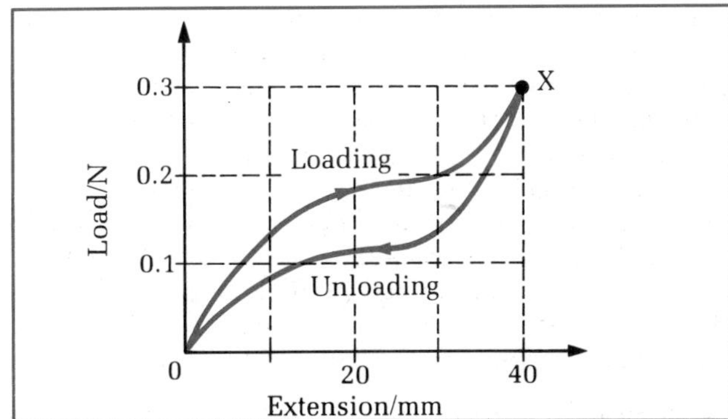

Figure 7.46

From the graph, estimate
a) the work done to extend the strip from its natural length to point X,
b) the gain of internal energy by the strip when it is loaded to X then unloaded.

7.10 The diagram below shows a typical stress-strain curve for steel. Describe the chief features of the curve and explain them in atomic terms. From the curve, estimate the total work done per unit volume to fracture a sample of steel.

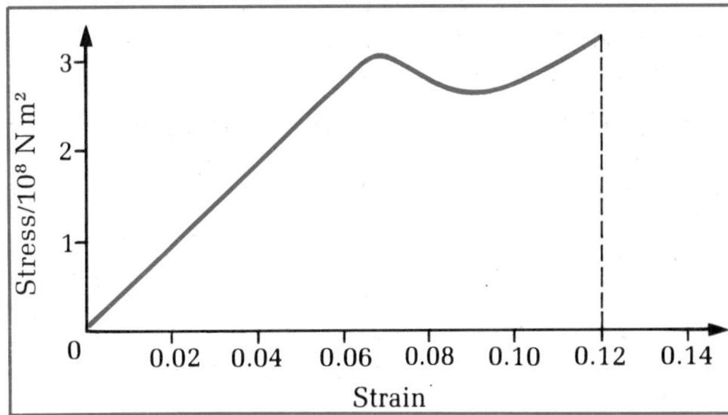

Figure 7.47

7.11 In an X-ray diffraction experiment, a beam of X-rays of wavelength 0.15 nm was directed at a metal wire specimen, as below. A strip of film was used as shown to record the position of the diffracted beams, dark lines on the film at P and Q being found after developing it. Calculate the spacing between the layers of atoms causing these dark lines.

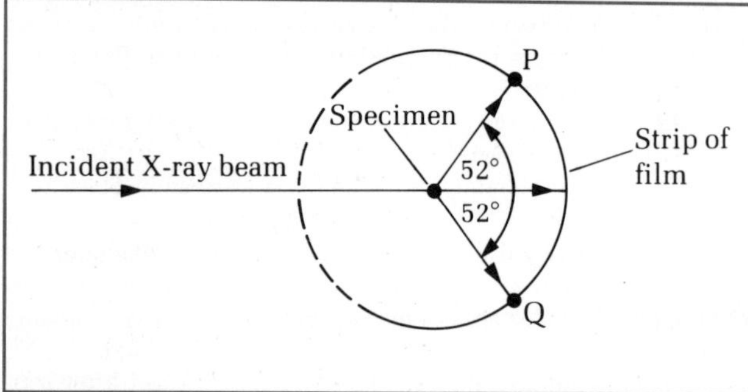

Figure 7.48

7.12 Describe how rubber differs from steel in terms of its mechanical properties. Explain two of the differences in terms of the structure of each material.

7.13 Sketch stress-strain curves on the same axes for copper, steel, glass and rubber. List the four materials in order of increasing

a) stiffness,

b) brittleness.

7.14

a) What is the difference between elastic and plastic behaviour? Explain the difference in simple molecular terms.

b) Explain why a metal bar snaps easily after being bent rapidly back and forth a few times.

7.15

a) What is the difference between a thermosetting and a thermoplastic polymer? Give an example of each type, and state one of its uses.

b) Sketch a typical stress-strain curve for a strip of polythene. Explain the curve in simple molecular terms.

8

Fluids

8.1 Fluids at rest

Pressure in fluids

Fluids are substances which can flow, so any gas or any liquid is a fluid. Fluids exert force and can be acted on by force. Try squeezing one of those long inflated balloons at one end. The other end pops out a little further because the squeeze force increased the pressure in the balloon.

Consider a small area of cross-section in a fluid. The force per unit area at right angles (i.e. normally) to the area is the **pressure** in the fluid.

$$P = \frac{F}{A}$$

where P = pressure,
F = force normal to area,
A = area on which force acts.

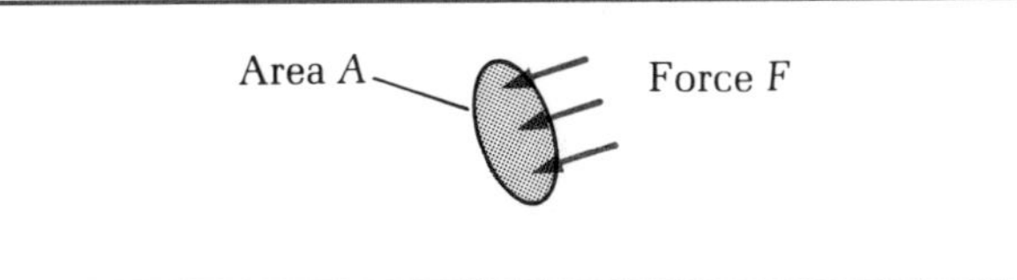

Figure 8.1 *Pressure*

The unit of pressure is the **pascal** (Pa), which equals $1\,N\,m^{-2}$.

For any fluid at rest,

a) the pressure at any point acts equally in all directions,
b) the pressure increases with depth.

Figure 8.2 shows two cylindrical water tanks connected by a pipe and valve. Both tanks are open to the atmosphere. Suppose the valve is initially closed with a higher level in the narrow tank. Then the pressure on the valve from the narrow tank is greater than from the wide tank. If the valve is opened, water flows from the narrow tank to the wide tank because of the pressure difference; but as the levels become more equal, the pressure difference lessens so the flow lessens. Eventually the levels become equal and the flow stops.

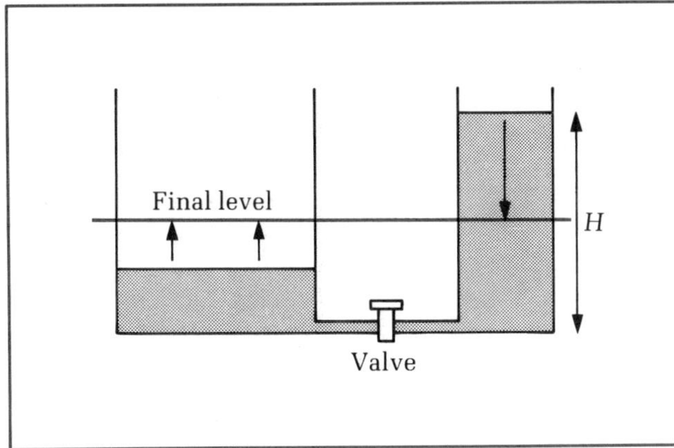

Figure 8.2 *Pressure difference*

Pressure increases with depth because of the weight of the fluid. Consider the narrow tank before the levels become equal. The volume of water is given by base area $A \times$ height H between the base and the surface of the water. So the mass of water m is equal to $AH\rho$ where ρ is the water density (since mass = volume × density). Therefore the weight of water, mg, is equal to $AH\rho g$. So the water pressure on the base (= weight of water/area of base) is equal to $AH\rho g/A$ which equals $H\rho g$.

$$P = H\rho g$$

where P = pressure in a fluid at rest,
H = depth beneath surface,
ρ = fluid density,
g = acceleration due to gravity.

Increase of pressure with depth is used in domestic water supplies. The mains water supply feeds an open tank in the loft of the house. The water from the tank feeds the hot water system. In a two-storey house, the hot water pressure on the ground floor is higher than the hot water pressure upstairs, because the downstairs taps are much further below the open water surface of the loft tank.

Measuring pressure

Lots of people need to measure pressure. Blood pressure, gas pressure, tyre pressure, atmospheric pressure are a few examples of different pressure measurements often made. There are different types of instrument for measuring pressure, the type chosen in any given situation determined by the range and accuracy required.

The U-tube manometer can be used to measure domestic gas pressure. The difference in height between the two levels is measured, as in Figure 8.3. Then the equation $P = H\rho g$ may be used to calculate the pressure difference between the two levels. With one limb open to the atmosphere and the other connected to the pressure source, the gas pressure is equal to $H\rho g$ + atmospheric pressure. (ρ is the manometer fluid density.)

The Bourdon gauge outlined in Figure 8.4 (on the next page) is a more robust instrument than the U-tube manometer. The pressure to be measured makes a flexible metal tube uncurl, causing a pointer to move over a scale.

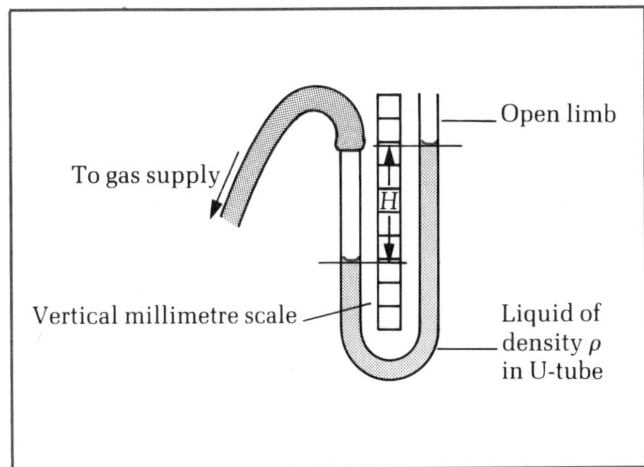

Figure 8.3 *U-tube manometer*

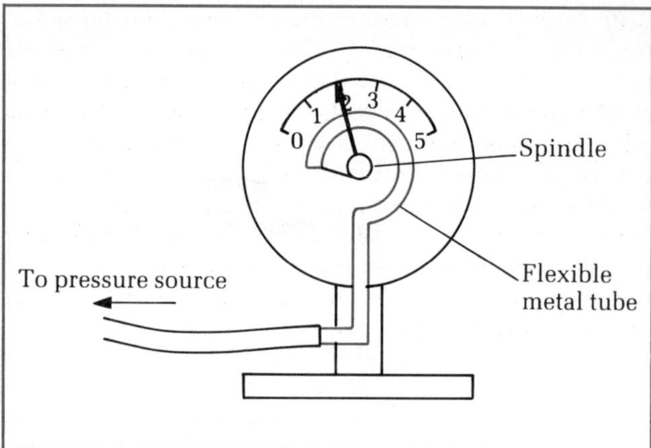

Figure 8.4 *A pressure gauge*

The scale can be calibrated (i.e. marked) directly in pascals. The measured pressure is the excess pressure of the pressure source compared with atmospheric pressure.

Electrical gauges produce a p.d. which varies with pressure. The capacitance-type gauge contains two closely spaced metal plates forming a capacitor. The pressure to be measured is exerted on the outer side of one of the metal plates, as in Figure 8.5. The plate is flexible enough to move nearer to the other plate, making the capacitance increase (see p. 168). As a result, charge flows onto the plates from a battery connected to the plates. The charge flow alters the p.d. across a capacitor C connected in series with the battery. So by measuring the p.d. the pressure can be monitored or recorded with a data recorder.

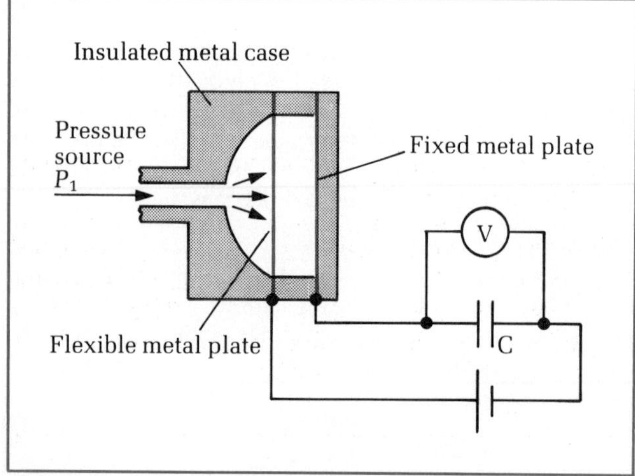

Figure 8.5 *Capacitance-type pressure gauge*

Barometers are used to measure atmospheric pressure. A Fortin (or mercury) barometer has a column of mercury supported by atmospheric pressure. The mercury is in an inverted tube, as in Figure 8.6. The lower end of the tube is in a reservoir of mercury which is open to the atmosphere.

No air is present at the top of the tube above the mercury, so the pressure in the mercury in the tube, level with the open surface, is equal to atmospheric pressure. Hence atmospheric pressure may be calculated from $H\rho g$, where H is the height between the mercury level in the tube and the reservoir and ρ is the density of the mercury. Atmospheric pressure does vary slightly from day to day due to weather conditions; if the mercury falls, you will need an umbrella outdoors!

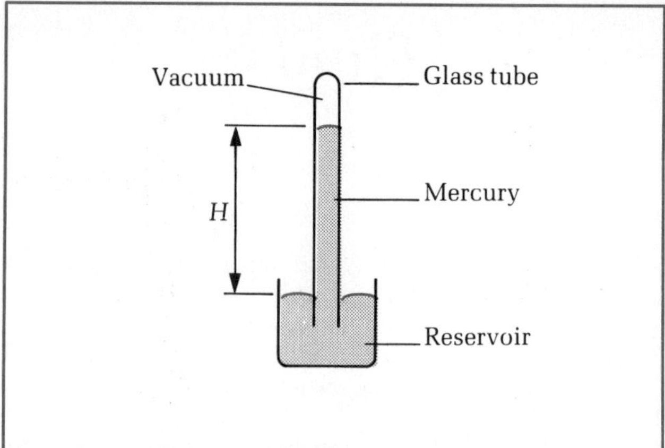

Figure 8.6 *Fortin's barometer*

Standard pressure is defined as the pressure exerted by a mercury column exactly 760 mm high. This is the average value for atmospheric pressure. Since the density of mercury is 13 600 kg m^{-3}, then standard pressure is equal to $0.760 \times 13\,600 \times 9.81$ Pa which equals 101 kPa.

Archimedes' principle

An object in a fluid experiences an upthrust from the fluid equal to the weight of fluid displaced.

The discovery of this principle was a result of Archimedes' efforts for his king. 'Find out if my new crown is solid gold, without cutting it', said the king to Archimedes. History records that Archimedes realized the displacement of water caused an upthrust when he entered his bath one day. Whether or not he greeted his discovery by running through the streets shouting 'Eureka!' is another matter. But his use of the principle is perhaps the earliest known example of non-destructive testing.

Suppose a spring balance is used to weigh a small object, first in air then with the object in water, as in Figure 8.7. The reading is smaller when the object is in water because there is an **upthrust** on the object from the water. Why? The cause is water pressure. Imagine a vertical cylinder as the object. The pressure at its base is $H_1\rho g$ where H_1 is the depth of the base and ρ is the water density. So the force on the base upwards is $H_1\rho gA$. The force on the top of the cylinder due to pressure is $H_2\rho gA$ where H_2 is the depth of the top. So the overall force is upwards, and equals $H_1\rho gA - H_2\rho gA = (H_1 - H_2)\rho gA$. But $(H_1 - H_2)A$ is the volume of the cylinder, so $(H_1 - H_2)A\rho g$ is the weight of the fluid displaced. Hence the upthrust = the weight of fluid displaced.

Worked example A glass stopper is weighed in air then immersed wholly in water and reweighed. The readings obtained are 2.4 N in air and 2.0 N in water. Given that the density of water is 1000 kg m^{-3}, calculate the density of the stopper.

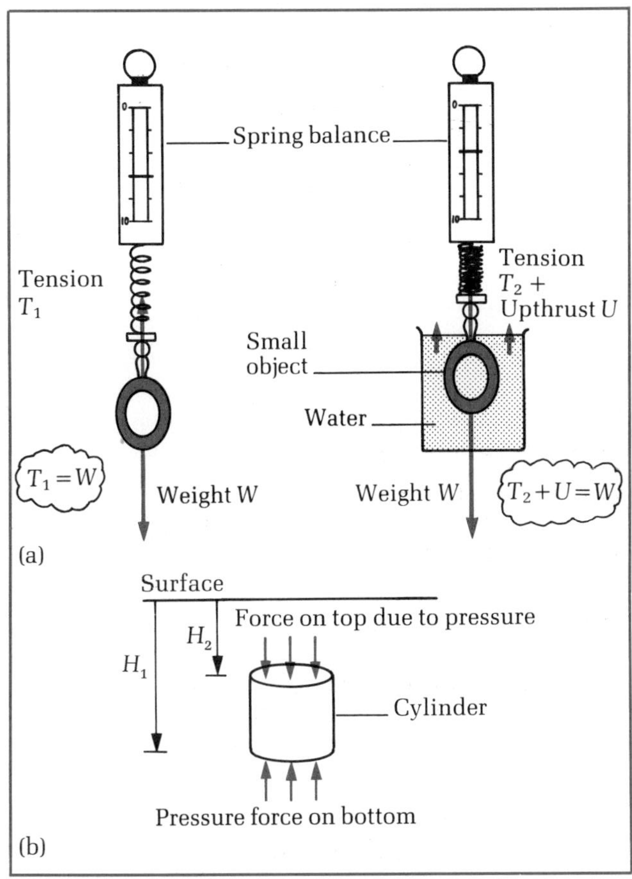

Figure 8.7 **a** *Upthrust* **b** *Cause of upthrust*

Solution The weight of water displaced = 0.4 N since the upthrust in water is 0.4 N.
So the mass of water displaced = 0.04 kg, assuming $g = 10\ \text{m s}^{-2}$.
Therefore the volume of water displaced = mass/density of water = 0.04/1000 m^3.
So the volume of the stopper = 0.04/1000 m^3, since it is wholly immersed.
The mass of the stopper = 0.24 kg since its weight is 2.4 N.
Therefore the density of the stopper = mass/volume = 0.24/(0.04/1000) = 6000 kg m^{-3}.

Hydrometers are used to measure liquid densities. In use, a hydrometer floats upright with part of its stem above the liquid, as in Figure 8.8.

For any floating object, the upthrust must be equal to its weight. A hydrometer floats higher in alcohol than in water because alcohol has a greater density. So the hydrometer does not need to sink in as far in alcohol to give the same upthrust as in water. In other words, the mass of alcohol displaced equals the mass of water displaced when the hydrometer floats in each, in turn. Because the density of alcohol is greater than the density of water, the volume displaced is less for alcohol than for water.

Usually the stem is calibrated using liquids of known density, so when the hydrometer is placed in a liquid of unknown density, the density is given directly from the liquid level against the scale on the stem.

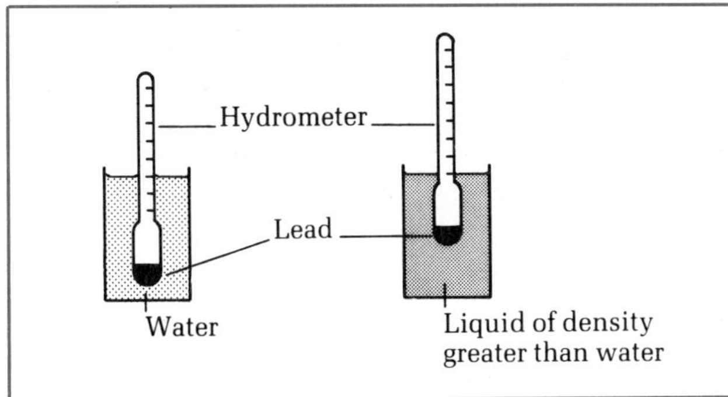

Figure 8.8 *Hydrometers*

Hydraulics

Hydraulic brakes and presses exert much greater forces than could be applied directly. Figure 8.9 shows a force F_1 applied to a narrow piston containing oil. The pressure on the oil, given by F_1/A_1, is transmitted through the pipes to the wide piston. So the force of the fluid on the wide piston is given by

$$F_2 = PA_2 = F_1A_2/A_1$$

Since the area A_2 of the wide piston is greater than the area A_1 of the narrow piston, then F_2 is greater than F_1. If, for example, $A_2 = 1000A_1$, then $F_2 = 1000F_1$. So a small effort F_1 can be used to move a much greater load.

However, the effort must travel further than the load when the forces move and do work. Push piston 1 in by a distance d_1 and piston 2 moves out a maximum distance d_2 equal to d_1A_1/A_2. We assume here that the volume of fluid d_1A_1 which leaves the narrow piston is equal to the volume d_2A_2 which enters the wide piston.

The work done by the effort is F_1d_1. The maximum work done on the load is given by

$$F_2d_2 = \frac{F_1A_2}{A_1} \times \frac{d_1A_1}{A_2}$$

which equals F_1d_1.

Provided the volume is unchanged the work done by the effort is transmitted completely to the load. If, however, the volume is reduced (i.e. the fluid is compressed), then the load does not move the maximum possible distance, so not all the work done by the effort reaches the load. Some is used to compress the fluid. If air gets into a hydraulic system then the fluid becomes compressible, and the system becomes less effective.

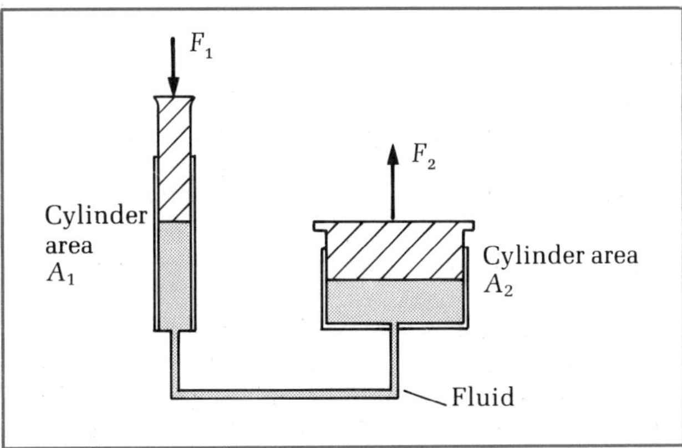

Figure 8.9 *A hydraulic press*

8.2 Surface tension

Some common effects due to surface tension are shown in Figure 8.10. For example, a clean needle can be made to float on water by first placing the needle on tissue paper then placing the paper and needle on water. The paper sinks, leaving the needle floating on a clean water surface. Try it; you ought to notice that the water surface curves under the needle, as if there is a skin across the surface. Weak attractive forces link molecules near the surface, creating surface tension which supports the needle. Another example is when a vertical microscope slide suspended by a thread is lowered until it is partly in water. As soon as the slide comes into contact with water, the pull of surface tension acts on it. To lift the slide out of the water, extra force needs to be used to overcome surface tension.

Surface tension is an important factor in a variety of situations. For example washing clothes is more effective if a powder is used to reduce surface tension. Then dirt particles can be removed more easily from the fibres of the clothing.

The **coefficient of surface tension**, γ, of a liquid is defined as the force per unit length acting at right angles to a length in the surface.

$$\gamma = \frac{F}{l}$$

where F = surface tension force,
l = length in the surface,
γ = surface tension coefficient.

The unit of γ is $\mathrm{N\,m^{-1}}$. The value of γ varies from one liquid to another, and it changes with temperature. For water at 20°C, the value of γ is $0.07\,\mathrm{N\,m^{-1}}$ compared with $0.05\,\mathrm{N\,m^{-1}}$ at 100°C.

Meniscus effects are due to surface tension. Any water surface in contact with a vertical glass wall curves up to meet the wall, as in Figure 8.13(a). Mercury in contact with a vertical glass wall curves down to meet the glass, as in Figure 8.13(b). To explain these meniscus

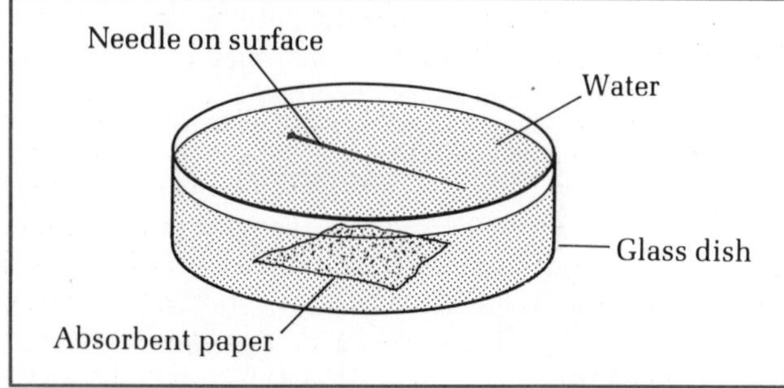

Figure 8.10 *A floating needle*

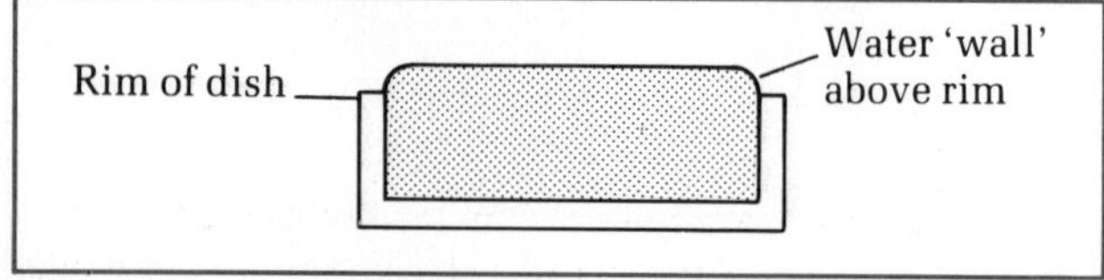

Figure 8.11 *Water walls*

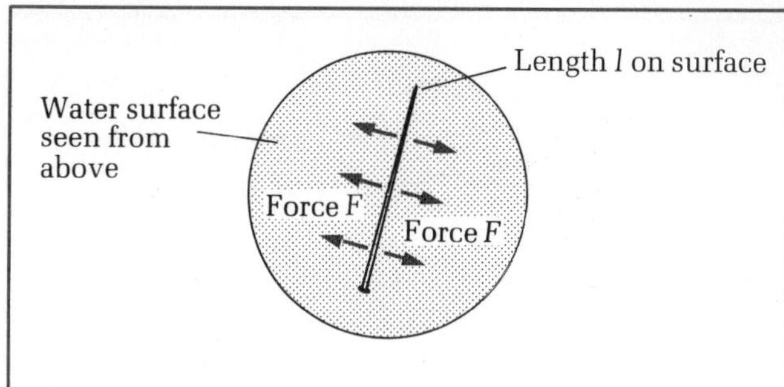

Figure 8.12 *Surface tension*

effects we need to look at the cause of surface tension once again. Molecules in a liquid are capable of exerting weak attractive forces on one another. Away from the surface inside the liquid, the molecules are, on average, at equilibrium spacing. So interior molecules exert no overall force on each other. But at and near the surface, the molecules are more widely spaced than inside the liquid, so they exert weak attractive forces on one another (see p. 71). These weak attractive forces link the surface molecules, causing surface tension.

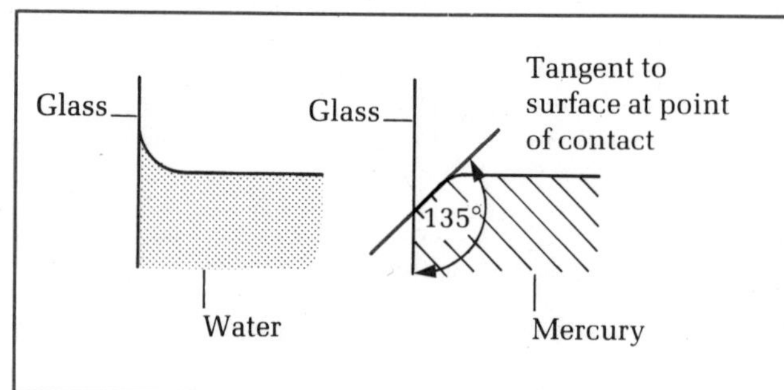

Figure 8.13 **a** *A water meniscus* **b** *A mercury meniscus*

Consider now what happens when water is placed in a glass beaker. The glass molecules pull on nearby water molecules, causing the water surface at the glass to curve up. Near the glass, the water surface changes from being horizontal, becoming parallel to the glass. At the point where the water surface is parallel to the glass, any water molecule is acted on by two equal and opposite forces. One force is due to neighbouring water molecules. The other force is due to glass molecules in the opposite direction. The angle between the two forces is 180° for water on clean glass. The result is that the angle between the water surface and the glass at the point of contact is zero, although this is not true for other liquids.

The angle of contact is defined as the angle between the liquid surface and the solid surface at the point of contact.

For mercury in contact with glass, the angle of contact is 135° which means that a mercury meniscus is almost like an inverted water meniscus. The force between mercury molecules is much greater than the force between a mercury and a glass molecule, so mercury near the glass is pulled down.

Capillary action is due to surface tension. When a vertical capillary tube, open at each end, is lowered partly into water, water rises up the tube. The level of water in

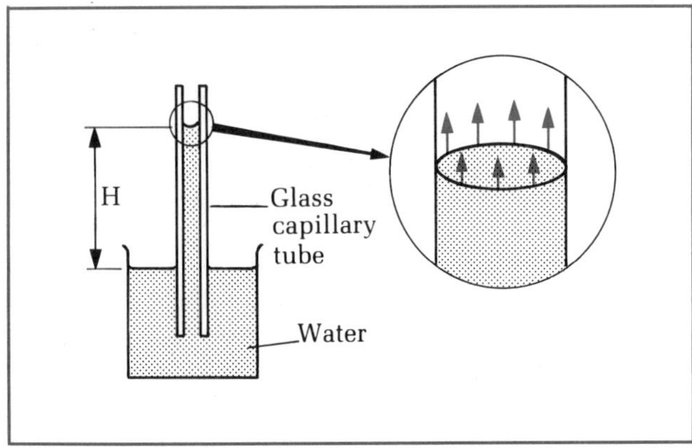

Figure 8.14 *Capillary action*

the tube becomes higher than the level outside. Figure 8.14 shows the effect. The water rises up the tube to a level which can just be supported by surface tension. The water in the tube in contact with the glass is acted on by the force of the glass molecules which is equal to the surface tension force. Contact is round the inside of the tube, a distance equal to $2\pi r$ where r is the tube radius. Since the force per unit length is equal to γ, then the pull due to the glass equals $\gamma \times 2\pi r$. But the pull supports a water column in the tube of height H. So the weight of water supported equals $\pi r^2 H\rho g$ where ρ is the density of water. Hence

$$\pi r^2 H\rho g = 2\pi r\gamma$$

which gives

$$H = \frac{2\gamma}{\rho g r}$$

where H = height rise,
γ = surface tension coefficient,
r = internal radius of tube,
ρ = liquid density,
g = acceleration due to gravity.

Bubbles are held together by surface tension. A small soap bubble in air forms a sphere. Surface tension in the film acts to reduce the bubble size so its internal pressure must be greater than the pressure outside.

Imagine the bubble in two equal halves, as in Figure 8.15. The pressure force of one half acting on the other half is $\Delta P \times \pi r^2$ where ΔP is the excess pressure inside, and r is the bubble radius. However a surface tension force acts where they are joined to keep them together.

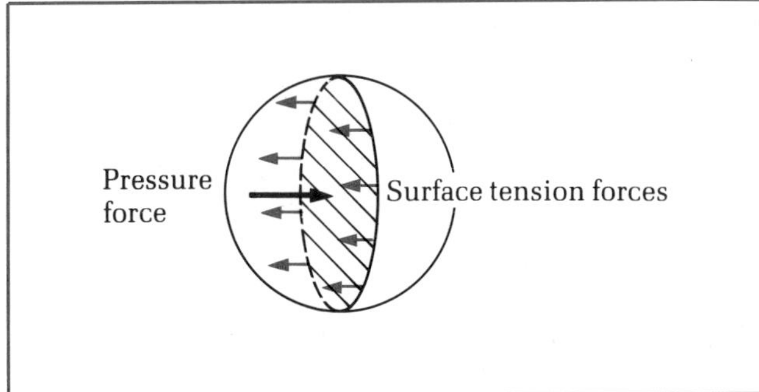

Figure 8.15 *Bubbles*

For an air bubble in water, the surface tension force equals $\gamma \times 2\pi r$. The two forces are equal and opposite, so $\gamma \times 2\pi r = \Delta P \times \pi r^2$.
Excess pressure (of an air bubble in liquid) is given by

$$\Delta P = \frac{2\gamma}{r}$$

where r = bubble radius,
γ = surface tension.

For a soap bubble in air, the bubble has two surfaces, one inside and the other outside. So the surface tension force is twice that above. Hence the excess pressure in a soap bubble is $4\gamma/r$.

Surface energy

Molecules in a liquid are, on average, separated from one another by a distance, measured centre to centre, which we call the **equilibrium spacing.** If they move closer together they repel one another. If they move away from one another, they attract one another. See p. 70 if necessary. To move two molecules away from one another, work needs to be done on the pair. When the surface area of a liquid is increased, molecules are moved from inside the liquid to its surface. Each molecule moved is pulled away from its neighbours inside the liquid and taken to the surface where it has neighbours on one side only. For example, if each molecule inside a given liquid has twelve 'nearest neighbour' molecules, then at the surface each molecule only has six nearest neighbours. So molecules moved to the surface need to be supplied with energy to allow each molecule to break away from neighbouring molecules.

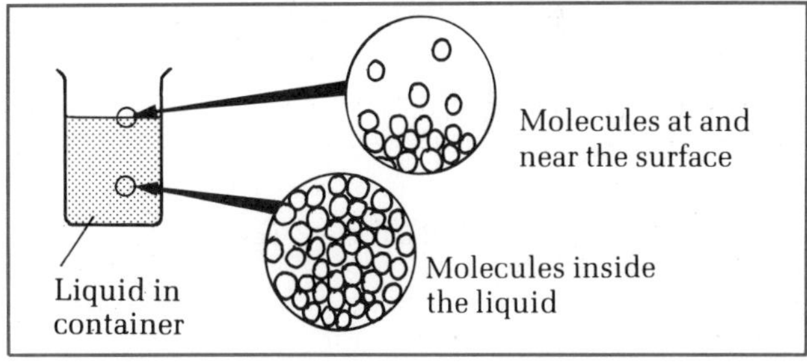

Figure 8.16 *Neighbours*

Bond energy is the energy needed to separate completely a pair of molecules from equilibrium spacing. Each molecule of the pair is therefore given energy $\varepsilon/2$ when they are separated. A molecule with n neighbours loses half its neighbours when moved to the surface. So to move a single molecule to the surface, it needs to be given energy equal to $n/2 \times \varepsilon/2$. If the surface has on average A molecules per unit area, then the energy needed to create unit surface area is approximately $A \times n\varepsilon/4$.
Surface energy per unit area, σ, is given by

$$\sigma \approx \frac{An\varepsilon}{4}$$

where A = number of molecules per m^2,
n = number of neighbours per molecule,
ε = bond energy.

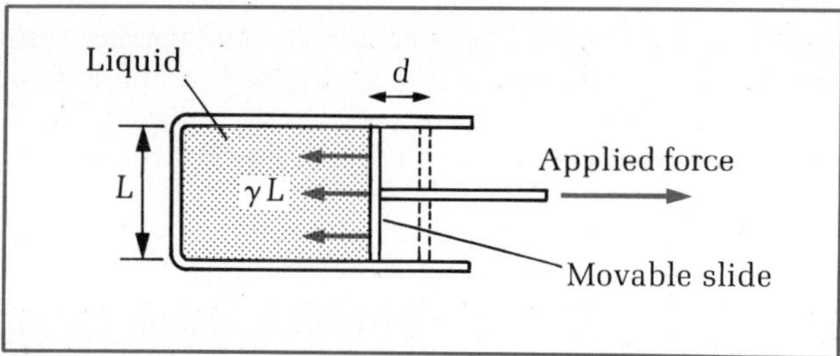

Figure 8.17 *Top view of tank*

Suppose the liquid is in a tank with a movable side AB, as shown in Figure 8.17. The surface tension force on the movable slide is equal to γL where L is the length of the slide. To extend the liquid surface, AB is pulled away from the opposite side. If AB is moved by a distance d, the work done against the force is γLd (= force × distance). But Ld is the extra surface area created. So the work done to create unit area of extra surface is $\gamma Ld/Ld$ which equals γ. Provided the temperature is unchanged, the work done equals the gain of surface energy.

Surface energy per unit area $\sigma = \gamma$.

Worked example Estimate the bond energy of water molecules, assuming each water molecule has, on average, twelve neighbours and that the equilibrium spacing for water molecules is about 10^{-10} m. The surface tension coefficient for water is $0.07\,\mathrm{N\,m^{-1}}$ at 20°C.

Solution $n = 12$

$\gamma = 0.07\,\mathrm{N\,m^{-1}}$

Each molecule at the surface is about 10^{-10} m from its nearest neighbour. So each surface molecule takes up an area of about $10^{-10} \times 10^{-10}\,\mathrm{m^2} = 10^{-20}\,\mathrm{m^2}$.

Therefore the number of molecules per m², $A = 10^{20}$.

Since $\gamma \approx \frac{An\varepsilon}{4}$ then the bond energy may be calculated.

$$\varepsilon \approx \frac{4\gamma}{An} = \frac{4 \times 0.07}{10^{20} \times 12} = 2 \times 10^{-22}\,\mathrm{J}.$$

Measuring surface tension

Figure 8.18 shows the arrangement. Use a travelling microscope to measure the height of the water in a vertical capillary tube above the water outside. To measure the water level in the beaker, a bent pin attached to the tube is lowered until it just touches the water. Then when the water is removed, the level of the tip of the pin can be measured easily. Direct measurement of the level in the beaker is difficult because of the meniscus at the edge.

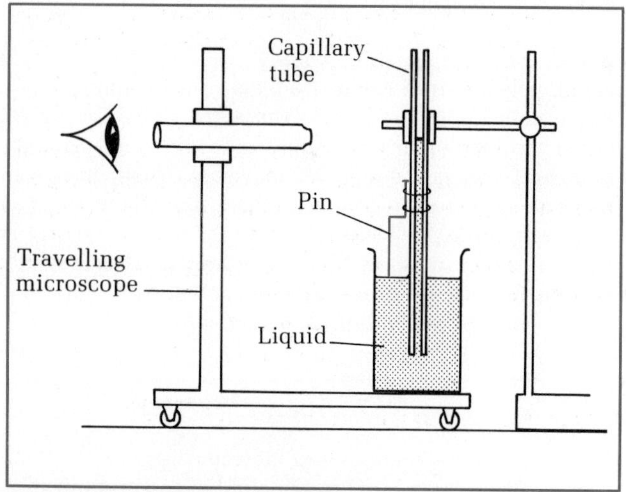

Figure 8.18 *Capillary tube method*

With the pin touching the water, the microscope is focussed on the water level in the tube and its reading taken. Then the water is removed and the microscope is readjusted to focus on the pin. The new reading is taken, and the height H of water in the tube is calculated from the difference of the readings.

The microscope can also be used to measure the internal diameter of the tube. Measurements are made across two perpendicular diameters to give an average value for the radius r. Assuming values of $1000\,\mathrm{kg\,m^{-3}}$ for ρ_{water} and $9.8\,\mathrm{m\,s^{-2}}$ for g, γ can be calculated using the capillary rise equation.

8.3 Fluid flow

Flow patterns

Oil, water and air are all fluids but their flower properties are very different. Oil flows with difficulty compared with water. Air easily becomes turbulent. To predict the motion of a fluid, we must consider the forces acting in the fluid. We must also take account of the presence of solid surfaces which affect the flow.

A simple flow tank using water is shown in Figure 8.19. Ink can be used to show the pattern of flow as the water drains down the tank. Flat objects can be placed in the flow to give different flow patterns. Tilting the tray more increases the flow rate. Some flow patterns are shown in Figure 8.20. Each flow pattern shows how the fluid motion changes with position or time.

A fluid element is the smallest volume of a fluid which follows the flow. If any smaller, molecular motion would make its path haphazard so it is just large enough for motion of individual molecules in it to balance out. If any larger, it might be pulled in different directions by the flow. So each fluid element follows the overall drift of the fluid motion.

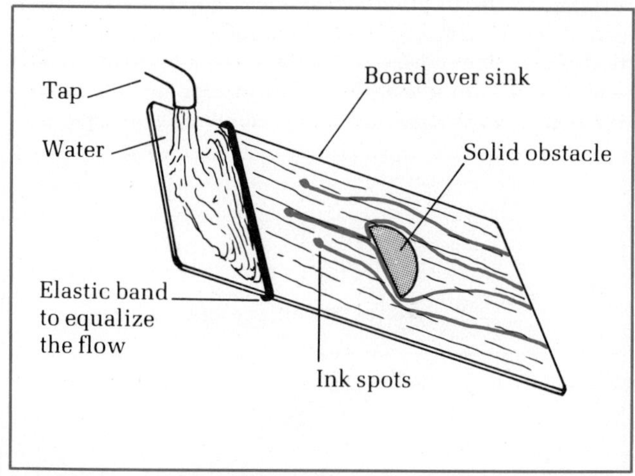

Figure 8.19 *A simple flow tank*

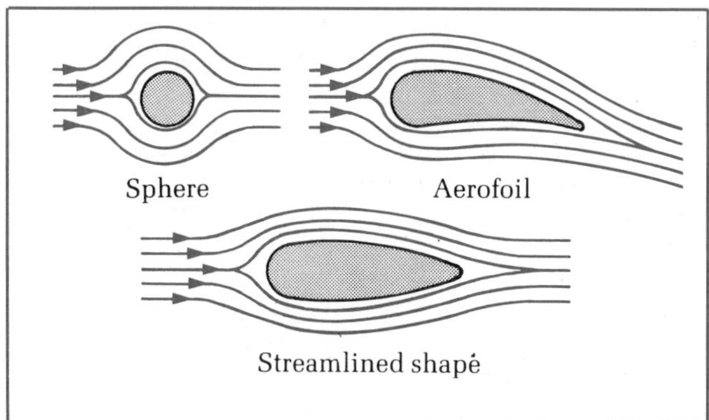

Figure 8.20 *Flow patterns*

A flowline is the path which an individual fluid element describes. In the flow tank of Figure 8.19, a small ink spot produces a coloured flowline as the ink is carried along by the flow. A **stable** flow pattern is one where the flowlines do not move about. Sometimes, the same flow pattern shows stable flowlines changing to unstable flow downstream from an obstacle. A stable flowline is one in which every element along the line follows the same path. Such a flowline is sometimes called a **streamline**, and some examples are given in Figure 8.20. So streamlined flow is stable flow where the velocity at any point does not change with time. The direction of a streamline at a point gives the velocity direction at that point. A given fluid element moving along a streamline may change its speed and direction as it moves along the streamline. But at any given fixed point on a streamline, the velocity of the fluid is fixed. A streamline may be thought of as a stable flowline.

In streamlined flow, fluid elements do not cross streamlines. Any given element moves along the same streamline so cannot cross to another. Where streamlines are parallel to one another, the flow is sometimes called **laminar**. Figure 8.21 shows the flow pattern produced by steady flow of a fluid round a plate fixed at right angles to the flow.

Turbulent flow occurs where the flow becomes mixed up. Fluid elements passing a given point do not follow the same path. In other words, the flowlines are unstable and keep changing. Figure 8.21 shows streamlined flow which becomes turbulent in the wake of the obstacle. Various factors determine whether or not fluid flow is turbulent; the flow speed, the shape and size of solid surfaces in the flow, the fluid density and fluid friction all contribute. For example, the force on a ball moving at low speed through a fluid is proportional to its speed, assuming the flow is laminar. But at high speed, the flow becomes turbulent and the force becomes proportional to the square of the speed.

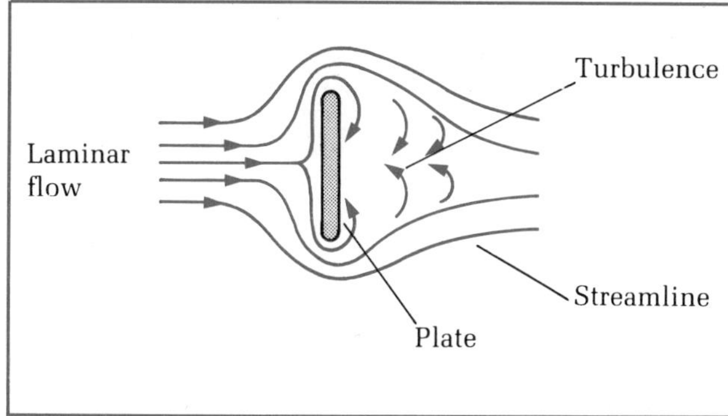

Figure 8.21 *Turbulence*

Continuity

Consider steady flow along a pipe which widens as shown in Figure 8.22. The flow becomes slower as the pipe widens out. The reason is that the mass of fluid per second entering the pipe is equal to the mass per second leaving, and since it leaves through a wider exit, it does not move as fast. In a small interval of time Δt, the volume of fluid passing point X in the pipe is $vA\Delta t$ (which is the volume of pipe section of area A and length $v\Delta t$). So the mass passing X in time Δt is $\rho vA\Delta t$. Hence the mass per second passing X is $\rho vA\Delta t/\Delta t$ which equals ρvA. Since X could be anywhere along the pipe, it follows that ρvA is the same for all points along the pipe. This is known as the **continuity equation**, and holds for all points along a pipe (or stream tube).

$$\rho vA = \text{constant}$$

where ρ = fluid density,
v = fluid speed,
A = area of cross-section.

The same equation applies to a tube of flow defined by streamlines. The tube can be thought of as a pipe since fluid does not cross streamlines. So the fluid in the tube stays in the tube as it moves along.

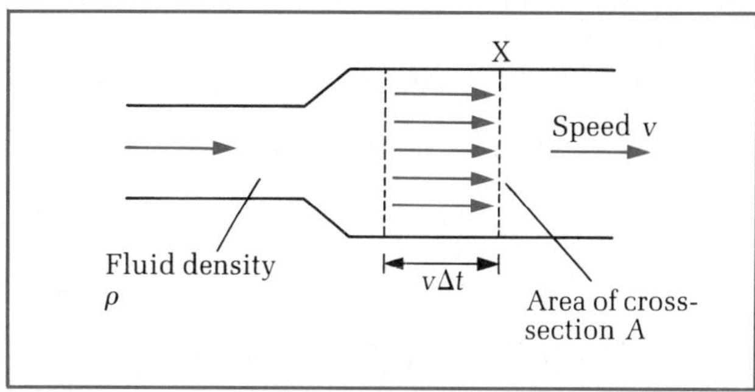

Figure 8.22 *Continuity*

An incompressible fluid is a fluid which has a constant density. Water is almost incompressible. Air can easily be compressed. For an incompressible fluid, the continuity equation gives vA = constant because ρ is constant.

Worked example A water hose with an internal diameter of 20 mm at the outlet discharges 30 kg of water in 60 s. Calculate the water speed at the outlet. Assume the density of water is 1000 kg m^{-3}.

Solution Mass flow/second = 30/60 = 0.5 kg s^{-1}

Cross-sectional area $A = \pi d^2/4$ (where d = diameter).

Therefore $A = \pi(20\times10^{-3})^2/4$.

The continuity equation gives $v = \dfrac{\text{mass/second}}{A\rho}$.

Hence $v = \dfrac{0.5\times4}{\pi(20\times10^{-3})^2\times1000} = 1.59\ \text{m s}^{-1}$.

8.4 Viscosity

Some liquids flow more easily than others. For example, water runs more easily than syrup. Gases flow much more freely than liquids. Why do some fluids flow more easily than others? Fluid flow generally involves different parts of a fluid moving at different velocities. Different parts of a fluid slide past each other, as if in layers. The layers slide past each other with ease if the fluid flows easily, as if there is little friction between the layers. Where flow is difficult to maintain, the layers slide over each other with difficulty. A **viscous** fluid is one like syrup, where flow is difficult to maintain. Water is much less viscous than syrup because it flows more easily than syrup.

The **viscosity** of a fluid is a measure of its resistance to flow. In other words, it is a measure of how viscous the fluid is.

Consider the flow of a viscous fluid through a pipe, as in Figure 8.23. The fluid layers near the pipe wall move slowly because of friction between the pipe and wall. In fact the fluid next to the wall almost sticks to the wall. But the fluid near the centre of the pipe moves faster. So the flow is in layers, parallel to the pipe wall. The layers move at different speeds, from zero at the wall to a maximum at the centre. Each layer slides past the next layer closer to the wall. Friction between the layers means that the pressure at the inlet must be greater than at the outlet to keep the layers moving. A pressure difference is essential to overcome the fluid friction due to viscosity.

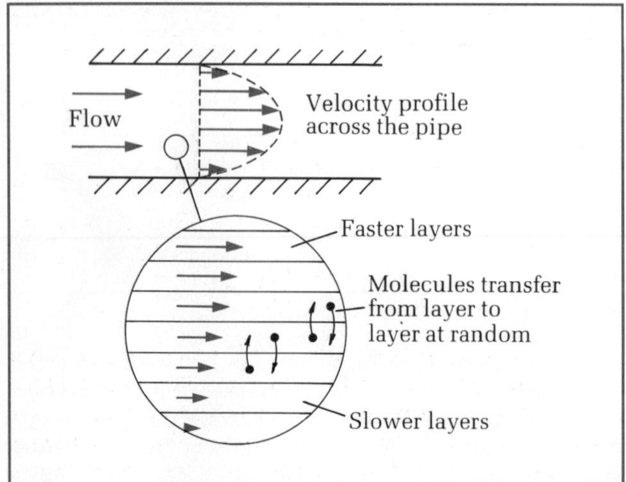

Figure 8.23 *Velocity gradient*

Molecular motion is the cause of viscosity. The molecules of a fluid at rest move about at random, continually colliding with each other. In a moving fluid, the molecules still move about at random, but a drift motion occurs in addition to the random motion. The drift motion takes the molecules along with the flow. Molecules move from layer to layer due to their random motion. But for every molecule which changes layers in one direction, another molecule will change in the opposite direction. Consider the transfer of molecules from a high speed layer to an adjacent layer moving at lower speed. The transfer of molecules is accompanied by transfer of momentum. Each transferred molecule takes momentum from the high-speed layer to the low-speed layer. So the high-speed layer loses momentum to its low-speed neighbour because high momentum molecules leave it and are replaced by low-speed molecules. Newton's second law tells us that loss of momentum is due to a force opposing the motion. So loss of momentum from a high-speed layer is the cause of the friction force on the layer due to its low-speed neighbour. The low-speed neighbour is pulled along by the high-speed layer, but in turn an even lower speed layer drags on the low-speed neighbour.

The velocity gradient across the pipe is defined as the change of velocity per unit distance along the pipe radius. If the velocity at a given point P is $0.20\,\mathrm{m\,s^{-1}}$ and at a point 1 mm away from P along the radius through P, the velocity is $0.18\,\mathrm{m\,s^{-1}}$, then the velocity gradient at P is $(0.20-0.18)/(1\times10^{-3})\,\mathrm{s^{-1}}$ which equals $200\,\mathrm{s^{-1}}$. In general, if the velocity at a point P is v and the velocity at a nearby point Q is $v+\Delta v$, then the average velocity gradient between P and Q is $\frac{\Delta v}{\Delta r}$ where Δr is the distance from P to Q. For P and Q very close together, we write the velocity gradient as $\frac{dv}{dr}$ where $\frac{d}{dr}$ means change per unit distance.

Figure 8.23(a) shows how the velocity varies across the pipe. The slope of the curve gives the velocity gradient. The velocity gradient is greatest at the edge, and is zero at the centre.

The tangential stress T between two adjacent fluid layers is the viscous force per unit area acting along the boundary between the two layers. As we saw earlier, the force is due to the motion of molecules between the two layers. Where two adjacent layers move at the same velocity, there is no tangential stress between them. The unit of tangential stress is the pascal (Pa), which is equal to $1\,\mathrm{N\,m^{-2}}$.

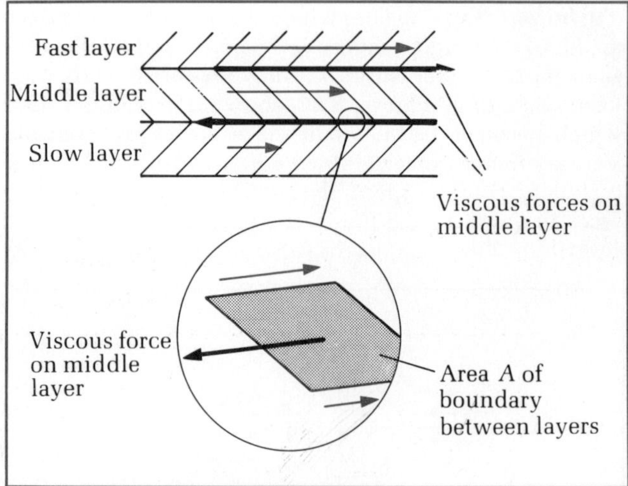

Figure 8.24 *Tangential stress* = F/A

A newtonian fluid is one for which the velocity gradient at any point is proportional to the tangential stress. The constant of proportionality is called the **coefficient of viscosity** η (said as 'eta').

$$T = \eta \frac{dv}{dr}$$

where T = tangential stress,
$\frac{dv}{dr}$ = velocity gradient.

The unit of η is N s m^{-2} since T has a unit of N m^{-2} and $\frac{dv}{dr}$ is in s^{-1}. A large value for η means a lot of friction between fluid layers, i.e. a very viscous fluid. Some values for different fluids are given below.

	η/N s m^{-2}	
	at 20°C	at 40°C
Air	1.8×10^{-5}	1.9×10^{-5}
Water	1.0×10^{-3}	0.66×10^{-3}
Castor oil	9.9	2.13

Most liquids become less viscous as they become warmer, In winter, cars are more difficult to start from cold because the engine oil is much more viscous than when warm. Viscostatic oil is specially made with the same viscosity when cold as when hot.

Some liquids like paint become less viscous the more they are stirred. Such liquids are **non-newtonian fluids.**

Poiseuille's equation for pipe flow

Consider steady flow of a viscous fluid through a uniform horizontal pipe. Assume the flow is laminar. To maintain flow, the pressure at the inlet must be greater than the pressure at the outlet since the fluid is viscous.

The flow is steady so the volume per second passing through the pipe is constant. If volume V passes through in time t, the volume per second is V/t. The volume per second is determined by the forces acting. The pressure force pushes the fluid through the pipe against the resistance of the viscous force. So the volume per second depends on

a) the pressure difference P between the inlet and outlet,
b) the pipe length L,
c) the pipe radius r,
d) the fluid viscosity η.

The volume per second is increased by either a greater pressure difference or a shorter length. The pressure gradient P/L is the factor that is important. So the volume per second is determined by P/L, r and η.

The method of dimensions is useful here to work out the link between the volume per second and the factors that determine its value. We express the link using an index for each factor, the index to be determined. Then each factor is written in terms of the basic dimensions (i.e. mass, length and time). Finally, we balance the dimensions on each side of the equation to give the index values. (See p. 531.) Let's see how it works here!

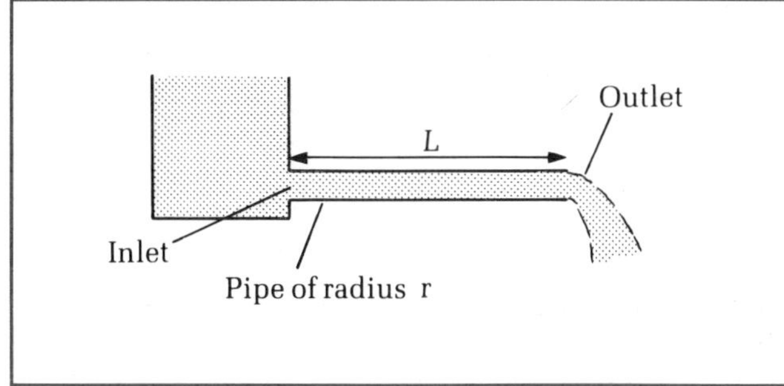

Figure 8.25 *Viscous flow in a pipe*

Stage 1 Assume $V/t = k\eta^x r^y (P/L)^z$ where x, y and z are to be determined, and k is a numerical constant.

Stage 2 We use M, L and T for the base dimensions of mass, length and time.

$[V/t] = L^3T^{-1}$
$[\eta] = [\text{newtons}]\ T^1L^{-2} = M^1T^{-1}L^{-1}$
since $[\text{newtons}] = MLT^{-2}$
$[r] = L$
$[P/L] = [\text{newtons/metre}^2]\ L^{-1} = M^1T^{-2}L^{-2}$

(The square brackets mean 'dimensions of'.)

Stage 3 Balance the dimensions of the equation.

$[V/t] = [\eta]^x[r]^y[(P/L)]^z$
so $L^3T^{-1} = (M^xT^{-x}L^{-x})(L^y)(M^zT^{-2z}L^{-2z})$

a) The indices for M give $0 = x + z$.
b) The indices for T give $-1 = -x - 2z$.
c) The indices for L give $3 = -x + y - 2z$.

These three equations give $x = -1$, $y = 4$ and $z = 1$. Hence we can now write the equation for V/t as follows.

$$V/t = \frac{kr^4(P/L)}{\eta}$$

The equation can be derived from $T = \eta \frac{dv}{dr}$, using a more complicated approach involving integration. This gives $\pi/8$ for the numerical constant k. The equation is known as **Poiseuille's equation** after its discoverer.

$$V/t = \frac{\pi r^4 P}{8\eta L}$$

where V/t = volume per second,
r = tube radius,
P = pressure difference between the two ends,
L = tube length,
η = fluid viscosity.

The flow rate (i.e. volume/second) depends on r^4 so if the tube radius is halved (by squeezing it, assuming flexible walls), the flow rate is reduced to 1/16 of the previous rate. An example of the consequence of this dependence is within each of us. Arteries are known to narrow due to deposits on the inside walls. The narrower the arteries become, the greater the pressure to keep the blood flowing through them. So the heart, the pumping station of the blood system, must work harder.

Poiseuille's equation is not unlike Ohm's law if we compare flow rate with current. Just as an electric current requires a p.d., so flow rate needs pressure difference. So pressure difference is like potential difference.

Ohm's law	**Poiseuille's equation**
$I = \frac{1}{R} \times V$	$V/t = \frac{(\pi r^4)}{8\eta L} \times P$

The tube 'resistance' is therefore $\frac{8\eta L}{\pi r^4}$

High viscosity or a narrow tube means a high resistance to the flow. For a non-viscous fluid or one with very low viscosity, there is no resistance to flow since $\eta=0$. So no pressure difference is needed to keep a non-viscous fluid moving. Can you think of a 'zero resistance' situation in electricity? No p.d. is needed to maintain the current if the resistance is zero. Certain metals and alloys at very low temperatures have zero electrical resistance. They are called **superconductors**.

Viscous drag

Any object moving through a viscous fluid is acted on by friction due to the fluid. The object drags fluid along near its surfaces, causing fluid layers to move at different speeds near the object. So the object experiences viscous drag due to the fluid. The force due to fluid drag increases with speed. For viscous drag, the force is in proportion to the object's speed. **Stokes' law** gives the exact link between the force and the speed for a sphere.

$$F=6\pi\eta Rv$$

where F = force due to viscous drag,
η = viscosity of fluid,
R = sphere radius,
v = sphere speed.

The terminal velocity of an object moving through a fluid is reached when the drag force balances out the other forces on the object. The resultant force on the object is then zero, so the object moves at constant velocity.

Consider a ball bearing released from rest in a fluid such as water. The viscous force may be written as $F=kv$ where k ($=6\pi\eta R$ from Stokes' law) is a constant. The resultant force on the object is equal to its weight mg − the viscous force. So when its speed is v, its acceleration is given by $(mg-kv)/m$ since acceleration = resultant force/mass.

$$\text{Acceleration } a=g-\frac{k}{m}v$$

Initially, $v=0$ so $a=g$. As the object speeds up, the acceleration becomes smaller according to the above equation. So its speed builds up at a slower rate. Gradually its acceleration falls to zero as its speed builds up to a level where $mg=kv$. The speed therefore reaches a constant value equal to mg/k. Figure 8.27 shows how the speed increases with time. The acceleration is given by the gradient of the curve.

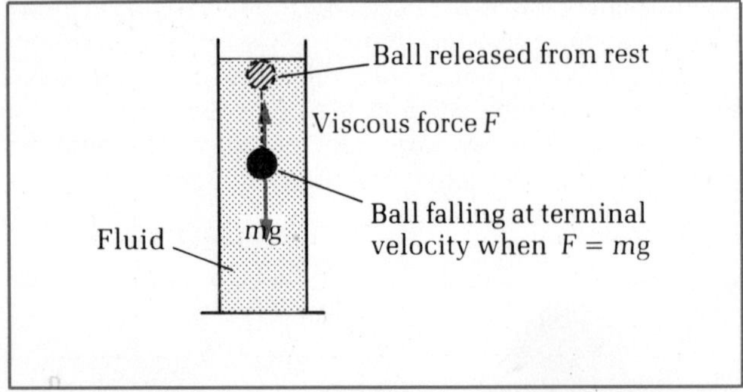

Figure 8.26 *Terminal velocity*

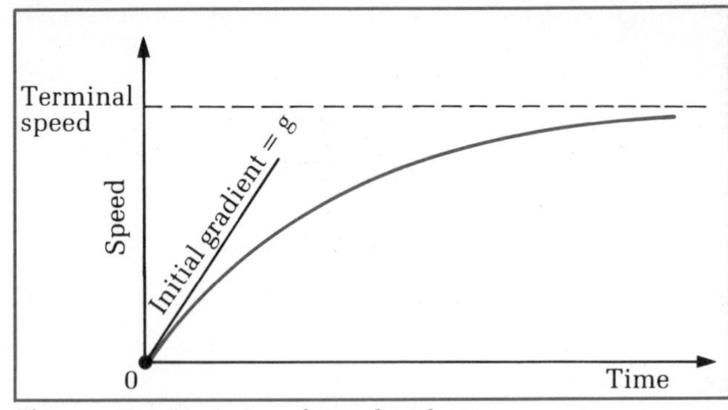

Figure 8.27 *Variation of speed with time*

Measuring viscosity

A steel ball bearing of known diameter is released from rest just below the surface of a liquid in a long vertical tube. The ball reaches terminal velocity quickly so by timing its fall over a measured distance, the terminal velocity may be calculated. The diameter can be measured accurately with a micrometer. The ball mass can be determined using a mass balance. Assuming Stokes' law applies, the viscosity can be calculated from $6\pi\eta Rv=mg$.

The ball diameter should be much smaller than the tube diameter, otherwise friction on the sides of the tube becomes significant. Two light gates may be used to time the ball. When the ball breaks the first light gate, an electronic timer is started. The timer is stopped when the ball breaks the second gate. Without too much difficulty, you can interface the light gates to a microcomputer and write a program to calculate η from the timing recorded.

How quickly does the ball reach terminal velocity? You could time the ball over different distances from rest. If the timings are not in proportion to the distances, then clearly the ball has not kept the same speed on average in each run.

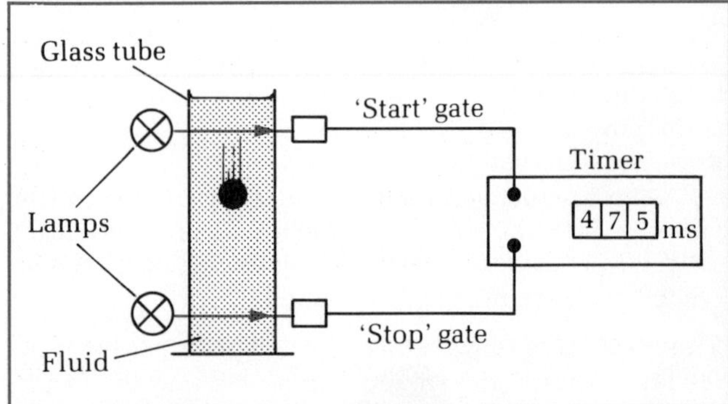

Figure 8.28 *Measuring viscosity*

8.5 Non-viscous flow

Energy

Forces do work when they cause movement along the force direction. Pressure is force per unit area so fluid pressure does work when the fluid moves. If the pressure changes along the flow direction, then the work done by pressure alters the energy of the fluid. Fluid energy can be kinetic energy, potential energy or internal energy.

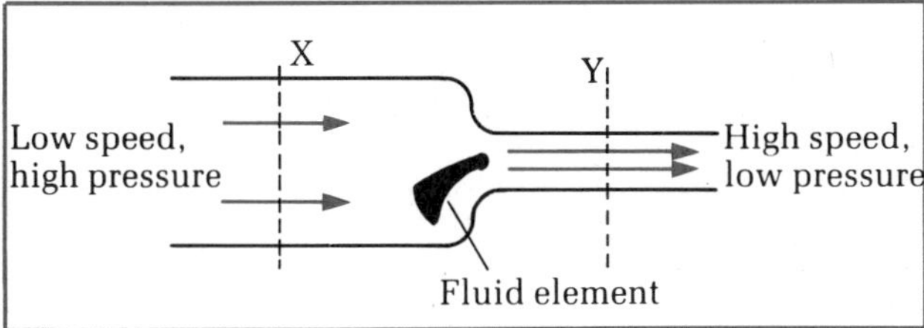

Figure 8.29 *Pressure and speed*

Viscous forces are responsible for increasing the internal energy of a fluid just as friction between moving solid surfaces increases the internal energy of the solids. In a non-viscous fluid, no internal energy can be produced from pressure changes so work done produces changes of P.E. and K.E.

For a non-viscous fluid moving at steady speed without change of height, the K.E. and P.E. of the fluid elements stay constant as the fluid elements move along the flowlines. Therefore, no work is done on the fluid elements so there is no change of pressure along the flow.

Where speed or height does change along the flow direction, there must be a change of pressure. Change of speed or height means change of K.E. or P.E. so work must be done. For a non-viscous fluid, work done by pressure changes must equal the changes of K.E. and P.E. of the fluid elements. For example, consider the flow of a non-viscous fluid along a horizontal pipe which narrows as shown in Figure 8.29.

The flow in the narrow section is faster than in the wide section because the mass per second leaving the narrow part must equal the mass per second entering the wide part. The continuity equation gives the ratio of the two speeds, $v_x/v_y = A_y\rho_y/A_x\rho_x$. The increase of speed of each fluid element as it passes from the wide to the narrow section means that its K.E. increases. So each fluid element has work done on it by pressure forces. Therefore the pressure in the wide section must be greater than the pressure in the narrow section. Imagine the fluid element just as it crosses from the wide section to the narrow section. The pressure on it from the wide section is bigger than the pressure on it from the narrow section, so the fluid element is speeded up.

For horizontal flow, where the speed is high the pressure is low

A very simple demonstration of this effect is given by blowing gently between two sheets of paper hanging side by side, as in Figure 8.30. The sheets move together when you blow between them. The reason is the low pressure caused by the airflow between the sheets. The higher pressure outside pushes the sheets together. Another example is well known to ball players! A spinning ball moving through the air veers to one side; this is called the 'magnus' effect, and happens because the airflow round one side of the ball is faster than round the other side. Figure 8.31 shows the idea. As a result, the higher pressure on the low speed side pushes the ball to one side.

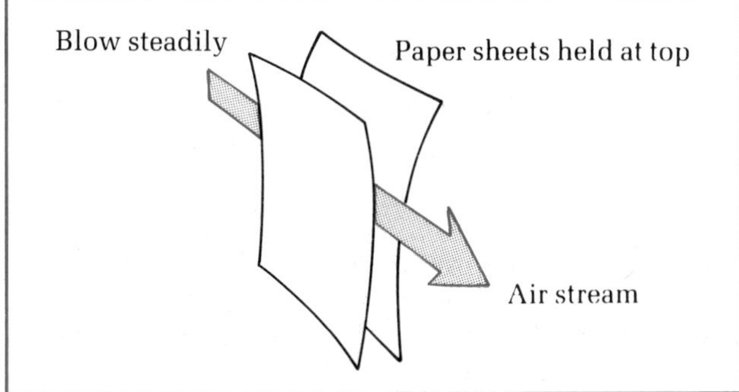

Figure 8.30 *Creating low pressure*

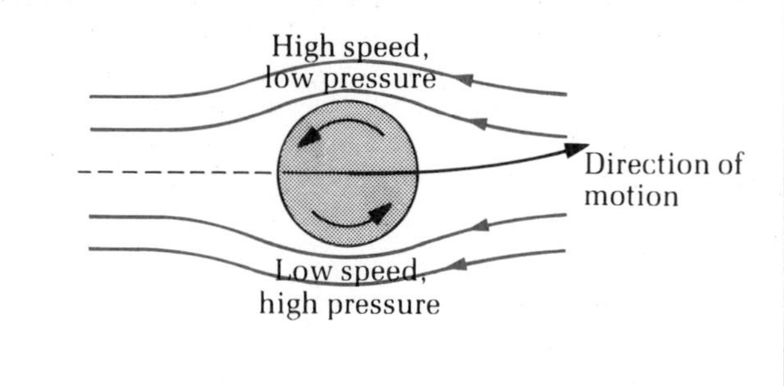

Figure 8.31 *The Magnus effect*

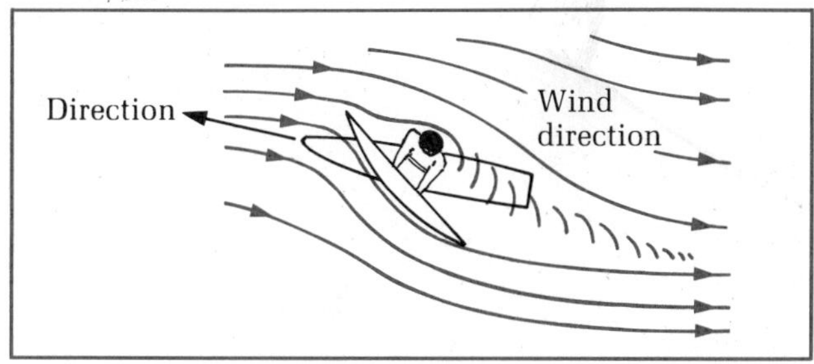

Figure 8.32 *Windsurfing*

Windsurfers use the principle too. Figure 8.32 shows an overhead view of a windsurfer moving into the wind. Can you see why the pressure on the trailing side of the sail is higher than on the leading side? All these effects are examples of a wider principle, the **Bernouilli principle**, which includes P.E. changes as well as K.E. changes.

The Bernouilli principle

Consider streamlined flow of an incompressible, non-viscous fluid, as shown in Figure 8.33. All the fluid passing through area A_x passes along a streamtube through

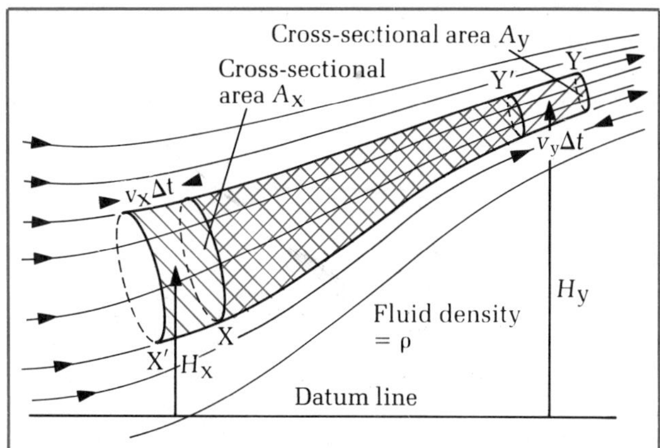

Figure 8.33 *Bernouilli's equation*

area A_y. The streamtube is defined by the streamlines through the edge of area A_x.

In a small interval of time Δt, the volume of fluid passing through A_x is given by $A_xv_x\Delta t$. Since the fluid is incompressible, an equal volume $A_yv_y\Delta t$ passes through A_y.

In effect, the fluid in X'Y' moves along to XY in time Δt. The overall result is the same as if XX' was replaced by YY', section XY' staying unchanged.

1 The change of K.E. is therefore given by K.E. of YY' − K.E. of XX'.
Hence change of K.E. $= \frac{1}{2}\rho Vv_y^2 - \frac{1}{2}\rho Vv_x^2$.

2 The change of P.E. is given by P.E. of YY' − P.E. of XX' hence change of P.E. $= \rho VgH_y - \rho VgH_x$
where H_y and H_x are the respective heights of YY' and XX' measured from the same 'datum' line.

3 The work done at end X' to move it to X is given by $(p_xA_x)(v_x\Delta t)$ since p_xA_x is the force on area A_x due to pressure p_x. This force moves a distance equal to $v_x\Delta t$. The work done by end Y' pushing against pressure p_y to reach Y is given by $(p_yA_y)(v_y\Delta t)$.
So the total work done

$$= \text{work done on X'} - \text{work done by end Y'}$$

$$= p_xA_xv_x\Delta t - p_yA_yv_y\Delta t$$

$$= p_xV - p_yV$$

since $V = A_xv_x\Delta t = A_yv_y\Delta t$.

Since no viscous forces act, the total work done is the sum of the change of P.E. and the change of K.E., which gives

$$p_xV - p_yV = (\rho VgH_y - \rho VgH_x) + (\tfrac{1}{2}\rho Vv_y^2 - \tfrac{1}{2}\rho Vv_x^2)$$

Rearranging, we obtain

$$p_xV + \tfrac{1}{2}\rho Vv_x^2 + \rho VgH_x = p_yV + \tfrac{1}{2}\rho Vv_y^2 + \rho VgH_y$$

We can cancel V from every term, and since X and Y could be any two positions along the streamtube, we then obtain **Bernouilli's equation**. This states that for all points along the same streamline

$$p + \tfrac{1}{2}\rho v^2 + \rho gH = \text{constant},$$

where p = pressure,
v = speed,
H = height,
at any point.

For **horizontal flow,** H is constant, so Bernouilli's equation becomes

$$p + \tfrac{1}{2}\rho v^2 = \text{constant}$$

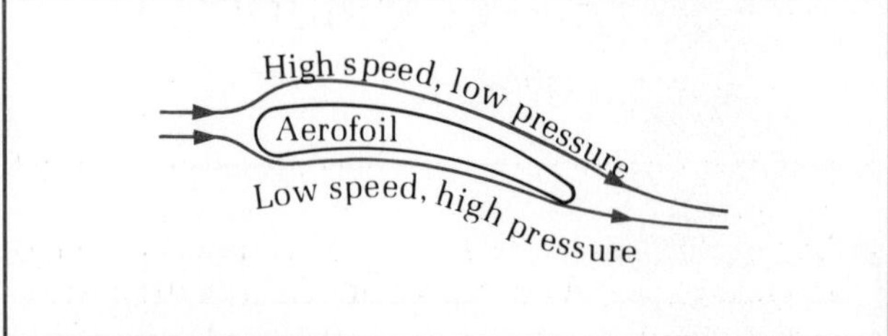

Figure 8.34 *Lift*

If a fluid element speeds up as it moves along a horizontal streamline, the pressure falls because $p + \frac{1}{2}\rho v^2$ is constant. The flow of air round an aerofoil is effectively horizontal flow since height changes are small. The aerofoil is shaped so that the air flows faster over the top than underneath, as in Figure 8.34. So the pressure underneath is higher than the pressure above the aerofoil. Hence the aerofoil experiences a lift force. Aeroplane wings are shaped to produce a lift force.

The Venturi duct is another example of Bernouilli's equation at work. To measure the flow speed of a fluid along a pipe, the pipe is constricted by a duct at one point along its length. The shape of the duct, shown in Figure 8.35, is to eliminate turbulence. The flow through the duct is faster than through the rest of the pipe, so the fluid pressure at the duct is less than in the rest of the pipe. A pressure gauge is used to measure the pressure difference Δp between the duct and a position upstream. Given the pipe diameters at the two points, the flow speed can be calculated.

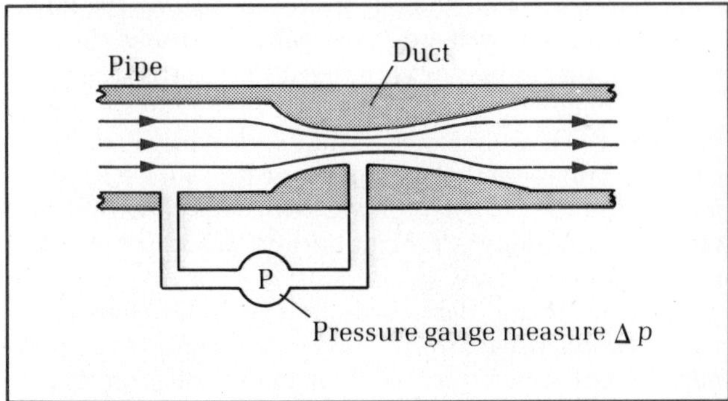

Figure 8.35 *The Venturi duct*

Assume the pipe is horizontal, the fluid is non-viscous and incompressible, and the flow is laminar. Suppose point X is at the duct and point Y is some point upstream. Since the flow is horizontal, we can write

$$p_x + \tfrac{1}{2}\rho v_x^2 = p_y + \tfrac{1}{2}\rho v_y^2$$

Also the continuity equation gives $\rho_xv_xA_x = \rho_yv_yA_y$.

Since $\rho_x = \rho_y$, then $v_x = \dfrac{A_y}{A_x}v_y$.

We can now obtain an expression for the speed v_y.

$$p_x + \tfrac{1}{2}\rho\left(\frac{A_yv_y}{A_x}\right)^2 = p_y + \tfrac{1}{2}\rho v_y^2$$

Rearranged, this gives

$$\Delta p = p_y - p_x = \tfrac{1}{2}\rho v_y^2\left(\frac{A_y^2}{A_x^2} - 1\right).$$

Hence v_y can be calculated if Δp is measured.

The carburretor of a car engine uses a venturi duct to feed the correct mix of air and petrol to the cylinders. Air is drawn through the duct and along a pipe to the cylinders. A tiny inlet at the side of the duct is fed with petrol. The air through the duct moves very fast, creating low pressure in the duct which draws petrol vapour into the air stream.

Worked example A non-viscous, incompressible fluid passes along a horizontal pipe of diameter 0.1 m. A venturi duct in the pipe has diameter of 0.05 m. A pressure gauge is connected to measure the pressure difference between the duct and a point upstream. Calculate the flow speed in the pipe when the gauge reading is 1.5 kPa, and determine the volume per second of fluid passing down the pipe at that speed. The fluid density is 800 kg m^{-3}.

Solution Let X be at the duct and Y a point upstream.

The area ratio A_y/A_x equals $\frac{\pi}{4}d_y^2 \Big/ \frac{\pi}{4}d_x^2 = d_y^2/d_x^2$

Hence $A_y/A_x = 4$.

Therefore the pressure difference $\Delta p = \frac{1}{2}\rho v_y(4^2 - 1)$, which gives $\Delta p = \frac{1}{2} \times 800 \times v_y^2 \times 15$.

$$\text{Hence } v_y^2 = \frac{2\Delta p}{800 \times 15} = \frac{2 \times 1500}{800 \times 15} = 0.25\ \text{m}^2\ \text{s}^{-2}$$

$v_y = 0.5$ m s^{-1}.

The volume per second passing along the pipe is given by $v_y A_y$.

Hence volume per second $= 0.5 \times \pi \times 0.1^2/4$

$= 3.9 \times 10^{-3}$ m^3 s^{-1}.

The pitot-static tube

Consider how fluid speed is measured in open fluids which are not confined to pipes. For example, suppose we wish to know the flow speed at different depths of a river. Or we may wish to measure the speed of a boat moving through a fluid.

The total pressure p_T at a point in a fluid is the pressure on a small surface at rest placed head-on to the flow.

The static pressure p_S at a point is the pressure on a small surface parallel to the flow.

At any given point, the total pressure is greater than the static pressure because of the impact of the fluid on any head-on surface in its path. The difference between the total and the static pressure depends on the fluid speed. By measuring the difference we can therefore determine the fluid speed.

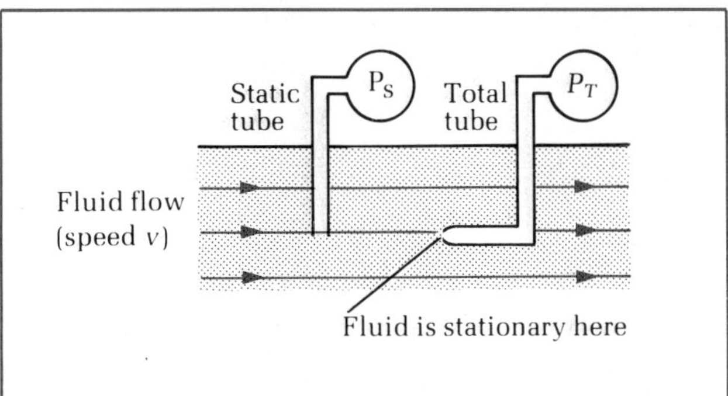

Figure 8.36 *Measuring total and static pressures*

Figure 8.36 shows how each type of pressure can be measured. The two pressure tubes are along the same streamline at the same level. The fluid at the tip of the 'total' tube T is at rest so the fluid speed at the tip, $v_T = 0$.

The fluid passes the static tube at speed $v_S = v$ where v is the undisturbed flow speed. Using Bernouilli's equation applied to horizontal flow

$$p_S + \tfrac{1}{2}\rho v_S^2 = p_T + \tfrac{1}{2}\rho v_T^2$$

Hence

$$p_S + \tfrac{1}{2}\rho v^2 = p_T,$$

So

$$v = \sqrt{\frac{2(p_T - p_S)}{\rho}}$$

The dynamic pressure is defined as the pressure due to the fluid's motion. Hence $\frac{1}{2}\rho v^2$ is the dynamic pressure, equal to (total pressure − static pressure).

Figure 8.37 shows a more convenient way of measuring the pressure difference. The two tubes of Figure 8.36 are combined as a **pitot-static** tube where the total tube (or pitot tube) passes down the centre of the static tube. A pressure gauge may be used to measure the pressure difference between the total and the static parts. Hence the fluid speed may be calculated using the above equation.

Boat speeds or aeroplane speeds are measured relative to the fluid in this way. Usually the gauge is calibrated (i.e. marked) directly in terms of speed but the scale would not be linear.

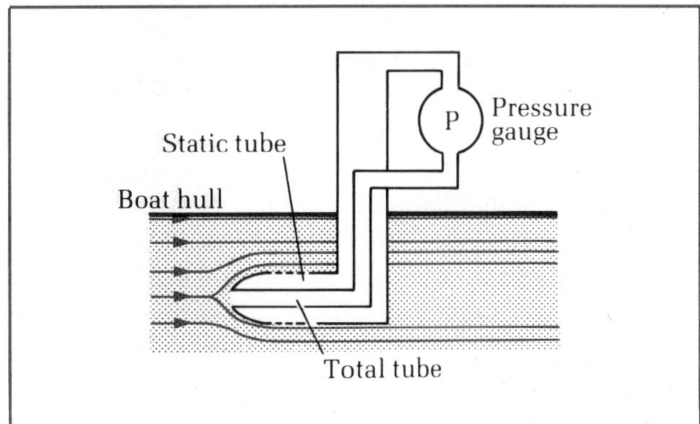

Figure 8.37 *A pitot-static tube*

Worked example A pitot-static tube used to measure boat speed is to be fitted with a pressure gauge. Given that the boat speed will not exceed 10 m s^{-1} and the density of sea water is 1050 kg m^{-3}, calculate the maximum pressure on the gauge.

Solution For $v = 10$ m s^{-1} and $\rho = 1050$ kg m^{-3},

$\frac{1}{2}\rho v^2 = \frac{1}{2} \times 1050 \times 100 = 52\,500$ Pa.

Hence the maximum pressure on the gauge will be 52.5 kPa.

8.6 Summary

Definitions

1 *Pressure* is force per unit area (acting at right angles to the area).
2 *Surface tension coefficient* γ is force per unit length, acting at right angles to a small length in the surface.
3 *Coefficient of viscosity* is tangential stress/velocity gradient.

Equations

1 Pressure due to depth $= H\rho g$
2 Height gain due to capillary rise $= 2\gamma/\rho g r$
3 Excess pressure inside a bubble $= 2\gamma/r$ (for a bubble in a liquid)
4 Surface energy per unit area $\sigma = \dfrac{An\varepsilon}{4}$
5 Continuity equation $\rho v A = \text{constant}$
6 Poiseuille's equation $\dfrac{V}{t} = \dfrac{\pi r^4 P}{8\eta L}$
7 Stokes' law $F = 6\pi\eta R v$
8 Bernouilli's equation $p + \frac{1}{2}\rho v^2 + \rho g H = \text{constant}$

Short questions

8.1 What would be the height of the Earth's atmosphere if it had uniform density 1.2 kg m^{-3} giving pressure at sea level equal to 10^5 Pa, the same as the average pressure at sea level? Is your estimate reasonable? If not, why not?

8.2 A test tube of length 160 mm is used as a hydrometer by weighting it with a small quantity of lead shot so that the tube floats upright. When placed in water (density 1000 kg m^{-3}), the test tube floats with 40 mm of its total length above water. When placed in salt solution, only 32 mm of its total length floats above the liquid level. The test tube has a uniform external diameter of 15 mm. Calculate
a) the volume and weight of water displaced when the tube floats in water,
b) the mass of the tube and contents,
c) the density of the salt solution. Assume $g = 10$ m s^{-2}.

8.3 A glass stopper is suspended in air by a thread attached to a spring balance which reads 26.0 g. When the stopper is fully immersed in water, the reading on the spring balance changes to 16.0 g. Calculate
a) the volume of the stopper,
b) its density.
Assume the density of water is 1000 kg m^{-3} and $g = 10$ m s^{-2}.

8.4 An underwater sonar detector floats 500 m beneath the ocean surface, attached to the upper end of a cable anchored on the sea bed, as in the diagram above.

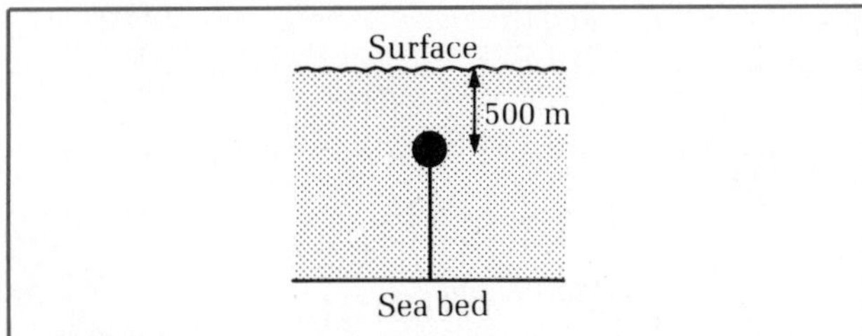

Figure 8.38

The detector is in a watertight drum, partly air-filled for buoyancy. The drum weight is 500 N and its volume is 0.25 m^3. Assuming the density of sea water is 1050 kg m^{-3} and $g = 10$ m s^{-2}, calculate
a) the pressure on the drum due to the sea water,
b) the upthrust on the drum
c) the tension in the cable, assuming it is vertical.

8.5
a) The pressure p of an air jet is given by $p = \frac{1}{2}\rho v^2$ where ρ is the density of air and v is the air speed. Estimate the minimum air speed to create a soap bubble from a soap film on a circular wire frame of diameter 30 mm. Assume the surface tension of the soap solution is 0.03 N m^{-1}, and the density of air is 1.2 kg m^{-3}.
b) Estimate the extra surface area created when the soap bubble in (a) is formed. Hence estimate the work done to blow such a bubble. Assume the bubble diameter is the same as that of the circular frame, and that the bubble is created at constant temperature.

8.6 A clean glass capillary tube of internal diameter 0.40 mm is dipped into a beaker of water. The water rises up the tube to a height of 75 mm above the water level in the beaker when the tube is vertical. Calculate the surface tension of water.
Assume the density of water = 1000 kg m^{-3}.

8.7 The average spacing between adjacent molecules in water is 0.31 nm. Estimate
a) the number of molecules per unit area at the surface of water,
b) the bond energy of water molecules, assuming each water molecule has twelve nearest neighbours on average. Assume the surface tension of water is 0.07 N m^{-1}.

8.8 Oil flows steadily through a uniform pipe of length 2 m and internal diameter 1.0 cm. If the flow rate is 8.5×10^{-6} m^3 s^{-1}, calculate the pressure difference between the ends of the pipe. Assume the oil viscosity is 0.01 N s m^{-2}, and its density is 900 kg m^{-3}.

8.9 Investigations show that fluid flow along a pipe changes from laminar to turbulent flow if the mean flow speed v is greater than $\dfrac{1000\,\eta}{\rho r}$, where η is the oil viscosity, ρ is the oil density and r is the pipe radius. Is the flow in **8.8** laminar or turbulent?

8.10 A steel ball bearing of diameter 8.0 mm is timed as it falls through oil at steady speed. Over a vertical distance of 0.20 m, it takes 0.56 s. Assuming the density of steel = 7800 kg m^{-3} and of the oil = 900 kg m^{-3}, calculate
a) the weight of the ball,
b) the upthrust on the ball,
c) the viscosity of the oil.

8.11 In an experiment to demonstrate the Bernouilli principle, two flow tubes fitted with manometers are used, as in the diagram. One tube, Y, has a narrow mid-section as shown.

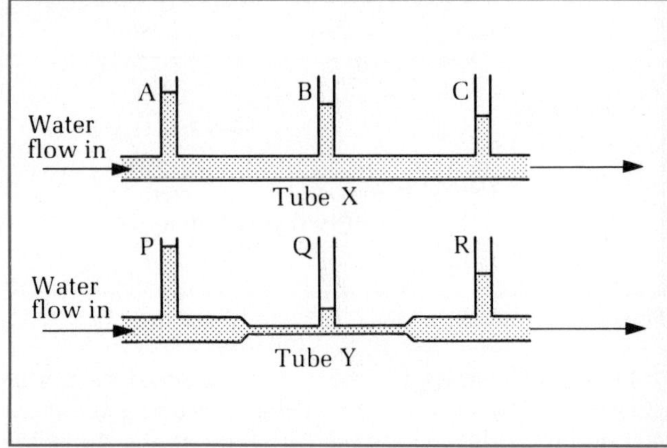

Figure 8.39

a) Explain why the manometer levels in the uniform tube show a steady pressure gradient along the tube.
b) Why is the level of manometer Q less than the average of manometers P and R on either side of Q?

8.12 An open water tank has a small hole of cross-sectional area 0.5 mm² in its wall at a depth of 0.60 m below the open water surface in the tank. Calculate
a) the speed at which water flows through the hole,
b) the volume of water per second leaving the tank.
Assume viscous effects are negligible and thet the Bernouilli principle applies.

8.13 A pitot-static tube is used to measure the speed of a boat in sea water. A pressure gauge connected between the total and the static tube gives a reading of 1870 Pa. Given the density of sea water is 1050 kg m^{-3}, calculate the boat speed.

8.14 Explain, with the aid of a diagram, why air flow over the wings of an aircraft causes lift.
Air flows over the upper surface of an aircraft's wings at a speed of 135 m s^{-1} and under the lower surface at a speed of 120 m s^{-1}. Calculate the pressure difference due to the flow, and so determine the lift force if the total wing area is 28 m². Assume the density of air is 1.2 kg m^{-3}.

8.15 A horizontal pipeline of diameter 50 mm contains liquid of density 800 kg m^{-3} and negligible viscosity. The pipeline has a constriction in the form of a duct at one position along its length, and its internal diameter narrows to 35 mm at the duct. A pressure gauge connected as shown in the diagram is used to measure the pressure difference between the constriction and a position upstream. For a pressure gauge reading of 3.8 kPa, calculate
a) the flow speed upstream and downstream of the duct,
b) the volume per second of liquid passing along the pipeline.

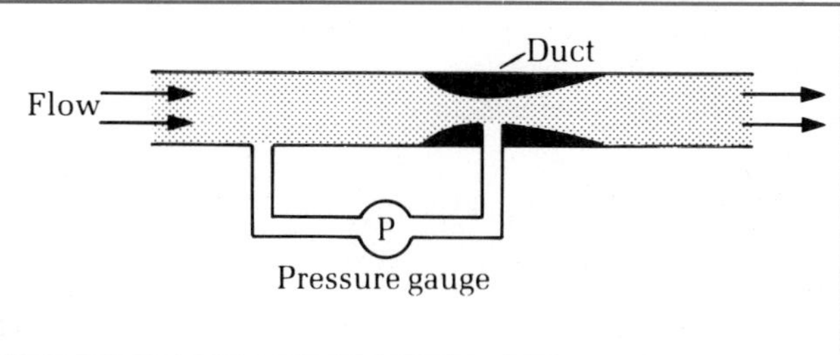

Figure 8.40

9

Thermal properties of materials

9.1 Temperature

Heat

Heat is energy transferred due to temperature difference. Ideas about heat are often misunderstood. For example, you may have heard phrases like, 'close the door or the cold will get in!' or, 'the heat of the building is unbearable'. Both are wrong in terms of physics. 'Letting the cold in' should really be 'letting energy out'; objects contain internal energy, *not* heat. Heat is energy *on the move* due to temperature difference, not stored energy. So we must think of heat in the same way as we think of work. Work is energy transferred by forces when they move. Heat is energy transferred due to temperature difference.

Internal energy is the energy contained in an object due to the K.E. and P.E. of its molecules. Suppose an object gains energy due to heat transfer and work done on it. Then its internal energy must increase by an amount equal to the transfer of heat + the work done.

> **Change of internal energy of the object ΔU = heat transferred to the object Q + work done on the object W.**

The equation above restates the **principle of conservation of energy**, and in the form above, is known as the **first law of thermodynamics**, discussed in more detail in Chapter 11. We may use the equation to define heat without involving temperature in our definition. All we need to state is that heat is energy transferred other than by applied forces. So if the internal energy of a body increases by ΔU and the work done on it is W, the heat transferred Q is $\Delta U - W$.

Let's look at an example using these ideas. Consider the energy changes when an electric drill is used to bore a hole into a wall. If you touch the drill bit immediately after use, you will notice it is very hot. Touch it again after letting it cool for a few minutes and it may still be warm. When in use, frictional forces (between the tip of the drill bit and the wall) do work on the drill bit. The drill bit gains internal energy so its temperature rises. Some heat transfer from the drill bit occurs, but when in use work is done on it at a faster rate, so the bit becomes hot. After a few minutes out of use, the bit feels cooler because its internal energy has decreased due to heat being transferred from it.

Temperature

Temperature is a measure of the degree of hotness of an object. The hotter an object is, the more internal energy it possesses. An ice cube in your hand feels cold because heat transfer from your hand to the cube reduces the internal energy of the skin. Place your hand in warm water and it gains internal energy due to heat transfer. If the water is at the same temperature as your hand, then the internal energy of your hand stays constant. There is no overall heat transfer between two objects at the same temperature, so **thermal equilibrium** exists. In other words, any two objects in thermal contact are in thermal equilibrium if there is no overall heat transfer between them.

Two objects in thermal contact at different temperatures cannot be in thermal equilibrium. Heat transfer occurs, and the hotter object loses internal energy. Assuming no work is done, the loss of internal energy from the hot object each second is equal to the gain of internal energy by the cold object each second. In simpler terms, heat transfer is always from hot to cold, provided no work is done.

Fixed points are used to define temperature scales. A fixed point is a standard degree of hotness which can be accurately reproduced. For example, the melting point of pure ice can be used as a fixed point. At least two fixed points are needed to define a scale of temperatures, with each fixed point given an exact numerical value. Any thermometer can be used to measure temperature on a given scale by calibrating it at the fixed points, then dividing the distance between the fixed point marks into an equal number of intervals or *degrees*.

The thermometric property of a thermometer is the physical property which is measured to determine the temperature. Examples of thermometric properties are:

1. **Length** for a liquid-in-glass thermometer, the length of the liquid thread in its stem is measured,
2. **Resistance** for a resistance thermometer, the electrical resistance is measured,
3. **Pressure** for a gas thermometer, the gas pressure is measured,
4. **Voltage** for a thermocouple thermometer, the e.m.f. of a thermocouple is measured.

Centigrade scales are often used for inexpensive thermometers. Two fixed points define a centigrade scale.

1. **Ice point**, defined as 0°, which is the temperature of pure melting ice.
2. **Steam point**, defined as 100°, which is the temperature of steam at standard pressure.

To calibrate a thermometer on the centigrade scale, its

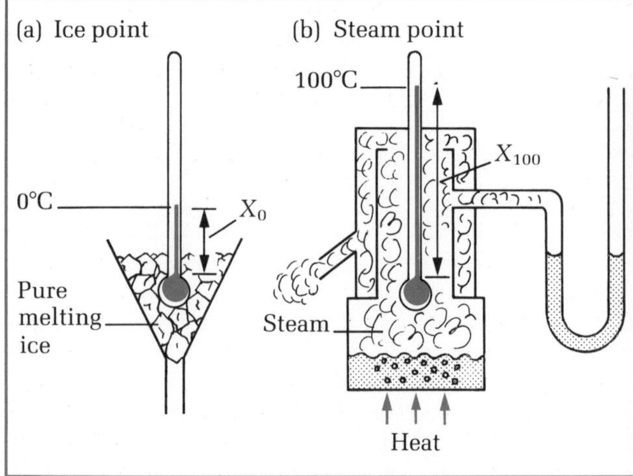

Figure 9.1 *Calibrating a thermometer*

thermometric property is measured when the thermometer is at ice point and when it is at steam point. Each degree of temperature then corresponds to $(X_{100}-X_0)/100$ where X_{100} and X_0 are the values of the thermometric property at steam point and ice point respectively.

After a thermometer has been calibrated, it can be used to measure unknown temperatures. If the thermometric property at some unknown temperature has a value X, then the unknown temperature is given by the following equation.

$$\theta=\frac{(X-X_0)}{(X_{100}-X_0)}\times 100^\circ$$

where θ is the unknown temperature in terms of the thermometer's centigrade scale.

Of course most thermometers, such as the liquid-in-glass thermometer of Figure 9.1, are simply marked with 100 equal degrees between the two fixed point readings. The temperature is then read off directly. However, you cannot do that with a resistance thermometer, for example. You have to measure the resistance at ice point, steam point and the unknown temperature. Then you must use the equation to calculate the unknown temperature.

Worked example The readings of a resistance thermometer are 20.0 ohms at ice point, 28.2 ohms at steam point, and 23.1 ohms at an unknown temperature. Calculate the unknown temperature on the centigrade scale of the thermometer.

Solution $X=23.1$ ohms, $X_0=20.0$ ohms, $X_{100}=28.2$ ohms.

Unknown temperature

$$\theta=\frac{X-X_0}{X_{100}-X_0}\times 100=\frac{3.1\times 100}{8.2}=37.8^\circ.$$

Different thermometers often give different values for the same unknown temperature. For example, a mercury thermometer might show a room temperature reading of 18° when an electrical thermometer next to it shows a reading of 20°. Even two thermometers of the same type might not agree. Which one is correct? We need a standard which other thermometers can be compared with. The problem of choosing the standard from practical thermometers is avoided by using the laws of thermodynamics. These are discussed in detail in Chapter 11, but all that is needed here is to know that the laws provide a theoretical scale known as the thermodynamic scale of temperatures. The properties of an ideal gas can be used to make measurements on this scale.

The thermodynamic scale (sometimes called the **absolute** scale) is defined in terms of two fixed points.

1 **Absolute zero**, which is the lowest possible temperature, is given the value 0 for its temperature.

2 **The triple point of water**, which is the temperature at which ice, water and water vapour are in thermal equilibrium. The triple point is given the value of 273.16 for its temperature. This choice seems strange at first but the reason is to make the temperature difference between ice point and steam point exactly 100 on this scale as well.

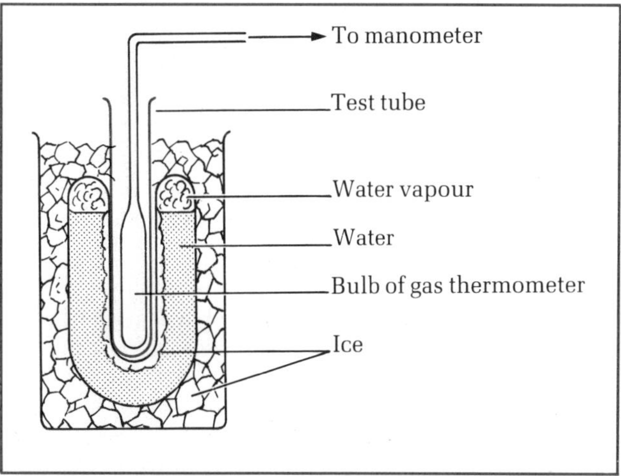

Figure 9.2 *The triple point of water*

The kelvin K is the unit of thermodynamic temperature. So absolute zero has a thermodynamic temperature defined as 0 K. The triple point of water is defined as 273.16 K. Figure 9.2 shows how the triple point is achieved in practice.

To measure temperatures on the absolute scale, we can use the properties of an ideal gas. An ideal gas has zero pressure at absolute zero, so we measure the pressure and volume of an ideal gas as our thermometric property. Measurements are made at the triple point of water and at the unknown temperature. The absolute temperature is then calculated using the following equation.

Absolute (or thermodynamic temperature)

$$T=\frac{(pV)}{(pV)_{Tr}}\times 273.16$$

where p = pressure,
V = volume,
$(pV)_{Tr}$ = pressure × volume at the triple point, (for a fixed mass of ideal gas).

The practical details of a gas thermometer are shown in Figure 9.5. In practice an ideal gas means one which obeys Boyle's law (pV = constant), so the gas used in the gas thermometer must obey Boyle's law. The thermodynamic temperature of ice point can be measured with a gas thermometer; its value is 273.15 K. Steam point has a thermodynamic temperature of 373.15 K because the difference between ice point and steam point is defined as 100 K.

The Celsius scale of temperature is defined from the thermodynamic scale according to the equation

$$t=T-273.15$$

where t = temperature on Celsius scale,
T = thermodynamic temperature.

The unit is the degree Celsius, °C, so ice point is 0°C and steam point is 100°C.

A gas thermometer with a centigrade scale is the only type of thermometer which gives readings directly in degrees Celsius. Suppose the pressure readings of a constant volume gas thermometer at ice point and steam

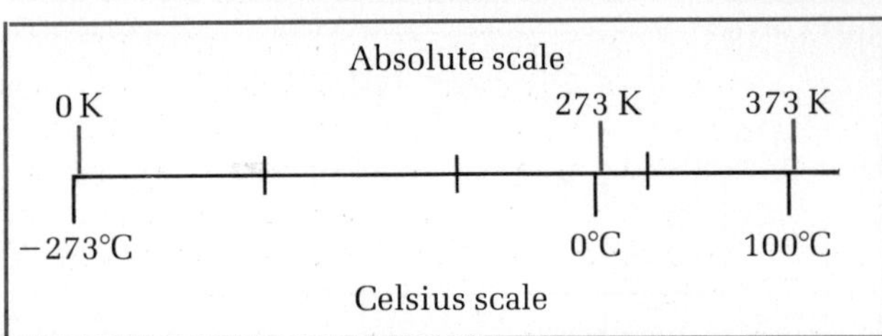

Figure 9.3 *Temperature scales*

point are p_0 and p_{100}. For a pressure reading p at an unknown temperature, then the unknown temperature on the centigrade scale of the thermometer is given by

$$\theta = \frac{p - p_0}{p_{100} - p_0} \times 100$$

However, the thermodynamic temperature is given by

$$T = \frac{(pV)}{(pV)_{Tr}} \times 273.16$$

which may be rearranged as

$p = cT$ where c is a constant equal to $p_{Tr}/273.16$

(assuming the volume V is constant).

So $p_0 = 273.15c$ and $p_{100} = 373.15c$. Putting these values, in terms of c, back into the equation for θ gives

$$\theta = \frac{cT - 273.15c}{373.15c - 273.15c} \times 100$$

After cancelling c, we obtain the following equation:

$$\theta = T - 273.15 = t$$

So the gas thermometer with a centigrade scale gives direct readings in degrees Celsius. Other types of thermometer with centigrade scales do not give direct values on the Celsius scale so their readings need to be corrected. Only after correction can the reading be given as °C. Before correction its reading should be given as 'degrees on its centigrade scale'.

Worked example A given liquid-in-glass thermometer marked with a centigrade scale is found to have a thread length L which varies with Celsius temperature t as shown by Figure 9.4. When the thermometer reads 50° on its centigrade scale, what is its temperature on the Celsius scale?

Solution From Figure 9.4, the length L at ice point is 65 mm. At steam point, the length is 248 mm.
So at 50° on its centigrade scale, the length is midway between ice point and steam point, i.e. 156.5 mm.
We can now read from Figure 9.4 what the Celsius temperature is for a length of 156.5 mm.
Hence $t = 58$°C.

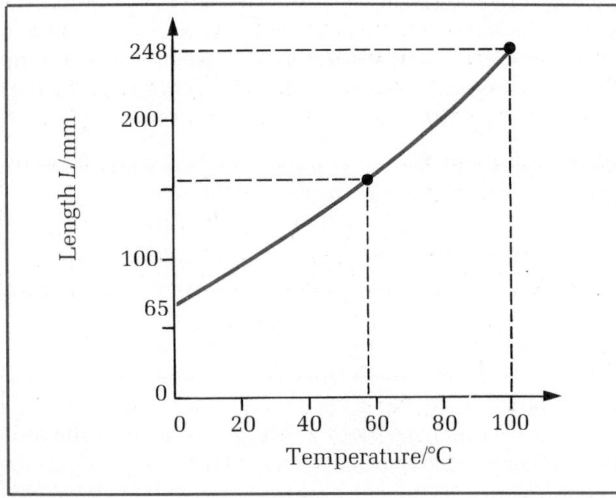

Figure 9.4 *A correction curve*

Types of thermometer

A practical thermometer must have a thermometric property which varies smoothly with temperature. Accurate measurement of the thermometric property is usually required so the thermometer must be sensitive to small changes of temperature. In some situations a thermometer must respond quickly to changes of temperature, and it may be necessary to read the thermometer with remotely placed measuring equipment. For example, the temperature inside a nuclear reactor is read from instruments in one building connected to the temperature probe in the reactor in a different building.

The constant volume gas thermometer uses the pressure of a fixed mass of gas at constant volume as its thermometric property (see Figure 9.5). Dry air or nitrogen is used since their behaviour is almost ideal at low pressures. Temperature readings made with a gas thermometer are sometimes called **ideal gas temperatures.** If calibrated at ice point and steam point, it gives readings on the Celsius scale. If calibrated at the triple point, it gives readings on the thermodynamic scale. It cannot be used for very low temperatures because the gas liquefies. Nor can it be used for very high temperatures: even a steel bulb melts if its temperature exceeds 2500 K.

How then can we measure very high or very low temperatures? For very low temperatures we can use a gas such as helium which remains a gas right down to a few degrees above absolute zero. For very high temperatures, pyrometers are used. See p. 109.

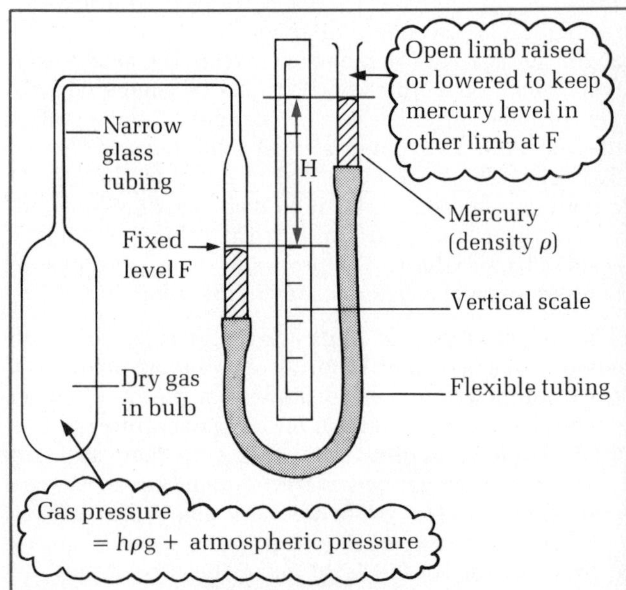

Figure 9.5 *A constant volume gas themometer*

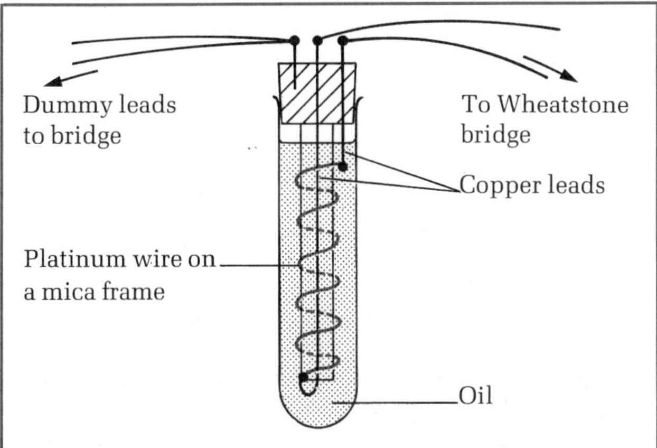

Figure 9.6 *A platinum resistance thermometer*

Resistance thermometers are more convenient than gas thermometers for accurate work. The measured resistance is that of a platinum wire in a tube, as in Figure 9.6. A Wheatstone bridge (see p. 000) is used to measure its resistance. For very accurate measurements, the resistance of the leads between the bridge and the platinum wire must be compensated for. This can be done by connecting dummy leads in the other arm of the bridge, as in Figure 9.6.

To use a resistance thermometer instead of a gas thermometer, the two thermometers must first be compared. A calibration curve like the one in Figure 9.4 is obtained by measuring the resistance at different temperatures measured with the gas thermometer. Then the resistance thermometer can be used instead of the gas thermometer; its resistance is measured at the unknown temperature, then the calibration curve is used to give the gas thermometer temperature.

Thermocouple thermometers have a much faster response than gas thermometers, so they are used to measure rapidly varying temperatures. The thermometric property of a thermocouple thermometer is the potential difference between two different metals in contact. A simple thermocouple can be made by twisting one end of a copper wire together with one end of an iron wire, as in Figure 9.7.

When two different metals are brought into contact, free electrons pass between them at the contact points. But because the electrons can leave one metal more easily than the other, a p.d. develops across the junction.

Thermocouple p.d.s vary smoothly with temperature so thermocouples make good thermometers. For accurate measurements, three wires are used to form two opposing thermocouples, as Figure 9.7. The free ends are then joined to a voltmeter or potentiometer. When the two junctions are at different temperatures they do not balance one another, so a net p.d. is produced. In practice, one junction is kept at ice point and the other is used as the temperature probe.

Thermocouple p.d.s of the order of millivolts so are best measured with a specially adapted potentiometer, shown in Figure 9.7. A resistor R of large resistance is connected in series with the driver cell D and the potentiometer wire AB. The thermocouple p.d. is then balanced against the p.d. along wire AB. To do this, contact C is moved along the wire until the meter shows no deflection. The length of wire from A to C, L_{AC}, is proportional to the thermocouple p.d. V since the p.d. per unit length along AB is constant.

With the probe junction at steam point, the length AC is measured after finding the balance point. This length can then be divided into 100 equal intervals to give a centigrade scale which can be read directly. Resistance R is chosen to keep length AC at steam point as long as possible. When the probe is at an unknown temperature, the potentiometer is rebalanced. The new position of contact C gives the temperature (on its centigrade scale) directly.

For example, suppose the steam point balance length was 780 mm and a balance length of 300 mm is measured at some unknown temperature. The unknown temperature is then $\dfrac{300 \times 100}{780} = 38.5°$ on its centigrade scale.

As with the resistance thermometer, a thermocouple thermometer can be calibrated against a gas thermometer, then used instead of the gas thermometer.

Liquid-in-glass thermometers are convenient and easy to use. Thermal expansion of the liquid in the thermometer makes the liquid thread in the stem increase in length. They need to be checked periodically because the initial calibration can become inaccurate after frequent use. The glass expands and contracts minutely each time a liquid-in-glass thermometer is used. Frequent use might cause the expansion of the glass to become permanent, so affecting its readings.

Pyrometers are used to measure very high temperatures, for example the temperature inside a furnance. The thermal radiation from the furnace is compared in terms of colour with the thermal radiation from a lamp filament in the pyrometer. Very hot objects emit light as well as infra-red radiation, and the colour changes with

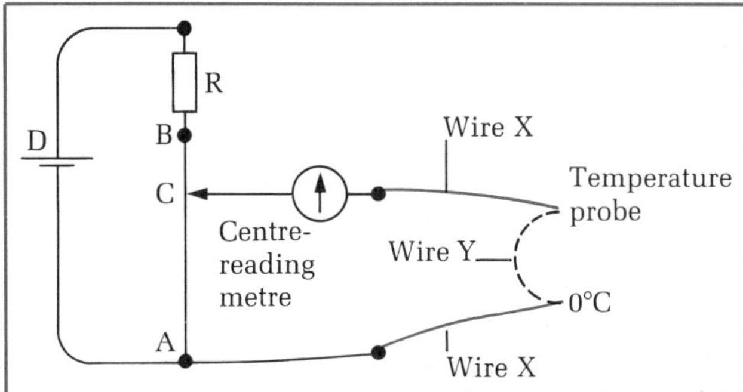

Figure 9.7 *A thermocouple thermometer*

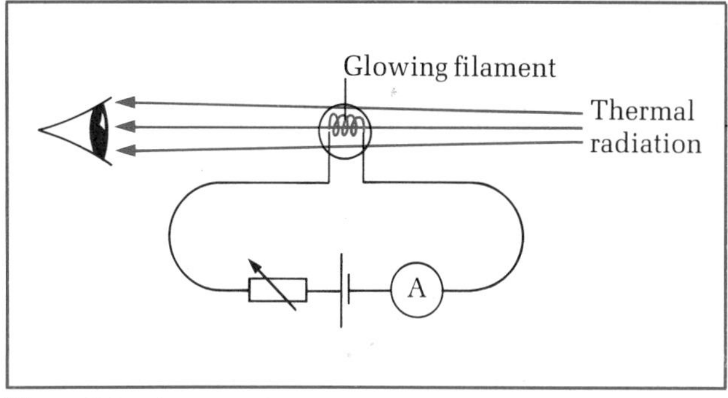

Figure 9.8 *A pyrometer*

increased temperature. Heated metal reddens as its temperature increases, and further heating cause the metal to glow orange. The temperature of a glowing object can therefore be estimated from its colour. The laws of thermal radiation are well-established, and may be used to determine the temperature accurately. See p. 115.

When a pyrometer is used, a hot wire filament inside the pyrometer is viewed against the glowing object. The filament current is increased from zero until it makes the filament exactly the same colour as the glowing object; when this happens, the filament 'blends' into the background formed by the glowing object. A meter in series with the filament can then be calibrated directly in terms of source temperatures, known using the laws of radiation.

The international practical temperature scale specifies which type of thermometer may be used as an alternative standard to the gas thermometer for different temperatures. Gas thermometers are often inconvenient due to slow response or size. More fixed points are chosen, their temperatures being measured using a gas thermometer. Then an alternative type of thermometer is chosen as the practical standard between successive fixed points. The alternative thermometer must first be calibrated against a gas thermometer. Then it can be used to give ideal gas temperatures.

9.2 Heat capacities

Specific heat capacity

Nowadays many buildings are heated by hot water piped from a boiler to radiators. When the boiler is switched off, the radiators become cold quite rapidly. Electric storage heaters operate differently. Electricity supplies them with internal energy which is released, by heat transfer, slowly to the surroundings. Storage heaters stay warm long after they are switched off. The reason is that they are made of material which needs a lot more internal energy to warm up than the metal and water of a water radiator. Also they contain more mass. So a storage radiator at the same temperature as a water radiator contains much more internal energy.

The specific heat capacity c of a material is defined as the heat energy required to increase the temperature of 1 kg by 1 K. The unit of c is $\mathrm{J\,kg^{-1}\,K^{-1}}$. Materials with low specific heat capacity like aluminium are easily heated and cooled. To raise the temperature of a mass m by $\Delta\theta$, the heat energy required is given by

$$Q = mc\Delta\theta$$

where Q = heat supplied,
c = specific heat capacity,
m = mass,
$\Delta\theta$ = temperature rise.

The heat capacity C of an object is defined as the heat energy required to increase the object's temperature by 1 K. The unit of C is $\mathrm{J\,K^{-1}}$. If the object is a single mass m of the same material, then its heat capacity equals its mass × its specific heat capacity.

Figure 9.9 *A comparison of types of thermometer*

Type of thermometer	Thermometric property	Advantages	Disadvantages	Use on IPTS scale
Mercury-in-glass (234 – 723 K)	Length of mercury thread	1 Convenient 2 Portable	1 Fragile 2 Limited range	From 234 to 723 K
Constant volume gas-thermometer (3 – 1750 K)	Pressure of a fixed mass of gas at constant volume	1 Wide range 2 Accurate 3 Gives ideal gas temperatures directly	1 Bulky, inconvenient 2 Slow to respond	Below 90 K
Resistance thermometer (90 – 1400 K)	Electrical resistance of a platinum coil	1 Wide range 2 Accurate 3 Best type for small temperature differences	1 Slow to respond	From 90 to 903 K
Thermocouple thermometer (25 – 1400 K)	e.m.f. across the junction of two dissimilar metals	1 Fast response 2 Wide range 3 Readings can be read remotely, stored or fed into a microcomputer	1 For accuracy, a high resistance millivoltmeter or potentiometer must be used	From 903 to 1336 K
Pyrometer (above 1250 K)	Colour of thermal radiation from hot object	1 No contact with hot object 2 Can be portable	1 Not highly accurate	Above 1336 K

For example, suppose an aluminium electric kettle has a mass of 1.5 kg when empty. The specific heat capacity of aluminium is $900\,J\,kg^{-1}K^{-1}$, so the heat capacity of the empty kettle is $900 \times 1.5\,J\,K^{-1} = 1350\,J\,K^{-1}$ (assuming non-aluminium parts are insignificant). If the kettle is then filled with 2.0 kg of water (specific heat capacity $4200\,J\,kg^{-1}K^{-1}$), its heat capacity increases by $2 \times 4200\,J\,K^{-1} = 8400\,J\,K^{-1}$. So the total heat capacity of the filled kettle is $1350 + 8400 = 9750\,J\,K^{-1}$. To increase the temperature of the water and kettle from 20°C to 100°C, the minimum amount of heat energy required is $80 \times 9750\,J = 780\,kJ$. In practice, heat losses to the surroundings means that more heat energy must be supplied.

Measuring specific heat capacities

An electrical method suitable for a liquid involves using an insulated container, called a calorimeter, as in Figure 9.10. The calorimeter's mass when empty is first measured. Its specific heat capacity must be known. The initial temperature of a measured mass of liquid is determined. Then the current through the heater is switched on for a measured time. The rheostat is used to maintain a constant current reading on the ammeter. The voltmeter reading is also recorded. After the measured time, the current is switched off and the highest temperature of the liquid is measured after stirring. The following measurements are made.

1 Mass of calorimeter m'
2 Mass of liquid m
3 Initial temperature of liquid θ_1
4 Final temperature of liquid θ_2
5 Heating time t
6 Current I
7 Heater p.d. V.

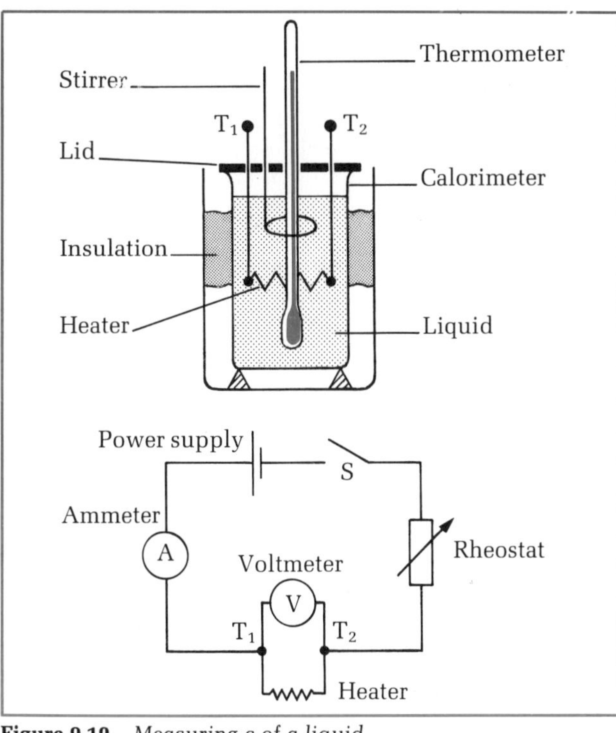

Figure 9.10 *Measuring c of a liquid*

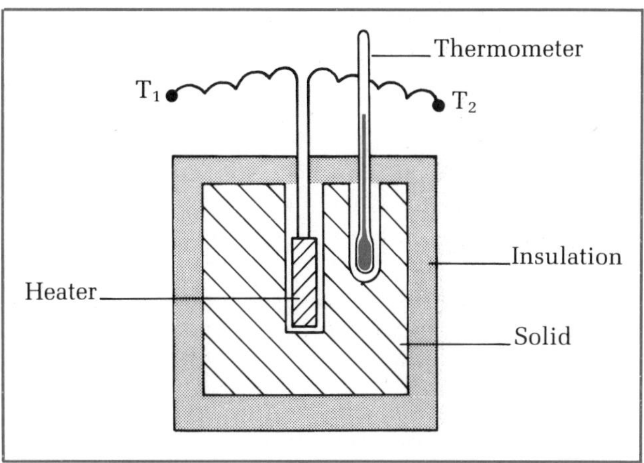

Figure 9.11 *Measuring c of a solid*

The electrical energy supplied $= IVt$ (see p. 314 for this equation).

The heat energy transferred to the liquid $= mc(\theta_2 - \theta_1)$.

The heat energy transferred to the calorimeter $= m'c'(\theta_2 - \theta_1)$ where c' is the specific heat capacity of the calorimeter.

No internal energy is lost from the liquid and calorimeter since we assume perfect insulation of the calorimeter. So the electrical energy supplied equals the heat transferred to the liquid + the heat transferred to the calorimeter.

$$IVt = mc(\theta_2 - \theta_1) + m'c'(\theta_2 - \theta_1)$$

The equation may be used to calculate c, given a value for c' and all the necessary measurements.

The specific heat capacity of a solid can be determined using the arrangement of Figure 9.12, on the next page. The solid is large enough for a hole to be made in it so that a small heater can be inserted. Another hole is made for a thermometer.

A continuous flow method is used for gases or liquids. A steady flow of the liquid or gas is passed along a pipe containing an electric heater, as in Figure 9.12. The inlet and outlet temperature is measured for a known flow rate and heater power. For a liquid, the flow rate is easily determined by collecting the liquid in a beaker for a measured time, then measuring the mass of liquid collected.

Assuming the temperatures are steady when measured, the electrical energy supplied is either carried away by the liquid or lost to the surroundings.

Electrical energy supplied in time t = heat transferred to liquid + heat loss to surroundings

$$IVt = mc(\theta_2 - \theta_1) + H$$

where m = the mass of liquid which passes through in time t,
H = the heat loss to the surroundings in time t,
θ_1 and θ_2 = the inlet and outlet temperatures respectively.

To calculate c from the above equation, heat loss H must be eliminated since it is not known what its value is. We could insulate the apparatus, but a better approach is

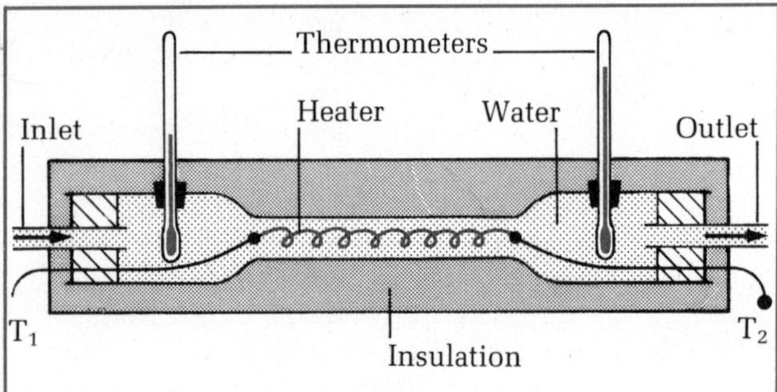

Figure 9.12 *Continuous flow*

to repeat the measurements at a different flow rate. To keep H the same, the heater power must be adjusted to maintain θ_1 and θ_2 at the same values. The rheostat is used to adjust the heater current, and once the new flow rate has achieved the same temperatures as before, the heater current and p.d. are remeasured. The new flow rate must also be determined. For the new readings

$$I'V't = m'c(\theta_2 - \theta_1) + H$$

where m' = the mass of liquid collected at the outlet in the same time as before.

Now we can calculate c by subtracting one equation from the other to eliminate H.

$$IVt - I'V't = mc(\theta_2 - \theta_1) - m'c(\theta_2 - \theta_1)$$

When the measured values are put into this equation, the only unknown quantity, c, can then be calculated.

Why doesn't the heat capacity of the tube appear in the equations? The reason is that the flow is at steady temperatures, so energy supplied electrically is not used to warm the tube (since the tube is at steady temperature).

Worked example A steady stream of air is drawn through a tube which contains a 500 W heater. Given that the density of the air is $1.2\,\mathrm{kg\,m^{-3}}$ and its specific heat capacity (at constant pressure) is $1000\,\mathrm{J\,kg^{-1}\,K^{-1}}$, calculate the maximum temperature rise of the air for a flow rate of $0.4\,\mathrm{m^3\,s^{-1}}$.

Solution In one second, the mass of air which passes over the heater = 0.48 kg since the volume/second is $0.4\,\mathrm{m^3\,s^{-1}}$ and the density is 1.2 kg. Also, the maximum heat transfer to the air is 500 J in one second.
$Q = 500\,\mathrm{J}$, $m = 0.48\,\mathrm{kg}$, $c = 1000\,\mathrm{J\,kg^{-1}\,K^{-1}}$.

$$\text{Temperature rise } \Delta\theta = \frac{Q}{mc} = \frac{500}{0.48 \times 1000} = 1.0\,\mathrm{K}.$$

Latent heat

When a solid is heated, its temperature rises due to the increase of the internal energy of the solid. The increased internal energy is in the form of increased K.E. of the molecules in the solid. They vibrate more when the solid is heated and its temperature rises. Continued heating increases the temperature to the melting point. At the melting point, further heating makes the solid liquefy without change of temperature. Once it is completely liquefied, the temperature continues to rise if more energy is supplied. To change the state of the material from solid to liquid, energy must be supplied to the material. To reverse the process (i.e. from liquid to solid), energy must be taken from the material. For example, when hypo crystals are heated in a test tube, the crystals liquefy. If the liquid is then allowed to cool, crystal formation can be observed when the temperature falls to the melting point temperature; the tube becomes warm when crystals form from the liquid because energy is released due to the crystals forming. Energy transferred to or from a material as a result of change of state is called **latent heat**. Latent heat must be supplied for changes from

a) solid to liquid (i.e. melting)
b) liquid to vapour (i.e. vaporization),
c) solid to vapour directly (which happens to some materials – the process is called **sublimation**).

Latent heat must be removed for the reverse processes, for example when a vapour condenses to form a liquid or when a liquid solidifies.

The specific latent heat of fusion (or vaporization) l of a material is defined as the heat energy required to change the state of 1 kg of material from solid to liquid (or liquid to vapour) without change of temperature. The unit of l is $\mathrm{J\,kg^{-1}}$ or sometimes $\mathrm{J\,mol^{-1}}$.

To change the state of mass m of material, the heat energy required is given by

$$Q = ml$$

where Q = heat transferred,
m = mass,
l = specific latent heat of fusion or vaporization.

Worked example 0.5 kg of water at 20°C in a plastic beaker is placed in a freezer which converts it to ice in 5 minutes 30 s. Calculate the rate of transfer of heat from the water. The specific heat capacity of water is $4200\,\mathrm{J\,kg^{-1}\,K^{-1}}$ and the specific latent heat of ice is $336\,000\,\mathrm{J\,kg^{-1}}$.

Solution Assume the heat capacity of the beaker is negligible.
To convert 0.5 kg of water at 20°C to water at 0°C, heat transfer Q_1 from the water is given by $Q_1 = mc\Delta\theta = 0.5 \times 4200 \times 20 = 42\,000\,\mathrm{J}$.
To convert 0.5 kg of water at 0°C to ice at 0°C, the heat transfer Q_2 must be given by $Q_2 = ml = 0.5 \times 336\,000 = 168\,000\,\mathrm{J}$.
Hence the total heat transfer $Q = Q_1 + Q_2 = 210\,000\,\mathrm{J}$, and this takes place in 330 s.
Therefore the heat transfer per second

$$= \frac{210\,000}{330} = 636\,\mathrm{J\,s^{-1}}.$$

Measuring specific latent heat of vaporization

The liquid in the inner flask shown in Figure 9.13 is heated electrically so that it boils at a steady rate. The vapour passes through holes in the neck of the inner flask into the outer flask. Condensation takes place on the outer wall and in the condenser. The condensed vapour runs down the tube into a collecting flask.

The rheostat is used to ensure a steady heater current.

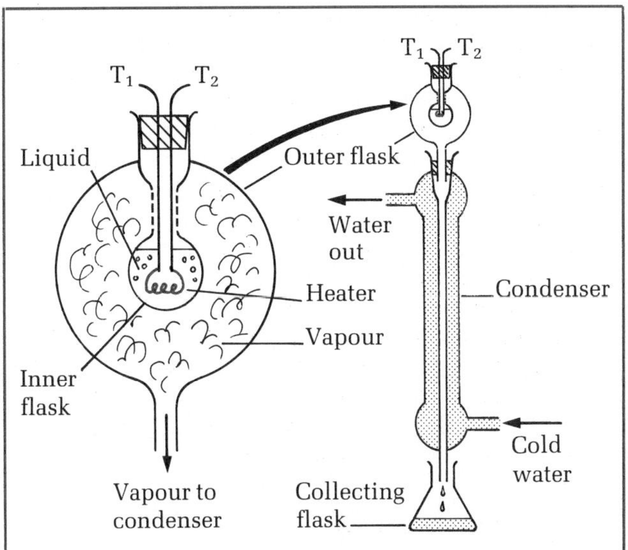

Figure 9.13 *Measuring l*

When liquid is running steadily into a collecting beaker, the beaker is replaced by a flask of known mass for a measured time. The flask is then weighed to give the mass of liquid condensed in the measured time. For steady flow, this gives the mass vaporized/second in the inner flask. The heater current and p.d. are also measured.

The electrical energy supplied in time t supplies latent heat and heat loss to the surroundings. Thus

$$IVt = ml + H$$

where I = the current,
V = the p.d.,
t = the time taken to collect mass m,
H = the heat loss in that time.

To calculate the specific latent heat l, we need to eliminate H from the above equation. The best way to do this is to run the experiment again at a different heater power. The rate of vaporization is different now, but the heat loss H in time t is the same, because the heat loss depends only on the liquid temperature, which stays at boiling point. So the new heater current and p.d. I' and V' are measured, and the mass collected in time t is remeasured.

Hence

$$I'V't = m'l + H$$

where m' is the mass collected in time t at the new rate.

To calculate l, we can eliminate H by subtracting one equation from the other.

$$IVt - I'V't = ml - m'l$$

When the measured values are put into the equation, l is the only quantity not known. Hence l can be calculated.

Molecular interpretation of latent heat

Heat must be supplied to melt a solid. The heat transfer increases the energy of the molecules, enabling them to break away from each other. In a solid the molecules are joined to one another by relatively strong bonds. To change the solid into a liquid, these relatively strong bonds must be broken. The molecules must do work to break the bonds which keep them in place, so they must be supplied with energy.

Suppose each molecule in the solid is joined by bonds to n neighbouring molecules, on average. Let the energy needed to break each bond (i.e. the *bond energy*) be ε. To break a bond, each of the two molecules which form the bond must be supplied with energy $\varepsilon/2$. So to remove one molecule completely from its neighbours, that molecule must be supplied with energy $n\varepsilon/2$. Hence, to remove N molecules from each other, the total energy required is $Nn\varepsilon/2$. This energy is supplied as latent heat.

$$l = \tfrac{1}{2}Nn\varepsilon$$

where l = specific latent heat,
N = number of molecules per unit mass,
n = average number of neighbours,
ε = bond energy.

Strictly, the equation applies to sublimation only, since the solid to vapour change does remove molecules completely from each other. Otherwise, it is only approximate.

The refrigerator

To cool an object below room temperature, its internal energy must be reduced. When a relatively warm object is put into a refrigerator, heat transfer from the object to the refrigerator takes place because the inside of the refrigerator is much cooler. So the object's internal energy falls. But what happens to the heat transferred to the refrigerator from the object? Does the refrigerator warm up? The answer can be found by putting your hand behind a refrigerator. You should find there is a warm panel there, outside the main refrigerator box but connected to it by pipes. The panel is used to transfer heat to the room from the refrigerator. Energy from an object placed in a refrigerator is therefore used to warm up the surroundings.

This is achieved by pumping a suitable fluid, the **refrigerant**, round a circuit containing an evaporator inside the refrigerator and a condenser panel behind the refrigerator. Valves ensure a one-way flow. Liquid in the evaporator at low pressure is vaporized by heat transfer from warm objects in the refrigerator. The vapour passes through a valve into a compressor which forces it under high pressure into the condenser panel where it liquefies, releasing internal energy. The liquid then passes into a storage tank from where it is fed through a valve to the evaporator. So the refrigerant transfers energy from any

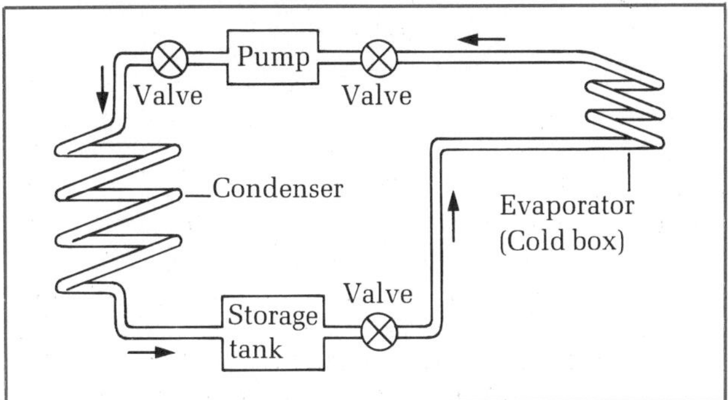

Figure 9.14 *A refrigerator*

warm object in the refrigerator to the room in which the refrigerator is placed. To drive the fluid round the circuit, the compressor must do work. So the heat transferred to the room is equal to the heat transferred from inside the refrigerator plus the work done by the compressor. Examples of refrigerants include ammonia, freon and sulphur dioxide.

9.3 Thermal expansion

Thermal cut-out switches used in electric heaters use thermal expansion of metals. When the temperature of a metal rises, the metal expands. Increased molecular vibrations cause the expansion, as explained on p. 71. Different metals expand at different rates. **A bimetallic strip** consists of two different metals bonded together. The strip bends when heated because the two metals expand at differing rates. Figure 9.15 shows a bimetallic strip used in a thermal switch.

Thermal expansion must be allowed for when buildings, bridges and other structures are designed. Temperatures can change by as much as 50°C from winter to summer so expansion gaps are necessary in structures to allow for increase of length due to thermal expansion. Expansion can be calculated using the following equation.

$$\text{Expansion of length } \Delta L = \alpha \times L \times \text{temperature rise } \Delta\theta$$

where α is the **coefficient of linear expansion** of the material. The unit of α is K^{-1}.

Let's look at an example of the use of this equation. Suppose steel girders of length 20 m are used to form the frame of a building. What gap should there be between girders end-to-end to allow for thermal expansion? The coefficient of linear expansion of steel is $1.2 \times 10^{-5} K^{-1}$. If we work on the basis of a maximum temperature rise of 50 K, the expansion of a single girder is given by

$$\text{Expansion} = \alpha L \Delta\theta$$
$$= 1.2 \times 10^{-5} \times 20 \times 50 = 1.2 \times 10^{-2} \text{ m}.$$

Most liquids expand when heated. An exception is water which contracts when its temperature rises from 0°C to 4°C. Above 4°C, it expands. This unusual thermal behaviour of water is the reason why water pipes sometimes burst in winter. If the water temperature in a pipe falls from 4°C to 0°C, the water expands, possibly bursting the pipe. Flooding results when the pipe unfreezes.

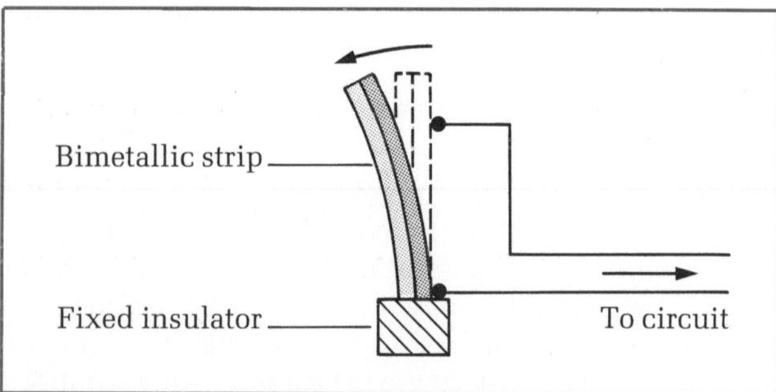

Figure 9.15 *A thermal switch*

Figure 9.16 *Thermal expansion gap in a railway line*

9.4 Thermal radiation

All objects emit thermal radiation. The hotter an object is, the more energy per second is carried away from it by thermal radiation. Thermal radiation consists of electromagnetic waves (see p. 291) with a range of wavelengths covering the infra-red and visible regions of the electromagnetic spectrum. When thermal radiation is directed at a surface, some of the radiation is absorbed by the surface and some is reflected. Some of the radiation may pass through the surface and be transmitted right through the material to pass out through the other side. Most materials do not transmit thermal radiation though, so thermal radiation incident on a surface is usually either reflected or absorbed.

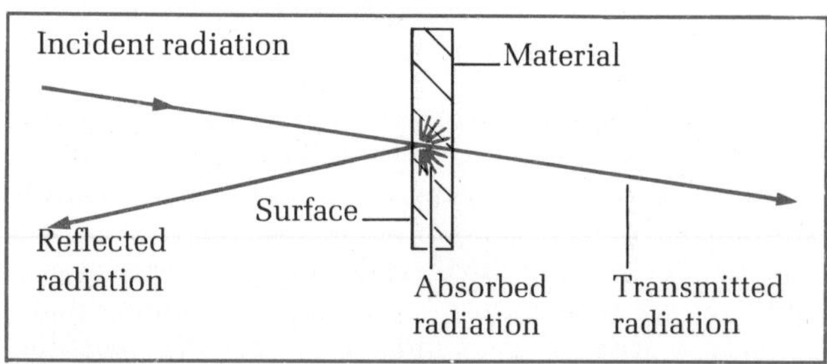

Figure 9.17 *Thermal radiation*

Shiny silvered surfaces are the best reflectors, whereas matt black surfaces are very good absorbers. Surfaces which appear black in daylight do not reflect any light so they must absorb all the light which falls on them. Such surfaces are good absorbers of thermal radiation as well.

When thermal radiation is absorbed by a surface at constant temperature, the surface loses energy by some means to balance the energy gained by thermal radiation. Possible ways of losing energy from the surface are either by thermal convection (see p. 119) to the fluid above the surface, by thermal conduction (see p. 116) or by re-radiating the energy. Each way makes a contribution, but if the surface is a good absorber of thermal radiation, it will also be a good emitter so it will re-radiate well. Figure 9.18 shows a simple test. Two aluminium beakers containing equal volumes of water are to be placed near a thermal radiator. One of the beakers is painted matt black, the other is given a shiny silvered surface. With a

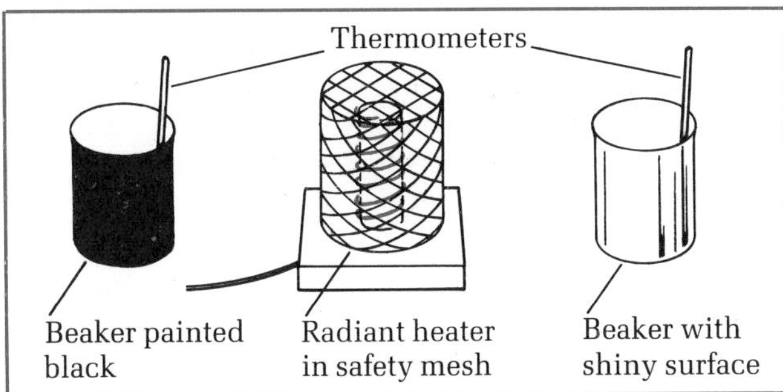

Figure 9.18 *Investigating surfaces*

thermometer in each, how would you compare their absorbing and emitting powers?

Absorption Position the beakers so they reach the same steady temperature. Which is closer to the heater for the same final steady temperature of the water? The shiny surface should be closer because it is a poor absorber and a good reflector compared with the black surface.

Emission With the beakers warmed to the same steady temperature, switch the heater off and remove it. Which beaker cools faster initially? The black surface cools faster because it radiates away heat at a faster rate. Thermal convection and conduction are the same in both cases because the initial temperatures are the same. So thermal radiation accounts for the difference.

A surface which is a good absorber of thermal radiation is also a good emitter of thermal radiation.

Black body radiation

An object with surfaces which absorb thermal radiation of all wavelengths is called a **black body** because it reflects no light.

Imagine the remnants of a star after it has used up all its energy. The remnants become invisible if unable to reflect light from other stars. What then is seen if the remnants become very hot again? The surface emits thermal radiation, and may become visible if hot enough. Because the surface is capable of absorbing thermal radiation of any wavelength, then that surface must also be capable of emitting any wavelength. So thermal radiation emitted by a black body contains all wavelengths.

Another example of a black body radiator is a small hole in the door of a furnace. Any thermal radiation which enters the hole from outside passes into the furnace and is absorbed by the inside walls. But the walls inside the furnace are very hot so they emit thermal radiation strongly. The inside of the furnace is therefore filled with thermal radiation passing from wall to wall. Some of this radiation passes out through the hole. The emitted radiation from the hole contains all wavelengths.

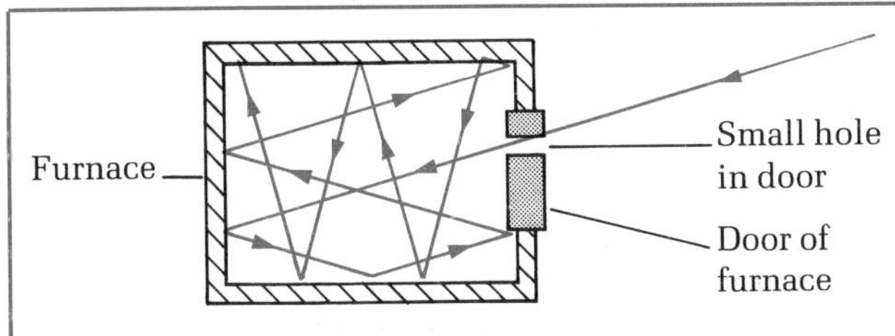

Figure 9.19 *A black body*

Stars too are black body radiators! Light directed at a star will not be reflected. So a star is a perfect absorber of any thermal radiation directed at it. Because its surface is capable of absorbing all wavelengths directed at it, then it must also be capable of emitting all wavelengths, which it does because it is very hot.

Black body radiation from a source at constant temperature contains a continuous range of wavelengths. The energy carried by the radiation is not distributed evenly across the wavelength range. Also the distribution changes if the source temperature alters. The proportion of energy carried by shorter wavelengths increases as the source temperature increases. This explains why a piece of steel, when heated, first becomes dull red then orange-red as it becomes hotter. Figure 9.20 shows how the energy is distributed over the wavelength range for several values of source temperature. At each value of temperature, the distribution curve has a peak. However the peak moves to shorter wavelengths as the temperature rises. So at any one temperature, the proportion of energy carried away is greatest for one value of wavelength. This peak wavelength becomes shorter as the temperature increases according to an equation known as **Wien's law**.

$$\lambda_p \times T = \text{constant}$$

where T = surface temperature of source, in K,
λ_p = peak wavelength.

The total energy emitted per second increases with surface temperature of the source. Increased area also increases the total energy per second emitted. The link between the radiated power (i.e. energy per second) from a black body radiator at constant temperature was first established in 1884. It is known as **Stefan's law**, after its discoverer.

$$Q/t = \sigma A T^4$$

where Q/t = energy/second radiated away,
A = surface area,
T = surface temperature, in kelvins,
σ = Stefan's constant.

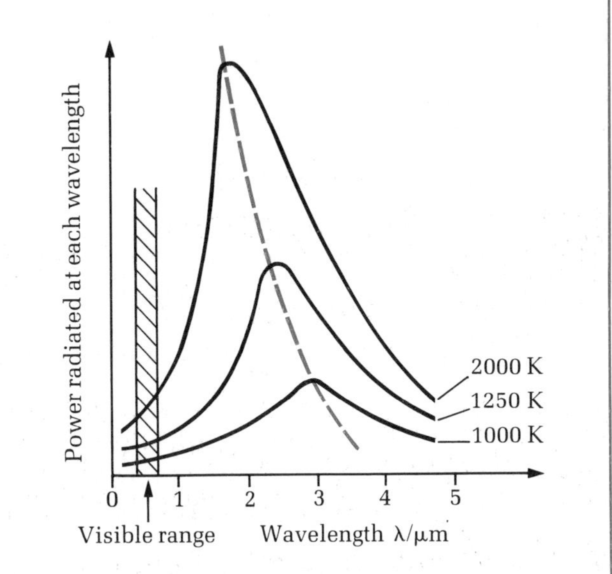

Figure 9.20 *Black body radiation curves*

Measurements give the value of σ as $5.7 \times 10^{-8}\,\mathrm{W\,m^{-2}\,K^{-4}}$. Wien's law and Stefan's law were established as a result of careful experiments. The explanation of these laws was not achieved until Einstein developed the photon theory of radiation (see p. 429). The laws are used in many situations. Astronomers use them to determine star temperatures. For example, *Betelgeuse* is a red-orange star in the constellation of *Orion*. The peak wavelength emitted by *Betelgeuse* is about 700 nm. The constant in Wien's law is known from laboratory measurements to have a value of 0.0029 K m. So the surface temperature of *Betelgeuse* is, using Wien's law, $0.0029/(700 \times 10^{-9})$ K which equals about 4000 K.

An entirely different example of the use of thermal radiation is in thermography, used in medical physics. A thermograph is a thermal image of the body, obtained by mapping the thermal radiation from lots of small areas on the body surface. Hot spots at the surface can be detected because such spots emit more thermal radiation than the surrounding surface. Hot spots or patches on the body surface are often due to disorders just beneath the surface, so thermographs can give useful information.

9.5 Thermal conduction

Understanding thermal conduction

Metals are good conductors of heat and electricity. The reason is the presence of free electrons inside the metal. At absolute zero, these electrons are attached to individual atoms in the outer shells. When the metal is supplied with energy to raise its temperature above absolute zero, the outer electrons of each atom break free as the atoms vibrate. So any metal above absolute zero contains lots of free electrons moving about inside the metal in a haphazard way. Free electrons collide with one another and with the atoms of the metal as they move about. When a p.d. is applied across the metal, the free electrons in the metal are attracted towards the positive terminal. Free electrons drift towards the positive terminal, so an electric current passes through the metal. When a metal is heated at one point, for example a metal bar heated at one end, the atoms in the heated part vibrate more. Free electrons colliding with these atoms gain kinetic energy so the free electrons in the heated part move about faster. These high-speed, free electrons collide with other free electrons from the colder parts, causing transfer of kinetic energy to the other electrons. An individual electron may only travel a short distance along the bar before colliding and giving up its extra kinetic energy to another electron. This other electron rebounds with extra kinetic energy and in turn may only travel a short distance before colliding and giving some of its kinetic energy to a slower electron. So energy is transmitted through the metal by movement of free electrons, even though the electrons only travel short distances.

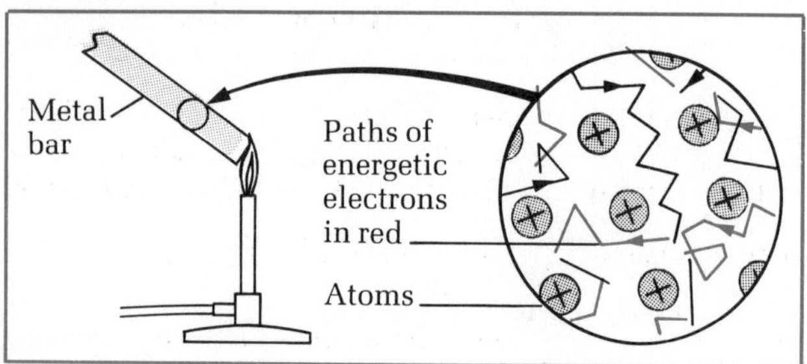

Figure 9.21 *Thermal conduction in a metal*

Insulators are poor thermal conductors because they have no free electrons. All the electrons are firmly attached to individual atoms, even at high temperatures. So insulators cannot conduct electricity at all. Yet they do conduct thermally, even if very poorly. Insulation to prevent heat loss from houses does not cut out heat losses altogether. The reason is that the vibrations of the atoms transmit energy through the material. When part of the material is heated, the atoms in that part vibrate more. The increased vibrations make neighbouring atoms vibrate more which in turn make other atoms further away vibrate more. So energy passes to other parts of the material. This mechanism is present in metals but energy transfer due to electrons is much greater than that due to atomic vibrations.

Gases and liquids conduct heat energy. When heated, molecules gain extra kinetic energy which is transferred to other molecules as a result of collisions between molecules. The mechanism is not unlike what happens in metals due to free electrons, but here the molecules, rather than free electrons, carry the energy. Gases and liquids are poor thermal conductors, in general, compared with metals. The reason is that molecules in a gas or liquid move much more slowly than the free electrons in a metal. So energy transfer in a liquid or gas is much slower. One type of liquid, though, is an excellent thermal conductor: liquid metals such as mercury. The reason is the presence of free electrons.

Thermal conductivity

Consider a uniform bar which is insulated along its sides. Suppose one end is at a constant temperature θ_1 and the other end at a lower constant temperature θ_2. Heat is conducted from the hotter end to the other end at a steady rate. Because the sides are perfectly insulated, the heat energy per second reaching the cold end is equal to the heat energy per second from the hot end. In other words, the heat energy per second passing any position along the bar is the same. The temperature varies along the bar as shown in Figure 9.22. The temperature fall per unit length along the bar is the same from one end to the other.

The temperature gradient at any position along the bar is defined as the change of temperature per unit distance along the bar. For Figure 9.22, the temperature gradient is constant, and is given by

$$\frac{(\theta_1 - \theta_2)}{L}$$

where L is the bar length.

Now consider the same bar without insulation along its sides. Assume the same end temperatures θ_1 and θ_2 as before. Heat energy is lost from its sides. So the heat energy per second reaching the colder end is less than the heat energy per second which flows from the hot end.

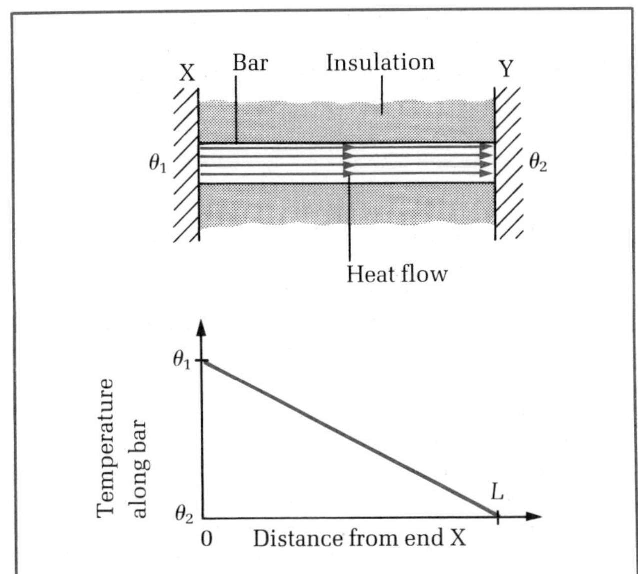

Figure 9.22 *Heat flow along a uniform insulated bar*

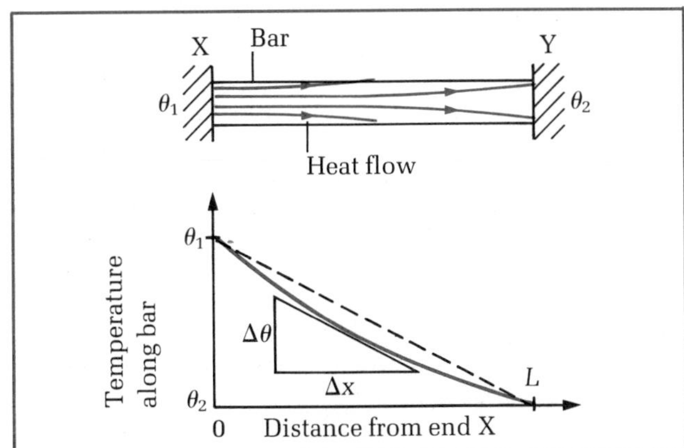

Figure 9.23 *Heat flow along a uniform non-insulated bar*

Figure 9.23 shows heat flow lines which spread out from the hot end. The temperature does not change uniformly with distance in this case. For two positions along the bar separated by distance Δx, the average temperature gradient between the two positions is $\frac{\Delta\theta}{\Delta x}$ where $\Delta\theta$ is the temperature difference between the two positions.

For two very close positions, Δx is very small and $\frac{\Delta\theta}{\Delta x}$ is now the gradient of the temperature curve shown in Figure 9.23. The curve is steepest at the hot end, so the temperature fall is steepest at that end.

The heat flow along an insulated uniform bar depends on

a) the temperature gradient $(\theta_1 - \theta_2)/L$ along the bar,
b) the area of cross-section of the bar,
c) the material of the bar.

To measure heat flow, the heat energy Q conducted along the bar in time t must be measured. The heat energy per second passing along the bar (i.e. the heat flow) is then given by Q/t. We measure water flow in the same way; for volume V of water passing along a pipe in time t, the flow rate is V/t. Figure 9.24 shows heat flow along several uniform insulated bars of the same material. Suppose bar 1 has length L, area of cross-section A and temperature difference 100 K (=100°C) across its ends. Let the heat flow be represented by Q/t for bar 1. Bar 2 has the same area and temperature difference, but its length is 2L. So the temperature gradient of bar 2 is 0.5 × the temperature gradient of bar 1. What is the heat flow along bar 2? With half the temperature gradient of bar 1, the heat flow is half that of bar 1, i.e. $Q/2t$. Bar 3 has twice the area of bar 1, other factors being equal. So bar 3 is equivalent to two 'bar 1s' side by side. The heat flow along bar 3 is therefore twice the heat flow along bar 1. What about bar 4's heat flow? With a temperature difference across its ends of 200 K (=200°C), area 2A and length 2L, show that the heat flow is just twice that of bar 1.

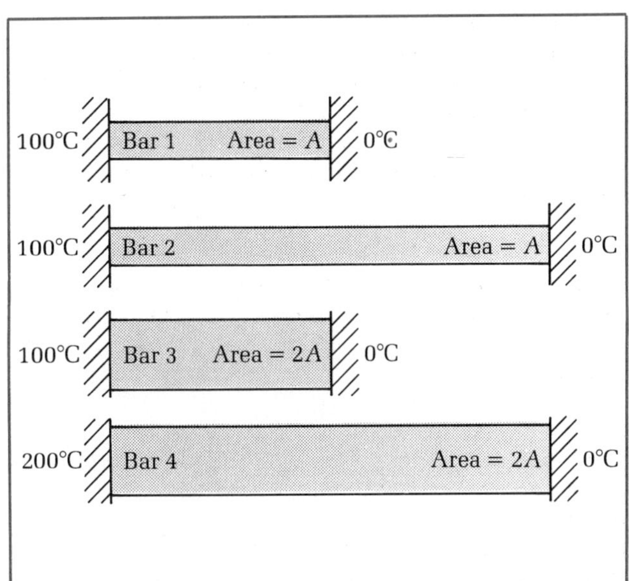

Figure 9.24 *Factors affecting heat flow*

From the ideas above, you ought to be able to see that the heat flow along an insulated uniform bar is proportional to

a) the temperature gradient,
b) the area of cross-section.

Therefore we can write an equation for the heat flow in terms of a constant of proportionality, k, which is called the **thermal conductivity** of the material.

$$\frac{Q}{t} = kA\frac{(\theta_1 - \theta_2)}{L}$$

where Q = heat energy conducted in time t,
$\theta_1 - \theta_2$ = temperature difference between the ends of the bar ($\theta_1 > \theta_2$),
A = cross-sectional area,
L = length of bar.

The equation is sometimes referred to as Fourier's law. The unit of k can be deduced from the equation. Q/t is in watts, A is in square metres, $(\theta_1 - \theta_2)$ in kelvins and L is in metres. Since $k = \frac{Q/t}{A(\theta_1 - \theta_2)/L}$, its unit is $\frac{\text{watts}}{\text{metre}^2 \times \text{kelvins/metres}}$ which simplifies to become $\text{W m}^{-1}\text{K}^{-1}$.

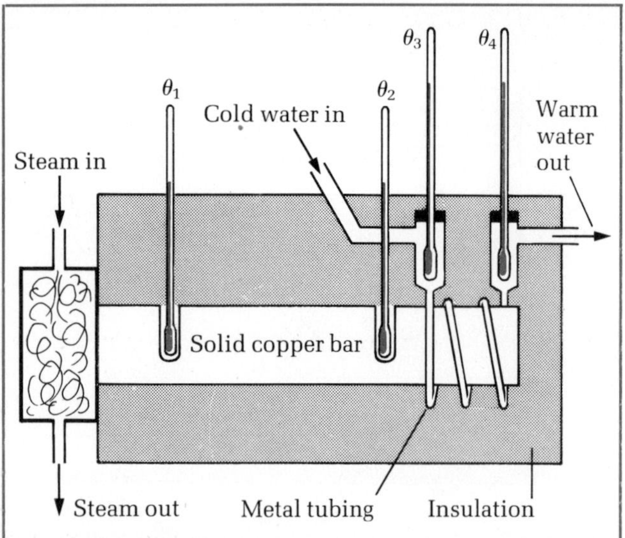

Figure 9.25 *Measuring* k

Worked example Searle's apparatus for measuring the thermal conductivity of copper is shown in Figure 9.25. One end of an insulated solid copper bar is in steam at 100°C. The other end is cooled by water piped through metal tubing wrapped round that end. Water flows steadily through the metal tubing, removing heat conducted along the bar. The temperature of the water at the inlet and at the outlet to the pipes is measured. The temperature gradient along the bar is determined by measuring the temperatures at two positions along the bar. At each position, a thermometer is placed in a small hole drilled into the bar.

The following measurements were made when the temperatures had become steady.
Temperatures along the bar: $\theta_1 = 74°C$, $\theta_2 = 55°C$
Water temperatures: at the inlet $\theta_3 = 16°C$ and at the outlet $\theta_4 = 25°C$
Mass of water collected at the outlet in 60 s = 0.150 kg
Bar diameter = 50 mm
Distance from θ_1 to θ_2 = 150 mm

Use this information to calculate the thermal conductivity of copper. The specific heat capacity of water is 4200 J kg^{-1} K^{-1}.

Solution The temperature gradient along the bar

$$= \frac{(\theta_1 - \theta_2)}{L} = \frac{74 - 55}{0.15} = 127\ \mathrm{K\,m^{-1}}$$

The area of cross-section $= \dfrac{\pi(\text{diameter})^2}{4}$

$$= \frac{\pi \times (5 \times 10^{-2})^2}{4} = 1.96 \times 10^{-3}\ \mathrm{m^2}$$

The heat energy/second conducted = heat gained per second by the water. That is, $Q/t = \dfrac{m}{t}c(\theta_4 - \theta_3)$ where m is the mass of water flowing in time t through the cooling pipes.

Therefore $Q/t = \dfrac{0.150}{60} \times 4200 \times (25 - 16) = 94.5\ \mathrm{W}$.

Using $Q/t = kA(\theta_1 - \theta_2)/L$ gives the value of k.

$$94.5 = k \times 1.96 \times 10^{-3} \times 127$$
$$k = 380\ \mathrm{W\,m^{-1}\,K^{-1}}.$$

Thermal resistance

The equation for k can be likened to $V = IR$ for electricity. Heat flow requires a temperature difference. An electric current requires a potential difference. To calculate heat flow we use $Q/t = kA\ (\theta_1 - \theta_2)/L$. To calculate electric current we use $I = V/R$. Clearly Q/t and I are both flow quantities, produced by $(\theta_1 - \theta_2)$ and V respectively, which drive the flow. So the quantity kA/L is the thermal equivalent of $1/R$. In other words we may define the *thermal resistance* of a conductor from the comparison. Thermal resistance of a thermal conductor, R_{th}, is given by

$$R_{th} = \frac{L}{kA}$$

where L = conductor length,
A = area of cross-section,
k = thermal conductivity.

The unit of thermal resistance is K W^{-1} since $R_{th} = \dfrac{(\theta_1 - \theta_2)}{(Q/t)}$ just as $R = \dfrac{V}{I}$.

Equivalent terms for thermal and electrical conduction are shown below.

Thermal conduction	**Electrical conduction**
Heat energy/second conducted Q/t	Current I
Temperature difference $(\theta_1 - \theta_2)$	Potential difference V
Thermal resistance L/kA	Electrical resistance R
$Q/t = kA(\theta_1 - \theta_2)/L$	$I = V/R$

Worked example Calculate the heat energy per second conducted through a closed window of height 1.0 m and width 1.5 m containing a single glass pane of uniform thickness 4.0 mm, with a 20 K temperature difference across the glass. The thermal conductivity of glass is 0.12 W m^{-1} K^{-1}.

Solution

Temperature gradient $\dfrac{(\theta_1 - \theta_2)}{L} = \dfrac{20}{4 \times 10^{-3}}$
$= 5 \times 10^3\ \mathrm{K\,m^{-1}}$.

Window area $A = 1 \times 1.5 = 1.5\ \mathrm{m^2}$.

So heat energy/second conducted through the window is given by

$$Q/t = kA\frac{(\theta_1 - \theta_2)}{L} = 0.12 \times 1.5 \times 5 \times 10^3 = 900\ \mathrm{W}.$$

Thermal conductors in series are two or more materials which are in contact so that heat energy passes through one then through the next. So the heat flow passing through them is the same. The same idea applies to electrical conductors in series; two such conductors pass the same current. Using thermal resistances, we can treat thermal conductors in series just as we treat electrical conductors in series. Consider the following examples.

Electrical circuit The circuit in Figure 9.26 shows a 6 ohm resistor in series with a 4 ohm resistor and a 10 V battery of negligible internal resistance.

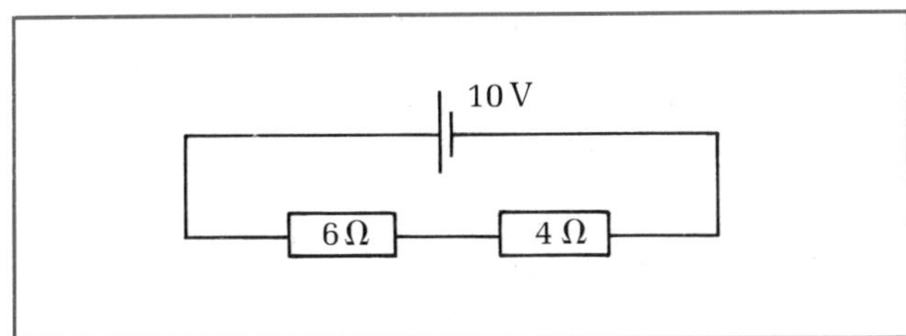

Figure 9.26 *Electrical resistors in series*

Total circuit resistance $=6+4=10$ ohms.

$$\text{Current}=\frac{V}{R}=\frac{10}{10}=1\text{ A}.$$

Thermal circuit The diagram below shows two insulated metal conductors joined end to end. The area of cross-section of each is 1 cm^2. For conductor X, $k=300\text{ W m}^{-1}\text{K}^{-1}$ and $L=1.8$ m. For conductor Y, $k=400\text{ W m}^{-1}\text{K}^{-1}$ and $L=1.6$ m. A temperature difference of 100 K is maintained across the free ends.

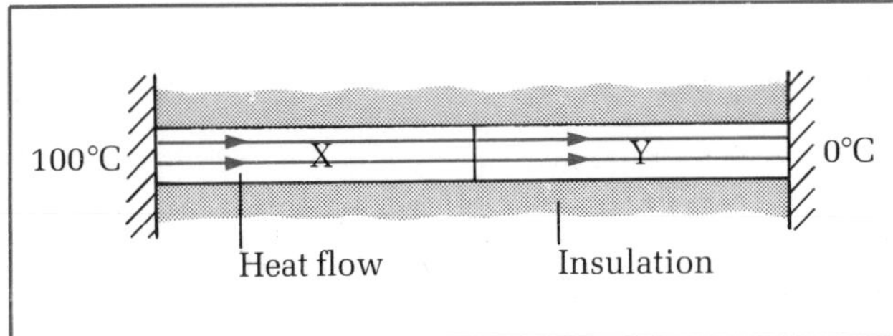

Figure 9.27 *Thermal resistors in series*

$$\text{Thermal resistance of X}=\frac{1.8}{300\times1\times10^{-4}}=60\text{ K W}^{-1}.$$

$$\text{Thermal resistance of Y}=\frac{1.6}{400\times1\times10^{-4}}=40\text{ K W}^{-1}.$$

Total thermal resistance $=60+40=100\text{ K W}^{-1}$.

$$\text{Heat energy/second } Q/t=\frac{\text{temperature difference}}{\text{total thermal resistance}}$$

$$=\frac{100}{100}=1\text{ W}.$$

Worked example A window of height 1.0 m and width 1.5 m contains a double glazed unit consisting of two single glass panes, each of thickness 4.0 mm separated by an air gap of 70 mm. Calculate the heat energy per second conducted through the window when the temperature difference across the unit is 20 K. The thermal conductivity of glass is $0.12\text{ W m}^{-1}\text{K}^{-1}$, and of air is $0.025\text{ W m}^{-1}\text{K}^{-1}$.

Solution The glass panes and the air gap are thermal resistors in series. The two glass panes can be treated as a single pane of thickness 8.0 mm for the purpose of thermal resistance calculations.

Thermal resistance of glass panes

$$=\frac{L}{kA}=\frac{8\times10^{-3}}{0.12\times1.5}=4.44\times10^{-2}\text{ K W}^{-1}.$$

Thermal resistance of the air gap

$$=\frac{L}{kA}=\frac{70\times10^{-3}}{0.025\times1.5}=1.87\text{ K W}^{-1}.$$

Total thermal resistance

$R_{th}=1.87+0.044=1.91\text{ K W}^{-1}$.

Heat energy/second conducted through the window is given by

$$Q/t=\frac{\text{temperature difference}}{\text{total thermal resistance}}=\frac{20}{1.91}=10.5\text{ W}.$$

In practice, thermal convection in the air gap and heat conduction through the frame would increase the heat flow. Nevertheless, the overall result is still a greatly reduced heat flow compared with a single pane.

9.6 Heating and insulation in the home

Fuel to heat your home costs money. One thing we can be quite sure of is that fuel costs will rise if fuel becomes scarce. Most of the fuel used in the home is for heating purposes. In winter, fuel bills are high because we use more fuel to keep warm. What happens to all the energy from the fuel? If homes were perfectly insulated, then fuel would only be needed to warm them up from cold. Once warmed up, no more fuel would be needed since there would be no heat losses. Alas, such an ideal home would be very expensive to construct even if it were technically possible. In real homes in winter, heat losses to the outside do occur, so we must use fuel to replace energy lost through heat transfer. However, heat transfer can be cut down by a variety of methods such as loft insulation, cavity wall insulation, double glazing or even closing the curtains!

Heat transfer mechanisms

There are three ways by which heat transfer occurs.

a) Conduction (See section 9.5 for details.)

b) Radiation (See section 9.4 for details.)

c) Convection occurs because hot fluid rises to displace cold fluid. So convection currents in the fluid carry energy away from the point of heating. Convection is produced when any fluid is heated because hot fluid has a lower density than cold fluid, due to thermal expansion. So hot fluid is pushed upwards by the surrounding cold fluid, causing convection currents. For example, hot air balloonists use a gas burner to create a rising current of hot air which fills the balloon, so making it float upwards.

Heat losses from buildings are caused in general by all three mechanisms. For example, most houses nowadays are built with cavity walls. A cavity wall has two layers of brick separated by a cavity, as in Figure 9.28. The inside surface layer is usually plastered as well. Heat transfer takes place through the cavity, through the brick and through the plaster. Conduction is responsible for heat

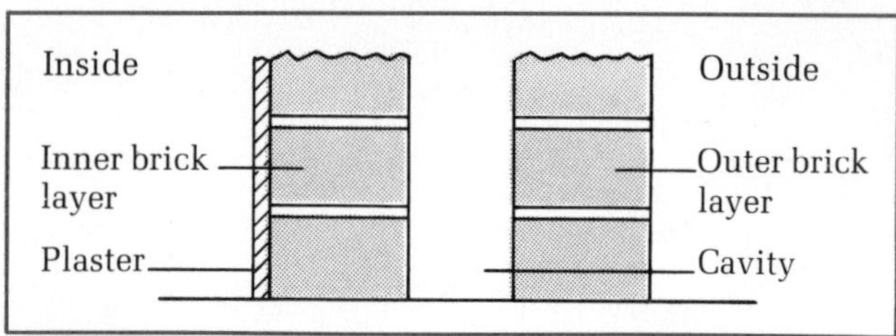

Figure 9.28 *Cavity wall*

transfer through the brick and plaster. The cavity acts as a thermal barrier reasonably effectively but conduction, convection and radiation all contribute to heat transfer across it. By filling the cavity with suitable insulating material radiation and convection can be eliminated, but conduction still occurs. The air near the inside and outside surfaces of the wall acts as a partly-insulating layer too! The outside surface is usually a little warmer than the surrounding air, so radiation and convection are responsible for heat transfer from the outside surface.

Double glazing also reduces heat losses. A double-glazed window consisting of two glass panes with an air layer between has a much greater thermal resistance than a single pane of glass. Figure 9.29 shows the variation of temperature from the inside to the outside across a double-glazed unit. The air near the inside and outside surfaces increases the effective thermal resistance since it lessens the temperature drop across the actual unit.

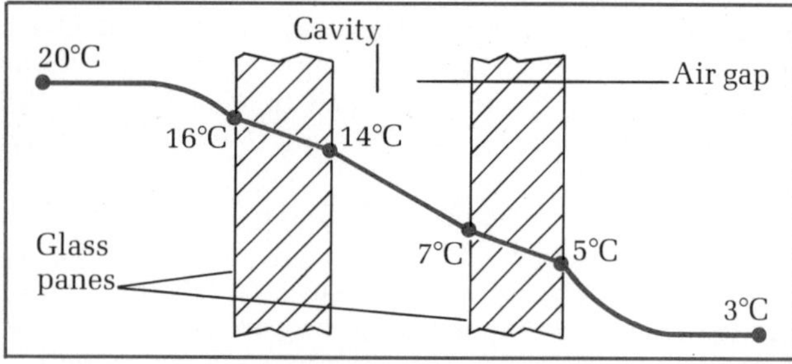

Figure 9.29 *A double-glazed unit*

Heat loss calculations

Because heat loss from buildings involves all three heat transfer mechanisms, heat loss calculations can be very complicated. Not only must thermal resistance due to the actual materials be taken into account, but so must the presence of cavities and the types of surfaces. Engineers use standard values for heat losses for each type of wall, window or roof. Floors too must be included. These standard values are worked out using known values of thermal conductivity or by direct measurements for convection and radiation. Values are always given for an area of 1 m² and a temperature difference of 1 K. So if the area and the temperature difference for any given surface are known, the heat loss per second can be calculated.

The thermal resistance coefficient of a material is the thermal resistance of unit area of the material. For material of thickness L and thermal conductivity k, its thermal resistance coefficient is L/k. This is because its thermal resistance is L/kA and since $A=1$ for unit area, then the thermal resistance of unit area is L/k. Its unit is therefore $\mathrm{m/(W\,m^{-1}\,K^{-1})}$ which equals $\mathrm{m^2\,K\,W^{-1}}$.

Consider the example of a single brick wall 96 mm thick. The thermal conductivity of brick is $0.84\,\mathrm{W\,m^{-1}\,K^{-1}}$. So the thermal resistance coefficient of the wall is $96\times10^{-3}/0.84=0.11\,\mathrm{m^2\,K\,W^{-1}}$.

Surfaces and cavities may also need to be taken into account. As Figure 9.29 shows, the air temperature near a surface varies with distance from the surface. So we need to know the equivalent thermal resistance of each surface. Likewise, a cavity acts as a partial thermal barrier so we need to know the equivalent thermal resistance of each cavity. For example, the surface resistance coefficient of a house wall is typically $0.13\,\mathrm{m^2\,K\,W^{-1}}$ for the inside surface, and $0.06\,\mathrm{m^2\,K\,W^{-1}}$ for the outside surface. Therefore a house wall of single brick has a total thermal resistance coefficient of 0.11 (brick) + 0.13 (inside) + 0.06 (outside) $=0.30\,\mathrm{m^2\,K\,W^{-1}}$. The thermal resistance of 1 m² of the wall is therefore $0.30\,\mathrm{K\,W^{-1}}$. The individual resistances are added together because they are in series with one another. They all pass the same heat flow. So a temperature difference across the wall of 10 K produces a heat flow equal to $10/0.30=33$ W for 1 m². Figure 9.30 compares the heat flow through a single brick wall with the heat flow through a cavity wall.

For the cavity wall of Figure 9.30, we need to know the cavity resistance coefficient as well as the other values. A typical cavity resistance coefficient is $0.25\,\mathrm{m^2\,K\,W^{-1}}$. So the total thermal resistance coefficient is 0.13 (inside) + 0.11 (inner brick layer) + 0.25 (cavity) + 0.11 (outer brick layer) + 0.06 (outside) $=0.66\,\mathrm{m^2\,K\,W^{-1}}$. So the thermal resistance of 1 m² of the wall is $0.66\,\mathrm{K\,W^{-1}}$. A 10 K temperature difference produces a heat flow equal to $10/0.66=15$ W for each m² of wall.

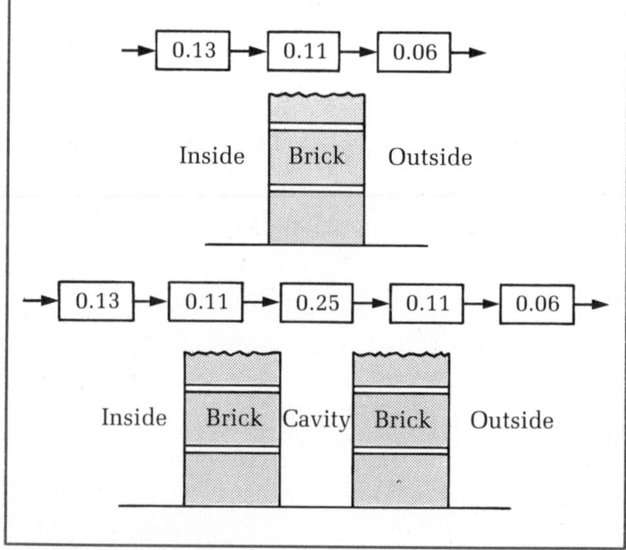

Figure 9.30 *Thermal resistances*

The U-value or thermal transmittance of a wall (or window or roof or floor) is the heat flow per m² produced by a temperature difference of 1 K. The U-value of the cavity wall of Figure 9.30 is therefore $1.5\,\mathrm{W\,m^{-2}\,K^{-1}}$. Given the U-value and area of each type of wall of a building, the total heat flow for any known temperature difference can be calculated. U-values for most types of walls, windows, roofs and floors are listed in reference books.

	Cavity brick wall at No. 1	at No. 3	Tiled roof at No. 1	at No. 3	Windows at No. 1	at No. 3	Floor
U-value/ $W\,m^{-2}\,K^{-1}$	0.6	1.6	0.6	1.9	3.2	4.3	0.5
Area/m^2 (of each)	136		60		12		48
Heat loss/ second/W for a 1 K temperature difference		218		116		51	24

Figure 9.31 *Home insulation*

Let's look at the example shown in Figure 9.31. The occupants of No. 1 have the benefit of loft insulation, cavity wall insulation and double glazing. None of these benefits is installed at No. 3. The heat losses can be compared using the figures above.

See if you can work out for yourself the heat loss/second from No. 1 for a 1 K temperature difference. The figures for No. 3 are worked out for you. The total heat loss/second from No. 3 for a 1 K temperature difference is $24+218+116+51=409$ W. So in winter, if the temperature difference between the inside and outside is 10 K, No. 3 loses $409\times10=4090$ W. Over a whole day, No. 3 would need to use 4.09×24 units of electricity ($=4.09$ kW $\times$ 24 hours) if heated electrically. At 5p per unit, No. 3 would pay £4.91 per day just to keep warm in winter. How much would No. 1 pay? Perhaps you can make an estimate for your own home.

9.7 Summary

Definitions

1 *Specific heat capacity* is the heat energy required to raise the temperature of 1 kg of material by 1 K.
2 *Specific latent heat capacity* is the heat energy to change the state of 1 kg of material without change of temperature.
3 *Thermal conductivity* is the heat per second conducted through an area of 1 m^2 by a temperature gradient of 1 K m^{-1} normal to the area.

Equations

1 Temperature on a centigrade scale
$$\theta=\left(\frac{X-X_0}{X_{100}-X_0}\right)\times100^\circ$$
2 Absolute temperature $T=\frac{(pV)}{(pV)_{Tr}}\times273.16$ K
3 Celsius temperature $t=T-273.15$°C
4 Heat supplied to raise temperature of mass m by $\Delta\theta$, $Q=mc\Delta\theta$
5 Heat supplied to change state of mass m, $Q=ml$
6 Energy required to change state of N molecules $=\frac{1}{2}Nn\varepsilon$
7 Thermal expansion of length $L=\alpha L\Delta\theta$
8 Wien's law: $\lambda_p\times T=$ constant
9 Stefan's law: energy/second radiated from a black body$=\sigma AT^4$
10 Fourier's law: $Q/t=kA(\theta_1-\theta_2)/L$
11 Thermal resistance$=L/kA$
12 Thermal resistance coefficient = thermal resistance of 1 m^2
13 U-value (or thermal transmittance) = heat flow/m^2 produced by a temperature difference of 1 K

Short questions

9.1 A constant volume gas thermometer and a thermocouple thermometer are used together to measure the boiling point of a certain liquid X. The readings of the two thermometers at ice point, steam point and the boiling point of X are given below.

	Gas thermometer	Thermocouple thermometer
Ice point	101 kPa	0 mV
Steam point	138 kPa	5.4 mV
Boiling point of X	124 kPa	3.4 mV

Calculate
a) the boiling point of X on the Celsius scale,
b) its boiling point on the thermocouple's centigrade scale.

9.2 A resistance thermometer has a resistance of 25.40 ohms at ice point, 27.34 ohms at steam point and 26.95 ohms at the melting point of a certain solid. Calculate the temperature of its melting point on the centigrade scale of the resistance thermometer.

9.3 State, with reasons, the type of thermometer you would choose to measure the temperature of
a) air as it is exhaled,
b) the engine block of a car,
c) a liquid at its boiling point, which is about 110°C.

9.4 The resistance of a thermistor over a limited range of temperature is given by the equation $R = c/(T - 203)$ where c is a constant and T is the absolute temperature. What would be the temperature on the centigrade scale of the thermistor at absolute temperature $T = 300$ K?

9.5 The brake linings of the wheels of a car have total mass 4.8 kg and specific heat capacity 1200 J kg^{-1} K^{-1}. Calculate the maximum possible temperature rise of the brake linings when the car (of mass 800 kg) travelling at 15 m s^{-1} is brought to rest by applying the brakes.

9.6 Calculate the heat energy required to convert 0.10 kg of ice at −10°C into steam at 100°C. The specific heat capacities of ice and water are 2100 and 4200 J kg^{-1} K^{-1} respectively; the specific latent heat of fusion of ice is 3.25×10^5 J kg^{-1} and the specific latent heat of vaporization of water is 2.25×10^6 J kg^{-1}.

9.7

a) A 3 kW electric kettle is filled with 1.6 kg of water at 20°C. Calculate the least possible time for the water to be brought to the boil. Use the value of specific heat capacity given in question **9.6**.

b) If the kettle in **a)** requires 0.5 kg of water to cover the heating element, estimate the time it would take to boil away enough water at 100°C to expose the element to air. Use the value of specific latent heat capacity given in question **9.6**.

9.8 A certain solid X is heated in an insulated container at a constant rate. The temperature of the material varies as shown in the diagram. If the specific heat capacity of the solid is 1800 J kg^{-1} K^{-1}, calculate

a) the specific latent heat of fusion of the solid,

b) the specific heat capacity of liquid X.

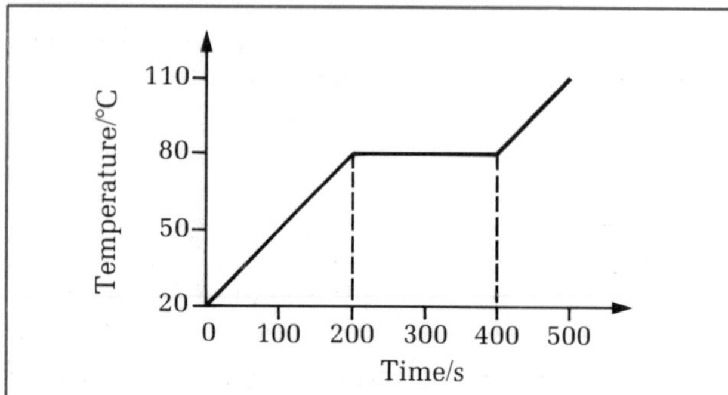

Figure 9.32

9.9 A hot water radiator has an inlet temperature of 60°C and an outlet temperature of 48°C maintained by water flowing at a rate of 1.1×10^{-5} m^3 s^{-1} through the radiator. Calculate the heat loss per second from the radiator, assuming it is at steady temperature. The density of water is 1000 kg m^{-3}. The specific heat capacity of water is 4200 J kg^{-1} K^{-1}.

9.10 A well-lagged calorimeter of mass 120 g contains 200 g of water and 50 g of ice, initially at 0°C. A jet of steam is blown through the water until the water temperature reaches 30°C. Calculate the mass of steam that must have condensed. The specific heat capacity of copper is 380 J kg^{-1} K^{-1}. Other values are as given in question **9.6**.

9.11 An undersoil heating system at a football stadium uses electric heating cables below ground to stop the pitch freezing in winter. Suppose the pitch temperature suddenly falls to −5°C to a depth of 0.04 m overnight when the heating system is off. Estimate the heat energy required to heat to 0°C each m^2 of pitch, assuming the specific heat capacity and density of soil values are 1500 J kg^{-1} K^{-1} and 2500 kg m^{-3} respectively.

9.12 The filament of a 1 kW electric heater is made using 20 m of tungsten wire of diameter 0.7 mm wrapped into many turns. Calculate

a) the surface area of the filament,

b) its surface temperature when operating normally, assuming it is a black body radiator.
(The Stefan constant $= 5.7 \times 10^{-8}$ W m^{-2} K^{-4}.)

9.13 A student working in a supermarket goes into the cold room, where the temperature is −20°C, to obtain some goods. Estimate the heat loss per second from the student after just entering the cold store from an outer room where the temperature is 15°C. The value of the Stefan constant is given in question **9.12**.

9.14 Determine the heat loss per hour through the brick wall of a house; the wall is of thickness 30 cm and area 20 m^2 and the outside and inside temperatures are 10°C and 0°C respectively.
(The thermal conductivity of brick = 0.4 W m^{-1} K^{-1}.)

9.15 If the wall of question **9.14** is covered with a layer of plaster 10 mm thick on the inside surface, what will the heat loss per hour now be for the same temperature difference?
(The thermal conductivity of plaster = 0.0040 W m^{-1} K^{-1}.)

9.16 From questions **9.14** and **9.15**, calculate the thermal resistance of the brick and of the plaster. Hence calculate the interface temperature.

9.17 A composite bar of diameter 10 mm is insulated along its sides as shown. Calculate the thermal resistance of each section, and hence determine the interface temperature. The thermal conductivity values are: iron, 40 W m^{-1} K^{-1}; copper, 360 W m $^{-1}$k^{-1}.

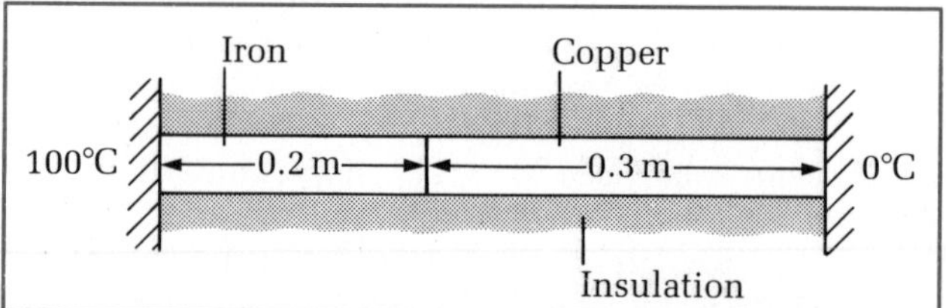

Figure 9.33

9.18 The ice on a pond is 10 mm thick when the air immediately above the ice is at a temperature of 263 K. Calculate

a) the rate of heat transfer through the ice (assume the water immediately below the ice is at 273 K),

b) the rate at which the ice thickness increases.
The thermal conductivity of ice is 2.3 W m^{-1} K^{-1}, the density of water is 1000 kg m^{-3}, and the specific latent heat of fusion of water is 3.25×10^5 J kg^{-1}.

9.19

a) Calculate the thermal resistance of a window pane of thickness 4 mm and size 1.5×2.0 m, given the thermal conductivity of glass is 1 W m^{-1} K^{-1}. What is the value of the thermal resistance coefficient of the window glass?

b) Given the internal and external surface resistance coefficients of the window glass are 0.06 and 0.13 m^2 K W^{-1}, calculate the U-value of the window.

9.20 A double-glazed window unit of size 1.5×2.0 m uses 2 glass panes, each of thickness 4 mm, separated by a cavity with resistance coefficient 0.14 m^2 K W^{-1}. Calculate the U-value of the unit.

10
Gases

10.1 Ideal gases

Boyle's law

For a fixed mass of gas at constant temperature, its pressure is inversely proportional to its volume. Robert Boyle, in the 17th century, was the first person to establish the link between gas pressure and volume. Imagine a gas-filled cylinder with a piston capable of compressing the gas, as in Figure 10.1. When the gas volume is reduced (i.e. the gas is compressed) its pressure increases. Reduce the volume of the gas to half its initial value and the pressure doubles. Boyle's law is usually stated as 'pressure × volume = constant' for a fixed mass of gas at constant temperature. So pressure $= \frac{\text{constant}}{\text{volume}}$ which is an equation for the statement that pressure is inversely proportional to the volume.

$$pV = \text{constant}$$

where p = gas pressure,
V = volume of fixed mass at constant temperature.

Boyle established the law by experiment. The measurements can be plotted to give curves of pressure against volume as in Figure 10.1. Each curve is given by measurements at constant temperature. Constant temperature curves are called **isothermals.** The temperature corresponding to each isotherm increases with distance from the origin. The measurements for any one isotherm of Figure 10.1 may be replotted on a graph of pressure against 1/volume to show a straight line passing through the origin. Figure 10.2 shows several isothermals of p against $1/V$. For a fixed mass of gas, the gradient of each

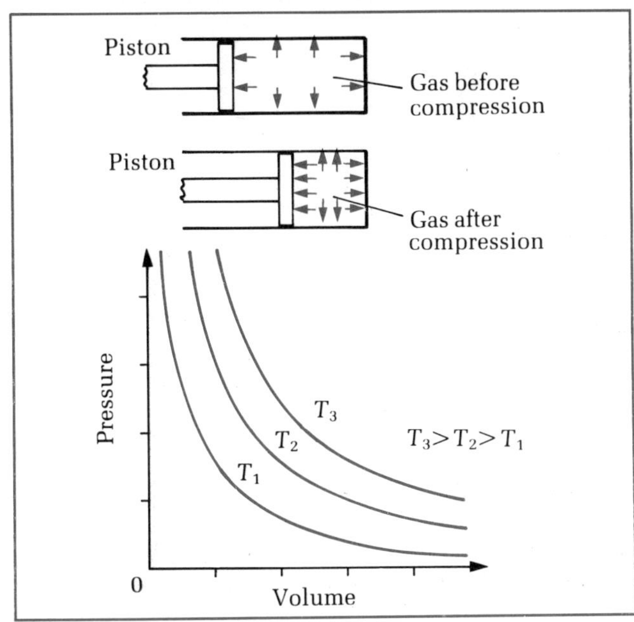

Figure 10.1 *Boyle's law*

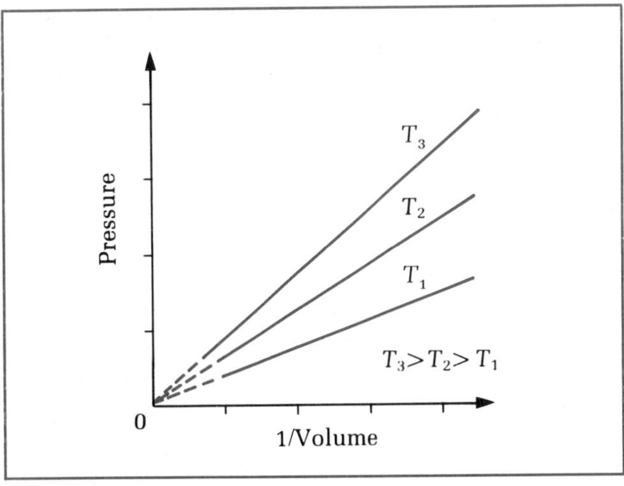

Figure 10.2 *Isotherms of p against 1/V*

isothermal increases with temperature. Each line of Figure 10.2 is represented by an equation of the form $p = \frac{\text{constant}}{V}$. The constant depends on mass and temperature in a way which is explained later.

Boyle's law is not obeyed by all gases or at all temperatures. Air is a mixture of oxygen and nitrogen, and both obey Boyle's law near room temperature, so air does too. But carbon dioxide gas near room temperature can be liquefied, so it clearly does not obey Boyle's law. Gases which obey Boyle's law are called **ideal gases**.

Charles' law

For a fixed mass of gas at constant pressure, its volume is proportional to its absolute temperature.

$$V = \text{constant} \times T$$

or $$\frac{V}{T} = \text{constant}$$

where T = absolute temperature,
V = volume of fixed mass at constant pressure.

The link between gas volume and temperature can be investigated using the arrangement shown in Figure 10.6. Measurements can be plotted on a graph of volume against temperature in °C. The graph is a straight line, as shown by Figure 10.3. The line intercepts the volume axis at V_0.

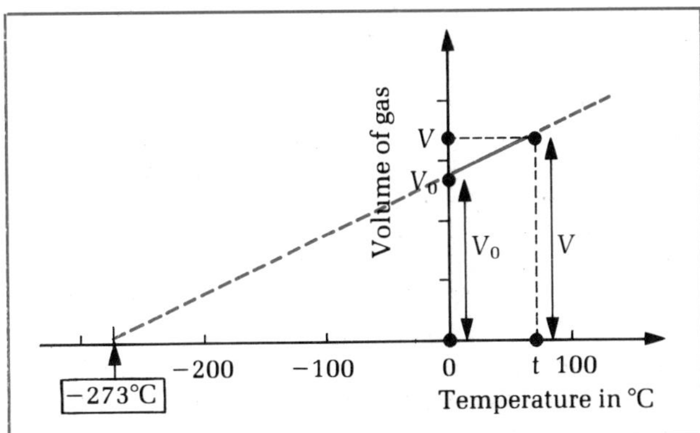

Figure 10.3 *Charles' law*

Charles discovered that all ideal gases intercept the temperature axis at the same point, which is $-273.15°C$ (see Figure 10.3). Clearly measurements on actual gases can be made only over a limited range of temperature. But the measurements give a straight line, and the line always points back to $-273.15°C$. The line for any ideal gas can therefore be represented by the equation

$$\frac{V}{(t+273)}=\frac{V_0}{273} \quad \text{where } t \text{ is the temperature in °C.}$$

Since absolute temperature T equals Celsius temperature $t+273$, then the equation may be written as

$$V/T=V_0/T_0$$

where $T_0 \approx 273$ K (273.15K, in fact).

Hence V/T always has the same value (for fixed mass at constant pressure).

Investigating the gas laws

Boyle's law The gas to be investigated is trapped at the top of a uniform vertical tube, as shown in Figure 10.4. Oil is used to trap the gas. To compress the gas, the valve is opened then a footpump is used to increase the pressure in the tube. The pressure is measured directly from a pressure gauge. Because the tube is uniform, the length of the trapped gas column can be used as a measure of its volume. A water bath round the tube keeps the gas temperature constant. The pressure is increased in steps, and at each step the volume is measured.

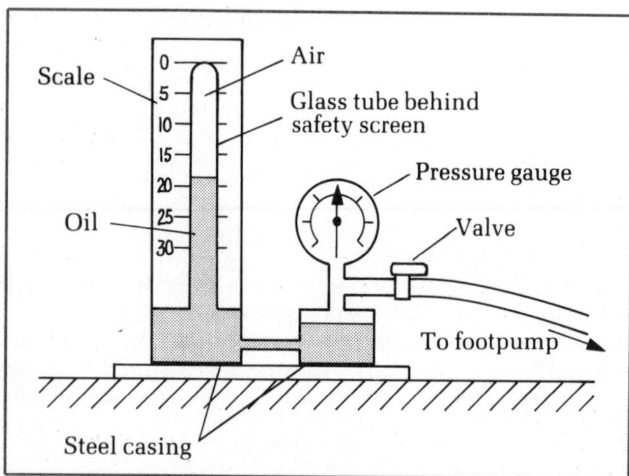

Figure 10.4 *Investigating Boyle's law*

The pressure law To investigate the link between gas pressure and temperature, a fixed mass of gas is heated up at constant volume. The gas, which must be dry, is in a glass bulb connected to a U-tube manometer or pressure gauge. The bulb is placed in a water bath which is heated to different temperatures. At each temperature, the gas pressure is measured from the gauge. A thermometer in the water is used to measure the water temperature which is assumed to be equal to the gas temperature. A graph of pressure against temperature in °C is a straight line, as shown by Figure 10.5. The line cuts the temperature axis at $-273°C$.

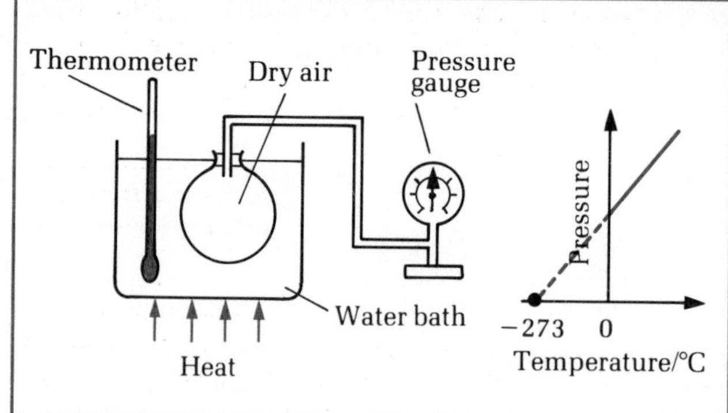

Figure 10.5 *Investigating the pressure law*

Charles' law Dry air is trapped in a uniform capillary tube by a thread of concentrated sulphuric acid. The acid keeps the air dry throughout. The tube is heated in a water bath. The trapped air temperature is assumed to be the same as the water temperature. Because the tube is uniform, the volume of the trapped air is proportional to the length of the air column. So the air column length is measured for different temperatures. The pressure of the trapped air stays constant throughout since the acid thread is acted on by constant pressure (i.e. atmospheric pressure) on its top side. The measurements can then be directly plotted as length of air column, l, against temperature.

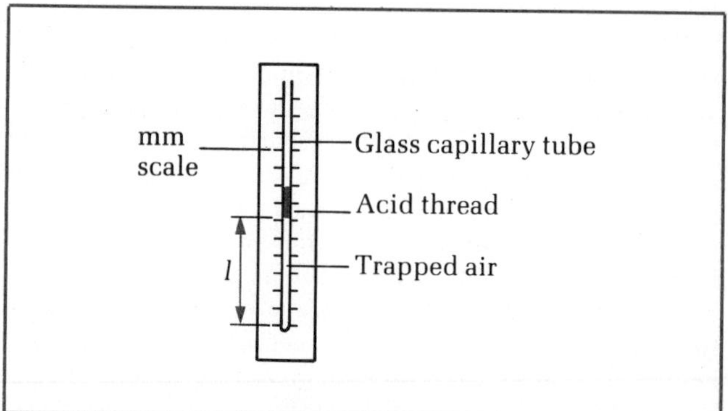

Figure 10.6 *Investigating Charles' law*

The ideal gas equation

The link between pressure and temperature for an ideal gas can be written as $\frac{p}{T}=$ constant for a fixed mass of gas at constant volume. So now we have three separate gas laws, each one to be used only in restricted circumstances (e.g. the temperature must be constant for Boyle's law to be used). The three laws are

Boyle's law $pV=$ constant
(For constant mass and temperature.)

Charles' law $\frac{V}{T}=$ constant

(For constant mass and pressure.)

Pressure law $\frac{p}{T}=$ constant

(For constant mass and volume.)

The combined gas law is a single equation which covers the separate gas laws. It is much more useful though because it can be used where pressure, volume and temperature all change, given that the mass of the gas remains constant.

$$\frac{pV}{T} = \text{constant}$$

If the temperature is constant, then 'pV = constant' results from the combined gas law. If the pressure is constant instead, then we have V/T is constant, which is Charles' law. If the volume is kept constant, then p/T is constant. So we can recover all three experimental gas laws from the combined gas law, and we can tackle problems where only the mass is constant.

Experiments show that the molar volume (i.e. volume of 1 mole) of any gas at standard pressure (= 101 kPa) at 0°C is always equal to 0.0224 m³. So for 1 mole of any gas, we can write the combined gas law as

$$\frac{pV}{T} = \frac{p_0V_0}{T_0}$$

where p_0 is standard pressure, $T_0 = 273$ K, and V_0 is always equal to 0.0224 m³. The value of p_0V_0/T_0 is therefore equal to $101 \times 10^3 \times 0.0224/273$ which equals 8.3 Pa m³ K⁻¹, provided the mass is equal to the molar mass. This value is called the **molar gas constant**, R. Its unit is usually given as J mol⁻¹ K⁻¹ (since 1 Pa = 1 N m⁻² giving 1 Pa m⁻³ = 1 N m = 1 J). So for 1 mole of any gas, we can write the combined gas law in the form $\frac{pV_m}{T} = R$, where V_m is the molar volume at pressure p and temperature T. In this form, the equation is called the **Ideal Gas equation**.

$$pV_m = RT$$

$$pV = nRT$$

where p = pressure,
T = absolute temperature,
V_m = volume of 1 mole,
V = volume of n moles,
R = molar gas constant.

For n moles of gas, the value of p_0V_0/T_0 is $8.3\,n$ J K⁻¹. For example, two moles of gas at standard pressure and 0°C would occupy a volume of 2×0.0224 m³.
So a calculation of p_0V_0/T_0 for 2 moles would give the value 2×8.3 J K⁻¹.

Worked example Calculate the density of air at 100°C and 200 kPa, given its density at 0°C and 101 kPa is 1.29 kg m⁻³.

Solution Let the number of moles in 1 kg of air = n.
The volume of 1 kg of air at 0°C = mass/density at 0°C = 1/1.29 = 0.78 m³.

At 0°C and 101 kPa, $p_0 = 101$ kPa,
$V_0 = 0.78$ m³ for 1 kg of air,
$T_0 = 273$ K.

At 100°C and 200 kPa, $p = 200$ kPa,
$T = 373$ K

V is to be calculated for a mass of 1 kg.

Using $\frac{pV}{T} = \frac{p_0V_0}{T_0}$ gives

$$\frac{200 \times 10^3 \times V}{373} = \frac{101 \times 10^3 \times 0.78}{273}$$

So $V = \frac{101 \times 373 \times 0.78}{200 \times 273} = 0.54$ m³.

Therefore, the density of air at 100°C and 200 kPa is mass/volume = 1/0.54 = 1.86 kg m⁻³.

Worked example Calculate the density of hydrogen gas at 20°C and 101 kPa, given the molar mass of hydrogen molecules is 2×10^{-3} kg. Assume the molar gas constant $R = 8.3$ J mol⁻¹ K⁻¹.

Solution Use $pV_m = RT$ to calculate the molar volume of hydrogen gas at 20°C and 101 kPa.
Hence $V_m = RT/p = 8.3 \times 293/101 \times 10^3 = 0.0241$ m³.

Therefore the density = mass/volume

$$\text{Density} = \frac{0.002}{0.0241} = 0.083 \text{ kg m}^{-3}.$$

10.2 The kinetic theory of gases

Molecules in gases

Gases are composed of molecules. If we could see the molecules in a gas, we would see that they are separated from one another by distances which are large compared with their actual sizes. Their speeds differ and they move about in random directions. They continually collide with one another and with the container walls. Of course, we cannot see gas molecules directly because they are far too small. But if we observe the **Brownian motion** of tiny smoke particles (see p. 64) we can see the effects of their collisions. Each smoke particle moves about at random because it is subjected to continual random impacts by molecules. Molecular speeds must be very high compared with the smoke particle speeds because the molecules are much less massive than the smoke particles.

A model of the molecules of a gas is shown in Figure 10.7. The picture shows a 'snapshot' of magnetic pucks moving about on a frictionless, level surface. With a magnetic strip round the sides, the pucks rebound off each other or off the sides without loss of kinetic energy. So the collisions are **elastic** because there is no loss of K.E.

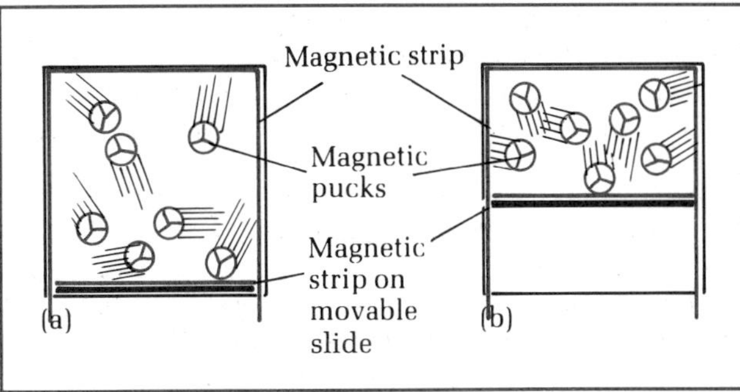

Figure 10.7 *A model of gas molecules*

Transfer of K.E. from one puck to another takes place when they collide, but the total K.E. is constant. The molecules in a gas move about in the same sort of way, making elastic collisions with one another and the container walls. Suppose we make the pucks move faster. The impacts on the sides become more forceful, and they happen more often. So the pressure on the sides increases. How can we make the molecules of a gas move faster? The answer is to raise the gas temperature; for a fixed volume, the pressure increases.

Boyle's law can also be explained. Suppose one of the sides could be moved inwards to reduce the space in which the pucks move. Because the average distance across the space is now shorter, the time taken for a puck to cross the space is less, so impacts on the sides are more frequent. Therefore the pressure increases as a result of reducing the amount of space.

The kinetic theory equation

To develop the explanation of the gas laws in molecular terms, we need to use some basic principles of mechanics. The pressure of a gas is due to lots of molecular impacts. Each impact of a molecule with a solid surface exerts a tiny force on the surface. Lots of impacts all over the surface average out to give a steady force on the surface. Hence a steady pressure is exerted on the surface. Assume the solid surface is smooth and flat, and consider the impact of one molecule with the surface. The molecule does not lose any K.E. as a result of hitting the surface so its speed after impact is unchanged. However its direction changes. Figure 10.8 shows an impact at angle θ to the normal. For speed U, the velocity before impact has components $U \sin\theta$ parallel to the surface and $U \cos\theta$ at right angles towards the surface. After the impact, the components are $U \sin\theta$ parallel to the surface (as before) and $U \cos\theta$ at right angles away from the surface. So the molecule's momentum keeps the same component parallel to the surface. But its component at right angles to the surface changes from $mU \cos\theta$ towards the surface to $mU \cos\theta$ away from the surface. Therefore the change of the molecule's momentum is $2mU \cos\theta$. This is because we must use a direction code '− for towards: + for away'. So the change of momentum is $mU \cos\theta - (-mU \cos\theta) = 2mU \cos\theta$, at right angles to the surface. To change the momentum, a force is needed. So we can use the change of momentum to give an expression for the force.

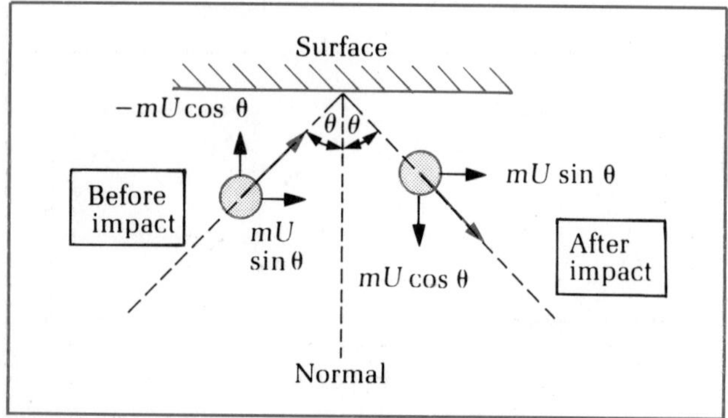

Figure 10.8 *Momentum changes*

To take the ideas further, we need to make some more assumptions. A list of assumptions is helpful here.

Point molecules This avoids the need to consider the size of different types of molecules. In practice, if the volume of the space is much greater than the actual volume of the molecules, then the molecules are effectively points in the space.

No intermolecular attractions If intermolecular attractions are significant, then impacts on solid surfaces are reduced in force. The molecules away from the surface hold back any molecules moving towards the surface, so the impact force is reduced. Provided the gas is not at high density or low temperature, then the assumption is valid.

Random motion The average of many random impacts gives a smooth pressure.

Elastic collisions If the collisions were inelastic, the K.E. of the molecules would be converted into other forms of energy. The gas pressure would fall. Since this does not happen, we can be confident that the collisions are elastic.

Impact time Force equals change of momentum/time taken. How long does an impact last? Provided it is much shorter than the time between impacts, then change of momentum/time between impacts gives the average force.

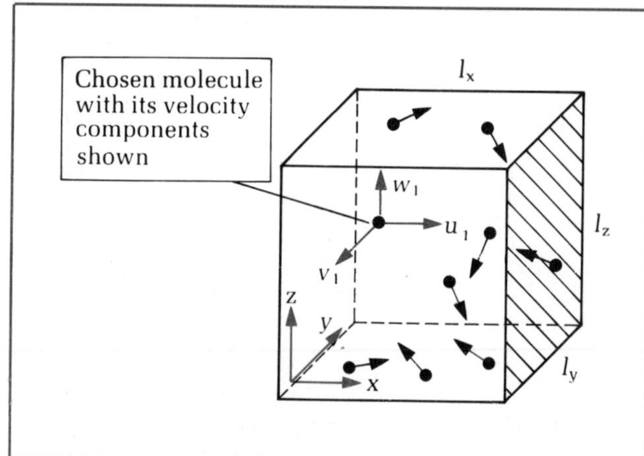

Figure 10.9 *Molecules in a box*

Now consider N gas molecules in a rectangular box with sides of length l_x, l_y, l_z as shown in figure 10.9. With the assumptions above, let's suppose that each molecule bounces from side to side, with little chance of hitting another gas molecule. The path of one molecule is shown in the figure. Let its velocity components in the x, y and z directions be u_1, v_1 and w_1 respectively. So its speed c_1 is given by the rule for adding perpendicular components.

$$c_1 = \sqrt{u_1^2 + v_1^2 + w_1^2}$$

Our chosen molecule hits the shaded face every time it has travelled a distance $2l_x$ along the x-direction (i.e. the distance back and forth). Its velocity component along the x-direction is $+u_1$ one way and $-u_1$ the opposite way. Its progress in the x-direction is at speed u_1, so the time between successive impacts on the shaded face is $t = \text{distance/speed} = 2l_x/u_1$.

Each impact of our chosen molecule on the shaded face causes only the x-component of momentum to change. It changes from $-mu_1$ to $+mu_1$, where m is the mass of the molecule. So the change of momentum is $2mu_1$, and this occurs every time the molecule passes back and forth. Since force = change of momentum/time taken, the average impact force on the shaded face is $F_1 = \frac{2mu_1}{(2l_x/u_1)} = \frac{mu_1^2}{l_x}$.

The average pressure of our chosen molecule on the shaded face is given by $p_1 = F_1/(l_y l_z)$ since the face area is $l_y l_z$. Hence $p_1 = \frac{mu_1^2}{l_x l_y l_z} = \frac{mu_1^2}{V}$ where V, the gas volume, equals $l_x l_y l_z$.

We can now choose each molecule in turn and write an equation for the pressure of each one on the shaded face. Each equation is slightly different because the x-components of velocity differ.

Pressure of first molecule $p_1 = mu_1^2/V$
Pressure of second molecule $p_2 = mu_2^2/V$
⋮
Pressure of Nth molecule $p_N = mu_N^2/V$

assuming each molecule's mass is m.

The total pressure of all N molecules on the shaded face is therefore given by $p = p_1 + p_2 + p_3 + \ldots + p_N$.
Hence

$$p = \frac{mu_1^2}{V} + \frac{mu_2^2}{V} + \frac{mu_3^2}{V} + \ldots + \frac{mu_N^2}{V}$$

$$= \frac{m}{V}(u_1^2 + u_2^2 + u_3^2 + \ldots + u_N^2).$$

We can write the equation for p in a more convenient form as follows:

$$p = \frac{Nm}{V}\langle u^2\rangle \text{ where } \langle u^2\rangle = \frac{u_1^2 + u_2^2 + \ldots + u_N^2}{N}$$

$\langle u^2\rangle$ is the mean value of (the x-components of velocity)^2. So we have an equation for the pressure p in terms of $\langle u^2\rangle$. Because the motion is random we could have obtained the equation for p in terms of $\langle v^2\rangle$ or $\langle w^2\rangle$.

$$\text{So } p = \frac{Nm}{V}\langle u^2\rangle = \frac{Nm}{V}\langle v^2\rangle = \frac{Nm}{V}\langle w^2\rangle$$

In other words $\langle u^2\rangle = \langle v^2\rangle = \langle w^2\rangle$.
But $\langle c^2\rangle = \langle u^2\rangle + \langle v^2\rangle + \langle w^2\rangle$
since the speed of each molecule is given by the sum of the squares of its velocity components. Therefore

$$\langle u^2\rangle = \langle v^2\rangle = \langle w^2\rangle = \tfrac{1}{3}\langle c^2\rangle.$$

So the equation for pressure becomes

$$p = \frac{1}{3}\frac{Nm}{V}\langle c^2\rangle = \tfrac{1}{3}\rho\langle c^2\rangle$$

The gas density ρ = mass of all the molecules Nm/volume V. The equation for pressure in terms of $\langle c^2\rangle$ is called the **kinetic theory equation.**

$$pV = \tfrac{1}{3}Nm\langle c^2\rangle$$
$$p = \tfrac{1}{3}\rho\langle c^2\rangle$$

where p = gas pressure,
V = gas volume,
N = number of molecules,
m = mass of each molecule,
ρ = gas density (= Nm/V),
$\langle c^2\rangle$ = mean square speed of the gas molecules.

The root mean square (r.m.s.) speed of the gas molecules, $c_{r.m.s.}$, is defined as the square root of the mean square speed $\langle c^2\rangle$. For N molecules with individual speeds $c_1, c_2, c_3, \ldots c_N$, the r.m.s. speed of the molecules is therefore given by the equation

$$c_{r.m.s.} = \sqrt{\langle c^2\rangle} = \sqrt{\left\{\frac{c_1^2 + c_2^2 + \ldots + c_N^2}{N}\right\}}$$

For example, supose a box contains just five molecules, with speeds 300, 350, 420, 500 and 580 m s^{-1}. The mean square speed $\langle c^2\rangle$

$$= (300^2 + 350^2 + 420^2 + 500^2 + 580^2)/5 = 195\,060\ \text{m}^2\ \text{s}^{-2}.$$

So the r.m.s. speed is given by

$$c_{r.m.s.} = \sqrt{\langle c^2\rangle} = \sqrt{195\,060} = 441\ \text{m s}^{-1}.$$

The r.m.s. speed is not the same as the mean speed. Using the example of the box with five molecules, above, the mean speed of the five molecules is equal to (300 + 350 + 420 + 500 + 580)/5 = 430 m s^{-1}. The r.m.s. speed is important because the kinetic theory equation for pressure is in terms of $\langle c^2\rangle$ which is the same as $c_{r.m.s.}^2$. So the kinetic theory equation can be written with $c_{r.m.s.}^2$ in place of $\langle c^2\rangle$ (i.e. $p = \frac{1}{3}\rho c_{r.m.s.}^2$).

Worked example Calculate the r.m.s. speed of oxygen gas molecules at a pressure of 100 kPa when the density is 1.43 kg m^{-3}.

Solution Use $p = \frac{1}{3}\rho\langle c^2\rangle$ to calculate the mean square speed $\langle c^2\rangle$.
The r.m.s. speed is then given by $\sqrt{\langle c^2\rangle}$.
Hence $\langle c^2\rangle = \frac{3p}{\rho} = \frac{3 \times 100 \times 10^3}{1.43} = 2.10 \times 10^5\ \text{m}^2\ \text{s}^{-2}$.
Therefore $c_{r.m.s.} = \sqrt{\langle c^2\rangle} = 458\ \text{m s}^{-1}$.

The distribution of speeds is continuous for many molecules. Figure 10.10 shows the number of molecules in each speed interval. Each interval covers the same range of speeds. You could obtain a similar distribution curve

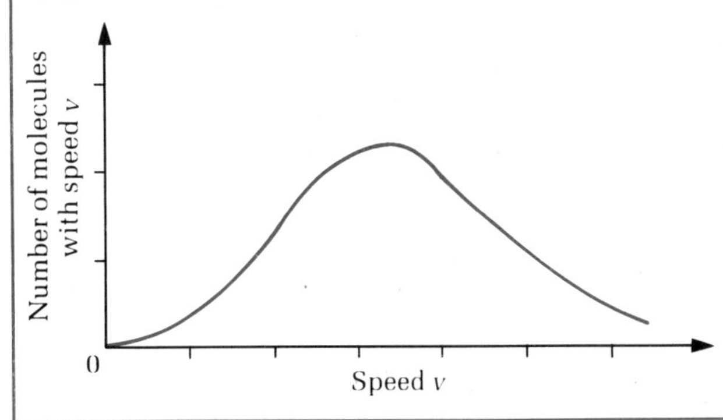

Figure 10.10 *Distribution of molecular speeds*

by measuring the speeds of lots of vehicles on a motorway, then count up the number of vehicles with speeds in each interval of say 5 mph. The first interval would be from 0 to 5 mph, the next from 5 to 10 mph, and so on. You would probably find the peak interval is 65 to 70 mph. If it is greater, inform the traffic police! The molecular distribution curve reaches a peak smoothly, then falls away at higher speeds smoothly. The r.m.s. speed is further along the speed axis than the peak because high-speed molecules contribute more to the r.m.s. value than low-speed molecules.

Kinetic energy

For N point molecules, each of mass m, the total kinetic energy is given by adding the K.E. of all N molecules. If the molecular speeds are $c_1, c_2, \ldots c_N$ then the molecular K.E.s are $\frac{1}{2}mc_1^2, \frac{1}{2}mc_2^2, \ldots \frac{1}{2}mc_N^2$. So the total kinetic energy is given by

$$\text{Total K.E.} = \tfrac{1}{2}mc_1^2 + \tfrac{1}{2}mc_2^2 + \ldots + \tfrac{1}{2}mc_N^2$$

$$= \tfrac{1}{2}Nm\left\{\frac{c_1^2 + c_2^2 + \ldots + c_N^2}{N}\right\}$$

$$= \tfrac{1}{2}Nm\langle c^2\rangle$$

Explanation of the ideal gas equation is now possible if we bring temperature into our scheme. The ideal gas equation is $pV = nRT$. The kinetic theory equation is $pV = \frac{1}{3}Nm\langle c^2\rangle$. If we assume that $Nm\langle c^2\rangle$ is proportional to the absolute temperature T (i.e. $Nm\langle c^2\rangle =$ constant $\times T$), then the kinetic theory equation becomes $pV =$ a constant $\times T$. The result is almost the same as the ideal gas equation. We can make it exactly the same by assuming $\frac{1}{3}Nm\langle c^2\rangle$ equals nRT. What does this assumption mean? The total K.E. of N gas molecules is $\frac{1}{2}Nm\langle c^2\rangle$ which is therefore $3nRT/2$. So the total K.E. is proportional to the absolute temperature.
The total K.E. of N molecules $\frac{1}{2}Nm\langle c^2\rangle = 3nRT/2$.
The mean K.E. of a gas molecule

$$= \frac{\text{total K.E. } (\frac{1}{2}Nm\langle c^2\rangle)}{\text{number of molecules } (N)}$$

$$= \tfrac{1}{2}m\langle c^2\rangle = \frac{3nRT/2}{N}$$

For n moles of gas, the total number of molecules N is equal to $n \times L$, where L is the Avogadro constant. Hence the mean K.E. of a gas molecule is therefore $\dfrac{3nRT/2}{nL}$ which equals $\dfrac{3RT}{2L}$. This expression can be written as $\frac{3}{2}kT$, where k $(= R/L)$ is called the Boltzmann constant. The value of k is $1.38 \times 10^{-23}\,\text{J K}^{-1}$ since $R = 8.3\,\text{J mol}^{-1}\text{K}^{-1}$ and $L = 6.02 \times 10^{23}\,\text{mol}^{-1}$.

$$\text{Mean K.E. of a gas molecule} = \tfrac{3}{2}kT$$

where k = Boltzmann constant,
T = absolute temperate.

In effect, we have 'proved' the ideal gas equation $pV = nRT$ from the kinetic theory equation $pV = \frac{1}{3}Nm\langle c^2\rangle$ by assuming the mean K.E. of a gas molecule $\frac{1}{2}m\langle c^2\rangle$ equals $\frac{3}{2}kT$. The ideal gas equation is based on experiments. The kinetic theory has been used above to derive the ideal gas equation. So our list of assumptions, on which the theory is based, tell us how molecules in an ideal gas behave. For a gas which does not obey the ideal gas equation, we can look again at the assumptions to see why they do not apply. Using the kinetic theory to derive the ideal gas equation is a very powerful example of the link between theory and experiment. Boltzmann developed the kinetic theory of gases and used it to explain and predict many properties of gases.

Avogadro's hypothesis can be explained using the kinetic theory equation. The hypothesis states that equal volumes of different gases at the same temperature and pressure contain equal number of molecules.

For equal volumes of two gases, A and B, at the same pressure

$$\tfrac{1}{3}N_A m_A\langle c_A^2\rangle = \tfrac{1}{3}N_B m_B\langle c_B^2\rangle$$

where subscripts A and B indicate the number of molecules, mass of each type of molecule, and mean square speed.

Since the temperature of A is the same as the temperature of B, then

$$\tfrac{1}{2}m_A\langle c_A^2\rangle = \tfrac{1}{2}m_B\langle c_B^2\rangle$$

Hence $N_A = N_B$. In other words, the number of molecules in A is equal to the number of molecules in B if equal volumes of each gas are at the same temperature and pressure.

Collisions between gas molecules do take place but have no effect on the pressure. When two gas molecules collide, the total K.E. of the two molecules after the collision is the same as before. K.E. is transferred from one molecule to the other as a result of the collison, but the total K.E. is not changed by the collision. The gas pressure is given by

$$p = \frac{Nm\langle c^2\rangle}{3V} = \frac{2 \times \text{the total K.E.}}{3V}$$

So the gas pressure is unchanged as a result of collisions between gas molecules because the total K.E. is not changed.

10.3 Diffusion

Gas molecules move at high speeds. Using $p = \frac{1}{3}\rho\langle c^2\rangle$ for air at room temperature, we can show that the r.m.s. speed of air molecules at room temperature is about $460\,\text{m s}^{-1}$. But if one gas is released into another gas, they mix very slowly. Figure 10.11 shows an arrangement for releasing bromine vapour into air. Bromine vapour is used because it is brown so its progress can be seen. A glass phial containing bromine liquid is placed in a sealed rubber tube connected to a large glass diffusion tube. The phial is cracked so that bromine liquid runs into the bottom of the diffusion tube when the tap is opened. Bromine vapour from the liquid spreads slowly into the air space

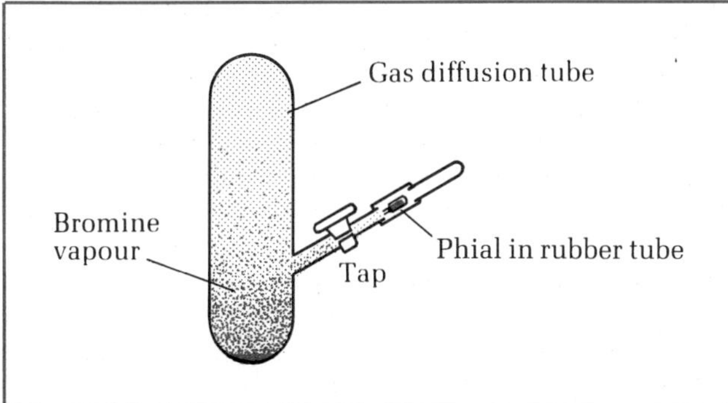

Figure 10.11 *Diffusion of bromine*

of the tube. The spreading process is an example of **diffusion**. Why is it so slow to spread when the molecules move at high speeds? If the air in the tube had been removed before starting, the vapour would have filled the tube in a fraction of a second. With air present, the process is much slower because the bromine molecules keep bumping into air molecules. If we could observe the path of a single bromine molecule, it would be a series of haphazard steps, as shown by Figure 10.12. The haphazard motion is due to the random motion of the air molecules. The bromine molecule collides, rebounds, then travels to hit an air molecule, then another, then another, etc.

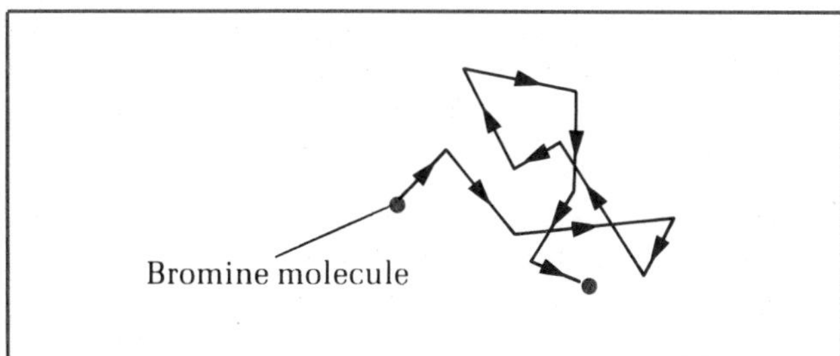

Figure 10.12 *Path of a bromine molecule*

The mean free path is the average distance, l which a molecule travels between successive collisions. The greater the density of molecules, the shorter the mean free path. For a gas containing N_v molecules per unit volume, each molecule has an average volume $1/N_v$ to move in. Now the volume swept out by each molecule between successive impacts, shown in Figure 10.13, is approximately $l\pi d^2$ where d is the molecular diameter. No other molecule must come into contact with the molecule which sweeps out the tube volume. So the tube diameter is $2d$. Now this 'swept-out volume', $l\pi d^2$, contains only the one molecule. So it equals $1/N_v$.

$$\text{Hence } l \approx \frac{1}{\pi d^2 N_v}$$

We can use this equation to give an estimate of the mean free path for air molecules at room temperature. The volume of L molecules at room temperature and pressure is about 0.022 m^3 (i.e. the molar volume). Hence the number of molecules per unit volume, N_v, is about $6 \times 10^{23}/0.022 = 2.7 \times 10^{25}\,\text{m}^{-3}$. Now the size of an air molecule, d, is about 3×10^{-10} m (see p. 67). So the mean free path is about 10^{-7} m, using the formula for l above.

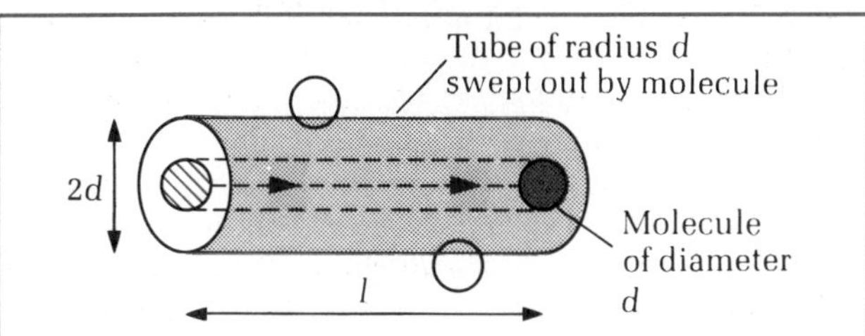

Figure 10.13 *Mean free path*

The random walk model is useful here. Find a chess board, a counter and a die. Start with the counter on a square near the middle. The rules are simple: you can only move the counter one square at a time, and not diagonally. So there are four possible moves, as in Figure 10.14, labelled 1 to 4. Throw the die to choose at random which direction to move. If the die throws up a 5 or a 6, just repeat the throw. Figure 10.14 shows a possible sequence for 25 moves. In theory, the counter could end up 25 squares away from the start but that is very unlikely because each move is at random. Statistics theory shows that for a large number of moves, N, you finish $\sqrt{N}$ squares from the start. Try it for $N = 100$ using a piece of graph paper instead of a chess board. If l is the length of the side of each square, then the distance from start to finish is $l\sqrt{N}$.

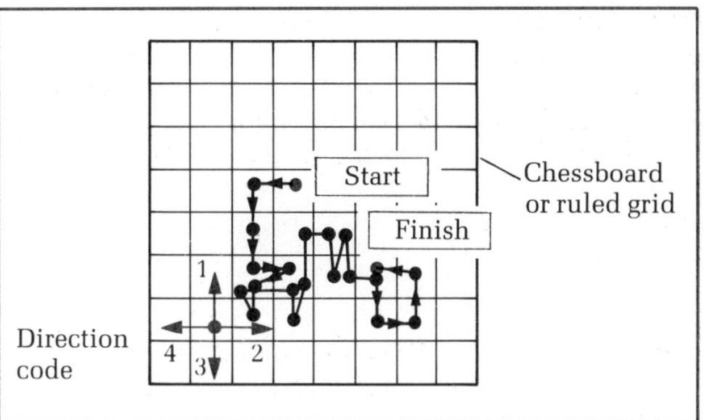

Figure 10.14 *Random walk*

Let's apply the ideas of the random walk model to bromine diffusion. The mean free path is the distance from one square to the next, so each molecule can change direction after each step. Of course real molecules can move in any direction after each collision, and the distance l is only the average distance from one collision to the next. Nevertheless, the $\sqrt{N}$ factor still applies.

The diffusion tube filled with air takes about 500 s to half-fill with bromine vapour. This is the time taken for the middle to become half-brown compared with deep brown near the bottom and light brown near the top. The tubes are about 0.4 m long, so the diffusion speed is $0.2/500 = 4 \times 10^{-4}\,\text{m}\,\text{s}^{-1}$. Compare this speed with the r.m.s. speed of bromine molecules at room temperature. Bromine has a molar mass M of 0.16 kg. Since $\frac{1}{3}Lm\langle c^2 \rangle = RT$ for 1 mole (using the kinetic theory equation and the ideal gas equation), then $\langle c^2 \rangle = 3RT/mL = 3RT/M$. So the r.m.s. speed $= 3 \times 8.3 \times 300/0.16 = 222\,\text{m}\,\text{s}^{-1}$. Bromine molecules move at speeds of the order of $220\,\text{m}\,\text{s}^{-1}$, but they diffuse about half-a-million times more slowly.

Suppose that each bromine molecule makes N collisions over the 500 s, then the total distance travelled for all N steps was Nl where l is the mean free path. This is half-a-million times more than the distance from start to finish $l\sqrt{N}$. So $Nl = 5 \times 10^5\ l\sqrt{N}$. Hence $\sqrt{N}$ is about 5×10^5, giving 2.5×10^{11} for N. The mean free path l can now be calculated since the distance from start to finish was about 0.2 m. Hence $l = 0.2/\sqrt{N} = 4 \times 10^{-7}$ m. This figure is in rough agreement for the value worked out for air molecules at room temperature. Although our estimate is based on a simple model, more advanced theory shows that our results are of the right order of magnitude.

Worked example Given the mean free path and r.m.s. speed of bromine molecules at a certain temperature and pressure are 4×10^{-7} m and 220 m s^{-1} respectively, how long would it take for bromine molecules to diffuse 10 m?
Solution Let N be the number of collisions over 10 m.

The distance from start to finish, $l\sqrt{N} = 10$ m, hence $\sqrt{N} = 10/l$ giving $N = 6.25 \times 10^{14}$.
The total distance travelled by a molecule is $Nl = 6.25 \times 10^{14} \times 4 \times 10^{-7} = 2.5 \times 10^8$ m for all N steps.
This total distance is travelled at a speed of 220 m s^{-1}. Hence

$$\text{Time taken} = \frac{\text{distance}}{\text{speed}} = \frac{2.5 \times 10^8}{220} = 1.1 \times 10^6 \text{ s}$$

Since the method is based on estimates, the time taken is therefore about 10^6 s. It would be nonsensical to give an answer with more significant figures since the approach is based on approximations.

10.4 Real gases

Vapours

Real gases liquefy when cooled and compressed. Nitrogen is an ideal gas at room temperature. However, if nitrogen gas is cooled below 127 K (= −146°C) then it can be liquefied if compressed sufficiently. Carbon dioxide can be liquefied if compressed sufficiently, provided its temperature is less than 31°C. At much greater temperatures, carbon dioxide behaves like an ideal gas. If cooled sufficiently, all gases can be liquefied if compressed enough. So at high pressure and low temperature, gases behave non-ideally. A gas behaves ideally if it obeys Boyle's law. We use the term **real gas** to mean a gas which behaves non-ideally. Real gas behaviour occurs because real gas molecules are not point molecules with zero volume. Real gas molecules attract one another too. If a gas is cooled and compressed enough, weak intermolecular attractions cause liquefaction of the gas.

Experiments have shown that gases can only be liquefied by compression if cooled to a temperature at or below a certain value, referred to as the **critical temperature**. The critical temperature of carbon dioxide is 31°C, so no amount of pressure on carbon dioxide above 31°C can liquefy it. Nitrogen has a critical temperature of −146°C. A gas which is capable of being liquefied by compression must therefore be at a temperature at or below its critical temperature. **Vapours** are gases which can be liquefied so a gas above its critical temperature cannot be referred to as a vapour. A vapour is material in the gaseous state at or below the critical temperature.

Any liquid produces vapour. The molecules of a liquid are in contact with one another, and they move about at random. Their kinetic energies vary, so at any one time some molecules will have much more energy than others. These molecules are capable of breaking away from their slower moving neighbours if near the surface. The more energetic molecules of a liquid can leave the liquid to pass into the vapour state. So if you raise the temperature of a liquid, more molecules pass into the vapour state: exactly what happens when water is warmed.

Suppose a beaker of liquid is placed in a closed room. Vapour is produced from the liquid. If only a small amount of liquid is used, all the liquid might become vapour. For example, water in a saucer left in a closed room for a day or so will probably evaporate and the room will therefore fill with water vapour. If no liquid is present, the vapour is said to be **unsaturated**. All the molecules of the substance are in the gaseous state.

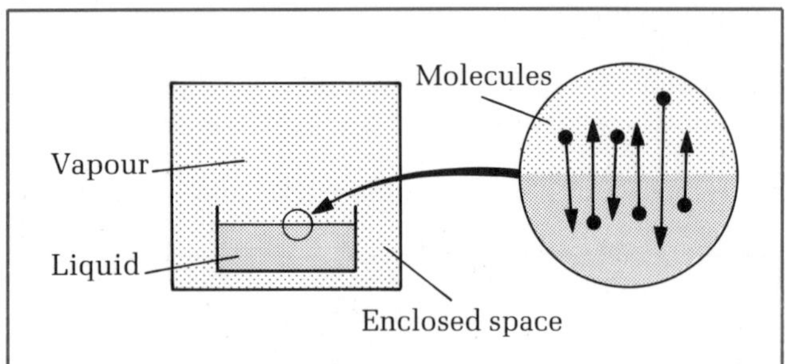

Figure 10.15 *Dynamic equilibrium*

A saturated vapour is one that is in thermal equilibrium with a liquid. For example, when a bucket of water is placed in a closed room, some of the water will evaporate to give water vapour. But with a bucketful, it is unlikely that all the water will evaporate – unless the room is a sauna! The water vapour in the room builds up to a certain level, which depends on temperature. Molecules leave the liquid to become vapour molecules, but they can also re-enter the liquid if their motion in the vapour state brings them back to the liquid surface. The number of molecules in the vapour state builds up until the number of molecules re-entering the liquid balances the number of molecules leaving the liquid in any given time. **Dynamic equilibrium** is the term we use to describe this situation.

The pressure of a saturated vapour increases with temperature but not in the way described by the pressure law. Figure 10.16 shows the variation for water vapour. The pressure increases much more than ideal gas pressure because increased temperature not only makes the vapour molecules move faster, it also increases the number of vapour molecules. When the temperature of the liquid rises, molecules gain K.E. so more of them pass into the vapour state. So, the number of molecules in the vapour state increases, but this also increases the re-entry rate until dynamic equilibrium is re-established.

A liquid boils when its saturated vapour pressure is equal to the external pressure on the liquid surface. Water at standard pressure (101 kPa) boils at 100°C so the saturated vapour pressure of water at 100°C is

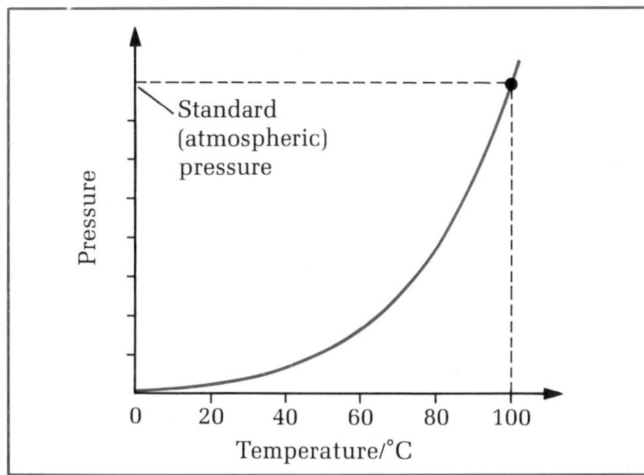

Figure 10.16 *Variation of SVP with temperature for water vapour*

101 kPa. The boiling point of water can be increased from 100°C by increasing the pressure on the water surface. Lower pressure gives a lower boiling point.

Saturated vapour pressure does not depend on volume If a saturated vapour is compressed (i.e. reduced in volume), the vapour molecules come closer together, so the rate of re-entry to the liquid increases. The transfer rate from the liquid stays constant (assuming constant temperature). Therefore, the number of molecules in the vapour state decreases so the re-entry rate falls. Eventually the re-entry rate regains its value before compression took place, so the pressure returns to its initial value. This happens when the number of vapour molecules per unit volume returns to its initial value.

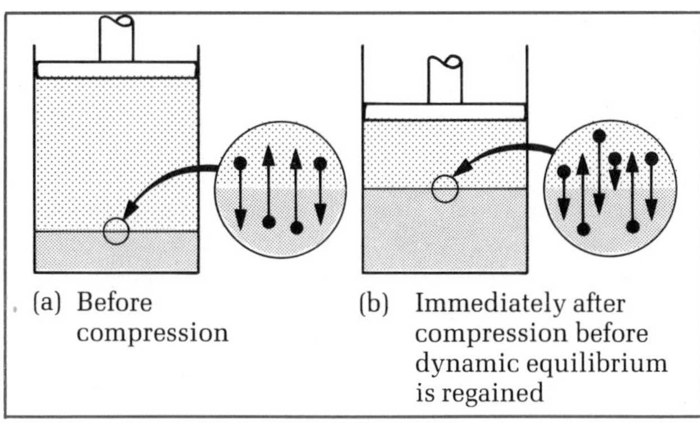

(a) Before compression
(b) Immediately after compression before dynamic equilibrium is regained

Figure 10.17 *SVP and volume*

Andrews' experiments

Liquefaction of gases was thoroughly investigated by Thomas Andrews in 1863. He devised apparatus capable of applying measured pressures up to 200 times atmospheric pressure to a gas. Figure 10.18 shows a cross-section of the apparatus.

Two capillary tubes sealed at the upper ends project vertically from a steel casing. The top part of one of the tubes contained the test gas, trapped by a thread of mercury. The top of the other tube contained dry air as a control gas. The mercury in each tube was held in by the water pressure in the casing. The pressure could be increased by turning the screws. Each tube was in its own

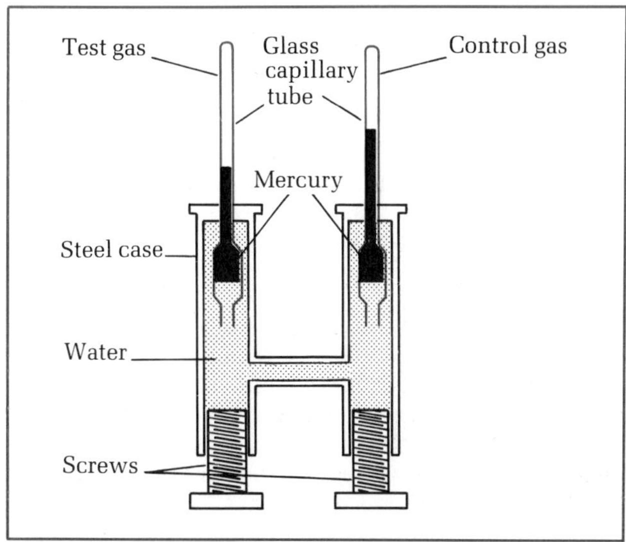

Figure 10.18 *Andrews' experiment*

water bath, the control bath at room temperature and the test bath at a temperature which could be changed. Both capillary tubes were previously checked for uniform diameter, and the volume of gas per mm length was measured before installing the tubes in the casing.

With the test gas at constant temperature, the pressure was increased in steps. At each step, the length of each gas column was measured when steady. Hence the volume of each gas could be determined. Since the control gas obeys Boyle's law its pressure could be calculated from its volume (using $pV = p_0V_0$, p and V being the pressure and volume of the control gas at any step, p_0 and V_0 being the pressure and volume of the control gas at atmospheric pressure). Because the test gas was at the same pressure as the control gas at each step, then the pressure of the test gas could be found. So Andrews was able to obtain readings of pressure and volume at different constant temperatures for a fixed mass of test gas.

Figure 10.19 shows the results for CO_2 as the test gas, with pressure plotted on the vertical axis and volume on the horizontal axis. Each curve is at different constant temperature. The curves are referred to as **isotherms** of p against V. The state of the CO_2 depends on its pressure, volume and temperature. The liquid state can only exist at or below the critical temperature.

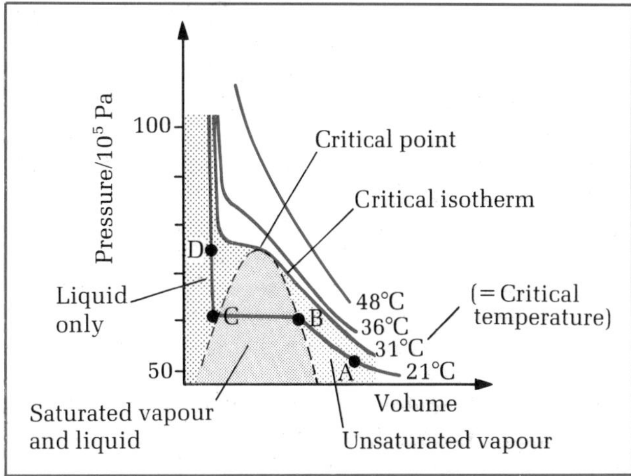

Figure 10.19 *CO isotherms of* p *against* V

The critical temperature T_c is the temperature above which no liquid can form, no matter how great the pressure of the gas is.

The critical isotherm is the isotherm at the critical temperature. It is flat at one point only. Isotherms below the critical temperature are flat over a range of volume corresponding to CO_2 as saturated vapour and liquid. The point where the critical isotherm is flat is the highest temperature where liquid is present. This point is called the **critical point**.

Consider the isotherm labelled ABCD. At A, the CO_2 is an unsaturated vapour since no liquid is present. If the volume is now reduced, the pressure rises to point B. At B, the CO_2 vapour is just saturated. Further reduction of the volume causes the saturated vapour to liquefy, the percentage of liquid increasing from B to C. The saturated vapour pressure stays constant. At C, the CO_2 is entirely liquid. Further reduction of volume is now difficult to achieve because liquids are almost incompressible. So the pressure rises sharply from C to D. All the other isotherms below the critical isotherm are like ABCD in general, each with a liquid-only section, a saturated vapour and liquid section, and an unsaturated section. Above the critical isotherm, no liquid is present so there are no flat parts on the curves.

Other gases give p against V curves with the same general shape as the CO_2 isotherms of Figure 10.19, but the critical temperature of each gas differs, and so too does the pressure at the critical point. Andrews' work on the liquefaction of gases led others to design and operate machines capable of liquefying the so-called *permanent gases*, nitrogen, oxygen and hydrogen etc. Very low temperatures are achieved when these gases are liquefied. Oxygen, for example, liquefies at 90 K at standard pressure. Its critical temperature is 155 K. Helium-3 (i.e. ^3_2He atoms) liquefies at 3.2 K at standard pressure. The usual way of supplying oxygen for industrial or medical use is in high pressure cylinders.

Explaining real gas behaviour

For an ideal gas at constant temperature, pV = constant. Real gas behaviour can be compared readily with ideal gas behaviour by plotting isotherms of pV against p. Figure 10.20 shows the 'pV against p' isotherms for air. The graph also shows ideal gas isotherms of pV against p. Since pV is constant for an ideal gas at constant temperature, the ideal isotherms are parallel to the horizontal (p) axis. The real gas isotherms are flat like the ideal isotherms at one particular temperature only, the **Boyle temperature** T_b. Even at T_b, the isotherm is flat at low pressures only.

a) For temperatures above T_b, the pV curve rises as p is increased.

b) For temperatures less than T_b, the pV curve falls as p is increased from low pressure, then the curve rises with further increase of p. If the temperature is less than T_c (the critical temperature) the pV curve drops sharply at constant pressure where liquefaction occurs. Once the 'gas' is completely liquid, pV rises as p is increased further.

To explain the pattern of the pV against p isotherms, we must look again at the kinetic theory assumptions: in particular, the assumptions that the gas molecules are point molecules with no attractive forces acting between them.

For temperatures above T_b the molecules become closer when the pressure is increased. So the actual volume of the molecules becomes more important compared with the measured volume of the gas. The available volume in which the molecules can move is less than the measured volume. We can write $V-b$ for the available volume where V is the measured volume; b is a constant, the **co-volume**, which represents the actual volume taken by the molecules themselves. So pV rises because the measured volume increasingly over-estimates the available volume at higher and higher pressures.

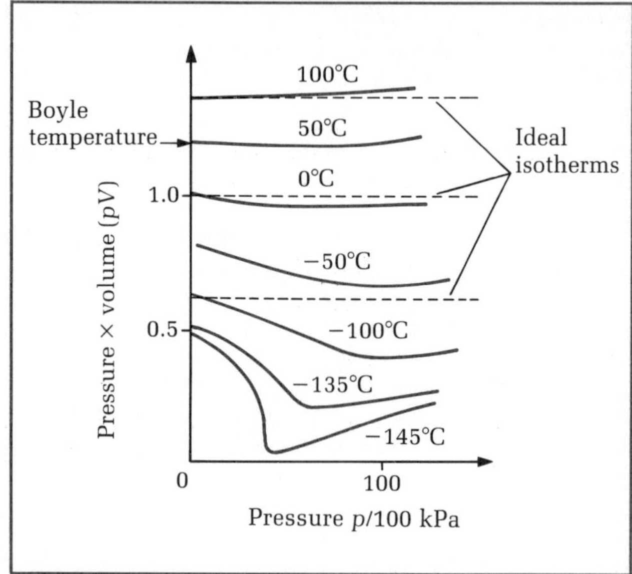

Figure 10.20 *pV against p isotherms for air*

For temperatures below T_b at low temperatures, the molecules move more slowly than they do at higher temperatures. So the weak attractive forces between the molecules have longer to act when molecules approach one another. The weak attractive forces pull on any molecules moving towards the container walls, so reducing the impact. Hence the pressure is reduced compared with the 'ideal gas' pressure. So pV falls compared with the ideal gas value. But as the pressure increases further, the volume of the molecules becomes more and more important. So pV stops falling and starts rising as the pressure is increased more and more.

The pressure reduction is due to attraction between molecules near the container wall and molecules in the bulk of the gas. The size of the reduction depends on the number of molecules per unit volume in each part. If we assume that the reduction is proportional to the number per unit volume at the surface × the number per unit volume in the bulk, we can write $\frac{a}{V^2}$ (a = constant) for the pressure reduction. So the ideal gas pressure can be written as $\left(p+\frac{a}{V^2}\right)$ where p is the measured pressure.

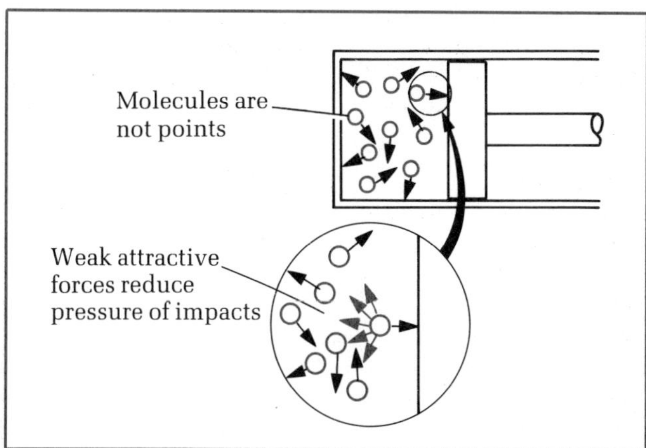

Figure 10.21 *Real gases*

Instead of writing $pV = nRT$, we can therefore write

$$\left(p + \frac{a}{V^2}\right)(V - b) = nRT$$

for real gas behaviour. This equation is known as the **Van der Waals equation**. The equation describes real gas behaviour reasonably accurately but not with complete success.

10.5 Summary

Definitions

1 *The root mean square speed* of the molecules of a gas is the square root of the mean value of (molecular speed)^2.
2 *Ideal gases* are gases which obey Boyle's law.
3 *The mean free path* of a gas molecule is the average distance it travels between successive collisions.
4 *The critical temperature* of a gas is the temperature above which it cannot be liquefied by increasing the pressure only.

Equations

1 The ideal gas equation for n moles is $pV = nRT$
2 The kinetic theory equation is $pV = \frac{1}{3}Nm\langle c^2\rangle$ or ($p = \frac{1}{3}\rho\langle c^2\rangle$ in an alternative form)
3 The mean K.E. of a gas molecule is $3kT/2$ for an ideal gas
4 The mean free path of a gas molecule $\approx 1/(\pi d^2 N_v)$

Short questions

Assume $R = 8.3\ \mathrm{J\ mol^{-1}\ K^{-1}}$,
$L = 6.02 \times 10^{23}\ \mathrm{mol^{-1}}$

10.1 An ideal gas in a sealed container, initially at a temperature of 17°C and 100 kPa pressure, is heated at constant volume until its temperature is 100°C. The container valve is then opened to allow gas to escape until the pressure falls back to 100 kPa at 100°C. Calculate
a) the pressure at 100°C before the valve was opened,
b) the fraction of the initial mass of gas lost as a result of opening the valve.

10.2 The molar mass of oxygen molecules is 0.032 kg. Calculate the density of oxygen gas at
a) 0°C,
b) 100°C when the gas pressure is 101 kPa

10.3 Two sealed identical glass bulbs are connected together by means of tubing and a valve. With the valve open, the two bulbs are placed in melting ice and the gas pressure throughout equalizes to initial pressure p_0. The connecting valve is then closed and one bulb is transferred to boiling water at 100°C. Calculate the new pressure in terms of p_0 in the hotter bulb. If the valve is opened to allow the pressures to equalize, calculate the final pressure in terms of p_0 with one bulb at 100°C and the other one at 0°C.

10.4 Two identical gas cylinders each contain 20 kg of compressed air at 1000 kPa pressure and 275 K. One of the cylinders is fitted with a safety valve which releases air from the cylinder into the atmosphere if the pressure in the cylinder rises above 1100 kPa. The cylinders are moved to a room where the temperature is 310 K. Calculate
a) the pressure in the cylinder which is not fitted with a safety valve,
b) the mass of gas lost from the other cylinder.

10.5
a) For 1 mole of an ideal gas, sketch a graph to show how its pressure × its volume (pV) varies with the gas temperature (t) in °C. Give an equation linking pV to t, and relate the equation to your graph.
b) Repeat (a) for 2 moles of ideal gas. Use the same axes.

10.6 A certain vacuum pump is capable of reducing the gas pressure in a sealed container to 0.001 × standard pressure if the temperature is maintained at 300 K. Calculate the number of molecules per m^3 in the container at this pressure and temperature. Standard pressure = 101 kPa.

10.7 Estimate the number of air molecules that enter your lungs when you inhale deeply. Assume the molar mass of air is 0.029 kg and its density at 0°C and 100 kPa pressure is 1.3 kg m^{-3}.

10.8
a) Describe and account for the motion of smoke particles suspended in air, viewed using a microscope and illuminated by a beam of light.
b) State how and explain why the motion would differ if (i) larger smoke particles had been used, (ii) the air temperature is increased.

10.9 Explain in molecular terms why the pressure of a fixed mass of gas increases when
a) the temperature of the gas is increased at constant volume,
b) the volume of the gas is reduced at constant temperature.

10.10
a) Calculate the r.m.s. speed of the molecules of oxygen gas at (i) 0°C, (ii) 100°C. The molar mass of oxygen molecules is 0.032 kg.
b) Calculate the total kinetic energy of 1 mole of oxygen gas at (i) 0°C, (ii) 100°C.

10.11
a) Calculate the r.m.s. speed of the molecules of nitrogen gas at 20°C. The molar mass of nitrogen molecules is 0.028 kg.
b) Explain why ammonia gas released at one end of a room spreads slowly throughout the room even though the air molecules move at very high speeds.

10.12 Smoke particles in air move about at random, continually changing direction, with speeds of the order of mm s^{-1}. Air molecules at the same temperature move about with speeds of the order of hundreds of metres per second. Use this information to estimate the ratio of the mass of a smoke particle to the mass of an air molecule.

10.13 Free electrons in a metal move about in continual random motion, like the molecules of an ideal gas. Using this comparison, the K.E. of 1 mole of free electrons in a metal at temperature T may be written as $3RT/2$. Calculate
a) the K.E. of 1 mole of free electrons at 20°C,
b) the mean K.E. of a free electron at (i) 20°C, (ii) 1500°C.

10.14 Liquid air has a density of 900 kg m^{-3} compared with 1.2 kg m^{-3} for gaseous air. Assuming that the molecules of liquid air are in contact with each other, estimate
a) the size of an air molecule,
b) the average spacing between molecules in ordinary air. Assume the molar mass of air is 0.029 kg.

10.15 The molar volume of oxygen gas at 0°C and 101 kPa pressure is 0.0224 m^3. Calculate the number of molecules per m^3 under these conditions. Assuming the size of an oxygen molecule is 3.5×10^{-10} m, estimate the mean free path of an oxygen molecule in these conditions.

10.16
a) Sketch isotherms for carbon dioxide to show how its pressure depends on its volume. Mark on your sketch the critical isotherm, and indicate where the CO_2 exists in the liquid state and where it is a vapour.
b) Unsaturated carbon dioxide vapour in an otherwise empty tube is compressed at a constant temperature that is less than its critical temperature. Show on your graph how the pressure changes as the volume is reduced.

10.17 Dalton's law of partial pressures states that, for a mixture of gases in a container, the total pressure is equal to the sum of the partial pressures (i.e. the pressure each gas would exert if alone in the container). Consider two gases, X and Y, in a container of volume V. Suppose there are N_x and N_y molecules of each gas, with r.m.s. speeds c_x and c_y.
a) Give an equation for the partial pressure of each gas in terms of its number of molecules, its r.m.s. speed and the volume V.
b) Both gases are at the same temperature so their molecules have the same mean K.E. Write an equation for this.
c) Use your equations to show that the total pressure is the sum of the partial pressures.

10.18 A mercury barometer tube of length 1000 mm above the reservoir level contains a small quantity of dry air in the space above the mercury at the top of the tube. On a day when the atmospheric pressure is 101 kPa and the temperature is 20°C, the barometer height is 740 mm. The next day when the temperature is 15°C, the barometer height is 750 mm. Calculate the atmospheric pressure on the second day. (A column of mercury of height 760 mm gives a pressure of 101 kPa at its base.)

10.19
a) Explain in molecular terms why the pressure of a saturated vapour at constant temperature is independent of its volume.
b) An anaesthetic machine pumps air through a volatile liquid, as shown in Figure 10.22. Explain why the liquid temperature falls as air is forced through. How does the fall of temperature affect the concentration of vapour in the air stream?

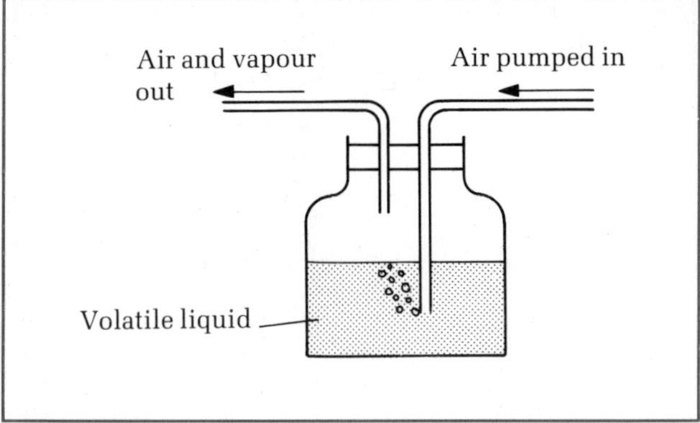

Figure 10.22

10.20
a) What do you understand by the terms (i) ideal gas, (ii) real gas?
b) Sketch isotherms of pressure × volume against pressure for (i) an ideal gas, (ii) a real gas.
c) Explain in molecular terms the shape of real gas isotherms you have drawn.

11

Thermodynamics

11.1 The first law of thermodynamics

Heat is energy transferred due to temperature difference. Heat transferred to an object either causes the internal energy of the object to rise or allows the object to do work, or both. Energy transformations often involve heating the surroundings. For example, a falling weight loses potential energy. What happens to the P.E. which it loses? If the weight is accelerating as it falls, it speeds up so it gains kinetic energy. The loss of P.E. supplies gain of K.E. If the object then hits the floor, what happens to its K.E.? Sound energy is produced and the atoms of the floor at the impact point gain some energy as well. So the floor becomes just a little warmer as a result of the impact.

Another example is when a battery is used to run an electric motor to raise a weight, as in Figure 11.1. The battery supplies electrical energy which is partly converted into potential energy of the raised weight. Some of the electrical energy is used for heating the motor windings because the windings have resistance. Also, frictional forces cause heating at the motor bearings. So not all the electrical energy is converted to potential energy. The difference is lost through heating of the motor and its surroundings. The surroundings gain energy through heat transfer.

Why does heat transfer always seem to make energy spread out? Is it possible to draw energy from the surroundings and concentrate it in some form? Thermodynamics is the study of energy transformations, and it gives the answers to these important questions. The laws of thermodynamics were established by scientists in the 19th century. These laws are based on experiments involving energy transformations, and scientists put great faith in them because no experiment has ever disproved them. To understand the reasons for the laws we must look at energy in atomic terms. Here we find the rules of chance come into play. However, with large numbers of atoms, the rules become very predictable. You have met the rules of chance if you own one or more premium bonds. What are the chances of the owner of a single bond ever winning? But the owner of a million bonds would probably win every month!

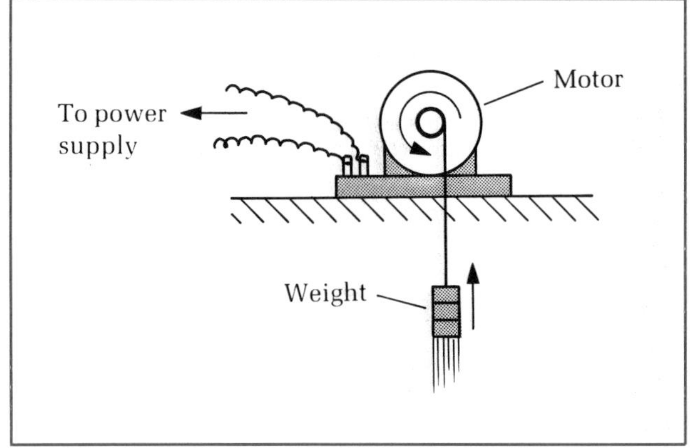

Figure 11.1 *Energy on the move*

The first law of thermodynamics states that heat energy supplied to a system either increases the internal energy of the system or enables it to do work, or both.

$$Q = \Delta U - W$$

where Q = heat transferred to system,
ΔU = gain of internal energy,
W = work done on the system.

If the internal energy of the object increases by ΔU, and the work done on the system is W, then heat transferred into the system must equal $\Delta U - W$. For example, if a system's internal energy rises by 20 J and the work done on it is 14 J, the heat transferred into the system must be 6 J.

The first law is based on conservation of energy. There are two ways of changing the internal energy of any system. One way is to use forces to do work. The other way is by heat transfer. So the first law tells us that the gain of internal energy equals the heat transfer to the system + the work done on the system. So we can write $\Delta U = Q + W$ which gives $Q = \Delta U - W$.

When the equation for the first law is used, sometimes the values are negative. A negative value for ΔU means the internal energy has decreased. A negative value for either Q or W means that energy is transferred out of the system. For example, if the work done on the system is 28 J but its gain of internal energy was only 20 J, then the equation gives $Q = 20 - 28 = -8$ J. So heat transferred *from* the system is 8 J.

Stretching a rubber band provides a simple example of using thermodynamics.

1. Stretch a rubber band quickly then hold it, still stretched, against your lower lip. It feels warm. The internal energy of the rubber increases because you have done work on the rubber band when you stretched it. The stretching was so rapid that heat transfer was negligible. An energy transfer where no heat transfer occurs is called an **adiabatic** change.
2. Keep the rubber band stretched against your lip and you should find that it cools to 'lip' temperature. Its internal energy falls due to heat transfer to the surroundings. No work is done because there is no movement.
3. Now release the rubber band quickly, and put it unstretched against your lip again. You ought to find that it becomes colder than 'lip' temperature as a result of unstretching. Its internal energy has fallen because internal forces in the rubber do work as it unstretches. No heat transfer occurs because the change is rapid.
4. Finally the unstretched rubber warms up to 'lip' temperature because heat transfer from the room increases its internal energy.

Compressing a gas provides another example. Figure 11.2, on the next page, shows a simple arrangement of a bicycle pump with a blocked outlet. A thermocouple inserted through the blocked outlet allows the air temperature inside the pump to be monitored.

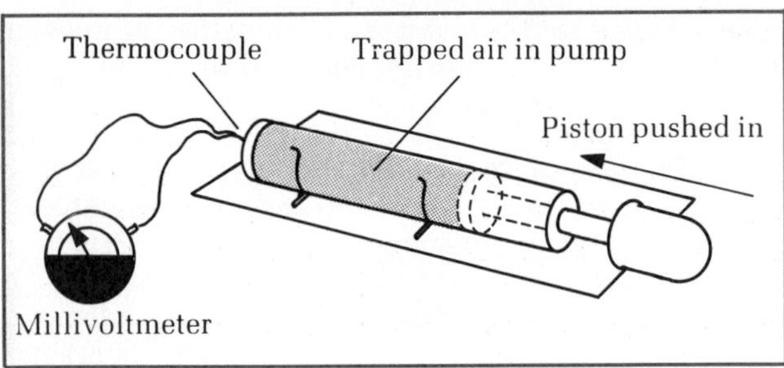

Figure 11.2 *Compressing a gas*

1 Push the piston in rapidly and hold it in position. The thermocouple thermometer ought to show a temperature rise due to increase of internal energy of the air. The push force does work on the air rapidly, so no heat transfer from the air occurs. Hence the work done increases the internal energy.
2 Return the piston to its initial position and allow the temperature to return to its initial value as well. Then push the piston in very gradually. No temperature rise occurs this time. Why? You still do work on the air as you push the piston in. However, any slight rise of temperature causes heat transfer from the air so the temperature returns to the initial value. In other words, heat transfer from the air occurs at the same rate as work is done on it. So no rise of internal energy occurs.

11.2 Thermodynamics of ideal gases

Heat capacities of an ideal gas

The internal energy U of an ideal gas is due to the kinetic energy of its molecules. No internal forces of attraction exist between 'ideal gas' molecules so an ideal gas does not possess molecular potential energy. To change the temperature of a gas, its internal energy U must be changed. To do this, work must be done or heat must be transferred to or from the gas.

Work is done on a gas when its volume is reduced (i.e. it is compressed). At constant volume, no work is done because the forces due to pressure in the gas do not move the container walls. When a gas is compressed, work is done on it by the applied forces which push the pressure forces back. Try compressing the air in a bicycle pump by blocking the outlet and pushing the piston in rapidly.

When a gas expands, the forces due to its pressure push back the container walls, as in Figure 11.3. So the gas does work when it expands. For example, when the valve on a CO_2 cylinder is released, the cylinder outlet releases a high-pressure jet of CO_2 gas into the air. The gas does work to expand and push back the air. The CO_2 uses up so much of its internal energy to expand that it cools and solidifies. You may have seen solid CO_2 produced in this way for use in cloud chambers (see p. 441).

Consider a tube of gas at pressure p trapped by a piston, as in Figure 11.3. The force due to pressure on the piston is pA, where A is the area of cross-section of the tube. To hold the piston in place, an equal and opposite force must be applied to it.

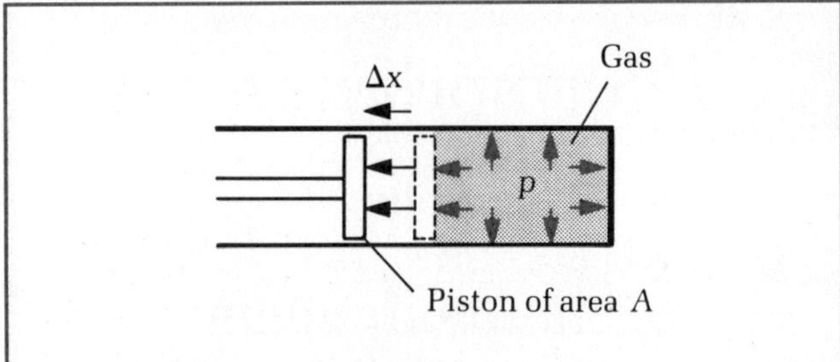

Figure 11.3 *Work done by a gas*

Suppose the gas is allowed to expand a little without any significant change of pressure. The piston moves out by a distance Δx. So the change of gas volume is $A\Delta x$. The work done by the gas is given by force × distance moved. Work done by gas $= F\Delta x = pA\Delta x = p\Delta V$ since $A\Delta x =$ the volume change, ΔV.

For example, if the volume increases from $0.100\,m^3$ to $0.102\,m^3$ at a pressure of 10^5Pa, the work done by the gas is $10^5 \times 0.002 = 200\,J$. If the volume had *decreased* from $0.100\,m^3$ to $0.098\,m^3$ at 10^5 Pa, the volume change would have been $-0.002\,m^3$, giving work done by the gas equal to $-200\,J$. In this case, the $-$ sign indicates energy transferred *to* the gas.

Work done by a gas $= p\Delta V$

where $\Delta V =$ change of volume at constant pressure p.

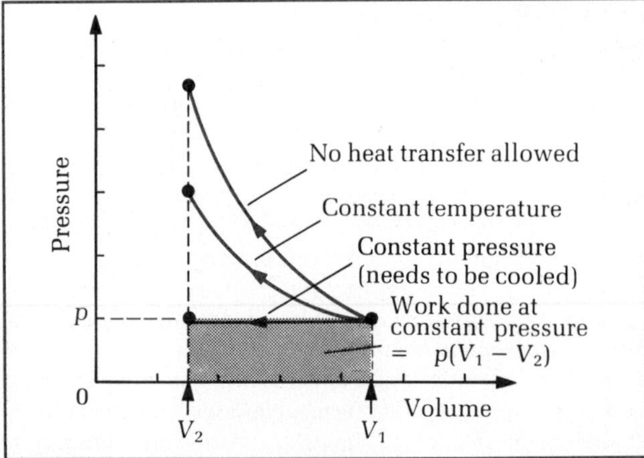

Figure 11.4 *Area under p-V curves*

Pressure against volume curves are often used to show changes of a gas. For a small change of volume ΔV at constant pressure p, $p\Delta V$ is represented on such graphs as the area under the curve from V to $V+\Delta V$. The area forms a strip of height represented by p and of width represented by ΔV. Since $p\Delta V$ is work done, then the area under the curve represents the work done. Figure 11.4 shows changes of volume of a fixed mass from V_1 to a smaller volume V_2 in different ways. Each way involves work done on the gas. Which way involves most work being done?

Heat transfer Q to a gas to change its temperature must increase its internal energy. But if the gas expands when it is heated, it uses energy to do work. So the heat transfer Q to the gas is equal to its increase of internal energy + the work done by the gas to expand.

$Q = \Delta U + p\Delta V$

where ΔV = change of volume at pressure p,
Q = heat transfer to gas,
ΔU = change of its internal energy.

The molar heat capacity at constant volume, C_v is defined as the heat transfer required to raise the temperature of 1 mole by 1 K at constant volume. If 1 mole of gas is heated at constant volume so that its temperature rises by ΔT, the heat transfer must equal $C_v\Delta T$. Since $\Delta V = 0$, then the rise of internal energy $\Delta U = C_v\Delta T$.

The molar heat capacity at constant pressure C_p is defined as the heat transfer to the gas required to raise the temperature of 1 mole by 1 K at constant pressure.

To raise the temperature of 1 mole by ΔT at constant pressure, the heat transfer must be $C_p\Delta T$. The internal energy increases by the same amount as at constant volume because the temperature increase is the same. So $\Delta U = C_v\Delta T$. However, the gas expands to keep the pressure constant. So it does work equal to $p\Delta V$ where ΔV is the increase of volume.

$$Q = C_p\Delta T$$
$$\Delta U = C_v\Delta T$$
$$W = p\Delta V$$

So $\quad C_p\Delta T = C_v\Delta T + p\Delta V$

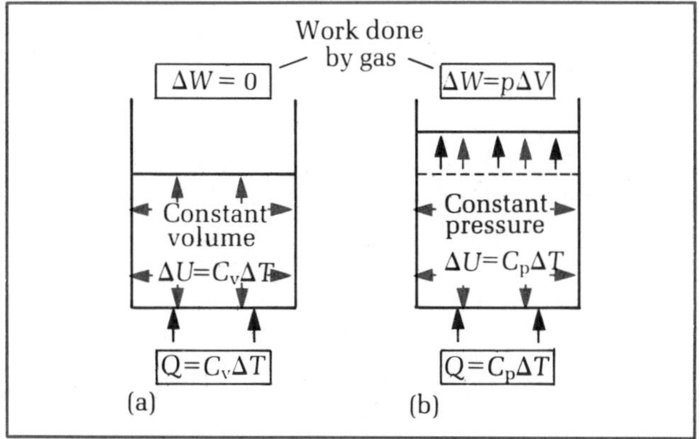

Figure 11.5 *Heat capacities*

For an ideal gas, $pV = RT$ for 1 mole. After heating at constant pressure, its volume increases from V to $V + \Delta V$, and its temperature from T to $T + \Delta T$. So $p(V + \Delta V) = R(T + \Delta T)$. Hence $p\Delta V = R\Delta T$. So the work done is $R\Delta T$. Hence

$$C_p\Delta T = C_v\Delta T + R\Delta T$$

So $\quad C_p = C_v + R$

A monatomic gas such as helium, is one with only one atom in each molecule. For a monatomic gas, we can use the kinetic theory equation to give its internal energy. From $pV = \frac{1}{3}Nm\langle c^2\rangle$, the total K.E. of the molecules is $\frac{3}{2}RT$ for 1 mole. Its internal energy U is entirely in the form of K.E. due to molecular speeds. Hence $U = \frac{3}{2}RT$. So if its temperature rises by ΔT, its internal energy rises by $\frac{3}{2}R\Delta T$. Since this is equal to the heat transfer at constant volume, $C_v\Delta T$,

$$C_v = 3R/2$$

Also, since $C_p = C_v + R$,

$$C_p = 5R/2$$

These values do not apply to molecules with more than one atom each. Such molecules also possess K.E. of vibration and K.E. of rotation. So C_p and C_v do not in general have the values above, but their difference, $C_p - C_v$ is always equal to R for an ideal gas. The actual values depend on the number of atoms per molecule.

Adiabatic changes of an ideal gas

An adiabatic change is defined as one which occurs without heat transfer. A gas can be compressed adiabatically by reducing its volume rapidly. The change is then too fast for measurable heat transfer. Another way is to insulate the gas container so heat transfer cannot occur.

Consider the adiabatic compression of an ideal gas in a little more detail. Work is done on the gas to reduce its volume, but heat transfer does not occur. So the internal energy increases, and therefore the temperature rises. Suppose a small change of volume ΔV causes the temperature to change by ΔT for 1 mole of gas. From $Q = \Delta U + p\Delta V$, we can write the following equation.

$$0 = C_v\Delta T + p\Delta V \quad \text{since } Q = 0 \text{ and } \Delta U = C_v\Delta T \qquad (1)$$

Before the change, the pressure and volume are linked by $pV = RT$.

After the change, the pressure changes to $p + \Delta p$ and the volume to $V + \Delta V$, so the new values are linked by $(p + \Delta p)(V + \Delta V) = R(T + \Delta T)$.

Now multiply out the left hand side of the new equation, but ignore $\Delta p\Delta V$ because it represents an exceedingly small change! Also $pV = RT$, so we are left with

$$p\Delta V + V\Delta p = R\Delta T$$

So $V\Delta p = R\Delta T + C_v\Delta T$ since $p\Delta V = -C_v\Delta T$ from (1).

Hence $V\Delta p = C_p\Delta T$ since $C_p = C_v + R$. Therefore

$$V\Delta p = C_p\Delta T \qquad (2)$$

If we now combine equation 2 with equation 1, we can eliminate ΔT.

$$V\Delta p = C_p\Delta T$$
$$p\Delta V = -C_v\Delta T$$

Hence $\dfrac{V\Delta p}{p\Delta V} = -\dfrac{C_p}{C_v}$

The ratio C_p/C_v is usually written as the symbol γ. So the pressure and volume changes are linked by the equation

$$\frac{\Delta p}{\Delta V} = -\gamma\frac{p}{V}$$

This equation for small changes can be solved by integration to give an equation for all changes, called **the adiabatic equation for an ideal gas.**

$$pV^{\gamma} = \text{constant}$$

Adiabatic changes of an ideal gas can be plotted on a graph of pressure against volume. The curves are steeper than isotherms, as Figure 11.6 shows. For example, if an

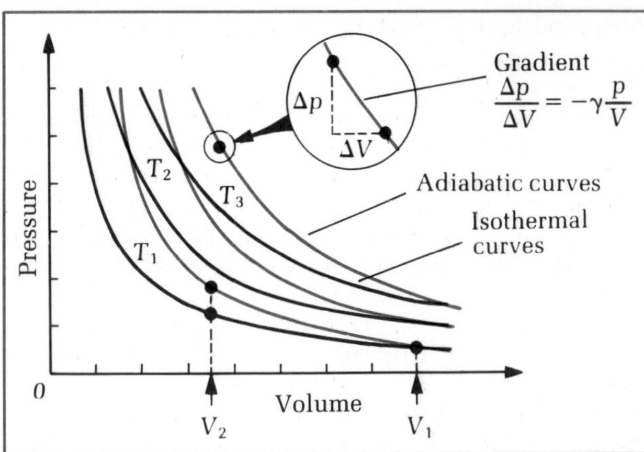

Figure 11.6 *Adiabatic curves*

ideal gas is compressed from volume V_1 to V_2 adiabatically, the final temperature is higher than if it is compressed isothermally. So the final pressure is higher.

11.3 The second law of thermodynamics

Engines

Engines are designed to do useful work. To do so, they must be supplied with energy, usually as a result of heat transfer. A steam engine or a car engine does work as long as it has fuel. Heat transfer from the fuel to the engine takes place, allowing the engine to do work. The internal energy of the engine does not change, provided its temperature is steady. So heat transfer to the engine is equal to the work done by the engine + the heat losses to the surroundings.

$$Q_1 = W + Q_2$$

where Q_1 = heat transfer to the engine,
W = work done by the engine,
Q_2 = heat losses to surroundings.

Consider the example of a steam turbine. Jets of high pressure steam etc., from a boiler are directed at the

Figure 11.7 *Heating the surroundings*

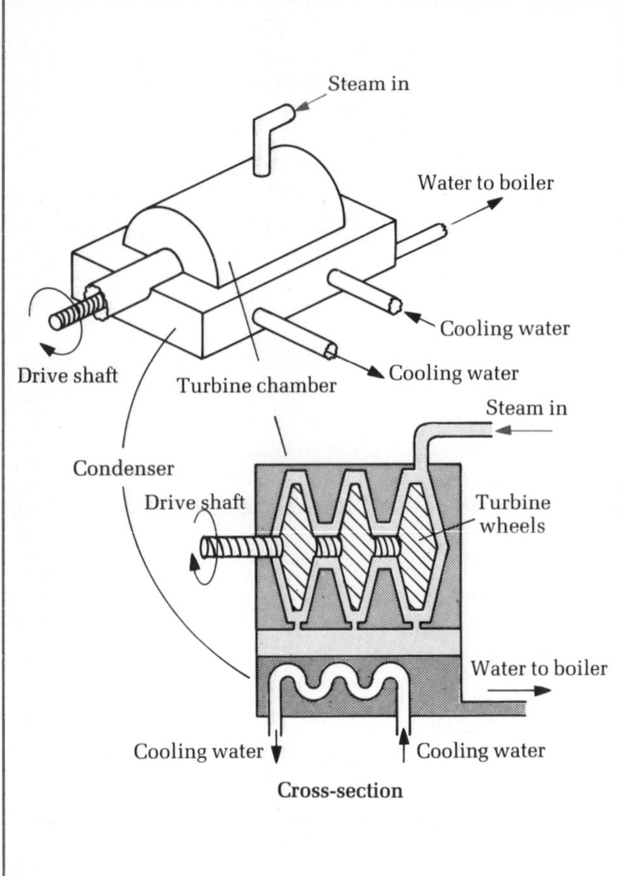

Figure 11.8 *Steam turbines*

blades of a turbine wheel. The force of the steam makes the turbine wheel turn. After impact, the steam is at much lower pressure. It passes into a condenser where it turns back to water which is returned to the boiler. So heat energy from the boiler produces steam which does useful work. Then the steam transfers heat energy to the condenser at a much lower temperature. Figure 11.8 shows the process. The energy gained by the condenser is waste energy in the sense that it has not been used for work.

Is it possible to design an engine where no energy is wasted? In the turbine of Figure 11.8, what would happen if the condenser were bypassed? The low pressure after the turbines would rise and stop the flow of steam into the turbines. So the condenser, a low temperature sink, is essential if heat transfer is to be produced by a high temperature source. This is true of any heat engine.

Another example is the turbojet engine shown in Figure 11.9. The air intake is compressed then heated strongly, so a jet of very hot air is forced from the exhaust; this pushes the engine in the opposite direction. The fuel supplies the energy for the work to be done by the engine. But the hot air from the exhaust transfers some of that energy to the atmosphere. The fuel burning supplies a high temperature source. The atmosphere provides a low temperature sink.

For an engine to do work as a result of heat transfer from a high temperature source, we need a low temperature sink to draw the energy from the source. So the work done can never be equal to the heat transfer from the source.

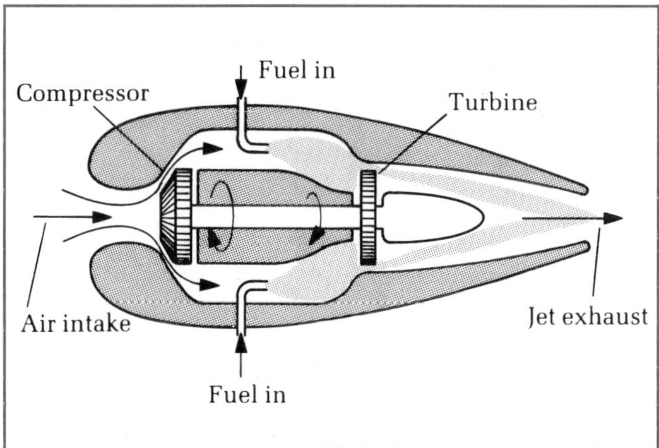

Figure 11.9 *A turbojet engine*

The second law of thermodynamics states that it is impossible for heat transfer from a high temperature source to produce an equal amount of work.

The first law tells us that the energy put into a machine is equal to the energy from the machine. You can't get more energy out than you put in. The second law goes further; the useful energy from any engine is always less than the energy put into the machine.

The efficiency of an engine is defined as

$$\frac{\text{the work done by the engine}}{\text{energy supplied to the engine}} \times 100\%.$$

So a heat engine has an efficiency of $(W/Q_1) \times 100\%$ where W is the work done by the engine, and Q_1 is the heat transferred to it. Since W is always less than Q_1, no engine can have an efficiency of 100% or more. The most efficient type of engine is a theoretical engine called a **reversible** engine. Its 'inventor' was a 19th century scientist called Carnot. A reversible engine is defined as one which can be operated in reverse, so that work W done on it causes maximum heat transfer from cold to hot, as in Figure 11.10. The heat transfer to the high temperature reservoir is given by $Q_1 = W + Q_2$. In other words, a reversible heat engine doesn't waste energy.

If you could design an even more efficient engine, you could disprove the second law. Your engine could do work on a reversible engine to make heat transfer from cold to hot without any help! Unfortunately, such an engine is impossible. Heat always tends to spread out.

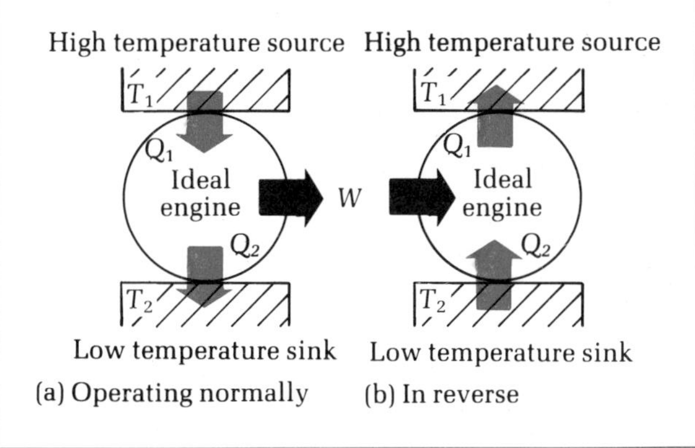

Figure 11.10 *A reversible engine*

Investigating heat engines

An engine at work due to heat transfer is shown in Figure 11.11. The thermopile unit is a series of thermocouples (see p. 109) arranged so that they produce a voltage when one side is hot and the other side is cold. Hot water is placed in one container and cold water is placed in the other to make the motor connected to the thermopile run. Swap the containers round and the motor reverses. The thermopile unit is a heat engine which accepts energy Q_1 from the hot water to do work W on the motor. To keep the energy moving, one side must stay colder than the thermopile. So the thermopile loses energy Q_2 to the cold water. See if you can estimate the efficiency of the system. A reversible engine operation between 373 K and 273 K has an efficiency of 27% (see later).

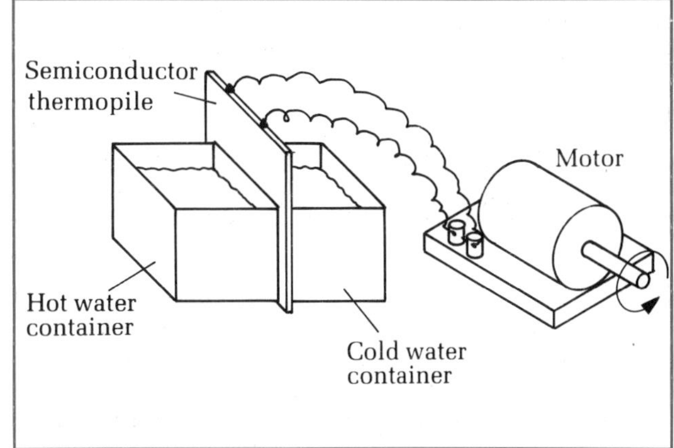

Figure 11.11 *Investigating heat engines*

An engine in reverse is sometimes called a heat pump. By passing current through the thermopile, you can make it pump heat energy from cold to hot. Replace the water in the containers with water at room temperature, then pass a current through the thermopile. You will need to measure and control the current, to ensure you do not exceed its current rating. With sensitive thermometers in the water in each container, you ought to find that one container warms up and the other cools. Although operating in reverse, the thermopile, etc., is not a reversible engine. Its efficiency is much less than 27% so it still wastes energy.

11.4 Entropy

The thermodynamic scale of temperatures is defined using a reversible engine. We define the source temperature T_1 and the sink temperature T_2 as in the ratio Q_1/Q_2, as shown in Figure 11.12 on the next page.

$$\frac{T_1}{T_2} = \frac{Q_1}{Q_2}$$

where Q_1 = heat transfer from the source at temperature T_1,
Q_2 = heat transfer to the sink at temperature T_2.

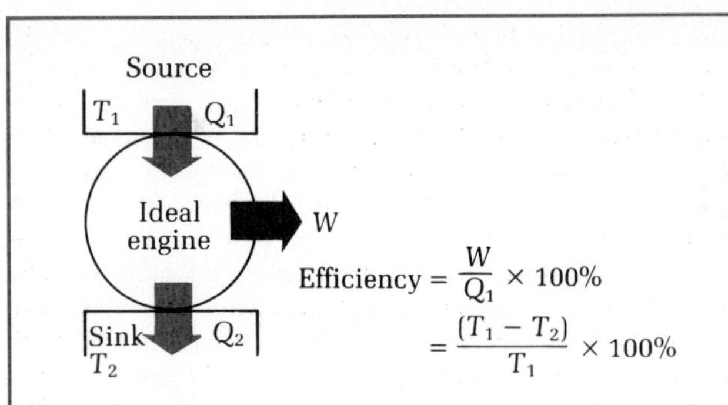

Figure 11.12 *Efficiency*

Since the efficiency of a heat engine is $(W/Q_1) \times 100\%$, and W is equal to $Q_1 - Q_2$, then the efficiency formula can be written as

$$\text{Efficiency} = \frac{(Q_1 - Q_2)}{Q_1} \times 100\% = \left(1 - \frac{Q_2}{Q_1}\right) \times 100\%$$

However, the most efficient type of heat engine is a reversible engine, and Q_2/Q_1 is equal to T_2/T_1 for a reversible engine. So the efficiency of a reversible engine is given by

$$\text{Maximum possible efficiency} = \left(1 - \frac{T_2}{T_1}\right) \times 100\%$$

$$= \frac{T_1 - T_2}{T_1} \times 100\%$$

Now you can see why the thermopile engine operating between 373 K and 273 K can never have an efficiency greater than 27%. The only way any engine could be made 100% efficient would be to make the sink at absolute zero (i.e. $T_2 = 0$), and that is impossible.

Spreading of energy so that it becomes less and less useful occurs because real engines operate at less than maximum efficiency. A reversible engine does work W by taking energy Q_1 as heat transfer from a high temperature source T_1. The engine has to supply energy Q_2 to a low temperature sink T_2 to enable it to do work. Suppose we store the energy produced by work. For example, the motor driven by the thermopile could be used to lift a weight. Then we use that stored energy to do work W *on* the heat engine, operating it in reverse. We simply transfer energy Q_1 back to T_1, taking energy Q_2 from T_2. Our operations have taken us back to the starting point, as if nothing had happened. No energy has been wasted. This happens only with a reversible engine for which

$$\frac{Q_1}{T_1} = \frac{Q_2}{T_2}$$

Now suppose we use a non-reversible engine instead. It does less work for the same heat transfer Q_1 from the high temperature source. Its efficiency is less than the reversible engine. We can write this as an inequality. The inequalities between a 'real' and an 'ideal' engine are set out at the top of the next column.

Efficiency of a 'real' engine < Efficiency of an 'ideal' engine

$$\left(1 - \frac{Q_2}{Q_1}\right) \times 100\% < \left(1 - \frac{T_2}{T_1}\right) \times 100\%$$

Hence $$\frac{-Q_2}{Q_1} < \frac{-T_2}{T_1}$$

So $$\frac{Q_1}{T_1} < \frac{Q_2}{T_2}$$

Entropy measures the way energy spreads out to become less useful. The quantity $\frac{Q}{T}$ is defined as the change of entropy when heat transfer occurs at constant temperature. For the reversible engine, no overall entropy increase occurs because the low temperature sink gains entropy equal to the loss from the high temperature source. But the 'real' engine does cause overall increase of entropy. The gain of entropy by the sink is more than the loss from the source. So the 'real' engine spreads energy out to make it less useful!

$$\text{Change of entropy } \Delta S = \frac{Q}{T}$$

where Q = heat transfer,
T = temperature of source or sink.

The unit of entropy is joules per kelvin (J K^{-1}). Because energy always tends to spread out to become less useful, entropy always tends to increase.

Worked example A thermopile engine operating between 373 K and 273 K has an efficiency of 10%. Calculate the increase of entropy when it does 100 J of work.

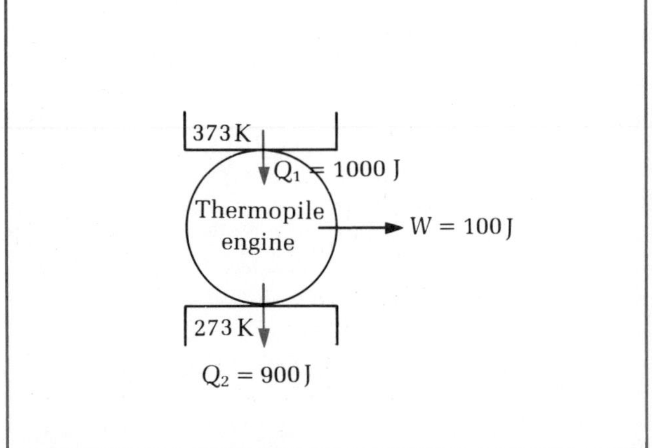

Figure 11.13 *Increasing entropy*

Solution The heat transfer to the engine $Q_1 = 1000$ J because the engine only does 100 J of work at 10% efficiency. So the loss of entropy from the source T_1 is equal to $Q_1/T_1 = 1000/373 = 2.68\ \text{J K}^{-1}$.
The heat transfer to the sink from the engine $Q_2 = Q_1 - W = 900$ J.
So the gain of entropy by the sink T_2 is equal to $Q_2/T_2 = 900/273 = 3.30\ \text{J K}^{-1}$.
Therefore, the overall gain of entropy $S = 3.30 - 2.68 = 0.62\ \text{J K}^{-1}$.

11.5 Chance and change

Spreading energy

Does energy always tend to spread out? A diver jumping into water obviously spreads energy out. Imagine a film of a diver jumping into water run in reverse. Conservation of energy doesn't forbid the diver leaping out of the water as a result of energy gained from the water, but thermodynamics does!

Another example of spreading is when you add milk to a cup of coffee or tea. Even without stirring, the milk spreads throughout the cup. The reverse process of the milk being concentrated after having spread out never happens. Lots of processes are 'one-way only', with no reverse possible.

The expansion of an ideal gas allows molecules to spread out. Let's consider such an expansion first in terms of heat transfer, and then in terms of the molecules.

1 Suppose we start with 1 mole of ideal gas molecules confined to one half of a box by a movable partition, as in Figure 11.14. Let V_0 be the volume of one half.

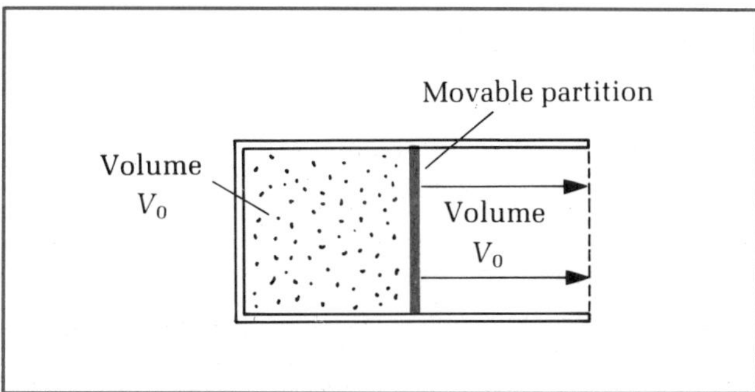

Figure 11.14 *Expansion of an ideal gas*

Then we move the partition back slowly until the volume is doubled. The slowness of the process allows heat transfer into the gas to make up for work done by the gas. So its internal energy stays the same, hence its temperature is constant. At any stage, the gas pressure is given by $pV = RT$ (as $n = 1$).

The work done by the gas in each small expansion $= p\Delta V$, where ΔV is the increase of volume.

Hence the work done at each stage $= RT\dfrac{\Delta V}{V}$.

The total work done to double the volume is given by adding the work done at each stage. This is achieved by integrating.

Total work done is given by

$$W = RT\int_{V_0}^{2V_0} \frac{dV}{V} = RT\ln\left(\frac{2V_0}{V_0}\right) = RT\ln 2$$

We have made use here of an integration formula explained on p. 507.

So the heat transfer to the gas $Q = RT\ln 2$.

Hence the increase of entropy of the gas $\Delta S = \dfrac{Q}{T} = R\ln 2$

2 In terms of molecules, how many ways are there of distributing the molecules between the two halves of the box? Initially they are all in one half, so to start with there is only one way. Finally, they are distributed at random, so there is a 1 in 2 chance of any one molecule being in one half. For L molecules (where L = the Avogadro constant), there are 2^L ways of distributing the molecules between the two halves. In terms of an equation, $W = 2^L$ gives the numbers of ways W. But $\ln W = \ln(2^L) = L\ln 2$.

Entropy is a measure of the number of ways in which energy can be spread. We define the entropy S of the molecules as $k\ln W$, where k is Boltzmann's constant R/L. In the above example, $\ln W = L\ln 2$ so the entropy change ΔS is therefore $kL\ln 2$ which equals $R\ln 2$. By defining entropy is this way, we obtain agreement with the 'heat transfer' approach. In addition, we can now see exactly what entropy means in molecular terms.

$$S = k\ln W$$

where W = number of ways of distributing the energy of a system,
S = the entropy of the system,
k = the Boltzmann constant.

This equation was first derived by Boltzmann, and it applies to all 'energy systems'. Using this equation, we can understand why energy always tends to spread out.

Thermal equilibrium

Energy is quantized By this statement, we mean that particles can have only certain amounts of energy. The minimum amount allowed is so small that we don't notice it on a large scale. For example, the energy of an atomic electron can take only certain allowed values. When an electron moves from a high energy level to a lower level, a photon is emitted by the electron. The photon carries away the energy lost by the electron. Since the electron is only allowed certain energy values, the photon can only have certain energy values. See p. 456.

Vibrating atoms in a solid are only allowed energy which is a multiple of a basic amount. An atom can have energy ε, 2ε, 3ε etc., but not $1\frac{1}{2}\varepsilon$ or any other 'in-between' value. The energy values of a vibrating atom can be pictured like the rungs of a vertical ladder; the atom can be on any rung, but not in between. When vibrating atoms gain or lose energy, they do so in units of ε. So energy transfer in a solid can be pictured as lots of energy units or 'quanta' jumping from one atom to another all through

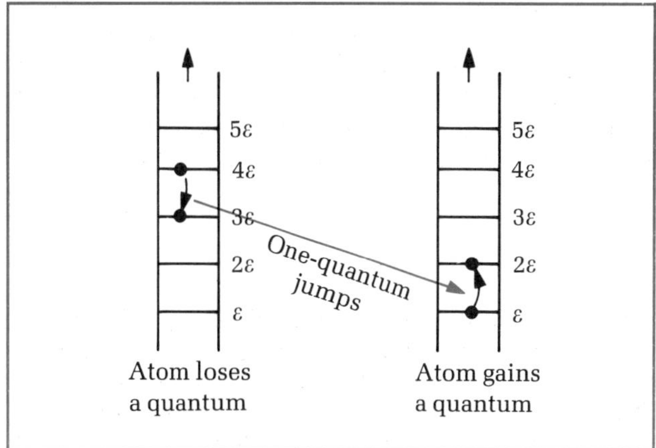

Figure 11.15 *Atoms in an Einstein solid*

the solid. At any one time, a given atom can be anywhere on its ladder, provided it is on a rung. In other words, a given atom might have any number of quanta at any time. But that number would continually change as quanta 'hop' from atom to atom. The model was devised by Einstein, and it can be used to explain thermal properties of solids.

The quantum shuffle uses the Einstein model of a solid to explain thermal equilibrium. Imagine a chess board where each square represents an atom, and counters are used to represent quanta. Suppose we start with an even distribution, i.e. the same number of counters on each square. Then the counters are moved at random from square to square. Each move is made by taking a counter off a randomly-chosen square. Then a square is chosen at random to receive the counter. To keep the model simple, we allow the counters to move from one square to any other square. A computer is helpful here, and can be used for much larger numbers of atoms and quanta. Nevertheless, you could use two dice and a six by six grid: one die could select a row, and the other a column, so you could use the two dice to choose one square from thirty-six at random. Figure 11.16 shows the results from a 20 × 20 grid. The initial distribution is shown by (a), then the picture after a given number of moves from the start is shown in (b), (c) and (d).

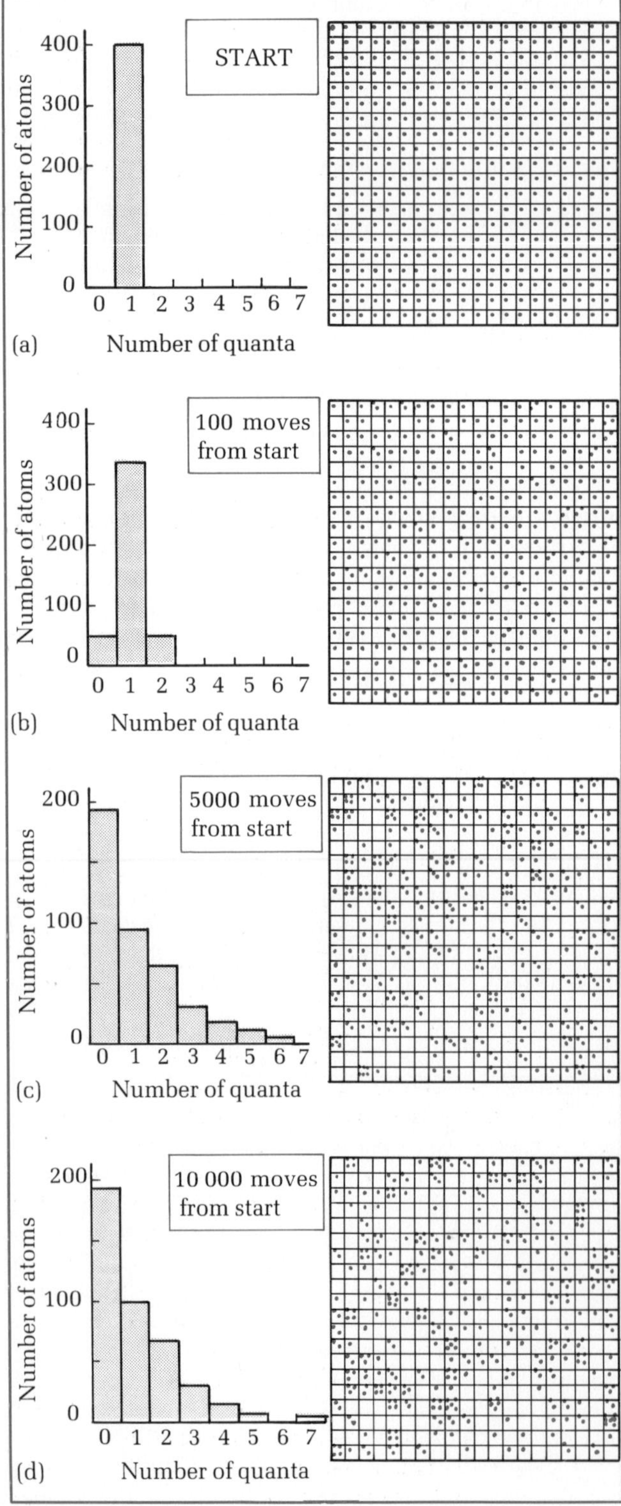

Figure 11.16 *The quantum shuffle*

Figure 11.16 shows that the distribution becomes uneven. It is difficult to see any meaning in the grid pictures. But if a **distribution curve** is plotted for each picture, we can see what equilibrium means for this model. Each curve shows the number of atoms with no quanta, with one quantum, with two quanta, etc., in each grid picture. The eventual distribution is an exponential decrease curve. The curve represents a distribution pattern where the ratio

$$\frac{\text{number of atoms with } q \text{ quanta}}{\text{number of atoms with } q-1 \text{ quanta}}$$

is the same for any value of q.

Why should one particular pattern be the eventual pattern? There are many ways to distribute the quanta. Because the quanta shuffle about at random, each way is just as likely to give the grid picture as any other way. Figure 11.17 shows the number of ways of distributing four quanta among three atoms. There are 15 ways in total, but one particular distribution pattern occurs more often. So this is the pattern most likely to be pictured. For large numbers of atoms and quanta, there is a huge number of ways to distribute the quanta, but one particular pattern occurs far, far more frequently than any other. This is the **equilibrium pattern**.

The equilibrium pattern is the one with the **greatest number of ways of distributing the quanta**. In other words, it is the one with the **greatest entropy**.

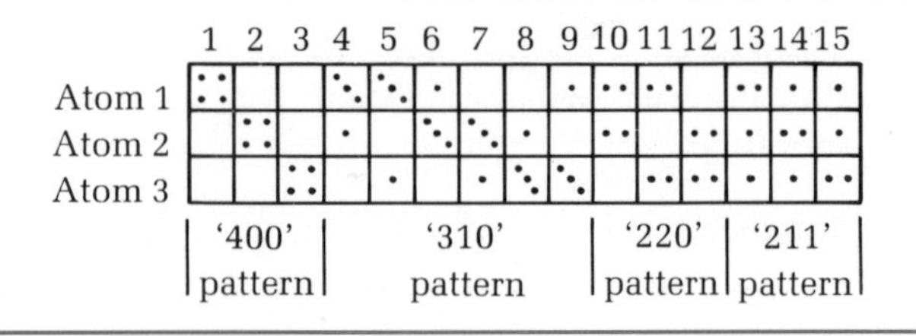

Figure 11.17 *Number of ways*

Heat transfer

Heating a solid involves supplying more energy quanta to the solid. With more quanta to be distributed among the same number of atoms, the number of ways to distribute the quanta increases. So the entropy of the solid increases.

Suppose before the transfer there are q quanta distributed amongst the N atoms of the solid. Let W_0 be the number of ways of distributing the quanta among the atoms. Probability theory tells us that one quantum gained by the solid increases the number of ways by a factor $\left(1+\frac{N}{q}\right)$. Let's use ω for this factor so $\omega = 1 + N/q$.

So the first quantum to be transferred increases the number of ways from W_0 to W_1 where $W_1 = \omega \times W_0$. The entropy changes from $k \ln W_0$ to $k \ln W_1$. So the increase of entropy is $k \ln W_1 - k \ln W_0 = k \ln(W_1/W_0) = k \ln\omega$. Each quantum transferred to the solid increases its entropy by $k \ln\omega$. Transfer in the opposite direction decreases the entropy by $k \ln\omega$ or $k \ln\left(1+\frac{N}{q}\right)$.

The factor $1+\frac{N}{q}$ can be proved from probability theory, based on random distribution of quanta among atoms.

Achieving equilibrium is a process where quanta transfer from hot to cold. Why? Consider two solids, X and Y, with equal numbers of atoms. Let's make the number of atoms in each 1000. Start with X and Y separated, with 700 quanta in X and 300 in Y. So X is initially hotter than Y.

X	Y
1000 atoms	1000 atoms
700 quanta	300 quanta

Figure 11.18 *Achieving equilibrium*

Now bring X and Y into thermal contact so quanta can transfer between them. The transfer factor for X is $1+\frac{1000}{700} = 2.43$. For Y, the transfer factor for Y is $1+\frac{1000}{300} = 4.33$.

1 Suppose the first transfer between X and Y is from X to Y.
So the number of ways of distributing the energy in Y, W_y, goes up by a factor of 4.33.
However, the number of ways of distributing the energy in X, W_x, goes down by a factor of $\frac{1}{2.43}$.
Therefore the total number of ways of distributing the energy in both X and Y changes by $\frac{4.33}{2.43}$ which equals a factor of 1.78.

2 Now suppose the first transfer had been from Y to X. So W_y changes by a factor of $\frac{1}{4.33}$ whereas W_x changes by a factor of 2.43. So the total number of ways changes by $\frac{2.43}{4.33}$ which equals a factor of 0.56.

Compare the two transfers. The number of ways of distributing the quanta is greatest if the transfer is from X to Y. So that is the most probable direction of transfer. For large numbers of quanta, the chance of transfer from X to Y far exceeds the chance of the reverse. So heat transfer is from hot to cold because that is the most likely change.

Entropy increases because the number of ways of distributing the energy increases. In the previous example, transfer from X to Y occurs because that is most likely. X loses a quantum of energy so its entropy falls by $k \ln 2.43$ which equals $0.88k$. Y gains a quantum of energy so its entropy rises by $k \ln 4.33$ which equals $1.46k$. So the total entropy rises by $1.46k - 0.88k$. Increasing the number of ways of distributing energy means increasing the disorder. **So entropy is a measure of the disorder of a system.** Entropy always tends to increase for an isolated system. Sometimes a system does become more ordered, so its entropy decreases. For example, when a battery is charged, positive ions are deposited in one part of the battery from the battery solution, and negative ions from the solution are deposited in a different part. Energy is stored in the battery, but to do this, work must be done on the battery. The work is done by a generator somewhere. Its entropy increases more than the entropy fall of the battery. So total entropy increases for any isolated system.

Boltzmann's tomb has the epitaph '$S = k \log W$'. The consequences of this little equation are indeed universal, for the Universe is an isolated system. Its entropy is increasing, and it is gradually becoming more and more disordered.

Activation processes

When an isolated system at constant temperature gains energy, its entropy increases because the number of ways of spreading the energy quanta in the system increases. If the system loses energy, then its entropy decreases, although the energy transfer causes an equal or greater entropy increase elsewhere. If the number of ways of spreading the energy decreases by a factor ω, then the entropy decrease ΔS equals $k \ln\omega$. Hence the energy loss from the system equals $-kT \ln\omega$ where the $-$ sign represents the loss. Thus the energy loss $E = -kT \ln\omega$; rearranged, this gives ω proportional to $e^{-E/kT}$. The factor $e^{-E/kT}$ is called the **Boltzmann factor**.

Consider how these ideas apply to processes where energy is distributed among many particles and where each particle needs a certain amount of energy to undergo a particular process. These are called **activation processes** because each particle needs a certain amount of energy to go through the process. For example, water molecules need a certain amount of energy to change state from liquid to vapour. Suppose the energy transfer when a molecule goes from one state to the other is E. The chance of transfer of a molecule from the liquid to the vapour state is governed by the Boltzmann factor $e^{-E/kT}$. This is because the chance of transfer is governed by how the energy is distributed. So the concentration of molecules in the vapour state above a liquid is proportional to the Boltzmann factor. Assuming the vapour pressure p is proportional to the concentration of vapour

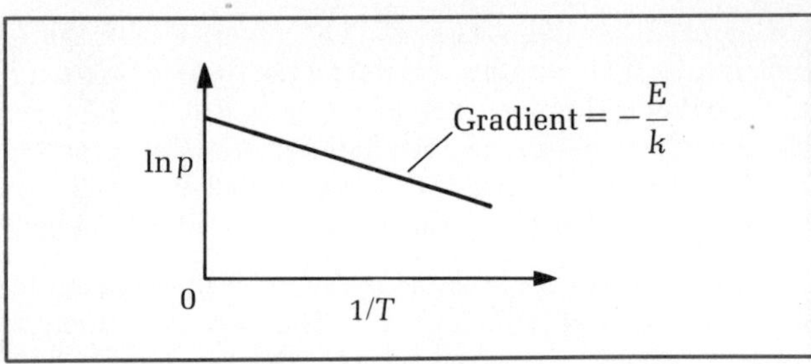

Figure 11.19 *Activation*

molecules, then p is proportional to $e^{-E/kT}$. The graph of lnp against $1/T$ should give a straight line with gradient $-E/k$, as in Figure 11.19. Measurement of the vapour pressure of water at different temperatures gives results in close agreement with Figure 11.19. The higher the temperature, the greater the rate of transfer of molecules.

Many processes and reactions occur at a rate proportional to the Boltzmann factor $e^{-E/kT}$. In each case, E is the energy needed to make the process happen. Increasing the temperature makes the process much more likely to happen. So the rate of reaction increases considerably. With many particles present, there will be a spread of energies among the particles. Some particles will have more than the required energy for the process. As the temperature increases, the number of particles with more than energy E increases. Hence the rate at which particles go through the process increases as the temperature is raised.

11.6 Summary

Laws of thermodynamics

1st law $Q = \Delta U - W$

2nd law No engine can convert heat transferred into it into an equal amount of work.

Definitions

1 *Molar heat capacity* C_p is the heat energy to raise the temperature 1 mole by 1 K at constant pressure.
2 *Molar heat capacity* C_V is the heat energy to raise the temperature of 1 mole by 1 K at constant volume.
3 *Entropy* $S = k \ln W$
4 *Efficiency* of a heat engine $= W/Q_1 \times 100\%$

Equations

1 For an ideal gas of point molecules $U = 3nRT/2$, $C_p = 5R/2$, $C_V = 3R/2$, $\gamma = 5/3$
2 For an adiabatic change of an ideal gas $pV^\gamma =$ constant
3 For an ideal heat engine, efficiency $= \dfrac{T_1 - T_2}{T_1} \times 100\%$
4 Entropy change due to heat transfer $\Delta S = \dfrac{Q}{T}$
5 The Boltzmann factor governing activation processes is $e^{-E/kT}$.

Short questions

11.1 When water at 100°C and a pressure of 101 kPa changes to steam under the same conditions, its volume increases by a factor of 1670. Given the density of water is 960 kg m^{-3} at 100°C and 101 kPa, and its specific latent heat of vaporization is 2.26×10^6 J kg^{-1}, calculate
a) the heat supplied to convert 1 kg of water at 100°C to steam at the same temperature,
b) the work done when 1 kg of water turns to steam at 101 kPa pressure,
c) the increase of internal energy.

11.2 A fixed mass of ideal gas is contained in a cylinder. The cylinder volume can be varied by moving a piston in or out. The gas has an initial volume 0.010 m^3 at 100 kPa pressure and its temperature is initially 300 K. The gas is cooled at constant pressure until its volume is 0.006 m^3. Sketch a pressure v volume graph to show the change. Calculate
a) the final temperature of the gas,
b) the work done on the gas,
c) the number of moles of gas, (assume $R = 8.3$ J mol^{-1} K^{-1}),
d) the change of internal energy of the gas,
e) the heat transfer from the gas.

11.3 Suppose the gas in question **11.2** had been compressed from its initial conditions ($V = 0.0100$ m^3, $p = 100$ kPa, $T = 300$ K) at constant temperature until its volume was 0.0060 m^3. Calculate the final pressure of the gas, and sketch a curve representing the change on your pressure against volume graph. From the curve, estimate the work done on the gas and compare your estimate with the work done in question **11.2**.

11.4 What is meant by an adiabatic change?
Two identical cylinders X and Y contain equal volumes of ideal gas at the same temperature and pressure. The volume of each cylinder can be varied by moving a piston in or out of the cylinder. The gas in each cylinder is then compressed to half its initial volume; X is compressed isothermally whereas Y is compressed adiabatically. Show the changes on a pressure against volume diagram, and compare the energy changes for the two gases.

11.5 Explain why the molar heat capacity of an ideal gas at constant pressure is greater than the molar heat capacity at constant volume. Prove that the difference is equal to the molar gas constant R.

11.6 Write down an equation for the pressure of an ideal gas in terms of its density ρ and the r.m.s. speed c of its molecules. Use this equation and the ideal gas equation to prove that the internal energy of one mole of an ideal gas is $3RT/2$ where T is the gas temperature.

11.7 Calculate the efficiency of an 'ideal' heat engine operating between temperatures of 500 K and 300 K for the hot and cold reservoirs respectively. If the engine does work at a rate of 5 kW, how much heat energy is wasted each second?

11.8 A heat engine operating between reservoirs at 700 K and 300 K is used to drive a turbine. The heat engine accepts heat energy from the hot reservoir at a rate of 5000 J s^{-1} and does work at a rate of 1.5 kW. Calculate
a) the heat energy wasted,
b) the engine's efficiency,
c) the change of entropy in 1 s,
d) the efficiency of an 'ideal' engine operating between the two reservoirs.

11.9 A heat pump is a device which does work to pump heat energy from one reservoir to a hotter reservoir. The coefficient of performance of a heat pump is the ratio of heat transferred from the cooler reservoir/the work done to cause the transfer. A heat pump with a coefficient of performance equal to 5 is used to maintain the temperature in a room at 20°C when the outside temperature is 5°C. With an electric heater instead of the heat pump, the heater uses 1.2 kW of electrical power to keep the temperature at 20°C in the room. When the heat pump was used, what was
a) its heat output to the room,
b) the electrical power used by the heat pump?

11.10 Entropy is a measure of disorder. Increased entropy means more ways of spreading energy about (i.e. more disorder). A system may be said to have achieved equilibrium when its entropy has stopped rising and is constant. Discuss the entropy changes, using the above terms, which take place when
a) hot and cold water are mixed,
b) coal is burned to heat a room,
c) milk is added to a cup of black coffee.

B

Multiple choice questions

BM.1 An oil droplet of diameter 0.4 mm is placed on a clean water surface and spreads out to form a circular patch of diameter 294 mm. The water surface is then cleaned and a second oil drop of diameter 0.3 mm is placed on the water surface to form a circular patch. The thickness, in nm, and diameter, in mm, of the second patch is

	A	B	C	D	E
Thickness	0.5	0.5	4.0	4.0	8.0
Diameter	150	190	220	250	220

BM.2 Gases have densities of about one thousandth of those of solids. If the distance between neighbouring molecules of a solid is *d*, the approximate average distance between neighbouring molecules of a gas is

A $10d$ **B** 10^2d **C** 10^3d **D** 10^6d **E** 10^9d

(N. Ireland)

BM.3 In an experiment to demonstrate Brownian motion in a gas, a cell containing smoke (carbon particles suspended in a gas) is brightly illuminated from the side and is viewed under a microscope. Bright specks of light are seen to be undergoing random motion. Which of the following statements is (are) correct?

1 The specks of light are produced by light which is scattered (or reflected) from the carbon particles.
2 The motion is random because random fluctuations occur in the number of gas molecules bombarding the carbon particles from different directions.
3 The average speed of the carbon particles is much less than that of the gas molecules.

A 1 only **B** 1 and 2 only **C** 1 and 3 only
D 2 and 3 only **E** 1, 2 and 3.

(N. Ireland)

BM.4 Which of the following statements about the force and P.E. of two molecules is *not* correct?

A The P.E. is least at equilibrium separation.
B The force is zero at equilibrium separation.
C The P.E. and force are both zero at infinite separation.
D The P.E. can be greater than zero if the molecules are very close together.
E The P.E. always falls as the molecules are moved closer together.

BM.5 A steel wire of length 2.00 m is stretched by a load of 60.0 N resulting in strain of 7.50×10^{-4} within the Hooke's law region. The energy stored in the wire is

A 4.5×10^{-2} J **B** 9.00×10^{-2} J **C** 3.75×10^{4} J
D 7.50×10^{4} J **E** 8.00×10^{4} J

(London)

BM.6 The following data was obtained when a wire was stretched within the elastic region.

Force applied to wire	100 N
Area of cross-section of wire	10^{-6} m^2
Extension of wire	2×10^{-3} m
Original length of wire	2 m

Which of the following deductions can be correctly made from this data?

1 The value of the Young modulus is 10^{11} N m^{-2}.
2 The strain produced is 10^{-3}.
3 The energy stored in the wire when the load is applied is 10 J.

A 1, 2 and 3 **B** 1 and 2 only **C** 2 and 3 only
D 1 only **E** 3 only

BM.7 The diagram shows the force *v* extension lines for two wires of the same material with the same initial length. When the force is 60 N, the extension of wire P is 3.4 mm and the extension of wire Q is 0.85 mm.

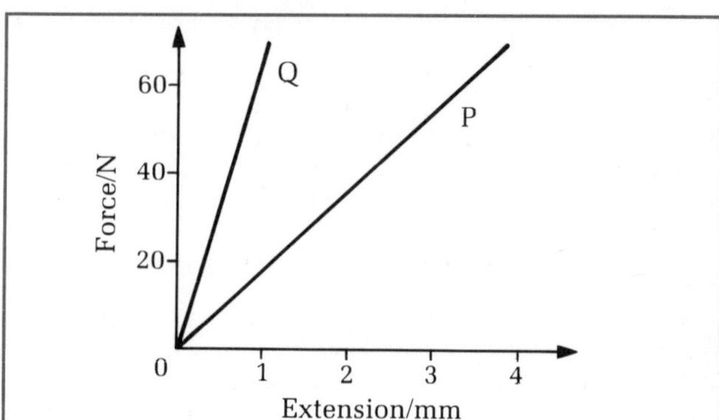

Figure BM1

The ratio of the diameter of P to that of Q must be

A 4 **B** 2 **C** 1
D 0.5 **E** 0.25

BM.8 For the two wires of question **BM.7**, the energy stored when P is extended by 1 mm from its initial length is equal to the energy stored in Q when Q is given an extension of

A 0.25 mm **B** 0.5 mm **C** 1.0 mm
D 2.0 mm **E** 4.0 mm

BM.9 The diagram below shows the flow of water at a steady rate through a uniform pipe with a narrow section. The water level in each manometer X, Y and Z indicates the pressure in each section. The level in Y is not shown.

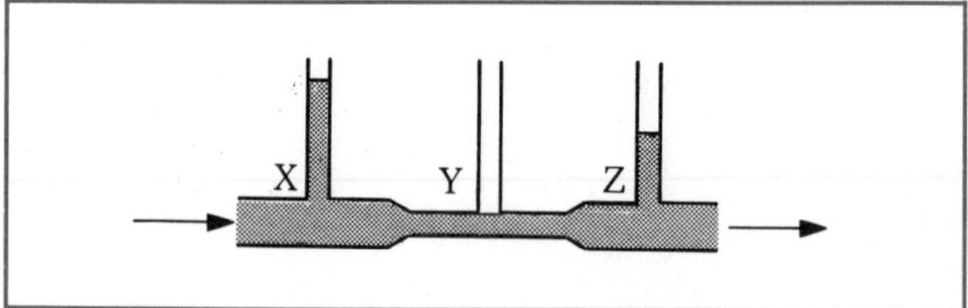

Figure BM2

Which of the following statements is (are) correct?

1 The water speed in the narrow section is greater than the speed in the other sections.
2 The water speed in section Z is less than in section X.
3 The water level in manometer Y is higher than the level in either X or Z.

A 1, 2 and 3 **B** 1 and 2 only **C** 2 and 3 only
D 1 only **E** 3 only

BM.10 Which of the following properties of a gas thermometer could *not* be regarded as a reason for its adoption as a standard?

A A wide range of temperatures can be covered.
B Expansion of the containing envelope is relatively small.
C It is capable of high accuracy.
D Thermometers using different gases give good agreement if the pressures are low.
E Use of a pressure gauge as an indicator makes it direct reading.

(London)

BM.11 The temperature of a hot liquid in a container of negligible thermal capacity falls at a rate of $2\,K\,min^{-1}$ just before it begins to solidify. The temperature then remains steady for 20 min by which time the liquid has all solidified.

The quantity $\dfrac{\text{specific heat capacity of liquid}}{\text{specific latent heat of fusion}}$, in K^{-1}, is

A 1/40 **B** 1/10 **C** 1 **D** 10 **E** 40

(AEB June 83)

BM.12 Two copper spheres of the same size are suspended in an evacuated enclosure which is maintained at 273 K. One sphere has a blackened surface and the other sphere has a polished surface. Both spheres are initially at a temperature of 300 K. Which of the following statements is correct?

A Both spheres will initially emit radiation at the same rate.
B Initially the blackened sphere's temperature will fall and the polished sphere's temperature will rise.
C Initially the blackened sphere's temperature will rise and the polished sphere's temperature will fall.
D The spheres will finally be at the same temperature as each other and the same temperature as the enclosure.
E The spheres will finally be the same temperature as each other but at a different temperature from the enclosure.

(London)

BM.13 The diagram shows two rods of different metals but of the same cross-sectional area welded together at Y.

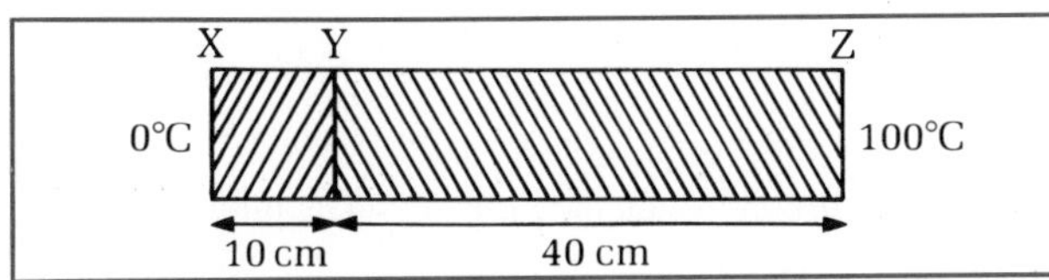

Figure BM3

XY is made of a metal whose thermal conductivity is four times that of YZ. The ends X and Z are maintained at temperatures of 0°C and 100°C respectively and the rods are well-lagged to prevent heat loss. When steady conditions obtain the temperature at Y (to the nearest degree) is

A 6°C **B** 20°C **C** 50°C **D** 80°C **E** 94°C

(London)

BM.14 In making a comparison between conduction of electricity along a uniform wire and conduction of heat along a lagged bar, which of the following statements is (are) correct?

1 The p.d. across the ends of the wire is analogous to the temperature difference between the ends of the lagged bar.
2 The electric current in the wire is analogous to the rate of flow of heat along the lagged bar.
3 The electrical resistance is analogous to 1/thermal conductivity.

A 1, 2 and 3 **B** 1 and 2 only **C** 2 and 3 only
D 1 only **E** 3 only

BM.15 Equal volumes of several gases are maintained under the same conditions of temperature and pressure. Which of the following statements about the gases under these conditions is (are) correct?

1 The number of molecules in each sample of gas is the same.
2 The ratio of the masses of the samples of any two gases is equal to the ratio of the molecular masses of the gases.
3 The ratio of the densities of any two gases is equal to the ratio of the molecular masses of the gases.

A 1, 2 and 3 **B** 1 and 2 only **C** 2 and 3 only
D 1 only **E** 3 only

BM.16 An ideal gas at pressure p and temperature T has N molecules per unit volume. If the pressure of the gas is changed to $0.5\,p$ and the temperature is changed to $2\,T$, the number of molecules per unit volume becomes

A $0.25N$ **B** $0.5N$ **C** N **D** $2N$ **E** $4N$

BM.17 The relative molecular masses of helium and oxygen are 4 and 32 respectively. At a given temperature, the ratio

$\dfrac{\text{r.m.s. speed of helium molecules}}{\text{r.m.s. speed of oxygen molecules}}$ is equal to

A $\sqrt{2}$ **B** $2\sqrt{2}$ **C** 4 **D** $4\sqrt{2}$ **E** 8

(AEB Nov. 84)

BM.18 The first law of thermodynamics may be written in the form of the equation $\Delta Q = \Delta U + \Delta W$ where ΔQ is the energy supplied to a gas, ΔU is the increase in its internal energy and ΔW is the work done by it in expanding. When a *real* gas undergoes a change at constant pressure which one of the following statements is correct?

A ΔQ is necessarily zero.
B ΔU is necessarily zero.
C ΔW is necessarily zero.
D None of ΔQ, ΔU, or ΔW is necessarily zero.
E The equation cannot be applied because it is true for an ideal gas only.

(London)

BM.19 A certain reaction proceeds at a rate which depends on temperature, in kelvins, according to the equation $A = e^{-E/kT}$, where A is a constant, k is the Boltzmann constant and E is the energy required to make the process occur. The rate is doubled when the temperature is increased from 300 K to 310 K. Which of the following statements is (are) correct?

1 The rate is doubled again when the temperature is increased from 310 K to 320 K.
2 A graph of ln R (where R is the rate) against $1/T$ gives a straight line.
3 E/k is equal to 9300 ln 2.

A 1, 2 and 3 **B** 1 and 2 only **C** 2 and 3 only
D 1 only **E** 3 only

BM.20 Energy will always flow from body X to body Y when they are in thermal contact if

1 X has more quanta than Y,
2 the average number of quanta per atom in X is greater than in Y,
3 X has more atoms than Y.

A 1 only **B** 2 only **C** 1 and 2 only
D 2 and 3 only **E** 1, 2 and 3

B

Long questions

BL.1

a) Draw a graph to show how the mutual potential energy $V(x)$ of two atoms separated by a distance x varies with x. Using the same axes, draw a graph to show how the mutual force $F(x)$ between the atoms varies with x. Explain how the two graphs are related.
Use your graphs:
i) to explain why solids expand when heated;
ii) to predict the extent to which solids have elastic properties.

b) Describe how you would attempt to measure the Young modulus of rubber. State what apparatus you would use, the form and dimensions of the specimen, what readings you would take, and how the result would be calculated.

c) A steel bar of length l and cross-sectional area A is clamped rigidly at its ends when it is at 0°C. The temperature of the bar is then lowered through $\Delta\theta$ by immersion in a cold liquid. Given that the linear expansivity α and the Young modulus E of the steel are constant over the temperature range involved, derive expressions for:
i) the tension in the cooled bar;
ii) the elastic energy stored in the bar.

(Oxford)

BL.2

a) Sketch a graph showing how the force between two atoms varies with the distance between the atoms. On the same axes, sketch a second curve showing how the potential energy of the two atoms varies with distance apart.
Use your graph to explain
i) why latent heat is needed to change that state of a solid,
ii) why most solids obey Hooke's law for small extensions.

b) The specific latent heat of sublimation of lead is $9.74 \times 10^5\,\text{J kg}^{-1}$. The density of lead is $11 \times 10^3\,\text{kg m}^{-3}$. Estimate
i) the average separation of adjacent lead atoms in the solid state,
ii) the energy needed to remove each lead atom, on average, from the solid state.
(The Avogadro constant is $6.02 \times 10^{23}\,\text{mol}^{-1}$; the molar mass of lead is 0.207 kg.)

BL.3 This question is about mechanical properties of materials. You may find the data at the top of the next column useful.

a) Compare the mechanical properties of steel and glass-reinforced plastic (e.g. Fibreglass). Credit will be given for the correct use of technical terms such as *toughness* and *strength*, and explanations of their meanings.

b) Explain, in terms of the structures of the two materials, why steel and Fibreglass have the mechanical properties that they do.
Include an explanation of how each material yields.

c) Choose two different applications of strong materials, one particularly suited to steel, and one particularly suited to Fibreglass.

Explain how the properties of the material are put to good use in these applications, and why the other material is less suitable.

	Steel	**Fibreglass**
The Young modulus	$21 \times 10^{10}\,\text{N m}^{-2}$	$3.5 \times 10^{10}\,\text{N m}^{-2}$
Ultimate tensile strength	$10^9\,\text{N m}^{-2}$	$10^9\,\text{N m}^{-2}$
Density	$7800\,\text{kg m}^{-3}$	$1850\,\text{kg m}^{-3}$
Cost per kilogram	15p–20p	£1.00–£1.50
Cost per cubic metre	£1170–£1560	£1850–£2775

(O and C Nuffield)

BL.4

a) Below is a graph showing how the extension of a steel wire of length 1.2 m and area of cross-section 0.012 mm² alters as a stretching force is applied.

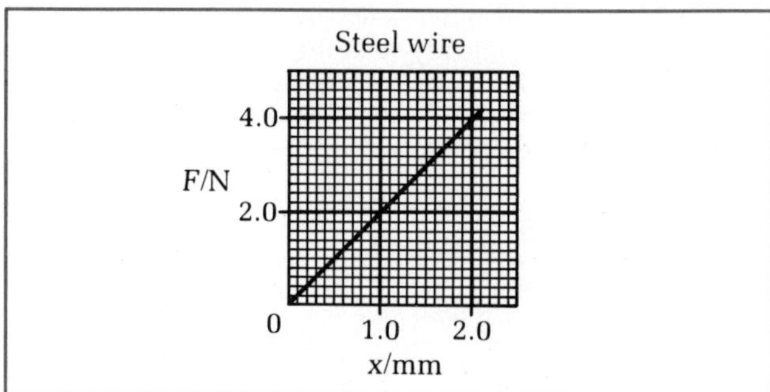

Figure BL1

i) Use the graph to calculate the Young modulus for steel.
ii) Draw a labelled diagram of an experimental arrangement suitable for obtaining such a set of results.

b) This graph shows the results of a similar experiment done with a copper wire. In this case the wire has been stretched until it breaks.

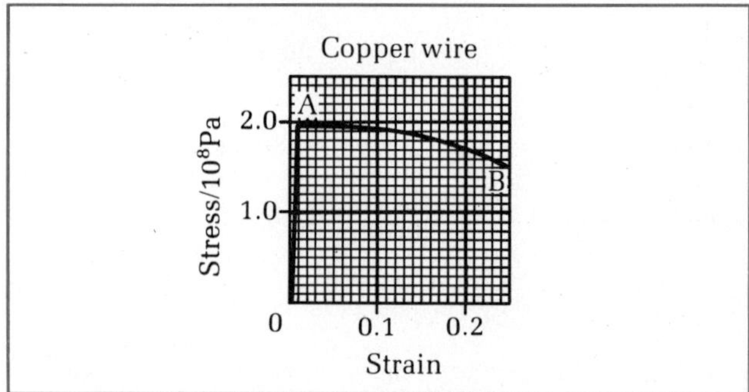

Figure BL2

i) The graph drawn in this instance is a stress-strain curve. Explain *one* advantage of representing the results in this way.
ii) Account in molecular terms for the behaviour of the wire as it is stretched from A to B.
iii) The copper wire used was 2.0 m long and 0.25 mm² in cross-section. Calculate the tension in the wire at A and an approximate value for the work done in producing a strain of 0.1.

c) A length of rubber cord is suspended from a rigid support and stretched by means of weights attached to its lower end.
  i) Sketch a stress-strain curve to represent the behaviour of such a cord as it is first loaded then unloaded.
  ii) Suppose the cord were continuously stretched and relaxed at a rapid rate. What might you notice? How would this be explained by the stress-strain graph?

(London)

BL.5 A copper wire and a length of rubber are each subjected to linear stress until they break. Sketch labelled graphs of stress against strain to show the behaviour of each material. Write brief notes on the important differences between the behaviour of these two materials. The table below shows how the extension e of a 10 m length of a certain nylon climbing rope depends on the applied force F.

e/m	0	1.9	2.8	3.4	3.8	4.1	4.3
F/kN	0	2.0	4.0	6.0	8.0	10.0	12.0

a) Draw a graph of applied force against extension. A climber of mass 70 kg, attached to a 10 m length of this rope, can withstand a force from the rope of no more than 6.5 kN without the risk of serious injury.

b) Read off from your graph the extension which would be produced in the rope for a force of 6.5 kN.

c) Use the graph to find the energy stored in the rope if it were stretched by this amount.

d) If the upper end of the rope were securely anchored, through what vertical distance could the climber fall freely (before the rope started to stretch) without risk of injury from the force of the rope when his fall was arrested?

(Cambridge)

BL.6 Sketch a typical stress-strain graph for the stretching to breaking point of a malleable metal such as copper. Mark on your graph the limit of proportionality, elastic limit, yield point and breaking stress (or tensile strength). Explain the meanings of each of these terms.
Account briefly for each region of your graph in terms of the microscopic structure of the metal.
Sketch typical stress-strain curves, for stretching to breaking, of
a) a brittle material such as glass,
b) polyethylene.
Label the regions of interest.
A mass of 20 kg is suspended from a length of copper wire of radius 1 mm. If the wire suddenly breaks, find the resulting change in temperature.
(Data for copper: Young modulus $= 12 \times 10^{10}$ Pa; density $= 9 \times 10^3$ kg m^{-3}; specific heat capacity $= 420$ J kg^{-1}K^{-1})

(WJEC)

BL.7 Give an account of the elastic properties of rubber and explain these as far as you are able in terms of the microscopic structure of rubber.
Why is elastic hysteresis in rubber of great importance in relation to the performance of car tyres?

(WJEC)

BL.8

a) i) State and explain the first law of thermodynamics.
  ii) Explain the meaning of the concept of *internal energy* as applied to monatomic gases, diatomic gases, and crystalline solids.
  iii) Discuss what happens to the heat supplied to water at 100°C to make it boil at normal atmospheric pressure.

b) Poiseuille's equation for the volume of liquid V flowing in time t under laminar-flow conditions through a uniform capillary tube of radius r and length l with a pressure difference p between its ends is

$$\frac{V}{t} = \frac{\pi r^4 p}{8 l \eta},$$

where η is the coefficient of viscosity of the liquid. In an experiment with a liquid at room temperature, the following readings are obtained:
$r = 0.80 \pm 0.05$ mm,
$p = 2000 \pm 20$ Pa,
$l = 0.300 \pm 0.001$ m,
rate of flow $= 53.6 \pm 0.2$ cm^3 per minute.
  i) Explain what is meant by *laminar flow*.
  ii) Given that the dimensions of η are $ML^{-1}T^{-1}$, show that the equation is dimensionally consistent.
  iii) Calculate the value of η for the liquid from this experiment and the maximum value of the error which may be expected.
  iv) How might a significantly high value for η in an experiment of this kind be explained?

(Oxford)

BL.9

a) What do you understand by the *equation of continuity* as applied to a fluid in motion?

b) Derive Bernoulli's equation for an incompressible fluid.

c) A simple syringe used to produce a jet of water consists of a piston of area 4.0 cm^2 which moves in a horizontal cylinder which has a small hole of area 4.0 mm^2 at its end. If the force on the piston is 50 N calculate a value for the speed at which the water is forced out of the small hole, assuming the speed of the piston is negligible. The density of water is 1.00×10^3 kg m^{-3}.

d) Explain why the speed of the piston may be ignored.

(JMB)

BL.10

a) Distinguish between *static pressure, dynamic pressure* and *total pressure* when applied to streamline (laminar) fluid flow and write down expressions for these three pressures at a point in the fluid in terms of the flow velocity v, the fluid density ρ, pressure p, and the height h, of the point with respect to a datum.

b) Describe, with the aid of a labelled diagram, the Pitot static tube and explain how it may be used to determine the flow velocity of an incompressible, non-viscous fluid.

c) The static pressure in a horizontal pipeline is 4.3×10^4 Pa, the total pressure is 4.7×10^4 Pa, and the area of cross-section is 20 cm^2. The fluid may be considered to be incompressible and non-viscous and has a density of 10^3 kg m^{-3}. Calculate
  i) the flow velocity in the pipeline
  ii) the volume flow rate in the pipeline.

(JMB)

BL.11

a) i) Explain the terms *specific heat capacity, specific latent heat* and *internal energy.*

ii) Why is a distinction between the specific heat capacity at constant pressure and that at constant volume important for gases but less important for solids and liquids?

b) Describe an accurate method to measure the specific latent heat of vaporization of a liquid boiling at atmospheric pressure.

c) Calculate the external work done and the internal energy gained when 1.0 kg of water at 100°C and 1.0×10^5 Pa pressure is converted to steam. Take the density of steam under these conditions to be 0.58 kg m^{-3}, the specific latent heat of vaporization of water to be 2.3×10^6 J kg^{-1}, and the density of water to be 1000 kg m^{-3}.

d) The specific latent heats of fusion of substances have markedly different values from those of their specific latent heats of vaporization at the same pressure. Explain in general terms how these differences arise.

(Oxford)

BL.12

a) A temperature scale can be set up by using fixed points in conjunction with some property which varies continuously and reproducibly with temperature.

i) What is meant by the term 'fixed points'? Give two examples of such points.

ii) One such property is the resistance of a platinum coil. How is temperature defined on such a scale?

b) Two thermometers are based on different properties but they are calibrated using the same fixed points.
To what extent are the thermometers likely to agree when used to measure a temperature

i) near one of the fixed points;

ii) midway between the two fixed points?

Justify your answers.

c) The diagram shows the main parts of the apparatus used in the determination of the specific heat capacity of a liquid by continuous flow.

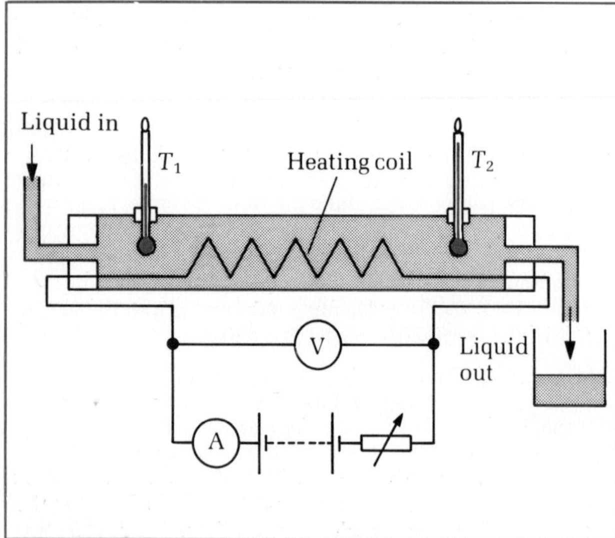

Figure BL3

i) What are the principal advantages of this method compared to other simple laboratory methods?

ii) When using the apparatus to determine the specific heat capacity of water, a student first adjusted the flow rate of the water so that 50 g was collected in 1 minute. The rheostat was then adjusted until the voltmeter read 12.0 V when the ammeter reading was 2.50 A. The readings on thermometers T_1 and T_2 eventually reached steady values of 20.0°C and 28.0°C respectively.
The student then reduced the flow rate to 25 g min^{-1} and adjusted the rheostat until the voltmeter read 8.8 V. The ammeter reading was 1.85 A, but both thermometer readings remained unchanged.
Use the results to calculate a value for the specific heat capacity of water.

(AEB June 84)

BL.13 Explain what is meant by a *scale of temperature*, discussing how, in principle, a scale would be established.

The following table gives an example of a thermometer, the associated physical property and the measurement taken. Make a list giving the entries for the blank spaces (*a*)–(*d*) in the table.

Instrument	Physical property	Measurement taken
Constant volume gas thermometer	Pressure change of gas	Difference in manometer levels supported by the pressure of a fixed volume of nitrogen
Liquid-in-glass thermometer	(*a*)	(*b*)
Thermocouple	(*c*)	(*d*)

Why is it *not* convenient to use a constant volume gas thermometer in most practical situations? To what use is it actually put?

The masses of hydrogen and oxygen atoms are 1.66×10^{-27} kg and 2.66×10^{-26} kg respectively. What is the ratio of the 'average' speed of hydrogen and oxygen atoms at the same temperature? What is usually meant by the term 'average' speed in this case?

For the thermocouple give, with reasons, a situation for which it would be particularly suitable.

(London)

BL.14

a) The kinetic theory of gases predicts that the root mean square (r.m.s.) speed of the molecules of an ideal gas is given by the expression $(3p/\rho)^{\frac{1}{2}}$, where p is thepressure and $\rho\pi$ is the density of the gas.
The graphs show how the pressure of oxygen gas temperatures, T and 300 K.

i) Use the graph at the top of the next page to calculate a value for the r.m.s. speed of the oxygen molecules at 300 K. Explain your working.

ii) Is the temperature T higher or lower than 300 K? Explain your reasoning.

iii) The graphs are based upon experimental results. What conclusion can you draw from them about the behaviour of oxygen?

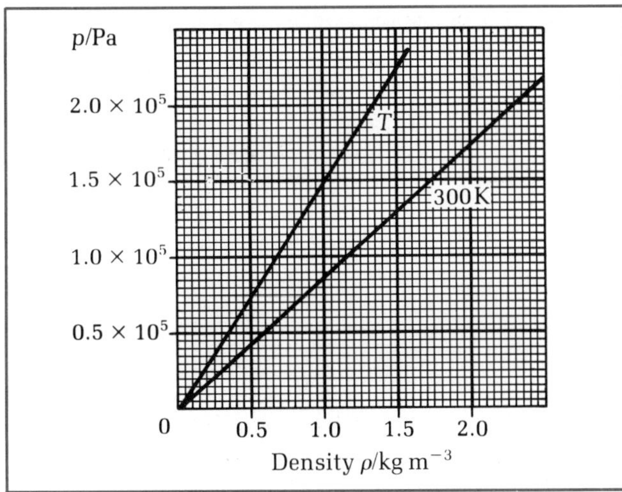

Figure BL4

iv) Outline a simple experimental procedure for investigating how the pressure of a known mass of air varies as its density changes at room temperature. Include a labelled diagram of the apparatus used.

b) The graphs below show how the speeds of the molecules in an ideal gas are distributed at wo temperatures. Use them to help you answer the following questions.

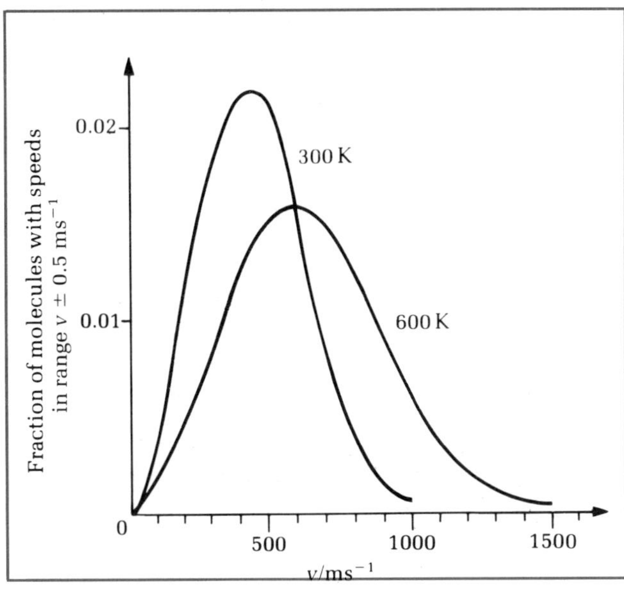

Figure BL5

i) In what two main ways does the temperature appear to affect the distribution of speeds?

ii) What is the value of v for which the fraction of molecules with speeds in the range $v \pm 0.5\ \mathrm{m\,s^{-1}}$ is a maximum at a temperature of 300 K? How does this value compare with the r.m.s. speed calculated in (a)(i) above?

(London)

BL.15

a) Explain what is meant by an ideal gas. State two respects in which a real gas differs from an ideal gas. Under what conditions do most gases approach the ideal closely enough for many purposes?

b) Using a simple kinetic theory it may be shown that, for an ideal gas, $p = \frac{1}{3}\rho\overline{c^2}$ where p is the pressure of gas, ρ its density and $\overline{c^2}$ the mean square speed of the molecules. Use this formula to calculate the root mean square speed of hydrogen molecules at s.t.p. if standard pressure is 1.01×10^5 Pa and the density of hydrogen at s.t.p. is $0.090\ \mathrm{kg\,m^{-3}}$. Sketch graphs, on the same axes, to indicate how the molecular speeds are distributed at two different temperatures. State clearly the meanings of the quantities to which the axes of your graph correspond. Mark which curve corresponds to the higher temperature.

c) Show how the equation $p = \frac{1}{3}\rho\overline{c^2}$ may be related to the equation of state for an ideal gas, $pV = nRT$, where p is the pressure, V the volume of n moles, R the molar gas constant and T the absolute temperature. Justify any assumption which you have to make.

d) The mass of one mole of oxygen molecules is 32.0 g and their root mean square speed is $460\ \mathrm{m\,s^{-1}}$ at 0°C. If the Avogadro number is $6.02 \times 10^{23}\ \mathrm{mol^{-1}}$, obtain a value for the Boltzmann constant.

(AEB Nov. 83)

BL.16

a) Explain what is meant by a saturated vapour.

b) Draw graphs to show the relationship between pressure and temperature (i) for an ideal gas, (ii) for saturated water vapour and (iii) for a mixture of the ideal gas and the saturated water vapour. The curves should be drawn on axes extending from 0–100°C (x-axis) and $0\text{–}1.2 \times 10^5$ Pa (y-axis) and should be reasonably accurate. The initial pressure of the ideal gas at 0°C is to be taken as 0.2×10^5 Pa.
(Normal atmospheric pressure is *approximately* 10^5 Pa.)

c) Describe how you would measure the saturated vapour pressure of water at various temperatures between 10°C and 80°C.

d) A mass of saturated water vapour at 50°C occupies a volume of $1.00 \times 10^{-2}\ \mathrm{m^3}$. When converted to steam at 100°C and pressure 1.01×10^5 Pa it occupies $1.60 \times 10^{-3}\ \mathrm{m^3}$. Assuming that, up to saturation point, the vapour behaves like an ideal gas, calculate the saturated vapour pressure of water at 50°C.

(AEB June 83)

BL.17 State the first law of thermodynamics in the form of an equation, explaining clearly the meaning of each term in your equation.

Starting from the first law, and stating and explaining any assumptions you make, prove that the difference between the principal heat capacities for 1 mole of an ideal gas is given by

$$C_P - C_V = R$$

where R is the gas constant for 1 mole.

The piston of a bicycle pump is slowly moved in until the volume of air enclosed is one fifth of the total volume of the pump and is at room temperature (290 K). The outlet is then sealed and the piston suddenly drawn out to full extension. No air passes the piston. Find the temperature of the air in the pump immediately after withdrawing the piston assuming that air is a perfect gas with $\gamma = 1.4$.

Sketch a graph showing how the internal energy of the enclosed mass of air varies with time, starting from the instant that the outlet was sealed and ending when the air in the pump has returned once more to room temperature.

(WJEC)

BL.18

a) Explain what is meant by *thermal conductivity* and use your explanation to derive the unit of thermal conductivity.

b) The ends of a vertical column of liquid are maintained at different steady temperatures. Describe how heat is transferred through the column when the hotter end is (i) at the top, (ii) at the bottom of the column. In which case is the rate of flow of heat the greater? Give the reason for your answer.

c)

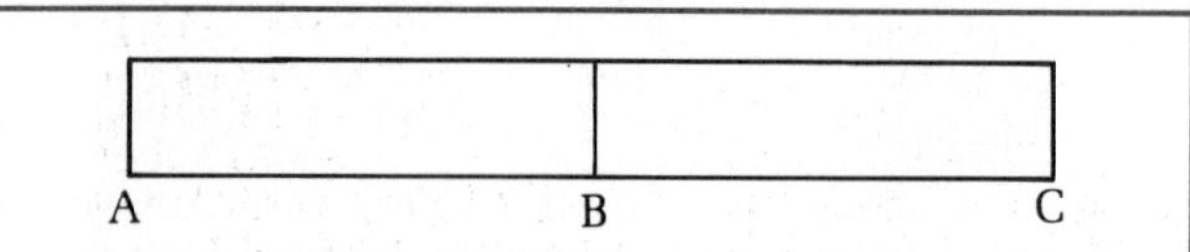

Figure BL6

The diagram shows two bars, AB and BC, of identical dimensions joined to form a composite bar. The thermal conductivities of AB and BC are k and $\frac{1}{2}k$ respectively. End A is maintained at 60°C and end C at 30°C.

i) Sketch the lines of heat flow through the bar when the bar is ideally lagged and when it is unlagged.

ii) For the lagged bar, calculate the temperatures at B and the mid-point of AB. If end C is then maintained at 60°C and end A at 30°C show that the magnitude of the rate of flow of heat is unchanged.

(JMB)

BL.19

a) Define thermal conductivity. What are its dimensions and in what unit is it measured?

b) Describe how you would measure the thermal conductivity of a bad conductor.
Why is the shape of the specimen of material which you would use for this experiment different from that which you would use for a good conductor?

c) A long cylindrical metal bar is heated at one end and kept at room temperature at the other end, the temperatures being maintained constant. Draw sketch graphs to show the temperature distribution along the length of the bar if it is (i) unlagged, (ii) perfectly lagged. In each case explain the shape of your graph.

(AEB June 82)

BL.20 Define (*a*) the *Stefan-Boltzmann constant* and (*b*) the *thermal conductivity* of a substance. Show that their units are $W\,m^{-2}K^{-4}$ and $W\,m^{-1}K^{-1}$ respectively.

The normal flux of radiant energy from the Sun at the distance of the Earth is 1360 $W\,m^{-2}$. Stating any necessary assumptions, calculate

i) the total power emitted by the Sun

ii) the temperature of its surface

iii) the rate of loss of mass by the Sun, given that the mass m_E of energy E is given by $m_E = E/c^2$, where c is the speed of electromagnetic waves in vacuum.

Explain how heat is conducted through a poor thermal conducting material. Does the mechanism you have described also apply for a good conductor such as silver? Why is silver so very much better as a conductor than glass?

Give two practical reasons why it is necessary to use a relatively thin specimen of large cross-sectional area in determinations of thermal conductivity of a poorly-conducting material.

(Sun–Earth distance $= 150 \times 10^6$ km
Sun's radius $= 0.7 \times 10^6$ km.)

(SUJB)

SECTION C

Fields

Gravitational fields · Electric fields · Magnetic fields · Electromagnetic induction

A space shuttle ascending into orbit above the Earth. Designed to be reusable, each shuttle can lift 29 tonnes of cargo into orbit, and can take on board damaged satellites for repair.

A field is an invaluable 'thinking model' to explain interactions between objects not in direct contact. There are four basic forms of interaction in physics, namely gravitational, electrical, magnetic and nuclear. Any type of force is due to one of these four basic interactions. One of the triumphs of 19th century physics was to show that electricity and magnetism are different aspects of the same form of interaction. The link is not easy to demonstrate in theory, but its consequences in the form of electromagnetic waves are profound.

The ideas of force and energy are used extensively to develop the description of fields in terms of field strength and potential. To explain interactions, we consider that an object creates a field around itself, and the field exerts a force on any similar object that is in the field. For example, a charged object creates an electric field around itself that affects any other charged object close to it. Moving charges create magnetic fields, and masses create gravitational fields. The force of interaction between two similar objects is measured by the field strength and the energy is measured by the potential. So the ideas of field strength and potential, developed from force and energy respectively, are features of all field descriptions.

12

Gravitational fields

12.1 Force and potential

Gravitational field strength

'What goes up must come down' – or must it? Throw a ball into the air and it returns to you because of the Earth's gravity. The force of gravity on the ball pulls it back to Earth. The force of attraction between the ball and the Earth is an example of gravitational attraction which exists between any two masses. It isn't obvious that there is a force of attraction between you and *any* object near you, but it is true. Any two masses exert a gravitational pull on one another. But the force is usually too weak to notice unless at least one of the masses is very large.

The mass of an object creates a force field around itself. Any other mass placed in the field is attracted towards the object. The second mass also has a force field round itself, and this pulls on the first object with an equal force. The force field round a mass is called a **gravitational** field. Suppose a small mass is placed close to a massive body. The small mass and the body attract each other, but the force is too small to move the massive body noticeably. The small mass, assuming it is free to move, is pulled by the force towards the massive body. The path which the smaller mass would follow is called a **field line**, or sometimes a **line of force**. Figure 12.1 shows the field lines near a planet. The lines are directed to the centre of the planet, so the field is a radial field.

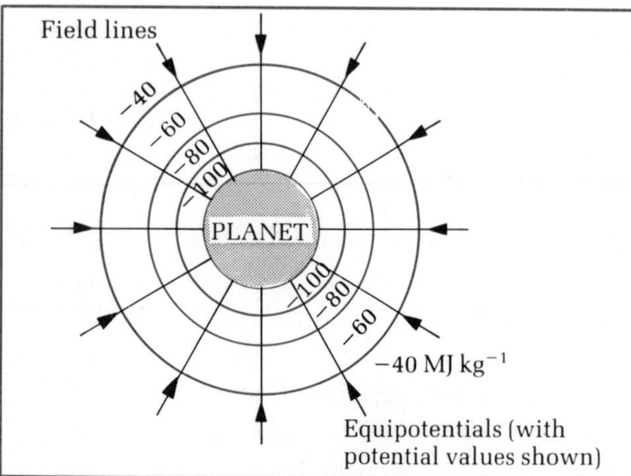

Figure 12.1 *Equipotentials near a planet*

The strength of a gravitational field, g is the force per unit mass on a small test mass placed in the field. In general the force varies from one position to another so g varies too. The unit of g is $N\,kg^{-1}$.

$$g = \frac{F}{m}$$

where F = force on a small test mass m placed in the field,
g = gravitation field strength where m is placed.

The test mass must be small, otherwise it might pull so much on the object producing the field that it might alter the field.

Consider a small mass m in free fall just above the Earth. The mass accelerates due to the pull of gravity. The force on m is mg, where g is the gravitational field strength. From Newton's second law, the acceleration equals $\frac{\text{force}}{\text{mass}}$. Hence the acceleration $= \frac{mg}{m} = g$. In other words the acceleration of free fall is equal to the gravitational field strength. So the gravitational field strength near the Earth's surface is $9.8\,N\,kg^{-1}$ because the acceleration of free fall is $9.8\,m\,s^{-2}$ just above the Earth.

A uniform field is one in which the field strength has the same value at all points in the field. In a uniform field, a test mass would experience the same force wherever it was placed. Is this the case for the Earth's field? Far from the Earth, its pull is much weaker, so clearly the Earth's gravitational field strength falls with distance from Earth. However, over small distances, the change is insignificant. For example, the measured value of g is the same 100 m above the Earth as on the surface. In theory, g is greater at the surface than higher up, but the difference is too small to be noticeable – provided we don't go too high! Only over distances which are small compared with the Earth's radius can the Earth's field be considered uniform.

Gravitational potential

Imagine you're in a space rocket about to blast off from the surface of a planet. The planet's gravitational field extends far out into space, although it becomes weaker with distance. To escape from the planet's pull, the rocket must do work against the pull of gravity from the planet. If the rocket fuel doesn't provide enough energy to escape, you are doomed to return! The planet's field is a gravitational trap which the rocket must climb out from to escape. As the rocket climbs, its potential energy rises. If it falls back, its potential energy drops. It's a bit like the situation of a bowling green which dips down in the middle. To make a ball reach the edge from the middle, it must be rolled with sufficient kinetic energy to start with. As it rolls up to the edge, its potential energy rises and its kinetic energy falls. If it stops before it reaches the edge, then the slope makes it roll back to the centre.

Potential energy is energy due to **position**. The position for zero P.E. is at infinity – in other words, so far away that the gravitational pull on a test mass is negligible. Our rocket climbing out of the planet's field needs to increase its P.E. to zero to escape completely. At the surface, its P.E. was negative so it needs to do work to climb out of the field to zero P.E.

The gravitational potential V at a point in a gravitational field is the P.E. per unit mass of a small test mass placed at that point. The unit of gravitational potential is $J\,kg^{-1}$.

Suppose our rocket has a mass of 1000 kg, and the planet has surface potential equal to $-100\,MJ\,kg^{-1}$. Assume the fuel is used quickly to boost the rocket to high speed, then the fuel stage is jettisoned, leaving the 1000 kg payload escaping from the planet. For the rocket to escape completely, the P.E. of the 1000 kg payload must increase from -100×1000 MJ to zero. So the work done on the payload must be at least 100 000 MJ to escape.

The work done to remove a test mass from potential V to infinity must be numerically equal to mV, where m is the mass being moved. So the potential at a point in a gravitational field is equal to the work done per unit mass to remove a test mass from that point to infinity. If the test mass is moved from potential V_1 to a different potential V_2, its change of P.E. is equal to $m(V_2 - V_1)$. If the rocket payload is only given 40 000 MJ of K.E. from the fuel, then it can only increase its P.E. by 40 000 MJ. So it can only reach a distance where the potential is -60 MJ kg^{-1}.

Equipotentials are lines of constant potential. Hill-walkers ought to know all about equipotentials since a map contour line is a line of constant potential. A contour line joins points of equal height above sea level. So a hillwalker following a contour line has constant potential energy. Sensible hillwalkers take great care where the contour lines are very close to one another. One slip and their P.E. might fall dramatically!

The equipotentials near the Earth are circles like Figure 12.1. Near the surface, the equipotentials are uniformly spaced. A 1 kg mass raised from the surface by 1 m gains 9.8 J of P.E.; if raised another 1 m, it gains another 9.8 J of P.E. So the P.E. rises by 9.8 J for each metre of height gained near the surface. However, much higher up, the gravitational field strength becomes weaker, so the gain of P.E. per metre of height gain becomes less. In other words, away from the Earth's surface, the equipotentials become more spaced out.

The potential gradient is the change of potential per metre. Near the Earth's surface, the potential changes by 9.8 J kg^{-1} for every metre of height gained. So the potential gradient near the surface of the Earth is constant, and equal to 9.8 J kg^{-1} m^{-1}. However, further from the Earth's surface, the potential gradient becomes less and less.

Consider a test mass m being moved out of a gravitational field, as shown in Figure 12.2. To move the mass by a small distance Δr in the opposite direction to the gravitational force on it, its P.E. must be increased by an equal and opposite force F. The work done by F is equal to $F\Delta r$. So the increase of P.E. of the test mass is equal to $F\Delta r$.

Hence the change of potential ΔV of the test mass$=\dfrac{F\Delta r}{m}$, since the change of potential equals the change of P.E. per unit mass.

$$\text{So} \quad \Delta V = \frac{F\Delta r}{m}$$

$$\text{giving} \quad F = m\frac{\Delta V}{\Delta r}.$$

Since the gravitational force is equal and opposite to the applied force, then the gravitational force$= -m\dfrac{\Delta V}{\Delta r}$.

The gravitational field strength is given by

$$g = \frac{\text{gravitational force}}{\text{mass}}$$

$$= -\frac{\Delta V}{\Delta r}.$$

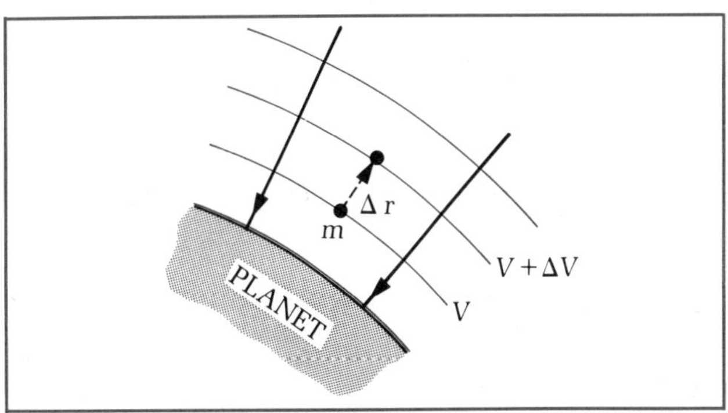

Figure 12.2 *Potential gradients*

Now suppose we let $\Delta r \rightarrow 0$, so $\dfrac{\Delta V}{\Delta r}$ is now written as $\dfrac{dV}{dr}$ where $\dfrac{d}{dr}$ means 'change per unit distance'. Hence $\dfrac{dV}{dr}$, the change of potential per metre (i.e. the potential gradient), equals $-g$.

Gravitational field strength, g, is the negative of the potential gradient. Therefore

$$g = -\frac{dV}{dr}$$

Potential gradients are like contour gradients on maps. The closer together the contours are on a map, the steeper the hill. Likewise, the closer the equipotentials are, the stronger the field is. Where the equipotentials are evenly spaced, the potential gradient is constant. Hence the gravitational field strength is constant, and the field is uniform.

A non-uniform gravitational field occurs where the gravitational field strength varies from one position to any other position in the field. Figure 12.3 shows a non-uniform field. The equipotentials at X are spaced very closely, giving a potential gradient of $\dfrac{5 \times 10^3 \text{ J kg}^{-1}}{1000 \text{ m}}$ which equals 5 J kg^{-1} m^{-1}. So the gravitational field strength at

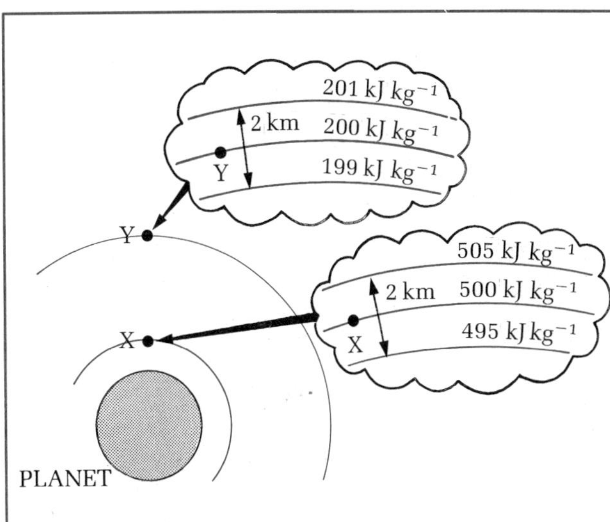

Figure 12.3 *Calculating g*

X is $-5\,N\,kg^{-1}$. The $-$ sign tells us that the gravitational force on a test mass at X is inwards. Work out for yourself the potential gradient at Y. Hence show that the gravitational field strength at Y is $-1\,N\,kg^{-1}$.

12.2 Newton's theory of gravitation

We owe our understanding of gravitation to Isaac Newton. 'The notion of gravity was occasioned by the fall of an apple!', said Newton when asked what made him develop the idea of gravity. Newton's theory of gravitation was an enormous leap forward because it explains events from the 'down-to-earth' falling apple to the motion of planets. And like any good theory, it can be used to make predictions; for example, the return of a comet and its exact path can be calculated using the law of gravitation.

Newton realized that gravitation is universal. Any two masses exert a force of attraction on each other. He knew about the careful measurements of planetary motion made by astronomers like Johannes Kepler. Forty or more years before Newton established the theory of gravitation, Kepler had shown that the motion of the planets was governed by a set of laws. Kepler had measured the motion of each of the planets, and showed that each planet orbits the Sun. The measurements that he made for each planet were its time period T (i.e. the time taken for one complete orbit) and the average radius r of its orbit. He showed that the value of r^3/T^2 was the same for all the planets. This is known as **Kepler's third law**.

	Mercury	Venus	Earth	Mars	Jupiter	Saturn
Average radius r of orbit/10^{10} m	6	11	15	23	78	143
Time T for one orbit/10^7 s	0.8	1.95	3.2	5.9	37.4	93.0
r^3/T^2 /10^{16} $m^3 s^{-2}$	337	350	330	349	340	338

Figure 12.4 *Kepler's third law*

To explain Kepler's third law, Newton started by assuming that the planets and the Sun were point masses. A scale model of the Solar System with the Sun represented by a 1p piece would put the Earth about a metre away, represented by a grain of sand! Newton then assumed that the force of gravitation between a planet and the Sun varied inversely with the square of their distance apart. Using this inverse-square law of force, he was able to prove that r^3/T^2 was the same for all the planets. (The actual proof is given in section 12.4.) He then went on to use the inverse-square law of force to explain and make predictions for many other events involving gravity.

The theory of gravitation assumes that the gravitational force between any two point objects is
a) always an attractive force,
b) proportional to the mass of each object,
c) proportional to $1/r^2$, where r is their distance apart.
These three requirements can be summarized as

$$F = -G\frac{m_1 m_2}{r^2}$$

where m_1 and m_2 = masses of the two objects,
r = distance apart,
F = gravitational force.

The constant of proportionality, G, in the above equation is called the **universal constant of gravitation.** The unit of G can be worked out from the equation above; rearranged, the equation gives $G = -Fr^2/m_1m_2$. So G can be given in units of $N\,m^2\,kg^{-2}$. The value of G is $6.7 \times 10^{-11}\,N\,m^2\,kg^{-2}$. The $-$ sign in the equation is because the force is always an attractive force.

Work out for yourself the gravitational force between two point masses, each of mass 10 kg at 0.1 m apart. The values for m_1, m_2 and r must be put into the equation in units of kilograms and metres. You should obtain a force value which is far too small to notice in a real situation. Only if one or more of the masses is very large does the force become noticeable, unless special techniques are used.

Measurement of G was first made in a laboratory by Henry Cavendish in 1798. He devised a torsion balance made of two small lead balls at either end of a rod. The rod was suspended horizontally by a torsion wire, as in Figure 12.5. The wire was calibrated by measuring the couple required to twist it, per degree. Then, with the rod at rest in equilibrium, two massive lead balls were brought near the torsion balance to make the wire twist. By measuring the angle which it twisted through, the force of attraction between each massive lead ball and the small ball nearest to it was calculated. The distance between the centres of the small and large masses was also measured. Then G was calculated using the equation for the law of gravitation.

Figure 12.5 *Cavendish's experiment*

12.3 Planetary fields

Field strength and potential

The law of gravitation can be used to determine the field strength or potential at any point in the field produced by a planet or any other spherical mass. Newton showed that the field of a spherical mass is the same as if the mass were concentrated at its centre. The field lines of a spherical mass are always directed towards the centre, so the field pattern is just the same as for a point mass, as shown in Figure 12.6. Newton used the law of gravitation, which is for point masses, to show that the field strength and the potential are the same too, just as if the field were produced by an equal point mass at the centre.

The gravitational field strength due to a planet of mass M can be calculated by treating the planet as a point mass M at its centre. Let the planet radius be R. Suppose a test mass, m, is placed at distance r from the centre of the planet. Treating the planet as a point mass M at a distance r from the test mass, Newton's law of gravitation gives the force on the test mass as

$$F = \frac{GMm}{r^2} \quad \text{(towards the centre of the planet)}$$

The gravitational field strength, g, at the test mass position is defined as the force per unit mass on the test mass. Hence

$$g = \frac{F}{m} = \frac{GM}{r^2}$$

1 Away from the planet, the gravitational field strength follows an inverse-square law. At the surface, distance r equals the planet radius R. So the surface gravity g_s is equal to GM/R^2. Rearranged, this gives $GM = g_sR^2$. So above the surface, we can write $g = \frac{g_sR^2}{r^2}$ for the gravitational field strength g at distance r from the centre.

We can use this equation to calculate g in terms of the surface gravity g_s for different distances from the planet.

Distance *r* from centre	R	$2R$	$3R$	$4R$
Gravitational field strength	g_s	$g_s/4$	$g_s/9$	$g_s/16$

The Moon's surface gravity is known to be about $1.6\,\text{N kg}^{-1}$. Its radius is about 1600 km. So here on Earth, the pull of the Moon is very weak because we are about 380 000 km away from the Moon. In fact, the gravitational field strength due to the Moon at this distance is given by

$$g_M = \frac{1.6 \times (1600 \times 10^3)^2}{(380\,000 \times 10^3)^2} = 2.8 \times 10^{-5}\,\text{N kg}^{-1}.$$

Not too much to worry about, you might think! But the Moon's gravity is the cause of ocean tides which can create great problems in coastal areas.

2 Inside a planet or any other spherical mass, gravitational field strength falls to zero at the centre. From the equation $g = GM/r^2$, you might think that g becomes ever larger and larger as r becomes smaller and smaller. However, inside the planet, only the mass in the sphere of radius r 'counts'. The rest of the mass outside radius r up to the surface gives no resultant force. So as r becomes smaller, the mass M which contributes to g becomes smaller too. At the centre, the mass which contributes is zero; hence g is zero at the centre.

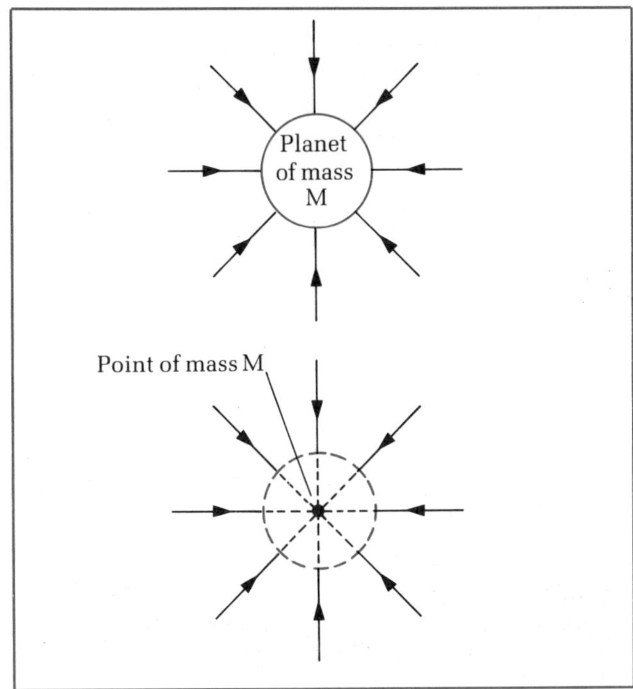

Figure 12.6 *Comparing fields*

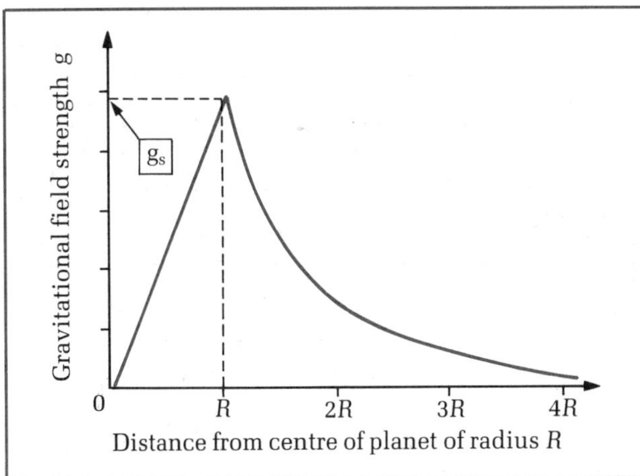

Figure 12.7 *Gravitational field strength*

The gravitational potential V of a planet can also be calculated from Newton's law of gravitation. Gravitational potential is the P.E. per unit mass of a small test mass in the field. We saw in section 12.1 that the gravitational potential at a point gives the work done per unit mass to move a test mass to infinity from that point. The gravitational force on the test mass varies with position, as shown by Figure 12.8. The force curve is described by the equation $F = -GMm/r^2$. To move the test mass away, an equal and opposite force $F'(= -F)$ is required.

Consider one small step in moving the test mass m from

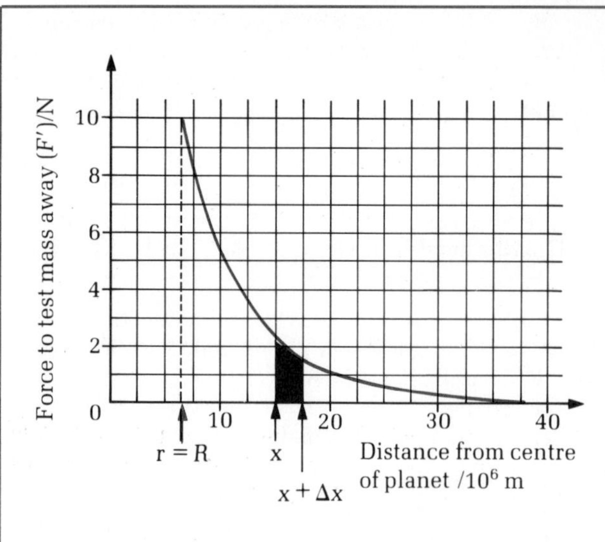

Figure 12.8 *Work done*

the distance r to infinity. Suppose in that small step the distance from the planet's centre increases from x to x+Δx. The average force F' over that step is GMm/x^2.

The work done on m to move it by a distance Δx is equal to $F'\Delta x$. This amount of work done is represented on Figure 12.8 by the area of the strip of width Δx under the curve.

So the work done ΔW for one small step away is given by

$$F'\Delta x = \frac{GMm}{x^2}\Delta x$$

The total work done W is given by adding up the work done in each small step from r to infinity. You can do this by estimating the total area under the curve from r to infinity; count the blocks of area under the curve. Part blocks over half-size can be counted as whole blocks; part blocks under half size can be ignored. In this way, you can obtain a reasonable estimate for the total area of all the strips from r to infinity. Then use the scales to determine how much work each block represents. The scales in Figure 12.8 are for a 1 kg test mass moved away from the Earth. Each block corresponds to a force of 1 N moved away by a distance of 2500 km. So each block represents 2.5 MJ. How much work is done to move a 1 kg mass from the Earth's surface to infinity? See if you can work it out from the 'area' method.

The mathematical method for determining the total work done uses integration. For this, we need to make the steps very small. Then we add the work done in all the steps by integrating.

So the total work done

$$W = \int_{x=r}^{\infty} \Delta W = \int_{x=r}^{\infty} \frac{GMm}{x^2} dx$$

where we use dx to mean a very small distance. The addition or integration is from x = r to x = infinity (∞).

Hence the total work done is given by

$$W = GMm\int_{x=r}^{\infty} \frac{dx}{x^2} = GMm\left[\frac{-1}{x}\right]_{x=r}^{\infty} = \frac{GMm}{r}.$$

See p. 506 for a further discussion of integration if necessary.

The work done to move a test mass m from r to infinity is therefore equal to GMm/r, where M is the planet's mass. We can use this to calculate the work done to remove a 1 kg mass from the Earth's surface to infinity. The Earth's radius $R = 6400$ km. We know that $GM = g_s R^2$ (see the previous section). So the work done for $m = 1$ kg is $g_s R^2/R$, since the starting point is $r = R$. Hence the work done $= g_s R = 9.8 \times 6400 \times 10^3$ J for 1 kg.

How does this value of 63 MJ kg^{-1} compare with the 'block counting' method? The block method is not too accurate unless you make each block very small, which is what the integration process does.

The work done to remove a test mass m from r to infinity $= \frac{GMm}{r}$. So when a test mass is moved from r to infinity, its P.E. increases by an amount equal to GMm/r. Since its P.E. at infinity is zero, then its P.E. at r must be equal to $-GMm/r$. Therefore potential at r (= P.E. per unit mass) is equal to $-\frac{GMm/r}{m}$ which equals $-GM/r$.

The gravitational potential V near a planet of mass M is given by

$$V = -\frac{GM}{r}$$

where r = distance from the planet's centre.

Figure 12.9 shows how the gravitational potential near a planet or other spherical mass varies with distance from its centre. The scales are marked with values worked out for the Earth's field, but they could be changed to apply to any spherical mass. The potential at the surface is equal to $-GM/R$, where M is the mass of the sphere and R is its radius. So if values of M and R are known, the surface potential can be calculated. Since the surface gravity $g_s = GM/R^2$, the surface potential is equal to $-g_s R^2/R$, which equals $-g_s R$. Given values of g_s and R, the surface potential can be determined.

The curve of Figure 12.9 is a $1/r$ curve, not an inverse square curve. So the potential at distance $2R$ from the centre is 0.5 × the surface potential, the potential $3R$ from the centre is 0.33 × the surface potential etc.

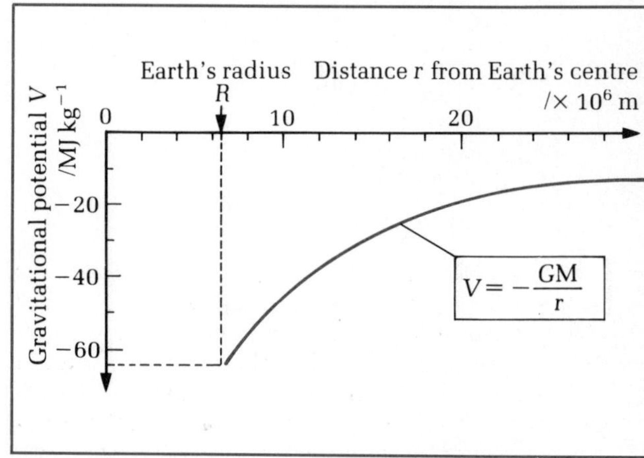

Figure 12.9 *Gravitational potential near Earth*

Escape velocity

To remove a 1 kg mass from the Earth's surface to infinity, 63 MJ of work must be done on the 1 kg mass. To cause it to escape by projecting it, it must be given 63 MJ of kinetic energy by launching it at high speed. What speed must it be launched with? We can use the K.E. formula $\frac{1}{2}mv^2$; so for $m=1$ kg,

$$\tfrac{1}{2}\times 1\times v^2 = 63\times 10^6\ \text{J}$$

Hence $v=\sqrt{2\times 63\times 10^6}=11\times 10^3\ \text{m s}^{-1}$.

The escape velocity from a point in a gravitational field is the minimum velocity of projection for any small mass to escape from the field to infinity.

Consider the escape velocity of a small mass near a planet. To escape from a position at distance r from the planet's centre, the P.E. of the small mass must increase from $-GMm/r$ to 0.

If projected, the small mass will escape if its initial K.E. is sufficient to increase its P.E. by GMm/r. Therefore, for speed of projection v, it will escape if $\frac{1}{2}mv^2$ is greater than or equal to GMm/r,

$$\text{i.e. } \tfrac{1}{2}mv^2 \geqslant \frac{GMm}{r}$$

$$\text{or} \quad v^2 \geqslant \frac{2GM}{r}$$

$$\text{So the escape velocity} = \sqrt{\frac{2GM}{r}}$$

Since the gravitational field strength g at the point of projection equals GM/r^2, then $2GM/r$ equals $2gr$. Hence the escape velocity is equal to $\sqrt{2gr}$.

Escape velocity from distance r from a planet of mass M is given by

$$V_{esc} = \sqrt{\frac{2GM}{r}} = \sqrt{2gr}$$

where r = distance of point of launch measured from the planet's centre.

Worked example The Moon's radius is 1600 km and its surface gravity g is $1.6\ \text{N kg}^{-1}$. Calculate the escape velocity from the lunar surface.

Solution $g=1.6\ \text{N kg}^{-1}$, $r=1600\times 10^3$ m.

To calculate the escape velocity, use $v_{esc}=\sqrt{2gr}$.

Hence the escape velocity is

$v_{esc}=\sqrt{2\times 1.6\times 1600\times 10^3}=2.3\times 10^3\ \text{m s}^{-1}$.

Black holes are 'cosmic crushers' with gravitational fields so strong that not even light can escape from them. Anything which falls into a black hole would never be seen again, even if it could withstand the enormous forces. If no light can escape from them, how do we know that they exist? They were 'invented' in the first place by theoreticians, but astronomers reckon they have discovered one or two possible candidates for 'black hole' status. One of these is a 'mystery' companion to a supergiant star in Cygnus. The mystery companion is a strong source of X-rays, thought to be produced as a result of drawing in mass from the supergiant.

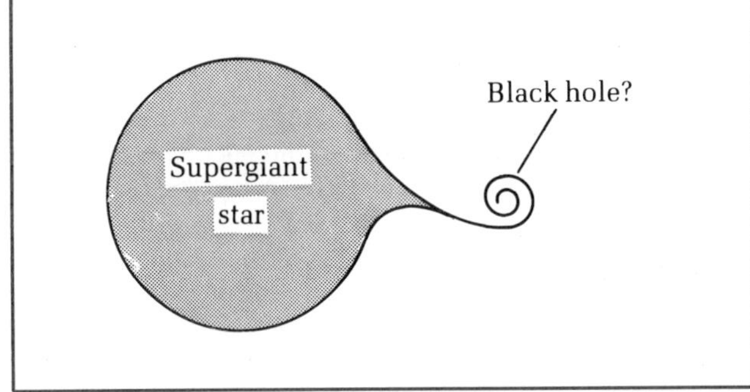

Figure 12.10 *Black holes*

Change of P.E. over small distances

A small mass m at a distance r from the centre of a much larger mass M has P.E. equal to $-GMm/r$. Suppose the small mass is moved away from the centre of the larger mass by a distance h, as in Figure 12.11. The P.E. of the small mass is therefore increased to $-GMm/(r+h)$.

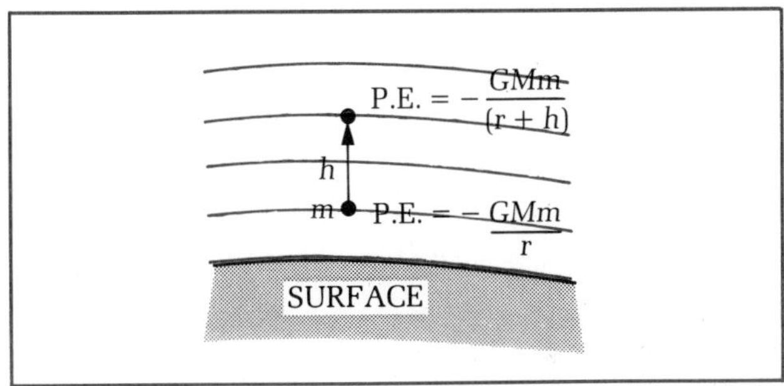

Figure 12.11 ΔP.E. = mgh

The change of P.E. is the P.E. at final position − P.E. at initial position, and is given by

$$\text{Change of P.E.} = \left[-\frac{GMm}{(r+h)}\right]-\left[-\frac{GMm}{r}\right]$$

$$= GMm\left(\frac{1}{r}-\frac{1}{(r+h)}\right) = GMm\left[\frac{(r+h)-r}{r(r+h)}\right]$$

$$= GMm\frac{h}{r(r+h)}$$

This complicated formula becomes very familiar if we assume h is much smaller than r. Then we can write r^2 in place of $r(r+h)$. Using $g=GM/r^2$,

$$\text{Change of P.E.} = \frac{GM}{r^2}mh = gmh = mgh$$

So we can only use the 'mgh' formula for P.E. change if h is much less than r. This amounts to the same as assuming g does not change significantly over distance h.

12.4 Satellite motion

On any clear night you ought to be able to see satellites passing overhead in the night sky. Although they are only pinpoints of light, they are noticeable because they move steadily through the constellations. Newspapers supply information to enable you to identify some of them from their directions. However, satellite motion is not confined to artificial satellites orbiting the Earth. Any small mass which orbits a much larger mass is a satellite. The Moon is the Earth's only natural satellite. Mars has two moons, *Phobos* and *Deimos*. Jupiter has at least fourteen satellites! The planets themselves are satellites in orbit round the Sun.

Kepler's laws describe the orbital motion of the planets round the Sun. Newton showed that the same laws apply to the motion of any satellite.

1 Kepler's first law Each planet moves in an elliptical orbit with the Sun at one of the two focal points of the ellipse.

In fact some of the planets move in orbits which are almost circular, with the Sun at the centre. A circle may be considered as a special case of an ellipse where the two focal points are at the same position.

2 Kepler's second law The line from the Sun to each planet sweeps out equal areas in equal times.

A planet in an elliptical (i.e. non-circular) orbit moves fastest when it is at the closest distance of approach to the Sun. Figure 12.12 shows the distance moved by a planet from P_1 to P_2 in a given time when it is near the Sun. The same diagram shows the planet when it is furthest from the Sun; it moves from P_3 to P_4 in the same time as it takes to move from P_1 to P_2. The area SP_1P_2 is equal to the area SP_3P_4. Newton's theory gives a complete explanation of this law.

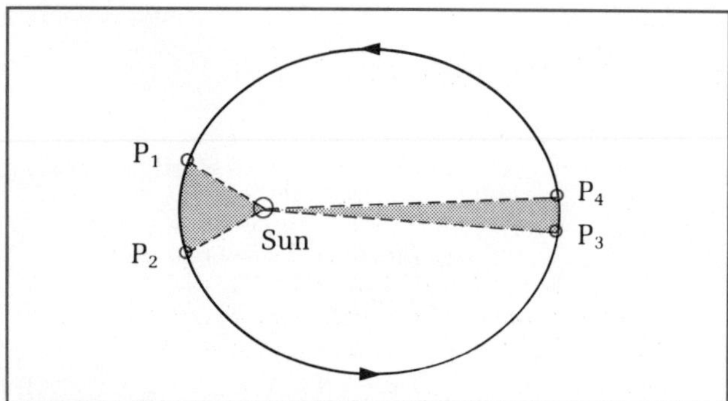

Figure 12.12 *An elliptical orbit*

3 Kepler's third law The ratio r^3/T^2 is the same for all the planets. T is the time taken for one complete orbit of average radius r.

The proof of Kepler's third law is easiest for circular orbits. Consider a planet of mass m moving in a circular orbit of radius r about the Sun. Let M represent the mass of the Sun.

At any position in the orbit, the velocity of the planet is at right angles to the radial line from the planet to the Sun. The force on the planet is always directed inwards towards the Sun. So the force is at right angles to the velocity. The planet therefore moves round at steady speed. Let its speed be v.

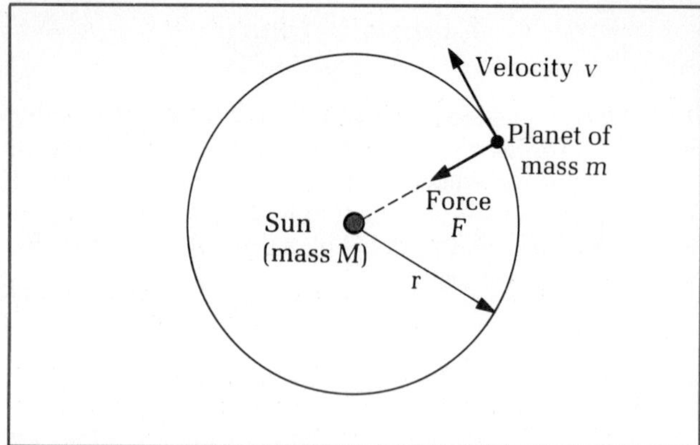

Figure 12.13 *Kepler's third law*

Since the planet follows a circular orbit of radius r at steady speed v, the centripetal acceleration of the planet $= v^2/r$.

Therefore the centripetal force $= mv^2/r$.

The centripetal force is provided by the force of gravitation, GMm/r^2.

Hence $$\frac{GMm}{r^2} = \frac{mv^2}{r}$$

So $$v^2 = \frac{GM}{r}$$

For steady speed, the time, T, to make one complete orbit is given by the equation $v = \dfrac{2\pi r}{T}$.

Therefore $$\frac{(2\pi r)^2}{T^2} = \frac{GM}{r}$$

Giving $$\frac{r^3}{T^2} = \frac{GM}{4\pi^2}$$

which is Kepler's third law, since $GM/4\pi^2$ is the same for all the planets. The value of $GM/4\pi^2$ is determined by the mass M of the central body round which the satellite orbits.

Newton's theory not only explains Kepler's laws, but it also allows the mass M to be calculated if G's value is known. Figure 12.4 shows that the value of r^3/T^2 for any planet is $3.35 \times 10^{18}\ \mathrm{m^3\,s^{-2}}$. Given $G = 6.7 \times 10^{-11}\ \mathrm{N\,m^2\,kg^{-2}}$, we can calculate the Sun's mass in kilograms. Work it out for yourself.

The motion of any satellite in orbit around a more massive body can be determined using Kepler's laws if the mass of the central body is known. For Earth's satellites, we can write Kepler's third law as follows.

$$\frac{r^3}{T^2} = \frac{GM}{4\pi^2} = \frac{g_s R^2}{4\pi^2}$$

where g_s is the surface gravity,
R is the radius of the planet.

Satellite calculations: worked examples

1 The Moon has an orbital time period of 27.3 days. Calculate its average distance from Earth. Use $g_s = 9.8\ \mathrm{N\,kg^{-1}}$ and $R = 6400$ km for the Earth's surface gravity and radius respectively.

Solution Let r = the average distance from Earth to Moon.

$$\frac{r^3}{T^2}=\frac{g_sR^2}{4\pi^2}=\frac{9.8\times(6400\times1000)^2}{4\pi^2}=1.0\times10^{13}\,\text{m}^3\,\text{s}^{-2}.$$

Hence $r^3 = 1.0\times10^{13}\times T^2$.
Since $T = 27.3$ days $= 27.3\times24\times3600$ s
$= 2.36\times10^6$ s
we can calculate r. Complete the calculation yourself to show that the required distance is 380 000 km.

2 The first artificial satellite to orbit the Earth was *Sputnik 1* which had a time period of 96 minutes. Calculate its average height above the Earth. Use the same values for g and R as in the previous example.
Solution $T = 96$ minutes $= 96\times60$ s $= 5760$ s
$\frac{r^3}{T^2} = 1.0\times10^{13}\,\text{m}^3\,\text{s}^{-2}$, from the previous example.
Hence $r^3 = 1.0\times10^{13}\times5760^2 = 3.32\times10^{20}\,\text{m}^3$.
So $r = 6923$ km.
So the first artificial satellite orbited the Earth at a height of about 500 km.

3 Communication satellites orbit the Earth with a time period of 24 hours, in orbits which make them appear stationary when seen from the Earth's surface. Their orbits are called **synchronous** orbits because they move round at the same rate as the Earth. We can use Kepler's third law to calculate the height of a communication satellite.
$T = 24$ hours $= 24\times3600$ s
Hence $r^3 = 1.0\times10^{13}T^2 = 1.0\times10^{13}\times(24\times3600)^2$
$= 7.5\times10^{22}\,\text{m}^3$.
So $r = 42\,100$ km.
The height above Earth is therefore 35 700 km, which is about $5\frac{1}{2}$ Earth radii.

4 Weather satellites orbit over the Earth's poles several times each day. They transmit detailed pictures of cloud patterns which weather forecasters use. Figure 12.14 shows a picture from a typical weather satellite as it passes over Europe from south to north. The picture is built up line by line as the satellite scans in strips from west to east. From the information given with Figure 12.14, see if you can work out the satellite's time period and height.

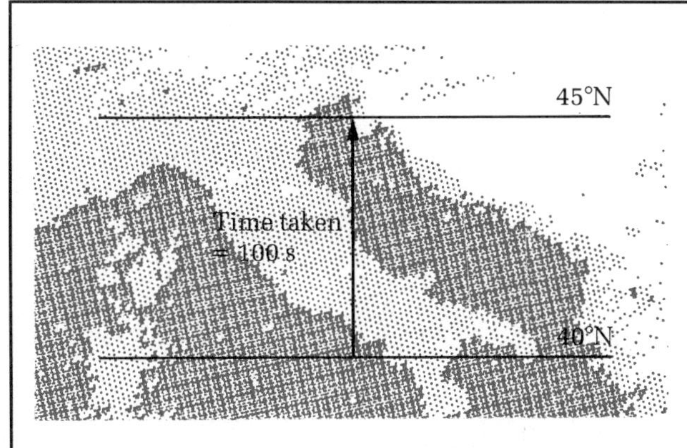

Figure 12.14 *Weather satellites*

12.5 Summary

Definitions

1 *Gravitational field strength* g at a point is the force per unit mass on a test mass placed at that point.
2 *Gravitational potential* V at a point is the P.E. per unit mass of a test mass placed at that point. The position of zero P.E. is at infinity.
3 *A uniform field* is one where the gravitational field strength is the same at all points.
4 *Potential gradient* is the change of potential per metre.

Equations

1 Gravitational field strength g

$$=\frac{F}{m}=-\text{potential gradient}, =-\frac{dV}{dr}$$

2 Near a large spherical mass M $g=\frac{GM}{r^2}$

$$V=-\frac{GM}{r}$$

3 Escape velocity $v_e=\sqrt{2gr}$

4 Satellite orbits $\frac{r^3}{T^2}=\frac{GM}{4\pi^2}$

Laws

1 Newton's law of gravitation $F=\frac{Gm_1m_2}{r^2}$

2 Kepler's third law r^3/T^2 is a constant

Short questions

Assume $G = 6.67\times10^{-11}\,\text{N m}^2\,\text{kg}^{-2}$

12.1 Use the information below to calculate
a) the Earth's gravitational field strength at the Moon,
b) the distance from the Earth's surface to the point between the Earth and Moon where the overall gravitational field strength due to Earth and Moon is zero. The Earth–Moon distance is 3.8×10^8 m.

	Surface gravitational field strength/N kg^{-1}	**Radius**/km
Earth	9.8	6400
Moon	1.6	1600

12.2 Figure 12.15, on the next page, shows the equipotentials near a certain non-spherical body. Calculate the P.E. of a 0.10 kg mass m at
a) P,
b) Q,
c) R.
What is the work that must be done on m to move it from
d) P to Q,
e) Q to R?

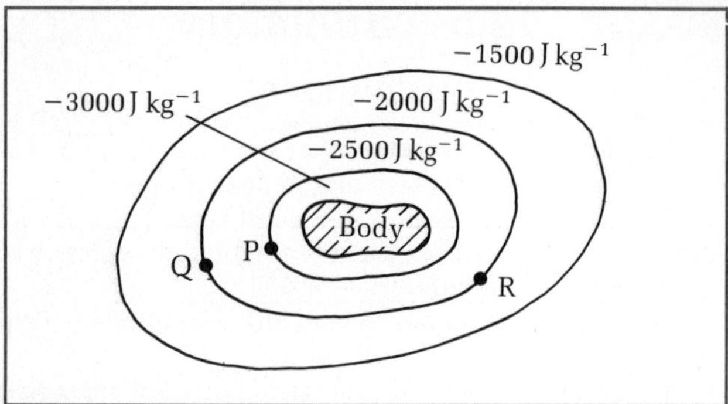

Figure 12.15

12.3 The variation of gravitational potential near a certain planet is shown by the graph. Use the graph to determine

a) the gravitational potential at X, a distance of 5000 km from the centre of the planet,

b) the change of P.E. of a 100 kg mass moved from point X to point Y, a distance of 4000 km from the centre of the planet,

c) the gravitational field strength at X.

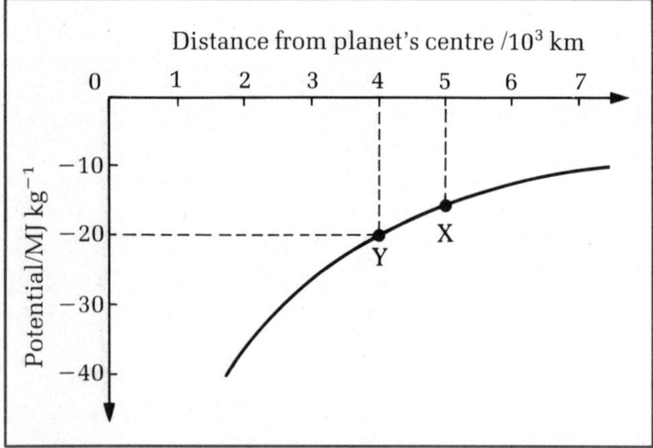

Figure 12.16

12.4

a) Write down an expression for the gravitational potential at distance *d* above the surface of a spherical planet of mass *M* and radius *R*.

b) Use your expression in **a)** to show that the escape velocity for a projectile from the planet's surface is $\sqrt{2g_sR}$ where g_s is the surface gravitational field strength.

12.5 Use the information given in question **12.1** and the value of G given to calculate

a) the force of gravitational attraction between the Earth and the Moon,

b) the time taken for the Moon to orbit the Earth once.

12.6 Astronomers reckon that the Milky Way Galaxy is rotating, and that the Sun takes 200 million years to make one complete orbit. The Sun is in one of the spiral arms of the galaxy, at a distance of 5×10^{21} m from the galactic centre.

a) Calculate the speed at which the Sun orbits the galactic centre.

b) Assuming most of the galactic mass is at its centre, calculate the mass of the galaxy.

c) Estimate the number of stars in the galaxy, given the mass of a typical star is 2×10^{30} kg.

12.7 Jupiter is the largest planet of the Solar System; its mass is 0.0011 × the Sun's mass. The distance from the Sun to Jupiter is 8×10^{11} m on average. Calculate

a) the distance from the Sun to the point along the line between the Sun and Jupiter at which their combined gravitational field strength is zero,

b) the gravitational potential at this point.

12.8 A certain weather satellite is in a circular orbit with a time period of two hours. Calculate its height above the Earth's surface, given the surface gravitational field strength is 9.8 N kg^{-1} and the Earth's radius is 6400 km.

12.9 Communications satellites orbit the Earth with a time period of 24 hours exactly so they stay in the same position relative to the Earth's surface. Calculate the height of such a satellite above the Earth's surface, given the surface gravity and radius of the Earth as in question **12.8**.

12.10 Calculate the total energy of a communications satellite of mass 100 kg in a stable circular orbit with a time period of 24 hours.

13

Electric fields

13.1 Field patterns

Static electricity

Most plastic materials can be charged quite easily by rubbing with a dry cloth. When charged, they can usually pick up small bits of paper. Try it with a plastic comb which you can probably charge by combing your hair! The bits of paper are attracted to the charged piece of plastic. Do charged pieces of plastic material attract one another? Figure 13.1 shows an arrangement to test for attraction. A charged perspex ruler will attract a charged polythene comb, but two charged rods of the same material always repel one another.

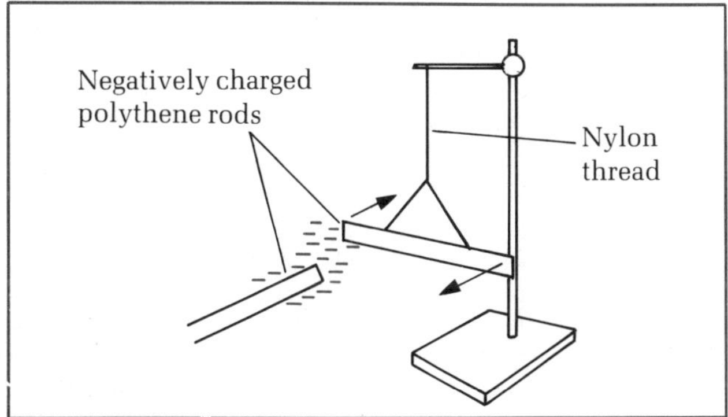

Figure 13.1 *Electrostatic forces*

Charging was first investigated scientifically in the early 17th century using natural materials such as ebonite and amber. The word 'electricity' dates from that time, derived from the Greek word for amber. It was realized that there were only two types of charge, and since they were found to neutralize each other, they were simply called positive and negative charge. Perspex is an example of a material which charges positively; polythene charges negatively. Tests such as that shown in Figure 13.1 show that objects with like charge always repel, and objects with unlike charge always attract.

Electrons are responsible for charging in most situations. As explained in Chapter 6, an atom contains a positively charged nucleus surrounded by electrons. The positive charge of the nucleus is due to protons, which each carry an equal and opposite charge to that of an electron. So an uncharged atom contains equal numbers of electrons and protons. Add electrons to an uncharged atom and it becomes a negatively charged ion. Remove electrons from an uncharged atom and it becomes a positively charged ion. An uncharged solid contains the same number of electrons and protons. To make it negatively charged, electrons must be added to it. Remove electrons from an uncharged solid and it becomes positively charged. When a perspex rod is charged by rubbing it with a dry cloth, electrons are transferred from the rod to the cloth, so the rod becomes positively charged, and the cloth becomes negatively charged.

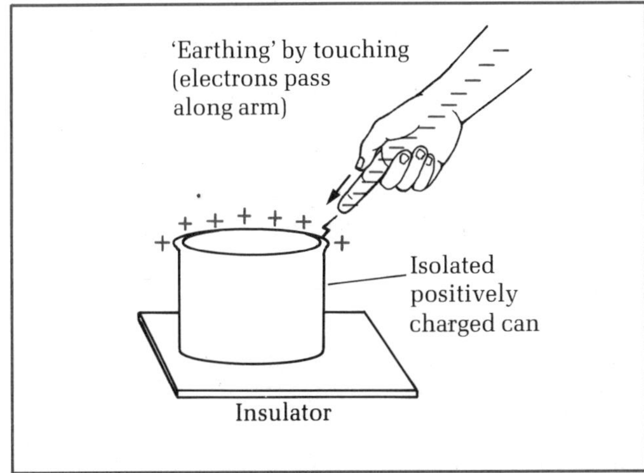

Figure 13.2 *Discharge to Earth*

Conducting materials such as metals contain lots of free electrons. These are electrons which move about inside the metal, and are not attached to any one atom. To charge a metal up, it must first be isolated from the Earth. Otherwise any charge which it is given is neutralized by electrons flowing to or from the Earth. Then the isolated metal can be charged by direct contact with any charged object. If an isolated metal is charged positively then 'earthed', electrons flow onto the metal from the Earth to neutralize or discharge it. Insulating materials do not contain free electrons. All the electrons in an insulator are attached to individual atoms. Some insulators, such as perspex or polythene, are easy to charge because surface atoms easily gain or lose electrons.

The shuttling ball experiment shows that an electric current is caused by a flow of charge. A conducting ball is suspended by an insulating thread between two vertical plates, as in Figure 13.3. When a high voltage is applied across the two plates, the ball bounces back and forth between the two plates. Why? Each time the ball touches the negative plate it gains some electrons, so the ball becomes negatively charged; it is then repelled by the negative plate. It is then pulled onto the positive plate, and when contact is made, electrons on the ball transfer to the plate. So the ball now becomes positively charged and is repelled back to the negative plate to repeat the

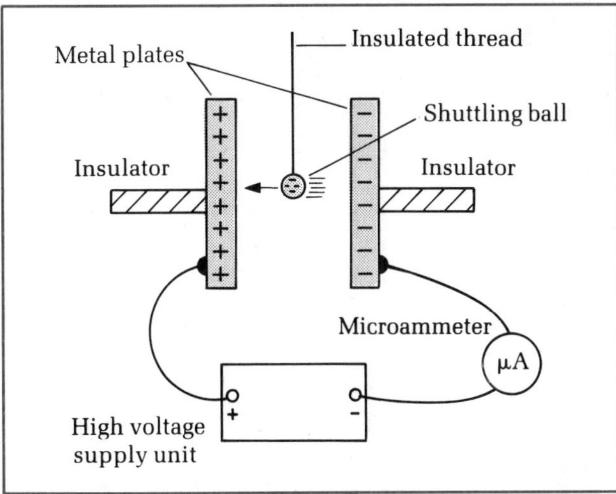

Figure 13.3 *The shuttling ball*

cycle. Therefore electrons from the high voltage supply pass along the wire to the negative plate. There, they are ferried across to the other plate by the ball. Then they pass along the wire back to the supply. A microammeter in series with the plates shows that the shuttling ball causes a current round the circuit.

What would you expect to observe if the two plates were moved closer together? The ball shuttles back and forth even more rapidly when the plates are brought closer. As a result, the microammeter reading increases because charge is ferried across at a faster rate.

The unit of charge is the **coulomb** (C), defined as the charge passing a point each second when the current is 1 A.

In the shuttling ball experiment, suppose the charge ferried across the gap each cycle is Q. If each cycle takes time t, then the current in the circuit is Q/t on average. See p. 313 for a further discussion of current and charge.

Each electron carries a charge of 1.6×10^{-19} C. To make a total charge of 1 C, 6.2×10^{18} electrons are needed. So a 0.3 A torch bulb has about 2 million, million, million electrons passing through it every second! Charge is often measured in microcoulombs; 1 microcoulomb (1 μC) equals 10^{-6} C. How many electrons are needed for 1 μC of charge? Very sensitive charge measuring devices called electrometers can measure charges as small as 10^{-12} C. Gold leaf electroscopes can also be used to detect small amounts of charge.

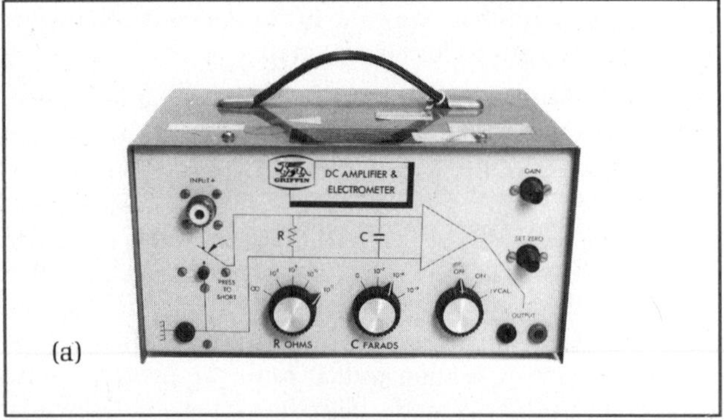

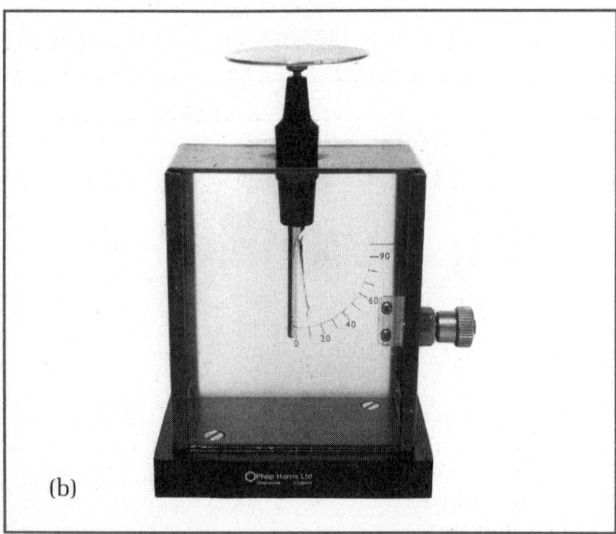

Figure 13.4 *Measuring charge* **a** *The electrometer* **b** *The gold leaf electroscope*

Electric field patterns

Suppose a small positive charge is placed as a test charge near a body with a much bigger charge, which is also positive. The test charge will be repelled by the larger charge. An electric field of force surrounds the bigger charge. Any test charge placed in the field experiences a force. Unless the test charge is much smaller than the large charge, it might exert a large enough force on the large charge to move it and so alter the field. So we assume that the test charge is too small to alter the field of the large charge. If the test charge is free to move, it follows a path away from the large charge. The path is called a field line, or line of force.

Field lines are the lines which positive test charges follow. Figure 13.5 shows the patterns of field lines around different charged objects. Each pattern is produced by semolina grains sprinkled on oil. An electric field across the surface of the oil is set up by connecting two metal conductors in the oil to opposite poles of a high voltage supply unit. Different shaped metal conductors give different field patterns. The grains line up along the field lines, like plotting compasses in a magnetic field.

1 Oppositely charged point objects create a field as shown in Figure 13.5(a). The field lines become concentrated at the points. A positive test charge released from an off-centre position would follow a curved path to the negative point charge.
2 A point object near an oppositely charged flat plate produces a field as shown in Figure 13.5(b). The field lines are concentrated at the point object, but they are at right angles to the plate where they meet the plate.
3 Two oppositely charged plates create a field as shown in Figure 13.5(c). The field lines run parallel from one plate to the other, meeting the plates at right angles. The field is uniform as the field lines do not spread out.

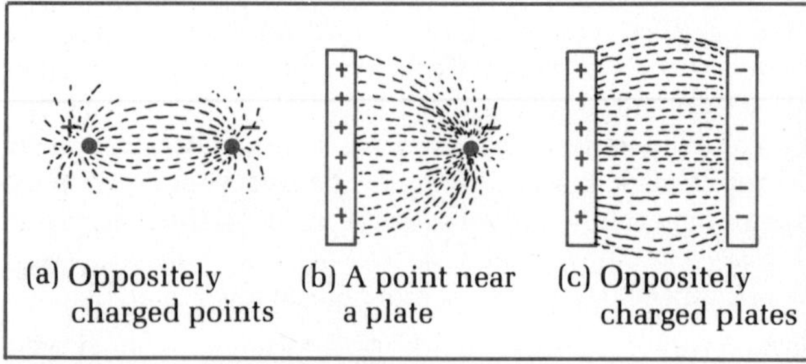

Figure 13.5 *Electric field patterns*

Force and energy

The electric field strength E at a point in a field is defined as the force per unit charge on a positive test charge placed at that point. The unit of E is N C^{-1}.

Since E is the force per unit charge, then the force on a test charge q is equal to qE. In other words, electric field strength equals force/charge.

$$E = \frac{F}{q}$$

where F = force on a test charge q,
E = electric field strength where the test charge is placed.

A uniform field occurs where the electric field strength is the same at all positions in the field. As shown in Figure 13.5(c), the field lines of a uniform field are parallel to one another. The force on a test charge in a uniform field is the same wherever the test charge is placed. So if a test charge is moved about in a uniform field, work is done on it unless it is moved at right angles to the field lines. The situation is like that of a mass moved about near the Earth's surface; the force mg is constant provided the mass is not moved too far from the surface. Work is done on the mass whenever it is moved up or down to change its potential energy. Move it horizontally and its P.E. does not change.

In a non-uniform field, the force on a test charge varies from one position to any other position. If a test charge is moved from outside the field to some position in the field, then its potential energy is changed.

The electric potential V at a point in a field is defined as the potential energy per unit charge of a positive test charge placed at that point. The position of zero P.E. is defined as infinity. The unit of V is the **volt** (V) which is equal to $1\,\text{J}\,\text{C}^{-1}$.

Suppose a test charge, $q = +1\,\mu\text{C}$, is moved into an electric field from infinity to reach a certain position where the electric potential is -1000 V. The P.E. of the test charge in the field is therefore -1000×10^{-6} J, which equals -10^{-3} J. To remove the test charge from the field, it must be given 10^{-3} J of energy; in other words, 10^{-3} J of work must be done on q to remove it from the field.

P.E. of a test charge q at potential $V = qV$

Equipotentials are lines of constant potential. A test charge moving along an equipotential has constant potential energy. No work is done on a test charge moving along an equipotential because the force due to the field is at right angles to the equipotential. In other words, the field lines (i.e. lines of force) cross the equipotential lines at right angles. Equipotentials for electric fields are like equipotentials for gravitational fields; both are lines of constant potential energy for the appropriate test object, in one case charge and in the other case mass.

Equipotentials are like map contours; a walker following a map contour round a hillside has constant potential energy. Figure 13.6 shows the equipotentials near two positive point charges. The pattern is like the contours round two hills next to each other. A test charge released from point X near to one of the point charges is pushed by the field along a path which takes it through point Y to

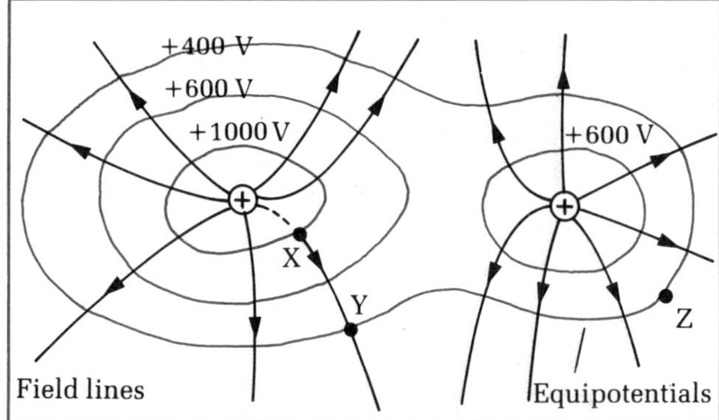

Figure 13.6 *Equipotentials*

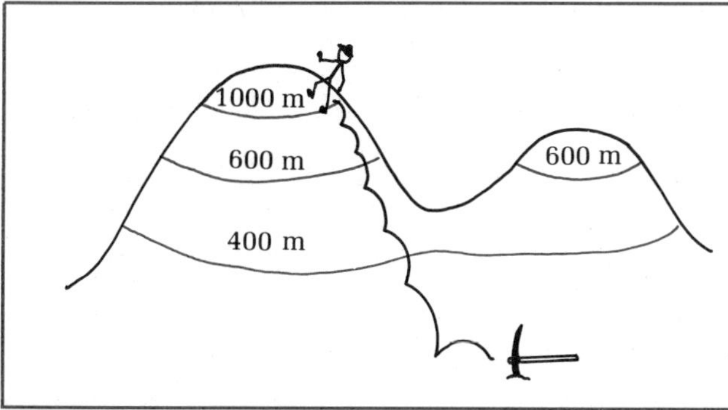

Figure 13.7 *Potential hills*

infinity. At every point along the path, the force on the test charge is at right angles to the equipotential line through that point. So the path from X through Y and beyond is a field line (i.e. line of force). Other field lines are shown, each one starting at one of the two positive point charges.

Suppose Figure 13.6 represented map contours; the hills would be shaped as in Figure 13.7. A ball released near the top of either hill would follow a curved path represented by a field line. The potential hills give a useful picture of the field near the two charges. We can use the same idea to deal with negatively charged bodies and their fields; however, we need to invert the potential hill making a valley in which the P.E. of a positive test charge would be less than zero.

Suppose a $+2\,\mu\text{C}$ test charge is moved from point X to point Y in Figure 13.6. The potential at X is $+1000$ V, so the test charge has P.E. equal to $+2 \times 10^{-6} \times 1000$ J, which equals $+2 \times 10^{-3}$ J. The potential at Y is $+400$ V, so the test charge has P.E. at Y equal to $+2 \times 10^{-6} \times 400$ J which equals $+8 \times 10^{-4}$ J. So moving the test charge from X to Y lowers its P.E. by 1.2×10^{-3} J. If the test charge is then moved along the equipotential through Y to point Z, its P.E. remains constant.

Plotting equipotentials

Conducting paper is used on top of carbon copy paper on top of a white sheet of paper. The conducting paper is pinned down, and two metal plates are placed on it at either end. Figure 13.8 shows the idea, with rectangular plates used. A battery is connected across the plates, then

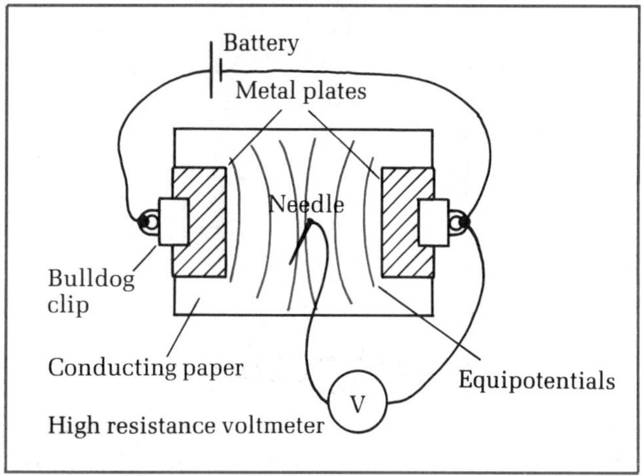

Figure 13.8 *Plotting equipotentials*

a contact needle connected to a voltmeter as shown is used to measure the potential on the conducting paper. The other terminal of the voltmeter is connected to the negative pole of the battery. Contact on the paper with the needle gives a voltmeter reading equal to the potential at the point of contact. A voltmeter with high resistance or an oscilloscope should be used.

Start by using the needle to outline the shape of the metal plates. Measure the potential difference across the plates. Then use the needle to trace equipotentials between the plates. Write the potential of each equipotential next to the trace. To trace each equipotential, you need to move the needle across the conducting paper so that the voltmeter reading stays constant. In this way, trace out several equipotentials at equal intervals of potential. When the white sheet of paper is removed, you should be able to see clearly the outlines of the two plates and the equipotentials which you traced. Try different shaped plates to give different patterns.

On each pattern, you can sketch the field lines because they always cross the equipotentials at right angles. Start each field line near the positive plate and sketch it across the equipotentials to the negative plate. Figure 13.8 shows the expected pattern for two parallel rectangular plates.

13.2 Uniform electric fields

Investigating uniform fields

Any charged body creates an electric field in the space around itself. Two oppositely charged parallel plates produce an electric field in the space between them. Figure 13.9 shows one way to measure the potential in the space between the plates. A flame probe is used to make the air near the probe conduct. The probe is a tiny gas flame which creates ions in the air. The probe itself is a hypodermic needle connected to an electroscope or electrostatic voltmeter. Wherever the probe tip is held between the two plates, the ions charge it to the potential at that point. By moving the probe about between the plates, the equipotentials can be mapped out.

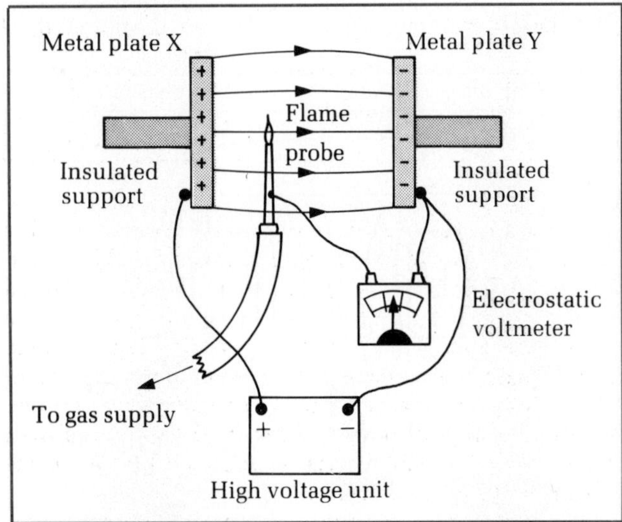

Figure 13.9 *A uniform field*

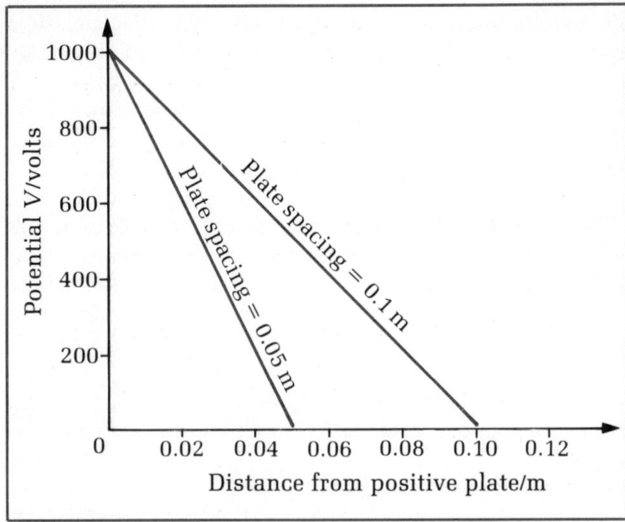

Figure 13.10 *Potential gradients*

Suppose the probe is moved directly from the positive plate across to the earthed plate. The potential decreases steadily with distance. The potential drops by equal steps for every centimetre the probe is moved. Figure 13.10 shows the variation of potential between the plates, along a direct line from the positive plate to the earthed plate.

Now suppose the plates are moved closer together; the shuttling ball experiment showed that the field is made stronger by doing this. The flame probe shows that the drop of potential is greater for each centimetre the probe is moved. An alternative to a flame probe is a home-made electroscope; it consists of a small piece of metal foil sellotaped to a charged perspex rule. When the rule is held vertical, the foil stands out from the rule, as shown in Figure 13.11.

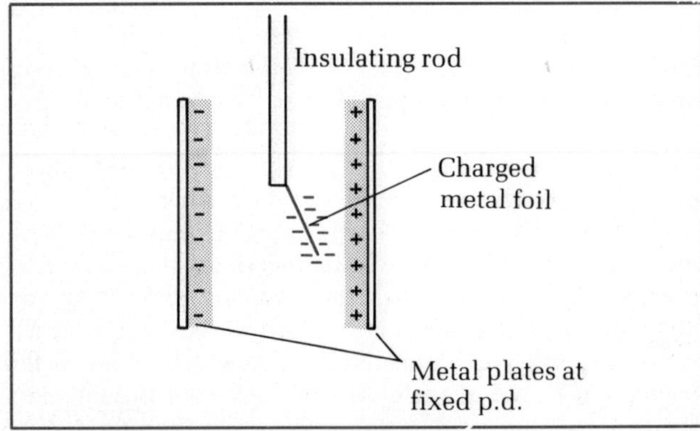

Figure 13.11 *A home-made electroscope*

Does the angle of the foil to the rule change as the rule is moved in the field? Does the angle change when the plates are moved together? You ought to find no difference is made by moving the rule about, but moving the plates closer together increases the angle. The field between the plates has the same strength at all points (i.e. a uniform field). Moving the plates together makes the field stronger.

Potential gradients

Between two oppositely charged parallel plates, the potential changes steadily from one plate to the other. The change of potential per unit distance along the line between the plates is constant. This is an example of a constant potential gradient, like a constant incline. Near a charged sphere the potential does *not* change steadily with distance, so the potential gradient near a charged sphere is not constant. Figure 13.7 shows a situation where the potential gradient changes with position.

The potential gradient at a point in a field is the change of potential per unit distance at that point. The unit of potential gradient is $\mathrm{V\,m^{-1}}$.

For any uniform field the potential gradient can easily be calculated. For Figure 13.10, the potential gradient with the plates at 0.1 m spacing is equal to $\dfrac{2000\ \mathrm{V}}{0.1\ \mathrm{m}}$, which equals $2\times10^4\ \mathrm{V\,m^{-1}}$. With the plates at 0.05 m spacing, the potential gradient is equal to $4\times10^4\ \mathrm{V\,m^{-1}}$. In general, if the potential difference between two points is V, and their distance apart is d, then the potential gradient is equal to V/d. The two points should be along the same field line.

$$\text{Potential gradient between two parallel plates} = \frac{V}{d}$$

where V = p.d. between plates,
d = distance between plates.

Suppose a test charge $+q$ is moved from the negative plate across to the positive plate of a parallel plate arrangement. The work done on the test charge is equal to qV, where V is the p.d. between the plates. Since work done equals force × distance, then the force on the test charge is given by qV/d where d is the plate separation. We have made use of the fact that the force on the test charge is constant since the field is uniform. Hence

$$F = qV/d$$

Since the electric field strength E is defined as F/q (i.e. force per unit charge), then the electric field strength is equal to V/d.

$$\text{Electric field strength between two parallel plates} = \frac{V}{d}$$

The unit for E can therefore be given as $\mathrm{V\,m^{-1}}$ as an alternative to $\mathrm{N\,C^{-1}}$. The direction of E is **opposite** to the potential gradient. The potential increases from the negative to the positive plate but the field lines are from positive to negative.

For any non-uniform field we must consider small distances only to work out potential gradients. Suppose a test charge $+q$ is moved through a small distance Δr against the electric field force, as shown in Figure 13.12. The force on the test charge due to the field is qE, where E is the electric field strength. So the force F' to move the test charge against the electric field force is $-qE$ (i.e. an equal and opposite force).

The work done by F' is equal to $F'\Delta r$ when the test charge is moved by a small distance. Hence the work done per unit charge is $F'\Delta r/q$ which equals $-qE\Delta r/q$ (since $F' = -qE$). In other words, the work done per unit charge by the applied force is $-E\Delta r$.

However, the work done changes the potential energy.

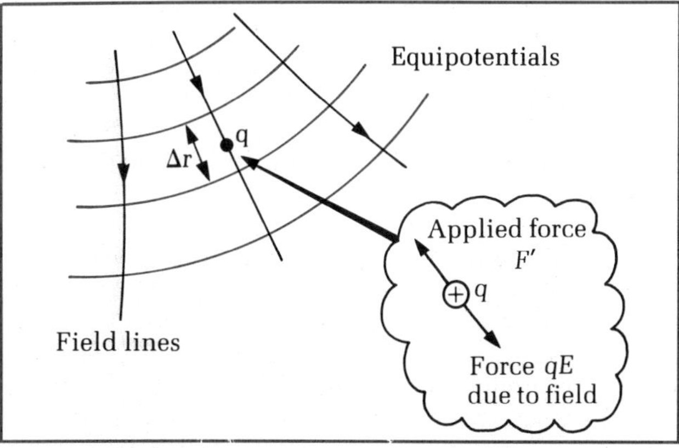

Figure 13.12 *A non-uniform field*

The work done per unit charge is equal to the change of P.E. per unit charge (i.e. the change of potential). Therefore, the change of potential ΔV is given by

$$\Delta V = -E\Delta r$$

Hence $$E = -\frac{\Delta V}{\Delta r}$$

Electric field strength = −potential gradient (at any point)

Factors affecting the field between parallel plates

To increase the strength of the field between parallel plates, we could move the plates closer together. Provided the plates are at fixed potential difference, the field strength is increased by moving them closer. The field could also be made stronger by charging the plates up more, or by using bigger plates. Figure 13.13 shows an arrangement to investigate the factors which determine the field strength.

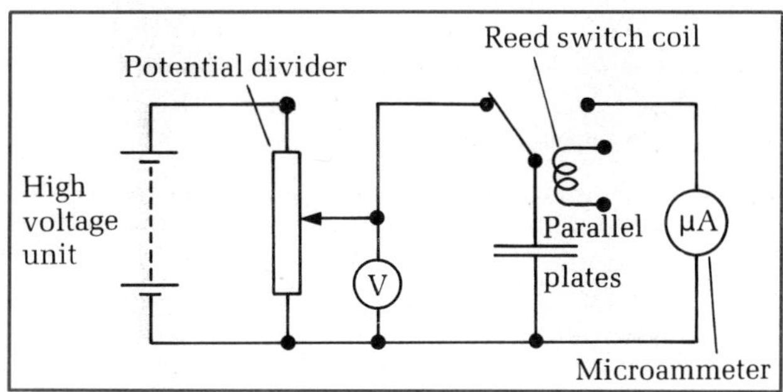

Figure 13.13 *Field factors*

A reed switch is used to repeatedly charge and discharge a pair of parallel plates. The switch operates from an alternating current supply of known frequency so the switch charges and discharges the plates at the same frequency. The plates charge up from a high voltage supply when the switch is connected to the supply. Then the switch reconnects to the microammeter, and the plates discharge through the meter. The switch is then moved back to the high voltage supply for the next cycle of charging and discharging.

By measuring the discharge current I and the a.c. frequency, we can calculate the charge on the plates each cycle.

Each cycle takes $1/f$ seconds. So the charge Q passing through the meter in that time is given by

$$Q = \text{current} \times \text{time} = I \times 1/f = I/f.$$

At constant frequency, the charge is therefore proportional to the current. Since the plates discharge at the same rate as they charge, the charge on the plates is equal to I/f.

1 Measure Q for different plate potential differences (p.d.s). The separation and plate area are kept constant. A potential divider is used to apply different plate p.d.s V, measured using a voltmeter. A graph of Q against V ought to show that Q is proportional to V.
2 Measure Q for different plate separations d. The p.d. and plate area are kept constant. A graph of Q against $1/d$ should show that Q is proportional to $1/d$.
3 Measure Q for different plate areas A. The p.d. and separation are kept constant. To change the plate area, the plates can be slid apart to give different overlap areas. A graph of Q against A should show that Q is proportional to A.

The results show that

1 $Q \propto V$
2 $Q \propto 1/d$
3 $Q \propto A$

so we can write

$$Q \propto \frac{VA}{d}$$

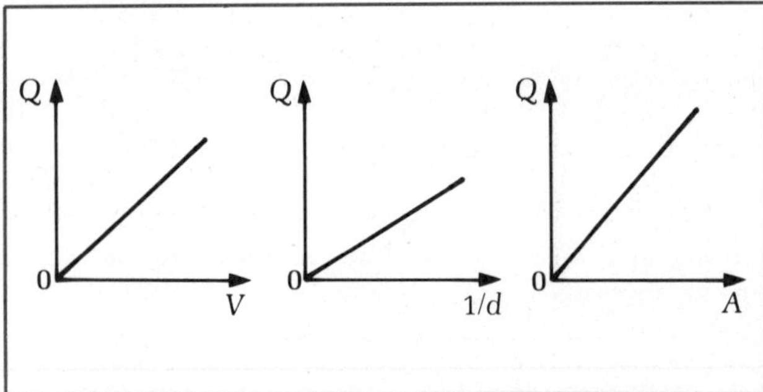

Figure 13.14 *Field factor graphs*

Using a constant of proportionality, we can therefore write an equation for Q.

The constant of proportionality is called the **electric field constant** ε_0 or the **permittivity of free space**.

$$Q = \frac{AV\varepsilon_0}{d}$$

where Q = charge on each plate,
A = plate area,
d = plate spacing,
V = plate p.d.

Electric field strength is given by

$$E = \frac{V}{d} = \frac{Q}{A\varepsilon_0}$$

The unit of ε_0 is F m^{-1}, where 1 farad (F) is defined as 1 coulomb per volt.

Measuring ε_0

The apparatus shown in Figure 13.13 may be used. The discharge current is measured for different plate potentials. The current I is measured using the microammeter. The plate potential V is measured using the voltmeter. The plate separation d and the area A are kept fixed. To measure d, a travelling microscope can be used at different positions round the edges. Use a millimetre rule to measure the plate length and width, and so calculate A. The frequency of the reed switch can be measured using a frequency meter.

To calculate ε_0, use the equation above. Rearranged, we obtain

$$I = \frac{AV\varepsilon_0}{d} \times f = \frac{A\varepsilon_0 f}{d} \times V$$

since $Q = I/f$.

Plot a graph of I against V. The gradient of the graph is equal to $A\varepsilon_0 f/d$. Hence ε_0 can be calculated if the gradient is measured. The accepted value of ε_0 is $8.85 \times 10^{-12}\ \text{F m}^{-1}$.

13.3 Parallel plate capacitors

Dielectrics

Suppose a pair of parallel plates are charged to a potential difference V. Each plate gains an amount of charge equal and opposite to the other plate. One plate loses a certain number of electrons, and the other plate gains an equal number of electrons. The charge Q on each plate is equal to $AV\varepsilon_0/d$, where A is the plate area and d is the plate spacing. We've assumed the space between the plates is empty. The charge stored Q can be altered by changing A or V or d. For example, if we move the plates closer together at fixed p.d., Q increases because d is made smaller.

Another way to increase the charge stored is to insert a **dielectric** between the plates. Dielectrics are insulating materials which increase the capacity of the plates to store charge. Polythene and waxed paper are examples of dielectrics.

Let's consider what happens when a dielectric is placed between charged parallel plates. Each molecule of the dielectric becomes **polarized**. This means that its electrons are pulled a little towards the positive plate, as shown in Figure 13.15. So the surface of the dielectric sheet facing the positive plate gains a layer of negative charge due to polarization. The other surface of the dielectric loses negative charge, so it gains a layer of positive charge. For fixed p.d. across the plates, the plates must be connected to a battery. The positive dielectric charge, which faces the negative plate, attracts more electrons onto the plate from the battery. The negative dielectric charge, facing the positive plate, pushes electrons off the plate back to the battery. So the negative plate gains more negative charge, and the positive plate becomes even more positive. In other words, the dielectric has increased the charge stored on the plates for a fixed plate p.d.

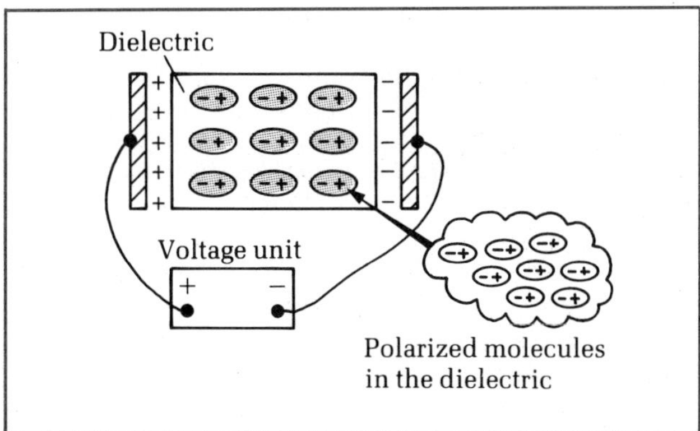

Figure 13.15 *Dielectrics*

The relative permittivity ε_r or dielectric constant, may be defined as the ratio of

$$\frac{\text{charge } Q \text{ stored with dielectric}}{\text{charge } Q_0 \text{ stored without dielectric}}$$

with fixed p.d. and the space either filled completely with a dielectric or completely empty.

Hence, the charge stored with a dielectric completely filling the space is given by

$$Q = \varepsilon_r Q_0 = \frac{A\varepsilon_0\varepsilon_r V}{d}$$

To measure ε_r, we can use the arrangement shown in Figure 13.13. The plates are kept at fixed p.d. and spacing. The microammeter reading I_0 without dielectric is measured. Then the dielectric is inserted into the space between the plates to fill the space. The new reading of the microammeter, I, is measured. Since the charge Q is equal to current × frequency of the switch, then

$$\varepsilon_r = \frac{Q}{Q_0} = \frac{I}{I_0}$$

Typical values for ε_r are 2.3 for polythene, 2.5 for waxed paper, 7 for mica and 81 for water! Water molecules are easily polarized.

Capacitance of parallel plates

Any device which can store charge is a capacitor. Insulate yourself from the Earth and you become a capacitor. Two oppositely charged parallel plates form a capacitor. Any capacitor consists of two conductors insulated from one another. If you insulate any object from the Earth, then that object and the Earth act as the two plates of a capacitor. The symbol for any capacitor is a pair of parallel plates, as in Figure 13.16, even though the actual capacitor might be different.

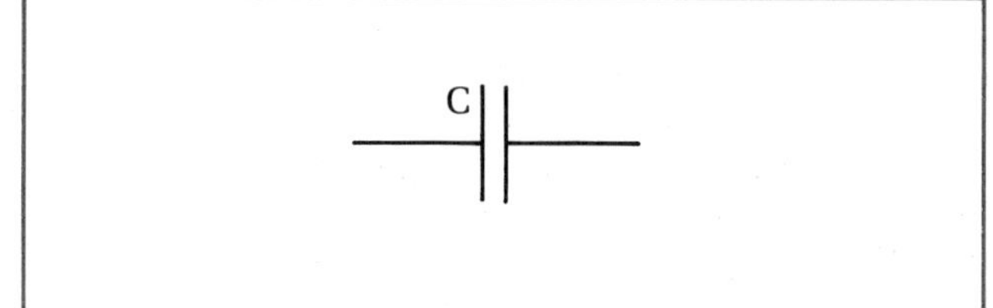

Figure 13.16 *Capacitor symbol*

The capacitance C of a capacitor is defined as the charge stored per unit p.d. applied across its plates.

$$C = \frac{Q}{V}$$

where C = capacitance,
V = p.d. across plates,
Q = charge stored.

The unit of capacitance is the **farad** (F), defined as 1 coulomb per volt.

Capacitance values are frequently given in microfarads ($1\,\mu F = 10^{-6}\,F$) or picofarads ($1\,pF = 10^{-12}\,F$). Capacitors in use in circuits are dealt with in detail in Chapter 22.

For a pair of parallel plates, charge stored is given by

$$Q = \frac{A\varepsilon_0\varepsilon_r}{d}V.$$

The capacitance of a pair of parallel plates is given by

$$C = \frac{Q}{V} = \frac{A\varepsilon_0\varepsilon_r}{d}$$

where A = plate area,
d = plate spacing,
ε_0 = electric field constant,
ε_r = relative permittivity.

The equation shows that a large capacitance can be achieved by

1 making the area A as large as possible,
2 making the plate spacing d as small as possible,
3 using a dielectric between the plates.

Worked example Calculate the capacitance of a pair of parallel plates 0.1 m by 0.1 m with an air gap of 5 mm. Assume $\varepsilon_0 = 8.85 \times 10^{-12}\,F\,m^{-1}$.

Solution Use the equation $C = \frac{A\varepsilon_0\varepsilon_r}{d}$

$A = 0.1 \times 0.1 = 10^{-2}\,m^2$
$d = 5 \times 10^{-3}\,m$.
$\varepsilon_r = 1$ since the plates are air-filled.
Hence

$$C = \frac{10^{-2} \times 8.85 \times 10^{-12} \times 1}{5 \times 10^{-3}} = 1.77 \times 10^{-11}\,F = 17.7\,pF.$$

Practical capacitors

Capacitors used in circuits range from capacitances of about 1 pF to more than 1 F, a factor of 10^{12} or more. Most practical capacitors are designed using two strips of aluminium foil as plates. A dielectric separates the two pieces of foil, and the arrangement is usually rolled up, as in Figure 13.17. In this way, a compact unit gives a large capacitance. The area can be large, the spacing can be small, and the whole device is a convenient size, unlike real parallel plates. Dielectrics such as mica or polyester are used for capacitances up to about 10 μF.

Electrolytic capacitors are designed to have much larger capacitances than 10 μF. They are used in power supply circuits to even out voltage changes (p. 348). High capacitance is achieved by a very thin layer of dielectric.

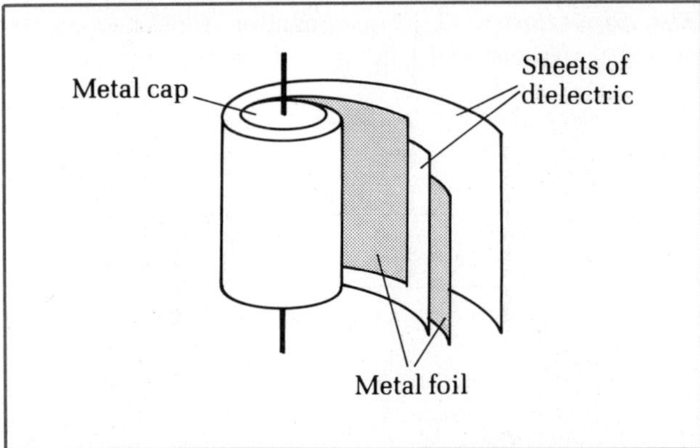

Figure 13.17 *A practical capacitor*

The capacitor is made with an electrolyte-soaked paper between the two aluminium strips. The electrolyte used is a solution of aluminium borate. When charged up, a layer of aluminium oxide forms against the positive plate. The oxide layer is a dielectric, so the actual capacitance is due to the oxide layer sandwiched between the positive aluminium foil plate and the electrolyte-soaked paper as the negative plate. The electrolytic action creates a very thin layer of dielectric, so the capacitance is very high. Electrolytic capacitors must be connected with correct polarity otherwise the dielectric layer may be damaged. Also, the maximum working voltage should not be exceeded or the dielectric layer becomes conducting.

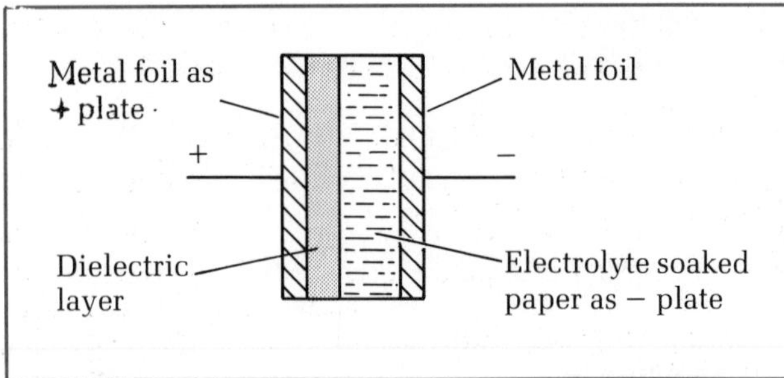

Figure 13.18 *An electrolytic capacitor*

Tuning capacitors are used in radio receivers to select different radio stations. A tuning capacitor consists of two sets of metal plates, interleaved as in Figure 13.19. When the tuning dial is turned, the overlap of the plates alters so the effective area changes. Hence the capacitance changes.

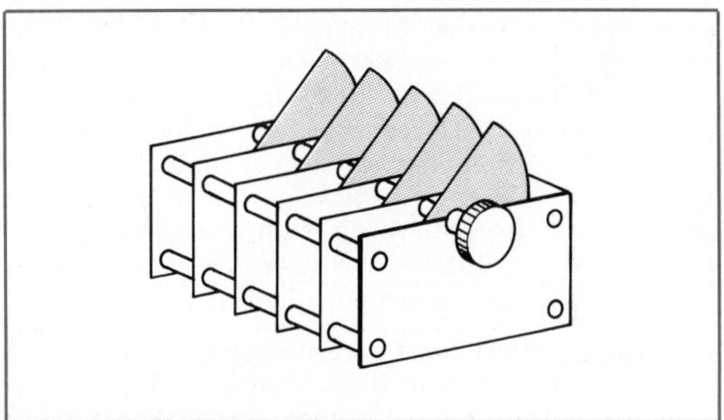

Figure 13.19 *A tuning capacitor*

Energy stored in a charged capacitor

When a capacitor is charged, energy is stored in it. A charged capacitor discharged across a torch bulb will release its energy in a brief flash of light from the bulb, as long as the capacitor has been charged to a large enough p.d. in the first place. When discharged through the torch bulb, charge flow is rapid enough to give a large enough current to light the bulb, but only for a brief time.

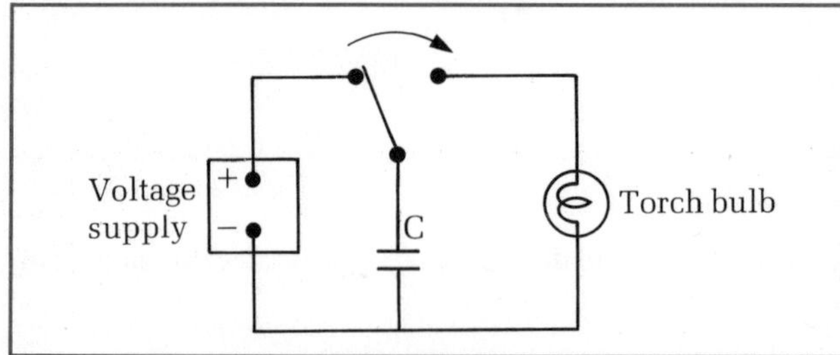

Figure 13.20 *Releasing stored energy*

Consider the charging process. The charge on the plates builds up to a final value Q_0 which depends on the p.d. of the battery used. At any stage during the charging process, the p.d. across the plates is given by

$$V = Q/C$$

where Q is the charge on the plates at that stage, C is the capacitance. Figure 13.21 shows how the plate p.d. varies with charge during the charging process.

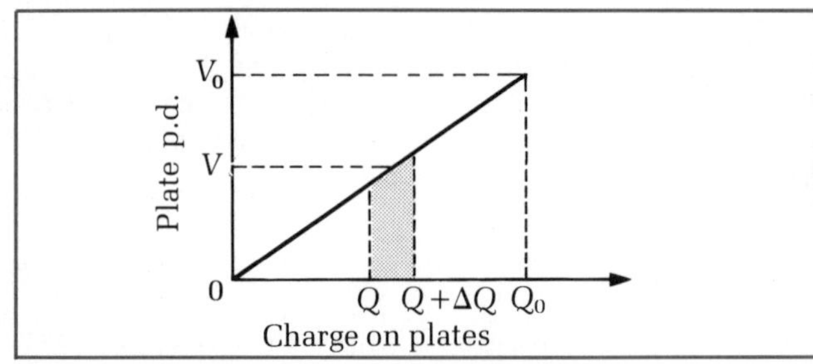

Figure 13.21 *Energy stored in a capacitor*

Let's look at one step of the charging process when the charge increases from Q to $Q+\Delta Q$. This involves putting an extra charge Q on the plates at p.d. V. Figure 13.21 shows the step. How much work must be done to do this? V is the energy per unit charge, so $V\Delta Q$ is the energy supplied to put the extra charge on the plates. Figure 13.21 shows that $V\Delta Q$ is represented by the area of the shaded strip: the strip width represents ΔQ and its height V.

The energy supplied in each step is represented on Figure 13.21 by the area of each strip. So the total energy supplied is represented by the total area of all the strips; in other words, the area under the line from O to final charge Q_0. The total area is therefore that of the triangle with height V_0 and base Q_0. Hence the total energy supplied $= \frac{1}{2}Q_0V_0$ ($= \frac{1}{2}$ base × height). This is the energy stored in the capacitor.

Energy stored in a charged capacitor is given by

$$E = \tfrac{1}{2}Q_0V_0 = \tfrac{1}{2}CV_0^2 = \tfrac{1}{2}Q_0^2/C$$

where Q_0 = final charge,
V_0 = final p.d.

We make use of the capacitance equation $C = Q_0/V_0$ to give two alternative forms of the equation; i.e. $\frac{1}{2}CV_0^2$ and $\frac{1}{2}Q_0^2/C$.

13.4 The electrical inverse square law

Coulomb's investigations

Like charges repel and unlike charges attract. The force between two charged objects depends on how close they are to each other. The exact link was first established by Charles Coulomb in 1784. He devised a very sensitive torsion balance to measure the force between charged pith balls. Figure 13.22 shows the arrangement. A needle with a pith ball at one end and a counterweight at the other end was suspended horizontally by a vertical wire. Another pith ball on the end of a thin vertical rod could be placed in contact with the first ball.

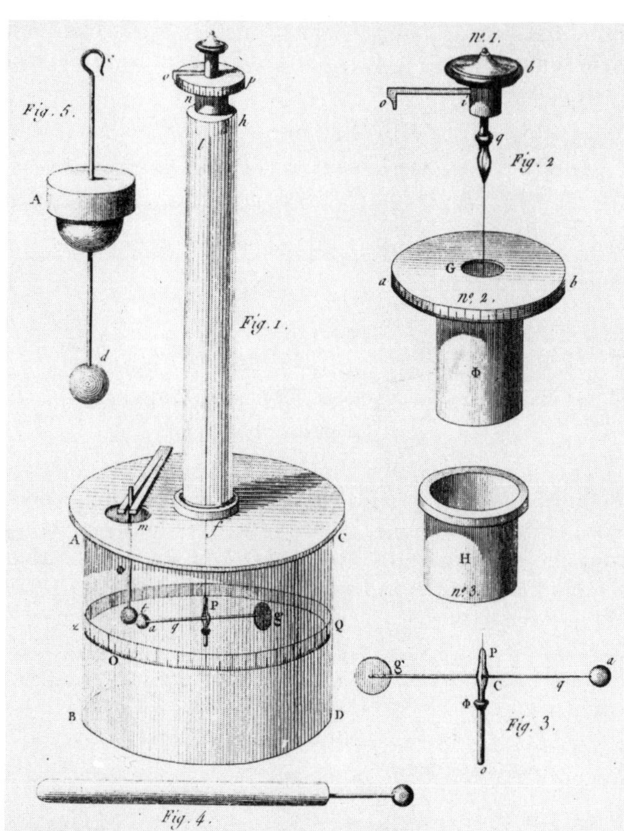

Figure 13.22 *Coulomb's torsion balance*

The pith balls were effectively point charges. The ball on the rod was charged and then placed in contact with the other ball. The contact between them charged up the second ball, which was then repelled by the ball fixed on the rod. This caused the wire to twist until the electrical repulsion was balanced by the twist built up in the wire. By turning the torsion head at the top of the wire, the distance between the two balls could be set at any required value. The amount of turning needed to achieve that distance gave the force. Some of Coulomb's many measurements are below.

Distance *r*	36	18	$8\frac{1}{2}$
Force *F*	36	144	567

Measurements for both variables were actually made in degrees, so the above values are in relative units. Can you make out a pattern for the measurements? Halving the distance from 36 to 18 makes the force increase by a factor of about 4. Halving the distance again from 18 to $8\frac{1}{2}$ (near enough 9)increases the force again by a factor of about 4. The measurements fit the link that F is proportional to $1/r^2$. All the other measurements which Coulomb made fitted the same link. Since the force is also proportional to the charge of each point object, the following equation, known as Coulomb's law, can be written for F.

$$F = k\frac{Q_1Q_2}{r^2}$$

The constant of proportionality k can be shown to be equal to $\frac{1}{4\pi\varepsilon_0}$. The equation can then be written

$$F = \frac{1}{4\pi\varepsilon_0}\frac{Q_1Q_2}{r^2}$$

where r = distance between the two point charges Q_1 and Q_2.

To prove that $k = 1/4\pi\varepsilon_0$, consider a test charge $+q$ between two oppositely charged parallel plates. The test charge is attracted by the negative plate and repelled from the positive plate. Every patch of charge on each plate exerts a force on the test charge. Coulomb's law gives the force due to each patch of charge, provided each patch is small enough to be considered effectively as a point. Then we can obtain the total force by adding the forces from the individual patches together. Integration is used for this. The result in terms of k for the total force F is

$$F = \frac{4\pi kqQ}{A}$$

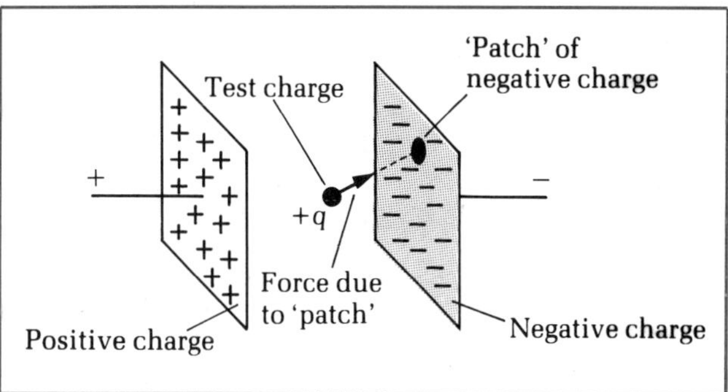

Figure 13.23 *Using Coulomb's law*

However, we can also use the parallel plate equation for the plate p.d. V to obtain an expression for the force on

q in terms of ε_0. Then we can compare the two force expressions. Since $V = \frac{Qd}{A\varepsilon_0}$, then the electric field strength $E = V/d = Q/A\varepsilon_0$.

The force on the test charge $F = qE = \frac{qQ}{\varepsilon_0 A}$.

Comparing the two expressions, $4\pi k$ is equal to $\frac{1}{\varepsilon_0}$. So the constant of proportionality k equals $1/4\pi\varepsilon_0$.

Investigating Coulomb's law

You can check Coulomb's law by using a metallized polystyrene ball suspended by nylon thread. The ball can be charged by direct contact with a charged polythene rod. Another charged ball fixed to the end of an insulating rod is used to repel the suspended ball. A lamp can be used to cast shadows of the balls and thread onto a graph paper screen.

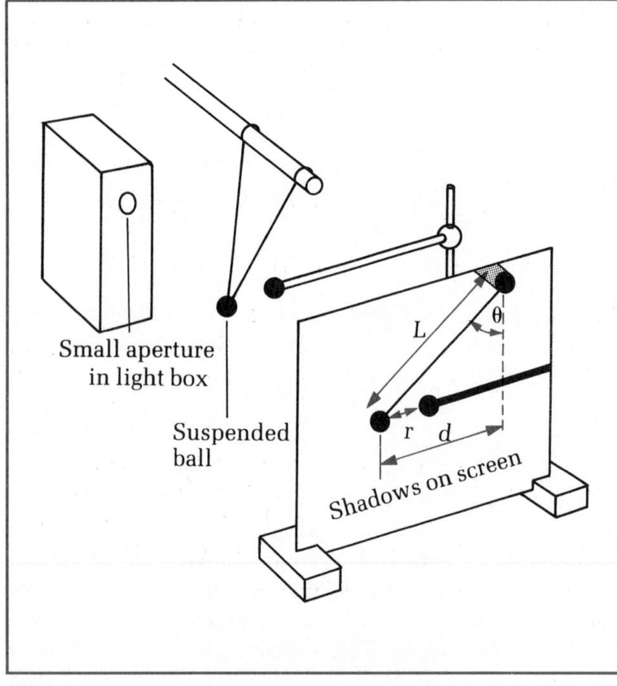

Figure 13.24 *Investigating Coulomb's law*

From the shadows on the screen, the distance r between the two point charges can be measured. The displacement d of the suspended ball from its undisturbed position can then be measured. This measurement is repeated for different values of r.

At each value of r, the suspended ball is in equilibrium caused by three different forces:

a) its weight W,
b) the tension T in the thread,
c) the electrostatic force of repulsion F.

The force diagram in Figure 13.25 shows the forces acting on the ball. From the diagram,

$T\cos\theta = W$ (balancing vertical components)
$T\sin\theta = F$ (balancing horizontal components)

where θ is the angle of the thread to the vertical.

Hence $F = W\tan\theta$

However $\tan\theta = d/L$ where L is the thread length.

So $F = Wd/L$

In other words, the force is proportional to displacement d. So a graph of d against $1/r^2$ represents F against $1/r^2$. The graph ought to be a straight line through the origin.

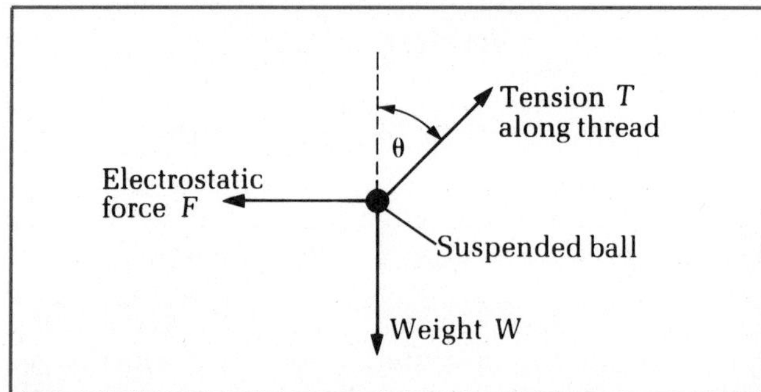

Figure 13.25 *Resolving forces*

The electric field near a point charge

We can use Coulomb's law to work out the strength or potential near a point charge. Consider a test charge q near a much larger point charge Q. The force on the test charge is given by

$$F = \frac{qQ}{4\pi\varepsilon_0 r^2}$$

where r is their separation.

The electric field strength E at the test charge is defined as F/q. Hence

$$E = \frac{Q}{4\pi\varepsilon_0 r^2}$$

where E = electric field strength at distance r from point charge Q.

The equation shows that the strength of an electric field produced by a point charge varies with distance according to the inverse square law. Double the distance and the field strength drops to 1/4 of its initial value. Treble the distance and the field strength drops to 1/9 of its initial value.

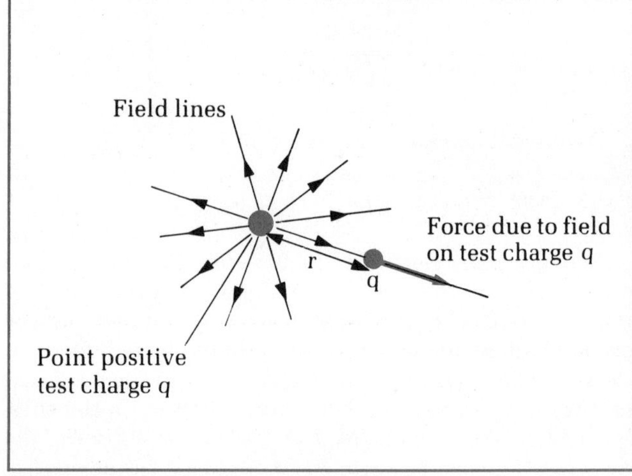

Figure 13.26 *Field near a point charge q*

The electric potential V at the test charge is equal to the work done per unit charge to move it to that position from infinity. To obtain an expression for V we can use the same approach as for gravitational potential. (See p. 54.) Newton's law of gravitation, $F = -GMm/r^2$, gives the force on a test mass m near a much larger mass. From that, we can obtain the expression for gravitational potential $-GM/r$. Compare the two force laws.

Coulomb's law	**Newton's law**
$F = \frac{1}{4\pi\varepsilon_0}\frac{Qq}{r^2}$	$F = -G\frac{Mm}{r^2}$

Both force laws are inverse square law forces, so we should expect the energy changes which they produce to be similar. The constants $1/4\pi\varepsilon_0$ and G play a similar role in the two equations. Q and q play the same role as M and m. So the gravitational potential formula $-GM/r$ has its electrical counterpart as $V = \frac{1}{4\pi\varepsilon_0}\frac{Q}{r}$.

Electrical potential	**Gravitational potential**
$V_E = \frac{Q}{4\pi\varepsilon_0 r}$	$V_G = -\frac{GM}{r}$

To prove the formula for electrical potential in detail, consider the test charge q moved in steps to distance r from infinity. Figure 13.27 shows the idea. The applied force F' to move q is equal and opposite to the electrostatic force. So the work done on q to move it to distance x from distance $x + \Delta x$ is given by

$$\Delta W = F'\Delta x$$

provided the step Δx is small.

Hence the total work done

$$W = \int_{\infty}^{r} \Delta W = \int_{\infty}^{r} -\frac{Qq}{4\pi\varepsilon_0 x^2}\Delta x = +\frac{Qq}{4\pi\varepsilon_0 r}$$

Since the initial P.E. (at infinity) is zero, the final P.E. (at r) must have been equal to $+\frac{Qq}{4\pi\varepsilon_0 r}$.

Hence the potential at r, equal to P.E./q is equal to $Q/4\pi\varepsilon_0 r$.

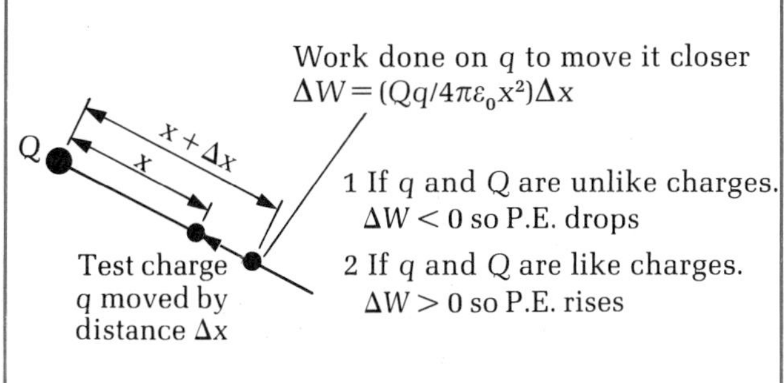

Figure 13.27 *Potential near a point charge*

Worked example Each electron carries a charge of -1.6×10^{-19} C. Calculate the electric field strength and electric potential at a distance of 10^{-10} m from an electron. Assume $\varepsilon_0 = 8.85 \times 10^{-12}$ F m^{-1}.

Solution $r = 10^{-10}$ m, $Q = -1.6 \times 10^{-19}$ C

To calculate E, use $E = \frac{Q}{4\pi\varepsilon_0 r^2}$

Hence

$$E = \frac{-1.6 \times 10^{-19}}{4\pi \times 8.85 \times 10^{-12} \times (10^{-10})^2} = 1.4 \times 10^{11}\ \text{V m}^{-1}.$$

To calculate V, use $V = \frac{Q}{4\pi\varepsilon_0 r}$

$$\text{Hence } V = \frac{-1.6 \times 10^{-19}}{4\pi \times 8.85 \times 10^{-12} \times 10^{-10}} = 14\ \text{V}.$$

Comment the calculations above apply to an electron in a vacuum since we have assumed the relative permittivity is 1. Air has a relative permittivity of 1.0005 so the calculations are nearly correct for an electron in air. However water has a relative permittivity of 81, so the values for E and V are much smaller in water than in air. The electric force between any two charges is made much weaker by putting them in water. This is the reason why, for example, when salt crystals are put into water, they dissolve. The water makes the electric forces between the ions of the salt crystal much weaker, so the ions break away from one another to form the solution.

13.5 Charged spheres

The Van der Graaf generator

Very high potentials up to a million volts or more can be produced using a Van der Graaf generator. A rubber belt driven by a motor carries charge to a large metal dome from a high voltage supply unit. One of the supply terminals is earthed. The other terminal is connected to a metal brush held against the rubber belt. Charge from the brush is carried by the belt to another brush in contact with the inside of the hollow dome. When the belt is driven at steady speed, charge builds up on the dome.

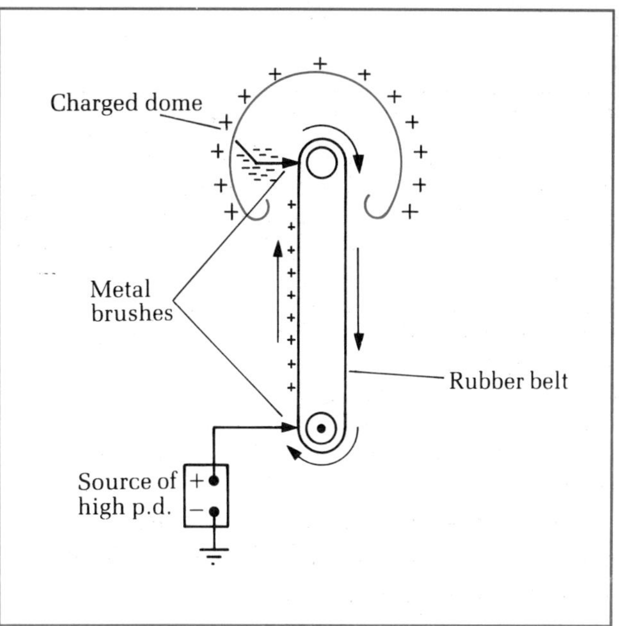

Figure 13.28 *Outline of a van der Graaf generator*

The dome potential can become large enough to cause 'lightning' sparks to jump across to any nearby earthed object. Bench-top Van der Graaf generators can easily produce sparks several centimetres in length, in air. The potential gradient to create a spark in air is about 30 000 V m^{-1}. Work out for yourself the potential needed to produce a spark of length 5 cm. Approach Van der Graaf generators with care!

Discharge from points can be demonstrated by fixing an upturned drawing pin to the dome (when uncharged). Sparks can no longer be produced when the generator is switched on because the dome no longer charges up enough. The reason is that the charge concentrates where the surface is most curved, at the tip of the pin, so an intense electric field is produced near the pin tip. Air molecules entering this area become ionized. The pin attracts oppositely charged ions, so the dome charge becomes neutralized. Lightning conductors work on this principle. Charged thunderclouds above a lightning conductor attract charge from the Earth to the tip of the conductor, so a strong electric field is created in the air above the conductor, ionizing the air molecules. The cloud is discharged by oppositely charged ions being attracted to it.

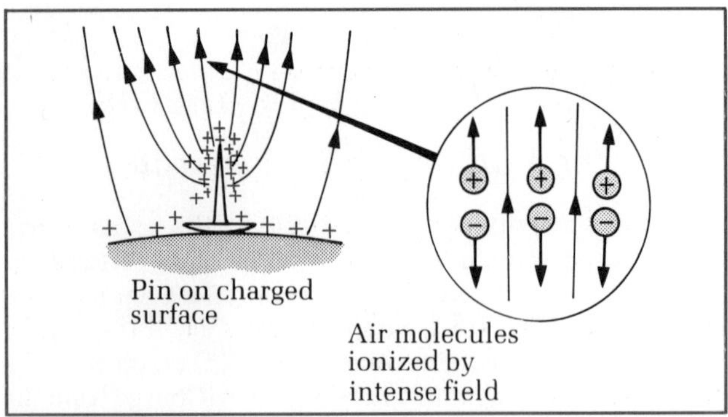

Figure 13.29 *Discharging from points*

Charging by induction can be demonstrated with the Van der Graaf generator too. A metal object on an insulated support is placed near the charged dome of the generator. Suppose the dome is positively charged. Free electrons in the metal object are attracted to the end nearer the dome, so that end becomes negatively charged, and the other end positively charged. The object is then connected to the Earth, and electrons from the Earth flow onto it to neutralize its positive end. The earth wire is then removed, and the sphere is left with a negative charge. Any charged body can charge an insulated metal object in this way.

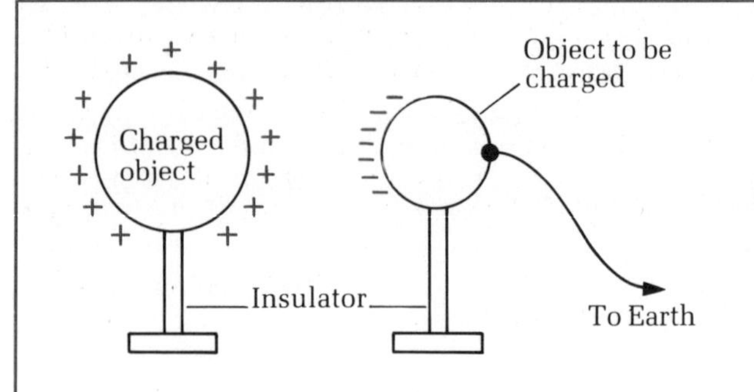

Figure 13.30 *Charging by induction*

The electric field near a charged sphere

The charged dome of a Van der Graaf generator produces a strong electric field near the dome. Figure 13.31 shows the pattern of field lines and equipotentials produced by a positively charged sphere. The field lines spread out radially, as if from the centre of the sphere. A test charge in the field experiences the same force as if all the charge were concentrated at the centre of the sphere. Outside the sphere, the field strength and potential are the same as if produced by an equal charge at the centre.

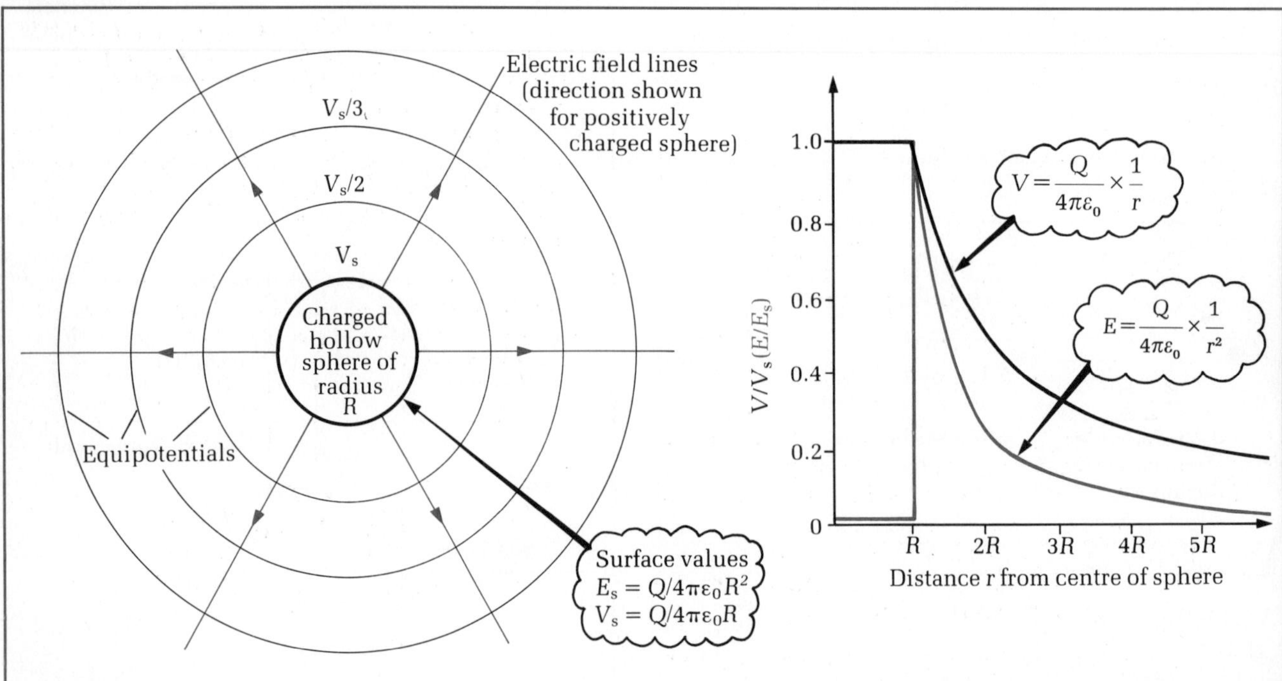

Figure 13.31 *Field near a charge sphere*

For a conducting sphere of radius R carrying charge Q we can make the following observations.

1 Outside the sphere at distance r from its centre

$$E = \frac{Q}{4\pi\varepsilon_0 r^2} \quad \text{for the electrical field strength } E,$$

$$V = \frac{Q}{4\pi\varepsilon_0 r} \quad \text{for the electric potential } V.$$

2 At the sphere surface, the above equations apply with $r = R$, so the potential of the sphere $V_s = \dfrac{Q}{4\pi\varepsilon_0 R}$.

3 Inside the sphere, there is no electric field. A test charge inside the sphere experiences zero overall force. All the sphere charge is at its surface. No matter where the test charge is placed, it is pulled equally in all directions. So no overall force results.

Since there is no electric field inside the charged sphere, no work is done on a test charge moved about inside. So the P.E. of a test charge inside is constant. Hence the potential inside the sphere is constant, equal to the surface potential.

Investigating the potential near a charged sphere

A flame probe connected to a gold leaf electroscope can be used to measure the potential near a charged sphere. The electroscope must first be calibrated by connecting it to a high voltage supply unit and voltmeter. The output from the supply unit is adjusted in steps; at each step the angle of the leaf to the vertical is measured together with the corresponding output p.d. A lamp and screen can be used to cast a shadow of the leaf onto the screen.

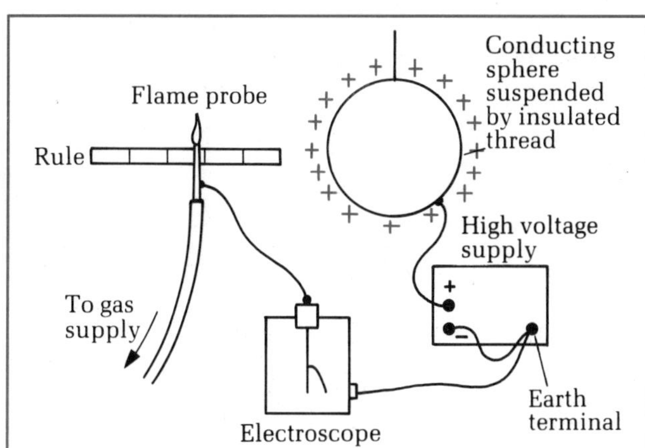

Figure 13.32 *Investigating potential*

The flame probe is positioned at different distances from the sphere centre. At each place, the angle of the leaf to the vertical is measured to give a value for the potential V. The position of the probe tip along a millimetre rule pointing to the centre is noted. After taking the measurements, the distance from each position along the rule to the sphere centre can be measured accurately.

Since $V = \dfrac{Q}{4\pi\varepsilon_0 r}$, a graph of V against $1/r$ should give a straight line through the origin.

13.6 Summary

Definitions

1 *Electric field strength* E at a point is the force per unit charge on a small positive charge at that point.

2 *Electric potential* V at a point is the P.E. per unit charge of a small positive charge at that point. The position of zero P.E. is at infinity.

3 *The potential gradient* at a point is the change of potential per unit distance along the field line through that point.

4 *Capacitance* C is defined as charge stored per volt.

Equations

1 Force on a test charge $F = qE$

2 P.E. of a test charge $= qV$

3 Potential gradient between parallel plates $= V/d$

4 Charge stored by parallel plates $Q = \dfrac{AV\varepsilon_0\varepsilon_r}{d}$

5 Energy stored in a charged capacitor $= \frac{1}{2}CV^2$

6 Coulomb's law $F = \dfrac{Q_1Q_2}{4\pi\varepsilon_0 r^2}$

7 Field formulae outside a charged sphere or near a point charge,

$$E = \frac{Q}{4\pi\varepsilon_0 r^2} \quad \text{and} \quad V = \frac{Q}{4\pi\varepsilon_0 r}$$

Short questions

Assume $\varepsilon_0 = 8.85 \times 10^{-12}\,\mathrm{F\,m^{-1}}$.

13.1

a) Explain with the aid of diagrams how an insulated metal object can be charged by induction using a negatively charged rod. Give your explanation in terms of movement of electrons.

b) A charged metal ball on an insulated thread is suspended near an earthed metal plate. Explain why the ball experiences a force of attraction towards the metal plate, irrespective of whether the ball is positively or negatively charged.

13.2 The diagram below shows the equipotentials of the electric field between two oppositely charged conductors. Copy the diagram and show the path taken by a positive test charge released from rest at X. Calculate

a) the P.E. of a +2 nC point charge at X,

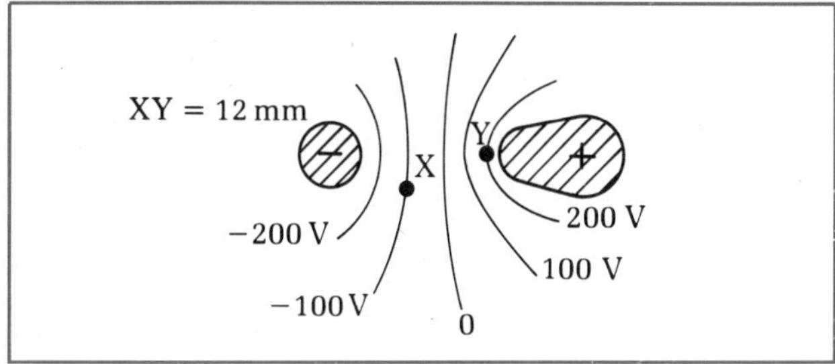

Figure 13.33

b) the work done to move the +2 nC charge from X to Y,
c) the average value of the electric field strength between X and Y.

13.3 Two horizontal metal plates A and B are spaced 50 mm apart, one on top of the other. A high voltage unit is used to maintain a constant p.d. of 1200 V between the plates, the lower plate A being earthed. Sketch a graph to show how the potential varies along a vertical line from the midpoint of the earthed plate to plate B. Calculate the electric field strength between the plates, and explain how the value can be determined from the graph. A tiny oil droplet of mass 3.9×10^{-16} kg is held stationary between the two plates by the electric field. Calculate the charge on the droplet, given $g = 10\,\mathrm{N\,kg^{-1}}$.

13.4
a) For question **13.3**, if the area of each plate is $8.0 \times 10^{-5}\,\mathrm{m^2}$, calculate (i) the charge on each plate, (ii) the charge per unit area on each plate.
b) If the plates are disconnected from the voltage supply unit and then moved to a spacing of 20 mm without loss of charge, calculate (i) the new electric field strength, (ii) the new p.d. between the plates.

13.5 A 4700 μF capacitor is charged to a p.d. of 10 V and is then discharged across the terminals of a small electric motor. The motor has a thread attached to its spindle, and the thread supports a 0.1 N weight. When the capacitor is discharged across the motor terminals, the motor raises the weight by 0.12 m. Calculate
a) the energy stored in the capacitor initially,
b) the work done on the weight when it is raised,
c) the efficiency of the energy transfer.
Account for the difference between **a)** and **b)**.

13.6 Two point charges X (= +5 nC) and Y (= −3 nC) are placed 40.0 mm apart. Calculate
a) the force between the charges,
b) the combined electric field strength and potential at (i) their midpoint, (ii) the point that is 100 mm from X and 60 mm from Y.

13.7 Equal and opposite 10 nC charges are situated at the vertices A and B of an equilateral triangle ABC where AB = 20 mm. Calculate
a) the electric field strength at C,
b) the electric potential at C.

13.8 In an experiment to determine the charge on a small insulated charged ball, the ball is suspended by an insulated thread between oppositely charged parallel plates as shown below. The plates are 80 mm apart at a p.d. of 600 V, and the ball's mass is 0.05×10^{-3} kg. If the thread lies at 5° to the vertical, calculate
a) the tension in the thread,
b) the electric force on the ball,
c) the charge on the ball.

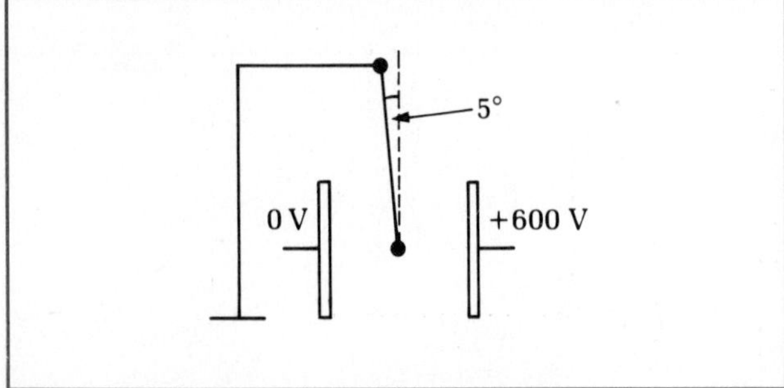

Figure 13.34

13.9 A flame probe connected to an electrostatic voltmeter is used to measure the electric potential at X, a distance of 0.20 m from the surface of a charged metal sphere of radius 0.10 m. With fixed charge on the sphere, the voltmeter reads 3000 V. Plot a graph to show how the electric potential varies from the surface of the sphere to a distance 0.5 m from its surface. What feature of the graph gives the electric field strength at any point? Hence determine the electric field strength at X.

13.10
a) In an experiment to determine ε_0, a pair of insulated parallel metal plates of area $0.15\,\mathrm{m^2}$ and spacing 1.2 mm was repeatedly charged and discharged 50 times each second using a reed switch, as below. The switch was used to charge the plates up to 60 V and then to discharge the plates through a microammeter. The reading on the microammeter was 3.32 μA. Calculate (i) the charge stored on the plates each cycle, (ii) the capacitance of the plates, (iii) the value of ε_0.
b) The spacing between the plates is completely filled with a dielectric material, causing the current to increase to 10.2 μA. Calculate the relative permittivity of the dielectric. What would the microammeter reading be if the space between the plates was only half filled with the dielectric material, as shown in the diagram?

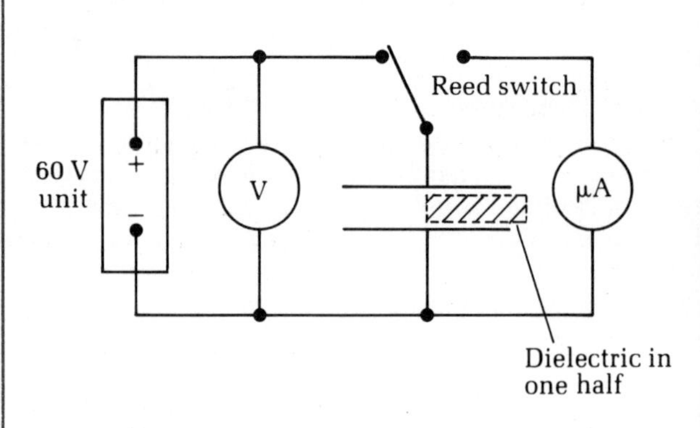

Figure 13.35

14

Magnetic fields

14.1 Magnetic field patterns

Magnetic fields are produced either by current-carrying conductors or by permanent magnets. The magnetic field near a bar magnet becomes evident if the bar magnet is placed under a sheet of paper with iron filings on top of it. The iron filings form a pattern of lines, called field lines, as shown in Figure 14.1 The lines concentrate at each end or pole of the bar magnet.

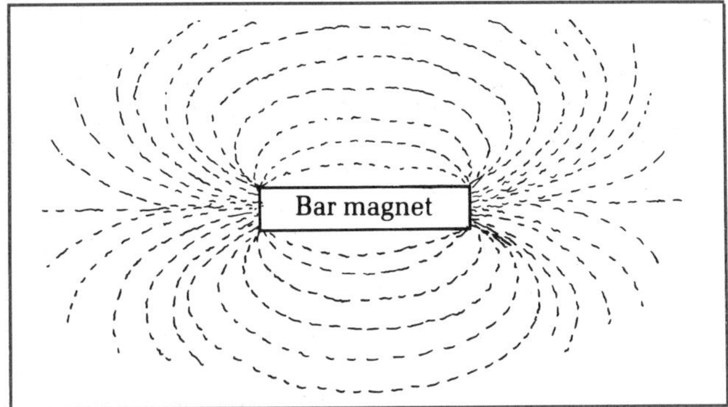

Figure 14.1 *The magnetic field near a bar magnet*

The Earth's magnetic field is used by map readers for direction-finding. If a bar magnet is suspended horizontally from its centre, it lines up along the north–south direction. The end pointing north is called the **north-seeking** or simply the **north pole**; the other end is the south-seeking or simply the **south pole**. The bar magnet, free to turn, lines up along the field lines of the Earth's magnetic field. The field lines of the Earth form a pattern as shown in Figure 14.2. The lines are concentrated near the Earth's geographical poles, as if the Earth contained a giant bar magnet. The magnetic poles are where the field lines pass through the Earth's surface vertically; they do not coincide with the geographical poles, and they are known to move gradually. Map-readers with a magnetic compass can locate the direction of magnetic north using the compass. Then to determine true north they need to know the **angle of variation** for their location. This is the angle between true and magnetic north from their position.

Except near the Equator, the field lines of the Earth's magnetic field are at an angle to the Earth's surface. So at any point on the surface, the Earth's field has a vertical and a horizontal component. In the UK, the field lines are at about 70° to the horizontal.

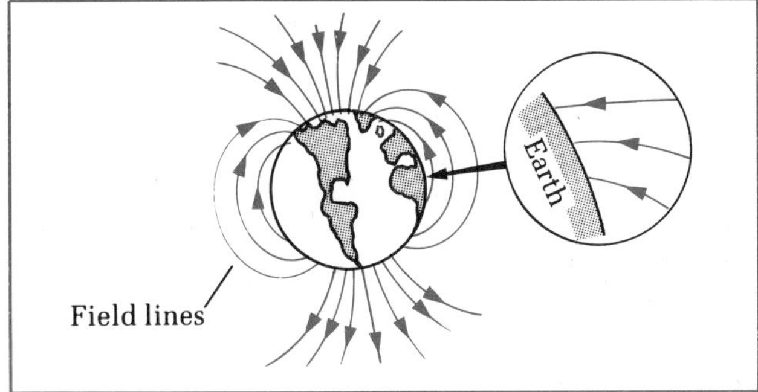

Figure 14.2 *The Earth's magnetic field*

The Earth's magnetic poles have gradually moved over many centuries. Clearly, the simple picture of a giant bar magnet fixed inside the Earth is far removed from the real situation. The study of rocks magnetized by the Earth's field when they solidified has given scientists some ideas and facts about the past behaviour of the field. It is thought that the Earth's field may even have reversed its polarity several times in the past. Studying the Earth's magnetic field gives useful information in the search for minerals. Satellites fitted with magnetometers can transmit information about the Earth's magnetic field high above the surface.

The field lines of a magnetic field are the lines which a tiny bar magnet would lie along. The direction of the field line is the direction which the north-seeking pole of the bar magnet points towards. Field lines give a useful picture of a magnetic field, but they are no more than a way of describing the field, just as gravitational or electric field lines give pictures.

An example of another magnetic field pattern is shown in Figure 14.3. Two bar magnets are placed end-to-end, so that they repel one another. The bar magnets exert equal and opposite forces on one another. The magnetic field of each one pushes on the other one. The overall magnetic field pattern is the combined effect of the two magnetic fields. Compare the pattern with the one produced by two bar magnets lined up to attract each other. Figure 14.4 shows the pattern; the field lines from one bar magnet link up with the lines of the other magnet. It isn't difficult to prove with two bar magnets that **unlike poles always attract** and **like poles always repel**. So Figure 14.3 is produced by two like poles facing one another.

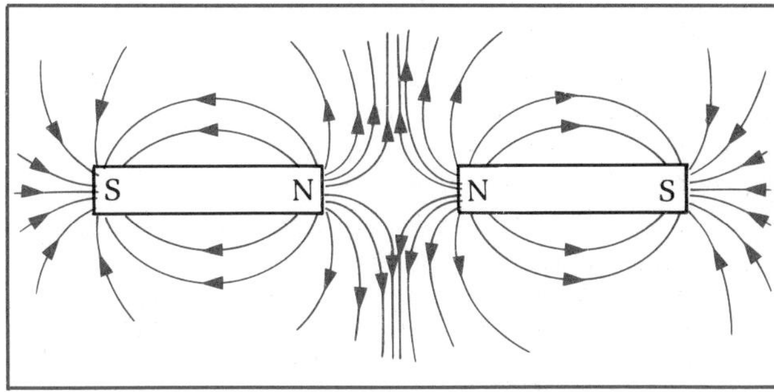

Figure 14.3 *Magnets repelling*

Figure 14.4 is produced by two unlike poles. The field pattern produced by a horseshoe magnet is like that between two bar magnets lined up to attract.

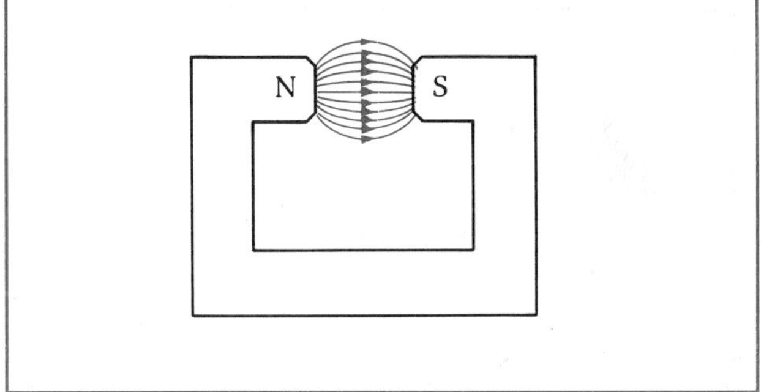

Figure 14.4 *Magnetic field of a horseshoe magnet*

Current-carrying wires produce magnetic fields which can be made stronger by increasing the current.

1 The magnetic field pattern near a long straight wire carrying a steady current is shown in Figure 14.5. The field lines are concentric circles around the wire. The direction of the field lines depends on the current direction. One simple way to remember the field line direction from the current direction is the **corkscrew rule**. Imagine driving a corkscrew into a cork. Its rotation is in the same direction as the field lines of a wire carrying current in the driving direction. In other words, a wire carrying current away from you has field lines which are clockwise round the wire as you see them.

2 The magnetic field produced by a solenoid carrying a steady current is like the field of a bar magnet. Outside the solenoid, the field lines loop round from one end to the other. However, unlike a bar magnet, the field lines pass straight down the middle of the solenoid's inside.

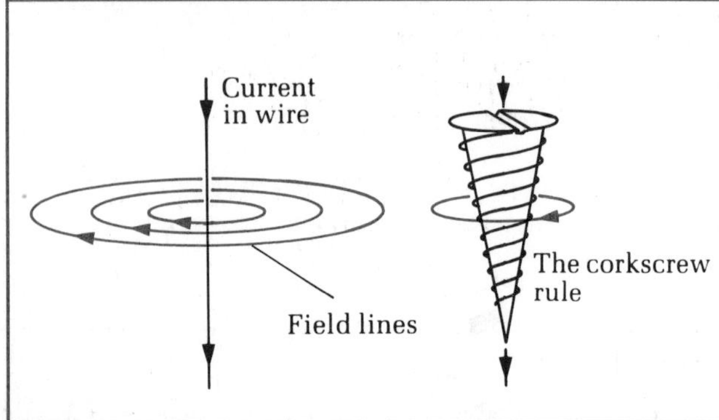

Figure 14.5 *The magnetic field near a wire*

The field lines along the inside of a solenoid are parallel to one another. So the field inside a solenoid is a **uniform field**.

Because the field of a bar magnet is like that of a solenoid, we can say that a solenoid has poles like a bar magnet. The field lines of a bar magnet spread out from the N-pole, loop round, then converge on the S-pole. A small plotting compass on one of its field lines lies with its N-pole pointing to the opposite pole of the bar magnet. The end of a solenoid from which field lines emerge is therefore like the N-pole of a bar magnet. A view of the current direction around that end would show an anticlockwise current. The field lines loop round the outside of the solenoid to re-enter at the other end, the S-pole. A view of that end shows a clockwise current. A simple way to remember the polarity of a solenoid, given the current direction, is the **solenoid rule**, shown in Figure 14.6.

Inside a solenoid, the magnetic field is uniform and its strength can be increased by increasing the current. Figure 14.7 shows a large solenoid which uses superconducting windings. Superconductors have zero electrical resistance, so superconducting windings can pass huge currents without over-heating. The result is to give very strong magnetic fields – much stronger than the strongest permanent magnets. The solenoid shown is used in a body scanner, to make atoms in the body emit tiny radio signals. The signals are picked up by detectors, and then processed by a computer. The computer is linked to a TV monitor which gives an image of a cross-section of any part of the body; all this is done without pain or damage to the patient!

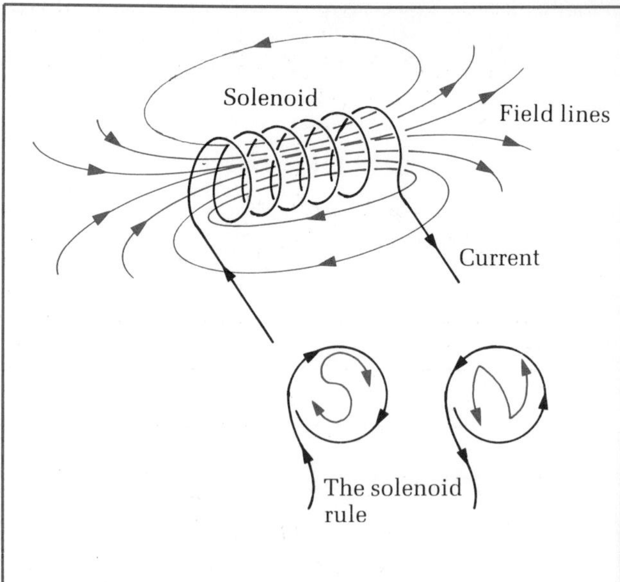

Figure 14.6 *The magnetic field of a solenoid*

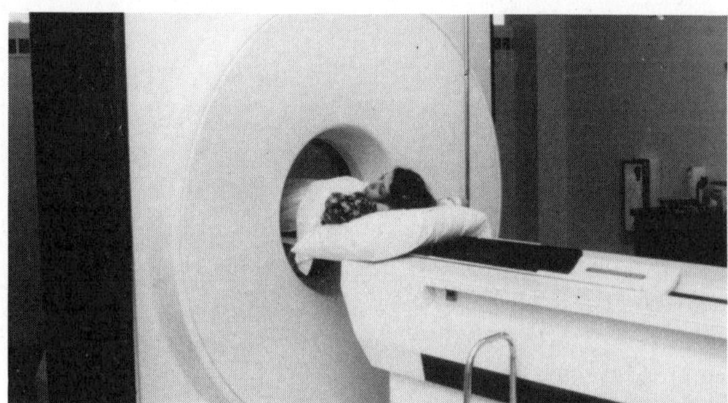

Figure 14.7 *A body scanner. The patient is partly in the solenoid*

14.2 Magnetic field strength

Investigating the force on a current-carrying conductor in a magnetic field

Hold a current-carrying wire between the poles of a horseshoe magnet. With a steady current in the wire, a force due to the magnet acts on the wire, strongest when the wire is at right angles to the field lines, as shown in Figure 14.8. Turn the wire from a line at right angles to the field lines until it is parallel to the field lines. The force becomes weaker as the wire is turned, until it becomes zero when the wire is parallel to the field.

Why should the wire be acted on by a force due to the magnet? The wire must carry a current, and it must lie at right angles to the field to experience the full effect. The overall magnetic field pattern is shown in Figure 14.9.

The field is stronger on one side of the wire than the other. The magnet's own field is combined with the field due to the wire to produce the overall field pattern. The wire is pushed out of the magnet's field in the direction

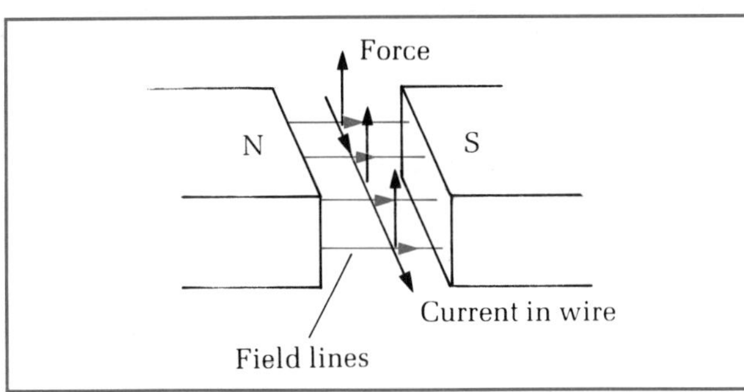

Figure 14.8 *Force on a conductor*

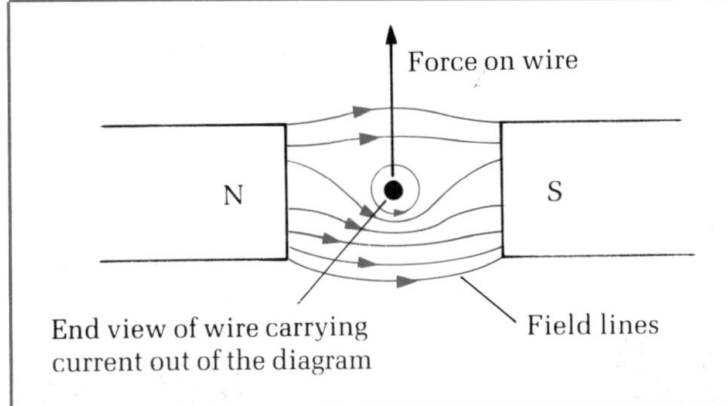

Figure 14.9 *Field pattern*

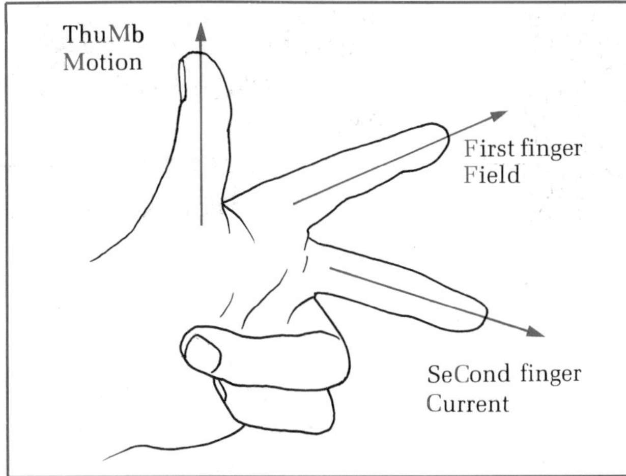

Figure 14.10 *The left-hand rule*

shown. A simple way to remember the force direction, given the current and field directions, is given by the **left hand rule.** See Figure 14.10.

Current-carrying wires produce magnetic fields and can be pushed about by magnetic fields. We measure the strength of a magnetic field in terms of the force it exerts on a current-carrying wire. This is similar to the way we measure electric or gravitational field strength. An electric field is the region near a charged object; electric charges placed near a charged object experience forces which we explain by saying that the electric field of the charged body acts on the charges. We define the strength of an electric field as the force per unit charge on a test charge. Similar ideas apply to gravitational fields. Gravitational field strength is defined as the force per unit mass on a test mass. For magnetic field strength, we use a current-carrying wire as the test object, but the wire must be at right angles to the field lines.

The force on a current-carrying wire at right angles to a uniform magnetic field is

a) proportional to the current I,

b) proportional to the length L of the wire.

An arrangement to test these statements is shown in Figure 14.11. The top pan balance is used to measure the force on the test wire. The wire is connected in series with a rheostat, an ammeter and a d.c. power supply.

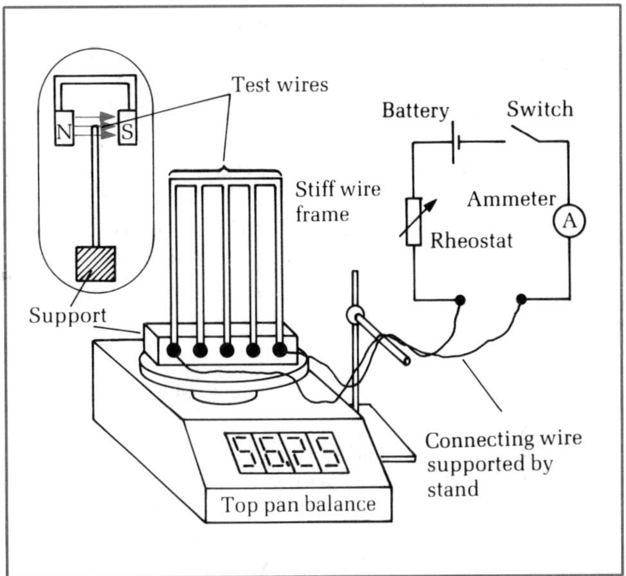

Figure 14.11 *Measuring* B

For **a)**, the rheostat is used to alter the current in steps. At each step the readings of the top pan balance and the ammeter are noted. The test wire is kept in the same magnetic field throughout, using the same length of test wire at each step. The measurements ought to show that the force on the test wire is proportional to the current.

For **b)**, the rheostat is used to keep the current the same throughout. The length of the test wire can be altered by reconnecting the wires to the frame. The same magnetic field is used throughout. The force is measured for the same current in each length of test wire in turn. The measurements ought to show that the force is proportional to the length.

Since the force is proportional to

1 the current I,

2 the length L,

then we can define the strength of the magnetic field as the force per unit length per unit current.

Magnetic field strength B is defined as the force per unit length per unit current on a current-carrying conductor at right angles to the field lines. The unit of magnetic field strength is the tesla (T), named after Nikola Tesla who invented a variety of electromagnetic machines including the widely-used induction motor. See p. 198.

$$B = \frac{F}{IL}$$

where F = force on wire at right angles to field lines,
I = current,
L = length of wire,
B = magnetic field strength.

For a straight wire at an angle to the field lines other than 90°, the force is less than if it were at 90°. If the wire is parallel to the field lines, there is no force on the wire. The magnetic field has direction and magnitude and it is the perpendicular component to the wire which exerts a force on the wire. Suppose the angle between the wire and the field lines is θ. Let the magnetic field strength be B. The field component parallel to the wire is $B\cos\theta$; the field component perpendicular to the wire is $B\sin\theta$. So the force on the wire is given by

$$F = B\sin\theta\, IL = BIL\sin\theta$$

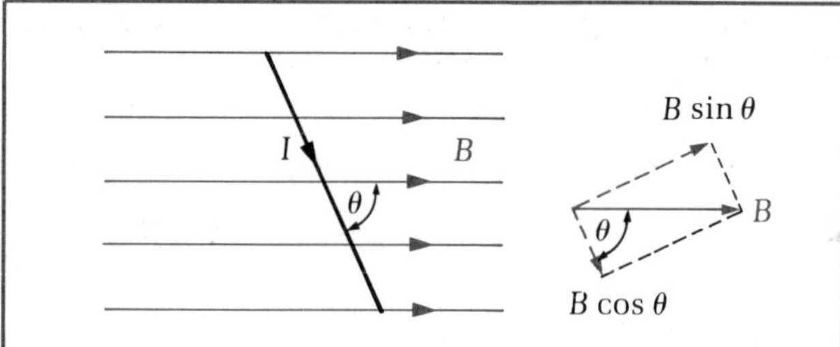

Figure 14.12 $F = BIL\sin\theta$

The direction of the force is still given by the left hand rule, but the first finger points in the direction of the perpendicular component of the field. The parallel component has no effect on the current-carrying wire.

Measuring magnetic field strength

We could use the arrangement of Figure 14.11 to measure the magnetic field strength of a horseshoe magnet. The length of the test wire is chosen according to the pole width of the magnet. To do this, the connecting wires from the circuit are connected to the appropriate sockets at the base of the frame. With a measured steady current passing through the test wire as shown in Figure 14.11, the reading of the top pan balance is noted. The reading is then noted without current in the test wire. The difference in readings gives the magnetic force on the test wire. The length of the test wire must also be measured. Then the magnetic field strength can be calculated using $B = F/IL$.

Worked example A test wire of length 0.05 m was placed at right angles to the field lines between the poles of a horseshoe magnet. When a current of 2.5 A passed through the test wire, the wire experienced a force of 0.025 N. Calculate the magnetic field strength of the magnet.

Solution

$F = 0.025\ \text{N}$, $I = 2.5\ \text{A}$, $L = 0.05\ \text{m}$

$$B = \frac{F}{IL} = \frac{0.025}{2.5 \times 0.05} = 0.20\ \text{T}$$

14.3 Motors and meters

The turning effect on a current-carrying coil

1 Consider a rectangular coil placed **parallel to** a magnetic field, as in Figure 14.13. The plane of the coil is parallel to the field lines; each side (i.e. WX and YZ) of the coil is at right angles to the field lines. When a steady current passes round the coil, a magnetic force acts on each side of the coil. The left hand rule can be used to determine the force directions. Each side is acted on by a force which is at right angles to the field lines and the wire. The force on one side is in the opposite direction to the force on the other side because the current along each side is in opposite directions. So the two forces provide a torque (i.e. a turning effect) on the coil. In other words, the coil is acted on by a **couple** due to the magnetic forces on the sides.

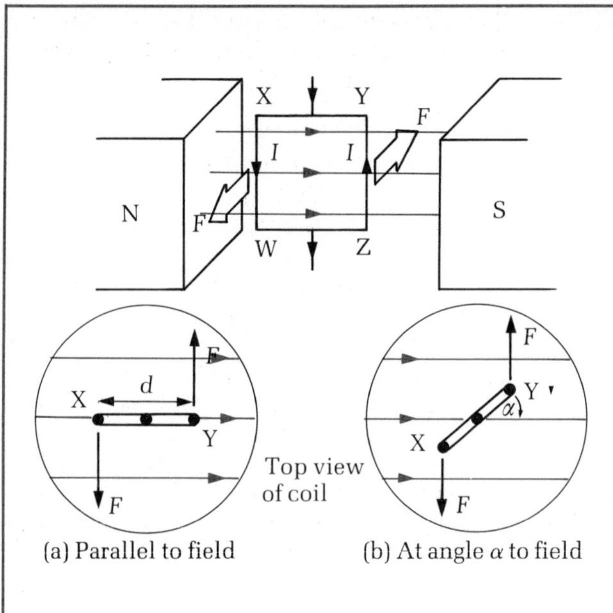

Figure 14.13 *Couple on a coil*

If the coil is pivoted as shown, the couple can turn the coil. This is the principle of moving coil meters and motors. The moving coil meter has a coil which turns against a hairspring; the motor has a coil with brush contacts so that it can rotate continuously. The principle of operation is the same for both, though. When current passes through a coil parallel to the lines of a magnetic field, a couple due to the field acts on the coil.

Suppose the length of the sides of the coil is l, and that the coil has N turns. When current I passes round the coil, the effective current along each side is NI since each side has N windings with current I in each winding.

Since each side is at 90° to the field lines, then the force on each side is given by

$$F = BNIl\sin 90° = BNIl$$

The couple C on the coil due to the magnetic forces is equal to the force $F \times$ the distance d across the coil. Hence

$$C = Fd = BNIld = BNIA$$

where A is the coil area $(= ld)$.

2 When the coil plane is **at an angle** to the field lines, the couple is less than when the coil is parallel to the field. The force on each side is exactly the same in size and direction, but the distance used to calculate the couple is less. The perpendicular distance between the force lines is equal to $d \cos \alpha$ where α is the angle between the coil and the field. So the couple equals $Fd \cos \alpha$.

$$\text{Couple } C = BNIA \cos \alpha$$

where B = magnetic field strength,
N = number of turns,
I = current,
A = coil area,
α = angle between coil and field.

The moving coil meter

The coil is pivoted between the poles of a horseshoe magnet (see Figure 14.14). Current enters and leaves the coil via the hairsprings. When current passes round the coil, the couple due to the field turns the coil. The turning movement is opposed by the tightening of the hairsprings, so the coil reaches a position where the couple due to the field is equally opposed by the hairspring couple.

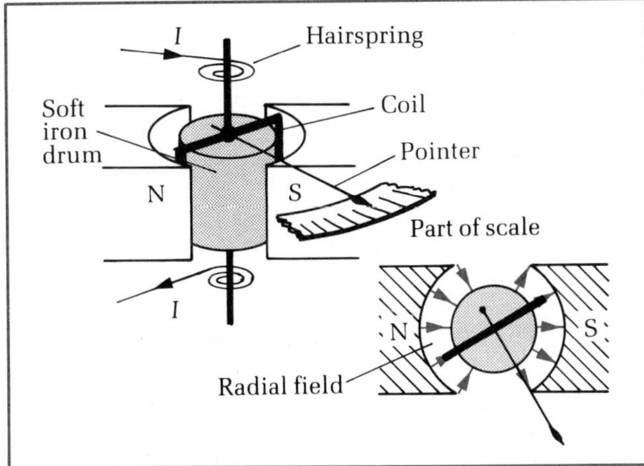

Figure 14.14 *Moving coil meter*

By including a soft iron cylinder in the coil, the field lines are made **radial**. This means that the field lines point towards or away from the centre of the cylinder. The result is that the coil is always parallel to the field lines, no matter what position it turns to – provided it doesn't turn too far. In practice the pointer reaches the end of the scale before the coil can turn too far.

The hairspring couple C is proportional to the angle which the coil turns through. So we can write $C = k\theta$, where θ is the angle turned by the coil and k is the **stiffness constant** for the hairspring.

The final position which the coil reaches when a steady current I passes through is given by

Couple due to field = hairspring couple

$$BNIA = k\theta$$

Hence
$$\theta = \frac{BNA}{k} I$$

The angle θ can be made larger by
a) using a stronger magnet (i.e. increasing B),
b) using a coil with a bigger area A,
c) increasing the number of turns N,
d) using a weaker hairspring (i.e. reducing k).

The suspension-type meter is much more sensitive than the pivoted meter. The moving coil is suspended by a fine wire called the **torsion wire.** Current enters and leaves the coil via the torsion wire and the hairspring. Instead of a pointer, a mirror attached to the torsion wire is used to reflect a light beam onto a scale. When the coil turns, the spot of light produced by the beam on the scale moves across the scale. Using a light beam pointer doubles the sensitivity because the spot moves through 2° for every degree the coil turns through. (The reflected ray moves through twice the angle the mirror turns through.) Suspension-type meters can measure currents of less than a microampere, but they must be handled very carefully or the suspension will be damaged.

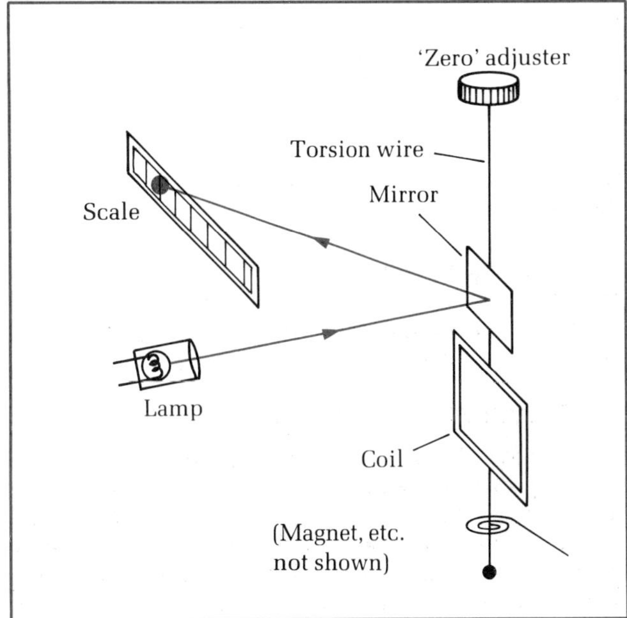

Figure 14.15 *Suspension-type meter*

Motors

To make a coil turn continuously in a magnetic field, the connections to the coil must allow for continuous contact. This is achieved using graphite brushes pressed against metal slip rings on the coil spindle. Figure 14.16 shows the idea. When current is passed through the coil, the coil turns until it faces the magnet poles. Further turning is only possible if the current in the coil changes direction.

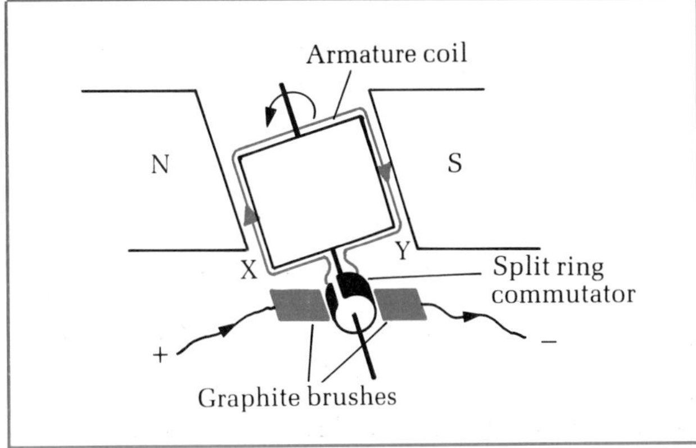

Figure 14.16 *A simple d.c. motor*

The d.c. motor has a 'split-ring commutator', as shown in Figure 14.16 on the previous page. Each time the coil turns through the position at right angles to the field, the coil connections become reversed, so the current changes direction in the coil, and the coil is pulled round for another half-turn, etc. Practical d.c. motors usually have several coils on the same spindle, spaced at equal angles. The commutator has a corresponding number of segments, each pair connected to a coil. In this way, the brushes are at any time in contact with one of the coils. So a more even turning effect is produced.

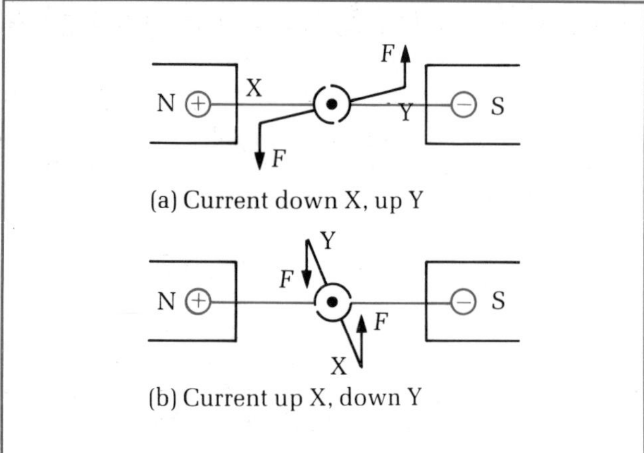

Figure 14.17 *End view of a commutator*

An electromagnet is often used instead of a permanent magnet. The electromagnet's current is taken from the same d.c. supply as the coil current. The windings of the electromagnet are usually referred to as the **field coil**, and the rotating coil is usually called the **armature coil**. The field coil may be in series or in parallel with the armature coil.

a) A series-wound motor has the field coil in series with the armature coil. This gives a powerful starting torque, because the initial current when the armature is at rest is large, so the field is initially very strong. Hence the armature is given a strong push to start with by the field.

b) A shunt-wound motor has the field coil and the armature coil in parallel, as shown in Figure 14.19. This gives a steady speed when the load being moved by the motor varies.

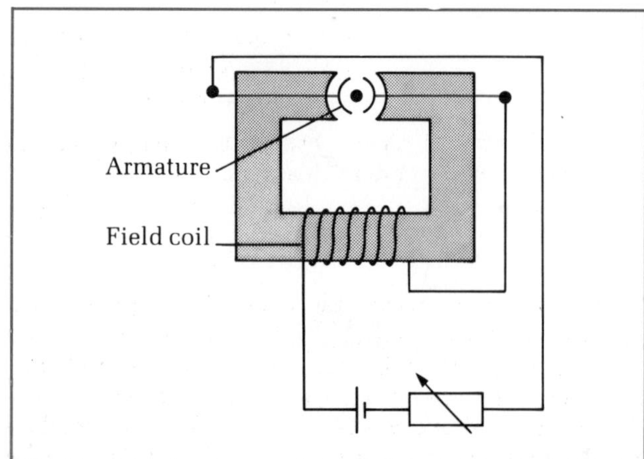

Figure 14.18 *Series-wound motor*

The a.c. motor is the same as the series wound d.c. motor, as shown in Figure 14.19. When the a.c. supply reverses, the current and the magnetic field both reverse, so the effect is the same as d.c.

A back e.m.f. is created when a motor armature turns in a magnetic field. The reason is that the armature windings cut across the field lines, so an e.m.f. is induced in the windings. This is the principle of electromagnetic induction, discussed in detail in Chapter 15. The induced e.m.f. acts against the applied p.d. to oppose the motion. The size of the induced e.m.f. is proportional to the motor speed. When a motor is switched on the induced e.m.f. is initially zero, so there is no opposition at first to the applied p.d. Hence a large current passes through the motor at first, but as the motor speeds up the back e.m.f. increases, so the opposition to the applied p.d. increases. Therefore the current becomes smaller as the motor speeds up. Connect an ammeter in series with a d.c. motor; you should find that the current surges through the motor at first, then decreases as the motor speeds up.

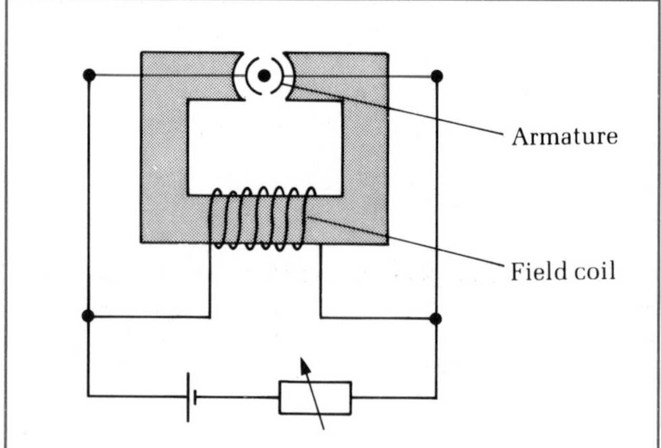

Figure 14.19 *Shunt-wound motor*

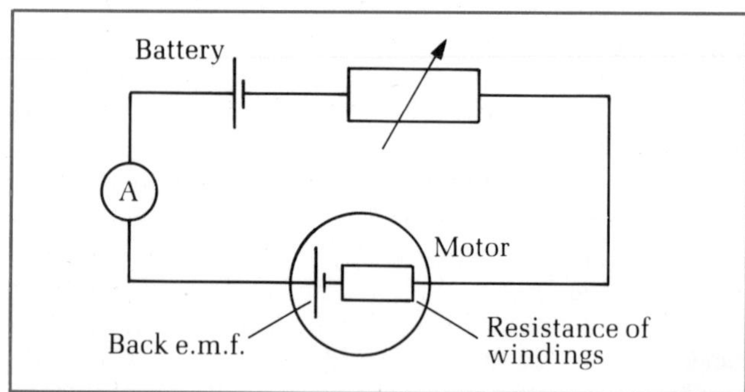

Figure 14.20 *Back e.m.f.*

14.4 Charged particles in magnetic fields

Electron beams

To produce a beam of electrons, a vacuum tube containing an electron gun must be used. The gun is essentially a heated metal filament and a nearby anode plate. The anode plate is made positive with respect to the filament.

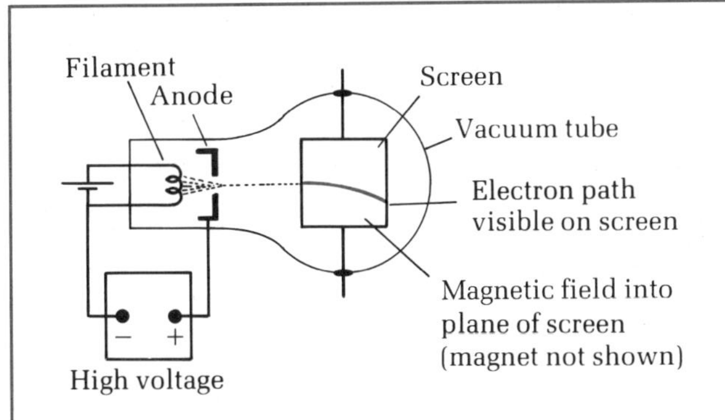

Figure 14.21 *Deflection tube*

The filament is heated by an electric current. Electrons from the hot filament are pulled towards the anode. Some of these electrons pass through a small hole in the anode to form an electron beam. This process, called **thermionic emission**, is discussed in more detail on p. 419.

To make the beam visible, a screen coated with fluorescent paint is used. The screen can be placed in line with the beam so a visible trace across the screen is produced by the beam. If the tube is then placed in a magnetic field the effect of the field on the beam can be studied.

With the field lines at right angles to the screen, the beam is pushed to one side of the screen. Each electron in the beam experiences a force due to the magnetic field. If the electrons had been moving inside a wire, the whole wire would have been pushed to one side, but because the electrons are in a beam rather than a wire, the beam is forced into a curve by the field. In a wire, the current direction is opposite to the electron direction because of the convention that current passes from positive to negative round a circuit. So the equivalent current for the electron beam is back towards the filament. We can use the left hand rule to work out the direction of the force on the electrons, but we must remember that the current direction is opposite to the direction of the electrons in the beam. Perhaps it's simpler to remember that electrons moving from left to right in a vertical plane are pushed down by a magnetic field directed into the plane. See Figure 14.22.

The reason why current-carrying wires in a magnetic field experience forces is that the electrons moving in the wire are pushed to one side by the field. You can see in Figure 14.22 electrons pushed by a magnetic field: if those electrons had been confined to a wire, the whole wire would have been pushed to one side.

All charged particles moving across the lines of a magnetic field are acted on by a force due to the field. Positively charged particles such as protons are pushed in the opposite direction to negatively charged particles like electrons. Figure 14.23 shows a pair of charged particles curving across a magnetic field. The particles are created by a nuclear collision. They curve in opposite directions because they carry opposite charge.

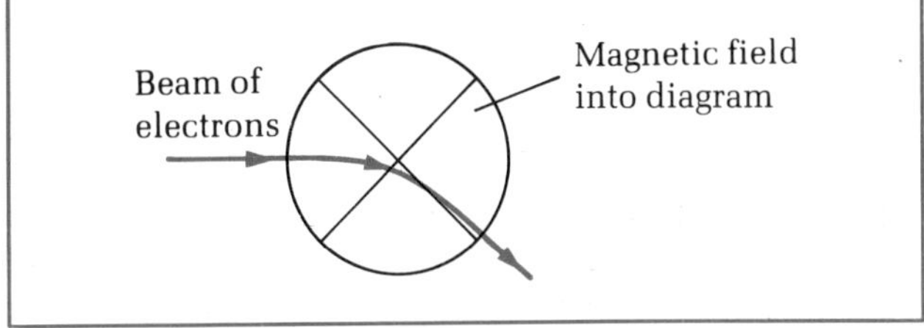

Figure 14.22 *Electrons in a magnetic field*

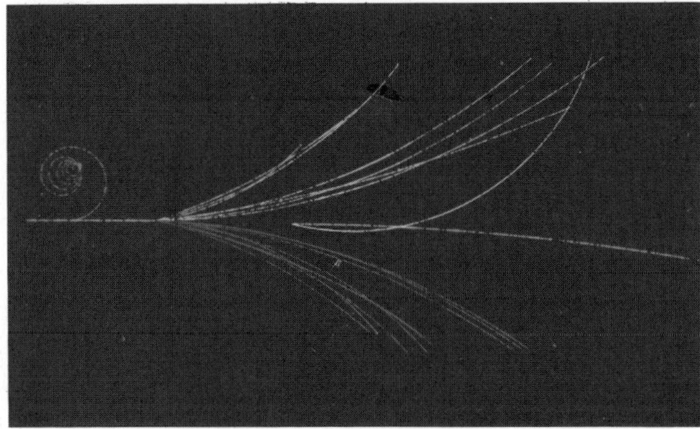

Figure 14.23 *Charged particles in a magnetic field*

The force on a moving charge

An electric current in a wire is due to the passage of electrons along the wire. The electrons carry charge along the wire. The charge passing a point along the wire in a time interval t is given by It, where I is the current.

A beam of charged particles in a vacuum tube is an electric current across the tube. Suppose each particle in the beam has charge q and speed v. In a time interval t, each particle travels a distance vt. Its passage is equivalent to current q/t along a wire of length vt.

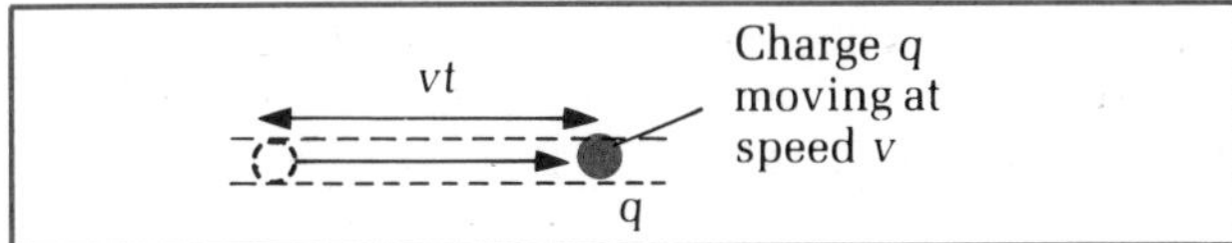

Figure 14.24 *Force on a moving charge*

At right angles to the field lines of a magnetic field of strength B, the charged particles experience a force F due to the field. If in a wire, the force would be given by $F = BIL$. For a moving charge, the same equation applies, but I is the equivalent current q/t and L is the length vt. Hence we can write $F = BIL = B(q/t)vt = Bqv$ for the force.

$$F = Bqv$$

where F = force,
q = charge carried by particle,
B = magnetic field strength,
v = speed of the particle.

The deflection of electron beams by magnetic fields is discussed in more detail in Chapter 27.

The Hall effect

Electrons moving across the lines of a magnetic field experience a force due to the field. If the electrons are in a beam in a vacuum tube, the field makes them change direction. If the electrons are moving along inside a conductor, the field pushes them to one edge of the conductor. Because the electrons are in a solid, they can only be pushed as far as the edge of the conductor, so that edge of

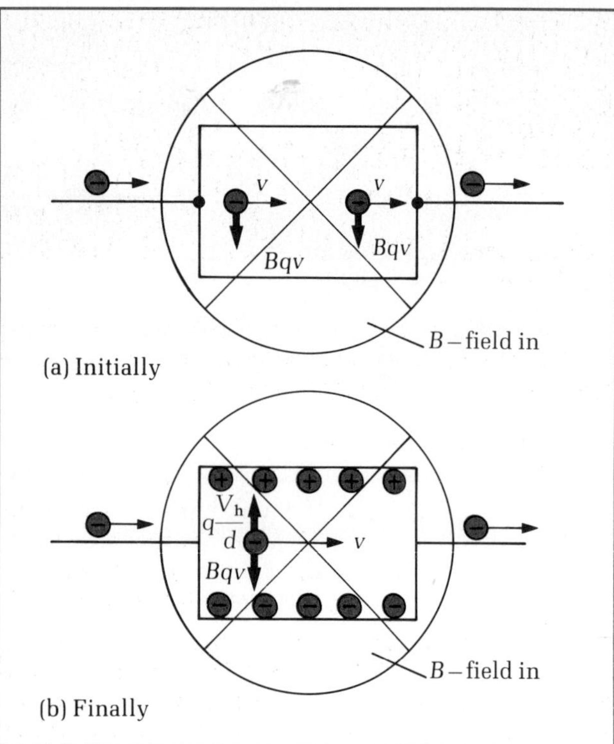

Figure 14.25 *The Hall effect*

the conductor ought to become negatively charged, and the opposite edge positively charged. This is in fact exactly what happens, but it is most noticeable in thin, wide conductors and semiconductors such as silicon. It is called the **Hall effect** after its discoverer.

Consider the passage of free electrons through a conductor, as in Figure 14.25. Suppose a magnetic field is applied to it, the field lines being directed into the conductor as shown. When the field is first applied, free electrons are forced by the field to the lower edge, so the lower edge becomes negatively charged and the upper edge positively charged. But the charge build-up at the edges acts to prevent further free electrons being pushed downwards by the magnetic field. The charge builds up to a level where the magnetic force on each electron moving across is balanced by the electric field due to the charge built up. So each incoming free electron passes straight through. In other words, the p.d. between the edges builds up until it has created an electric field strong enough to prevent the magnetic field pushing any more electrons down. The p.d. between the edges is called the **Hall p.d.**

The factors which determine the Hall p.d. include the strength of the magnetic field. The stronger the field, the greater the p.d. To determine the exact link, consider the situation when the Hall p.d. has built up, so charge carriers (i.e. electrons in metals) pass through the material undeflected because the magnetic force is balanced by the electric field force.

The magnetic force $= Bqv$ where B is the magnetic field strength, q is the charge of each particle, v is the speed of each particle.

The electric force $= qV_h/d$, where V_h is the Hall p.d. which develops across distance d between the two edges. See p. 167 if necessary. The electric field strength between the two edges is V_h/d so the force equals charge × electric field strength.

Since electric field force = magnetic field force, we can write

$$qV_h/d = Bqv$$

so $$V_h = Bvd$$

The Hall p.d. is proportional to the magnetic field strength; so by measuring the Hall p.d. we can determine the strength of the magnetic field B. This is the basis of the Hall probe, discussed later.

The speed v of the charge carriers depends on the current. Suppose each charge carrier takes time t to cross the conductor. If the number of charge carriers per unit volume is n, then the total number of charge carriers is nV where V is the conductor volume. However, the conductor volume equals the area of cross-section A × its length L. Since $L = vt$, then $V = Avt$. So the total number of charge carriers in the conductor is $nAvt$. If the charge of each carrier is q, then the total charge in the conductor is $nAvtq$. All this charge leaves the conductor in time t, so the current I through the conductor is given by

$$I = nAvtq/t = nAvq$$

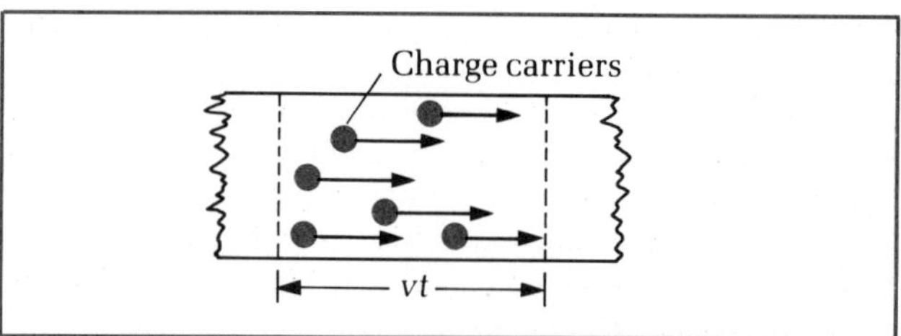

Figure 14.26 *Charge carriers*

The speed of the charge carriers v is therefore equal to I/nAq. We can use this expression to show how the Hall p.d. depends on the current.

$$V_h = Bvd = \frac{BId}{nAq} = \frac{BI}{nxq}$$

since the area of cross-section A is equal to the thickness x × the distance d across the conductor.

$$V_h = \frac{BI}{nxq}$$

where B = magnetic field strength,
I = current,
n = number of charge carriers per m^3,
x = thickness,
q = charge of each carrier.

Worked example A strip of metal foil of thickness 0.1 mm carries a current of 5.0 A. When a magnetic field of strength 0.2 T is directed into the foil at right angles to its plane, a p.d. across its edges is produced due to the Hall effect. Measurement of the p.d. gives a value of 6.25 μV for a current of 5 A. Calculate the number of charge carriers per unit volume, given the charge of each carrier is 1.6×10^{-19} C.

Solution

$B = 0.2$ T, $I = 5.0$ A, $x = 1 \times 10^{-4}$ m,
$q = 1.6 \times 10^{-19}$ C, $V_h = 6.25 \times 10^{-6}$ V.

To calculate n, rearrange the equation $V_h = BI/nxq$ to give $n = \frac{BI}{xqV_h}$. Therefore

$$n = \frac{0.2 \times 5.0}{1 \times 10^{-4} \times 1.6 \times 10^{-19} \times 6.25 \times 10^{-6}} = 1 \times 10^{28}\ \text{m}^{-3}.$$

The charge carriers in a metal such as aluminium are free electrons. Certain metals and semiconductors have positive charge carriers, called **holes**, instead of free electrons. When placed in a magnetic field, these materials give a Hall p.d. which has reversed polarity compared with free electron conductors or semiconductors. A hole is a vacancy in the bonds linking atoms together; the vacancy is caused by a missing electron. An electron from a nearby atom can jump across to fill the vacancy, but the electron leaves behind a vacancy. Effectively, the vacancy or hole moves in the opposite direction, so it is equivalent to a positive electron. Semiconductors such as silicon can be doped with other atoms to make the material conduct either by free electrons or by holes as the charge carriers.

For hole conduction, the added atoms each have one electron less than a silicon atom. So when the added atoms form bonds with silicon atoms, each added atom has a vacancy. For free electron conduction, the added atoms each have one electron more than a silicon atom. So when each added atom forms bonds with silicon atoms, an extra electron is left over as a free electron. Semiconductors which are doped are called **p-type** or **n-type** according to whether the conduction is by holes (for p-type) or by electrons (for n-type). See p. 203 for a fuller discussion of semiconductors.

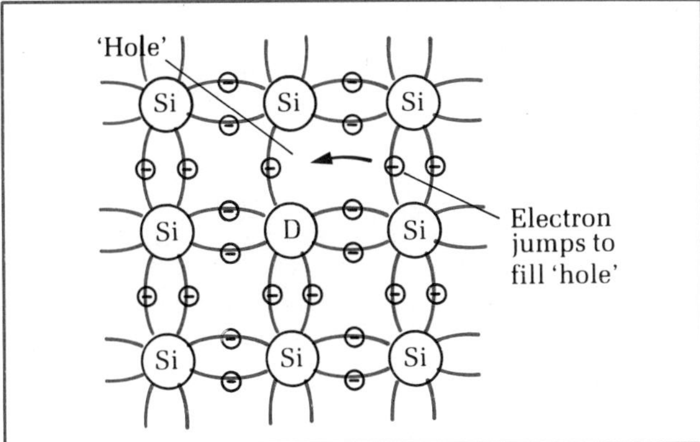

Figure 14.27 *Conduction by holes*

Hall probes are used to measure magnetic field strength. A Hall probe is a slice of doped semiconductor. Such material can be doped to have a much lower concentration of charge carriers than in metals. So they produce much bigger Hall voltages than metals. Two connecting wires to either end of the slice supply a constant current. Another two wires connected across the edges of the slice allow the Hall p.d. to be measured. The slice must be positioned so that the field lines of the magnetic field are at right angles to its plane. With a constant current, the Hall p.d. is proportional to the magnetic field strength. The probe can be calibrated using a magnetic field of known strength. This would simply involve noting the voltmeter reading with the probe in the known field. Then

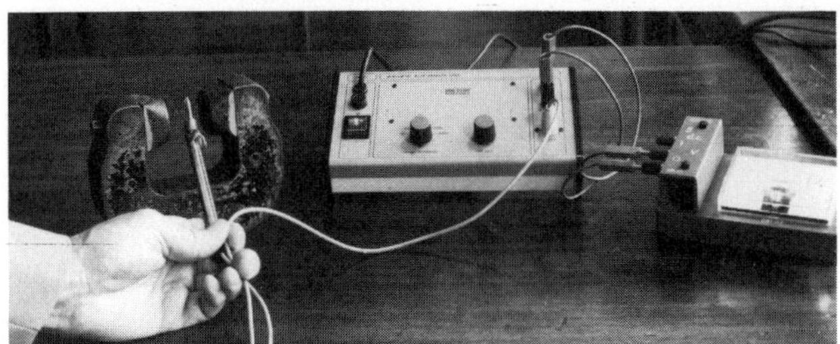

Figure 14.28 *The Hall probe*

the probe is put into the unknown field and the voltmeter reading is noted once more. The field strength can then be calculated since it is proportional to the Hall p.d.

$$B = \frac{B_0}{V_0} V$$

where B = 'unknown' field strength,
V = Hall p.d. produced by B,
B_0 = known field strength,
V_0 = Hall p.d. produced by B_0.

What are the advantages of a Hall probe over the current balance shown in Figure 14.11 for measuring magnetic field strengths? The current balance is easier to use since it does not need to be calibrated by a known field. It gives the magnetic field strength directly from the force, current and length of the test wire. However, the Hall probe is much easier to use in practical situations. For example, satellite magnetometers use Hall probes to measure the strength of the Earth's magnetic field high up in Space. At the other extreme, right down in the ground, mining engineers use Hall probes to check that hydraulic pit props are unmagnetized – otherwise particles of iron stick to the hydraulics and prevent them operating correctly! So Hall probes, if correctly calibrated, are much more convenient to use. In addition, they can measure very weak magnetic fields.

14.5 Field formulae for current-carrying wires

Investigating the magnetic field of a solenoid

A slinky coil is an excellent adjustable solenoid. It can be stretched or compressed as required to alter its coil spacing and length. Figure 14.29 (on page 186) shows a circuit used to supply a slinky solenoid with a steady current. A Hall probe is used to measure the magnetic field strength at different positions in and out of the solenoid.

The variation of magnetic field strength across and along the solenoid To measure this, start by placing the Hall probe into the solenoid through the spacing of the windings. The slice of the probe must be at right angles to the solenoid axis since the field lines in the solenoid are parallel to its axis. The rheostat and ammeter are used to keep the current constant. Measurements of the magnetic field strength can be made at different positions across the solenoid to show that the field strength

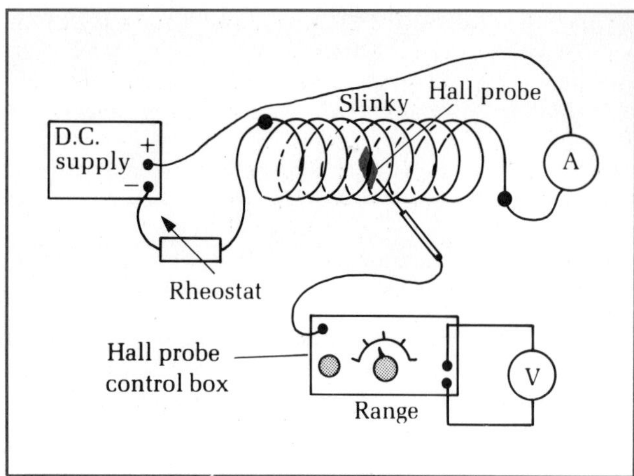

Figure 14.29 *Investigating magnetic fields*

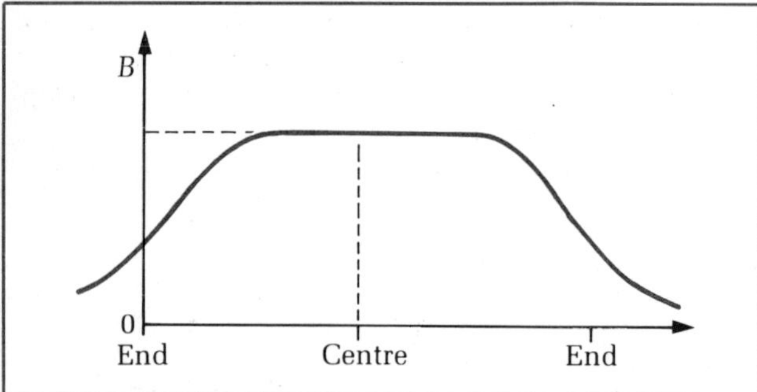

Figure 14.30 *Solenoid field*

is reasonably constant across any diameter. Measurements at different positions along the solenoid can also be made. A graph of magnetic field strength against distance from one end ought to be like Figure 14.30. Away from the ends, the field strength is constant.

The variation of magnetic field strength with current To measure this the Hall probe is kept in one position inside the solenoid. The rheostat is used to alter the solenoid current in steps. At each step, the current and field strength are measured. The results can be plotted as a graph of magnetic field strength against current. Figure 14.31 shows such a graph. The magnetic field strength is proportional to the current.

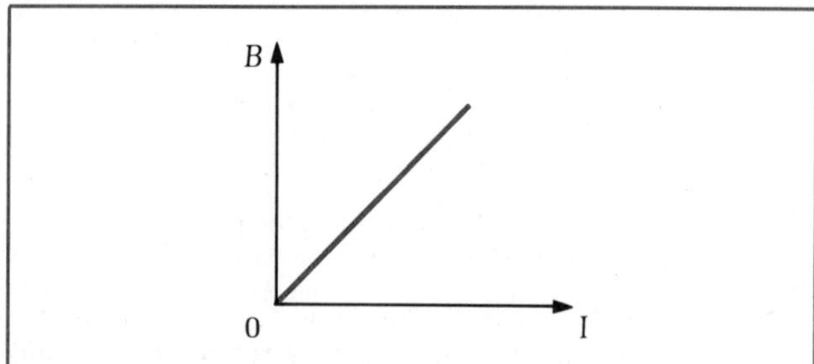

Figure 14.31 B v I *for a solenoid*

The variation of magnetic field strength with the number of turns per metre To measure this, start with the windings as close as possible but not touching each other. Count the number of turns per metre, then pass a measured current through the solenoid and measure the magnetic field strength at its centre. Then pull the windings out a little, and repeat the procedure. Do this for several winding spacings. A graph of magnetic field strength against number of turns per metre ought to give a straight line through the origin. The magnetic field strength is proportional to the number of turns per metre.

The variation of magnetic field strength with solenoid diameter To measure this, you will need to use several metres of insulated wire to make solenoids of different diameters. Each solenoid has the same length and number of turns. Pass the same current through each solenoid, and measure the magnetic field strength inside. Your measurements ought to show that the diameter of the solenoid does not affect the magnetic field strength inside it.

Field formulae

In a solenoid The results of the above experiments show that the magnetic field strength inside a solenoid is

a) proportional to the current I,

b) proportional to the number of turns per metre.

We can therefore write an equation for the magnetic field strength in terms of a constant of proportionality, called the **magnetic field constant**, μ_0.

$$B = \mu_0 \frac{N}{l} I$$

where B = magnetic field strength in the solenoid,
N = number of turns,
l = length,
I = current,
μ_0 = magnetic field constant.

The magnetic field constant is sometimes called the **permeability of free space.** Its value is set at $4\pi \times 10^{-7}\,\mathrm{H\,m^{-1}}$ because of the way the ampere is defined. See p. 188. The unit μ_0 is $\mathrm{H\,m^{-1}}$ where the henry (H) is the unit of inductance of a coil. See p. 505.

The solenoid field can be used to define the magnetic field constant in the same sort of way that the electric field constant is defined by parallel plates. In both cases, we looked at the factors which determine the field strength. Then a constant of proportionality is introduced to give an equation for the field strength.

Worked example A solenoid with 1000 turns and of length 0.1 m is lined up so that its axis is parallel to the Earth's field lines, as in Figure 14.32. The strength of the

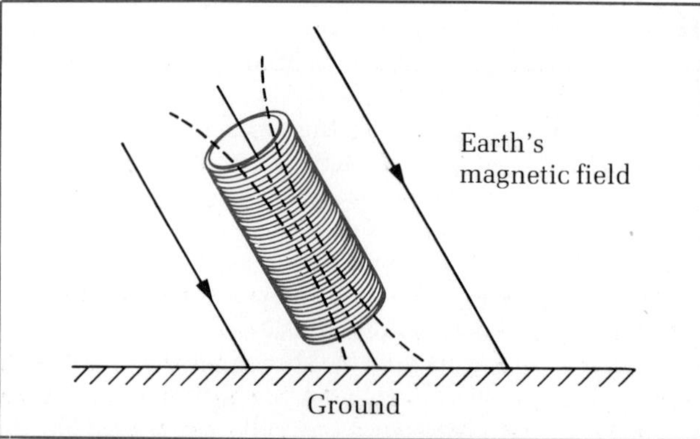

Figure 14.32 *Diagram for worked example*

Earth's magnetic field is 70 μT at the position of the solenoid. Calculate the current in the solenoid which will create a magnetic field of the same strength as the Earth's field. Assume $\mu_0 = 4\pi \times 10^{-7}\,\mathrm{H\,m^{-1}}$.

Solution $N = 1000$ turns, $l = 0.1$ m, $B = 70 \times 10^{-6}$ T

Rearrange the equation $B = \mu_0 NI/l$ to give $I = Bl/\mu_0 N$.

$$\text{Hence } I = \frac{70 \times 10^{-6} \times 0.1}{4\pi \times 10^{-7} \times 1000} = 5.6 \times 10^{-3}\,\text{A}.$$

Near a long straight wire A Hall probe can be used to investigate the magnetic field strength near a long straight wire. The wire carrying a steady current produces magnetic field lines which are circles round the wire. Does the magnetic field strength vary with position round the wire? How does the magnetic field strength vary with current or with distance from the wire? Graphs based on measurements using a Hall probe are shown in Figure 14.33. To produce the measurements for each graph, the relevant factor was altered in steps. At each step the magnetic field strength was measured. For example, Figure 14.33(a) shows that the current was altered to give a set of readings of magnetic field strength and current in the wire. The probe itself was kept in the same position throughout.

The graphs show that the magnetic field strength B is

a) proportional to the current I,

b) proportional to $1/r$, where r is the distance from the probe to the centre of the wire.

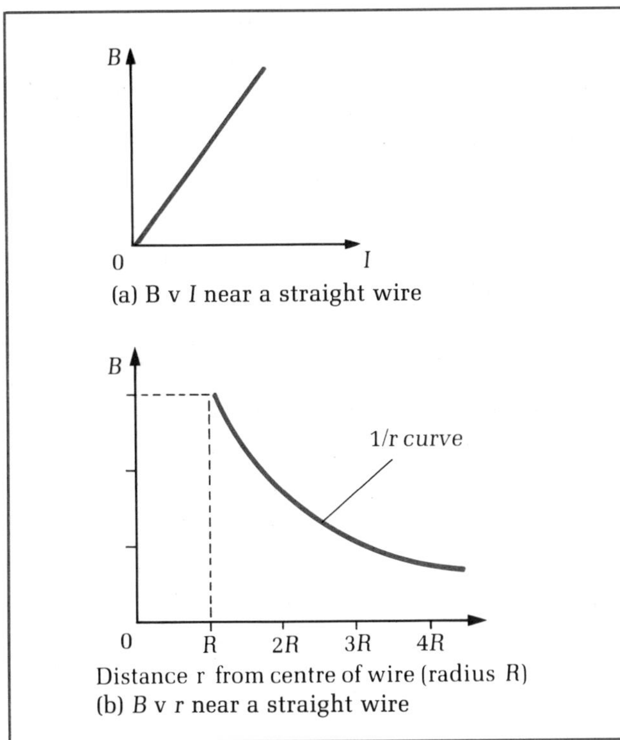

Figure 14.33

The theory of magnetic fields produced by current-carrying wires was first developed by André Ampere in the early 19th century. He realized that the field lines produced by currents were always continuous loops. He considered a solenoid with its ends bent round to fit together (i.e. a toroidal solenoid), and showed that its magnetic field was uniform all the way round. The magnetic field strength in this arrangement is given by the equation $B = \mu_0 NI/l$. Ampere used this to develop a more general rule; the magnetic field strength × the distance round the loop equals μ_0 × the total current through the loop's plane. In other words, $Bl = \mu_0(NI)$.

Ampere applied the same approach to the field round a straight wire. The loop is a circle of radius r, so l equals $2\pi r$. Hence $B(2\pi r) = \mu_0 I$.

$$B = \frac{\mu_0 I}{2\pi r}$$

where B = magnetic field strength at distance r from the wire centre,
I = current,
μ_0 = magnetic field constant.

These two situations are shown in parts (a) and (b) of Figure 14.34. The rule for the strength of a magnetic field is much more complicated when the field strength changes along the loop.

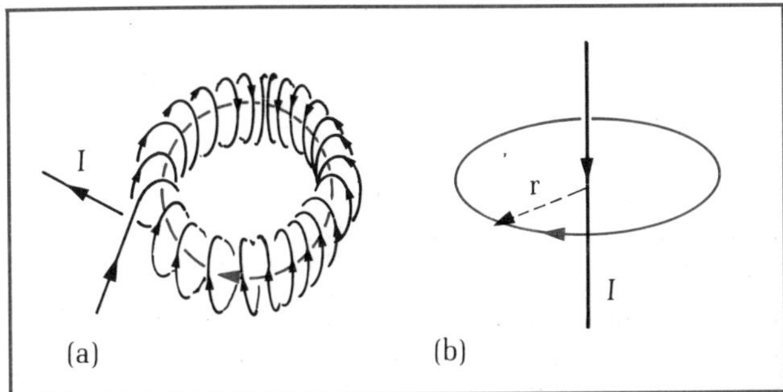

Figure 14.34 *Ampere's rule*

Worked example Two long straight wires are lined up parallel to one another at a separation of 0.10 m. Currents of 5.0 A and 3.0 A pass along each wire in the same direction. Calculate the magnetic field strength due to the two wires at a point on the mid-line between the two wires. Assume $\mu_0 = 4\pi \times 10^{-7}\,\mathrm{H\,m^{-1}}$.

Solution To determine the magnetic field strength at the midpoint, the contribution from each wire must first be calculated, then the contributors must be combined, taking account of directions.

For the 5 A wire, use $B = \mu_0 I/2\pi r$ where $r = 0.05$ m.

$$\text{Hence } B = \frac{4\pi \times 10^{-7} \times 5}{2\pi \times 0.05} = 2.0 \times 10^{-5}\,\text{T}$$

For the 3.0 A wire, $r = 0.05$ m.

$$\text{Hence } B = \frac{4\pi \times 10^{-7} \times 3}{2\pi \times 0.05} = 1.2 \times 10^{-5}\,\text{T}$$

The two field contributions are in opposite directions. So the resultant magnetic field strength at the midpoint $= 2.0 \times 10^{-5} - 1.2 \times 10^{-5} = 0.8 \times 10^{-5}$ T.

The definition of the ampere

There is a magnetic field around a current-carrying wire. Place two current-carrying wires parallel to one another, and they exert a force on one another. Figure 14.35 shows an arrangement to demonstrate the effect.

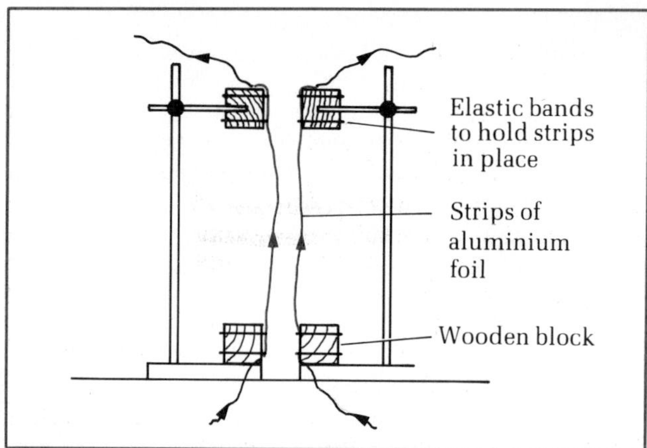

Figure 14.35 *Force between current-carrying conductors*

The two wires attract when the currents are in the same direction; they repel when the currents are in opposite directions.

The current in each wire produces a magnetic field which acts on the other wire. Figure 14.36 shows that the wires exert equal and opposite forces on one another. The field lines from one wire pass through the other wire at right angles. Using the left hand rule, you can show that the wires are forced towards one another.

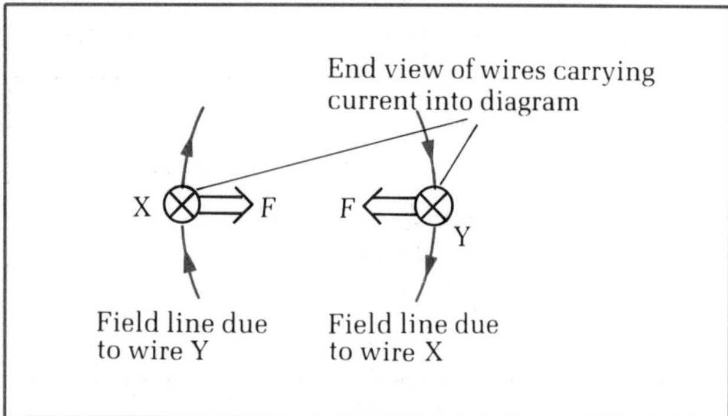

Figure 14.36 *Using the left-hand rule*

Consider the force on wire X due to wire Y, as in Figure 14.36. The magnetic field strength due to Y at X is given by

$$B=\frac{\mu_0 I_y}{2\pi d}$$

where I_y is the current in Y and d is the distance between the two wires.

Hence the force on length L of wire X is given by

$$F=BI_xL=\frac{\mu_0 I_y I_x L}{2\pi d}$$

where I_x is the current in X.

One ampere is defined as the current in two infinitely long straight wires 1 m apart in a vacuum which produces a force of $2\times10^{-7}\,\mathrm{N\,m^{-1}}$ on each wire.

You can now see why the value of μ_0 is $4\pi\times10^{-7}\,\mathrm{H\,m^{-1}}$. The formula for the force is given above. For $I_x=I_y=1$ A and $d=1$ m, the force is 2×10^{-7} N per metre length of each wire. The formula for the force gives

$$\text{Force per unit length } F/L=\frac{\mu_0 I_x I_y}{2\pi d}$$

Hence $2\times10^{-7}=\dfrac{\mu_0\times1\times1}{2\pi\times1}$ which gives

$$\mu_0=4\pi\times10^{-7}\,\mathrm{H\,m^{-1}}$$

Using a current balance as an ammeter

How do you know than an ammeter is accurate? Moving coil meters can become inaccurate, perhaps due to the magnet becoming gradually weaker or the hairspring being damaged. Digital meters too can become inaccurate if internal components fail. Ammeters need to be checked from time to time to make sure they remain accurate. A **current balance** connected in series with a suspect ammeter can be used to check for accuracy. Figure 14.37 shows the arrangement. The current balance and the suspect ammeter are in series so they take the same current. The current balance is used to determine the current, and the value can be used to check the ammeter reading.

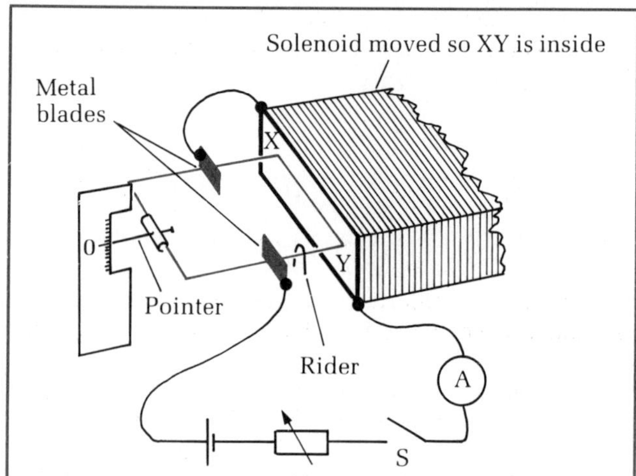

Figure 14.37 *A current balance*

A metal frame is balanced level on two sharp metal blades. The frame is positioned so that end XY is inside a suitable solenoid. The other end of the frame has a pointer against a fixed scale. A rider of known weight is moved along the side of the frame until the pointer is exactly against the scale zero.

The solenoid is connected in series with the frame wire XY, using the two metal blades as contacts. Then the solenoid and test wire XY are connected into the circuit shown in Figure 14.37. The suspect ammeter is included in the circuit.

When the circuit switch is closed, the frame becomes unbalanced. The reason is that the current creates a magnetic field along the solenoid axis. The current also passes through the wire XY which is at right angles to the magnetic field, so a magnetic force acts on wire XY to unbalance the frame. The frame is rebalanced level by repositioning the rider. The distance x moved by the rider is then measured. The length L of wire XY, and the distance d from the blades to XY must also be measured.

1 To calculate the magnetic force F on XY, the principle of moments is used. To rebalance the frame,

The moment due to force F = the moment due to the shift of the rider

So $Fd = mgx$

where mg is the weight of the rider. Hence

$$F = mgx/d$$

2 To calculate the current I, use the formula for the magnetic field strength in the solenoid.

$$B = \mu_0 nI$$

where n is the number of turns per metre of the solenoid.

Since $F = BIL$, then we can obtain an expression for F in terms of I.

$$F = (\mu_0 nI)IL = \mu_0 nI^2L$$

$$\text{Hence } I = \sqrt{\frac{F}{\mu_0 nL}}$$

Given $\mu_0 = 4\pi \times 10^{-7}\,\text{H m}^{-1}$ with values of F, n and L determined from the experiment, the current can be calculated. The calculated value can then be compared with the reading of the ammeter.

14.6 Magnetic materials

Ferromagnetism

Ferromagnetic materials are materials which can be permanently magnetized, such as iron, steel, cobalt and nickel. The magnetic field produced by a d.c. solenoid becomes many times stronger when a bar of iron or steel is placed inside the solenoid. When the solenoid current is switched off, the bar retains its magnetism.

The cause of ferromagnetism is the electrons in the outer shells of each atom. Electrons moving round the nucleus create tiny magnetic fields because they are moving charges, and moving charges form electric currents. In ferromagnetic atoms, the electrons of each atom produce a resultant magnetic field; in all other types of atoms, the electrons produce zero resultant magnetic field. So ferromagnetic atoms are like tiny magnets. When a bar of iron is placed in a d.c. solenoid the atomic magnets line up along the solenoid field, all pointing in the same direction. The combined effect then is to give a very strong overall magnetic field. When the current in the solenoid is switched off, the atomic magnets stay partly lined up.

Heating a bar magnet causes it to lose its magnetism. The effect is to give the atomic magnets enough energy to turn in random directions. If iron is heated to a temperature of 800°C or more, it loses its magnetism completely.

To investigate ferromagnetism, the arrangement shown in Figure 14.38 may be used. A bar of the material to be investigated is placed in an a.c. solenoid. The solenoid is in series with a resistor and the a.c. supply. The alternating p.d. across the resistor is connected to the X-input of an oscilloscope, so the spot on the oscilloscope screen moves back and forth at the same frequency as the alternating current. The end of the solenoid is positioned pointing to the oscilloscope screen as shown. The oscilloscope tube must be unscreened so that the electron beam in the tube is deflected by the solenoid field. The

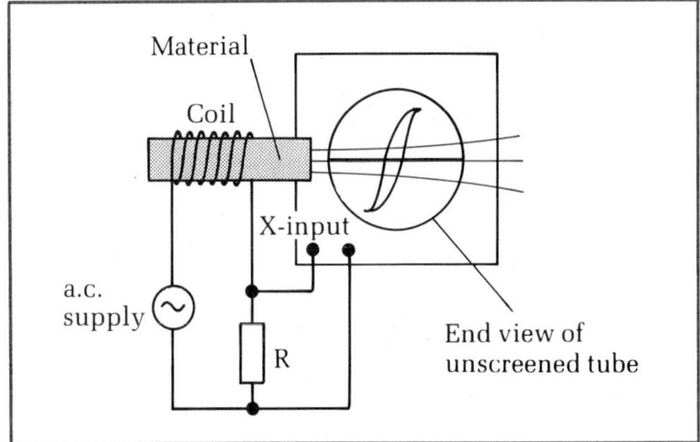

Figure 14.38 *Investigating ferromagnetism*

solenoid is positioned so that its magnetic field forces the beam up and down. The X-input should be disconnected to check that the field deflects the beam correctly. When the X-input is reconnected, the beam traces out a loop on the screen each cycle.

The loop on the screen is called a **hysteresis loop**. The x-direction across the screen represents the magnetizing current in the solenoid. The y-direction on the screen represents the magnetic field strength produced by the material in the solenoid. So the trace on the screen shows how the magnetic field strength varies with magnetizing current. Figure 14.39 shows loops for different materials.

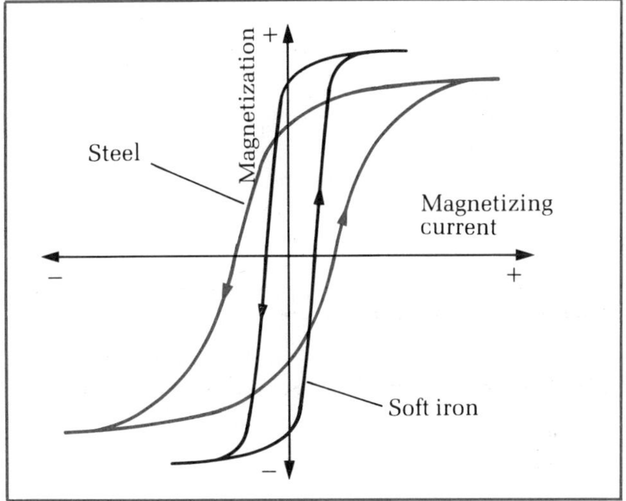

Figure 14.39 *Hysteresis loop*

The loops show the same general shape. The chief features are as follows.

1. **Saturation** The magnetic field strength reaches a maximum value which cannot be increased by increasing the current. The material is said to be magnetically saturated when it gives maximum magnetic field strength. The atomic magnets are then all lined up.
2. **Remanence** When the current is zero, the magnetic field strength is non-zero, so when the current is switched off, the material remains magnetized. The atomic magnets interact to stay partly in line, even though the solenoid current is zero.
3. **The loop area** This is a measure of the energy needed to magnetize and demagnetize the material each cycle. **Hard** magnetic materials like steel cannot be easily

magnetized or demagnetized, so they have a large loop area compared with **soft** magnetic materials such as iron which can easily be magnetized.

The relative permeability μ_r of the material is

$$\frac{\text{The magnetic field strength with the material}}{\text{Magnetic field strength without the material}}$$

in a toroidal (i.e. endless) solenoid. The value of μ_r for iron is about 2000.

Without material in the solenoid, the magnetic field strength is given by

$$B_0 = \mu_0 nI$$

where n is the number of turns per metre.

With material in the solenoid, the field strength B is equal to $\mu_r B_0$. Hence

$$B = \mu_r \mu_0 nI$$

where μ_r is the relative permeability of the material.

To calculate magnetic field strengths, care must be taken using values of μ_r from reference books. All magnetic materials become saturated in a strong magnetic field; once saturated, the field strength produced by the material is no longer $\mu_r \times$ the field strength without the material. For example, an air-filled solenoid might have a magnetic field strength of 0.1 mT along its axis; filled with iron, the magnetic field strength would increase to 0.2 T if μ_r were 2000. But iron saturates at about 1 T or more, so increasing the solenoid current by ten times would not increase the field strength to 2 T when filled with iron.

Permanent magnets are made from steel because it is hard to demagnetize. Strong permanent magnets must also have high remanence. One way to destroy the magnetism of a permanent magnet is to heat it. As already discussed, this causes the atomic magnets to point in different directions. Another way to destroy its magnetism is to withdraw it slowly from an alternating current solenoid. The solenoid produces an alternating field which continually changes the direction of the atomic magnets. By withdrawing the material slowly from the field, the atomic magnets are left pointing in different directions, so the material loses its overall magnetization.

When unmagnetized, the atomic magnets line up in small regions, called **domains**, as shown in Figure 14.40. Each domain has its atomic magnets pointing in the same direction, but the direction differs from one domain to another at random; so no overall magnetic field is produced. The pattern of domains can be observed using a microscope. The surface to be viewed must be very smooth. Then fine iron dust sprinkled on the surface sticks to the domain boundaries so each domain is apparent.

Electromagnets used for lifting iron objects should be easy to magnetize and demagnetize. They must lose their magnetism when the current is switched off, so electromagnets are made from materials with low remanence. A simple electromagnet can be made using two iron C-cores, suitable wire and a low voltage power unit. Even with this arrangement, you need to pull with considerable force to break apart the two C-cores. The two C-cores channel the field lines round, as if in a magnetic circuit.

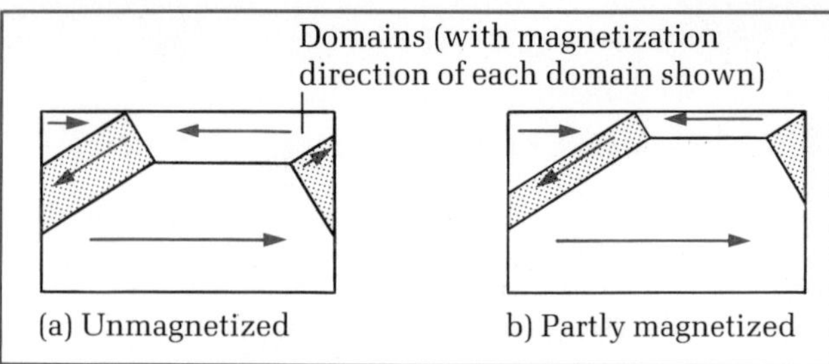

Figure 14.40 *Domains*

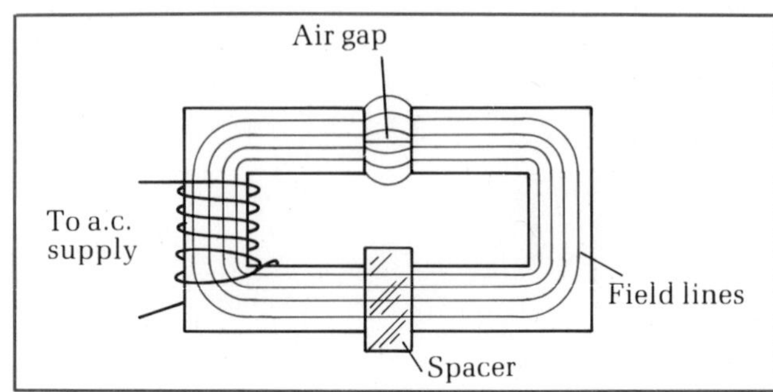

Figure 14.41 *A magnetic circuit*

Suppose a steel spacer is fitted on one side so the two cores don't quite fit together on the other side. Figure 14.41 shows the idea. The force of attraction between the two cores is now much less. Use a Hall probe to search for field lines near the cores. The field lines cross the air gap, but the presence of the air gap weakens the field.

Electromagnets are used in relay switches. A relay switch is actuated (i.e. opened or closed) by passing current through the relay coil. Figure 26.59 shows a simplified relay. When current passes through the coil, the iron core is magnetized. As a result, the hinged flap near one end is pulled onto the core because the flap is made of iron. The movement of the flap opens or closes the relay switches according to the design of the relay. So electrical machines can be switched on or off by small currents passed through the relay coil.

14.7 Summary

Definitions

1. *Magnetic field strength* is the force per unit length per unit current on a current-carrying conductor at right angles to the field lines.
2. One *ampere* is the current in two infinitely long parallel wires 1 m apart which causes a force of $2 \times 10^{-7}\,\text{N m}^{-1}$ on each wire.

Equations

1. Force on a current carrying conductor $F = BIL \sin\theta$
2. Force on a moving charge $F = Bqv \sin\theta$
3. Couple on a coil $C = BNIA \cos\alpha$
4. Deflection of a meter coil $\theta = \dfrac{BNA}{k} I$
5. Hall p.d. $V_h = Bvd = \dfrac{BI}{nxq}$

6 Magnetic field strength is a solenoid $B = \frac{\mu_0 NI}{l}$

7 Magnetic field strength near a straight wire $B = \frac{\mu_0 I}{2\pi r}$

Short questions

Assume $\mu_0 = 4\pi \times 10^{-7}\,\text{H m}^{-1}$

14.1 The Earth's magnetic field strength at a certain position P on its surface has a horizontal component of 18 μT due north and a downwards vertical component of 55 μT. Calculate
a) the magnitude and direction of the Earth's magnetic field strength at P,
b) the force on a 0.5 m length of straight wire carrying a steady current of 4.0 A when the wire is (i) vertical with the current passing downwards, (ii) horizontal so its current passes from east to west.

14.2 A current balance consists of a rectangular wire frame ABCD balanced on pivots midway along AB and CD as shown. A small rider of mass $m = 4.5 \times 10^{-4}$ kg is positioned along AB to balance the frame horizontally. End AD is positioned at the end of a solenoid at right angles to the solenoid axis. When a steady current of 4.5 A is passed through the solenoid and along end AD in series with one another, the rider must be moved 34 mm towards the pivot to regain the balance. Given AB = CD = 240 mm and AD = BC = 100 mm, calculate
a) the force on end AD due to the solenoid field,
b) the magnetic field strength at the end of the solenoid,
c) the number of turns per metre, n, of the solenoid, given that the field strength at the end is equal to $\frac{1}{2}\mu_0 nI$ where I is its current. Assume $g = 10\,\text{N kg}^{-1}$.

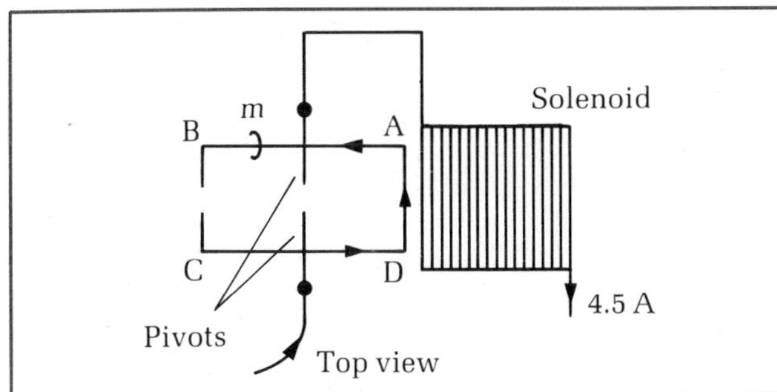

Figure 14.42

14.3 A rectangular coil of area $3.0 \times 10^{-4}\,\text{m}^2$ and 100 turns is placed in a uniform magnetic field of strength 55 mT with the coil axis perpendicular to the field lines. A steady current of 0.85 A is passed through the coil windings. Calculate the couple exerted on the coil when it is
a) positioned with its plane parallel to the field lines,
b) positioned with its plane at 60° to the field lines.

14.4 A long straight vertical wire of diameter 2.0 mm carries a steady current upwards. At 30 mm from the centre of the wire, the magnetic field strength is 0.10 mT. Sketch a graph to show the variation of magnetic field strength from the wire surface to a distance of 40 mm from the centre of the wire. Calculate the magnetic field strength at
a) the surface of the wire,
b) 40 mm from the centre of the wire.

14.5 A second long straight wire is positioned parallel to the wire of question **14.4** at a distance of 20 mm from the wire. Calculate the force on each metre length of the second wire when a steady current of 4.0 A is passed upwards along it. The current in the first wire is as in question **14.4**.

14.6 For the arrangement of question **14.5**, calculate the resultant magnetic field strength at
a) the midpoint between the two wires,
b) a point that is 20 mm from the centre of each wire.

14.7 A long solenoid has 1500 turns m^{-1} along its length. It is placed with its axis parallel to the field lines of a uniform magnetic field of strength 24 mT so the field lines pass along the solenoid axis.
a) Calculate the steady current that must be passed through the solenoid to give zero resultant magnetic field strength at its centre. Sketch the arrangement and show the directions of the current and the external field.
b) If the solenoid current is changed to 3.5 A, calculate the magnitude and direction of the resultant field in the solenoid.

14.8 A rectangular coil 60 mm × 40 mm with 25 turns is positioned with its longer sides parallel to a long straight vertical wire carrying a current of 6.8 A. The coil is placed so that its long sides and the vertical wire are in the same vertical plane, with the wire 40 mm away from the nearest long side of the coil. A current of 2.0 A is passed round the coil.
a) Calculate the force on each long side of the coil due to the magnetic field created by the vertical wire.
b) Explain why the top and bottom sides of the coil experience equal and opposite forces.
c) Calculate the overall force on the coil.

14.9
a) A rectangular slice of n-type semiconducting material, 10 mm × 6 mm × 1.5 mm, takes a steady current of 65 mA when connected into a circuit as shown in the diagram. When a uniform magnetic field of strength 90 mT is applied across the slice as shown, a p.d. of 3.5 mV develops between the sides parallel to the field lines as shown. (i) Explain the cause of this p.d. (ii) Calculate the number of free electrons per m^3 in the slice. (Assume $e = 1.6 \times 10^{-19}$ C.)

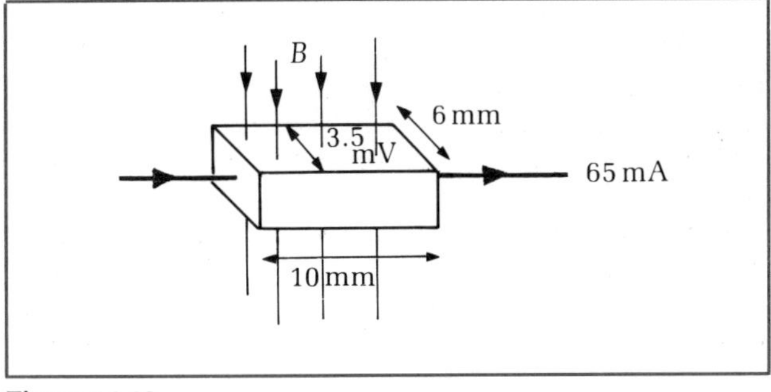

Figure 14.43

b) To demonstrate the Hall effect in a metal, it is necessary to use a thin metal foil carrying a current of the order of 5 A or more. What can you deduce about the number of free electrons per m^3 in a metal compared with a semiconductor such as in **a)**?

14.10 Use the ideas of the domain theory of magnetism to
a) explain why the strength of an electromagnet cannot exceed a certain value no matter how much current is passed through its windings,
b) explain why permanent magnets lose their strength if repeatedly struck,
c) explain why an iron bar removed from a d.c. solenoid retains some magnetism.

15

Electromagnetic induction

15.1 The principles of electromagnetic induction

Making a dynamo

Dynamos generate electricity. To make a dynamo, all you need is a length of wire and a magnet. Connect the wire to a sensitive meter to detect current. To generate a current, the wire must be moved relative to the magnet. It doesn't matter if the magnet is stationary while the wire is moved or the wire is kept fixed while the magnet is moved. Whenever there is relative motion between a wire and a magnet, an e.m.f. is **induced** in the wire. So if the wire is connected to a meter, the induced e.m.f. drives a current round the circuit. In other words, moving a magnet near a wire which is part of a circuit drives electrons round the circuit. The wire is therefore a source of electrical energy, just as a battery is. The induced e.m.f. is the electrical energy produced per coulomb as a result of moving the wire near the magnet.

Make the wire into a coil and move the magnet in and out of the coil. The induced e.m.f. is much greater with many turns in the coil, so a meter connected to the coil responds much more than if it is connected to a single wire. A bicycle dynamo is a very simple dynamo in which a magnet is made to spin round near a fixed coil. The relative motion causes an induced e.m.f. in the coil which drives current through any lamp connected to the dynamo. In other words, the moving magnet drives electrons round the circuit. The passage of electrons round the circuit is an induced current. If the lamp is disconnected, the moving magnet still produces an induced e.m.f. in the coil, but no induced current passes through the coil or lamp because there is a break in the circuit as a result of disconnecting the lamp.

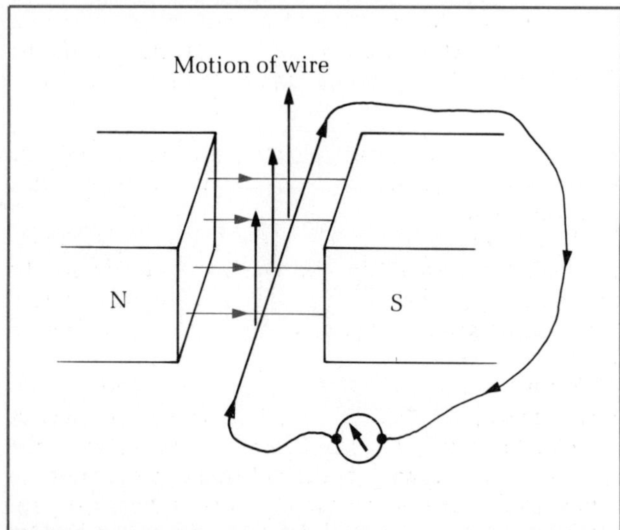

Figure 15.1 *Generating an electric current*

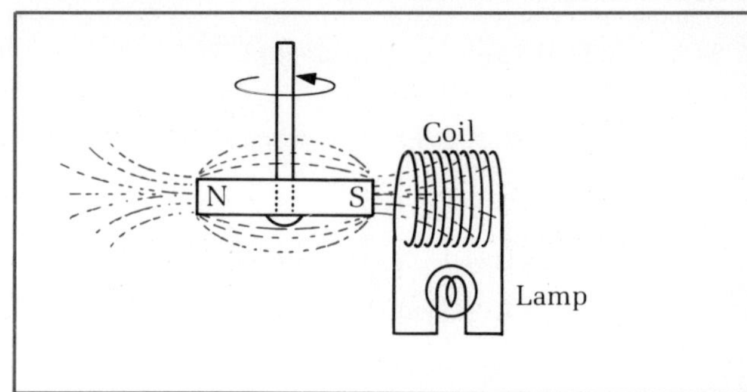

Figure 15.2 *A home-made dynamo*

Another way to generate electricity is to use a motor in reverse. If the motor does not have a permanent magnet, it is necessary to connect its field windings to a battery.

When the spindle is turned, an e.m.f. is induced in the armature windings. The reason is that the windings cut through the lines of the magnetic field of the motor magnet, so an e.m.f. is induced in the windings of the armature. The faster the spindle is turned, the larger the induced e.m.f. This is a general rule; try moving a single wire connected to a meter across a magnetic field at different speeds. At low speed, there is hardly any deflection of the meter, but the deflection is much greater if the speed is increased considerably.

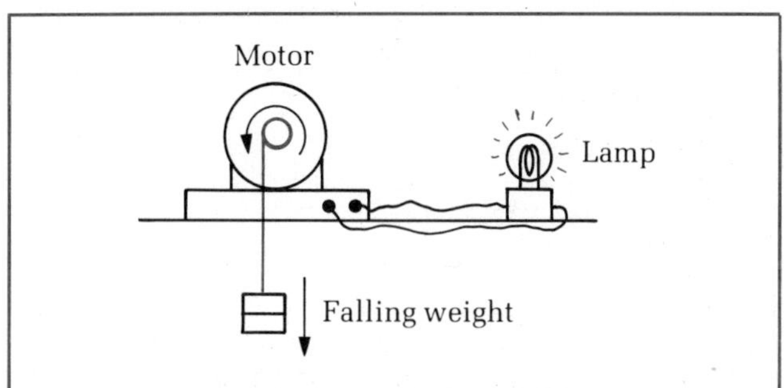

Figure 15.3 *A motor as a generator*

Understanding electromagnetic induction

When a beam of electrons is directed across a magnetic field, the electrons are deflected by a magnetic force, as shown in Figure 15.4(a). Each electron experiences a force at right angles to its direction of motion and to the field direction.

A metal rod is like a tube containing lots of free electrons. Suppose a metal rod is moved across a magnetic field as in Figure 15.4(b). All the electrons in the rod are forced to one end of the rod; the same force acts as if the electrons are beam electrons. So one end of the rod becomes negative (because the field pushes electrons to that end) and the other end becomes positive (because electrons are pushed away from that end). Hence an e.m.f. is induced across the ends of the rod. Connect the moving rod into a circuit and the e.m.f. across its ends drives current round the circuit.

Whenever electrons are directed across a magnetic field, the electrons are pushed to one side by the field. It makes no difference to the electrons whether they are in a

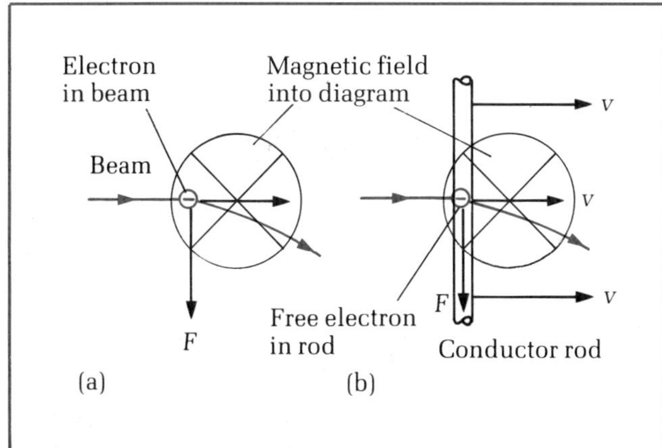

Figure 15.4 *Deflection of electrons in a magnetic field*

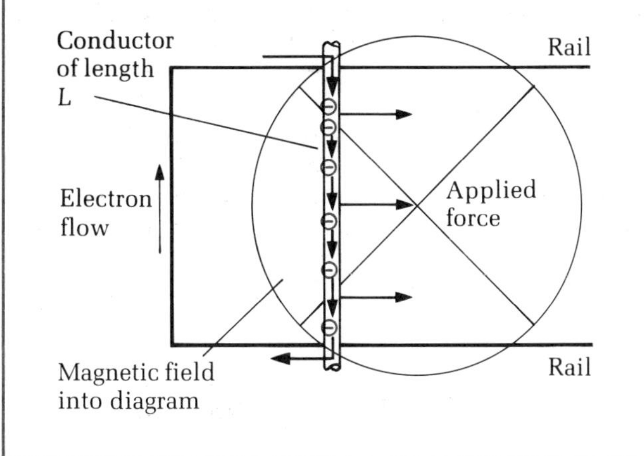

Figure 15.5 *Induced e.m.f. in a rod*

beam in a vacuum tube, in a Hall effect device or in a conducting rod or wire cutting across the field.

Consider the arrangement shown in Figure 15.5. The rod is on a pair of rails which are part of a complete circuit. Moving the rod along the rails as shown involves cutting across magnetic field lines, so the magnetic field forces the electrons in the rod to one end and then round the circuit. In other words, an induced current passes round the circuit. To keep the current steady, the rod must be kept moving at steady speed across the field.

However, the rod is now a current-carrying conductor in a magnetic field, so it experiences a magnetic force which opposes its motion. The left hand rule can be used to show that the current and field directions are such as to direct the magnetic force opposite to the motion of the rod. To keep the rod moving, the applied force F must be equal and opposite to the magnetic force BIL. Hence

$$F = BIL$$

The rod, moving at speed v, moves a distance of $v\Delta t$ in time Δt. So the work done by the applied force is given by

$$\text{Work done} = \text{force } F \times \text{distance moved } v\Delta t = BILv\Delta t$$

Electrical energy is produced from the work done. When a source of e.m.f. ε produces a current I for time Δt, the electrical energy produced equals $I\varepsilon\Delta t$.

Hence $\varepsilon I\Delta t = BILv\Delta t$

So $\varepsilon = BLv$

The laws of electromagnetic induction

Suppose a coil is connected to a centre-reading meter, as in Figure 15.6. When a bar magnet is pushed in and out of the coil, the meter deflects. An e.m.f. induced in the coil drives current through the meter. What determines the direction of the induced current? Consider a bar magnet approaching end X of the coil of Figure 15.6, with the N-pole leading. Current flow round the coil creates a magnetic field due to the coil. The coil field must oppose the incoming N-pole, otherwise it would pull the N-pole in faster, making the induced current bigger, pulling the N-pole in even faster still, etc. Clearly, the conservation of energy tells us that the coil field must oppose the incoming N-pole. Otherwise, we would produce electrical and kinetic energy from nowhere! So the induced current creates a magnetic field in the coil which opposes the incoming N-pole. The field due to the coil must therefore have field lines emerging from end X; in other words, the induced polarity of end X is a N-pole, which tries to repel the incoming N-pole. The solenoid rule tells us that the current flow round end X is therefore anticlockwise.

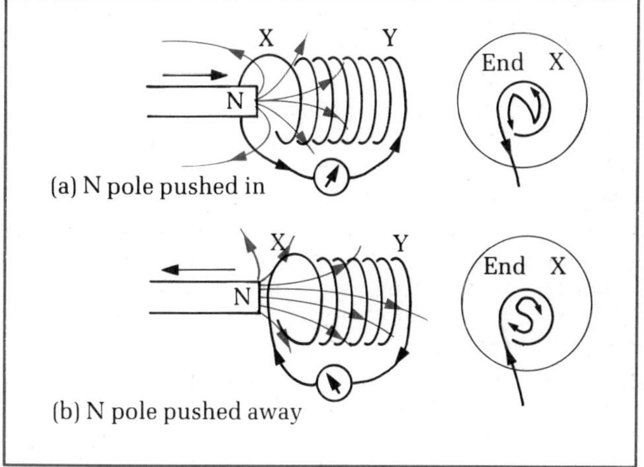

Figure 15.6 *Lenz's law*

What if the magnet is removed? If the N-pole of the bar magnet is taken from the coil, the induced polarity of end X is a S-pole. So the induced current must pass in a clockwise direction. Once again, the direction of the induced current is such as to oppose the cause of the induced current.

Lenz's law states that the direction of the induced current is always so as to oppose the change which causes the current.

The explanation of Lenz's law is that energy is never created or destroyed. The induced current could never be in a direction to help the change which causes it; that would mean producing electrical energy out of nowhere, which is forbidden!

Two applications of Lenz's law are given here.

1 **Electromagnetic brakes** are often fitted to coaches. A metal disc driven by the wheel axle turns between the poles of a horseshoe magnet. When the disc rotates the field pushes free electrons in the disc to the rim or centre according to the field direction. So an e.m.f. is induced between the rim and the centre of the disc. If a

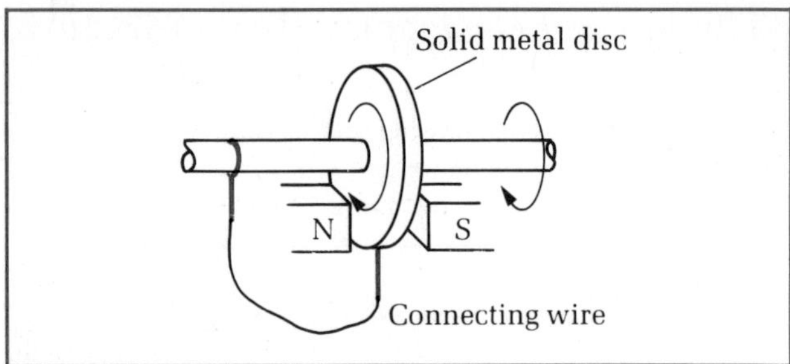

Figure 15.7 *Braking*

wire is connected between the rim and centre, the induced e.m.f. drives current through the wire and disc. The disc is now a current-carrying conductor, so a magnetic force acts against its motion. The magnetic force therefore acts as a braking force.

2 **Eddy currents** help to protect meter coils when the meter is moved. Suspension-type meters are very sensitive but easily damaged by careless handling. To protect the meter when not in use, the meter terminals should be connected together. Then if the meter is moved carelessly, the motion of its coil in the magnetic field creates induced current in the coil. The induced current causes magnetic forces to act on the coil to oppose its motion relative to the field. So movement of the sensitive coil relative to the magnet is greatly reduced by connecting the terminals together. If the coil is wound on a metal frame, current is induced in the frame too, so that also helps to prevent the coil swinging too freely. Also, when in use, induced current in the frame serves to slow the frame down, which stops the meter pointer from over-shooting. Such currents are sometimes called eddy currents.

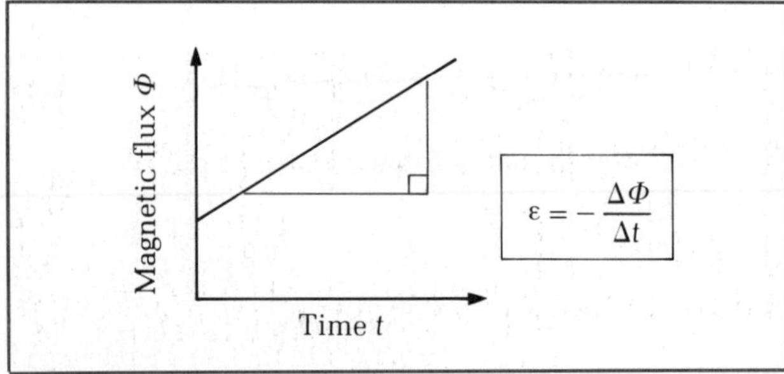

Figure 15.8 *Steady flux change*

Faraday's law Electromagnetic induction was discovered in 1831 by Michael Faraday. He showed how to calculate induced e.m.f.s in any situation. We have seen in the previous section that the e.m.f. induced in a rod moving at right angles to a uniform magnetic field is given by the equation

$$\varepsilon = BLv$$

In a short time interval Δt, the rod moves a distance $v\Delta t$. So it sweeps out an area A equal to $Lv\Delta t$. We can write the equation for ε as

$$\varepsilon = \frac{BLv\Delta t}{\Delta t} = \frac{BA}{\Delta t}$$

The product of field strength and area, BA, is called the **magnetic flux** passing through the area A at right angles to the field lines. The idea of magnetic flux is very useful for calculating induced e.m.f.s. The example of the rod moving across the magnetic field shows that the induced e.m.f. is equal to the magnetic flux swept out by the conductor each second. Faraday was the first person to show how induced e.m.f.s could be calculated from magnetic flux changes.

Magnetic flux $\varphi = BA$

Flux linkage Φ through coil of n turns $= n\varphi = BAn$

where B = magnetic field strength normal to area A.

The unit of magnetic flux and of flux linkage is the **weber** (Wb), equal to 1 T m^2.

In the circuit of Figure 15.5, the moving rods sweeps out area A in time t. So the change of magnetic flux through the circuit in that time is equal to BA. Therefore the induced e.m.f. is equal to the change of magnetic flux per second.

Faraday's law of electromagnetic induction states that the induced e.m.f. in a circuit is equal to the rate of change of flux linkage through the circuit.

$$\varepsilon = -\frac{d\Phi}{dt}$$

where ε = induced e.m.f.,

$\dfrac{d\Phi}{dt}$ = change of flux linkage per second.

The equation is the mathematical way of writing Faraday's law, in which $\dfrac{d}{dt}$ means change per unit time, so $\dfrac{d\Phi}{dt}$ means change of flux linkage per second. The $-$ sign is because the induced e.m.f. acts to oppose the change. Figure 15.8 shows how the flux changes with time in the circuit of Figure 15.4(a). The steady motion of the rod across the magnetic field means that the flux through the circuit increases steadily with time. So Figure 15.8 is a straight line with a positive gradient. The induced e.m.f. is given by the gradient of the line. The gradient triangle drawn has a height represented by change of flux linkage $\Delta\Phi$, and its base is represented by the time taken Δt. So the gradient is equal to $\Delta\Phi/\Delta t$.

Where the flux does not change at a steady rate, the induced e.m.f. varies. For example, if the rod of Figure 15.4(a) is accelerated across the field, the flux increases as shown in Figure 15.9. The induced e.m.f. at any point is given by the gradient of the tangent to the curve at that point.

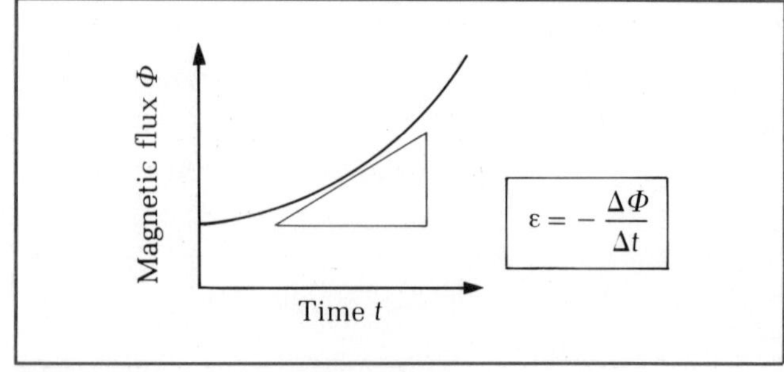

Figure 15.9 *Increasing flux change*

Measuring magnetic field strength using a search coil

A search coil is a small coil with a known number of turns and area. It can be used to measure magnetic field strength by placing it at right angles to the field lines. Then the coil is suddenly withdrawn from the field. The change of flux through the coil causes an e.m.f. to be induced in the coil. By connecting the coil to a suitable instrument, to allow measurements to be made when it is pulled out of the field, the magnetic flux in the coil initially can be determined. Then the magnetic field strength at its initial position can be calculated.

A ballistic galvanometer can be used to measure the change of magnetic flux. It is a moving coil meter with a light-beam pointer. The meter has a coil with a plastic frame so it can swing freely without eddy currents opposing its motion. The search coil is connected in series with the meter and a resistance box. When the search coil is removed from the magnetic field to be measured, an e.m.f. is induced in the search coil. The induced e.m.f. only lasts for as long as it takes to remove the search coil from the field, so a brief flow of charge passes round the circuit. This brief current sets the meter coil oscillating, so the light spot on the scale of the meter swings over to one side of the zero, then it swings back to the other side. It continues to swing from side to side long after the charge flow has stopped. The initial deflection θ (or first throw) of the spot is proportional to the charge passed Q. If the meter is calibrated so that its charge sensitivity is known, the charge passed by the search coil can be calculated.

$$Q = k\theta$$

where Q = induced charge,
θ = first throw,
k = charge per unit deflection, (i.e. the charge sensitivity).

To calculate the magnetic field strength B from the induced charge, we must use Faraday's law. The induced e.m.f. is given by

$$\varepsilon = \frac{\text{change of flux linkage through the coil}}{\text{time taken for the change}}$$

The initial flux through each turn of the coil is $B \times A$, where A is the coil area. So the initial flux linkage through the coil is BAn. The final flux linkage through the coil is zero. So the change of flux linkage is equal to BAn. Hence

$$\varepsilon = BAn/t$$

where t is the time taken.

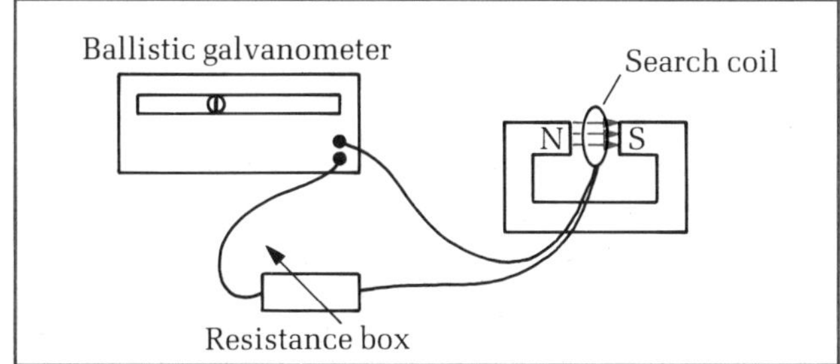

Figure 15.10 *Using a search coil*

The induced current $I = \dfrac{\text{induced e.m.f. } \varepsilon}{\text{circuit resistance } R}$

So $I = BAn/Rt$

However, the induced charge $Q = It$

Therefore $Q = BAn/R$

The equation can be used to calculate the magnetic field strength B since values of A, n and R are known. Q is measured, as explained above. The resistance R is the total resistance of the circuit. In practice, the resistance box in the circuit must be set at a high value to allow the meter coil to swing freely. Otherwise, the current opposes its motion too much.

$$B = \frac{QR}{An}$$

where B = magnetic field strength,
Q = induced charge,
R = circuit resistance,
A = area of search coil,
n = number of turns of the search coil.

A data recorder can be used to measure the induced e.m.f. as it changes. Data recorders are electronic devices which measure and record voltages at regular intervals. For example, some data recorders can take over 1000 readings at any one of several possible rates. Figure 15.11 shows a data recorder set to take readings at a rate of one per millisecond. Resetting its switches would allow it to take readings at other possible rates. As set, it is very suitable for measuring the e.m.f. induced in a search coil when the coil is withdrawn from the field to be measured. The run button on the recorder is released at the instant the coil is pulled out of the field, so the data recorder measures the e.m.f. as it changes every millisecond.

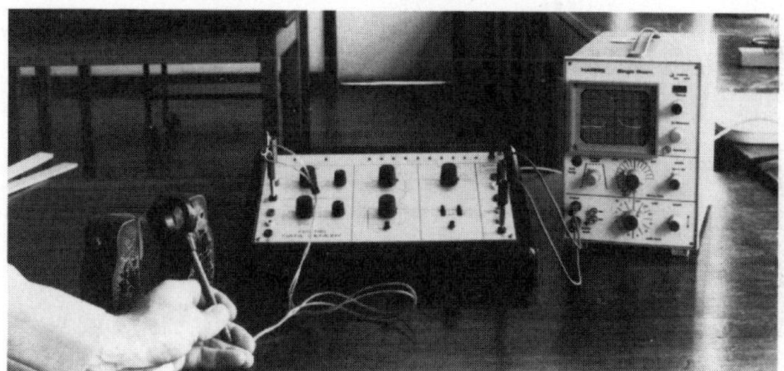

Figure 15.11 *Recording fast-changing e.m.f.s*

The recorded measurements can be displayed directly on an oscilloscope screen. Figure 15.12 shows a typical read-out trace. The y-axis represents the induced e.m.f. and the x-axis represents the time from the start. The trace shows that the induced e.m.f. reaches a peak then falls to zero. The ballistic galvanometer theory given in the previous section assumes a steady e.m.f.; this represents the average e.m.f. over the time the pulse lasts for.

The induced e.m.f. at any instant, ε, is equal to the rate of change of flux linkage, $\dfrac{d\Phi}{dt}$. So in as short time interval Δt, the change of flux linkage is equal to $\varepsilon\Delta t$. However, $\varepsilon\Delta t$ is represented on the graph by the area of the strip of width Δt under the curve. So the total area under the curve represents the total change of flux linkage. The link

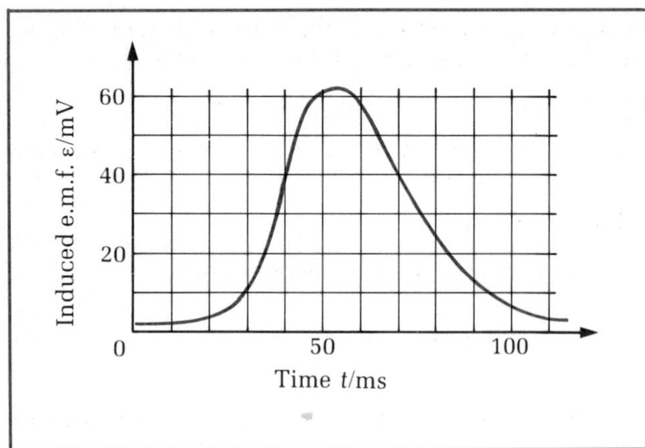

Figure 15.12 *E.M.F. v time*

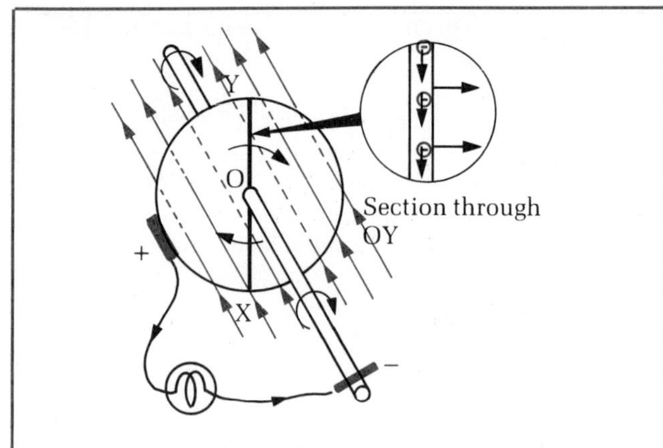

Figure 15.13 *Disc generator*

between ε and Φ is like the link between speed and distance. Just as the area under a speed v time curve represents distance moved, so the area under the e.m.f. v time curve represents flux linkage change. The settings of the data recorder ought to allow the scales to be worked out for the oscilloscope trace. So the area under the trace can be estimated, and then the change of flux linkage can be determined. Since the change of flux linkage equals BAn, then B can be determined if the coil area A and number of turns n are known.

For example, suppose the y-axis for Figure 15.12 is marked in intervals, each corresponding to 10 mV of e.m.f. Also, suppose the x-scale intervals each represent 10 ms of time. Each block of Figure 15.12 represents a flux linkage change of 10 mV × 10 ms which equals 1×10^{-4} Wb. If the total area under the curve is 26 blocks, then the total flux linkage change must have been 2.6×10^{-3} Wb.

Worked example A search coil has 100 turns and a diameter of 34 mm. It is connected into a circuit in series with a resistance box and ballistic meter. The total resistance of the circuit is 5000 ohms. The charge sensitivity of the meter is 0.02 μC mm^{-1}. The search coil is placed between the poles of a horseshoe magnet with the field lines at right angles to the coil. When the search coil is rapidly withdrawn from the field, a first throw of 85 mm is measured. Calculate the magnetic field strength between the poles of the horseshoe magnet.

Solution To calculate the induced charge Q, use $Q = k\theta$.

$k = 2 \times 10^{-8}$ C mm^{-1}, $\theta = 85$ mm

Hence $Q = 2 \times 10^{-8} \times 85 = 1.7 \times 10^{-6}$ C

To calculate B, use $B = QR/An$.

$R = 5000$ ohms, $n = 100$ turns, $A = \pi(\text{diameter})^2/4$

$$\text{Hence } B = \frac{1.7 \times 10^{-6} \times 5000}{\pi/4 \times (34 \times 10^{-3})^2 \times 100} = 9.4 \times 10^{-2}\ \text{T}.$$

15.2 Generators

The disc generator

When a disc is turned in a magnetic field, as in Figure 15.13, an e.m.f. is induced between its centre and rim. We have already met the idea in electromagnetic brakes. Imagine the disc as a wheel with lots of spokes. As it turns, the spokes cut across the field lines, so the free electrons in each spoke are pushed to one end of the spoke. Spoke OX in Figure 15.13 is cutting across the field lines from right to left, so its electrons are pushed towards end O by the field. Spoke OY is cutting the field from left to right, so its electrons, like electrons in a beam moving from left to right, are pushed down, towards O. Electrons pile up at the centre, so the centre becomes negative with respect to the rim. Brush contacts at the centre and the rim can be connected to a voltmeter to measure the induced e.m.f. ε.

The size of the induced e.m.f. can be calculated using Faraday's law. Each spoke sweeps out an area πR^2 in one turn, where R is the disc radius. So the flux swept out per turn is $B\pi R^2$. At f turns per second, the flux swept out each second is $B\pi R^2 f$. From Faraday's law, the induced e.m.f. is equal to the flux swept out each second. Hence the induced e.m.f. is given by

$$\varepsilon = B\pi R^2 f$$

where B = magnetic field strength,
R = disc radius,
f = frequency.

The speedometer of a car uses a disc generator. A metal disc on a spindle is turned by a link from the driving wheels. The disc turns between the poles of a horseshoe magnet, so a voltmeter connected via contact brushes at the rim and spindle gives a reading in proportion to the disc speed. The voltmeter is therefore an indicator of the speed of the car.

The alternator

An alternator or a.c. generator is essentially a rectangular coil which is turned between the poles of a horseshoe magnet, as in Figure 15.14. Each end of the insulated coil windings is connected to a slip ring round the spindle. The contact brushes pressed against the slip rings give continuous contact between the alternator coil and the circuit which it supplies current to.

Consider the coil turning at a steady rate. As a result, the flux through the coil continually changes to give an alternating e.m.f. across the coil terminals. Figure 15.15

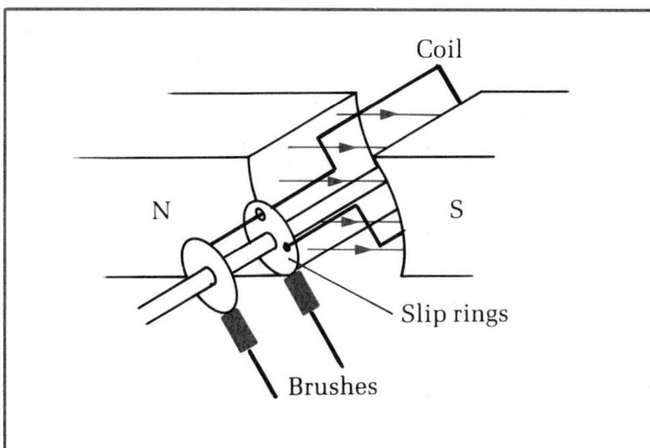

Figure 15.14 *A.C. generator*

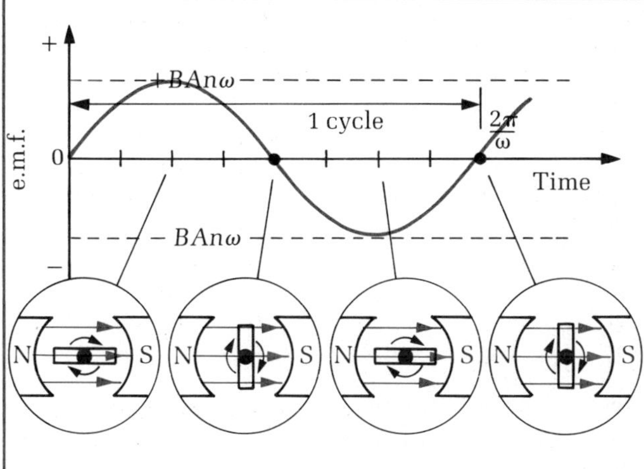

Figure 15.15 *e.m.f. v time for an a.c. generator*

shows how the e.m.f. varies with time. Each cycle of the e.m.f. corresponds to one turn of the coil in the field.

The induced e.m.f. is at a peak when its sides cut directly across the field lines. When the coil sides move parallel to the field lines, the induced e.m.f. is zero. The sides must cut across the field lines if the field is to push electrons round the windings. No cutting of the field lines takes place when the sides move parallel to the field lines.

We can use Faraday's law to obtain an expression for the induced e.m.f. Consider Figure 15.16, which shows the coil at angle θ to the mid-line between the poles. Assume the coil is turning so that angle θ increases at a steady rate.

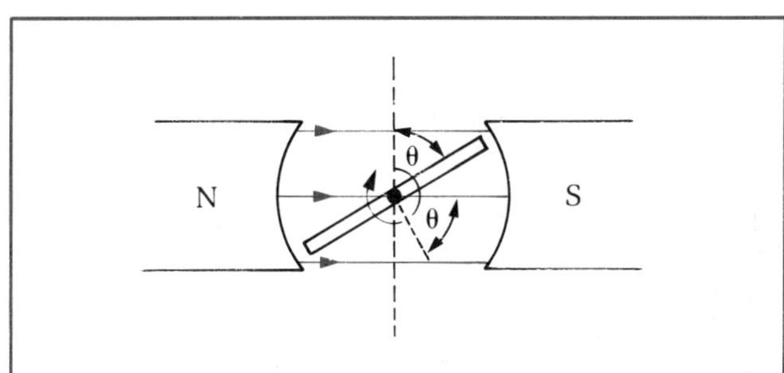

Figure 15.16 *Producing sine waves*

The magnetic field has a component $B\cos\theta$ perpendicular to the coil plane. So the flux through each turn of the coil is given by $BA\cos\theta$, where A is the coil area. Therefore the flux linkage through the coil is given by

$$\Phi = BAn\cos\theta$$

For steady rotation, $\theta = \omega t$ where ω is the angular speed of the coil. Hence

$$\Phi = BAn\cos\omega t$$

Faraday's law gives the induced e.m.f.

$$\varepsilon = -\frac{d\Phi}{dt} = -\frac{d}{dt}(BAn\cos\omega t)$$

Therefore since BAn is constant and $\frac{d}{dt}(\cos\omega t)$ is equal to $-\omega\sin\omega t$ we can write

$$\varepsilon = BAn\omega\sin\omega t$$

where ε = induced e.m.f.,
B = magnetic field strength,
A = area of coil,
n = number of turns,
ω = angular speed = $2\pi \times$ frequency.

The induced e.m.f. is therefore a sine waveform with a peak value equal to $BAn\omega$. The faster the alternator is turned, the greater is the peak value.

A d.c. generator can be made by replacing the two slip rings by a split-ring, as shown in Figure 15.17. The ends of the coil windings are connected to the split-ring. Turning the coil at a steady speed produces an induced e.m.f. which never changes direction.

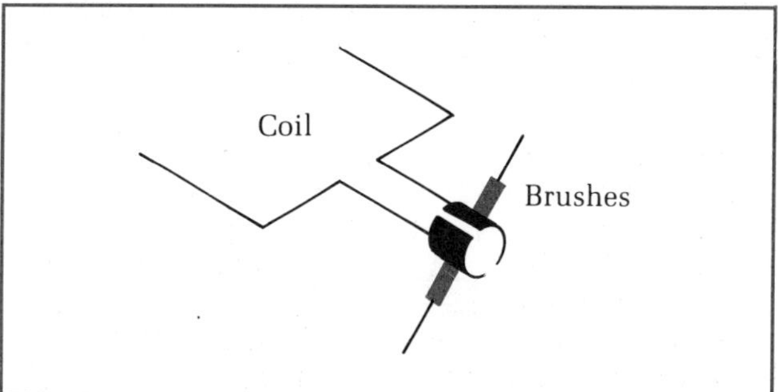

Figure 15.17 *The split ring commutator*

The induced e.m.f. varies with time as shown by Figure 15.18. In an alternator, the induced e.m.f. changes direction every half-cycle as the coil turns through the mid-line position. But in the d.c. generator, the coil connections to the brushes swap over each half-cycle, so the induced e.m.f. does not reverse.

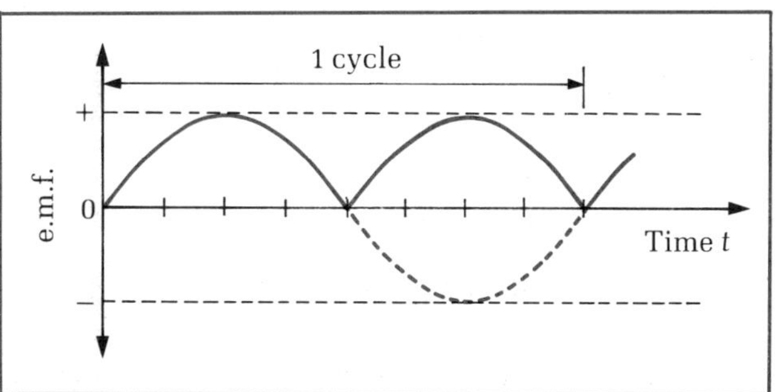

Figure 15.18 *E.M.F. v time for a d.c. generator*

Power station alternators are designed with three sets of coils at 120° to one another. Each set of coils produces an alternating e.m.f. out of phase by 120° with each of the other two e.m.f.s. The coils are called the 'stators' because they are stationary, and an electromagnet called the 'rotor' spins between them.

The electromagnet is supplied with current from a d.c. generator. So the turning of the rotor between the stator coils induces an alternating e.m.f. in each set of stator coils. The three phases are distributed via transformers and power lines to factories and local sub-stations. Your local sub-station supplies each home in the neighbourhood with electricity from one phase only. Some homes will be on one phase, some on another, and the rest on the third phase. This is why your home can sometimes suffer a blackout when other homes nearby still have electricity. This sometimes happens when a fault in the local sub-station cuts out one phase but not the others.

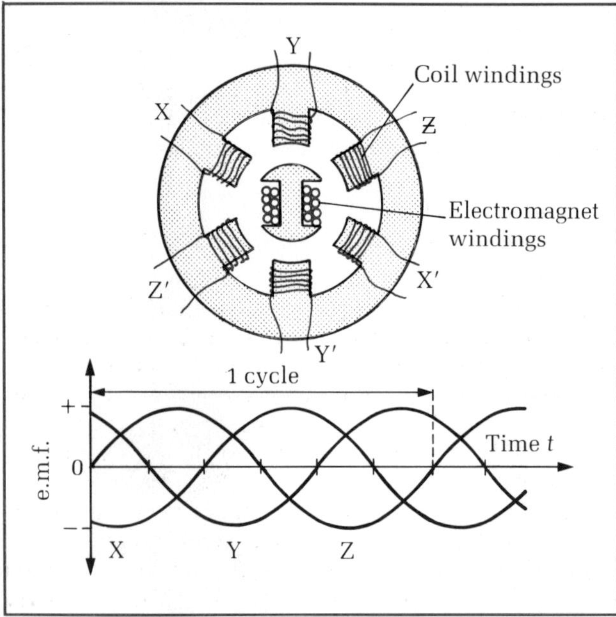

Figure 15.19 *Three-phase alternator*

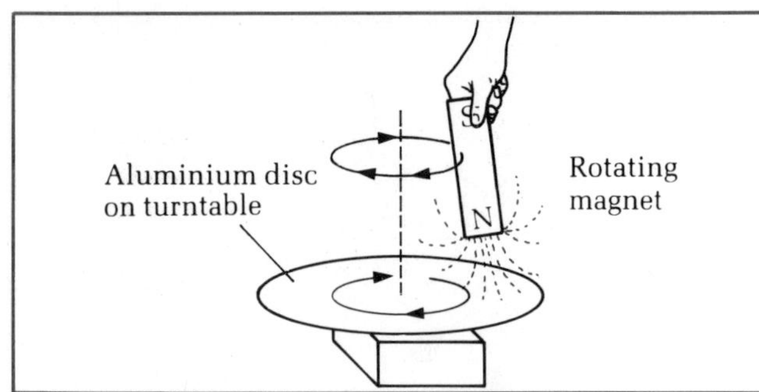

Figure 15.20 *Rotation by inducing current*

A rotating magnetic field can be produced without moving magnets round and round, by using two or more electromagnets. Each electromagnet is supplied with alternating current to produce an alternating magnetic field. However, each current is a fraction of a cycle out of phase with the current in the next electromagnet, so each electromagnet reaches peak strength at a different time to the one next to it. With the electromagnets round the edge of a circle, the peak strength therefore moves round the circle just as if a magnet is moved round and round. Figure 15.21 shows two electromagnets used to turn a metal can on a pivot. The field of one electromagnet peaks first to produce eddy currents in the can, then the other electromagnet's field peaks and pushes the can round a little. Before the cap stops, the peak field has moved back to the first electromagnet. So the can is given another push to keep it going.

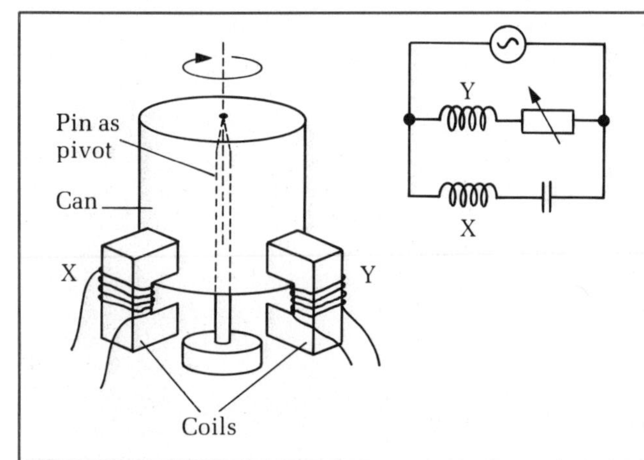

Figure 15.21 *A model induction motor*

The circuit to supply the electromagnets is also shown. One of the electromagnet coils is in series with a capacitor; the capacitor delays the peak current in the coil, so the field in the other electromagnet reaches its peak first.

15.3 Induction motors

Making an induction motor

When a magnet is moved round and round near a metal disc, the disc tries to keep up with the magnet. Figure 15.20 shows a simple demonstration, using an aluminium disc on a frictionless turntable. When the magnet is turned round and round above the disc, the disc starts to turn and to follow the magnet. Why? The motion of the magnet relative to the disc induces eddy currents in the disc. In other words, the moving magnetic field pushes the free electrons inside the disc around. The eddy currents oppose their cause, which is the relative motion between the magnet and the disc. So the currents are acted on by magnetic forces, which try to make the disc catch up with the magnet. It's not unlike stirring a bowl of water on a turntable round and round! The water currents created by stirring drag the bowl round.

Induction motors at work

Three-phase induction motors are the workhorses of industry. As explained previously, electricity generators which supply the National Grid produce three-phase alternating current. Each phase is carried by a separate cable in the network of power lines and transformers which make up the National Grid. Factories are supplied with all three phases for their 3-phase induction motors.

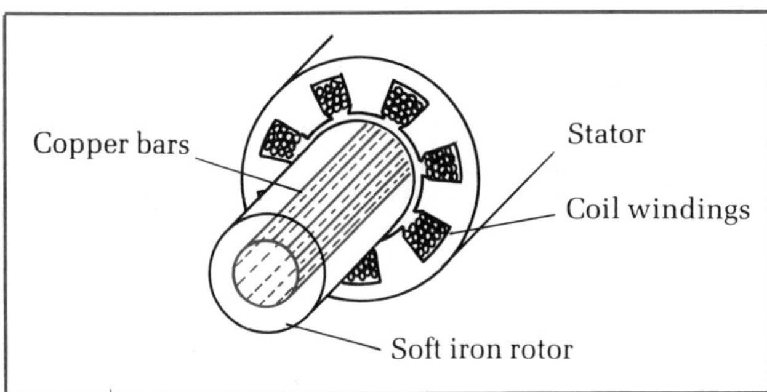

Figure 15.22 *A three-phase induction motor*

A 3-phase induction motor has at least three sets of coils, each supplied with a phase different from those supplied to the coils either side. Figure 15.22 shows an arrangement with just three sets of coils. The rotor is an iron cylinder with copper rods along the sides. When in use, the magnetic field rotates as the field of each coil peaks out of phase with the next coil. So the rotor is pulled round by the rotating field. The changing field of each coil induces current in the nearby copper rods. The adjacent coil then reaches its peak field and pushes the current-carrying rods round. Unlike the moving coil motor, induction motors are brushless; the rotor current is induced by the changing magnetic fields which then force the current-carrying conductors round. The rotor is designed with copper rods set into an iron core. Why? Copper is a good conductor of electricity, and iron makes the magnetic field much stronger, so the rotor gives a much stronger turning effect than if it were made only from iron or copper.

Shaded-pole induction motors are designed to operate with single phase alternating current. Domestic consumers receive single phase a.c. via the local sub-station. Figure 15.23 shows a simple model of a shaded pole motor. A metal plate covers half the area of one of the pole faces of the electromagnet. Without the shading plate, the rotor will not turn, but as soon as the plate is in position, the electromagnet pulls the rotor round. The reason is that the shading plate delays the magnetic field, so the phase of the shaded field lags behind the unshaded field. The situation is then like the two out-of-phase coils shown in Figure 15.21. The rotor is pulled towards the shaded field from the unshaded field.

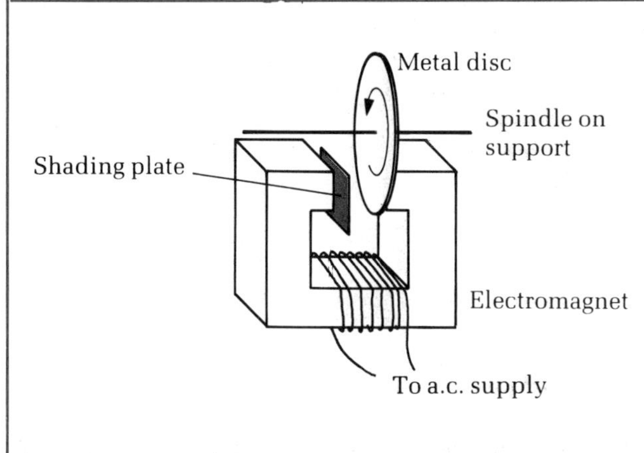

Figure 15.23 *A shading plate*

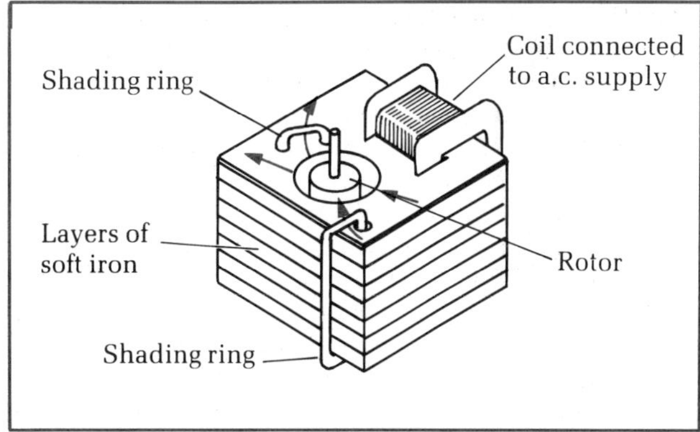

Figure 15.24 *A single-phase induction motor*

Shading rings can be used instead of a shading plate. Figure 15.24 shows a turntable motor with two shading rings. The field through the rings is delayed by the induced current in the rings. In effect, the two rings produce a magnetic field out of phase and at an angle to the field outside the rings, so the rotor is pushed round. Cut through the shading rings and the motor no longer works. Record disc turntables are driven by this type of motor.

15.4 Transformers

Suppose a bar magnet is repeatedly pushed in and out of a coil connected to a centre-reading meter. Induced current repeatedly passes through the meter, first in one direction then in the opposite direction, each cycle. The changing magnetic field caused by moving the bar magnet forces the free electrons in the coil to move round the coil.

Instead of moving a bar magnet in and out, an electromagnet supplied with alternating current could be used. If the a.c. frequency is less than about 5 Hz, the meter pointer deflects repeatedly in opposite directions. If the frequency is greater, then the meter should be replaced with an oscilloscope to monitor the changing e.m.f. Figure 15.25 shows the arrangement, which is an example of a simple transformer. The a.c. electromagnet is usually called the primary coil; the other coil is called the secondary coil.

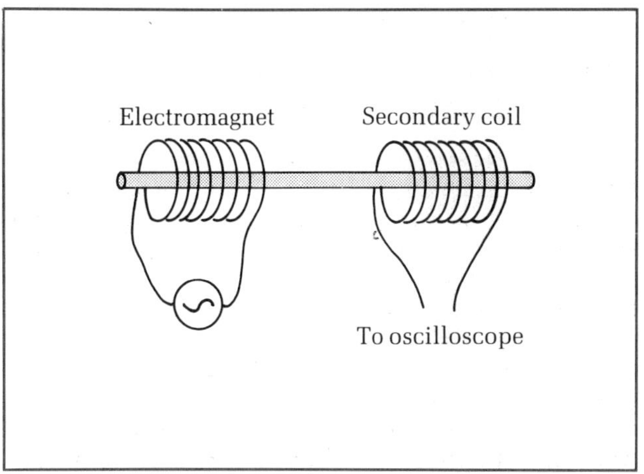

Figure 15.25 *A simple transformer*

Transformers are used to change an alternating p.d. to a different peak value. Any transformer has two coils, called the primary coil and the secondary coil. The two coils have the same soft-iron core. When the primary coil is connected to an alternating p.d., an alternating magnetic field is produced in the core. The field passes through the secondary coil. So an alternating e.m.f. is induced in the secondary coil by the changing magnetic field through it. The symbol for a transformer is shown in Figure 15.27.

Investigating transformers

Use two C-cores, plenty of insulated wire, a low voltage a.c. supply and an oscilloscope. Use the wire to make a primary coil wound round one side of a C-core. Connect the ends of the wire to the low-voltage a.c. supply. Then make a secondary coil with the same number of turns, but wide enough to slip on and off the core. Connect the secondary coil to the Y-input of an oscilloscope. The secondary coil can now be used as a 'flux-finder' to search for magnetic flux from the primary. With the oscilloscope time base off, any alternating e.m.f. induced in the secondary will make a vertical trace on the screen of the oscilloscope. The induced e.m.f. will make the spot on the screen go up and down so fast that a line is seen on the screen. The line length is a measure of the magnetic flux through the secondary coil.

Place the secondary coil on the same side of the core as the primary. Note the trace length. Now try the secondary coil at different positions round the core and in the gap across the ends of the core. Is the flux the same wherever the coil is placed? Magnetic flux is defined as $B \times A$, where A is the area of a coil placed at right angles to a magnetic field of strength B. Where the field is strongest, the magnetic flux is greatest. A helpful picture of magnetic flux is given by patterns of field lines. Lots of field lines through a coil on a diagram indicate that the coil is in a strong field, so a large amount of magnetic flux passes through it. The flux-finder coil used here ought to show that the flux in the core is the same wherever the flux-finder is placed on the core. However, the field spreads out in the air gap between the ends of the core, so the flux through the secondary coil in the air gap is less.

Suppose the other C-core is fitted on top of the first C-core, with the secondary coil still in the air gap. The trace length is now much shorter. The reason is that the magnetic flux is now confined to the cores. Slide the cores apart a little, and you may be able to detect 'flux leakage' where the cores join.

Now slide the secondary coil onto the other core, and fit the two cores together as in Figure 15.26(c). The trace length is now much greater than at the start, in Figure 15.26(a). Without an air gap, the flux in the core is much greater because the magnetic field strength is much greater. Alter the number of turns on the secondary coil. What difference is made to the trace length? You ought to find that the trace length is made longer by increasing the number of secondary turns.

The transformer rule

Transformers are designed so that all the magnetic flux produced by the primary coil passes through the secondary coil. Figure 15.27 shows a practical arrangement where the soft iron core channels the flux round from one end of the primary to the other end. The core is laminated so that layers of soft iron are separated from one another by layers of insulator. In this way, induced currents (i.e. eddy currents) in the core are reduced so the flux from the primary is as high as possible.

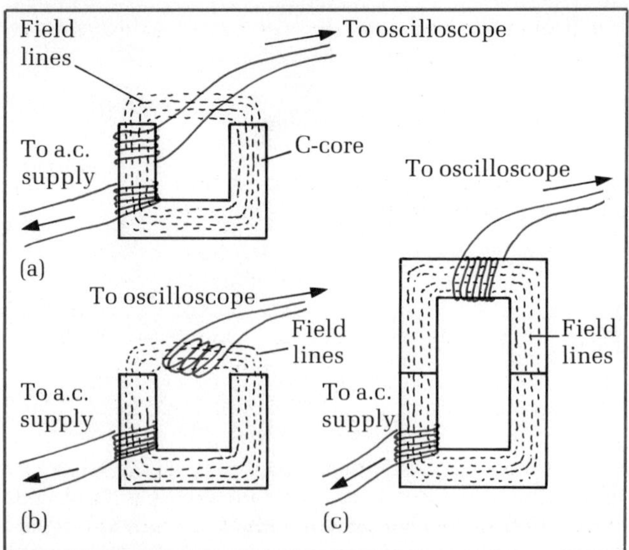

Figure 15.26 *Investigating transformers*

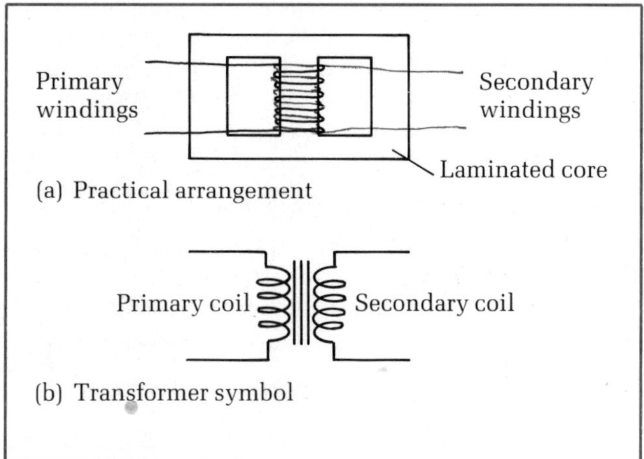

Figure 15.27 *Transformers*

Let φ = the flux in the core passing through each turn.

Hence the flux linkage through the secondary coil = $N_s\varphi$, where N_s is the number of turns of the secondary coil.

From Faraday's law, the induced e.m.f. V_s in the secondary is given by

$$V_s = -\frac{d}{dt}(N_s\varphi) = -N_s\frac{d\varphi}{dt}$$

The changing flux through the primary coil induces a **back e.m.f.** in the primary coil. The back e.m.f. opposes the applied p.d. So the applied p.d. V_p must equal the back e.m.f. to keep the flux changing. Faraday's law gives the back e.m.f. as $-\frac{d}{dt}(N_p\varphi)$ since the flux linkage through the primary is equal to the number of primary turns $N \times$ the flux φ through each turn.

Hence $V_p = -\frac{d}{dt}(N_p\varphi) = -N_p\frac{d\varphi}{dt}$

By dividing the equation for V_s by the equation for V_p, we obtain **the turns rule** for a transformer.

$$\frac{V_s}{V_p} = \frac{N_s}{N_p}$$

where V_s = induced e.m.f. in secondary,
V_p = p.d. applied to primary,
N_s = number of turns of secondary,
N_p = number of turns of primary.

The theory given above is made simpler by assuming the coils have negligible resistance, so the input power is equal to the output power when the transformer delivers current. Since electrical power is calculated from current × p.d., then the following equation applies to a transformer with 100% efficiency.

$$I_sV_s = I_pV_p$$

where I_s = secondary current,
I_p = primary current.

In practice, transformers are almost 100% efficient because they are designed with laminated soft-iron cores and low resistance windings.

Step-up transformers are designed with more turns in the secondary coil than in the primary coil. So the secondary e.m.f. is stepped up compared with the p.d. applied to the primary. For example, if $N_s = 20N_p$, then V_s is $20V_p$. However, the current is stepped down because I_sV_s cannot be greater than I_pV_p.

Step-down transformers are designed with the number of secondary turns lower than the number of primary turns, so the secondary e.m.f. is stepped down compared with the p.d. applied to the primary. The current in the secondary is greater than current in the primary current.

Figure 15.28 *Area control centre*

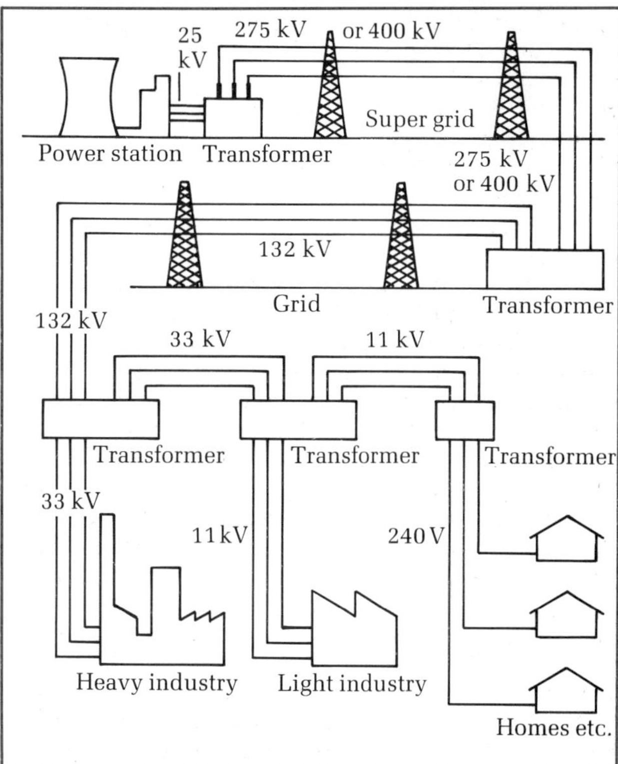

Figure 15.29 *The grid system*

The National Grid

Electricity from power stations in the UK is fed into the National Grid system which supplies electricity to most parts of the mainland. The National Grid is a network of cables, underground and on pylons, which covers all regions of the country. Each power station generates alternating current at a precise frequency of 50 Hz at about 25 kV.

Step-up transformers at the power station increase the p.d. to 400 kV or more for long-distance transmission via the grid system. Step-down transformers in each area are used to supply electricity from the grid system to local users. The step-down transformers operate in stages, as shown in Figure 15.29. Factories are supplied with all three phases at either 33 kV or 11 kV. Homes are supplied via a local transformer sub-station with single phase a.c. at 240 V.

Before the grid system was established, each locality had its own electricity power station. Some produced direct current, others produced alternating current. Voltage levels, too, often differed from one locality to another. However the economic benefits of a.c. transmission soon became obvious. By the time the National Grid was set up, most areas were supplied with a.c. at 50 Hz. The grid system links areas together so that sudden demand for power in one area can be met from other areas.

Transmission using alternating current is more efficient than d.c. transmission. This is because high voltage transmission is more efficient than low voltage transmission. Because a.c. can be stepped up and down much more easily than d.c., a.c. is more efficient.

A simple demonstration is shown in Figures 15.30 and 15.31, on the next page. In Figure 15.30 transformers are not used, so the transmission is at low voltage (12 V). The low voltage lamp at the end of the 'power line' does not light up. However, in Figure 15.31 a step-up transformer is used to increase the power line p.d. to 240 V, then a step-down transformer is used at the end of the power line to supply the low voltage lamp. The lamp at the end of

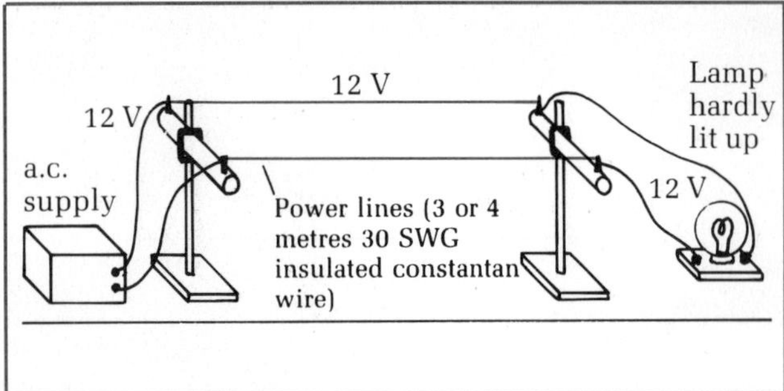

Figure 15.30 *Inefficient low-voltage transmission*

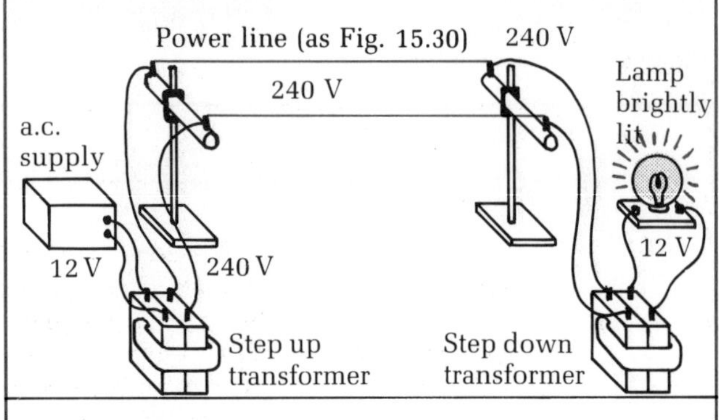

Figure 15.31 *Efficient high-voltage transmission demonstration*

the power line lights up brightly this time. This is because the current along the power line is much smaller. The step-up transformer increases the p.d. but reduces the current, so less power is wasted heating up the power lines because the current is smaller.

To deliver power W at voltage V, the current required is given by $W = IV$. Hence $I = W/V$. If the resistance of the cables is R, the power wasted through heating the cables is I^2R. So if the voltage V is high, the current I is low. Hence the wasted power is low. The wasted power $= W^2R/V^2$.

Mutual inductance

Wherever two separate electrical circuits are near to each other, changes in the magnetic flux produced by one circuit can affect the other circuit. This is used as the basic principle of the transformer.

Consider two coils X and Y near one another, as in Figure 15.32. Coil Y is connected to a lamp. Coil X is connected to an a.c. supply. The coils are in separate electrical circuits. Nevertheless, the lamp will light if the coils are close enough. If an iron core is placed through the two coils, the lamp becomes even brighter. The changing magnetic flux from coil X induces an e.m.f. in coil Y.

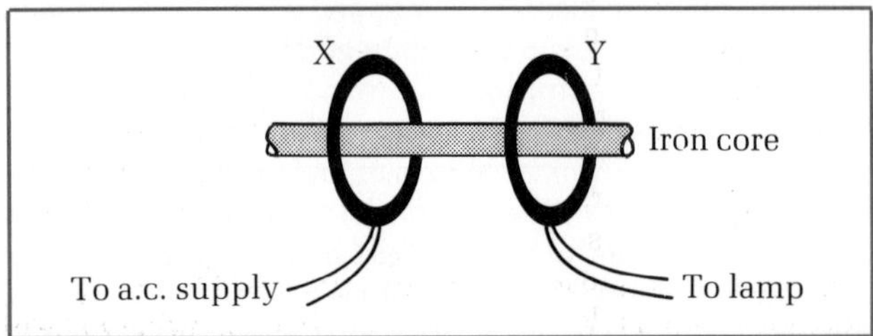

Figure 15.32 *Mutual inductance*

Does the effect work if the coils are swapped over? Coil X is now connected to the lamp, and Y is connected to the a.c. supply. Try it and you ought to find that it does work. The effect is said to be mutual because changing magnetic flux due to either coil affects the other coil.

The induced e.m.f. in one coil is proportional to the rate of change of flux through that coil from the other one. In other words, the induced e.m.f. in the secondary coil is proportional to the rate of change of flux linkage from the primary. However, the flux from the primary is proportional to the primary current. So the induced e.m.f. in the secondary is proportional to the rate of change of primary current.

In other words,

$$\Phi_s = MI_p$$

where Φ_s = flux linkage in the secondary produced by the primary,
I_p = primary current,
M = a constant of proportionality.

Since the secondary e.m.f. is given by $V_s = -\dfrac{d\Phi_s}{dt}$ from Faraday's law, then

$$V_s = -M\frac{dI_p}{dt}$$

The mutual inductance M of the two coils is defined as

$$\frac{\text{Induced e.m.f. in one coil}}{\text{Change of current per second in the other coil}}$$

$$M = \frac{-V_s}{\left(\dfrac{dI_p}{dt}\right)}$$

The unit of M is the **henry** (H), named after Joseph Henry who first developed the idea of mutual inductance. The unit of $\left(\dfrac{dI_p}{dt}\right)$ is $A\,s^{-1}$, so the unit of M is the same as $\dfrac{V}{(A\,s^{-1})}$ or $V\,A^{-1}\,s$. However, VA^{-1} is the unit of electrical resistance, the ohm (Ω). So the henry is the same as the ohm second. Whichever of the two coils is used as the primary, the value of M is the same.

Worked example A mutual inductor is supplied with current to one of its coils. The current rises and falls at a steady rate, as shown in Figure 15.33. In each cycle, the

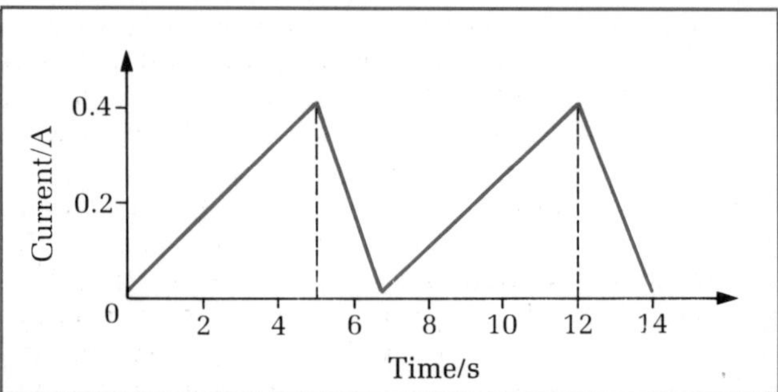

Figure 15.33

current rises from 0 to 0.4 A in 5 s, then falls back to zero in the next 2 s. An oscilloscope is used to measure the induced e.m.f. across the secondary coil. When the current is rising, a steady e.m.f. of −40 mV is measured. Calculate

a) the mutual inductance of the coils,

b) the induced e.m.f. across the secondary when the current falls.

Solution

a) For rising current, $\frac{dI_p}{dt}=\frac{0.4}{5}=8\times10^{-2}\,\text{A s}^{-1}$.

$$\text{Hence } M=\frac{-V_s}{\left(\frac{dI_p}{dt}\right)}=\frac{-(-0.04)}{8\times10^{-2}}=0.5\,\text{H}$$

b) When the current falls, $\frac{dI_p}{dt}=\frac{-0.4}{2}=-0.2\,\text{A s}^{-1}$

$$\text{So } V_s=-M\left(\frac{dI_p}{dt}\right)=-0.5\times-0.2=+0.1\,\text{V}$$

Figure 15.34 shows how the secondary e.m.f. varies with time.

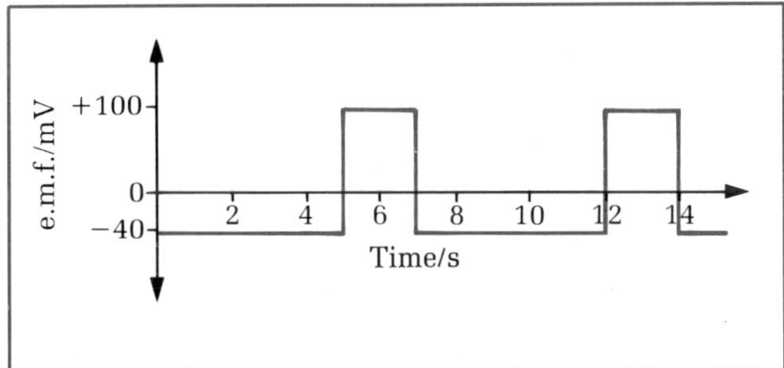

Figure 15.34

15.5 Self inductance

Consider the circuit shown in Figure 15.35. A coil is connected in series with lamp X. A second lamp Y in series with a rheostat is connected in parallel with the coil and X. The rheostat is adjusted so that both lamps become equally bright when connected to the battery. When the switch is closed, lamp Y lights up straight away but X is delayed. Why? Interchanging the lamps makes no difference. The lamp in series with the coil always lights up after the other lamp.

The coil delays the growth of current through its lamp and itself. When the switch is first closed, the growth of current in the coil causes growth of magnetic flux in its core. But the growth of magnetic flux induces a back e.m.f. in the coil. The back e.m.f. opposes the battery p.d. The current in the coil grows more slowly than in the rheostat, so lamp X lights up after lamp Y.

The e.m.f. induced in the coil due to its own changing flux is called a **self-induced e.m.f.** The magnetic flux change produced by turns in any one part of the coil induces e.m.f.s in all the other parts of the same coil. The overall effect is the self-induced e.m.f., which acts against the battery. So self-induced e.m.f.s are sometimes called **back e.m.f.s.**

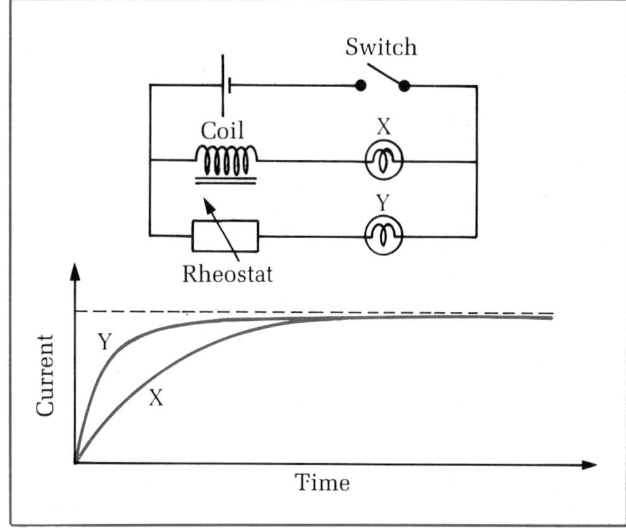

Figure 15.35 *Self-inductance*

Without a soft iron core in the coil, the delay between the two lamps is much shorter. The soft iron core makes the self-induced e.m.f. much greater because the magnetic field in the core is much stronger. Coils designed to produce large self-induced e.m.f.s are called **inductors** or chokes. In d.c. circuits they are used to slow the growth of current; for example, a choke in series with a motor prevents the motor from being overloaded when first switched on. But they also tend to keep decreasing currents going. Disconnect a wire from a coil carelessly and you could receive a hefty shock! The sudden drop of current as a result of the disconnection causes the field in the coil to collapse, so a huge e.m.f. is induced between the wire and the coil.

In a.c. circuits, coils react against changing currents, so they can be used to limit currents without wasting power. They are also used in tuning circuits. Slowly changing currents pass through coils more easily than fast changing currents. This is because fast changes create large induced e.m.f.s which oppose the changes, so high frequency signals do not pass through coils as easily as low frequency signals. See p. 351.

The magnetic flux linkage Φ in a coil is proportional to the current I in the coil. So we can write $\Phi=LI$, where L is the constant of proportionality for that coil.

Faraday's Law gives the induced e.m.f. in the coil, $V=-\frac{d\Phi}{dt}$ Hence

$$V=-\frac{d}{dt}(LI)=-L\frac{dI}{dt}$$

The self-inductance L of a coil is defined as

$$\frac{\text{Self-induced e.m.f.}}{\text{Change of current per second}}$$

$$L=\frac{V}{\left(\frac{dI}{dt}\right)}$$

The unit of L is the henry, the same as for mutual inductance. Sometimes L is referred to as the **inductance** of a coil.

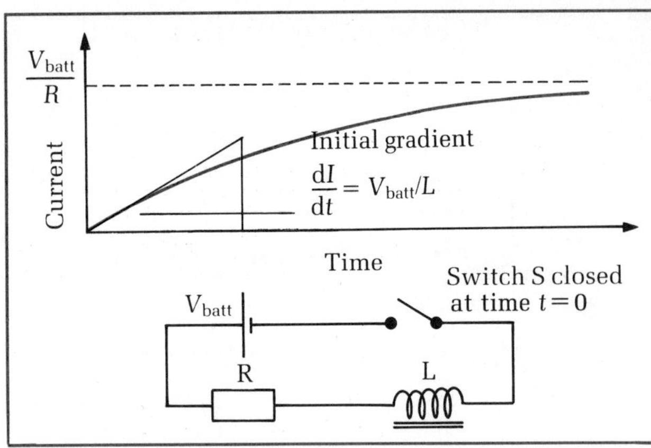

Figure 15.36 *Growth curve*

The growth curve for the current taken by an inductor when switched into a d.c. circuit is shown in more detail in Figure 15.36. The inductance L is in series with a resistance R and a battery which supplies an e.m.f. V_{batt}. When switch S is closed, the current in the circuit increases from zero and builds up to a final steady value.

Consider the circuit at time t after the switch is closed, when the current equals I.

$$\text{Battery e.m.f. } (V_{batt}) + \text{self-induced e.m.f. due to the coil} \left(-L\frac{dI}{dt}\right) = \text{resistor p.d. } (IR)$$

$$V_{batt} - L\frac{dI}{dt} = IR$$

In other words, the self-induced e.m.f. acts against the battery e.m.f. so the total e.m.f. $V_{batt} - L\dfrac{dI}{dt}$ is equal to the resistor p.d.

1 At time $t=0$, the current is zero so the resistor p.d. is zero initially. Therefore, $L\dfrac{dI}{dt} = V_{batt}$ at the start. So the initial growth of current $\left(\dfrac{dI}{dt}\right)_{t=0}$ is equal to V_{batt}/L.

2 As the current grows from zero, the resistor p.d. grows. So the self-induced e.m.f. $(= V_{batt} - IR)$ falls; hence the rate of growth of current falls. The growth curve flattens out more and more as time goes on because the rate of growth decreases and decreases.

3 As time $t \to \infty$, the rate of growth of current $\dfrac{dI}{dt} \to 0$. So the resistor p.d. rises to equal the battery e.m.f. Hence the current rises to become equal to V_{batt}/R as $t \to \infty$.

Initial rate of growth of current $= V_{batt}/L$

Final current $= V_{batt}/R$

Worked example An inductor of inductance 0.2 H and negligible resistance is connected in series with a 9 Ω resistor and a cell and switch. The cell has an e.m.f. of 1.5 V and its internal resistance is 1 ohm. Calculate

a) the initial rate of growth of current,

b) the final current,

c) the rate of growth of current when the current is 0.1 A.

Solution

a) Initially $I=0$ so the initial rate of growth of current $= V_{batt}/L = 1.5/0.2 = 7.5\ \text{A s}^{-1}$.

b) Finally, $\dfrac{dI}{dt} = 0$ so the final current $= V_{batt}/R = 1.5/(9+1) = 0.15$ A. (The total resistance in the circuit is 10 ohms.)

c) When $I = 0.1$ A, the p.d. across the total resistance $= IR = 0.1 \times 10 = 1.0$ V. So the self-induced e.m.f. is equal to 0.5 V $(= V_{batt} - IR)$. Hence $L\dfrac{dI}{dt} = 0.5$ V so $\dfrac{dI}{dt} = 0.5/L = 0.5/0.2 = 2.5\ \text{A s}^{-1}$. So the rate of growth of current $= 2.5\ \text{A s}^{-1}$ when $I = 0.1$ A.

Measuring self-inductance

Use a coil with low resistance. Connect the coil in series with a resistance box and a low voltage a.c. supply. Connect the leads from a double beam oscilloscope as in Figure 15.37. Input Y_1 therefore measures the supply p.d.

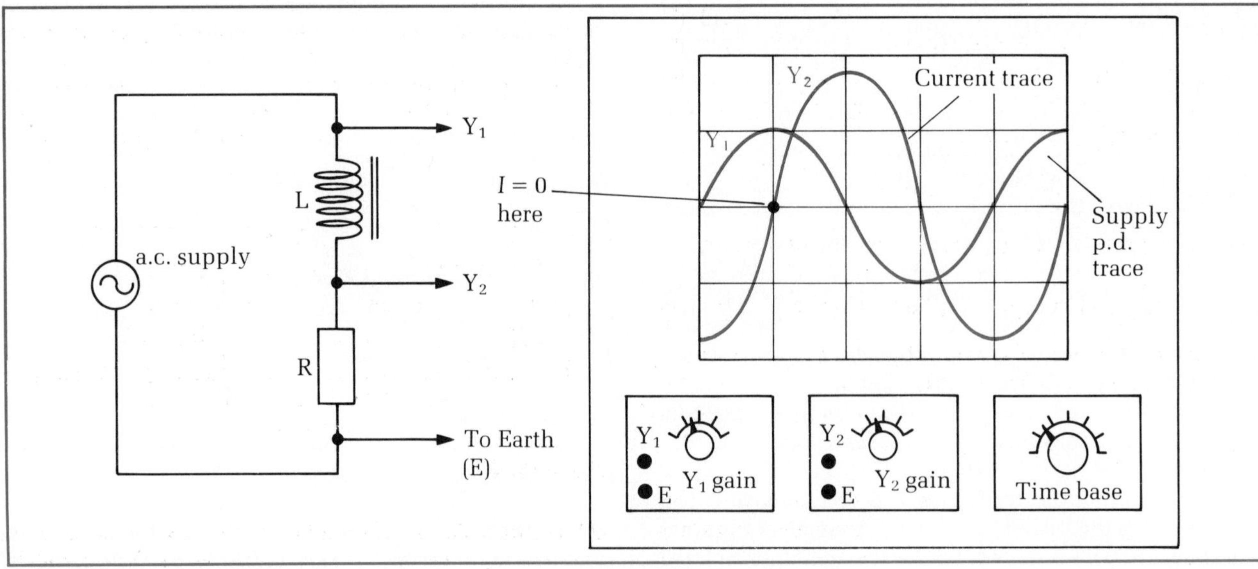

Figure 15.37 *Measuring self-inductance*

Input Y_2 measures the resistor voltage which is proportional to the current. So Y_2 gives a trace which represents the current.

The oscilloscope shows two traces, one for the current and the other for the supply p.d. Locate a point on the current trace where the current is zero. This is at the mid-point between a crest and a trough on the current trace. Estimate the gradient of the current trace at that point. The gradient is the change of resistor p.d. per second, $\frac{dV_r}{dt}$. Since the current I equals $\frac{\text{resistor p.d.}}{\text{resistance } R}$ then $\frac{dI}{dt}$ is equal to the gradient $\frac{dV_R}{dt}\Big/$ resistance R. So the change of current per second $\frac{dI}{dt}$ at that point can be determined.

The supply voltage V at the chosen point must also be measured off the other trace. Then L can be calculated from $V\Big/\left(\frac{dI}{dt}\right)$.

Energy stored in an inductor

When the current in an inductor is suddenly switched off, a spark may be produced across the switch gap. The inductor tries to keep the current going as its magnetic field collapses, so a large e.m.f. is induced across the switch gap and a spark may be produced. The energy stored in the magnetic field is suddenly released.

How much energy is stored in an inductor? Suppose the current in an inductor increases from 0 to final value I. Consider a short interval Δt when the current increases during the build up from i to $i+\Delta i$.

The induced e.m.f. $V=L\frac{\Delta i}{\Delta t}$ where L is the self-inductance of the coil.

The electrical energy supplied = current × p.d. × time $= iV\Delta t$

The electrical energy supplied $= iL\frac{\Delta i}{\Delta t}\Delta t = iL\Delta i$

To determine the total energy supplied, we add the energy supplied in all the time intervals. Integration is the mathematical process of doing this. Δi must be considered as a very small increase of current. So the total energy supplied is written as

$$\text{Total energy supplied} = \int_0^I Li\,di$$

where di represents a very small increase of current.

$$\text{Hence the total energy supplied} = \left[\tfrac{1}{2}Li^2\right]_0^I = \tfrac{1}{2}LI^2 \text{ since}$$

$\int i\,di$ equals $\frac{1}{2}i^2$. Since this is equal to the energy stored in the inductor, then the total energy stored can be written as

$$\text{Energy stored in an inductor} = \tfrac{1}{2}LI^2$$

where I = current,
L = self-inductance.

15.6 Summary

Definitions

1 *Mutal inductance M* is
$$\frac{\text{Induced e.m.f. in the secondary}}{\text{Change of current per second in the primary}}$$
2 *Self inductance L* is induced e.m.f./change of current per second

Laws

1 *Lenz's law* Induced current always flows in a direction so as to oppose the change which causes it.
2 *Faraday's law* The induced e.m.f. is equal to the rate of change of magnetic flux.

Equations

1 E.M.F. induced in
a) a rod: $\varepsilon = BLv$
b) a rotating disc: $\varepsilon = Bf\pi R^2$
c) a rotating coil: $\varepsilon = BAn\omega \sin \omega t$
2 Faraday's law $\varepsilon = -\frac{d\Phi}{dt}$
3 Magnetic flux $= BA$, Flux linkage $\Phi = BAn$
4 Search coil equation $Q = BAn/R$
5 Transformer equation $V_s/V_p = N_s/N_p$
6 Energy stored in an inductor $= \frac{1}{2}LI^2$

Short questions

15.1 A horizontal rod of length 400 mm lying along a line from east to west is released from rest above the Earth's surface. The Earth's magnetic field strength at this position has components 18 μT due north in the horizontal direction and 55 μT vertically downwards. Calculate the e.m.f. induced across the ends of the rod 2.0 s after its release. Assume $g = 10\,\text{N kg}^{-1}$.

15.2 A circular search coil of diameter 34 mm with 100 turns is connected to a data recorder. The coil is then positioned at rest between the poles of a horseshoe magnet with the plane of the coil parallel to the magnet's pole faces. The coil is then rapidly removed from the magnet and the data recorder is used to measure the induced e.m.f. at intervals of 10 ms. A graph of the measurements is shown below.

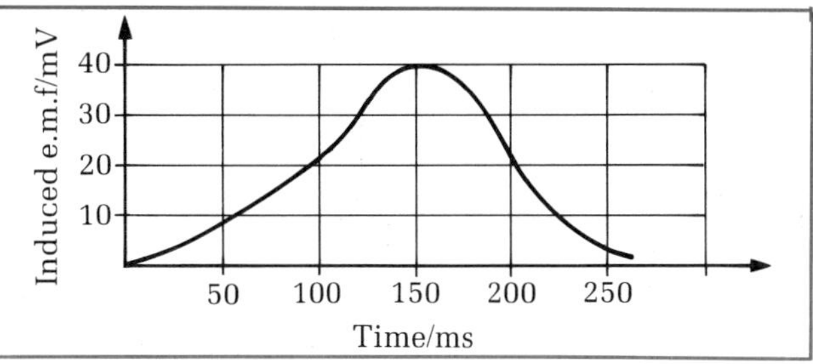

Figure 15.38

a) Use Faraday's Law to explain why the area under the curve of the graph represents the change of flux linkage through the coil.

b) From the graph estimate the change of flux linkage when the coil is removed from the magnetic field.
c) Hence calculate the magnetic field strength between the pole pieces.

15.3 The search coil of question **15.2** is now connected in series with a resistance box and a light beam galvanometer. The coil is then placed between the poles of a horseshoe magnet. The total resistance of the circuit is 6500 ohms. After the coil is rapidly withdrawn, the withdrawal makes the spot of the galvanometer swing from side to side of its scale with an amplitude of 78 mm. The charge sensitivity of the galvanometer is then determined by discharging a 0.40 μF capacitor charged to 1.50 V across the meter's terminals; as a result, the spot is set into oscillation with an amplitude of 55 mm. Calculate
a) the charge sensitivity of the galvanometer,
b) the magnetic field strength between the poles of the second horseshoe magnet.

15.4 The speedometer system of a certain make of car uses a metal disc of diameter 80 mm which is made to rotate when the wheels are turning. The disc turns between the poles of a horseshoe magnet so that the field lines are at right angles to the disc plane. The field strength of the magnet between its poles is 110 mT; a voltmeter connected between the rim and the centre of the disc gives a reading of 16 mV when the car's speed is 30 km h^{-1}. Calculate
a) the frequency of rotation of the disc at this speed,
b) the voltage range of the speedometer meter if its scale is to register speeds up to 140 km h^{-1}.

15.5 An a.c. generator consists of a square coil of area 2.0×10^{-3} m^2 with 200 turns. The coil is set turning at a rate of rotation of 30 Hz in a uniform magnetic field of strength 85 mT, with the coil axis at right angles to the magnetic field lines.
a) Calculate the peak e.m.f. induced across the generator's terminals.
b) Sketch a graph to show how the induced e.m.f. varies over one cycle; show on your graph the points in the cycle corresponding to the coil plane parallel to the field lines.
c) If the rate of rotation is reduced to 15 Hz, sketch on the same axes the new waveform for the induced voltage.

15.6
a) A bar magnet is suspended at rest horizontally above a copper disc on a stationary turntable. The centre of the bar magnet is directly above the centre of the disc. When the turntable is set into steady rotation, the bar magnet starts turning too. Explain in simple terms why the bar magnet turns even though it is not in contact with the disc.
b) Explain the principle of operation of a shaded-pole induction motor.

15.7 Two coils X and Y are positioned facing one another a few centimetres apart. When the current in coil X is increased at a steady rate of 5 A s^{-1}, a constant e.m.f. of 35 mV is induced across the terminals of coil Y.
a) Calculate the mutual inductance of the arrangement.
b) Coil Y is now connected in series with a rheostat, an ammeter and a battery. A voltmeter is connected across the terminals of coil X. When the rheostat is used to increase the current in Y from 0.5 A to 3.0 A at a steady rate the voltmeter reads 42 mV. How long did the change take?
c) The coils are now moved to new positions much closer together, and then the current in Y is reduced from 3.0 A to 0.5 A in the same time as the increase took. Would you expect the induced e.m.f. in coil X to be greater or smaller than 42 mV?

15.8 The circuit below shows a 6 V battery of negligible internal resistance connected across a coil in series with a lamp bulb X. A rheostat and a second lamp bulb Y are connected in parallel with the coil and X. The coil has a resistance of 9.0 Ω and the rheostat is adjusted until its resistance in circuit is 9.0 Ω.

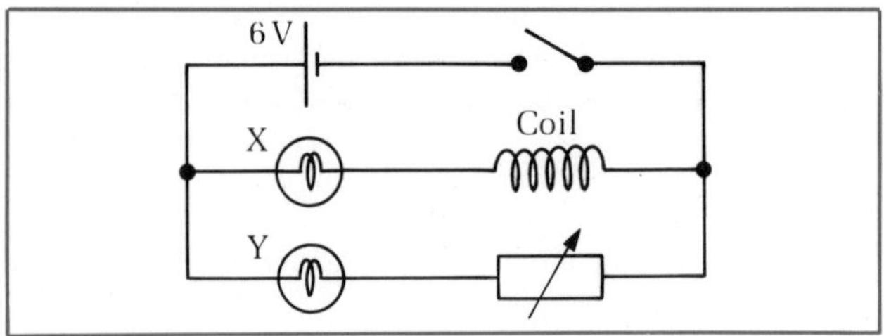

Figure 15.39

a) Explain why X lights up after Y when the switch S is closed.
b) If the coil's inductance is 0.4 H, calculate (i) the initial rate of growth of current through the coil, (ii) the maximum current if each bulb has 3 ohms resistance when lit, (iii) the energy stored in the coil at maximum current.

15.9 An iron ring has area of cross-section 8.5×10^{-5} m^2 and its average circumference is 320 mm. An insulated wire is wrapped around the ring to make a coil with 500 evenly spaced turns as shown in the diagram. Then another length of insulated wire is used to form a second coil with 800 evenly spaced turns over the first coil. When the coils are tested for mutual inductance, the current in the first coil A is increased steadily from 0 to 0.2 A in 10 s to give an induced e.m.f. of 6.0 mV across the terminals of the other coil B. Calculate
a) the mutual inductance of the two coils,
b) the magnetic flux in the ring when the current in A is 0.2 A and B is on 'open-circuit',
c) the self-inductance of A.

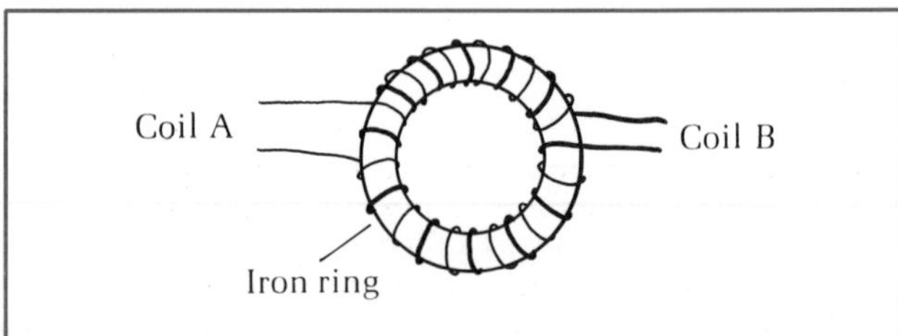

Figure 15.40

15.10
a) When a permanent magnet-type electric motor supplied by a battery is slowed down due to increased load, the current taken by the motor increases. Explain why this happens.
b) Such a motor has an armature of resistance 2.5 Ω and is used with a 12 V battery of internal resistance 0.5 Ω. When the motor is on load and turning at a steady speed of 25 Hz, its current is 2.5 A. Calculate (i) the back e.m.f. at this speed, (ii) the power supplied to the motor, (iii) the output power of the motor.
Account for the difference between (ii) and (iii).
c) If the load is increased and the motor speed falls to 20 Hz, what would the current become?

C

Multiple choice questions

CM.1 The mass and diameter of a planet are both four times those of the Earth. Assuming that each body is a sphere of uniform density then the ratio of the acceleration of free fall on the planet's surface to that on the Earth's surface is
A 16:1 **B** 4:1 **C** 1:1 **D** 1:4 **E** 1:16
(AEB June 82)

CM.2 A particle moves from a point X to a point Y in a gravitational field. No external forces other than gravitational forces act on it. Which of the following statements about the change in gravitational potential energy of the particle is (are) correct?
1 The change of potential energy is independent of the path between X and Y.
2 The change in potential energy is equal to the difference between the kinetic energy of the particle at Y and that at X.
3 The change in potential energy is independent of the mass of the particle.
A 1 only **B** 2 only **C** 1 and 2 only
D 2 and 3 only **E** 1, 2 and 3
(N. Ireland)

CM.3 A spacecraft is coasting towards the Earth under the influence of the Earth's gravitational field. Which of the following quantities is (are) changing with distance according to an *inverse square* law?
1 The gravitational potential energy of the spacecraft
2 The kinetic energy of the spacecraft
3 The gravitational field acting on the spacecraft
A 1 only **B** 2 only **C** 3 only
D 1 and 2 only **E** 1, 2 and 3
(N. Ireland)

CM.4 A body of mass 1 kg has a weight of 10 N at the surface of the Earth, which may be assumed to be a uniform sphere of radius R. When moved to a point at a distance $2R$ from the centre of the Earth its weight, in newtons, would be
A 0 **B** 1 **C** 2.5 **D** 5 **E** 10
(N. Ireland)

CM.5 Two satellites P and Q describe circular orbits of radii 7000 km and 8000 km respectively about the centre of the Earth. If P and Q complete one full orbit of the Earth in times T_P and T_Q, then the ratio T_P/T_Q is
A 7/8 **B** $(7/8)^{2/3}$ **C** $(7/8)^{3/2}$ **D** $(8/7)^{2/3}$ **E** $(8/7)^{3/2}$
(N. Ireland)

CM.6 An insulated metal plate which is parallel and close to a similar earthed plate is given an electric charge by being connected briefly to a d.c. supply. The separation of the plates is now increased using a non-conducting handle. Which, if any, of the following will remain constant?
1 The charge
2 The capacitance
3 The potential difference between the plates
A 1, 2 and 3 **B** 1 and 2 only **C** 2 and 3 only
D 1 only **E** 2 only
(Scottish H.)

CM.7 A charge of 2.0 μC is moved through a uniform electrostatic field of strength 10 V m^{-1} so that it travels 1.0 m along a path parallel to the field and then 1.0 m along a path perpendicular to the field. The change in potential energy, expressed in μJ, associated with this movement is
A 2.0 **B** 10 **C** 20 **D** $20\sqrt{2}$ **E** 40
(AEB Nov. 84)

CM.8 Two isolated metal spheres X and Y, of different radii, are positively charged while they are in contact. When the spheres are separated
A they have equal charges
B they have equal potentials
C a spark passes between the spheres
D a charge would pass from Y to X if the spheres were now connected by a conductor
E the charges would be neutralized if the spheres were now connected by a conductor
(AEB June 85)

CM.9 An electric charge is given to an insulated metal plate which is parallel and close to a similar earthed plate. The separation of the plates is now increased (using a non-conducting handle). Which of the following statements is (are) correct?
1 Work must be done to separate the plates.
2 The electric field strength between the plates increases.
3 Charge flows onto the earthed plate.
A 1 only **B** 2 only **C** 3 only
D 1 and 2 only **E** 1, 2 and 3

CM.10 The plates of a capacitor P, Q are connected to a d.c. supply and a slab of material S is introduced between the plates. The value of a quantity Y is found to vary with distance d across the system as shown below.
It is most probable that
A S is a metal and Y is electric field strength
B S is an insulator and Y is electric field strength
C S is an insulator and Y is potential gradient
D S is a metal and Y is potential
E S is an insulator and Y is potential
(London)

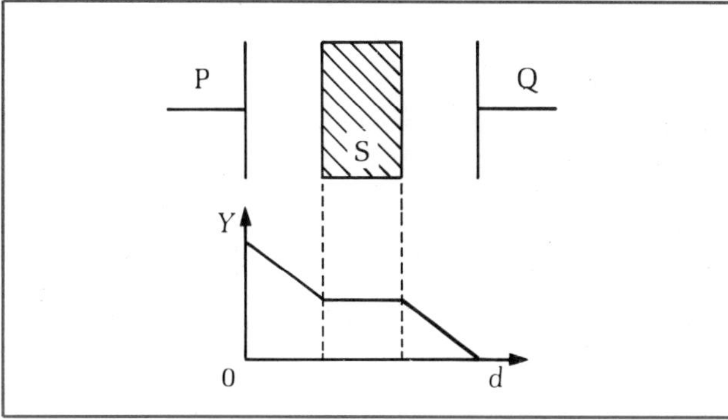

Figure CM1

CM.11 The force between two similar long thin parallel wires each carrying the same current is F. If the current in each wire is doubled and the distance between the wires is trebled the new force between them is
A $2F/9$ **B** $4F/9$ **C** $2F/3$ **D** $4F/3$ **E** $6F$
(AEB Nov 84)

CM.12 A small rectangular coil carrying constant current is placed on the axis and at one end of a horizontal solenoid. The initial position of the coil plane is vertical, parallel to the coil axis, as in Figure CM2 on the next page.

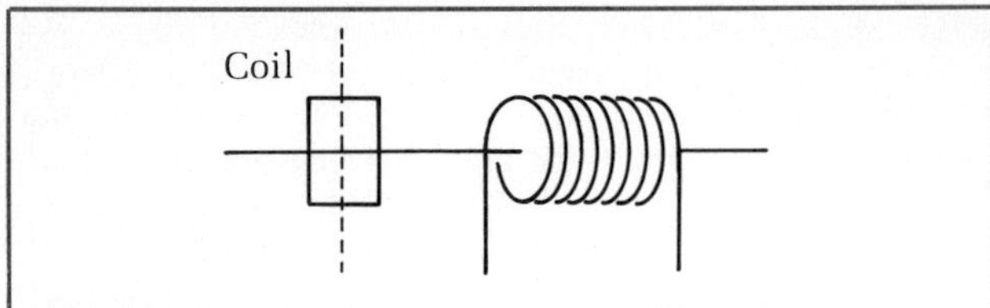

Figure CM2

When a steady current is passed through the solenoid,
A the coil is pulled into the solenoid
B the coil is pushed out of the solenoid
C the coil experiences a torque about a horizontal axis
D the coil experiences a torque about a vertical axis
E the coil experiences no resultant force or torque

CM.13 The current sensitivity of a moving coil galvanometer is increased by
1 increasing the number of turns on the coil
2 reducing the restoring couple per unit deflection
3 reducing the area of the coil
A 1, 2, and 3 are all correct **B** 1 and 2 only are correct
C 2 and 3 only are correct **D** 1 only is correct
E 3 only is correct

(London)

CM.14 An open spring with none of its turns touching each other has a length of 0.4 m. A constant current is passed through the spring which gives rise to a magnetic flux density B_1 at the geometrical centre of the spring. When the spring is stretched to a length of 0.9 m the same current gives rise to a magnetic flux density B_2 at the geometrical centre. The ratio B_1/B_2 is
A 81/16 **B** 9/4 **C** 4/9 **D** 3/2 **E** 2/3

(AEB Nov 83)

CM.15 A short, straight piece of copper wire is placed at the centre of a solenoid so that it is perpendicular to the axis of the solenoid. A circuit is arranged to pass current through the solenoid and the copper wire, which are connected in series. The electromagnetic force on the copper wire is
1 directed along the axis of the solenoid
2 directly proportional to the square of the current
3 directly proportional to the area of cross-section of the solenoid.
Which of these statements is (are) correct?
A 1 only **B** 2 only **C** 3 only
D 1 and 2 only **E** 1 and 3 only

(N. Ireland)

CM16 In Figure CM3 at the top of the next column, ABCD is a plate of n-type semiconductor. A magnetic field is applied normal to the plate, vertically into the page. In the plate
1 electrons are flowing from C to A
2 the current density is greatest near to side B
3 a potential difference is developed between B and D

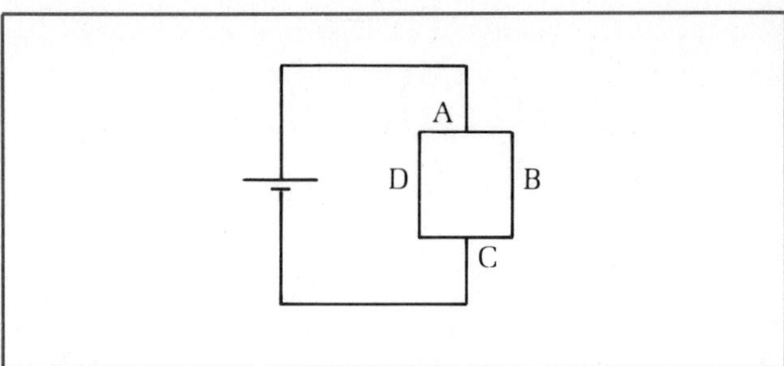

Figure CM3

A 1, 2, 3 correct **B** 1, 2 only correct
C 2, 3 only correct **D** 1 only correct
E 3 only correct

(AEB June 85)

CM.17 A metal disc spins in a uniform magnetic field perpendicular to the disc axis. Using contact brushes and a voltmeter, the p.d. between the axis and the rim is measured at 3.2 mV for a certain disc speed, U.
1 If the speed is halved, the p.d. between the rim and the axis becomes 1.6 mV.
2 If the contact brushes are placed on the rim directly opposite each other, the p.d. between the brushes becomes 6.4 mV for speed U.
3 At speed U with one contact on the axis and the other halfway between the centre and the rim, the p.d. is 1.6 mV.
A 1, 2, 3 correct **B** 1, 2 only correct
C 2, 3 only correct **D** 1 only correct
E 3 only correct

CM.18 A circular coil with its plane horizontal is held firmly in a clamp and its terminals are connected to a data recorder. A bar magnet with its axis along the coil axis is released from rest from a position above the coil so it drops through the coil. The data recorder is used to measure the induced e.m.f. at millisecond intervals. Which graph in Figure CM4 shown below best represents the variation of induced e.m.f. ε with time?

CM.19 A circuit contains a battery, a switch, a resistor and an inductor. The battery e.m.f. is E, the total circuit resistance is R and the inductance is L. Which of the following statements is *not* correct?
A When the switch is closed, an induced e.m.f. initially acts against the battery e.m.f.
B When the switch is closed, the initial rate of rise of current is E/L.
C When the switch is closed, the maximum current is E/R.
D Opening the switch can cause a spark to jump across the switch gap.
E Opening the switch causes an induced e.m.f. which acts against the battery e.m.f.

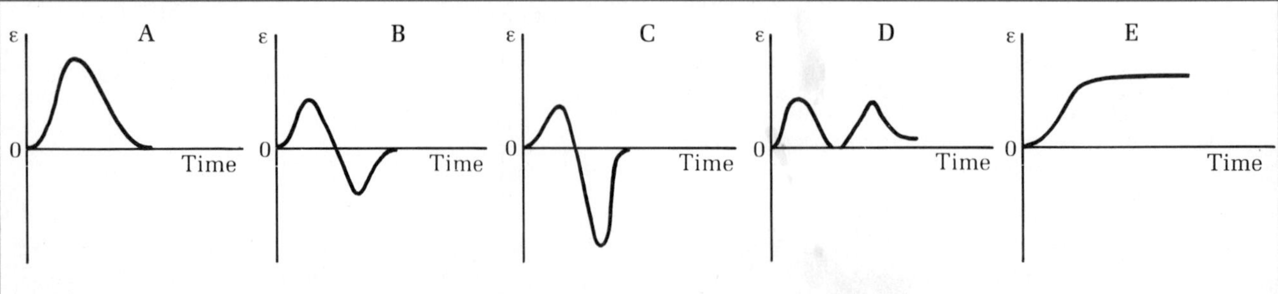

Figure CM4

CM.20 A coil P is placed next to an identical coil Q. When the current in Q changes at a rate of $2\,A\,s^{-1}$, an e.m.f. of 1.5 mV is induced in P. Which of the following statements is (are) correct?

1 Arranging for the current in Q to flow in the opposite direction to the current in P will reduce the interaction.

2 When the current in P changes by $2\,A\,s^{-1}$, an e.m.f. of 2 mV will be induced in Q.

3 Increasing the physical separation of the circuits will reduce the interaction.

A 1, 2, 3 correct **B** 1, 2 only correct **C** 2, 3 only correct
D 1 only correct **E** 3 only correct

(London)

C

Long questions

CL.1 State the law describing the gravitational force between two point masses M and m a distance r apart.

Two alternative units for gravitational field strength are $N\,kg^{-1}$ and $m\,s^{-2}$. Use the method of dimensions to show that they are equivalent.

State the general relationship between the field strength at a point in a field of force and the potential gradient at that point. Write down an expression for the gravitational potential at a point distant r from a mass M. Distinguish between *gravitational potential* and *gravitational potential energy*.

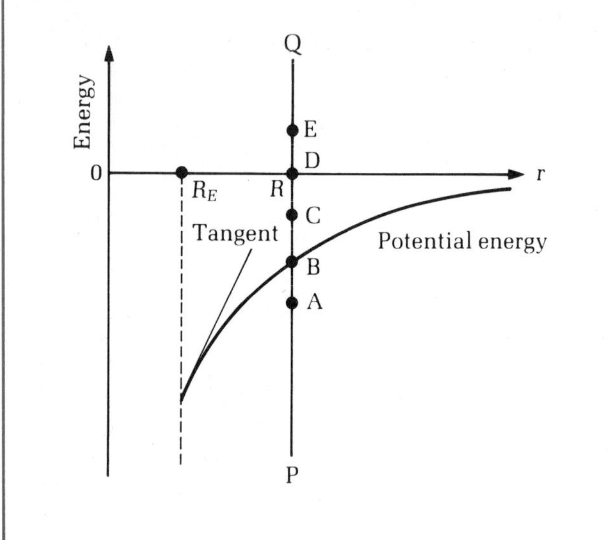

Figure CL1

The curve in the diagram above shows the way in which the gravitational potential energy of a body of mass m in the field of the Earth depends on r, the distance from the centre of the Earth, for values of r greater than the Earth's radius R_E. What does the gradient of the tangent to the curve at $r = R_E$ represent?

The body referred to above is a rocket which is projected vertically upwards from the Earth. At a certain distance R from the centre of the Earth, the *total* energy of the rocket (i.e., its gravitational potential energy plus its kinetic energy) may be represented by a point on the line PQ. Five points A, B, C, D, E have been marked on this line. Which point (or points) could represent the total energy of the rocket

a) if it were momentarily at rest at the top of its trajectory,
b) if it were falling towards the Earth,
c) if it were moving away from the Earth, with sufficient energy to reach an infinite distance?

In each case, explain briefly how you arrive at your answer.

(Cambridge)

CL.2

a) i) State Newton's law of gravitation.
ii) Describe an experiment by which the gravitational constant G may be determined.

b) A space station of mass m is in a circular orbit at a height h above the Earth's surface. In terms of these quantities together with the radius r of the Earth and the acceleration due to gravity g_0 at the Earth's surface, derive expressions for the space station's:
i) potential energy,
ii) kinetic energy,
iii) total energy.
The space station descends from a height of 300 km to 299 km in one orbit. If $m = 8.0 \times 10^4$ kg, $r = 6400$ km and $g_0 = 9.8\ m\,s^{-2}$, calculate:
iv) the energy lost by the space station,
v) the retarding force acting on it.

(Oxford)

CL.3 This question is about gravity and the similarities between gravity and electric and magnetic effects.

The passage below sets out three sets of ideas about gravity. For each of the sections (i) to (iii) you are asked to write a more complete explanation of the ideas: your explanations may include

– quantitative calculations to illustrate the ideas,
– fuller explanations of the theoretical ideas,
– discussion of possible experiments.

You should pay particular attention to the words and phrases that are underlined in each section.

Passage

i) There is something peculiar about gravity: it is such a small force that if we didn't live on a big lump of matter called the Earth we might not notice that it affected man-size objects at all. In fact the simplest calculations can show that it is very hard to demonstrate that the effect exists between all pieces of matter.
ii) There is a close analogy between the theoretical ideas involved in electricity and in gravity, and this can be of great value in discussing such abstract ideas as field and potential. Thus problems such as the scattering of alpha-particles by a nucleus and the path of a spaceship round the moon have many similarities though there are also important differences.
iii) However, electrical and magnetic effects are so much bigger, for man-size experiments, that they swamp all effects of gravity. The fact that when we come to matter on an astronomical scale, gravity is by far the most important force is then hard to explain – it must be due to electrical neutrality of big objects.

(O and C Nuffield)

CL.4

a) State Newton's law of gravitation and explain how this law was established.

b) Use Newton's law to deduce expressions for:
i) the period T of a satellite in circular orbit of radius r about the Earth in terms of the mass m_E of the Earth and the gravitational constant G;
ii) the gravitational intensity g at this orbit in terms of the orbital radius r, the gravitational intensity g_0 at the Earth's surface, and the radius r_E of the Earth (assumed to be a uniform sphere).

c) A satellite of mass 600 kg is in a circular orbit at a height of 2000 km above the Earth's surface.
(Take the radius of the Earth to be 6400 km, and the value of g_0 to be $10\ N\,kg^{-1}$.)

Calculate the satellite's:
i) orbital speed,
ii) kinetic energy,
iii) gravitational potential energy.

d) Explain why any resistance to the forward motion of an artificial satellite in space results in an increase in its forward speed.

(Oxford)

CL.5 This question is about a theory that the inner planets of the Solar System could have grown by collecting very small planets (asteroids) which collided with them and did not escape.

i) The idea of the theory is that asteroids collide with a planet and cannot escape after the collision. The main theoretical result to be used is that the gravitational potential at the surface of a planet of mass M and radius R is $-GM/R$, where G is the gravitational constant. Show from this that the lowest escape velocity, v_{esc}, which an asteroid of mass m must have at the surface if it is to escape from the planet is given by

$$v_{esc} = \sqrt{\frac{2GM}{R}}$$

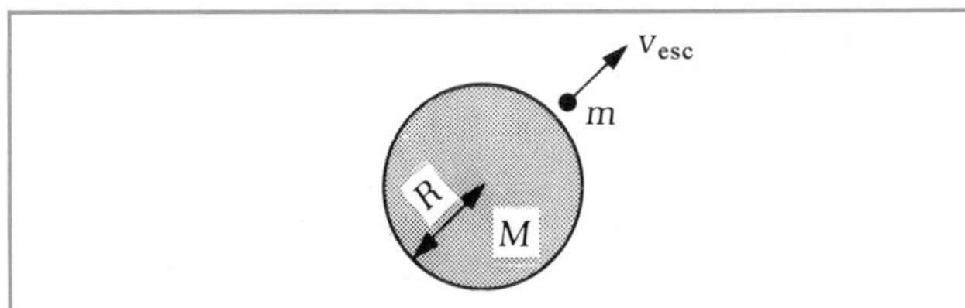

Figure CL2

ii) Show that for planets of different radii R but the same density, v_{esc} is proportional to R. Calculate the values of v_{esc} for the planet Mercury (radius $R = 2.4 \times 10^6$ m, density $= 5.4 \times 10^3$ kg m^{-3}) and for a very small planet with $R = 10^3$ m and the same density (volume of a sphere is $\frac{4}{3}\pi R^3$).

iii) If an asteroid A approaches a planet P and has a velocity v_{far} when a very long way off, the velocity of collision v_{col} will be greater than v_{far}. Explain why this is so.
Show that

$$v_{col}^2 = v_{far}^2 + v_{esc}^2.$$

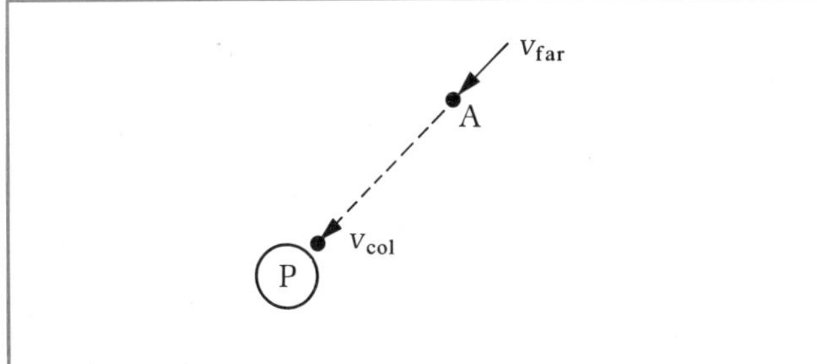

Figure CL3

iv) Asteroids might make such collisions if they are in orbits close to the planet's own orbit as shown in the diagram. Referring to Figure CL3, if P is a planet in a circular orbit, explain why the asteroid A will have a higher speed than P if they meet at the point C.

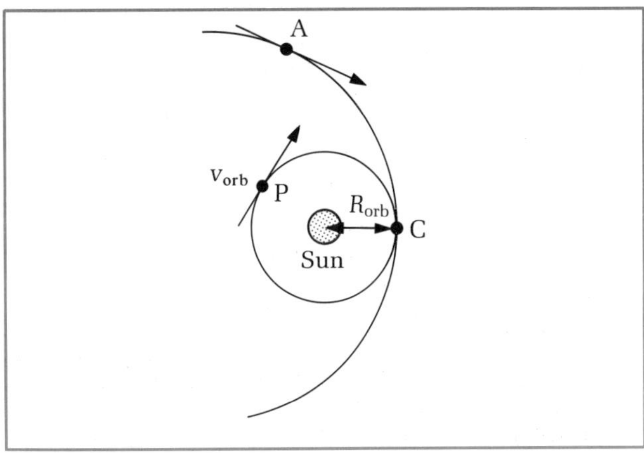

Figure CL4

v) In order to estimate collision speeds it is useful to estimate the value of the orbital speed, v_{orb}, for the planet P. Show from the force attracting the planet to the sun that

$$v_{orb} = \sqrt{\frac{GM_{sun}}{R_{orbit}}}$$

vi) In order to explore how planets might form and grow, start by assuming that asteroids and planets such as A and P approach at C with relative speeds v_{far} of about 10% of v_{orb}.
Because the collisions are inelastic, the colliding particles will usually coalesce for all values of v_{col} up to $3v_{esc}$. Given that v_{orb} for Mercury is about 5×10^4 m s^{-1}, use the ideas and results from parts (i) to (v) to show

- that an asteroid could stick to Mercury on collision,
- that an asteroid would not stick to a very small planet in the same orbit and with the same density as Mercury, but with a radius of 10^3 m.

vii) A theory of this type has been proposed to explain how planets might grow in the first place out of swarms of tiny asteroids. Discuss how the assumptions in (vi) might be altered for this purpose and outline how the theory might then give an explanation.

(O and C Nuffield)

CL.6

a) A conductor carrying a negative charge has an insulating handle. Describe how you would use it to charge (i) negatively, (ii) positively a thin spherical conducting shell which is isolated and initially uncharged. In each case explain why the procedure you describe produces the desired result.

b) For one of these cases, sketch graphs showing how the electric field strength and the electric potential vary along a line outwards from the centre of the shell, when the charging device has been removed.

c) The diagram below shows an arrangement of two point charges in air, Q being 0.30 μC.

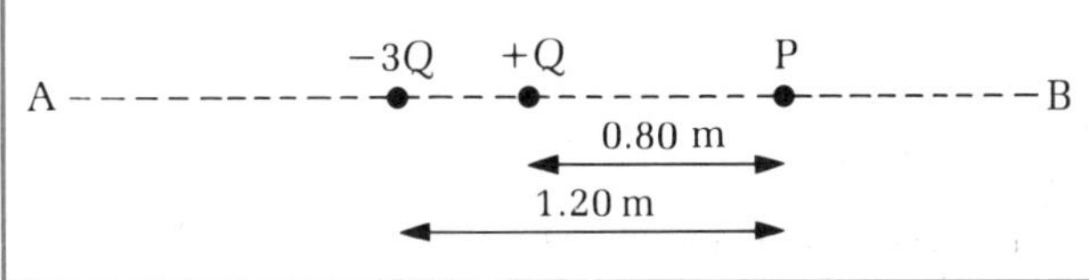

Figure CL5

i) Find the electric field strength and the electric potential at P.
ii) Find the point on AB between the two charges at which the electric potential is zero.
iii) Explain why the potential on AB to the left of the $-3Q$ charge is always negative.

Take $\varepsilon_0 = 8.8 \times 10^{-12}\,\text{F m}^{-1}$ or $= \dfrac{1}{36\pi} \times 10^{-9}\,\text{F m}^{-1}$.

(JMB)

CL.7

a) The electric field intensity E at the distance r from a point charge Q_1 is given by

$$E = \frac{Q_1}{4\pi\varepsilon_0 r^2}$$

Use this to derive the expression for the electrostatic potential energy of a point charge Q_2 at a distance r from the point charge Q_1.

b) i) Draw an arrangement of point charges which gives a point in space where the electric potential is zero and the electric field intensity is non-zero.
ii) Draw an arrangement of point charges which gives a point in space where the electric potential is non-zero and the electric field intensity is zero.
In each arrangement indicate the point concerned.

c) Two protons are initially separated by a very large distance. They are projected towards each other along the same straight line with equal speeds.
i) Find the initial speed of the protons if their minimum separation is 5.0×10^{-15} m. State any principle used in your calculation.
ii) Describe the motion of the protons after they reach their minimum separation.
iii) Calculate the potential difference required to accelerate protons in a linear electrostatic accelerator in order to achieve the speed determined in **c**)(i).

(Scottish SYS)

CL.8

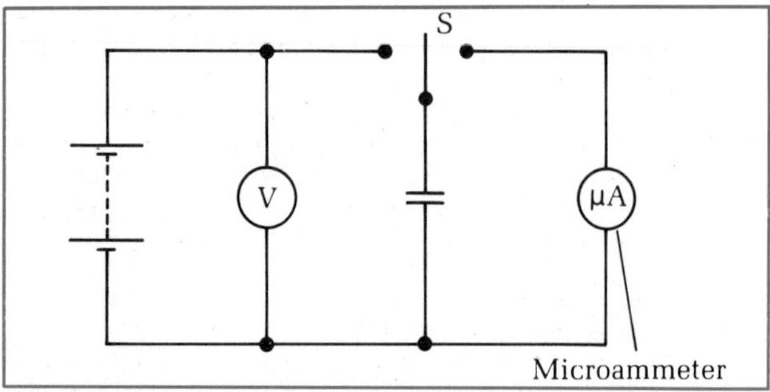

Figure CL6

In the circuit shown above, S is a vibrating reed switch and the capacitor consists of two flat metal plates parallel to each other and separated by a small air-gap. When the number of vibrations per second of S is n and the potential difference between the battery terminals is V, a steady current I is registered on the microammeter.

a) Explain this and show that $I = nCV$, where C is the capacitance of the parallel plate arrangement.

b) Describe how you would use the apparatus to determine how the capacitance C depends upon (i) the area of overlap of the plates, (ii) their separation, and show how you would use your results to demonstrate the relationships graphically.

c) Explain how you could use the measurements made in **b**) to obtain a value for the permittivity of air.

d) In the above arrangement, the microammeter records a current I when S is vibrating. A slab of dielectric having the same thickness as the air-gap is slid between the plates so that one-third of the volume is filled with dielectric. The current is now observed to be $2I$. Ignoring edge effects, calculate the relative permittivity of the dielectric.

(JMB)

CL.9

a) Define (i) electric field strength, (ii) electric potential at a point. Show that the electric field strength at a point is equal to the negative potential gradient at that point.

b) Why is the capacitance of a single isolated metal plate less than the capacitance of an arrangement consisting of the same plate with a similar earthed metal plate placed close to and parallel with it?

c) State three factors which affect the capacitance of a parallel plate capacitor and describe how the effect of each of these factors may be investigated experimentally. Compare the energy stored in a 100 μF capacitor, charged to a p.d. of 400 V, with that stored in a 12 volt 40 ampere-hour car battery.

(AEB June 82)

CL.10

a) Two small, identical, conducting spheres, each of weight 4.0×10^{-3} N, are suspended by non-conducting threads each 1.0 m long.

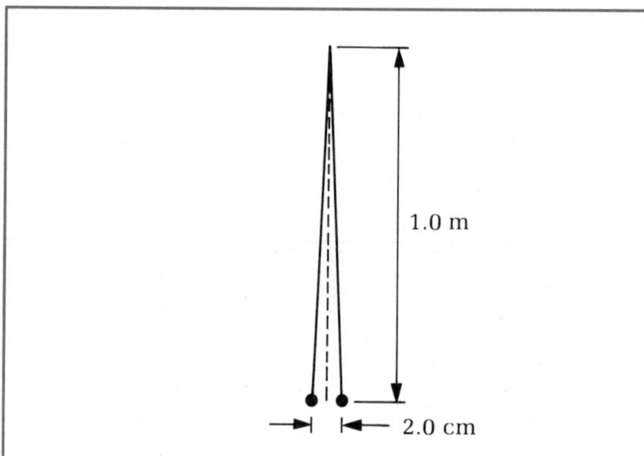

Figure CL7

The spheres are given equal negative charges and are found to remain stationary when their centres are 2.0 cm apart.
i) How would you attempt to charge each sphere to ensure that both have an equal charge?
ii) Calculate the charge on each sphere.
iii) An attempt is made to hold a ruler near the spheres to measure the separation. Explain why this could affect the separation. Describe a more satisfactory way of measuring this separation.

b) The diagram at the top of the next page, shows two positive point charges with X having four times the charge of Y.
Draw this figure and for the area represented show

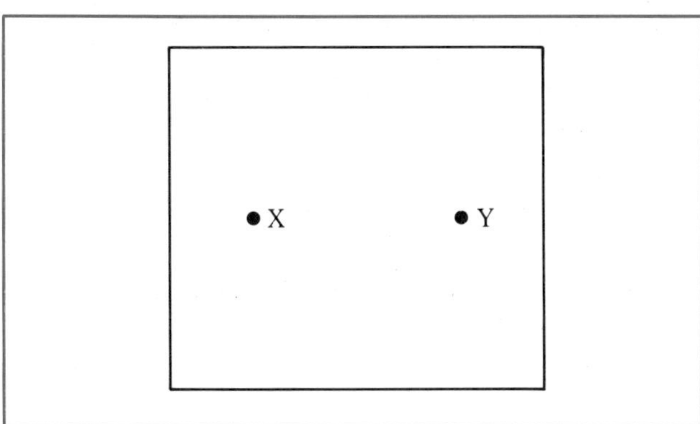

Figure CL8

i) the position between X and Y of the point where the electric field intensity is zero;
ii) lines which represent the electric field intensity;
iii) lines which represent the equipotentials.

(Scottish SYS)

CL.11

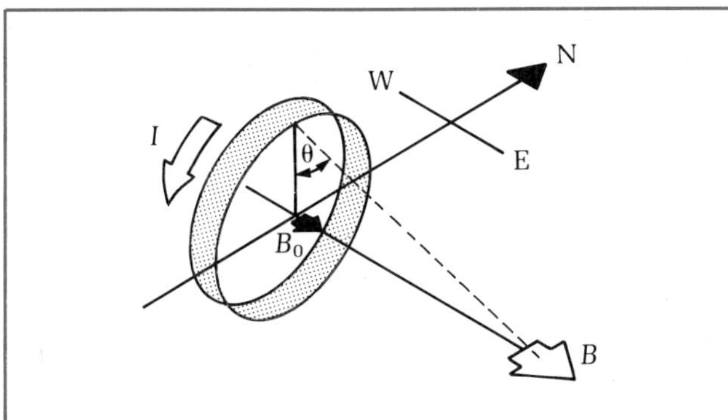

Figure CL9

a) The diagram above shows a conducting circular coil of radius 10 cm mounted in a north-south vertical plane. A current in the coil generates a magnetic flux density B_0 in an easterly direction at its centre and a magnetic flux density B at a point along the line perpendicular to the plane of the coil and passing through its centre. If θ is the angle shown in the diagram it can be shown that

$$B = B_0 \cos^3\theta.$$

Explain how you would verify this relation experimentally. Give details of the apparatus you would use and the measurements you would make.

b) Write down an expression for the force F on a straight wire of length l and which carries current I in a uniform magnetic field of flux density B.
Use this expression to derive a relation for the couple on a rectangular coil with sides of length a and b, with N turns and carrying current I, mounted with its plane parallel to a uniform magnetic field of flux density B.

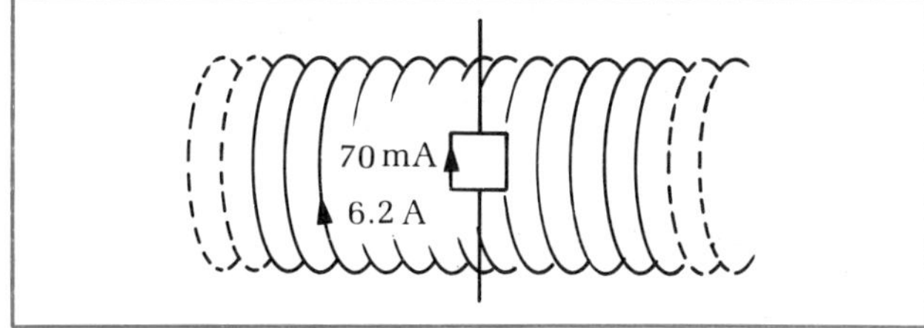

Figure CL10

A square coil of side 1.2 cm and with 20 turns of fine wire is mounted centrally inside, and with its plane parallel to the axis of, a long solenoid which has 50 turns per cm. The current in the coil is 70 mA and the current in the solenoid is 6.2 A, as shown at the bottom of the last column.
Calculate
i) the magnetic flux density in the solenoid,
ii) the couple on the square coil.
(Permeability of vacuum $= 4\pi \times 10^{-7}\,\mathrm{H\,m^{-1}}$.)

(London)

CL.12 Write down an expression for the force between two long parallel current-carrying conductors. Define the *ampere*. Show how this definition is equivalent to fixing the value of μ_0.

Draw a labelled diagram to show the essential features of a moving-coil meter. Explain how a linear scale may be achieved.

Deduce an expression for the angular current sensitivity, θ/I, of such a meter in terms of the flux density B, the number of turns N, and their mean area A, given that a torque (or couple) C is needed to rotate the coil through an angle θ of one radian.

Two such meters, P and Q, differ from each other only in that P has eight times as many turns as Q and the resistance of P is twenty times that of Q. They are connected in turn to a thermocouple of negligible resistance to detect its e.m.f. Determine the ratio of the meter deflections, θ_P/θ_Q.

Discuss which of these instruments is the more suitable for measuring a small e.m.f. from a source of **a)** low resistance, **b)** high resistance.

(Cambridge)

CL.13
a) Two long thin parallel wires of length l, separated by a distance d in a vacuum, carry equal currents I in opposite directions.
i) Sketch a diagram showing the magnetic field in a plane perpendicular to the two wires. Indicate clearly the directions of the currents, the flux lines and the force between the wires.
ii) Obtain an expression for the force F between the wires in terms of I, l, d and μ_0, the permeability of free space.
(The magnetic flux density at a point a distance r from a long straight wire carrying a current I in vacuum is $\mu_0 I/2\pi r$.)

b) i) A rectangular galvanometer coil of 200 turns, each 12 mm wide by 15 mm long, is situated in a uniform horizontal magnetic field of flux density 0.10 T. In the equilibrium position, when the current in the coil is zero, the plane of the coil is parallel to the direction of the uniform field. A current of 10 μA passed through the coil causes it to deflect 40° from the equilibrium position. Calculate the restoring torque exerted on the coil by the control spring of the galvanometer. Hence find the restoring torque per degree of deflection.
ii) In many galvanometers, the direction of the magnetic field is radial to the axis of rotation. How is such a field achieved? If a radial field, with a flux density of 0.10 T at the vertical conductors of the coil, replaced the uniform field in the galvanometer described in **b)**(i) above, calculate the angle of deflection from the

equilibrium position which would occur for a current of 10 μA, all other factors remaining the same.

iii) Sketch a graph of deflection θ against current I for the galvanometer with the radial field. Explain how the current sensitivity (the deflection per unit current) is obtained from this graph.

c) Explain *two* different ways in which the design of a given galvanometer with a radial field could be altered to obtain higher current sensitivity.

d) When deflected the coil of a galvanometer comes to rest with little or no oscillation. During movement, damping forces act on the coil. State what is meant by *damping forces* and suggest *two* possible sources of damping in this instance.

(N. Ireland)

CL.14

a) i) Sketch a graph which shows the relationship between the flux density and the strength of the magnetizing field in a material such as steel, over a complete magnetizing cycle, assuming that the specimen has been through several such cycles.

ii) Referring to the graph, explain the meaning of the terms *remanence* and *coercivity*.

iii) Describe *two* important differences between the graph you have sketched for steel and the one you would expect for soft iron.

iv) State, with reasons, whether steel or soft iron would be the more suitable material to use for the core of a transformer.

b) The diagram shows two pole pieces A and B between which there is a magnetic field maintained by a current flowing in coils (not shown).

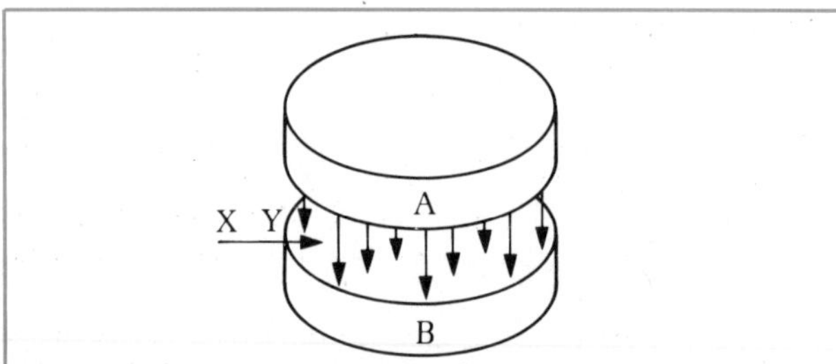

Figure CL11

i) Describe briefly how you would use a Hall probe to investigate whether the field between the poles is uniform.

ii) Outline how a search coil could be used to measure the magnetic flux density. Show clearly how you would use the observations to obtain a value for the flux density.

iii) A positively charged ion enters the region between the pole pieces travelling in the direction XY. Assuming that the field is uniform, describe and explain the path followed by the ion, first as it moves through the field and then after it has emerged from the field.

(AEB Nov 84)

CL.15 Describe, with the aid of a diagram, the construction of a ballistic galvanometer.

A search coil consists of 10 turns and is of area 5 cm². The coil is connected to a ballistic galvanometer, and the circuit containing the coil and the galvanometer has resistance 100 Ω. The coil is originally placed between the pole pieces of a permanent magnet, with its plane normal to the field lines, and when it is withdrawn quickly, a deflexion of 5 divisions on the galvanometer scale is noted. It is known that 1 division corresponds to a circulation of 5 μC. Find the flux density between the poles of the magnet, explaining each step in your argument.

Describe, with the aid of a circuit diagram, how you would calibrate a Hall probe to measure magnetic flux density, referring where appropriate to the earlier part of your answer.

(WJEC)

CL.16

a) Explain the terms *magnetic flux, magnetic flux density*. State the units in which these quantities are measured.

b) Explain how you would use a simple form of current balance to measure a current of about 5 A. Give a diagram of the apparatus, list the readings you would need to take, and show how the result is calculated. Discuss the results obtained were your apparatus used with an alternating current of frequency 50 Hz.

c) A 5.0 μF capacitor is charged to a potential difference of 100 V and discharged through a ballistic galvanometer of coil resistance 3 Ω. An undamped ballistic throw of 85 mm is observed.
The galvanometer is then connected to a search coil with 600 turns of mean area 4.0×10^{-4} m² and resistance 30 Ω. The search coil is placed with its plane perpendicular to the field between the poles of a permanent horseshoe magnet, and is then quickly withdrawn from the field; the undamped ballistic throw is observed to be 62 mm.
Calculate:

i) the charge sensitivity of the galvanometer;

ii) the magnetic flux density in the field.

(Oxford)

CL.17

a) A long vertical straight wire carries a steady electric current supplied by a battery of cells. Sketch the magnetic flux pattern in a horizontal plane around the wire

i) when there is no other magnetic flux, and

ii) when there is also a uniform horizontal magnetic flux.

In case ii) indicate the direction of any force experienced by the wire.

If the wire were allowed to move in the direction of this force would the current in the wire change? Explain your answer.

b) An expression for the magnetic flux density at a distance r from a straight wire carrying a current I is $\frac{\mu_0 I}{2\pi r}$, where $\mu_0 = 4\pi \times 10^{-7}$ Hm⁻¹. In the situation ii) above it is found that when the horizontal flux density is 2.0×10^{-5} T there is no resultant flux density at a point 15 cm from the wire. What is the value of the current in the wire?

What would be the new value of the uniform horizontal flux density if the point of no resultant flux density were to be moved to 5.0 cm from the wire?

c)

The rectangular coil shown at the top of the next page is in a vertical plane and is connected to the Y plates of an oscilloscope as shown. The coil is moved in a uniform horizontal magnetic flux

i) by taking it sideways in a vertical plane across the flux, without rotation, and

ii) by rotating it about the vertical axis AB at 10 rev s⁻¹.

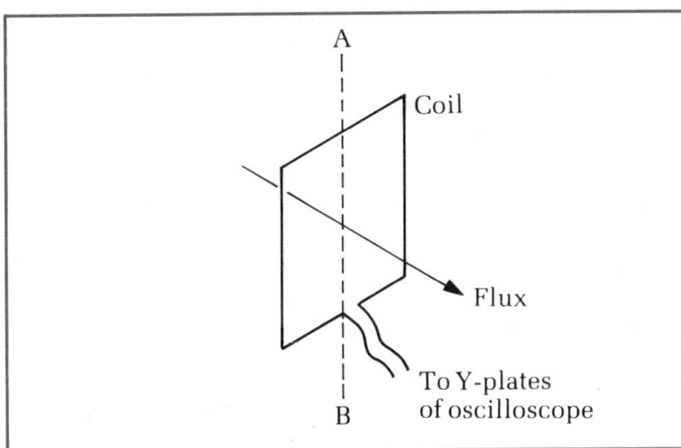

Figure CL12

In each case describe and explain the appearance of the oscilloscope trace if a time base frequency of 5 Hz were applied to the X plates.

What would happen to the trace if the frequency of rotation of the coil were doubled to 20 rev s^{-1}?

(London)

CL.18

In the diagram a uniform magnetic field of 2.0 T exists, normal to the paper, in the shaded region. Outside this region the magnetic field is zero. X, Y and Z are three loops of wire.

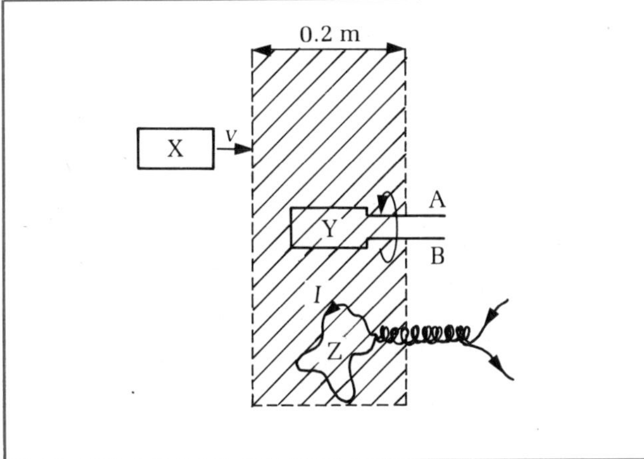

Figure CL13

a) Loop X is a rigid rectangle in the plane of the paper with dimensions 50 × 100 mm and resistance 0.5 Ω. It is pulled through the field at a constant velocity of 20 mm s^{-1} as shown. Its leading edge enters the field at time $t=0$. Sketch graphs, giving explanations, of how the following vary with time from $t=0$ to $t=20$ s:

i) the magnetic flux linking loop X,
ii) the induced e.m.f. around the loop,
iii) the current in the loop.

Neglect any effect of self-inductance.

b) Loop Y of the same dimensions as X is mounted on a shaft within the field region. It is rotated at a constant frequency f to provide a source of alternating e.m.f. At $t=0$ the loop is in the plane of the paper.

i) Give an expression for the e.m.f. between the terminals A and B at any instant.
ii) Calculate the frequency of rotation required to give an output of 3 V r.m.s.

c) Loop Z, made of a fixed length 0.3 m of flexible wire, rests on a smooth surface in the plane of the paper within the field region. A direct current I in the wire causes it to take up a circular shape.

i) Explain this observation.
ii) What would you expect to happen if the current I was reversed in direction?

(Oxford and Cambridge)

CL.19 State the *laws of electromagnetic induction*. Describe, and explain in terms of these laws, what happens when a powerful electromagnet is switched off, with special reference to the e.m.f.s and the current flow.

What is the origin of the back e.m.f. in an electric motor? Describe and explain what happens when a simple permanent magnet d.c. motor, connected to a source of constant e.m.f., is switched on

a) with very little friction or mechanical loading,
b) with considerable mechanical loading.

Illustrate your answers with *four* sketch graphs, using similar time scales, to show, in each case, how (i) the current, (ii) the speed, vary with time.

A toy car, driven by such a motor, runs steadily on level ground at 0.5 m s^{-1} and takes 1 A from a 12 V supply. When it runs into an obstruction and the wheels stop, the current taken is 3 A. What current will it take when running uphill at 0.4 m s^{-1}?

Steam engines are often fitted with a device to limit their top speed. Why is no equivalent device generally necessary for electric motors?

(Cambridge)

CL.20

a) Describe an experimental method of measuring the magnetic induction between the poles of a permanent magnet. You should include,

i) a sketch of the apparatus;
ii) an outline of the physical principles of the measurement.

b) In the figure the rod XY is of length 20 mm and is moved at a constant velocity of 6.0 m s^{-1} in a direction perpendicular to a magnetic field of induction 1.5 T. The ends of the stationary conductors are joined to complete a circuit of total resistance 4.0 Ω.

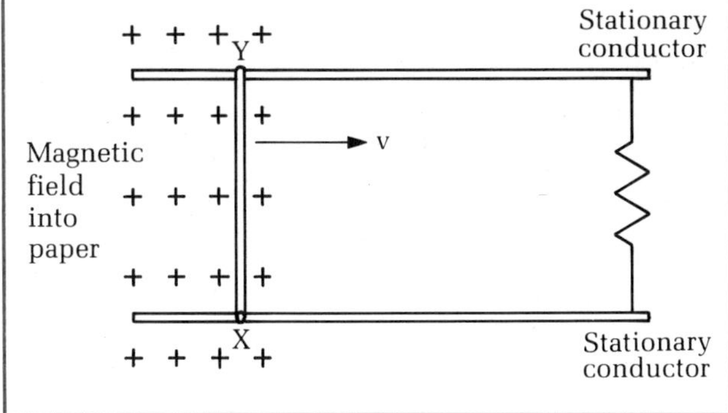

Figure CL14

i) Explain why, in the absence of friction, a constant external force is required to keep the rod moving with constant velocity.
ii) Calculate the magnitude of this external force.
iii) Calculate the magnitude and direction of the induced e.m.f. between the ends of the rod XY.
iv) Show that the mechanical power required to keep the rod moving with constant speed equals the electrical power dissipation in the circuit.

(Scottish SYS)

SECTION D

Waves

Oscillations · Wave motion · Sound · Physical optics · Optical instruments · Electromagnetic waves

Figure D1 *The Tidal Bore of the Severn. At mid-spring or mid-autumn near New or Full Moon, tides are particularly strong. In the Severn Estuary, the advancing tide is funnelled into the narrow river to create a wall of water or 'bore' up to 5m high that travels upstream for as far as 30 km.*

Most branches of physics involve the study of some aspect of wave motion. For example, materials scientists use X-rays to determine the structure of materials. Electrical engineers need to study wave motion to understand alternating current theory and radio communications. The link between force and motion is used in the study of oscillations to understand the effect of vibrations on bodies. Where oscillating bodies interact, energy can be transferred from one to another, giving rise to wave motion if many oscillating bodies are linked together.

Different forms of waves include electromagnetic waves, sound waves, water waves and mechanical waves; each waveform is due to a different form of interaction but they all share common properties. The study of wave motion involves describing and understanding these common properties, and showing how they apply to different waveforms.

16

Oscillations

16.1 Describing oscillations

Have you ever tried jumping up and down on a trampoline? With practice, you can make yourself bob up and down on the trampoline sheet. The trampoline sheet is said to **oscillate** when it moves up and down. A child on a playground swing moving to-and-fro may also be said to oscillate. Any object which moves to-and-fro repeatedly is said to oscillate. Each to-and-fro movement is **one complete cycle** of oscillation. The child on the swing would move from one side to the other side and back again in one complete cycle.

The time period T of an oscillating object is the time it takes to go through one complete cycle of oscillation. The unit of time period is the **second**.

The frequency f of the oscillations is the number of complete cycles per second made by the oscillating object. The unit of frequency is the **hertz** (Hz), equal to one cycle per second.

$$f=\frac{1}{T}$$

If a pendulum takes 25 s to complete 10 cycles exactly, then the time period is 2.5 s and the frequency is 0.4 Hz (=1/2.5).

There are many examples of oscillating motion in everyday life. For example, when a car travels over a bump, the suspension springs bounce the car up and down for a short time afterwards. Strike a piano key and a hammer inside the piano is made to hit a wire under tension. The wire vibrates, and every point on it oscillates to-and-fro making the air vibrate, so sound waves are created. Electronic oscillators are widely used. Microcomputers have an internal clock which is driven by an electronic oscillator circuit. From the moment your alarm clock, electronic or not, goes off you are surrounded by oscillations of one kind or another.

***Simple harmonic motion* (s.h.m.)** is a special kind of oscillating motion. An example of s.h.m. is the motion of a weight bobbing up and down on a spring. A pendulum swinging to-and-fro is another example. The **displacement** of the oscillating object from its equilibrium position changes with time. The object passes from one side of equilibrium to the other side and back each cycle. Its motion is s.h.m. if its displacement changes **sinusoidally** with time. Sinusoidal changes are described by sine waves, although they do not need to start with zero displacement.

A direct way of investigating the oscillating motion of a pendulum is to attach a fine paintbrush to the bob. A heavy bob is best because it keeps oscillating longer than a light bob. Figure 16.2 shows such an arrangement. When the bob oscillates, the paintbrush traces a fine line on the paper beneath. The bob is left to oscillate, and as it does so the paper is pulled steadily out from under the pendulum. Provided the paper is pulled out in a direction at right angles to the pendulum oscillations, the brush

Figure 16.1 *Surrounded by oscillations*

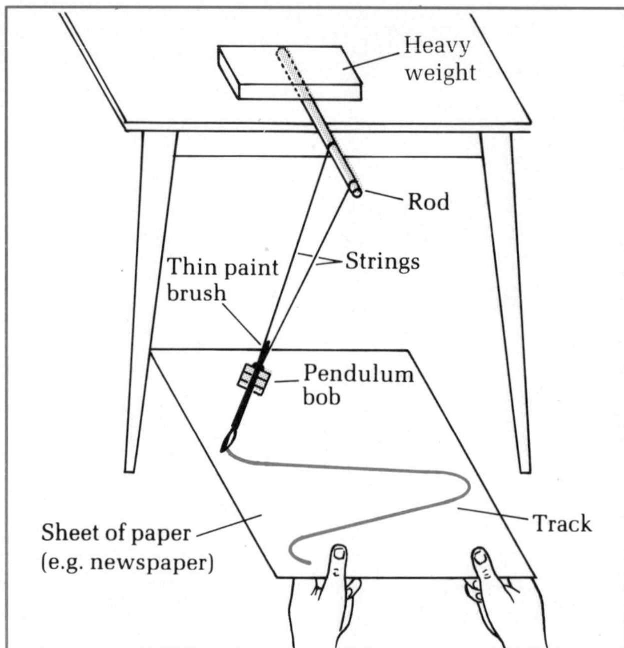

Figure 16.2 *Investigating oscillations*

ought to trace out a sine wave on the paper. Try it!

A frictionless pendulum would give a displacement-time graph as shown in Figure 16.5 (on page 220) if the pendulum was started by pulling it to one side then releasing it. Because the curve starts at one side of zero displacement, then strictly it is a cosine wave. The term sinusoidal is still used though, whatever its starting point.

The **amplitude** of the oscillations is the maximum displacement from the equilibrium position. During one complete cycle, the displacement changes from maximum on one side of equilibrium, to zero, to maximum on the other side of equilibrium, back to zero, then to maximum on the first side.

The **angular frequency** ω of the oscillations is defined as $\frac{2\pi}{T}$. Its unit is rad s^{-1}. Just as the time period T is a constant for any given object moving with s.h.m., so too is the angular frequency. We need to know the angular frequency if we wish to calculate the displacement at any given time during the motion, as explained below.

Linking s.h.m. and circular motion

Using the arrangement shown in Figure 16.3, a turntable with a ball stuck on its rim rotates underneath a simple pendulum. The pendulum is set oscillating by releasing it from rest directly over the rim when the ball passes underneath it. The turntable speed is adjusted so that the rotation frequency equals the pendulum frequency. A light beam from a projector is used to cast a shadow of the pendulum bob and the ball onto a screen. With careful adjustment of the speed of the turntable, it ought to be possible to release the pendulum so that its shadow moves exactly in phase with the ball's shadow, so the two shadows have the same motion. Now the motion of the ball's shadow is the component of the ball's motion parallel to the screen, which is sinusoidal. See Figure 16.3. Therefore the pendulum's motion is also sinusoidal.

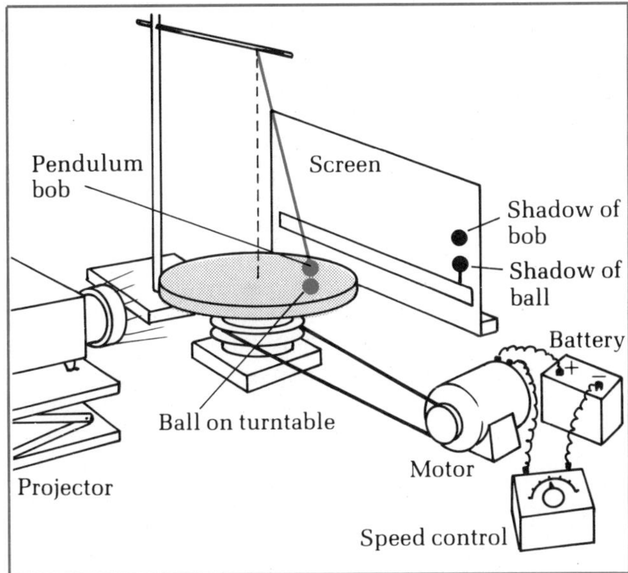

Figure 16.3 *Comparing s.h.m. with circular motion*

Now we know that the rotating ball moves round a circle of radius r with a time period T. Suppose at time $t=0$ the ball passes through a fixed point F on the circle, as shown in Figure 16.4. After time t, the line from the circle centre O to the ball must have turned through an angle θ given by

$$\theta = \frac{2\pi}{T}t = \omega t$$

where $\omega = 2\pi/T$

So the displacement of the ball along OF from O is therefore given by $r\cos\theta$ which equals $r\cos\omega t$.

Because the pendulum bob is exactly in phase with the ball, its displacement from equilibrium, s is given by the same expression. So the displacement of the bob is given by

Displacement $s = r\cos\omega t$ (Assuming it is released as above.)

We can make further use of the link between s.h.m. and circular motion to consider the pendulum's velocity. The ball moves round the circle with a steady speed equal to

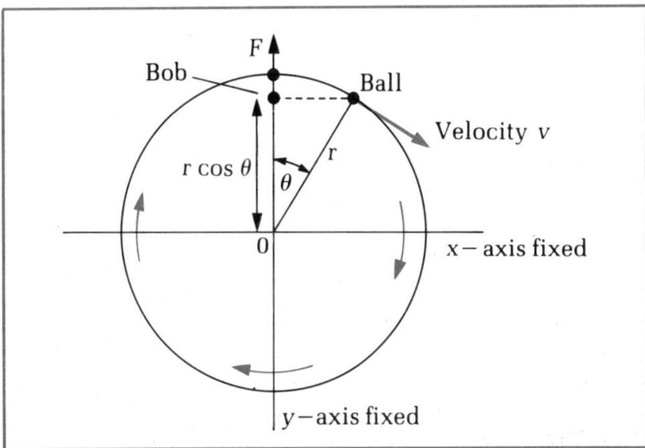

Figure 16.4 *Top view of turntable of Figure 16.3*

$\omega r(=2\pi r/T)$. So at the point on the circle shown in Figure 16.4, its velocity is ωr along the tangent to the circle. Hence the component of its velocity parallel to the line OF is $-\omega r\sin\omega t$ (negative sign (−) indicates opposite direction to the displacement). So the ball's shadow on the screen has a velocity given by $-\omega r\sin\omega t$.

Since the pendulum bob keeps up with the shadow, its velocity is therefore given by

velocity $v = -\omega r\sin\omega t$ (Assuming it is released as above).

Calculus 'buffs' should note that $\frac{ds}{dt}\left(=\frac{d}{dt}(r\cos\omega t)\right)$ gives v, to be expected since velocity equals rate of change of displacement $\left(\frac{ds}{dt}\right)$.

An equation for speed in terms of displacement is often useful, and we can now develop such an equation from the equations above. (We shall need to use the rule $\sin^2\theta+\cos^2\theta=1$.)

Since displacement $s=r\cos\omega t$, then $\omega^2s^2=\omega^2r^2\cos^2\omega t$
Also velocity $v=-\omega r\sin\omega t$, so $v^2=\omega^2r^2\sin^2\omega t$
Hence $v^2+\omega^2s^2=\omega^2r^2\cos^2\omega t+\omega^2r^2\sin^2\omega t$
$=\omega^2r^2(\cos^2\omega t+\sin^2\omega t)$
$=\omega^2r^2$ because $\cos^2\omega t+\sin^2\omega t=1$
giving $v^2=\omega^2r^2-\omega^2s^2=\omega^2(r^2-s^2)$
and finally $v=\pm\omega\sqrt{(r^2-s^2)}$.

This useful equation enables the speed to be calculated for any value of displacement. The calculation gives a positive value and a negative value (the two square roots) which correspond to the two possible directions.

Worked example A small mass is hung from the lower end of a vertical spring. The upper end of the spring is fixed. When the mass is displaced and then released, it oscillates with s.h.m. with an amplitude of 40 mm and a time period of 0.35 s. Calculate the speed when the displacement is **a)** zero, **b)** +20 mm.

Solution First, calculate the angular frequency $\omega\ (=2\pi/T)$.
$\omega = 2\pi/0.35 = 17.95\ \text{rad s}^{-1}$.

a) For the speed at zero displacement, $s=0$.
Using $v^2=\omega^2(r^2-s^2)$ gives
speed $v=\omega r=40\times10^{-3}\times17.95=0.718\ \text{m s}^{-1}$.

b) For the speed at displacement $s = +20$ mm,
$v^2 = \omega(r^2 - s^2) = 17.95^2 \times (0.04^2 - 0.02^2)\ \mathrm{m^2\ s^{-2}}$
which gives $v = 0.622\ \mathrm{m\ s^{-1}}$.

Note that the speed was asked for, not the velocity, so directions do not need to be considered.

16.2 Principles of simple harmonic motion

We have seen in the preceding section that simple harmonic motion can be described using sine wave graphs, although not necessarily starting at zero displacement. Now we want to try to understand s.h.m., rather than simply describing it. Let's return to the frictionless pendulum which gave a displacement-time curve as in Figure 16.5. The pendulum was released from one side of equilibrium with zero initial speed. It reached maximum speed passing through the middle, then slowed down as it moved up the other side. At the highest point on the other side its speed was zero for an instant, then it swung back to the centre and returned to the same side as it started from, ready for a further cycle. Its velocity when it swings back is in the opposite direction to when it swings forward. Figure 16.6 shows how the velocity varies during the cycle. Comparing with Figure 16.5, the velocity is zero when the displacement is greatest, and greatest when the displacement is zero.

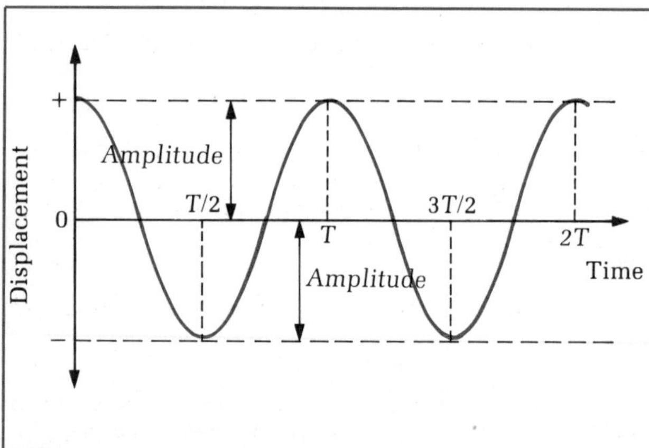

Figure 16.5 *Displacement-time curve for an object moving with s.h.m.*

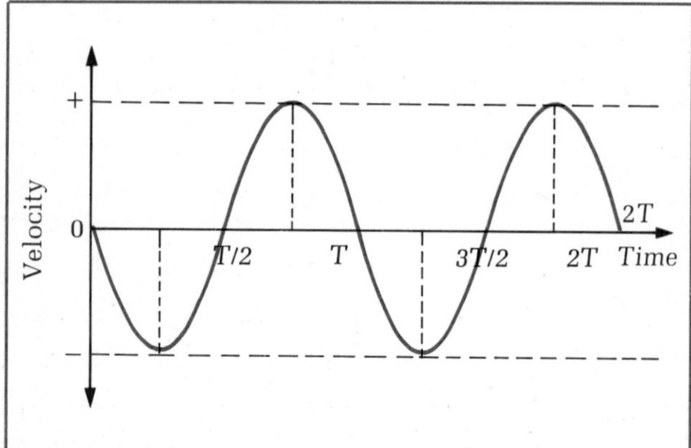

Figure 16.6 *Velocity-time curve for an object moving with s.h.m.*

The **velocity** is given by the gradient of the displacement-time graph. The gradient of the displacement-time curve of Figure 16.5 is zero initially, but then becomes more and more negative for $\frac{1}{4}$ cycle. Then, the gradient becomes less and less negative over the next quarter cycle, becoming zero at half a cycle from the start. This, of course, is the way that the velocity changes. The direction of the velocity is indicated by the sign of the gradient. If we choose the initial displacement as the positive direction, then the velocity during the first half cycle is in the negative direction. So, given that the displacement-time curve is a cosine wave (like Figure 16.5), then the velocity-time curve is an inverted sine wave (like Figure 16.6).

The **acceleration** is given by the gradient of the velocity-time graph. The gradient of Figure 16.6 therefore gives the pendulum's acceleration. The gradient of Figure 16.6 is negative at first, becoming zero at quarter-cycle, then reaching a maximum positive value at half-cycle. Thus the acceleration was initially in the negative direction, becoming zero at quarter-cycle, then changing direction when the object passed through equilibrium. Therefore the acceleration varies with time as shown by Figure 16.7. The acceleration curve is an inverted cosine wave (i.e. an inverted displacement-time curve). Whenever the displacement is positive, the acceleration is negative; whenever the displacement is negative, the acceleration is positive. Even more important, the acceleration is proportional to the displacement because the acceleration-time curve has its maximum values at the same times as the displacement-time curve. The link between acceleration and displacement is used to define what we mean by simple harmonic motion.

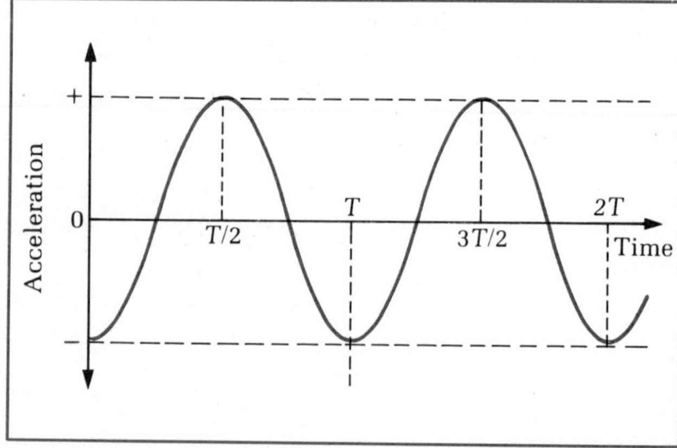

Figure 16.7 *Acceleration-time curve for an object moving with s.h.m.*

Simple harmonic motion of an object is oscillating motion where the acceleration of the object is
a) always directed towards equilibrium,
b) proportional to the displacement from equilibrium.

During s.h.m., when the object is on one side of equilibrium, the acceleration is towards the other side. When the object moves from one side to the other, the acceleration changes direction. So the acceleration is always directed towards equilibrium. We can make an equation to represent the link stated above between acceleration and displacement. The equation would be in the form 'acceleration = − constant × displacement'. The negative sign shows that the acceleration is in the opposite direction to the displacement direction (and hence towards the middle because displacement is always measured outwards from the middle). The constant shows that the acceleration is proportional to the displacement. A large value for the constant of proportionality would give large acceleration values, so giving rapid oscillations. Clearly, the constant of proportionality is linked to the frequency of the oscillations.

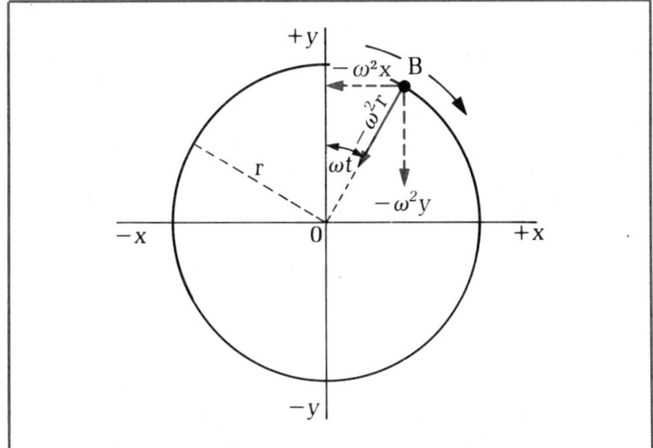

Figure 16.8 *Resolving centripetal acceleration*

To see what the link is, let's consider the circle of Figure 16.8. Point B moves round the circle at angular speed ω since radius OB turns through 2π radians in time T. Now the acceleration of B is centripetal (i.e. towards the centre), and is given by $a = -\omega^2 r$. The y-component of this acceleration is $a \cos \omega t$ which equals $-\omega^2 r \cos \omega t$ which in turn equals $-\omega^2 \times$ (the y-displacement). Since the y-displacement varies sinusoidally, we see that the constant of proportionality linking acceleration and displacement for s.h.m. is in fact ω^2.

Equation for s.h.m.

$$a = -\omega^2 s$$

where a = acceleration,
s = displacement,
$\omega = 2\pi/T$ = angular frequency.

Forces and s.h.m.

The motion of an object is determined by the forces acting on it, so when an object moves with s.h.m., the forces on it are the cause of that motion. Simple harmonic motion is defined as oscillating motion where the acceleration is given by $a = -\omega^2 s$.

The resultant force F on any object causes an acceleration given by $F = ma$, so for s.h.m., the resultant force must be given by

$$F = -m\omega^2 s$$

where F = resultant force,
m = mass.

The forces acting on the object must therefore give a resultant force F which is

a) always directed towards equilibrium (i.e. so as to try to restore the object to equilibrium),

b) always proportional to the displacement.

If we want to know whether or not an object will oscillate with s.h.m., it is only necessary to consider the forces acting. If the forces always give a resultant which obeys the two conditions above, then the object must move with s.h.m. when it oscillates.

16.3 Oscillations of loaded springs

When a spiral spring is stretched, the tension in it increases. The tension is proportional to the extension (i.e. amount of stretching) of the spring. This is true for all spiral springs, provided they are not overstretched. This fact was first discovered by Robert Hooke in 1676, and so it is referred to as **Hooke's law**. For example, if a spring requires a pull of 1.0 N to stretch it by 0.05 m, then a pull of 2.0 N will stretch it by 0.10 m. The tension T in this spring can be calculated from the equation $T = 20x$, where x is the extension of the spring. More generally, the tension in any spring is given by an equation of the form,

$$T = kx$$

where k is called the **spring constant**,
T = tension,
x = extension.

Because the equation is true for all springs, then any 'springy' system has a restoring force which is proportional to its displacement from equilibrium. So springy systems oscillate with s.h.m.

Investigating the oscillations of loaded springs

A trolley is held at rest by two stretched springs as shown in Figure 16.9. When the trolley is displaced from equilibrium then released, it oscillates about equilibrium.

The combined effect of the two springs is to try to restore the trolley to its equilibrium position. Each time the trolley is pulled back to equilibrium, it gains momentum which takes it through equilibrium to the other side. The springs then act to pull it back once again, but once again the trolley overshoots to the other side of equilibrium. In this way, the trolley is made to oscillate about equilibrium.

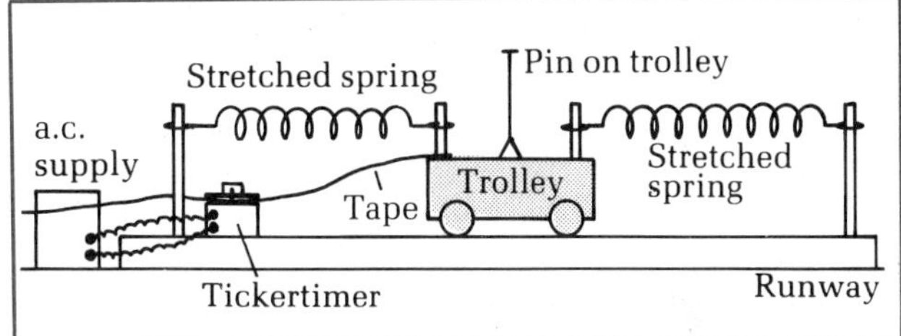

Figure 16.9 *Investigating oscillations*

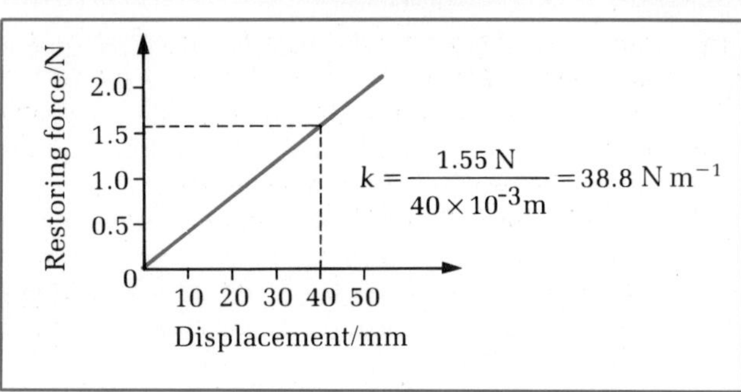

Figure 16.10 *Force v displacement graphs*

To investigate the restoring force, a spring balance may be used to pull the trolley from equilibrium. If the trolley is pulled to a constant displacement, the spring balance gives a reading equal to the restoring force at that displacement, so the restoring force at different displacement values can be measured. Then the spring constant k for the system can be determined. The best way to do this is to plot a graph of restoring force against displacement, as in Figure 16.10. The spring constant k is equal to the gradient of the graph.

To show directly that the motion is s.h.m., a tickertape can be attached to the trolley as shown in Figure 16.9. The tickertape is passed through a tickertimer. The trolley is displaced and held near the tickertimer end of the runway with both springs taut. The tickertimer is switched on, and the trolley is released. The trolley pulls the tape through the tickertimer until it reaches the maximum displacement on the other side. Then the trolley returns without returning the tape. The tape records only the first half-cycle of the motion, but that is nevertheless useful. Each dot on the tape corresponds to 1/50 second (assuming the tickertimer operates at 50 dots/second). So by measuring the distance of each dot from the midpoint of the dotted section of the tape, a graph of displacement against time may be plotted for the first half-cycle. The midpoint between the dots at either end corresponds to the equilibrium position of the trolley. If the 'dotted' section of the tape is short enough, it can be stuck alongside the displacement axis of the graph, as in Figure 16.11. The

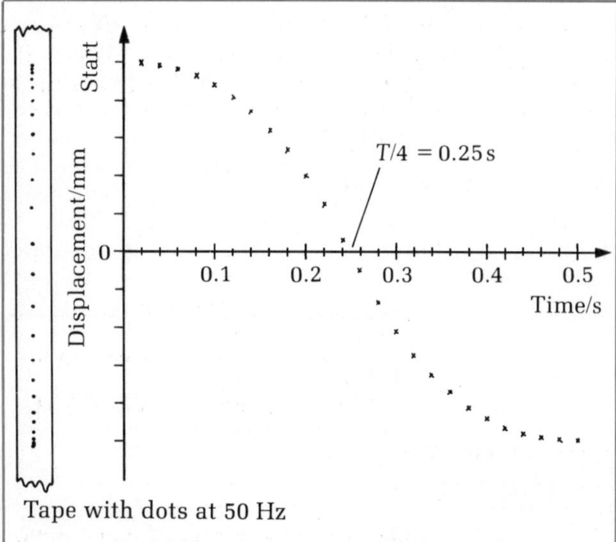

Figure 16.11 *Displacement-time curve from a tickertape*

time taken for the first half cycle to be completed can be read from the graph to give the time period T of the motion. The displacement-time curve is a cosine wave with amplitude equal to the initial displacement.

If the apparatus is then dismantled, the mass M of the trolley can then be measured directly. Theory shows that the time period T is given by $T = 2\pi\sqrt{\frac{M}{k}}$ so the time period may be calculated and compared with the measured value.

Theory of loaded spring oscillations

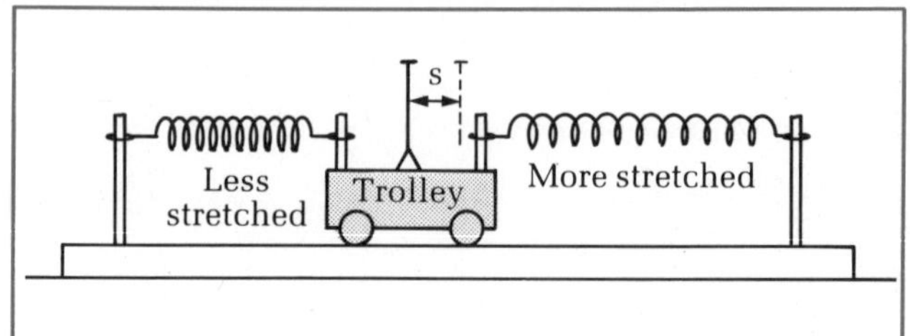

Figure 16.12 *Restoring forces*

Consider the trolley in Figure 16.12 at some point during the motion. Let its displacement from equilibrium be s at that point. One of the two springs has been extended by s, and the other has been shortened by s. So compared with equilibrium, one spring has extra tension k_1s; the other spring has its tension reduced by k_2s. The spring constants of the individual springs are k_1 and k_2 respectively. The extra tension from one spring combines with the reduced tension from the other to give a restoring force of $k_1s + k_2s$. So the restoring force may be written

$$F = -ks$$

where $k = k_1 + k_2$ and the $-$ sign indicates it acts towards equilibrium.

Thus the acceleration of the trolley a is given by

$$a = -\frac{k}{M}s$$

This is the s.h.m. equation in disguise! If we write ω^2 in place of $\frac{k}{M}$, the equation becomes recognizable as

$$a = -\omega^2 s$$

where $\omega^2 = \frac{k}{M}$.

So we have proved that the oscillations are s.h.m., and from the proof, we can obtain a useful formula for the time period T.

$$T = \frac{2\pi}{\omega} = 2\pi\sqrt{\frac{M}{k}}$$

since $\omega^2 = \frac{k}{M}$.

The same equation for T applies to any springy system. The simplest system is a vertical spring supported from a fixed point with a mass M hanging from its lower end, as

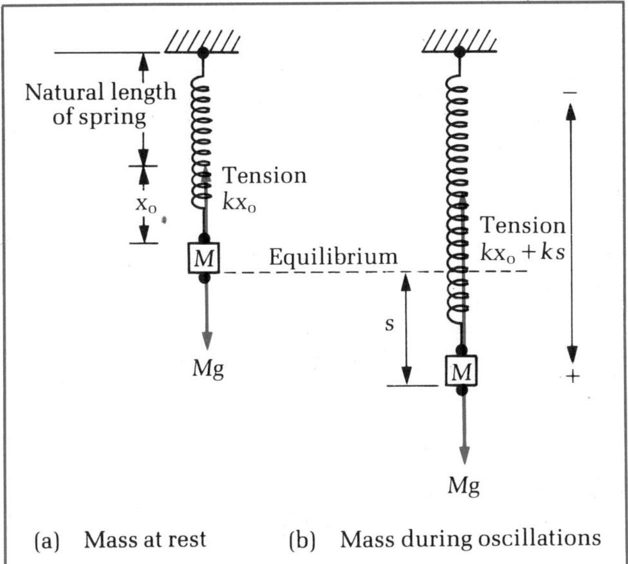

Figure 16.13 *Loaded spring*

shown in Figure 16.13. In equilibrium (i.e. at rest), the spring extension x_0 is given by

$$kx_0 = Mg$$

because the tension balances the weight; k is the spring constant.

During oscillations when M is at displacement s from equilibrium, the spring extension is $(s+x_0)$. The tension pulling on M is $k(s+x_0)$, so the resultant force on M (= tension − weight) is

$$k(s+x_0) - Mg = ks \text{ upwards}$$

Therefore, the restoring (i.e. resultant) force is given by $F = -ks$ once again.

Worked example A spiral spring is hung vertically from a fixed point. Its length when unloaded is 240 mm. When a mass of 0.15 kg is hung from its lower end, its length becomes 300 mm with the mass at rest. Calculate **a)** the spring constant of the spring **b)** the time period of the oscillations when the mass is pulled down by a small amount and then released.
Assume $g = 10\ \text{m s}^{-2}$.
Solution

a) In equilibrium with the mass on the lower end, the extension of the spring is 60 mm.

Hence $k = \dfrac{0.05 \times 10}{60 \times 10^{-3}} = 8.67\ \text{N m}^{-1}$

(Using $kx_0 = Mg$.)

b) Using $T = 2\pi\sqrt{\dfrac{M}{k}}$ gives $T = 2\pi\sqrt{\dfrac{0.15}{8.67}} = 0.826$ s.

Measuring g is possible, using a spiral spring, as top-watch, a metre rule and several different masses. The spring is hung vertically from a fixed point, and its length when unloaded is measured. Then one of the masses is hung from its lower end. With the mass at rest, the new length of the spring is measured. The equilibrium extension x_0 can then be calculated. The mass is then displaced slightly from equilibrium and then released. By timing ten complete cycles, the time period of its oscillations can be determined. The procedure is then repeated for the other masses.

The theory is based upon two equations.

a) At equilibrium, for mass M, $Mg = kx_0$ so $M/k = x_0/g$.

b) In oscillation, its time period is given by $T = 2\pi\sqrt{\dfrac{M}{k}}$.

Therefore $T^2 = 4\pi^2\dfrac{M}{k} = 4\pi^2\dfrac{x_0}{g}$, so a graph of T^2 against x_0 has a gradient of $4\pi^2/g$.

By using the results to plot such a graph, g can be measured.

16.4 The simple pendulum

To understand why a simple pendulum oscillates with s.h.m., consider a snapshot of a simple pendulum which is oscillating. Figure 16.14 shows the pendulum at a point somewhere between zero and maximum displacement in the positive direction. The bob is acted on by two forces which are its weight mg and the tension in the string T.

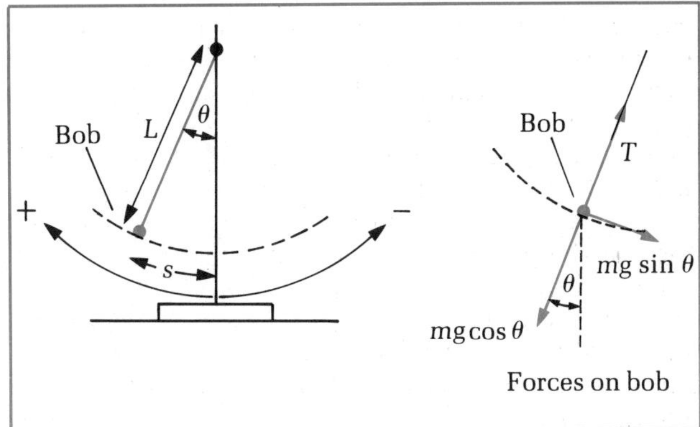

Figure 16.14 *The simple pendulum*

The bob moves along a circular arc, and the tension is always at right angles to the arc. So the tension cannot accelerate the bob along its path. The component of the weight acting along the arc provides the force causing the motion. In more detail, resolve the forces parallel and perpendicular to the path.

Perpendicular to the path:

$$T - mg\cos\theta = m\frac{v^2}{L}$$

where v = speed of the bob,
$\dfrac{v^2}{L}$ = its centripetal acceleration,
θ = angle of string to vertical.

Parallel to the path:

$$-mg\sin\theta = ma$$

where a = acceleration *along* its path.

So the acceleration along its path $a = -g\sin\theta$.

Now provided θ does not exceed about 10°, then $\sin\theta = s/L$ where s is the displacement, and L is the length of the string. Hence the acceleration, a, can be written as

$$a = -\frac{g}{L}s$$

which may be written as:

$$a = -\omega^2 s$$

by writing ω^2 for g/L.

So the s.h.m. equation is obtained provided θ does not exceed about 10°. Also, the time period is given by

$$T = \frac{2\pi}{\omega} = 2\pi\sqrt{\frac{L}{g}}$$

Using the simple pendulum to measure g

Adjust the string so that the length is about 1 m, then measure the time period by timing 20 oscillations. Measure the length of the pendulum accurately to the middle of the bob. Repeat for several shorter lengths. A graph of T^2 against L ought to give a straight line through the origin. From the equation for T, the gradient of the graph is equal to $4\pi^2/g$. Hence determine g.

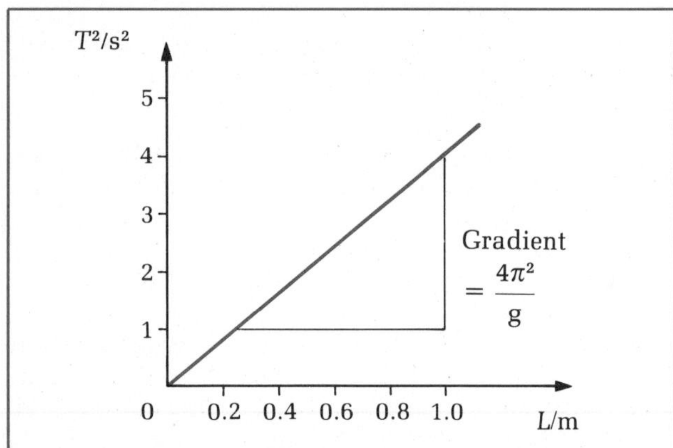

Figure 16.15 *T² v L graph*

16.5 Energy of oscillating systems

Free oscillations occur where the total energy of an oscillating system stays constant. The energy of the system changes from P.E. to K.E. and back every half-cycle. But the total energy (=P.E.+K.E.) remains constant. For example, the simple pendulum moved to the side is given P.E. When released, its P.E. falls and its K.E. rises. But at any stage, the sum of the P.E. and K.E. is equal to the initial P.E. As it swings through the centre, it has maximum K.E. because its speed is greatest passing through the centre. As it swings towards maximum displacement, its K.E. changes to P.E. as it slows down to reach the highest point of the swing.

In practice, the amplitude of the oscillations gradually becomes smaller and smaller unless the pendulum is supplied with energy. Air resistance is one reason why the amplitude decreases gradually. Friction due to motion through the air gradually converts the energy of the system into heat energy. Unless the system is designed with a mechanism to feed energy in at the same rate as it is lost, the energy decreases, so the amplitude becomes smaller and smaller; therefore the system does not oscillate freely.

Let's consider the energy involved in free oscillations in a little more detail. The K.E. is given by $\frac{1}{2}mv^2$. Now we know that $v^2 = \omega^2(r^2 - s^2)$, where s is the displacement, r the amplitude, and ω is the angular frequency. Hence the K.E. at displacement s is equal to $\frac{1}{2}m\omega^2(r^2 - s^2)$. In fact, when $s = 0$, we see that the K.E. equals $\frac{1}{2}m\omega^2 r^2$ which is therefore the maximum K.E. A graph of K.E. against displacement s is shown in Figure 16.16.

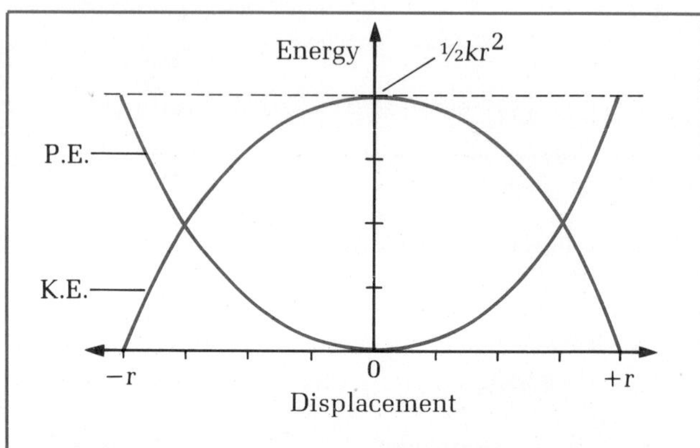

Figure 16.16 *Energy of oscillations*

For an oscillating mass on a spring, we know that $\omega^2 = k/m$ so the expression for K.E. simplifies to become

$$\text{K.E.} = \tfrac{1}{2}m\omega^2(r^2 - s^2) = \tfrac{1}{2}k(r^2 - s^2)$$

The potential energy (P.E.) at displacement s is equal to $\frac{1}{2}ks^2$. The reason is that the restoring force equals ks. So when the system is displaced from 0 to s, the average force from 0 to s is $\frac{1}{2}ks$. Therefore the work done from 0 to s is the average force × displacement which equals $\frac{1}{2}ks^2$. See p. 18 if necessary.

$$\text{P.E.} = \tfrac{1}{2}ks^2$$

The total energy E is equal to the K.E.+P.E., which means that

$$E = \tfrac{1}{2}k(r^2 - s^2) + \tfrac{1}{2}ks^2$$

using the two expressions above. So the total energy is just $\frac{1}{2}kr^2$ which is constant because k and r are constants. The total energy does not change during the motion; when the mass moves away from its equilibrium position, it loses K.E. but the loss of K.E. is balanced by the gain of P.E. When it moves inwards towards equilibrium, the K.E. increases but is balanced by the fall of P.E. So at any point during the motion, the total energy is always the same. (See Figure 16.16.)

In more general terms, $\text{K.E.} = \frac{1}{2}m\omega^2(r^2 - s^2)$ and $\text{P.E.} = \frac{1}{2}m\omega^2 s^2$, so the total energy is $\frac{1}{2}m\omega^2 r^2$.

Damped oscillations are oscillations where the amplitude becomes smaller and smaller. The simple pendulum, in practice, gives an example of damped oscillations, although many cycles may pass before it stops. In

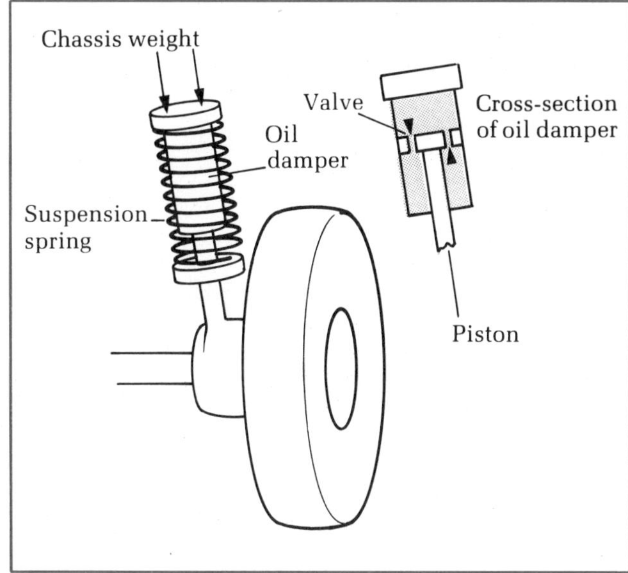

Figure 16.17 *Car suspension*

this instance, the damping force is friction due to air resistance. Like friction, damping is sometimes a nuisance but it can sometimes be useful. Try taking a ride along a bumpy road in a springy car! If it is oversprung, you'll realize in a practical way why damping is an important feature of car suspension systems.

Where the oscillations gradually die away, the damping is said to be **light**. The simple pendulum provides an example of a lightly damped system, so light that the oscillations are almost free. When the damping is more noticeable, the displacement-time graph is as shown in Figure 16.18.

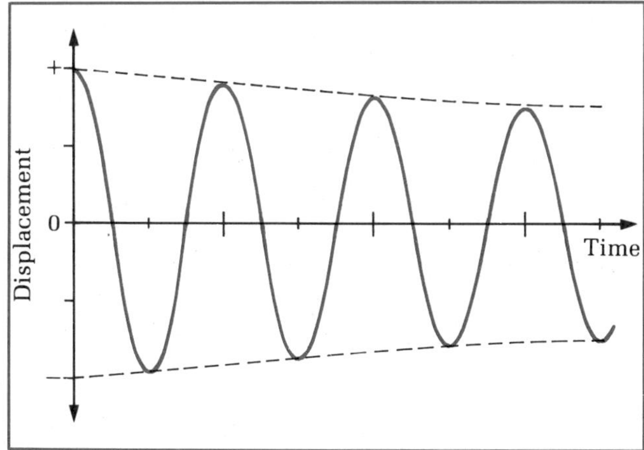

Figure 16.18 *Lightly damped oscillations*

A car suspension system with light damping would not be very satisfactory. After passing over a bump, the occupants would be thrown up and down until the oscillations died away. One method of damping the springs of a suspension system of a car is to use oil dampers, as in Figure 16.17. The motion is slowed down by the flow of oil through valves in the piston dampers. If very thick oil were used, the damping would be too great; when displaced from equilibrium and released, the system would be too slow in returning to equilibrium. The damping is then said to be **heavy**, or the system said to be **over-**

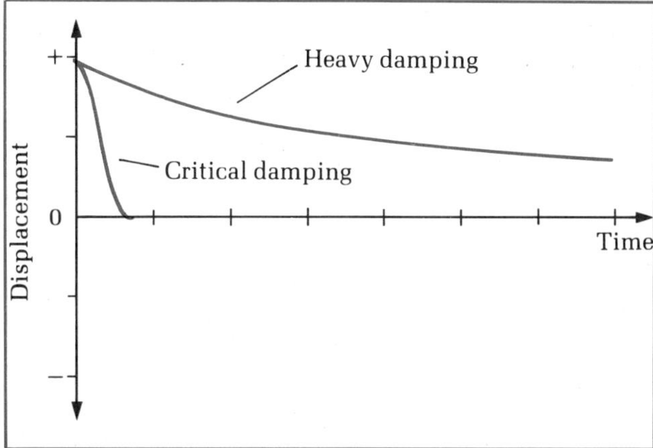

Figure 16.19 *Critical and heavy damping*

damped. Figure 16.19 shows how the displacement changes with time where the system is overdamped.

If a system, when displaced and released, returns to equilibrium without overshooting *and* in the shortest possible time, the system is said to be **critically damped**. Its displacement-time curve is also shown in Figure 16.19. Critical damping is an important feature of moving coil meters which are used to measure current and voltage. When the reading changes, it's no use if the pointer oscillates for a while before settling down to the new reading. You need to make the new reading quickly in case it changes again. If the coil is critically damped, the pointer moves to its new reading in the shortest possible time without oscillating.

All oscillating systems have three features in common.

Inertia due to mass in a mechanical system, which gives the system its K.E.

Restoring forces which give the system its P.E. In mechanical systems, restoring forces are due to springiness in the system.

Damping which converts the total energy of a system into heat energy. The forces responsible for damping are often called **dissipative** forces because they dissipate (i.e. waste) the system's energy.

Figure 16.20 shows some examples of oscillating systems not previously discussed. Try to identify in each system which part is responsible for each of the above features.

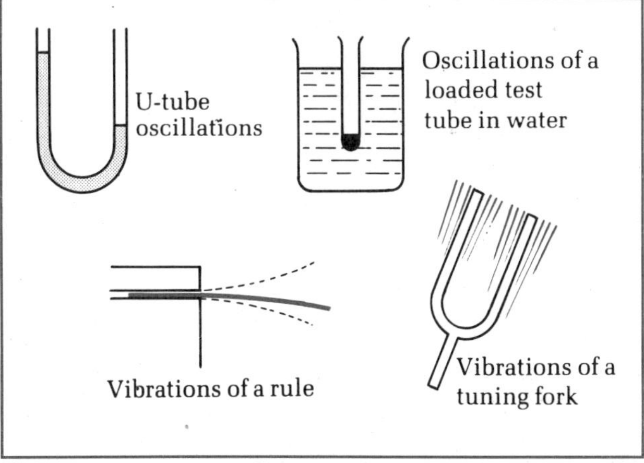

Figure 16.20 *Some oscillating systems*

16.6 Forced oscillations and resonance

Imagine pushing a child on a swing. If each push is timed suitably, the child swings higher and higher. The pushes are a simple example of a **periodic** force, a force applied at regular intervals. When a periodic force is applied to an oscillating system, the response depends on the frequency of the periodic force. When the system oscillates without any force applied, its frequency is called its **natural frequency**. If the frequency of the periodic force is equal to the natural frequency of the system, then each push can build the amplitude up further. If however the frequencies are not equal, the periodic force will build the amplitude up for a while, then it will slow the motion down for a while, then build up again etc. Whether or not the frequencies are equal, the oscillations are said to be **forced** when a periodic force acts. The periodic force provides a means of supplying energy to the system. Only if the frequency of the periodic force is matched (i.e. equal) to the natural frequency can energy be taken in over many cycles. When not matched, energy is fed in for a few cycles then taken out again for a few cycles, etc.

When the force frequency equals the natural frequency, energy is fed in continuously so the amplitude can build up more and more. The system is then said to be in **resonance**. Each push supplies energy just at the most suitable point in each cycle.

At resonance, the force frequency is equal to the natural frequency of the system.

How far can the amplitude build up at resonance? The damping forces convert the system's energy into waste energy. So the amplitude builds up until the rate at which the periodic force supplies energy is balanced by the rate at which the damping forces waste energy. At first the amplitude builds up because more energy is supplied (by the periodic force) than is lost (by the damping force). As the amplitude becomes bigger, so does the damping. Eventually, the amplitude reaches a constant value. Then, the damping removes energy at the same rate as the periodic force supplies it, so the system's energy stays constant.

Investigating resonance of a mechanical system

To make the arrangement shown in Figure 16.21, the upper spring is attached to the top of a tall stand and a suitable mass is hung from its lower end. The lower spring is attached to the weight, and is pulled to set both springs under tension. The lower end of the lower spring is then firmly fixed to a vibrator to be driven by a signal generator. The mass between the two springs can now be made to bob up and down and, provided its amplitude of oscillation is not too large, it moves along a vertical line. Use a stopwatch to determine its time period for free oscillations; time twenty oscillations and hence calculate its time period. Then its natural frequency f_0 (=1/time period) can be calculated.

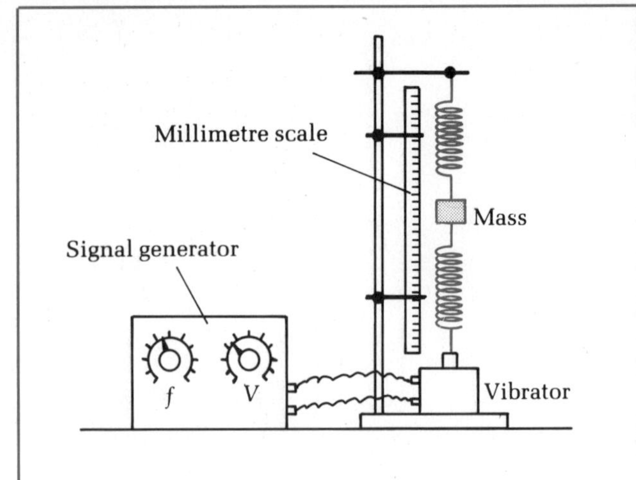

Figure 16.21 *Investigating resonance*

When the vibrator is operated using the signal generator, the mass responds most to the applied force when the forcing frequency is equal to the natural frequency f_0. The mass-spring system is then in **resonance**, and the energy supplied by the vibrator makes the amplitude build up.

For several different values of frequency on each side of resonance, measure the maximum amplitude. Note the frequency of the applied force from the signal generator dial, or use a CRO to measure the frequency of the signal voltage. The results can then be plotted to give a **resonance curve** as in Figure 16.22. The curve shows that the amplitude reaches a sharp peak, the resonance peak, at a single value of the applied (i.e. force) frequency. This value should equal the natural frequency of the system.

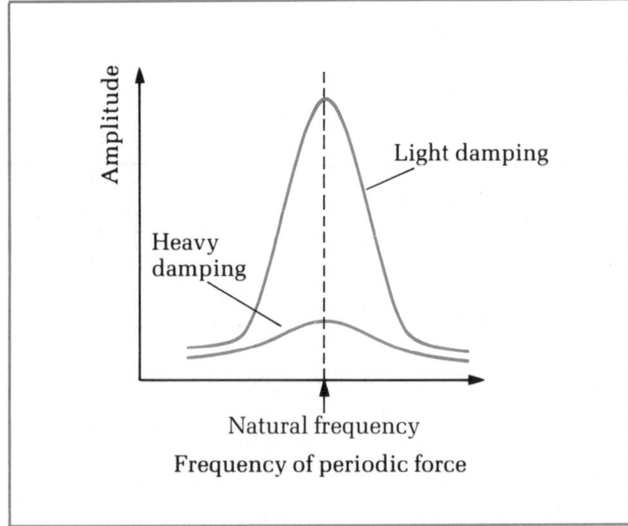

Figure 16.22 *Resonance curve*

Extra damping can be applied by attaching a vane, in the form of a piece of card, to the mass. Then, when it oscillates, the air resistance is much greater than without the vane. So even at resonance the amplitude does not build up as greatly or as sharply as without damping. With the vane the damping force removes energy at a greater rate than without, so the amplitude does not build up to the high level possible without damping.

Some other examples of resonance

There are many examples around us. Any system capable of oscillating can be set into resonance, sometimes with disastrous results.

1 **Shattering a wine glass** It has been known for high-pitched sound waves to shatter fragile objects; for example an opera singer hitting a top note may shatter a wine glass. Perhaps less dramatic is the effect of blowing gently over the top of a bottle. The air in the bottle creates a resonant sound at the right speed of blowing. Try it!

Figure 16.23 *A wine glass is shattered by sound*

2 **Tuning a guitar** To tune a guitar string to a piano note or to another guitar string, the 'paper rider' method is helpful. A small paper rider is balanced on the string to be tuned. Then, when the piano note is sounded, the rider will fall off if the string is tuned correctly. The reason is that the piano creates sound waves which pass over the guitar string. If the sound waves have the same frequency as the natural frequency of the string, then the string resonates and the rider falls off. Try it if you have a guitar.

3 **Barton's pendulums** The driver pendulum D (see Figure 16.24) is pulled aside and then released. The oscillating motion from it is transmitted along the supporting thread. So the other pendulums are subjected to a periodic force from D. In fact, only R oscillates much at all. The reason is that D and R are the same length, so they have the same time period. Therefore, the forcing frequency (that of D) is equal to the natural frequency of pendulum R only.

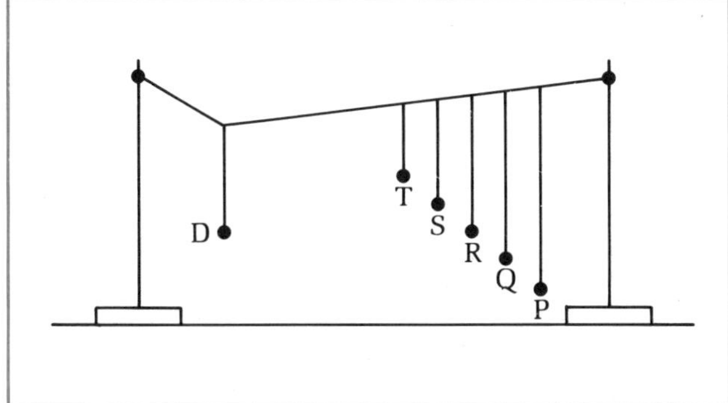

Figure 16.24 *Barton's pendulums*

4 **The collapse of the Tacoma Narrows Bridge in 1940** The dramatic collapse of this bridge was actually filmed. The long span of a bridge has a natural frequency of oscillation, a bit like a springboard. Usually, damping makes any oscillations die away quickly. However, on the day of the disaster, the wind caused the bridge to resonate. This is because a crosswind forms eddies on the downwind side of a bridge – and each time an eddy forms, the bridge is given a little sideways push. The disaster arose because the wind speed was such as to produce eddies at the same frequency as the natural frequency of the bridge. With not enough damping in this particular bridge, the bridge span oscillated at ever-increasing amplitude until it collapsed. You will be glad to know that since then bridges and buildings have been designed to avoid such effects.

Figure 16.25 *Collapse of Tacoma Narrows bridge*

16.7 Summary

Definitions

1 Angular frequency, $\omega = 2\pi/T = 2\pi f$
2 Amplitude is the maximum displacement.
3 S.h.m. is defined as oscillating motion where the acceleration is
a) always directed towards equilibrium,
b) proportional to the displacement from equilibrium.
4 Critical damping is where the system, when displaced and released, returns to equilibrium, without overshooting, as quickly as possible.
5 At resonance, the frequency of the periodic force is equal to the natural frequency of the system.

Equations

1 $a = -\omega^2 s$ (the defining equation for s.h.m.)
2 $v^2 = \omega^2(r^2 - s^2)$ (for speed v at displacement s, r = amplitude)

3 $T = 2\pi\sqrt{\frac{m}{k}}$ (mass on a spring)

4 $T = 2\pi\sqrt{\frac{L}{g}}$ (simple pendulum)

Short questions

Assume $g = 10\,\text{N kg}^{-1}$.

16.1 A small mass attached to a spring oscillates with simple harmonic motion with amplitude of 35 mm, taking 6.5 s to make 20 complete oscillations. Calculate
a) its angular frequency,
b) its maximum speed,
c) its maximum acceleration.

16.2 For the small mass of question **16.1**, calculate its speed when its displacement from equilibrium is
a) 10 mm,
b) 20 mm.

16.3 The displacement s of a point object oscillating in simple harmonic motion is described by the equation s (in mm) = 15 cos $10\pi t$ (in radians) where t is the time (in s) since passing a fixed point P.
a) What is the amplitude and the time period of the motion?
b) Calculate the displacement and speed when $t = 0.040$ s.

16.4 The pistons of a 4-cylinder car engine when idling move approximately in simple harmonic motion with an amplitude of 50 mm at a frequency of 110 Hz. Calculate
a) the maximum speed of the pistons,
b) the maximum acceleration of the pistons.

16.5 A spiral spring of natural length 300 mm is suspended vertically from a fixed point with its upper end fixed. A mass of 0.150 kg is suspended at rest from the lower end of the spring to increase its length to 355 mm. The mass is then pulled down a further distance of 30 mm and released from rest so it oscillates about equilibrium. Calculate
a) the stiffness (i.e. spring constant) k of the spring,
b) the time period of oscillations,
c) the maximum speed and the maximum K.E. of the mass,
d) the maximum and minimum tension in the spring.

16.6 When a student pushes down with a force of 200 N on the wing of a car above one of its front wheels, the wing is displaced downwards by 12 mm.
a) Calculate the stiffness (i.e. spring constant) k of the suspension spring at that wheel.
b) Assuming each wheel is fitted with an identical spring and the mass of the car (with driver) is 600 kg, calculate its natural frequency of oscillation.
c) Estimate the speed at which the car would resonate when travelling over regularly spaced bumps 15 m apart.

16.7 A trolley of mass 0.9 kg on a horizontal runway has springs attached as shown in the diagram at the top of the next column. To displace the trolley 50 mm from its equilibrium position, a force of 2.5 N is required to hold the trolley.
a) Calculate the stiffness (i.e. spring constant) k of the system.

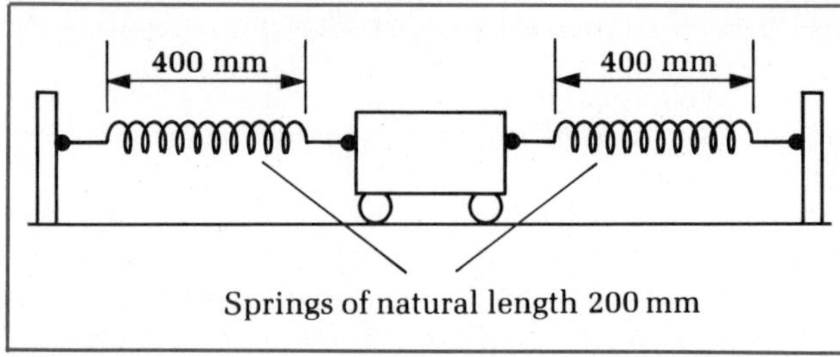

Figure 16.26

b) The trolley is released from rest at 50 mm displacement. How long does it take to reach maximum displacement on the other side of equilibrium, and what is its speed as it passes through equilibrium?
c) If its initial displacement in (b) had been 25 mm, what would your answers for (b) be now?

16.8 What additional mass should be loaded onto the trolley in question **16.7** to double the time period?

16.9 A simple pendulum consists of a spherical bob of mass 0.024 kg suspended by a light thread of fixed length 1.5 m. The bob is displaced from its equilibrium position with the thread taut until the centre of the bob has risen 20 mm above its equilibrium position. Then the bob is released from rest. Calculate
a) the time period of oscillations,
b) the maximum speed of the bob,
c) the maximum tension in the thread.
What is the angle between the thread and the vertical at maximum tension?

16.10 Calculate the K.E. and the P.E. of the bob in question **16.9** when it is
a) at maximum height,
b) passing through equilibrium,
c) at a height of 10 mm above equilibrium.

16.11 What length should the thread be changed to, in question **16.9**, in order to make the time period exactly 1.0 s?

16.12 Two identical simple pendulums are suspended from a taut horizontal string, as shown in the diagram below. Bob X is displaced from equilibrium in a direction at right angles to the string. With its thread taut, bob X is then released. Describe the subsequent motion of the pendulum bobs.

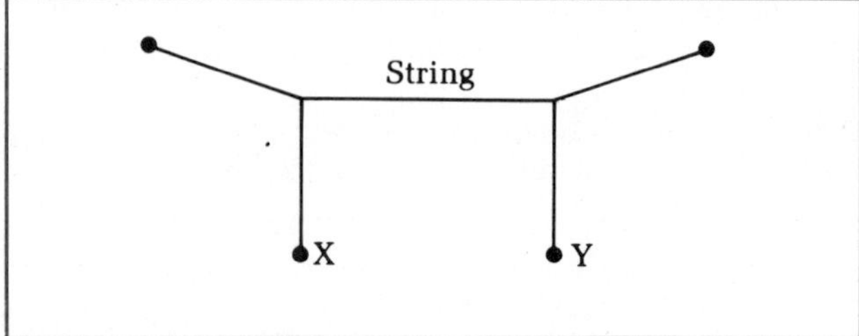

Figure 16.27

16.13 To investigate the vibrations of a building, a student designs a detector consisting of a light steel strip, fixed at one end, with different masses to be attached to the other end. The idea is that vibrations make the mass at the end oscillate noticeably. To determine the stiffness (i.e. spring constant) k of the strip, the student noted that a mass of 0.050 kg attached when the strip was horizontal made the free end drop by 2.5 mm.
a) Calculate the stiffness of the strip.

b) The student discovered strong vibrations of the strip when a mass of 0.080 kg was fitted. Calculate the frequency of these vibrations.

16.14 When sound waves fall on the eardrum, the drum vibrates in approximate s.h.m. transmitting vibrations to the bones of the middle ear. At a sound frequency of 3000 Hz for an amplitude of oscillation of the eardrum of 0.1 μm, calculate

a) the maximum speed of the eardrum,

b) its maximum acceleration.

Estimate the maximum pressure difference between the outer and middle ear (i.e. across the eardrum), given the thickness × the density of the drum tissue is approximately 0.1 kg m^{-2}.

16.15 A small mass is placed on a scale pan supported by a spring. When the scale pan is displaced downwards then released, it oscillates with s.h.m. Tests show that if the amplitude of oscillation is greater than a certain value, the mass leaves the scale pan at the highest point. If the time period of the oscillations is 0.4 s, what is the maximum amplitude for the mass to remain in contact with the pan at all times?

17 Wave motion

17.1 Progressive waves

Waves carry energy without carrying matter. To understand wave motion, we need to begin by studying simple wave systems such as water waves in a ripple tank or waves on a vibrating string. Only by learning from these simple examples can we hope to understand more complicated types of wave motion.

Drop an object into water and ripples spread out across the water; the ripples are examples of progressive waves because they carry energy across the water surface. We meet other examples of progressive waves in many branches of physics, such as in the study of sound or in the study of optics, but all types of progressive waves have the common feature of disturbances in motion carrying energy.

Waves may be divided into two forms which are:

Transverse where the disturbances are at **right angles** to the wave direction. Figure 17.1 shows a transverse wave produced on a Slinky spring by a quick flick of one end up and down, and the same wave is shown a little later further along. Other examples of transverse waveforms are waves on a string (see Figure 17.30 on page 237), water waves and electromagnetic waves.

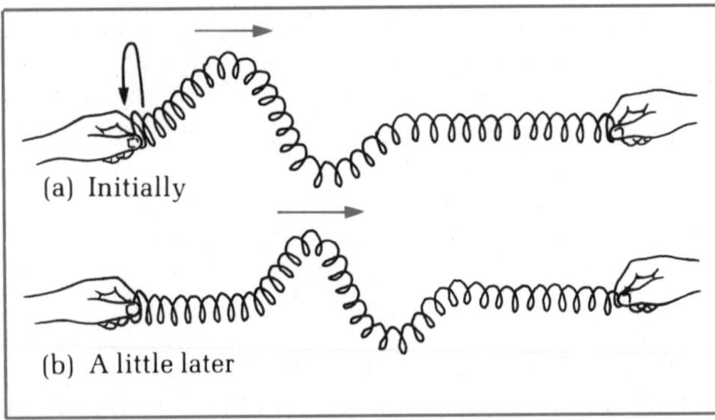

Figure 17.1 *A transverse wave pulse*

Longitudinal where the disturbances are **parallel** to the line of the wave direction. Figure 17.2 shows a Slinky spring carrying a longitudinal wave, produced by a quick to-and-fro movement of the end of the Slinky. The same wave is shown a little further along. Sound waves are longitudinal; a vibrating loudspeaker pushes backwards and forwards repeatedly on the surrounding air, and this causes pressure waves which travel through the air. The

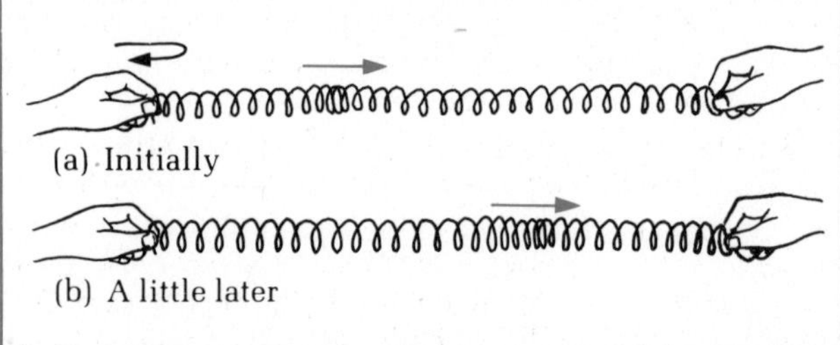

Figure 17.2 *A longitudinal wave pulse*

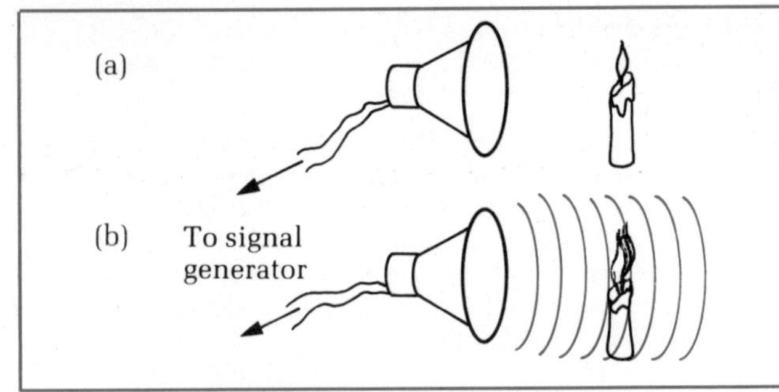

Figure 17.3 *Sound waves*

effect of sound waves on a candle flame causes the flame to flicker, as illustrated by Figure 17.3.

Longitudinal waves are not as easy to picture as transverse waves, but if the undisturbed Slinky is sketched alongside the same slinky carrying longitudinal waves, then the displacement of each coil from its undisturbed position can readily be seen, as shown in Figure 17.4. In fact, a graph of displacement against position along the slinky can then be drawn, but remember that the graph does not represent a transverse waveform here!

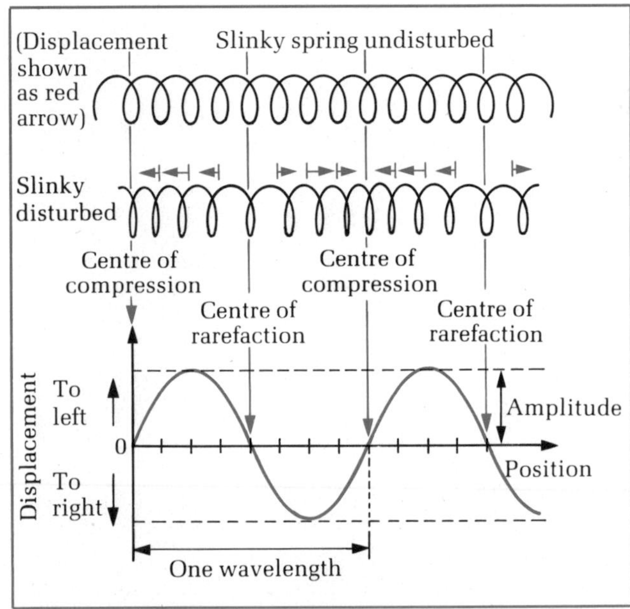

Figure 17.4 *Representing longitudinal waves*

17.2 Measuring waves

Imagine being on board a ship in a storm. Each wave throws the ship up, then the ship crashes down, only to be thrown up again by the next wave.

Each up and down movement as the ship rises from wave trough to wave crest, and then falls back to a trough is a complete **cycle of oscillation**.

The wave machine pictures of Figure 17.6 show how the crests and troughs progress as the individual particles of the medium, like corks on water, bob up and down. Each picture is a 'snapshot' of the wave motion, as waves travel from left to right. The first picture, part (a), shows particle P on the crest of a wave. A little later, in part (b), P has moved down, but a nearby particle Q has

Figure 17.5 *Water waves. The highest wave ever measured was recorded in the Atlantic Ocean on 30 Dec 1972 by the ship* Weather Reporter. *It measured 26 m from crest to trough.*

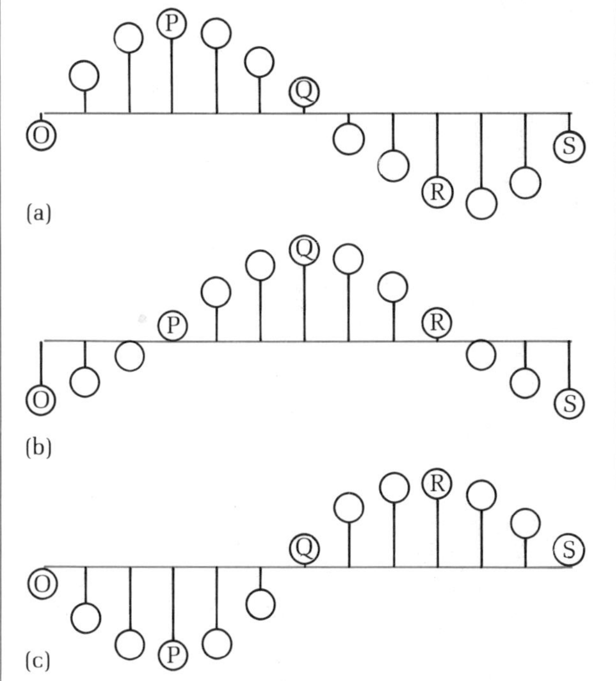

Figure 17.6 *Progressive waves*

moved up to the wave crest. A little later again, in part (c), Q has moved down and the wave crest has moved along to where particle R is. It follows that every particle on a wave crest makes one complete cycle of oscillation before it is next on a wave crest, so the wave crests and wave troughs travel along but the particles of the medium just oscillate about their undisturbed positions.

The **frequency** f of the wave motion is the number of wave crests passing a given position each second. The unit of frequency is the **hertz** (Hz). From the preceding paragraph, it ought to be clear that the frequency of oscillation of any particle is the same as the wave motion frequency. Also, because f wave crests pass a given point each second, then the time between one wave crest and the next arriving at that point, the **time period** T, is equal to $1/f$.

$$T = 1/f$$

The **amplitude** of a wave is the maximum displacement of a particle from its undisturbed position. In simple terms, for a water wave, it is the height of the wave crest measured from the **midpoint**, so the height from wave trough to wave crest is twice the amplitude.

Heinrich Hertz was the discoverer of electromagnetic (e-m) waves, predicted 20 years earlier by James Clerk Maxwell. Hertz went on to show that e-m waves had all the usual wave properties, including polarization. To demonstrate refraction he used a 'pitch' prism 1.5 m in height.

Figure 17.7 *Heinrich Hertz (1857–94)*

The **phase difference** φ between two particles along the wave is the fraction of a cycle by which one moves behind the other. Two particles which are in phase oscillate together so in Figure 17.6, O and S move in phase because they reach the wave crest at the same times; their phase differences is zero. However, P and Q have a phase difference of a quarter-cycle because Q is at the wave crest a quarter of a cycle after P. Sometimes, phase difference is expressed as an angle in radians or degrees; to convert from fraction of a cycle to an angle, the fraction must be multiplied by 2π (for radians) or 360 (for degrees).

1 cycle corresponds to 2π radians or 360°.

The **wavelength** (λ) of the waves is the distance from one particle to the next particle in phase with it. In other words, it is equal to the distance between adjacent wave crests, as shown in Figure 17.6. If the two particles are at a distance x apart, then their phase difference is $2\pi x/\lambda$ because they are (x/λ) wavelengths apart; in fact, x could be greater than one wavelength so the phase difference is the fraction of a cycle left over from dividing x by λ. For instance, suppose $\lambda = 0.20$ m and two points are 0.85 m apart, then the phase difference between the two points is a quarter of a cycle because their separation is $4\frac{1}{4}$ ($= 0.85/0.20$) wavelengths. Clearly, two points separated by a whole number of wavelengths must be in phase, so for $4\frac{1}{4}$ wavelengths it is the extra $\frac{1}{4}$ wavelength which determines the phase difference.

The speed of a progressive wave, its **wave speed**, is the distance travelled per second by its wave crests. To denote wave speed c is often used. Figure 17.8 shows straight wave crests at a given instant, then the same wave crests are shown again exactly one cycle later. One wave crest, marked W, is labelled so that its progress may be followed.

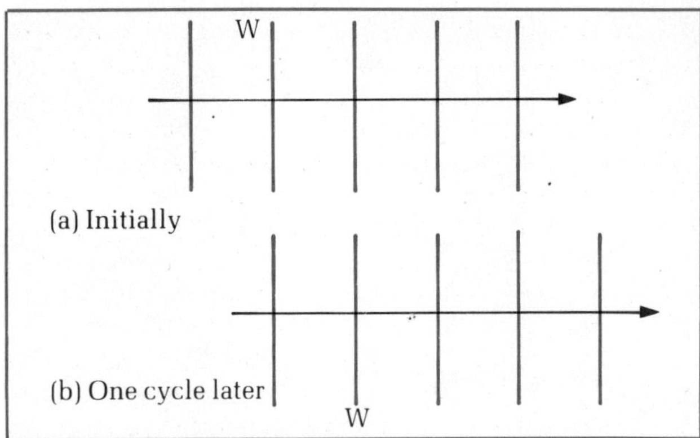

Figure 17.8 *Wave speed*

In the time taken for one complete cycle, W travels a distance of one wavelength, and takes time equal to 1/frequency. Therefore, the speed of W(= distance/time) is given by $\lambda/(1/f)$ which equals λf.

$$c = \lambda f$$

Measuring electrical waves

Connect the output from a signal generator to the Y-input terminals of an oscilloscope. Set the frequency dial on the signal generator to a convenient value, and adjust the

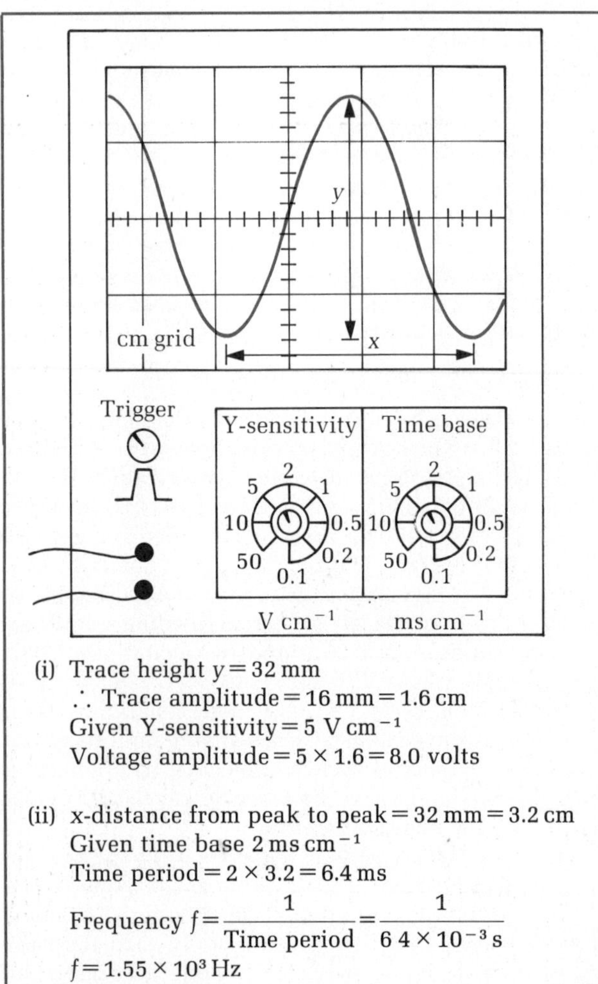

y
x
cm grid
Trigger
Y-sensitivity
Time base
5 2 1
10 0.5
50 0.2
0.1
V cm^{-1}
ms cm^{-1}
(i) Trace height y = 32 mm
∴ Trace amplitude = 16 mm = 1.6 cm
Given Y-sensitivity = 5 V cm^{-1}
Voltage amplitude = 5 × 1.6 = 8.0 volts
(ii) x-distance from peak to peak = 32 mm = 3.2 cm
Given time base 2 ms cm^{-1}
Time period = 2 × 3.2 = 6.4 ms
Frequency $f = \frac{1}{\text{Time period}} = \frac{1}{6.4 \times 10^{-3}\,\text{s}}$
$f = 1.55 \times 10^3$ Hz

Figure 17.9 *Measuring waveforms*

output voltage to display the waveform on the oscilloscope screen. It is probably necessary to adjust the sensitivity control for the Y-input, and also perhaps the time base control. As a rule, it is usually best to start with least sensitivity on the Y-input, then increase the sensitivity until a suitable level is reached. Also, the trigger control is to freeze the display, so if the displayed waves move across the screen, adjust the trigger control.

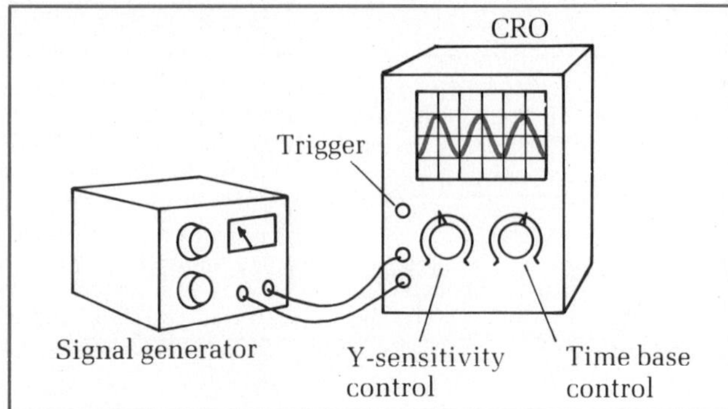

Figure 17.10 *Investigating waveforms*

Try to arrange the display as in Figure 17.10 and then measure the distance from one wave crest to the next; note the time-base setting and calculate the time period T and frequency f of the wave motion. You can compare your calculated value of f with the value on the dial of the signal generator.

To measure the signal amplitude, measure the distance from crest to trough along the Y-axis of the display, then note the sensitivity setting of the Y-input (often in volts/cm) and calculate the amplitude.

Figure 17.11 Using $c = f\lambda$

Using a ripple tank

A good way of studying waves is to use a ripple tank and a stroboscope. A vibrator on the water surface, as in Figure 17.12, produces a steady train of waves, and a suitable lamp placed above should cast shadows of the waves on the white screen below the tank. The waves

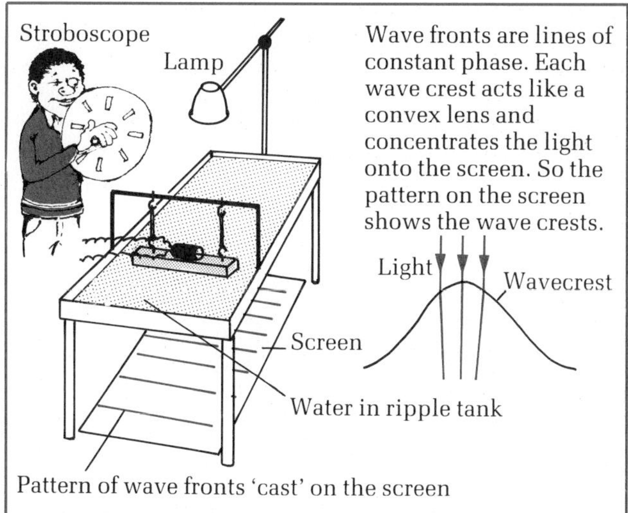

Figure 17.12 *The ripple tank*

often travel too fast to measure so a stroboscope can be used to freeze the pattern on the screen. As shown in Figure 17.12, the strobe has a number of slits so when it is turned at just the right speed for the waves, the viewer sees the shadows on the screen fixed in position; the reason is that the wave crests all look the same, so if a glimpse through the strobe is obtained every time a wave crest arrives at a given position, then the viewer sees a stationary pattern on the screen.

Some interesting experiments to try are as follows.

Measure the frequency of the waves Turn the strobe at the slowest possible speed which will freeze the motion. Enlist a friend to count the number of turns you make in 30 s, then from knowing the number of slits in the disc, work out the number of glimpses seen per second. This will be the frequency of the waves.

Measure the wavelength then calculate the wave speed. Place a metre rule on the screen along the wave velocity direction. Use the strobe to freeze the motion. Measure the distance along the rule of the spacing across a given number of wave fronts. So calculate the wavelength (that is, the distance between adjacent wave fronts). Use your value of f from above to calculate the wave speed from $c = f\lambda$.

Make the pattern move backwards Turn the disc just a little faster than the 'right' speed. The wave crests appear to move backwards because glimpses are seen too soon for wave crests to arrive at the correct positions for freezing.

Try putting different shaped reflectors in the tank, and observe the shape and direction of the reflected waves. Reflection of waves is discussed in detail in section 17.3.

17.3 Wave properties

Wave motion occurs in many different branches of physics. Whatever form the wave motion takes, there are a number of wave properties which often arise. For example, reflection of waves features in radio astronomy, in ultrasonic testing, and of course in optics; by studying reflection of waves in a ripple tank, we can understand reflection better when it arises in these other branches of physics. Other examples of useful wave properties to study are refraction, dispersion, diffraction, interference and polarization. All these are discussed in more detail below.

To explain many wave properties, **Christian Huygens** in the 17th century devised a theory for light which supposes that every point on a wave front emits wavelets in the forward direction. In this way, a new wave front is formed, and every point on the new wave front sends out further wavelets as shown in Figure 17.13.

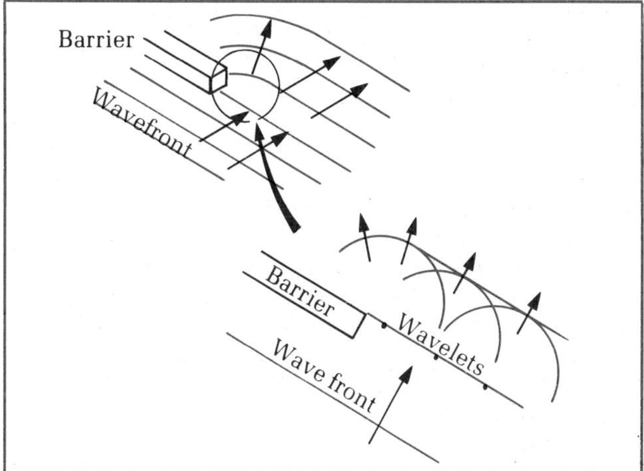

Figure 17.13 *Diffraction at an edge*

Reflection

The simplest example is when a straight (sometimes called **plane**) wave front reflects off a straight reflector. The reflected wave front is at the same angle to the reflector as the incident wave front, as shown in Figure 17.14. Huygens explained this by choosing two points on

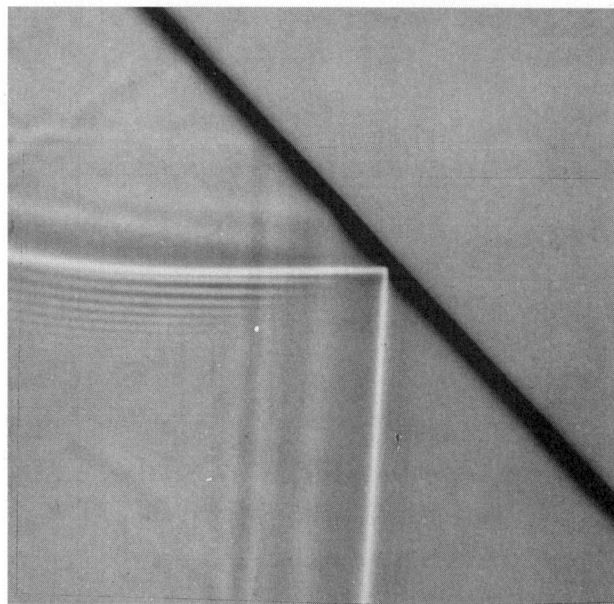

Figure 17.14 *Reflection of plane waves*

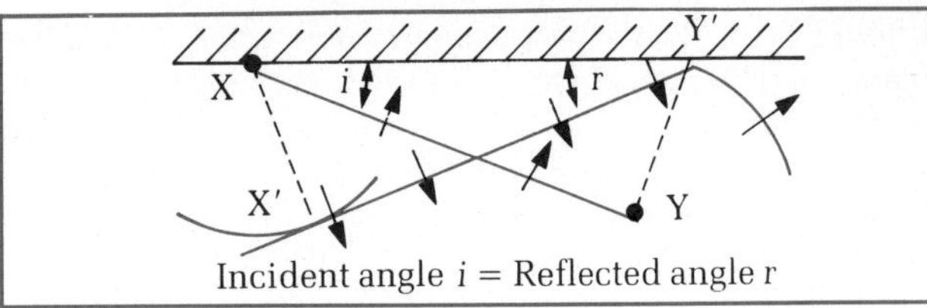

Figure 17.15 *Explaining reflection*

the incident wave front, X and Y, point X being where the wave front meets the reflector. Both points emit wavelets at the same time, as shown in Figure 17.15. The wavelets reach X′ and Y′ respectively at the same time so the reflected wave front X′Y′ is the tangent from X′ to Y′ at the reflector, which is at the same angle to the reflector as the incident wave front.

If either the incident wave front or the reflector is curved, the reflected waves in general will not be straight. Figure 17.16 shows circular wave fronts reflected off a straight reflector; the reflected waves appear to come from an image point behind the reflector.

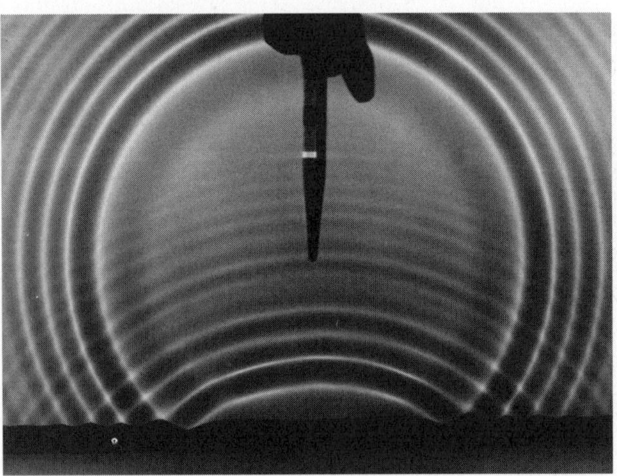

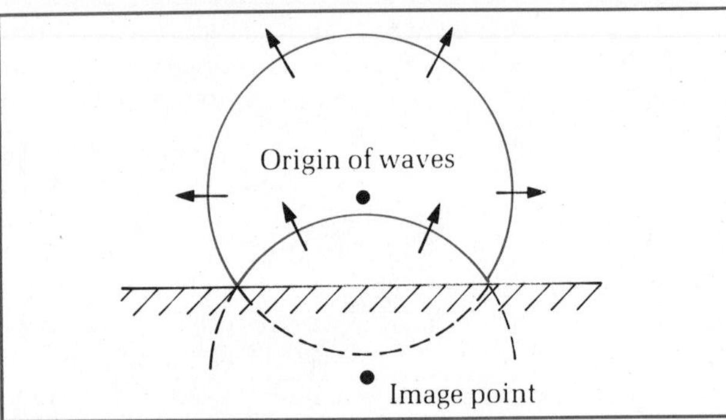

Figure 17.16 *Reflection of circular waves*

Refraction

Straight wave fronts change direction when they travel at an angle across a boundary between deep and shallow water. The waves have been **refracted** at the boundary because the wave speed in deep water is greater than in shallow water. When a wave front travels from deep to shallow water, its direction becomes closer to the **normal** (i.e. perpendicular to the boundary) because the waves move more slowly in the shallow water, as in Figure 17.17.

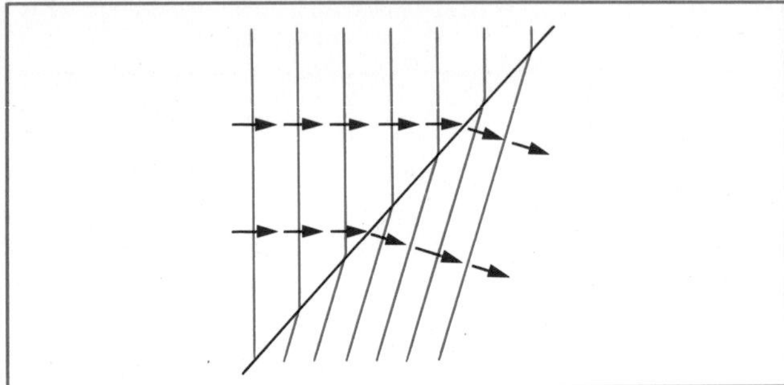

Figure 17.17 *Refraction*

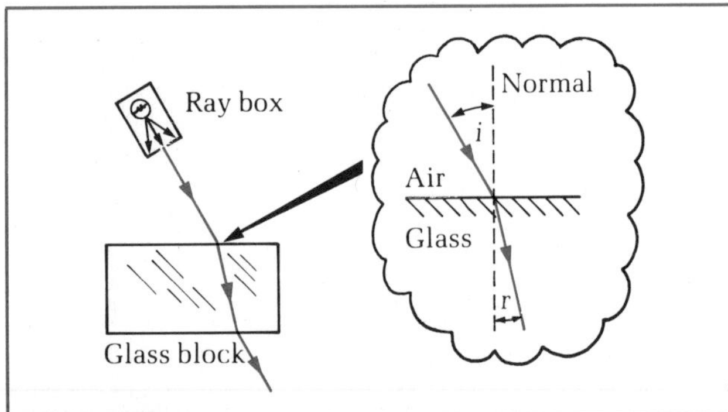

Figure 17.18 *Refraction of light*

Refraction of light can easily be demonstrated by directing a ray of light at a glass block, as in Figure 17.18. By experiment, it is reasonably easy to check the relationship, known as Snell's law, which is

$$\frac{\sin i}{\sin r} = \text{a constant}$$

where i is the angle between the incident ray and the normal,
r is the angle between the refracted ray and the normal,
and the constant is called the **refractive index** n.

Huygens' wave theory explains refraction by considering a wave front XY, as in Figure 17.19 about to cross a straight boundary from one medium into another where the wave speed is smaller.

Each point along XY emits a wavelet at the same time, but the wavelet from X travels more slowly than the one

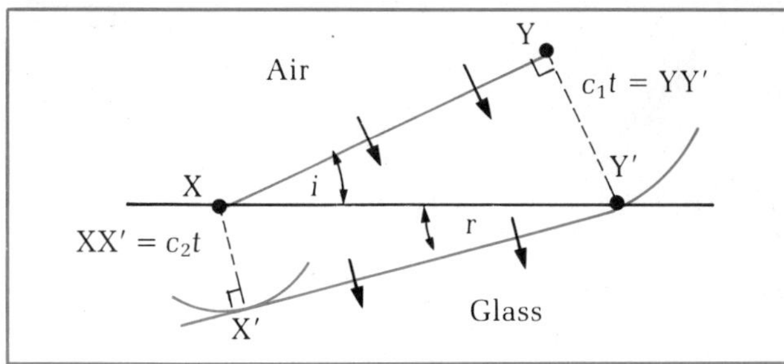

Figure 17.19 *Explaining refraction*

from Y because it is over the boundary. By the time Y's wavelet reaches Y', X's wavelet has only travelled as far as X', so the new wave front X'Y' is at a different angle than XY and its direction differs.

In more detail, let c_1 and c_2 be the wave speeds of the incident and refracted wave fronts respectively. Suppose the wave front takes time t to travel from XY to X'Y'; then $YY'=c_1t$ so $XY'=c_1t/\sin i$ because the wave front is at angle i to the boundary. Also $XX'=c_2t$ so $XY'=c_2t/\sin r$ since r is the angle between the refracted wave front and the boundary. Therefore

$$\frac{c_1t}{\sin i}=\frac{c_2t}{\sin r} \text{ which gives } \frac{\sin i}{\sin r}=\frac{c_1}{c_2}$$

The refractive index n from Snell's law is now seen to equal the wave speed ratio c_1/c_2; in fact, because the frequency does not change, the wave speed ratio is equal to the wavelength ratio λ_1/λ_2 since $c_1=\lambda_1 f$ and $c_2=\lambda_2 f$.

Refractive index from medium 1 to medium 2

$$n=\frac{c_1}{c_2}=\frac{\lambda_1}{\lambda_2}$$

For light waves, n is always given in terms of air for medium 1.

Dispersion

Dispersion of white light is the splitting of a beam of white light into colours when the beam is refracted by a transparent material such as glass. Figure 17.20 shows dispersion of white light by a prism; the white light beam consists of all the colours of the spectrum together, so it has a range of frequencies corresponding to those colours. Now the prism splits the beam into colours because it refracts higher frequencies more (e.g. blue light) that it refracts lower frequencies (e.g. red light). This in turn is due to the wave speed in glass of blue light being less than for red light; in other words, the wave speed of light in glass depends on the light frequency, so the refractive index for blue light is larger than for red

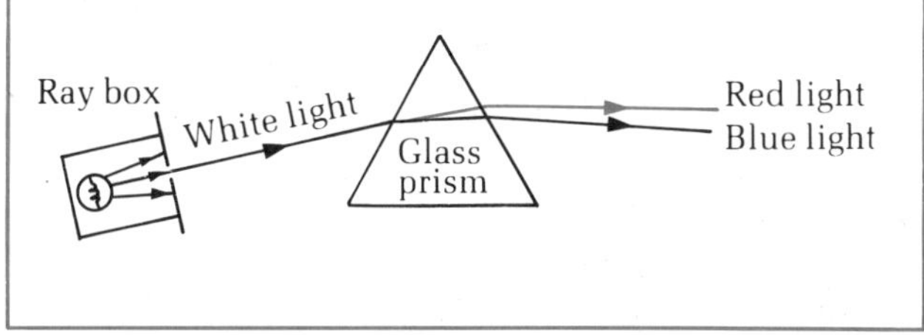

Figure 17.20 *Dispersion of light*

light. Dispersion of waves happens wherever wave speed depends on frequency.

Diffraction

Waves spread out when they pass by an edge or through a gap. This effect is called **diffraction**. For straight wave fronts travelling towards a gap. Figure 17.21 shows an important point – the narrower the gap, the more the waves spread out.

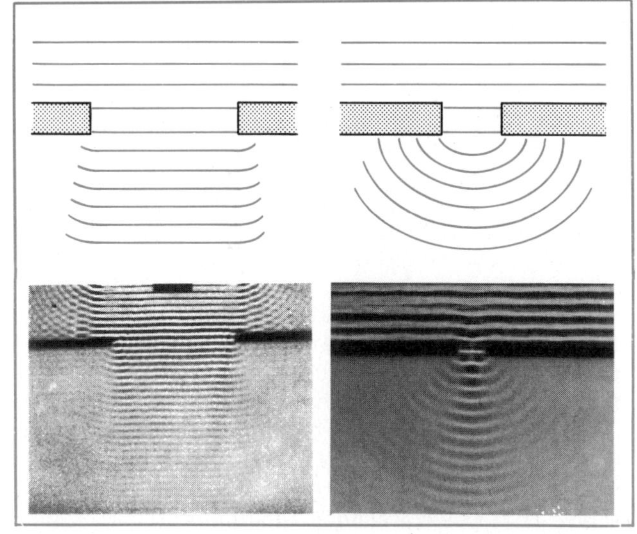

Figure 17.21 *Effect of gap width*

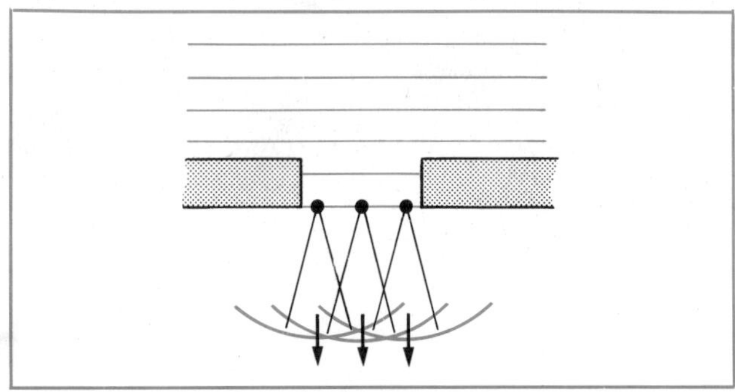

Figure 17.22 *Explaining diffraction*

Using Huygens' wave theory, we can explain diffraction; consider points along the wave front at the gap. Each point emits a wavelet in the 'forward' direction, and the new wavelets combine to form a new wave front spreading beyond the gap. Where the gap is very wide compared with the wavelength, diffraction hardly takes place and this is why diffraction of light waves is hard to demonstrate; in practice, gap widths need to be of the order of 0.1 mm.

Interference

Where two sets of waves pass through one another, the wave crests from one set reinforce the wave crests from the other set to form supercrests when they meet. This is an illustration of the **principle of superposition** which states that:

> **The total displacement at a point = the sum of the individual displacements at that point.**

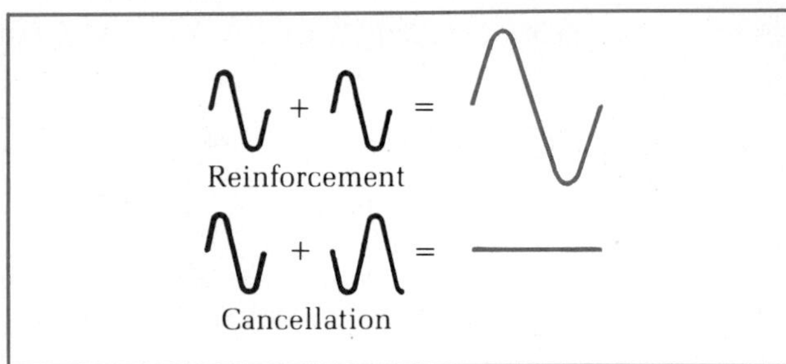

Figure 17.23 *Interference*

When a wave crest meets a wave trough, the result is that they cancel one another out.

Crest + crest = supercrest } = reinforcement
Trough + trough = supertrough }
Crest + trough = zero } = cancellation

Provided a constant phase difference is maintained between the sources of the waves, the points of cancellation and of reinforcement do not move about so the two sets of waves are said to have produced an **interference pattern**. Interference of water waves may be demonstrated using a ripple tank with two vibrating dippers, as shown in Figure 17.24. The two dippers are supported by a vibrating bar so they produce waves of the same frequency which spread out from each dipper. The two sets of waves pass through one another to produce easily observed points of cancellation and reinforcement. A snapshot of the two sets of waves would show points of cancellation where wave crests from one dipper cross wave troughs from the other dipper, as shown in Figure 17.25.

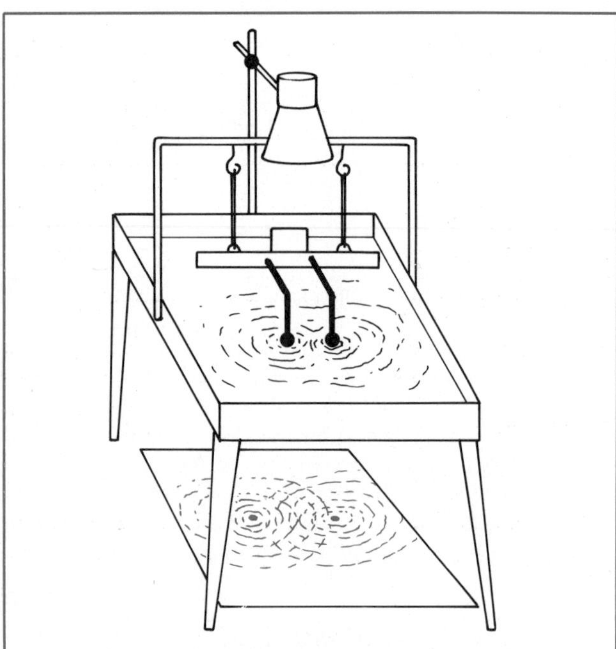

Figure 17.24 *Interference of water waves*

Polarization

If you wish to make waves travel along a rope, you need to flick the rope from side to side across the rope direction; the waves produced are transverse waves. The disturbances can be in any direction at right angles to the rope direction, and they will always be transverse because they *are* at right angles. However, there are an infinite number of directions at right angles to the rope, just as there are an infinite number of directions horizontal to a vertical line. If the rope is flicked from side to side along a fixed line, the disturbances will be parallel to that line only, as illustrated by Figure 17.26(a). The waves are then said to be **polarized**, and may be compared with the **unpolarized** waves of Figure 17.26(b) produced by continually changing the direction in which the rope is moved.

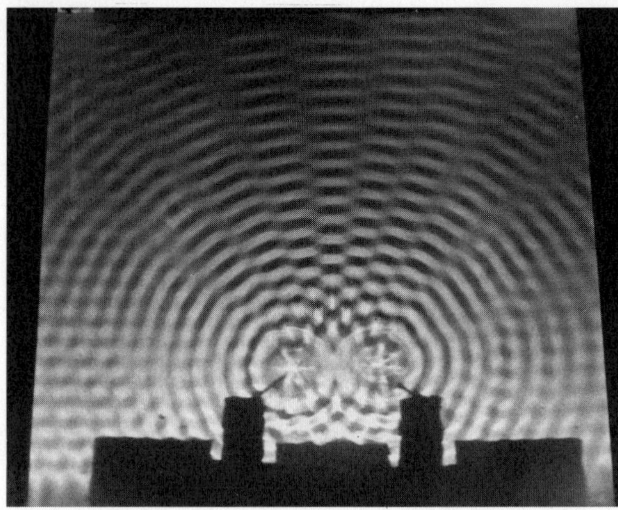

Figure 17.25 *An interference pattern*

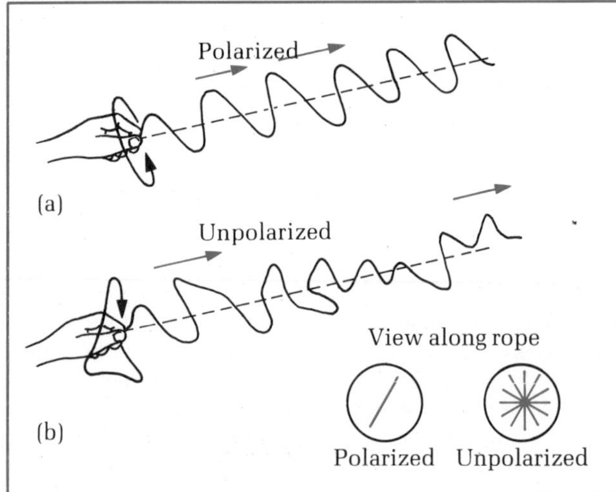

Figure 17.26 *Polarization*

Only transverse wave forms can be polarized because longitudinal waves oscillate *along* the direction in which the wave is travelling.

Light is a transverse wave form, and can easily be polarized by passing it through a **polaroid** filter. The filter only allows through light waves which oscillate in a preferred direction determined by filter itself. If the polarized light from such a filter is then directed at a second filter, the light will not be allowed through the second filter if its preferred direction is at right angles to that of the first filter, as illustrated by Figure 17.27. The second filter is sometimes called the **analyser.**

To understand in simple terms how the analyser

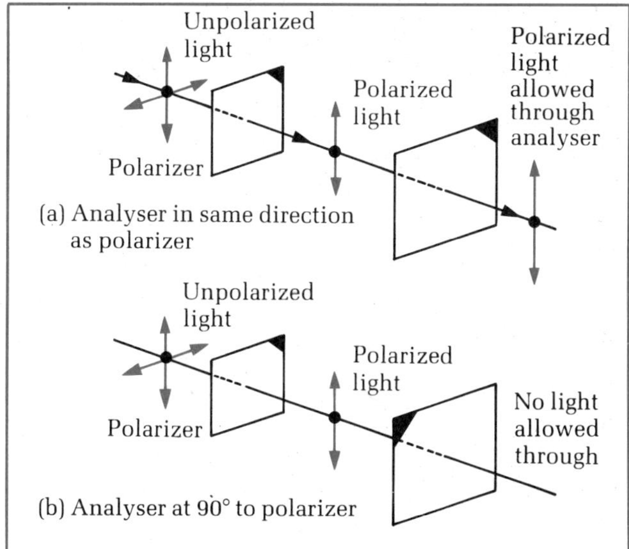

Figure 17.27 *Explaining polarization of light*

works, a board with a slit across the middle may be used as a polarizer for rope waves. The rope passes through the board as shown in Figure 17.28, and the board only allows oscillations in one direction through. A second similar board, the analyser, placed further along the rope only allows the polarized waves through if its slit is in the same direction as the first slit.

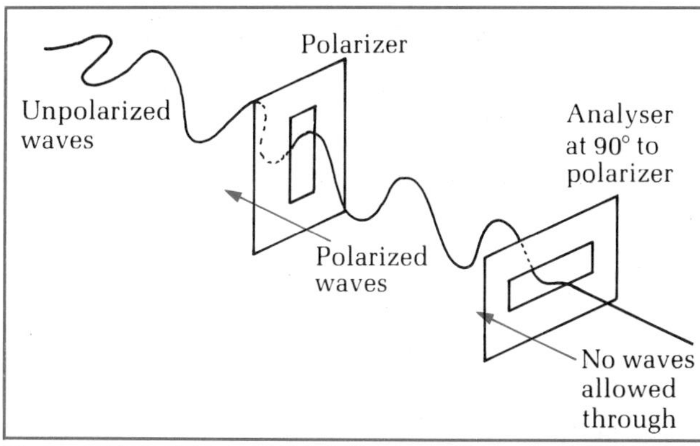

Figure 17.28 *Crossed analyser*

17.4 Stationary waves

Waves carry energy. In certain conditions, wave energy can be localized which means it can be stopped from moving through the medium. When this happens, the waves are said to be **stationary** or **standing** waves. You can make stationary waves on a rope or string with the aid of a friend. To do this, one person sends waves continuously along the rope from one end by moving the end from side to side. The other person does the same at the other end, producing waves of the same frequency moving in the opposite direction. By changing the frequency at which the waves are created, you should be able to create stationary waves at just the right frequency. It is easy to tell when the stationary wave pattern is set up because the rope then looks like Figure 17.29. In fact, this figure shows just one of the possible patterns. Every stationary

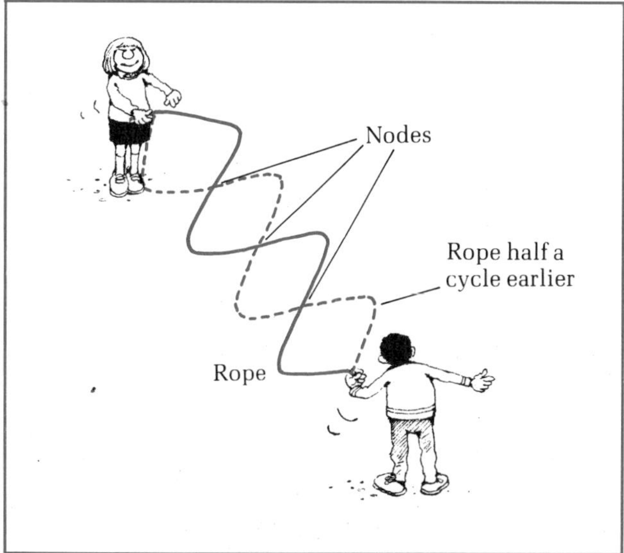

Figure 17.29 *Making stationary waves*

wave pattern has points where there is never any displacement. These points are called **nodes**.

Another way of making stationary waves on a rope or string is to tie one end to a fixed point such as a door handle. Then by waving the other end from side to side, waves can be sent along the rope to reflect back at the fixed point. So the reflected waves travel back towards you, and pass through waves sent out later, moving to the fixed end. Try it.

Investigating stationary waves on a string

Using a light string rather than a heavy rope, much higher frequencies are involved. So we need to use a vibrator connected to a frequency generator, as in Figure 17.30. One end of the string is tied to the vibrator. The other end

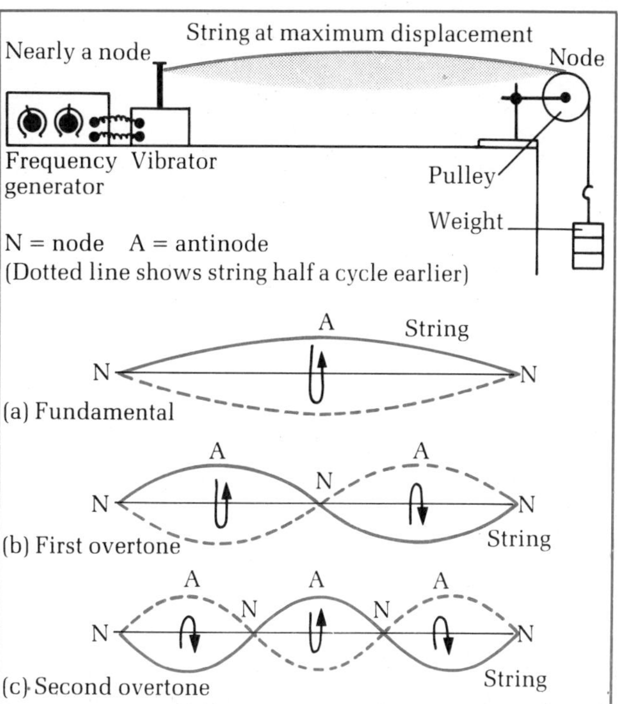

Figure 17.30 *Stationary waves on a string*

passes over a pulley, and supports a weight (two or three newtons is suitable if the string length is of the order of a metre).

When the amplitude of the vibrator is turned up, the string may or may not show a stationary wave pattern. However, by adjusting the frequency of the vibrator, any of the stationary wave patterns of Figure 17.30 can be set up on the string. The appearance of such a pattern often astonishes those unprepared for it! Even better, by using a stroboscope, the string can be frozen so that instead of seeing a blur at maximum displacement, it is possible to see the string at maximum displacement.

The pattern which appears at the lowest possible frequency has a node at either end, with maximum displacement at the mid point. Where the displacement is a maximum, the term **antinode** is used. This pattern is called the **fundamental** pattern. At higher frequencies, other patterns appear; these are referred to as **overtones**. The first overtone is where there are two loops on the string (compared with one loop for the fundamental). The second overtone has three loops on the string, etc. In fact, the frequency of the first overtone is 2 × the fundamental frequency, the frequency of the second overtone is 3 × the fundamental frequency etc, so in general,

the frequency of any pattern is equal to the number of loops × the fundamental frequency.

Describing stationary waves

The wave peaks of a stationary wave do not move along. The wave energy stays at the same positions along the string, unlike in progressive waves. Figure 17.29 shows that there are positions where there is never any displacement. These positions are the *nodes*. Midway between the nodes, the string vibrates with its greatest amplitude; these positions are the *antinodes*. The positions of the nodes and antinodes along the string do not change, hence the use of the term stationary wave. Each point along the string vibrates with constant amplitude, but the amplitude varies with position along the rope or string.

Nodes: positions of zero amplitude
Antinodes: positions of greatest amplitude

During one full cycle, a stationary wave pattern on a string changes as shown in Figure 17.31. At each antinode, the string moves from maximum amplitude to zero, to maximum amplitude the other way, back to zero, then back to the first displacement. Figure 17.31 shows that all points between two adjacent nodes move in phase with one another. They all reach maximum displacement together, and all pass through zero displacement together. However, any two points separated by just one node move out of phase by 180° or π radians: when one such point is at maximum positive displacement, the other is at maximum negative displacement. More generally, any two points have a phase difference equal to $n\pi$ where n is the number of nodes between them.

To create stationary waves, two sets of progressive waves of the same frequency must pass through one another in opposite directions. Let's consider why stationary waves are formed when two sets of waves, as described, pass through each other. The principle of superposition

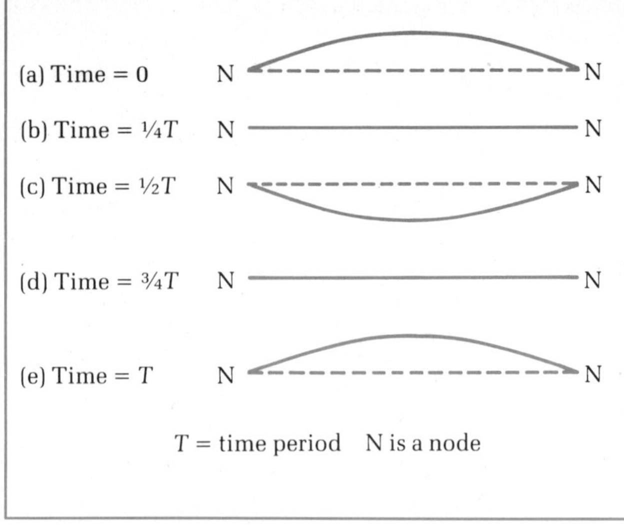

Figure 17.31 *A stationary wave pattern*

must be used. This is explained in more detail on p. 235 but essentially it tells us that

When two waves meet, the total displacement is equal to the sum of the individual displacements.

Of course, we do not see the individual sets of waves in Figure 17.30. We only see their combined effect which is the stationary wave pattern. What we can do however is to imagine where the individual sets of waves would be if we could see them. Then we can combine them to give the stationary wave pattern.

When the stationary wave pattern is at maximum displacement, the two sets of progressive waves are positioned so that they reinforce one another.

When the stationary wave pattern is at zero displacement everywhere, the two sets of progressive waves are positioned so that they cancel one another out.

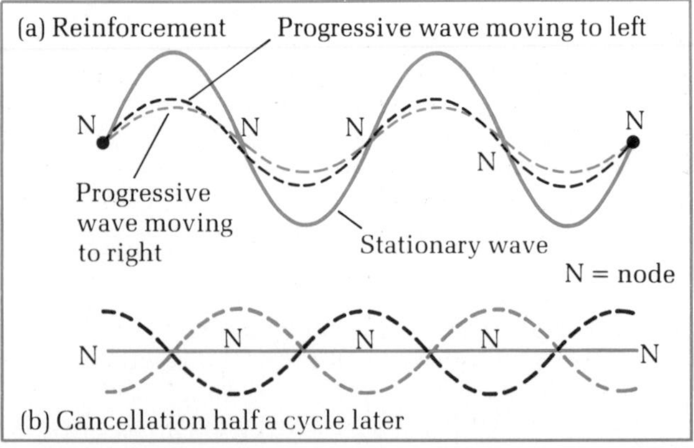

Figure 17.32 *Explaining stationary waves*

In one half of a complete cycle, the stationary wave pattern on the string swings from maximum displacement one way to maximum displacement in the opposite direction. Suppose at a particular antinode in a given half-cycle, the string swings from positive to negative displacement. In the same half-cycle, at the next antinode, the string swings the other way, i.e. from negative to positive. So in one half-cycle, wave peaks are replaced by wave troughs at the antinodes or vice versa. Therefore, the distance between one antinode and the next antinode is exactly one half-wavelength of the progressive waves which form the pattern. So from one node to the next is one half-wavelength

Distance between adjacent antinodes = one half-wavelength.

Now we can see the reason for the frequency relationship of stationary waves on the string. The fundamental pattern of Figure 17.32 is when the string length L is one half-wavelength. The fundamental wavelength λ_0 is therefore $2L$. Given the wave speed of the progressive waves on the string is denoted by c, then the frequency of the fundamental pattern is

$$f_0 = c/\lambda_0 = c/2L.$$

Now the first overtone has just two loops along its length, as in Figure 17.32(b), so the wavelength λ_1 here is L. Hence the frequency of the first overtone must be c/L which is $2f_0$. The second overtone has three loops along its length. So its wavelength λ_2 is $\frac{2}{3}L$, giving its frequency f_2 as $c/(\frac{2}{3}L)$ which is $3f_0$.

Measuring the wavelength of microwaves

Microwaves are short-wave radio waves, and can be produced from specially-made transmitters. Figure 17.33 shows such a transmitter, the microwaves being emitted from the horn of the transmitter. A flat metal plate is used to reflect the microwaves back towards the transmitter. A microwave detector diode connected to a microammeter is moved along the line between the plate and transmitter. The diode and plate are adjusted in position until the meter reading is at a non-zero value. Then the plate is gradually moved along the line until the reading is highest. When the diode is moved along the line (plate fixed), the reading drops to zero at each node and reaches its greatest value at each antinode. In this way, the distance between adjacent nodes can be determined to give one half-wavelength exactly. In fact, it is best to measure across several nodes. For example, if the distance from one node to the fifth node along is 80 mm, then five half-wavelengths measure 80 mm. So the value of the wavelength is 80/2.5 mm which equals 32 mm.

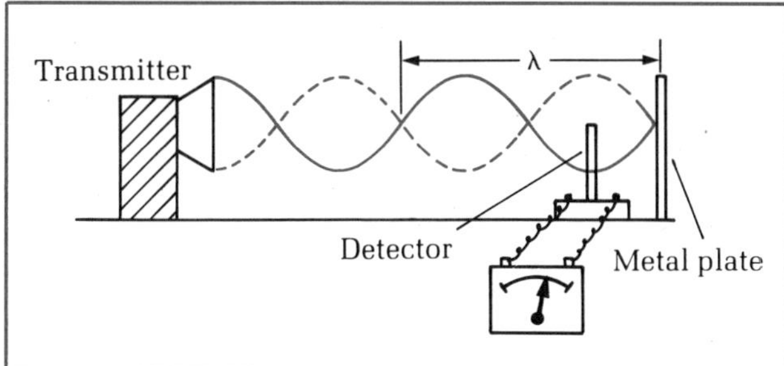

Figure 17.33 *Measuring λ for microwaves*

Acoustic resonance is another situation involving stationary waves. When sound waves are created near one end of a tube, the waves travel to the other end and reflect back. Inside the tube, there will be two sets of sound waves of the same frequency passing through one another. Stationary sound waves can thus be set up in the tube, and at certain frequencies the tube resonates with sound. The situation is then like the 'waves on the string' demonstration of Figure 17.30 (except that sound waves are longitudinal). Resonance happens only if each wave peak returns to the sound source just as the source is creating another wave peak, so the wave energy builds up and up to resonance. Clearly, the source frequency for resonance depends on how far the waves travel from the source to the far end of the tube and back. Figure 17.34 shows a small loudspeaker connected to a signal generator, sending sound into a tube. By sprinkling some very light powder along the length of the tube, acoustic resonance can actually be seen. By adjusting the frequency at the higher range, resonance can be detected by ear and by eye; the powder is pushed into a series of regularly-spaced heaps, each heap being at a node. The powder is pushed away from the antinodes and settles at the nodes in heaps. Acoustic resonance is discussed further on p. 249.

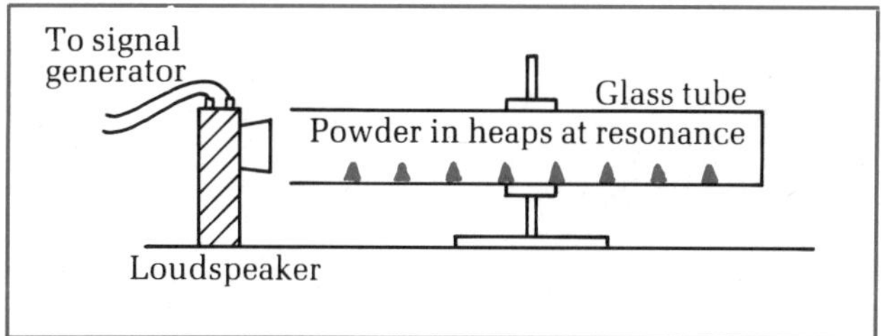

Figure 17.34 *Acoustic resonance*

17.5 Mechanical waves and resonance

A single mass on a spring has its own natural frequency of oscillation. If a periodic force acts on the mass, then resonance occurs if the force frequency is equal to the natural frequency of the mass-spring system). See p. 226 if necessary.) What if there are a large number of masses, all interconnected by springs? For example, consider a system of trolleys and springs connected as in Figure 17.35. If the end trolley is moved back and forth its oscillations pull on the next trolley. This trolley then starts to move back and forth, and pulls on the next one, etc. So waves of oscillations are sent down the line of trolleys. The system is a bit like the Slinky spring in Figure 17.1. Each time the end trolley is moved back and forth, a pulse of waves travels to the far end, reflects, and returns. If the returning waves are in phase with new waves being created, then the energy of oscillations builds up and up.

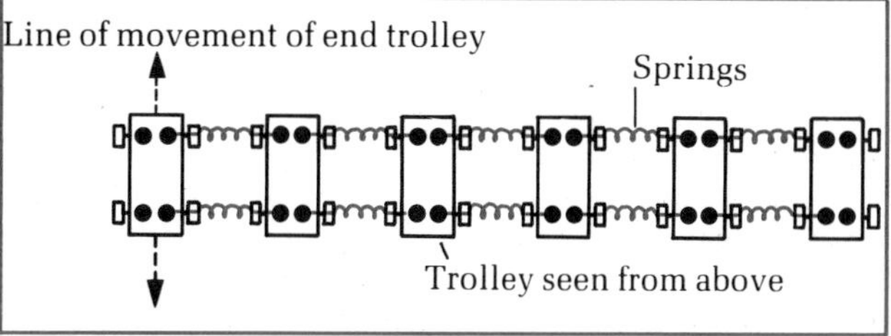

Figure 17.35 *Transverse waves along a line of trolleys*

The system can be made to resonate, provided the frequency of moving the end trolley back and forth is suitable. The lowest frequency for resonance is called the **fundamental**, and corresponds to the displacement pattern shown in figure 18.16(a). Resonances at higher frequencies (i.e. overtones) are possible, each with an antinode at either end, but differing numbers of nodes between. Where there are lots of masses and springs, the system resonates at more than one frequency value. These values depend on the speed at which waves travel through the system, and on the length of the system.

The speed of mechanical waves depends on the masses of the oscillating bodies as well as on the stiffness of the springs. Figure 17.36 shows an arrangement to measure the speed of a pulse of waves which pass along a line of spring-loaded trolleys.

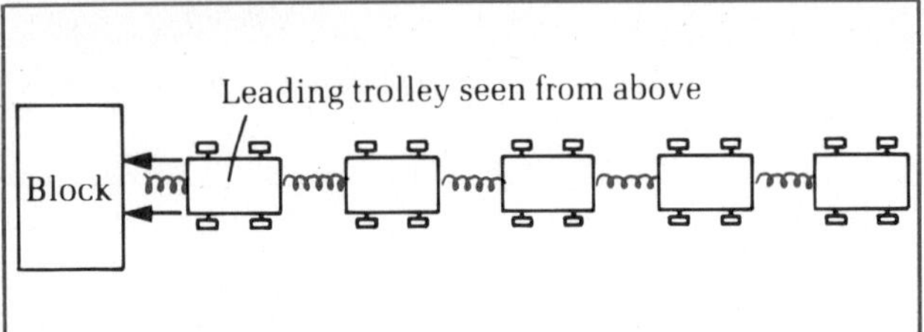

Figure 17.36 *Transmitting an impact pulse*

The line of trolleys is given a push so that the leading trolley travels a short distance before its spring hits a heavy block. The impact causes the other trolleys to push together, and a compression wave passes down the line to the last trolley. Then the last trolley rebounds and pulls all the others back in turn. So an expansion wave passes back towards the block, and finally pulls the leading trolley away from the block. Using an electrical timer, the speed of the pulse can be determined by timing the contact between the block and the spring which is pushed into it. This timing is equal to the time taken for the pulse to travel the length of the line and back. So, by measuring the distance as well, the pulse speed can be calculated.

Do you think the pulse speed would be greater if the springs were stronger? Stronger springs would transmit the oscillations down the line more quickly. Do you think that the pulse speed would be greater if the trolleys were each of greater mass? The trolleys would take longer to respond if more massive, so the pulse speed would be less. The atoms of a solid are a bit like the trolleys in Figure 17.36. Atoms in solids are connected to one another by bonds. (See p. 68.) The bonds are strong enough to stop the solid from being deformed unless large forces are applied. But the bonds can be stretched or squeezed like tiny springs between the atoms, so when an atom oscillates, its vibrations are passed on to neighbouring atoms.

Suppose a solid bar is struck with a hammer at one end. The atoms at that end are set into oscillation, and the bonds transmit the oscillations along the bar. So a pulse of energy is sent along the bar by the hammer. As the pulse passes each position along the bar, the atoms there are set into oscillation. So the sound of the impact travels along the bar in the form of a pulse transmitted by atoms passing their oscillations on.

Measuring the speed of sound in a solid

A metal rod is suspended by means of threads (see Figure 17.37). The end of the rod is struck sharply with a hammer. The electric circuit is connected to measure the contact time between hammer and rod on impact. The atoms of the rod at the impact end transmit a pulse down the rod via the atoms along the bar. The pulse is reflected at the other end and returns. When it reaches the impact end, the contact is broken. By measuring the bar's length L and the contact time t, the pulse speed may be calculated from speed of pulse $= 2L/t$.

Try using a stronger bar. Choose one made of a material with a greater modulus of elasticity (see p. 76). For example, steel has a greater modulus than brass, so steel is stronger. Does the pulse travel faster in a stronger bar? It ought to because the atomic springs are stronger.

Try using a denser bar. What conclusion would you draw if the speed were greater?

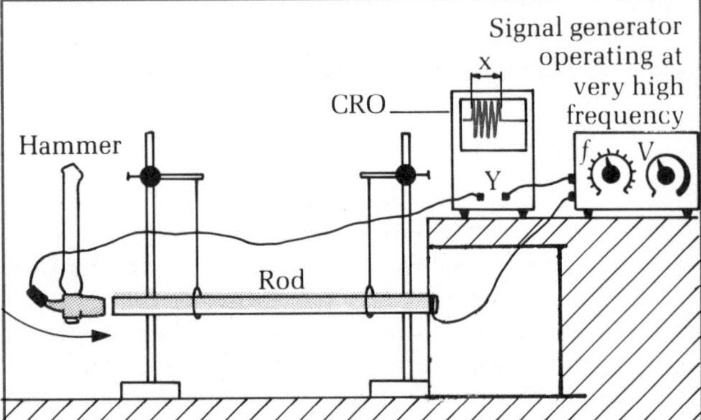

When the rod and hammer are in contact, the signal from the generator is displayed on the CRO. To measure the contact time (t), estimate the length (x) of the wave trace on the screen. Then use the time base calibration to determine t.

Figure 17.37 *Speed of sound in a solid*

Resonance of a solid bar can be demonstrated by attaching one end of the bar to a vibrator driven by a signal generator. A bar in the form of a long thin rod is best. The vibrator sends waves along the bar to the far end, where they reflect and return. Starting with a very low frequ-

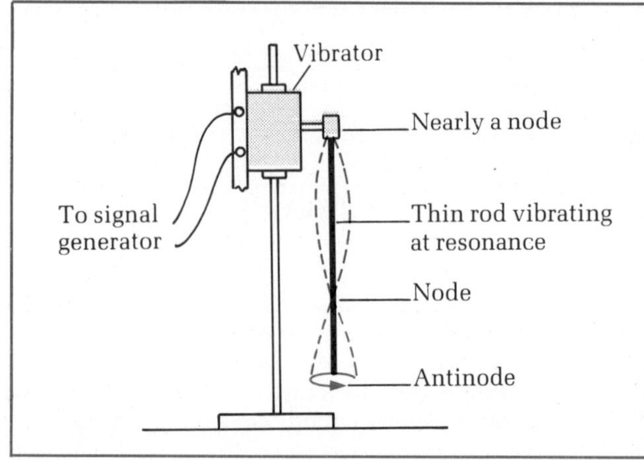

Figure 17.38 *Resonance of a rod*

ency of vibration, the frequency is increased until the bar is seen to resonate. At the fundamental frequency, there will be an antinode at either end and a node in the middle, as shown in Figure 17.38. The fundamental frequency is the frequency value where the time for the waves to travel the length of the bar and back is exactly equal to the time for one cycle. So wave peaks from the vibrator travel to the far end and return at the same time as a new wave peak is being created. The wave energy therefore builds up.

The bar also resonates at higher frequencies, each one with an antinode at either end but a different number of nodes between. At the fundamental frequency f_0, there is exactly one half-wavelength along the bar. So the wavelength $= 2L$ where L is the bar length. Hence the fundamental frequency $f_0 = c/2L$ where c is the speed of sound along the bar. You ought to be able to show that resonances also occur at $2f_0$, $3f_0$, $4f_0$ etc. (See *Tube open at the far end* on p. 250.)

Chladni's plate is a good example of resonance in a metal sheet. The plate is fixed horizontally to a vibrator, as shown in Figure 17.39. Sand is sprinkled evenly in a thin layer over the plate. The vibrator frequency is changed until the sand can be seen forming a pattern. The pattern is caused by waves travelling from the vibrator to the edge, reflecting at the edge, and returning. So sets of waves pass through each other. At just the right frequency, the amplitude of the stationary wave pattern becomes large enough to shift the sand to nodes. In fact, the nodes are lines for a two-dimensional pattern. There are a number of frequencies at which the plate resonates. Each gives a different pattern.

Avoiding resonance of metal plates or panels is important in the design of machines. A car bonnet which resonates when the engine is turning at a certain speed is quite obviously undesirable!

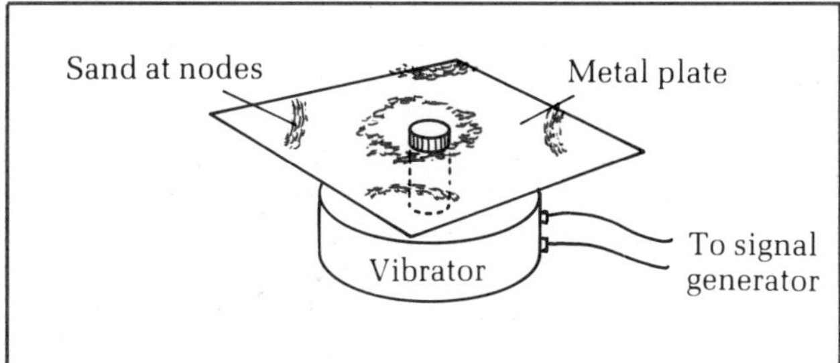

Figure 17.39 *Chladni's plate*

17.6 Summary

Definitions

1 Transverse waves occur where the displacement is at right angles to the wave direction.
2 Longitudinal waves occur where the displacement is along the line of the wave direction.
3 Frequency of the wave motion is the number of wave crests per second passing a given point.
4 Amplitude is the maximum displacement from equilibrium.
5 Phase difference between two particles along the wave is the fraction of a cycle by which one is behind the other.
6 Wavelength is the distance from one particle to the next particle in phase with it.

Equations

1 $c = \lambda f$

where c = wavespeed,
f = frequency,
λ = wavelength.

2 Snell's law of refraction

$$\frac{\sin i}{\sin r} = \frac{c_i}{c_r} = \frac{\lambda_i}{\lambda_r} = n$$

where i = incident angle,
r = angle of refraction,
n = refractive index (for light passing from incident to refracted medium),
c_i = wave speed in incident medium,
λ_i = wavelength in incident medium,
c_r = wave speed in refracted medium,
λ_r = wavelength in refracted medium.

Laws and rules

Principle of superposition When two or more waves meet, the total displacement is equal to the sum of the individual displacements.

Conditions for stationary waves to be formed Two sets of progressive waves of the same frequency, speed and amplitude must pass through one another in opposite directions.

Short questions

17.1 One end of a long straight rope lying along a corridor is picked up by a student who waves the end from side to side to make waves travel along the rope. The student notices that the wave amplitude decreases along the rope. Sketch graphs to show how the displacement of the rope
a) at a given position along the rope varies with time,
b) at a given instant varies along the rope.
Represent the wavelength and frequency on your graphs.

17.2 Two boats X and Y, 100 m apart on a lake, each contain a fisherman. Waves from a passing speedboat travelling in the direction XY pass boat X first then pass boat Y. The fisherman in Y notes that when his boat is on a crest, boat X is in a trough with one crest only between the two boats. He also notes that each wave crest takes 4.0 s to travel from X to Y. Calculate the wavelength, the wave speed and the frequency of the waves.

17.3 What is the difference between longitudinal and transverse waves? How would you show that
a) sound waves are longitudinal,
b) light waves are transverse.

17.4 The diagram at the top of the next page shows wave fronts on the water surface of a ripple tank 0.1 s afer being created at position P. Copy each diagram and sketch the wave fronts' position **a)** 0.1 s later, **b)** 0.2 s later.

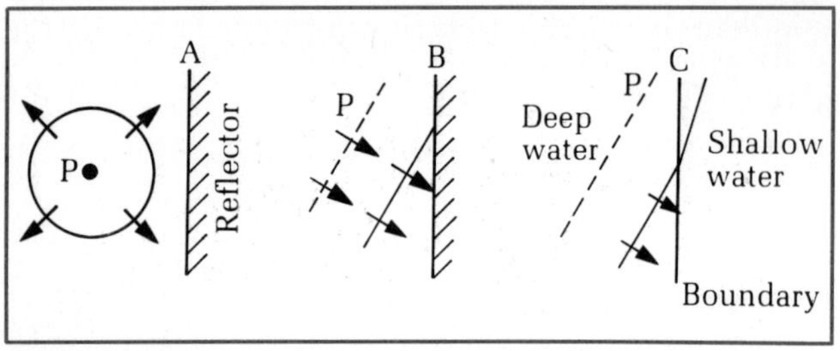

Figure 17.40

17.5 Plane waves on the water surface of a ripple tank travelling at a speed of $0.10\,m\,s^{-1}$ pass over a straight boundary from deep to shallow water where the wave speed is $0.15\,m\,s^{-1}$. If the incident wave fronts are at 30° to the boundary, calculate the angle between the boundary and the refracted wave fronts. Give a sketch of such a wave front half in deep and half in shallow water, and show the direction of each section of the wave front.

17.6 A cine camera is used to take a film of waves on a rope against a 1 m rule as a scale. When the film is projected onto a screen, the waves appear frozen as in the diagram below.

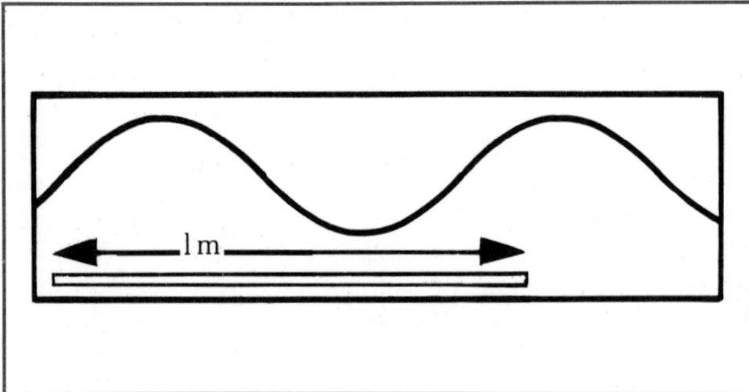

Figure 17.41

a) Can you tell if the waves were stationary or travelling waves from the picture?
b) The camera took the pictures at 20 frames per second. Assuming the waves were travelling (not stationary), estimate the least possible value for the wave speed.
c) Had the waves been stationary waves, how would the film differ if the cine camera had operated slightly more slowly?

17.7 Alternating current is passed along a 1.5 m length of wire stretched at constant tension between two fixed points. The wire passes between the poles of a horseshoe magnet at its centre. When the frequency of the alternating current is adjusted to 85 Hz, the wire vibrates with three antinodes along its length.
a) Sketch the pattern seen.
b) How does the phase vary along the wire compared with the midpoint?
c) Calculate the wavelength and the wave speed of the waves on the wire.

17.8 A student near a portable TV fitted with a loop aerial notices that moving 0.4 m nearer the set makes the picture fade, then return to its original quality as she moves away. Explain why this happens and calculate
a) the wavelength of the TV signal carrier waves,
b) the wave frequency.
(The speed of electromagnetic waves in air $= 3 \times 10^8\,m\,s^{-1}$.)

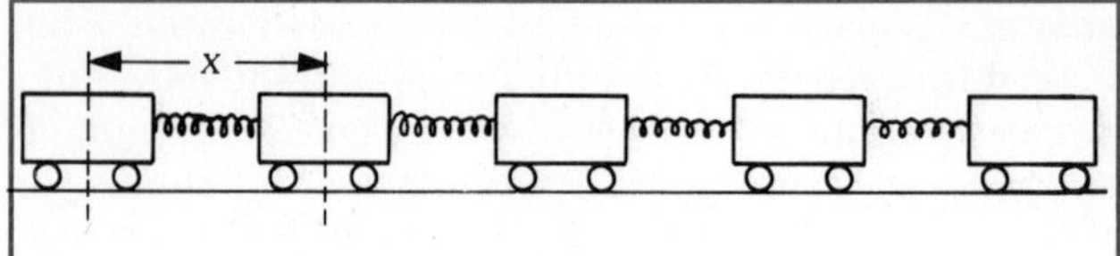

Figure 17.42

17.9
a) For a row of trolleys, each of mass m, linked by springs of stiffness (i.e. spring constant) k, the speed c of a compression pulse along the row is given by $c = x\sqrt{(k/m)}$ where x is the distance from the centre of one trolley to the centre of the next adjacent trolley when at equilibrium as in the diagram at the top of the next column. How long would it take a compression pulse to travel along a row of ten such trolleys, each of mass 0.8 kg, at spacing $x = 0.35$ m, given the stiffness $k = 25\,N\,m^{-1}$?
b) Now consider a row of atoms linked by atomic bonds as springs. The atoms belong to a solid of density ρ. (i) Show that the density of the solid $= m/x^3$ where m is the mass of an atom and x is the equilibrium spacing between adjacent atoms. (ii) Hence show that the wave speed of a pulse travelling through the solid is given by $c = \sqrt{k/\rho x}$. (iii) Show that the Young modulus for the solid $E = k/x$ and hence derive an expression for the wave speed c in terms of E and ρ. (iv) Calculate the wave speed of a pulse through aluminium, given $E = 0.7 \times 10^{11}$ Pa and $\rho = 2700\,kg\,m^{-3}$.

17.10 A Slinky spring is allowed to hang vertically with its lower end fixed.
a) Explain why the coils are more spaced out near the top of the Slinky compared with lower down.
b) When a transverse pulse is sent down the Slinky, why does it travel more slowly as it progresses downwards?
c) Sketch the shape of the Slinky at maximum displacement when oscillating in its fundamental mode due to its top end being moved from side to side at the right frequency.

18

Sound

18.1 Nature of sound waves

Producing sound waves

Most of us start our lives by producing sound waves! We spend much of our life surrounded by objects which produce sound waves. Any vibrating object in air produces sound waves which spread out through the surrounding air. Most machines in use vibrate so the only sure way to silence them would be to put them in vacuum chambers. In a vacuum there would be no surrounding medium for the vibrating surfaces of the machine to push against, hence no sound waves. Sound waves require a medium, which may be any solid, liquid or gas.

An example of sound travelling through a solid is when someone taps on heating pipes with, for example, a coin. The tapping noise can usually be heard by anyone in a different room if the pipes pass into it. Another example is to listen to your own voice first with then without a tape recorder. Played back from a tape recorder, your voice sounds different to you than it does when you speak. The reason is that when you speak your ears do not only hear sound waves arriving through the air; they also detect sound waves which pass through the inside of your head. The internal sound waves pass through bone and tissues which muffle the sound, causing the difference. Try it, if you have a tape cassette recorder available, then try to describe the difference in simple terms.

Sonar detection systems make use of the fact that sound waves pass through liquids. Submarine propellers produce sound waves in the surrounding sea water, and the sound can be detected by special underwater microphones called hydrophones. By trailing a hydrophone in the sea from a helicopter, the presence of submarines up to several kilometres away can be detected. See Figure 18.1.

Figure 18.1 *A helicopter trailing a sonar detector in the sea*

To understand the nature of sound waves, consider how they are produced by a loudspeaker connected to a signal generator, as in Figure 18.2. The alternating current supplied by the signal generator makes the loudspeaker coil vibrate to and fro, so the loudspeaker diaphragm connected to the coil is pushed to and fro. Each time the diaphragm is pushed forward, it compresses the surrounding air a little. The push gives momentum to surrounding air molecules which therefore move away. But they travel only a short distance before bumping into other air molecules, and they then pass their momentum on. So the compression moves through the air away from the diaphragm. In other words, a **compression** wave travels through the air. Meanwhile, the diaphragm is pulled back by the reversal of the alternating current. The effect is to give the surrounding air molecules more space. Nearby molecules move to fill the space, so the expansion moves away from the diaphragm. Molecules further away then move to fill the space created by nearby molecules, so the molecules further away leave a space, to be filled by molecules yet further away. In this wave, an expansion wave called a **rarefaction** travels away from the diaphragm. So the continuous motion of the diaphragm to and fro causes a continuous train of compressions and rarefactions to pass through the surrounding air. At any fixed position in front of the diaphragm, the molecules there bunch up as each compression passes. Then they space out as each rarefaction passes before the next compression. The pressure increases above the undisturbed level as each compression passes, then the pressure falls below the undisturbed level as a rarefaction follows each compression.

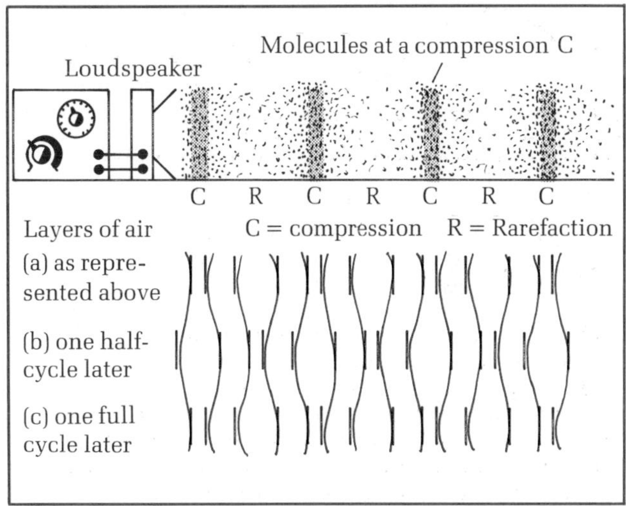

Figure 18.2 *Displacement of air layers*

Of course gas molecules move about at random. Diffusion experiments show that the molecules in a gas make little progress, and the reason is that they are continually colliding with one another. Each molecule bounces off one neighbour, and moves only a short distance before hitting another neighbour, then rebounding to hit another. With molecules all around, a molecule doesn't move very far unless the gas is at very low density. So when sound waves pass, compressions cause the molecules to push forward onto their neighbours which then push forward onto *their* neighbours, etc. Each rarefaction causes them to move backwards, as they move in to fill space, so leaving space for molecules further away to fill, etc. Each layer of molecules is alternately pushed forwards then backwards when sound waves pass. Sound waves are therefore **longitudinal waves** because the displacement is always along the line in which the wave travels. Figure 18.2 represents the movement of the layers of air when a sound wave passes.

All layers have equal amplitude. The phase of each layer depends on its position, one layer a little out of phase with the next, which is out of phase by the same amount with the next, and so on. So two layers separated by one complete wavelength have a phase difference of 2π radians, which means they are back in phase. Maximum pressure is at the centre of a compression where the layers become crowded together. Minimum pressure is where they are most spaced out. However, at the centre of each compression or rarefaction, each layer is at zero displacement. So the displacement zeros are at the same positions as the pressure extremes, and vice versa, at any given instant.

The passage of sound waves through a solid is easier to understand than through a gas. Atoms in a solid are held together by bonds. There are different kinds of interatomic bonds, and some of these are described on p. 68. They all have in common the fact that they lock the atoms together. The atoms do not move from their positions in the solid because of the bonds, but they vibrate about their own positions without changing position with neighbouring atoms. When sound waves pass through a solid, each atom is displaced first one way then in the opposite direction, then back again, etc. The situation is like a row of trolleys joined by stretched springs. If the end trolley is moved to and fro, the movement is passed along the row.

Describing sound waves

There is a huge difference between the sound of a *747* Jumbo Jet taking off and the sound from a well-played flute. Not only does the loudness differ, but the pitch and quality are very different too.

Loudness is determined by the amplitude of the sound waves. The sound from the loudspeaker of Figure 18.2 becomes louder when the signal generator voltage is increased. A bigger voltage makes the diaphragm of the loudspeaker vibrate with a bigger amplitude, so the amplitude of the sound waves is increased, and the sound is louder.

Pitch depends on the frequency of the sound waves. In Figure 18.2, if the frequency of the signal generator voltage is increased, the pitch of the sound becomes greater. In fact, most sounds are not pure sine wave notes like the sound from the loudspeaker of Figure 18.2. To investigate sounds from different sources, a microphone connected to an oscilloscope may be used. The oscilloscope trace can show the wave form of the sound waves as they are produced. Figure 18.3(a) shows the wave form of the sound from a loudspeaker connected to a signal generator. The wave form of the note from a flute, shown in Figure 18.3(b) is very different. Because Figure 18.3(b) is not a pure sine wave like Figure 18.3(a), its **quality** is not as high.

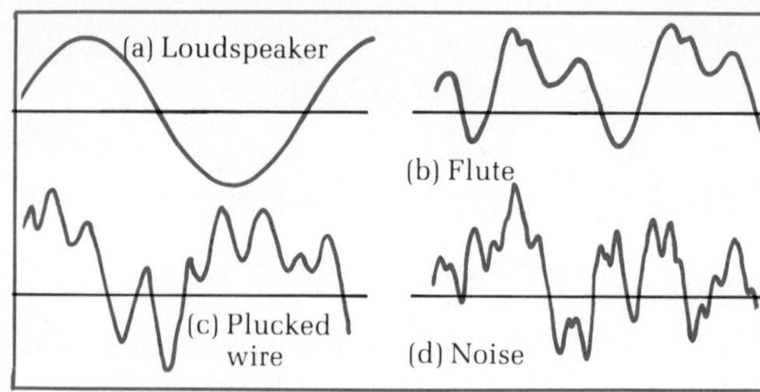

Figure 18.3 *Intensity*

Quality depends on how the sound is produced. For example, the sound from a vibrating wire depends on the mode (i.e. pattern) of vibration of the wire. Under special circumstances, the wire can be made to vibrate in one of its overtone modes or in its fundamental mode. See Figure 17.30 if necessary. Each mode has a fixed frequency of vibration, so emits sound waves of that frequency. When the wire is plucked it vibrates in a more complicated way, each mode making a contribution, so the sound from it contains several different frequencies, giving a complicated wave form like Figure 18.3(c).

A sound can be classed from its wave form as either a noise or a note.

Noise occurs where the wave form shows irregular changes, as in Figure 18.3(d), so is unpleasant and unwanted.

Notes have wave forms with changes which repeat regularly, as Figure 18.3(c), so notes are easier to listen to, and a high-quality note contains fewer frequencies than a low-quality note.

Decibels

Sound waves from a loudspeaker carry energy away from the loudspeaker. The intensity of the sound becomes smaller further from the loudspeaker because the sound waves spread out. So a listener far from the speaker receives less sound energy per second than one closer to the speaker.

Intensity of sound is defined as the sound energy per second arriving at unit area, the area being placed at right angles to the sound direction. The unit of intensity is $J\,s^{-1}\,m^{-2}$ (or $W\,m^{-2}$).

The average human ear can detect sounds at frequencies up to about 18 000 Hz, and is at its most sensitive at about 3000 Hz. The least intensity of sound which the ear can detect is an incredibly low $10^{-12}\,W\,m^{-2}$ (= 1 picowatt/m^2). At the other extreme, sound intensities much over $1\,W\,m^{-2}$ causes pain, with permanent loss of hearing at higher intensities. The ear therefore has an enormous range of sensitivity from 10^{-12} to $1\,W\,m^{-2}$, a factor of a million million. Yet judgement of loudness is much more limited. Experiments show that the ear's response (i.e. the subject's judgement of loudness) varies **logarithmically**

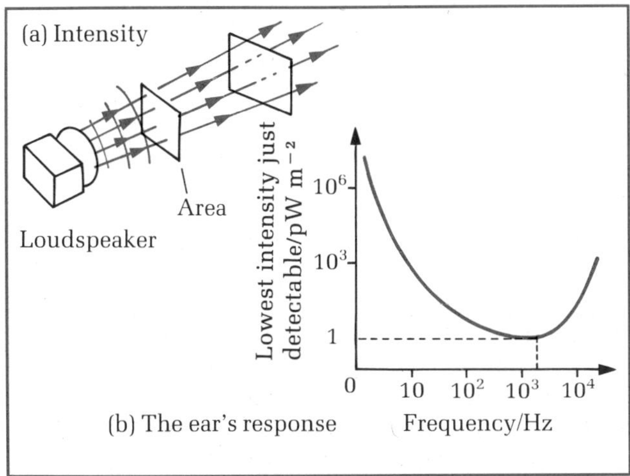

Figure 18.4 *The ear's response*

with intensity, as shown in Figure 18.4(b). So for a note of constant frequency, increase of loudness in equal steps is caused by increasing the intensity by a constant factor for each step (e.g. × 10 each step).

Intensity level (IL) of a sound is defined as

$$\text{IL} = \log_{10}\frac{I}{I_0}$$

where I = the intensity of the sound,
I_0 = 1 picowatt (the minimum detectable intensity).

The **bel** (B) is the unit of intensity level, although the **decibel** (dB) is more commonly used. One bel equals ten decibels. The table in Figure 18.5 shows intensity levels calculated for various intensity values.

Because intensity level is defined as a logarithmic scale, then it closely matches our judgement of loudness. a **decibel meter** may be used to measure loudness. For example, if such a meter reads 40 dB in a quiet room, when a radio is turned on in the room the reading increases. Suppose the radio is turned up gradually until the meter reading is 80 dB. Every 10 dB increase corresponds to the same increase of loudness.

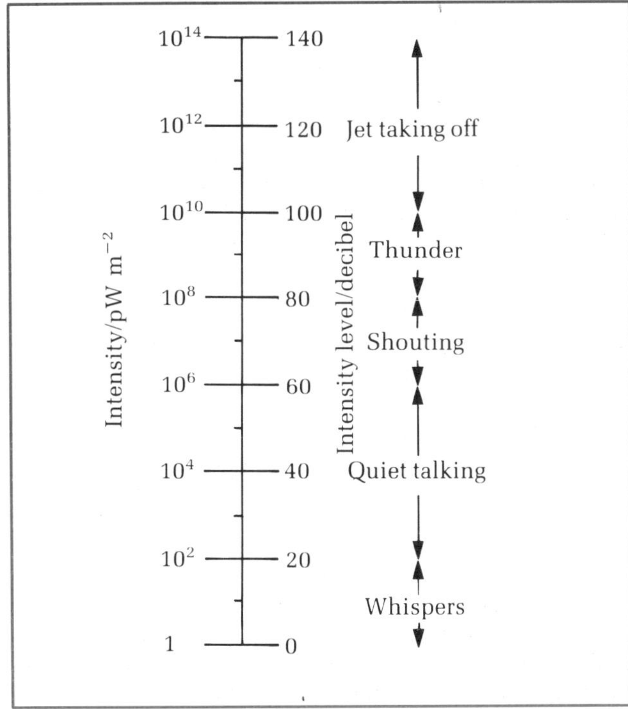

Figure 18.5 *The decibel scale*

Ultrasonics

High frequency sound waves beyond the range of the human ear are called **ultrasonics**. So ultrasonics cover a frequency range from 18 000 Hz upwards. Compared with sonics (i.e. sound waves we can hear) ultrasonics have shorter wavelengths because their frequencies are higher. For example, ultrasonics of frequency 40 kHz are used for industrial cleaning. In air where the speed of sound is approximately 330 m s^{-1} the wavelength of these ultrasonic waves is 8.25 mm ($= c/f = 330/40\,000$ m). By comparison, sound waves of frequency 1000 Hz in air have wavelength of 0.33 m. Equipment to be cleaned by ultrasonics is placed in a tank of water which ultrasonics pass through. The ultrasonic waves pass through the water, agitating and loosening particles of dirt and grease.

Another use of ultrasonics is in medical imaging. For example, they are used in prenatal care to give an image of a baby inside the womb. Ultrasonic scans do not harm the baby, and are much safer than X-ray scans in this situation. To produce the ultrasonic waves, an ultrasonic **transducer** is used. This converts electrical energy into ultrasonic energy by applying an alternating voltage across a quartz crystal. Quartz is used because it changes length slightly when a voltage is applied across it. So an alternating voltage makes the quartz vibrate. By making the applied frequency equal to the natural frequency of vibration of the crystal, the vibrations become very large; the crystal resonates. So the crystal produces ultrasonic waves of frequency equal to its natural frequency. The

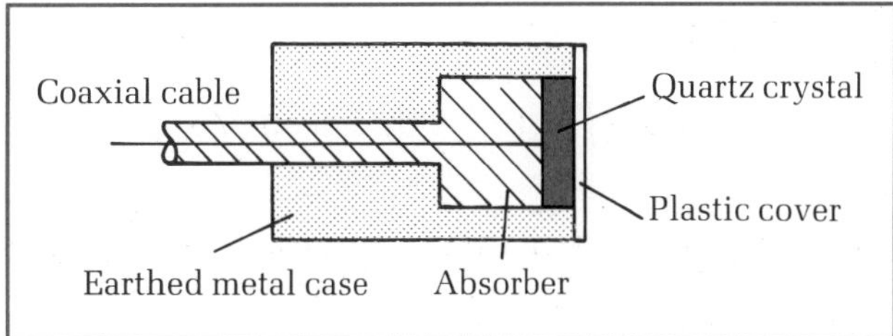

Figure 18.6 *An ultrasonic transducer*

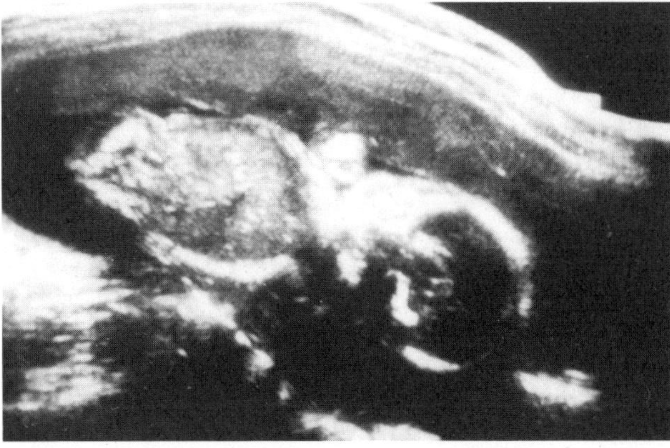

Figure 18.7 *An ultrasonic scan picture of a foetus in a womb*

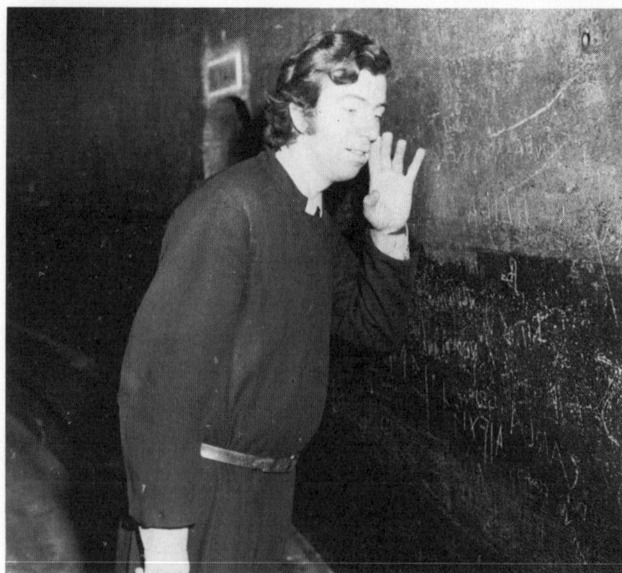

Figure 18.8 *The Whispering Gallery, St Paul's Cathedral*

frequency is in the megahertz (MHz) range, so the ultrasonics pass directly into the body when the transducer is placed on the body surface. Inside the body, tissue boundaries reflect part of the incoming ultrasonic energy, and a second transducer can be used to detect the reflected ultrasonics. The second transducer converts ultrasonic energy back to electrical energy, so enabling an image on an oscilloscope to be built up showing the internal boundaries. Linked with modern computers, ultrasonic scanning has brought great benefits to medicine.

18.2 Properties of sound

Reflection

In a sports hall or a gymnasium, sounds do not die away as quickly as in a library. The bare walls of a sports hall reflect sound waves much better than the book-lined walls of a library. Sound waves obey the rules of reflection just as any other type of wave. A smooth bare wall is a good reflector of sound waves for two main reasons:

a) Most of the incident sound wave energy is reflected from a smooth bare wall.

b) A smooth wall reflects the sound waves evenly. A rough wall would scatter the reflected waves.

Figure 18.8 (above) shows the famous Whispering Gallery in St Paul's Cathedral where faint sounds made on one side can be heard on the other side more than 30 m away. The reason is that the concave walls reflect and concentrate sound waves.

Echoes can easily be created in a large hall such as a gymnasium. Sound reflects off one wall, travels to another wall and reflects, and so on. Someone at the centre of the floor who makes a single hand clap may well hear several echoes. The hall is said to **reverberate** with sound because the sounds live on. Concert halls such as the one shown in Figure 18.9 are specially designed to prevent undue reverberation. The walls are draped with suitable absorbing materials like curtains or soft wallpaper.

Figure 18.9 *A concert hall*

However, if there is too much absorption and hardly any reflection, then the audience hears deadened sounds. A well-designed concert hall even takes account of the audience as absorbers of sound, just like the curtains and upholstery, so a concert in a full hall may indeed sound better than in a half-empty hall! To test sound equipment, special rooms with absorbing surfaces all over the floor, walls and ceiling are used. Voices are very different in such rooms because there are no reflections.

Refraction

In air, sound travels faster where the temperature is higher. At night, air temperatures near the ground are often lower than in the air higher up, and this causes sound to travel further at night. When waves change speed at a boundary, the wave direction changes if they arrive at an angle to the boundary. Figure 17.17 (on page 234) shows the change of direction of water waves in a ripple tank when moving from deep to shallow water. At the boundary, the waves slow down so they change direction. When sound waves pass from cold air to warmer air, their direction changes. Because they speed up as they move into warmer air, their direction changes away from the vertical. The vertical is the direction of the normal to

Figure 18.10 *Refraction of sound waves*

the boundary between cold and warm air here. Thus, sounds created at the surface are refracted. Instead of travelling upwards as they would in the daytime, they travel further horizontally.

Interference and diffraction

Sound waves diffract (i.e. spread) round corners and obstacles. Conversation in a corridor can be heard in a room with a door open to the corridor, even if the conversation is not taking place right outside the door. The sound waves reaching the doorway spread out into the room. The extent of the diffraction (i.e. spreading) depends on the sound wavelength in relation to the doorway width. Figure 17.21 shows how different gap widths with the same wavelength produce different amounts of diffraction. With a small wavelength in comparison with the door width, there is hardly any diffraction at all.

To demonstrate interference of sound waves, a signal generator and two loudspeakers may be used. The two speakers are connected in series with one another, then connected to the signal generator, as in Figure 18.11.

The two speakers operate in phase because they are connected in series. They are coherent emitters of sound

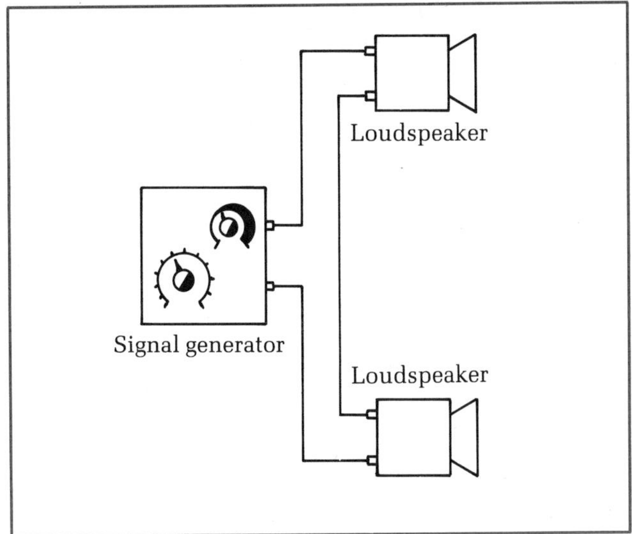

Figure 18.11 *Interference of sound waves*

waves because the phase difference does not change. By walking slowly in front of the two speakers, a careful listener ought to be able to locate points of cancellation and reinforcement.

Suppose we use a frequency of 2000 Hz. What should the spacing between the two loudspeakers be? With a listener about 2 m in front of the speakers, we can calculate the speaker spacing using the double slits equation of p. 260. The equation may be written as

$$\frac{y}{X} = \frac{\lambda}{d}$$

where y = fringe spacing,
X = slit-screen spacing,
d = distance between the double slits,
λ = wavelength.

We want to apply this equation to the twin speaker interference arrangement of Figure 18.11. Given the speed of sound in air is $330\,\mathrm{m\,s^{-1}}$, the wavelength may be calculated. The listener detects points of cancellation by moving parallel to the line between the speakers, so the fringe spacing y is represented by the distance between adjacent zeros. A reasonable value for y for sound interference would be a little larger than the size of a normal head, on the basis of sharp minima not too close for the ear to detect. Finally, the slit-screen spacing X is represented by the distance from the listener to the speakers, say 2 m. With your estimates and calculations for y, X and λ, calculate the value of d which represents the distance between the speakers.

To prove that the points of cancellation are indeed caused by interference try short-circuiting across the terminals of one of the loudspeakers. Reduce the sound level first though or the other speaker becomes too loud. Before shorting out, locate by ear a point of cancellation. Then ask a friend to short out one of the two speakers. The point of cancellation will then cease to exist because waves from one speaker only are produced.

With both speakers operating, what difference would you expect if the speakers are moved closer together? Look to the double slits equation again. With λ and X fixed, how does y change when d is reduced? The points of cancellation become more widely spaced when the speakers are moved together.

Beats

With a guitar, you can produce beats by first tuning two strings almost to the same note. When the two strings are plucked at the same time, the sound from them comes and goes continuously. In other words, the intensity rises and falls continuously. The same effect can be demonstrated with two loudspeakers, each operated by a different signal generator. The listener ought to be able to detect when the frequencies are close in value. Beats can then be heard if the two sounds are equally loud. Adjusting one signal generator only to make the two frequencies even closer causes the beats to become more spaced out in time.

A beat of sound is heard when the two sets of sound waves arrive in phase at the ear. The principle of superposition tells us that the waves reinforce one another when they arrive in phase, so maximum intensity is heard.

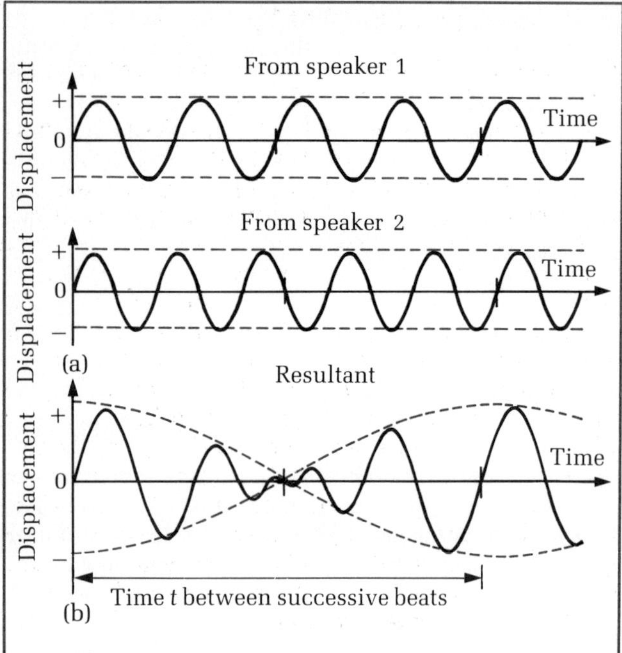

Figure 18.12 *Beats*

Because the two frequencies are different, a little later the waves arrive out of phase by 180°, so they cancel one another out and the sound fades. A little later again, and the two sets of waves are back in phase, hence another beat; and so it continues.

Figure 18.12 shows how the displacement of each sound waves varies with time at the position of the listener. One set of waves has a slightly higher frequency than the other. Using the principle of superposition, the resultant wave form can be established from the two individual sets of waves. Figure 18.12(b) shows the resultant wave form of the two sets of waves in Figure 18.12(a). The amplitude of the resultant wave form varies from zero to a maximum. Each beat is caused by maximum amplitude arriving at the ear.

To determine the time between successive beats, suppose the higher frequency is f_1 and the lower frequency is f_2. Let T be the time between successive beats. From one beat to the next, the higher frequency source emits f_1T cycles, and the lower frequency source emits f_2T cycles. Because the two sets of waves are back in phase, then the higher frequency source must have emitted **exactly one more** cycle in that time than the other source.

So $f_1T - f_2T = 1$ which gives $T = \frac{1}{(f_1 - f_2)}$. Therefore the beat frequency f_B is given by

$$f_B = 1/T = (f_1 - f_2)$$

$$f_B = f_1 - f_2$$

Beats are produced with a frequency equal to the difference between the frequencies of the two sets of waves which combine.

Measuring the speed of sound in air

A loudspeaker connected to a signal generator is used to produce sound waves. A microphone is used to detect the sound waves from the loudspeaker, as shown in Figure 18.13. The microphone is connected to one of the Y-inputs of a double beam oscilloscope. The oscilloscope controls must be adjusted to enable the sound waves to create a trace on the oscilloscope screen. It may be necessary to adjust the time-base control and the Y-input sensitivity control. Connections are also made from the other Y-input to the loudspeaker terminals.

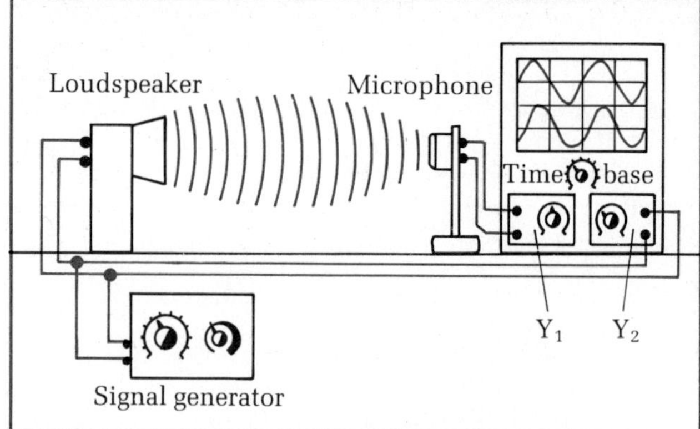

Figure 18.13 *Speed of sound in air*

The screen ought to show two sets of sine waves, and it ought to be possible to arrange one set above the other, using the Y-shift controls. One set of sine waves is due to the direct signal from the loudspeaker to the oscilloscope; the other set is via the microphone. Move the microphone away from the speaker and the trace from the microphone ought to move relative to the other trace. The reason is that the waves from the speaker take longer to reach the microphone when the microphone is moved away, so the oscilloscope spot, tracing the waves from left to right, forms the microphone display further and further to the right as the movement continues. With the two sets of waves initially in phase on the screen, moving the microphone makes the two sets move out of phase. But continued movement in the same direction brings the two sets back in phase; the wave peaks of one set have then caught up with the adjacent wave peaks of the other waveset. The microphone has then been moved through exactly one wavelength. By moving the microphone through a fixed number of complete cycles (i.e. wavelengths), its total displacement may be measured. Then the wavelength can be calculated.

The frequency of the sound waves may be measured from the oscilloscope, provided the time-base is calibrated. The horizontal distance across the screen for several cycles of one of the traces is measured. Using the time-base calibration, the time for one cycle (i.e. time period) can then be determined. Hence the frequency may be calculated. Finally, the speed of sound in air c may be calculated using the equation $c = f\lambda$.

Other possibilities with this experiment are:

a) to investigate if the speed of sound in air varies with frequency,

b) to investigate if the speed of sound in tubes of air differs from the value in open air. A small microphone to fit inside the tube would be needed here. Also, tubes of different diameters could be tried if available.

A single beam oscilloscope may be used for this experiment. The X-input is used instead of the second Y-input.

With the time-base off, the screen trace shows an ellipse. When the microphone is moved away from the speaker, the ellipse becomes a line then a circle before reforming as an ellipse. Separation by one complete wavelength takes the pattern through one complete cycle (e.g. line with positive slope – ellipse – circle – ellipse – line with negative slope – ellipse – circle – ellipse – back to line with positive slope). (See Lissajous' figures, p. 425.)

18.3 Resonance of air columns

When a loudspeaker producing sound is placed near the end of a hollow tube, the tube resonates with sound at certain frequencies. Sound waves travel along the tube, reflect at the far end and return to the loudspeaker end. So stationary waves are set up inside the tube because two sets of waves of the same frequency pass through one another. See p. 239 if necessary.

Why should the tube resonate with sound at certain frequencies only? At resonance, the loudspeaker emits wave peaks and wave troughs towards the far end of the tube. The waves reflect at the far end and return. If each returning wave peak reaches the loudspeaker just as a new wave peak is being produced, the wave peak is given a little more energy than otherwise. So the amplitude builds up until the tube resonates with sound. The best position for the loudspeaker is a little beyond the end of the tube, so at that position, that is an **antinode** (i.e. maximum amplitude). There may be other antinodes along the tube, and at these positions, the two sets of waves always reinforce one another. Where the two sets of waves always cancel one another, the amplitude is always zero. These positions are the **nodes**.

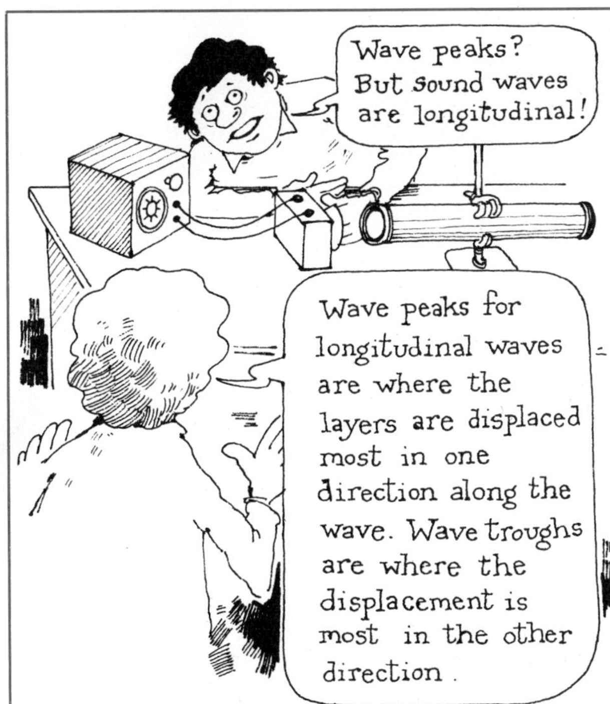

Figure 18.14 *Sound wave peaks*

Tube closed at the far end

At a closed end, there can be no displacement of the layers of air because the end is rigid. The waves must change phase by 180° when they reflect at a closed end. To illustrate this, tie one end of a string to a fixed point. Flick the other end to make a wave peak pass along to the fixed end. The wave reflects, but returns as a wave trough!

So a wave peak sent out returns as a wave trough and resonance occurs if a new wave trough is being created when the returning wave trough reaches the loudspeaker. Therefore, the time from a wave peak being created to a wave trough being created must equal the time for sound to pass along the tube and back. From wave peak to wave trough can be half a cycle, or it can be $1\frac{1}{2}$ cycles, or $2\frac{1}{2}$ cycles, or any whole number plus a half cycle. At frequency f, the time should therefore be $1/2f$ or $3/2f$ or $5/2f$ etc. For a tube of length L, the time for sound to travel along and back is given by distance/speed $= 2(L+e)/c$ where e is the distance from the speaker to the end of the tube. The speed of sound in the tube is represented by c.

At the **fundamental resonance**

$$1/2f = 2(L+e)/c$$

Since the wavelength λ of sound in the tube equals c/f then it follows that

$$\lambda = 4(L+e)$$

This result gives one-quarter of a wavelength from the loudspeaker to the closed end. So it corresponds to an antinode at the loudspeaker and a node at the closed end, with no other nodes or antinodes between.

Overtones are the resonances at higher frequencies. For the tube closed at one end, there is always an antinode at the open end and a node at the closed end. But overtones have intervening nodes and antinodes. Figure 18.15(b) shows the displacement pattern for several overtones. There is always an odd number of quarter wavelengths between the loudspeaker and the closed

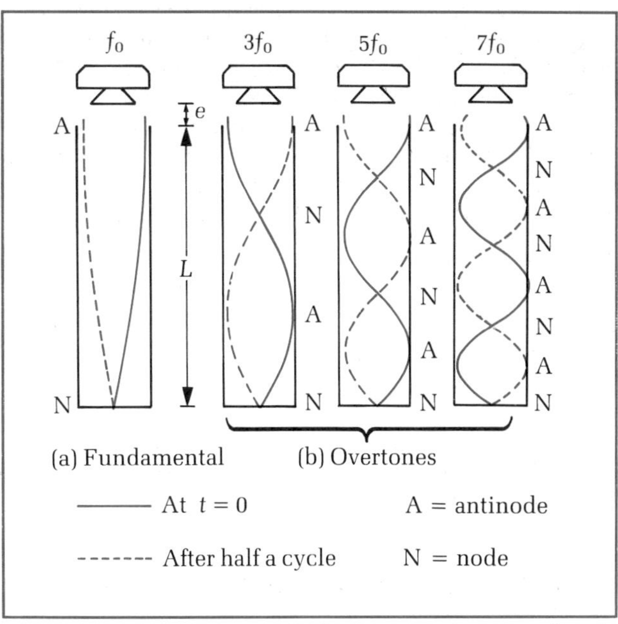

Figure 18.15 **a** Fundamental **b** Overtones in a pipe closed at one end

end. You can fit a quarter of a wavelength (for the fundamental) or three-quarters of a wavelength (for the first overtone) or five-quarters of a wavelength (for the second overtone), etc. Each extra half-wavelength fitted in makes an extra loop in the pattern. So the waves travelling along the tube take an extra half cycle in each direction, giving an extra full cycle altogether.

Tube open at the far end

Sound waves passing along an open-ended resonance tube (i.e. open at both ends) are **partly reflected** when they reach the open end. Most of the sound energy passes to the outside at the open end, but some sound energy returns along the inside, carried by reflected sound waves. However, there is no phase reversal on reflection at the open end. This is not unlike waves along a rope if the far end is not tied down. Flick the near end (held in the hand) and a wave peak passes along the rope to the far end. At the far end, the disturbance reflects and returns on the same side (i.e. as a wave peak). So, at the far end there is an antinode. In fact, in the open tube example, the reflection occurs just a little beyond the actual end of the tube. This extra distance e is the end-correction.

For resonance, the returning wave peak must reach the sound source (e.g. loudspeaker) just as the source produces a new wave peak. So the time for the wave peak to travel the tube length and back must be a complete number of cycles.

The **fundamental resonance** occurs when the trip time (i.e. time for wave peak to travel to the other end and back) is exactly equal to one complete cycle.

$$\frac{1}{f_0} = \frac{2(L+2e)}{c}$$

So the wavelength is given by

$$\lambda_0 = c/f_0 = 2(L+2e)$$

where L = tube length,
e = end correction at either end,
c = speed of sound in the tube.

The distance from the speaker to the position of reflection at the far end, $L+2e$, is exactly one half-wavelength, so it corresponds to an antinode at the speaker and another antinode just beyond the far end, with only one node between.

Overtone resonances must always have an antinode at either end too, if open at both ends. Figure 18.16(b) shows the displacement pattern for several overtones. There is always a complete number of half-wavelengths from one end to the other. For example, the first overtone has exactly two half-wavelengths in the tube; so its wavelength λ_1 is exactly equal to $(L+2e)$. Therefore, the frequency of the first overtone f_1 is exactly $2f_0$. This corresponds to the trip time equal to exactly two cycles. Other overtone resonances occur at $3f_0$, $4f_0$, $5f_0$ etc.

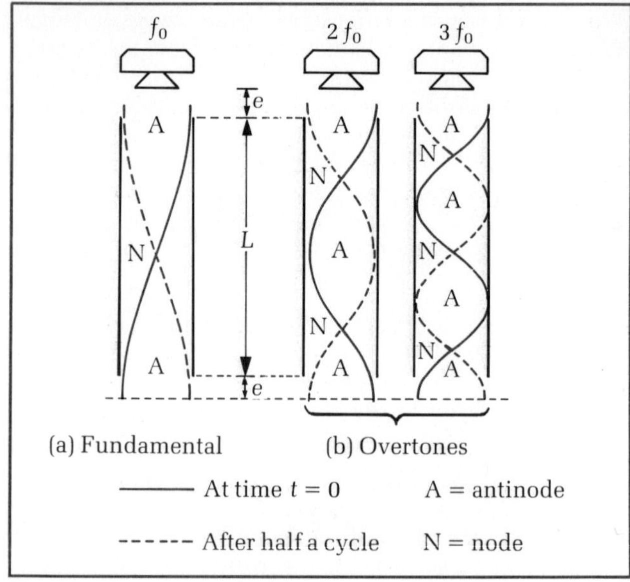

Figure 18.16 **a** *Fundamentals* **b** *Overtones in a pipe open at both ends*

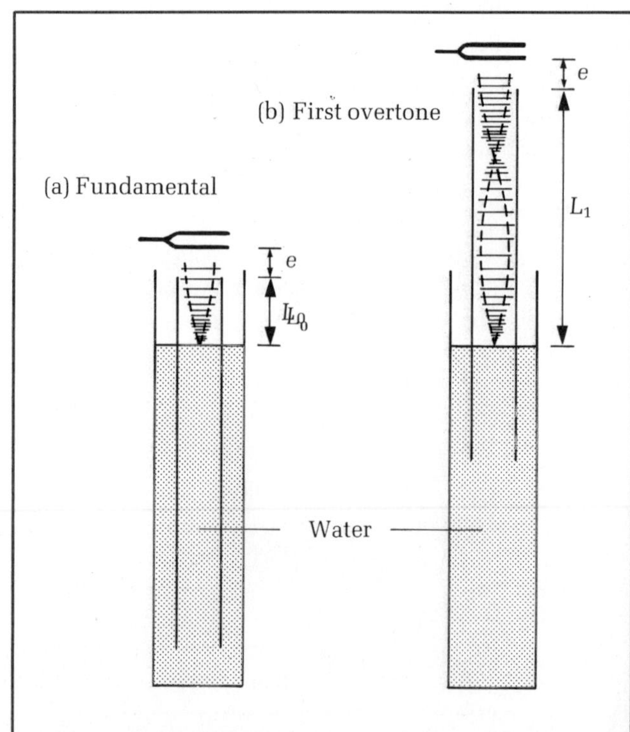

Figure 18.17 *Investigating resonance*

Investigating resonance in air columns

Use a tuning fork of known frequency to obtain resonances in an air column of variable length. One arrangement for varying the length is shown in Figure 18.17. The inner glass tube is moved up and down, with its lower end beneath the water level in the wide tube. The length of the air column in the inner tube may be measured using a metre rule. The best way to proceed is to start with the air column length almost zero. Hold the vibrating tuning fork as in Figure 18.17(a), and gradually raise the inner tube. The tube ought to resonate with sound at a certain length. Assuming this is the shortest length for resonance, then it is the fundamental resonance. Measure the resonant length L_0 accurately.

In the same way it ought to be possible to locate resonant lengths for the same tuning fork at greater lengths than L_0. Measure the resonant length L_1 for the first

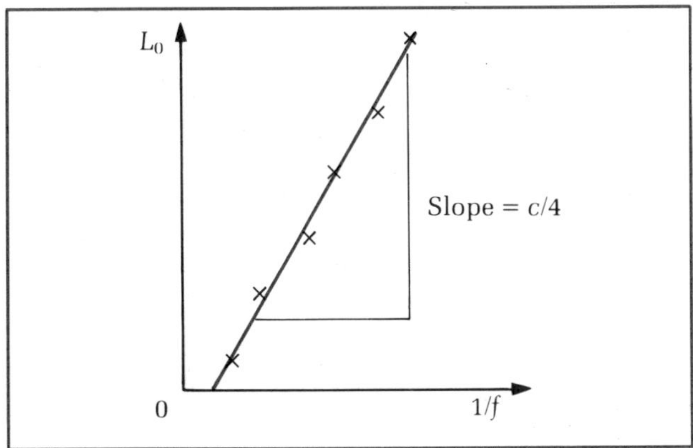

Figure 18.18 *The graph method*

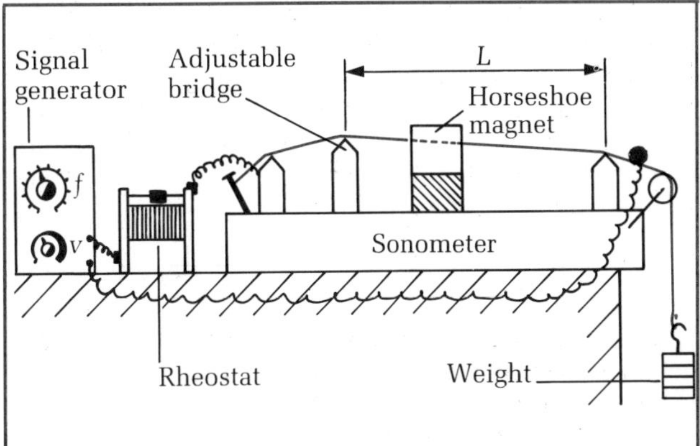

Figure 18.19 *The sonometer*

overtone. Since the tube is closed at its lower end, the resonant lengths are given by

$L_0 + e = \lambda/4$ for the fundamental resonance

and

$L_1 + e = 3\lambda/4$ for the first overtone

where λ is the wavelength of sound produced by the tuning fork.

Hence, the wavelength may be calculated from

$$\lambda = 2(L_1 - L_0)$$

If the frequency of the tuning fork is known, then the wave speed of sound in the tube may be calculated from $c = \lambda f$. If a number of tuning forks of known frequencies are available, the speed of sound may be calculated for each frequency. You ought to find that the speed of sound in air does not vary with frequency.

A graph method for determining c is to use only the fundamental lengths for each tuning fork. Each length L_0 is given by

$$L_0 + e = \lambda/4 = c/4f$$

where f = frequency of each tuning fork, e = end correction.

So a graph of L_0 against $1/f$ ought to give a straight line. See Figure 18.18. By comparison with $y = mx + k$ (where m = slope, k = y-intercept), the above equation for L_0 shows that the slope of L_0 against $1/f$ is $c/4$. Hence, the value of c can be determined from the slope of the graph.

18.4 Vibrations of strings and wires

To set up stationary waves on a stretched string, the arrangement shown in Figure 17.30 may be used. A metal wire under tension can be made to vibrate in the same fashion if arranged as shown in Figure 18.19.

The wire is part of a **sonometer**. The wire passes over a pulley and supports a set of weights as shown. In this way, the wire is put under tension. The length of the vibrating section may be altered by moving the adjustable bridge away or towards the fixed bridge. The wire is connected in series with a variable frequency signal generator and a rheostat. A horseshoe magnet is placed across the middle of the wire.

When a current passes along the wire, the wire experiences a force due to the magnetic field. When the frequency of the signal generator is increased from a low value, the wire resonates when the signal frequency equals the fundamental frequency of the wire. The vibration pattern at the fundamental frequency has a node at either end, and an antinode at the middle, as explained on p. 238. If the signal frequency is increased further, the fundamental pattern is lost. However, overtone patterns appear at signal frequencies of $2f_0$, $3f_0$, $4f_0$, etc. In each case for the best effect, the magnet should be positioned at an antinode.

What causes each pattern? Consider what happens at the fundamental frequency. The wire is pushed first to one side then to the other side, then back again, etc., for each cycle of the alternating current. Each push creates wave peaks that travel to either end. The wave peaks reflect at either end (i.e. at the bridges), return as wave troughs, and pass through each other. So stationary waves can be set up. For resonance, wave peaks or troughs must return to the centre exactly when new wave peaks or troughs are being produced at the centre. So the trip time (centre to end and back) must be exactly half a cycle. Then each wave peak sent out from the centre reflects at the end and returns as a wave trough, reaching the centre exactly as the next wave trough is being produced. Remember that on reflection at a node, the phase is reversed. (See p. 237 if necessary.) So the returning wave trough adds its energy to the new wave trough and the vibrations build up to resonance.

Since the trip time is L/c, where L is the wire length and c is the wave speed, then it follows that the fundamental frequency f is given by

$$\frac{L}{c} = \frac{1}{2f_2} \quad (= \text{time for half a cycle})$$

so the fundamental wavelength $\lambda_0 = c/f_0$, is given by

$$\lambda_0 = 2L$$

This agrees with the simple picture of nodes at either end with one antinode at the centre. The wire length is exactly half a wavelength at the fundamental resonance.

The overtone frequencies correspond to the trip time being 1 cycle (for the first overtone) or $1\frac{1}{2}$ cycles (for the

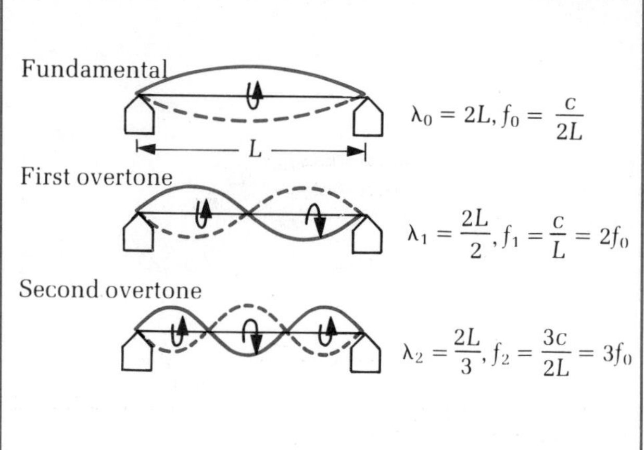

Figure 18.20 *Resonances of a vibrating wire*

second overtone), etc. So the first overtone occurs at a frequency equal to $2f_0$, corresponding to 1 wavelength along the wire as shown in Figure 18.20. The second overtone occurs at a frequency equal to $3f_0$, corresponding to $1\frac{1}{2}$ wavelengths on the wire, etc.

When the wire vibrates, it creates sound waves which travel through the surrounding air. The frequency of the sound waves will be equal to that of the wire. However, the wire produces a stationary wave pattern only in special circumstances, like the arrangement of Figure 18.19. When simply plucked at the centre, it produces sound waves with a mixture of frequencies composed of the fundamental and overtones. Try to tune a wire to the same note as a tuning fork. When the wire is tuned, the tuning fork frequency equals the fundamental frequency of the wire. Yet the two notes differ because the wire produces overtones as well. It takes a trained ear to pick out the fundamental frequency from the mixture produced by the wire.

Factors that determine the fundamental frequency

The note from a guitar string can be pitched higher by tightening the string or by making the string shorter. Also, a light string produces a note of higher pitch than a heavier string with the same tension and length. So the fundamental frequency of a vibrating wire depends on

a) its tension T,
b) its mass M,
c) its length L.

One way to find out how the frequency depends on these factors is to use the method of 'dimensional analysis' as described on p. 512. The idea is to assume that the fundamental frequency f_0 is given by an equation of the form

$$f_0 = kT^xM^yL^z$$

where k is a constant, and x, y and z are to be determined as follows.

Each quantity is expressed in terms of the three basic dimensions which are **M** (for mass), **L** (for length) and **T** (for time). Using the square bracket to mean 'the dimensions of', we can use the dimensions in the equation above to work out the values of x, y and z.

The dimensions of $f_0 = [f_0] = \mathbf{T}^{-1}$
$[T] = \mathbf{MLT}^{-2}$
$[M] = \mathbf{M}$
$[L] = \mathbf{L}$

From the equation above,

$$f_0 = kT^xM^yL^z$$

which gives $\mathrm{T}^{-1} = [\mathbf{MLT}^{-2}]^x[\mathbf{M}]^y[\mathbf{L}]^z$

so $\mathrm{T}^{-1} = \mathbf{M}^{(x+y)}\mathbf{L}^{(x+z)}\mathbf{T}^{-2x}$

and comparing powers of each dimension on either side,

$0 = x + y$ (from the powers of M)
$0 = x + z$ (from the powers of L)
$-1 = -2x$ (from the powers of T)

which gives $x = \frac{1}{2}$, $y = -\frac{1}{2}$, $z = -\frac{1}{2}$.
So the original equation for f may now be written as

$$f_0 = kT^{\frac{1}{2}}M^{-\frac{1}{2}}L^{-\frac{1}{2}} = k\sqrt{\frac{T}{ML}}$$

Now the wavelength $\lambda_0 = 2L$, so the speed of waves on the wire is given by wave speed c, where

$$c = \lambda_0 f_0 = 2Lk\sqrt{\frac{T}{ML}} = 2k\sqrt{\frac{T}{M/L}} = 2k\sqrt{\frac{T}{\mu}}$$

where μ is the mass per unit length M/L.

In fact, a more detailed analysis shows that the wave speed is given by

$$c = \sqrt{\frac{T}{\mu}}$$

therefore the constant k must have a value of 0.5. So we can state the equation for the fundamental frequency as

$$f_0 = \frac{1}{2L}\sqrt{\frac{T}{\mu}}$$

where T = tension,
μ = mass per unit length,
L = length of vibrating wire.

Investigating notes from a plucked wire

The fundamental frequency depends on the tension, length and mass of the wire. A sonometer may be used to investigate the relationship by experiment.

At constant tension The wire is set under constant tension by hanging a weight from its end. A set of tuning forks of known frequency is used. The wire length is adjusted by sliding the movable bridge along under it. For each tuning fork, the wire length is adjusted until its note

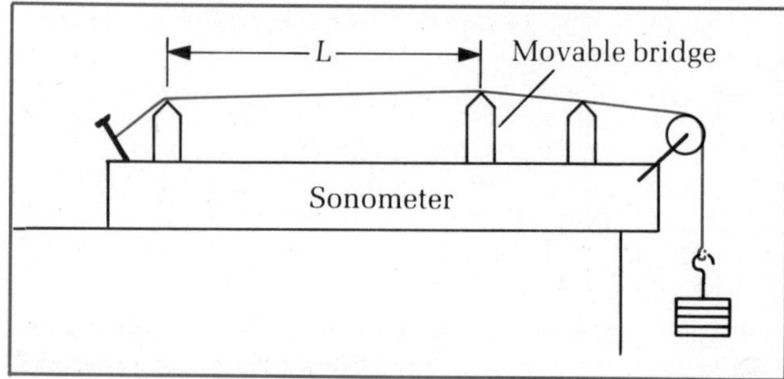

Figure 18.21 *Tuning a sonometer*

is the same as that of the tuning fork. When matched, a paper rider balanced on the wire ought to fall off when the tuning fork is sounded. The sound waves from the tuning fork cause the wire to resonate when matched. The procedure is repeated for each tuning fork, to give a set of results for frequency and length. A graph of frequency against 1/length ought to give a straight line through the origin. At constant tension using the same wire, the equation for the fundamental frequency is written as

$$f=\frac{1}{2}\sqrt{\frac{T}{\mu}}\times\frac{1}{L}=\frac{\text{constant}}{L}$$

At constant length The tension is changed by altering the weight attached to the end of the wire. A loudspeaker and signal generator may be used to give a note of variable frequency. For fixed length, the weight is increased in steps. At each step, the wire is plucked and the loudspeaker note is tuned in to the wire. When matched, the signal frequency is read from the dial of the signal generator, and the frequency and weight are recorded. In this way a set of readings of frequency and tension (= weight) can be obtained.

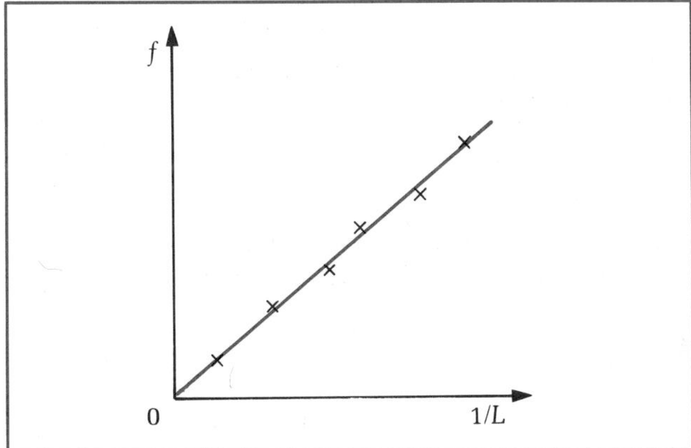

Figure 18.22 *Frequency v l/length*

Suppose you did not know what link to expect between f and T. Your results may be used to determine the link by plotting a log graph. Assume the link is in the form

$$f=kT^{x}$$

where k is a constant, and x is to be determined by experiment.

Now we take logs on each side of the equation to give

$$\log f=\log k+x\log T$$

(See p. 503 if necessary.)

So by plotting a graph of $\log f$ against $\log T$, ($\log f$ up the vertical axis) we ought to obtain a straight line of slope x, as shown in Figure 18.23. By measuring the slope from the graph, the value of x can be determined. Of course the full equation for f shows that $f=\text{constant}\times T^{\frac{1}{2}}$, so the value of x is 0.5. A graph of f against $T^{\frac{1}{2}}$ could have been plotted instead to give a straight line. This would have confirmed the link, but the log graph is much better because it enables us to find the link if it was not known.

Worked example A steel wire hangs vertically from a fixed point, supporting a weight of 80 N at its lower end. The length of the wire from the fixed point to the weight is

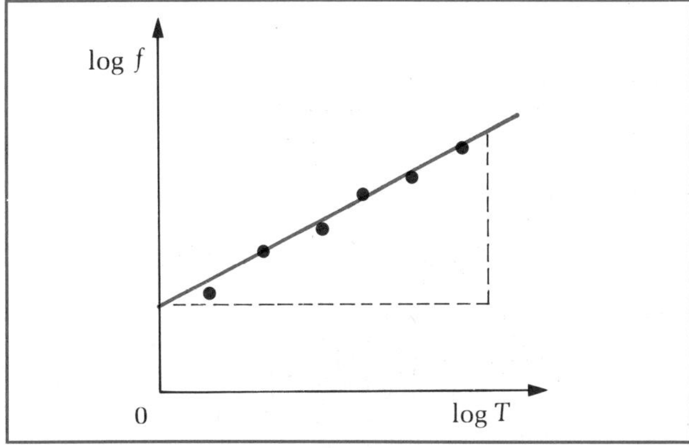

Figure 18.23 *log f v log T*

1.5 m. Given the density of steel is 7800 kg m^{-3}, calculate the fundamental frequency emitted by the wire when it is plucked, if its diameter is 0.50 mm.

Solution First the mass per unit length must be calculated. The wire should be considered as a solid cylinder of length 1.5 m and diameter 0.50 mm.

The volume of the wire = length × area of cross-section

mass = volume × density and

the mass of wire = length × area of cross-section × density.

Therefore

So the mass per unit length is given by

$$\mu=\frac{\text{mass}}{\text{length}}=\text{density}\times\text{area of cross-section}$$

which gives

$$\mu=7800\times\frac{\pi(0.5\times10^{-3})^2}{4}\ \text{kg m}^{-1}$$

Hence m = 1.53 × 10^{-3} kg m^{-1}.

Finally, we use the equation $f_0=\frac{1}{2L}\sqrt{\frac{T}{\mu}}$ to calculate f_0.

$$\text{So } f_0=\frac{1}{2\times1.5}\times\sqrt{\frac{80}{1.53\times10^{-3}}}$$

$$=76\ \text{Hz}.$$

18.5 The Doppler effect

When a car sounding its horn speeds past, the note heard by a bystander changes pitch. When approaching, the pitch is higher; when moving away, the pitch is lower. This is an example of the Doppler effect: when there is relative motion between a source of sound and an observer, the observed frequency is different from the emitted frequency.

A simple demonstration involves connecting a small loudspeaker to a signal generator using very long connecting wires. A string tied to the loudspeaker can be used to whirl it round in a circle. The wires need to be held with the string, and the string must be firmly attached to the speaker.

Observers listening outside the circle hear the pitch rise as the speaker moves towards them. Then as it moves

Figure 18.24 *Doppler shift*

away the pitch falls. The person at the centre of the circle who is whirling the speaker round hears a constant pitch, provided the string doesn't break! When the speaker moves round on its circular path, there is no relative motion between the observer at the centre and the speaker.

The Doppler effect can also be noticed by an observer moving either towards or away from a stationary source. Provided there is relative motion between observer and source along the line between them, then the effect can be detected. Any form of wave motion is capable of giving the effect, but it is most readily noticed with sound waves. To understand the cause, consider first the effect of the source moving, then the effect of the observer moving.

Source moving Assume that the source moves at steady speed v_s along the line between the observer and source. Suppose the source emits sound at a constant frequency f_s. Let the speed of sound be represented by c. In one second, the source emits f_s waves and travels a distance equal to v_s.

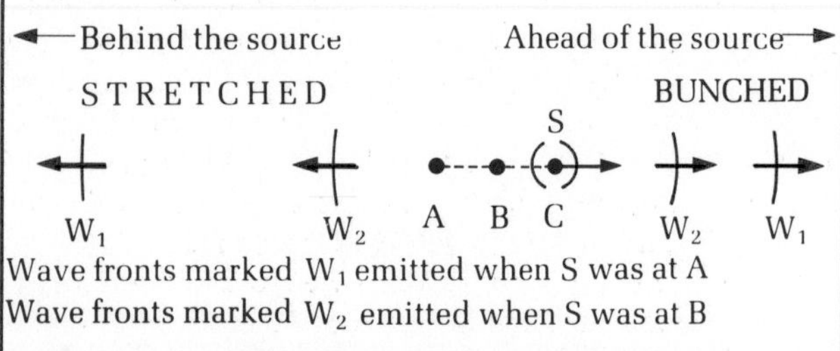

Figure 18.25 *The effect of a moving source*

Ahead of the source the waves are bunched up because the source is moving in the same direction as the waves. Suppose at time $t = 0$ a wave peak is emitted. Then one second later, that wave peak has travelled a distance equal to c. But the waves emitted in that second will be spaced over a distance of $(c - v_s)$ because of the motion of the source. So f_s waves are spaced over a distance of $(c - v_s)$ ahead of the source. The wavelength ahead of the source is given by

$$\lambda = \frac{(c - v_s)}{f_s}$$

Behind the source the waves are stretched out because the waves move in the opposite direction to the source. So f_s waves are spaced over a distance of $(c + v_s)$ because the source moves away from the waves. The wavelength behind the source is therefore given by

$$\lambda = \frac{(c + v_s)}{f_s}$$

Observer moving Suppose the observer is moving towards the waves at speed v_0. The relative speed of the waves to the observer is $(c + v_0)$. So the observed frequency f_0 is given by

$$f_0 = \frac{(c + v_0)}{\lambda} \quad \text{(Observer moving } \textit{towards} \text{ source.)}$$

This is equivalent to the viewpoint that the observer moves a distance v_0 in one second, so detects an extra number of waves v_0/λ in that second. So the observed frequency is $\frac{c}{\lambda} + \frac{v_0}{\lambda}$.

For an observer moving away from the source, the relative speed of the waves is $(c - v_0)$, so the observed frequency is

$$f_0 = \frac{(c - v_0)}{\lambda} \quad \text{(Observer moving } \textit{away} \text{ from source.)}$$

We can now use these formulae to calculate the observed frequency in any given situation.

Example 1 Source moving at steady speed v towards a stationary observer. The waves emitted towards the observer are bunched up because the source moves towards the observer. So the wavelength of the sound waves is given by

$$\lambda = \frac{(c - v_s)}{f_s}$$

where f_s = the source frequency.

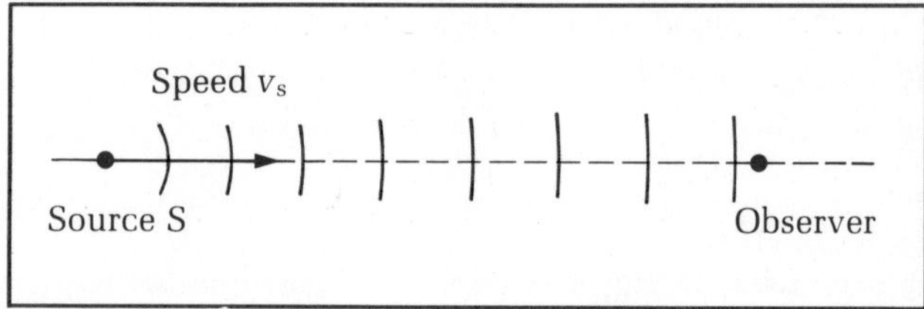

Figure 18.26

Since the observer is stationary, the observed frequency is given by

$$f_0 = \frac{c}{\lambda} = \frac{c}{(c - v_s)} \times f_s$$

Because $\frac{c}{(c - v_s)}$ is greater than 1, then f_0 is greater than f_s. The observed frequency is greater than the source frequency.

Example 2 Source and observer move at different speeds in the same direction along the same line, the source behind the observer.

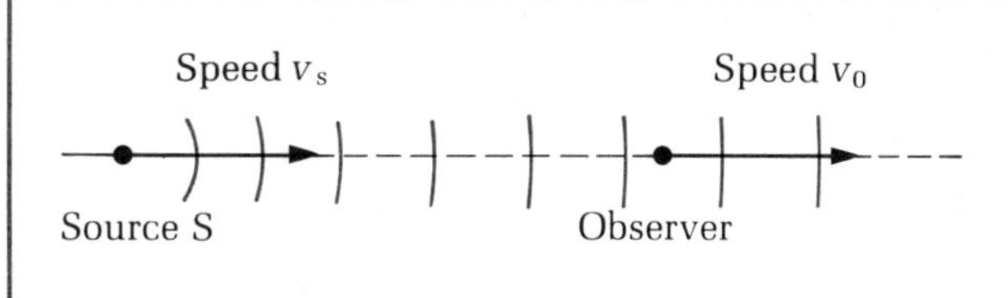

Figure 18.27

Because the source moves towards the observer, the waves emitted towards the observer are bunched up. The wavelength is given by

$$\lambda = \frac{(c - v_s)}{f_s}$$

The observer is moving away from the source, so the observed frequency is given by

$$f_0 = \frac{(c - v_0)}{\lambda} = \frac{(c - v_0)}{(c - v_s)} \times f_s$$

If $v_0 = v_s$, then the above equation gives $f_0 = f_s$. This is to be expected because when $v_0 = v_s$, there is no relative motion between observer and source.

The equation from Example 2 may be modified for other situations. For instance, if the source and observer move in opposite directions, then we change v_s to $-v_s$. So the equation becomes

$$f_0 = \frac{(c - v_0)}{(c + v_s)} \times f_s$$

Consider another example; suppose the observer moves at steady speed away from the stationary source. Then, using the equation of Example 2, we set $v_s = 0$ since the source is stationary. So the observed frequency is given by

$$f_0 = \frac{(c - v_0)}{c} \times f_s$$

The Doppler shift Δf is the difference between the observed frequency and the source frequency. So for Example 1, the Doppler shift is given by

$$\Delta f = f_0 - f_s = \left[\left(\frac{c}{c - v_s}\right) - 1\right] f_s = \frac{v_s}{c - v_s} \times f_s$$

If the source speed is much less than the wave speed, the Doppler shift is very small. For sound waves, the wave speed in air is about 340 m s^{-1}, so sound sources moving as slow as 10 m s^{-1} give a noticeable Doppler shift. However, for electromagnetic waves, the wave speed is 3×10^8 m s^{-1}, so source speeds need to be very much greater for electromagnetic waves to give a noticeable effect.

For electromagnetic waves only the relative velocity needs to be considered. The observer can be considered at rest, and the source is assumed to have a speed v (which is the relative speed along the line between source and observer). So the wavelength emitted by the source is given by

$$\lambda = \frac{(c \pm v)}{f_s}$$ (− if source moves towards the observer, + if source moves away.)

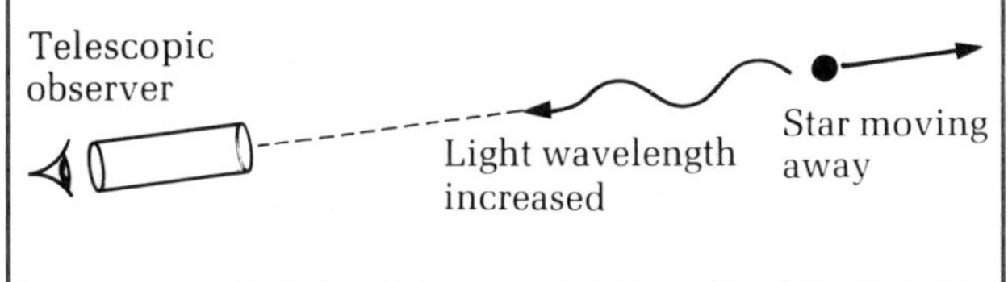

Figure 18.28 *Red shift*

Since a stationary source would emit a wavelength $\lambda_0 = \frac{c}{f_s}$, then the change of wavelength due to the relative motion is given by

$$\Delta\lambda = \frac{(c \pm v)}{f_s} - \frac{c}{f_s} = \pm\frac{v}{f_s} = \pm\frac{v}{c}\lambda_0$$

Astronomers use the Doppler effect to calculate speeds of distant stars and galaxies. By comparing the line spectrum of light from the star with light from a laboratory source, the Doppler shift of the star's light can be measured. Then the speed of the star can be calculated.

Stars moving *towards* Earth show a **blue shift**. This is because the wavelengths of light emitted by the star are shorter than if the star had been at rest. So the spectrum is shifted towards shorter wavelengths, i.e. to the blue end of the spectrum.

Stars moving *away* from the Earth show a **red shift**. The emitted waves have a greater wavelength than if the star had been at rest, so the spectrum is shifted towards longer wavelengths, i.e. towards the red end of the spectrum. Astronomers have discovered that all the distant galaxies are moving away from us, and by measuring their red shifts, they have estimated their speeds. The furthermost galaxies have been estimated to have speeds approaching the speed of light.

An important application of the Doppler shift using electromagnetic waves is the **radar speed trap**. Microwaves are emitted from a transmitter in short bursts. Each burst reflects off any obstacle in the path of the microwaves. In between sending out bursts, the transmitter is open to detect reflected microwaves. If the reflection is caused by a moving obstacle, the reflected microwaves are Doppler-shifted. By measuring the Doppler shift, the speed at which the obstacle moves (along the line between it and the transmitter/receiver) can be calculated. Figure 18.29 shows the arrangement.

If the transmitter emits microwaves at frequency f_s, and the obstacle is moving at steady speed v towards the transmitter then the transmitter emits waves of wavelength $\lambda = c/f_s$. However, the reflector makes the wavelength shorter, so the received frequency (of the reflected microwaves) is higher. The Doppler shift Δf, which is the difference between the received and the transmitted frequency, is given by

$$\Delta f = \frac{2v}{c} f_s$$

where c = speed of electromagnetic waves.

The factor of 2 is because the reflected waves reflect off as if from an image transmitter behind the reflector, like the image of an object viewed in a plane mirror. Because the image is twice as far from the object transmitter as the

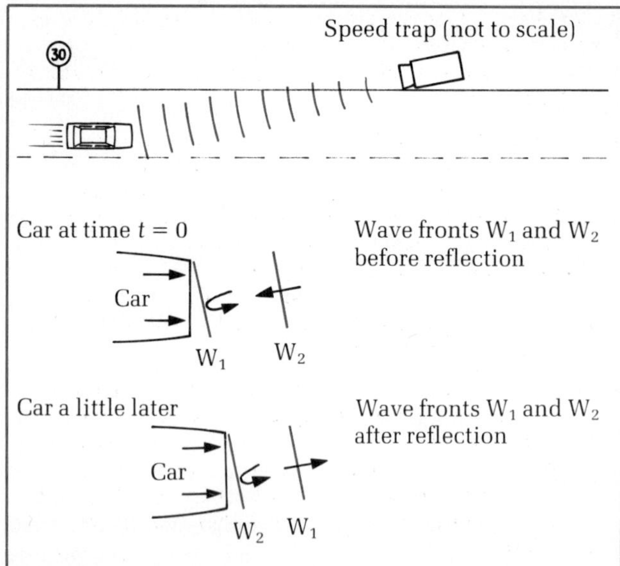

Figure 18.29 *Radar speed trap*

reflector the image speed is twice the reflector speed. The value of c is 3×10^8 m s^{-1}; what would be a typical value in m s^{-1} for v? Use your value to estimate the order of magnitude for $\Delta f/f_s$. Typical transmitter frequencies for microwaves are of the order of 10^{10} Hz (=10 GHz), so the Doppler shift Δf produced by a car moving at 10 m s^{-1} is 667 Hz for a 1 GHz transmitter. In a radar speed trap, the received (i.e. reflected) signal is mixed with the transmitted signal to produce a beats signal. The beats signal frequency is equal to the difference between the received and transmitted frequencies (i.e. $=\Delta f$). So an electronic filter is used to pass the beats signal to a measuring unit.

18.6 Summary

Definitions

1 *Beat frequency* is the difference between the two frequencies producing the beats.
2 *Speed of sound in air* is proportional to $\sqrt{T}$, where T is the air temperature in kelvins.
3 *For a resonance tube closed at one end:*
Fundamental note has a wavelength $\lambda_0 = 4(L+e)$, and frequency $f_0 = c/\lambda_0$ where L is the tube length, e is the end correction, c is the speed of sound in the tube.
Overtones occur at frequencies $3f_0$, $5f_0$, $7f_0$, etc.
4 *For a resonance tube open at both ends*
Fundamental note has a wavelength $\lambda_0 = 2(L+2e)$, and frequency $f_0 = c/\lambda_0$
Overtones occur at frequencies $2f_0$, $3f_0$, $4f_0$, etc.

Equations

1 Vibrating wire fundamental frequency: $f_0 = \dfrac{1}{2L}\sqrt{\dfrac{T}{\mu}}$

where T = tension,
L = length,
μ = mass per unit length.

Overtones occur at frequencies $2f_0$, $3f_0$, $4f_0$ etc.

2 For the Doppler effect:
Motion of the source gives a wavelength

$$\lambda = \frac{c \pm v_s}{f_s} \qquad \begin{array}{l} -\text{ for ahead} \\ +\text{ for behind.} \end{array}$$

Motion of the observer gives observed frequency

$$f_0 = \frac{c \pm v_0}{\lambda} \qquad \begin{array}{l} +\text{ if towards S,} \\ -\text{ if away from S.} \end{array}$$

Short questions

Assume the speed of sound in air at 20°C = 340 m s^{-1}

18.1 The human ear is most sensitive at a frequency of 3000 Hz when it can just detect sound waves of intensity 10^{-12} W m^{-2}. Calculate
a) the sound energy incident each second on an eardrum of area 20 mm^2 at this intensity,
b) the wavelength of the sound waves in air at 3 kHz. Comment on this value for wavelength in relation to possible resonances in the tube leading to the eardrum.

18.2 By definition, the intensity level of 0 dB is where the intensity is 10^{-12} W m^{-2}. A decibel meter at a distance of 10 m from a small source of sound gives a reading of 70 dB.
a) What is the intensity of the sound waves at the meter?
b) What would be the dB reading on the meter at 30 m from the source of sound?

18.3 A microphone connected to an oscilloscope is used to display sound wave forms. The wave form produced by a tuning fork note is shown below. Copy the sketch and on the same axes show the wave form you would expect for
a) the same tuning fork sounded louder,
b) a different tuning fork of higher pitch but at the same loudness as **a)**.

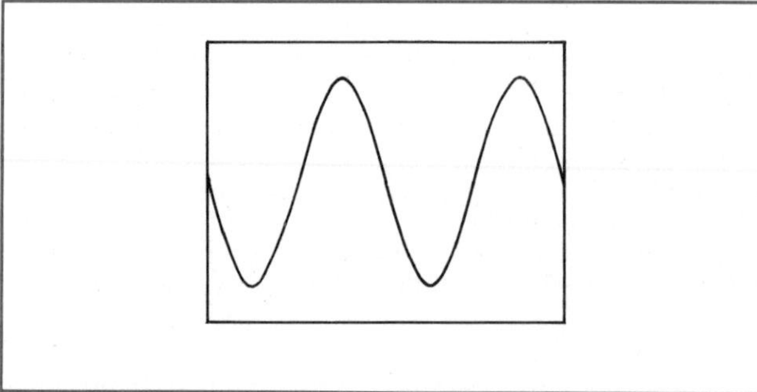

Figure 18.30

18.4 A tuning fork of frequency 256 Hz is used to tune a sonometer wire of length 0.850 m. The vibrating length of the wire is then shortened to 0.800 m.
a) What would be the new frequency of the string when plucked?
b) What would be the beat frequency heard when the tuning fork and the shortened wire are sounded together?

18.5 Two tuning forks A and B are sounded together to produce beats at 8 Hz. A has a known frequency of 512 Hz. When B is loaded with a small piece of plasticene, beats at 2 Hz are heard when two tuning forks are sounded together. Calculate the frequency of B when unloaded.

18.6 A small loudspeaker is placed near the open end of a pipe of length 400 mm closed at its other end. The minimum

frequency at which the pipe resonates is 215 Hz.
a) Estimate the speed of sound in the pipe.
b) Calculate the next highest frequency for resonance.

18.7 A church organ consists of open-ended pipes of differing lengths. The minimum length is 30 mm and the longest is 4.0 m. Given the speed of sound in air is $340\,\text{m s}^{-1}$, estimate the frequency range of the fundamental notes.

18.8 A vertical pipe that is open-ended at its upper end is connected to a water tap at its lower end. The length of the air column in the pipe can be varied by raising or lowering the level of water in the pipe. A tuning fork of frequency 342 Hz is sounded and then held with its vibrating tips over the open end of the pipe. At the same time, the water level is gradually lowered from the top. The air column in the tube resonates when its length is 245 mm then again when its length is 744 mm. Calculate
a) the speed of sound in the tube,
b) the 'end-error' correction.

18.9 A vertical steel wire of length 1.25 m and diameter 0.36 mm supports a weight of 50 N from its lower end. The density of steel is $7800\,\text{kg m}^{-3}$. Use this information to calculate
a) the mass per unit length of the wire,
b) the frequency of fundamental vibrations of the wire.

18.10 A uniform metal strip of length 760 mm is attached at one end to a vibrator connected to a variable frequency generator. As the frequency of the generator is increased from zero, the strip first vibrates very strongly when the frequency is 15 Hz. Calculate
a) the wavelength of the transverse waves on the strip,
b) the speed of the transverse waves.

18.11 An ultrasonic transducer emits short pulses of ultrasonic waves of frequency 2.0 MHz at a pulse rate of 1 kHz. The transducer acts as a receiver when it is not transmitting, so pulses reflected back to the transducer by obstacles can be detected. By measuring the time interval between a pulse being transmitted and returning, the distance to the obstacle can be calculated. Figure 18.31 shows an oscilloscope trace of the transducer signal when the transducer was placed on top of a solid block of steel of thickness 40 mm. When the transducer was placed on a second steel block of the same thickness, an extra pulse was observed on the trace.
a) Calculate the speed of ultrasonic waves in steel.
b) Explain the likely cause of the extra pulse in the second block.

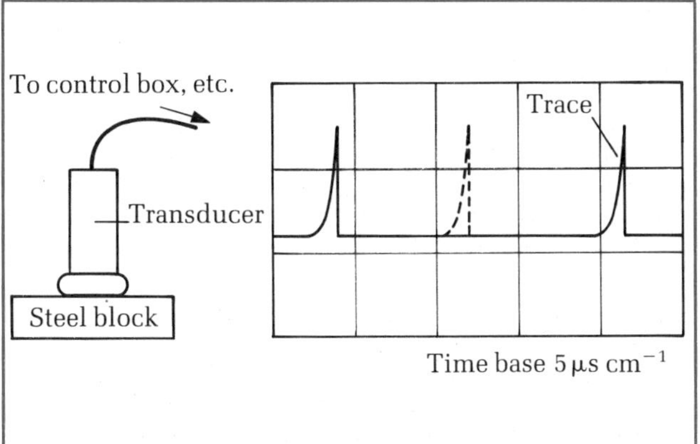

Figure 18.31

18.12 The silencer box of a certain car is open at either end and has a length of 0.8 m. Given the speed of sound is $340\,\text{m s}^{-1}$, calculate the minimum frequency for resonance in the box.

18.13 A car sounds its horn as it travels at a steady speed of $15\,\text{m s}^{-1}$ along a straight road between two stationary observers X and Y. The observer at X hears a frequency of 538 Hz whilst the observer at Y hears a lower frequency.
a) Is the car travelling towards X or towards Y?
b) What would be the frequency heard from the car by either X or Y if the car stopped and sounded its horn?
c) What frequency does Y hear when X hears 538 Hz? The speed of sound in air $= 340\,\text{m s}^{-1}$.

18.14 Two cars P and Q are travelling along a motorway in the same direction. The leading car P travels at a steady speed of $12\,\text{m s}^{-1}$; the other car Q, travelling at a steady speed of $20\,\text{m s}^{-1}$, sounds its horn to emit a steady note which P's driver estimates has a frequency of 830 Hz. What frequency does Q's own driver hear? (Speed of sound in air $= 340\,\text{m s}^{-1}$.)

18.15 A train waiting in a station sounds its horn before it sets off and an observer waiting on the platform estimates its frequency at 1200 Hz. The train then moves off and accelerates steadily to reach a speed of $25\,\text{m s}^{-1}$. Fifty seconds after departure, the driver sounds the horn again and the platform observer estimates the frequency at 1140 Hz. Calculate the train speed 50 s after departure. How far from the station is the train after 50 s? The speed of sound in air $= 340\,\text{m s}^{-1}$.

19 Physical optics

19.1 The wave nature of light

Young's double slits experiment

It isn't obvious that light consists of waves because although reflection and refraction of light are easily demonstrated, other wave properties such as interference and diffraction are much more difficult to observe. Even a great scientist like Sir Isaac Newton preferred to think that light was made up of tiny particles which he called 'corpuscles'; reflection of light by a mirror was, according to Newton, because each corpuscle of light bounced off the mirror like a tennis ball bouncing off a wall! There seemed no reason for scientists at the time to accept Huygens' theory that light is composed of waves. They reasoned, 'If light does consist of waves, then why are sharp shadows formed? The light ought to spread round the edges of any obstacle creating a shadow, so the shadow ought not to be sharp![1] Yet we now know that if the so-called sharp shadow is looked at closely, evidence of diffraction is in fact present. So poor old Huygens did not get very far in his efforts to persuade his fellow scientists. As far as they were concerned, Newton, as ever, was right! Don't be too harsh, though, on the scientists of those times; it must be very hard to demonstrate diffraction or interference of light using a *candle* as a light source!

The first experimental evidence for the wave nature of light came from an experiment devised by Thomas Young in 1801. He observed the interference of light waves using a double slit arrangement as shown in Figure 19.3.

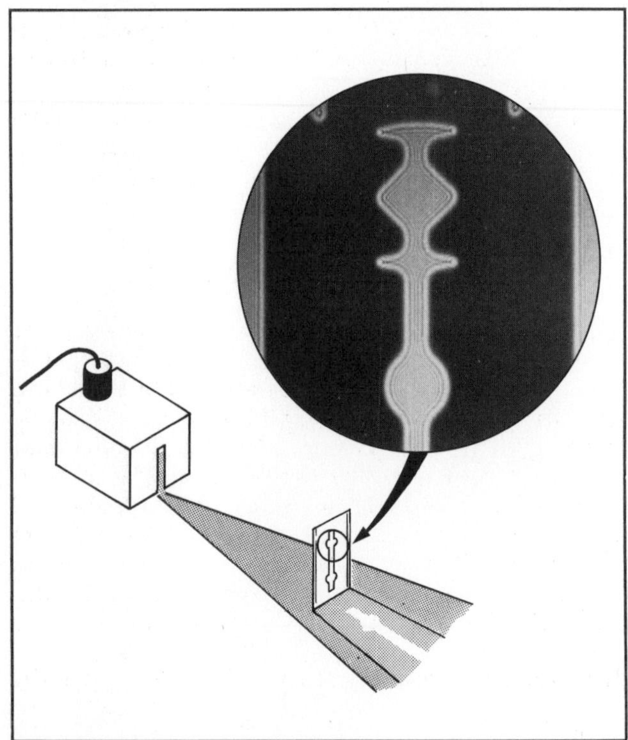

Figure 19.1 *Diffraction at an edge*

Figure 19.2 *Thomas Young, 1773–1829*

The development of our understanding of light is a good example of the link between theory and experiment in science. Newton's corpuscular theory was accepted until Young's experiments gave results that could only be explained by Huygens' wave theory.

*In 1890, experiments on photoelectricity led to a revolutionary new way of thinking about light. The results of these experiments could not be explained by wave theory, and it was not until 1905 that Einstein devised an acceptable theory – using the idea of wave packets called **photons**.*

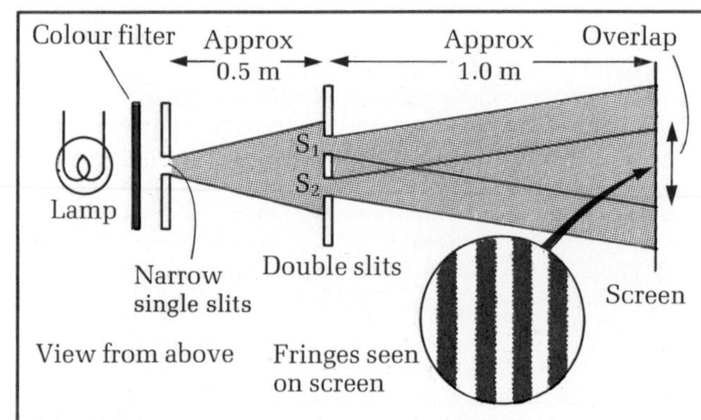

Figure 19.3 *Young's double slit experiment*

The narrow single slit S transmits light from the lamp, and the light is allowed to fall on the double slit arrangement. Each of the double slits transmits light towards the screen so that light from one of the double slits overlaps light from the other. Bright and dark fringes are seen on the screen in the overlap area. These fringes are formed by interference of the light waves from the double slits.

The use of double slits makes sure that the two sets of light waves are emitted with a constant phase difference (e.g. when one slit emits a crest, the other always emits a crest a set time later). The two slits are then said to be **coherent sources**. As a result, points of cancellation and reinforcement on the screen stay fixed in position.

Dark fringes on the screen are formed at positions where waves from one slit arrive out of phase by 180° with the waves from the other slit; in other words, wave crests from one slit arrive at the same time as wave troughs from the other, so they cancel one another out.

Bright fringes occur where waves from one slit reinforce waves from the other slit. In other words, the waves from one slit arrive in phase with the waves from the other slit.

By using different coloured filters between the lamp box and the double slits, different colours of light can be used in turn to form the fringe pattern. Each colour gives a fringe pattern of equally spaced bright and dark fringes, parallel to the double slits. However, the spacing of the fringes does depend on the light colour; in fact, red light gives a greater spacing than blue light.

The **wavelength** λ of the light used may be calculated from the following equation.

$$\lambda = \frac{yd}{X}$$

where y = the fringe spacing,
d = distance between the double slits (from centre to centre),
X = distance from the double slits to the screen.

By measuring the fringe spacing y for each colour, as well as values for d and X, the wavelength of light of each colour may be determined. Accurate measurements ought to show that the wavelength of red light is about 650 nm compared with the wavelength of blue light which is about 400 nm.

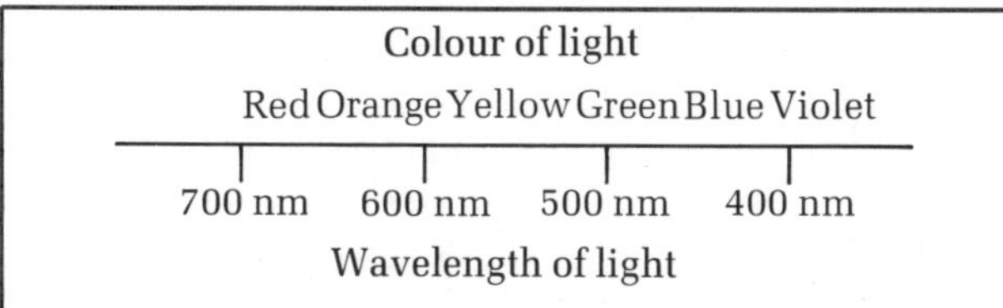

Figure 19.4 *Wavelength and colour*

Measurements

1 Distance from the double slits to the screen X: use a metre rule marked in millimetres.
2 Distance between the double slits d: use a travelling microscope to measure from centre to centre. Another method is to use a lens to throw a magnified image of the slits onto a screen, measure the spacing of the image slits on the screen then divide that value by the magnification (v/u) of the lens to give the actual slit spacing d. Figure 19.5 shows the principle of this method.
3 Fringe spacing y: if the fringes are formed on a screen, then measure across as many fringes as possible and divide the measurement by the number of fringe spacings, as in Figure 19.6.

A millimetre rule would give the measurement to the nearest 0.5 mm but greater accuracy is possible using a travelling microscope to measure across the fringes. If a travelling microscope is used instead of a screen to

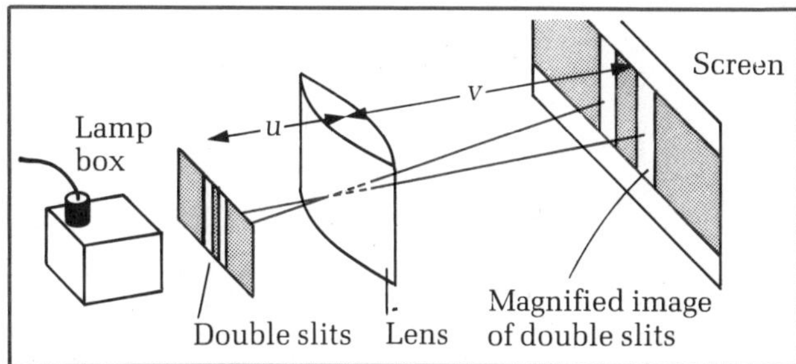

Figure 19.5 *Measuring the slit spacing*

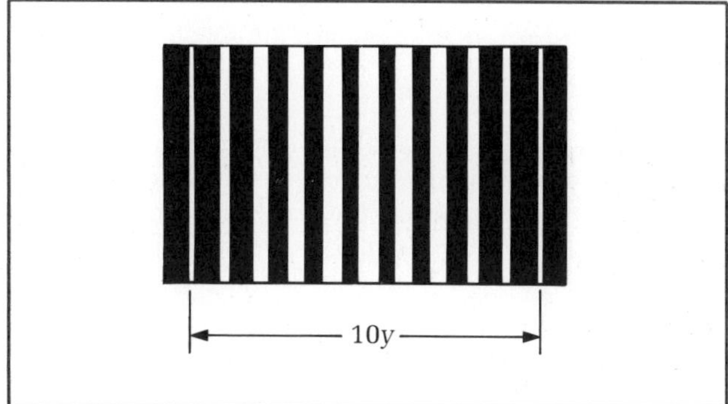

Figure 19.6 *Double slits fringe pattern*

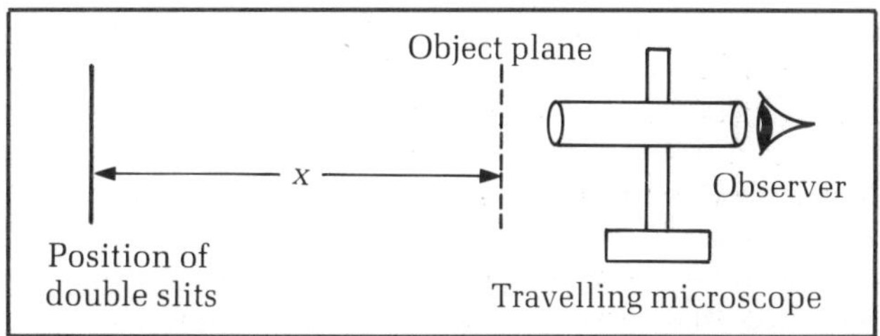

Figure 19.7 *Measuring the fringes*

observe the fringes, then it is very important to measure distance X from the double slits to the object plane of the microscope. This is because the fringes seen through the microscope are those formed in its object plane. To locate the object plane, move a pinhead to and fro between the double slits and the microscope; the pinhead will be in focus when in the microscope's object plane.

Theory of the double slits equation

Suppose the two slits S_1 and S_2 emit wave crests at the same time. At point P on the screen, the wave crest from S_1 arrives later than the wave crest from S_2. In fact, P can only be a point of reinforcement if the wave crest from S_1 arrives a complete number of cycles later than the wave crest from S_2. If so, then the wave crest from S_1 is reinforced at P by a later wave crest from S_2. This will only happen if light waves from S_1 have to travel an extra distance equal to a whole number of wavelengths compared with the light waves from S_2. This extra distance $S_1P - S_2P$ is called the **path difference.**

For reinforcement at P, the path difference is given by

$$S_1P - S_2P = m\lambda$$

where $m = 0, 1$ or $2 \ldots$ etc.

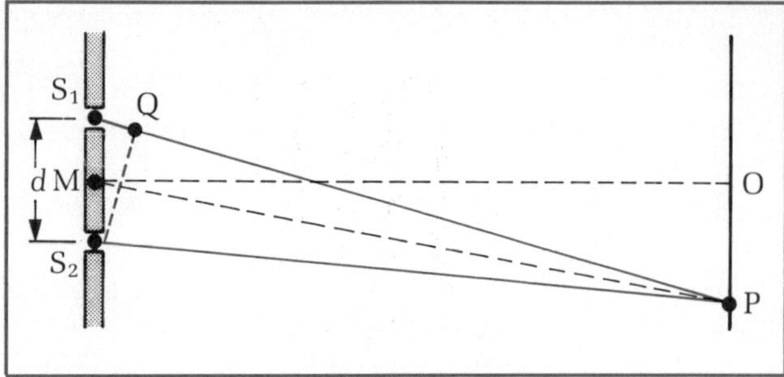

Figure 19.8 *The theory of the double slits*

For cancellation at P, the path difference is given by

$S_1P - S_2P = (m + \frac{1}{2})\lambda$

The condition for cancellation is because a wave crest from one slit can only be cancelled out if a wave trough from the other slit arrives at the same time. This will only be so if the light waves from S_1 travel an extra distance equal to $(m + \frac{1}{2})\lambda$ compared with the light waves from the other slit.

Now consider the geometry, as shown in Figure 19.8. A point Q has been marked along the line S_1P at a position such that distance QP equals S_2P. Hence

$$S_1Q = S_1P - S_2P$$

Next, compare the shape of triangles MOP and S_1S_2Q; they are similar triangles (i.e. have the same shape) – near enough – provided OP is much smaller than OM; this is true in practice because OP is of the order of millimetres compared with OM which is up to 1 metre. Because the two triangles have the same shape, the ratio S_1Q/S_1S_2 (i.e. short side/long side) is equal to OP/OM.

$$\frac{S_1Q}{S_1S_2} = \frac{OP}{OM} \text{ so } OP = \frac{OM \times S_1Q}{S_1S_2}$$

For the bright fringes only,

$$S_1Q = S_1P - S_2P = m\lambda$$

from above. Also OM is the distance X and S_1S_2 is the slit spacing *d*. Therefore,

$$OP = \frac{m\lambda X}{d}$$

where $m = 0, 1, 2, \ldots$ etc.

In fact, $m = 0$ corresponds to point P at the central bright fringe at O, so imagine point P moving outwards away from O. The next bright fringe will be when OP equals $\lambda X/d$ so it corresponds to $m = 1$. In other words, S_1 is one wavelength further from P at this position compared with S_2. Moving further from O, the next bright fringe is where OP equals $2\lambda X/d$ so this one corresponds to $m = 2$. It is at the position where S_1 is two wavelengths further than S_2. In general, *m* is called the **order number**, and is used to label the *m*th bright fringe from the centre, as shown in Figure 19.9.

Clearly, the fringe spacing between two adjacent bright fringes *y* is therefore given by

$$y = \frac{\lambda X}{d}$$

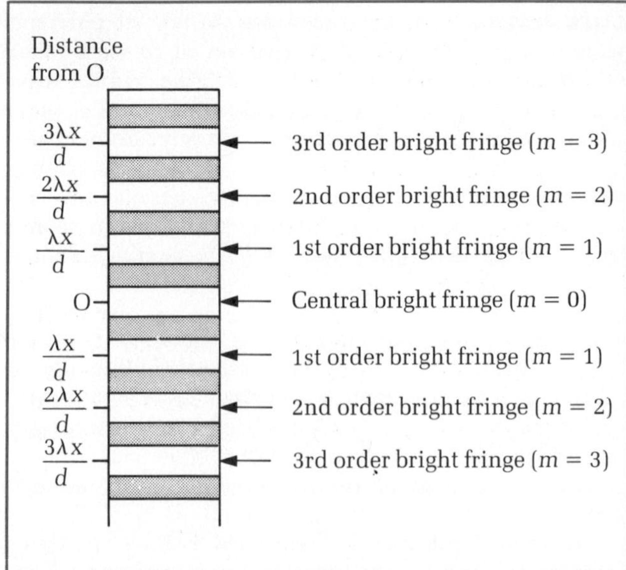

Figure 19.9 *Fringe positions*

The essential points in the above theory are as follow.

1. For the first bright fringe from the centre, $S_1P - S_2P = \lambda$.
2. Mark point Q along S_1P so that $PQ = PS_2$. Therefore $QS_1 = \lambda$.
3. Triangles MOP and S_1S_2Q are similar in shape, so $OP/OM = QS_1/S_2S_2$
4. Insert distance symbols d ($= S_1S_2$), λ ($= QS_1$), X ($=$ OM) and y ($=$ OP) to give

$$\frac{y}{X} = \frac{\lambda}{d}$$

White light fringes

White light consists of all the colours of the spectrum with wavelengths ranging from 700 nm (deep red) to 400 nm (blue). If a red filter is placed between the lamp box and the double slits, only red light is allowed through the filter. All the other colours are absorbed by the filter, and the fringe pattern is therefore made up of red and dark fringes. Now a blue filter would give blue and dark fringes closer together than the red fringes were (because the wavelength of blue light is less than of red light), but the central fringe would still be in the same place, as shown in Figure 19.10.

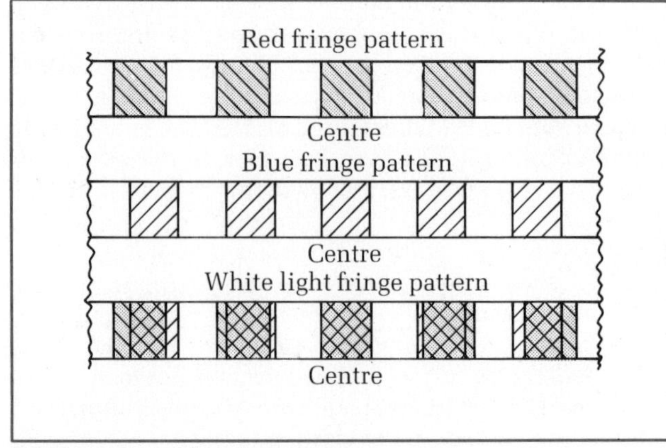

Figure 19.10 *White light fringes*

If no filter is used, each colour of the white light beam gives a fringe pattern so all the patterns are formed on top of one another. However, because all the central fringes are in the same place, a white fringe is seen at the centre because all the colours contribute. Away from the centre, the fringes will partly overlap so that first order fringes will appear tinged with blue on the side towards the centre and tinged with red on the outer side. The reason is that the first order blue fringe is closer to the central fringe than the first order red fringe.

As shown in Figure 19.10, the inner fringes apart from the central fringe will appear white with blue and red tinges. Further away from the centre, the overlap increases much more between the different colours of adjacent fringes, and as a result the whole pattern appears white away from the centre.

Figure 19.11 *Coherence*

Lloyd's mirror

A different way of demonstrating interference of light was devised by Dr Lloyd in 1834. A plane mirror is positioned near a narrow single slit, parallel to the mirror, in such a way that the reflected light overlaps with the direct light from the slit. Figure 19.12 shows the essential arrangement.

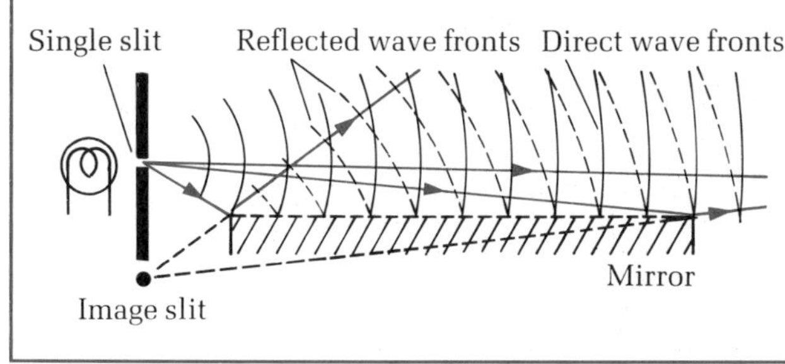

Figure 19.12 *Lloyd's mirror*

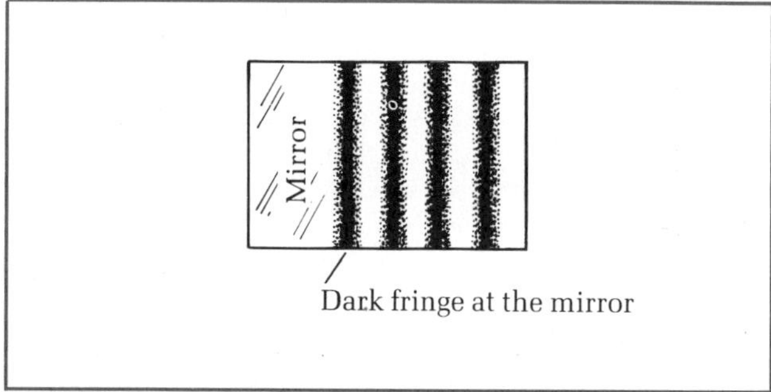

Figure 19.13 *Lloyd's mirror fringes*

The reflected light appears to come from an image slit I so that the image slit and the actual slit act as coherent sources. Interference fringes are therefore seen where the reflected light overlaps with the direct light. The appearance of the fringes is almost the same as Young's fringes, except they are formed on the front side of the mirror only. The most significant difference between Lloyd's fringes and Young's fringes is, however, best seen if a white light source is used; there is a **dark** fringe at the mirror surface. In fact, the pattern is reversed compared with the Young's fringes in the sense that the bright fringes of Young's experiment are formed in the place of the dark fringes of the Lloyd's experiment (and *vice versa*). The reason for the difference is that there is a **phase reversal** whenever light is reflected off glass (or at any denser medium). Therefore, at the surface where the reflected light has travelled the same distance as the direct light, the reflected light cancels the direct light to give a dark fringe.

19.2 Interference by thin films

Coherence

A ripple tank may be used to demonstrate interference of water waves, as shown in Figure 17.12. Each dipper produces a continuous sequence of waves which pass through the waves from the other dipper. The overlapping waves interfere to give points of cancellation and reinforcement. Figure 17.12 shows an arrangement to make the two dippers go up and down exactly out of phase with one another. Will an interference pattern still be seen? The answer is yes, because the new arrangement still gives points of cancellation and reinforcement, though the points are not at the same positions as before. Whatever the phase difference between the two dippers, an interference pattern will be seen, provided the phase difference does not vary. If it does vary, the positions of cancellation and reinforcement keep changing; if the changes are too fast to follow by eye, then no interference pattern will be seen. So to observe an interference pattern, the dippers must maintain a constant phase difference. They are then said to act as **coherent** sources of waves.

Coherent sources of waves are sources which maintain a constant phase difference.

The Young's double slits experiment on p. 258 needs great care to produce an interference pattern with light waves. The double slits give sets of overlapping waves from each slit. The two sets of waves are produced with a constant phase difference because every wave front from the single slit (behind the double slits) produces a wave front from each of the double slits. So the double slits act as coherent emitters of waves. At any position of the interference pattern, the waves from one slit maintain a constant phase difference with the waves from the other slit. At points of reinforcement, the two sets of waves arrive in phase. At points of cancellation, they arrive out of phase exactly.

Two sets of waves are coherent if they maintain a constant phase difference.

When an atom emits light, it does so in a short burst which lasts no more than a few nanoseconds (i.e. of the order of 10^{-9} s). Light travels at a speed of $3 \times 10^8 \text{ m s}^{-1}$, so each short burst produces a train of light waves up to 0.3 m long. The wavelength of light is approximately 600 nm, so each burst of light from an atom can contain up to half a million cycles. But if an emitting atom bumps into another atom, the impact causes the phase of the wave train to change suddenly. In a lamp, atoms are continually bumping into one another, so an emitted wave train of light has many phase changes along its length. In practice, the average length of a section of wave train without jumps of phase can be less than 1 mm or so. The average length between two sudden phase changes is called the **coherence length** of the waves.

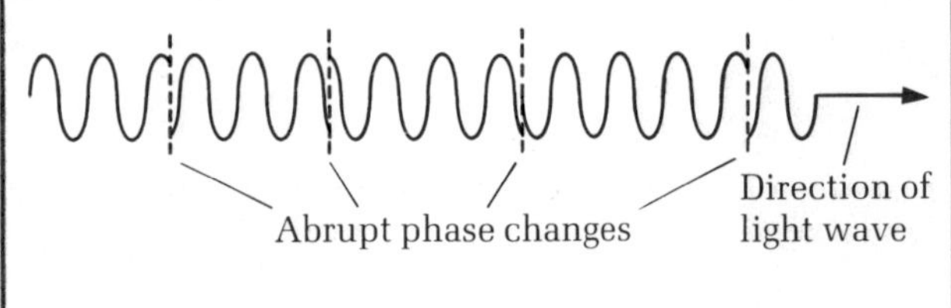

Figure 19.14 *A wave train of light*

Interference is only possible if the two sets of overlapping waves are coherent. In other words, they must have arrived at the observing screen with a constant phase difference. In general, one set of waves will have travelled further than the other set, both emitted by the source at the same time. The extra distance which one set of waves travels compared with the other is called **the path difference**. If the path difference is too great, interference will not be seen no matter how good the equipment. In the double slits experiment, each burst of light from the lamp is divided into two sets of waves by the double slits. If the path difference is too great, one set of waves will have passed the observing position before the other one arrives. So the two sets of waves arriving at the same time will not then have a constant phase link. The path difference should not be greater than the coherence length.

In general, there are two ways of dividing the waves from a lamp to produce two sets of waves. One way is to **divide the wave fronts** as in the double slits experiment and Lloyd's mirror. One set of waves is produced from one part of the wave fronts, the other set from a different part. The other way is to **divide the amplitude**; this happens when light passes through glass. Some of the light is reflected, and some passes through (i.e. is transmitted). When you look at a pane of glass, you see your own reflection; light from you to the window is partly reflected (the rest is transmitted). All along each incident wave front, there is some reflection and some transmission. Figure 19.15 shows the idea.

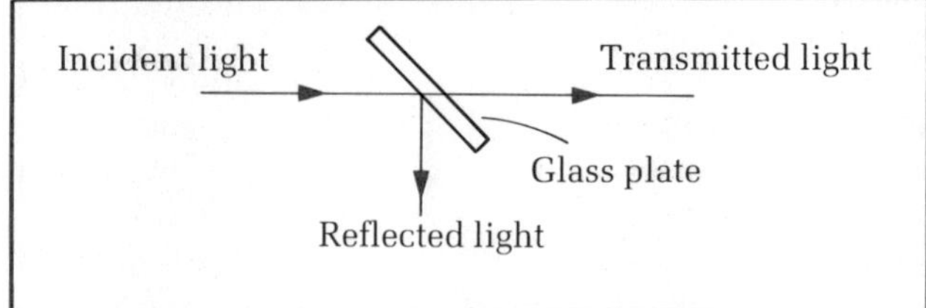

Figure 19.15 *Partial reflection*

So the reflected waves and the transmitted waves are both from the same burst of waves produced by the source. If the two sets of waves can be made to overlap, then interference can be produced.

Lasers

Few areas of science have caught the popular imagination more than the laser. Space fiction films often depict laser beams as all-powerful death rays, yet laser light can be reflected like sunlight, so a weapon which can be turned back on its user isn't much use. In fact, there are many uses to which lasers are nowadays applied. For example, in medicine, there is a well-established method of removing unwanted skin blemishes using laser energy. Lasers are used in engineering, for surveying and guidance. In entertainment, their use in compact discs has been promoted as an alternative to tape cassettes. The word LASER stands for **L**ight **A**mplification by **S**timulated **E**mission of **R**adiation. Each burst of light from the atoms inside a laser tube causes (i.e. stimulates) other atoms to

Figure 19.16 *Using lasers*

emit light. In the gas laser, the tube is sealed at each end by parallel mirrors which reflect light back into the tube. The reflected light waves stimulate other atoms to emit light waves in phase. So as the light passes back and forth along the tube, the energy of the light waves builds up and up. So how does the light get out? In fact, one of the mirrors transmits a small fraction of the light through, so out comes the laser beam. The beam is almost perfectly parallel, and is much more coherent than ordinary light; the coherence length is much greater. There is only one value of wavelength in the beam so it is monochromatic (i.e. a single colour).

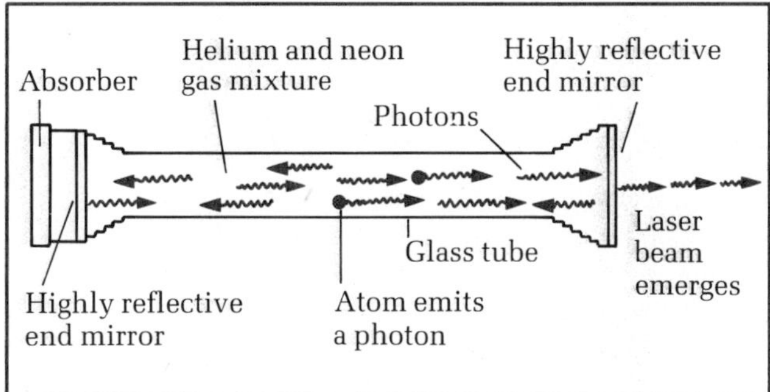

Figure 19.17 *Inside a laser tube*

Because the beam is almost perfectly parallel and monochromatic, a convex lens can focus it to a very fine spot. The beam power is then concentrated in a very small area. Hence the danger of laser beams entering the eye. The eye lens would focus the laser beam onto a tiny spot on the retina of the eye, and the intense concentration of light at that spot would burn the retina!

Always wear safety goggles in the presence of laser beams.

Never look along the beam, even after reflection.

Thin films with parallel sides

When light passes into a transparent medium such as glass, from air, some of the light is reflected at the boundary. The rest of the light energy is transmitted into the medium. When light passes from the medium into air, partial reflection also takes place so only part of the light energy emerges from the medium. When light is directed at a thin film such as a soap film, partial reflection occurs twice, once at the air-film boundary, and again at the film-air boundary. So two reflected beams return back into the air on the incident side of the film, as in Figure 19.18.

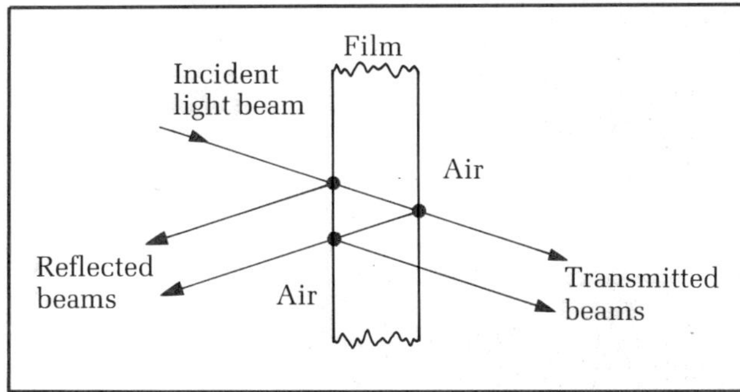

Figure 19.18 *Thin film interference*

Provided the film is not too thick, the two reflected beams produce an interference effect.

You may have noticed such effects when looking at a good-quality camera lens. The lens reflects very little so it appears dark. Why? The outer lens surface is coated with a thin transparent film. The reflected light from each surface of the film interferes destructively (i.e. cancels), so the overall effect is that the film, viewed from the same side as the light source, appears dark. What happens to the light which falls on the film? Figure 19.18 shows that there is more than one transmitted beam. The transmitted beams reinforce one another, so all the light energy passes into the camera to give a better image than with an uncoated lens. The process of coating the film on the lens is called 'blooming', and it is a carefully controlled process. The film must be uniform and of certain thickness. For the reflection from the second surface to cancel the reflection from the first surface, they must be out of phase by 180°. Because there is a phase reversal on reflection at each surface, then the path difference between the two reflected beams for cancellation must equal an odd number of half wavelengths. The reason for the phase reversal at each surface is that the refractive index of the film is between that of air and glass, so both reflections count as external reflections. For normal incidence, the path difference is twice the film thickness. Therefore by making the film thickness equal to one quarter the wavelength of yellow light, cancellation of reflected yellow light occurs. Since yellow is near the middle of the white light spectrum, then most of the light falling on the lens passes through the lens into the camera.

For cancellation of reflected light,

$$2t = \tfrac{1}{2}\left(\frac{\lambda_0}{n}\right)$$

where t = film thickness,
λ_0 = wavelength of yellow light in air,
n = refractive index of film.

Wedge-shaped films

Hold a soap film up and view it in reflected light. Alternate bright and dark fringes ought to be seen, each fringe horizontal. The film forms a wedge, thin at the top and wide along its lower edge. Each fringe is a contour of equal film thickness, so the fringes are horizontal. The position of each dark fringe is given by the equation $2t = \frac{m\lambda_0}{n}$, as explained in Figure 19.20.

Allow the film to drain and you ought to observe that the fringes move down and space out. Just before the film breaks, it becomes dark at the top. This is where the film is so thin that the two reflected beams have almost no path difference; because one beam has its phase reversed upon reflection, the two beams cancel out completely. So it appears dark just before it breaks at the top. Try it. With white light you should observe a spectrum of colours in each bright fringe. Does each bright fringe have red or blue at the top? Can you explain why? Remember that blue light has a smaller wavelength than red light.

A more controlled way of observing wedge fringes is to use two flat glass slides, placed one on top of the other

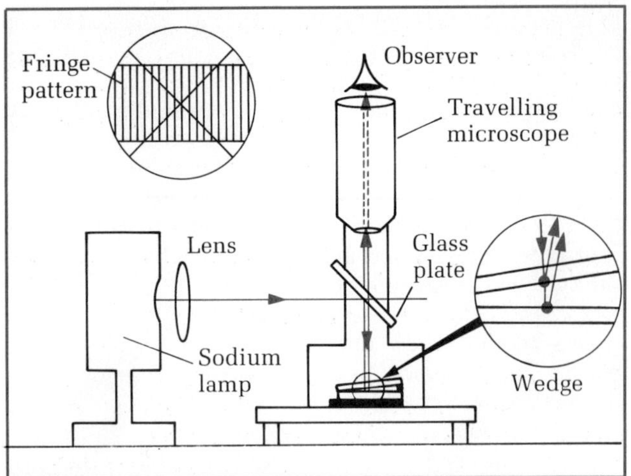

Figure 19.19 *Wedge fringes*

with a spacer at one end. A thin wire is suitable as a spacer. Figure 19.19 shows how the fringes can be illuminated and observed in reflected light. The glass plate is used to direct light onto the wedge, and to allow the reflected light to pass up to the viewing microscope. Using monochromatic (i.e. a single value of wavelength) light, the fringes are equally spaced bright and dark fringes. They are contours of equal wedge thickness, parallel to the thin end of the wedge. At each dark fringe, the thickness is given by $2t = m\lambda_0$ (for an air wedge), or $t = m\lambda_0/2$. From one dark fringe to the next, the wedge thickness increases by $\lambda_0/2$. So the fringe spacing y (from one dark fringe to the next) is given by

$$y = \frac{(\lambda_0/2)}{\tan\theta}$$

where θ is the angle of the wedge.

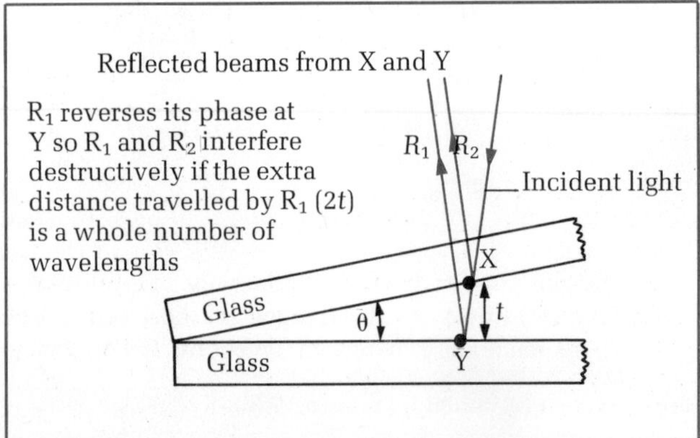

Figure 19.20 *The wedge angle*

$$\text{Fringe spacing } y = \frac{\lambda_0}{2\tan\theta}$$

The wavelength of the light used may be determined by measuring the fringes. A travelling microscope may be used to measure across several fringe spacings to give an accurate value of y. The microscope can also be used to measure the distance from the spacer to the end of the wedge (X). Then measurement of the spacer width (W) with a micrometer allows the wedge angle to be calculated from $\tan\theta = W/X$.

Worked example A wedge of air is formed using two flat glass plates of length 120 mm in contact at one end and separated by a wire at the other end. Using monochromatic light of wavelength 590 nm, the wedge is viewed in reflected light. A distance of 1.97 mm was measured across 20 fringe spacings. Calculate the wire diameter.

Solution First, calculate the angle of the wedge using the fringe spacing equation,

$$\tan\theta = \frac{\lambda_0}{2y} = \frac{590 \times 10^{-9}}{2 \times (1.97 \times 10^{-3}/20)} = 2.99 \times 10^{-3}$$

Then use $W = X\tan\theta$ to calculate the wire diameter W, X being 120 mm.

Hence $W = 120 \times 10^{-3} \times 2.99 \times 10^{-3} = 3.59 \times 10^{-4}$ m.

Newton's rings

Another way of making a thin film is to place a convex lens on a flat glass plate. There is a very thin film between the lens and the plate. If the lens and plate are used in the arrangement in Figure 19.19 instead of the wedge, then a pattern of concentric bright and dark rings is seen. The rings are centred on the point of contact between lens and plate. A dark spot is seen at the centre if the lens is in contact with the plate. Sir Isaac Newton was the first person to describe the rings.

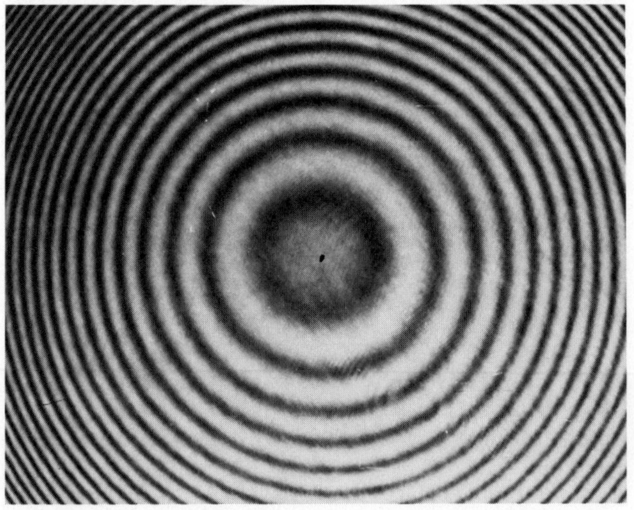

Figure 19.21 *Newton's rings*

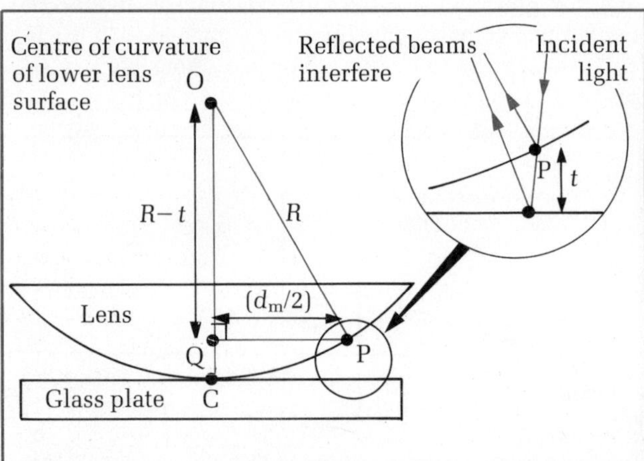

Figure 19.22 *Newton's rings – the theory*

Light is directed down onto the lens and plate by means of the partial reflector plate at 45° to the vertical. The light passes through the lens and the plate, but at each boundary some light is reflected. Light is reflected at the underside of the lens and at the top surface of the plate. These two reflected sets of light waves produce the interference pattern. Each **dark** fringe is formed where the film thickness is given by $2t=\frac{m\lambda_0}{n}$. The reasons for this equation are:

a) there is a phase reversal upon reflection at the top surface of the plate,

b) the path difference, $2t$, between the two beams must therefore be a whole number of film wavelengths,

c) the overall effect is that the two reflected beams cancel one another where $2t=m\lambda_0/n$.

So each fringe is a contour of equal film thickness, which is why they are circles. The diameter of each dark fringe can be written in terms of film thickness using the geometry of Figure 19.22. For the mth dark ring from the centre, diameter d_m, where the film thickness is t, triangle OPQ gives

$$\frac{d_m^2}{4}+(R-t)^2=R^2$$

where R is the radius of curvature of the lower lens surface. This gives

$$\frac{d_m^2}{4}=2Rt$$

if t^2 is neglected because it is much smaller than the other terms.

Since $2t=\frac{m\lambda_0}{n}$, then

$$d_m^2=4mR\frac{\lambda_0}{n}$$

where d_m = diameter of mth dark ring from centre,
n = refractive index of film,
λ_0 = wavelength of light in air.

Using a sodium lamp to give monochromatic light, the rings can be measured with a travelling microscope. For a number of the inner dark rings, the diameter of each is measured from one side of the ring, through the centre, to the opposite side. The ring number from the centre must be noted. A graph of d_m^2 against m may then be plotted to give a straight line of gradient $4R\lambda_0/n$. For an air film, $n=1$ so if the value of R is known, λ_0 may be calculated.

Using white light, each bright ring is tinged with blue at the inside and with red at the outer side. The reason is the longer wavelength of red light compared with blue light. With a red filter, red rings only would be seen; with a blue filter, only blue rings would be seen; but the mth red ring would be slightly larger than the mth blue ring. So with white light, which has all the colours of the spectrum, the mth ring is red at its outer edge and blue at its inner edge. At the centre though, the pattern ought to have a black spot if the lens and plate are in contact. The path difference is zero at the centre for all wavelengths so with the phase reversal, all wavelengths cancel.

The bright rings are at positions where the path difference $2t$ is a whole number of wavelengths plus a half-wavelength. So with the effect of the phase reversal, the two reflected beams reinforce if $2t=(m+\frac{1}{2})\frac{\lambda_0}{n}$.

19.3 Diffraction by slits and obstacles

Waves spread round corners. Light waves are no exception but you need to look very carefully for the evidence. Figure 19.1 shows a photograph of the pattern of light behind a razor edge. Light directed at the edge diffracts (i.e. spreads) after passing the edge.

Single slit diffraction

Fringes are also seen when a point source of light is viewed through a narrow slit. Figure 19.23 shows a possible arrangement. Try it. The fringe pattern formed by a single slit consists of alternate bright and dark fringes, each fringe parallel to the slit. Although the fringes fade away from the centre, it ought to be possible to see the central fringe and several fringes either side.

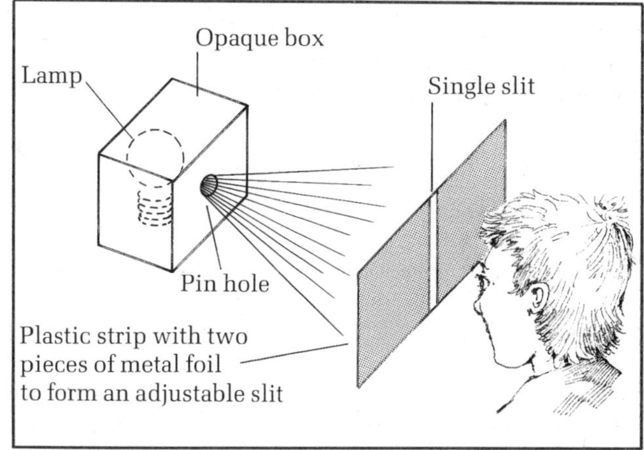

Figure 19.23 *Viewing single slit fringes*

The fringes can be photographed by allowing the diffracted light to fall on a photographic film. Figure 19.24 shows an enlarged view of such a photograph. The photograph can be used to obtain a graph of fringe brightness against position, as shown in Figure 19.25 (on the next page).

The graph shows that the central fringe is much brighter than the others. Also, the central fringe is twice the width of each of the other fringes. The intensity (i.e. brightness) is least at certain positions along the screen, each position being that of a dark fringe.

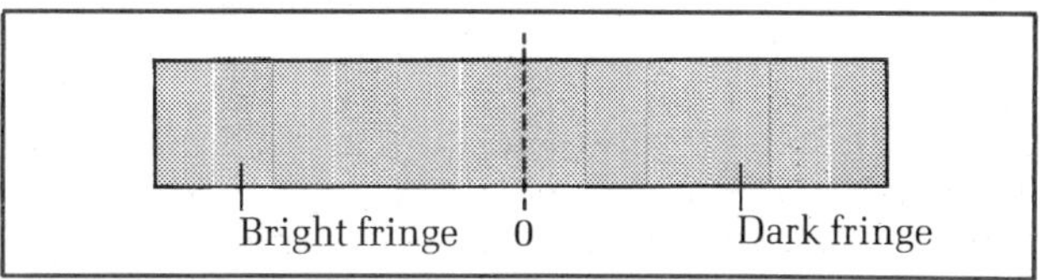

Figure 19.24 *Single slit fringes*

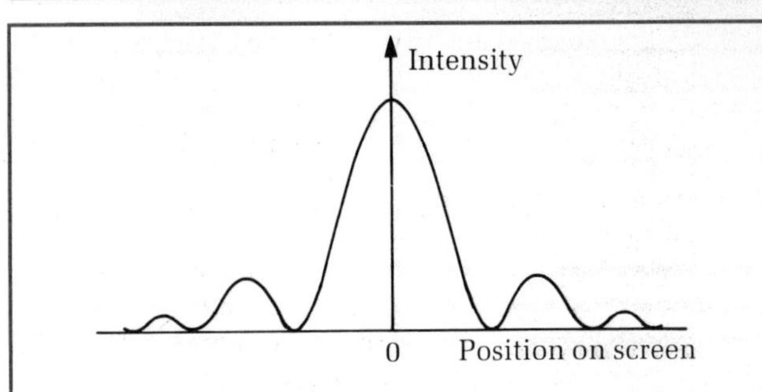

Figure 19.25 *Intensity graph*

What if a narrower slit is used? The fringes are even more widely spaced out with a narrower slit. The whole pattern is stretched out more. Try it!

The narrower the slit, the greater the diffraction (for constant wavelength).

Try using one or two colour filters, each in turn, over the light source. You ought to observe that wider fringes are obtained with red light compared with blue light. Red light has a longer wavelength than blue light.

The longer the wavelength, the greater the diffraction (for constant slit width).

Huygens' wave theory can be used to explain diffraction effects. The theory supposes that each point on a wave front emits secondary wavelets which form a new wave front. See p. 23 if necessary. So each section of wave front which passes through the slit spreads out. All the points on the section passing through contribute to the total intensity. At a dark fringe, the contributions all cancel one another out. Figure 19.26 shows fringes formed (by a single slit) on a screen. Each point along the wave front at the slit contributes to the pattern. At any given position on the screen, the contributions differ in phase. This is because the distance from each point on the wave front to P differs. So if each point emits a wavelet at the same time, the wavelet from the nearest point to P arrives first. Each of the other wavelets arrive in turn, the wavelet from the furthermost point arriving last. So at any one time at P, the contributions differ because of their phase difference.

Consider the position at the first dark fringe. The contributions cancel one another out so darkness results. To understand this, divide the slit gap into two halves. Each point in one half gives a contribution which is exactly cancelled by the corresponding point in the other half.

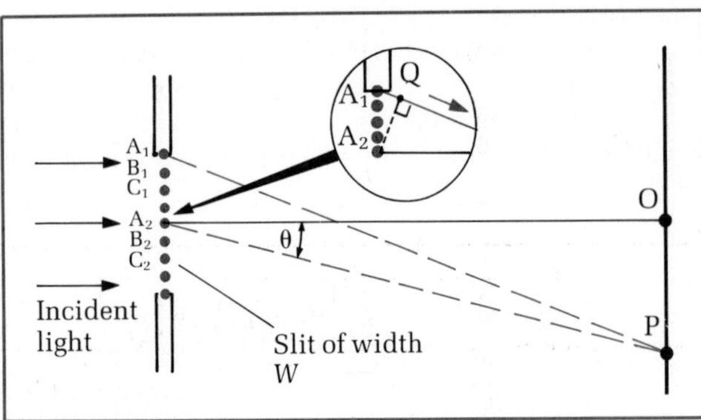

Figure 19.26 *Single slit equation*

Figure 19.26 shows the idea. Point A_1 gives a contribution cancelled by the contribution from point A_2. Point B_1 cancels point B_2, and so on. So the path difference from corresponding points to the dark fringe must be exactly half a wavelength. In other words, wavelets from A_1 travel an extra half a wavelength of distance to the first dark fringe, compared with wavelets from A_2. Likewise for all pairs of corresponding points.

For the first dark fringe, $A_1P - A_2P = \lambda/2$.

Now suppose we choose a position along line A_1P, to be labelled Q, where the distance A_1Q is exactly half a wavelength. So $QP = A_2P$. See Figure 19.26. The angle θ between the first dark fringe and the central bright fringe to the slit centre is shown as angle PA_2O. But angle QA_1A_2 is the same (because in practice OP is much less than OA_2). Also, angles POA_2 and A_1QA_2 are right angles. So triangles A_1A_2Q and OPA_2 are similar triangles. Therefore,

$$\frac{OP}{PA_2} = \frac{QA_1}{A_1A_2}$$

But $\sin\theta = \dfrac{OP}{PA_2}$ and $QA_1 = \dfrac{\lambda}{2}$ and $A_1A_2 = \dfrac{W}{2}$, where W is the slit width. Hence,

$$\sin\theta = \frac{\lambda/2}{W/2} = \frac{\lambda}{W}$$

For example, with a slit of width 0.4 mm, using light of wavelength 500 nm, the first dark fringe is at angle θ given by

$$\sin\theta = 500 \times 10^{-9}/0.4 \times 10^{-3}$$
$$\theta = 0.072°$$

so on a screen placed 1 m beyond the single slit, the distance from the first dark fringe to the central fringe would be $1 \times \tan 0.072°$ m = 1.26 mm.

More generally, the position of each dark fringe is given by

$$\sin\theta = \frac{m\lambda}{W}$$

where θ = angle subtended to the centre,
λ = wavelength used,
W = slit width,
m = 1, 2, 3, ... etc.

Multiple slits

Using two slits instead of a single slit gives the fringe pattern first observed by Thomas Young. Alternate bright and dark interference fringes are seen. However the intensity of the bright fringes changes from one to the next. Compare Figure 19.25 (single slit pattern) with Figure 19.27 which shows the double slits pattern.

You ought to be able to pick out the single slit pattern in the double slit pattern. The fringes of the double slits pattern fade away from the centre and disappear at the single slit minimum. Then the fringes reappear further out, fading at each single slit minimum. So the double slit fringes are modulated by the single slit pattern of each of the slits.

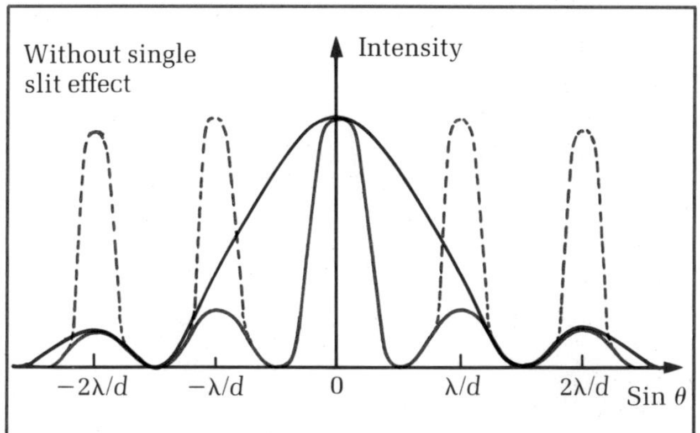

Figure 19.27 *Double slit pattern*

For three equally spaced slits the fringe pattern is more complicated. Consider the arrangement shown in Figure 19.28 where there are three slits, X, Y and Z.

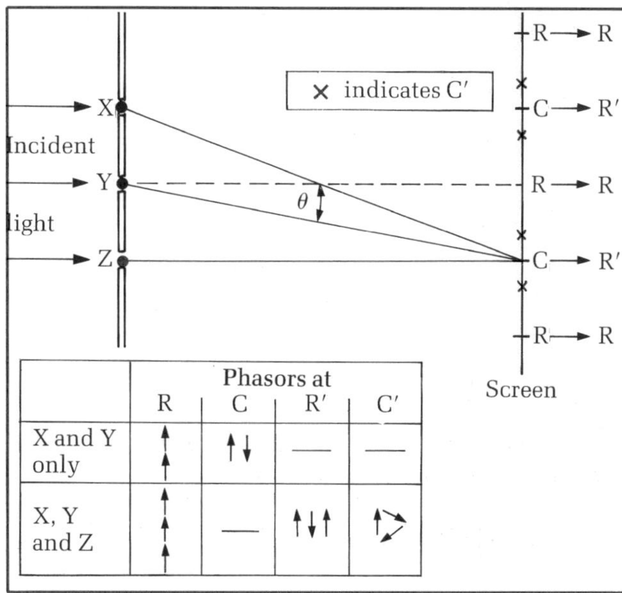

Figure 19.28 *Three-slit system*

1 If slit Z is blocked off, the fringe pattern is the double slit pattern of alternate bright and dark fringes. Suppose the contribution from each slit is represented by a phasor arrow, as in Figure 19.28. At each bright fringe, the contribution from X is in phase with the contribution from Y so the phasor for X is in the same direction as the phasor for Y. The total contribution is then represented by the two phasors end-to-end. At each dark fringe, the contribution from X is 180° out of phase with the contribution from Y. So the phasor for X points in the opposite direction to the phasor for Y. The total contribution is then zero since the phasors cancel each other out. The use of phasors is not essential here but they can be very helpful. The phase angle between two contributions is the angle between the phasors.

2 Now suppose slit Z is unblocked. Reinforcement still takes place at the same positions as before since Z's phasor is in phase with Y's phasor which is in phase with X's phasor at such positions. But cancellation no longer occurs where X's phasor is 180° out of phase with Y's phasor; the reason is that Z now supplies a contribution too, so the total contribution is not now zero at such positions. Reinforcement, not as strong as where X, Y and Z are all in phase, takes place at these positions. Figure 19.28 shows positions of reinforcement R where X, Y and Z are all in phase; positions where Z is 180° out of phase with Y which is 180° out of phase with X are labelled R′.

Cancellation does take place between adjacent maxima. At each point of cancellation C′ on the screen, the three contributions give zero resultant, as shown in Figure 19.28. The contribution from Z is 120° out of phase with the contribution from Y which is 120° out of phase with the contribution from X. So the phasors form an equilateral triangle, hence there is zero resultant.

The intensity distribution for three slits is shown by Figure 19.29. Comparison with the double slits pattern shows that there is a subsidiary maximum (corresponding to R′) between the double slit maxima (corresponding to R). So the three-slit fringe pattern shows a subsidiary bright fringe with a dark fringe either side between adjacent double slits fringes. The double slit fringes become narrower and hence sharper.

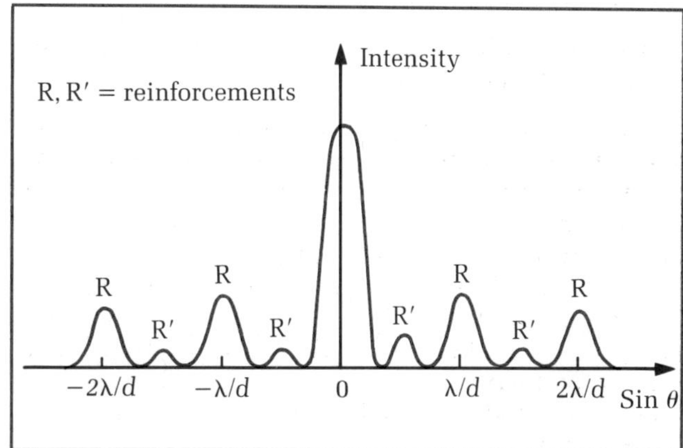

Figure 19.29 *Three-slit pattern*

Single slit diffraction modulates the intensity distribution in practice as shown in Figure 19.29. As with the double slits pattern, Figure 19.27, the three slit pattern fades away from the centre.

For more than three equally spaced fringes each extra slit in excess of two gives an extra subsidiary fringe between adjacent double slit fringes. So for four equally spaced slits, there will be two subsidiary maxima between adjacent double slit fringes. Again, the double slit

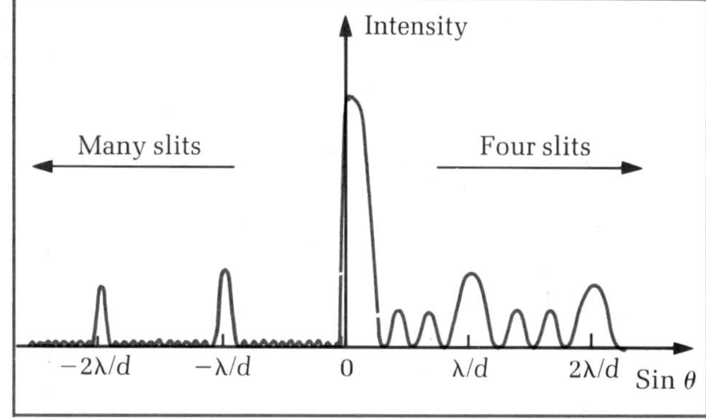

Figure 19.30 *More than three slits*

fringes become sharper as the number of slits is increased. The intensity distribution for four slits is shown in Figure 19.30 (on the previous page). In each case the distribution pattern is modulated by the single slit diffraction pattern. Also, the subsidiary maxima become less and less significant as the number of slits is increased. So the double slit fringes become sharper and sharper.

For many slits the subsidiary maxima are too weak to detect and the double slit fringes become exceedingly sharp; in fact, they appear as thin lines. Now imagine the slits could be made much closer so the arrangement becomes a **diffraction grating**. As explained on p. 269, the position of each line is given by the equation

$$\sin\theta = m\lambda/d$$

where d is the distance between the centres of adjacent slits.

The smaller the value of d, the larger the value of θ becomes. A typical diffraction grating has 600 lines per mm so d is very small indeed. The result is that the fringes are widely spaced, bright, sharp lines with darkness between them. Such gratings are used to measure wavelengths accurately. The single slit diffraction pattern is always superimposed on the interference pattern due to the slits; the effect may be to cause some of the sharp lines to be much weaker in intensity than other lines.

Diffraction by a circular gap

The diffraction pattern produced by a rectangular hole is shown in Figure 19.31. There are two sets of fringes. One set corresponds to the width, the other set to the height of the hole. Because the width is less than the height of the hole, the fringes due to the width are more widely spaced out than the height fringes. Narrower gaps produce more diffraction.

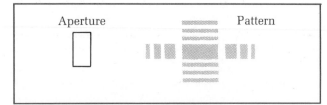

Figure 19.31 *Diffraction by a rectangular aperture*

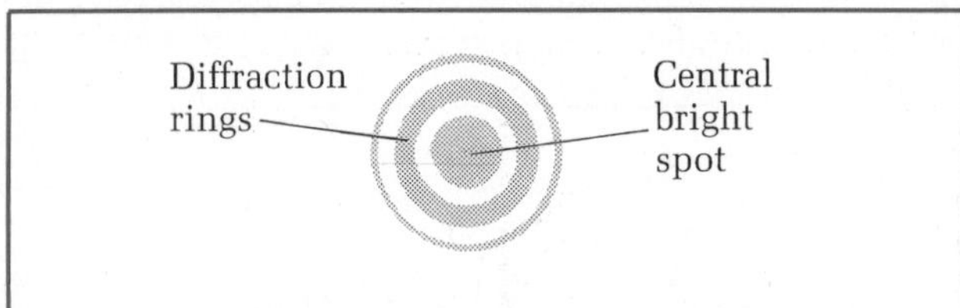

Figure 19.32 *Diffraction by a circular aperture*

Now consider the diffraction pattern produced by a circular hole, as in Figure 19.32. The pattern has a bright central spot, surrounded by bright and dark rings. It can be proved that the angle of the first dark ring is given by

$$\sin\theta = \frac{1.22\lambda}{W}$$

where W = hole diameter,
λ = wavelength,
θ = angle subtended by ring radius to hole.

The theory is more complicated than for the single slit; the basic principle is the same though. At a dark ring, the contributions from every point on a wave front at the hole cancel out. The theory is beyond the scope of A-level physics, but just as for the single slit equation, you can see that the diffraction is greater the smaller the hole or the longer the wavelength.

Optical instruments are designed to take account of the diffraction of light passing through the aperture hole into the instrument. Even where diffraction is small because the hole is large, the spreading could be enough to spoil the use of the instrument. Imagine yourself in the place of an astronomer who intends to design a telescope to study the surface of a planet such as Mars. The entrance hole of a refracting telescope is the objective lens (or mirror if a reflector) so the entrance hole width W is the objective lens width. Now the magnifying power of a telescope is equal to (focal length of objective/focal length of eyepiece). So would the astronomer be correct in assuming that a long focus objective allows all the details of the planet to be easily seen? In fact even the most powerful telescope would not allow details less than a certain size to be seen. Each point on the object (i.e. planet's surface) gives an image point which is spread due to diffraction by the telescope objective. If two point objects are too close together, their images overlap and cannot be seen separately. By using a *wide* objective lens, the spread of each point image is made smaller, so the image of close objects can be seen separately. In other words, the ability of the telescope to **resolve** detail is improved by using a wide objective lens.

The Rayleigh criterion for resolving two points (i.e. being able to tell them apart) is that their angular separation θ must be greater than the value given by

$$\sin\theta = \frac{1.22\lambda}{W}$$

where W is the 'objective' width (i.e. diameter).

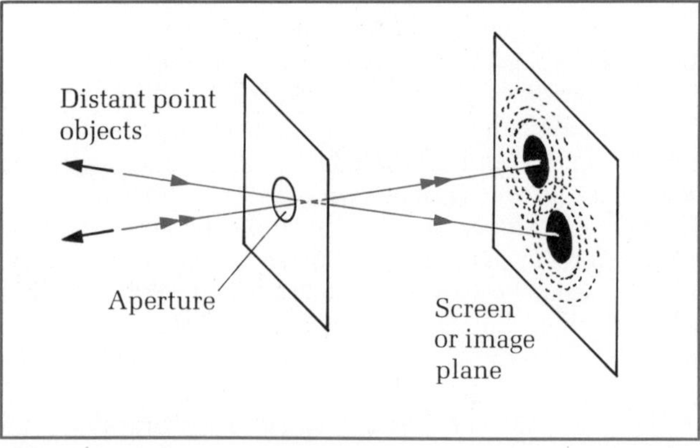

Figure 19.33 *Resolving power*

This corresponds to the centre of one of the diffracted images being at the same position as the first dark fringe of the other image.

The same general idea can be applied to microscopes, radio telescopes, spy satellite cameras, etc. Wider objective lenses give more detail in the image. The wider the objective lens (or dish in the case of the radio telescope), the less the diffraction of the light passing through, so the images are less smeared out.

Also, by using filters to view objects using blue light, more detail can be made out than with white light. This is because white light contains a range of wavelengths from 700 nm for red to 400 nm for blue. By filtering out the red light, only the short wavelength part of the visible spectrum is used. So the minimum angle θ between two point objects which can just be distinguished can be made smaller using blue light only. Blue light, in other words, is less smeared out by the objective gap than red light.

Holograms

These amazing pictures enable viewers to see round the sides of images in them. A 3-D effect is obtained because the images in the picture can be seen at different angles simply by changing the position of viewing. It is almost as if the objects themselves were present. How does a hologram work? To make a hologram, lenses are used to widen a laser beam. Then the beam is divided by a glass plate acting as a partial reflector. One of the two beams from the glass plate is directed at the object to be holographed. The light from the beam is scattered off the object onto a special photographic film. The other beam from the plate is also directed at the film but none of its light is scattered. The film records the interference pattern produced by the scattered light and the direct light.

The interference pattern formed on the film is the hologram. Viewed under suitable conditions, a 3-D image of the original object is seen in the hologram.

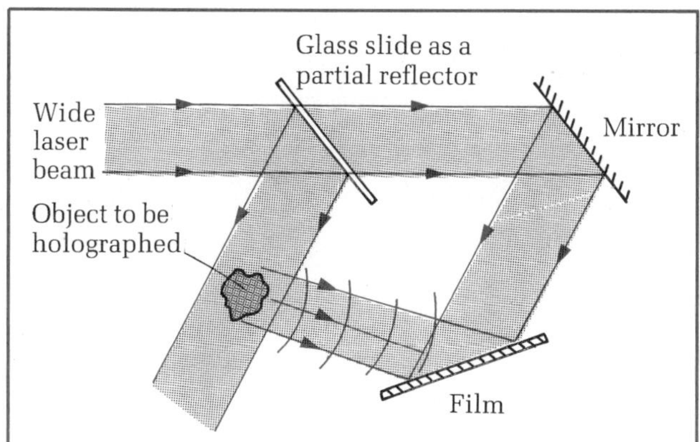

Figure 19.34 *Making a hologram*

19.4 Diffraction gratings and spectra

Diffraction gratings

In the double slits experiment, the light from one slit interferes with the light from the other slit. The result is a pattern of alternate bright and dark fringes. If the double slits are replaced by multiple slits, fewer but sharper fringes are seen. If there are many closely spaced slits,

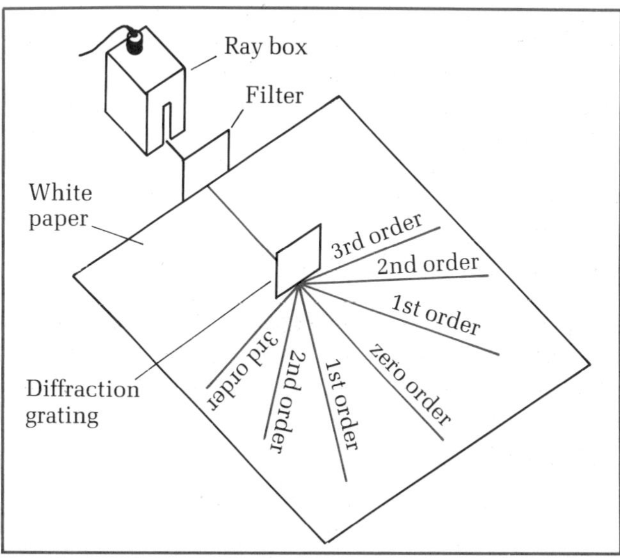

Figure 19.35 *Diffraction gratings*

reinforcement (i.e. positive interference) only happens in a limited number of directions. Each of the many slits transmits light waves, but they cancel one another out in most directions. A plate with many close, regularly spaced slits is called a **diffraction grating**. Figure 19.35 shows the effect of a diffraction grating on a parallel beam of monochromatic light. The transmitted light is channelled only in certain directions by the grating. Each slit produces a diffracted set of light waves, but the different sets of waves reinforce one another only in certain directions.

One of these directions is always the same as the direction of the incident light. This central direction is called the zero order of diffraction. The other beams from the grating are numbered outwards from the zero order beam, the same numbers used either side of the centre. Each diffracted beam on one side is matched by a beam at the same angle to the zero order on the other side of the zero order.

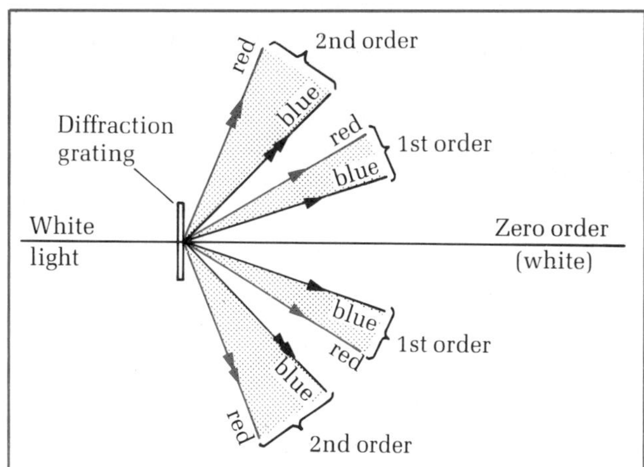

Figure 19.36 *White light directed at a diffraction grating*

The closer the slits, the more widely spaced are the diffracted beams.

Using several diffraction gratings in turn, each with a different spacing, you can show that the grating with the most slits per millimetre produces the fewest diffracted beams.

The longer the wavelength of light used, the more widely spaced are the diffracted beams.

Use a white light source with coloured filters. You should find that red light gives greater angles between each diffracted beam and the zero order. Does red light have a greater wavelength than blue light? See p. 259 if necessary.

White light gives a mini-spectrum of colours in each diffracted beam.

White light consists of all the colours of the spectrum from red to blue, with a continuous range of wavelengths. The red light component is diffracted more than the blue light component, other colours falling in between. So each diffracted beam has red at its outer edge and blue at its inner edge, with the other colours between. The central order is white because all the colours in it travel in the same direction as the incident beam, so there is no separation of colour in the zero order (i.e. central) beam.

So far, only **transmission gratings** have been considered. A transmission grating consists of fine grooves cut into a glass plate. When a parallel beam of light is directed at a transmission grating, the grooves scatter the light in all directions. The slits are the spaces between the grooves, and these spaces allow light to pass through. Each space produces a set of diffracted waves and the different sets of diffracted waves reinforce in certain directions only. Plastic replica gratings may be made from glass gratings by using the glass grating as a mould.

Reflection gratings may be made by cutting regularly spaced grooves on a polished metal surface. The gaps between the grooves reflect the incident light, each gap producing a diffracted set of light waves. So the principle is the same as for the transmission grating. A record disc can be used as a very simple reflection grating. Its grooves are regularly spaced, although not as close as a proper reflection grating. Hold the disc at an angle to a point source of light and it ought to be possible to see colours due to diffracted orders of white light. Another way of making a reflection grating is to stick a plastic replica grating onto a concave mirror. Concave reflection gratings channel most of the diffracted light into a single diffracted beam. This makes the study of light from low intensity sources easier because the light energy would otherwise be channelled into several diffracted beams. Diffraction of ultra-violet radiation is also possible with reflection gratings.

The diffraction grating equation

Consider an enlarged cross-section of a transmission grating, as in Figure 19.37. Light waves arrive continuously at the grating so each slit continuously transmits diffracted waves. The diffracted waves reinforce one another in certain directions only. For example, consider the way in which the first order diffracted beam is produced (i.e. the one nearest to the central order). Its direction is precisely where waves from one slit line up exactly with waves emitted one cycle earlier from the next slit. So the

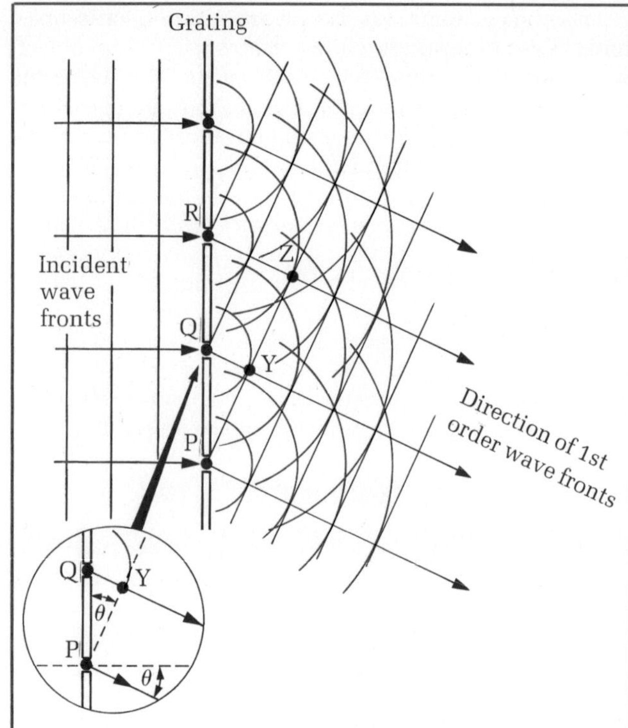

Figure 19.37 *Formation of the first order wave front*

wave fronts reinforce. Figure 19.37 shows the formation of a first order diffracted wave front, labelled PYZ. It is formed by the wave front at P lining up with the wave front from Q emitted one cycle earlier which lines up with the wave front from R emitted two cycles earlier, etc.

From △PQY, distance QY is exactly one wavelength. Also distance PQ is equal to the slit spacing, d. Also $Q\hat{Y}P$ is a right angle. Therefore

$$\sin Q\hat{P}Y = \frac{QY}{PQ} = \frac{\lambda}{d}$$

However, $Q\hat{P}Y$ is equal to the angle of diffraction θ $(= O\hat{Q}Y)$ so the angle of diffraction for the first order beam is given by

$$\sin \theta_1 = \frac{\lambda}{d}$$

For the mth order of diffraction, its direction is such that wave fronts from one slit reinforce wave fronts emitted m cycles earlier by the adjacent slit. So distance QY is equal to $m\lambda$ now. Its angle of diffraction is given by

$$\sin \theta_m = \frac{m\lambda}{d}$$

where θ_m = angle of diffraction of the mth order,
λ = wavelength,
d = slit spacing,
m = order number of the beam.

The slit spacing is sometimes given in terms of the number of slits per metre. For a grating with N slits per metre, its slit spacing in metres is $1/N$.

The number of diffracted beams can be worked out if the grating spacing d and the wavelength λ are known.

Because $\sin\theta$ can never be greater than 1, then $m\lambda/d$ ($=\sin\theta_m$) can never be greater than 1.

$$\frac{m\lambda}{d} \leqslant 1 \qquad (\leqslant \text{ means 'is less than or equal to'})$$

So the order number m can never be greater than d/λ.

$$m \leqslant \frac{d}{\lambda}$$

To calculate the number of diffracted beams, the value of d/λ is first determined. The highest order number is then given by the value of d/λ rounded down to the nearest whole number. The number of diffracted beams (excluding the zero order) is given by the highest order number.

Worked example A parallel beam of monochromatic light of wavelength 650 nm is directed normally at a diffraction grating which has 600 lines per mm. Determine the number of diffracted beams and the angle of diffraction of the highest order.

Solution First, calculate the value of the slit spacing d.

Since the grating has 600 lines per mm, then $d=\frac{1}{600}\,\text{mm}=1.67\times10^{-6}\,\text{m}$.

Then calculate the value of d/λ.

$$\frac{d}{\lambda}=\frac{1.67\times10^{-6}}{650\times10^{-9}}$$

Which gives $d/\lambda=2.57$.

So the highest order number is $m=2$.

For $m=2$, the angle of diffraction θ_2 is given by

$$\sin\theta_2=\frac{m\lambda}{d}=\frac{2\times650\times10^{-9}}{1.67\times10^{-6}}=0.778$$

So $\theta_2=51.08°=51°5'$.

Missing orders may be caused because of the single slit diffraction pattern. Each slit produces a set of diffracted waves, but in certain directions, the light intensity is zero. These directions are the dark fringes of the single slit pattern, and may cause missing orders in the light from a grating. Even though wave fronts from one slit may be reinforced by wave fronts from other slits, if each diffracted wave front is weak in that direction, then the reinforced wave front will be weak – or even missing altogether in that direction. See p. 268.

In the worked example above, suppose the slits and grooves of the grating are the same width. Since the spacing from one slit to the next slit (d) is known ($=1.67\times10^{-6}\,\text{m}$), then the width of each slit is $d/2$ and the width of each groove is $d/2$. Now the single slit diffraction pattern has minima of intensity at angles given by

$$\sin\theta=\frac{m\lambda}{W}$$

where W, the slit width, is equal to $d/2$ in this example.

So the first dark fringe of the single slit pattern is given by $m=1$. Hence

$$\sin\theta_1=\frac{1\times650\times10^{-9}}{(1.67\times10^{-6}/2)}=0.778$$

So the first minimum is at angle $\theta=51.08°$. This is the same angle as the second order diffracted beam, so the second order beam would be missing if the grating has slits of width W equal to half the slit spacing (i.e. if $W=d/2$). Therefore a grating should not have a slit width which causes missing orders. Concave reflection gratings have a great advantage over transmission gratings because the central peak of the single slit pattern can be channelled down one of the diffracted orders.

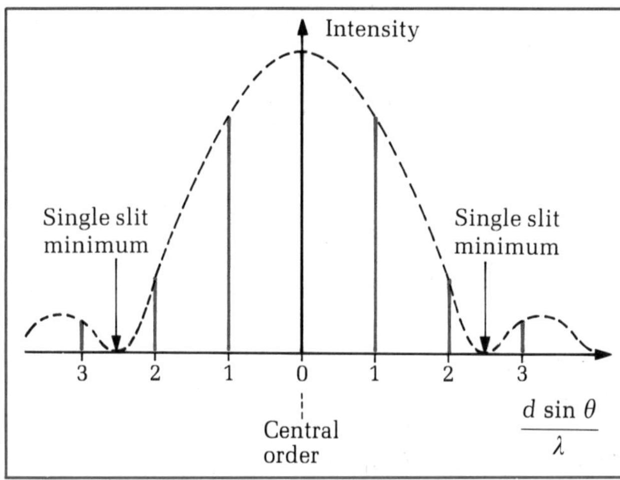

Figure 19.38 *Missing orders*

Using a diffraction grating to measure light wavelengths

A spectrometer is used to enable angles to be measured very accurately. First, though, the spectrometer must be

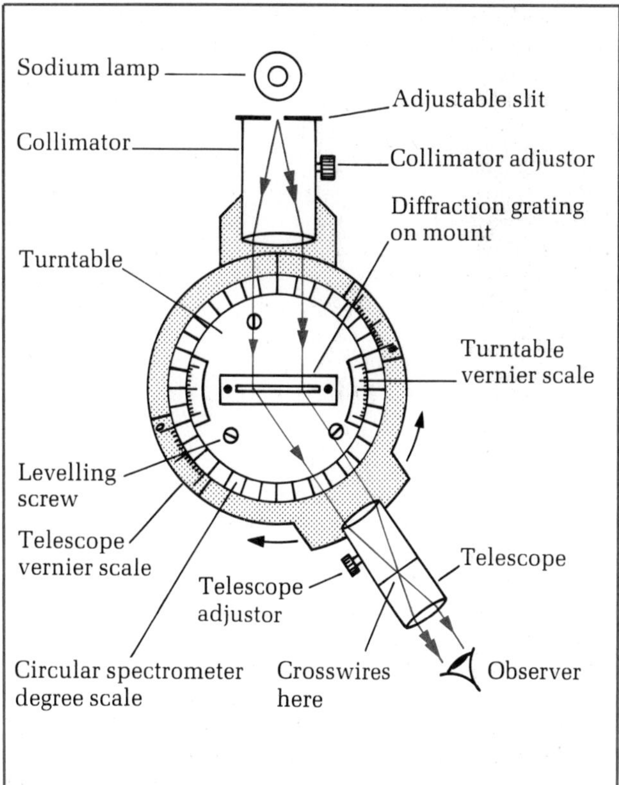

Figure 19.39 *Diffraction grating used on a spectrometer*

correctly adjusted. To do this, the telescope is turned to view a distant object. When the telescope is focused on the object, the telescope is thus set correctly to receive parallel light. Then the telescope is turned to face the collimator. The slit of the collimator viewed through the telescope is then brought into focus by adjusting the collimator. The collimator is thus set to produce parallel light. The slit is then adjusted to a narrow line, and the sodium lamp is placed next to the slit. This set up is shown in Figure 19.39 (on the previous page).

Now we need to make sure that the turntable which the grating is to be placed on is horizontal (assuming the bench is horizontal!), otherwise the diffracted beams miss the telescope altogether. A spirit level is used, placed along the line between two of the three levelling screws. The two screws are adjusted until the line is horizontal, judged by the spirit level. Then the spirit level is placed at right angles to the line so it passes over the third levelling screw. This screw is then adjusted to make the spirit level once again horizontal. The whole turntable is now horizontal, and the grating can be placed in its mounting which is now vertical.

Can we be sure that the grating is at right angles to the light from the collimator? To be exact, line the telescope up with the collimator by viewing the slit and positioning its image at the centre of the telescope crosswires. Note the telescope's position from the vernier scale. Then turn the telescope to one side by 90° exactly. Now turn the grating on the turntable until an image of the slit by reflection from the grating is on the centre of the crosswires. Then the grating is at 45° to the light from the collimator. Finally, turn the turntable and grating back by exactly 45° to get the grating at exactly 90° to the incident light.

Before making any measurements, scan across with the telescope to observe each diffracted order. For a monochromatic source of light, each order will appear as a line image of the slit. In fact, most vapour lamps emit several values of wavelength, so each order consists of several lines, each of different colour corresponding to one of the values of wavelength. The pattern of lines makes up the spectrum of light from the lamp. For a sodium lamp, the brightest lines of the spectrum are a pair of very close orange-yellow lines, called the *sodium D-lines.*

By turning the telescope, each line of each order can be centred on the crosswires, and so the position of each line is measured. Its colour is also noted. The procedure is carried out for measurements on both sides of the central order. Figure 19.40 shows typical measurements taken using a sodium lamp and a grating with 600 lines per mm. The angle of diffraction for each line may be calculated from the measurements taken for that line on either side of the centre. The angle of diffraction is half the difference between the two measurements. The colour of each line may be used to identify its order number. Hence the wavelength of each line can be calculated using the diffraction grating equation.

	Position on left	right	Angle of diffraction
	θ_L	θ_R	$\theta = \frac{(\theta_R - \theta_L)}{2}$
1st order yellow	78°0′	119°41′	20°51′
2nd order yellow	53°28′	144°12′	45°22′

Number of lines per metre = 600 000

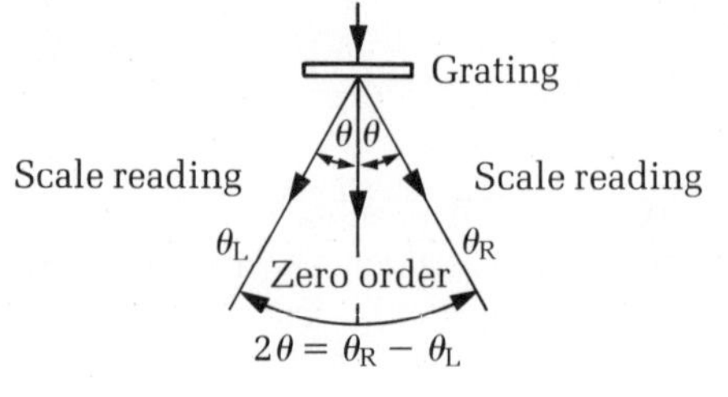

Figure 19.40 *Typical results*

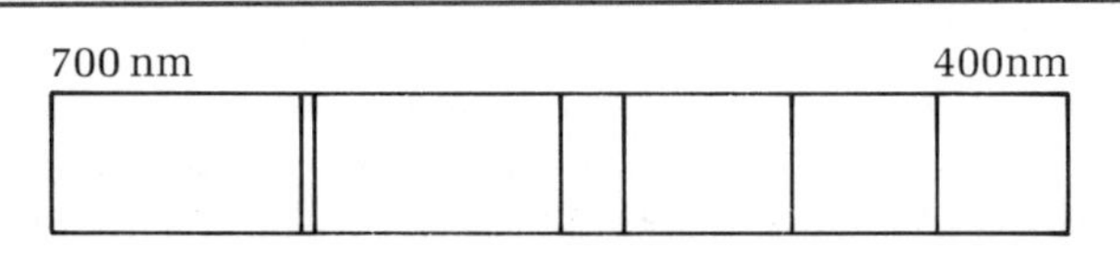

Figure 19.41 *Line emission spectrum*

Types of spectra

Line emission spectra for example the sodium spectrum, consist of thin vertical lines of different colours, set against a dark background. Each line corresponds to one value of wavelength emitted by the source. Vapour lamps and discharge tubes emit line spectra. The atoms of the light-emitting material emit **photons** of certain energies only. Each photon is emitted when an electron of an atom moves from one energy level to a lower energy level, as explained in detail on p. 456. The photon carries away an amount of energy equal to the difference between the energies of the two levels. Since the energy of a photon is given by hf where f is the frequency and h is a constant (called the **Planck constant**), then only certain values of frequency are produced, so only certain photon wavelengths are emitted.

The pattern of a line emission spectrum can be used to identify the type of atom producing the spectrum. In other words, it can be used to identify the element producing the light. Each pattern is characteristic of the element which produces the light; hence the energy levels of each types of atom are unique to that type.

Continuous spectra for example the white light spectrum, consist of a continuous range of colours from deep red to deep blue. A prism can be used to produce a continuous spectrum of white light, as shown on p. 235, as an alternative to using a diffraction grating. White light may be produced from any filament lamp. The electric current passing through the filament heats up the filament, and the atoms of the filament vibrate so much that their energy levels become spread out. The atoms emit a continuous range of photon energies and hence wavelengths. Sunlight gives a continuous spectrum, first demonstrated by Sir Isaac Newton who showed that sunlight (and white light produced by candles) can be split into the colours of the spectrum.

A continuous spectrum cannot be used to identify the type of atom which emits the light, but it can be used to determine the temperature of the source. The most intense part of the spectrum changes with temperature. If the temperature is increased, the brightest part of the spectrum changes from red to orange to yellow, etc. The wavelength of the brightest part of the spectrum becomes shorter as the temperature increases. See p. 115. Astronomers use this fact to deduce the surface temperatures of stars. For example, the star *Betelgeuse* in the constellation of *Orion* (the top left-hand star of *Orion* for northern hemisphere observers) is orange-red in colour. Its surface temperature has been estimated, from its colour, to be 3000 K. The Sun is yellow in colour and its surface temperature has been estimated at 6000 K. Bluish stars have surface temperatures in the range 20 000 to 30 000 K.

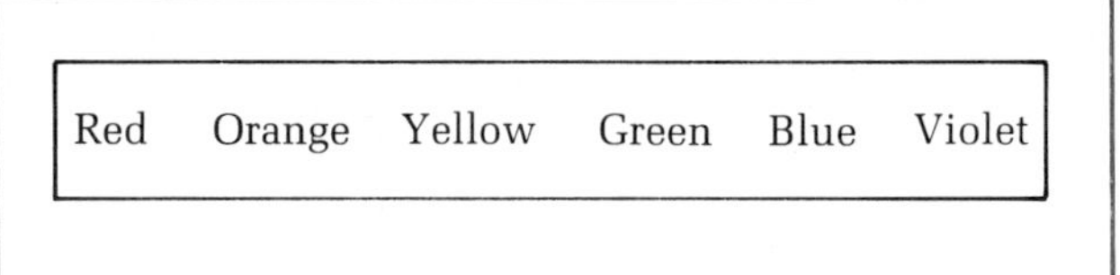

Figure 19.42 *Continuous spectrum*

Band spectra are produced by molecules rather than single atoms. A band spectrum consists of one or more groups of spectral lines, as shown in Figure 19.43. The atoms of a molecule are joined to one another by 'bonds', the bonds formed by electrons. The energy levels of the bond electrons are regularly spaced, so the emitted photons produce groups of closely spaced lines.

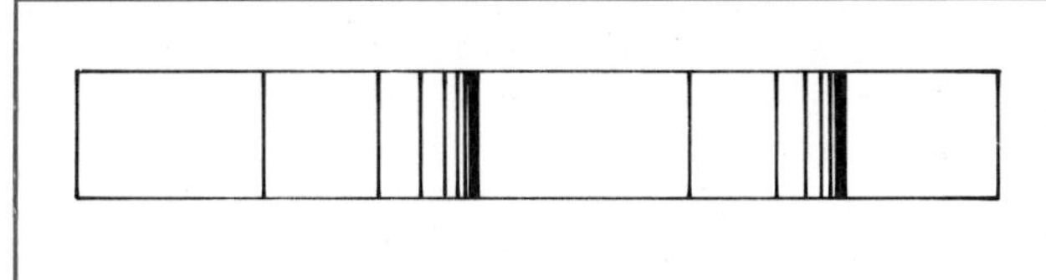

Figure 19.43 *Band spectrum*

Absorption spectra consist of dark vertical lines set against a background of the continuous spectrum. If the spectrum of sunlight is carefully studied, it is possible to observe dark vertical lines against the continuous spectrum. Another way to produce an absorption spectrum is to pass white light through a sodium flame, made by heating common salt in a bunsen burner flame. If the light then passes into a spectrometer, the white light spectrum will be seen to contain thin dark vertical lines. Switch the white light source off and the continuous spectrum is replaced by the line emission spectrum of the sodium flame, and the lines are at the same positions as the dark lines of the absorption spectrum. The absorption spectrum is like the negative of the emission spectrum in the sense that colour in the absorption spectrum is replaced by darkness in the emission spectrum, and the dark lines are replaced by coloured lines.

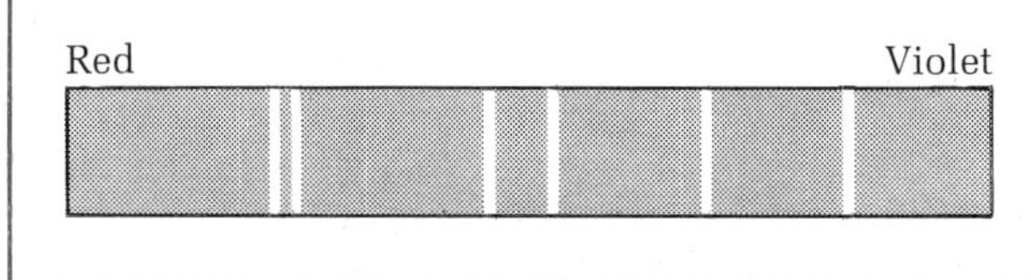

Figure 19.44 *Line absorption spectrum*

A sodium atom emits a photon of light when one of its electrons moves from one energy level to a lower energy level. The reverse process happens when white light passes through a sodium flame. An electron moves from a low energy level to a higher level as a result of absorbing a photon, but the photon must have just the right amount of energy, equal to the difference of the two energy levels. White light contains a continuous spread of photon energies, so the atoms of the sodium flame absorb certain photon energies only, corresponding to the possible energy level differences. The white light passes through the flame into the spectrometer short of certain photon energies, so the white light spectrum has certain wavelengths missing, each dark line produced by photons of a particular energy (and hence wavelength) having been absorbed. Thus the dark lines are the negative of the line emission spectrum. The pattern of the dark lines of an absorption spectrum can therefore be used to identify the type of atom which absorbed the light.

19.5 Summary

Equations

1 Young's double slits equation

$$y=\frac{\lambda X}{d}$$

where y = fringe spacing (centre to centre),
d = slit spacing (centre to centre),
X = slit-screen distance,
λ = wavelength.

2 Parallel-sided thin films are dark when viewed by reflection if

$$2t=\frac{m\lambda_0}{n}$$

where t = film thickness,
λ_0 = wavelength in air,
n = refractive index of film,
m = 0.1, 2, 3, . . .

3 Wedge fringes have a spacing given by

$$y=\frac{\lambda_0}{2\tan\theta}$$

where y = fringe spacing (centre to centre),
θ = wedge angle.

4 Newton's rings; the diameter of the mth dark ring from the centre is given by

$$d_m{}^2=4mR\lambda$$

where d_m = ring diameter,
R = radius of curvature of the lower surface of the lens.

5 Single slit diffraction gives minimum intensities at positions given by

$$\sin\theta_m = \frac{m\lambda}{W}$$

where θ_m = angle subtended,
W = slit width,
m = order number of dark fringe from the centre.

6 Diffraction pattern of a circular hole has the first dark ring at an angle given by

$$\sin\theta = \frac{1.22\lambda}{W}$$

where W = hole diameter.

7 The diffraction grating equation

$$d\sin\theta_m = m\lambda$$

where d = slit spacing = 1/number of slits per metre,
θ_m = angle of diffraction of the mth order.

8 Photon energy $E = hf$

where h = Planck constant,
f = frequency.

Definitions and rules

Coherence two sets of waves which overlap are coherent if they maintain a constant phase relationship.
Resolution of two point objects by an optical instrument is possible if the angle subtended by the two points to the instrument is more than a value given by

$$\sin\theta = \frac{1.22\lambda}{W}$$

where W = width of the objective.

Short questions

19.1 To demonstrate interference of light waves, Thomas Young allowed light from an illuminated single slit to fall on a pair of narrow closely-spaced slits. Explain
a) why light from two separate lamp bulbs can never produce an interference pattern,
b) why double slits used as above do give an interference pattern.
How would the interference pattern differ if (i) the double slits had been closer together, (ii) the single slit had been much wider?

19.2 Atoms emit light in short bursts lasting of the order of nano-seconds. Given the speed of light in air is $3 \times 10^8\,\mathrm{m\,s^{-1}}$, calculate
a) the length of a typical wave train of light in air,
b) the number of cycles in such a wave train if the wavelength in air is 500 nm.

19.3
a) In a double slit experiment, a screen is positioned at a distance of 0.80 m from the two slits, as shown in the diagram. Light of wavelength 590 nm is directed at the slits normally to give an interference pattern on the screen with five fringes spaced over a distance of 4.8 mm. Calculate the slit spacing.
b) How would the pattern in **a)** differ if white light had been used?

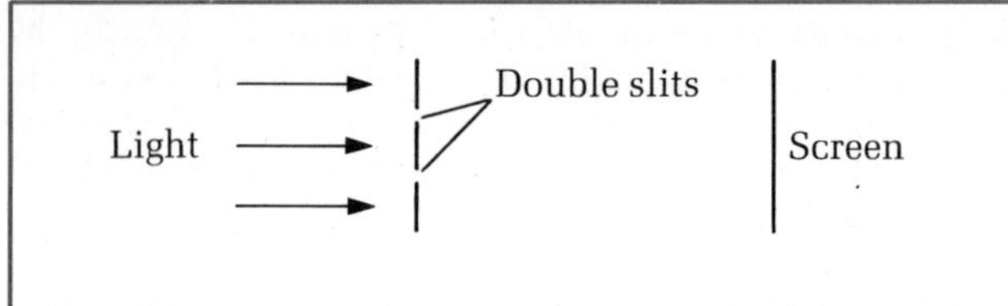

Figure 19.45

19.4 A flat glass surface is coated with a thin uniform layer of transparent material of refractive index 1.65 and thickness 250 nm. White light is passed through the coated glass surface and the transmitted light is viewed using a spectrometer. The observed spectrum shows a dark band set against the colours of the white light spectrum.
a) Explain the cause of the dark band.
b) Calculate the wavelength of the missing light that caused the dark band.

19.5
a) Explain why a bloomed lens observed by reflection in daylight appears purple.
b) What must be the minimum thickness of the coating on such a lens if it is to eliminate reflection of light of wavelength 560 nm?
The refractive index of the coating is 1.38 and of the glass is 1.53.

19.6 Give a labelled diagram of an arrangement you could use to observe the interference of monochromatic light reflected from an air wedge. Describe and explain the observed pattern of fringes.

19.7 An air wedge is formed between two flat glass plates of length 45 mm by using a spacer at one end. Wedge fringes spaced 0.19 mm apart are observed by reflection when light of wavelength 430 nm is directed normally at the wedge. Calculate the angle of the wedge and the thickness of the spacer.

19.8 What would the fringe spacing in question **19.7** become if the wedge was filled with water? (Refractive index = 1.33.)

19.9
a) Explain the formation of Newton's rings observed when monochromatic light is reflected from a planoconvex lens in contact with a flat glass plate.
b) Explain why a black spot is seen at the centre of the ring pattern provided the lens is in contact with the plate.

19.10 A parallel beam of light of wavelength 650 nm is directed normally at a single slit of width 0.14 mm in a darkened room. A screen is placed 1.50 m from the single slit as shown in the diagram.
a) Sketch a graph to show how the intensity of light falling on the screen varies with position across the screen.

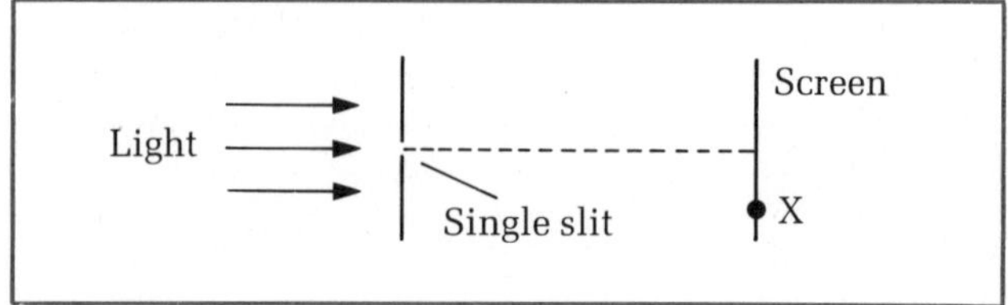

Figure 19.46

b) Calculate the width of the central fringe.
c) Suppose the slit could be made wider. Sketch a graph to show how the intensity at X, a distance of 5 mm from the central maximum, varies with slit width as the slit is widened.

19.11
a) Estimate the size of craters on the lunar surface that can just be resolved using a telescope fitted with an objective of diameter 100 mm. Assume the wavelength of light observed is 600 nm, and that the distance from the Earth to the Moon is 380 000 km.
b) Estimate the minimum width of the lens fitted to a satellite 'spy camera' for a satellite that orbits at a height of 1600 km above the Earth.

19.12 Monochromatic light of wavelength 600 nm is directed normally at four parallel narrow slits spaced at 0.1 mm apart. The two outer slits are blocked off. The transmitted light falls onto a screen 1.8 m from the slits, the screen being placed at right angles to the direction of the incident beam. The diagram below shows the variation of intensity with position across the screen.

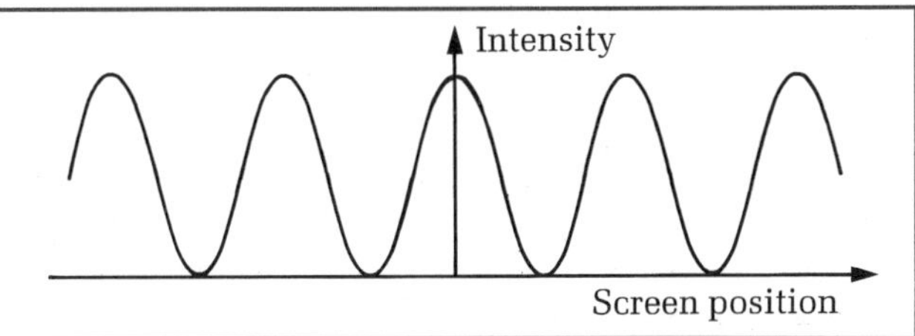

Figure 19.47

a) Calculate the fringe spacing.
b) Copy the diagram and sketch on your copy the variation of intensity with position for (i) the two outer slits only used, (ii) all four slits used. Explain the pattern of (ii) in general terms.

19.13 Light of wavelength 430 nm is directed normally at a diffraction grating. The first-order transmitted beams are at 28° to the zero-order beam. Calculate
a) the number of slits per mm on the grating,
b) the angle of diffraction for each of the other diffracted orders of the transmitted light.

19.14 Light directed normally at a diffraction grating contains wavelengths of 586 and 580 nm only. The grating has 600 lines per mm.
a) How many diffracted orders are observed in the transmitted light?
b) For the highest order, calculate the angle between the two diffracted beams.

19.15 A diffraction grating is designed with each gap of the grating of width 0.83 μm. When used in a spectrometer to view light of wavelength 430 nm, diffracted beams are observed at angles of 14°55′ and 50°40′ to the zero-order beam.
a) Assuming the low-angle diffracted beam is the first-order beam, calculate the number of lines per mm on the grating.
b) Explain why there is no diffracted beam between the two observed. What is the order number for the beam at 50°40′?

20

Optical instruments

20.1 Mirrors

Waves and rays

The wave theory of light devised by Huygens (see p. 233) can be used to explain reflection and refraction. Yet when we consider the effect of lenses or mirrors on the path of light, we usually prefer to draw diagrams using light rays. Light rays represent the direction of travel of wave fronts. Imagine light spreading out from a point source such as a small lamp bulb. The waves spread out in all directions; we can represent this by light rays directed outwards from the bulb. Light rays are straight lines, provided the direction does not change. There are several ways of changing the direction of light, such as reflection off a mirror or refraction at a glass block. So when we draw diagrams to show the path of light through a lens or off a mirror, the rays are straight lines which change direction only at the point of reflection or refraction.

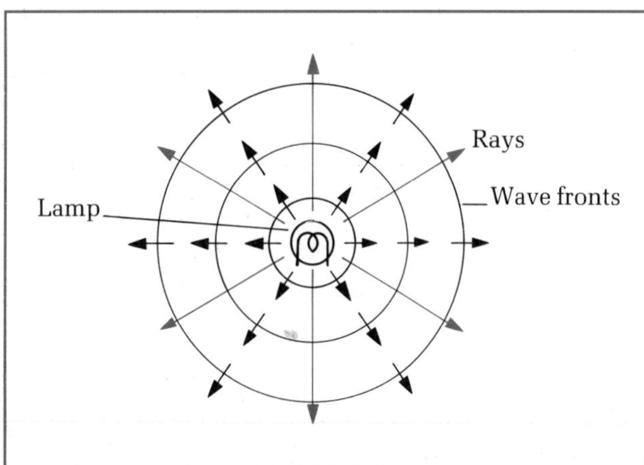

Figure 20.1 *Rays and waves*

Diffraction of light by gaps or obstacles changes the direction of the light, but the change of direction is not an abrupt change like at a mirror surface, and it can only be understood using the wave theory. So in situations where diffraction is likely to be significant, the idea of rays may be used, but with great care. In practice, apertures and other features which cause diffraction are usually wide enough so that diffraction effects are too small to be noticeable. An exception is when an optical instrument is used to inspect an object for detail. Then diffraction effects, usually not noticed, set a limit to the amount of detail which can be seen.

So in most circumstances we can think in terms of light rays when we attempt to understand how optical instruments work. This approach, sometimes called **geometrical optics**, is a good deal easier to follow than visualizing wave fronts passing through optical instruments. Provided we remember that the use of light rays has limits, as explained above, we can use the ray approach.

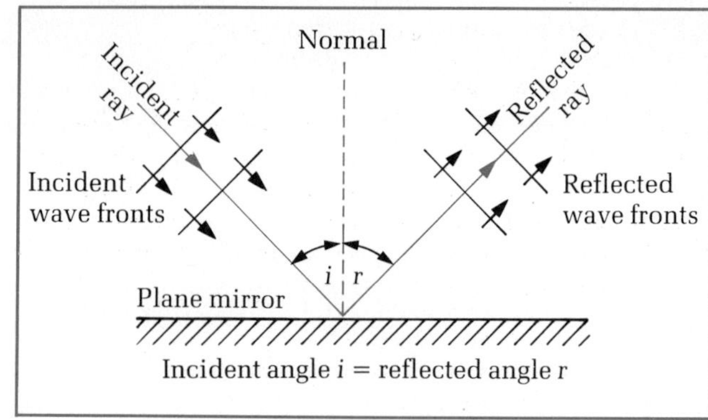

Figure 20.2 *Reflection by a plane mirror*

Plane mirrors

All types of mirrors obey the **law of reflection** which is illustrated by Figure 20.2. The angle between the incident ray and the normal is always equal to the angle between the reflected ray and the normal. The normal is the line perpendicular to the mirror at the point of incidence. Look in a plane mirror and you will see your own image behind the mirror. The image is at the same distance behind the mirror as the object is in front. The image is also **laterally inverted**, as shown by Figure 20.3, which is the reason why there are 'mirror signs' on the front of some police cars, ambulances and fire engines.

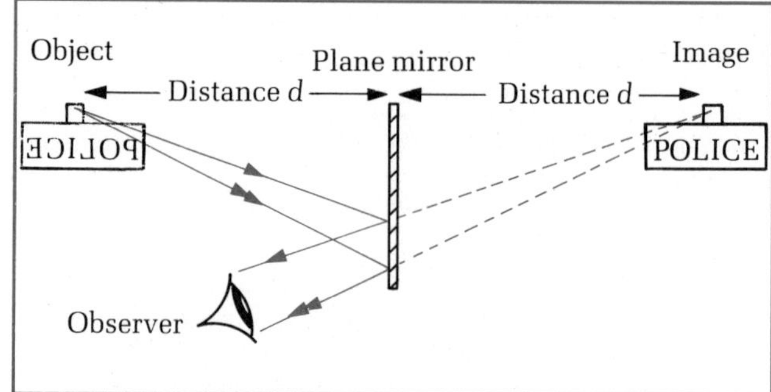

Figure 20.3 *Mirror images*

Ammeters and voltmeters used for accurate measurements usually have a strip of mirror along the scale. The reason is to ensure that the reading is made directly above the pointer. Otherwise, the pointer reading will be incorrect because the pointer will not be in line with the correct reading.

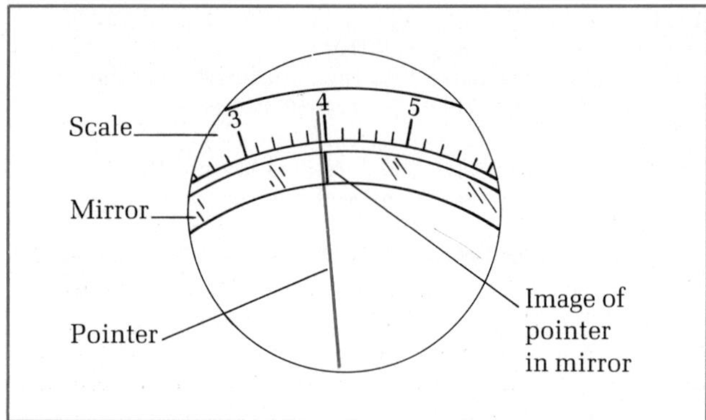

Figure 20.4 *Reading a meter*

Concave mirrors

A spherical concave mirror is one with a surface which is part of a sphere. The sphere radius is called the **radius of curvature** of the mirror. The straight line from the centre of curvature to the centre of the mirror is called the **axis**. The **focal point** of a concave mirror is where parallel light along the axis is focused after reflection by the mirror.

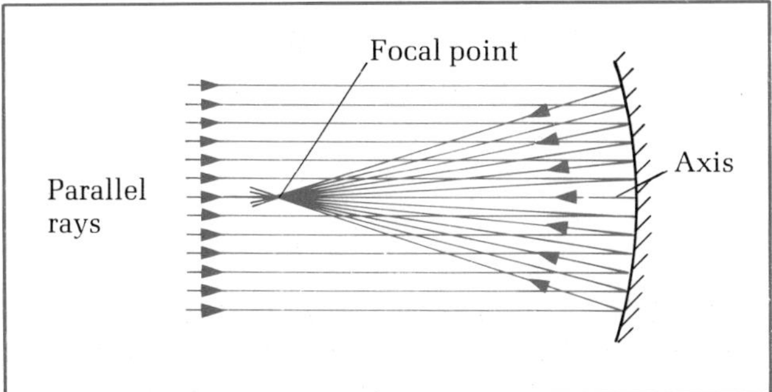

Figure 20.5 *The concave mirror*

The radius of curvature is usually much greater than the width of the mirror. If it were not, parallel light could not be focused to a single point. The outer rays of a parallel beam would focus nearer the mirror than the inner rays of the beam. Figure 20.6 shows the situation of a highly curved mirror with a width comparable with its radius of curvature. Each incident ray reflects off in accordance with the law of reflection, but because the mirror is curved, the direction of the normal differs across the mirror. The reflected rays from the parallel beam do not reflect to the same point for a spherical mirror, but they define a **cusp** as in Figure 20.6. By cutting off the outer rays of the parallel beam, the cusp is narrowed down to a point, the focal point. Another way to reduce the cusp to a point is to use mirrors with width much less than the radius of curvature, which is what you would have if only the central part of a much wider mirror were used.

The **focal length** of a concave mirror is the distance from the focal point to the centre of the mirror.

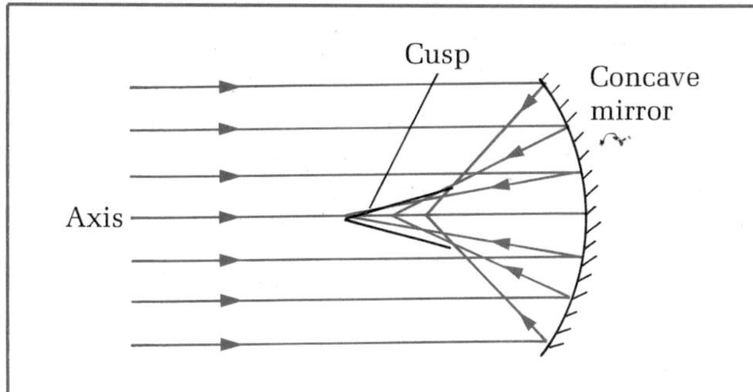

Figure 20.6 *A highly-curved mirror*

Image formation by a concave mirror depends on how far away the object is placed from the mirror. Consider the light rays from a point object which reflect off the mirror. We consider only mirrors which have a radius of curvature much greater than the width, so the reflected rays form a point image. To locate the point image, consider three key rays.

1 The ray to the lens centre reflects off at an equal angle.

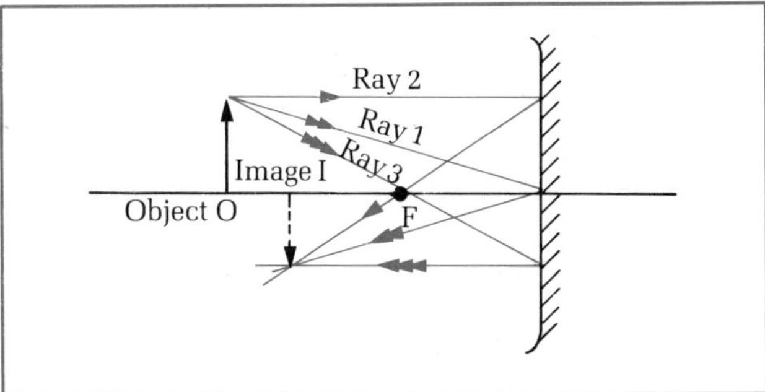

Figure 20.7 *Image formation*

2 The ray parallel to the axis reflects off through the focal point.

3 The ray through the focal point reflects off parallel to the axis, (the reverse of **2**).

Figure 20.7 shows the situation where the three key rays meet (i.e. converge) after reflection to form a clear point image. All the other rays from the point object which reflect off the mirror must pass through where the three key rays meet.

A second point object on the axis in line with the first object would give a corresponding point image on the axis, in line with the first image. So an extended object (in other words, a real object) between the two point objects would give an extended image between the two point images.

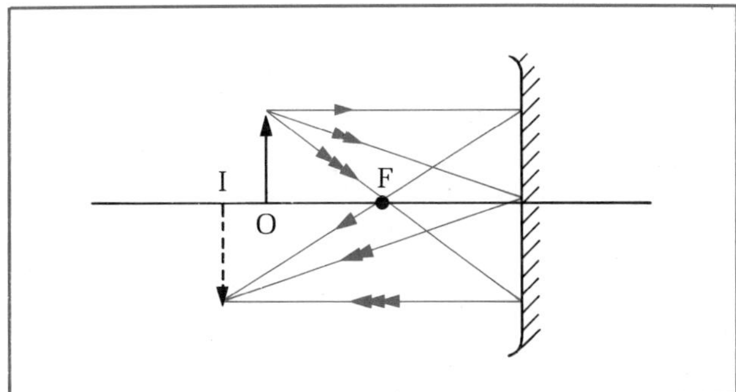

Figure 20.8 *Object closer than 2F*

The image position and size depend on how far the object is placed from the mirror. Figures 20.7 and 20.8 show that the nearer the object is to the mirror, the further away the image is formed, provided the object is no nearer than the focal point.

If the object is placed between the focal point and the mirror, the reflected rays diverge from one another after reflection, as shown in Figure 20.9. An image is still formed in this situation, but it can only be seen by looking

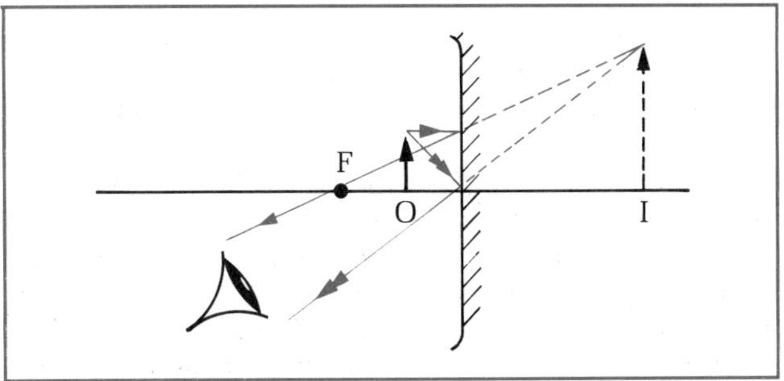

Figure 20.9 *A virtual image*

into the mirror. It is a **virtual** image, formed where the rays appear to come from. A virtual image cannot be formed on a screen or photographic film, unlike the images formed in Figures 20.7 and 20.8. These images are **real** images, formed where the reflected rays meet. The virtual image formed in Figure 20.9 is enlarged compared with the object; a face mirror is usually a concave mirror, so the user sees an enlarged upright virtual image behind the mirror.

The concave mirror formula may be used to calculate the image position if the object position and focal length values are known.

$$\frac{1}{u}+\frac{1}{v}=\frac{1}{f}$$

where u = object distance from mirror,
v = image distance from mirror,
f = focal length.

To prove this formula, consider object O placed at distance u from a mirror of focal length f as shown in Figure 20.10.

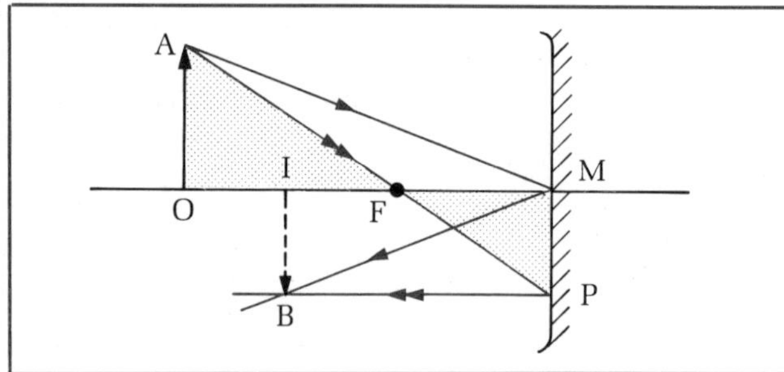

Figure 20.10 *The mirror formula*

The ray from A to the middle of the mirror, M, reflects off at an equal angle to the axis. So triangles AMO and BMI are similar. Hence

$$\frac{OA}{IB}=\frac{OM}{IM}$$

The ray from A through F strikes the mirror at P. Triangles AFO and MFP are similar (because angle AFO = angle MFP, and each triangle has a right angle). Hence

$$\frac{AO}{MP}=\frac{OF}{FM}$$

However, because the ray reflected at P is parallel to the axis, MP must equal IB. Hence

$$\frac{OM}{IM}=\frac{OF}{FM}$$

and since OM = u, IM = v (image distance), FM = f and OF = $u-f$ then

$$\frac{u}{v}=\frac{u-f}{f}=\frac{u}{f}-1$$

which gives

$$\frac{1}{v}=\frac{1}{f}-\frac{1}{u}$$ and hence the required equation.

The linear magnification M is defined as $\dfrac{\text{image height (IB)}}{\text{object height (OA)}}$, which equals $\dfrac{v}{u}$.

If the formula is applied to Figure 20.9, the value of v is negative. For example, suppose the object distance is 0.20 m and the focal length is 0.25 m, then the image distance is given by $1/v = 1/0.25 - 1/0.2 = -1.0$. So the image distance is -1.0 m. The negative value for v must be interpreted as indicating a virtual image is formed. This interpretation is called a **sign convention**.

Real is positive, virtual is negative.

Measuring the focal length of a concave mirror

Method 1 Arrange the illuminated object facing the mirror (see Figure 20.11). The concave mirror is then moved along the line between it and the object. When the object is exactly at the centre of curvature of the mirror, a clear image is seen beside the object. The light from the object then strikes the mirror normally at every point so it is reflected back to where it started. The distance from the centre of the mirror to the object is equal to its radius of curvature R. Since its focal length equals $\frac{1}{2}R$, measurement of the distance from object to mirror allows f to be calculated.

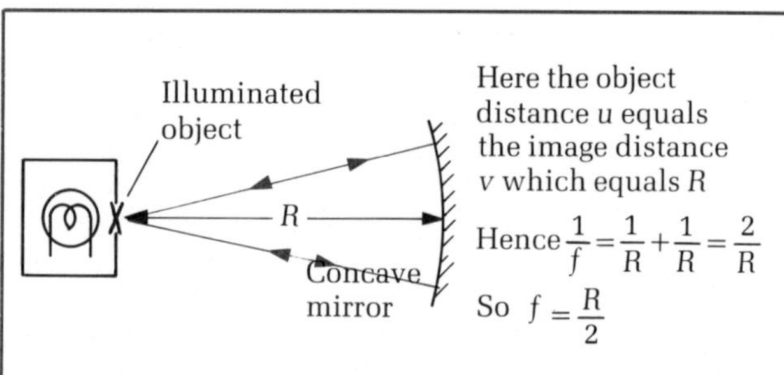

Figure 20.11 *Method 1*

Method 2 Use a screen to locate the image (see Figure 20.12). For several values of object distance u, locate the image and measure the image distance v. Use object distances which give equal changes of $1/u$ (e.g. u = 0.50 m, 0.33 m, 0.25 m, 0.20 m, 0.165 m) so that the points on a graph of $1/u$ against $1/v$ are regularly spaced.

Since $1/u + 1/v = 1/f$, then the graph ought to be a straight line with a slope of -1. The intercepts on both axes should equal $1/f$. Use the average value of the intercepts to calculate f.

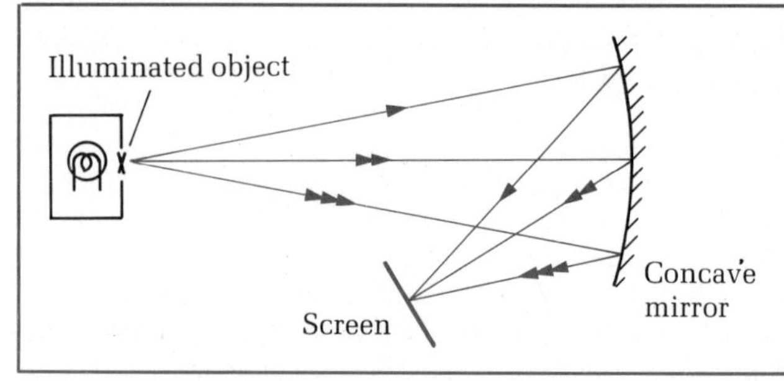

Figure 20.12 *Method 2*

Convex mirrors

If parallel rays are directed at a spherical convex mirror, the reflected rays diverge.

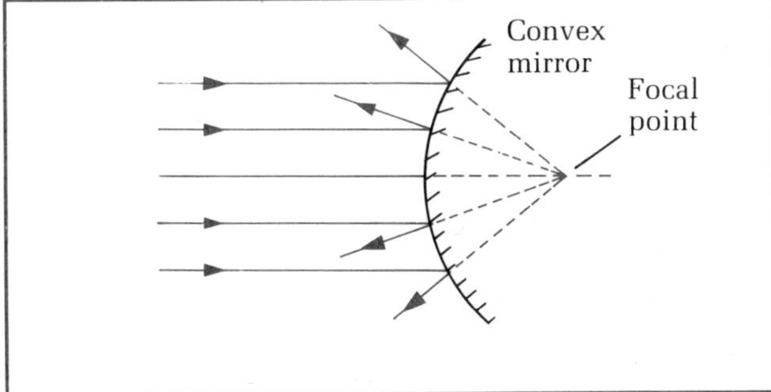

Figure 20.13 *The convex mirror*

The **focal length** of a convex mirror is the distance from the mirror to the point where parallel rays along the axis appear to diverge from after reflection. A convex mirror always forms a virtual image of a real object. The image is always smaller than the object, and nearer the mirror than the object.

The formula $1/u+1/v=1/f$ can be used for convex mirror calculations, provided the focal length is inserted as a negative value. The 'real is positive' convention must also be used.

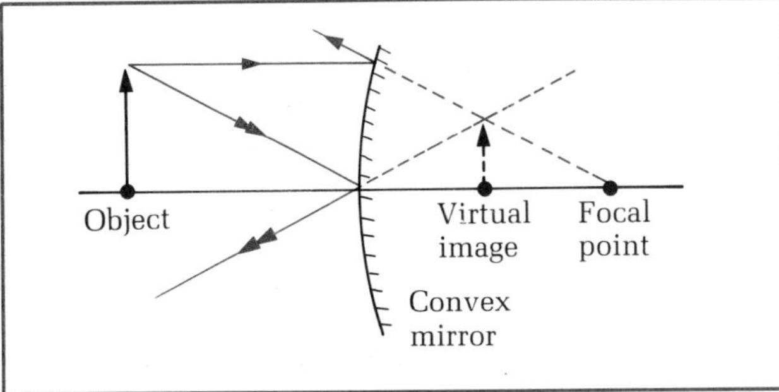

Figure 20.14 *Image formation*

Worked example A point object is placed on the axis of a convex mirror at a distance of 0.20 m from the mirror. Calculate the position of the image if the focal length is 0.15 m.

Solution $u=0.20$ m, $f=-0.15$ m.

To calculate f, use $1/f=1/u+1/v$.

So $1/-0.15=1/0.20+1/v$

which gives $-6.67=5+1/v$

hence $-1/v=5+6.67=11.67$

so $v=-1/11.67=-0.086$ m.

Driving mirrors are usually convex mirrors because they give a wide field of view behind the driver.

20.2 Lenses

Refraction

When a light ray passes from air into a transparent medium such as glass, the direction changes if it was not initially along the normal. The light ray in the glass is closer to the normal than the light ray in the air. The change of direction may be calculated using Snell's law.

$$\frac{\sin a}{\sin b}=n$$

where a = angle in air,
b = angle in the medium,
n = refractive index of the medium.

The refractive index is equal to the ratio of the wavelength in air to the wavelength in glass, as explained on p. 236.

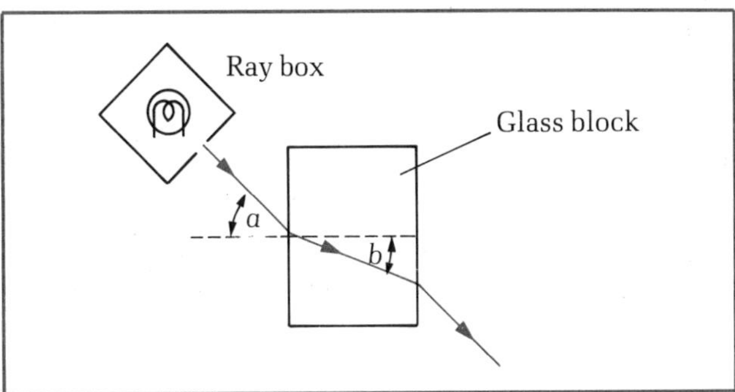

Figure 20.15 *Snell's law*

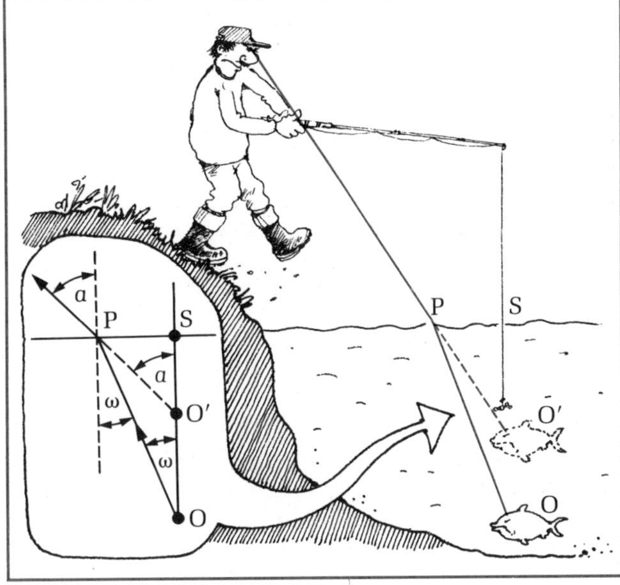

Figure 20.16 *Apparent depth*

Refraction of light causes the water in a swimming pool to appear shallower than it really is. An object on the floor of the swimming pool appears nearer the surface when viewed from above because the light from the object refracts away from the normal at the surface. Just as the light refracts towards the normal when it passes from air to water, so it refracts away from the normal when it passes out.

The apparent depth of the water is equal to the real depth/the refractive index, if the water is viewed from a near-normal position. In Figure 20.16, the light ray from O to the surface at point P refracts as if from O′. The angle of incidence between OP and the normal at P is labelled ω. From triangle OPS, $\sin\omega=SP/OP$. The angle of refraction is labelled a. From triangle O′PS, $\sin a=SP/O'P$.

Snell's law applied to refraction at P must be used with care because the ray passes from water to air. The refractive index is given by

$$n=\frac{\sin a}{\sin \omega}=\frac{(SP/O'P)}{(SP/OP)}=\frac{OP}{O'P}$$

Provided the ray from O to P is near the normal, then OP is the real depth and O'P is the apparent depth. Hence

$$n=\frac{\text{real depth}}{\text{apparent depth}}$$

Measuring the refractive index of a glass block

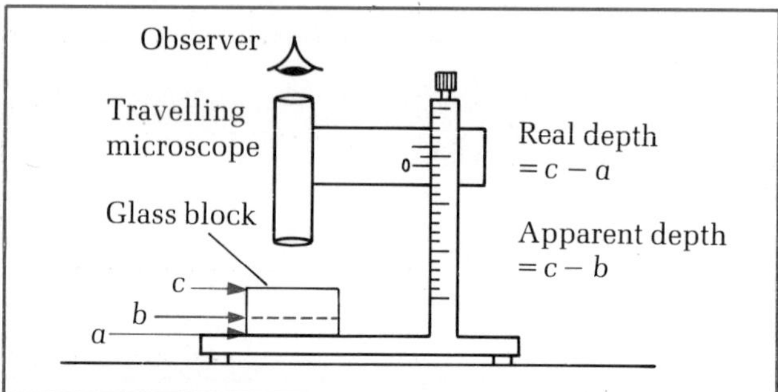

Figure 20.17 *Measuring refractive index*

Use a travelling microscope with a cross marked on paper as an object (see Figure 20.17). Focus the microscope on the object which should be sellotaped so it doesn't move. Take the reading a of the microscope. Place the glass block over the object and refocus the microscope on the image seen through the block. Take the new reading b which gives the image position. Finally, refocus the microscope on the top of the block to give a third reading c.

Total internal reflection

When light passes from a transparent medium into air, it bends away from the normal. If the angle of incidence exceeds a critical value, the light is totally reflected at the boundary so it does not pass into the air.

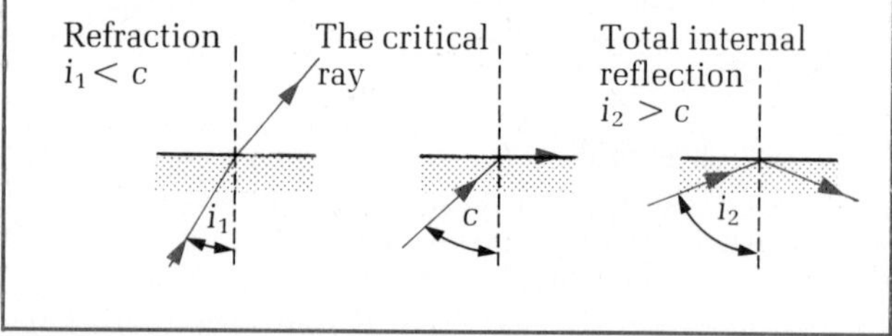

Figure 20.18 *Internal reflection*

Where the angle of incidence in the glass is equal to the **critical angle**, the refracted ray passes exactly along the boundary. So the angle of refraction in air is exactly 90°. Hence the refractive index $n=\sin 90°/\sin c$ where c is the critical angle. Since $\sin 90°=1$, then

$$n=\frac{1}{\sin c}$$

In the previous equation n is the refractive index and c the critical value.

Fibre optics make use of total internal reflection to guide light along transparent fibres, even though the fibres might be twisted and bent. At a bend in the fibre, the light rays reflect internally off the sides of the fibre as shown in Figure 20.19. So the light rays are guided along the fibre. Fibre optics are used in medicine to view the inside of the body. The individual fibres must be flexible and thin so they can be bent tightly without the light leaving the fibre at the bend. An image formed by a lens on the end of a bundle of fibres will be transmitted to the other end of the bundle. Provided the fibres keep the same relative positions at each end, the light transmits the image to the other end of the bundle.

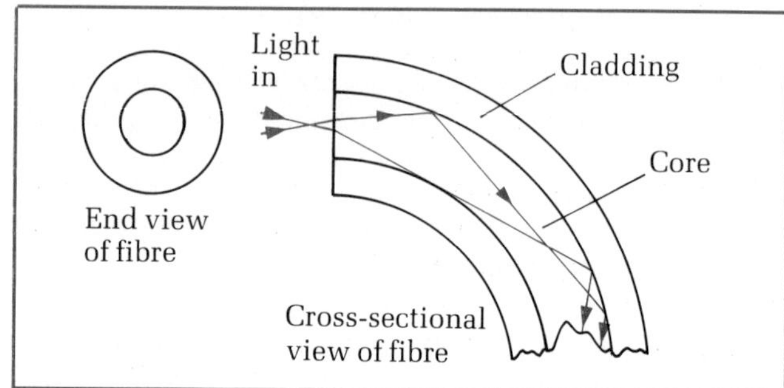

Figure 20.19 *Fibre optics*

To prevent light passing from one fibre to an adjoining fibre at points of contact, the fibres are clad with a medium of low refractive index. Total internal reflection then takes place at the boundary between the fibre and the cladding, so light ought not to reach the boundary with an adjacent fibre. Without cladding, light rays would pass between fibres at the points of contact; as a result, the fibres would become uniformly bright so no image could be seen.

Optical fibres are also used for telecommunications. Using light to carry information allows much more information to be carried than by using radio waves. This is explained on p. 294. The reason is that the frequency of light is much greater. Cables which carry optical fibres must use very clear material because of the long distances over which the cables are used. Even if each metre of cable absorbs only a tiny fraction of the light passing along it, after a distance of several kilometres the total amount of light absorbed can be considerable. Suppose a material is used for the fibres which only absorbs 1% of the light per metre of its length. How many metres long could the fibre be, if the amount of light transmitted through it is to be greater than 50%?

The first metre of its length absorbs 1% and transmits 99% into the next metre. The next metre transmits 99% of 99% of the light entering the fibre. The next metre transmits 99% of 99% of 99% of the light entering the fibre. How many times should you multiply 99% by itself before you obtain less than 50% as the answer? Try it! The answer gives the length of the fibre, and it works out to be between 60 and 80 m. So fibre optic cables for telecommunications must use very clear material, and even so need **repeater units** at intervals to boost the light signal.

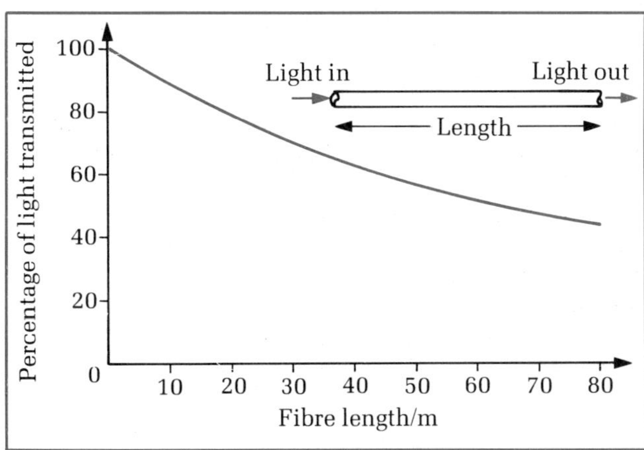

Figure 20.20 *Absorption*

Convex lenses (converging lenses)

Direct a beam of parallel light rays at a convex lens and the light is brought to a focus by the lens. If the parallel light is directed along the lens axis (the line through the centre of each surface), the rays are brought to a focus on the axis at the **focal point** of the lens.

The **focal length** *f* of a convex lens is the distance from the lens to the point where rays parallel to the axis are brought to a focus.

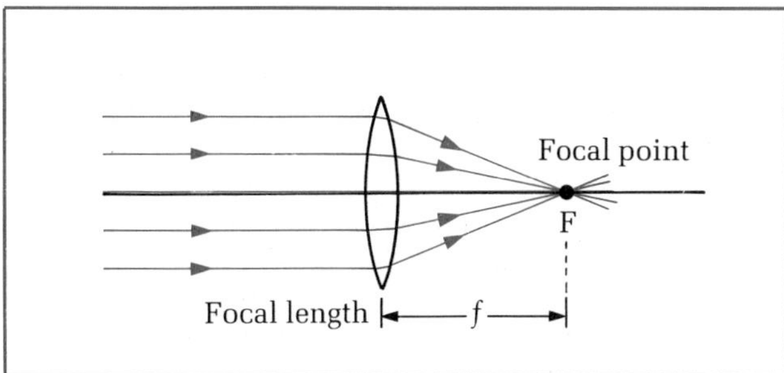

Figure 20.21 *The convex lens*

Convex lenses can have different shapes as shown in Figure 20.22. Provided the lens is thin, the focal length is the same on either side. Each light ray which passes through is refracted at the first surface then at the second surface. However the two points at which it is refracted are so close for a thin lens that lens diagrams usually show a single refraction only.

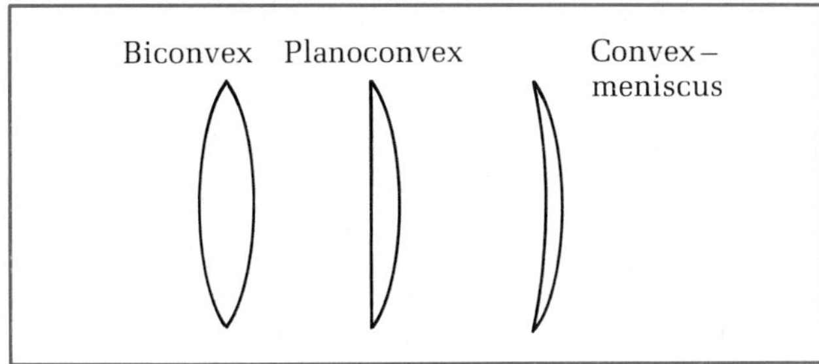

Figure 20.22 *Types of convex lens*

Ray diagrams are useful for showing how images are formed. For a convex lens used to form an image of a point object, three key rays need to be drawn.

1 The ray to the lens centre, which passes straight through.
2 The ray parallel to the axis, which refracts through the focal point.
3 The ray through the focal point, which refracts parallel to the axis.

Figure 20.23 shows the three key rays used to locate the image of a point object O above the axis. All other rays from the point object to the lens must refract through the point image formed. An extended object from O to the axis at O′ would give an extended image from I to the axis at I′.

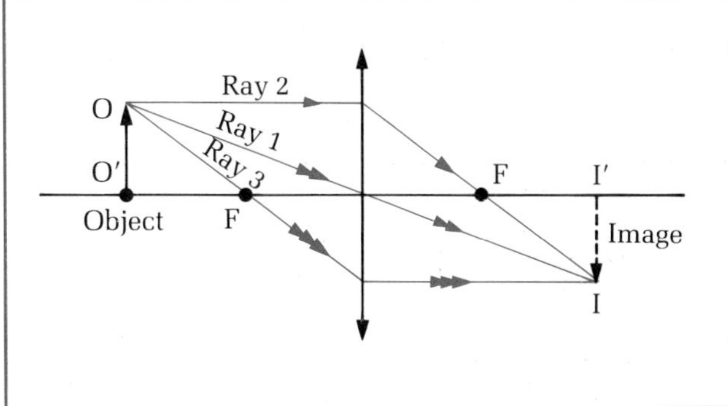

Figure 20.23 *Image formation*

The linear magnification is defined as the ratio of the $\frac{\text{image height}}{\text{object height}}$

and since $\frac{\text{image height}}{\text{object height}} = \frac{\text{image distance}}{\text{object distance}}$ then

$$\text{Linear magnification} = \frac{v}{u}$$

where u = object distance from lens,
v = image distance from lens.

The image size depends on how far the object is placed from the lens. Figure 20.24 shows that the nearer the object is to the lens, the further away the image is, provided the object is no nearer than the focal point.

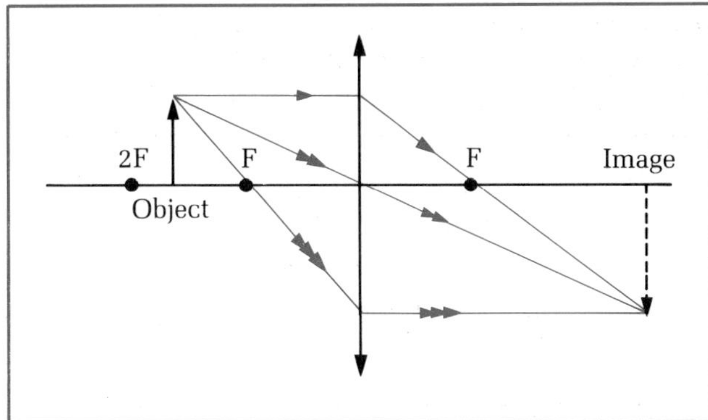

Figure 20.24 *Object closer than 2F*

If the object is placed between the focal point and the lens, the refracted rays cannot be brought to a focus. But an observer looking into the lens from the other side to the object can see an image formed where the rays appear to come from. The image is a virtual image, and it is magnified compared with the object. So a convex lens can be used as a magnifying glass by viewing as shown in Figure 20.25 (on the next page).

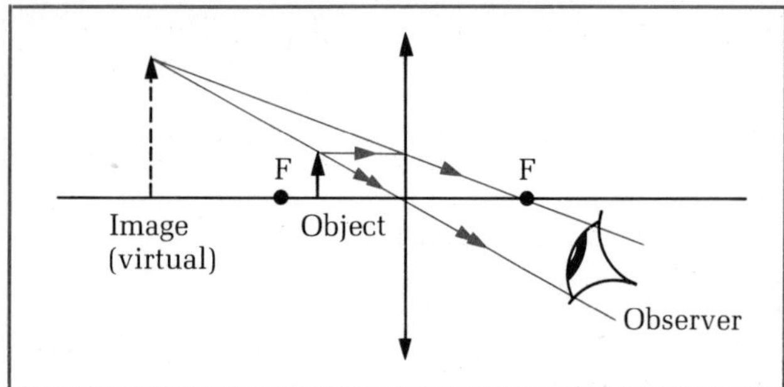

Figure 20.25 *The magnifying glass*

The lens formula can be used to calculate the position of an image, given the object position and focal length of the lens.

$$\frac{1}{u}+\frac{1}{v}=\frac{1}{f}$$

where u = object distance,
v = image distance,
f = focal length.

Where the object distance is less than the focal length, the image distance calculated by the formula has a negative value. For example, if $u=0.15$ m and $f=0.25$ m, then $1/v=1/0.25-1/0.15=-2.67=$ so $v=-0.375$ m. The negative value must be interpreted as indicating a virtual image, so we must again use the 'real is positive, virtual is negative' sign convention with the formula.

Measuring the focal length of a convex lens

Method 1 Arrange the plane mirror so it reflects light from the illuminated object back into the lens (see Figure 20.26). By adjusting the distance from the lens to the object, a clear image can be brought into focus beside the object. The object is then placed exactly at the focal point of the lens; light from the object is refracted into a parallel beam passing from the lens to the plane mirror. The mirror reflects the light back along its own path to give an image next to the object.

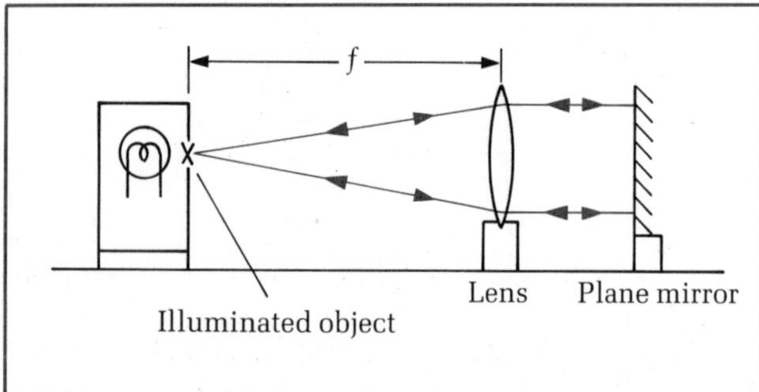

Figure 20.26 *Method 1*

Method 2 Locate the image on a screen (see Figure 20.27). Measure the image distance for several measured values of object distance. Choose object distances which give equal changes of $1/u$ so the points on a graph of $1/u$ against $1/v$ are equally spaced. Since $1/u+1/v=1/f$, the graph is a straight line with intercepts on both axes equal

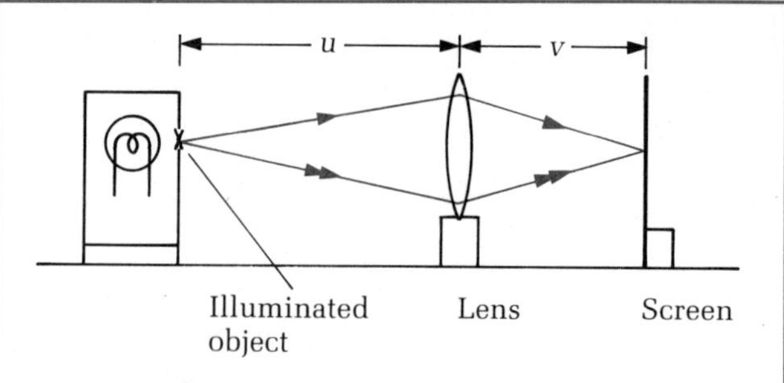

Figure 20.27 *Method 2*

to $1/f$. Check your value of f with the more direct value from method 1.

Concave lenses (diverging lenses)

When a parallel beam of light is directed at a concave lens, the light rays are made to diverge from the lens. The **focal length** of a concave lens is the distance from the lens to the point where parallel rays along the axis appear to diverge from.

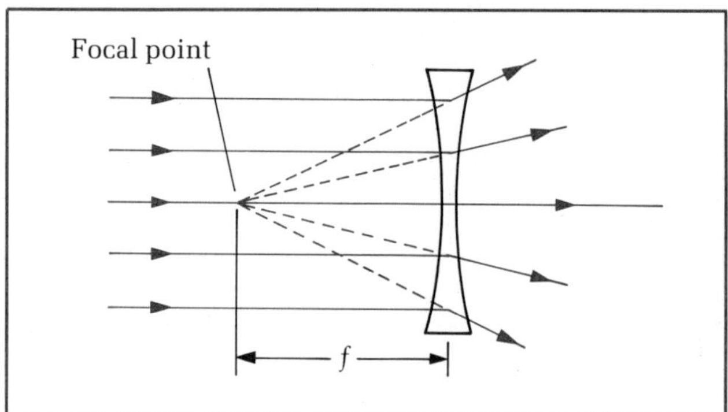

Figure 20.28 *The concave lens*

To determine the position of an image formed by a concave lens, we may use either the lens formula or the ray diagram approach. If the formula is used, we need to remember the 'real is positive' sign convention, and to make the focal length value negative. For example, if an object is placed 0.375 m in front of a concave lens of focal length 0.25 m, then the object distance $u=0.375$ m, and the focal length $f=-0.25$ m, so the image distance v is given by

$$1/v=1/f-1/u=(1/-0.25)-(1/0.375)=(-4)-(-2.67)$$
$$=-6.67$$

Hence $v=-1/6.67=-0.15$ m (a virtual image is formed).

To draw a ray diagram for a concave lens, the three key rays are:

1. The ray to the centre, which passes straight through.
2. The ray parallel to the axis which refracts as if from the focal point on the 'incident' side.
3. The ray directed at the focal point on the other side which refracts parallel to the axis.

Figure 20.29 shows the ray diagram for the problem above. An observer looking in at the lens would see a virtual image on the same side as the object. The image is upright and diminished in size but nearer to the lens than the object.

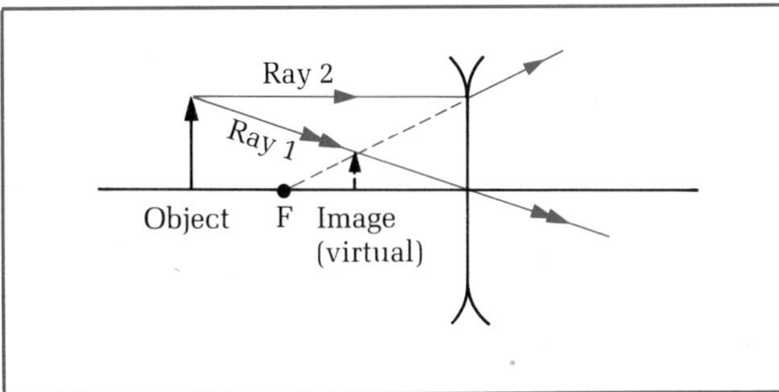

Figure 20.29 *Image formation*

Sight correction

The eye is capable of focussing objects at different distances. This is achieved by automatic adjustment of the thickness of the eye lens. To focus a distant object, the eye lens is made thinner, so less powerful. The light rays from a distant point object are brought to a focus on the retina by the eye lens. For near objects, the eye lens must be made thicker and hence more powerful. Then the rays from a near object can be brought to a focus on the retina as shown in Figure 20.30.

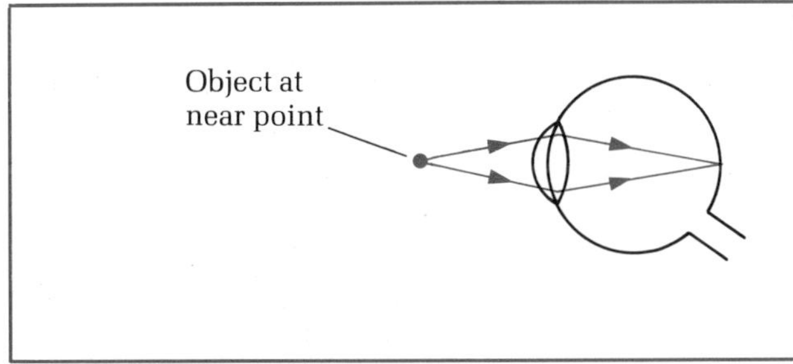

Figure 20.30 *Near point of normal eye*

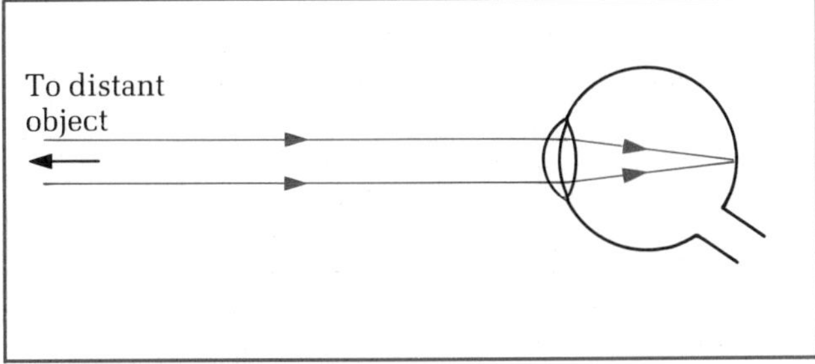

Figure 20.31 *Far point of normal eye*

A **normal** eye can see objects in focus in the range from infinity to 25 cm. This range is based upon the average human eye which has an age of 40 years. Young persons have a much wider range but the average 70-year-old has a much reduced range. The normal eye has a **near point** of 25 cm and a **far point** of infinity.

A **short-sighted eye** cannot focus on distant objects. Its far point is nearer than infinity because the eye muscles cannot make the lens thin enough. By using a suitable concave lens in front of the eye, distant objects can be brought into focus as shown in Figure 20.32. The concave lens makes parallel rays from a distant point diverge and the eye therefore sees an image (of the distant object) at its far point. The concave lens must therefore have a focal length equal to the uncorrected far point distance.

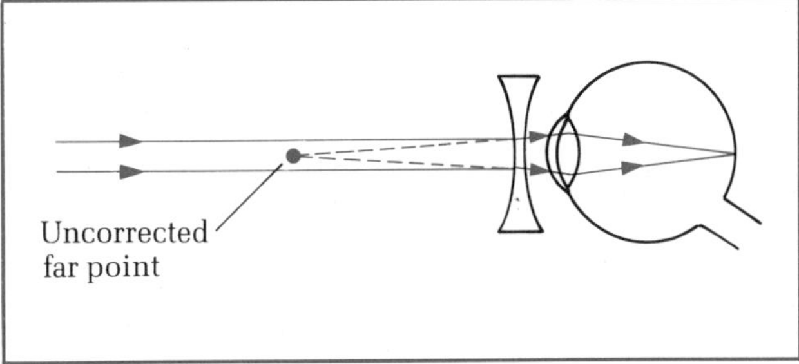

Figure 20.32 *Short sight correction*

A **long-sighted eye** cannot focus near objects, such as the print of a book. Its near point is further than 25 cm away because the eye lens cannot be made thick enough by the eye muscles. A suitable convex lens must be used to bring near objects into focus, as shown in Figure 20.33.

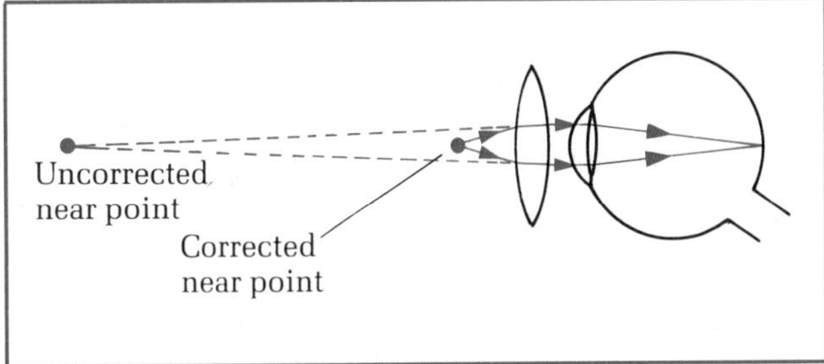

Figure 20.33 *Long sight correction*

The convex lens forms an image (of a near object) beyond its near point, so the eye therefore sees the image in focus. The convex lens acts like a magnifying glass, but it's not the magnification which is vital here; the key point is that the image is formed further away than the object. The focal length of the convex lens may be calculated using the lens formula, remembering that the image is virtual. So if an eye has a near point of 1.0 m, to correct it with a convex lens, an object placed 0.25 m from the lens (assumed right in front of the eye) must give a virtual image at its near point. Hence $u=0.25$ m, $v=-1.0$ m. So $1/f=1/u+1/v$ gives $1/f=1/0.25+1/-1=4-1=3$. Hence $f=0.33$ m.

The **power** of a lens is defined as 1/(its focal length in metres). The unit of power is the dioptre (D). A lens with a focal length of $+0.33$ m has a power of $+3$ D.

20.3 The camera

To photograph objects at different distances, the camera lens must be moved away from or towards the film to a set position for each object distance. For a scenic photograph, the lens is positioned so that the film is exactly at its focal point. Then a clear image of the scene is focused on the film when the shutter is pressed. For a close-up photograph, the lens must be moved further away to increase the image distance because the object distance is less.

A simple camera can be made using a single lens, but more expensive cameras use lens combinations to avoid **chromatic aberration**. A single lens would refract blue light more than red light, so the photograph would show

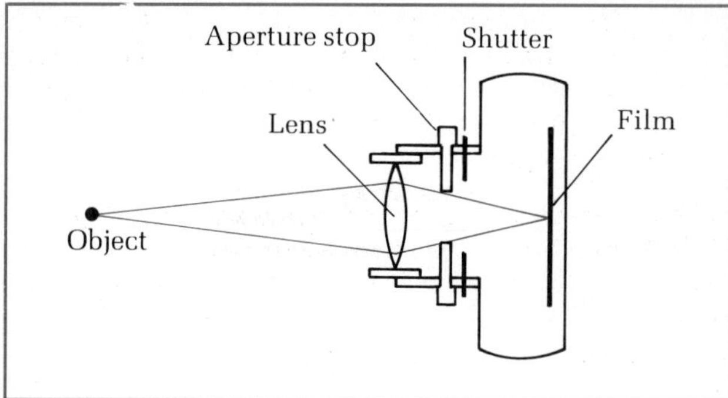

Figure 20.34 *A simple camera*

coloured tinges. An achromatic lens is a lens system which refracts all colours by the same amount. so it would give a clear photo without coloured effects spoiling it.

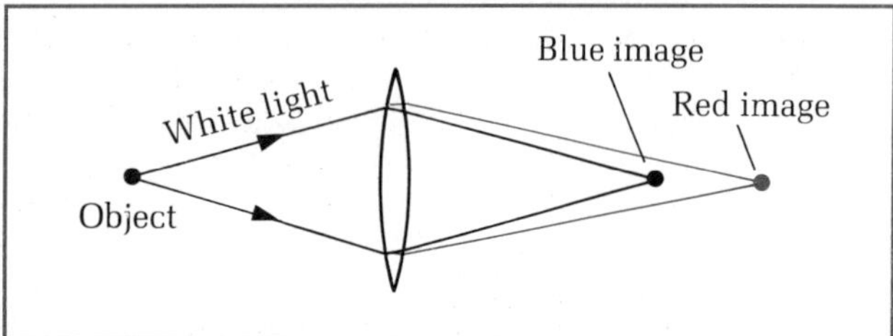

Figure 20.35 *Chromatic aberration*

An aperture stop is a feature of a good camera. It can be adjusted to allow more or less light through onto the film. For high speed photography, the shutter is opened for a very short time only, so the aperture needs to be wide open to let as much light through in that short time. Otherwise the image will be too faint. For a given shutter speed, the ***f*-number** setting controls the amount of light reaching the film. The *f*-number setting determines the area of the aperture on a scale such that the width equals the focal length/the *f*-number. To widen the aperture, the *f*-number should therefore be decreased.

The depth of field is affected by the aperture width. When an object is photographed, other objects in view will also be on the same photograph. The depth of field determines whether or not these other objects give sharp images, assuming the main image is sharp. Diffraction is the cause of depth of field, a narrower aperture giving a greater depth of field (because the amount of diffraction is greater for a narrower gap). A high speed photograph which requires a fast shutter speed and hence wide aperture will show the background out of focus; the wide aperture gives a small depth of field so the background will not be in focus. **Spherical aberration** results from a

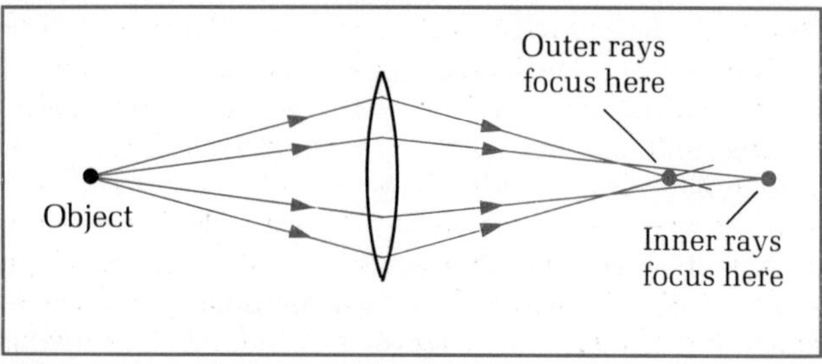

Figure 20.36 *Sherical aberration*

wide aperture because the lens refracts outer rays slightly more than inner rays. To ensure sharp images, the aperture is narrowed to stop outer rays reaching the film.

20.4 Microscopes

The simple microscope (or magnifying glass)

Hold a convex lens over some print and you will be able to see a magnified image of the print. The print must be placed between the lens and its focal point. If the lens is moved upwards off the print, the image becomes larger and more distant. Imagine you are inspecting a diamond with the aid of a magnifying glass. Hold the magnifying glass to the eye and move the diamond to and from the lens. What should be the best position to hold the diamond to inspect it as closely as possible? The best position is where its **image** lies at the eye's near point. The magnifying power is then highest. Try it; have a look at the details on a coin if you don't have a diamond.

The magnifying power (of a microscope) is defined as

$$\frac{\text{the angle subtended by the image to the lens}}{\text{the angle subtended by the object at the near point to the eye}}$$

For an object of height h_0 placed at distance u from the lens, the angle subtended by its image is given by $\tan\beta = \frac{h_0}{u}$, as shown by Figure 20.37.

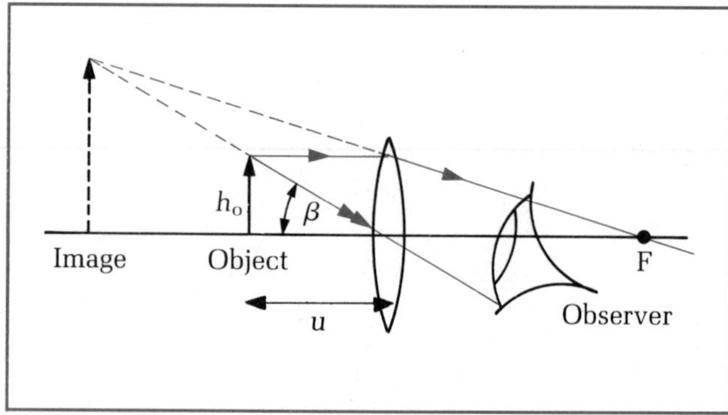

Figure 20.37 *The simple microscope*

For the object viewed unaided at the near point, distance D from the eye, the angle subtended by the object to the eye is given by $\tan\alpha = \frac{h_0}{D}$, as in Figure 20.38. The microscope makes the object appear bigger because β is greater than α. So the magnifying power for small angles is given by $\frac{h_0/u}{h_0/D}$ since the small angle rule can be used. (For small angles, the angle in radians equals the tangent of the angle.)

Hence the magnifying power $= \frac{D}{u}$

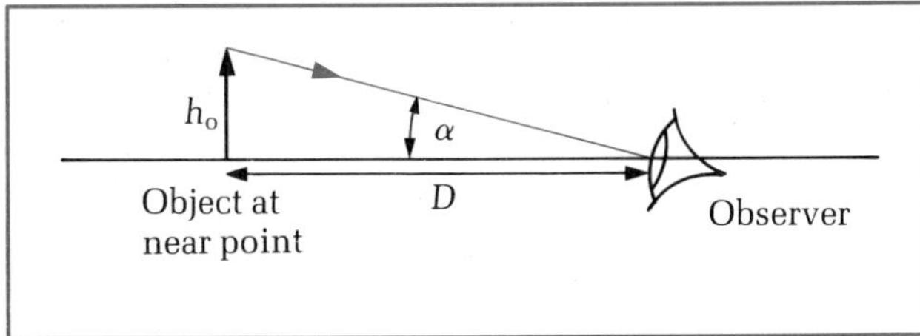

Figure 20.38 *Direct viewing.*

The object distance can take any value in the range from the focal point to the point where the image lies at the near point. There is no point in the object being any closer because it would form the image closer than the near point. So the greatest value of magnifying power is when the object distance is least, which is when the image is formed at the near point.

Maximum magnifying power occurs when the image is formed at the near point. The image distance, v, is then equal to $-D$ ($-$ for virtual). So $1/u=1/f-1/v$ gives $1/u=1/f-1/-D=1/f+1/D$.

$$\text{Hence the maximum magnifying power} = \frac{D}{f}+1$$

Minimum magnifying power occurs when the object is at the focal point so the viewer sees the image at infinity. Since the object distance is equal to the focal length here, then $D/u=D/f$.

$$\text{Hence the minimum magnifying power} = \frac{D}{f}$$

The compound microscope

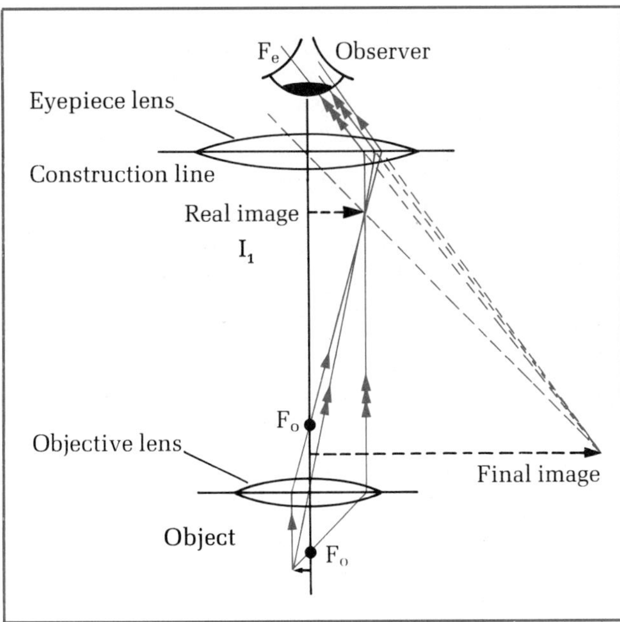

Figure 20.39 *The compound microscope*

Essentially a compound microscope consists of two convex lenses, referred to as the objective and the eyepiece. The object to be viewed is placed just outside the focal length of the objective lens. Light rays from the object which pass into the objective are brought to a focus inside the focal length of the eyepiece.

The objective forms a real image between the eyepiece and the eyepiece's focal point. So the viewer, looking through the eyepiece, sees a magnified virtual image of the picture formed by the objective, i.e. of the real image (as shown in Figure 20.39).

In normal adjustment the eyepiece is positioned so that the final image lies at the near point of the eye. The magnifying effect is produced partly by the objective and partly by the eyepiece. The final image is much less distorted than a similar-sized image formed by a simple microscope.

The magnifying power is defined as the ratio of

$$\frac{\text{the angle subtended by the final image to the eye}}{\text{the angle subtended by the object at the near point to the eye}}$$

If the final image has height h_2 and is formed at distance v_2 from the eyepiece, then the angle subtended by the final image to the eye (assumed at the eyepiece) is given by $\beta=h_2/v_2$. Assume β is small so $\tan\beta=\beta$.

When the object of height h_0 is placed at the near point (distance D) and viewed unaided, then the angle subtended by the object is given by $\alpha=h_0/D$.

Hence the magnifying power M is given by

$$M=\frac{(h_2/v_2)}{(h_0/D)}=\frac{h_2}{h_0}\times\frac{D}{v_2}$$

$$M=\text{linear magnification}\times\frac{D}{v_2}$$

since h_2/h_0 is the linear magnification.

In normal adjustment, the final image lies at the near point so $v_2=D$. Thus the magnifying power in normal adjustment is equal to the linear magnification h_2/h_0.

Since $\frac{h_2}{h_0}=\frac{h_2}{h_1}\times\frac{h_1}{h_0}$ where h_1 is the height of the real image formed by the objective, then the linear magnification is equal to the linear magnification of the objective × the linear magnification of the eyepiece. So if an eyepiece is chosen which has a linear magnification of 8, and the objective has a linear magnification of 3, then the linear magnification in total is 24. Hence the magnifying power in normal adjustment is 24.

Some compound microscopes are fitted with crosswires or an eyepiece scale. The crosswires (or eyepiece scale) are positioned where the real image I_1 is formed by the objective. So the viewer can see the final image in focus with the image of the crosswires.

The resolving power of a microscope is its ability to enable detail in the image to be made out. The limit to which a microscope can be used to resolve detail depends on the width of the objective lens. This is because of diffraction effects produced by the objective gap. This effect is discussed in more detail on p. 268. A wider objective produces less diffraction so points in an image under study can be resolved better. Use of short wavelength (i.e. blue) light also reduces diffraction. So viewing an image in blue light allows more detail to be made out.

Worked example An object of size 5 mm is placed 50 mm in front of an objective of a compound microscope. An eyepiece of focal length 160 mm is positioned so that a virtual image of the object can be seen 250 mm from the eye. Calculate the size of the final image, and determine the separation of the lenses, given that the focal length of the objective is 40 mm.

Solution First determine the position of the real image formed by the objective.

$u_1 = 50$ mm, $f_1 = 40$ mm.

Since v_1 is given by $\frac{1}{u_1} + \frac{1}{v_1} = \frac{1}{f_1}$

then $1/v_1 = 1/40 - 1/50 = 50 - 40/50 \times 40$

giving $v_1 = 200$ mm

Also, the linear magnification of the objective is $\frac{v_1}{u_1} = \frac{200}{40} = 5$.

Next determine the distance from the real image to the eyepiece. Treat the real image formed by the objective as an **object** for the eyepiece.

Calculate the object distance u_2 for the eyepiece, given its focal length f_2 and the image distance v_2.

Now $f_2 = 160$ mm and $v_2 = -250$ mm ($-$ since virtual)

so $1/u_2 = 1/f_2 - 1/v_2 = (1/160) - (1/-250)$

$= (250 + 160)/(250 \times 160)$

Hence $u_2 = 98$ mm.

The separation between the two lenses is equal to $v_1 + u_2$ which gives 298 mm for the separation.

The linear magnification of the eyepiece is given by $\frac{v_2}{u_2}$ which equals $-250/98$.

So the total linear magnification is $5 \times (-250/98)$ which gives -12.8.

The negative sign indicates a virtual image is formed. So the size of the final image is 12.8×5 mm which equals 64 mm.

20.5 Telescopes

The simple refracting telescope

To make a simple refracting telescope, two convex lenses are needed. The objective, the lens facing the object, needs to have a long focal length compared with the eyepiece lens. The objective is used to form a real image of the distant object being viewed. The real image lies in the focal plane of the objective, and the eyepiece is used to give a magnified view of the real image. The viewer, looking in at the eyepiece, sees a virtual image which is the magnified view of the real image. To appreciate how the lenses function in the telescope, make a home-made telescope using two suitable convex lenses on a slider bar as in Figure 20.41. View a distant lamp through the telescope, and while viewing, ask a friend to hold a piece of tracing paper where the real image is formed. Your friend can see the real image on the paper, and you can see a magnified view of it when looking through the eyepiece.

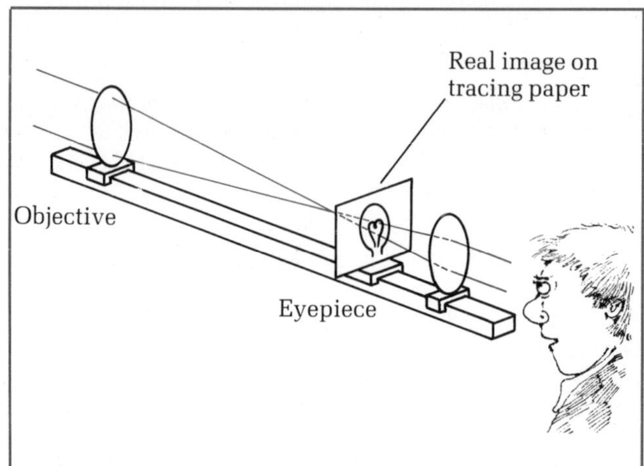

Figure 20.41 *A home-made telescope*

In normal adjustment the final image seen through the eyepiece is adjusted to lie at infinity. So when correctly adjusted, it ought to be possible to see the final image of a distant object with one eye at the eyepiece; at the same time, it ought to be possible to see the object with the other eye directly. Since the final image is at infinity, the real image must be at the eyepiece's focal point.

The magnifying power of a telescope is defined as the ratio of

$$\frac{\text{the angle subtended by the final image at infinity to the eye}}{\text{the angle subtended by the distant object to the eye}}$$

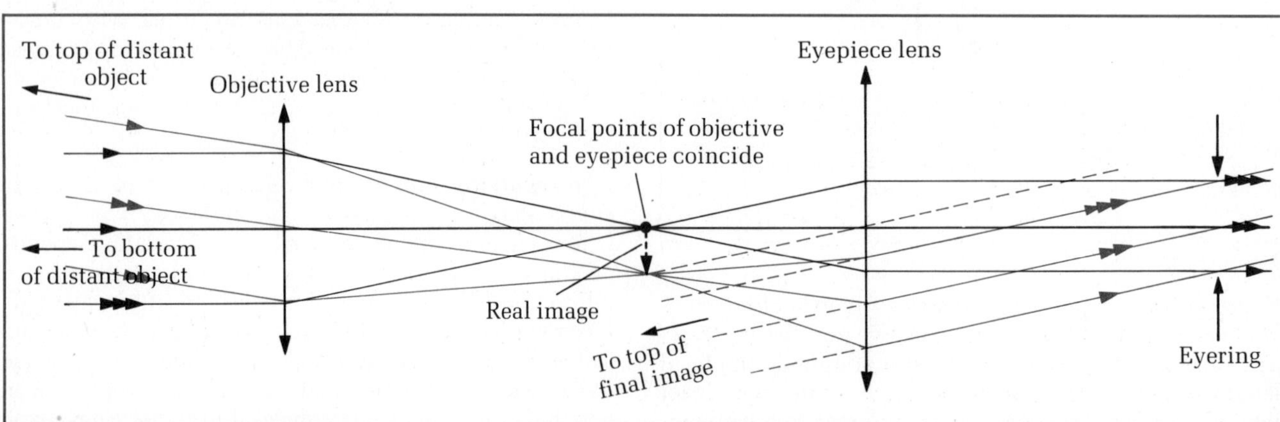

Figure 20.40 *Simple refracting telescope*

Suppose a telescope with a magnifying power of 20 is used to view the Moon which subtends an angle of about 0.5° to the eye. Then the angle subtended by the final image to the eye will be 10° (= 20 × 0.5°).

Magnifying power formula in normal adjustment, the magnifying power is given by the ratio of

$$\frac{\text{the focal length of the objective } (f_0)}{\text{the focal length of the eyepiece } (f_e)}$$

To prove this formula, consider part of Figure 20.40 as shown in Figure 20.42. The ray through the centre of the objective intercepts the tip of the real image, so the angle subtended by the object α is given by

$$\tan\alpha = \frac{h_i}{f_0}$$

where h_i is the height of the real image. Assuming small angles (i.e. $\tan\alpha = \alpha$), then the angle subtended by the object $\alpha = h_i/f_0$.

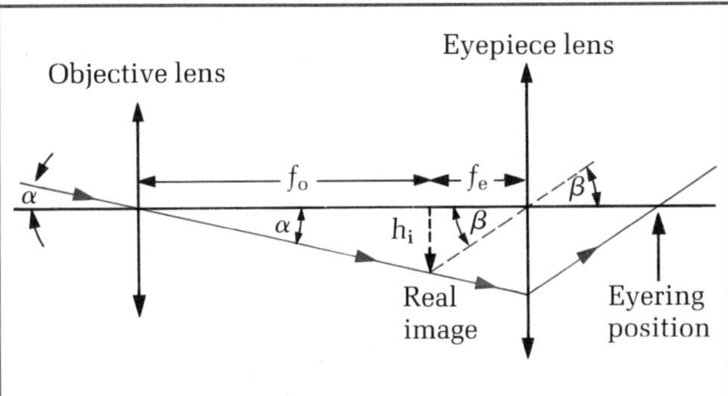

Figure 20.42 *Magnifying power*

Now consider the line from the tip of the real image through the centre of the eyepiece lens. The angle of this line to the axis is the angle subtended by the final image. So the angle subtended by the final image β is given by

$$\tan\beta = \frac{h}{f_e}$$

Assuming the angles are small gives

$$\beta = \frac{h}{f_e}$$

So the magnifying power is

$$\frac{(h/f_e)}{(h/f_0)} \text{ which is equal to } \frac{f_0}{f_e}$$

The **length** of a telescope in normal adjustment is equal to the sum of the two focal lengths ($f_0 + f_e$). The reason is that the real image is formed at the focal point of the objective, and the eyepiece is adjusted so that its focal point is at the real image. So the distance from the objective to the eyepiece is equal to $f_0 + f_e$. This fact caused great problems for 18th century astronomers who attempted to make ever more powerful telescopes by using objectives with long focal lengths. Imagine a telescope with an eyepiece of focal length 5 cm and a magnifying power of 60. It would need to have a focal length for its objective of 300 cm – giving a length of 3 m 5 cm. Reflecting telescopes do not need to be as long.

The eyering is the best position to place the eye, as shown in Figure 20.40. All the light from a distant object which passes through the telescope must pass through the eyering after leaving the telescope, so by placing the eye at the eyering, the viewer is able to see as much of the final image as possible. The geometry of Figure 20.40 shows that the eyering diameter is equal to the objective diameter/the magnifying power (assuming normal adjustment). Most telescopes for direct viewing are designed with an eyering diameter about the same size as the eye pupil in the dark: about 8 mm.

The resolving power is determined by the width of the objective lens. Suppose a telescope is used to view a pair of binary stars. The two stars have a very small angular separation. Because of diffraction by the objective gap, the image of each star will be smeared out, and may overlap with the other image. The limit of resolution is when the diffracted image of one star cannot be distinguished from the image of the other star; in other words, when the two images cannot be individually distinguished. The theory of diffraction is discussed in more detail on p. 268; the result of the theory is all that is needed here. Two stars can just be resolved if their angular separation is $1.22\,\lambda/W$ where λ is the light wavelength and W is the objective gap width. By designing a telescope with a wide objective, its resolving power as well as its light gathering power is increased.

Worked example A simple refracting telescope is constructed from two convex lenses of focal lengths 50 mm and 500 mm.

a) In normal adjustment what is the distance between the lenses, and what is the magnifying power?

b) To study the Sun, the telescope is adjusted to form a real image of the Sun on a screen 300 mm behind the eyepiece. Given that the Sun subtends an angle of 0.01 radians to Earth, what will be the diameter of its image on the screen?

Solution

a) The separation of the lenses is equal to sum of the two focal lengths (for normal adjustment). So the distance between the two lenses is 550 mm. The magnifying power in normal adjustment = objective focal length/eyepiece focal length, which gives 500/50 for the magnifying power.
Hence magnifying power = 10.

b) A telescope must *never* be used to study the Sun directly. By increasing the distance between the lenses, it is possible to form a real image of the Sun on a screen as shown by Figure 20.43.

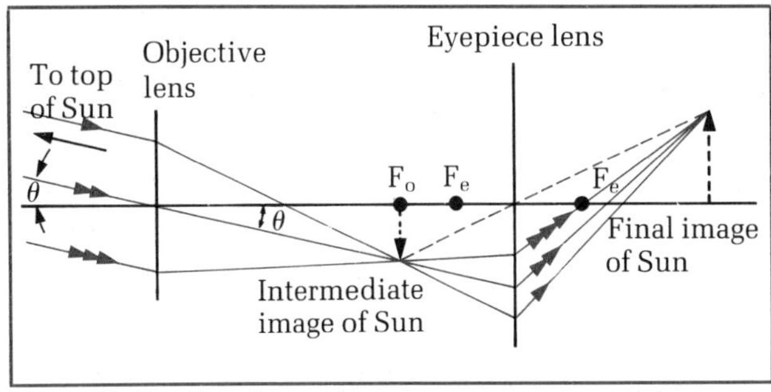

Figure 20.43 *Forming an image on a screen*

The diameter of the intermediate image formed by the objective is equal to the angle subtended by the Sun × the focal length of the objective.
The image formed by the objective lies in its focal plane.
So its diameter is $0.01 \times 500 = 5$ mm.
The linear magnification produced by the eyepiece is given by $\frac{\text{image distance}}{\text{object distance.}}$
We know that the image distance $v = 300$ mm, and the focal length of the eyepiece is 50 mm. Hence the object distance (which is the distance from the eyepiece to the intermediate image at F_0) is given by

$$1/u = 1/f_e - 1/v$$

where $f_e = 50$ mm, $v = 300$ mm.
So $1/u = 1/50 - 1/300 = 5/300$.
Hence $v/u = 300 \times 5/300 = 5$.
Therefore the diameter of the final image on the screen $= 5 \times$ the diameter of the intermediate image.
Hence the diameter of the screen image $= 5 \times 5 = 25$ mm.

The reflecting telescope

By using a concave mirror instead of a convex lens as the objective, telescopes can be made much more powerful in terms of magnifying power and resolving power. As well, the problem of chromatic aberration – a lens producing coloured images with white light, see p. 284 – is eliminated. Reflection does *not* split white light into colours, so the concave mirror does not produce chromatic aberration. If the mirror shape is parabolic instead of spherical, then spherical aberration can be eliminated (see p. 284). So there are several advantages in the use of reflecting telescopes rather than refractors.

The Newtonian reflecting telescope is the basis of most small telescopes commonly used by amateur astronomers. A small plane mirror is used to direct the light from the concave mirror into an eyepiece, usually near the top of the tube (as shown below).

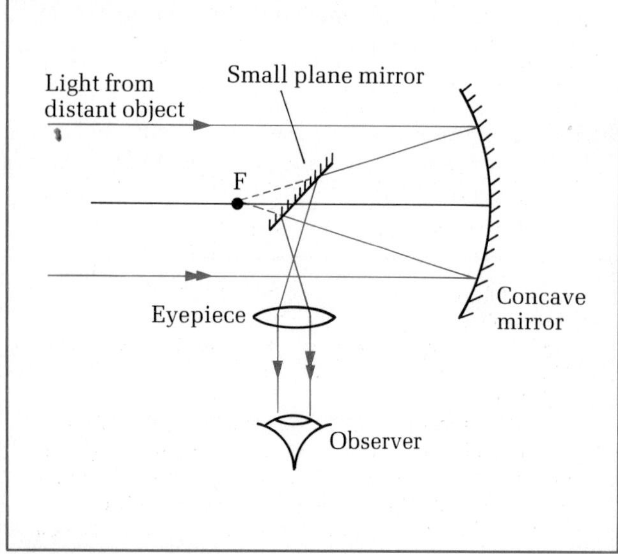

Figure 20.44 *The Newtonian reflector*

The Cassegrain reflecting telescope is the type used in most observatories. A small convex mirror is used to return the light back to the centre of the concave mirror where there is a small hole to allow the light through. So the large concave mirror and the small convex mirror used with it have an effective focal length of about twice that of the concave mirror alone.

Telescopes with wide objectives allow lots of light to be collected, so very faint stars and distant galaxies can be observed with such telescopes. By siting telescope observatories on mountain tops in relatively cloud-free countries such as Hawaii, astronomers can make best use of the instruments. Even then, the Earth's atmosphere limits the resolving power of telescopes because the atmosphere spreads the light out a little. So only telescopes with objective widths of less than about 50 cm can reach their theoretical limit for resolving power; in other words, bigger telescopes don't achieve their theoretical limit (as given by $\sin\theta = 1.22\lambda/W$; see p. 268) because the atmosphere smudges the images. Satellite-based telescopes offer exciting possibilities because they do not suffer the effects of the Earth's atmosphere, so much more detail is to be expected from them.

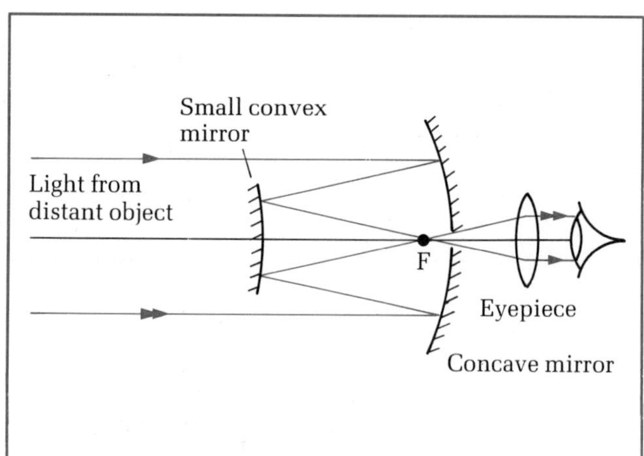

Figure 20.45 *The Cassegrain reflector*

20.6 Summary

1 *Mirrors*

a) Law of reflection   The angle between the incident ray and the normal is equal to the angle between the reflected ray and the normal.

b) Curved mirror formulae
Radius of curvature $R = 2 \times$ the focal length.

$$\frac{1}{u} + \frac{1}{v} = \frac{1}{f}$$

where u = object distance,
v = image distance,
f = focal length

Sign convention   real is positive, virtual is negative } for u and v
concave focal length is positive
convex focal length is negative

2 *Lenses*

a) Snell's law $\frac{\sin i}{\sin r} = n$

where i = incident angle,
r = refracted angle,
n = refractive index

b) Critical angle c is given by $\sin c = \frac{1}{n}$

c) Lens formulae $\frac{1}{u} + \frac{1}{v} = \frac{1}{f}$

as **1b)** including the sign convention but convex focal length is positive, concave negative.

d) Lens power in dioptres = 1/focal length in metres.

3 *Microscopes*

a) Magnifying power

$$= \frac{\text{angle subtended by the image}}{\text{angle subtended by the object at the near point to the eye}}$$

b) For a simple microscope, the magnifying power $= \frac{D}{u}$

where D = near point distance,
u = distance from object to lens.

4 *Telescopes*

a) Magnifying power in normal adjustment

$$= \frac{\text{angle subtended by the final image at infinity to the eye}}{\text{angle subtended by the distant object to the eye}}$$

b) In normal adjustment the magnifying power $= \frac{f_o}{f_e}$

where f_o = objective focal length,
f_e = eyepiece focal length.

c) Resolving power = $1.22\,\lambda/W$

where λ = wavelength,
W = objective width.

Short questions

20.1 Determine the position of the image formed by a concave mirror of radius of curvature 300 mm when a point object is placed on the mirror axis
a) 200 mm from the mirror's pole,
b) 400 mm from the pole.

20.2 A driving mirror consists of a convex mirror of radius of curvature 2.00 m. Determine the position of the image of an object
a) 10 m from the mirror,
b) 5 m from the mirror.

20.3 Calculate the position and linear magnification of the image formed by a convex lens of focal length 150 mm when the object is
a) 300 mm from the lens,
b) 100 mm from the lens.
Assume the object is a small object placed on the axis of the lens.

20.4 A convex lens of focal length 800 mm is used to form a real image of the Moon on a screen. If the Moon subtends an angle of 0.5° to an observer on Earth, calculate the size of its image formed on the screen.

20.5
a) A parallel beam of light is directed at the centre of one of the faces of an equilateral prism of refractive index 1.5 at an angle of incidence of 40°. Sketch the ray path through the prism and calculate the angle between the emergent beam and the incident beam.
b) The angle of incidence is gradually reduced from 40°. Total internal reflection occurs at the opposite face when the angle of incidence at the first face falls below a certain value. Calculate this value.

20.6 A convex lens is used to project the image of an illuminated slide of height 10 mm onto a screen placed 2.0 m from the lens. The slide must be placed 400 mm from the lens to give a clear image on the screen. Calculate
a) the focal length of the lens,
b) the height of the image on the screen.

20.7 An illuminated object of height 5 mm is placed 1.20 m from a screen. A convex lens is moved between the object and the screen until a magnified image is seen on the screen. The lens is then 0.35 m from the object.
a) Calculate the focal length of the lens.
b) The lens is now moved towards the screen to a new position which gives a clear image on the screen. Calculate (i) the distance the lens must have been moved to this new position, (ii) the height of the new image/the height of the first image.

20.8
a) A short-sighted eye has a far point of 4.00 m and a normal near point. Calculate the power of a suitable correcting lens. Give a diagram to show how the correcting lens works.
b) A long-sighted eye has a far point at infinity and its near point is 500 mm away. Calculate the power of a suitable correcting lens. Give a diagram to show how the lens works.

20.9 A student wishes to photograph a small plant on the bed of a shallow pool of clear water of depth 300 mm. A camera fitted with a convex lens of focal length 72 mm is used with a film 75 mm behind the lens. Calculate
a) the apparent depth of the pool when viewed normally,
b) the height of the camera lens above the water surface to give a clear image on the film. The refractive index of water is 1.33.

20.10 Discuss the *f*-number and film speed settings for question **20.9**. How would you change these settings to take a shot of a fish moving near the plant?

20.11
a) With the aid of a labelled diagram, show how a convex lens can be used as a magnifying glass. Explain how maximum magnifying power is achieved by adjusting the object position until the image lies at the viewer's near point.
b) A 120 mm focal length convex lens is used to form a magnified image of a small object. For a viewer with a near point of 250 mm, how far from the lens should the object be placed to achieve maximum magnification when (i) the lens is held to the viewer's eye, (ii) the lens is held 200 mm from the viewer's eye? Sketch a ray diagram for each situation.

20.12 With the aid of a labelled diagram, show that the magnifying power (i.e. angular magnification) of a compound microscope is equal to its linear magnification when the final image is at the viewer's near point.

20.13 A compound microscope consists of two convex lenses of focal lengths 8.0 mm and 50 mm. A small object is placed on the microscope axis at a distance of 10 mm from the objective. The distance between the two lenses is then adjusted so that a magnified virtual image is seen at a distance of 250 mm from the eyepiece. Calculate

a) the distance from the image formed by the objective to each lens,

b) the distance between the two lenses,

c) the magnifying power.

20.14 In question **20.13**, suppose the distance between the two lenses is altered to 80 mm by changing the position of the eyepiece. Calculate

a) the new position of the final image,

b) the magnifying power now.

20.15 In question **20.13**, a graticule consisting of a grid of lines spaced 1 mm apart ruled on a glass plate is positioned between the lenses so that the final image is seen against an image of the grid in focus. Where should the graticule be placed between the lenses?

20.16 A refracting telescope consists of two convex lenses of focal lengths 50 mm and 650 mm. The telescope is used in normal adjustment to view the lunar surface. Calculate

a) the separation of the lenses,

b) the angular size of the final image if the lunar disc subtends an angle of 0.5° to an observer on Earth.

20.17 What should be the distance between the lenses in question **20.16** if the final image of the lunar disc is to be formed 250 mm from the eyepiece at the near point of an observer?

20.18 The arrangement in question **20.16** is modified to photograph the lunar surface by placing a photographic film 750 mm from the eyepiece lens on the other side to the objective lens. Calculate

a) the required separation of the lenses,

b) the size of the image formed on the film.

21

Electromagnetic waves

21.1 The nature of electromagnetic waves

White light can be split into colours, using either a prism or a diffraction grating. The spectrum of colours produced by splitting white light is from red to blue, covering a wavelength range from 700 to 400 nm approximately. The colours change with position from red to orange to yellow to green to blue to violet.

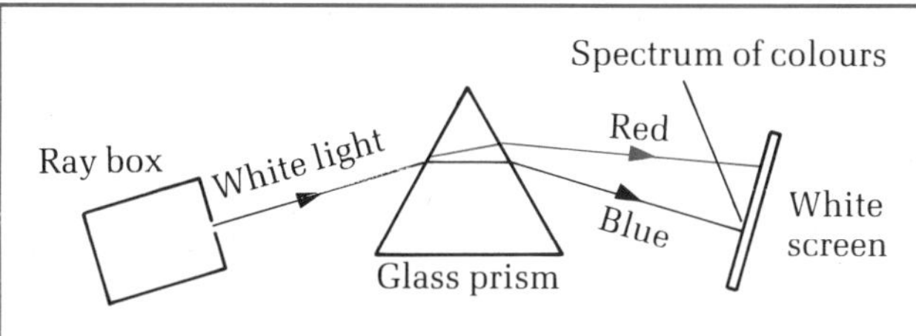

Figure 21.1 *The visible spectrum*

Beyond either end of the spectrum, the eye is unable to detect any light, yet a blackened thermometer bulb placed just outside the red part of the spectrum shows a temperature rise. The bulb absorbs **infra-red** radiation which the eye is unable to detect. Beyond the blue part of the spectrum, a strip of fluorescent paper glows due to **ultra-violet radiation** falling on it. If the eye could detect radiation outside the visible spectrum, the world about us would appear very different, as you can see from Figure 21.2.

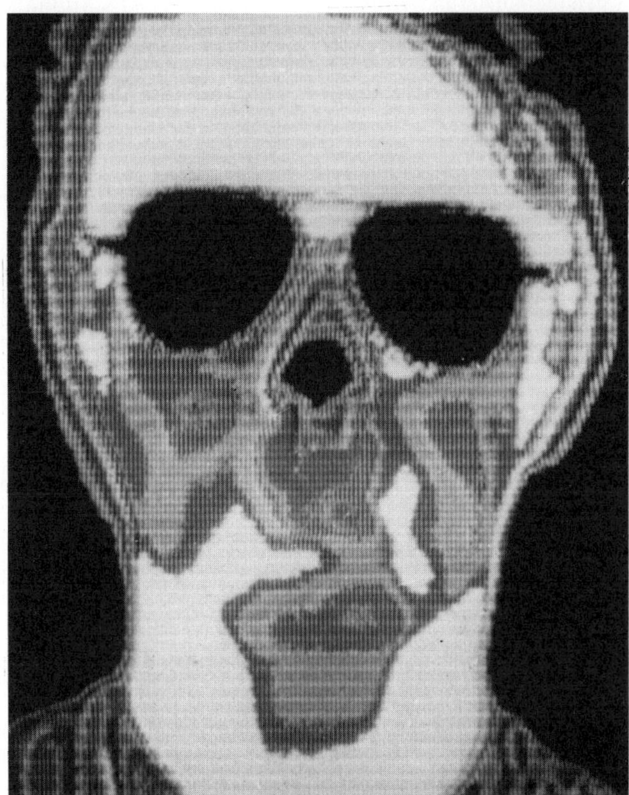

Figure 21.2 *An infra-red photograph of a human face with sunglasses*

Light, infra-red and ultra-violet radiation are part of the **electromagnetic spectrum**. Figure 21.3 shows the full extent of the electromagnetic spectrum. It covers a very wide range from radio waves, infra-red light, ultra-violet and X and γ (gamma) radiations. The wavelength range is enormous since it covers radio waves, with wavelengths of more than 1 km possible, to gamma wavelengths which are of the order of 10^{-15} m. Yet all electromagnetic radiations have the same fundamental nature.

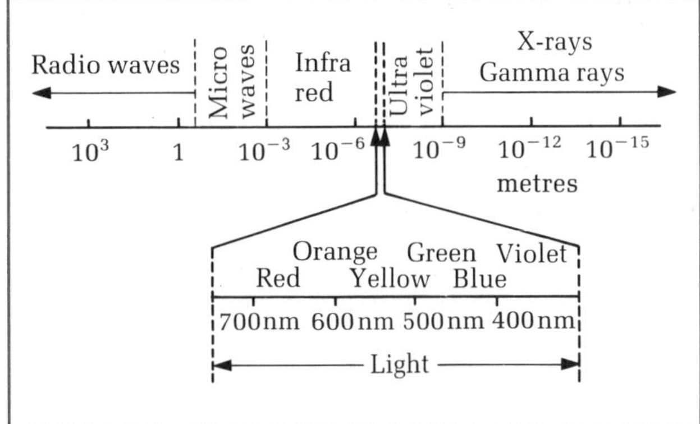

Figure 21.3 *The electromagnetic spectrum*

In 1862, James Clerk Maxwell devised a theory linking electricity and magnetism together. He used the earlier discoveries of Michael Faraday. Faraday knew that a magnetic field was produced when an electric current passed along a wire. Could a magnetic field produce an electric current? Faraday made the very important discovery that a **changing** magnetic field produced an electric current: the principle of electromagnetic induction, see p. 194. Maxwell used his enormous mathematical talent and his intuition to take the ideas further. He proved that a changing magnetic field could create a changing electric field, and that a changing electric field could create a changing magnetic field. Then he showed that a changing current in a wire creates a changing magnetic field that creates a changing electric field that creates a changing magnetic field further away that creates a changing electric field that creates ... In other words, he showed that a changing current in a wire creates **electromagnetic waves** that radiate outwards from the wire. Electromagnetic waves are joint electric and magnetic fields which travel through space without the need for a medium to carry them. If we could see the electric and magnetic fields of an electromagnetic wave, the picture would look like Figure 21.4.

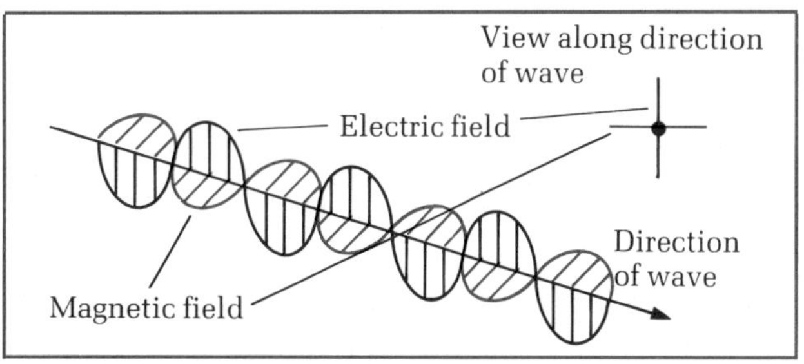

Figure 21.4 *Electromagnetic waves*

Both fields vary sinusoidally, and they are exactly in phase with one another. The electric field is, at any point, at right angles to the magnetic field, and both fields are at right angles to the direction in which the wave travels. So electromagnetic waves are transverse waves because the electric and the magnetic fields are at right angles to the direction in which the wave travels. An electric charge placed in the path of an electromagnetic wave would oscillate from side to side as the wave passed it. The charge would be pushed first one way by the electric field, then half a cycle later the electric field would push it in the opposite direction and so on.

The triumph of Maxwell's theory was an equation for the speed of electromagnetic waves c in terms of the electric field constant ε_0 (see p. 168) and the magnetic field constant μ_0 (see p. 186). He showed that

$$c = \frac{1}{\sqrt{\mu_0 \varepsilon_0}}$$

Maxwell used the known values of these constants to calculate c. The value of ε_0 is $8.85 \times 10^{-12}\,\mathrm{F\,m^{-1}}$; a method for measuring ε_0 is described on p. 168. The value of μ_0 is equal to $4\pi \times 10^{-7}\,\mathrm{H\,m^{-1}}$, as explained on p. 188. Use these values to calculate the speed of electromagnetic waves. Imagine Maxwell's amazement when he arrived at the same result as the measured speed of light! Maxwell had shown that light waves are electromagnetic waves, and the fact that the measured speed is equal to the calculated speed made his theory indisputable.

Of course all good theories not only explain but they predict too. Maxwell predicted that light was just one part of the wide spectrum of electromagnetic radiation. Other scientists had discovered by experiment the presence of infra-red and ultra-violet radiation just outside the visible spectrum, so he could explain the nature of their discoveries. He predicted that the range was even wider, yet the theory was such a huge advance that it was not until 1887 that radio waves were discovered and identified as long wavelength electromagnetic waves.

The means of producing and detecting radio waves was first discovered by Heinrich Hertz. He used his transmitter and detector to show that radio waves had all the properties to be expected of any type of wave. He proved by experiment that radio waves could be reflected, refracted or diffracted. He demonstrated interference of radio waves, and proved that they could be polarized. Hertz's skill as an experimenter allowed him to justify in full the predictions which Maxwell had made. His discoveries were eagerly taken up by other scientists and engineers such as Marconi who showed that radio communication across the Atlantic Ocean was possible.

21.2 Radio waves and microwaves

Hertz's experiments

Hertz showed that radio waves were produced when high voltage sparks jumped across an air gap. He used a wire loop with a small gap in it as a detector, as shown in Figure 21.5. The sparks produce radio waves which spread out from the spark gap and pass through the detector loop. Tiny sparks jump across the detector gap when radio waves pass through it because the radio waves induce a voltage in the detector loop and the voltage makes sparks jump across the detector gap.

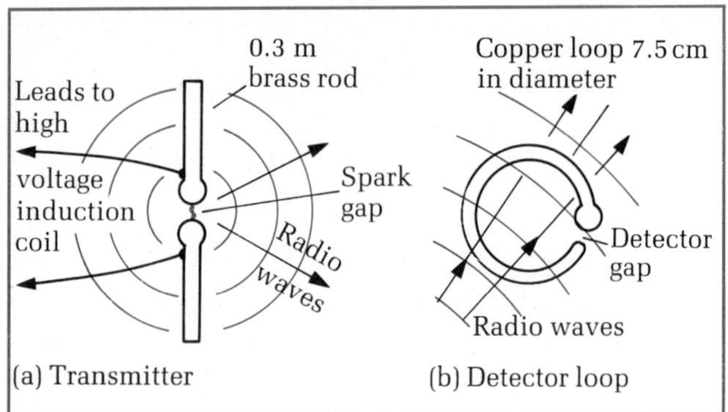

Figure 21.5 *Hertz's experiments*

By placing a metal sheet between the transmitter and the detector, Hertz showed that radio waves do not pass through metal but are reflected by metal. Insulators do not stop radio waves, as Hertz discovered, to his surprise, when he took his detector to a different room from the transmitter and found that the detector still gave sparks. He was able to strengthen the radio beam by using a concave metal reflector behind the transmitter. With the transmitter spark gap at the focal point of the concave reflector, the radio waves reflected off the concave reflector to give a stronger beam. The idea is much the same as positioning a lamp bulb at the focal point of a concave mirror to give a parallel beam of reflected light.

To measure the wavelength, Hertz produced stationary radio waves. To do this he used a flat metal sheet in front of the transmitter to reflect the waves back to the transmitter. So waves travelling towards the flat metal sheet pass through reflected waves off the metal sheet. When the detector was moved along the line between the flat metal sheet and the transmitter, the detector signal varied with position. No signal could be detected when the detector was 33 cm from the flat metal sheet, nor at 65 cm, nor at 98 cm. So the distance between adjacent nodes was

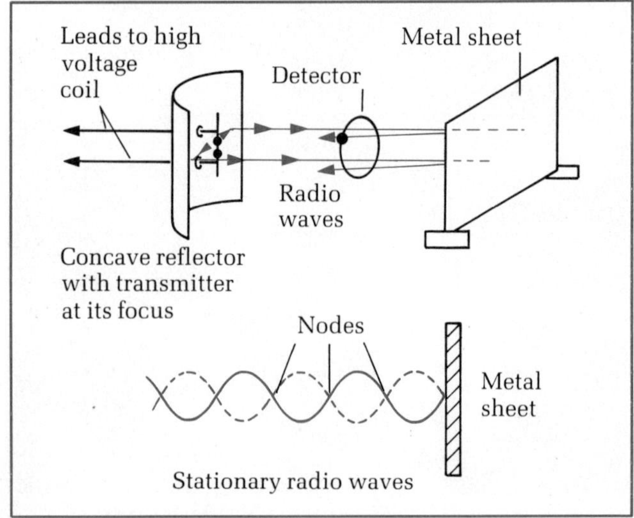

Figure 21.6 *Measuring the wavelength*

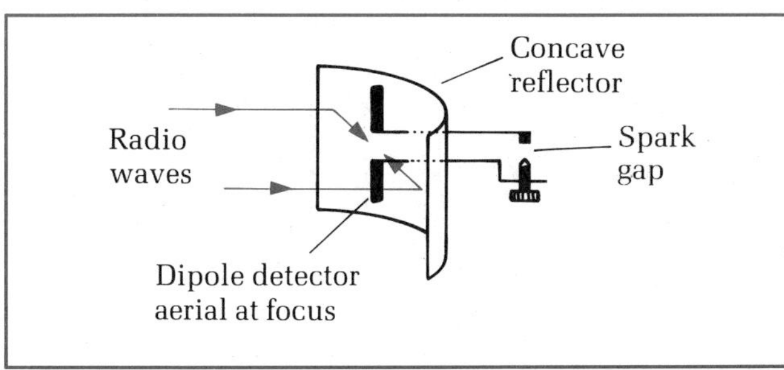

Figure 21.7 *An improved detector*

approximately 33 cm, which gave a radio wavelength of 66 cm. (The distance between adjacent nodes is one half-wavelength.)

The transverse nature of radio waves was demonstrated with an improved detector. By using a **dipole**, as in Figure 21.7, at the focus of another concave reflector, the detected signal from the dipole was passed along wires to a spark gap behind the reflector. When the reflector and dipole were parallel to the spark gap transmitter, a good signal was received. But when the reflector and dipole were turned through 90°, the signal became zero. So Hertz concluded that radio waves from the transmitter were polarized; to receive maximum signal the detector must be parallel to the polarization direction of the radio waves. You may have met polarization of waves in more general terms on p. 235.

Radio and TV communications

Nowadays radio waves are produced using transmitter aerials supplied with high frequency alternating current. The frequency of the radio waves is the same as the a.c. frequency. The simplest form of radio communication is to use a morse code key to switch the alternating current on and off. Pulses received at the detector would need to be decoded. By mixing an audio frequency signal with the high frequency **carrier**, the received pulses can be converted to sound pulses.

Pulse modulation is a modern form of morse code used to transmit information in digital code. Computers operate with electrical signals which are either 'high' (1) or 'low' (0) in terms of voltage. Data is converted and used in a computer in binary form, using 1s and 0s, as explained on p. 367. So data can be carried by radio waves by switching the transmitter current on and off as required, in much the same way as sending a signal by flashing a light on and off. But the electronic switching is much much faster, in pulses of the order of microseconds or less. Teletext is broadcast in this way using a carrier frequency of the order of 500 MHz.

Amplitude modulation occurs when signals from a microphone are used to vary (i.e. modulate) the amplitude of the high frequency carrier waves. In other words, the high frequency waves carry the microphone signals; when the microphone signal is strong, the carrier waves are produced with a large amplitude. When the microphone signal falls, the amplitude of the carrier waves falls. At the receiver, the carrier waves are filtered out so that only the microphone signal passes into the amplifier and loudspeaker. What would be the frequency of carrier waves produced by a local radio station operating at a (carrier) wavelength of 200 m? Use the value of 3×10 m s^{-1} for the speed of radio waves. By comparison, the frequencies of sound waves producing the microphone signal are in the range up to about 15 000 Hz. So even Figure 21.9 (on the next page) does not show how closely spaced the carrier waves are compared with the microphone signal.

Before Marconi first demonstrated that radio signals could be transmitted across the Atlantic Ocean, many scientists doubted that radio waves could travel round the Earth. They thought that radio waves would pass through the atmosphere into space. In fact, most simple types of aerial transmit radio waves in all directions except along the line of the aerial. The radio waves can reach a receiver in three possible ways.

1 The ground wave signal Radio waves transmitted along the ground are affected by the ground, and they

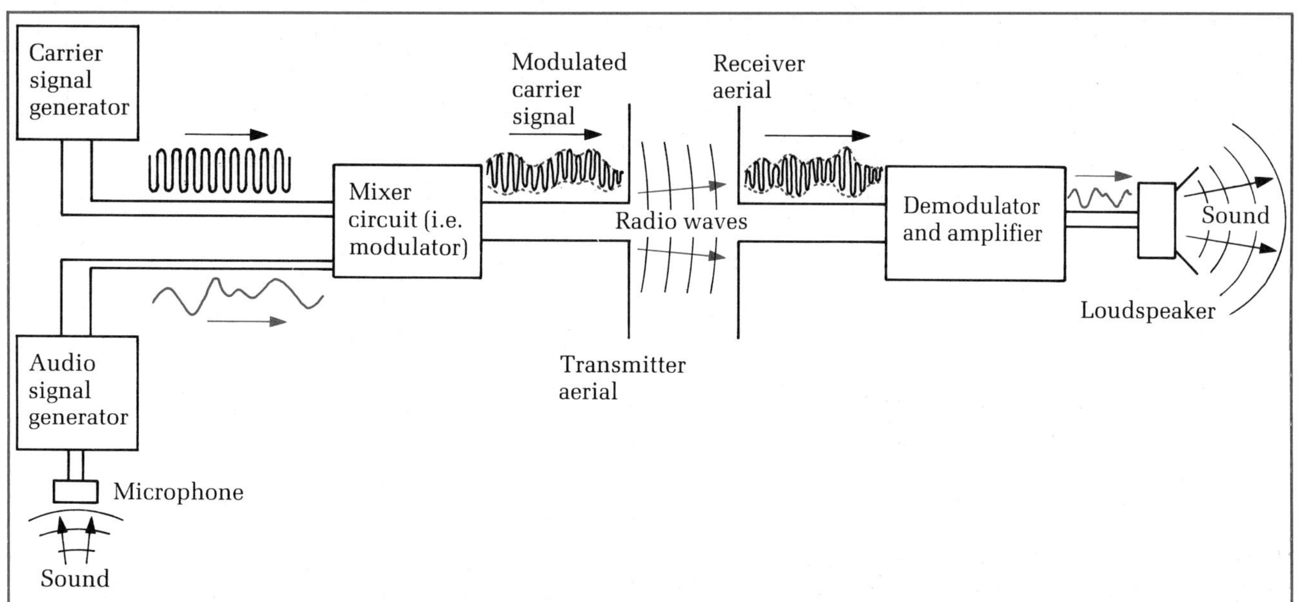

Figure 21.8 *Radio communication*

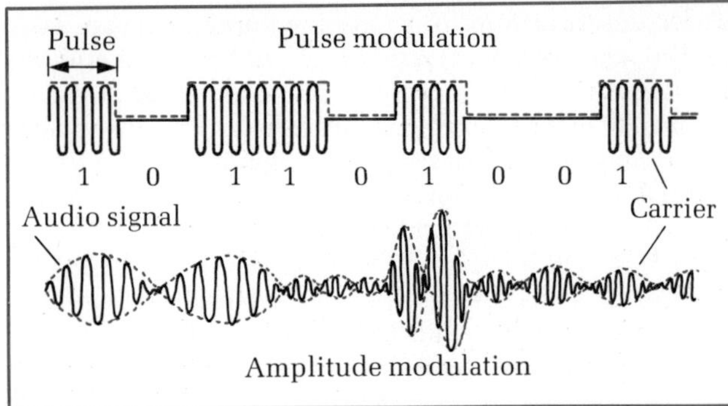

Figure 21.9 *Types of modulation*

follow the Earth's curvature round. The ground signal travels further over sea water than over dry land because sea water is conducting. Also, carrier waves of low frequency travel much further than high frequency carrier waves. At 20 to 30 kHz, the ground signal range is thousands of kilometres; at 50 MHz, it is only a few kilometres.

2 **The space wave signal** This travels to the receiver along the line of sight from the transmitter. The higher its frequency, the less is the energy absorbed from it by the atmosphere.

3 **The sky wave signal** Radio waves are reflected from the upper atmosphere if the frequency is less than about 30 MHz. The reflecting layers make up the **ionosphere** where ultra-violet radiation from the Sun ionizes the gas molecules. So the layers are electrically conducting, and act like metal reflectors. Low frequency waves are reflected weakly by layers about 50 km above the surface. Higher frequencies are reflected by layers up to 400 km above the surface – and much more effectively than low frequency signals lower down. However, if the frequency is greater than about 30 MHz the signal passes through into space. So sky waves can travel round the Earth by bouncing back and forth between the ionosphere and the surface of the Earth. Interference results when the sky wave from a transmitter meets its ground wave, sometimes causing TV programmes to fade out.

Radio waves are used to carry TV signals as well as radio programmes. In addition, satellite links make broadcasting of live events on TV possible across the world. Whatever type of signal is being carried, the term 'radio wave' is used for the carrier. But the frequency range for radio programmes is much smaller than for TV programmes, and satellite links use even higher frequencies. Higher frequencies are necessary where much more information is to be carried. For example, pulse modulation with a high frequency carrier can be used to send many more pulses per second than with a low frequency carrier, so TV programmes are transmitted at much higher frequencies than radio programmes; and because TV frequencies pass out through the ionosphere (being in the range 200 MHz to 900 MHz), satellite links are necessary to transmit TV programmes across the world. In fact, TV signals are beamed to satellites using microwaves as the carrier. Microwaves have frequencies of 3000 MHz or more, so because they have a much shorter wavelength than normal TV carrier frequencies, they can be directed in a narrow beam from Earth to satellite and back to another part of the Earth. Normal TV carrier frequencies pass into Space but spread out too much, so not enough energy would be received from them. In addition, microwave links can carry much more information because their frequency is higher.

Transmitter aerials need to be about the same length as the wavelength for efficient transmission. Marconi used a 122 m aerial to transmit from Great Britain to America. A common type of transmitter aerial is the **half-wave dipole**, as shown in Figure 21.11. Its length is exactly half a wavelength so the alternating voltage produces resonance of the current along its length. Thus the transmitter power radiated away is greatest at this length. The waves spread in all directions except along its length. By using two metal rods either side of the dipole, the radio waves can be beamed as shown by Figure 21.12. The

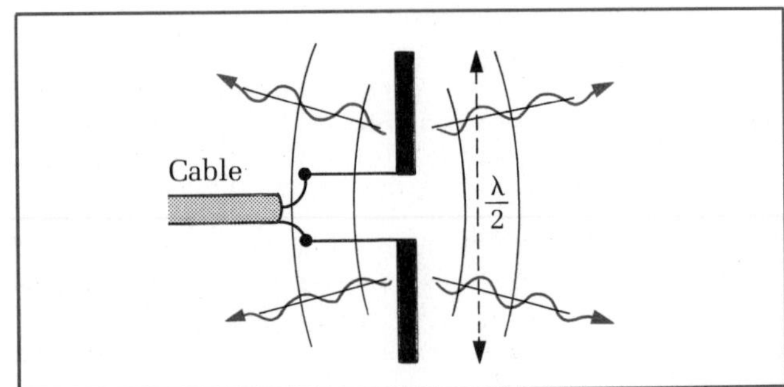

Figure 21.11 *A half-wave dipole*

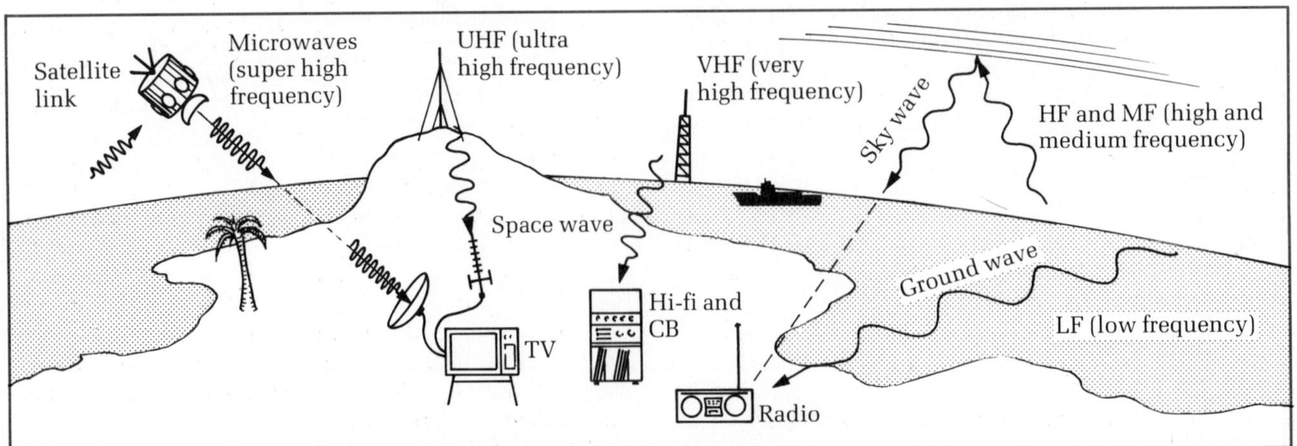

Figure 21.10 *Using radio waves*

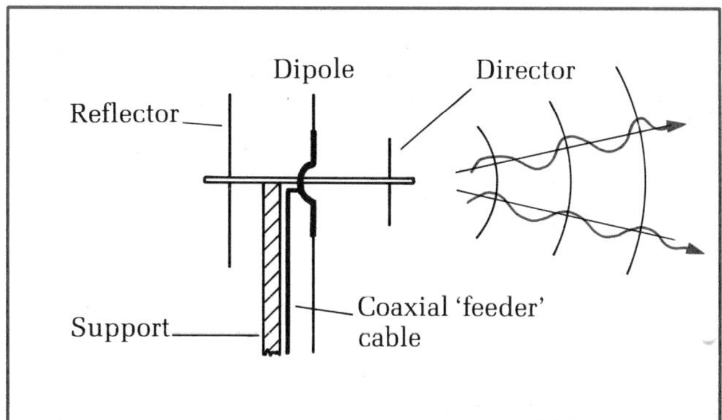

Figure 21.12 *Radio beam*

longer rod (the reflector), and the shorter rod (the director), are positioned one quarter of a wavelength from the dipole. Then the radio waves are beamed in the direction of the shorter rod.

A parabolic dish, with the dipole at its focus, is another way of directing radio waves. It is most effective for microwaves which only need a short dipole since the wavelength is small. A short dipole can be placed exactly at the focus so the reflected microwaves do not spread very much. A large dipole would produce a much wider beam because much of the dipole would not be at the focus. Even so, some spreading is caused by **diffraction** by the dish. The same idea applies here as to diffraction of light by gaps, with the dish acting as a reflecting gap. The extent of the spreading can be estimated using the diffraction equation on p. 268.

$$\sin\theta = \frac{1.22\,\lambda}{W}$$

where W is the width of the dish.

So for 3 cm microwaves and a dish width of 1.0 m, the angle of spread to one side, θ, is given by $\sin\theta = 1.22 \times 0.03/1$ and thus $\theta = 2°$, giving a beam spread of 4°. What would be the width of the beam 100 km away?

Receiving aerials are often made in the form of dipoles. A radio telescope uses a parabolic reflecting dish to focus radio waves onto the dipole. A large dish allows as much energy as possible to be collected. Also, the accuracy of locating radio sources in space is much greater for a larger dish. The dish reflects the radio waves to the focus, but because of diffraction the focus is smeared out, in the same sort of way as light is after passing through a pinhole. The equation of the previous paragraph applies. A dish of width 100 m operating at a wavelength of 21 cm would be able to pinpoint radio sources to within a fraction of a degree. More down-to-earth are UHF aerials which can be seen as TV aerials on many rooftops. The simple dipole is modified by the addition of a number of short bars to make it more sensitive when pointed towards the TV transmitter mast. Even more commonplace are ferrite rod aerials in transistor radios.

Using microwaves

Microwaves are produced by specially designed electronic valves. High power microwave sets each use a **magnetron valve** which can radiate kilowatts of power. Radar waves are microwaves produced by magnetron valves. Low power microwaves can be produced by **klystron valves**, with typical power outputs of a few milliwatts. A low power microwave set can be used to demonstrate wave properties as follows.

i) Reflection Direct the microwaves straight at a metal plate. Place the detector behind the transmitter; a signal should be received by the detector. Now remove the metal plate, and the signal should become zero. The metal plate reflects microwaves back towards the receiver. With the metal plate in the same position, see if the detector can pick up a signal behind the metal plate. A metal plate should reflect all the microwave energy directed at it.

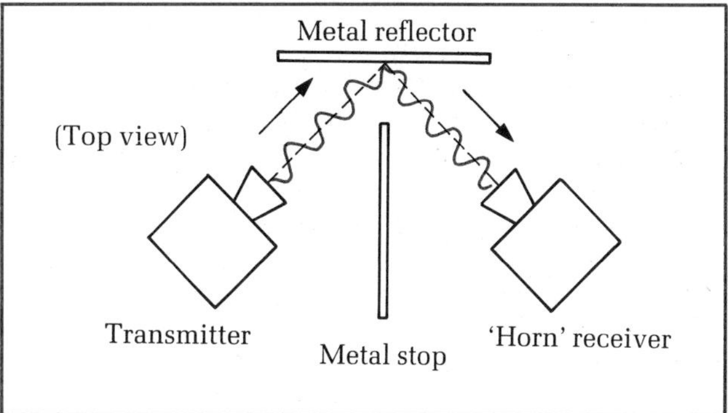

Figure 21.13 *Reflection*

Microwaves obey the law of reflection. Set up the microwave transmitter and detector as in Figure 21.13 with a metal plate between to stop direct microwaves. A metal reflector plate is then turned in front of the transmitter. The detector signal ought to reach a maximum when the angle of the reflected microwave beam to the plate is the same as the angle of the incident beam to the plate.

Try the experiment with a sheet of thin card instead of a metal reflector plate. You ought to find that some microwave energy passes through the card and some is reflected, so the reflected beam is not as strong as from a metal plate. But the law of reflection is still obeyed. Suppose you found that the reflected beam and the beam which passes through are both very weak, compared with the incident beam. What would you conclude? The microwave energy directed at the object (e.g. card) must have been absorbed, so the internal energy of the material of the object must have increased. In other words, the object's temperature is raised. Microwave ovens use this principle which is particularly effective with water molecules in food.

ii) Polarization Line the detector up facing the transmitter so a good signal is received (see Figure 21.14 at the top of the next page). Then turn the detector on its side. The signal should become zero. Turn the transmitter on its side as well and the signal returns. The reason is that the microwaves from the transmitter are polarized in one plane only. The detector must be lined up in that plane if it is to give a signal. At 90°, no signal can be detected because the electric field of the microwaves (see p. 298), which defines the plane of polarization, is at right angles to the detector aerial.

With transmitter and detector upright, place a metal grille between the two. When the grille is turned about

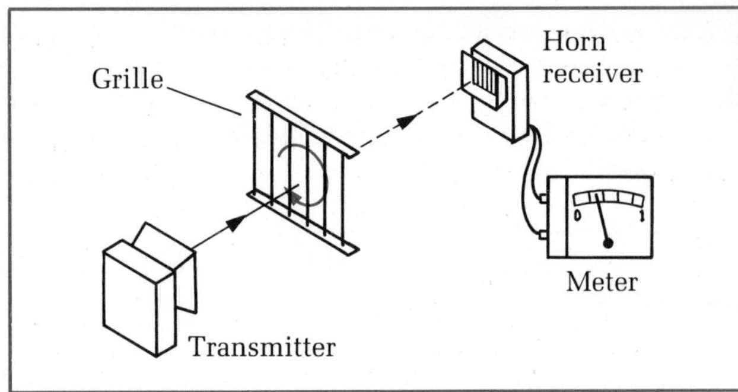

Figure 21.14 *Polarization*

the beam, the signal changes. The minimum signal is found when the grille bars are at right angles to the plane of polarization.

iii) Measuring the wavelength using stationary waves Arrange the apparatus as shown (Figure 21.15), and move the detector along the line between the transmitter and the plate. The signal varies with position, its strength changing from zero to a maximum according to where the detector is placed. The reason is that the reflected microwaves and the incident microwaves form a stationary wave pattern. From one zero to the next gives the spacing between two nodes, which is half a wavelength. Measure across several nodes to determine the wavelength more accurately. What is the microwave frequency? Assume the speed of the waves is $3 \times 10^8 \text{ m s}^{-1}$.

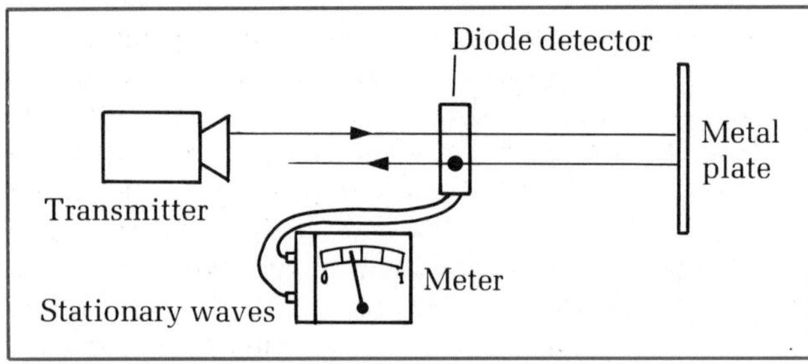

Figure 21.15 *Stationary waves*

iv) Diffraction Use metal plates to form a gap. Investigate the extent of the spreading of microwaves which pass through the gap. Try to locate the first minimum of the single slit diffraction pattern (see p. 266). Change the gap width and investigate how the diffraction changes.

v) Interference Form two gaps using metal plates (see Figure 21.16). Make the gaps each about 2 cm wide so that the waves from each gap spread out in all directions. Move the detector about and locate points of cancellation and reinforcement. Find a point of cancellation. Then block one gap. The signal ought to return because the waves are from the other gap only.

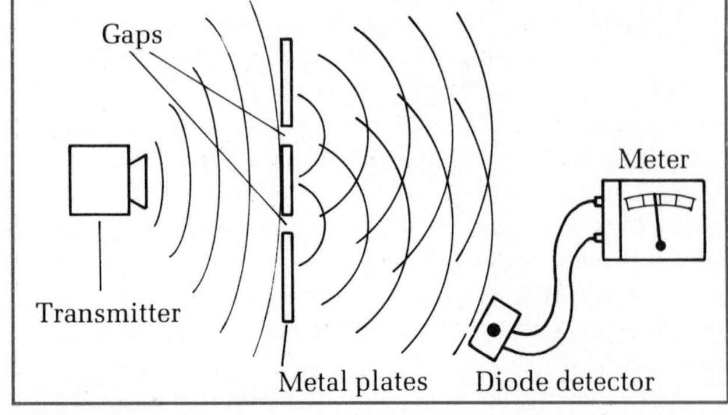

Figure 21.16 *Interference (top view)*

vi) Refraction Microwaves have a much greater wavelength than light, so lenses and prisms for microwaves need to be bigger than for light. Figure 21.17 shows a big wax lens used to focus microwaves. The lens refracts the microwaves because microwaves travel more slowly in wax than in air.

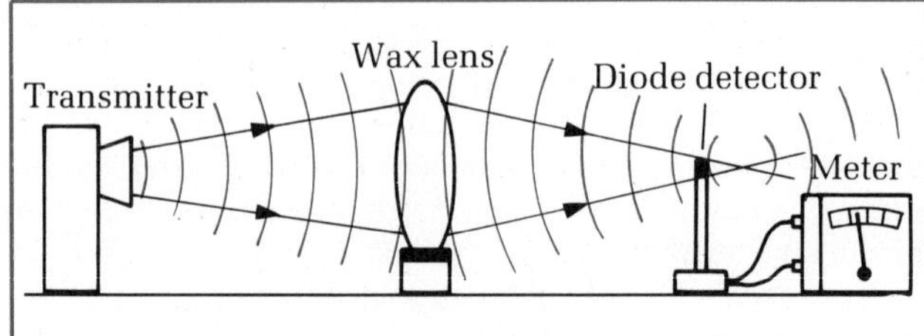

Figure 21.17 *Refraction (side view)*

21.3 Infra-red radiation

The wavelength range of infra-red radiation is from about 1 mm to about 750 nm, where the red part of the visible spectrum ends. All objects emit infra-red radiation; the hotter an object is, the greater is the energy per second carried away by infra-red radiation. So an infra-red photograph provides a map of 'hot spots' on an object's surface. Electromagnetic radiation from hot objects is sometimes called **thermal radiation**. Any object heated until it glows emits visible radiation as well as infra-red radiation, so the term 'thermal radiation' covers visible and infra-red radiation. Thermal radiation is discussed in more detail on p. 115.

To produce infra-red radiation in a controlled way, an electric heater filament may be used (as shown in Figure 21.18). Tungsten wire, wrapped round a suitable heat-resistant former, can be supplied with current from a low voltage unit if the length of wire is short enough. With the filament heater at the focus of a large metal concave reflector, a beam of infra-red radiation can be directed as required. A second metal concave reflector may be used to

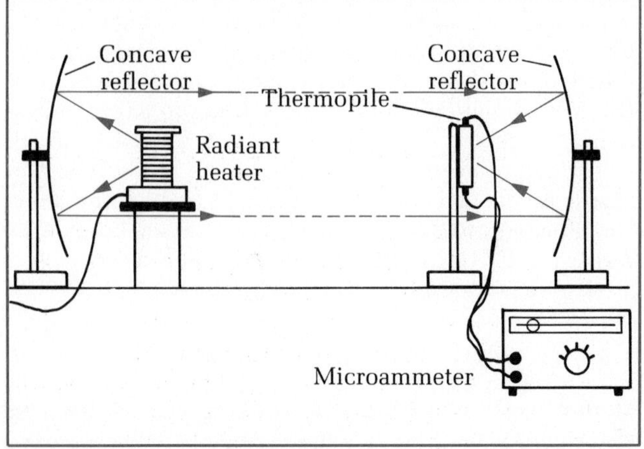

Figure 21.18 *Infra-red radiation*

focus the beam directed at it. So any object placed at the focus of the second reflector will be warmed up.

To detect infra-red radiation, the simplest way is to use a blackened thermometer bulb. Black surfaces are best at absorbing infra-red radiation; polished silvered surfaces are the best reflectors.

More accurate detectors

Thermopiles consist of thermocouple junctions fixed to a black surface. A thermocouple junction is formed whenever two different metals are joined. So a wire of copper joined at one end to a wire of iron forms a simple thermocouple. A potential difference is created across the junction because electrons can leave one metal more easily than the other. The p.d. changes with change of temperature. So when infra-red radiation is absorbed by the black surface of the thermopile, the surface temperature increases causing a change of the thermocouple p.d. By monitoring the thermocouple p.d., infra-red radiation directed at the thermopile can be detected.

Bolometers use temperature-sensitive resistors called **thermistors**. By connecting the thermistor into one of the arms of a Wheatstone bridge circuit, as in Figure 21.19, small changes of the resistance of the thermistor can be detected. In the absence of infra-red radiation, the bridge is balanced by altering the resistances in one or more of the other arms. Then infra-red radiation is directed at the bolometer, causing its thermistor to warm up. Thus the thermistor resistance alters and the bridge becomes unbalanced, so current passes through the centre-reading meter causing a reading off-centre. Some types of bolometer use thermistors cooled to within a few degrees of absolute zero to make them very sensitive.

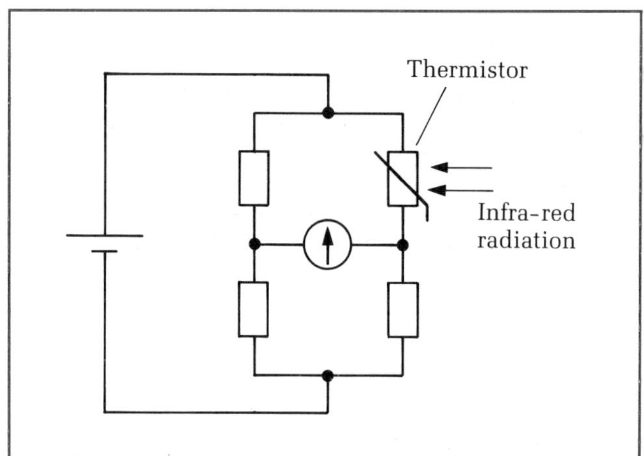

Figure 21.19 *Thermistor used to detect infra-red radiation*

Photographic film needs to be of a special type because ordinary film is only sensitive to light. Infra-red film has a special dye in the film emulsion. The dye absorbs infra-red radiation and transfers the absorbed energy to the film emulsion. Because infra-red film is also sensitive to light, a filter must be fitted to the camera. The filter only allows infra-red radiation to pass through to the film, so when the film is developed, the picture shows infra-red sources only.

Photoelectric detectors are usually made from semiconducting material such as silicon. Energy absorbed by the semiconductor from the infra-red radiation causes electrons to be freed. So by applying a p.d. which draws a current as a result of the freed electrons, the infra-red radiation can be detected and measured. This type of detector does not respond to infra-red wavelengths greater than about 1000 nm.

Glass absorbs infra-red radiation wavelengths greater than about 2500 nm so when a spectrum is produced by a prism (using light from a filament lamp) much of the infra-red radiation is absorbed by the prism. A small amount does pass through, and this would be detected just beyond the red part of the spectrum.

21.4 Ultra-violet radiation

Ultra-violet radiation can make cloth glow in the dark. The glow is caused by the presence of **fluorescent dyes** from washing powders which absorb ultra-violet radiation and then emit light. The presence of ultra-violet (UV) radiation beyond the blue part of the visible spectrum was first demonstrated using fluorescent paper. However, disco lovers and users of UV 'sun tan' lamps should beware! Ultra-violet radiation is harmful to the eyes. Fortunately, the Earth's atmosphere filters out the UV radiation from the Sun. Other types of detector include:

a) Photographic film which is sensitive to wavelengths down to 200 nm. For shorter wavelengths, specially prepared film must be used. Ordinary glass absorbs ultra-violet radiation so ultra-violet lenses must be made from transparent materials such as quartz. Even so, these materials do not transmit wavelengths of less than about 100 nm. Also, filters must be used to allow only the ultra-violet radiation to reach the film. Otherwise, light will produce much stronger images.

b) Photoelectric detectors such as vacuum photocells. In a vacuum photocell, there is a metal surface called the cathode. When ultra-violet radiation is directed at the cathode, electrons at the cathode surface gain energy from the radiation. Some of the electrons gain enough energy to leave the cathode and reach the anode. So a meter connected to the photocell would give a non-zero reading when ultra-violet radiation falls on the photocathode. Photocells can also detect light so a filter is required to eliminate light from the incident radiation.

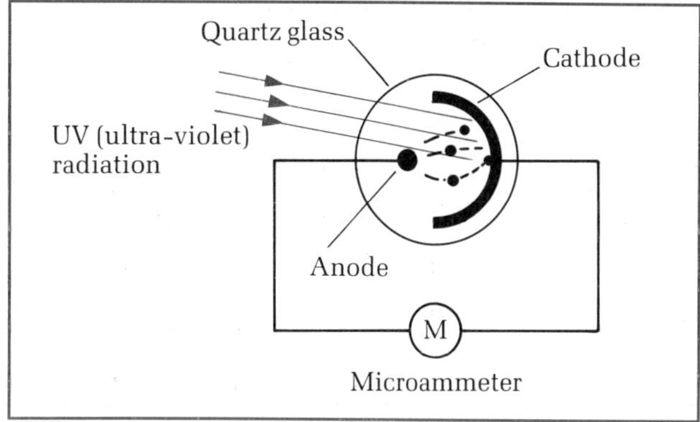

Figure 21.20 *Detecting ultra-violet radiation*

21.5 Polarization of electromagnetic waves

Any electromagnetic wave consists of an electric field component and a magnetic field component. As shown in Figure 21.4, the two components vary sinusoidally but they are always in phase and at right angles to one another and to the wave direction. Hertz's experiments using radio waves showed that they are polarized, and that the electric field component can be used to define the plane of polarization. Radio waves from a transmitter aerial are produced with the electric field component always parallel to the aerial. So the waves are polarized when they are produced. This is not, in general, the case for light. Light from a filament lamp is unpolarized, which means that each 'burst' of light from the lamp is polarized in a random direction. So at any given point, the polarization direction changes as one burst follows another.

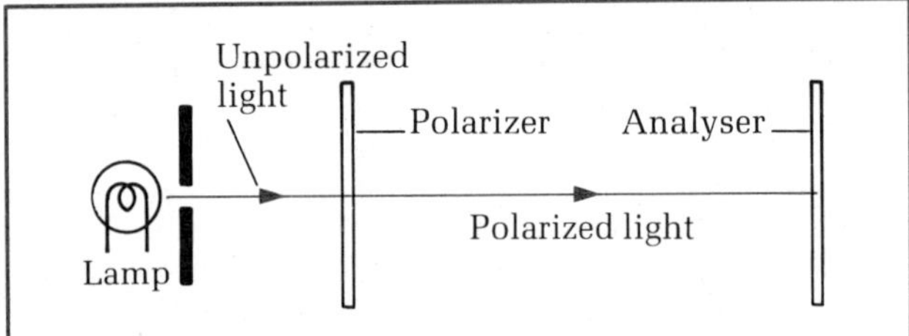

Figure 21.21 *Crossed polaroids*

Investigating polarized light

1) A beam of light from a ray box is directed through a polaroid filter. The polaroid only allows through light which is polarized in a certain plane so the beam becomes polarized after passing through the polaroid. The plane of polarization produced by the polaroid is determined by the line of the molecules in the polaroid. If a second polaroid, called the analyser, is placed in the path of the beam after the first filter, the intensity of the beam can be changed by turning the second filter. When the filters are lined up with one another, all the light from the first passes through the second. But when the two filters are 'crossed', no light emerges from the second one. The molecules of the second filter are at right angles to the plane of polarization of the light from the first filter. So no light can pass through the second filter.

2) Certain liquids are capable of rotating the plane of polarized light. Suppose two polaroids are 'crossed' so that no light can emerge through them, then a tube of liquid such as glucose solution is placed between the two polaroids. Light can then be seen through the analyser. The liquid has rotated the plane of polarization of the light from the first polaroid. So the analyser is no longer at right angles to the plane of polarization. By turning the analyser, the light can once again be blocked out. The angle that the analyser is turned through is equal to the angle which the plane of polarization is rotated by.

3) Light can be polarized by reflection off glass or any other transparent medium, as shown in Figure 21.22.

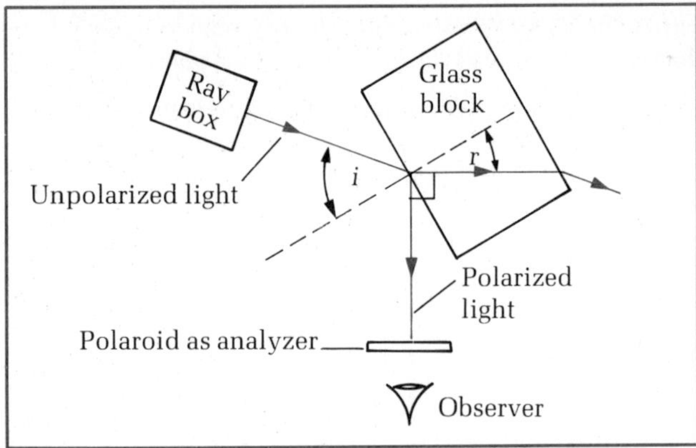

Figure 21.22 *Polarization by reflection*

Direct a beam of light at an angle to the surface of a glass block. View the reflected beam (off the surface of the block) through a polaroid. Turn the polaroid about the reflected beam, and the intensity ought to alter. At a particular angle of incidence, the polaroid can block the light altogether. So at this angle, the reflected beam is completely polarized in one plane only. This angle is called the **Brewster angle**. Try to measure it for glass.

At the Brewster angle, the reflected beam is exactly at right angles to the refracted beam, so the angle of refraction equals 90° – the angle of reflection. But the angle of incidence equals the angle of reflection. Hence the angle of refraction equals (90° – the angle of incidence).

$$r = 90° - i$$

From Snell's law $\dfrac{\sin i}{\sin r} = n$

where i = angle of incidence,
r = angle of refraction,
n = refractive index of the glass block.

Since $\sin r = \sin(90° - i) = \cos i$, then the Brewster angle i is given by

$$\frac{\sin i}{\cos i} = n$$

or more simply,

$$\tan i = n$$

Calculate the Brewster angle for a glass block of refractive index 1.5.

Applications

Stress analysis Place a perspex rod between crossed polaroids. When the rod is flexed coloured fringes may be seen in the parts of the rod under greatest stress, viewing through the analyser. Engineers often use this technique at the design stage when a model of the device under test is constructed using perspex, as is shown in Figure 21.23. Areas of high stress can be located by viewing the model between crossed polaroids.

Liquid crystals Displays for digital watches and calculators use liquid crystals. Try turning a piece of polaroid directly over a liquid crystal display (LCD). The contrast between the displayed information and the back-

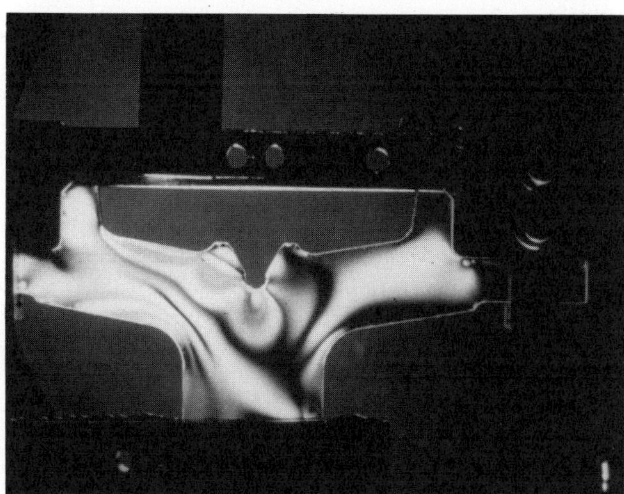

Figure 21.23 *Stress fringes*

ground changes. Liquid crystals are not rigid and can flow, but their molecules are in a regular order just like atoms in a solid crystal. Each LCD is a matrix of segments, each segment activated by a tiny electric current which makes it darker than its background. So by activating a suitable combination of segments, the LCD can be made to display any number required.

The reason why a segment becomes darker when activated is that it does not reflect light when it is activated. The segment is constructed from a cell of liquid crystal between crossed polaroids. The underside of the top piece of the cell is marked with lots of fine parallel lines along which the molecules of the liquid crystal line up. The bottom of the cell is also marked with fine parallel lines, but positioned at right angles to the top set, so the line up of molecules changes by 90° from the top to the bottom. As a result, light which passes through the cell has its plane of polarization turned through 90°. Therefore even though the cell is between crossed polaroids, light can pass through it. With no voltage across the cell, light from the room passes through it, reflects off a mirror underneath, and passes back out again; so the cell allows the light to be reflected. When a voltage is applied across the cell, the molecules line up along the electric field instead of the twisted arrangement explained above. The light cannot pass through since the new line does not rotate the plane of polarization. Hence the cell is dark when activated.

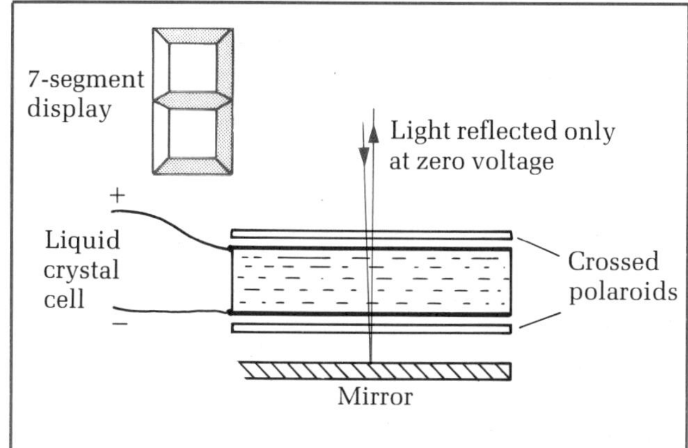

Figure 21.24 *Liquid crystal cell*

21.6 The speed of light

Accurate measurement of the speed of light is important not just for scientific purposes but also for applications such as aircraft navigation. During the Second World War, navigation errors arose when radar started to be used for guidance. The value for the speed of light was only known to an accuracy of one part in 30 000, so scientists devised even more accurate ways of measuring it. Calculations of distance involved multiplying the speed of light by the time taken for radio waves to travel from transmitter to aircraft. If the speed of light was not known accurately enough, the distance calculations would be in error. Astronomers use the same idea to measure very accurately the distance from Earth to the Moon. Radar pulses bounced off the Moon are timed from the instant the pulse is sent out to its return. The Moon–Earth distance is then calculated from (speed of pulse × timing)/2.

The main problem in measuring the speed of light is the timing of very short intervals. Even over a distance of 10 km or more, light takes less than 50 microseconds.

Fizeau's method used a rapidly rotating cog wheel with lots of teeth round the edge. A narrow beam of light was directed parallel to the axis of the wheel so that the light could pass exactly through a single gap between the cogs when the wheel was in the right position. The light was then allowed to travel a long distance to a reflecting mirror which returned the light to the same point on the wheel. With a gap present, the returning light could be observed as shown in Figure 21.25.

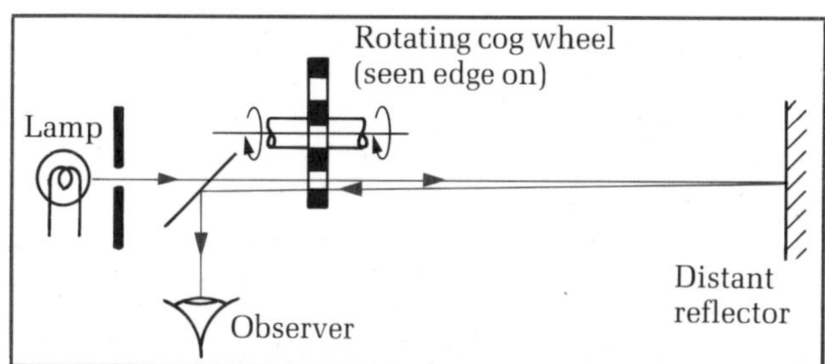

Figure 21.25 *Fizeau's method*

With the wheel turning, the returning light does not pass totally to the observer because the gap which allowed it through is replaced by a tooth. As the speed of the wheel is increased, a point is reached where the light is totally cut off. The gap which allowed light through has then been replaced by the next tooth in the time taken for the light to go and return. By measuring the frequency of rotation, f, the time taken can be calculated if the number of teeth on the wheel, N, is known. To replace one gap by the next tooth, the wheel must turn through $1/2N$ of 1 rotation. So the time taken must equal $\frac{1}{2Nf}$.

Light is cut off at rotation frequencies of $3f$, $5f$, $7f$, etc. each one corresponding to a gap being replaced by the next but one tooth, then the next again, etc.

Fizeau first performed the experiment in 1849. He used a cog wheel with 720 teeth, and allowed the light to travel a distance from gap to mirror and back of 17.266 m. The lowest frequency at which the light was cut off was 12.6 Hz. What value does this give for the speed of light?

Michelson's method used an octagonal (i.e. eight-sided) mirror made of polished steel. Light from an intense source was reflected off one face of the octagonal mirror, then directed at a distant concave reflector. The concave mirror reflected the light back to the opposite side of the octagonal mirror, so an observer could see an image of the light source by reflection off the octagonal mirror, as shown in Figure 21.26.

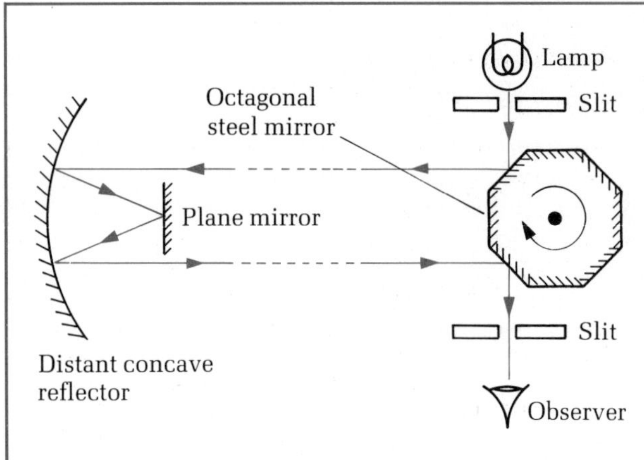

Figure 21.26 *Michelson's method*

The octagonal mirror was capable of being rotated at high speed. As the speed of rotation is increased from zero, the image first disappears, but then reappears at a certain frequency of rotation, f. Increased rotation makes the image disappear again, but it reappears at frequencies of $2f$, $3f$, $4f$, etc. The reason why the image reappears at frequency f is that in the time taken for the light to travel from one side to the other side the mirror makes exactly 1/8 of a rotation. So the second reflection off the mirror is exactly at the correct angle to pass to the observer. At $2f$, the mirror makes exactly 2/8 of a rotation in the trip time of the light, and so on. The trip time is exactly equal to $1/8f$. If the trip distance is D, then the speed of light is given by $D/(1/8f) = 8fD$.

Michelson made the distance as great as possible by setting the octagonal mirror on a mountain top with the concave reflector on another mountain 35 km away. His measurement gave a value for the speed of light of 299 774 km s^{-1} which is very close to the currently accepted value of 299 793 km s^{-1}.

The Kerr cell uses a special transparent cell between two crossed polaroids. When a voltage is applied across the cell, light is allowed through because the electric field rotates the plane of polarization of the light passing through. So an alternating voltage turns the cell on and off repeatedly. The cell can be used like a 'Fizeau cog wheel'. What is the fraction of a cycle between the cell being on, off, then on again? In that time, the alternating voltage changes from peak to zero and back to peak, corresponding to a gap then a tooth then a gap in the beam path. If the cell is operated at the lowest frequency to block the returning light, the time taken for the light to go and return is equal to the time for the voltage to change from peak to zero.

21.7 Waveguides

Where a radio or TV aerial on a mast is connected to a receiver circuit, the connecting wire is usually coaxial cable. This type of cable consists of a single copper wire along the centre of an insulating roll. A sheath of copper surrounds the roll, with an insulating layer over the sheath, as in Figure 21.27. The sheath prevents unwanted signals being picked up because it shields the inner wire.

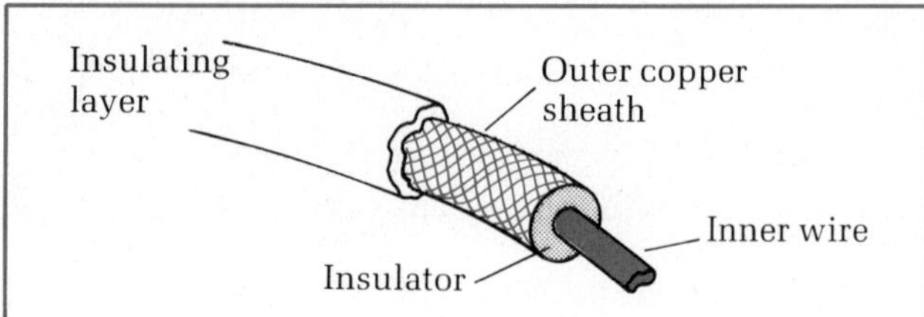

Figure 21.27 *A coaxial cable*

More important, it allows the signal from the aerial to be transmitted without energy loss. See the effect for yourself of using two separate wires to link an aerial to a receiver; then use coaxial cable and observe the difference. Ordinary wires act as transmitters of electromagnetic waves. Instead of feeding the signal energy from the aerial to the receiver, the ordinary wires radiate the energy away to the surroundings. The coaxial cable keeps the electromagnetic waves in the space between the inner wire and the outer sheath; the signal is guided in the form of electromagnetic waves from the aerial to the receiver instead of being radiated away. The coaxial cable is an example of a **waveguide** which transmits electromagnetic waves from one point to another without allowing the waves to spread out.

Waveguides can be very efficient at transmitting energy. Figure 21.28 shows a rectangular waveguide where opposite metal sides act like the inner and outer wires of the coaxial cable. This type of waveguide can transmit microwaves carrying large amounts of power with very little heating effect. For power transmission, waveguides operate most effectively if the wavelength is the same order as the 'width' of the guide. So in practice, microwaves are used.

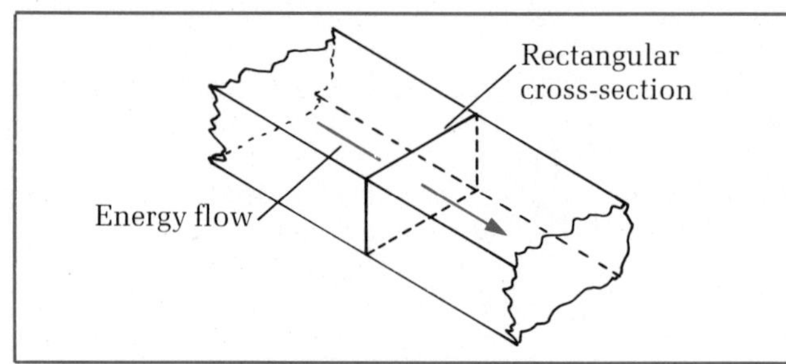

Figure 21.28 *A rectangular waveguide*

Measuring the wave speed of guided waves

A pulse generator operating at 150 kHz is used to apply voltage pulses to one end of a long coaxial cable (see

Figure 21.29). The cable length must be measured and needs to be at least 100 m.

A dual beam oscilloscope is used to display the voltage pulse. One beam displays the pulse when it is applied to one end (end A) of the cable. The other beam displays the pulse when it arrives at the other end (end B) of the cable. The time taken for the pulse to travel from A to B along the cable is determined by measuring the distance on the oscilloscope screen between the two pulse displays. The time-base setting must be known so that the time taken can be calculated. Then the wave-speed can be calculated from cable length/time taken. The resistance boxes should be set at values that prevent reflected pulses appearing on the screen. Try different values. If the boxes are incorrectly set, extra pulses appear on the screen due to reflections at the ends of the cable.

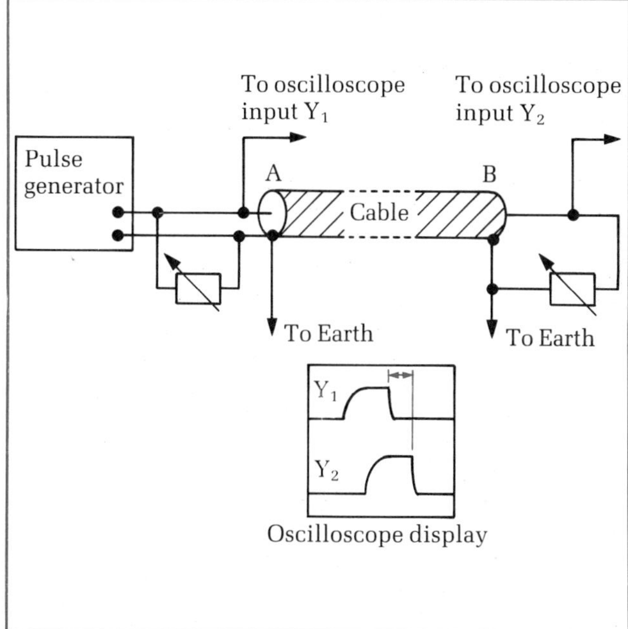

Figure 21.29 *Measuring the wave speed*

Estimating the wave speed

Consider a simplified waveguide made up of two long parallel metal plates as in Figure 21.30. One plate is earthed (plate X in the diagram) and the other plate (Y) is made positive when the battery switch is closed. Closing the switch causes a voltage pulse to run along Y from the battery end to the far end. The pulse travels along Y at steady speed v, charging each section positive as the pulse reaches and passes that section.

Suppose we could picture the pulse at the instant that section S in Figure 21.30 charges up. The charge flow onto S is a current and so creates a magnetic field between the plates. As S becomes fully charged, the current drops so the magnetic field strength falls. The falling magnetic field induces a voltage between the plates in the next section, so the next section receives the pulse. The pulse is caused by the changing magnetic field inducing a voltage and hence electric field. The changing electric field generates a changing magnetic field further along, and so on.

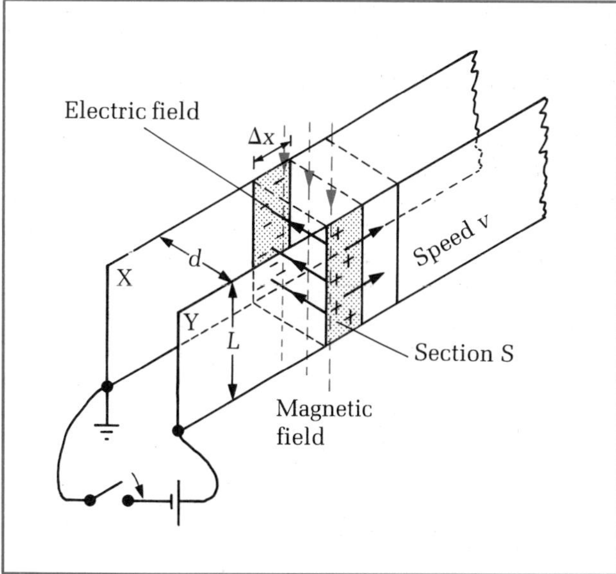

Figure 21.30 *Estimating the wave speed*

1 Section S charges up like a parallel plate capacitor. Its area is $L\Delta x$, and the plate spacing is d as shown in Figure 21.30. So the charge flow onto S is given by

$$\Delta Q = \text{capacitance} \times \text{p.d.} = \frac{L\Delta x\varepsilon_0 V}{d}$$

2 The flow of charge creates current $I = \Delta Q/\Delta t$

$$= \frac{L}{d}\varepsilon_0 V\Delta x/\Delta t$$

Since $\Delta x/\Delta t =$ pulse speed v, then

$$I = L\varepsilon_0 Vv/d$$

3 The current produces a magnetic field in the space between the plates. Consider the section as a one-turn solenoid of length L. The magnetic field strength is given by

$$B = \frac{\mu_0 I}{L}.$$

$$\text{Hence } B = \frac{\mu_0 L\varepsilon_0 Vv}{dL} = \frac{\mu_0\varepsilon_0 Vv}{d}$$

4 As section S becomes fully charged, the current decreases so the magnetic field strength falls. The falling magnetic field between the plates generates an induced voltage V across the plates further along. From electromagnetic induction, $V = Bvd$ gives the induced voltage.

Therefore

$$V = \frac{\mu_0\varepsilon_0 Vv^2 d}{d}$$

$$\text{So wave speed } v = \frac{1}{\sqrt{\mu_0\varepsilon_0}}$$

For the coaxial cable, v is less than above because the relative permittivity of the insulator is more than 1.

21.8 Summary

Common properties of all electromagnetic waves

1 All electromagnetic waves travel at the same speed in a vacuum.
2 All electromagnetic waves are transverse waves so can be polarized.
3 All electromagnetic waves consist of electric and magnetic fields at right angles to one another and to the direction of travel. The two fields vary sinusoidally but are always in phase with one another.

Short questions

Speed of light in a vacuum $= 3 \times 10^8\ \mathrm{m\,s^{-1}}$

21.1
a) With the aid of a diagram, describe the nature of electromagnetic waves in terms of the electric and magnetic fields that make up the waves.
b) The intensity of an electromagnetic wave in a vacuum is equal to $\frac{1}{2}\varepsilon_0 E_0^2 c$ where c is the speed of light, ε_0 is the electric field constant ($= 8.85 \times 10^{-12}\ \mathrm{F\,m^{-1}}$) and E_0 is the peak value of its electric field. A certain radio receiver is capable of detecting radio waves if the electric field strength of the radio waves is at least $1\ \mathrm{mV\,m^{-1}}$. Calculate the minimum intensity of radio waves that can just be detected by the receiver.

Differences between types of electromagnetic waves

Type	Wavelength range	Production	Detection	Properties
Radio	>0.1 m	Rapid acceleration and deceleration of electrons in aerials	1 Receiver aerials	1 Reflected from metals
Microwaves	0.1 m to 1 mm	Klystron valve or magnetron valve	1 Point contact diodes	1 Reflected by metals 2 Absorbed by non-metals partly
Infra-red	1 mm to 700 nm	Vibrations of atoms and molecules	1 Thermopiles 2 Bolometers 3 Infra-red photographic film	1 Polished silvered surfaces are the best reflectors 2 Matt black surfaces are the best absorbers
Light	700 nm to 400 nm	Atomic electrons emit light when they move from one energy level to a lower level.	1 The eye 2 Photocells 3 Photographic film	1 In white light, coloured surfaces only reflect their own colour of light. All other colours are absorbed
Ultra-violet	400 nm to 1 nm	Inner shell electrons moving from one energy level to a lower level	1 Photocells 2 Photographic film	1 Absorbed by glass
X-rays	<1 nm	X-ray tubes or Inner shell electrons	1 Photographic film 2 Geiger tube 3 Ionization chamber (X-rays and gamma rays)	1 Penetrate metal 2 Ionize gases (X-rays and gamma rays)
Gamma rays	<1 nm	Radioactive decay of the nucleus		

21.2 Calculate
a) the frequency of radio waves of wavelength 1500 m in air,
b) the wavelength of microwaves of frequency 1 GHz in a medium of refractive index 1.5.

21.3 State two similarities and two differences between radio waves and ultra-violet radiation.

21.4 Two radio masts 1000 m apart transmit continuously at the same frequency of 120 MHz with a constant phase difference. An aircraft flies parallel to the line joining the masts along a flight path that is 30 km from the line.
a) Explain why the radio signal received by the aircraft varies in strength as the aircraft moves along its flightpath.
b) If the received signal varies in strength at a frequency of 1.5 Hz, calculate the aircraft speed.

21.5 A radar speed trap consists of a microwave transmitter and receiver side-by-side pointing in the same direction, as in Figure 21.31. The receiver takes a direct signal from the transmitter via a link cable, and it also detects signals reflected by any obstacle in the path of the microwave beam.

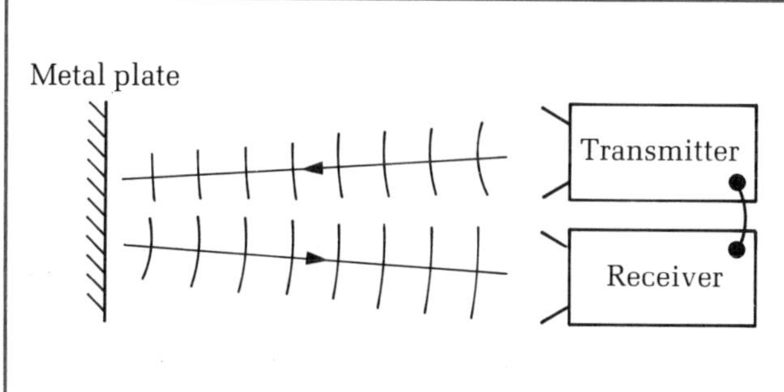

Figure 21.31

a) Explain why, with a metal plate as an obstacle, the receiver output rises and falls as the obstacle is moved towards the transmitter and receiver.
b) If the metal plate must be moved 15 mm as above for the receiver output to rise, fall, then rise again, calculate the wavelength and frequency of the microwaves in air.
c) In use, the beam is directed towards an approaching car which causes the receiver output to vary at a frequency of 720 Hz. Calculate the speed of the approaching car.

21.6 A transmitter of 3 cm microwaves is positioned at the focal point of a concave reflecting dish used to direct the microwaves at a receiver 10 km away. The reflecting dish has a diameter of 1.5 m. Use the ideas of resolving power to estimate the width of the beam at the receiver.

21.7 A small microwave receiver is placed in the beam from a microwave transmitter. A metal grille is then placed between the receiver and the transmitter with its plane at right angles to the beam. The grille consists of a set of regularly spaced parallel metal rods. When the grille is rotated by 180° about the beam, the receiver signal varies from maximum to minimum and back to maximum again as the grille is turned. Explain why the receiver signal varies in this way as the grille is turned.

21.8
a) Explain why the intensity of light passing through two pieces of polaroid varies when one piece of polaroid is rotated about the direction of the incident light. Sketch a graph to show how the intensity varies with angle as the polaroid is turned.
b) A beam of light can be polarized by reflecting the light off a glass plate. Calculate the angle of incidence for complete polarization of such a beam, given the refractive index of the glass is 1.5.

21.9 In a Michelson-type experiment using a 32-sided mirror M, a beam of light was reflected off one side of M then directed at a distant reflector R. The beam was then returned to M to reflect off the opposite side and then pass through a narrow gap to be observed. With M stationary, the gap was positioned so that light could be observed. Then M was rotated at increasing speed until once more light could be seen through the gap.
a) Explain why the light beam passed through the gap when the speed of rotation reached a certain value.
b) Further increasing the speed enabled the light to be seen through the gap at rotation frequencies of 500 and 750 Hz. What was the lowest speed at which the light could be seen?
c) Given the distance from R to M was 18.75 km, calculate a value for the speed of light.

21.10 A parallel beam of light is directed at the edge of a rotating cog wheel in a direction along the wheel axis. The beam travels to a distant mirror where it is reflected back along its own path to the edge of the cog wheel. The distance travelled by the beam from leaving the cog wheel to returning is 10.8 km. The minimum speed of rotation of the cog wheel at which the returning beam passes through the gaps at the edge of the cog wheel is 42.7 Hz. The cog wheel has 650 teeth. Calculate the speed of light from this information.

D

Multiple choice questions

DM.1 Which of the following is (are) simple harmonic oscillations?
1 a mass moving up and down on the end of a spring always in tension
2 a small ball-bearing rolling up and down on a saucer
3 a simple pendulum with an amplitude of 45°
A 1 only **B** 2 only **C** 1 and 3 only
D 2 and 3 only **E** 1, 2 and 3

DM.2 A body executes simple harmonic motion of amplitude 100 mm along the line PQ centred on O. When passing through O the body has 50 J of kinetic energy. When it is at a point 40 mm from O the values of its kinetic energy E_k and potential energy E_p are

	A	B	C	D	E
E_k/J	42	30	20	18	10
E_p/J	8	20	30	32	40

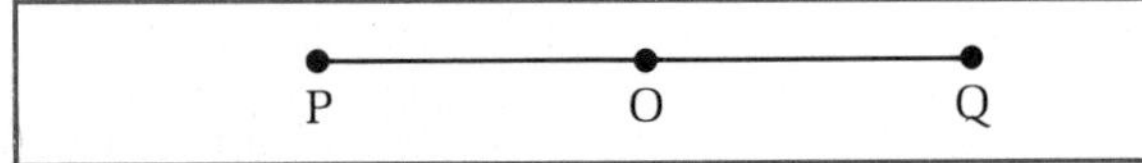

Figure DM.1

(N. Ireland)

DM.3 The following are quantities associated with a body performing simple harmonic motion.
1 The velocity of the body
2 The accelerating force on the body
3 The acceleration of the body
Which of these quantities are exactly in phase with each other?
A None of these **B** 1 and 2 only **C** 1 and 3 only
D 2 and 3 only **E** 1, 2 and 3

(N. Ireland)

DM.4 When mechanical resonance occurs in a lightly damped system the
1 amplitude of vibration is at a maximum
2 natural and forcing frequencies are equal
3 rate of energy transfer to the vibrating system is at a minimum
A 1, 2, 3 correct **B** 1, 2 only **C** 2, 3 only
D 1 only **E** 3 only

(London)

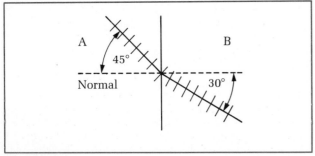

DM.5 Waves travel from medium A to medium B. In A, their direction is at 45° to the normal and in B it is at 30° to the normal.

Figure DM.2

If the velocity of the waves in medium A is 0.283 m s^{-1}, the velocity in medium B is
A 0.200 m s^{-1} **B** 0.231 m s^{-1} **C** 0.347 m s^{-1}
D 0.400 m s^{-1} **E** 0.425 m s^{-1}

(Scottish H.)

DM.6 When light passes from air into glass, which of the following quantities change?
1 frequency **2** wavelength **3** speed
A 1, 2, 3 correct **B** 1, 2 only **C** 2, 3 only
D 1 only **E** 3 only

DM.7 The diagram below shows a row of trolleys and springs used as a model to make predictions about the speeds of pulses in various solids. A student investigating the model discovers that compression pulses along the row travel more slowly if the mass of each trolley is increased or if weaker springs are used. From these investigations, the student makes three deductions about the atoms in two solids, X and Y. Which of these deductions *cannot* be supported by these investigations?

Figure DM.3

1 Pulses travel more slowly in solid X than in solid Y if the atomic mass of X is greater than Y.
2 Pulses travel more slowly in solid X than in solid Y if the atomic bonds in X are weaker than in Y.
3 If pulses in solid X travel at the same speed as in Y, and the atomic mass of X is greater than the atomic mass of Y, then the atomic bonds in X must be weaker than the bonds in Y.
A 1, 2, 3 correct **B** 1, 2 only **C** 2, 3 only
D 1 only **E** 3 only

DM.8 A wire 1.20 m long is fixed at each end under tension. A transverse wave of speed 300 m s^{-1} is propagated along the wire and forms a standing wave pattern by reflection at the ends. In a certain mode of vibration M it is found that the nodes are 0.40 m apart. Which of the following gives the frequency, in Hz, of the mode M and of *all* the possible *lower* resonant frequencies?

	A	B	C	D	E
Frequency of mode M	375	375	375	750	750
Lower resonant frequencies	125	187.5	125 250	187.5 375	250 500

(N. Ireland)

DM.9 A gramophone record is designed to rotate at $33\frac{1}{3}$ rev/min and carries a recording of a pure tone of 1.00 kHz. It is played on a record player whose turntable rotates incorrectly at 33 rev/min. At the same time, a constant tone of 1.00 kHz is sounded at about the same volume. The frequency of the resulting beats is about
A 1/3 Hz **B** 1 Hz **C** 10 Hz **D** 20 Hz **E** 67 Hz

(London)

DM.10 A narrow tube of adjustable length is open at both ends and is situated in air at a temperature such that the speed of sound is 340 m s^{-1}. A sound source of frequency 680 Hz is placed near the tube. Resonance will occur for tube lengths of about

1 125 mm **2** 375 mm **3** 500 mm

A 1, 2, 3 correct **B** 1, 2 only **C** 2, 3 only
D 1 only **E** 3 only

DM.11 A taut wire is set into resonance with a node at either end and a single node at the centre of the wire. Which one of the following statements is *not* correct?

A The wavelength of the waves on the wire is equal to the length of the wire.
B All points to one side of the centre vibrate in phase with one another.
C The amplitude of the vibrations is greatest midway between the centre and either end of the wire.
D Any two points either side of the centre have a phase difference of 90°.
E Two points equidistant from the centre on either side of the centre have the same amplitude of vibration.

DM.12 S_1 and S_2 are two sources of waves of equal amplitude and wavelength λ. The instantaneous positions of two wave crests from each source are shown. Which of the following is true?

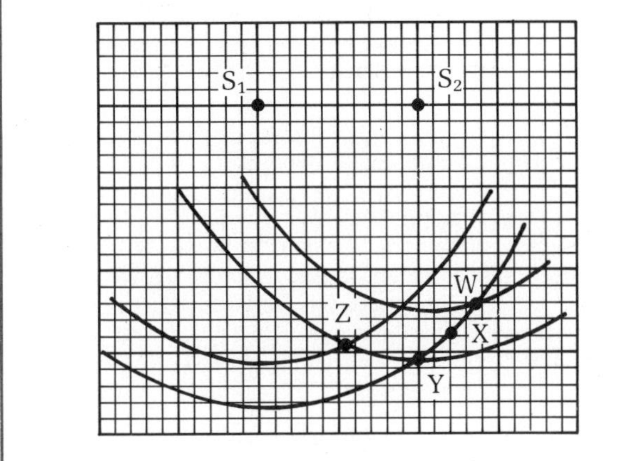

Figure DM.4

A X is a point of destructive interference.
B W is a point of destructive interference.
C $S_1Z - S_2Z = (2n-1)\lambda/2$ where n is an integer.
D $S_1Y - S_2Y = (2n-1)\lambda/2$ where n is an integer.
E For maximum interference $S_1S_2 = \lambda$.

(London)

DM.13 White light in a parallel beam is directed at two parallel slits spaced 0.30 mm apart, the direction of the beam at right angles to the slits. Interference fringes are observed on a screen placed 0.800 m from the slits, at right angles to the beam direction. The distance across three fringe spacings is measured at 4.0 mm. Which of the following statements is (are) correct?

1 The average value for the light wavelength is 500 nm.
2 A red filter placed in the path of the beam in front of the slits gives red fringes with a greater spacing than 1.33 mm.
3 If the slits are replaced by double slits at a spacing of 0.6 mm, the new fringe spacing for white light is 0.67 mm.

A 1, 2, 3 correct **B** 1, 2 only **C** 2, 3 only
D 1 only **E** 3 only

DM.14 A parallel beam of light of wavelength 600 nm is directed normally at a diffraction grating with 300 lines per mm. The number of transmitted beams, including the central order, is

A 5 **B** 6 **C** 11 **D** 51 **E** 61

DM.15 If each gap of the diffraction grating of question **DM.14** has a width equal to 0.2 × the line spacing of the grating, single slit diffraction will reduce the number of transmitted orders. The number of transmitted beams, including the central order, should be

A 4 **B** 5 **C** 6 **D** 9 **E** 41

DM.16 In an experiment to determine the focal length of a converging lens, the distance u from the object to the optic centre of the lens is varied and the corresponding distance v from the optic centre of the lens to the real image is measured. A straight line graph would be obtained by plotting

1 v against u
2 $1/v$ against $1/u$
3 uv against $(u+v)$

A 1, 2 and 3 correct **B** 1 and 2 only correct
C 2 and 3 only correct **D** 1 only correct
E 3 only correct

(AEB June 84)

DM.17 Which one of the following statements about a compound microscope consisting of two convex lenses is *not* correct?

A The final image is virtual and inverted.
B The image formed by the objective lens is real and inverted.
C The object's distance from the objective must be less than the focal length of the objective lens.
D The image formed by the objective lens must lie between the eyepiece and its focal point.
E The microscope graticule is positioned where the image formed by the objective lens is formed.

DM.18 The diagram below represents two thin lenses L_1 and L_2 placed coaxially 30 cm apart.

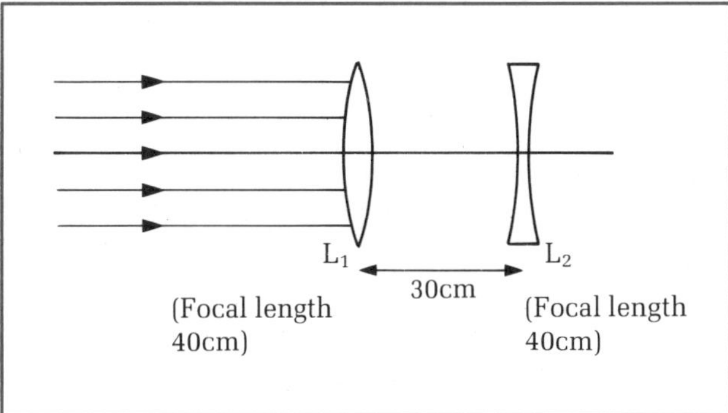

Figure DM.5

A beam of light parallel to the axis is incident on L_1. The final image formed by refraction through both lenses is

A real and between L_1 and L_2
B real and on the right of L_2
C virtual and on the left of L_1
D virtual and on the right of L_2
E at infinity

(London)

DM.19 A refracting telescope consisting of two convex lenses has an eyepiece lens of focal length 20 mm. In normal adjustment when the final image is at infinity, the separation between the lenses is 460 mm. The focal length F of the objective and the magnifying power M for this separation are given at the top of the next page.

	A	B	C	D	E
F/mm	240	240	440	440	460
M	12	23	23	22	23

DM.20 Dark lines in the sun's spectrum are caused by
A diffraction of light by dust particles in the air
B interference between light from different parts of the sun
C polarization of light by dust particles in the air
D strong absorption of particular frequencies by the sun's outer atmosphere
E strong radiation of particular frequencies by the sun's outer atmosphere

(London)

D

Long questions

DL.1 Define *simple harmonic motion*, and explain the meaning of the terms *amplitude, period, damping*. State the relation between the amplitude, the period and the maximum velocity of a particle moving with simple harmonic motion.
A weighted test-tube, which has an external cross-section of area 1.0 cm², floats in water of density 1000 kg m^{-3} with a length 5.0 cm submerged. What is the mass of the test-tube?
Show that if the floating test-tube is pushed vertically down a small distance and then is released, it will move up and down with simple harmonic motion. What is the period of this motion?

(SUJB)

DL.2 Define **a**) *displacement*, **b**) *amplitude*, **c**) *angular frequency*, of a simple harmonic motion and give an expression relating them, explaining all symbols used.

A student is under the impression that ω, the angular frequency of oscillation of a simple pendulum is dependent solely upon the length l of the pendulum and the mass m of its bob. Show, by dimensional analysis, that this cannot be correct. Derive from first principles the correct equation,

$$\omega^2 = g/l,$$

where g is the acceleration of free fall.

A small spherical mass is hung from the end of an elastic string of natural length 40.0 cm and when the pendulum so formed is set swinging with small amplitude, 20 oscillations are completed in 26.0 s. The bob is then replaced by one of the same size but of a different mass and the new time for 20 oscillations is 26.4 s. Account for this change and calculate the ratio of the masses.

(Cambridge)

DL.3 Find, stating clearly any assumptions or conditions, an expression for the period of oscillation of a simple pendulum.
Such a pendulum is of length 1 m; the bob of mass 0.2 kg, is drawn aside through an angle of 5° and released from rest. The subsequent motion is described by

$$x = a \sin(\omega t + \varepsilon)$$

where x is the displacement of the bob (in metres) and t the time (measured in seconds from the instant of release).
Find values for a, ω and ε.
What is the maximum velocity and maximum acceleration experienced by the bob?
What are the maximum and minimum values for the tension in the string, and where in the motion do these occur?
The angular amplitude reduces to 4° in 100 s. Find the mean loss of energy per cycle.

(WJEC)

DL.4
a) A simple pendulum moves with simple harmonic motion provided the angle of deflection is small.
i) Show that the period is given by $T = 2\pi\sqrt{l/g}$ where l is the length of the pendulum.
ii) If such a pendulum is constructed on the Earth to have a period of 0.5 s and is then taken to the Moon, what is its period there?
b) A mass released from point Q moves with simple

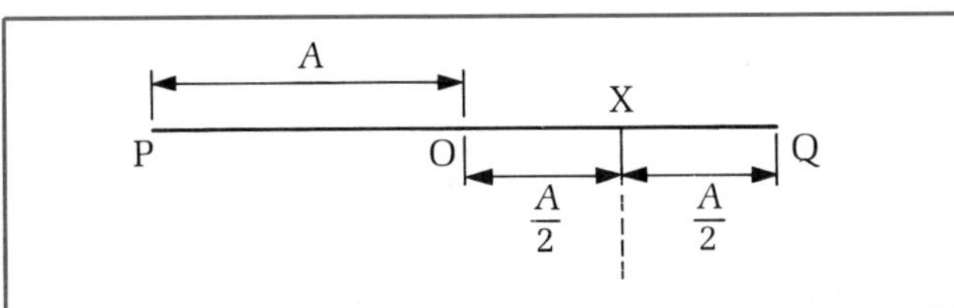

Figure DL.1

harmonic motion of amplitude A along the straight line PQ.
Prove that the time taken by the mass to move from Q to X is twice the time it takes to move from X to O.

c) A girl notices that a boat moored behind a sea wall is bobbing vertically up and down due to the action of the waves.

Figure DL.2

She sees that the lamp at the top of the mast is visible above the wall for a total time of 1.0 s and is hidden behind the wall for a total of 3.0 s. The maximum height of the lamp above the wall is 0.30 m.
Assuming that the motion is simple harmonic, calculate
i) the amplitude of the motion of the boat, and
ii) the speed of the lamp as it disappears behind the wall.

(Scottish SYS)

DL.5 This question is about the behaviour of water waves. Figure DL.3 shows a wave in water deep enough for the wavelength to be much less than the depth. As the wave moves to the right, the water at P acquires in succession the velocities v_1, v_2, v_3, v_4 and v_5 ($v_5 = v_1$) and it can be shown that it moves in a circle with constant speed, where the radius of the circle is equal to the wave amplitude A.

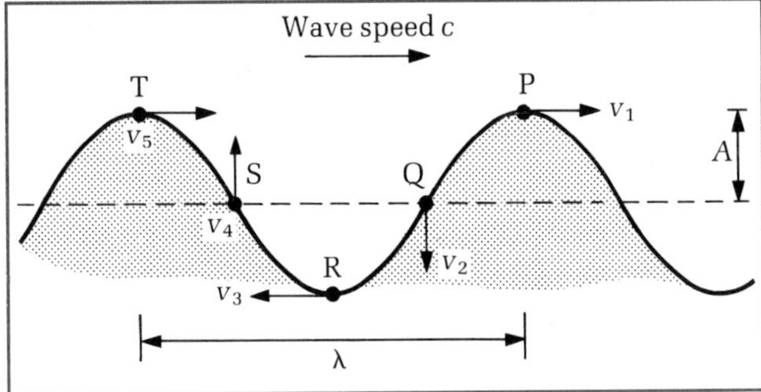

Figure DL.3

a) i) How would you find the time it takes for a water particle to go once round in a circle?
ii) If the water particles are moving in circles, how would you find the magnitudes, and what are the directions, of their accelerations at P, and at Q?

b) The speed c of deep water waves is given by $c^2 = \lambda g/2\pi$. Sketch graphs of (i) speed c, and (ii) frequency, against wavelength λ, for deep water waves in the range $\lambda = 1$ m to $\lambda = 100$ m. Indicate on your sketch the orders of magnitude of speed and frequency for wavelengths 1 m and 100 m. What would you plot in order to obtain straight line graphs relating wavelengths λ to (i) speed, (ii) frequency?

c) i) Suppose a storm at sea generates waves of wavelengths in the range 1 m to 100 m. What wavelengths of waves from the storm will be felt by a ship 100 km from the storm during the 24 hours following the onset of the storm?
ii) A small boat at sea has to ride such waves. What will be the speed of the water around circles in waves of wavelengths 100 m and amplitude 10 m? Describe the motion of such a boat (short compared to the wavelength) which rides such waves and also travels forward in the direction of travel of the waves at a mean speed of about 2 m s^{-1}.
iii) What will be the maximum vertical acceleration of the water in such waves? How will the maximum force on the yacht causing this acceleration compare with the weight of the yacht?

d) If the wavelength λ is larger than the depth d, the water no longer moves in circles and the wave speed for such shallow water waves is given by $c^2 = gd$.
Calculate the wave speed for waves of wavelengths 1 m and 1000 m in a sea of depth 100 m.
Combine these results and those from **b)** above to draw a rough sketch-graph to show how wave speed will vary with wavelength in the range $\lambda = 1$ m to $\lambda = 1000$ m for a sea of depth 100 m so that the waves are 'deep' water waves for short λ and 'shallow' waves for long λ. Label and explain the main features of your sketch.

(O and C Nuffield)

DL.6

a) i) Explain how stationary transverse waves form on a stretched string when it is plucked.
ii) State the factors that determine the frequency of the fundamental vibration of such a string and give the formula for the frequency in terms of these factors.

b) When two notes of equal amplitude but with slightly different frequencies f_1 and f_2 are sounded together, the combined sound rises and falls regularly.
i) Explain this, and draw a diagram of the resulting wave form.
ii) Show that the frequency of these variations of the combined sound is $|f_1 - f_2|$.

c) A car engine has four cylinders, each producing one firing stroke in two revolutions of the engine. The exhaust gases are led to the atmosphere by a pipe of length 3.0 m.
i) Assuming that vibration anti-nodes occur near each end of the pipe, calculate the lowest engine speed (in revolutions per minute) at which resonance of the gas column will occur.
ii) What may happen at higher speeds?
iii) Where, in the gases in the pipe, will the greatest fluctuations of pressure take place at resonance?
(Take the speed of sound in the hot exhaust gases to be 400 m s^{-1}.)

(Oxford)

DL.7

a) Describe the contrasting features of progressive (travelling) and stationary (standing) waves.
With the aid of a sketch, outline an experiment to demonstrate standing waves. Briefly describe the apparatus used and any adjustment that has to be made to obtain a satisfactory result.

b) A woodwind player in an orchestra tunes his instrument to a fundamental frequency of 256.0 Hz in a cold dressing-room where the mean air temperature in the instrument is 285 K. During the performance in a warm concert hall his instrument, playing what is intended to be the same note, produces a beat of frequency 2.5 Hz with a similar instrument which was tuned in the warmer air of the hall to 256.0 Hz. (Assume that the dimensions of the instrument do not change significantly, and that $c \propto \sqrt{T}$, where c is the speed of sound in air and T is the thermodynamic (absolute) temperature.)

i) Explain whether the first instrument produces a note of higher or lower frequency when played in the warmer air of the hall.

ii) Estimate the mean temperature of the air in the first instrument when played in the concert hall.

c) i) Explain what is meant by plane-polarized light.

ii) Describe how you would produce a parallel beam of plane polarized light, starting with a tungsten-filament lamp.

iii) How would you determine by a different technique the plane of polarization of the beam?

iv) With the aid of a diagram, explain how a photographer makes use of a Polaroid filter in front of his camera lens to photograph objects seen through a shop window from a brightly-lit street.

(Oxford)

DL.8

a) Explain what is meant by *resonance*.
Describe briefly three examples of resonance (each from a different field of physics) met in experimental work in a school laboratory.

b) Describe, illustrating the descriptions with suitable graphs:

i) the mutual interaction of two similar vibrating systems which are not quite in resonance with each other;

ii) the effects of progressively increased damping on a resonant vibration.

How would you investigate one of these effects? Describe the apparatus you would use and how the readings would be obtained.

c) A steel tube 3.0 m long, carrying a stream of gas in an industrial plant, is clamped at each end and subjected to transverse vibrations. The speed of transverse waves along the material of the tube is 400 m s^{-1}. Calculate the lowest frequency of standing wave that can form on the tube.
The transverse vibrations of the tube cause pressure fluctuations in the gas column forming standing waves with vibration antinodes at either end. If the speed of sound in the gas is 300 m s^{-1}, calculate the lowest frequency at which the vibrations of the tube and of the gas column would resonate.

(Oxford)

DL.9 Describe the Doppler effect and derive an equation for the apparent change in frequency when

i) the source is moving and the observer is stationary and

ii) the source is stationary and the observer is moving.

When do the equations not apply to electromagnetic waves?
A line of wavelength 590 nm in the sun's spectrum is observed first at one edge then at the opposite edge of the sun's disc on an equatorial diameter of the sun. A difference in wavelength of 0.0074 nm is found. The rotational period of the sun is 27 days. Calculate the sun's diameter.
(Speed of light = 3.00×10^8 m s^{-1}.)

(WJEC)

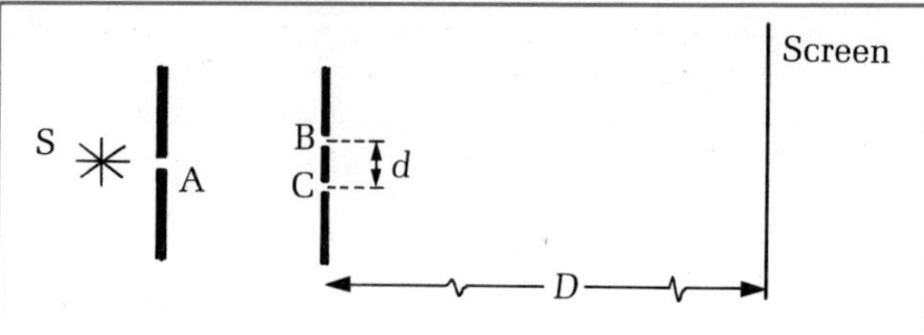

Figure DL.4

DL.10 What do you understand by **a)** *coherence*, **b)** *interference*, between two separate wave trains?
The figure above illustrates apparatus for an optical 'Young's slits' experiment. A source of light S illuminates a narrow slit A which acts as a source for the narrow slits B and C and produces fringes on the screen. With light of wavelength λ, bright fringes are formed on the screen with a separation s. Derive a relation between λ, s, d and D. Suggest suitable values for d and D.
Describe and explain what happens to the fringes if

i) both slits B and C are made narrower whilst keeping d constant,

ii) the light emerging from slit B is reduced in intensity to half that from slit C,

iii) a thin sheet of transparent plastic is inserted between slit B and the single slit A,

iv) slits B and C are both covered with sheets of polaroid and that in front of B is slowly rotated.

(Cambridge)

DL.11 Give the theory of the action of a diffraction grating on a parallel monochromatic beam of light incident normally on the grating.
A pure spectrum is one in which there is no overlapping of light of different wavelengths. Describe how you would set up a diffraction grating to display a pure spectrum on a screen. Explain the purpose of each optical component used.
When a diffraction grating is used on a spectrometer in normal adjustment, the light from a sodium lamp is diffracted through 16° in the first order.

a) What is the highest order of diffraction that can be seen, and what is the angle of diffraction in this order?

b) How would the angle of diffraction in the highest order be changed if the air between the grating and the telescope were replaced by a gas of refractive index (relative to air) of 1.002?

c) If the ruled lines of the diffraction grating are all 1 cm long, and the rulings extend over 2 cm, what is the minimum diameter of the object glass of the telescope if it is to be capable of receiving all the light diffracted in the highest order?

(SUJB)

DL.12 State *two* conditions necessary for the superposition of two waves to give rise to a well-defined interference pattern.
Draw labelled diagrams to illustrate the apparatus required for the production of

a) Newton's rings,

b) Young's two-slit interference fringes.

Explain clearly how the conditions for the production of a well-defined interference pattern can be met in each of these cases. What experimental evidence could confirm that Newton's rings lie approximately in the plane of contact of the two surfaces involved but that Young's fringes are not confined to any one plane?
A thin air-film is trapped between two glass plates inclined at a small angle to each other. Light of

wavelength 5×10^{-7} m is reflected normally from the film and interference fringes 0.6 mm apart are observed. When liquid of refractive index 1.3 replaces the air, the fringe separation changes. Explain this change.

Calculate

c) the angle between the plates,

d) the final fringe separation.

(*Cambridge*)

DL.13 State the relation between the wavelength, the spacing of the slits and the angles at which maxima are observed for light incident normally on a diffraction grating.

Describe, in as much detail as you can, the spectra you would observe when the source of light illuminating the diffraction grating was

a) a hot tungsten filament,

b) a discharge lamp containing mercury vapour,

c) a hot tungsten filament behind a blue liquid.

Light from a sodium lamp contains yellow light of two wavelengths. Explain the observation that, when viewed through a spectrometer, white light shone through a sodium flame shows a continuous spectrum with two dark lines at the positions where the sodium lamp gave two bright lines.

The wavelengths of the two sodium lines are 589.0 nm and 589.6 nm. Sodium light is incident normally on a grating with 500 lines per mm, placed in front of a camera lens which focuses the spectrum onto a film. Calculate the angular separation, and the separation of the images on the film, of the sodium lines in the second order spectrum with a lens of focal length 135 mm.

An experimenter wishes to use this camera and grating, without a collimator, to photograph the solar spectrum. The solar spectrum contains dark lines in the positions of the sodium lines. Neglecting diffraction effects and aberrations due to the lens, explain whether he will be able to resolve the dark lines on his photograph.

(The Sun subtends an angle of 10^{-2} radian at the Earth.)

(*Oxford and Cambridge*)

DL.14

a) Describe the phenomena of *diffraction* and *interference*. Discuss carefully how they contribute to the production of fringes by a diffraction grating illuminated by monochromatic light.

b) Parallel light of wavelength λ is incident normally on a single slit AB of width a. The diffraction pattern is viewed on a very distant screen S. Deduce the angle θ at which the first minimum in the diffraction pattern occurs.

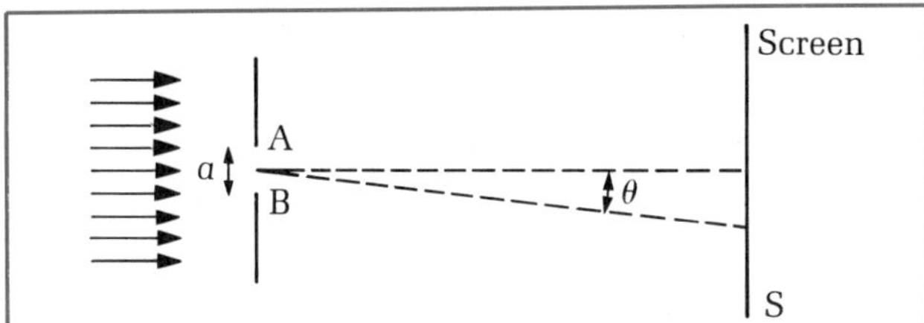

Figure DL.5

Sketch a graph showing the variation of light intensity seen on the screen as θ is varied.

c) The slit AB is then replaced by two very narrow parallel slits a distance a apart. Compare the variation of intensity seen on the screen with that produced by the single slit in **b)**.

d) A comb, of which the centres of the teeth are 3×10^{-3} m apart, is illuminated from behind with sodium light of wavelength 589 nm. It is viewed normally using a telescope having an objective lens of diameter 4×10^{-2} m. Estimate the maximum distance from the comb to the telescope at which it should be possible to distinguish separately the teeth of the comb.

(*Oxford and Cambridge*)

DL.15

a) Light of wavelength 5.9×10^{-7} m is incident normally on a slit of width 0.10 mm.

i) Sketch a graph to show how the intensity of the diffraction pattern produced by the slit varies with direction.

ii) Calculate the angle at which the first minima occur in the diffraction pattern.

iii)

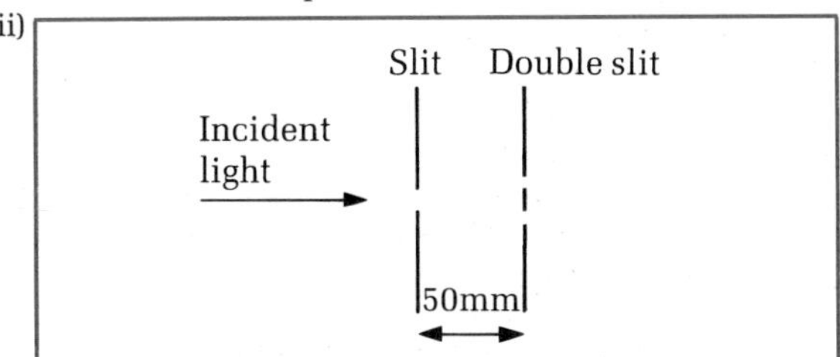

Figure DL.6

In an attempt to produce an interference pattern a double slit, with slit separation 1.0 mm, was placed 50 mm from the original single slit as shown in the diagram. Use your result in (ii) to suggest one reason why no interference fringes would be seen.

b) An interference pattern can be produced using a thin air film between two plane parallel glass plates. Draw a labelled diagram of the apparatus which could be used to produce and view the pattern. Describe the pattern and explain how it is formed.

(*JMB*)

DL.16

a) Draw ray diagrams to show what is meant by the *principal focus* and *focal length* of (i) a converging (convex) lens, and (ii) a diverging (convex) mirror.

b) Explain the terms *magnification* and *angular magnification* as applied to optical instruments.

c) The diagram shows parts of an astronomical telescope. F_0 is one of the principal foci of the objective lens. Parallel rays from the top of a distant object are incident on the objective, and a final image is formed a distance D from the eyepiece lens.

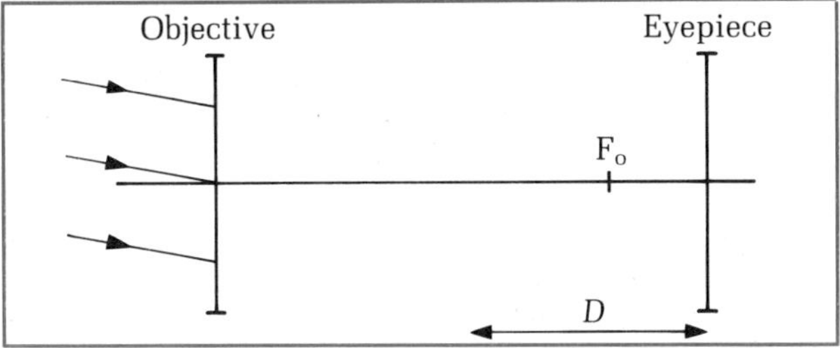

Figure DL.7

i) Copy the diagram and complete it by showing the paths of the rays through the two lenses. Show clearly the real image of the top of the object formed by the objective lens. Show also the final image produced by the telescope.

ii) If D is 250 mm and the focal lengths of the objective and eyepiece are 400 mm and 50 mm respectively, what is the required distance between the two lenses?

iii) The telescope is used to observe the Moon which subtends an angle of 9.0×10^{-3} rad at the objective. Calculate the angle subtended by the final image at the eyepiece, and determine the angular magnification of the telescope assuming that the telescope is adjusted as in (ii).

iv) On another occasion the telescope is adjusted to produce a real image of the Sun on a screen placed 500 mm beyond the eyepiece. Given that the Sun and Moon subtend equal angles at the Earth, calculate the diameter of the image of the Sun on the screen.

(AEB June 85)

DL.17

a) i) Draw a labelled diagram of a camera.

ii) When a sharp image of an object is formed on the film in a camera, what is the relationship between the distances of the object and the film from the camera lens and the focal length of the lens? Hence show that the linear magnification m due to the camera lens is given by $m = 1/(R-1)$, where R is the ratio of the object distance to the focal length of the lens.

iii) If the focal length of a camera lens is 50 mm and the *height* of the film is 25 mm, at what distance must an object 200 mm tall be placed from the camera lens so that the image occupies *half* the height of the film?

iv) If the distance of the object from the lens remains as in (iii) above, what is the focal length of the camera lens required to obtain a sharp image of the same object *equal* to the height of the film?

v) Explain with the aid of a labelled diagram the principle of a telephoto lens.

b) i) What is meant by the *f-number* of a camera lens? For a given light exposure to the film, how are *f*-number and exposure time related? Explain what advantage and disadvantage there might be in using a small aperture with a long exposure time.

ii) A simple camera lens may show a defect known as *chromatic aberration*. Describe this defect.

(N. Ireland)

DL.18

a) i) Define the term *magnifying power* for a compound microscope.

ii) Two lenses are arranged to form a compound microscope in order to examine a small object. Draw a ray diagram to show the passage through the microscope of two continuous rays from a non-axial point on the object, the final virtual image being formed at the near point. Distinguish any constructional rays you draw.

iii) Prove that the magnifying power of the microscope in this adjustment is the product of the linear magnification of the objective lens and the linear magnification of the eyepiece.

b) A grating which has 240 lines per mm is observed by a compound microscope. The distance between two adjacent lines in the final virtual image, formed at the near point (250 mm from the eyepiece), is 0.5 mm.

i) What is the magnifying power of the microscope?

ii) If the linear magnification produced by the eyepiece is 5.0, find its focal length.

iii) Find also the focal length of the objective lens, if the distance between objective lens and eyepiece is 150 mm.

iv) How far, and in what direction, must the eyepiece of this microscope be moved to form the final image at infinity?

v) What advantage and what disadvantage is there in having the final image at infinity?

(N. Ireland)

DL.19 Describe a terrestrial method for measuring the speed of light in air.
If it were suggested that there might be a time delay between light falling on the distant mirror and being re-emitted (reflected), how would you demonstrate that this is not the case?
Show how the refraction of light at a plane surface is explained by the wave theory of light, and derive a relation between refractive index and speed of light.

(SUJB)

DL.20 Most people do not realize that the space around us is filled with electromagnetic vibrations of a very wide range of frequencies, just a little of which we notice as 'heat' and 'light'.
Write a short essay about electromagnetic radiation, including

- evidence to show the common properties of this wide range of frequencies
- why the radiation is called electromagnetic
- some important applications of electromagnetic radiation of different frequencies.

(O and C Nuffield)

SECTION E

Electricity

Current and charge · Meters and bridges · Alternating current circuits · Digital electronics · Electronics in control

Figure E1 *Robots at work. The photograph shows a robot welding line making Austin Metros. Robots such as these are programmed to perform limited functions repeatedly for long periods. More advanced robots operate in hazardous situations, such as where explosive or radioactive materials are present.*

Our way of life has improved considerably over the past fifty years or so as a result of the development and use of electricity. Most activities depend in some way on electricity so this important branch of physics is perhaps best appreciated in terms of its applications. In this section, key concepts such as potential difference are first established, then they are applied to a wide range of situations from microelectronics to the National Grid. New uses for electricity are continually being discovered and new devices developed for such applications and uses. New devices make existing devices obsolete, and the user who understands key concepts can readily take advantage of the benefits of new devices. This is particularly important in the field of microelectronics where changes are taking place faster than ever before. Learning about the characteristics of a particular device is perhaps important if you are to use that device; but of much greater importance is the need to understand its general functions (and those of other devices) so that the potential of new devices can be fully exploited.

22

Current and charge

22.1 Electrical conduction

Understanding electricity

Without electricity, modern living would be impossible because we depend so much on it. Even apart from using electricity in our homes, almost all modern industrial processes need electricity. Our knowledge of electricity was gained by scientists such as Michael Faraday over 150 years ago. They could scarcely have imagined where their investigations would lead. Can you think of any part of your life which does not depend on electricity?

To understand electricity, we need to think in terms of atoms. As explained in Chapter 6, each atom has a positively charged nucleus. The nucleus is surrounded by electrons, each with a tiny negative charge. An uncharged atom has an exact balance between the charge of its nucleus and the charge of all its electrons. An atom becomes charged by changing its number of electrons. For example, an uncharged sodium atom has 11 electrons, each carrying a charge of -1.6×10^{-19} C. So its nucleus carries a positive charge of $11 \times 1.6 \times 10^{-19}$ C. If one electron is removed from the sodium atom, then its overall charge becomes $+1.6 \times 10^{-19}$ C. Charged atoms are called **ions**. Ions are created when electrons are removed or added to uncharged atoms. If electrons are added, the ions are negatively charged. If electrons are removed, the ions are positively charged.

An electric current in a material is the passage of charge through the material. In some materials such as a metal, the charge is carried by electrons. In other materials, such as salt solution, the charge is carried by ions. Insulating materials do not contain charge carriers at all; every electron in an insulator is fixed firmly to an atom so the electrons are not free to carry charge through the material. Figure 22.1 shows a simple test for conduction of electricity. The meter indicates a current whenever any conducting material is connected into the circuit. The battery forces the charge carriers in the conducting material to move round the circuit in one direction only. So a current is produced.

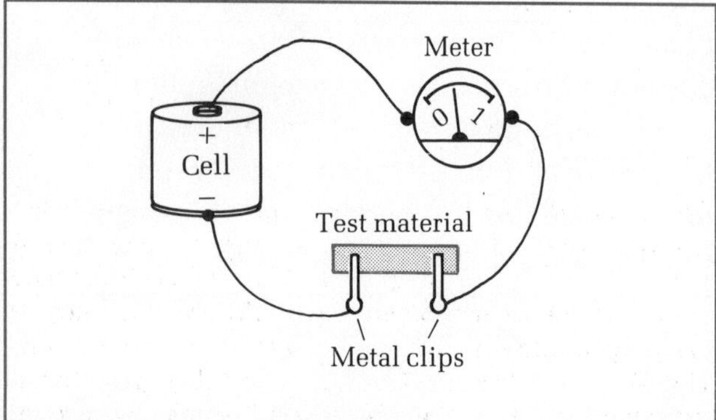

Figure 22.1 *Testing for conduction*

Free electrons are the charge carriers in a metal. At the absolute zero of temperature, these electrons are attached to individual atoms as outer electrons. Above absolute zero, the atoms vibrate and the outer electrons become detached from them. These electrons are free to move throughout the metal, so they are called **free** electrons.

Suppose in the circuit of Figure 22.1 a metal is placed into the circuit as the test material. The free electrons in the metal are attracted to the end connected to the positive terminal of the battery. These electrons leave the metal and pass round the circuit to enter the battery via its positive terminal. The electrons leaving the metal are replaced by other free electrons from the negative terminal of the battery. So a steady flow of electrons is produced round the circuit as the battery forces electrons out from its negative terminal to re-enter via its positive terminal. A break in the circuit at any point would stop the movement of electrons round the circuit.

Electrons always pass round any circuit from the negative terminal of the battery to its positive terminal. The scientists of 150 years ago did not know about electrons. They established the convention that the direction of electric current is from positive to negative because they imagined the charge carriers as positive charges. So current directions on circuit diagrams follow the convention 'from positive to negative round the circuit', even though free electrons move from negative to positive. Once established, conventions tend to stick. Compare driving on the left in the UK to driving anywhere else in Europe!

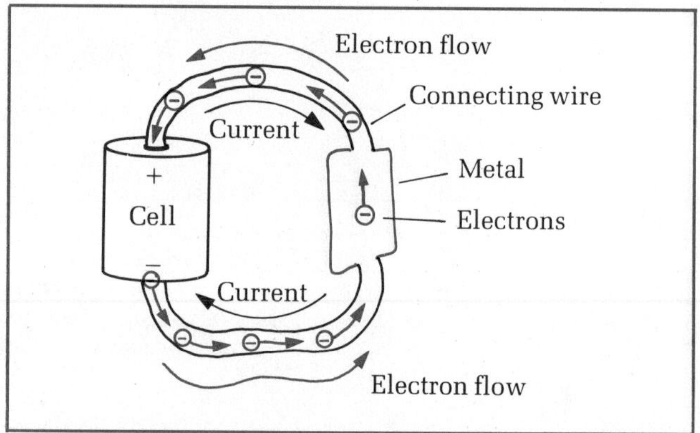

Figure 22.2 *Convention for current*

Units

The ampere **(A)** is the unit of current. It is defined in terms of the magnetic force between two current-carrying conductors (see p. 188). All other electrical units are based on the ampere. For small currents, we often use the milliampere (mA) or the microampere (μA). 1 mA = 10^{-3} A and 1 μA = 10^{-6} A.

The coulomb **(C)** is the unit of charge, defined as the charge passing a point in a wire each second when the current is exactly 1 A.

Each electron carries a charge of -1.6×10^{-19} C. So 6.2×10^{18} electrons are required for 1 coulomb of charge. When a wire conducts a current of 1 A, 6.2×10^{18} electrons pass along the wire each second.

For a steady current I the charge passed in time t is given by

$$Q = It$$

where Q = charge,
I = current,
t = time.

For example, if a wire conducts a current of 5 A for 60 s, then the total charge passed is 300 C. If a charge of 50 C passes along a wire in 20 s, then the average current is 50/20 = 2.5 A.

For changing currents the rate of flow of charge varies. In a short time Δt, the charge passed ΔQ is given by

$$\Delta Q = I\Delta t$$

where I is the current.

Hence $I = \dfrac{\Delta Q}{\Delta t}$.

By letting Δt become very small, we can write $\dfrac{\Delta Q}{\Delta t}$ as $\dfrac{dQ}{dt}$ where $\dfrac{d}{dt}$ means 'change per unit time, or second'. So the current I is equal to the charge passed per second $\dfrac{dQ}{dt}$.

$$I = \frac{dQ}{dt}$$

Drift velocity

When you switch on an electric light, the response is almost instant. Yet the free electrons pass through the lamp filament at 'snail's pace'. Figure 22.3 shows an arrangement to observe the speed of ions as charge carriers. A strip of filter paper is soaked in ammonia or salt solution and fixed to a microscope slide by clips at either end. A crystal of potassium permanganate is placed on the filter paper. When an electric field is set up from one end to the other using a high voltage unit as in Figure 22.3, a purple stain from the crystal moves towards the positive end of the slide. The moving stain is due to negatively charged permanganate ions drifting slowly to the oppositely charged end of the slide. As the ions are attracted towards the oppositely charged end of the slide, each ion repeatedly collides with other ions; so ions are prevented from moving rapidly across the filter. Each permanganate ion is a carrier of a tiny amount of charge moving at a speed of the order of 1 mm per second towards the oppositely charged end.

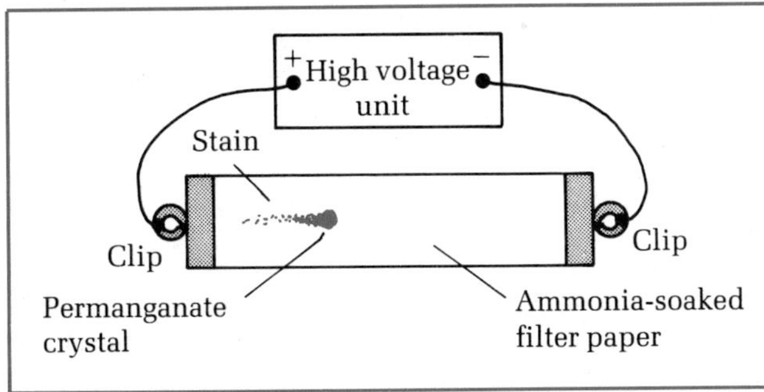

Figure 22.3 *Ions on the move*

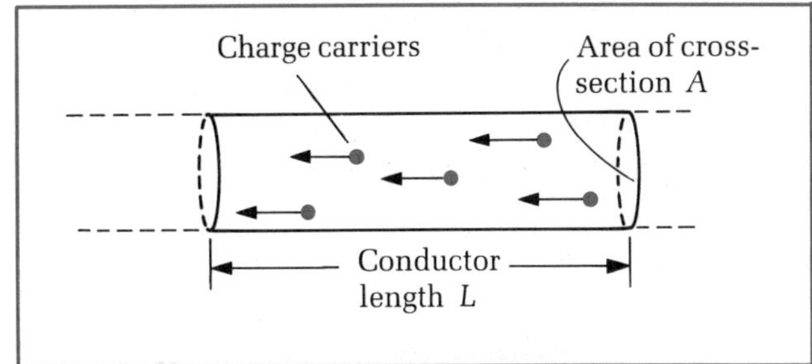

Figure 22.4 *Charge carriers*

Consider charge carriers moving through a conductor, as in Figure 22.4. Suppose the conductor has uniform cross-sectional area A, and that it contains n charge carriers per m³. Each carrier is assumed to move in the same direction towards one end at speed v. Let q be the charge of each carrier.

Suppose each carrier takes time t to move from one end of the conductor to the opposite end. Hence the length of the conductor $L = vt$. So the volume of the conductor is $vt \times A$.

With n carriers per unit volume in the conductor, the total number of carriers is given by $nvtA$. So the total charge of all the carriers in the conductor is given by $qnvtA$.

The current I is equal to the charge passed/time taken. In time t, all the charge carriers in the conductor leave from one end. So the charge passed in time t is given by $qnvtA$. Hence $I = qnvtA/t = qnvA$.

$$I = nvAq$$

where I = current,
n = number of charge carriers per m³,
v = velocity of carriers,
A = cross-sectional area,
q = charge of each carrier.

Worked example Estimate the speed of the free electrons in a copper wire of diameter 1 mm carrying a current of 5 A. The number of free electrons per m³ in copper at room temperature is about 10^{29}. The charge of each electron is 1.6×10^{-19} C.

Solution Area of cross-section $A = \pi(\text{diameter})^2/4$

$= \dfrac{\pi}{4} \times (1 \times 10^{-3})^2$ m².

Current $I = 5$ A.

Charge of each carrier $q = 1.6 \times 10^{-19}$ C.

Number of carriers per m³, $n = 10^{29}$.

From $I = nvAq$,

$$v = \frac{I}{nAq} = \frac{5}{10^{29} \times \frac{\pi}{4} \times (1 \times 10^{-3})^2 \times 1.6 \times 10^{-19}} \text{ m s}^{-1}.$$

Hence $v = 4.0 \times 10^{-4}$ m s⁻¹.

In a metal the worked example shows that free electrons drift along wires at speeds of the order of 1 mm s⁻¹. If we could actually see the free electrons moving about inside the metal, we would see that they repeatedly collide with one another and with the metal ions. The picture of free electrons in a metal is like that of molecules in a gas. When no current is passing through the metal, there is no overall movement of electrons though the metal. The electrons move about inside the metal at random. Now consider what happens when the metal is connected into a circuit so it passes a current. The free electrons are attracted towards the positive end of the metal, but the repeated collisions with other electrons slow their progress. The motion towards the positive end is in addition to the random motion. Each free electron accelerates briefly towards the positive end but is repeatedly stopped by impacts with other free electrons. Its motion is a gradual drift to the positive end superimposed on its haphazard motion.

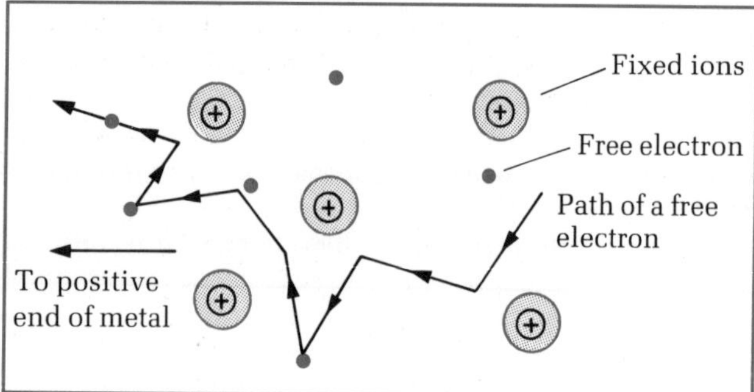

Figure 22.5 *Drift velocity*

22.2 Potential difference

Energy

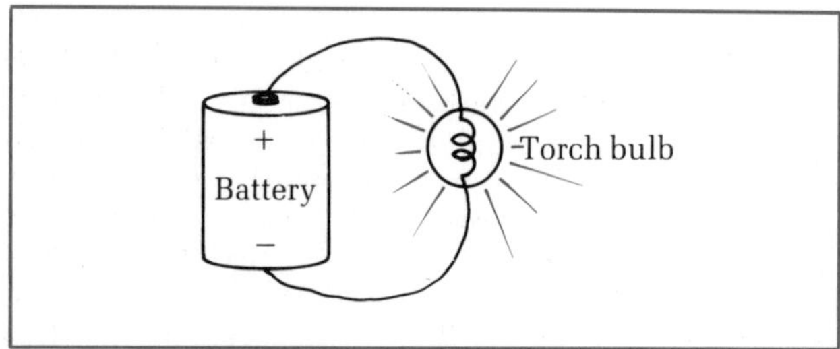

Figure 22.6 *Energy*

When a torch bulb is connected to a battery, electrons deliver energy from the battery to the torch bulb. Each electron which passes through the bulb takes a fixed amount of energy from the battery and delivers it to the bulb. After delivering energy to the bulb, each electron re-enters the battery via the positive terminal to be resupplied with energy to deliver once more to the bulb. The situation is like the so-called 'merry-go-round' trains which transport coal from pit to power station day and night. Figure 22.7 shows the idea. Each trainload of fuel is a fixed quantity of energy taken from the pit to the power station. The train then returns to the pit empty to be re-supplied. Suppose each trainload carries 1000 tonnes of fuel, and there are 10 trains per day. The quantity of fuel delivered each day is therefore 10 000 tonnes, equal to the quantity of fuel per trainload × the number of trainloads per day.

We can apply the same idea to electrical energy. The electrons leaving the battery are like wagons of a train leaving a pit. Each electron carries a fixed amount of energy on leaving the battery, to be delivered to the components which make up the circuit.

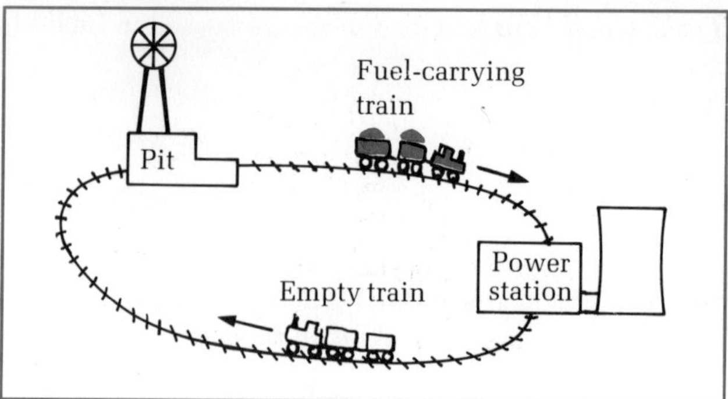

Figure 22.7 *A fuel circuit*

The potential difference (p.d.) between two points in a circuit is the electrical energy converted into other forms of energy when 1 coulomb of charge passes from one point to the other. The unit of p.d. is the volt (V), equal to $1\,\mathrm{J\,C^{-1}}$.

Suppose the p.d. across the terminals of an electric motor in operation is 6 V. Every coulomb of charge passing through the motor delivers 6 J of electrical energy which is converted to other forms of energy by the motor. Two coulombs of charge would deliver 12 J of electrical energy. To deliver 48 J of electrical energy, 8 coulombs of charge would need to pass through the motor. In general, where a charge Q delivers electrical energy W between two points of a circuit, the p.d. between the two points is W/Q.

$$V=\frac{W}{Q}$$

where W= electrical energy delivered,
Q= charge passed,
V= potential difference.

Consider the simple circuit of Figure 22.6. Suppose the p.d. aross the lamp is 2.5 V and the current passing through it is 0.5 A. In 1 second, 0.5 coulombs of charge passes through the lamp. Each coulomb delivers 2.5 J of electrical energy to the lamp. So each second, 1.25 J of electrical energy is delivered to the lamp to produce light and heat energy. In other words, the electrical power supplied to the lamp is 1.25 W ($=1.25\,\mathrm{J\,s^{-1}}$).

More generally, suppose some electrical device in a circuit has a p.d. V across its terminals when it takes a steady current I. In time t, the charge which passes through the device is given by It. Each coulomb of charge delivers energy V, so the electrical energy delivered to the device in time t is given by ItV. Hence the electrical power (i.e. energy per second) supplied to the device is given by $\frac{\text{energy } ItV}{\text{time } t}$ which equals IV.

Electrical power supplied to a device $=IV$

where V= p.d. across its terminals,
I= (steady) current.

We can use this equation to calculate fuse ratings for electrical appliances. For example, suppose we wish to fit a fuse to a 240 V, 3000 W appliance. Since the electrical power taken by the device is 3000 W, then the current for normal operation can be calculated. From the above

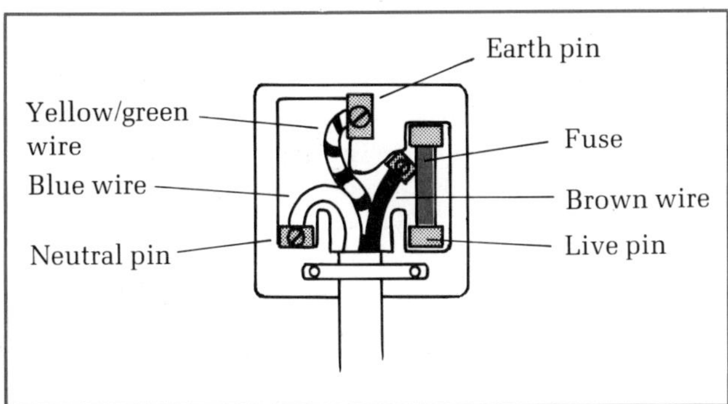

Figure 22.8 *A three-pin plug*

equation, $3000 = I \times 240$, so the current I must equal $3000/240 = 12.5$ A. Fuses are supplied by manufacturers with different ratings, including 13 A. The appliance should be fitted with a 13 A fuse so that the normal current of 12.5 A can pass. If, however, a fault develops in the appliance making the current rise, the fuse will blow (i.e. melt) at 13 A.

The kilowatt hour is a unit of electrical energy used to price electricity. 1 kilowatt hour is the electrical energy supplied to a 1 kilowatt heater in one hour. In other words, it is the amount of electrical energy supplied in 3600 s at a rate of $1000\,\mathrm{J\,s^{-1}}$. Hence 1 kilowatt hour is 3.6 MJ. An electricity bill which states the price of electricity as 5p per unit means that each kilowatt hour used costs the user 5p. Estimate the number of units used in your own home in 24 hours; for example, suppose the following appliances are used over a 24 hour period.

100 W (i.e. 0.1 kW) lamp for 6 hours uses 0.6 kilowatt hours
5 kW electric cooker for 0.5 hours uses 2.5 kilowatt hours
0.5 kW television for 2 hours uses 1.0 kilowatt hours
Hence the total energy used = 4.1 kilowatt hours.

Measuring p.d. without a voltmeter

Voltmeters measure the p.d. between two points in a circuit. A voltmeter which reads 5 V when connected to the terminals of a resistor in a circuit tells us that each coulomb of charge passing through the resistor delivers 5 J of electrical energy. But how is the voltmeter calibrated? We need to devise a way to measure p.d. without using a voltmeter. Then we can calibrate the scale of the voltmeter with known values of p.d.

To measure p.d. without a voltmeter we can compare heating by electricity with heating by mechanical work. Figure 22.9 shows a suitable arrangement in which a metal drum can be heated either electrically or by friction between the cord and the drum when the drum is turned.

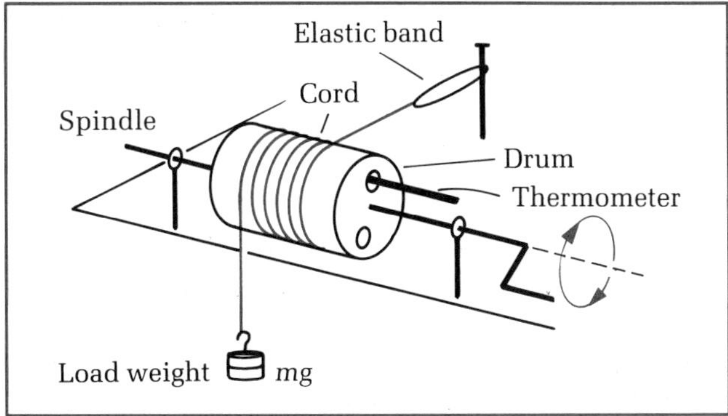

Figure 22.9 *Heating by mechanical work*

Electrical heating is by means of a heater inserted into a hole drilled into the solid metal drum (Figure 22.10). A thermometer inserted into a second hole as shown is used to measure the temperature of the drum. The heater is connected in series with a rheostat, switch, ammeter and power supply. The initial temperature of the block is noted, then the switch is closed and a constant current is supplied for a measured time until the temperature has risen by 10 K. The rheostat is used to keep the current at a fixed value, measured using the ammeter. The charge Q which passes through the heater during the heating time is given by current × time.

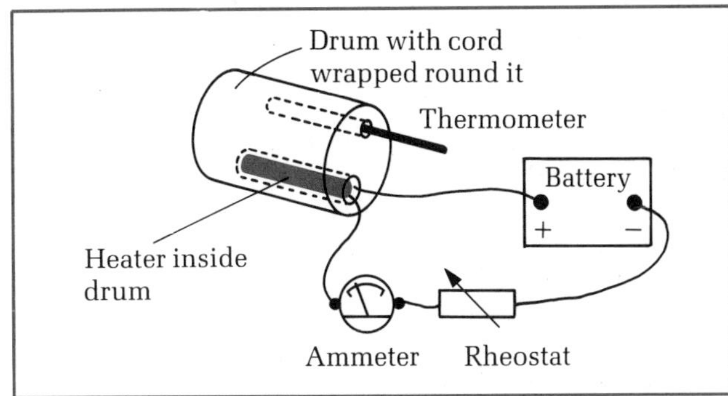

Figure 22.10 *Heating by electricity*

Mechanical work is now done to supply the same amount of energy. The drum is allowed to cool to its initial temperature. The nylon cord wrapped round the drum (Figure 22.9) supports a load weight at one end. The other end of the cord pulls on an elastic band when the drum is stationary. The drum is to be turned at a steady rate so that the rubber band just becomes slack. Then, the load weight is prevented from falling by the frictional force between the cord and the drum. In other words, the frictional force is then equal to the load weight. The drum is turned in this way until its temperature rises by 10 K as before. The number of turns of the drum (to produce this temperature rise) is measured.

To calculate the energy supplied, the circumference C of the drum (equal to $\pi \times$ diameter of drum) with the drum must be measured. Each turn of the drum therefore involves the frictional force moving through a distance C. So the work done per turn is equal to the load weight $Mg \times$ distance C. Hence, for n turns, the total work done is equal to $MgCn$.

Since the temperature rise is the same for both electrical and mechanical heating, the electrical energy supplied by the heater must have been equal to $MgCn$. The charge passed through the heater is given by It, where I is the heater current and t is the time for electrical heating.

Therefore, the p.d., V, across the heater terminals is given by

$$V = \frac{MgCn}{It}$$

since p.d. equals energy supplied/charge passed. Hence the p.d. V can be calculated.

To calibrate a voltmeter, the voltmeter must be connected across the heater terminals for the first part of the experiment. The scale reading of the voltmeter would be noted during heating. After the mechanical heating part of the experiment, the calculated value of p.d. could then be marked on the voltmeter scale at the appropriate position.

Electromotive force

Any circuit needs a supply of power to drive current round the circuit. The device which supplies the electrical power, the source, has one of its terminals positively charged and the other one negatively charged. In a circuit, the source forces electrons out from its negative terminal; if the current is steady, the electrons move round the circuit and re-enter the source via its positive terminal. Inside the source, electrical energy is produced from other forms of energy so that the source terminals stay charged. Some examples of different types of sources are given here.

i) Dynamos which convert mechanical work into electrical energy (see p. 192).

ii) Cells which convert chemical energy into electrical energy. A battery is a combination of two or more cells.

iii) Solar cells which convert light energy into electrical energy. Solar powered calculators each have a panel of solar cells. If you block out the light from its solar cells, the displayed numbers vanish.

The symbol for any source of direct current is shown in Figure 22.11. Such a source might be a direct current dynamo, or a battery, or any other device which can supply a direct current. We can use the symbol too for a mains-powered d.c. supply unit. But remember that such a unit is supplied with electrical energy so it does not actually produce electrical energy from other forms.

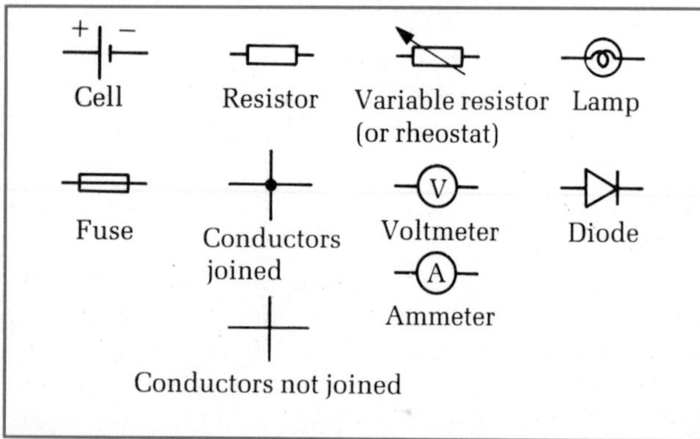

Figure 22.11 *Circuit symbols*

The electromotive force (e.m.f.) of a source is defined as the electrical energy produced per unit charge inside the source. The unit of e.m.f. is the volt since 1 V is equal to $1\,\mathrm{J\,C^{-1}}$.

For example, a battery with an e.m.f. of 12 V produces 12 J of electrical energy for every coulomb supplied by the battery to an external circuit. So if the battery is used to drive five coulombs of charge round a circuit, the amount of electrical energy produced inside the battery is $12 \times 5 = 60$ J. Some of the electrical energy produced may be wasted inside the battery instead of being delivered to the external circuit. This is because the battery may have **internal resistance** (see p. 322). When electrons pass through a battery with internal resistance, each electron is supplied with a fixed amount of electrical energy. However, some of the energy of each electron is used inside the battery to force its way through parts of the battery. So not all the energy converted into electrical energy inside the battery is delivered to the external circuit.

Consider the circuit of Figure 22.12 where two lamps X and Y in **parallel** are connected to a battery in **series** with another lamp Z and an ammeter. The battery, lamp Z and the ammeter are said to be 'in series' with one another because they pass the same current. The charge per second through the ammeter is the same as through Z which is the same as through the battery. So they each carry the same current. Lamps X and Y are said to be 'in parallel' because the p.d. across X is the same as across Y. Each coulomb of charge passing from point A to point B delivers a fixed amount of energy no matter whether the charge passes through X or through Y. The energy situation is a bit like a two-storey house with two staircases, one at the front and one at the back of the house. A person who walks downstairs from the upper floor to the ground floor loses the same amount of potential energy no matter which staircase is used.

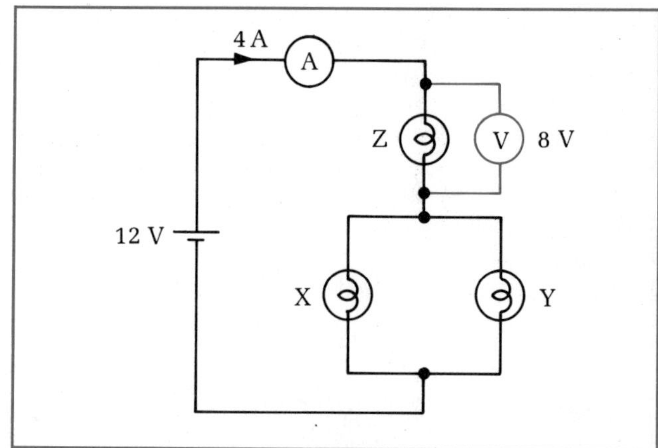

Figure 22.12 *Using electrical energy*

Suppose the battery in Figure 22.12 has an *e.m.f.* of 12 V and its internal resistance is zero. Suppose the ammeter reads 4 A and that the three lamps are identical. A voltmeter connected across lamp Z reads 8 V.

The battery is in series with the ammeter so the current supplied by the battery is 4 A. Therefore, in one second, 4 coulombs of charge leaves the battery. Each coulomb of charge takes 12 J of electrical energy from the battery (since the battery e.m.f. is 12 V). So 48 J of electrical energy leaves the battery each second.

Lamp Z is in series with the ammeter so it too passes 4 coulombs of charge each second. With a p.d. of 8 V across

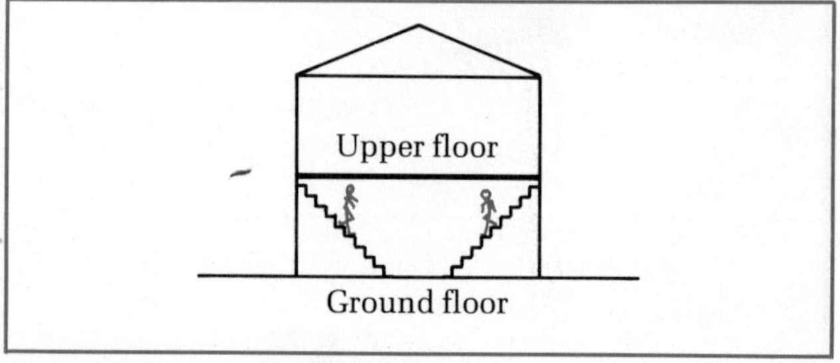

Figure 22.13 *Parallel staircases*

the terminals of lamp Z, each coulomb through lamp Z delivers 8 J of electrical energy which is converted to light and heat energy. So lamp Z uses 32 J (= 8 × 4) of electrical energy each second.

Lamps X and Y are identical so they take equal currents. The current from the battery divides into two equal branches from A to B. So 2 coulombs per second pass through X and 2 coulombs per second pass through Y. Each coulomb of charge is supplied with 12 J of electrical energy by the battery, and uses 8 J to pass through Z. So each coulomb delivers 4 J of electrical energy between A and B. Therefore, each lamp uses 8 J of electrical energy every second.

To sum up, lamp Z uses 32 J s^{-1},
lamp X uses 8 J s^{-1},
lamp Y uses 8 J s^{-1},
and the battery supplies 48 J s^{-1}.

So we can see how the electrical energy supplied by the battery is used. The example shows the use of basic circuit rules too.

For currents

a) Components in series take the same current.
b) At a junction in a circuit, the total current into the junction is equal to the total current leaving the junction.

Figure 22.14 shows an example of a junction between 4 wires A, B, C and D. If the currents in wires A, B and C are known, the current in wire D can be calculated. Suppose wires A and B carry currents of 1 A and 3 A towards the junction, and C carries 5 A from that junction. Then wire D must carry 1 A towards the junction.

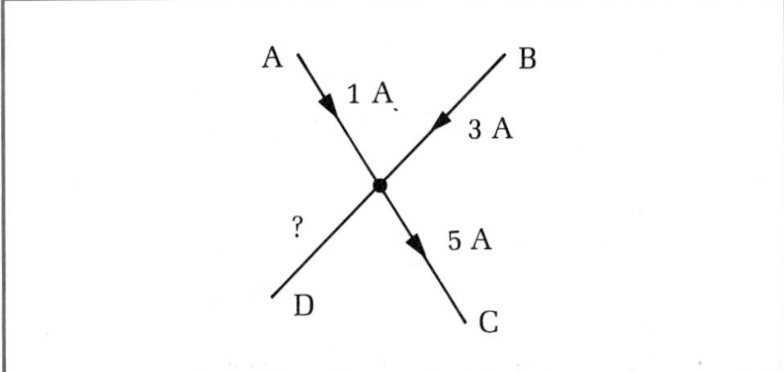

Figure 22.14 *Currents at a junction*

For potential differences

a) Components in parallel have the same p.d. across their terminals.
b) The total p.d. across any two components in series is equal to the sum of the p.d.'s across the individual components.

For example, Figure 22.15 shows two resistors P and Q in series. Each coulomb of charge which passes through P also passes through Q. The total energy delivered per coulomb to P and Q is equal to the energy per coulomb delivered to P + the energy per coulomb delivered to Q. So the p.d. across P and Q is equal to the p.d. across P + the p.d. across Q.

Cells and batteries

We use batteries for electrical power whenever we need a portable source not connected to a mains power point.

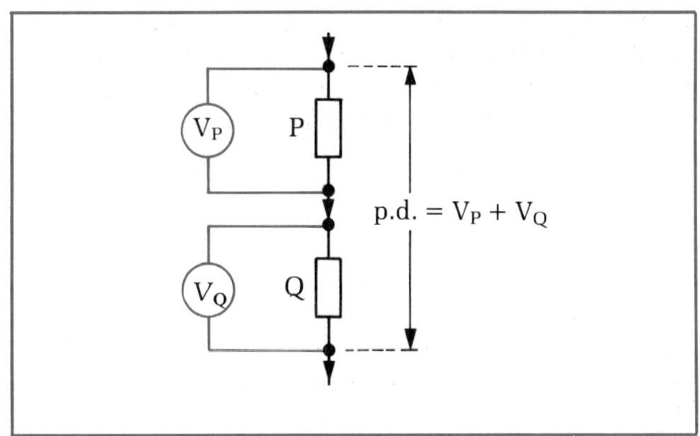

Figure 22.15 *Adding p.d.s*

The batteries used by a calculator are very different to the batteries used in a car. However, all batteries produce electrical energy. One of the first batteries to be invented was made by Alessandro Volta in 1800. He used discs of silver and zinc, stacked alternately as in Figure 22.16, with brine-soaked cardboard on top of each zinc disc. The result is a battery called a voltaic pile where the topmost silver plate is the positive terminal and the lowest zinc plate the negative terminal. Try it yourself using silver coins and aluminium foil in place of zinc.

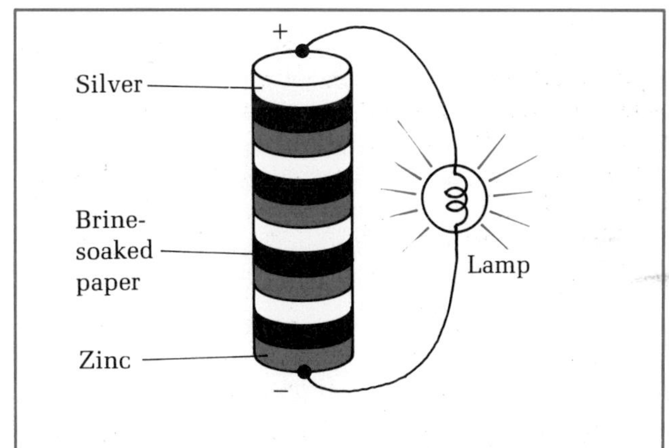

Figure 22.16 *A voltaic pile*

Batteries in continuous use run down once the chemicals inside have been used up. Electrical energy in a chemical battery is produced by different materials in the battery reacting together. Once all the material has been used up, no more electrical energy can be produced.

Primary cells cannot be recharged once they run down. They must be replaced by fresh cells once they can

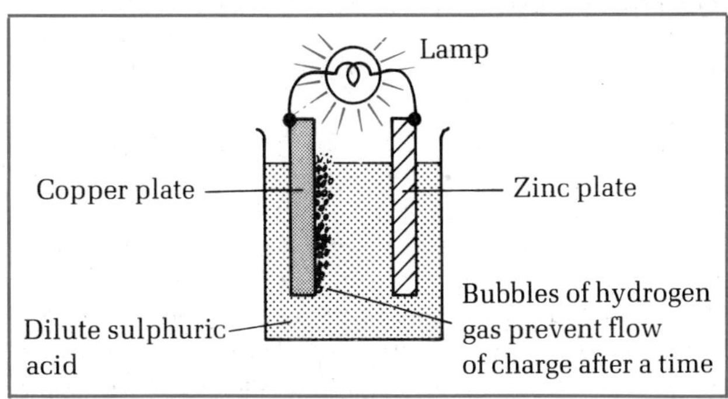

Figure 22.17 *A simple cell*

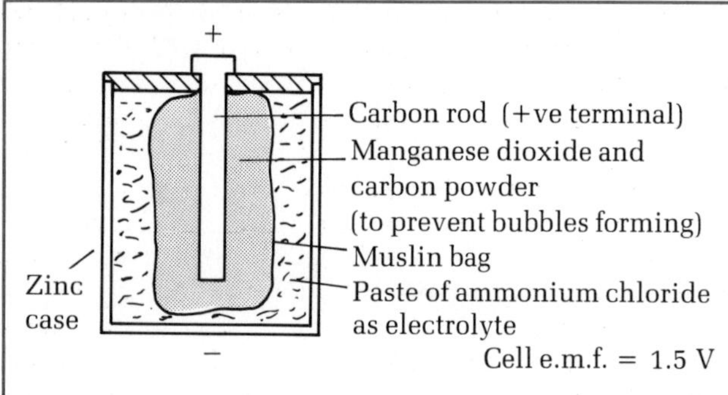

Figure 22.18 *A dry (Leclanche) cell*

no longer supply electrical energy. Torch batteries and non-rechargeable calculator batteries use primary cells.

Secondary cells are rechargeable. After a secondary cell has run down, it can be recharged from a suitable battery charger. The battery charger drives electrons *into* the negative terminal of the cell and out from the positive terminal. The electrons forced into the cell carry energy into the cell to enable the materials to be renewed. Most car batteries consist of six lead-acid cells in each battery. The cells are connected in series. In use to deliver current, the lead plates react with the acid to produce electrical energy from chemical energy. Gradually the reaction makes the lead dissolve in the acid. The acid becomes weaker as more and more lead ions pass into the solution. By recharging the battery, lead ions from the solution are driven back onto the plates. So, after recharging, the battery can be used once again to supply electrical energy. Electric vehicles such as milk floats use lead-acid batteries, recharged each day. Research laboratories in several countries are engaged in work to develop lighter, more powerful batteries for use in electric cars.

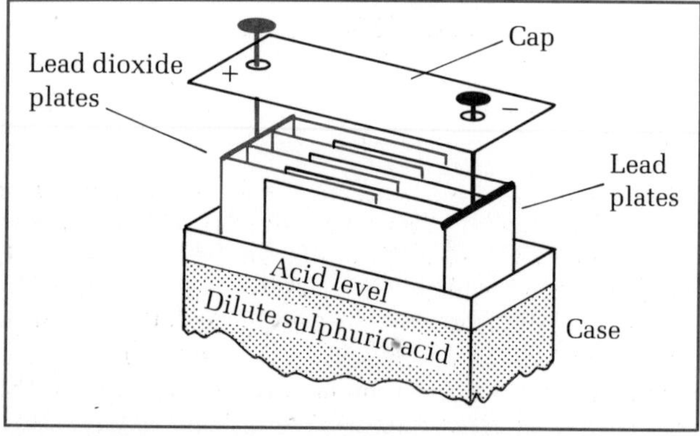

Figure 22.19 *A lead acid cell*

The capacity of a cell or battery is a measure of how long a cell can deliver current before it runs down. Capacity is usually given in ampere hours; for example, a cell with a capacity of 30 ampere hours at 2 A can supply that current for 15 hours before it runs down. The same cell could supply a steady current of 0.5 A for at least 60 hours.

Standard cells are sources of accurately-known e.m.f. The Weston standard cell has an e.m.f. of 1.018 V at 20°C. It becomes run down very easily so it can only be used to supply very small currents (less than 10 μA) for very short periods at a time. It can be used to check the accuracy of a voltmeter, as explained on p. 339.

22.3 Resistance

An electric circuit consisting of a battery connected across a lamp can be likened to a water circuit where water is pumped through pipes. Figure 22.20 shows the idea. The pump drives water through the pipes, causing a pressure difference across the ends of each pipe to drive the water through. In the simple circuit of Figure 22.20, the pipes are in series, because the same amount of water passes through each, per second. Compare the narrow pipe X with the wide pipe Y. To force water through the narrow pipe at the same flow rate as in the wide pipe, the narrow pipe needs a bigger pressure difference. The resistance of the narrow pipe to the flow of water is greater than the resistance of the wide pipe.

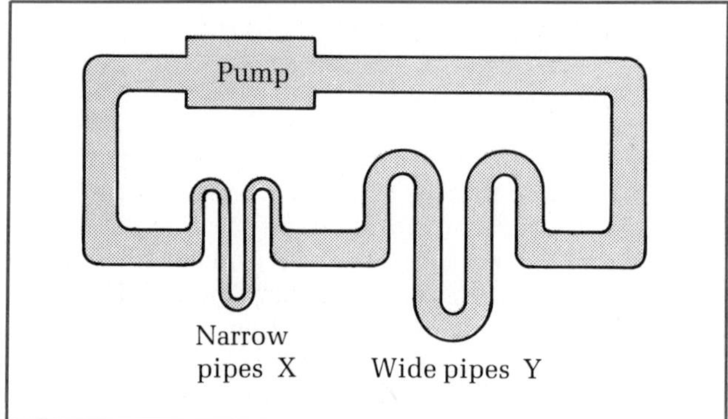

Figure 22.20 *A water circuit*

When a battery is connected across a conductor, the charge carriers in the conductor gradually move towards one end of the conductor. The charge carriers continually collide with one another and with the atoms of the conductor, as they gradually drift towards one end. So the passage of each charge carrier through the conductor is opposed by conductor atoms and other charge carriers. In other words, the passage of charge through the conductor is resisted by the conductor.

The resistance R of a conductor is defined by the equation

$$R = \frac{V}{I}$$

where R = resistance,
V = p.d. across the conductor,
I = current.

The unit of resistance is the **ohm** (Ω), equal to 1 V A^{-1}. **Conductance** is defined as 1/resistance.

Ohmic conductors have resistance which does not depend on current. Georg Ohm in 1826 was the first scientist to develop the idea of resistance. He showed that the current passed by a wire in a circuit was proportional to the p.d. across the wire. So for a wire, p.d./current is a constant provided the temperature stays the same. This statement is known as **Ohm's law**, and is the same as stating that the resistance of a wire does not depend on current, provided the temperature is unchanged. Any conductor with a resistance which changes with current is a non-ohmic conductor. For example, a diode is a component which allows current to pass easily in one direction but not in the opposite direction. So its resistance changes with current direction.

Investigating resistance

The circuit in Figure 22.21 is used to pass current through the device under test. The p.d. across the device is measured using a high resistance voltmeter. The current through the device is varied in steps, using the rheostat. At each step the p.d. is measured from the voltmeter and the current is measured from the ammeter. The measurements can be displayed as a graph of current against p.d.

Typical graphs for different test devices are shown in Figure 22.22.

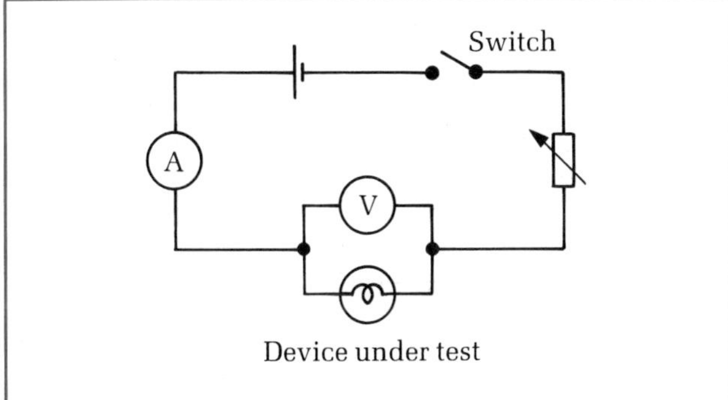

Figure 22.21 *Investigating resistance*

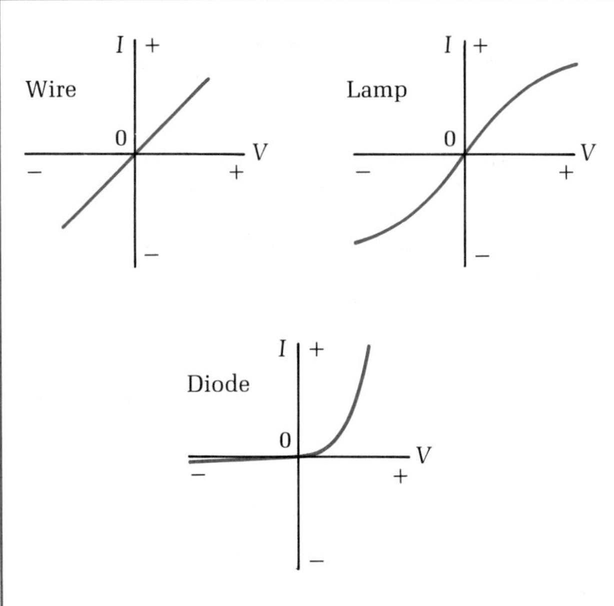

Figure 22.22 I v V *graphs*

i) A length of wire The graph is a straight line through the origin, so the current is proportional to the p.d. Hence the resistance ($=V/I$) is constant, showing that the wire is an ohmic conductor.

ii) A lamp filament The graph of I against V becomes less and less steep as the current increases from zero. So the value of V/I increases as the current increases. Hence the resistance of the filament increases with increasing temperature. This is because the filament becomes hotter as its current increases. Like any metal, the filament resistance increases with increased temperature. So its resistance increases with increased temperature.

iii) A diode The graph shows that the diode conducts in one direction only. This direction, called the forward direction, allows conduction much more easily than in the reverse direction. In the reverse direction, a diode has a very high resistance. The symbol for a diode is shown in Figure 22.11.

Resistivity

The circuit in Figure 22.21 may be used to investigate the factors that determine the resistance of a uniform metal wire.

i) Length of wire Different lengths of the same wire could be tested. For each length, the rheostat is adjusted to give the same current. The p.d. and current are then measured from the meters. Hence the resistance R of each length can be calculated. Figure 22.23 shows typical results plotted on a graph of resistance R against length L of wire. The graph shows that the resistance is proportional to the length. In other words, for the same wire, the resistance per unit length is the same all along the wire.

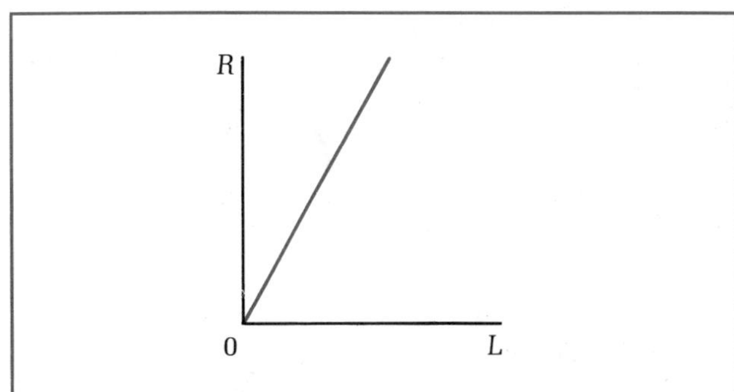

Figure 22.23 *Resistance v length*

ii) Diameter Wires of different thicknesses of the same material are used. The resistance per metre of each wire is determined as explained above. The diameter of each wire is measured with a micrometer. A graph of resistance per metre against 1/area of cross-section ought to give a straight line, as shown by Figure 22.24. For each wire, the area of cross-section A is calculated from the diameter d using the formula $A=\pi d^2/4$.

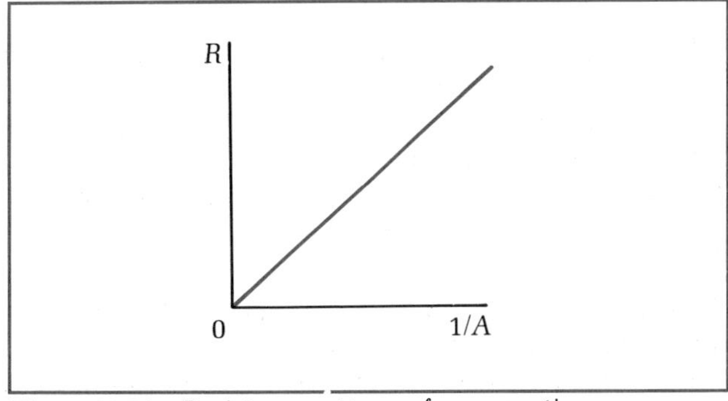

Figure 22.24 *Resistance v 1/area of cross-section*

For different wires of the same material, the resistance R of each wire is
a) proportional to its length L,
b) proportional to $1/A$, where A is the area of cross section of each wire.

So we can write the resistance of each wire in terms of a constant of proportionality called the **resistivity** of the material.

$$R=\frac{\rho L}{A}$$

where R = resistance,
L = length,
A = area of cross-section,
ρ = resistivity of the material.

The unit of resistivity is the ohm metre (Ωm). This can be seen by rearranging the equation above to give $\rho = RA/L$. The unit of ρ is the same as the unit of $R \times$ the unit of A/the unit of L (i.e. $\Omega m^2/m = \Omega m$). Some values of resistivity for different materials at room temperature are listed below.

Copper	$1.7 \times 10^{-8}\,\Omega m$
Constantan	$5.0 \times 10^{-7}\,\Omega m$
Carbon	$3 \times 10^{-5}\,\Omega m$
Silicon	$2300\,\Omega m$
PVC	about $10^{14}\,\Omega m$

Worked example What length of constantan wire of diameter 0.40 mm has a resistance of 10.0 ohms? Assume the resistivity of constantan is $5.0 \times 10^{-7}\,\Omega m$.

Solution

$R = 10.0$ ohms, $\rho = 5.0 \times 10^{-7}\,\Omega m$,

Area of cross-section $A = \pi\,(\text{diameter})^2/4$

$$= \frac{\pi}{4} \times (4 \times 10^{-4})^2\ m^2.$$

From $R = \rho L/A$, $L = \frac{RA}{\rho} = \frac{10.0 \times \pi \times (4 \times 10^{-4})^2}{4 \times 5 \times 10^{-7}}$

$= 2.52$ m.

Resistors in practice

A **resistor** is a device designed to have resistance. Some different types of resistors are given here:

Wire-wound resistors are used where accuracy is essential. For the highest possible precision, manganin wire is used because its resistivity increases less with temperature than any other metal. Wire wound resistors are also used where large currents are likely. The heating effect of a large current would damage other types of resistor. Resistance boxes used for accurate electrical measurements often contain wire-wound resistors. These boxes are designed with switches or brass plugs to select different values of resistance. For example, Figure 22.25 shows plugs removed from a resistance box to give a resistance of 8.0 ohms. Any resistance value up to 10 Ω in 1 Ω steps can be selected by taking out the appropriate plugs. A switch-type box is shown in Figure 22.26. The switches are set for a resistance of 5230 ohms. The box can be set at any resistance value up to 9990 ohms in 10 ohm steps.

Carbon resistors are the most common type of resistor, used where high precision is not essential and where currents are unlikely to cause damage through heating. The most accurate carbon resistors are made from a film of carbon although these are more expensive than pressed carbon resistors. Two factors are important when deciding which type of resistor to choose in any given situation.

a) The accuracy to which the resistance must be known: for example, a 100 ohm resistor accurate to 10% could have a resistance anywhere between 90 and 110 ohms.

b) The power likely to be supplied to the resistor: the power supplied to a resistor produces a heating effect. The power is said to be 'dissipated' since it cannot be returned to the battery. For example, suppose a 100 ohm resistor is likely to have currents up to 0.1 A passing through it. The maximum p.d. across the resistor is therefore given by $V = IR = 100 \times 0.1 = 10$ volts. So the maximum power supplied to the resistor will be 10×0.1 (= p.d. × current) = 1 watt. A resistor chosen with a power rating of 0.5 W would be damaged by the heating effect of the current. Figure 22.27 shows the colour code for resistors.

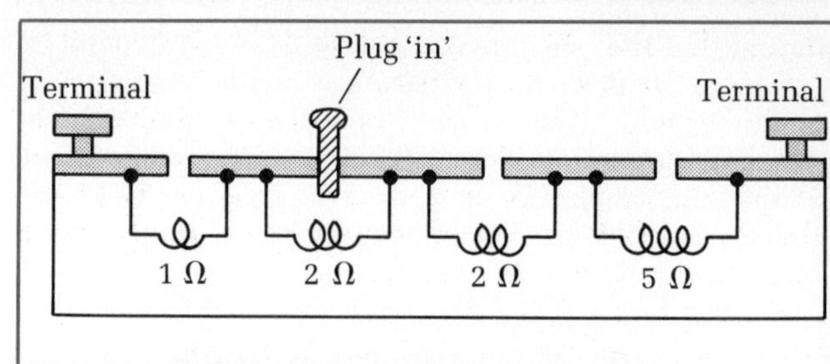

Figure 22.25 *A plug-type resistance box*

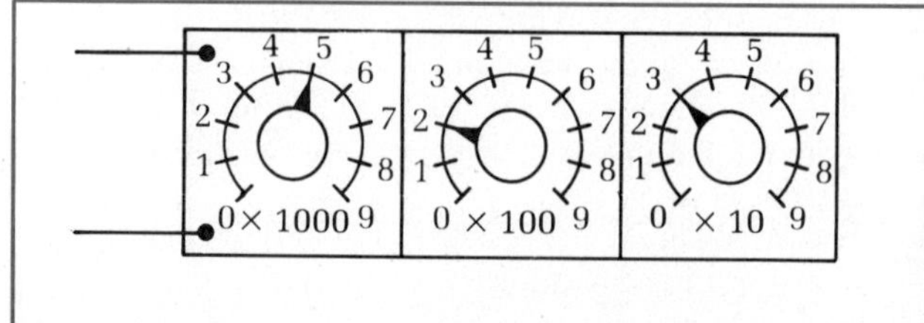

Figure 22.26 *A switch or dial-type resistance box*

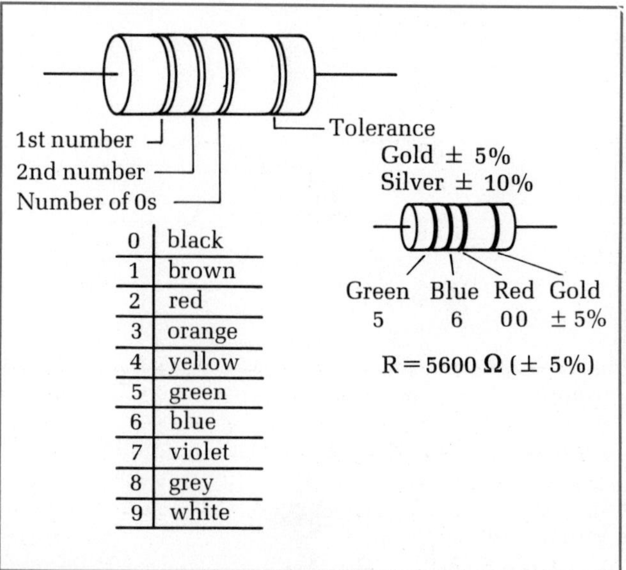

Figure 22.27 *The resistor colour code*

Variable resistors are used to control either current or p.d. There are two main types.

i) Rheostats as in Figure 22.28, are used to control currents. A wire-wound rheostat has one terminal at one end of the wire. The other terminal is connected to a sliding contact on the wire. The resistance between the two terminals varies when the sliding contact is moved along the wire.

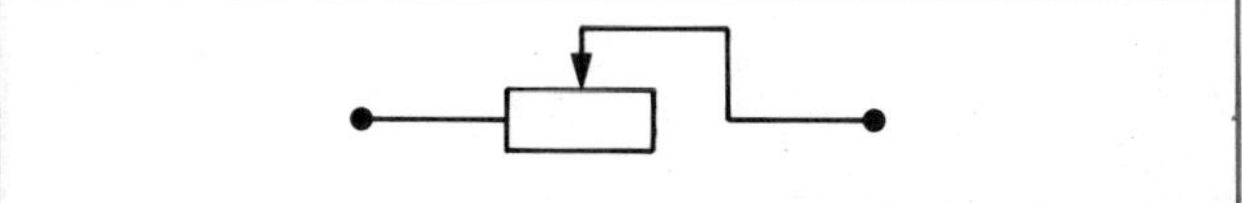

Figure 22.28 *A rheostat*

ii) Potentiometers as in Figure 22.29, are used to supply variable p.d. A fixed p.d. is connected across the ends of a conducting track. A sliding contact moved along the track gives a variable p.d. between the contact and one end of the track. See p. 337 for details of the slide-wire potentiometer.

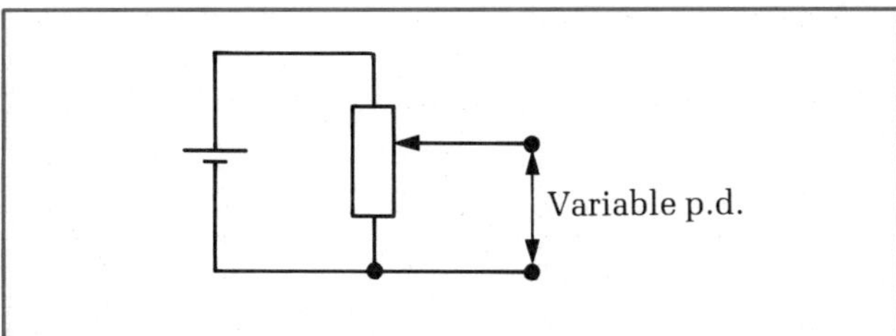

Figure 22.29 *A potentiometer*

Resistor combination rules

Resistors in series Resistors in series pass the same current. The reason is that equal amounts of charge pass through each resistor every second. For two resistors of resistance R_1 and R_2 in series, as in Figure 22.30,

the p.d. across R_1, $V_1 = IR_1$
the p.d. across R_2, $V_2 = IR_2$,

where I is the current.

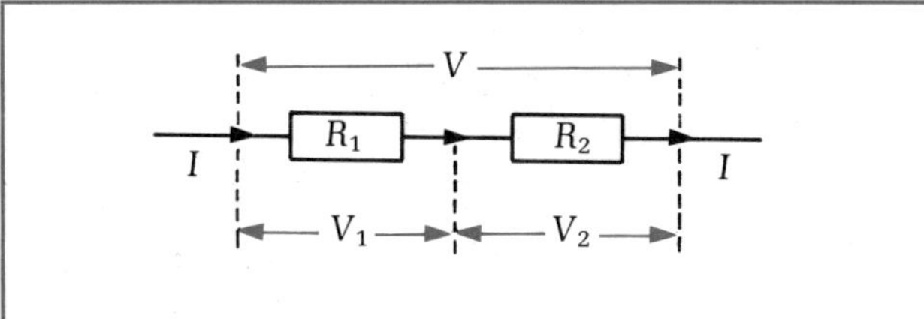

Figure 22.30 *Resistors in series*

The total p.d. V across the two resistors $= V_1 + V_2$. This is because each coulomb of charge passing through delivers energy V_1 to R_1 and V_2 to R_2. So the total energy delivered per coulomb is $V_1 + V_2$.
Hence $V = IR_1 + IR_2$
so the total resistance $R = V/I$ is given by

$$R = \frac{V}{I} = \frac{IR_1 + IR_2}{I} = R_1 + R_2$$

For more than two resistors in series, the theory can easily be extended to show that the total resistance equals the sum of the individual resistances.

$$R = R_1 + R_2 + R_3 + \ldots$$

Resistors in parallel Resistors in parallel have the same p.d. The reason is that each coulomb of charge passing through the combination delivers the same amount of energy whatever route the charge takes.

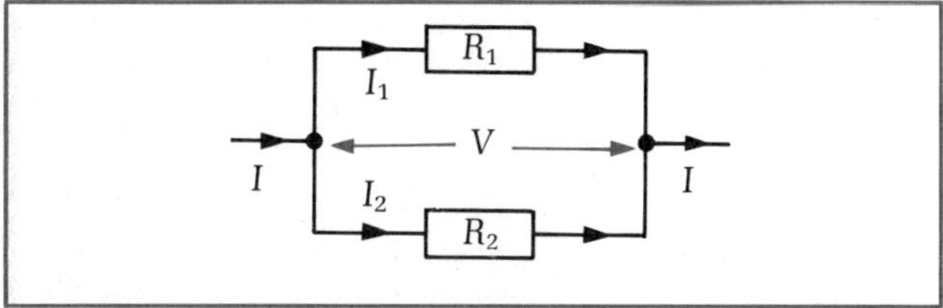

Figure 22.31 *Resistors in parallel*

For two resistors in parallel, as in Figure 22.31,

the current through R_1, $I_1 = V/R_1$
the current through R_2, $I_2 = V/R_2$

where V is the p.d. across the combination.

The total current I entering the combination $= I_1 + I_2$, since the charge flow per second entering is equal to the charge flow per second through R_1 + the charge flow per second through R_2.
Hence $I = V/R_1 + V/R_2$.
The total resistance $R = V/I$ so $1/R = I/V$.
Therefore

$$1/R = I/V = \frac{V/R_1 + V/R_2}{V} = 1/R_1 + 1/R_2$$

For more than two resistors in parallel, the total resistance R is given by the same form of equation as for two resistors in parallel.

$$1/R = 1/R_1 + 1/R_2 + 1/R_3 + \ldots$$

Worked example A battery with an e.m.f. of 12 V and internal resistance 2 Ω is connected to a wire-wound resistor of resistance 10 ohms.

a) Calculate the p.d. across the 10 ohm resistor.

b) What will the p.d. across the 10 ohm resistor become if a 15 ohm resistor is connected in parallel with the 10 ohm resistor?

Solution

a) Figure 22.32 shows the circuit for the battery and 10 ohm resistor.
The 10 ohm resistor is in series with the internal resistance of the battery since they carry the same current.

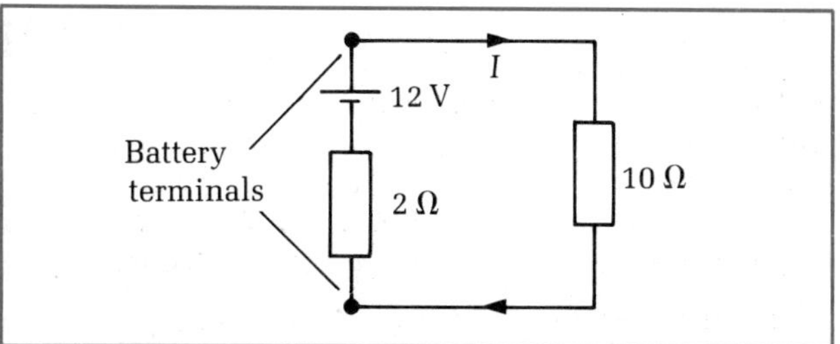

Figure 22.32

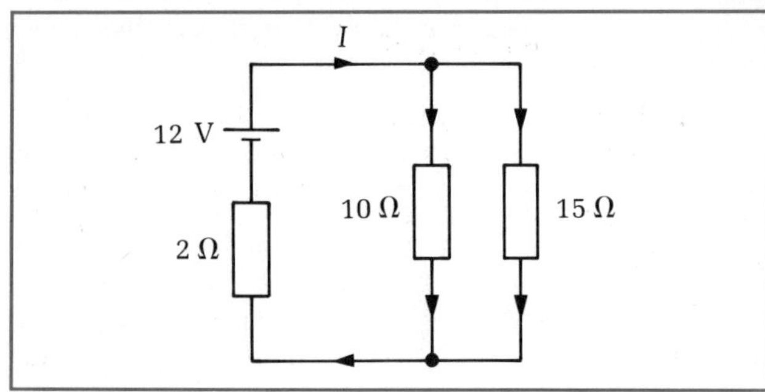

Figure 22.33

Hence the total resistance $R = 10 + 2 = 12$ ohms.
Therefore the current from the battery I = battery e.m.f./total resistance = 12/12 = 1 A.
Hence the p.d. across the 10 ohm resistor,
$V = I \times 10 = 1 \times 10 = 10$ V.

b) With the 10 ohm resistor now in parallel with a 15 Ω resistor, the circuit must be redrawn as in Figure 22.33. The combined resistance R of the 10 ohm resistor in parallel with the 15 ohm resistor is given by
$1/R = 1/10 + 1/15 = \frac{3+2}{30} = \frac{5}{30}$.
Hence the combined resistance R is 30/5 = 6 ohms.
The total resistance of the circuit is 2 + 6 = 8 ohms. This is because the parallel combination is in series with the internal resistance of the battery.
Hence the current from the battery, I = battery e.m.f./total resistance = 12/8 = 1.5 A.
Therefore the p.d. across the parallel combination = current × resistance of the combination = 1.5 × 6 = 9 V.
So the p.d. across the 10 ohm resistor = 9 V.

22.4 Circuits and cells

Internal resistance

A cell in a circuit supplies electrical energy to other components of the circuit. If the cell has internal resistance, some of the electrical energy produced is wasted due to heating inside the cell. The heating effect is due to the passage of current through the cell's internal resistance.

Consider the circuit of Figure 22.34(a). A cell of e.m.f. 2 V and internal resistance 1 ohm is connected to a 4 ohm resistor. The cell is represented by a source of e.m.f. *and* an internal resistance. The 4 ohm resistor is in series with the cell's internal resistance so the total circuit resistance is equal to 5 ohms (= 1 + 4).

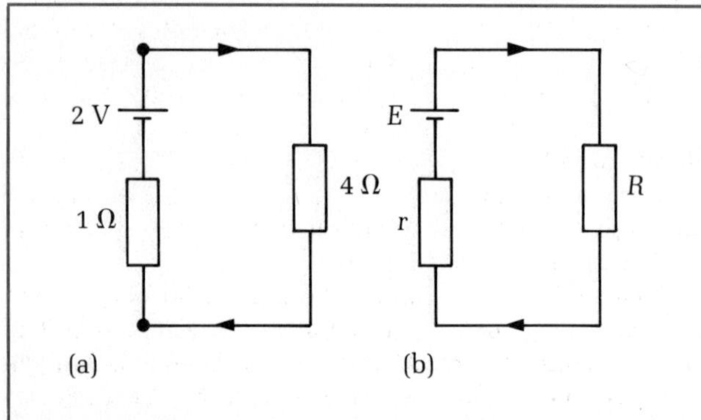

Figure 22.34 *Internal resistance*

The current I from the cell = cell e.m.f./total circuit resistance = 2/5 = 0.4 A.

Therefore the p.d. across the 4 ohm resistor = current × resistance = 0.4 × 4 = 1.6 V, and the p.d. across the 1 ohm internal resistance = 0.4 × 1 = 0.4 V.

In energy terms, each coulomb of charge is supplied with 2 J of electrical energy by the cell. However, 0.4 J is wasted by each coulomb inside the cell. So each coulomb delivers 1.6 J only to the external circuit (i.e. the 4 ohm resistor). The p.d. across the internal resistance is due to electrical energy being converted to heat energy inside the cell. So the p.d. across the cell's terminals is equal to the cell's e.m.f. less the lost p.d. across its internal resistance.

Suppose the 4 ohm resistor is replaced by a 3 ohm resistor. The total circuit resistance is now 3 + 1 = 4 ohms. So the current from the cell is 2 V/4 ohms = 0.5 A. Hence the lost p.d. (i.e. p.d. across the internal resistance) is equal to 0.5 × 1 = 0.5 V. The p.d. across the external resistor = 3 × 0.5 = 1.5 V. Once again, the cell's e.m.f. is equal to the p.d. across the external resistor + the lost p.d. across its internal resistance.

In general, suppose the cell is connected to a resistance R, as in Figure 22.34(b). The cell's e.m.f. is E and its internal resistance is r. The total resistance of the circuit is $R + r$. Hence the current from the cell, I = cell e.m.f./total circuit resistance = $E/(R + r)$. Rearranged, this equation becomes

$$E = I(R + r) = (IR) + (Ir)$$

The equation above tells us that each coulomb of charge from the cell is supplied with electrical energy E by the cell. However, each coulomb uses energy equal to Ir to pass through the cell. So the energy per coulomb delivered to the external circuit (i.e. to resistance R) is equal to IR. The p.d. across the cell terminals, equal to IR, is therefore equal to the cell e.m.f. E − the lost p.d. (Ir).

Electrical power is given by current × p.d., as explained on p. 315. Since the cell e.m.f. is given by

$$E = IR + Ir$$

then multiplying each term by I gives an equation in terms of power.

$$IE = I^2R + I^2r$$

The first term, IE, is the electrical energy produced per second by the cell. In other words, it is the power generated inside the cell.

The term I^2R is the heat energy produced per second in the resistance R. This is because the power used by the resistor = current (I) × p.d. (IR).

The term I^2r is the heat energy produced per second in the internal resistance of the cell.

So the power equation above tells us that the power generated by the cell is equal to the power supplied to resistor R + the power wasted inside the cell due to its internal resistance r.

Measuring internal resistance

The cell under test is connected in series with an ammeter, rheostat and lamp. A high resistance voltmeter is connected across the cell terminals (see Figure 22.35). The lamp is included to prevent an excessive current damaging the ammeter. A wire-wound resistor could be used in place of the lamp. The rheostat is used to alter the current in steps. At each step, the cell current and p.d. across the cell terminals are measured. The ammeter is used to measure the cell current, and the voltmeter to measure the cell p.d. As the current is increased, the p.d. across the cell terminals falls. This is because the lost p.d. increases when the current increases.

To determine the internal resistance r of the cell, a graph of cell p.d. against cell current is plotted. Figure 22.36 shows a typical graph.

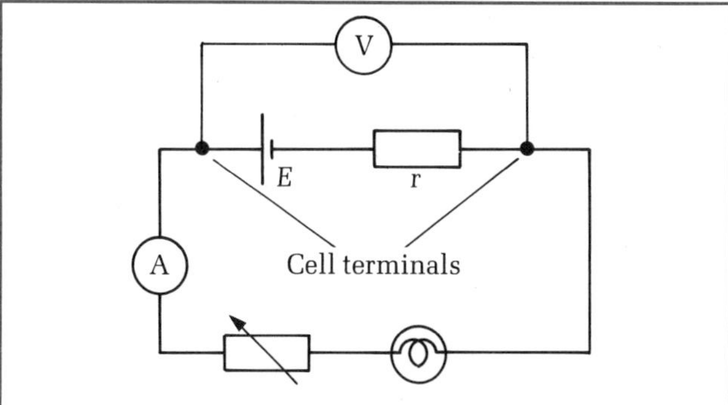

Figure 22.35 *Measuring internal resistance*

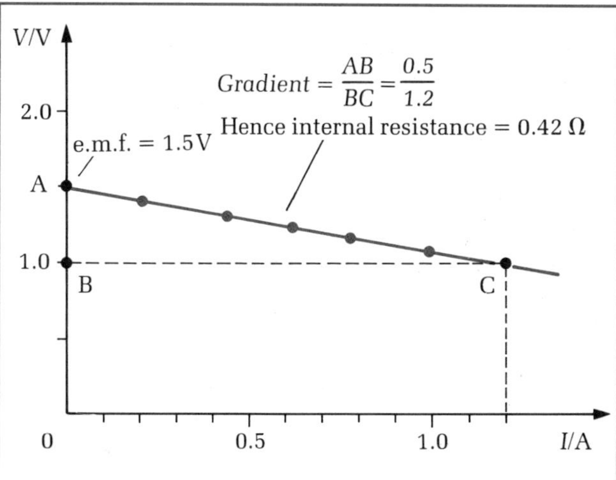

Figure 22.36 *Calculating internal resistance*

For zero current there is zero p.d. across the internal resistance, so the cell p.d. is equal to its e.m.f. when no current is supplied. Therefore, the intercept on the cell p.d. axis gives the cell's e.m.f.

For current I the lost p.d. is equal to Ir. So the cell p.d. equals $E-(Ir)$ where E is its e.m.f. So (Ir) is represented on the graph by the drop of the cell p.d. from the intercept on the vertical axis. In Figure 22.36, the height AB of the gradient triangle represents the lost p.d. Ir. The base BC of the triangle represents the current I. So the gradient of the line AC represents lost p.d./current which equals internal resistance r. Hence r can be calculated.

Kirchhoff's laws

As explained previously, components in series take the same current. Also, components in parallel have the same p.d. across their terminals. The rules for circuit calculations were first established by Gustav Kirchhoff in the 19th century.

Kirchhoff's first law At a junction, the total current entering the junction is equal to the total current leaving the junction.

Consider the junction shown in Figure 22.37. Suppose the current in wire 1 is 0.5 A towards the junction, and the current in wire 2 is 1.5 A from the junction. Then the third wire must carry a current of 1.0 A towards the junction. We can write the currents as $I_1=-0.5$ A, $I_2=+1.5$ A and $I_3=-1.0$ A, where currents entering the junction are given negative values. Currents leaving the junction are given positive values, so we can write

$$I_1+I_2+I_3=0$$

The reason for Kirchhoff's first law is that the total charge entering the junction is equal to the total charge leaving the junction. In Figure 22.37, 1.5 coulombs of charge leaves the junction along wire 2 each second. Wire 1 carries 0.5 coulombs each second into the junction, and wire 3 carries 1 coulomb each second into the junction. So the charge flow in is equal to the charge flow out.

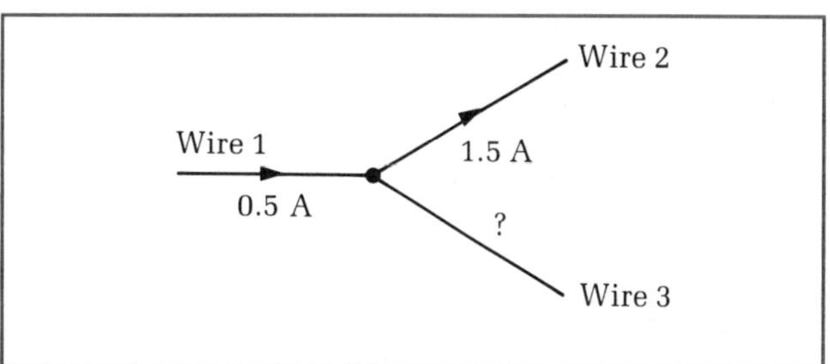

Figure 22.37 *Kirchhoff's first law*

Kirchhoff's second law The net e.m.f. round a circuit loop is equal to the sum of the p.d.s round the loop.

A circuit loop is any path round a circuit from a starting point back to the same point without changing direction. In other words a circuit loop is like a route round a one-way road system which brings you back to the start. The reason for Kirchhoff's second law is that e.m.f. and p.d. are terms used for energy per unit charge. A source of e.m.f. is a source of electrical energy; potential difference (p.d.) represents a sink of electrical energy where electrical energy is converted into other forms of energy. So Kirchhoff's second law amounts to a statement of conservation of energy. Imagine the journey of 1 coulomb of charge round a circuit loop; the total energy it delivers to the components in the loop (i.e. the sum of the p.d.s round the loop) is equal to the total electrical energy which is supplied to it (i.e. the net e.m.f.).

Consider the simple circuit of Figure 22.38 (at the top of the next page) where there is only one possible loop. The total resistance of the circuit is 5 ohms since the cells have zero internal resistance. The total e.m.f. is equal to $1.5+2=3.5$ V. Hence the current supplied by the battery of two cells is $3.5/5=0.7$ A.

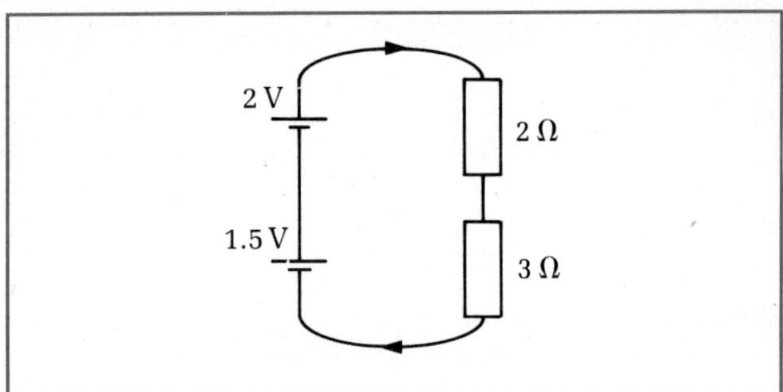

Figure 22.38 *Kirchhoff's second law*

The p.d. across the 3 ohm resistor = 3 × 0.7 = 2.1 V.

The p.d. across the 2 ohm resistor = 2 × 0.7 = 1.4 V.

Hence the sum of the p.d.s is equal to the total e.m.f. supplied by the two cells.

Suppose one of the two cells is reconnected in the opposite direction, so the two cells act against one another. The total e.m.f. is now 2 − 1.5 = 0.5 V. The total resistance is unchanged at 5 ohms. So the current supplied by the cells is now 0.5/5 = 0.1 A.

The p.d. across the 3 ohm resistor = 0.1 × 3 = 0.3 V.

The p.d. across the 2 ohm resistor = 0.1 × 2 = 0.2 V.

So the sum of the p.d.s is equal to the *net* e.m.f. of the two cells. We use the term 'net' because the 2 V cell supplies electrical energy but the 1.5 V cell uses electrical energy. So the overall effect is equivalent to a cell of e.m.f. 0.5 V.

Worked example Calculate the current, p.d. and power supplied to each resistor of the circuit shown in Figure 22.39. The cell has an e.m.f. of 4 V and its internal resistance is 1 ohm.

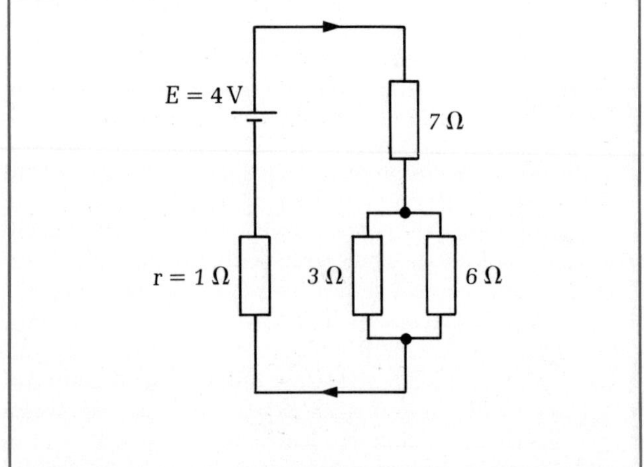

Figure 22.39

Solution Start by indicating the current direction in each branch of the circuit. There are two possible ways to determine the current in each resistor. Consider each way in turn.

1 The total circuit resistance can be calculated using the resistor combination rules. The combined resistance R of the 3 ohm and 6 ohm resistor in parallel is given by $1/R = 1/3 + 1/6$.
Hence $R = 2$ ohms.
So the total circuit resistance R_T is 2 + 7 + 1 = 10 ohms.

The current from the cell I = cell e.m.f./total circuit resistance = 4/10 = 0.4 A.
Hence the p.d. across the internal resistance = current × internal resistance = 0.4 V.
Also, the p.d. across the 7 ohm resistor = 0.4 × 7 = 2.8 V.
Lastly, the p.d. across the parallel combination = current × combined resistance = 0.4 × 2 = 0.8 V.
The current through each resistor of the parallel combination = p.d./resistance of each resistor.
So the current through the 3 ohm resistor = 0.8/3 = 0.267 A.
The current through the 6 ohm resistor = 0.8/6 = 0.133 A.

Resistor	Current/A	P.D./V	Power/W
1 Ω int. resistance	0.4	0.4	0.16
7 Ω resistor	0.4	2.8	1.12
3 Ω resistor	0.267	0.8	0.21
6 Ω resistor	0.133	0.8	0.11

2 Using Kirchhoff's laws:
let x = current through the 3 ohm resistor,
and y = current through the 6 ohm resistor.
From Kirchhoff's first law, the total current entering the parallel combination is equal to (x + y).
So the current through the cell and the 7 ohm resistor is (x + y).

The p.d. across the internal resistance = current × resistance = (x + y) × 1.

The p.d. across the 7 ohm resistor = (x + y) × 7.

The p.d. across the 3 ohm resistor = x × 3.

The p.d. across the 6 ohm resistor = y × 6.

Now apply Kirchhoff's second law to the circuit loop containing the cell, the 7 ohm resistor and the 3 ohm resistor, as in Figure 22.40.

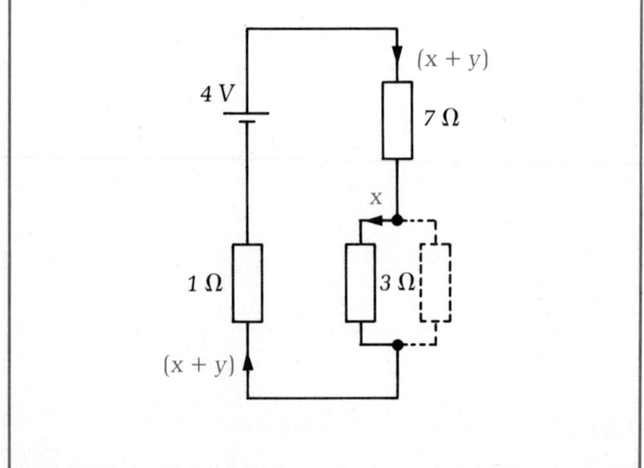

Figure 22.40

The net e.m.f., 4 V = the sum of the p.d.s
= p.d. across the internal resistance
+ the p.d. across the 7 ohm resistor
+ the p.d. across the 3 ohm resistor.

Hence 4 = (x + y) + 7(x + y) + 3x

giving 4 = 11x + 8y (1)

Next we apply the second law to the circuit loop shown in Figure 22.41.

The net e.m.f. 4 V = (x + y) + 7(x + y) + 6y

Hence 4 = 8x + 14y (2)

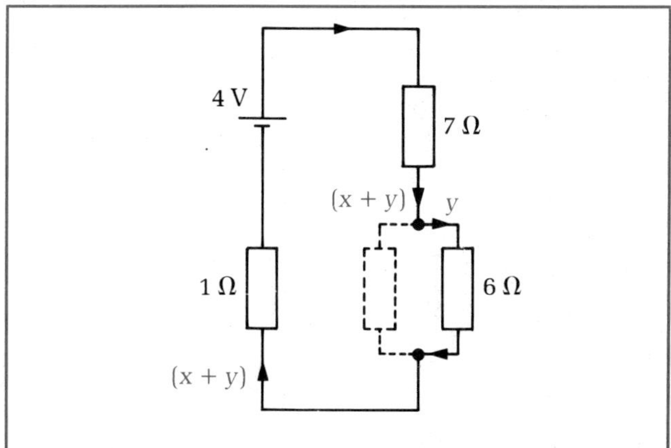

Figure 22.41

So we now have two equations with two unknowns, x and y. We can solve for x and y using the method of simultaneous equations.

From the first equation, $32 = 88x + 64y$.

From the second equation, $44 = 88x + 154y$.

Hence $44 - 32 = 88x + 154y - 88x - 64y$

giving $12 = 90y$ so $y = 12/90 = 0.133$ A.

The value of x can now be calculated from either equation, to give $x = 0.267$ A.

From the values of currents x and y, the current, p.d. and power in each resistor can be calculated.

Worked example A 3 ohm resistor is connected to the terminals of a 1.5 V cell in parallel with a 2 V cell. The internal resistance of the 1.5 V cell is 1 ohm, and the internal resistance of the 2 V cell is 2 ohms. Calculate the current supplied by each cell.

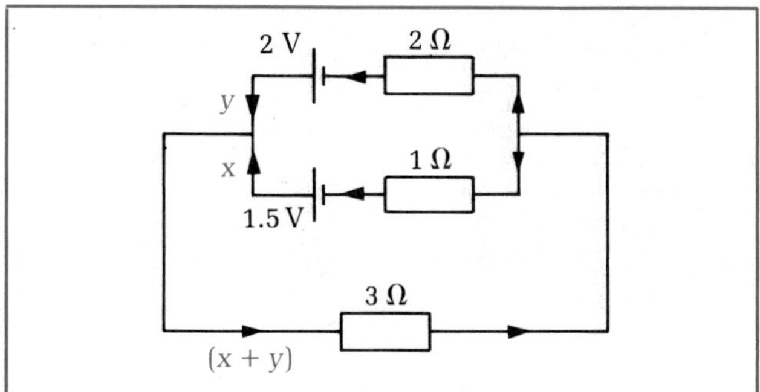

Figure 22.42

Solution The circuit is shown in Figure 22.42.

Let x = current supplied by the 1.5 V cell,

y = current supplied by the 2 V cell.

Hence current supplied to the 3 ohm resistor $= (x + y)$.

To determine x and y, we need to use Kirchhoff's laws applied to two circuit loops.

Loop 1 Consider the loop formed by the 1.5 V cell and the 3 ohm resistor, as in Figure 22.43.

The net e.m.f. in the loop = 1.5 V.

The p.d. across the 1 ohm internal resistance = current × resistance $= x \times 1$.

The p.d. across the 3 ohm resistor = current × resistance $= (x + y) \times 3$.

From Kirchhoff's second law, $1.5 = 1x + 3(x + y)$

Hence $1.5 = 4x + 3y$ **(1)**

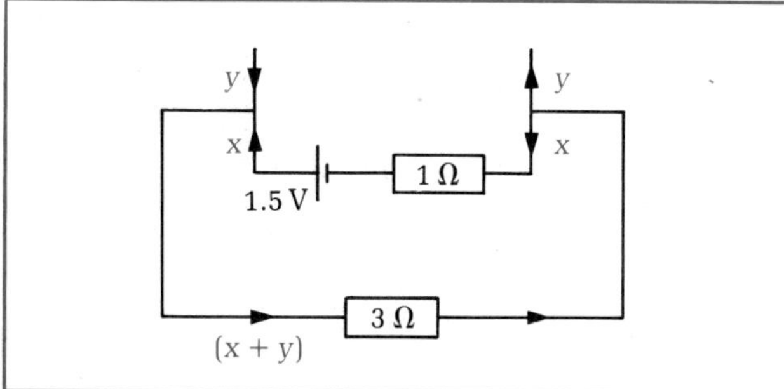

Figure 22.43

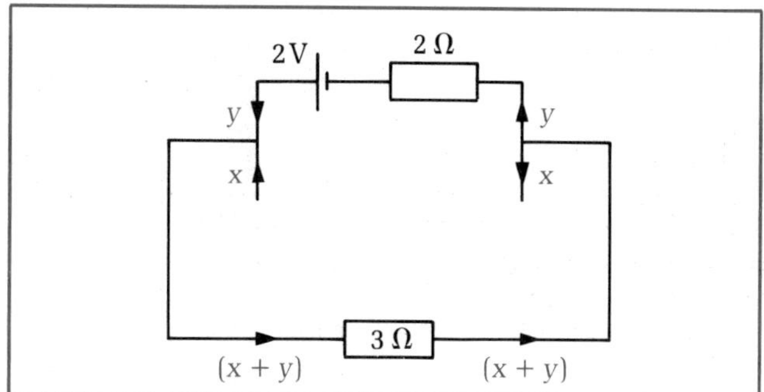

Figure 22.44

Loop 2 Now consider the loop formed by the 2 V cell and the 3 ohm resistor, as in Figure 22.44.

The net e.m.f. in the loop = 2 V.

The p.d. across the 2 ohm internal resistance = current × resistance $= y \times 2$.

The p.d. across the 3 ohm resistor $= 3(x + y)$, as before.

Hence, using Kirchhoff's second law, $2 = 2y + 3(x + y)$.

Thus $2 = 3x + 5y$. **(2)**

To solve for x and y, equation **(1)** is multiplied by 3 on both sides to give $4.5 = 12x + 9y$.

Equation (2) is multiplied by 4 on both sides to give $8 = 12x + 20y$.

Subtracting one equation from the other gives

$8 - 4.5 = 12x + 20y - 12x - 9y = 11y$.

Hence $y = 3.5/11 = 0.32$ A

Substituting in the first equation with the known value of y gives $1.5 = 4x + (3 \times 3.2)$.

Hence $4x = 1.5 - 0.96 = 0.54$, giving $x = 0.135$ A.

22.5 Capacitors in d.c. circuits

Capacitance

A capacitor is any device which can store charge. Two parallel metal plates placed near each other form a capacitor, as explained on p. 169. When a capacitor is connected to a battery, electrons from the negative terminal of the battery flow onto one of the plates. An equal number of electrons leave the other plate to return to the battery via its positive terminal. So each plate gains an equal and opposite charge.

Any capacitor is made up of two conductors insulated from one another. One of the conductors could be the Earth. When charged, one of the two conductors gains

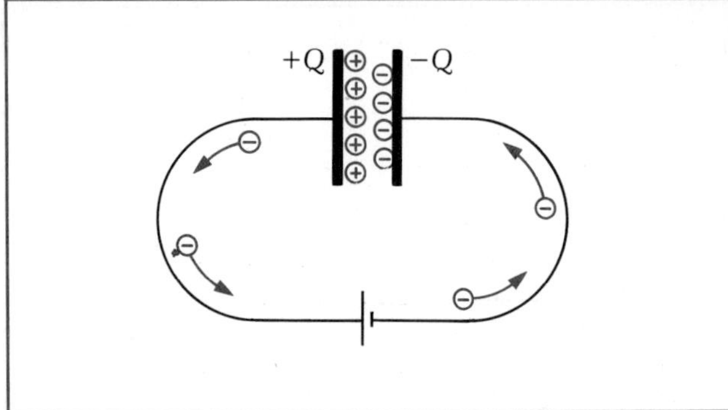

Figure 22.45 *Storing charge*

electrons and the other loses electrons. So the conductors gain equal and opposite charge. When we say that the charge stored by a capacitor is Q, we mean that one plate stores charge $+Q$ and the other plate stores charge $-Q$.

Charge stored by a capacitor can be investigated using the circuit of Figure 22.46. When the switch S is closed, electrons pass along the wires to charge the capacitor. The microammeter indicates a current due to the flow of charge. The capacitor p.d. is measured using a high-resistance voltmeter connected across the capacitor plates. The variable resistor is used to keep the current constant for as long as possible. When the switch is closed, a stopwatch is started and the voltmeter reading is noted at regular intervals. At the same time, the variable resistor is continually adjusted to keep the current constant until the capacitor becomes fully charged.

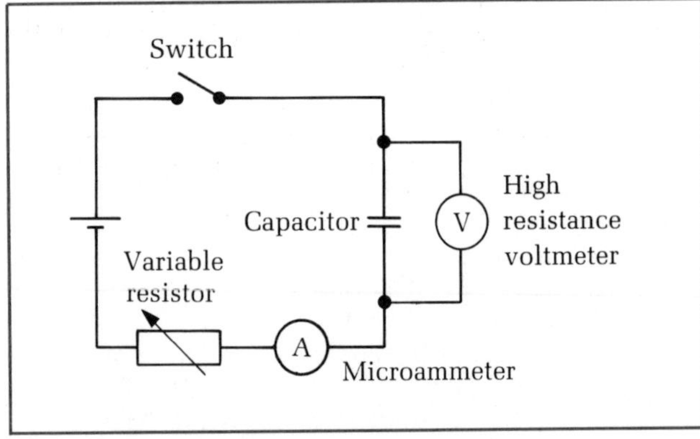

Figure 22.46 *Investigating capacitors*

Typical readings are shown in Figure 22.47. The charge stored after time *t* is given by

$$Q = It$$

where *I* is the charging current.

Time/s	0	20	40	60	80	100
p.d./volts	0	0.29	0.62	0.90	1.22	1.50
Current = 15 μA						

Figure 22.47 *Typical results*

Hence the charge stored for different values of p.d. can be calculated. A graph of charge stored Q against p.d. V, like that in Figure 22.48 shows that the charge stored is proportional to the p.d.

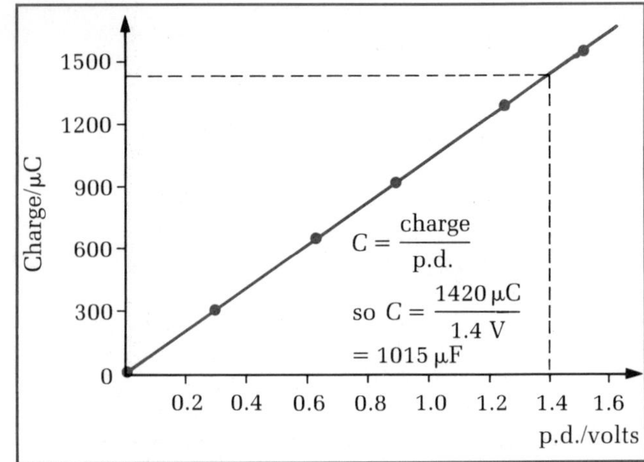

Figure 22.48 *Graph of results*

From this we can say that $\frac{\text{charged stored}}{\text{p.d. applied}}$ is constant for each capacitor.

The capacitance C of a capacitor is defined as the charge stored per unit p.d. applied to the capacitor. The unit of C is the **farad** (F), equal to 1 coulomb per volt. Note that 1 microfarad (μF) $= 10^{-6}$ F.

$$C = \frac{Q}{V}$$

where C = capacitance,
Q = charge stored,
V = applied p.d.

Capacitor combination rules

Capacitors in parallel Two capacitors in parallel have the same p.d. Suppose the p.d. across the two capacitors C_1 and C_2 of Figure 22.49 is V.

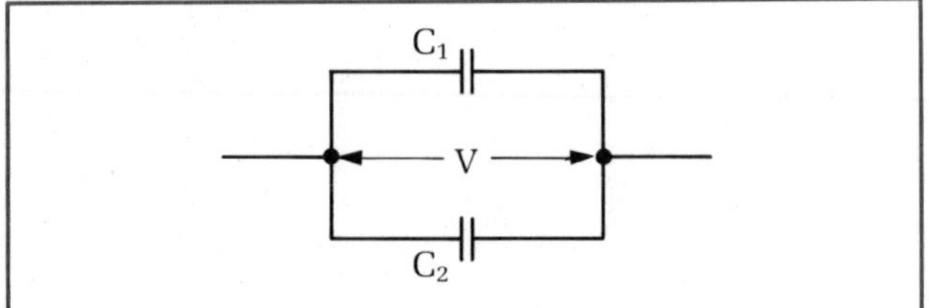

Figure 22.49 *Capacitors in parallel*

Hence the charge Q_1 stored by C_1, is given by $Q_1 = C_1V$
the charge Q_2 stored by C_2, is given by $Q_2 = C_2V$.
So the total charge stored $Q = Q_1 + Q_2 = C_1V + C_2V$.

Therefore the combined capacitance *C* is given by

$$C = \frac{\text{total charge stored } Q}{\text{applied p.d. } V}$$

$$\text{Hence } C = \frac{C_1V + C_2V}{V}$$

$$C = C_1 + C_2$$

Capacitors in series Two capacitors in series, as in Figure 22.50, store equal charge. When the capacitors C_1 and C_2 of Figure 22.50 are connected to a battery,

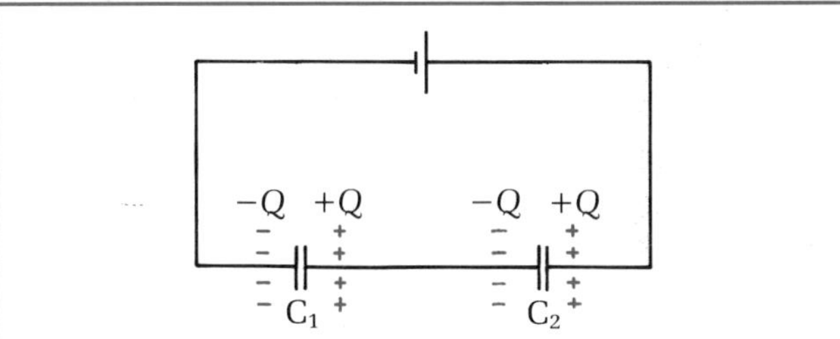

Figure 22.50 *Capacitors in series*

electrons from the negative terminal of the battery flow onto the left-hand plate of C_1. As a result, an equal number of electrons leave the right hand plate of C_1 and flow onto the left hand plate of C_2. This causes an equal number of electrons to leave the right hand plate of C_2 to return to the battery. Hence C_1 and C_2 store equal charge.

Suppose the charge stored $= Q$. Then the p.d. across each capacitor is equal to charge stored/capacitance.
The p.d. across $C_1 = Q/C_1$, and the p.d. across $C_2 = Q/C_2$.

Since the capacitors are in series, the total p.d.
$=$ p.d. across C_1 + p.d. across C_2.
Hence the total p.d., $V = Q/C_1 + Q/C_2$.

The combined capacitance C is given by
charge stored/total p.d. $= Q/V$.
Hence $1/C = V/Q$.
Since $V = Q/C_1 + Q/C_2$ then $1/C = V/Q = 1/C_1 + 1/C_2$. We can therefore write

$$1/C = 1/C_1 + 1/C_2$$

Figure 22.51 *Charged stored*

Worked example A 3 μF capacitor is connected in series with a 6 μF capacitor. The combination is then connected in parallel with a 1 μF capacitor to a 12 V battery, as in Figure 22.51. Calculate the charge stored by each capacitor.

Solution

For the 1 μF capacitor, the p.d. across its terminals is 12 V. So the charge stored = capacitance × p.d. = 1 × 12 = 12 μC.

For the 3 μF and 6 μF capacitor in series, the combined capacitance C is given by $1/C = 1/3 + 1/6$.

Hence $C = 2$ μF.

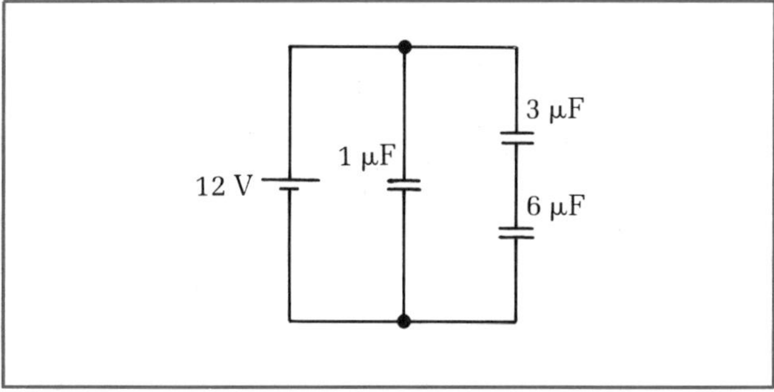

Figure 22.52

Therefore the charge stored by each capacitance = combined capacitance × p.d. = 2 × 12 = 24 μC.

So the 3 μF capacitor stores 24 μC, and the 6 μF capacitor stores the same amount of charge. Since they are in series, the charge stored by the 3 μF and 6 μF capacitor together is still 24 μC, *not* 48 μC.

The situation is not unlike two resistors in series; such resistors take the same current. The current passed is equal to the current through either resistor.

Energy stored in a charged capacitor

Consider the energy changes when a capacitor is charged from a battery then discharged through a lamp. Figure 22.53 shows a suitable circuit to do this. When the switch is connected to the battery, the capacitor C charges up so electrical energy is stored in capacitor C. When the switch is reconnected across to the lamp, the capacitor discharges through the lamp. So electrical energy in the capacitor is converted into light and heat energy by the lamp.

Suppose another capacitor identical to C is connected in parallel with C, as in Figure 22.54. Enough energy is

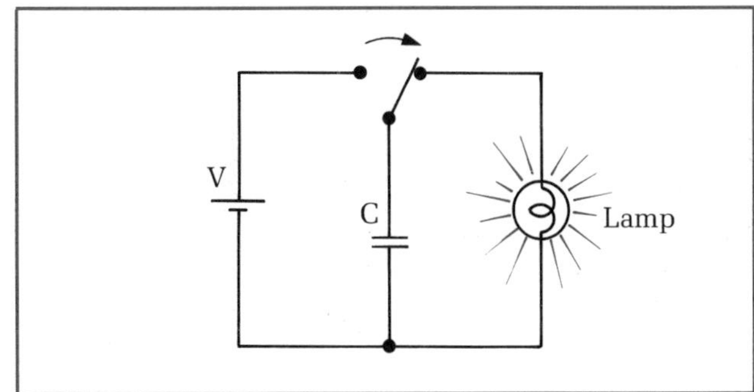

Figure 22.53 *Using stored energy*

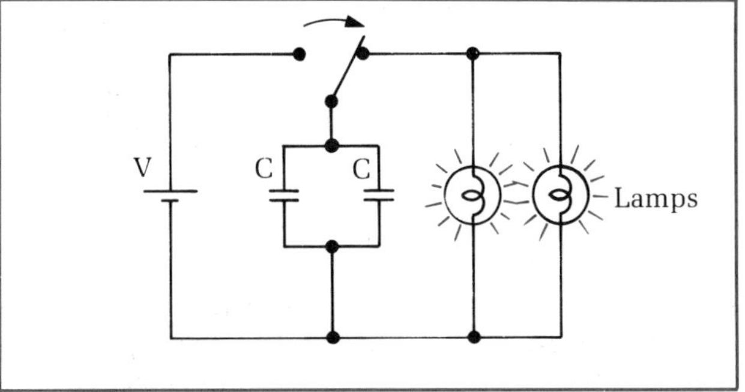

Figure 22.54 *Using two capacitors*

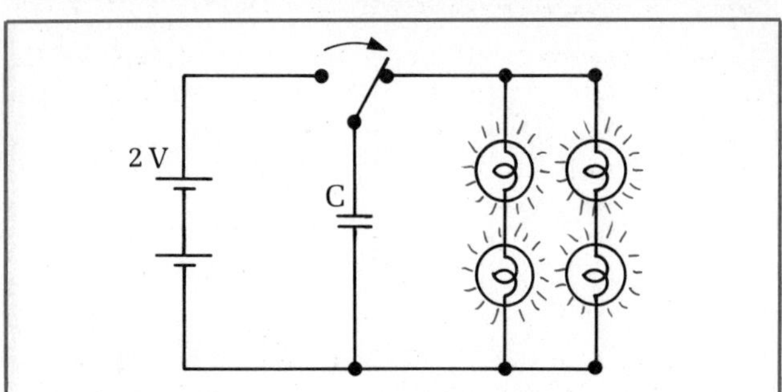

Figure 22.55 *Using two batteries*

now stored in the two capacitors to light briefly two lamps in parallel; so twice as much energy can be stored using two capacitors in parallel at the same p.d.

Now suppose the circuit of Figure 22.53 is changed by using two batteries in series, as in Figure 22.55. Capacitor C is now charged to twice the p.d. How many lamps can it light briefly now? In fact it will light four lamps briefly as in Figure 22.55 because four times as much energy is stored at twice the p.d.

As was explained on p. 170, the energy stored in a charged capacitor C is given by $\frac{1}{2}CV^2$, where V is the p.d. across the capacitor terminals. The charging process involves transferring charge Q from the battery to the capacitor. Since the battery p.d. is the electrical energy per unit charge, then the battery supplies energy QV to the circuit. Since $Q = CV$, the energy supplied by the battery is also equal to CV^2.

Energy supplied by battery $= CV^2$

Energy stored in capacitor $= \frac{1}{2}CV^2$

What happens to the rest of the energy supplied by the battery? Only 50% of the energy supplied is actually stored in the capacitor. The other 50% is wasted due to the heating effect of the charging current. The resistances in the circuit, including the internal resistance of the battery, cause a heating effect when current passes through them.

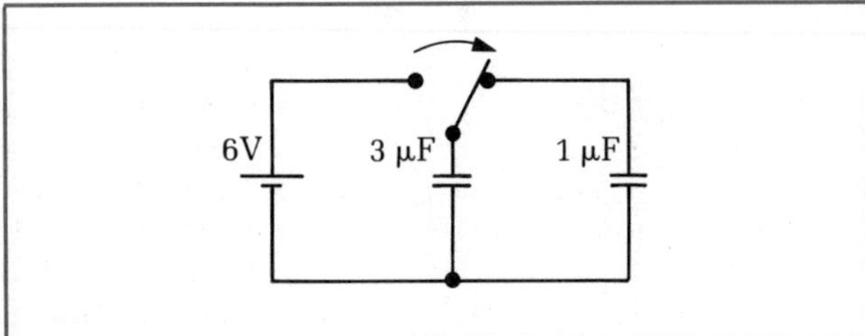

Figure 22.56

Worked example In the circuit shown in Figure 22.56, a 3 μF capacitor is charged from a 6 V battery by connecting the switch to terminal X. Then the switch is reconnected to terminal Y to charge a 1 μF capacitor from the 3 μF capacitor. Calculate

a) the charge and energy initially stored in the 3 μF capacitor,

b) the final charge and energy stored in each capacitor.

Solution

a) Initially, the 3 μF capacitor is charged to a p.d. of 6 V. Initial charge $Q = CV = 3 \times 6 = 18$ μC. Initial stored energy $= \frac{1}{2}CV^2 = \frac{1}{2} \times 3 \times 6^2 = 54$ μJ.

b) Finally, the 3 μF capacitor and the 1 μF capacitor have the same p.d. V. So the two capacitors are in parallel. The combined capacitance $C = 1 + 3 = 4$ μF.
Since the total final charge stored in the two capacitors = the initial charge, then $4\,V = 18$.
Hence $V = 18/4 = 4.5$ volts.
So the 3 μF capacitor discharges to a p.d. of 4.5 volts.
The energy stored in the 3 μF capacitor is now $\frac{1}{2} \times 3 \times 4.5^2 = 30.4$ μJ.
The charge stored in the 3 μF capacitor $= 3 \times 4.5 = 13.5$ μC.
The 1 μF capacitor becomes charged to a p.d. of 4.5 V due to the transfer of charge from the 3 μF capacitor.
So the charge stored in the 1 μF capacitor becomes $1 \times 4.5 = 4.5$ μC.
The energy stored in the 1 μF capacitor is now $\frac{1}{2} \times 1 \times 4.5^2 = 10.1$ μJ.

Note that the final total energy stored ($= 30.4 + 10.1$ μJ) is less than the energy initially stored in the 3 μF capacitor. This is because the charging current when the 1 μF capacitor is charged from the 3 μF capacitor causes a heating effect in the resistance of the connecting wires.

Charging and discharging a capacitor

Discharging through a fixed resistance allows electrons from the negative plate to pass through the resistance and move onto the positive plate. In this way, the plates become neutralized. Figure 22.57 shows a suitable circuit which allows a capacitor to be charged from a battery then discharged through a resistance.

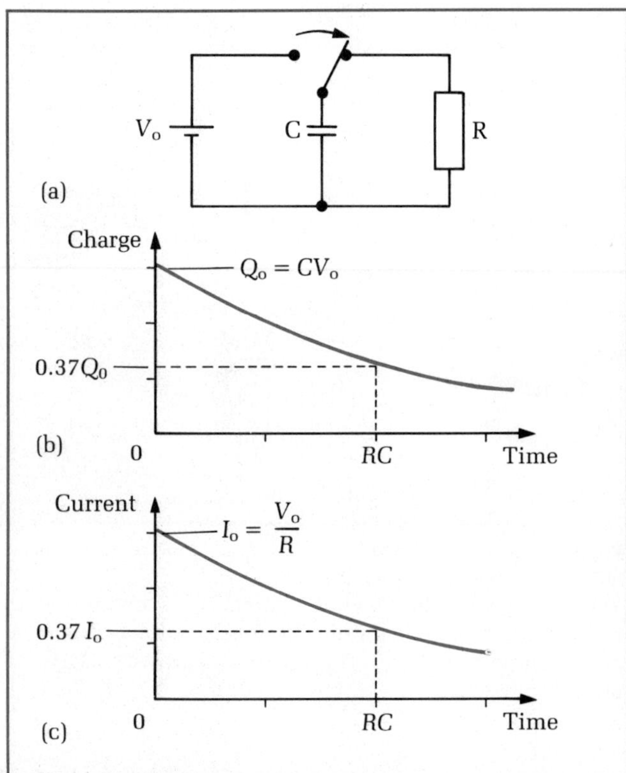

Figure 22.57 *Capacitor discharge*

Initially, the rate of discharge is high because electrons flow off the plates at a high rate at the start. However, as the p.d. across the plates lessens, the rate of discharge falls too. So the discharge current is initially high, but becomes less and less as the capacitor discharges. The situation is like the flow of water from a water tank. The pressure of the water at the outlet is high at the start because the level in the tank is high at the start, so the initial flow rate is large. But as the tank empties, the level falls, so the flow rate falls too.

Consider the discharge process in small steps. At time t after the start of discharge, suppose the charge still on the plates is Q. Let ΔQ be the charge flow off the plates in the short interval of time from t to $t+\Delta t$.

$$\text{The current } I=\frac{\text{p.d. across plates } V}{\text{resistance } R}$$

Since $V=Q/C$ we can write

$$I=\frac{Q}{CR}.$$

So the discharge current is proportional to the charge still on the plates.

However, for changing currents, the drop of charge $\Delta Q=-I\Delta t$ ($-$ for the drop of charge on the plates). Hence

$$\Delta Q=-\frac{Q\Delta t}{RC}, \text{ which gives}$$

$$\frac{\Delta Q}{Q}=-\frac{\Delta t}{CR}$$

The equation tells us that the fractional drop of charge, $\Delta Q/Q$, is the same for any short interval of time Δt during the process of discharge. For example, suppose $t=10$ s and the product $CR=100$. Then $\Delta Q/Q=-0.1$. In other words, every 10 seconds, the charge drops to 9/10 of its value at the start of the 10 s interval. So if the initial charge is Q_0, then the charge still on the plates after 10 s is $0.9\ Q_0$; after 20 s from the start, the charge will have fallen to $0.81\ Q_0$, etc.

Time from start	0	10 s	20 s	30 s	40 s . . .
Charge on plates	Q_0	$0.9Q_0$	$0.81Q_0$	$0.74Q_0$	$0.67Q_0$

The charge on the plates never quite becomes zero. Figure 22.57(b) shows how the charge falls with time. The curve is called an **exponential decrease** curve. Exponentials occur whenever the rate of change of a quantity is proportional to the quantity itself. See p. 507 for further details about exponentials in general.

Now suppose we consider very short intervals of time (i.e. $\Delta t \to 0$). Then the equation for the fractional charge $\Delta Q/Q=-\Delta t/CR$ may be written as

$$\frac{dQ}{dt}=-\frac{Q}{CR}$$

where $\frac{dQ}{dt}$ is equal to the current I.

The graphical solution of this equation is shown in Figure 22.57(b). The mathematical solution is written as

$$Q=Q_0e^{-t/RC}$$

where Q_0 = the initial charge in the capacitor,
e = the exponential function.

The value of $R\times C$ is called the **time constant** for the circuit. In a time t equal to $R\times C$, the charge falls to 0.37 ($=e^{-1}$) of its initial value.

Time constant $=RC$

where R = circuit resistance,
C = capacitance.

Since the discharge current $I=-Q/RC$, then the discharge current drops at the same rate as the charge. So the curve of current against time shown in Figure 22.57(c) is an exponential decrease too. Note that the gradient of the charge against time curve, $\frac{dQ}{dt}$, is equal to the current I.

Worked example A 3 μF capacitor charged to an initial p.d. of 12 V is discharged through a resistance of 0.1 MΩ. Calculate

a) the initial discharge current,

b) the p.d. across the capacitor plates after 0.3 s from the start of discharge,

c) the p.d. across the capacitor plates after 1.0 s from the start.

Solution

a) The initial current $=\frac{\text{initial p.d.}}{\text{resistance}}$

$$=\frac{12}{0.1\times 10^{+5}}=1.2\times 10^{-4}\text{ A}.$$

b) The time constant for the circuit $=RC$
$=3\times 10^{-6}\times 0.1\times 10^{6}=0.3$ s.
So in a time equal to 0.3 s from the start, the charge drops to 0.37 ($=e^{-1}$) of its initial value. Hence the p.d. drops to 0.37 of its initial value in that time.
So the p.d. after 0.3 s $=0.37\times 12=4.44$ V.

c) To calculate the p.d. after 1.0 s from the start, we need to use the equation $Q=Q_0e^{-t/RC}$.
Since $t=1.0$ s and $RC=0.3$ s, then $Q=Q_0\ e^{-1/0.3}$.
Hence $Q=Q_0\times 0.036$. Since the p.d. V is proportional to the charge Q, the p.d. falls to 0.036 of its initial value. Hence the p.d. $=0.43$ V.

Charging of a capacitor from a battery through a fixed resistance is represented in Figure 22.58(a) at the top of the next page. The resistance of the wires, circuit resistors and internal resistance of the battery are all lumped together in the resistance R. When switch S is closed, electrons flow from the battery's negative terminal onto the right hand plate of capacitor C. So electrons are forced off the left hand plate of C at the same rate, to return to the battery. Initially the plates are uncharged so the flow onto the plates is large. However, as the capacitor plates become more and more charged, the flow lessens. So the charging current becomes less and less as the capacitor charges up. Figure 22.58(b) shows how the charge builds up with time to a final value CV_0. The curve is called a **build-up exponential**.

The charging current I is equal to the rate of flow of charge onto the plates. So the gradient of the charge against time curve of Figure 22.58(b) gives the current. This initial current is high because the initial rate of flow of charge is high; so the initial gradient in Figure 22.58(b) is high. The gradient becomes less and less as the charge flow becomes smaller and smaller because the capacitor

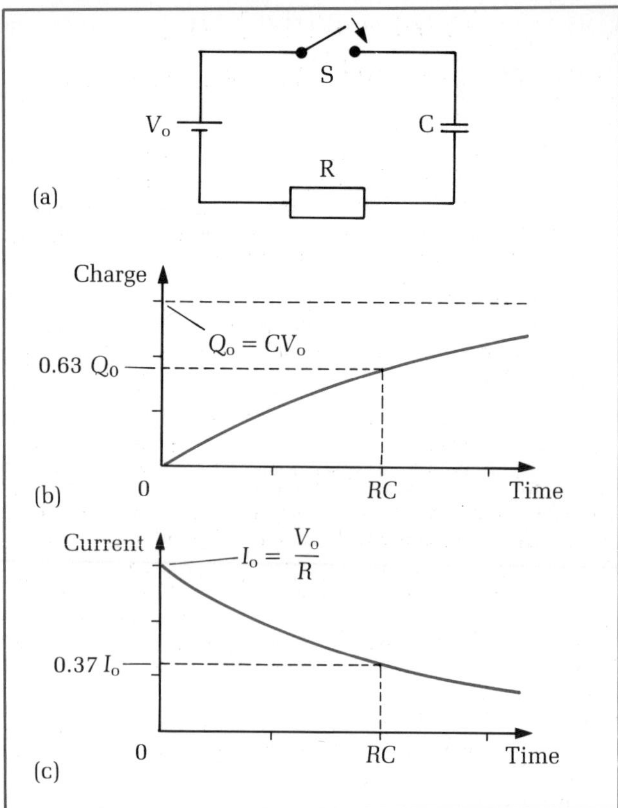

Figure 22.58 *Capacitor charging*

becomes fully charged. The charging current decreases exponentially so Figure 22.58(c) shows the way the current falls with time.

$$I = I_0 e^{-t/RC}$$

where RC = time constant,
I_0 = initial current = V_0/R,
t = time from start.

Investigating charging and discharging using an oscilloscope

Using a battery A capacitor is discharged through a large resistance using the circuit of Figure 22.59. An oscilloscope connected across the resistor allows the p.d. across the resistor to be measured at intervals. With the oscilloscope time-base off, a spot is seen on the screen.

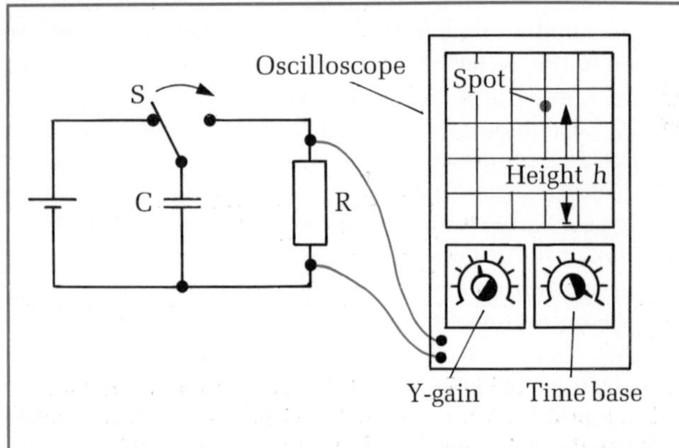

Figure 22.59 *Using a battery*

The position of the spot is adjusted to the bottom of the screen when the capacitor is fully discharged. Then the capacitor is charged from the battery using switch S. A stopwatch is started at the same time as the switch is reconnected to start the discharge. The spot on the screen jumps to near the top of the screen when the capacitor is charged. Then, as the capacitor discharges through the resistor, the spot gradually moves back to its zero position as the p.d. across the resistor falls. The position of the spot above its zero level is noted at regular intervals. The height of the spot above the zero level represents the p.d. across the resistor, and hence across the capacitor. So a graph of height against time ought to give an exponential decrease.

Using a square-wave generator A square-wave generator produces a p.d. which switches on and off regularly. Figure 22.61 shows the wave form from a square-wave generator. The generator is connected to a resistor in series with a capacitor (see Figure 22.60). An oscilloscope is used to display the changes of capacitor p.d. and resistor p.d. A dual-beam oscilloscope can be used to show one wave form above the other, as in Figure 22.62.

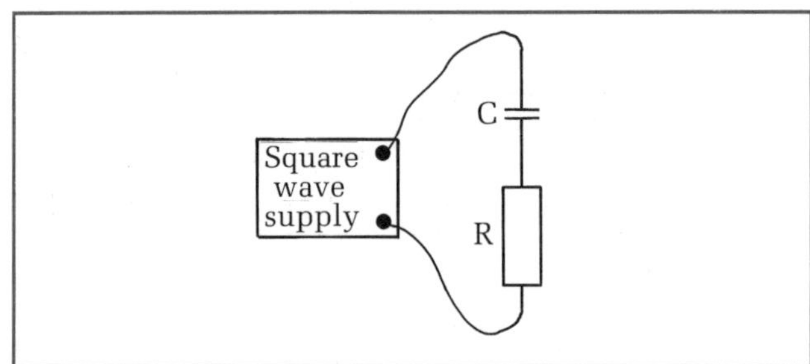

Figure 22.60 *Using square waves*

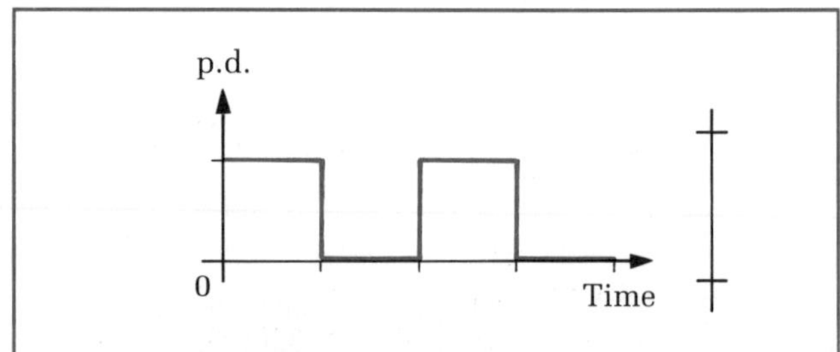

Figure 22.61 *Square wave form*

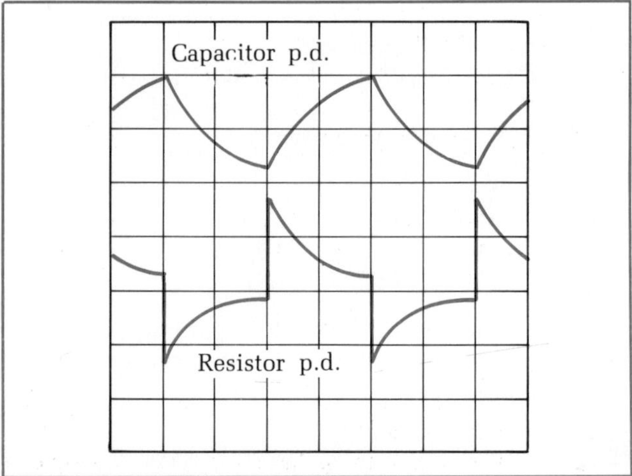

Figure 22.62 *Resistor and capacitor wave forms*

The capacitor p.d. repeatedly builds up then decreases, as the square-wave generator switches on and off repeatedly. Try different values of resistance and capacitance to see the effect on the wave form. If either R or C is substituted by a lower value, the time constant RC is reduced. So the wave form builds up and decreases more sharply, as in Figure 22.62.

The resistor p.d. is proportional to the current. The square-wave generator repeatedly charges and discharges the capacitor, so the current repeatedly decreases from an initially high value, first in one direction then in the opposite direction and so on. The resistor p.d. gives a wave form as in Figure 22.62. Again, substituting R or C with a lower value makes the wave form decrease more sharply.

22.6 Summary

Units and definitions

1 **The coulomb** is the unit of charge, equal to the charge passed when the current is 1 A for 1 s.
2 **The volt** is the unit of potential difference, equal to the electrical energy used per unit charge.
3 **The ohm** is the unit of resistance, equal to $1\ \mathrm{V\,A^{-1}}$.
4 **The farad** is the unit of capacitance, equal to $1\ \mathrm{C\,V^{-1}}$.
5 **The e.m.f.** of a cell is defined as the electrical energy per coulomb converted from other forms of energy inside the cell.

Equations

1 For steady currents, $Q = It$

2 For changing currents, $I = \dfrac{dQ}{dt}$

3 Potential difference $V = \dfrac{\text{electrical energy used } W}{\text{charge passed } Q}$

4 Electrical power $= IV$
5 Current through a conductor $I = nvAq$
where v is the drift velocity of charge carriers q
6 Resistance $R = V/I$
7 Resistivity $\rho = RA/L$
8 Resistor combination rules
Series: $R = R_1 + R_2$
Parallel: $1/R = 1/R_1 + 1/R_2$
9 Capacitance $C = Q/V$
10 Capacitor combination rules:
Series: $1/C = 1/C_1 + 1/C_2$
Parallel: $C = C_1 + C_2$
11 Energy stored in a charged capacitor $= \frac{1}{2}CV^2$
12 Time constant for a CR circuit $= RC$

Laws

Kirchhoff's laws
1st Law The total current entering a junction = the total current leaving the junction.
2nd Law The sum of the p.d.s round a closed loop of a circuit = the net e.m.f. round the loop.

Short questions

Circuits with steady currents

22.1 Calculate
a) the charge passed through a torch bulb in 5 minutes when the torch bulb carries a steady current of 0.3 A,
b) the number of electrons hitting the screen of a TV tube each second when the beam current is 1 mA. Assume the charge of an electron $= -1.6 \times 10^{-19}$ C.

22.2 Estimate the drift velocity of free electrons along a copper wire of diameter 0.4 mm carrying a steady current of 5.0 A. Assume that copper contains 1.0×10^{29} electrons per $\mathrm{m^3}$.

22.3 A 12 V battery is capable of delivering 1 A for 20 hours before it becomes flat. Estimate the total energy stored in the battery when fully charged.

22.4 In an electrolysis experiment, a steady current of 0.40 A is passed for 20 minutes through a cell containing copper sulphate solution. Both electrodes in the cell are copper plates. Copper ions from the solution are attracted to the cathode and neutralized by gaining two electrons per ion. Each copper atom leaving the anode goes into solution as a copper ion by releasing two electrons. Calculate
a) the total charge passed,
b) the number of copper atoms deposited on the cathode,
c) the mass deposited on the cathode, given each copper atom has a mass of 1.1×10^{-25} kg.

22.5 For a wire of uniform cross-section A carrying current I, the current density J is defined as I/A. Show that the p.d. per unit length along the wire is equal to J/σ where σ is the conductivity of the wire.

22.6 For each of the three circuits below, calculate
a) the total circuit resistance,
b) the cell current,
c) the p.d., current and power delivered to each resistor,
d) the electrical power delivered by the cell.

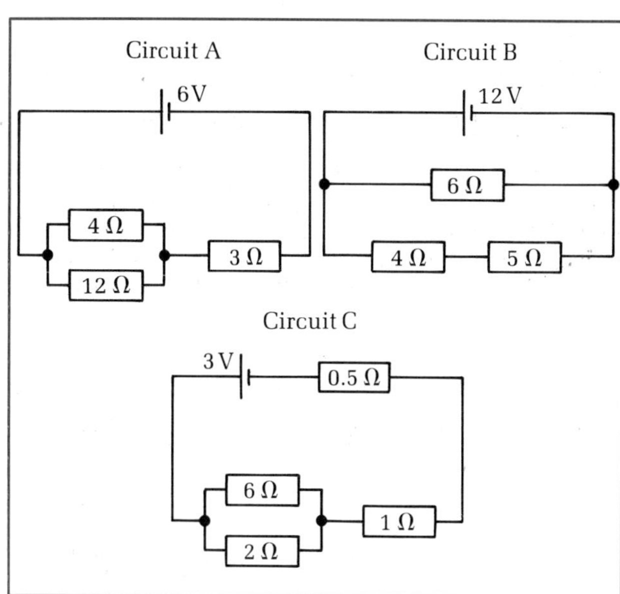

Figure 22.63

22.7 A torch bulb rated at 2.5 V, 0.3 A operates at normal brightness from two identical cells, each of e.m.f. 1.5 V and internal resistance r. Sketch the circuit and calculate
a) the internal resistance of each cell,

b) the electrical power supplied by the cells,
c) the power delivered to the torch bulb. Account for the difference between **b)** and **c)**.

22.8 In the previous question, suppose a second identical torch bulb is connected in parallel with the initial bulb when the initial bulb is operating at normal brightness. Explain why the initial bulb goes dimmer after the second bulb is connected into circuit.

22.9 A dry cell can deliver 0.15 A when connected to an 8.0 ohm resistor. When another 8.0 ohm resistor is connected in series with the first, the cell can only deliver 0.08 A to the combination. Use this information to calculate the cell's e.m.f. and internal resistance.

22.10 Suppose the two resistors in question **22.9** are reconnected in parallel with one another and then reconnected to the cell terminals. Calculate the current passing through each resistor now.

22.11 The circuit below shows a battery charger used to charge a 12 V car battery. The battery charger output is adjusted until it shows an output p.d. of 15 V and it supplies 2.0 A. Calculate
a) the charge and energy supplied by the charger in 1 hour,
b) the internal resistance of the battery,
c) the energy stored in the battery in 1 hour.

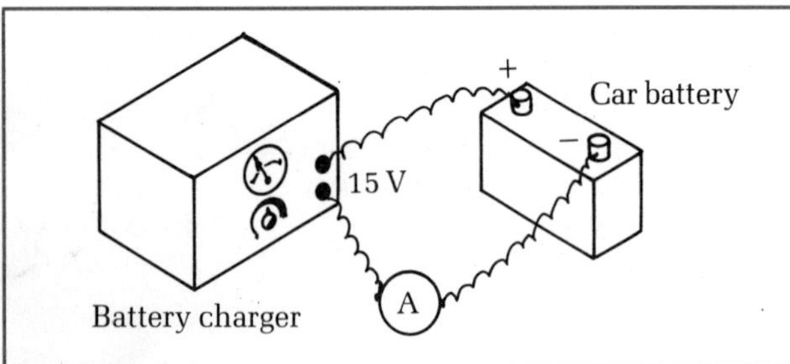

Figure 22.64

Account for the difference between the energy supplied and the energy stored.

22.12 In an experiment, the current and p.d. for a 12 V lamp bulb is measured at 2 V steps, giving results as follow:

p.d./volts	0	2.00	4.00	6.00	8.00	10.00	12.00
current/A	0	1.36	2.07	2.64	3.13	3.58	4.05

a) Plot a graph of p.d. against current,
b) From your graph, calculate the resistance of the bulb at (i) 1.0 A, (ii) 2.0 A, (iii) 4.0 A.

22.13 Puzzle boxes! An opaque box with four terminals gave the following measurements when it was tested by connecting its terminals in pairs in turn to a 1.5 V cell in series with a milliammeter.

		Negative terminal			
		1	2	3	4
Positive terminal	1	—	15 mA	7.5 mA	0
	2	15 mA	—	15 mA	0
	3	7.5 mA	15 mA	—	0
	4	5 mA	7.5 mA	15 mA	—

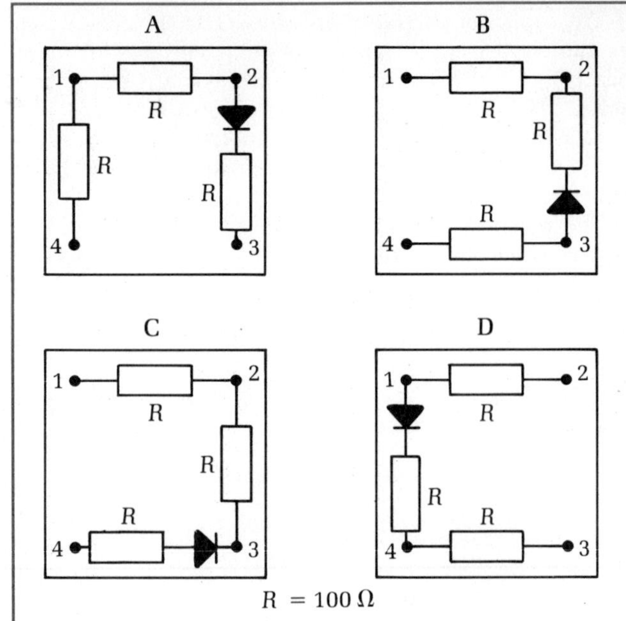

Figure 22.65

Given that the box's internal circuit is one of the four shown above, work out which one it contains.

22.14 Calculate the resistance per metre of constantan wire of diameter 0.35 mm, given its resistivity is $4.8 \times 10^{-7}\,\Omega$m. What length of this wire would be needed to make a 1.0 ohm resistor?

22.15
a) Calculate the resistance of a strand of aluminium wire of diameter 4.0 mm and length 1.0 m, given its resistivity is $2.5 \times 10^{-8}\,\Omega$ m. What would be the resistance per metre of 50 strands of this wire, using the strands in parallel as a cable?
b) An overhead cable made in this way has 50 strands of 4 mm aluminium wire surrounding a core of 8 steel strands, each of diameter 5.0 mm. Calculate the resistance per metre of the steel core, given its resistivity is $16 \times 10^{-8}\,\Omega$ m. What is the total resistance per metre of the cable? Why is the cable made in this way?

22.16 A battery of e.m.f. 6.0 V and internal resistance 3.0 ohms and another battery of e.m.f. 4.0 V and internal resistance 2.0 ohms are connected in parallel by joining like poles together. Then a 10 ohm resistor is connected from the positive to the negative terminals. Sketch the circuit and calculate
a) the current passing through each cell,
b) the current supplied to the 10 ohm resistor,
c) the p.d. across the 10 ohm resistor.

22.17 Four identical cells, each of e.m.f. 1.5 V and internal resistance 2 ohms are connected in parallel by joining like poles together. A 7 ohm resistor is then connected from the positive to the negative terminals. Sketch the circuit diagram, and calculate the current through the 7 ohm resistor.

22.18 In 24 hours, a certain household uses the following electrical devices:
a 3 kW kettle for 5 minutes on each of six occasions,
a 0.5 kW TV for 3 hours,
a 6 kW electric cooker for 2 hours,
a 100 W light bulb for 5 hours.
Calculate the total number of units (i.e. kilowatt hours) used. At 5p per unit, should the householder economize?

Capacitors in d.c. circuits

22.19 A 5000 μF capacitor is charged by supplying it with a steady current of 0.5 mA for 30 s using a variable resistor to keep the current steady. Calculate
a) the charge supplied to the capacitor,
b) the capacitor p.d. after 30 s.

22.20 For each of the capacitor combinations below, calculate
a) the charge stored and the p.d. across each capacitor,
b) the total charge and energy stored.

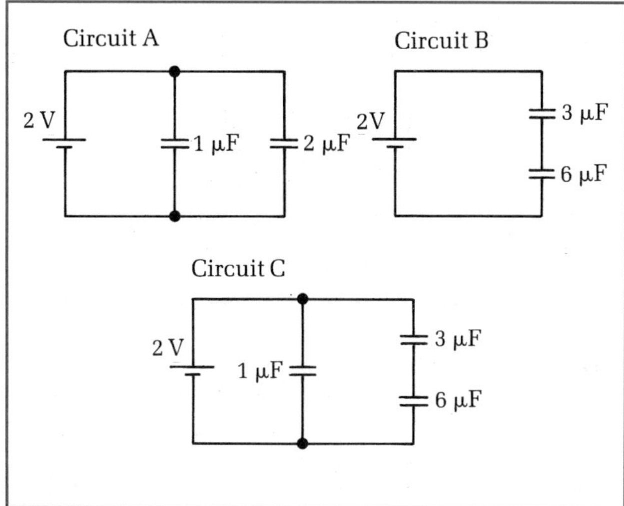

Figure 22.66

22.21 A 600 W flashlamp fitted to a camera is operated by discharging a 10 000 μF capacitor through the lamp. The capacitor is charged initially from a 6 V battery. Calculate
a) the energy stored in the capacitor when fully charged,
b) the duration of the flash.

22.22 Given six identical capacitors, each rated at 100 μF with a maximum working voltage of 6 V, give a circuit diagram to show how you would combine them to store maximum charge from a 12 V battery. Calculate the charge and energy stored in total for your arrangement.

22.23 A 2 μF capacitor is charged from a 6 V battery, and then the capacitor is discharged through a 0.5 MΩ resistor. Calculate
a) the energy stored in the capacitor before discharge,
b) the initial discharge current,
c) the p.d. across the capacitor at 1.0 s after the start of discharge.

22.24 A 1 kΩ resistor in series with a 0.1 μF capacitor is connected across the output terminals of a square wave generator operating at 2 kHz, as below. The voltage pulses from the square wave generator are 4 V high, each half-cycle being at 4 V followed by the next half-cycle at 0 V.

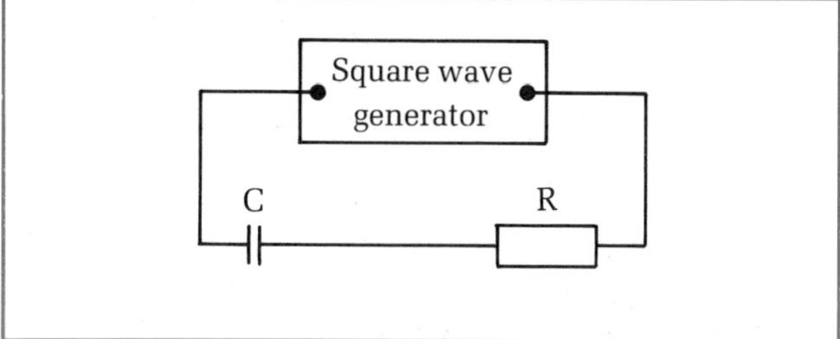

Figure 22.67

Sketch the wave forms for
a) the output of the signal generator,
b) the current,
c) the p.d. across the capacitor.

22.25 A 6 μF capacitor is charged up to a p.d. of 12 V then disconnected from the charging circuit. The capacitor is then connected across the terminals of an uncharged 4 μF capacitor. Calculate
a) the initial charge and energy stored in the 6 μF capacitor,
b) the final p.d. across the combination,
c) the energy stored finally in each capacitor.
Account for the difference between the total final energy and the initial energy stored.

23

Meters and bridges

23.1 Ammeters and voltmeters

Moving coil meters

Moving coil meters must take a current to give a non-zero reading. Most meters have linear scales, which means that equal intervals on the scale represent equal changes of current (or p.d.). Figure 23.1 shows a linear scale compared with a non-linear scale from a different type of meter. The maximum reading of a moving coil meter is called the **full-scale deflection** (f.s.d.). A meter which is described as having a full-scale deflection of 0.010 A will give the maximum scale reading when a current of 0.010 A passes through it. If the same meter has a resistance of 1000 ohms, then a p.d. of 10 V (= 0.010 A × 1000 Ω) applied to its terminals will give full-scale deflection. The construction and operation of moving coil meters are described on p. 181.

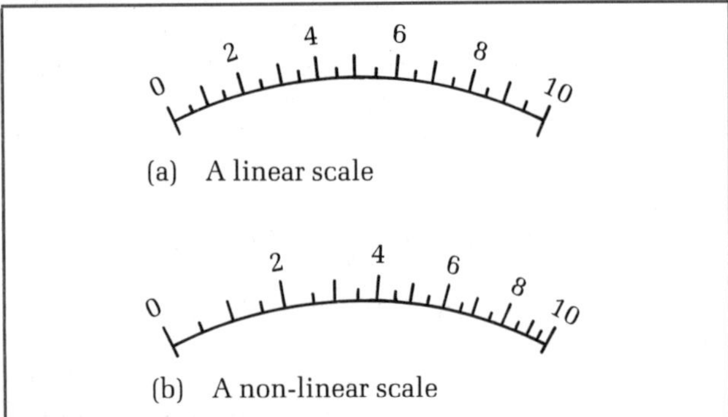

Figure 23.1 *Scales*

Moving coil ammeters A moving coil ammeter with a f.s.d. of 0.010 A would be damaged if the current through it exceeded 0.010 A. The meter can only be used to measure currents in the range from 0 to 0.010 A. To measure currents greater than 0.010 A, a **shunt** or **bypass resistor** must be connected in **parallel** with the meter. The shunt resistance is chosen according to the range of currents to be measured. For example, consider a meter with f.s.d. of 0.050 A and resistance of 10 ohms. Suppose it is to be used to measure currents up to 5.0 A. Figure 23.2 shows the meter fitted with a shunt resistor to allow measurements up to 5 A.

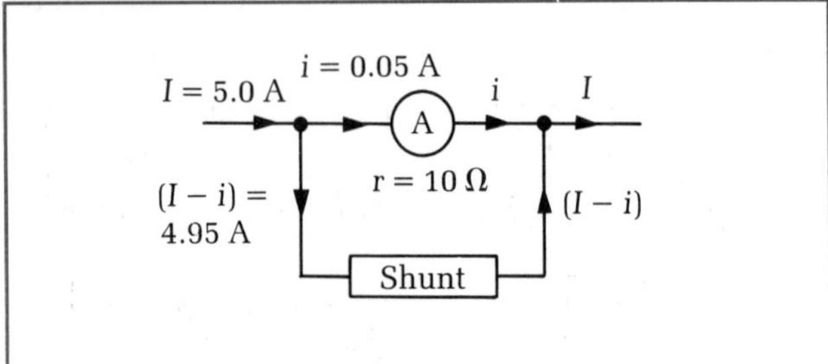

Figure 23.2 *Ammeter conversion*

To calculate the resistance of the shunt, we need to know the current which the shunt is to take and its p.d. when the meter shows full-scale deflection. The current which the shunt must take is the difference between the maximum current to be measured (5 A) and the maximum current through the meter (0.050 A). Hence, the current through the shunt must be 4.950 A. The p.d. across the shunt will be equal to the p.d. across the meter. For full-scale deflection, the meter p.d. = f.s.d. current × meter resistance = 0.050 × 10 = 0.50 V. So, the maximum p.d. across the shunt = 0.50 V. Hence the resistance of the shunt = p.d./current = 0.5/4.950 = 0.101 ohms.

More generally, suppose the meter resistance = r and its f.s.d. current = i. To measure maximum current I, the shunt must take current = $I - i$. The p.d. across the meter for full-scale deflection = f.s.d. current × meter resistance = ir. Since the shunt is in parallel with the meter, the shunt p.d. = ir. Hence the shunt resistance is given by

$$r_s = \frac{\text{p.d.}}{\text{current}} = \frac{ir}{(I-i)}$$

Some ammeters are dual-range meters, like the one shown in Figure 23.3. One of its terminals is always used; the other terminal is selected according to the current range to be measured. The choice of the second terminal connects different shunt resistors across the actual meter. For example, Figure 23.3 shows the 5 A terminal connected into a circuit. The meter scale is from 0 to 10 in steps of 0.5. As shown, the pointer is on the 3.5 mark on the scale. So the current is 1.75 A, because the full-scale deflection is 5 A.

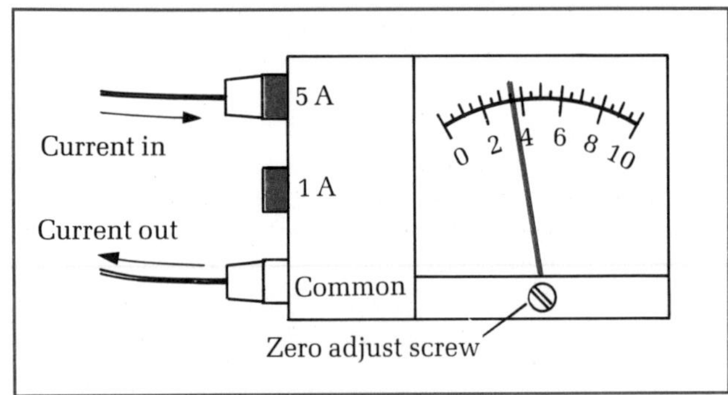

Figure 23.3 *A dual-range ammeter*

Moving coil voltmeters The potential difference across the terminals of a moving coil meter is proportional to the current through the meter. In other words, the meter has constant resistance. So the deflection of its pointer from zero is proportional to the p.d. applied to it. We can use any moving coil meter to measure p.d. provided the meter range is suitable. For example, a meter with a full-scale deflection current of 0.010 A and resistance 10 ohms can be used to measure p.d.s up to 0.1 V (= 0.01 A × 10 ohms). In other words, its full-scale deflection p.d. is 0.1 V.

To measure p.d.s greater than the full-scale value, a resistor must be connected in **series** with the meter. The series resistor is called a **multiplier**. The resistance of the multiplier is chosen according to the range of p.d.s to be measured. For example, suppose a meter with a full-scale deflection current of 0.010 A and resistance 10 ohms is to

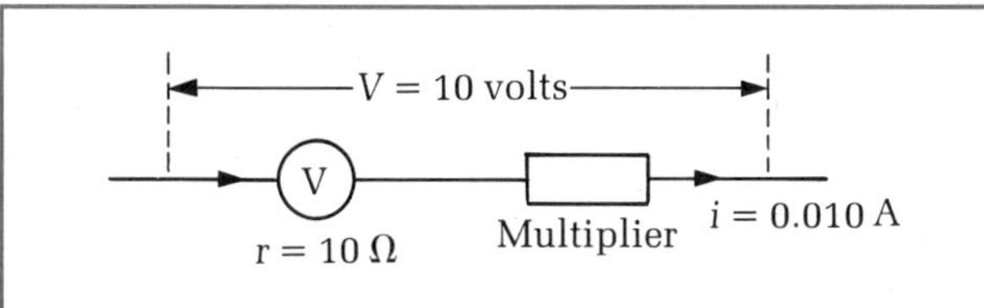

Figure 23.4 *Voltmeter conversion*

be used to measure p.d.s up to 10 V. The combined resistance of the meter and multiplier must only allow 0.010 A to pass when the p.d. across the meter and multiplier is 10 V. Figure 23.4 shows the idea. The resistance of the meter and multiplier $= \frac{\text{total p.d.}}{\text{current}} = \frac{10}{0.010} = 1000$ ohms.

The meter resistance is only 10 ohms. Therefore, the multiplier must have a resistance of 990 ohms, because it is in series with the meter.

More generally, suppose the f.s.d. current of the meter $= i$ and its resistance $= r$. To measure maximum p.d. V (assumed greater than ir), the combined resistance of meter and multiplier must equal V/i. Hence the multiplier must make up the difference in resistance between V/i and r.

$$\text{Multiplier resistance} = \frac{V}{i} - r$$

Multimeters These are designed to allow current or p.d. to be measured using the same instrument. A range switch is used to select one of several current or p.d. ranges. The switch connects the moving coil meter to an appropriate shunt or multiplier. Figure 23.5 shows a simplified multimeter with its range switch and meter scale. The internal circuit of shunts and multipliers is shown in Figure 23.6.

The basic meter has a resistance of 100 ohms and its full-scale deflection current is 10 mA. The range switch can be used to select any one of four resistors, or it can be set to the 'off' position. Each resistance value is chosen to give the required range.

Switch at P This allows the meter to be used to measure p.d.s up to 10 V. A 900 ohm resistor is connected in series with the meter when the switch is set at P. Use the multiplier formula above to check that this is right.

Switch at Q A 2400 ohm resistance is connected in series with the basic meter by setting the switch at Q. So p.d.s up to 25 V can be measured with the switch at Q.

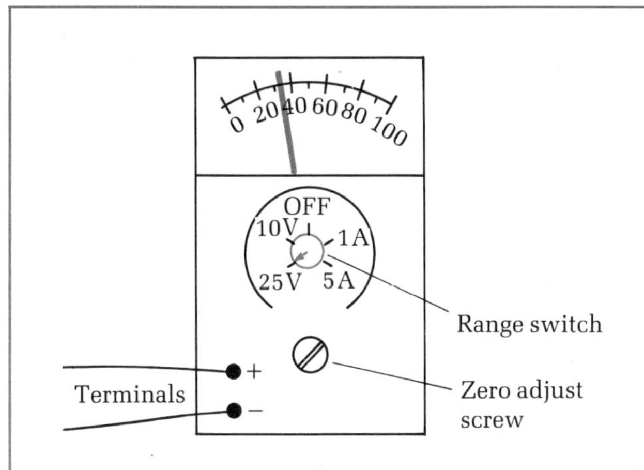

Figure 23.5 *A simplified multimeter*

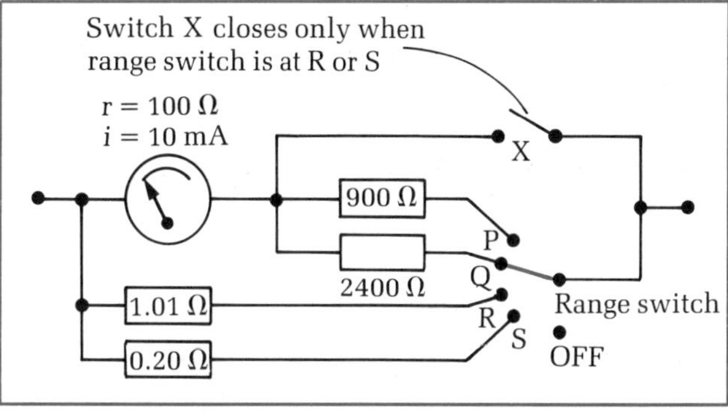

Figure 23.6 *Multimeter circuit*

Switch at R A shunt resistor is now connected to the basic meter, allowing currents up to 1.0 A to be measured. Use the shunt resistance formula to show that the shunt resistance must be 1.01 ohms.

Switch at S A different shunt resistor is now connected into the circuit, allowing currents up to 5.0 A to be measured. The resistance of the shunt must be 0.200 ohms. Check the value using the shunt resistance formula.

Consider the reading of the multimeter shown in Figure 23.5. The pointer reading is 35 on a scale up to 100. The range switch is set at Q for 25 V maximum. What is the meter reading? The scale reading of 100 corresponds to 25 V, so the scale reading of 35 corresponds to $35 \times 25/100 = 8.75$ V. In other words, the meter reading is 8.75 V.

Using moving coil meters

To measure resistance Compare the circuit of Figure 23.7(a) with that of Figure 23.7(b). Both circuits are to measure the resistance of identical resistors. In Figure 23.7(a), the voltmeter is connected directly across the resistor to be measured. The ammeter is in series with the test resistor and the voltmeter. So the ammeter measures the current through the test resistor *and* the current through the voltmeter. For accurate results, the ammeter current must equal the current through the test resistor only. The voltmeter must have a very high resistance so that negligible current passes through it. Then the test resistance can be calculated from the voltmeter reading/the ammeter reading.

In Figure 23.7(b), the voltmeter is connected across both the ammeter and the test resistor, so the ammeter current is the same as the current through the test resistor. However, the voltmeter reading is equal to the p.d. across the ammeter plus the p.d. across the test resistor. Hence the ammeter should have as low a resistance as

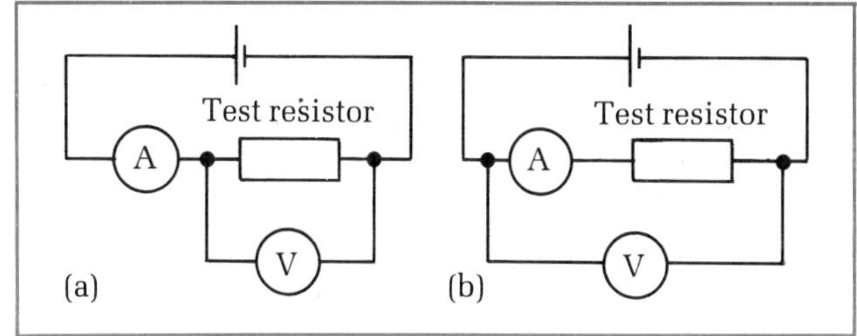

Figure 23.7 *Measuring resistance*

possible so the p.d. across it is negligible. Then the test resistance can be calculated from the voltmeter reading/the ammeter reading.

To measure the e.m.f. of a cell Suppose a voltmeter of resistance R is connected across the terminals of a cell of e.m.f. E and internal resistance r. Figure 23.8 shows the arrangement. No other connections are made to the cell so the circuit consists of an e.m.f. E in series with resistances r and R.
The current I through the cell is given by

$$I = \frac{\text{cell e.m.f. } E}{\text{total circuit resistance } (R+r)}$$

Hence the p.d. V across resistor R (i.e. across the voltmeter)

$$= \text{current } I \times \text{resistance } R = \frac{ER}{(R+r)}.$$

So the volmeter reading is given by

$$V = \frac{ER}{R+r}$$

Since $\frac{R}{R+r}$ is less than 1, then the voltmeter reading is always less than the cell e.m.f. E.

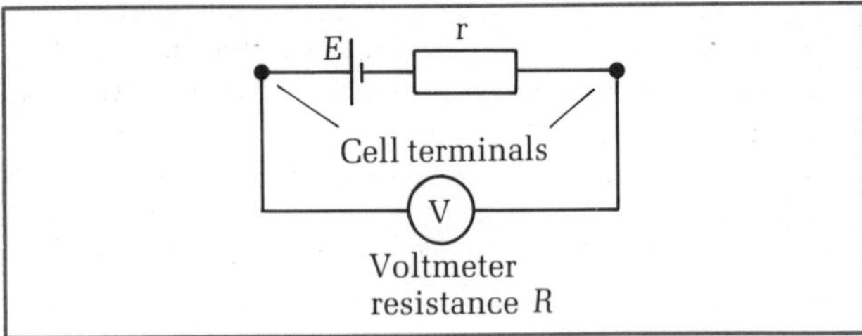

Figure 23.8 *Using a voltmeter*

Try it with numbers. Suppose $E = 12.0$ V, $r = 50$ ohms and the voltmeter resistance is 3000 ohms. Show that the voltmeter reading is 11.80 V.

Why is the voltmeter reading less than the cell's e.m.f.? Suppose the equation for the current I is rearranged to give

$$E = I(R+r) = (IR) + (Ir)$$

This rearranged equation tells us that the cell e.m.f. E is equal to the p.d. across the voltmeter (IR) + the p.d. across the internal resistance of the cell (Ir). So the voltmeter reading (IR) is less than the cell e.m.f. E because there is also a p.d. across the internal resistance (Ir).

Only by using a high resistance voltmeter can the cell e.m.f. be measured accurately. A high resistance voltmeter would take very little current from the cell, so the p.d. across the internal resistance would then be very small. Hence the voltmeter reading would measure the cell e.m.f. more accurately. Ideally, the voltmeter should draw no current from the cell under test; in theory, the voltmeter ought to have infinite resistance. In practice, this means its resistance must be much greater than the cell's internal resistance. Moving coil voltmeters do not give highly accurate results for measuring e.m.f.s because they must draw a current to operate. The typical resistance of a moving coil voltmeter is not great enough for measuring e.m.f.s accurately. An alternative to a moving coil voltmeter is a digital voltmeter which has a much greater resistance. A potentiometer can also be used to measure the e.m.f. of a cell (see p. 337).

Digital meters

Digital meters contain electronic circuits which measure p.d. The measured p.d. is displayed as a digital read-out. In use, most digital meters switch automatically to the most appropriate range. For example, suppose a digital meter has two ranges, from 0 to 200 mV and from 200 mV to 2.0 V. If the p.d. being measured suddenly falls from 1.5 V to 0.1 V, then the meter automatically switches from the upper to the lower range. Some digital meters have a manual range switch which needs to be set before use. So an auto-ranging DVM (that is, digital voltmeter) only needs to be set for current or p.d. before use.

When a DVM is set to measure current using its selector switch, as in Figure 23.9, the switch connects a resistor across the meter's terminals inside its casing. The resistor has a very accurate value of resistance so the current through it is proportional to the p.d. across it. The meter actually measures the p.d. across the resistor when used as an ammeter. For example, if the resistance is exactly 1 ohm, then the measured p.d. will equal the numerical value of current through the resistor.

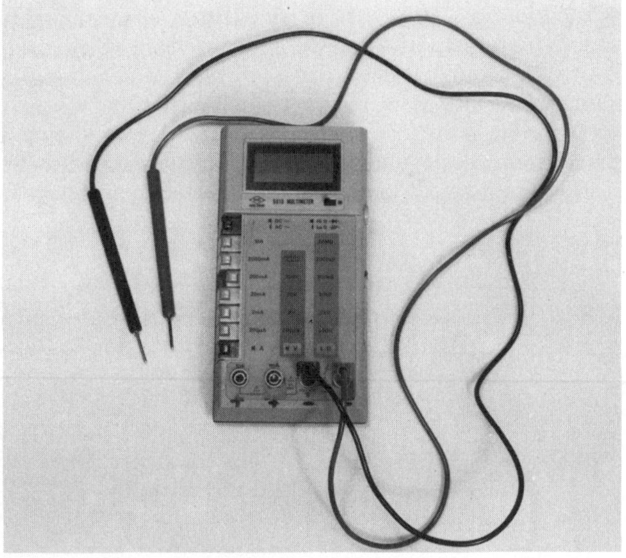

Figure 23.9 *A digital multimeter*

23.2 Potentiometers

The principle of the potentiometer

The potential divider is used to supply variable p.d. This is achieved by connecting a battery across a fixed length of wire of uniform resistance. A sliding contact can then be moved along the wire, as illustrated in Figure 23.10, giving a variable p.d. between the contact and one end of the wire.

Sometimes the resistance wire is coiled round a circular track, as in Figure 23.11(b). When the spindle is turned, the sliding contact moves round the track, so the p.d. across the output terminals varies with the position of the contact on the track.

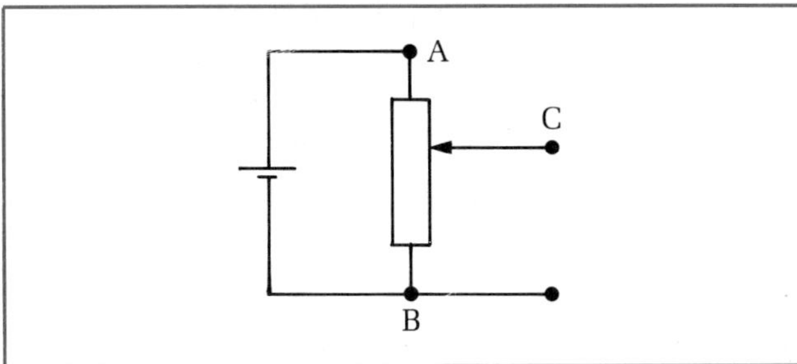

Figure 23.10 *Potential divider circuit symbol*

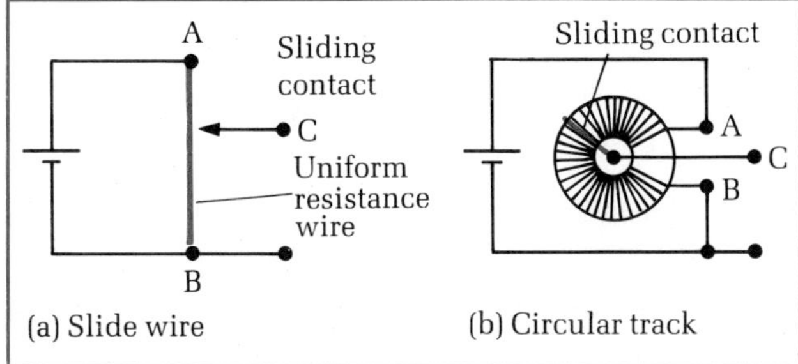

Figure 23.11 *Practical potential dividers*

The potentiometer is a potential divider used to measure e.m.f.s and potential differences. The slide-wire potentiometer consists of a fixed length of uniform resistance wire, as in Figure 23.11(a). A cell is used to maintain a constant p.d. across the ends of the wire. Since the wire is uniform all the way along its length, the p.d. across each centimetre of length is the same all along the wire. Hence the p.d. between the sliding contact and one end of the wire is proportional to the length of wire between the two points. The p.d. V from B to C is proportional to the length L from B to C.

$$V_{BC} = kL_{BC}$$

where k is a constant.

Suppose a test cell is connected in series with a centre-reading meter and then to points B and C of the potentiometer circuit, as in Figure 23.12. If the sliding contact is moved along the wire, the meter deflection will change. Current will be forced through the test cell if V_{BC} is greater than the cell's e.m.f. If, however, V_{BC} is less than the cell's e.m.f., then current will pass through the cell and meter in the opposite direction. At one position of the sliding contact, the meter deflection will be exactly zero (i.e. a null deflection). At this position, the cell's e.m.f. is exactly equal and opposite to the p.d. between B and C, V_{BC}. If the constant k is known, then by measuring the length from B to C, the value of V_{BC} can be calculated. Hence the cell e.m.f. can be determined.

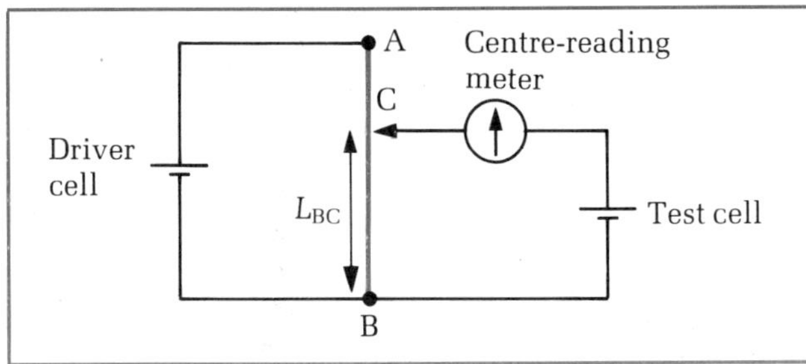

Figure 23.12 *Balancing a test cell*

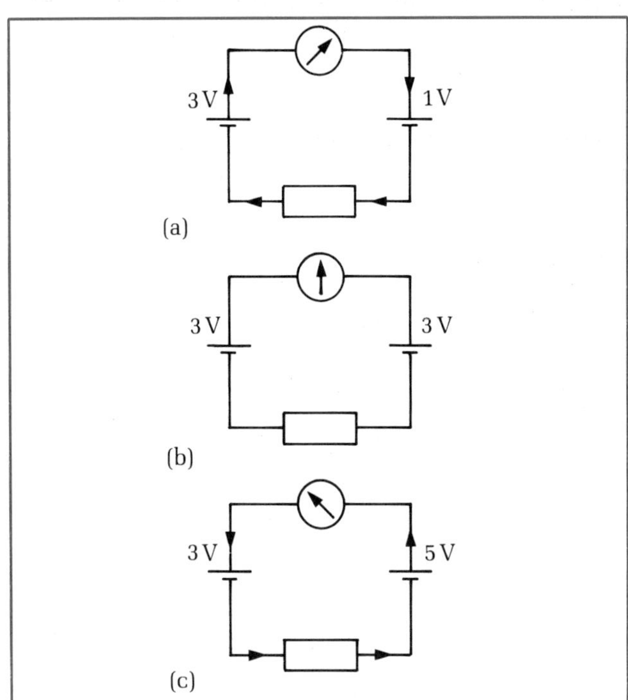

Figure 23.13 *Opposing cells*

The position of the sliding contact where the meter gives a null deflection is called the **balance position**. The length BC is then called the **balance length**. The term 'balance' is used because it indicates that the cell e.m.f. trying to push current one way is exactly opposed by the p.d. from B to C trying to push current in the opposite direction. The result is no current at all. The same applies to the circuit shown in Figure 23.13(b). The two cells in that circuit exactly balance one another out so there is no current. This is not the situation in the circuits of Figure 23.13(a) or (c). In part (a), the 3 V cell is opposed by the 1 V cell so the net e.m.f. is 2 V; hence current passes clockwise, as shown. In part (c), the 3 V cell is opposed by a 5 V cell giving a net e.m.f. of 2 V in the opposite direction to part (a), so the current is anticlockwise, as shown. In part (b), the 3 V cell is opposed by another 3 V cell so there is no current passing through the circuit.

Measuring the e.m.f. of a cell

The test cell is connected into the circuit shown in Figure 23.14. The rheostat is adjusted so that the test cell can be

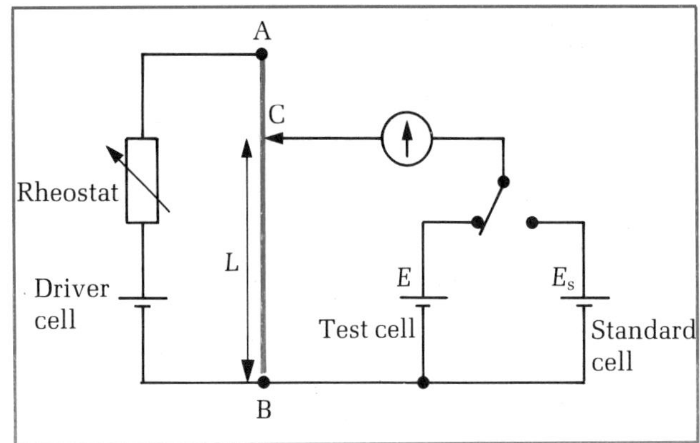

Figure 23.14 *Measuring the e.m.f. of a test cell*

balanced with the balance length as long as possible. The balance point for the sliding contact C is accurately located and the balance length L is then measured.

Without altering the rheostat, the standard cell is then connected in place of the test cell to calibrate the wire (i.e. to determine the p.d. per unit length, k). To do this, the balance point for the standard cell is located, then the balance length L_s for the standard cell is measured accurately. Given the e.m.f. of the standard cell, E_s, the p.d. per unit length k can be calculated from $k = E_s/L_s$. Then the e.m.f. of the test cell can be calculated from

$$E = kL = \frac{E_s}{L_s}L$$

where E = test cell's e.m.f.,
E_s = standard cell's e.m.f.,
L = test cell's balance length,
L_s = standard cell's balance length,
k = p.d. per unit length.

Consider the following points.

1 If the test cell cannot be balanced, it could be because it has been connected into the circuit with the wrong polarity. If so, the test cell's e.m.f. and the p.d. along the wire act in the same direction, so they cannot balance one another out. The remedy is to reconnect the test cell with the current polarity.

The driver cell supplies a p.d. across the full length of the wire. A balance point is only possible if the test cell's e.m.f. is not more than the p.d. across the wire. If the driver cell runs down during the experiment, the p.d. per unit length increases, so the calculation for the test cell's e.m.f. will not be accurate.

2 If the standard cell cannot be balanced after balancing the test cell, one possible reason may be that the standard cell's e.m.f. is greater than the p.d. across the wire. The rheostat must then be reset to give a balance point for the standard cell on the wire. Then the measurement with the test cell must be repeated. The rheostat setting determines the current through the potentiometer wire. Since the p.d. across the wire is proportional to the current, then the rheostat determines the p.d. across the wire.

3 At balance, no current passes through the test cell because there is a null reading on the meter, so there is no p.d. across the internal resistance of the test cell. Hence the p.d. across the cell terminals is equal to its e.m.f. The p.d. across the balance length is therefore equal to the test cell's e.m.f.

Compare using a potentiometer with using a moving coil voltmeter to measure the test cell's e.m.f. As explained on p. 336, a moving coil meter must pass current to operate. So the p.d. across the test cell's terminals will be less than its e.m.f., due to its internal resistance. The advantage of the potentiometer is that no current is passed through the test cell when it is at balance. Therefore the potentiometer gives a more accurate measurement of the test cell's e.m.f.

4 The internal resistance of the test cell can be determined using the potentiometer. First, the balance length of the test cell is measured. Then a known resistance R is connected across the test cell's terminals, as in Figure 23.15. The balance length is remeasured, to

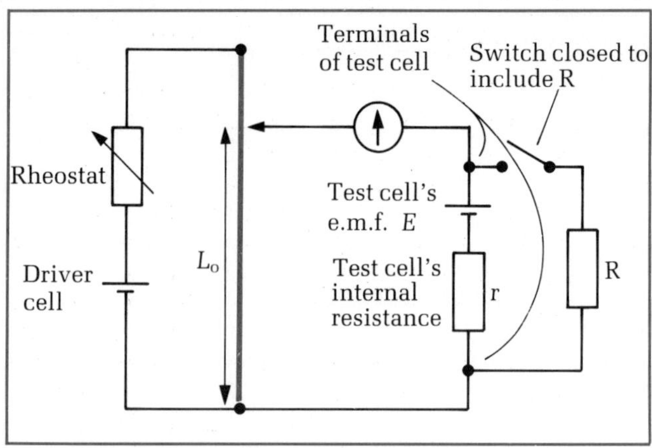

Figure 23.15 *Measuring internal resistance*

give a value less than the first measurement. The reason is that resistnce R allows current to pass through the cell, even at balance.

Suppose L_0 = the balance length without R, and L = the balance length with R, then the test cell's e.m.f. $E = kL_0$ (since no current passes through the test cell at balance without R). Therefore the p.d. V across its terminals with R in the circuit is given by $V = kL$, where k = p.d. per unit length.

From p. 336, $V = \dfrac{ER}{R+r}$

Hence $kL = \dfrac{kL_0R}{R+r}$

Which gives $r = R\dfrac{(L_0 - L)}{L}$

Worked example A potentiometer is to be used to measure the e.m.f. and internal resistance of a test cell. Without any resistor in parallel with the test cell the balance length is 740 mm. When a 5 ohm resistor is connected across the test cell, the balance length falls to 550 mm. When the test cell and resistor are replaced by a standard cell, the balance length is 535 mm. Given the e.m.f. of the standard cell is 1.1 V, calculate **a)** the e.m.f., **b)** the internal resistance of the test cell.

Solution The p.d. per unit length, k = standard cell e.m.f./standard balance length.

Hence $k = \dfrac{1.1\ \text{V}}{0.535\ \text{m}} = 2.06\ \text{V m}^{-1}$

a) For the e.m.f. of the test cell, $E = kL_0$, where L_0 is the balance length without the resistor.
Therefore $E = 2.06 \times 0.740 = 1.52$ V.

b) For the internal resistance of the test cell, r, the p.d. across its terminals with the 5 ohm resistor in circuit $= kL$, where L is the balance length.
Hence $V = 2.06 \times 0.550 = 1.13$ V.
Therefore the current through the 5 ohm resistor at balance = p.d./resistance = 1.13/5 = 0.227 A.
Hence the current from the cell = 0.227 A.
The p.d. lost across the internal resistance $= E - V = 1.52 - 1.13 = 0.39$ V.

So the internal resistance $= \dfrac{\text{lost p.d.}}{\text{current}} = 1.72$ ohms.

Calibrating a voltmeter

All voltmeters need to be checked from time to time to make sure that they are still accurate. Moving coil meters can become inaccurate if the restoring springs weaken or if the magnet weakens (see p. 181). Digital voltmeters can become inaccurate if electronic components in the meter fail. A quick check for accuracy is to connect a standard cell to the meter, but this only checks the meter at one point on its scale. Also, standard cells like the Weston standard cell (p. 318) must not be allowed to deliver more than 10 μA for about 10 s. Otherwise the standard cell itself becomes damaged.

A potentiometer can be used to check the accuracy of a voltmeter over the whole of its range. The potentiometer wire is first calibrated by measuring the balance length with a standard cell in the circuit, as in Figure 23.14. Hence the p.d. per unit length k can be calculated from the standard cell e.m.f./standard balance length.

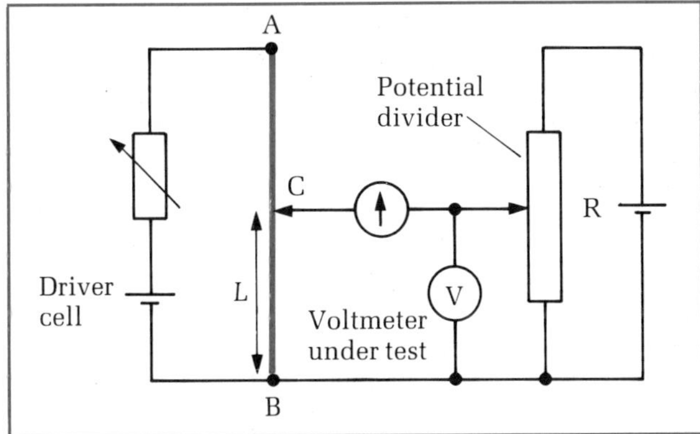

Figure 23.16 *Calibrating a voltmeter*

The voltmeter under test is then connected across the output terminals of a potential divider, as shown in Figure 23.17. The p.d. across the voltmeter is balanced against the p.d. across the potentiometer wire; the potential divider is altered in steps to give different voltmeter readings. At each step, the balance point is located and the balance length is measured. The p.d. V along the balance length L can be calculated from $V = kL$. The voltmeter reading V_m can then be compared with the calculated p.d. V for a range of values.

What if the voltmeter range is greater than the maximum p.d. across the wire? For example, suppose the p.d. across the whole length of the potentiometer wire were 2 V. How would you calibrate a 5 V voltmeter? You could easily check from 0 to 2 V, but what about from 2 to 5 V? Use two accurate resistance boxes R_1 and R_2 in series, connected to the voltmeter, as in Figure 23.17. Balance the p.d. across one of the resistances against the p.d. along the wire.

Suppose the p.d. across resistor $R_1 = V_1$.
Then the current through R_1, $I = V_1/R_1$.
Hence the p.d. across R_1 and R_2 in series

$$= I(R_1 + R_2) = V_1\frac{(R_1 + R_2)}{R_1}$$

So the p.d. across the voltmeter (= p.d. across R_1 and R_2 in series) can be calculated from the measured value of V_1 and the chosen values for R_1 and R_2.

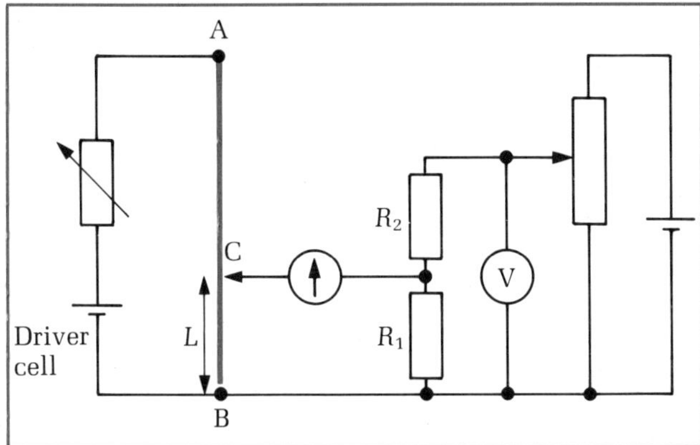

Figure 23.17 *Extending the calibration*

Measuring small e.m.f.s

Consider a potentiometer with a wire of length 1.00 m which has a p.d. across its ends of 2.0 V. Suppose we use it to measure a p.d. of 5 millivolts. The balance length for 5 mV is $\frac{0.005}{2.0} = 0.0025\,\text{m} = 2.5\,\text{mm}$. To measure such a small balance length would be impossible, given that the accuracy for locating a balance point is 1 mm or so. Clearly the straightforward potentiometer circuit of Figure 23.14 is quite unsuitable for measuring small p.d.s or e.m.f.s.

Measuring small p.d.s can be achieved by using a long potentiometer wire. Suppose the wire is 100 m long with a p.d. across its ends of 2.0 V. The p.d. across each metre is therefore 2.0/100 = 0.02 V. In other words, the p.d. per unit length is 0.02 V m^{-1} or 20 mV m^{-1}. So a small p.d. of 2.5 mV would be balanced across 0.125 m (= 2.5/20 m) of wire. Even better would be a potentiometer wire 500 m long; the balance length would then be 0.625 m.

In practice we only need to use 1 m of such a long wire, so the rest of the long wire can be replaced by an equivalent resistance, as in Figure 23.18. Suppose a 1 m poten-

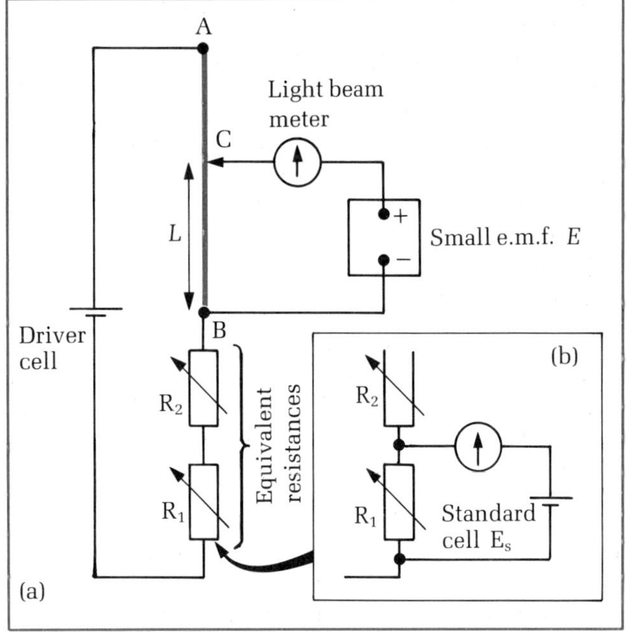

Figure 23.18 **a** *Measuring a small e.m.f.* **b** *Calibrating the wire*

tiometer wire has a resistance of 5 ohms. Then a resistance of 5 × 499 ohms (= 2495 ohms) is equivalent to 499 m of the same wire. So a resistance of 2495 ohms in series with the wire gives an equivalent total length of 500 m.

To calibrate the wire of Figure 23.18 (i.e. to determine the p.d. per unit length), we would need to use a standard cell. But standard cell e.m.f.s are in the range from 1 to 2 V depending on the type of cell, so it would not be possible to balance a standard cell along the 1 m length of the wire. Instead, the equivalent resistance is made up using two resistance boxes R_1 and R_2, as shown. The standard cell is balanced against the p.d. across R_1 by adjusting R_1 and R_2, keeping $R_1 + R_2$ constant. The enlarged part (b) of Figure 23.18 shows the circuit with the standard cell, in series with a centre-reading meter, connected across R_1.

To calculate the p.d. per unit length, k, consider the standard cell at balance across R_1. The current I along the wire is then equal to the current through R_1 and R_2. So the p.d. across $R_1 = IR_1$. Since this is the p.d. which balances the standard cell's e.m.f. E_s, then $E_s = IR_1$. Hence $I = E_s/R_1$.

Since current I passes through the potentiometer wire of resistance r, then the p.d. across the potentiometer wire $= Ir = E_s r/R_1$.

Hence the p.d. per unit length k is given by

$$k = \frac{\text{p.d. across wire}}{\text{length of wire } L_w} = \frac{E_s r}{L_w R_1}$$

Worked example A potentiometer used to measure the e.m.f. of a thermocouple has two resistance boxes, totalling 2000 ohms, in series with the wire. The wire resistance is 3 ohms and its length is 1.00 m. A standard cell of e.m.f. 1.018 V is balanced across one of the resistance boxes when its resistance is 1244 ohms and that of the other box is 756 ohms. A thermocouple of unknown e.m.f. is then balanced across 0.640 m of the potentiometer wire. Calculate the e.m.f. of the thermocouple.

Solution At balance, the current through the 1244 ohm

$$\text{resistance box} = \frac{1.018 \text{ V}}{1244 \text{ ohms}} = 8.18 \times 10^{-4} \text{ A}.$$

Hence the p.d. across the wire = current × wire resistance $= 3 \times 8.18 \times 10^{-4} = 2.455 \times 10^{-3}$ V.

Therefore the p.d. across 0.640 m of wire

$$= \frac{\text{p.d. across wire}}{\text{length of wire}} \times 0.640 = \frac{2.455 \times 10^{-3}\text{V}}{1.0\text{m}} \times 0.640$$

$$= 1.57 \times 10^{-3} \text{ V} = 1.57 \text{ mV}.$$

The thermocouple e.m.f. = p.d. across 0.640 m of wire, so the thermocouple e.m.f. is 1.57 mV.

23.3 The Wheatstone bridge

Comparing resistances

In the circuit shown in Figure 23.19, two fixed resistors R and S are connected in series with one another. Then the two resistors are connected in parallel with a fixed length of uniform resistance wire. A driver cell supplies a fixed p.d. across the wire and the two resistors.

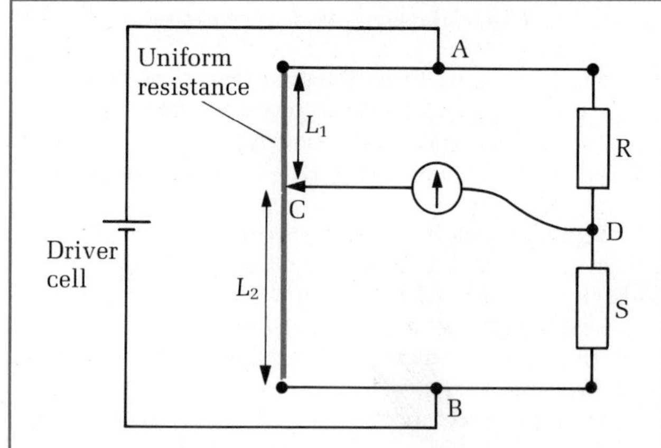

Figure 23.19 *A Wheatstone bridge*

The p.d. across resistor R is constant. Its value depends on the ratio of R and S. The p.d. across length L_1 of the wire can be varied by moving the sliding contact along the wire.

The sliding contact C is connected via a centre-reading meter to point D which is between R and S. By moving C along the wire, a position can be found where the meter shows a null reading. At this point, the p.d. across R is exactly equal to the p.d. across length L_1; so there is no p.d. between C and D. The circuit is said to be in balance because the two p.d.s exactly oppose one another.

The circuit shown in Figure 23.19 is known as a **Wheatstone bridge**, invented by Charles Wheatstone for measuring resistances very accurately. Suppose the value of S is known (i.e. S is a standard resistor) and we wish to determine R. At balance, the wire is divided into an upper section of length L_1 and a lower section of length L_2 $(= L_w - L_1)$, where L_w is the wire length. The resistance ratio R/S is equal to the length ratio L_1/L_2. So if L_1 and L_w are measured, and S is known, the value of R can be calculated.

Theory of the balanced Wheatstone bridge

Consider the Wheatstone bridge circuit of Figure 23.20. The wire of Figure 23.19 has been replaced by two sep-

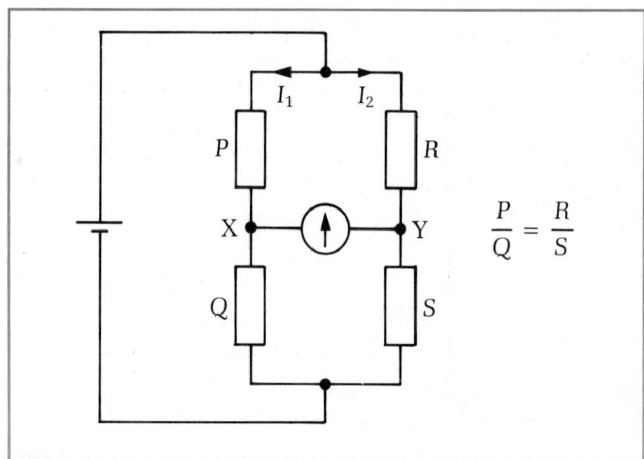

Figure 23.20 *The balanced bridge*

arate resistors P and Q. Suppose P, Q and R are fixed resistors, and S is a resistance box.

The resistance of S is adjusted until the meter shows a null reading. The bridge circuit is now balanced. There is no current through the meter so the current through P is the same as the current through Q. Also, the current through R is the same as that through S.

The p.d. across P, $V_p = I_1 P$.

The p.d. across Q, $V_q = I_1 Q$, hence $V_p/V_q = P/Q$.

The p.d. across R, $V_r = I_2 R$.

The p.d. across S, $V_s = I_2 S$, hence $V_r/V_s = R/S$.

Since there is no current through the meter, the potential at X is the same as the potential at Y. Hence $V_p/V_q = V_r/V_s$. Therefore $P/Q = R/S$.

$$\frac{P}{Q} = \frac{R}{S}$$

If P and Q form a wire of length L_W, and P corresponds to length L_1 of the wire then

$$\frac{R}{S} = \frac{L_1}{(L_W - L_1)}$$

Worked example In the Wheatstone bridge circuit of Figure 23.20, P and Q are 500 ohm resistors, R is a resistor of unknown resistance and S is a resistance box. The value of S can be adjusted in 1 ohm steps from 0 to 1000 ohms. The bridge is balanced when S is set at 345 ohms. **a)** Calculate the resistance of R. **b)** If a 200 ohm resistor is then connected in parallel with R, calculate the value of S necessary to rebalance the bridge.

Solution At balance $\frac{R}{S} = \frac{P}{Q}$

a) For $P = Q = 500$ ohms, $S = 345$ ohms,

$$\frac{R}{345} = \frac{500}{500}$$

Hence $R = 345$ ohms.

b) R is now replaced in Figure 23.20 by a parallel combination of R and a 200 ohm resistor, as in Figure 23.21.

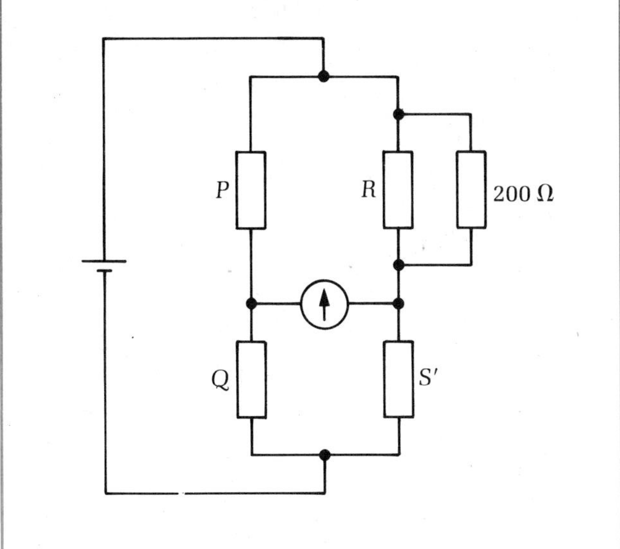

Figure 23.21

The equivalent resistance R' of the combination is given by

$$\frac{1}{R'} = \frac{1}{R} + \frac{1}{200} = \frac{1}{345} + \frac{1}{200}$$

Hence $R' = \dfrac{345 \times 200}{345 + 200} = 127$ ohms.

Since the condition for the balanced bridge is now

$$\frac{P}{Q} = \frac{R'}{S'}$$

where S' is the new value of the resistance box to be calculated, then $S' = R' \times \frac{Q}{P} = 127 \times \frac{500}{500} = 127$ ohms.

Hence $S' = 127$ ohms.

Using the Wheatstone bridge

To investigate the variation of resistance of a thermistor with temperature A thermistor is a resistor designed to have a resistance which changes with temperature. Its resistance can be measured by connecting it into one arm of a Wheatstone bridge circuit, as shown in Figure 23.22. A uniform wire AB with a sliding contact C forms resistances P and Q of the circuit. A resistance box of resistance S is used to complete the bridge circuit, chosen to give a balance point near the middle of the wire.

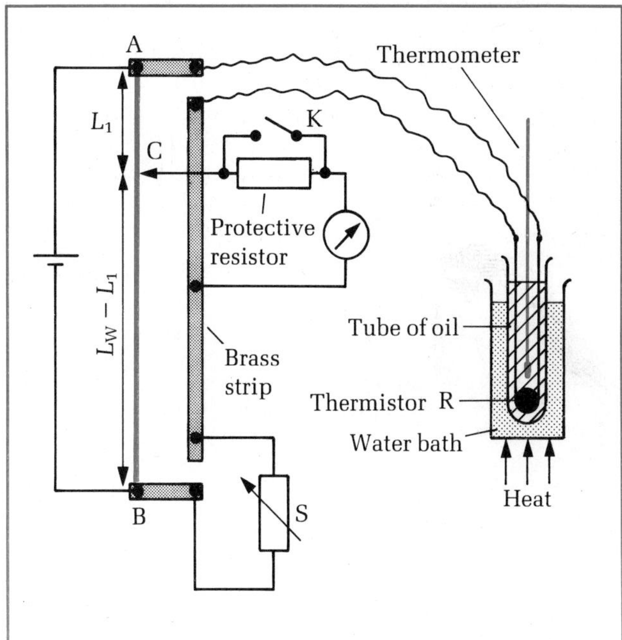

Figure 23.22 *Investigating thermistors*

The thermistor is contained in a tube of oil containing a thermometer. The oil is heated to different temperatures using a water bath, as shown. At each temperature, the bridge is balanced and the balance length L_1 and oil temperature θ are measured. A protective resistor in series with the meter is used to limit the current through the meter when it is off-balance. Then, after locating the approximate balance point, the protective resistor is short-circuited by closing switch K. The meter is now much more sensitive to changes in the position of the sliding contact. The balance point can be located much more accurately with K closed, but hunting for the balance position requires K to be open to limit the current through the meter.

At each temperature θ, the resistance of the thermistor can be calculated from the equation for the balanced bridge. $\frac{R}{S}=\frac{P}{Q}$. Since the wire is uniform, then P is proportional to length L_1, and Q is proportional to length $L_w - L_1$. Hence

$$\frac{P}{Q}=\frac{L_1}{L_w - L_1}$$

so

$$R=\frac{L_1}{L_w - L_1}\times S$$

where R = thermistor resistance,
S = resistance of resistance box,
L_1 = balance length (facing R),
L_w = length of the wire.

To measure the resistivity of a uniform wire The thermistor of Figure 23.22 is replaced by a measured length of the test wire. The bridge is balanced, choosing S to give a balance near the middle. The above equation is used to calculate the resistance R of the test wire. In the above equation, L_1 is the balance length, L_w is the **bridge wire** length, not to be confused with the test wire length.

The procedure is repeated for different lengths of the test wire from 0.2 m to 1.0 m in steps of 0.2 m. The oil bath used in Figure 23.22 is not needed here, but care must be taken to make sure the test wire does not twist round and short-circuit itself. Finally, the diameter of the test wire is measured using a micrometer at several positions along its length. Hence the average diameter d is calculated.

The resistance R of each length l of test wire is given by the resistivity equation.

$$R=\frac{\rho l}{A}$$

where A = area of cross-section = π (diameter)^2/4,
ρ = resistivity of test wire.

Hence a graph of R against l should give a straight line passing through the origin. The gradient of the graph is

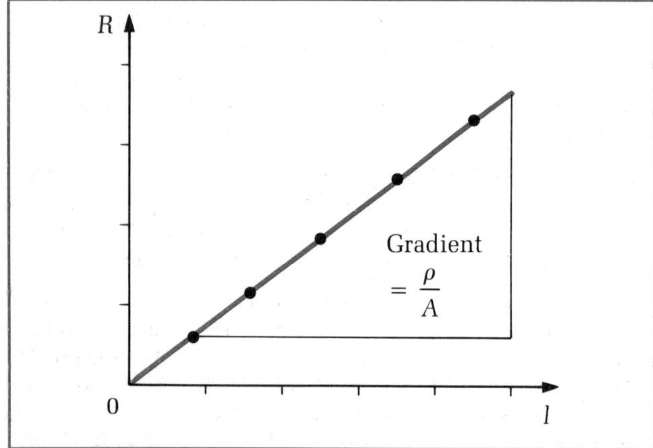

Figure 23.23 *Measuring resistivity*

equal to ρ/A, so the resistivity can be calculated from the measured value of gradient and the value of A.

Consider the following points.

1 The standard resistance S, chosen from a resistance box, is selected to give a balance point in the middle third of the bridge wire. The calculated value of R is then more accurate than if the balance point were near either end.

Suppose we can locate the balance point to within 5 mm. Compare the choices of S that give a balance point near the middle or near the end.

a) If S = 10 ohms gives a balance point at the middle, then R = 10 ohms. But assuming the bridge wire is 1000 mm long, the balance point could have been up to 5 mm either way from the middle. So the balance length could have been in the range 495 to 505 mm. For balance length L_1 = 495 mm, the calculated value of R would be 495 × 10/505 = 9.80 ohms. For balance length L_1 = 505 mm, the value of R would be 505 × 10/495 = 10.20 ohms. So the value of R is accurate to 0.2 ohms.

b) With the same resistance R, suppose we had chosen S = 2 ohms. The balance length L_1 is given by

$$\frac{L_1}{1000 - L_1}=\frac{R}{S}=\frac{10}{2}=5$$

Hence $L_1 = 5000 - 5L_1$, which gives $6L_1 = 5000$.
Hence L_1 = 5000/6 = 833 mm.

However, we can only locate the balance point to within 5 mm. So the measured value of L_1 would be in the range from 828 mm to 838 mm. For L_1 = 828 mm, the equation for R gives R = 828 × 2/172 = 9.63 ohms. For L_1 = 838 mm, the calculation for R gives R = 838 × 2/162 = 10.35 ohms. So the value of R is accurate to 0.36 ohms in this case. Clearly, a more accurate value is obtained by choosing S to give a balance point nearer the middle.

2 The bridge wire must be uniform. If not, the resistance of each section is no longer proportional to its length, so the equation for R becomes inaccurate.

3 The balance point does not alter if the driver cell runs down. The balance point depends only on the resistances P, Q, R and S.

The Wheatstone bridge as a detector

To detect temperature changes Suppose a thermistor R is connected into the Wheatstone bridge circuit shown in Figure 23.24. The circuit has fixed resistors P and Q, and a variable resistor S. With the thermistor at a set temperature, the value of S is chosen to balance the bridge. The meter reads zero. If the thermistor's temperature changes the bridge becomes unbalanced, because the thermistor's resistance depends on its temperature. So the meter no longer reads zero. We can use this arrangement to monitor the thermistor temperature. The meter reads either side of zero, depending on whether the thermistor temperature is more or less than the initial set value.

The bridge used in this way may be connected to an electronic circuit to operate an alarm (see p. 400).

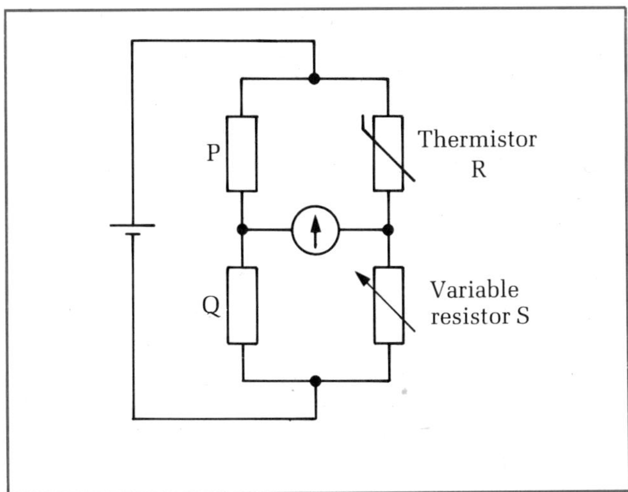

Figure 23.24 *Detecting temperature change*

To detect changes of light intensity An LDR (i.e. light-dependent resistor) has a resistance which falls when it is illuminated. If the thermistor of Figure 23.24 is replaced by an LDR, we can use the circuit to monitor light intensity levels. The resistance box is set so that the bridge is balanced when the LDR is illuminated at a set level (e.g. 1 m from a 100 W lamp in a darkened room). Change of illumination above or below that level makes the meter deflect off-zero in one direction or the other.

The circuit can be used to switch on an electric lamp automatically if the light intensity falls below a certain level. See p. 400 for further details.

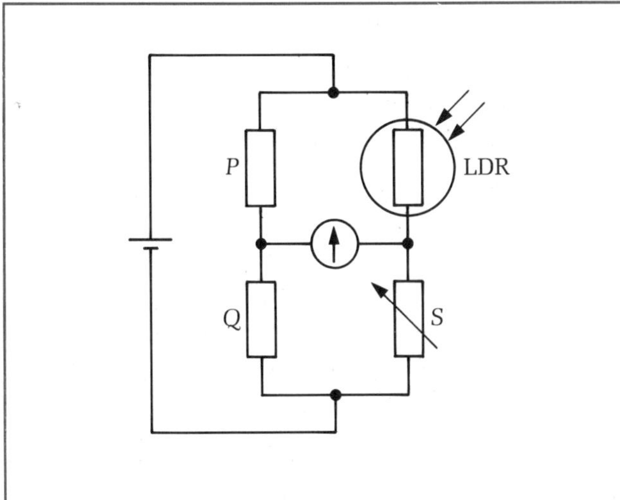

Figure 23.25 *Detecting changes of light intensity*

23.4 Summary

1 Shunts and multipliers:

a) Shunt resistance $= \dfrac{ir}{(I-i)}$ for a meter with resistance r to extend its current range from i to I.

b) Multiplier resistance $= \dfrac{V}{i} - r$

for a meter with resistance r and f.s.d. current i to extend its range to read up to p.d. V.

2 Potentiometer used to compare e.m.f.s,

$$\frac{E}{E_s} = \frac{L}{L_s}$$

3 Wheatstone bridge used to measure resistance

$$\frac{P}{Q} = \frac{R}{S}$$

Short questions

23.1 A meter has a resistance of 75 ohms and gives full-scale deflection when it passes a current of 1.0 mA. How would you adapt the meter to read
a) currents up to 1 A,
b) voltages up to (i) 5 V, (ii) 100 V?

23.2 The diagram below shows a circuit suitable for adapting a meter of resistance 50 ohms to read several ranges of current or voltage. The full-scale deflection current for the meter is 0.1 A. What position would you set the switches at P and Q to measure
a) currents up to 4 A,
b) voltages up to 12 V?

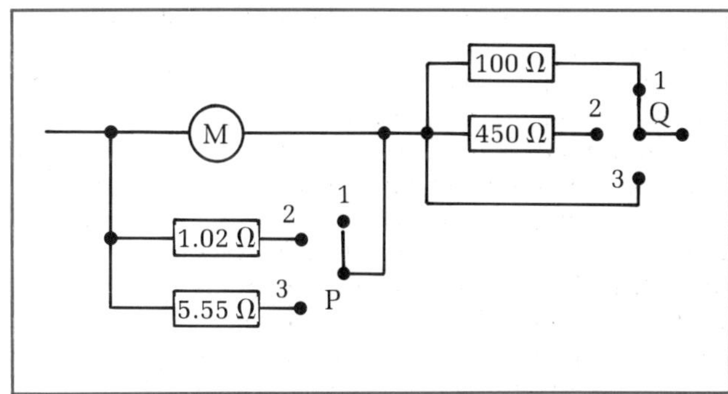

Figure 23.26

23.3 A voltmeter of resistance 1000 ohms is used to measure the e.m.f. of a battery. When connected across the battery terminals, the voltmeter reads 5.5 V. Then a 1000 ohm resistor is connected across the battery terminals in parallel with the meter which now reads 5.0 V. Sketch the circuit in each case, including the battery's internal resistance, and calculate the battery e.m.f. and its internal resistance.

23.4 A 2000 ohm resistor is connected in series with a 4000 ohm resistor and a 2 V cell of negligible internal resistance. A voltmeter of resistance 2000 ohms is then connected across the 2000 ohm resistor. Sketch the circuit and calculate the voltmeter reading. What would the voltmeter reading be if it is reconnected across the 4000 ohm resistor instead of across the 2000 ohm resistor?

23.5 A high voltage unit has an internal resistance of 50 MΩ. On open circuit, the p.d. across its output is 3000 V. When a resistor R is connected across the output, the p.d. at the output falls to 1000 V. Calculate the resistance of R.

23.6 A digital voltmeter with high internal resistance gives a maximum reading of 2.0 V. It is used to determine current by measuring the p.d. across a standard resistor S through which the current is passed. What should the resistance of S be to measure currents up to
a) 0.1 A,
b) 5 A?

23.7 The e.m.f. of a cell X is measured using a digital voltmeter which gives a reading of 1.53 V when connected directly across the terminals of X. When a 10 ohm resistor is connected across the cell terminals, the DVM reading falls to 1.48 V. Assuming the DVM takes negligible current, determine
a) the cell's e.m.f.,
b) the cell's internal resistance.

23.8 In a potentiometer experiment, a test cell X is compared with a standard cell S of e.m.f. 1.08 V. The balance length for cell X was found to be 630 mm whereas S gave a balance length of 425 mm. Calculate the e.m.f. of cell X.

23.9 In question **23.8**, the balance length for cell X is reduced when a 5 ohm resistor is connected across X's terminals. Explain why this happens. If the balance length for X is reduced to 335 mm, calculate the internal resistance of X.

23.10 Two cells A and B, connected in series, give a balance length of 758 mm along a potentiometer wire. When cell B is reversed, the balance length falls to 123 mm. Given the e.m.f. of cell A is 1.50 V, calculate the e.m.f. of cell B.

23.11 For the circuit shown below, calculate the meter current for each of the following situations.
a) X and Y are short-circuited.
b) A 3 kΩ resistor is connected between X and Y.
c) A 4 V battery is connected between X and Y with its positive pole at X.

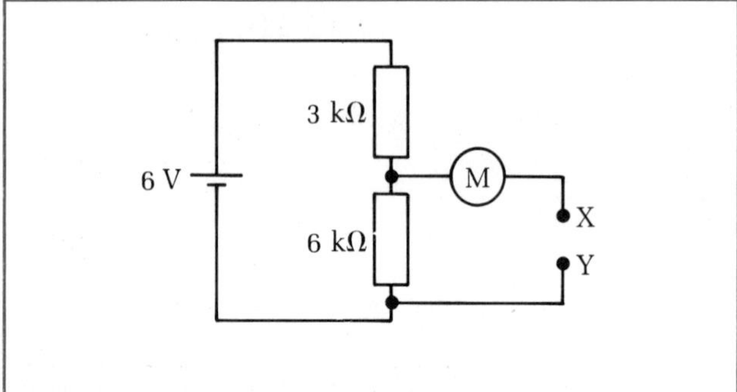

Figure 23.27

23.12 In an experiment to measure the e.m.f. of a thermocouple, a 2 V driver cell is connected in series with two resistance boxes and a potentiometer wire of length 1.00 m and resistance 3.2 ohms. The thermocouple e.m.f. is balanced across 754 mm of the potentiometer wire when the total resistance of the boxes is 1050 ohms. Then a standard cell of e.m.f. 1.08 V is balanced by the p.d. across one of the resistance boxes set at 510 ohms. The other box is set at 540 ohms. Calculate
a) the current in the potentiometer wire at balance point,
b) the p.d. across the potentiometer wire,
c) the thermocouple e.m.f.

23.13
a) For the Wheatstone bridge shown at the top of the next column, the bridge is balanced when $P=15$ ohms, $Q=8$ ohms and $S=12$ ohms. Calculate the resistance of X.
b) If P is then replaced by an 8 ohm resistor, what value of resistance connected in parallel with X is necessary to restore the balance condition?

23.14 In the Wheatstone bridge shown at the top of the next column, X is replaced by a light-dependent resistor (LDR), P and Q are replaced by identical 1000 ohm resistors and S is replaced by a resistance box. The bridge is balanced

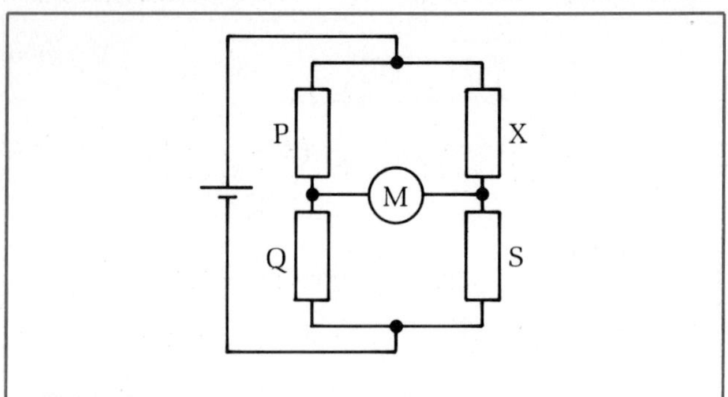

Figure 23.28

a) with the LDR in the dark by setting the resistance of S at 4950 ohms,
b) with the LDR exposed to daylight by setting S at 860 ohms.
Calculate the resistance of the LDR in the dark and in daylight. Explain why the LDR resistance changes.

23.15 In question **23.14**, P is replaced by an LDR (Y) identical to the LDR at X. S is then adjusted to give the balance conditon when X and Y are equally illuminated. Sketch the circuit, and determine the current direction through the meter when X is illuminated more than Y.

24

Alternating current circuits

24.1 a.c. measurements

Describing alternating current

Suppose an a.c. generator like the one described on p. 196 is connected to a suitable lamp. When the generator is turned at a steady rate, an alternating e.m.f. is induced in the rotating coil of the generator. The coil forms a complete circuit with the lamp, and the alternating e.m.f. forces the free electrons in the circuit repeatedly to change from one direction to the opposite direction. An individual free electron would be pushed one way, then the opposite way, then back again, and so on. Figure 24.2 shows how the current changes with time. In one full cycle, the current increases from zero to a peak in one direction, then it reverses to reach a peak in the opposite direction, then back to zero for the next cycle. Each full turn of the generator coil makes the current go through one full cycle.

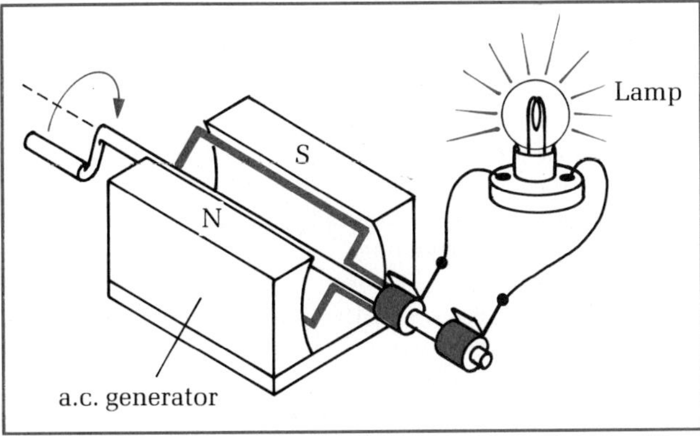

Figure 24.1 *Generating a.c.*

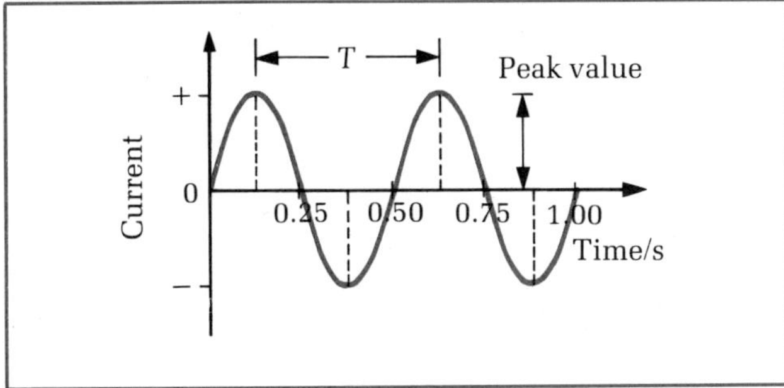

Figure 24.2 *Sinusoidal a.c.*

The time period T of an alternating current or p.d. is the time taken for one complete cycle.

The frequency f of an alternating current or p.d. is the number of complete cycles per second. The unit of frequency is the hertz (Hz), equal to one cycle per second.

$$T=\frac{1}{f}$$

In the example of Figure 24.2, the time period is equal to 0.50 s, and hence the frequency is 1/0.5 = 2 Hz. Another example is provided by the electricity in your home; electric heaters use alternating current at 50 Hz from the mains. The time period T is 1/50 = 0.020 s = 20 ms. So the current through a mains heater changes from one direction to the opposite direction and back again every 20 milliseconds.

The peak value of an alternating current or p.d. is the greatest value of the current or p.d. in a cycle. Alternating current with a peak value of 3 A changes from a peak of 3 A in one direction to a peak of 3 A in the opposite direction each half-cycle.

Alternating currents and p.d.s which are *not* sine waves can be produced by specially-designed electronic circuits. Figure 24.3 shows other possible wave forms.

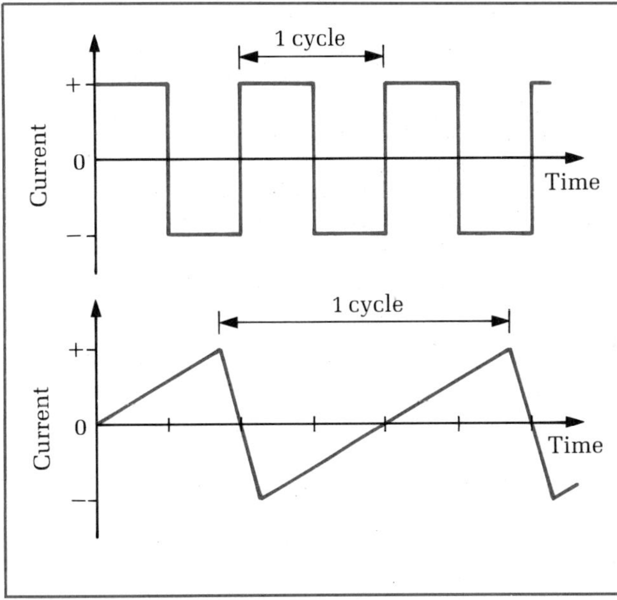

Figure 24.3 *Non-sinusoidal wave forms*

1. Square waves, where the current or p.d. is steady in one direction, then it abruptly reverses direction, stays steady, then abruptly reverses back again, etc.
2. Saw-tooth waves, where the current or p.d. changes at a steady rate then abruptly reverses before resuming its steady rate of change. This type of wave form is used for the time-base p.d. of an oscilloscope. See p. 348.

Unless otherwise stated, assume from now on that wave forms are **sinusoidal**, like Figure 24.2, for the rest of this chapter. So where the term alternating current is used, assume it has the same wave form as Figure 24.2 unless it is described as non-sinusoidal. We use the term sinusoidal because the same shape applies to the wave form, no matter whether it starts at zero, at a peak or anywhere between.

Comparing alternating current with direct current

Filament lamps supplied with alternating current usually work just as effectively as with direct current. For example, a 12 V car headlamp bulb will operate just as well on either a.c. or d.c. When electric lighting was first developed using a.c. it was realized that the frequency needed to be high enough to eliminate any flickering effects. European countries decided on 50 Hz as a suitable frequency, but in North America, the standard was set at 60 Hz. At either frequency, flickering is not a problem. However, dangers can arise due to stroboscopic effects. For example, an electric motor turning at 50 Hz viewed by a filament lamp operated at 50 Hz appears stationary. The reason is that the lamp flashes on each time the motor is in the same position, so the motor appears fixed, even though the eye cannot tell that the lamp brightness continually varies. The solution is to use strip lighting by means of fluorescent tubes, since these give a steady output of light.

Heaters work just as well with a.c. as with d.c. The flow of electrons through the heater filament causes heating whichever direction along the filament the electrons move. Figure 24.4(a) shows a heater in an a.c. circuit. Consider the heating effect of any resistor in an a.c. circuit. The power supplied to the resistor is equal to the current (I) × the p.d. (V). In a resistor, the power supplied to the heater is converted to energy of the surroundings. Even though the current and p.d. change continuously, we still use the same formula for the power. In a short time interval, Δt, the charge passed through the resistor is equal to $I \times \Delta t$. Since the energy supplied per unit charge is V, then the energy supplied in time Δt is $V \times I \times \Delta t$. Hence the energy per second supplied is $V \times I$. The difference with a.c. is that the power varies, unlike d.c. in which it is steady.

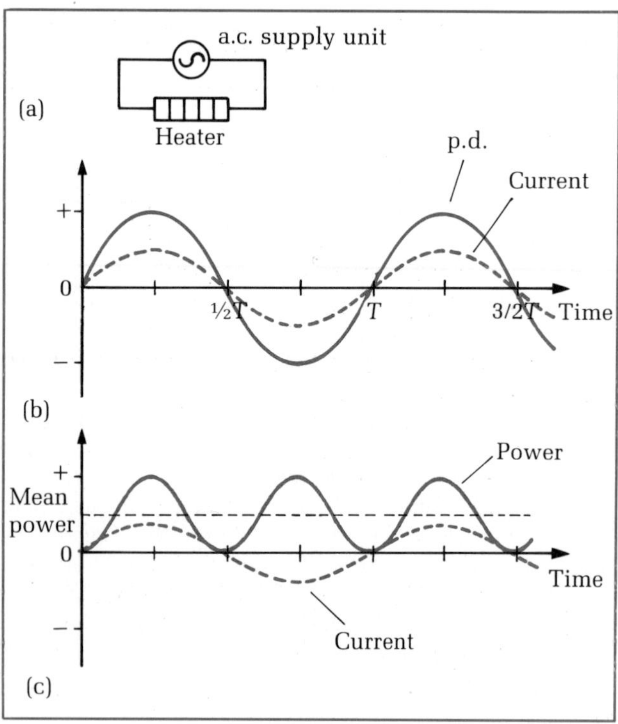

Figure 24.4 *Power used by a resistor*

The p.d. across a resistor is proportional to the current through it. We can use the link between p.d. and current with a.c. just as we can with d.c. The link tells us that the p.d. changes as the current changes. At any instant, p.d./current is always equal to the resistance. For example, suppose the resistance is 10 ohms and it passes an a.c. with a peak value of 2 A. The peak p.d. is $2 \times 10 = 20$ V when the current is 2 A. When the current is 0.5 A, the p.d. is $0.5 \times 10 = 5$ V. Figure 24.4(b) shows how the p.d. and current vary over several cycles.

Since the power supplied $= IV$ and the p.d. $V = IR$, then the power supplied $= I^2R$. Figure 24.4(c) shows how the power varies with time during several cycles. Power is still used by the resistor whichever direction the current is in. So whether the current is in the clockwise or anticlockwise direction, heat energy is always produced by the resistor.

Imagine a very low frequency a.c. generator used to supply a heater. Peak heating is when the current is at a peak in either direction; however, as the current passes through zero from one direction to the opposite direction, far less heat energy is produced. The power at any instant varies during the a.c. cycle. In the example above the peak current is 2 A and the peak p.d. is 20 V at the same time. So the peak p.d. is 40 W (= 2 A × 20 V). The power varies between 40 W and zero each half-cycle.

The mean power supplied to a resistor in an a.c. circuit is the average value of power over one complete cycle. In the power graph of Figure 24.4(c) the power varies from zero to a peak value. If the peak current is I_0, the peak power supplied is I_0^2R. Because Figure 24.4(c) is a graph of power against time, the area under the curve represents power × time, which is energy. So the area under one cycle of the power curve represents the heat energy produced in that time. The mean power is the value of steady power which will give the same energy over each cycle, so the mean power is represented on Figure 24.4(c) by the steady power level which has the same area underneath as the a.c. power curve has.

Over one full cycle, the area under the power curve is the same as the area under the midway level (i.e. the level at $\frac{1}{2}I_0^2R$). This can be proved mathematically but the key point is that the power curve itself is sinusoidal, with its mid-level at $\frac{1}{2}I_0^2R$. Hence the mean power is $\frac{1}{2}$ × the peak power.

$$\text{Mean power} = \tfrac{1}{2}I_0^2R$$

where I_0 = peak current,
R = resistance.

The root mean square (r.m.s.) or effective value of an alternating current or p.d. is the value of direct current or p.d. which would supply the same power in a given resistor.

Consider the example of a 2 ohm resistor which passes an alternating current with a peak value of 3 A. The peak power is $3^2 \times 2 = 18$ W, so the mean power is 9 W. What value of direct current through the 2 ohm resistor gives 9 W of power? Let the equivalent direct current $= I$.

We know that $I^2R = 9$, so $I^2 \times 2 = 9$. Thus $I = \sqrt{4.5} = 2.12$ A. In other words, a direct current of 2.12 A produces the same power as a 3 A alternating current. The effective or r.m.s. value of the alternating current is 2.12 A in this example.

In general, where the peak current is I_0 and the resistance is R, the peak power is I_0^2R. So the mean power is $\frac{1}{2}I_0^2R$. What is the r.m.s. value when the peak value of current is I_0?

Let the r.m.s. value of the current $= I_{r.m.s.}$; this is the value of direct current which gives power equal to the mean power $\frac{1}{2}I_0{}^2R$.

Hence $(I_{r.m.s.})^2R = \frac{1}{2}I_0{}^2R$

so $$I_{r.m.s.} = \frac{1}{\sqrt{2}}I_0$$

and $$V_{r.m.s.} = I_{r.m.s.}R = \frac{1}{\sqrt{2}}V_0$$

The r.m.s. value $= \frac{1}{\sqrt{2}} \times$ peak value (for sinusoidal wave forms).

For non-sinusoidal wave forms the r.m.s. value can be determined from the power curve. The power at any instant supplied to resistance R is equal to I^2R. The mean power can be estimated from the power curve. Then the r.m.s. value can be calculated from the mean power.

For example, consider a rectangular wave form as shown in figure 24.5. Suppose the resistance is 1 ohm. The time period of the wave form is 3 s.

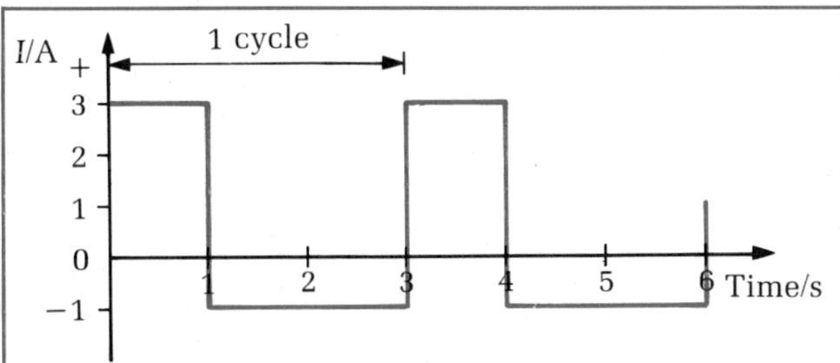

Figure 24.5

1 From 0 to 1 s, $I = 3$ A Hence the power supplied $= I^2R = 9$ watts. So the energy supplied $= 9$ J.
2 From 1 to 3 s, $I = -1$ A So the power supplied $= I^2R = 1$ watt. The energy supplied is thus 2 J.
The total energy supplied $= 9 + 2 = 11$ J and the time taken $= 3$ s.
Hence the mean power $=$ energy supplied/time taken $= 11/3 = 3.67$ W.

Let the r.m.s. current $= I_{r.m.s.}$, so this is the value of direct current which gives the same power.

Therefore $(I_{r.m.s.})^2R = 3.67$
so $(I_{r.m.s.})^2 = 3.67$ (since $R = 1$ ohm).

The r.m.s. current is thus 1.91 A $(= \sqrt{3.67})$.

Measuring alternating current

Using a meter Connect a centre-reading meter in series with a resistor and a low frequency a.c. generator. The meter pointer repeatedly swings from one side of the scale zero to the other side. Increasing the frequency makes the pointer oscillate even more rapidly. However, the meter cannot respond quickly enough to the changing current; at frequencies above 10 Hz, the pointer hardly moves at all, even though the peak current may be quite large.

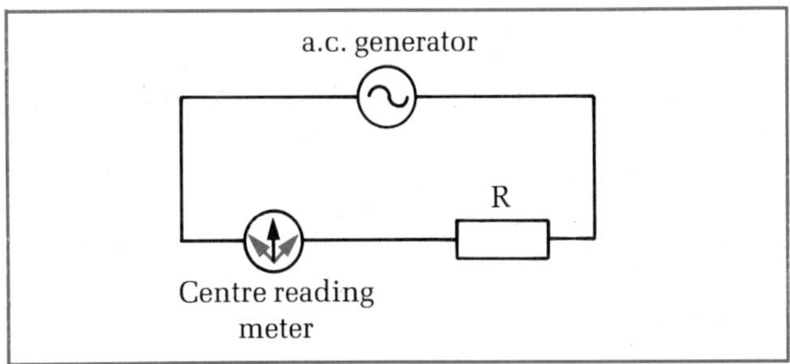

Figure 24.6 *Low-frequency a.c.*

It is possible to measure alternating current with a moving coil meter if the current is first converted into direct current. Converting alternating to direct current is called **rectifying** the current. **Diodes** are used for this purpose because they pass current in one direction only. Figure 24.7 shows the circuit of Figure 24.6 with a diode in series with the meter. The direction of current through the diode is called the **forward direction**; the opposite direction is called the **reverse direction**. The diode has low resistance in its forward direction, but its resistance in the reverse direction is high. The meter of Figure 24.7 deflects in one direction only; which is when the generator e.m.f. pushes current round the circuit in the forward direction of the diode. When the generator e.m.f. reverses polarity, it tries to push current round the circuit in the reverse direction; however, the diode resistance is very high in the reverse direction so almost no current passes. Therefore the meter deflects every other half-cycle, always in the same direction. Figure 24.8 shows how the generator e.m.f. and current vary with time; the current wave form is called a **half-wave** curve.

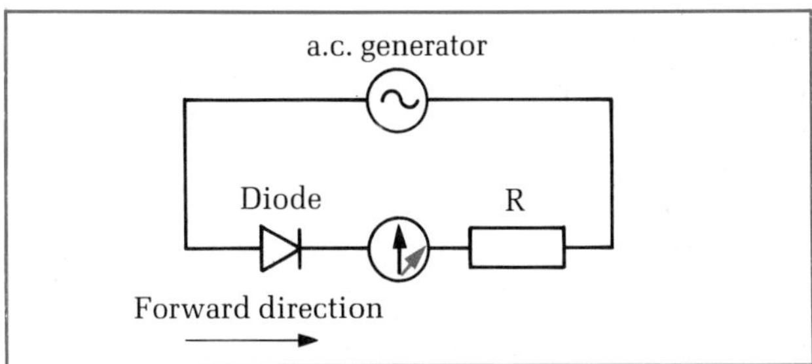

Figure 24.7 *Using a diode*

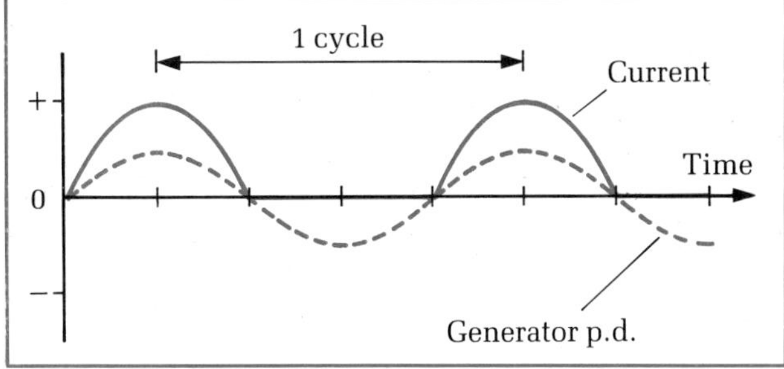

Figure 24.8 *Half-wave current*

A bridge rectifier used with a moving coil meter allows current to pass through the meter on both halves of each a.c. cycle. Four diodes are used to form a bridge rectifier, as in Figure 24.9, at the top of the next page. On one half of each a.c. cycle, diodes A and C conduct, so current passes,

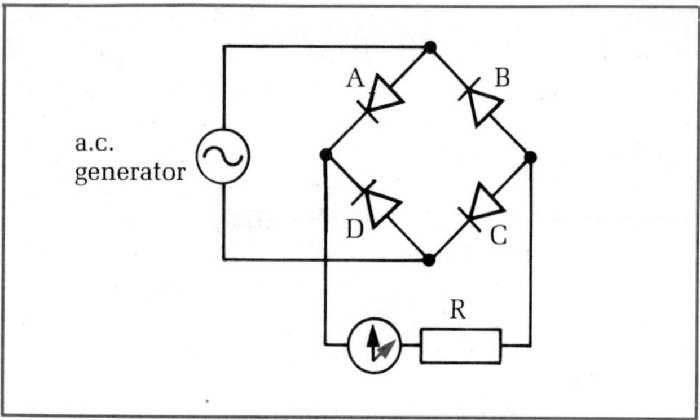

Figure 24.9 *A bridge rectifier*

left to right, through the meter. On the other half of the cycle, diodes B and D conduct, so once again, current passes left to right through the meter. Figure 24.10 shows how the current and generator e.m.f. vary with time. The curve is called a **full-wave rectified curve**. Where diodes are used with moving coil meters to measure alternating current, the meter is usually calibrated to read r.m.s. values directly from its scale. So in an a.c. circuit in which the frequency is more than 10 Hz, the meter pointer shows a steady reading which gives the r.m.s. value of the current.

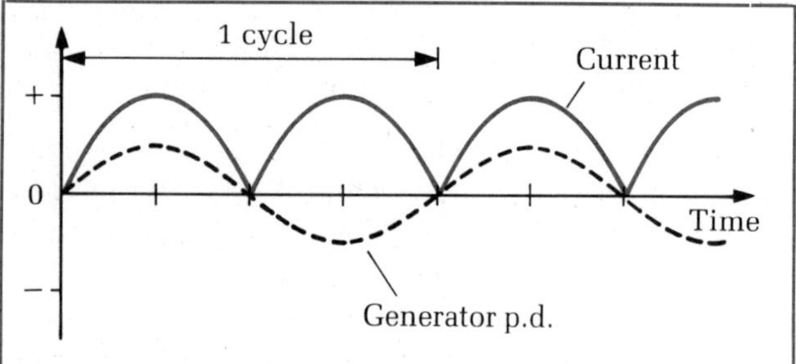

Figure 24.10 *Full-wave current*

Using an oscilloscope An oscilloscope consists of a specially-made electron tube and associated control circuits. In the tube, an electron gun at one end shoots electrons in a beam at a screen at the other end of the tube, as shown in Figure 24.11. Where the beam hits the screen, the impact causes light to be emitted from that point on the screen.

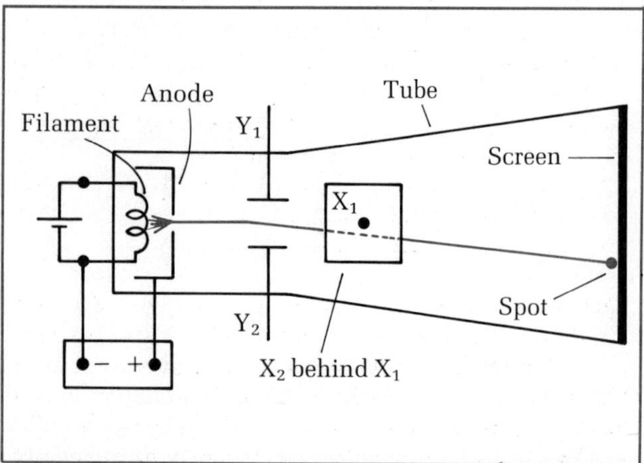

Figure 24.11 *An oscilloscope tube*

With no p.d. across either set of deflecting plates, the spot of light on the screen stays in the same position. A p.d. applied across the X-plates makes the spot deflect horizontally. A p.d. across the Y-plates makes the spot deflect vertically.

In use, the X-plates are connected to a circuit in the oscilloscope called the **time-base circuit**. The p.d. from this circuit repeatedly makes the spot move slowly, left to right, across the screen, then rapidly back again to repeat the slow left to right motion. The time-base circuit needs to produce a saw-tooth wave form as in Figure 24.3 to make the spot move in this way.

The p.d. to be investigated is applied to the Y-plates. So as the spot moves left to right, it moves up or down the screen according to the variation of the applied p.d. with time. Figure 24.12 shows the trace produced by the spot for a sinusoidal p.d. applied to the Y-plates.

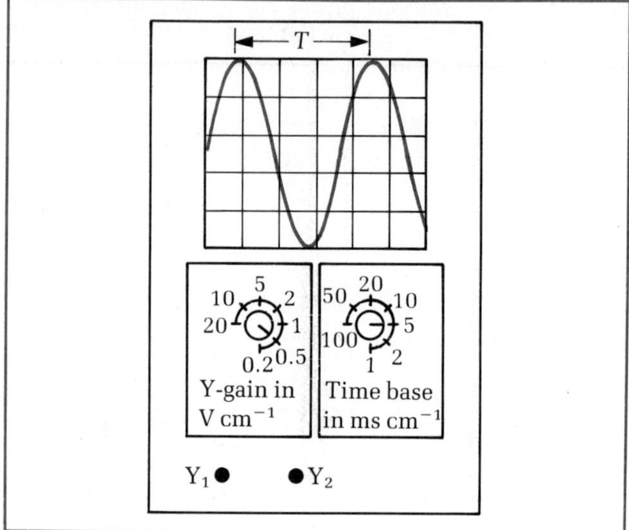

Figure 24.12 *Using an oscilloscope*

To measure the peak p.d. the peak-to-peak height of the trace on the screen is measured in millimetres. The Y-sensitivity control switch on the front of the oscilloscope gives the p.d. per cm used to deflect the spot vertically. The switch allows the applied p.d. to be amplified (i.e. made greater) by fixed amounts. Using the switch, the applied p.d. is 'routed' through an internal amplifier to enlarge the applied signal before supplying the Y-plates. Given the peak-to-peak height of the trace, the peak-to-peak p.d. can be calculated if the Y-sensitivity is known.

For example, Figure 24.12 shows a sine wave with a peak-to-peak height of 50 mm. The Y-sensitivity switch is set at 0.5 V cm^{-1}; so a deflection of 50 mm (5.0 cm) requires an applied p.d. of $0.5 \times 5 = 2.5$ V. Hence the peak-to-peak p.d. of the sine wave signal applied to the Y-input terminals is 2.5 V. So the peak voltage (i.e. peak p.d.) is 1.25 V.

To measure the time period *T* the distance along the x-axis for one full cycle is measured. The time-base control gives the time per cm for the spot to move horizontally. Hence the time for one cycle can be calculated.

For example, the horizontal distance for 1 cycle of the wave form of Figure 24.12 is 35 mm. With the time-base control at 5 ms cm^{-1}, the time period is $5 \times 3.5 = 17.5$ ms $= 0.0175$ s. Hence the frequency $= 1/0.0175 = 57$ Hz.

24.2 Reactance and resistance

Investigation phase differences

The phase difference between two oscillators is the fraction of a cycle by which one oscillator is behind the other. In any a.c. circuit, the current oscillates since it repeatedly changes direction. The p.d. across any device repeatedly alternates too when alternating current passes through it. In general, in an a.c. circuit with two or more devices, the p.d.s and currents are not necessarily in phase.

The circuit shown in Figure 24.13 may be used to investigate phase differences between the current and the p.d. for any individual device. An a.c. generator with a low frequency output p.d. is used. The centre-reading meter shows how the current taken by the test device varies; the p.d. across the test device terminals is applied to an oscilloscope as a voltmeter. In operation, the meter pointer oscillates about the scale zero (at the centre) at a frequency low enough to follow by eye. The oscilloscope time-base can be switched off so that the spot on its screen oscillates up and down in response to the p.d. applied to its Y-input terminals. By arranging the meter on its side above the oscilloscope screen, as in Figure 24.13, the phase link between the current and p.d. can be investigated for different test devices in turn, as below.

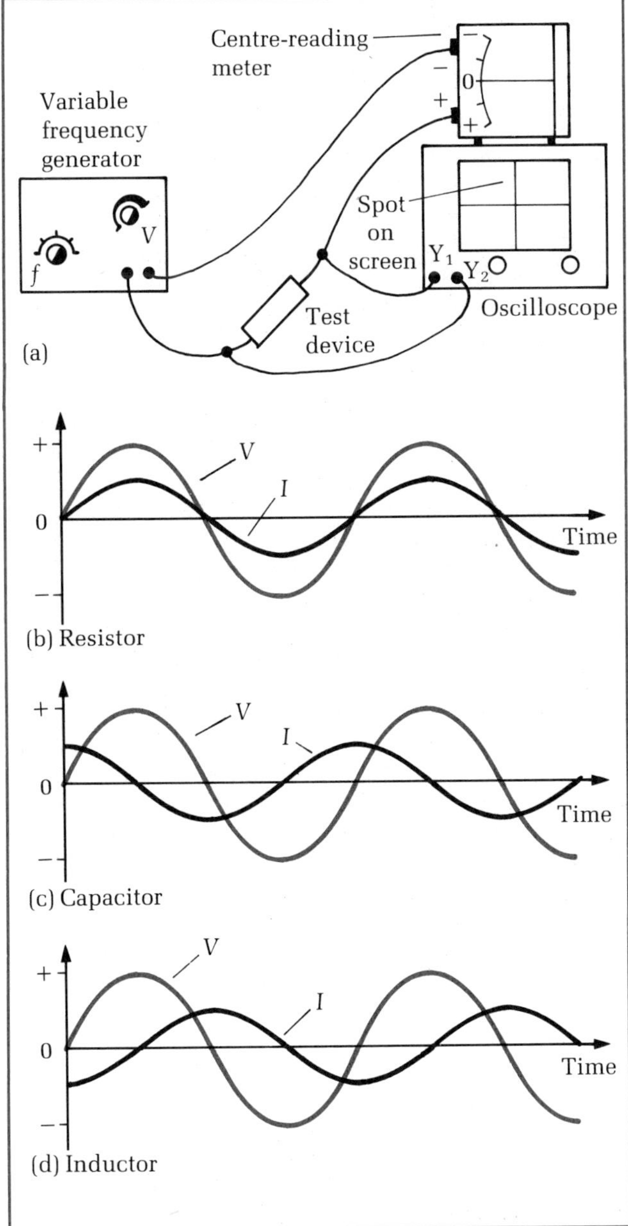

Figure 24.13 *Investigating phase differences*

A resistor The spot and pointer move exactly in phase, rising and falling together at the same rate. So the current is always in phase with the p.d. Figure 24.13(b) shows how the current and p.d. vary with time.

At any instant, the p.d. = current × resistance. In other words, p.d./current is a constant (= the resistance), provided the values are measured at the same time. So the usual resistance equation $V = IR$ applies.

A capacitor The pointer moves a fraction of a cycle ahead of the spot, so the current is ahead of the p.d. Perform a duet with a friend here by asking your friend to shout each time the spot reaches its highest position on the screen. You shout out each time the pointer passes through the zero position moving downwards. You ought to find that you and your friend shout at the same times because the highest position of the spot is when the pointer passes downwards through zero each cycle. The peak p.d. (i.e. spot deflection highest) is when the current is zero. The peak current was exactly one quarter-cycle earlier. At each shout, the current peaked one quarter of a cycle earlier. So the current through a capacitor is always one quarter of a cycle ahead of the applied p.d.

Why does this happen? At zero p.d. the capacitor is uncharged. As the p.d. rises from zero, the capacitor easily charges up so the current is high as the applied p.d. swings through zero. At peak p.d. the capacitor stops charging and begins discharging as the p.d. falls. So at peak p.d. the current is zero because it is changing from one direction to the opposite direction. Figure 24.13(c) shows how the p.d. and current for a capacitor vary with time.

An inductor Use a coil with as low a resistance as possible. This time the pointer lags behind the spot. So the current (from the pointer reading) is always a fraction of a cycle behind the applied p.d. Try a shouting duet again. You ought to discover that when the spot is at its highest position, the meter pointer is moving up through zero towards its highest position, so the current (i.e. pointer reading) reaches its peak one quarter of a full cycle after the p.d. (i.e. spot position) peaks. Figure 24.13(d) shows how the current and p.d. for an inductor vary with time. The current is always a quarter-cycle behind the p.d.

To understand why the current lags a quarter of a cycle behind the p.d. consider the ideas of self-inductance on p. 203 in a little more detail. The rate of change of current is highest when the current swings through zero; this is where the gradient of the current curve is greatest. The induced e.m.f. is proportional to the rate of change of current, so the induced e.m.f. is greatest when the current is zero, because then the rate of change of current is greatest. The applied p.d. must overcome the induced e.m.f., so the applied p.d. is highest when the rate of change of current is greatest. Hence the applied p.d. is highest when the current is swinging through zero.

Power

For steady currents, the power used by an electrical device is given by the power rule

Power = current × p.d.

For alternating currents, the same rule is used, even though the current and p.d. values change. So the power changes, unlike steady currents where the power is steady.

Resistors As explained in section 24.1, the power at any instant $= I^2R$, where R is the resistance and I is the current at that instant. Since the current changes, so too does the power. The power varies between zero and a peak value of $I_0{}^2R$, where I_0 is the peak current. The mean power over one full cycle is $\frac{1}{2}I_0{}^2R$.

The power supplied to a resistor is used to heat the surroundings. The energy lost to the surroundings can never be fully reconverted to electrical energy. We say that the power supplied is **dissipated** as energy to the surroundings.

Capacitors or inductors For either a capacitor or an inductor, the current and p.d. are a quarter of a cycle out of phase with one another. When the current is zero, the p.d. is at a peak in one direction or the other. When the p.d. is zero, the current is at a peak in one direction or the other. Since the power supplied to the device is given by current × p.d., the power is zero whenever either current or p.d. is zero. Consider power changes during one full cycle of the current for a capacitor.

1st quarter-cycle The current I falls from peak value I_0 to 0, the p.d. V increases from 0 to peak value V_0, so the power supplied rises from zero to a peak then falls back to zero.

2nd quarter-cycle The current falls from 0 to $-I_0$, the p.d. changes from V_0 to 0, so the power supplied changes from zero to a negative value then returns to zero.

The variation of power with time is shown by Figure 24.14, which also shows the variation of current and p.d. with time. What does a negative value of power mean? A positive value means that electrical energy is being converted into stored energy in the capacitor. A negative value means that the capacitor returns the energy to electrical energy in the circuit.

3rd quarter-cycle The current rises from $-I_0$, to 0, the p.d. changes from 0 to $-V_0$, so the power supplied changes from zero to a positive value (because I and V are both negative $I \times V$ is positive) then returns to zero.

4th quarter-cycle The current rises from 0 to I_0, the p.d. changes from $-V_0$ to 0, hence the power supplied changes from 0 to a negative value, then returns to zero.

To sum up
1st quarter: electrical energy converted to stored energy in the capacitor.
2nd quarter: stored energy is converted back to electrical energy in the circuit.

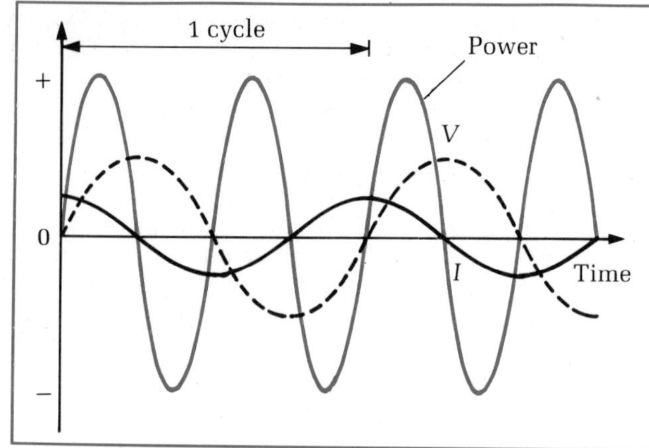

Figure 24.14 *Power curve for a capacitor*

3rd quarter: electrical energy converted to stored energy in the capacitor.
4th quarter stored energy is converted back to electrical energy in the circuit

Over one full cycle, all the energy supplied to the capacitor by the power supply is returned to the power supply. We assume that the circuit has zero resistance in the wires, so in total no energy is supplied to the capacitor.

Any device for which the current and p.d. are out of phase with one another by a quarter of a cycle gives the same effect. No power is taken overall from the circuit. So inductors as well as capacitors consume no power from an a.c. circuit. Because capacitors and inductors return energy to the power supply unit as explained above, they are called **reactive** components.

Reactive components in a.c. circuits store energy from the power supply unit in alternate quarter-cycles. The stored energy is returned to the power supply in the quarter-cycles between so no overall power is taken from the power supply unit.

Resistive components take energy from the power supply unit and dissipate it.

Reactance

Connect a variable frequency a.c. generator unit to a 10 μF capacitor in series with a torch bulb and an a.c. milliammeter. The torch bulb should give normal brightness when a current of 60 mA passes through it. The milliammeter should be capable of measuring currents up to 100 mA. Figure 24.15 shows the circuit.

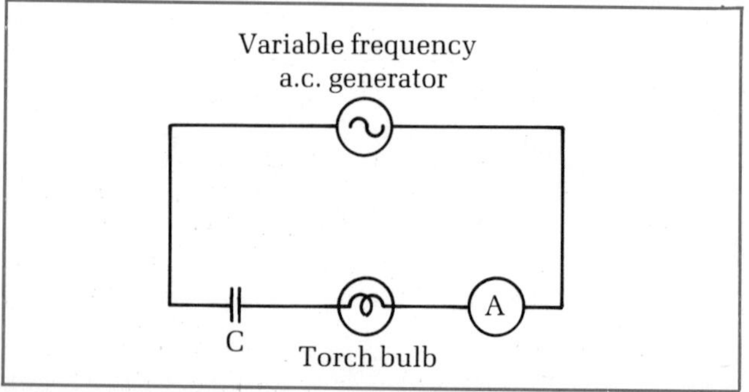

Figure 24.15 *Capacitor reactance*

With the output voltage from the signal generator set at 6 V, the bulb brightness can be varied by changing the frequency of the generator unit. At 50 Hz, the bulb is much dimmer than normal since the current is much less than 60 mA. But as the frequency is increased, the bulb brightness increases towards normal brightness, even though the output voltage stays constant. The capacitor opposes the passage of alternating current round the circuit; the lower the frequency, the greater the opposition. At very low frequencies, the current changes so slowly that it is almost like direct current; in fact, direct current is equivalent to a.c. at zero frequency. Capacitors block steady d.c., so as the a.c. frequency is adjusted towards zero the opposition becomes greater and greater.

Replace the capacitor by an inductor. This time, the opposition is greatest at high frequency, and so the bulb fades out as the frequency is increased.

Why should capacitors and inductors oppose the passage of alternating current? And why should the opposition depend on frequency in a different way for each?

For an inductor changing current induces an e.m.f. which opposes the applied p.d. The higher the frequency, the faster the rate of change of current. Faraday's law tells us that the opposition to the applied p.d. is even greater, so the current is smaller, the higher the frequency.

For a capacitor the higher the frequency, the faster the rate of charging and discharging the capacitor. So the current (that is, the rate of flow of charge) is larger, the higher the frequency.

Representing alternating currents and p.d.s by equations allows the effects of capacitors and inductors in a.c. circuits to be calculated. Consider the graph of p.d. against time shown in Figure 24.16. The wave form is a sine wave, since the p.d. is zero at time $t = 0$. There are three ways to determine the p.d. at any point in time.

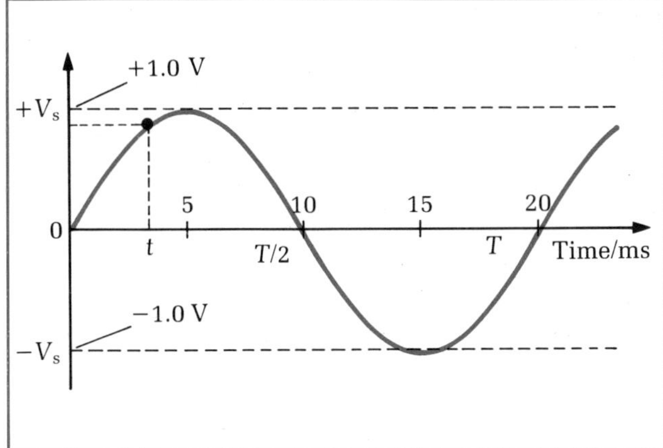

Figure 24.16 *Calculating sine values*

a) Graphical method We need to know the time period T to calibrate the time scale of Figure 24.16 in seconds. At any particular point in time, the p.d. at that instant can be read from the graph. For example, from Figure 24.16, the p.d. at 4.0 ms from time $t = 0$ is 0.95 V.

b) Using phasors A **phasor** is a rotating vector, as in Figure 24.17. Its length represents the peak value V_0, and it rotates about one end at a steady rate. Figure 24.17

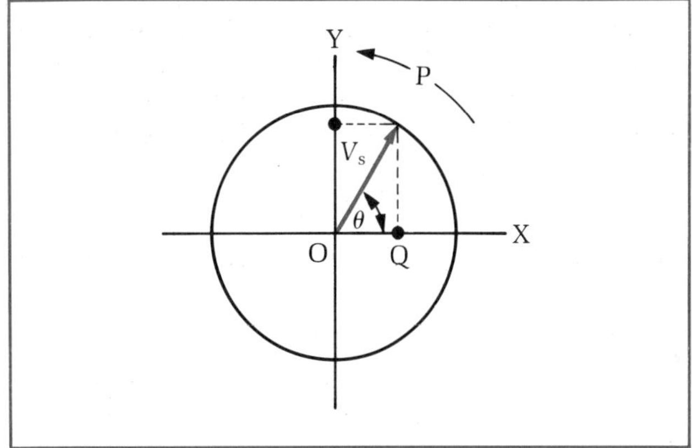

Figure 24.17 *Phasors*

shows a 'snapshot' of a phasor **OP** rotating about O. Point O is the origin of axes OY and OX.

Consider point Q which is directly beneath point P on the x-axis. The length PQ, called the **projection** of **OP** onto the y-axis, is equal to OP sin θ. As the vector rotates, θ changes at a steady rate. So PQ varies with time as shown by the graph in Figure 24.16. Hence PQ represents the value of the sine wave at any instant during the cycle. As the phasor rotates, the instant value changes sinusoidally.

Suppose phasor **OP** represents an alternating p.d., V. The length OP represents the peak value of the p.d. V_0. If we know the time period T, the time for the phasor to rotate once, then the phasor position can be determined at any particular point in time. Then we can draw a phasor diagram like Figure 24.17 to determine the value of V at that instant.

For example, suppose the peak p.d. is 1 volt and the time period T is 20 ms. Assume the start of each cycle is when **OP** lies along the + x-axis. The p.d. 4.0 ms later is worked out as follows.

OP rotates through 360° in 20 ms, hence in 4 ms **OP** rotates through $\theta = 360 \times 4/20 = 72°$. So the p.d. at 4 ms, $V = V_0 \sin\theta = 1 \times \sin 72° = 0.951$ volts.

c) Using equations the phasor method shows that the p.d. at any instant is given by $V = V_0 \sin\theta$. The phasor turns through angle θ in time t.

At frequency f, the time period T is equal to $1/f$. The phasor turns through an angle of $360° = 2\pi$ radians in time T. So it turns through $2\pi/T$ radians each second. In other words, the angular speed of the phasor is $2\pi/T$. This is usually called the **angular frequency** ω of the a.c. signal.

At time t, the phasor has turned through angle $\theta = (2\pi/T)t = \omega t$ where θ is in radians. So at time t, the p.d. at that instant is given by

$$V = V_0 \sin\frac{(2\pi t)}{T} = V_0 \sin\omega t$$

where t = time from start,
T = time period,
ω = angular frequency = $2\pi/T$,
V_0 = peak value.

For a capacitor in an a.c. circuit, as in Figure 24.18(a), the charge on its plates is proportional to the applied p.d.

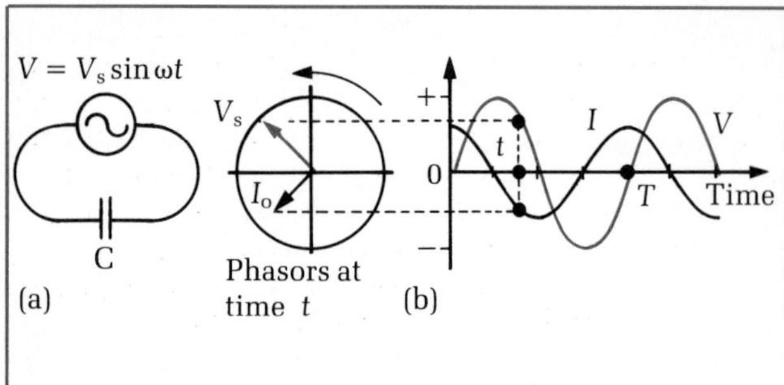

Figure 24.18 *Capacitors in a.c. circuits*

Suppose at time t the p.d. $V = V_0 \sin \omega t$.
Since the charge $Q = CV$, where C is the capacitance

$$Q = CV_0 \sin \omega t$$

The current in the circuit, $I = \dfrac{dQ}{dt}$

so $I = \dfrac{d}{dt}(CV_0 \sin \omega t) = CV_0 \dfrac{d}{dt}(\sin \omega t)$

As explained on p. 505, $\dfrac{d}{dt}(\sin \omega t) = \omega \cos \omega t$ giving

$I = CV_0 \times \omega \cos \omega t$

Now we can see the link between current and voltage precisely. The current is a cosine wave which is quarter of a cycle ahead of the sine wave for voltage. The theory supports the experiment of Figure 24.13(c); the current is always ahead of the voltage by a quarter-cycle (i.e. 90°). In addition, we can see how the peak current I_0 relates to the peak voltage V_0 because $\cos \omega t$ has a maximum value of +1. So the peak current $I_0 = CV_0 \omega \times 1 = \omega CV_0$.

The reactance of a capacitor is defined as $\dfrac{1}{\omega C}$. Since $I_0 = \omega CV_0$, then the reactance is equal to V_0/I_0. The unit of reactance is the **ohm**, so reactance is a measure of the opposition of the capacitor to the passage of alternating current. Because the reactance X is inversely proportional to ω, then high frequency gives low reactance and low frequency gives high reactance.

$$X_c = \frac{1}{\omega C}$$

where X_c = reactance of capacitor,
ω = angular frequency = $2\pi f$ (where f = frequency),
C = capacitance.

Worked example Calculate the reactance of a 1 μF capacitor at 50 Hz. What is the peak value and the r.m.s. value of the current passed by the capacitor when an alternating voltage of r.m.s. value 12 V is applied across its terminals?
Solution

Reactance $X_c = \dfrac{1}{\omega C} = \dfrac{1}{2\pi f C} = \dfrac{1}{2\pi \times 50 \times 1 \times 10^{-6}}$
Therefore X_c = 3180 ohms.

Peak voltage $V_0 = \sqrt{2} \times$ r.m.s. voltage = 17.0 V.
Since $V_0/I_0 = X_c$, then the peak current is given by

$$I_0 = \frac{V_0}{X_c} = \frac{17.0}{3180} = 5.35 \times 10^{-3}\ \text{A}.$$

Hence r.m.s. current =

$$I_0/\sqrt{2} = \frac{5.35 \times 10^{-3}}{\sqrt{2}} = 7.56 \times 10^{-3}\ \text{A}.$$

For an inductor in an a.c. circuit, as in Figure 24.19(a), the alternating voltage V from the a.c. generator causes a changing current through the coil. From Faraday's law (see p. 194), the rate of change of current $\dfrac{dI}{dt}$ is given by

$$L\frac{dI}{dt} = V$$

where L is the coil's self-inductance.

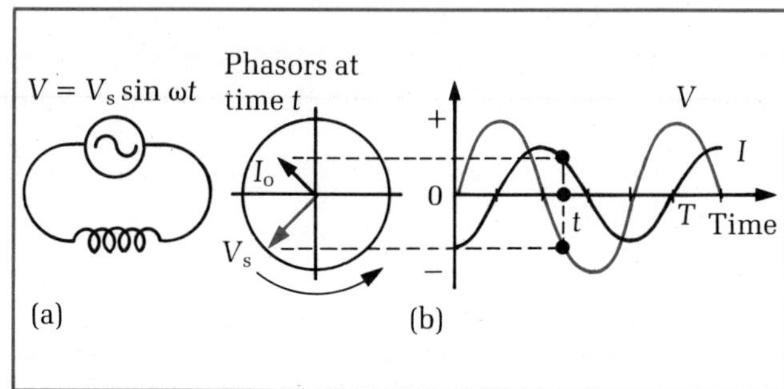

Figure 24.19 *Inductors in a.c. circuits*

Assuming $V = V_0 \sin \omega t$

Then $\quad L\dfrac{dI}{dt} = V_0 \sin \omega t$

Hence $\quad \dfrac{dI}{dt} = \dfrac{V_0}{L} \sin \omega t$

So $\quad I = \dfrac{-V_0}{\omega L} \cos \omega t$

Check: $\quad \dfrac{dI}{dt} = \dfrac{d}{dt}\left(\dfrac{-V_0}{\omega L} \cos \omega t\right) = \dfrac{-V_0}{\omega L}\dfrac{d}{dt}(\cos \omega t)$

$$= \frac{V_0}{\omega L}\,\omega \sin \omega t$$

$$= \frac{V_0}{L} \sin \omega t$$

The equations show the link between current and voltage again. The current is a negative cosine wave which is $\frac{1}{4}$-cycle behind the sine wave for voltage. The current is always $\frac{1}{4}$-cycle behind the voltage, in agreement with the experiment of Figure 24.13(d). Also, since $\cos \omega t$ has a maximum value of +1 and a minimum value of −1, then the current has a peak value I_0 equal to $V_0/\omega L$.

The reactance of an inductor is defined as ωL. Since $I_0 = V_0/\omega L$, then the reactance is equal to V_0/I_0, just as for a capacitor. This time though, the reactance is proportional to ω. So the higher the frequency, the higher the reactance. In other words, the opposition of an inductor to alternating current increases with frequency.

$$X_L = \omega L$$

Figure 24.20

In the previous equation on the inductance of an inductor, X_L refers to the inductor's reactance, L to its self-inductance and ω to its angular frequency.

24.3 Series circuits

Components in series have the same current passing through them. In a d.c. circuit, the current is usually constant; in an a.c. circuit the current is continually changing. But at any instant in an a.c. circuit, the current passing through components in series is the same.

CR circuits

Consider the circuit shown in Figure 24.21(a), in which an a.c. supply unit is connected in series with a resistor R and a capacitor C. Assume the circuit has no additional resistance, so all its resistance is represented by R. Suppose a voltmeter is connected across the terminals of the supply unit to measure the supply p.d. Then the voltmeter is reconnected across R to measure the p.d. across R. Finally, the voltmeter is reconnected across C to measure the p.d. across C. An example of the measurements might be:

supply p.d. = 9.2 V (r.m.s.) = 13 V (peak)
resistor p.d. = 3.5 V (r.m.s.) = 5 V (peak)
capacitor p.d. = 8.5 V (r.m.s.) = 12 V (peak)

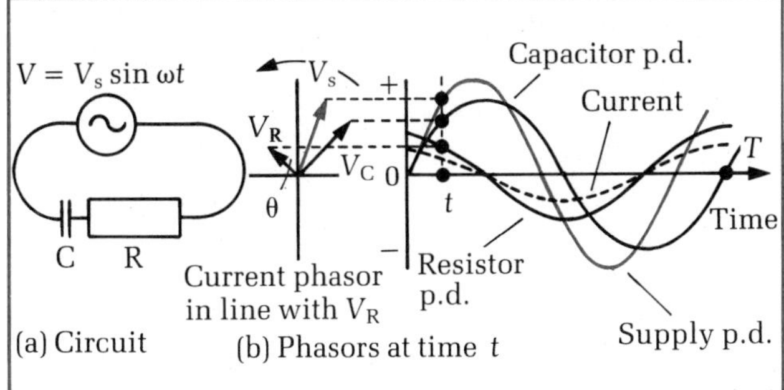

Figure 24.21 *CR series circuit*

The results show that the supply p.d. does not equal the sum of the resistor p.d. + the capacitor p.d. in this situation. Why? The voltmeter measures the r.m.s., value, so we could calculate the peak voltage for each of the three measurements. Do they add up now? All we have done is to multiply each measured value by $\sqrt{2}$, so they still dc not add up. The reason is that the resistor p.d. is in phase with the current, but the capacitor p.d. is $\frac{1}{4}$-cycle behind the current. So the peak value of the resistor p.d. is $\frac{1}{4}$-cycle ahead of the peak value of the capacitor p.d.

Figure 24.21(c) shows how the current and p.d.s vary with time. We cannot just add the peak capacitor p.d. to the peak resistor p.d. because the two peaks are at different times in each cycle. What we can do is to add the p.d.s at any one instant to give the supply p.d. at that instant. The result is shown in Figure 24.21(c). The resistor p.d. is a cosine wave with a peak value of 12 volts; the capacitor p.d. is a sine wave with a peak value of 5 volts. The two waves are added by taking lots of instants over each cycle. The result is a wave with a peak value of 13 V for the supply p.d. Check some instants yourself, even though it may be tedious.

Phasors are very useful here. Figure 24.21(b) shows the phasor diagram corresponding to one instant of Figure 24.21(c). The capacitor p.d. phasor is 90° behind the resistor p.d. phasor and the current phasor. The supply p.d. phasor is the vector sum of the resistor and capacitor p.d. phasors. Now you can see why 5 V across R and 12 V across C adds up to 13 V across the supply (and not 17 V). The resistor p.d. phasor is at 90° to the capacitor p.d. phasor; so they form adjacent sides of a rectangle. Their vector sum is the diagonal shown in Figure 24.21(b), so this represents the supply p.d. phasor. Hence the resistor p.d. phasor, the capacitor p.d. phasor and the supply p.d. phasor form three sides of a right-angled triangle. So, in the above example, the peak supply p.d. is equal to $\sqrt{5^2+12^2}=13$ volts (use Pythagoras' theorem).

$$V_S^2 = V_R^2 + V_C^2$$

where V_S = peak supply p.d.,
V_R = peak resistor p.d.,
V_C = peak capacitor p.d.

The equation applies to all series CR circuits because the resistor p.d. is always out of phase with the capacitor p.d. by 90°. For example, suppose the peak value of the resistor p.d. is 40 V and the peak value of the supply p.d. is 41 V. Then the peak p.d. across the capacitor must be 9 V because $9^2+40^2=41^2$.

Since the current is the same in each component, then

$$V_R = I_0 R \quad \text{and}$$

$$V_C = I_0 \frac{1}{\omega C}$$

where I_0 is the peak value of the current.

$$\text{Hence } V_S^2 = (I_0 R)^2 + \left(I_0 \frac{1}{\omega C}\right)^2 \quad \text{since } V_S^2 = V_R^2 + V_C^2$$

$$\text{So} \quad V_S^2 = I_0^2 \left\{R^2 + \left(\frac{1}{\omega C}\right)^2\right\}$$

$$\text{Giving } \frac{V_S}{I_0} = \sqrt{\left\{R^2 + \left(\frac{1}{\omega C}\right)^2\right\}}$$

The impedance Z of an a.c. circuit is defined as

$$\frac{\text{the peak p.d.}}{\text{the peak current}}.$$

The unit of impedance is the ohm.

The equation for a series CR circuit is given at the top of the next page.

$$Z = \frac{V_S}{I_0} = \sqrt{R^2 + \left(\frac{1}{\omega C}\right)^2}$$

where Z = impedance,
R = resistance,
C = capacitance,
ω = angular frequency,
V_S = peak supply p.d.
I_0 = peak current.

The phase angle φ between the supply p.d. and the current is the angle between the corresponding phasors, as shown in Figure 24.21(b). The effect of the capacitor is to make the supply phasor fall behind the current phasor; the effect of the resistor is to make it catch up. Figure 24.21(b) shows that the phase angle is given by

$$\tan\varphi = \frac{V_C}{V_R} = \frac{(1/\omega C)}{R} = \frac{1}{\omega CR}$$

Worked example A 2 µF capacitor and a 2 kΩ resistor are connected in series with an a.c. supply unit. The supply unit operates at 50 Hz, providing a peak p.d. of 12.0 V across the two components. Calculate
a) the peak current in the circuit,
b) the peak p.d. across each component.

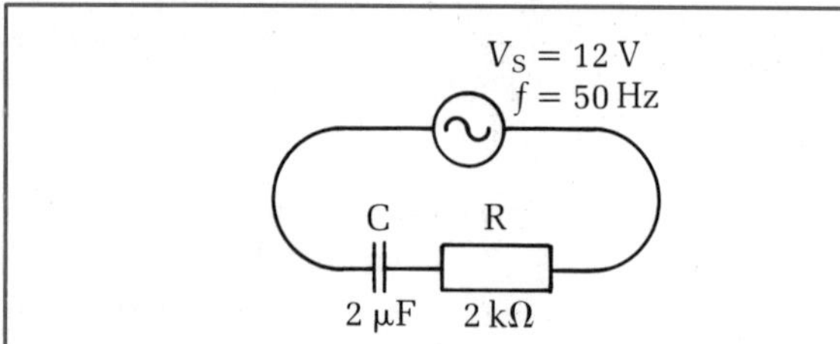

Figure 24.22

Solution
a) The reactance of the capacitor

$$= 1/\omega C = \frac{1}{2\pi \times 50 \times 2 \times 10^{-6}} = 1590\,\Omega$$

The circuit impedance Z

$$= \sqrt{R^2 + \left(\frac{1}{\omega C}\right)^2} = \sqrt{2000^2 + 1590^2} = 2560\,\Omega$$

Hence the peak current I_0 = peak supply p.d./Z
$= 12/2560 = 4.69 \times 10^{-3}\,\text{A} = 4.69\,\text{mA}$.

b) The peak p.d. across C

$$= I_0\frac{1}{\omega C} = 4.69 \times 10^{-3} \times 1590 = 7.46\,\text{V}$$

The peak p.d. across R
$= I_0R = 4.69 \times 10^{-3} \times 2000 = 9.38\,\text{V}$.

Check: does $V_S^2 = V_R^2 + V_C^2$? Try it for yourself.

LR circuits

Now consider an a.c. circuit with a resistor R in series with an inductor L, as in Figure 24.23(a). Once again, a voltmeter can be used to measure the p.d. across each component and across the supply unit. Just as for the series CR circuit, the p.d. measured across the supply is not equal to the sum of the p.d.s measured across the two

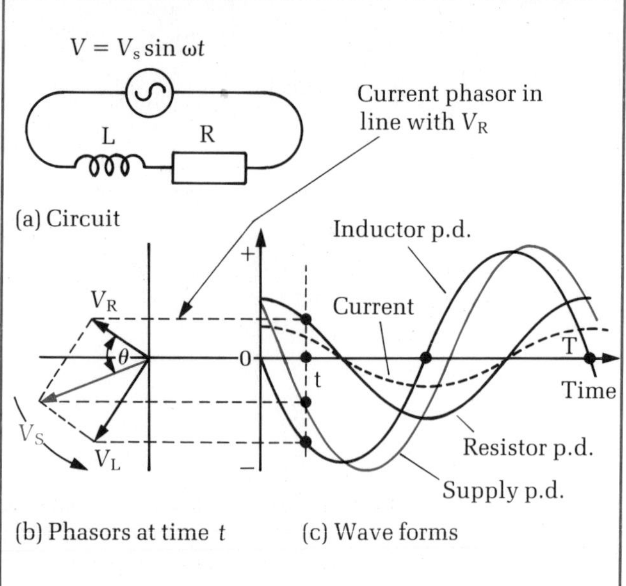

Figure 24.23 *LR series circuit*

components. The reason is that the p.d. across L is ¼-cycle ahead of the p.d. across R. The two components are in series so they pass the same current; the p.d. across R is in phase with the current, but the p.d. across L is ¼-cycle ahead. Figure 24.23(c) shows the wave forms for the p.d.s and the current.

At any instant, the supply p.d. is equal to the p.d. across R plus the p.d. across L. By considering lots of instants in each cycle, using Figure 24.23(c), we can therefore determine the wave form for the supply p.d. The result is shown in Figure 24.23(c).

Phasors make the circuit easier to understand. Figure 24.23(b) shows the phasors for the p.d.s and current at one instant. The inductor p.d. phasor is 90° ahead of the current phasor. The resistor p.d. phasor is in phase with the current phasor. The supply p.d. phasor is the vector sum of the resistor p.d. phasor and the inductor p.d. phasor. Using Pythagoras' theorem, as before, gives

$$V_S^2 = V_R^2 + V_L^2$$

where V_S = peak supply p.d.,
V_R = peak resistor p.d.,
V_L = peak inductor p.d.

Since the current at any instant is the same in each component, then

$$V_R = I_0R \text{ and}$$
$$V_L = I_0(\omega L)$$

where I_0 is the peak current.

Hence $V_S^2 = V_R^2 + V_L^2 = (I_0R)^2 + (I_0\omega L)^2 = I_0^2[R^2 + (\omega L)^2]$

so $$\frac{V_S}{I_0} = \sqrt{R^2 + (\omega L)^2}$$

The impedance Z of the circuit is defined as

$$\frac{\text{the peak p.d.}}{\text{the peak current}}.$$

For a series LR circuit the equation is as follows.

$$Z = \frac{V_S}{I_0} = \sqrt{R^2 + (\omega L)^2}$$

where Z = impedance,
V_S = peak supply p.d.,
I = peak current,
R = resistance,
L = inductance,
ω = angular frequency.

The phase angle φ between the supply p.d. and the current is the angle between the corresponding phasors, as shown in Figure 24.23(b).

$$\tan\varphi = \frac{\omega L}{R}\left(= \frac{V_L}{V_R}\right)$$

The effect of the inductor is to make the supply p.d. phasor ahead of the current phasor. The resistor has the effect of making the two phasors line up. So the inductor in this circuit has the opposite effect to the capacitor in the series CR circuit.

Coils have resistance as well as self-inductance. So a coil in an a.c. circuit is treated as a resistance in series with an inductor. Hence the peak current passed by a coil is equal to

$$\frac{\text{peak p.d. across its terminals}}{\text{coil impedance } Z}.$$

Worked example A coil with a resistance of 10 ohms and unknown self-inductance L is connected to an a.c. supply unit. The supply unit produces a peak p.d. of 10 V at 50 Hz across the coil terminals. An ammeter in series with the coil and supply unit gives an r.m.s. current of 0.40 A.

a) Calculate the self-inductance of the coil.
b) Determine the power dissipated in the coil.

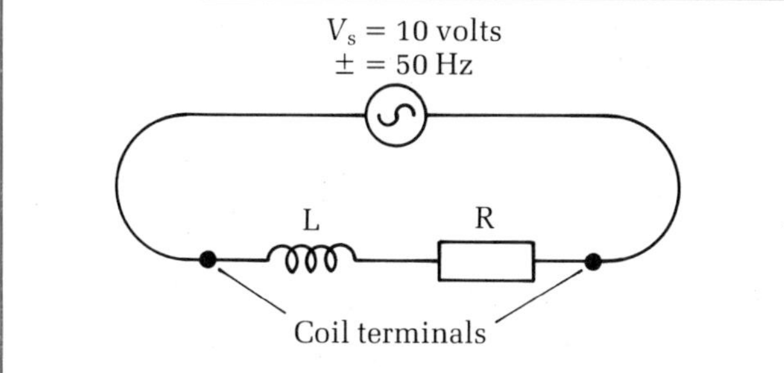

Figure 24.24

Solution

a) The peak current $I = \sqrt{2} \times$ the r.m.s. current $= \sqrt{2} \times 0.40 = 0.57$ A.
Since the impedance of the coil $Z = V_S/I_0$,

then $Z = \frac{10}{0.57} = 17.7$ ohms.

Using $Z = \sqrt{R^2 + (\omega L)^2}$ gives

$$17.7^2 = 10^2 + (2\pi \times 50 \times L)^2$$

Hence $(2\pi \times 50 \times L)^2 = 17.7^2 - 10^2 = 212$

So $2\pi \times 50 \times L = \sqrt{212} = 14.6$

Therefore $L = \frac{14.6}{2\pi \times 50} = 4.64 \times 10^{-2}$ H $= 46.4$ mH.

b) Power is dissipated only in the resistance of the coil.
Power dissipated $= \frac{1}{2}I_0^2R = \frac{1}{2} \times 0.57^2 \times 10 = 1.62$ W.

Measuring reactance

Capacitors The capacitor under test is connected into the circuit shown in Figure 24.25. The a.c. ammeter measures the r.m.s. value of the current. The oscilloscope can be used to measure the peak p.d. across the capacitor terminals. The oscilloscope can also be used to measure the time period of the alternating p.d.

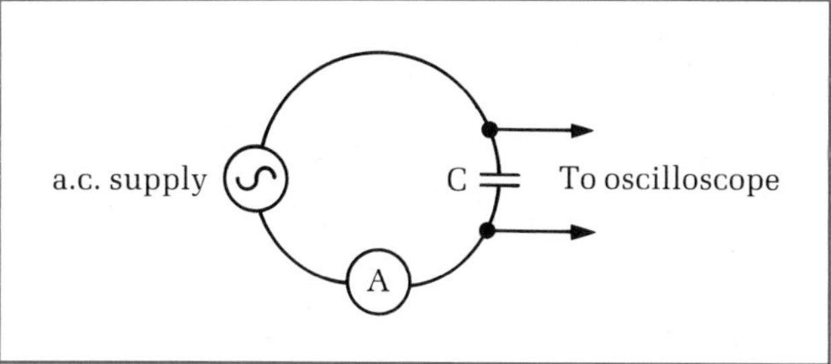

Figure 24.25 *Measuring reactance*

The peak current I is given by

$$I = \sqrt{2} \times \text{r.m.s. current}$$

Hence the reactance of the capacitor

$$= \frac{\text{peak p.d.}}{\text{peak current}} = \frac{\text{peak p.d.}}{\sqrt{2} \times \text{r.m.s. current}}$$

The reactance can be calculated from the measurements using the above equation. If possible, the supply p.d. can be varied in steps; at each step measurements of the peak p.d. and r.m.s. current are made. From the equation for reactance, a graph of y = peak p.d. against x = r.m.s. current ought to give a straight line passing through the origin.

$$y = \text{peak p.d.} = \sqrt{2} \times \text{reactance} \times \text{r.m.s. current}$$

so $$y = \sqrt{2} \times \text{reactance} \times x$$

This graph is shown in Figure 24.26. Its gradient is equal to $\sqrt{2} \times$ reactance. By measuring the gradient off the graph, the reactance can be calculated. If the angular frequency ω is calculated from the time period T (using $\omega = 2\pi/T$), then the capacitance C can also be calculated using reactance $= 1/\omega C$.

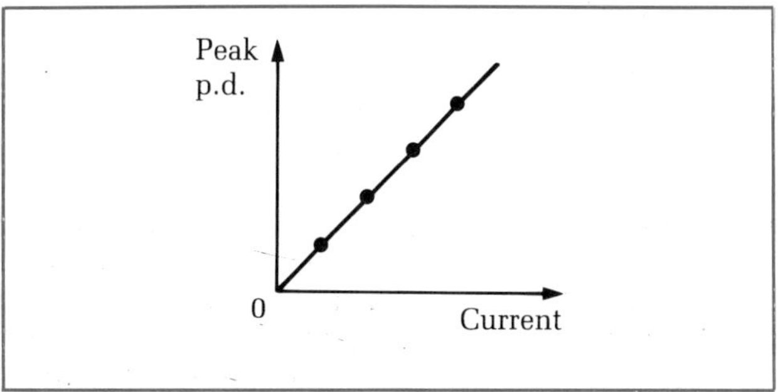

Figure 24.26 *p.d. v current for a capacitor*

Inductors Use the circuit of Figure 24.25, replacing the capacitor with the coil under test. The supply p.d. is varied in steps; at each step the ammeter reading is recorded. Also, the peak p.d. across the coil is measured using the oscilloscope. A graph of peak p.d. across coil against peak current should give a straight line passing through the origin. Since the coil impedance Z = peak p.d./peak current = gradient of graph, the impedance of the coil can be determined.

To determine the coil's self-inductance L, the a.c. frequency must be measured. This can be done using the oscilloscope to measure the time period $T = 1/\text{frequency}$. Then the coil resistance R must be measured, using the circuit of Figure 24.27. Since values of impedance Z, resistance R and frequency f have all been measured, the value of L can be determined using

$$Z^2 = R^2 + (\omega L)^2$$

where $\omega = 2\pi f$.

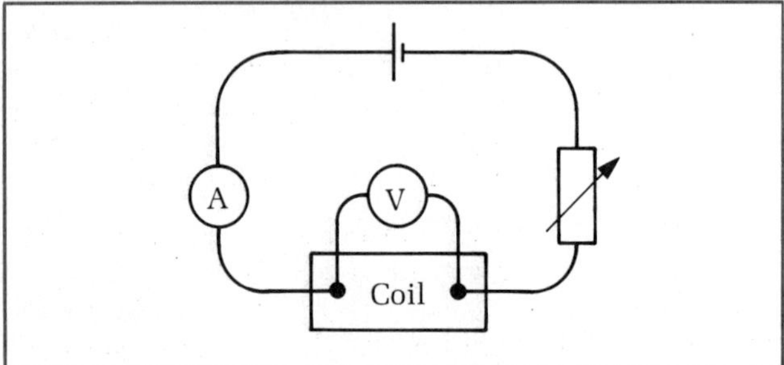

Figure 24.27 *Measuring the resistance of a coil*

24.4 Resonant circuits

Filters

Capacitors pass high frequency signals more easily than low frequency signals. Inductors pass low frequency signals more easily than high frequency signals.

Consider a source of alternating current which produces a complicated wave form, as in Figure 24.28. For example, a microphone and amplifier produce a signal like Figure 24.28 when a guitar string is plucked near the microphone. The signal consists of a mixture of sine waves of different frequencies and peak values. Suppose we want to separate the higher frequencies from the lower frequencies. This is done in hi-fi systems to achieve high-quality sound reproduction. Each part of the frequency range is supplied to a loudspeaker suitable for that range. Figure 24.28 shows a hi-fi speaker which contains two loudspeakers, the 'woofer' for low frequencies and the 'tweeter' for high frequencies. A series LC circuit is used to supply the two loudspeakers.

The series LC circuit is used because:

a) high frequency signals pass through the capacitor easily and build up across the inductor; so the tweeter is connected across the inductor to pick up the high frequency signals,

b) low frequency signals pass through the inductor easily but build up across the capacitor; hence the woofer is connected across the capacitor.

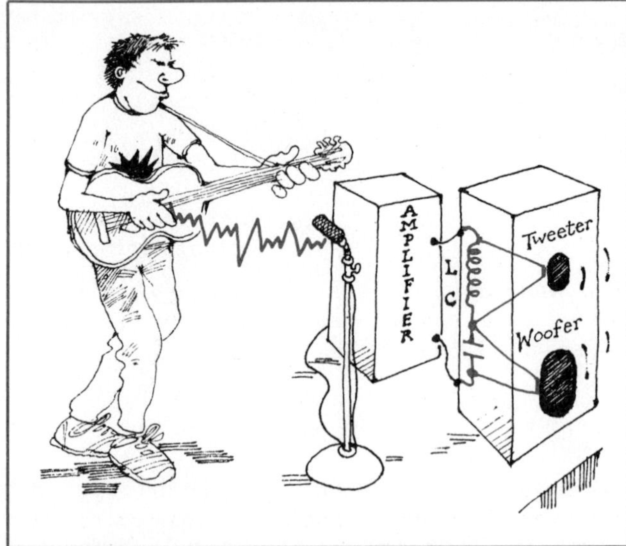

Figure 24.28 *Series LC circuit in action*

Series LCR circuits

The circuit of Figure 24.29 shows a capacitor C, an inductor L and a resistor R connected in series with one another, and with a variable frequency signal generator and an a.c. ammeter. A torch bulb is also connected in series with the ammeter.

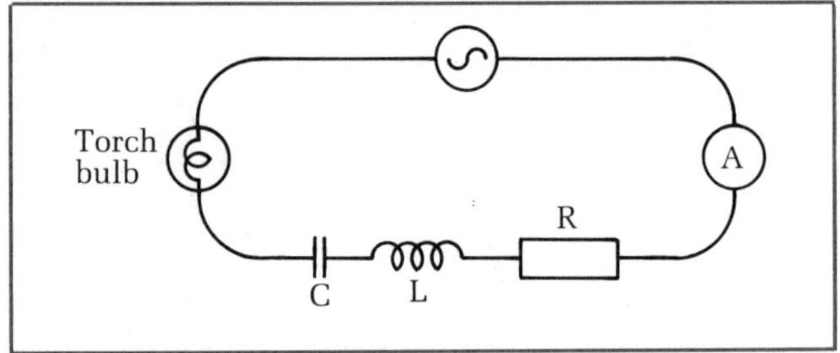

Figure 24.29 *Series LCR circuit*

Suppose the signal generator is set to give an output p.d. with a constant peak value. If the frequency of the generator is then gradually increased from low to high frequency, the current increases, then decreases again. The bulb is dim at low frequencies, then brightens as the frequency increases, and then becomes dim again at even higher frequencies. It ought to be possible to tune the generator frequency to give maximum brightness of the bulb at one position of the frequency dial. The current is then at its greatest. At higher or lower frequencies, the current is less than when the bulb is at maximum brightness. The reason is that at low frequencies, the capacitor provided most of the opposition to alternating current. At high frequencies, the inductor provides most of the opposition.

Phasors are useful here to understand the situation. All the components in Figure 24.29 take the same current because they are in series.

1 The resistor p.d. phasor V_R is in the same direction as the current phasor I_0.
2 The capacitor p.d. phasor V_C is 90° behind I_0.
3 The inductor p.d. phasor V_L is 90° ahead of I_0.

The phasor diagram shown in Figure 24.30 gives the phasor positions at an instant during one cycle. The phasors always maintain the same position relative to one another.

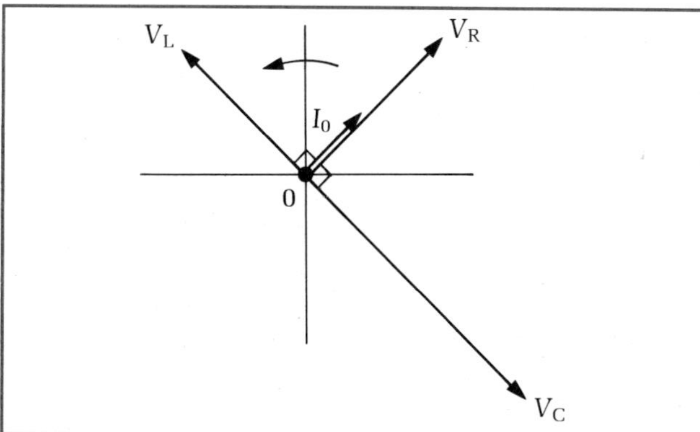

Figure 24.30 *Phasor diagram*

The supply p.d. phasor is the vector sum of the p.d. phasors for L, C and R. The diagram can be made simpler by adding the inductor and capacitor phasors, as in Figure 24.31. The supply p.d. phasor is the vector sum of V_R and $(V_C - V_L)$; this is because V_L and V_C are in opposite directions, so their vector sum is $(V_C - V_L)$.

Using Pythagoras' theorem gives the peak value of the supply p.d. V_S in terms of V_L, V_C and V_R.

$$V_S^2 = V_R^2 + (V_C - V_L)^2$$

Since $V_R = I_0 R$

$V_L = I_0(\omega L)$ and

$$V_C = I_0\left(\frac{1}{\omega C}\right)$$

$$\text{Then } V_S^2 = (I_0 R)^2 + \left(I_0 \omega L - \frac{I_0}{\omega C}\right)^2$$

$$\text{So} \quad V_S^2 = I_0^2 R^2 + I_0^2\left(\omega L - \frac{1}{\omega C}\right)^2$$

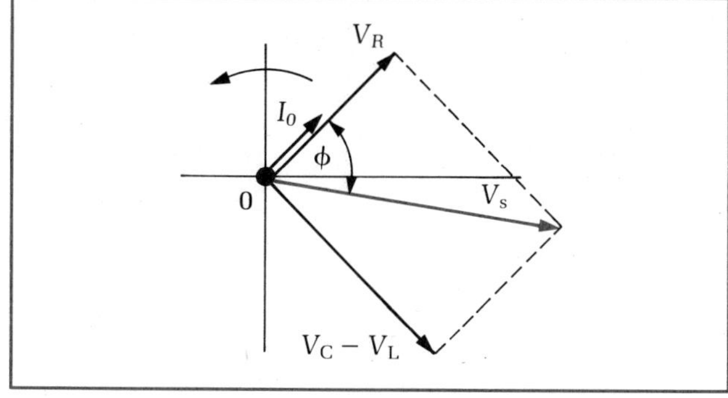

Figure 24.31 *Simplified phasor diagram*

The impedance Z of the circuit is defined as

$$\frac{\text{peak supply p.d.}}{\text{peak current}}.$$

$$Z = \frac{V_S}{I_0} = \sqrt{R^2 + \left(\omega L - \frac{1}{\omega C}\right)^2}$$

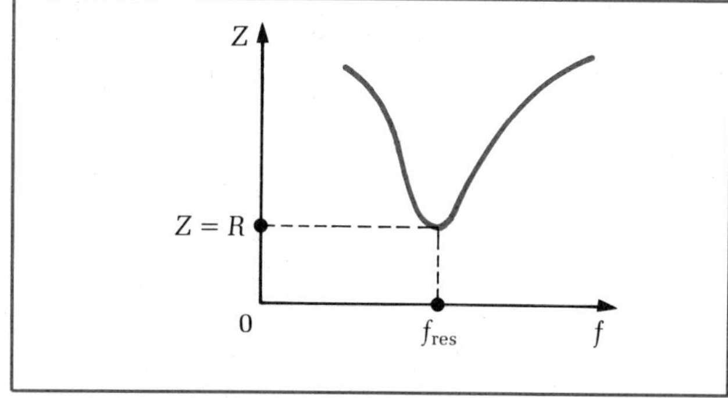

Figure 24.32 *Impedence v frequency*

The variation of Z with frequency is shown in Figure 24.32. At low frequencies, Z is large because $\frac{1}{\omega C}$ is large. At high frequencies, Z is again large because ωL is large. Somewhere between low and high frequencies, Z is a minimum because ωL is equal to $1/\omega C$.

Suppose the peak value of the supply p.d. V_S is kept constant. How does the current vary with the frequency? Since $I_0 = V_S/Z$, then the current is highest when the impedance is least. So when ωL is equal to $1/\omega C$, the current is a maximum. The circuit is then said to be at **resonance**.

The resonant frequency in a series LCR circuit occurs when the reactance of the capacitor $(1/\omega C)$ is equal to the reactance of the inductor (ωL). When this is so, V_L and V_C are exactly equal to one another. So, since they are 180° out of phase, they cancel each other out exactly. The circuit impedance is then equal to its resistance, and the supply p.d. is exactly in phase with the current.

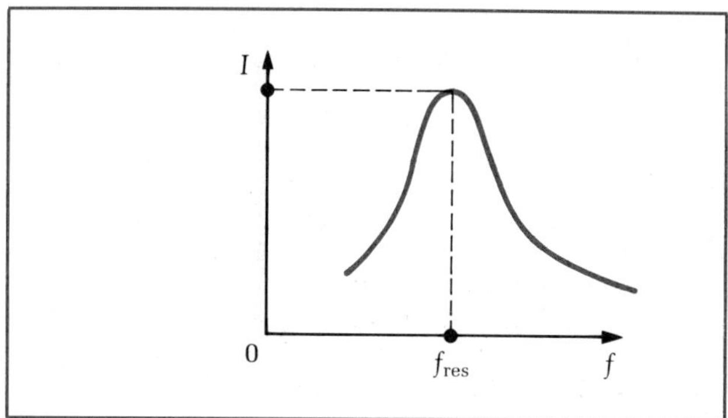

Figure 24.33 *Current v frequency*

$$\text{At resonance, } \omega L = \frac{1}{\omega C}$$

$$\text{Hence} \quad \omega^2 = \frac{1}{LC}$$

$$f_{res} = \frac{\omega}{2\pi} = \frac{1}{2\pi\sqrt{LC}}$$

where f_{res} = resonant frequency,
L = self-inductance,
C = capacitance.

Worked example A 1.0 μF capacitor is connected in series with a coil of resistance 50 ohms and self-inductance 49 mH and an a.c. ammeter. A current of r.m.s.

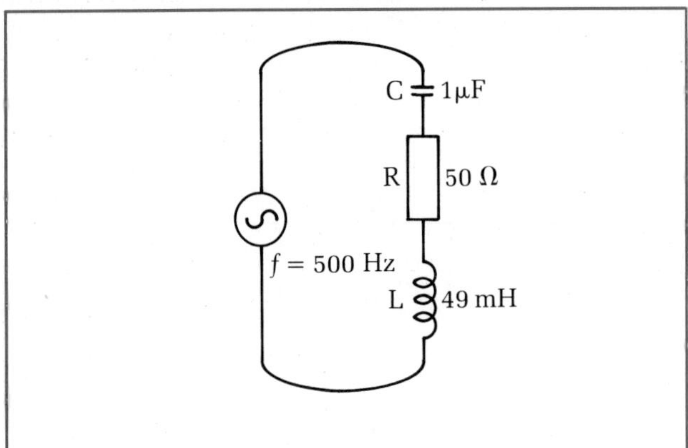

Figure 24.34

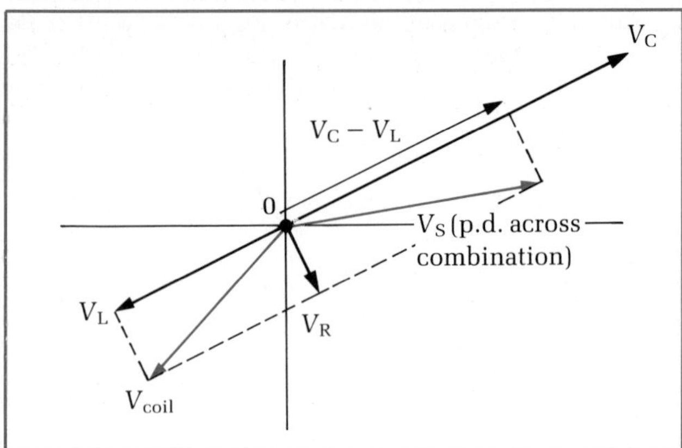

Figure 24.35 *Phasor diagram*

value 0.030 A and frequency 500 Hz is passed through the combination.

a) What is the peak p.d. across (i) the capacitor, (ii) the coil, (iii) the combination?

b) What is the resonant frequency of the circuit?

Solution

a) The peak current $I_0 = \sqrt{2} \times$ r.m.s. current $= 0.042$ A.

i) Hence peak p.d. across the capacitor

$$= I_0 \times \frac{1}{\omega C} = 0.042 \times \frac{1}{2\pi \times 500 \times 1 \times 10^{-6}}$$
$$= 13.5 \text{ V}$$

ii) The impedance of the coil, $Z_{coil} = \sqrt{R^2 + (\omega L)^2}$

$= \sqrt{50^2 + (2\pi \times 500 \times 0.049)^2}$ so $Z_{coil} = 161$ ohms.

Hence $V_{coil} = I_0 Z_{coil} = 6.80$ V.

iii) The impedance of the combination Z

$$= \sqrt{R^2 + \left(\omega L - \frac{1}{\omega C}\right)^2}$$

$$= \sqrt{50^2 + \left(2 \times 500 \times 0.049 - \frac{1}{2 \times 500 \times 1 \times 10^{-6}}\right)^2}$$

$$= \sqrt{2500 + (154 - 318)^2}$$
$$= 171 \text{ ohms.}$$

Hence $V_{combination} = I_0 Z = 0.042 \times 171 = 7.21$ V.

Comments not required for the answer:

$V_R = I_0 R = 0.042 \times 50 = 2.1$ volts.

$V_L = I_0 \omega L = 0.042 \times 2\pi \times 500 \times 0.049 = 6.47$ V.

$V_C = 13.5$ V (as calculated in (i)).

The phasor diagram is shown in Figure 24.35.

The coil p.d. $V_{coil} = \sqrt{V_R{}^2 + V_L{}^2}$

The p.d. across the combination $= \sqrt{V_R{}^2 + (V_L - V_C)^2}$

Check for yourself that the values of p.d. for R, L and C give the values of coil p.d., etc. as calculated above. You ought to obtain complete agreement because the method used for the solution involves calculating impedances and then p.d.s for the components. The other method involves calculating reactances to give the p.d. values for L, C and R separately. Then the p.d. values are combined to give the component p.d.s. So the two methods just involve slightly different approaches, leading to the same result.

b) The resonant frequency is given by

$$f_{res} = \frac{1}{2\pi\sqrt{LC}} = \frac{1}{2\pi \times \sqrt{0.049 \times 1 \times 10^{-6}}}$$

$f_{res} = 718$ Hz.

Tuning circuits

When you tune in to your favourite radio station, you alter a variable capacitance in the radio. The capacitance is in parallel with an inductor, as in Figure 24.36. The radio signals are picked up by the aerial. An aerial picks up many different frequencies, so it is essential to filter out all but the required signal frequency. Then the required signal is passed into an amplifier.

The capacitance and inductance in Figure 24.36 form a parallel LC circuit called a 'tuning circuit'. Low frequency signals pass to Earth through the inductor because inductors have low reactance at low frequency. High frequency signals pass to Earth through the capacitor because capacitors have low impedance at high frequencies. So there is a frequency between high and low where signals

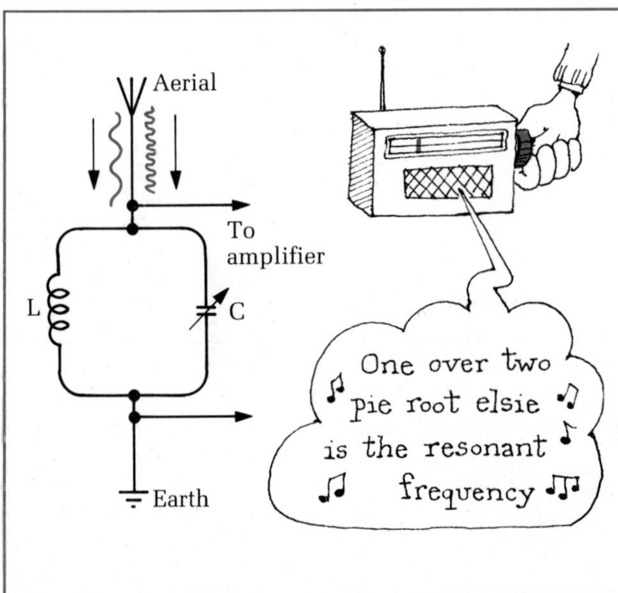

Figure 24.36 *Tuning circuits*

cannot pass to Earth through L or through C. The signals pass into the amplifier instead. At this frequency, the resonant frequency, the inductance and capacitance reactances are equal, that is, $\omega L = \frac{1}{\omega C}$.

As we have already seen, the resonant frequency is given by

$$f_{res} = \frac{\omega}{2\pi} = \frac{1}{2\pi\sqrt{LC}}$$

By altering the value of C, the resonant frequency is varied until it is equal to the carrier frequency of the radio station you want to listen to. See p. 293 for radio communications.

To understand the action of the tuning circuit, consider Figure 24.37, which shows an inductor and lamp in parallel with a capacitor and lamp. The parallel combination is in series with a third lamp and a variable-frequency supply unit.

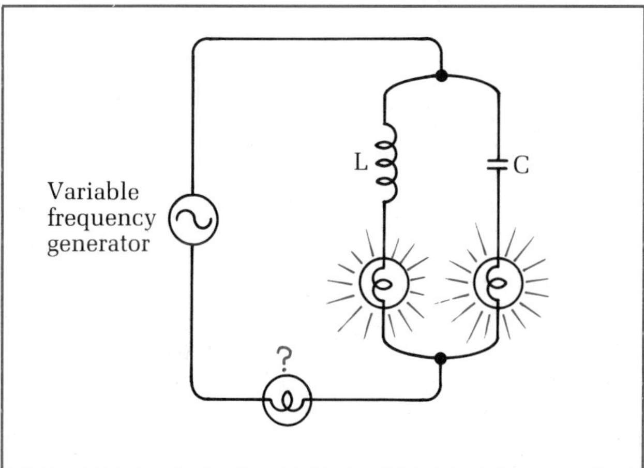

Figure 24.37 *A resonance puzzle*

At low frequencies, the inductor lamp lights up because low frequency signals pass through the inductor easily. The capacitor blocks low frequency signals.

At high frequencies, the capacitor lamp lights because the capacitor passes high frequency signals easily. But the inductor lamp is now out because the inductor blocks high frequencies. At both high and low frequencies, the third lamp lights as brightly as the lamp which is 'on' in the parallel combination.

Now adjust the frequency until the capacitor and inductor lamps are equally bright. This is the resonant frequency, as given by the previous equation. But the lamp in series with the combination is now *off*! Why? At resonance, electrons from the capacitor discharge through the inductor which acts to try to keep the falling current steady. So the inductor makes electrons from one plate of the capacitor pile up on the other plate. Then the electron flow reverses as the inductor makes the electrons pile up on the first plate again. So an alternating current passes round the LC combination with no current taken from the supply unit. The circuit is in resonance because the p.d. across L is exactly equal and opposite to the p.d. across C. So because the current in L is the same as in C, the reactance of L is equal to the reactance of C for resonance.

Where else have you met resonance. See p. 227 for some non-electrical examples. All resonance situations involve a component which tries to keep the motion going and another component which tries to restore the system to equilibrium. Provided energy losses from the system due to friction or resistance are small, the system oscillates with a large amplitude at resonance.

Components	**System**		
	LCR circuit	*Mass on a spring*	*Resonating wire*
Keeping it going	Self-inductance	Mass	Wire mass
Restoring it	Capacitance C	Spring	Tension
Wasting its energy	Resistance R	Tension	Friction

24.5 Summary

Definitions

1 The time period is the time taken for 1 complete cycle.
2 The frequency is the number of complete cycles per second.
3 The angular frequency is $2\pi/T$, where T is the time period.
4 The peak value is the maximum value of an alternating p.d. or current.
5 The root mean square value of an alternating current or p.d. is the value of direct current which would supply the same power in a given resistor.
6 The reactance of a capacitor C is defined as $1/\omega C$.
of an inductor L is defined as ωL.
7 The impedance of a circuit is peak p.d./peak current

Equations

1 Time period $T = 1/f$
2 Angular frequency $= 2\pi f = 2\pi/T$
3 r.m.s. value $=$ peak value$/\sqrt{2}$ (for sinusoidal a.c.)
4 Power dissipated $= (I_{r.m.s.})^2R = \frac{1}{2}I_0^2R$
5 Resonant frequency $= \frac{1}{2\pi\sqrt{LC}}$

Comparison of series circuits

A.C. supply unit connected across	Impedance Z	Phase angle φ between supply p.d. V_S and current I
R only	R	0
C only	$1/\omega C$	I is 90° ahead of V_S
L only	ωL	I is 90° behind V_S
C and R in series	$\sqrt{R^2+\left(\frac{1}{\omega C}\right)^2}$	I is at angle φ ahead of V where $\tan\varphi=\frac{1}{\omega CR}$
L and R in series	$\sqrt{R^2+(\omega L)^2}$	I is at angle φ behind V where $\tan\varphi=\omega L/R$
L, C and R in series	$\sqrt{R^2+\left(\omega L-\frac{1}{\omega C}\right)^2}$	I is ahead of V by angle φ where $\tan\varphi=\frac{1}{\omega CR}-\frac{\omega L}{R}$

Short questions

24.1 An oscilloscope is used to measure the voltage wave form across a 500 ohm resistor in an a.c. circuit. The wave form is shown in the diagram below. Given that the time base of the oscilloscope is set at 5 ms cm^{-1}, and its Y-gain is set at 0.5 V cm^{-1}, determine
a) the time period and hence the frequency,
b) the peak to peak voltage and hence the r.m.s. voltage,
c) the r.m.s. current through the resistor,
d) the mean power dissipated in the resistor.

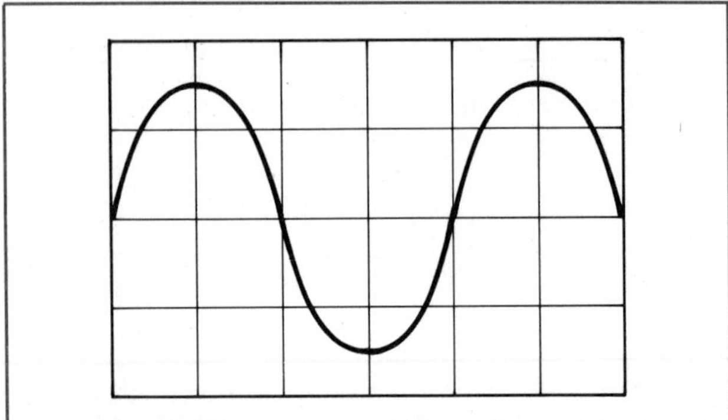

Figure 24.38

24.2 Determine the r.m.s. value for each of the following.
a) A sinusoidal current with a peak value of 2.0 A
b) A full-wave rectified current with a peak value of 3.0 A
c) A square-wave current with a frequency of 1 Hz, which is at 0.1 A for one-half cycle and −0.1 A for the next half-cycle
d) An uneven square wave voltage as below.

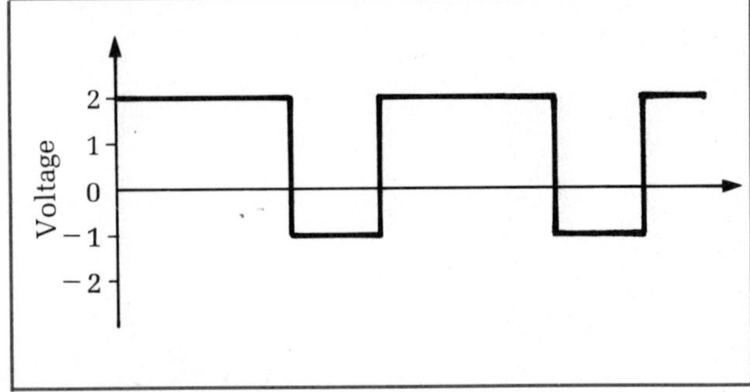

Figure 24.39

24.3 Sketch the wave form for the alternating p.d. between points X and Y in each of the circuits below.

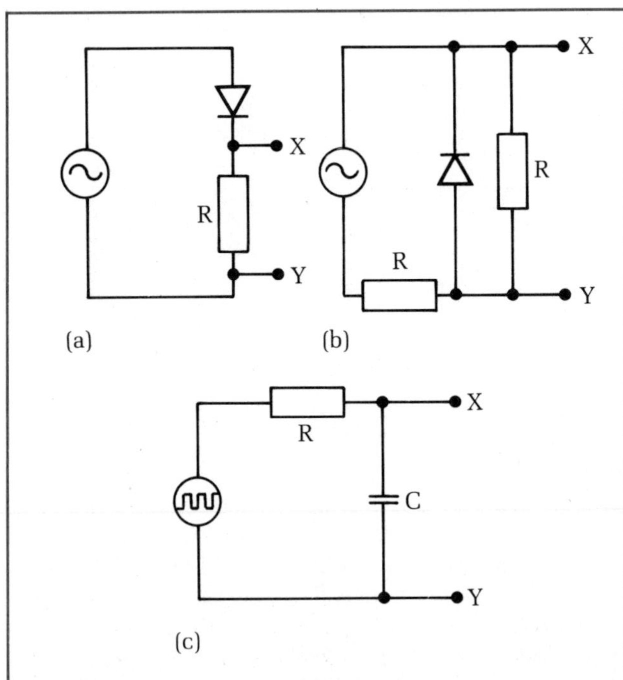

Figure 24.40

24.4 A sinusoidal current has a peak value of 5 A and its frequency is 50 Hz. Given that the current is represented by the equation

$I=I_0 \sin \omega t$,

use a phasor diagram to determine the current at time t equal to **a)** 4 ms, **b)** 12 ms, **c)** 22 ms.

24.5 Prove that the current through a capacitor is $\frac{1}{4}$-cycle ahead of the capacitor voltage. Show that the reactance of a capacitor is equal to $1/(2\pi fc)$, were f is the frequency and C is the capacitance. Use this result to explain why capacitors act as filters for blocking low frequency signals.

24.6 Explain, in terms of power, the difference between a resistive and a reactive component.
In an experiment to measure the reactance of a capacitor, the r.m.s. current is measured at 10 mA using an a.c.

milliammeter. The p.d. across the capacitor terminals is measured at the same time using an oscilloscope, giving a peak-to-peak voltage of 16 V. Calculate

a) the reactance of the capacitor,

b) the capacitance C, given that the frequency is 100 Hz.

24.7 Sketch curves on the same axes to show how the current through and the p.d. across an inductor vary with time when the inductor is in an a.c. circuit. Assume the inductor has negligible resistance. On the same axes, sketch the power curve for the inductor, and indicate the direction of energy transfer over one full cycle. Use your power curve to explain why an inductor does not dissipate power.

24.8 A 12 V lamp bulb is connected in series with an air-cored inductor and a low-voltage a.c. supply unit. The output from the supply unit is adjusted until the lamp is at normal brightness. Explain why the lamp brightness is reduced when a solid iron bar is inserted into the core of the inductor.

24.9 For the circuit shown below, the peak supply p.d. is kept fixed at 10 V but the supply frequency can be varied continuously over the range from 10 Hz to 1000 Hz. For each of the following components in turn connected between X and Y in the circuit, sketch a graph to show the variation of r.m.s. current with frequency over the full range.

a) A 100 ohm resistor connected between X and Y.

b) A 100 μF capacitor connected between X and Y.

c) A coil of resistance 8 ohms and inductance 5 mH connected between X and Y.

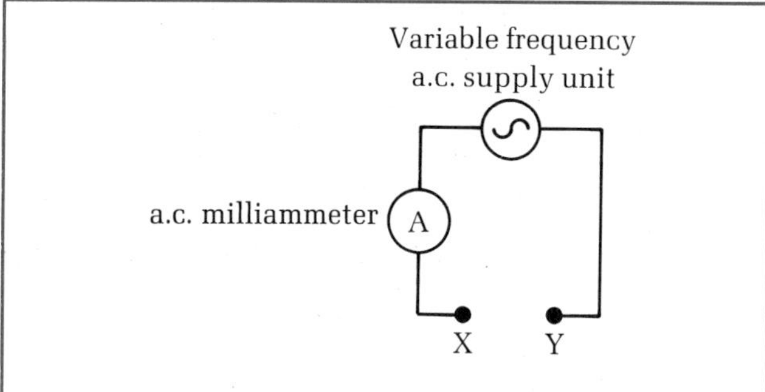

Figure 24.41

24.10 A student decides to make a filter to supply a suitable loudspeaker with high frequency signals only from an amplifier. He decides to use a resistor and a capacitor in series and considers two arrangements as shown below. Which one would you choose, and why?

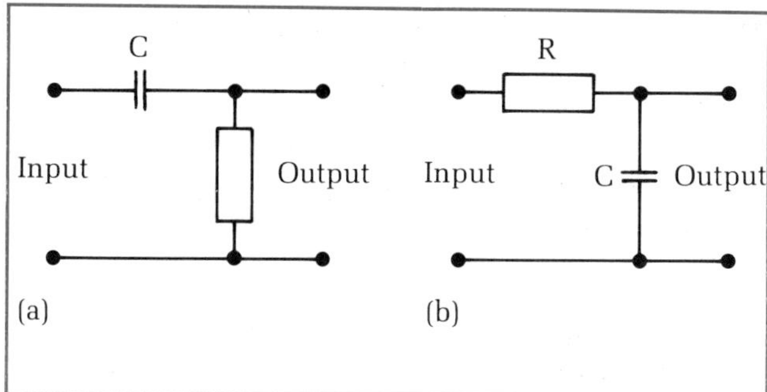

Figure 24.42

24.11 A coil X in a d.c. circuit passes a current of 1.0 A when the p.d. across its terminals is 9.0 V. The same coil in an a.c. circuit operating at 50 Hz passes an r.m.s. current of 0.1 A when the r.m.s. p.d. across its terminals is 4.1 V. Use this information to calculate

a) the resistance,

b) the inductance of the coil.

24.12 A resistor and capacitor are connected in series with one another and an a.c. milliammeter. The combination is then connected to the output terminals of a variable frequency generator. The output voltage of the generator is monitored using an a.c. voltmeter connected across the output terminals. At frequency $f = 50$ Hz, the milliammeter reads 0.025 A and the voltmeter reads 4.7 V; at $f = 100$ Hz, the voltmeter reads 3.2 V for the same current as before. Use this information to calculate the resistance and the capacitance.

24.13 A capacitor C, a resistor R and an inductor L are connected in series with one another, and then the combination is connected to the output of an a.c. supply unit. The resistance of R is known to be 120 Ω. A voltmeter reading r.m.s. values is then used to measure the p.d. across each component in turn. Its readings are 12 V across R, 15 V across L and 20 V across C.

a) Use the resistor voltage to calculate the r.m.s. current.

b) Calculate the r.m.s. value of the supply p.d. and sketch a phasor diagram to show how the voltage readings are related to one another.

c) Given the frequency is 50 Hz, calculate the capacitance of C and the inductance of L.

24.14 A coil has an inductance of 30 mH and a resistance of 5.0 Ω. it is to be connected into a circuit in series with a capacitor C and an a.c. supply unit with an output of 12 V r.m.s. at 40 Hz.

a) What value of capacitance for C would give resonance?

b) What would be the r.m.s. current at resonance?

c) What would be the r.m.s. p.d. across the coil at resonance?

24.15 The coil in question **24.14** is connected in parallel with a 0.01 μF capacitor. Sketch a graph to show how the impedance of the parallel combination varies with frequency.

25

Digital electronics

25.1 The systems approach

Using electronics

Electronics is a fast-changing subject. Few areas of our lives have escaped the sweeping changes caused by modern electronics. Today's devices soon become obsolete as engineers and scientists find new and better methods and materials. If cars had developed over the past 20 years at the same rate as electronics, today's new cars would cost less than £100 and would give more than 1000 miles per gallon!

Today's silicon chips used in any calculator each contain hundreds of transistors and resistors; twenty years ago, calculators were costly and heavy because the circuits were made using transistors and resistors, etc. as separate (i.e. discrete) components. In twenty years time, who knows?

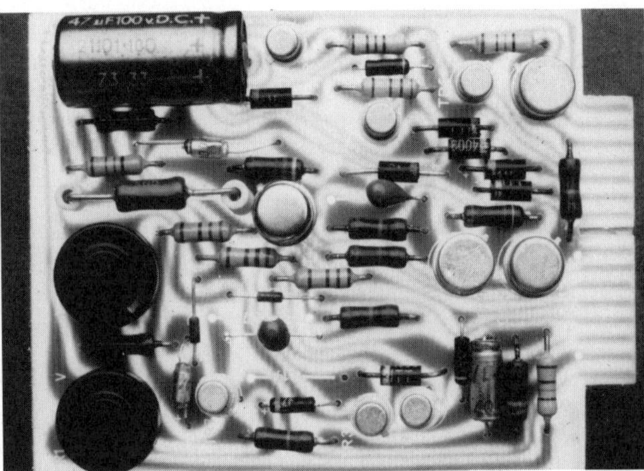

a) *Discrete components*

b) *Microchips*

Figure 25.1

Using electronics does not need detailed knowledge of how silicon chips are made or of their internal circuitry. Any electronic system can be considered as a set of units linked together. Each unit has a particular function, and what goes on inside the unit is of less importance to the user than what the unit actually does. Each unit is treated as a 'black box' which produces an output signal from the input signal supplied to it. To use each black box, there is no need to open it; all we need to know is how the input signal will affect the output signal.

A digital watch is an example of an electronic system. A quartz oscillator produces electrical pulses at a precise frequency. The pulses are 'fed' into a 'frequency divider' unit which produces output pulses at an exact rate of one per second. These pulses are counted, and the number of counts is displayed as the time in seconds. Each time the number of seconds displayed reaches 59, a pulse is sent to the 'minutes' counter so the number of minutes shown is increased (or 'incremented' to use the jargon) by one. In In the same way, each time the number of minutes reaches 59, a pulse is sent to the 'hours' counter. Figure 25.2 shows the system made up of 'black box' units. All that in a digital watch – amazing!

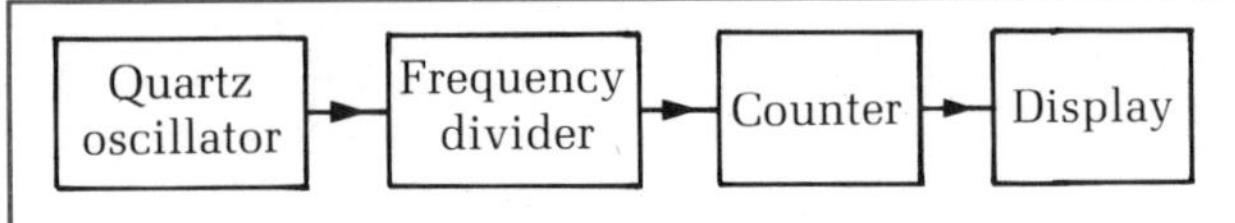

Figure 25.2 *A digital watch system*

Each pulse fed to the counter makes the input voltage jump from 0 volts (**low**) to a constant positive voltage (**high**) then back to zero. So we say that the input voltage jumps from a '0' (low) to '1' (high) then back to '0' for each pulse. The '0' represents low voltage at the input; the '1' represents constant high voltage at the input. Typically, a '0' is less than +0.5 V and a '1' is +5 V. No other voltage level occurs in the system, and that is why digital watches are so accurate. The quartz oscillator produces a precise frequency so pulses at exactly 1 Hz enter the counter from the divider unit.

Information handling has improved enormously as a result of digital electronics. You may well have had difficulty in the past chatting over the 'phone to your friends because of a 'poor line'. Yet long-distance calls as far as the other side of the Earth are usually 'as clear as a bell'. The telephone system is being developed to make increasing use of digital electronics, with international users the first to benefit. A poor line is usually due to background noise along the line, often caused by the random motion of electrons. Conversation can be difficult if it is accompanied by hissing due to background noise. Noise is added to the voice signals; amplifiers used to make the voice signals stronger make the noise signal stronger too, so the listener has difficulty picking out the conversation.

Digital systems used in the telephone network cut out much of the noise at the receiver. The voice signal is **digitized** and transmitted as a rapid sequence of pulses. The pulses pick up noise, but **regenerator** units at intervals along the line clean the pulses up. Each regenerator responds only to voltage changes from 'low' to 'high' or *vice versa*. The noise signal therefore does not affect the regenerator, so each regenerator sends 'clean' pulses down the line. Figure 25.4 shows the idea. Cleaned pulses are fed to a converter unit at the receiver end to produce the original voice signal without any noise present.

Figure 25.3 *Noise signals*

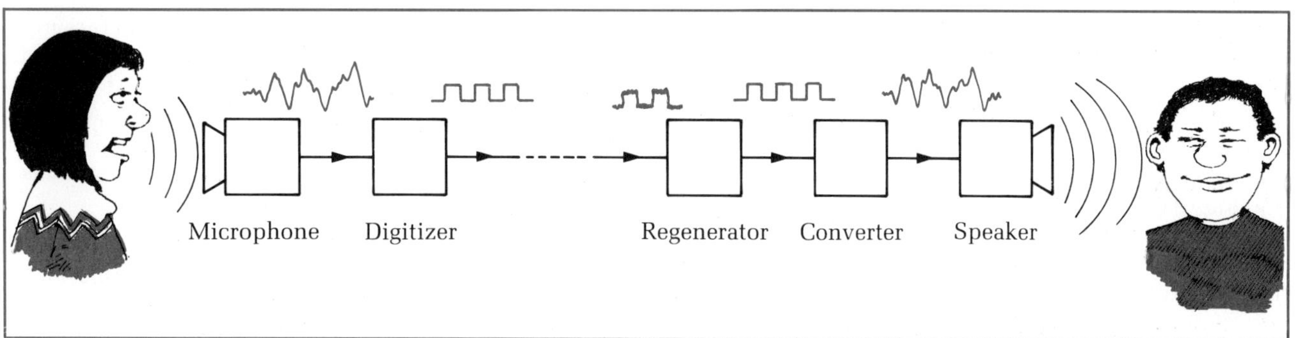

Figure 25.4 *Eliminating noise*

In electronic systems, we use the term **digital** where voltage levels can be either high or low; no other level is allowed. At any point in a digital system, the voltage is in one of two states – either high or low. Where voltages can be at any level, we use the terms **linear** or **analogue**. A **linear circuit** is one where the voltage at any point can have any value between an upper and a lower limit set by the power supply unit.

Logic is at the core of digital electronics. Signals are either low or high; like 'false' or 'true'. Consider the logic involved in the operation of a warning light for the two front seat belts of a car. The light will be on or off, depending on whether or not

1 the driver's seat belt is unfastened,
2 the brakes are off,
3 the passenger seat is occupied,
4 the passenger seat belt is unfastened.

If any of the above statements is true, represent it by a'1'; if false, then we shall use a '0' to represent it. Let's look at all the possibilities.

In fact, there are 16 possibilities in total because there are two possibilities for each statement. So in total there are $2 \times 2 \times 2 \times 2 = 16$ possibilities. What are the conditions for the light to be **on**? For the light to be **on**:

a) If statement **1** is true, then the light should come on only if statement **2** is true as well.

b) If statement **3** is true, then statement **4** must be true as well for the light to come on.

In other words, the light is on if 1 **AND** 2 are true **OR** if 3 **AND** 4 are true. The **truth table** at the top of the next column shows the conditions for the light to be on.

	Statements				Light
	1	**2**	**3**	**4**	
1st set of conditions	1	1	0	0	on
2nd set of conditions	0	0	1	1	on
3rd set of conditions	1	1	1	1	on

where '1' = true, and '0' = false.

A circuit with four switches, A, B, C and D will operate the warning light if it is connected as in Figure 25.5. A closed switch represents a true statement; an open switch represents a false statement. The light is on if

A AND B are both closed	**or**	C AND D are both closed	**or**	ALL are closed

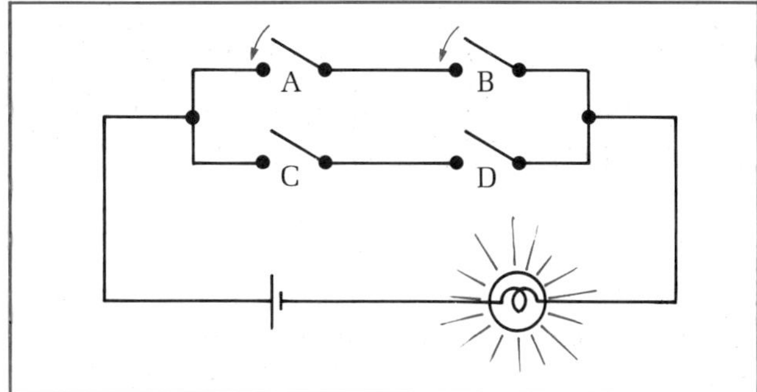

Figure 25.5 *Logic switches*

Digital circuits contain lots of electronic switches, not only in logic units but also in memory units, counters and registers, etc. The function of each unit is important to the user, not how the switches inside work. You will learn a little of what takes place inside such units in Chapter 26. Here, the emphasis is on using the units to design and build systems, just as engineers do.

25.2 Logic gates

Gates and truth tables

Electronic gates are units which each give '1' at the output when 'open' and '0' when 'closed'. Logic gates are opened or closed by input signals, always in digital form. There are different types of logic gate, the simpler ones are listed below. They can be combined to give **adders**, **registers**, **counters** and other electronic units.

The inverter or NOT gate This gate has one input and one output. In other words, the input voltage is applied between the input terminal and the 0 V line. The output voltage is produced between the output terminal and the 0 V line. The circuit symbol just shows the input and output terminals without the 0 V line or the positive power supply line. The voltage levels are either 'high' or 'low' relative to the 0 V line (which is 'low').

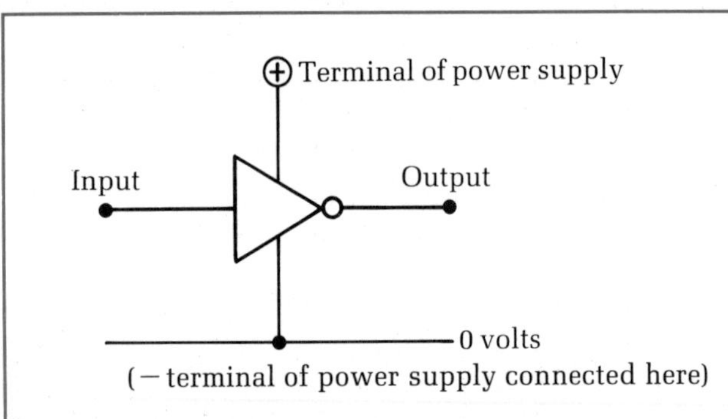

Figure 25.6 *An inverter*

If the input voltage is high, then the output voltage is low. If the input voltage is low, then the output voltage is high. So the unit is called an **inverter** because its output is always in the opposite state to the input. The link between the input voltage and the output voltage can be represented by a **truth table**, a table which shows the output state produced by each possible input state.

Input voltage or 'state'	Output voltage or 'state'
0	1
1	0

The fact that the output state is always the opposite state to the input is sometimes expressed by saying that the output is the **complement** of the input. In other words, the complement of 0 is 1, and the complement of 1 is 0.

The NAND gate With two or more inputs and one output, the output is NOT high only if each AND every input is high. NAND is short for NOT AND. The symbol for a two-input NAND gate is shown in Figure 25.7. The truth table is also shown in the diagram. The output state is '0' only if inputs A and B are **both** '1'. In other words, the output state is low only when inputs A and B are both high. If either input is low, then the output is high. If both inputs are low, the output is also high.

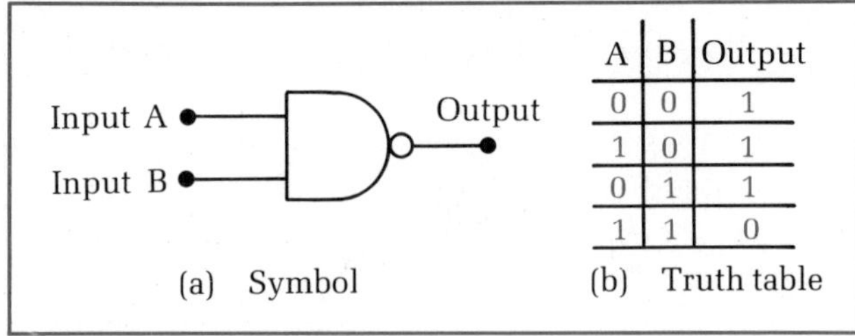

A	B	Output
0	0	1
1	0	1
0	1	1
1	1	0

(b) Truth table

Figure 25.7 *A NAND gate*

NAND gates are the easiest type to manufacture in integrated form on silicon chips. See p. 385 for brief details of the manufacturing process. Other types of gate, described below, can be made from NAND gates. An inverter can be made from a 2 i/p (i.e. two-input) NAND gate by connecting the two inputs together, and using them as a single input.

The AND gate The output is high only if each AND every input is high. The symbol and truth table for an AND gate are shown in Figure 25.8. A NAND gate and an inverter combined as in Figure 25.9 form an AND gate.

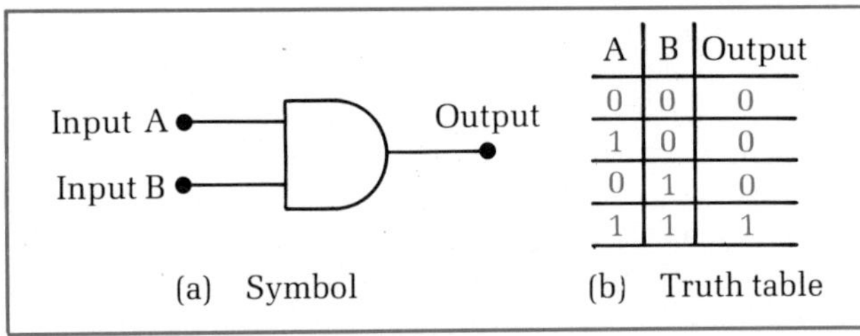

A	B	Output
0	0	0
1	0	0
0	1	0
1	1	1

(b) Truth table

Figure 25.8 *An AND gate*

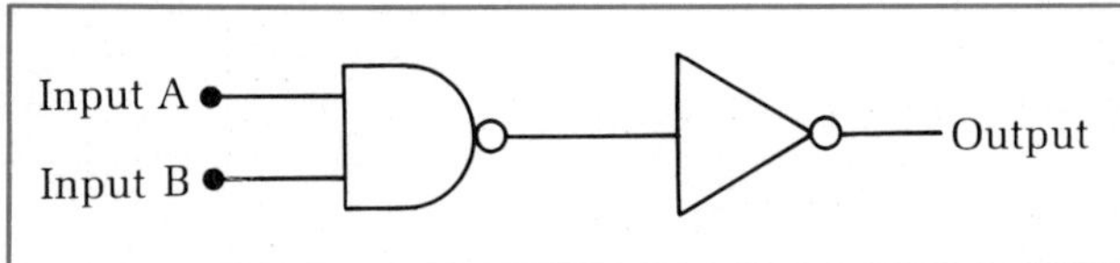

Figure 25.9 *Making an AND gate*

AND gates are used to control the progress of signals in a microprocessor. Figure 25.10 shows two AND gates

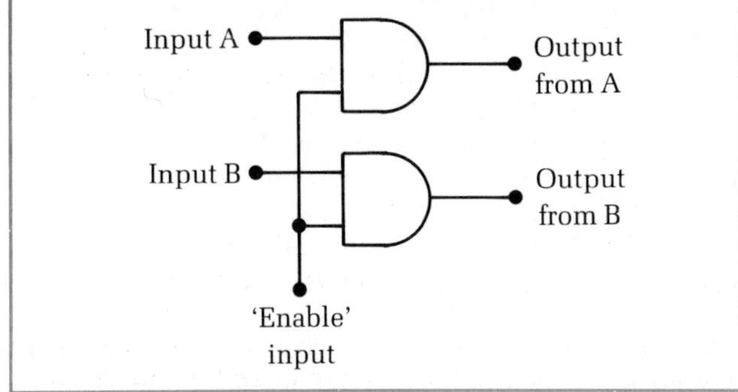

Figure 25.10 *Holding a signal*

used to 'hold' a signal at the gate inputs until a '1' is applied to the 'enable' line from the control circuit. Then the signal passes to the outputs.

The OR gate The output is high if one OR more inputs are high. Figure 25.11 shows the truth table and the symbol. As explained in the next section, OR gates are used in **adder** circuits. Figure 25.12 shows how to construct an OR gate from three NAND gates.

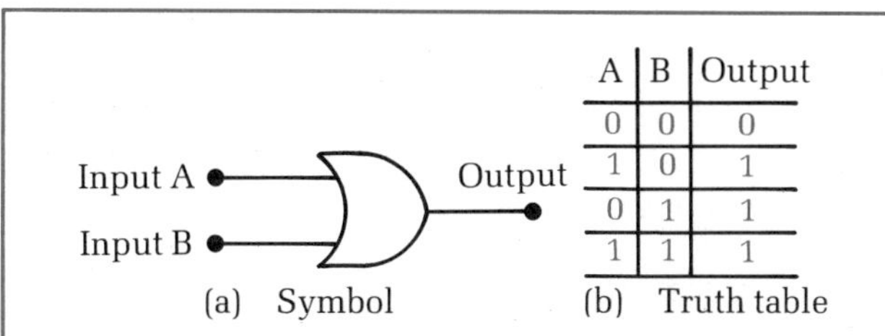

A	B	Output
0	0	0
1	0	1
0	1	1
1	1	1

Figure 25.11 *An OR gate*

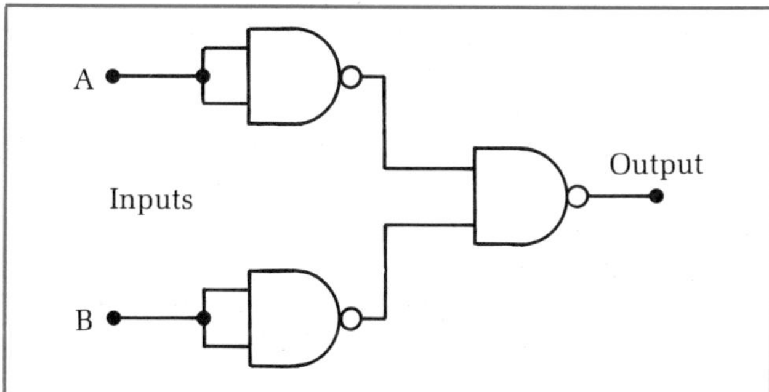

Figure 25.12 *Making an OR gate*

The NOR gate The output is NOT high if one OR more inputs are high. The symbol and truth table for a NOR gate are shown in Figure 25.13. A NOR gate can be made from four NAND gates. The NAND gates are used to make an OR gate and an inverter, as shown in Figure 25.14.

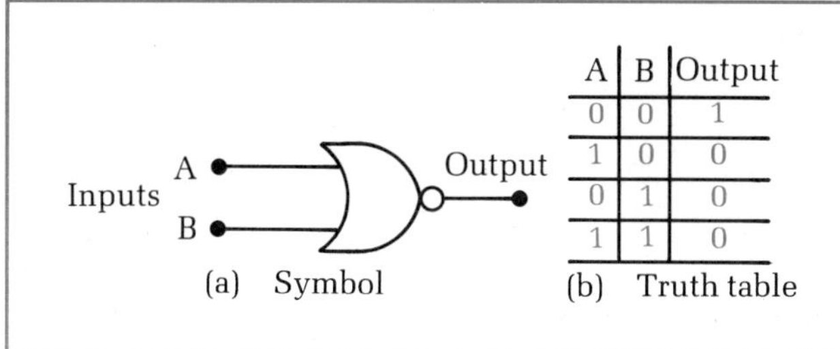

A	B	Output
0	0	1
1	0	0
0	1	0
1	1	0

Figure 25.13 *A NOR gate*

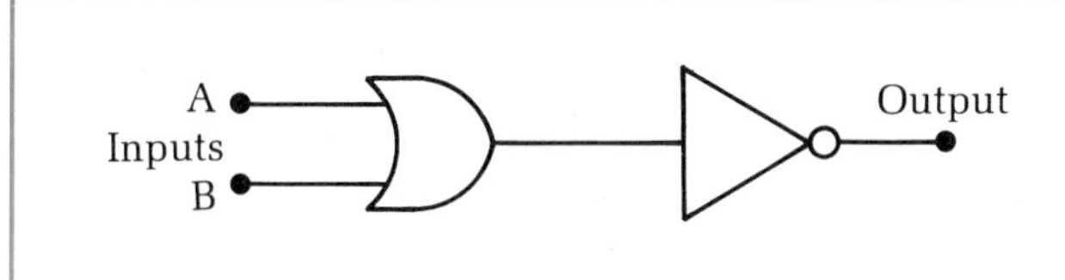

Figure 25.14 *Making a NOR gate*

Investigating gates

Practical work with gates can be made easier by using specially prepared circuit boards like the one shown in Figure 25.15. The pin connections for the power supply,

Figure 25.15 *A NAND board*

the inputs and the output terminal for each gate are clearly shown. Take care with such boards to ensure the correct value and polarity for the power supply is used. Otherwise, the logic chip may be damaged.

If a specially prepared board is not available, you will need the pin diagram for the chip under test. Figure 25.16 shows the pin diagram for a quad 2 i/p NAND chip; this type of chip has four NAND gates, each with two inputs. You also need to know the power supply voltage. The chip can be mounted and soldered to a suitable circuit board; better still, a suitable IC (i.e. chip) holder can be soldered to the board. Then, the chip can be inserted into the holder. Even simpler, an IC board can be used where the chip is inserted into the board.

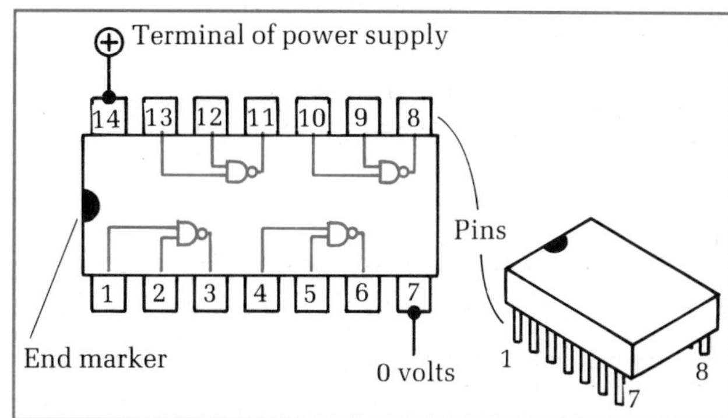

Figure 25.16 *A quad two-input NAND chip*

Logic levels Connect the inputs of a NAND gate together to make an inverter. When the input is high, the output is low. When the input is low, the output is high. The circuit of Figure 25.17 may be used to investigate the

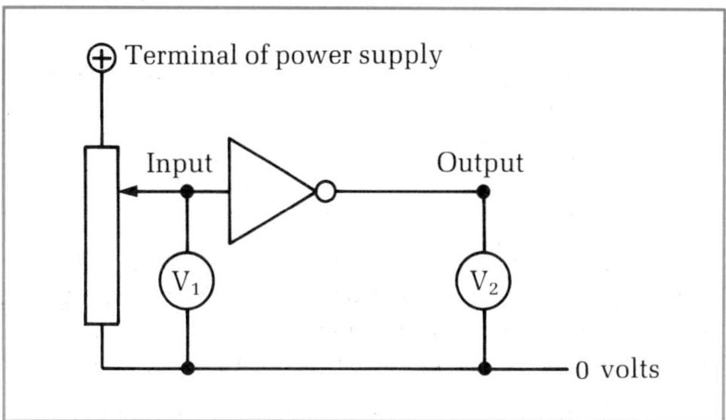

Figure 25.17 *Investigating logic levels*

variation of output voltage with input voltage. In theory, a '0' is exactly at zero volts, and a '1' is at a fixed positive voltage. In practice, the output is high if the input voltage is less than a fraction of one volt. In Figure 25.17, the potential divider is used to change the input voltage in small steps. At each step, the input and output voltages are measured from the voltmeters. A graph of output voltage against input voltage shows the high and low levels for the input and output voltages. The graph which you obtain will depend on the type of IC used. The two most common types are called TTL and CMOS. Each type is designed with different specifications, as described below. A fuller explanation of each type is given on p. 390. Figure 25.18 shows the output against input graph for a TTL NAND gate.

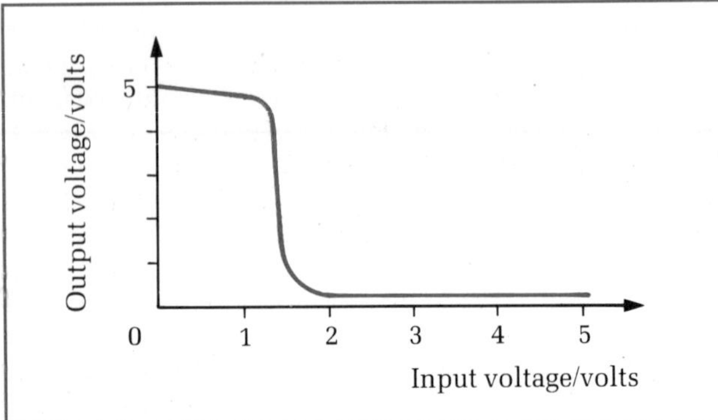

Figure 25.18 *A TTL inverter*

	TTL	CMOS
1 Supply voltage V_s	5 V ± 0.25 V d.c.	3 V to 15 V d.c.
2 Logic levels		
low	< 0.8 V	< 2 V
high	> 0.2 × V_S	> 0.8 × V_S
3 Power used	milliwatts	microwatts
4 Fan-out (i.e. number of inputs driven by one output)	about 10	about 50
5 Unused input pins	assume logic level 1	behave erratically so need to be set at 0 or 1 logic level

Testing logic gates Each of the gates described in the previous section can be tested using an LED indicator. An LED is a **light-emitting diode**; it conducts in one direction only, and when it does so, it emits light. The LED must be connected in series with a resistor to limit the current it passes when it conducts. Otherwise, it overheats and fails. Used as in Figure 25.19, the LED lights up when the voltage at its input terminal is high; when the voltage is low, the LED is off. So the state of the LED (on or off) indicates the voltage level at its input terminal.

Test each of the logic gates described in the previous section. If necessary, use NAND gates to make the other gates. The input(s) at each gate can be set high or low by connecting directly to the positive terminal of the power supply (for high) or to the 0 V line (for low). Remember

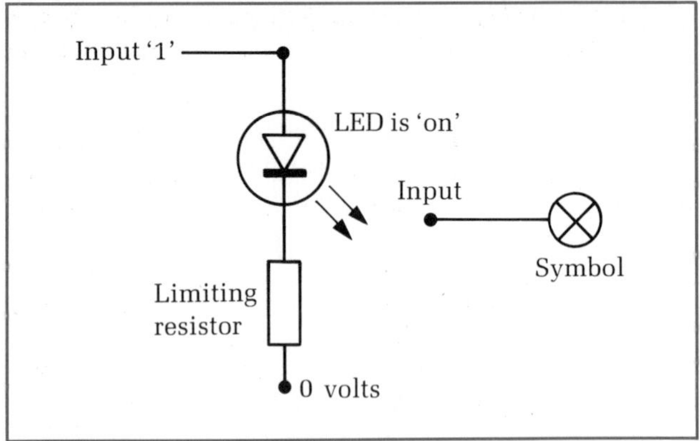

Figure 25.19 *A LED indicator unit*

that voltages are measured relative to the 0 V line, which is connected to the negative terminal of the power supply. The output of each gate is connected to the LED indicator circuit, as shown in Figure 25.19. For each gate in turn, its truth table can be checked. For example, Figure 25.20 shows an AND gate made from two NAND gates being checked. With both inputs high, the indicator light is on.

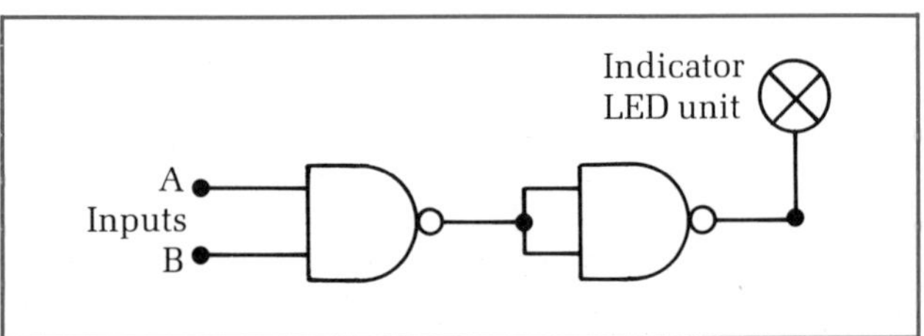

Figure 25.20 *Testing gates*

Design a three-input NAND gate using several two-input NAND gates If possible, construct and test your design. Figure 25.21 shows the truth table for a three-input NAND gate. One possible design is to start with an AND gate, as in Figure 25.22. An AND gate can be made

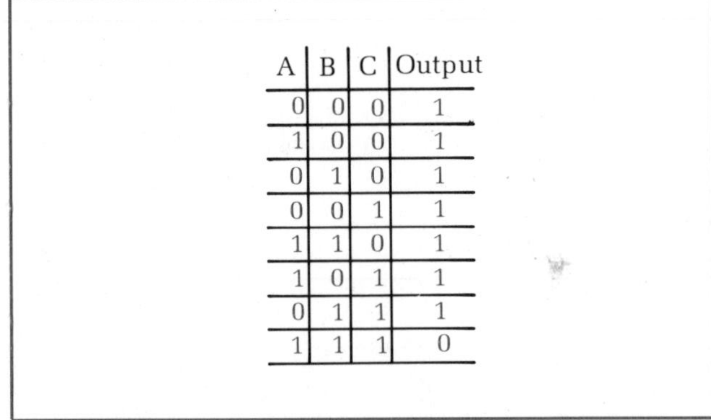

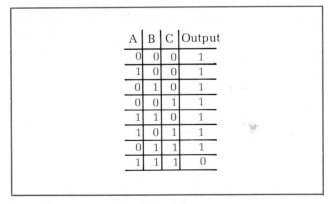

A	B	C	Output
0	0	0	1
1	0	0	1
0	1	0	1
0	0	1	1
1	1	0	1
1	0	1	1
0	1	1	1
1	1	1	0

Figure 25.21 *A truth table for a three-input NAND gate*

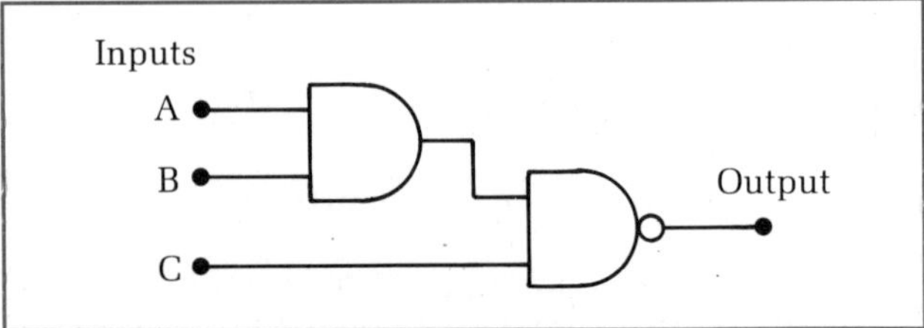

Figure 25.22 *A three-input NAND gate*

using two NAND gates, as explained earlier. So you can make a three-input NAND gate from several two-input NANDs.

Construct and test a light-operated switch Cameras often have a 'warning light' seen by the photographer in the viewfinder of the camera. The warning light switches on if the scene is too dim.

The warning light circuit is shown in Figure 25.23. In good lighting, the light-dependent resistor (LDR) has a low resistance, so the voltage at point X is high. Point X is connected to the input of an inverter, so the output of the inverter, Y, is low. Hence the indicator LED is off.

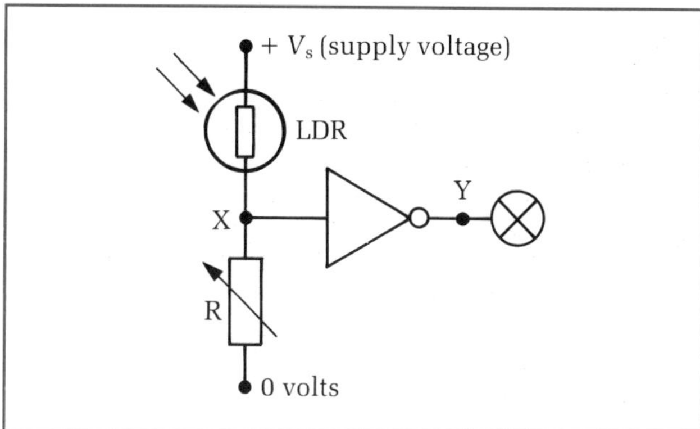

Figure 25.23 *A light level indicator*

In dim light, the LDR has high resistance, so the voltage at X is now low. Hence the voltage at Y is high, so the indicator LED is on.

Use a resistance box if necessary for resistor R in series with the LDR. Set the value of R so that the voltage levels at X will operate the indicator LED as required. If the value of R is too low, then in bright light, the voltage at X does not go high enough to turn the LED off. If the value of R is too high, then the LED stays off in dim light when it ought to be on. The voltage at X cannot go low enough if R is too high, so the value of R must be set correctly.

An exclusive OR gate (EOR) gives a 1 at its output only if the input voltages are different. Its truth table and circuit symbol are shown in Figure 25.24. A 1 at both inputs gives a 0 at the output. Construct and test the combination of NAND gates shown in Figure 25.25, and show that it acts as an EOR gate.

You can work out the output state for any input provided you know the truth table for a NAND gate; a two-

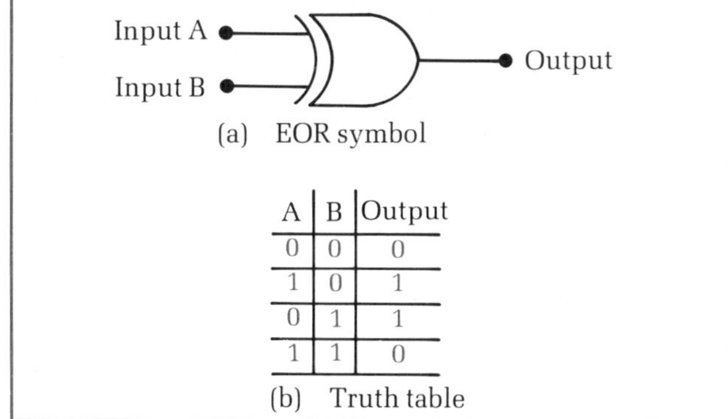

Figure 25.24 *The EOR gate*

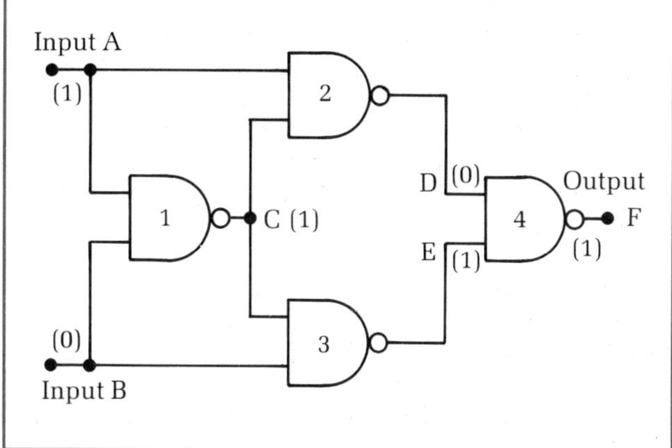

Figure 25.25 *Making an EOR gate*

input NAND gate gives 0 at its output only if there is a 1 at both inputs. Now consider Figure 25.25; suppose we apply a 1 to input A and a 0 to input B, so gate 1 gives a 1 at its output (i.e. A = 1, B = 0 gives C = 1). Gate 2 has both inputs high so its output (D) is low (i.e. D = 0). Gate 3 has one input high and the other low, so its output is high (i.e. E = 1). Hence gate 4, with inputs D = 0 and E = 1, gives a 1 at its output (F). So overall, A = 1 and B = 0, gives F = 1.

Input states					Output state
A	B	C	D	E	F
1	0	1	0	1	1

Work out for yourself the output state for each of the other three input conditions shown in the truth table.

25.3 Addition using logic gates

Binary numbers

With five digits on each human hand, we should not be too surprised at the evolution of our number system. The decimal system expresses any number in terms of powers of ten; so 345 means $(3 \times 10^2) + (4 \times 10^1) + (5 \times 10^0)$. The decimal system uses 10 as its base.

The binary system uses 2 as its base; any number can be made up in terms of powers of two. For example

$$13 = \begin{matrix} (1 \times 2^3) + (1 \times 2^2) + (0 \times 2^1) + (1 \times 2^0) \\ (8) \quad + \quad (4) \quad + \quad (0) \quad + \quad (1) \end{matrix}$$

A 1 or 0 is used to indicate the presence or absence of a particular power of two. So 13 written as a binary number is 1101. Each 1 or 0 is called a **bit**, so a binary number is a **string** of bits. The right hand bit, called the **least significant** bit (LSB), represents the presence or absence of 2^0 (= 1). So odd numbers always, in binary, have a 1 as the least significant bit. The next bit from the right represents the presence or absence of 2^1, the next bit 2^2 etc. Figure 25.26 (at the top of the next page) shows the binary numbers for decimal numbers 1 to 15.

	2^3	2^2	2^1	2^0
0	0	0	0	0
1	0	0	0	1
2	0	0	1	0
3	0	0	1	1
4	0	1	0	0
5	0	1	0	1
6	0	1	1	0
7	0	1	1	1
8	1	0	0	0
9	1	0	0	1
10	1	0	1	0
11	1	0	1	1
12	1	1	0	0
13	1	1	0	1
14	1	1	1	0
15	1	1	1	1

Figure 25.26 *Binary numbers*

Computers and electronic calculators operate using binary numbers. Digital electronics is used throughout, so the voltage at any point represents a single bit. A 4-bit binary number could therefore be carried on four wires in parallel, each wire carrying a single bit. Figures 25.27 shows how four switches could be used to display, on indicator lamps, any binary number up to 1111 (=15).

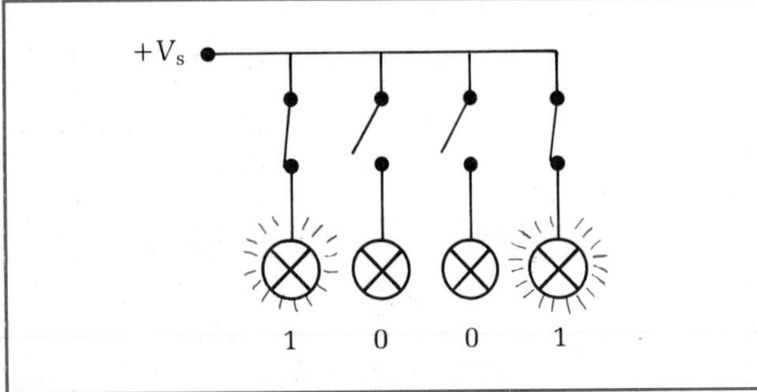

Figure 25.27 *Displaying 9 in binary code*

Binary addition is easy because you only need to count up to 1! Adding two bits, A and B, gives a **sum** and a **carry**, as shown below.

A	B	Carry	Sum
0	0	0	0
1	0	0	1
0	1	0	1
1	1	1	0

So A added to B gives a two-bit binary number, where the least significant bit is the sum. The other bit is the carry.

Compare the carry and sum columns above with the truth tables for the logic gates described in the previous section. The sum is identical to the output of an EOR gate. The carry is the same as the output of an AND gate. So we can use an EOR gate and an AND gate to add two bits to give a sum and a carry. Figure 25.28 shows how the two gates are connected together. The combination is called a

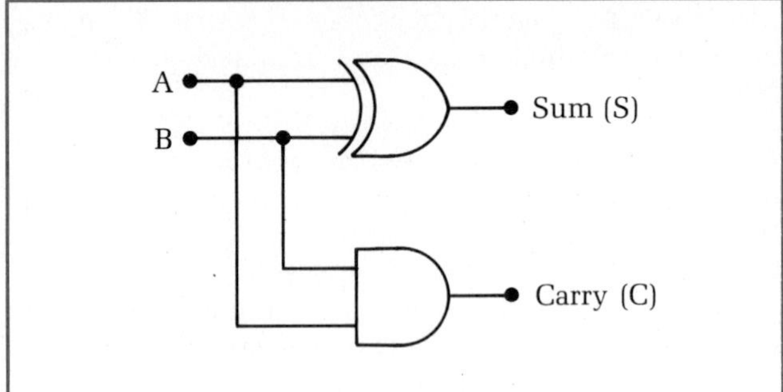

Figure 25.28 *A half-adder*

half-adder because it does not include an input for a 'carry forward' bit.

Consider the operations involved in adding two binary numbers, each with the same number of bits. We start with the two LSBs (i.e. least significant bits) and add them to give a sum and a carry. The carry is then added to the next bit from each of the two numbers, to give a new sum and carry. It's easiest to use a simple example.

	Decimal	**Binary**
A	9	1001
B	5	0101
A+B	14	1110

Let the LSB for A be represented by A_0, the next bit by A_1, the next bit by A_2, and the most significant bit by A_3. So $A_3A_2A_1A_0=1001$. The same idea is used for B; so $B_3B_2B_1B_0=0101$.

Step 1 $A_0+B_0=\text{Carry}+\text{Sum}$
1 1 1 0

Step 2 $A_1+B_1+\text{Carry from Step 1}=\text{Carry}+\text{Sum}$
0 0 1 0 1

Step 3 $A_2+B_2+\text{Carry from Step 2}=\text{Carry}+\text{Sum}$
0 1 0 0 1

Step 4 $A_3+B_3+\text{Carry from Step 3}=\text{Carry}+\text{Sum}$
1 0 0 0 1

The result is $C_3S_3S_2S_1S_0=01110$ (=14), where C_3 is the carry bit from Step 3, S_3 is the sum bit from Step 3, etc.

The half-adder can be used for Step 1 because there are two inputs only (A_0 and B_0) in Step 1. However, the other steps involve an input which is carried forward from the previous stage. So we need a unit which will add three bits (i.e. with three inputs) to give a sum and carry at its output. Such a unit is called a **full-adder**.

Full-adders

A full-adder has three inputs and two outputs. One of these inputs takes the bit carried forward from the 'previous' adder. Figure 25.29 shows three full-adders and a half-adder used to add two 4-bit binary numbers. Each full-adder adds one bit from one number to the corresponding bit from the other number, and a carry bit. The output from each full-adder gives a sum and a carry bit.

We have seen in the previous section how a half-adder can be made using an EOR gate and an AND gate. An EOR gate can be made using four NAND gates, one of which can be used for the AND gate. So a half-adder needs five NAND gates, as in Figure 25.29.

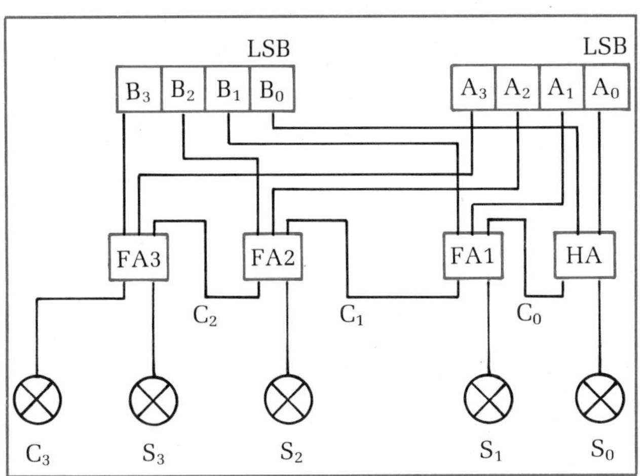

Figure 25.29 *Adding two binary numbers*

Now consider the design of a full-adder. The truth table is shown below.

Inputs			Outputs	
A	B	Carry-in	Carry-out	Sum
0	0	0	0	0
1	0	0	0	1
0	1	0	0	1
0	0	1	0	1
1	1	0	1	0
1	0	1	1	0
0	1	1	1	0
1	1	1	1	1

The truth table shows that A added to B added to the carry-in gives a two-bit binary number. The number's LSB is the sum; the other bit is the carry-out. We can generate the truth table using two half-adders and an OR gate, as in Figure 25.30. Bits A and B are supplied to a half-adder HA1. The sum bit from HA1 is then supplied to one input of a second half-adder HA2. The other input of HA2 is supplied with the carry-in bit. HA2 gives the final sum. The final carry-out is from a two-input OR gate, where the carry bits from HA1 and HA2 supply the two inputs.

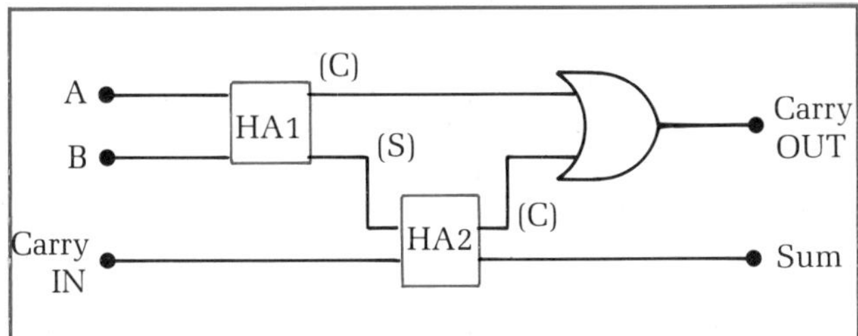

Figure 25.30 *A full-adder*

Let's consider how the full-adder circuit shown in Figure 25.30 works.

1 Carry-in = 0 The final carry-out is from HA1 via the OR gate.
The final sum is from HA2 supplied from HA1.

So the carry and sum from HA1 supply the final carry-out and sum as if HA2 and the OR gate were not present. Hence the four lines of the truth table for carry-in = 0 are the same as the truth table of a half-adder.

2 Carry-in = 1 Making the carry-in bit high causes the sum from HA2 to change its state. If HA1's sum output is high as well, then HA2 gives a '1' at its carry output. Hence the final carry-out goes high if A and B are different.

Hence, if the carry-in bit is low to start with, and A and B are different, then the full-adder gives a '1' at its sum output and a zero at its carry output. Now suppose the carry-in bit goes high; then the sum output from the full-adder changes to 0, but the final carry output goes high. So the output state of the full adder represents the binary number 10 (= 2).

Using adders

The half-adder Use the information in the previous pages to construct a half-adder using five NAND gates. Test your design by checking its truth table, comparing your results with the half-adder truth table on p. 368.

The full-adder Construct a full-adder from two half-adders and an OR gate, as shown in Figure 25.30. If necessary, make the half-adders and the OR gate from NAND gates. Figure 25.31 shows how nine NAND gates can be connected together to make a full-adder; why not thirteen NAND gates? (That is five for each half-adder and three for the OR gate.) Figure 25.31 shows that two pairs of NAND gates have no useful function. Each pair simply inverts then reinverts, so we can do without four of the thirteen, to give a cheaper full-adder.

Test your design by checking that its truth table agrees with the full-adder truth table.

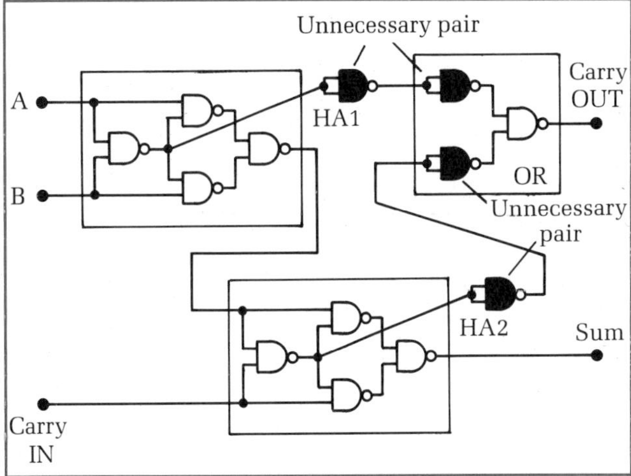

Figure 25.31 *Making a full-adder*

A 2-bit calculator Use a half-adder and a full-adder to add two 2-bit binary numbers. Connect the units as shown in Figure 25.32, on the next page. The half-adder adds the two LSBs to give a sum and carry. The full-adder is supplied with the other two bits and the carry bit from the half-adder. So a 3-bit binary number is produced.

The input states are set at '0' or '1' by connecting each input to the 0 volt line or to the positive supply voltage line. So each pair of inputs represents a 2-bit binary number. The output states are displayed using LEDs at each

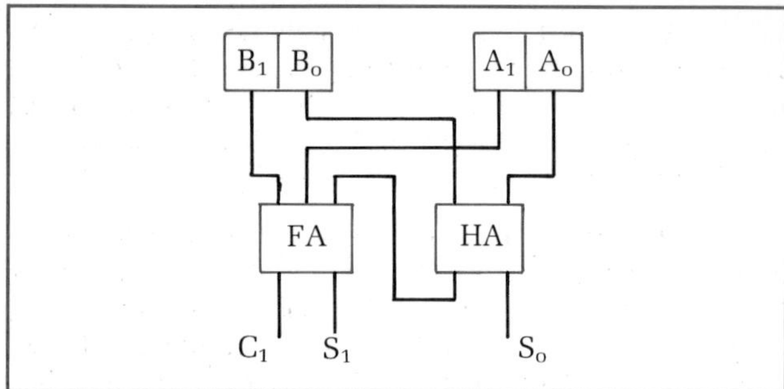

Figure 25.32 *A two-bit calculator*

output line. Try adding different numbers together using the system. Check your results from the LEDs agree with the calculated result.

25.4 Pulse-producing circuits

Multivibrators

Pulses of electricity can be used to operate such devices as warning indicators, alarm buzzers, pre-set timers, etc. A single pulse carried by a wire causes the voltage level to go from low to high to low ($0 \rightarrow 1 \rightarrow 0$) as in Figure 25.33. Pulse-producing circuits can be made using logic gates with capacitors and resistors.

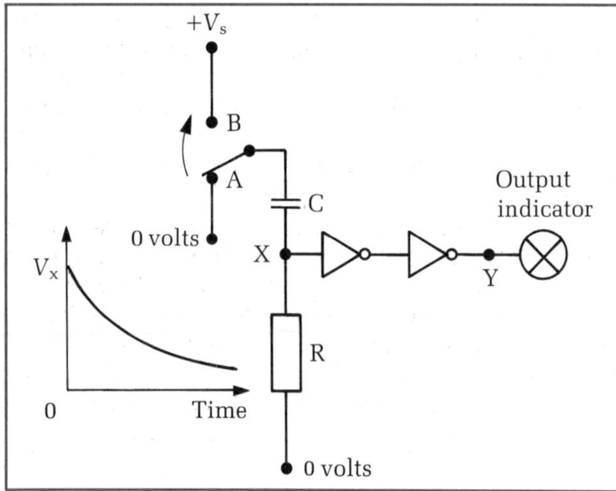

Figure 25.33 *A monostable multivibrator*

The one-shot (or monostable) multivibrator Imagine you are asked to design a circuit that will turn on a light for a given time at the press of a switch. After the switch is pressed, the circuit keeps the light on for a given time. The circuit is called a **one-shot** or **monostable** multivibrator. Pressing the switch produces one pulse only, but the length of the pulse depends on the resistance and capacitance values chosen.

A suitable circuit is shown in Figure 25.33. When the switch is reset from A to B the capacitor charges up through a resistor. The charging current is large initially but decreases exponentially as the capacitor charges up. So the voltage across R is initially large, but falls as shown in Figure 25.33. The resistor voltage provides the input to two inverters in series.

Whenever the voltage at X is above a fixed level (i.e. when V_x is high), the voltage at Y is high at a constant level. So the indicator light at Y switches on when the switch is reset from A to B; the light stays on until the voltage at X has fallen below the 'high' level. The time interval for which the indicator is on is determined by the time constant *RC* of the charging circuit. As explained on p. 328, the voltage at X drops to 37% of its initial value in a time interval *RC*. The bigger the value of either *R* or *C*, the longer the time which the light stays on for. When the voltage at X falls below the 'high' level, the voltage at Y becomes zero, so the light is turned off.

What would happen if only one inverter was used? Resetting the switch from A to B would turn the light off for a fixed time. Then the light would turn on automatically.

Astable multivibrators An astable multivibrator is a circuit which produces pulses at a steady rate. Such a circuit is needed to make an indicator light flash on and off repeatedly. Warning lights which flash on and off are more readily noticed than lights which stay on. To make an astable multivibrator, we can use resistors, capacitors and inverters once again. An astable multivibrator is sometimes called a **free-running** multivibrator because it produces a continuous train of pulses until it is disconnected from its power supply.

One way to make an astable multivibrator is shown in Figure 25.34. To understand how the circuit works, consider the sequence of events after indicator A switches on.

1 A goes high so the voltage at X must have gone high too. X going high makes C_2 charge up through R_2. (A on / B off)
The charging current through R_2 is initially large but then falls (see p. 331). So the voltage at Y goes high but drops as the current drops. With Y high, indicator B switches off because Z goes low.

2 After a fixed time, the voltage at Y drops low enough to make inverter 2 produce a high at Z, so switching B back on again. (B on / A off)
As soon as Z goes high, capacitor C_1 starts to charge up through R_1.
So now the voltage at W goes high and then falls gradually.
Hence indicator A now goes off as soon as B comes on until the voltage at W has fallen to a low enough level to switch A back on.

3 When A comes back on, the whole sequence repeats itself.

The sequence can be summed up by saying that when A is switched on, B is switched off for a while. Then when B switches back on again, it switches A off for a while. Then when A switches back on again, B switches off, etc. The two inverters behave like two punch-drunk boxers who thump only after being thumped. A thumps B who reacts after a delay by thumping back. A then reacts after a delay by thumping back, and so the slogging match goes on!

Astable multivibrators can be made using transistors as inverters, as explained on p. 386. Also, an operational amplifier (see p. 397) can be used with capacitors and resistors to make an astable multivibrator.

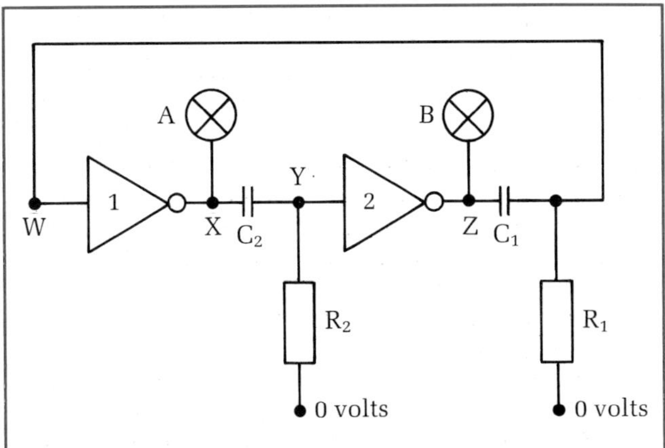

Figure 25.34 *An astable multivibrator*

The **frequency** of a multivibrator is the number of complete pulses produced per second. Each pulse produced in Figures 25.34 corresponds to one indicator being on. The time for which indicator B is on depends on the rate of charging of the capacitor C_2. Since C_2 charges up through resistor R_2, the time which B stays on for is given approximately by the time constant R_2C_2. So the time constant gives a rough value for the time period of the multivibrator. For example, suppose Figure 25.34 uses 1 μF capacitors and 10 MΩ resistors, then each indicator would stay on for approximately 10 s. So the time period of the pulses would be about 20 s. If 10 kΩ resistors were used instead, the time period would be about 20 ms (= $2RC$), giving a frequency of 50 Hz.

Using multivibrators

To produce equal pulses at any required frequency Equal pulses means the 1 lasts as long as the 0 for each cycle. Connect up the circuit shown in Figure 25.34, using different values of resistances and capacitances in turn to see the effect on the indicators. If the frequency is too high, use an oscilloscope to monitor the output. Figure 25.35 shows the variation of output voltages at each inverter. As explained in the previous section, the input voltage falls exponentially after going high. The time constant RC is an approximate measure of the 'on' time for each indicator.

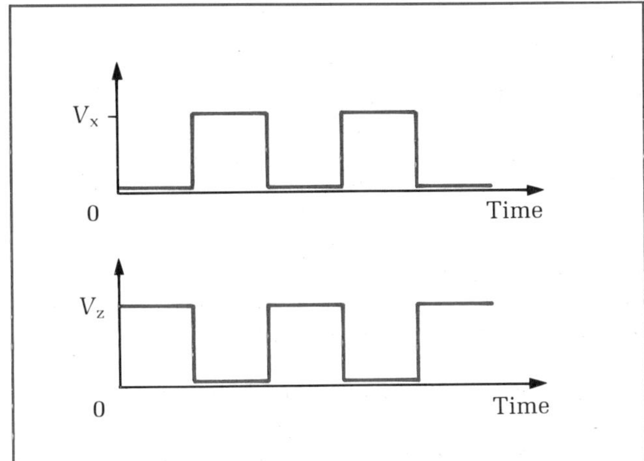

Figure 25.35 *Output voltage wave forms*

With equal values for the two resistances and for the two capacitances, the 'on' time for each indicator should be the same as the 'off' time for that indicator. Use the oscilloscope to measure the time period of the pulses, and check that it is about the same as $2RC$.

To make unequal pulses Unequal pulses from the circuit of Figure 25.34 mean that one lamp is on for longer than the other, each cycle. To make unequal pulses, values of resistances and capacitances should be chosen to give different values for R_1C_1 and R_2C_2. So the time constant for charging each capacitor differs. Try different values for the resistances and capacitances, using an oscilloscope to measure the time for which each lamp is on in each cycle.

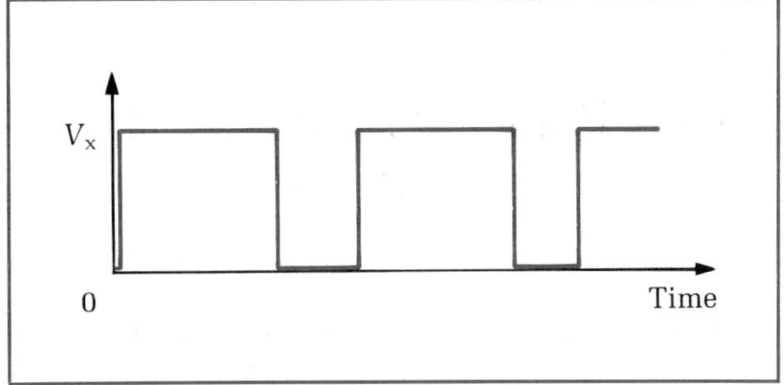

Figure 25.36 *Unequal pulses*

To make an alarm that switches off after a set time In addition to an astable multivibrator, another resistor and capacitor are needed. Also an AND gate and an earpiece are required. The resistance and capacitance values of the multivibrator circuit are chosen so that its frequency is in the audio range (e.g. about 1 kHz). With the earpiece connected to the output of the astable unit, a continuous note ought to be emitted by the earpiece.

How can the note be switched off automatically after being on for a given time? The multivibrator signals can be supplied to one of the inputs of an AND gate. The earpiece is then connected to the output of the AND gate, as in Figure 25.37. When the other input of the AND gate is high, the signals from the multivibrator pass through the AND gate to the earpiece so the note is heard.

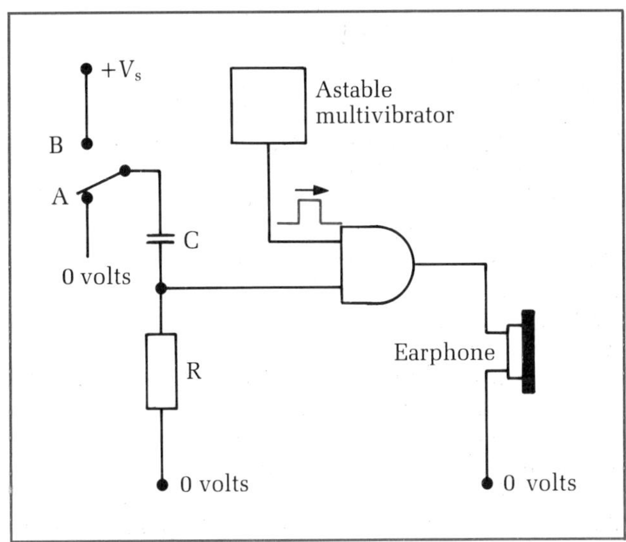

Figure 25.37 *An alarm*

However, if the other input of the AND gate is low, the AND gate does not pass the signals. Both inputs must be high to give a 1 at its output, otherwise there will always be a 0 at its output. So we need to make the other input go high for a set time.

Using the additional resistor and capacitor, as in Figure 25.37, enables this to be done. When switch S is reset from A to B, capacitor C charges up. So the voltage at X goes high, then decreases exponentially as C charges up. As a result, the AND gate passes the multivibrator signals as long as the voltage at X is above the minimum level for a 1. However, when the voltage at X falls below that level, the AND gate is switched off so the note stops.

25.5 Flip-flops

The bistable circuit

A bistable circuit has two stable states. At any time, it is in one of these two states. To make it change from one state to the other, it must be supplied with an input pulse. Compare the bistable unit with an astable unit; the astable unit repeatedly switches from one state to the other with no input pulses needed. The bistable unit needs an input pulse to make it 'flip' from one state to the other; the next input pulse makes it 'flop' back again.

Using inverters Connect two inverters as in Figure 25.38, with indicators at each output. Use a 'flying lead' to make the input of inverter A briefly high, so A's output is forced low, making B's output high. However, the output of B is fed back to the input of A, so even after the flying lead is removed, A's input stays high. Indicator X stays off and indicator Y stays on as a result of briefly making inverter A high.

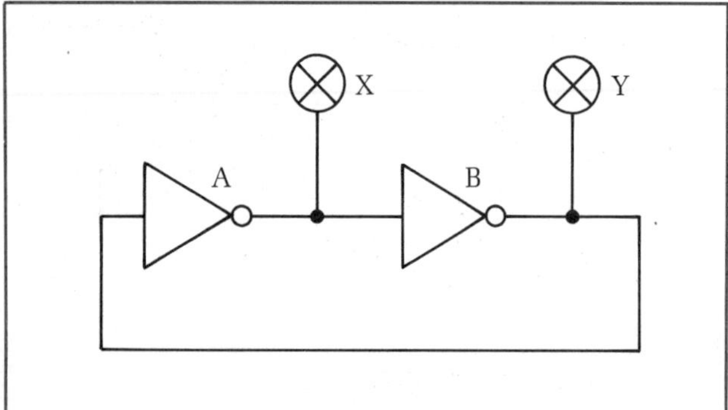

Figure 25.38 *A bistable circuit*

Now use the flying lead briefly to make the input of inverter B high. The indicators change state and stay in their new states. X is turned on and Y is turned off. The brief 1 at B's input forced its output (and hence A's input) to become low; so A's output is forced high to hold the circuit in its new state.

Using NAND gates A 0 on any input of a NAND gate always sets its output at 1. Two NAND gates interconnected as in Figure 25.39 can be used to make a bistable circuit. The truth table can be worked out with the results given at the top of the next column.

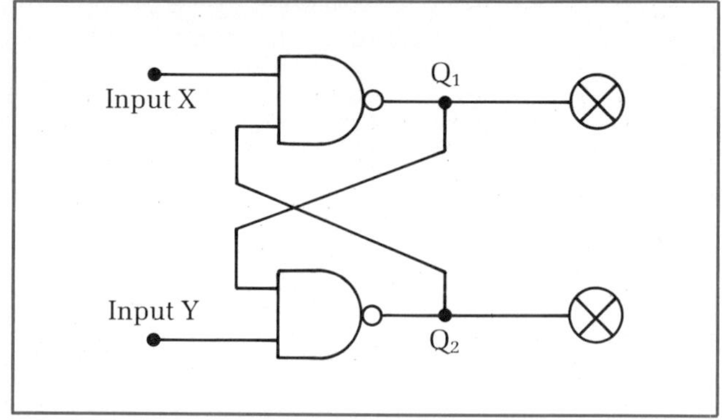

Figure 25.39 *Using NAND gates*

Inputs X	Inputs Y	Outputs Q_1	Outputs Q_2	Comment
0	0	1	1	A 0 on any input gives a 1 at the output
1	0	0	1	A 0 at Y makes $Q_2=1$ so $Q_1=0$
0	1	1	0	A 0 at X makes $Q_1=1$ so $Q_2=0$
1	1	1 or 0	0 or 1	Depending on which input is made 1 first

Provided X=0, Y=0 is not applied, then the outputs have two stable states.

a) $Q_1=1$, $Q_2=0$
b) $Q_1=0$, $Q_2=1$

If both inputs are connected high to start with, the circuit assumes (or takes on) one of its stable states. The circuit can be repeatedly switched from one stable state to the other by setting each input in turn briefly at 0. Try it!

Fans of TV panel games could develop the circuit for 'fast-response' questions. Whichever of two teams hits the button first would switch its indicator light on and the other team's off. Figure 25.39 shows how to do it. Team X's button supplies a 1 to input X when pressed. Team Y's supplies a 1 to input Y when pressed. Imagine both teams being asked a question; team X hit their button first, so their indicator light, from output Q_2, is switched on and the other team's light, from Q_1, is switched off. A 1 from team Y an instant later makes no difference provided the buttons are switches which hold in place after being pressed. Can you spot any problems with the system?

Unfortunately, both lights are on, at the start anyway. We can get round this problem with two AND gates, as shown in Figure 25.40. Both lights must now be off to start with because each button supplies a 0 to its AND gate. As soon as one team presses its button, the AND gate connected to the button receives 1s on both inputs, so that team's indicator is switched on.

Clocked flip-flops

The SR flip-flop Suppose we add an inverter to each input of Figure 25.39. Using NAND gates as inverters would give a circuit as in Figure 25.41. The logic state at X is the complement of the state at S; the state at Y is the

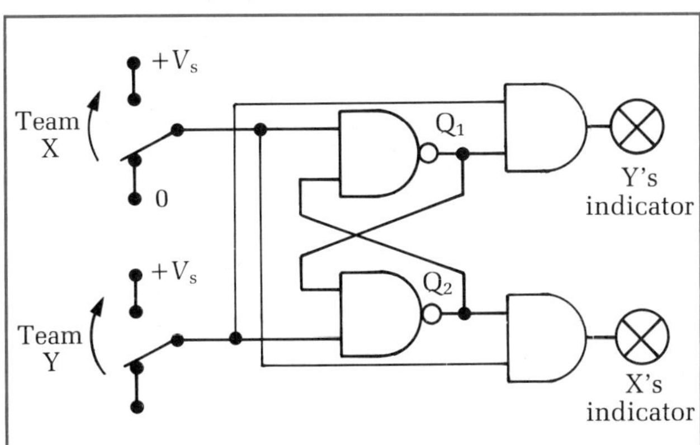

Figure 25.40 *Panel games*

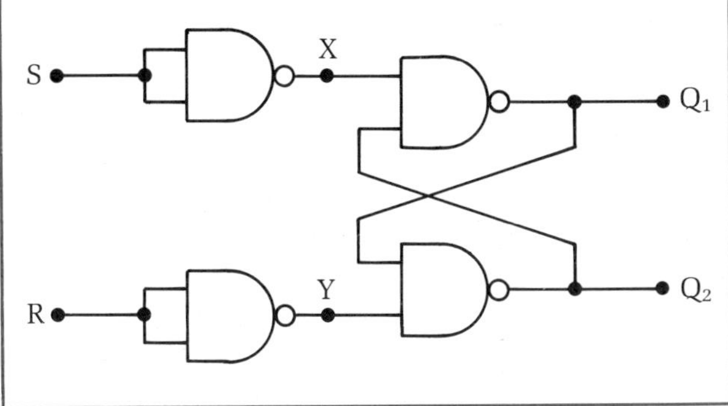

Figure 25.41 *An SR flip-flop*

complement of the state at R. In other words, whatever the state at S or R, X or Y will be in the other state. The circuit of Figure 25.41 is called an **SR flip-flop**. Its truth table is shown below.

Inputs				Outputs		
S	R	X	Y	Q_1		Q_2
0	0	1	1	1 0	or	0 1
1	0	0	1	1		0
0	1	1	0	0		1
1	1	0	0	1		1

Notice that the last three lines give the same states at the output as at the input. Provided '$S=1$, $R=1$' is not applied, the outputs are the same as the inputs (i.e. $Q_1 = S$; $Q_2 = R$). Input S is called the **set input** because $S=1$ sets Q_1 equal to 1. Input R is called the **reset input** because $R=1$ resets Q_1 to 0 (provided $S=0$).

We can use the SR flip-flop as a **latch** to remember input states, provided '$S=1$, $R=1$' does not occur. With $R=0$, then making $S=1$ briefly makes Q_1 go high and it stays high even after S drops to 0. So Q remembers the brief pulse at S. To clear Q_1 (i.e. to reset Q_1 to 0), a brief pulse applied to input R is needed. The circuit is called a latch because the input state at S is latched onto Q_1 and stored. So a bit at S is stored at Q_1.

Clocking the input states S and R through to the outputs allows the progress of bits to be controlled. Figure 25.42 shows the circuit for a clocked SR flip-flop. Compare it with Figure 25.41; two-input NAND gates

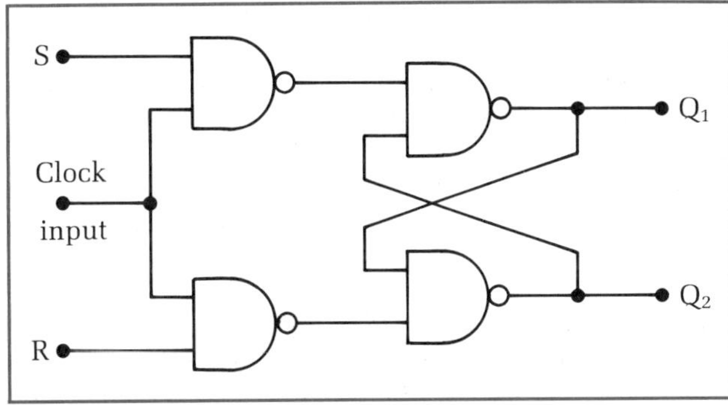

Figure 25.42 *Clocking in*

have been used in place of inverters. When a pulse is applied to the 'clock' input, the logic state at that input goes from 0 to 1 to 0. When the clock input goes high, the bits at S and R are 'clocked' through and stored at Q_1 and Q_2 respectively.

D-type flip-flops (or **data latches**) store a single bit when a clock pulse is applied. An inverter is used, as in Figure 25.43, to ensure that 'R = S' can never arise.

a) D = 0 (the logic state at the 'data' input)
Hence $S=0$ $R=1$
giving $Q_1=0$ $Q_2=1$ when the clock pulse is applied.

b) D = 1
Hence $S=1$ $R=0$
giving $Q_1=1$ $Q_2=0$ when the clock pulse is applied.
Since $Q_2=\bar{Q}_1$ in both cases, write Q for Q_1 and $\bar{Q}$ for Q_2, where Q can be 0 or 1.

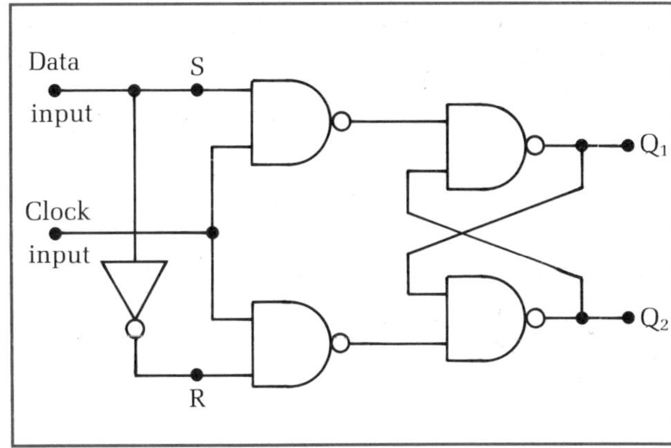

Figure 25.43 *A D-type flip-flop*

The circuit is called a data latch because the bit of data at D is latched onto and stored at output Q_1. The clock input ensures that the data is latched at the required time so the movement of the data bit is controlled. A data latch is a basic **memory unit** because it stores the data. If we wish to store a 4-bit binary number, we need four data latches operated from the same clock. When a clock pulse is applied, each latch stores one of the bits.

The four data latches shown in Figure 25.44 (at the top of the next page) form a memory system called a 'parallel shift register'. A single clock pulse 'shifts' the bit from each input and stores it at the output. The four bits are shifted 'in parallel' at the same time.

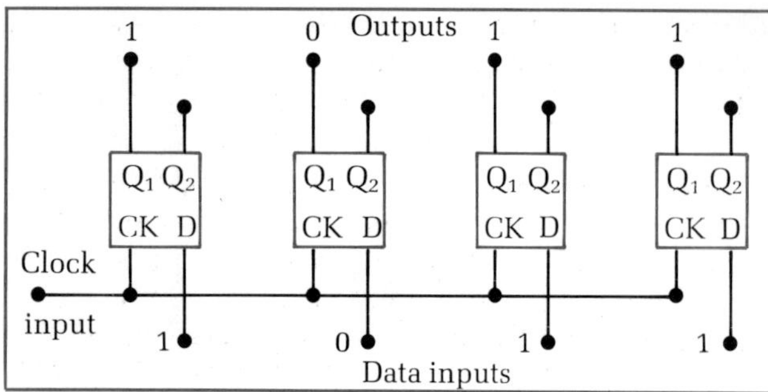

Figure 25.44 *Using data latches*

25.6 Counters and displays

Flip-flops in sequence

D-type flip-flops can be used to count brief pulses. Figure 25.45 shows the circuit for a single D-type flip-flop. Output $\bar{Q}$ is connected to the data input. A brief pulse at the clock input makes the flip-flop change state. The circuit is sometimes also called a **T-type** flip-flop because each input pulse 'toggles' (changes) the output.

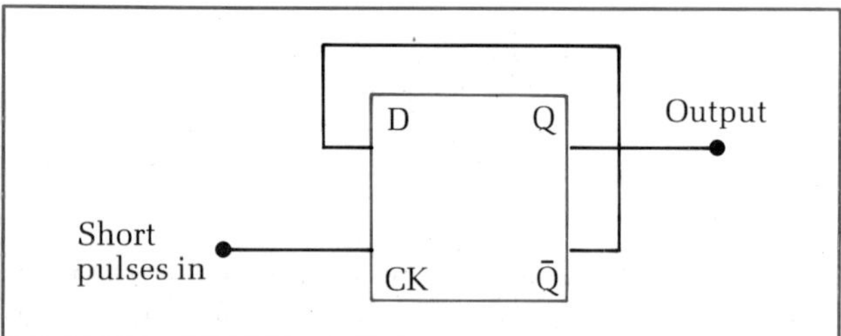

Figure 25.45 *A T-type flip-flop*

To understand why the circuit changes its state, suppose the outputs are $Q=1$, $\bar{Q}=0$ initially. A brief pulse at the clock input makes the output states change to $Q=0$, $\bar{Q}=1$. The clock pulse must be brief though, or else the new state of $\bar{Q}$ ($=1$) reaches the data input when the clock input is still high and the circuit changes state again to make $Q=1$, $\bar{Q}=0$ once more. It's as if a 1 'races' round from $\bar{Q}$ to the data input to make $\bar{Q}=0$ again; and if the clock pulse is still high, the 0 from $\bar{Q}$ races round to the data input to change the state yet again. So anything but a brief pulse at the clock input gives an unpredictable output state. Long pulses could be used as an electronic 'heads or tails' system.

The master-slave principle can be used to stop pulses racing round. Two clocked SR flip-flops in series are used, but the clock input of the second flip-flop is connected via an inverter to the clock pulse circuit.

1 Initially, suppose the bits at the inputs are *A* and *B*, where *A* and *B* represent logic states 0 or 1 without stating which. Assume $A=0$, $B=0$ is not applied.
2 The clock pulse goes high, so bits A and B are clocked through the first flip-flop (the 'master') but not through the second flip-flop (the 'slave'). The slave is inactive because its clock input is low.
3 The clock pulse now goes low, so *A* and *B* are clocked through to slave outputs Q_1 and Q_2 because the slave is now active. The master is inactive.

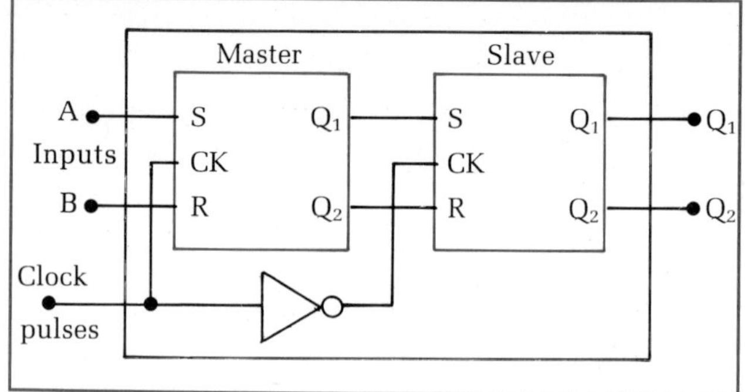

Figure 25.46 *A master-slave circuit*

So one complete pulse (low→high→low) clocks *A* and *B* to the outputs. If $A=0$, $B=0$, the master and slave outputs are unchanged by the clock pulse. So a 1 at either input is clocked through to the corresponding output.

The JK flip-flop, designed using the master-slave circuit and some more gates, can store data or count pulses. Output Q_1 is always the complement of Q_2. To store data, make $J=\bar{K}$ using an inverter from J to K. A data bit at J is stored at Q_1 when the clock pulse is applied. To count pulses, make $J=K=1$. One complete clock pulse makes the outputs change state.

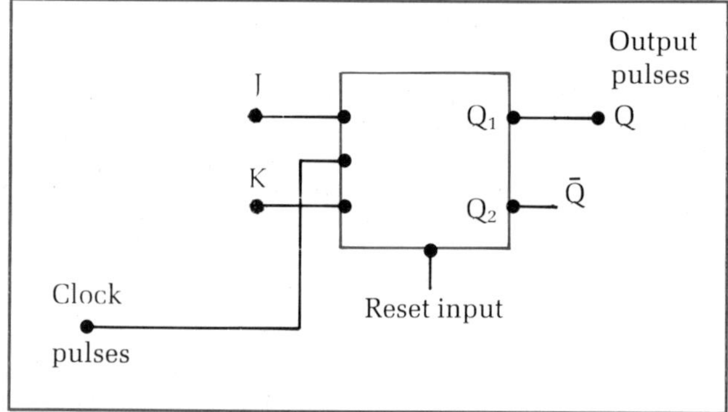

Figure 25.47 *A JK flip-flop*

The output Q_1 changes state when the clock pulse falls from 1 to 0. In other words, the 'trailing edge' of the clock pulse triggers the change of state. Some types of flip-flop are designed to trigger from the 'leading edge' of the clock pulse (i.e. when the clock pulse goes high).

If a JK flip-flop with $J=K=1$ is supplied with a continuous train of clock pulses, the output changes state every time the clock pulse falls from 1 to 0. Two clock input pulses in sequence change the output state and then change it back again. In other words, one output pulse is produced for every two clock input pulses. Therefore, the output pulses are at half the frequency of the clock input pulses. The circuit used in this way is called a **divide-by-two** (÷2) circuit.

A binary counter can be constructed by connecting ÷2 flip-flops in series, as shown in Figure 25.48. The Q output of the first flip-flop supplies the clock input of the second flip-flop; the Q output of the second flip-flop supplies the clock input for the third flip-flop, etc. All the Q output states can be reset at 0 by a brief 1 on the common reset line.

Suppose all the indicator LEDs are off to start with. The trailing edge of the first pulse at the counter input

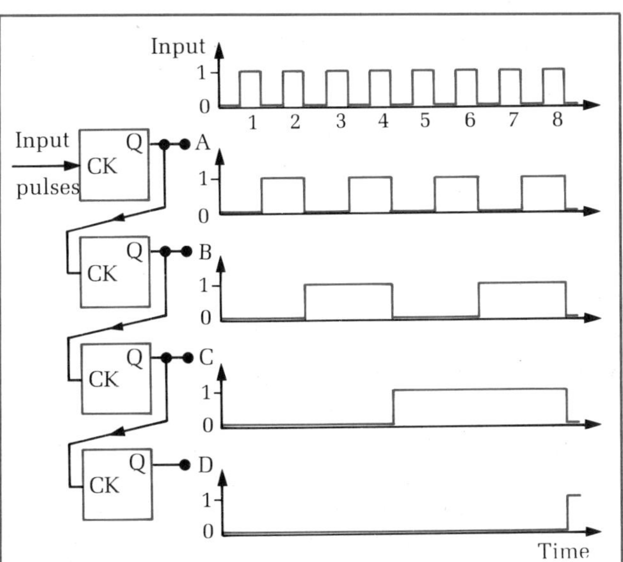

Figure 25.48 *A binary counter*

switches indicator A on because Q_A goes high. The trailing edge of the second pulse switches indicator A off, but indicator B goes on. This is because the input to the second flip-flop falls to 0 when indicator A goes off. The trailing edge of the third pulse switches A back on without affecting indicator B. So after three pulses, the indicator lamps B and A represent the binary number 11 (=3). The trailing edge of the fourth pulse switches indicator A off, which switches B off, which switches indicator C on. So lamps C, B and A now represent the binary number 100 (=4). Figure 25.48 shows the state of the indicators after *n* pulses. At any stage, the indicators show the binary number for *n*.

n	**Output states**			
	Q_D	Q_C	Q_B	Q_A
0	0	0	0	0
1	0	0	0	1
2	0	0	1	0
3	0	0	1	1
4	0	1	0	0
5	0	1	0	1
6	0	1	1	0
etc.				

A counter with four flip-flops will count up to 15 (=1111 in binary), then start again from 0 with the 16th pulse. In general, a counter with *N* flip-flops will count up to 2^N-1. Work out for yourself how many flip-flops are needed to count up to 1 million. The **modulo** of a counter is the number of pulses needed to reset its output states at 0, starting from 0. For example, a modulo-8 counter counts up to 7 then resets itself to zero.

Investigating counters

A single stage Use an astable multivibrator to supply a continuous train of pulses to a single stage. Use a master-slave flip-flop as the single stage. The input and output pulses can be displayed using a double-beam oscilloscope. The oscilloscope traces ought to show if the flip-flop is triggered by the trailing edge or by the leading edge of each pulse. Check that the flip-flop acts as a 'divide-by-two' circuit.

A modulo-8 counter Use three flip-flops as shown in Figure 25.49 to allow counting from 0 (=000) to 7 (=111 in binary). If the flip-flops are triggered by the leading edge of the input pulses, connect the $\bar{Q}$ output from each flip-flop to the clock input of the next stage. Supply low-frequency pulses from an astable multivibrator. Use a LED indicator to display the Q output of each flip-flop. Perhaps you could use a different coloured LED to display the input pulses. Note how the sequence of output LEDs changes as the pulses are counted. Provided the pulses are slow enough, it ought to be possible to check at any stage that the output LEDs are counting the number of input pulses in binary.

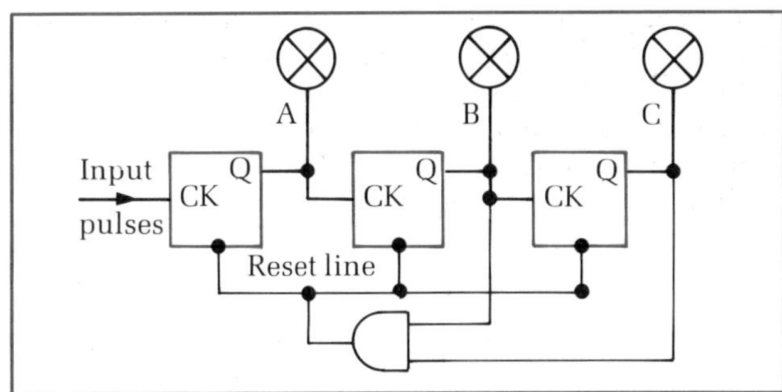

Figure 25.49 *Resetting*

A modulo-n counter Suppose you want a counter which will count up to 5. For example, one of the numerals on a digital watch resets after 5 (e.g. 11.59 changes to 12.00). Three flip-flops are needed because the binary number for 5 is 101, but they need to reset to zero after counting to 5. The reset line could be operated manually when 101 is displayed. Try it with the flip-flops of Figure 25.49.

Far better, the reset line could be operated by the output of an AND gate, as shown in Figure 25.49. The inputs to the AND gate are connected to those flip-flop outputs which go high at the sixth pulse. So the sixth pulse causes all the flip-flops to reset back to 000. The indicator LEDs do not show the sixth pulse as 110 because the reset is very rapid.

Can you design a counter which counts up to 9 then resets? The binary number for 9 is 1001, so a four-stage counter is needed. We want it to reset on the tenth (=1010) pulse, so the AND inputs must be connected to the outputs from the second and fourth stages.

Decoders and displays

So far we have seen how indicator LEDs can be used to display the output states of a counter. The display gives a 'read-out' in binary form which then has to be decoded into decimal form. It is much more convenient to use some more 'hardware' (i.e. electronics) to convert the binary read-out into a decimal display. To do so, the binary read-out must be decoded then displayed as a decimal number.

Two ways to display a decimal electronically are as follow.

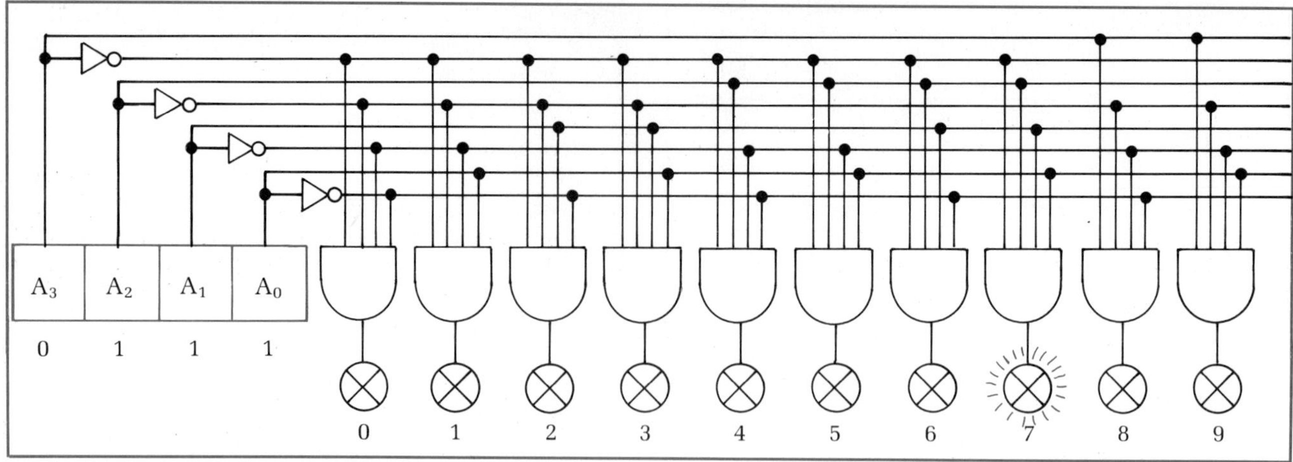

Figure 25.50 *A decimal decoder*

A decimal display which has 10 indicators, numbered from 0 to 9. Only one indicator must be lit at any time. The display itself is simple enough; ten LED indicators will do. However, a 'decoder' is necessary to convert the binary read-out from the counter into decimal form. The binary numbers for 0 to 9 are from 0000 to 1001. So the decoder must have four input terminals to accept a 4-bit binary number. Also it must have nine output terminals, one for each decimal number, as in Figure 25.50. So if the binary number 0111 is supplied to the inputs, then the output for 7 goes high and the indicator for 7 lights up.

The decoder circuit, usually on a single chip called a **decade counter**, contains a combination of AND and inverter gates between the input and output lines.

A 7-segment display which can show any number from 0 to 9, depending on which segments are activated. Digital watches and some types of calculator use 7-segment displays. If possible, have a close look at such a display. Figure 25.51 shows the layout of the seven segments. Each segment is activated by a 1 applied to its input terminal. The display shown in Figure 25.51 indicates 3 if inputs a, b, c, d and g are made high. The other inputs must be connected to 0 volts.

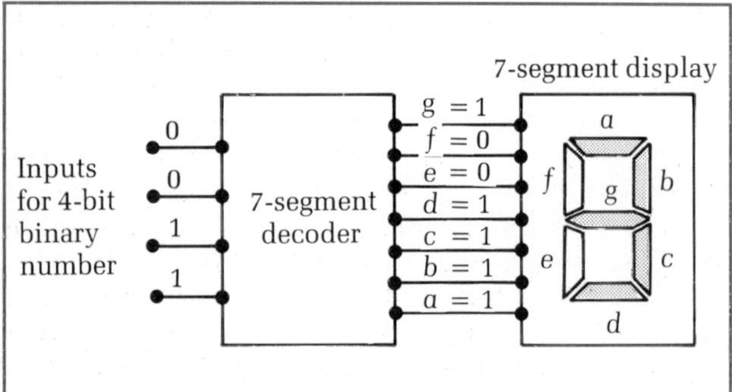

Figure 25.51 *Using a seven-segment display*

Two types of 7-segment displays in general use are:

a) LED displays, where each segment is an individual light-emitting diode (LED) of suitable shape. As explained on p. 384, an LED is a p-n junction diode which is made from a semiconductor called gallium arsenide phosphide. Like any other diode, it passes current in one direction only; however, when it conducts, light is emitted from the material. So a 1 applied to any segment of an LED display makes that segment bright.

b) Liquid crystal displays, where each segment is an individual liquid crystal 'cell'. As explained earlier, when a 1 is applied to a cell, the cell becomes dark against its background. LCDs can therefore only be used in reasonable lighting but they use much less power than LEDs.

To operate a 7-segment display, a decoder chip is essential to convert a 4-bit input to drive the seven segments. A **7-segment decoder**, considered as a black box, has four input lines and seven output lines. The circuit inside the decoder contains logic gates connected in such a way that each 4-bit binary number at the inputs activates the appropriate output lines. For example, suppose the binary number 0111 is supplied to the inputs of the decoder of Figure 25.51. As a result, output lines a, b and c go high; the other lines stay low so the display shows 7.

25.7 Designing digital systems

In the preceding sections you have met different types of digital circuits. The designer of a digital system uses these circuits as black boxes; knowledge of what each circuit can do is essential, but how each circuit works is not so important to the designer. Each circuit is literally a black box anyway since it is on a chip as an integrated circuit. Some examples to show how a designer works are discussed below. With each example, some technical details of relevant hardware is given in the hope that you will 'have a go'. The best way to learn about electronics is to design and make your own systems. Most electronics magazines list details and suppliers of chips, etc.

Traffic lights The simplest type of traffic lights is used at roadworks. The traffic flow is controlled by red/green lights at either end of the roadworks.

When one end shows a red light, the other end shows a green light, so traffic flow is in one direction only until the lights change. Then the flow is in the opposite direction.

Suppose you are asked to design an automatic system

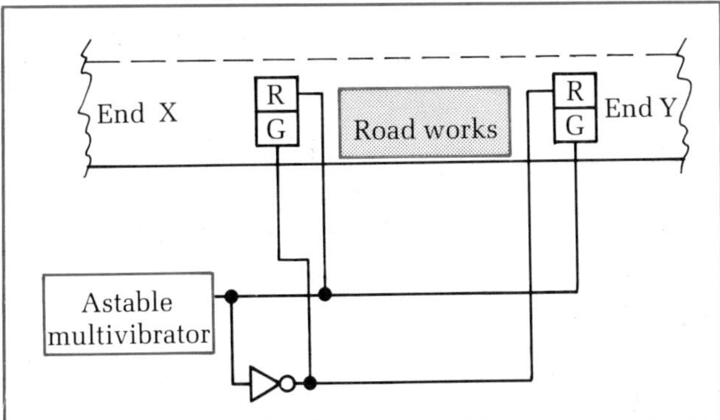

Figure 25.52 *Traffic lights*

to operate the lights. Consider one possible solution which allows equal time for traffic flow in each direction. A slow-running astable multivibrator can be used, as in Figure 25.52. Its output must operate four lights, red (R_X) and green (G_X) at end X and red (R_Y) and green (G_Y) at end Y.

R_X and G_Y must be on at the same time as R_Y and G_X are off. Then, R_Y and G_X must switch on together for a set time, and R_X and G_Y must go off. In other words, the system has two possible output states.

a) $R_X = G_Y = 1 \quad R_Y = G_X = 0$
b) $R_X = G_Y = 0 \quad R_Y = G_X = 1$

The multivibrator output can therefore be connected to R_X and G_Y directly. But the same output is connected to R_Y and G_X via inverter gates, as shown in Figure 25.52. Choose the resistance and capacitance values for the multivibrator to give suitable time intervals. Figure 25.53 shows a circuit using a special chip to make a multivibrator.

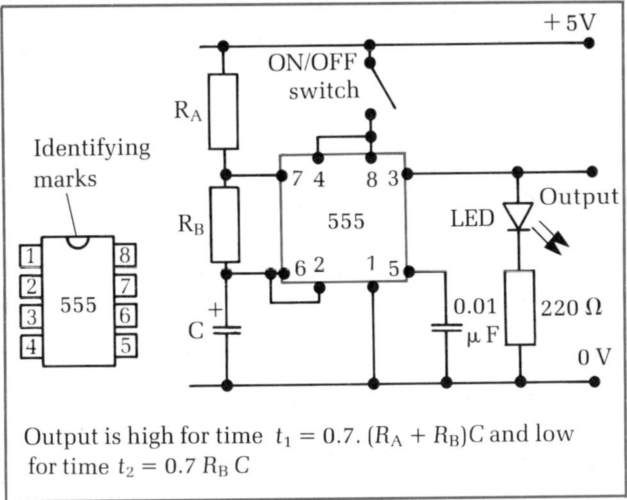

Figure 25.53 *A practical astable unit*

A reaction timer We want to measure the time taken for a person to react to an indicator light switching on. The 'supervisor' presses a switch to start the timing at the same instant as the light switches on. The 'guinea pig' responds by pressing a different switch which stops the timer.

To keep the system simple, we can use a binary counter as a timer. An astable multivibrator is used to produce pulses at a frequency of about 100 Hz. Figure 25.54 shows

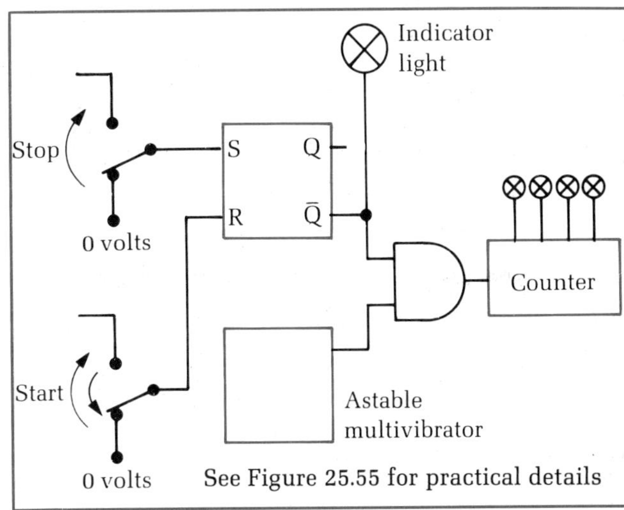

Figure 25.54 *A reaction timer*

one possible arrangement using an SR flip-flop, an AND gate and a binary counter with indicators.

When the 'start' switch is opened briefly, the reset line R goes high, so output Q from the bistable goes low. The other output $\bar{Q}$, goes high so the indicator light is connected to $\bar{Q}$. $\bar{Q}$ also supplies an input to the AND gate. The other input of the AND gate is supplied from the astable multivibrator. So opening the start switch enables the AND gate to pass the pulses from the astable unit to the counter. The person under test opens switch S as soon as possible after seeing the indicator light come on. Opening switch S makes $\bar{Q}$ go low. This disables the AND gate and stops pulses passing to the counter, so the timing is stopped. The reaction time can then be calculated by reading the state of the counter's indicator LEDs to give the number of pulses received. If the time period of the pulses is known, the reaction time is given by the number of pulses received × the time period of the pulses.

An electronic die Suppose you have a 7-segment display which is connected into a circuit to make it count repeatedly through the sequence 0 to 9 very fast. You would not be able to follow the changing display, but if counting is stopped at random, you will not be able to predict its display. You can use it to select any number from 0 to 9 at random.

The system could be as shown in Figures 25.55 and 25.56 (on the next page). A fast astable multivibrator supplies pulses to a binary counter. The counter drives a 7-segment decoder and display. How can the counter be stopped at the press of a switch? One way is to supply the astable unit's pulses to one input of a two-input AND gate. The other input is operated using a switch, as in Figure 25.56. When the switch is set at 0 volts, the AND gate is disabled so the astable's pulses no longer pass to the counter. Hence the display stops at the last pulse received. Lift the switch off 0 volts and the counting restarts.

Suppose you wanted the counter to count from 0 to 5 then reset itself to recount from 0 again. Another AND gate would be required for this purpose to operate the counter's reset line at the sixth pulse of each sequence. Where would you connect the AND gate's inputs? See p. 375 if necessary.

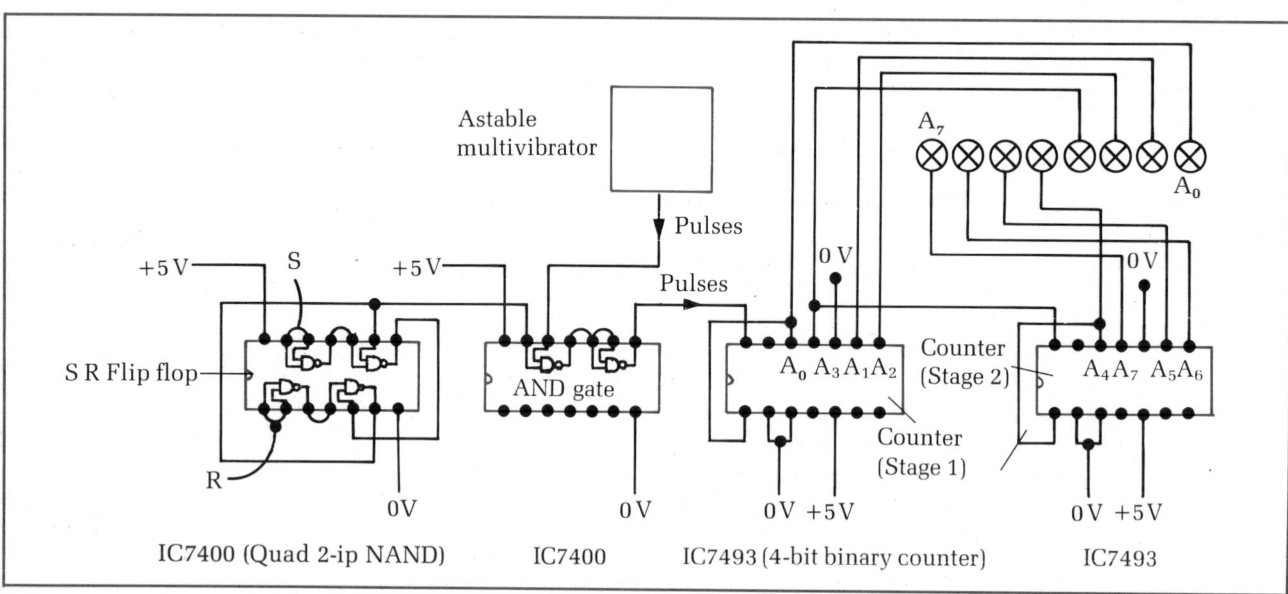

Figure 25.55 *A practical reaction timer with a modulo-255 counter*

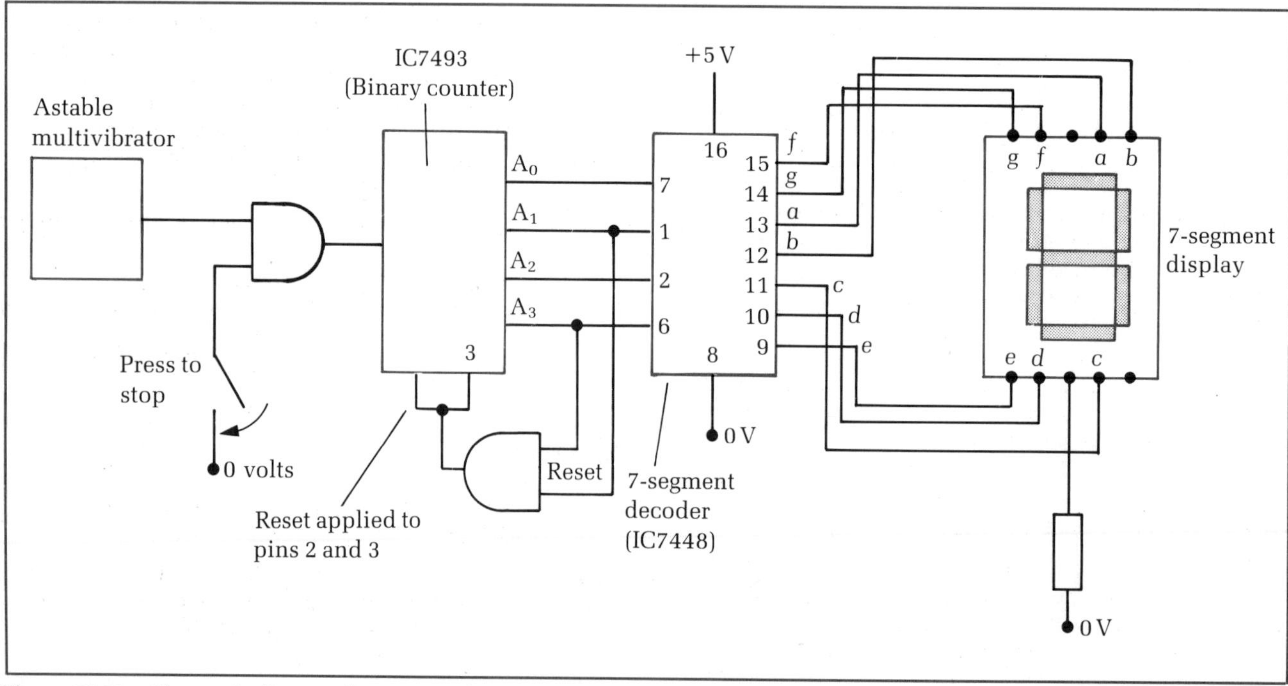

Figure 25.56 *Electronic dice (see figure 25.55 for AND gates and counter)*

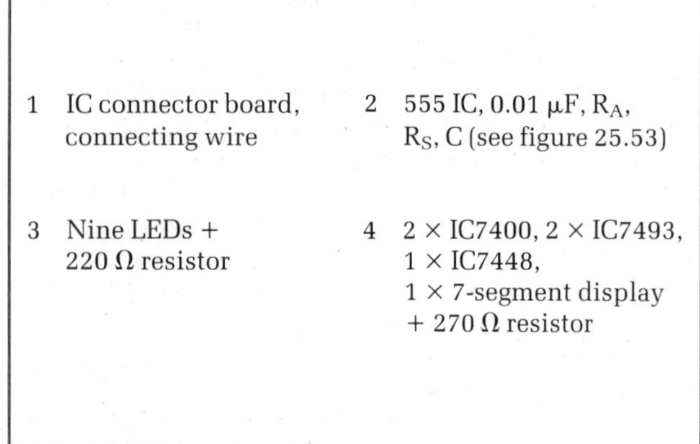

1	IC connector board, connecting wire	2	555 IC, 0.01 μF, R_A, R_S, C (see figure 25.53)
3	Nine LEDs + 220 Ω resistor	4	2 × IC7400, 2 × IC7493, 1 × IC7448, 1 × 7-segment display + 270 Ω resistor

Figure 25.57 *Components list for Figures 25.53, 25.55 and 25.56*

25.8 Summary

Definitions

1 *Logic level* '1' (i.e. high) occurs when the voltage is at or above a fixed positive value.
2 *Logic level* '0' (i.e. low) occurs when the voltage is at 0 volts.
3 *A digital system* occurs where voltage levels are either high or low.
4 *A linear system* occurs where the voltage can have any value between the limits set by the power supply.
5 *A modulo-n* counter counts from 0 to $n-1$ then resets back to zero automatically.

Types of gate

1 An inverter or NOT gate gives a 1 at its output when its input is 0, and *vice versa*.
2 An AND gate gives a 1 at its output only if all its inputs are 1.
3 A NAND gate gives a 0 at its output only if its inputs are all 1.
4 An OR gate gives a 1 at its output if any of its inputs is 1.
5 A NOR gate gives a 0 at its output if any of its inputs is 1.
6 An EOR gate gives a 1 at its output only if its two inputs differ.
7 A multivibrator is a circuit with two states.
a) A monostable multivibrator has one stable and one unstable state. When 'triggered' it goes into the unstable state for a fixed time then reverts to the stable state.
b) An astable multivibrator has two unstable states. It switches from one to the other repeatedly.
c) A bistable multivibrator (or flip-flop) has two stable states. When triggered it goes from one state to the other.

Short questions

25.1 Draw the circuit symbol and give the truth table for each of the following:
a) a 2-input AND gate
b) a 2-input NOR gate
c) a 3-input OR gate.

25.2 Draw a circuit to show how three NAND gates may be connected together to form a two-input OR gate. Indicate the logic states at the input and output lines of each NAND gate when one input of the OR gate is high and the other input is low.

25.3 Write down the truth table for an EOR gate and show how an EOR gate may be constructed from four NAND gates.

25.4 A comparator is a two-input logic unit which gives a '1' at the output if the two inputs are the same as one another. Determine the truth table for the combination of NAND gates shown below, and use your results to explain why a NOT gate at the output makes the combination into a comparator.

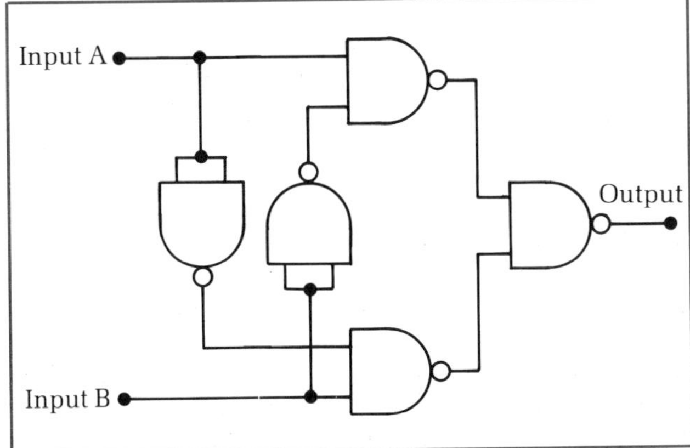

Figure 25.58

25.5 Determine the truth table and suggest an application for each of circuits shown below.

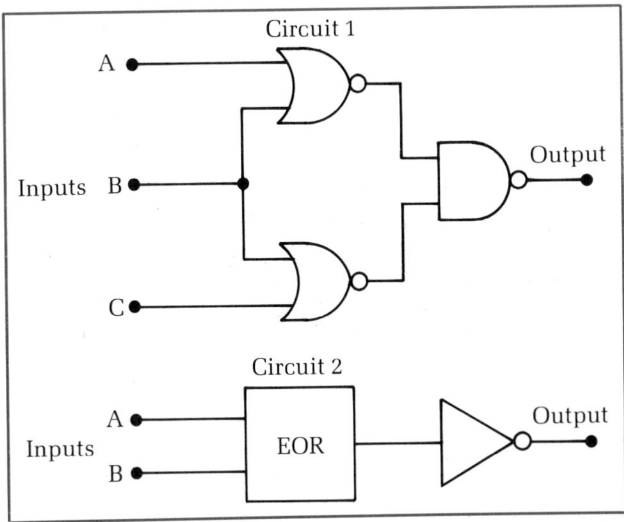

Figure 25.59

25.6 A self-service petrol pump offers a choice of 2-star or 4-star petrol by pushing either one of two buttons after the pump nozzle is lifted from its holder. The control unit, shown below, supplies a '1' to the 2-star valve or to the 4-star valve, according to the inputs supplied from the buttons and the nozzle. Design a suitable logic gate combination to use in the control unit.

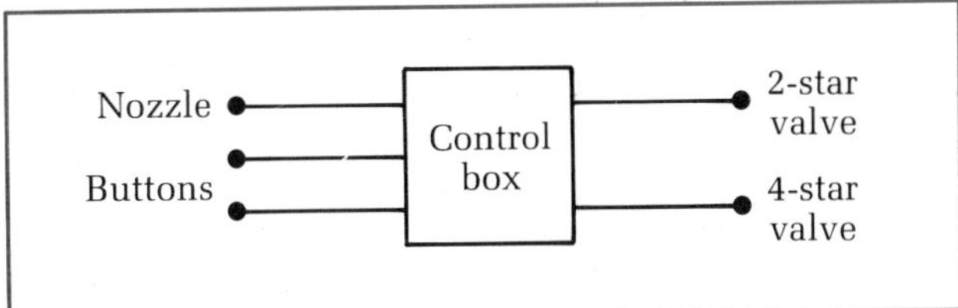

Figure 25.60

25.7 Show how a full-adder and a half-adder can be combined to construct a 2-bit binary adder.

25.8 Design a circuit to operate an indicator lamp so that the lamp, initially off, flashes on five times and then goes off. Use an astable multivibrator, a monostable multivibrator and an AND gate in your design.

25.9 The sequence for a set of traffic lights is shown below.

Red	on	on	off	off	on
Orange	off	on	off	on	off
Green	off	off	on	off	off

The circuit shown below is used to operate the red and orange lights. Copy the circuit and add to it a suitable combination of logic gates to operate the green light from the logic states of the red and orange lights.

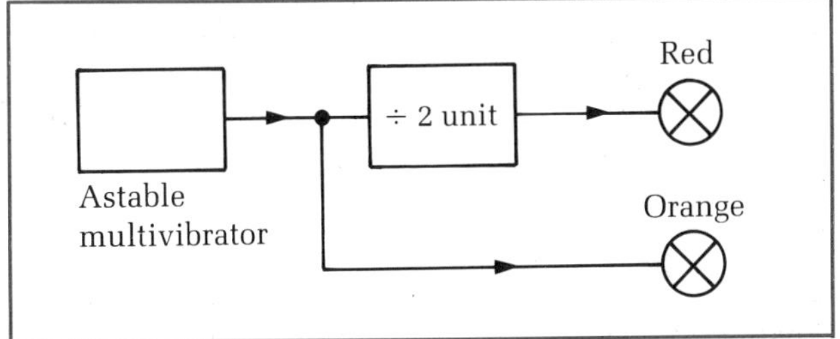

Figure 25.61

25.10 Show how a binary counter counting repeatedly from 0 to 5 can be designed using an astable multivibrator and three suitable flip-flops.

25.11 An LDR has a resistance of 500 Ω in daylight and 10 000 Ω in darkness. It is connected in series with a resistor X as shown below. Explain how the circuit can be used with an inverter to operate an indicator lamp as a 'darkness indicator'. If the inverter requires at least 2 V at its input for a '1' and less than 0.8 V for a '0', calculate a suitable value for X.

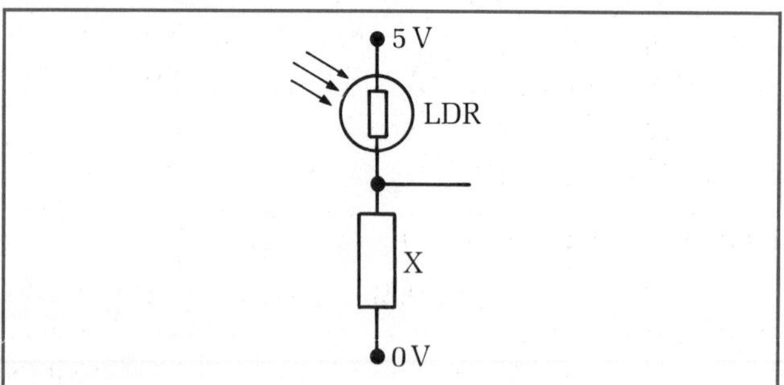

Figure 25.62

25.12 Design a 'door open' warning indicator lamp for a four-door car using several two-input OR gates and an indicator lamp. Assume each door is fitted with a switch which supplies a '1' to your unit unless the door is closed.

25.13 The circuit below is designed to supply a continuous train of pulses to either line X or line Y, according to the logic state of the 'select' line. Explain how it works.

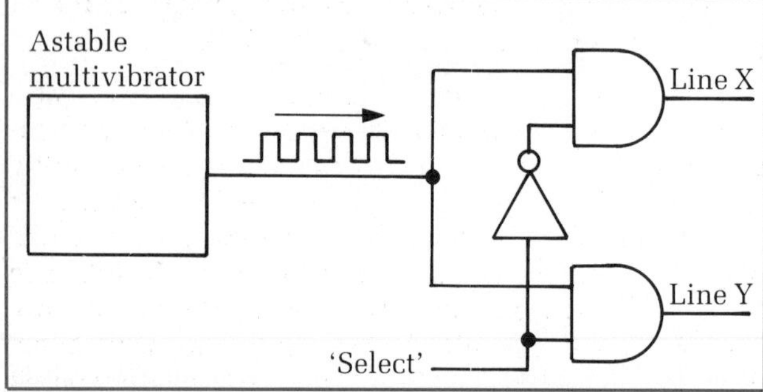

Figure 25.63

25.14 An LED indicator lights up noticeably if its current is more than 5 mA but it is damaged if the current exceeds 40 mA. What should the resistance of the resistor in series with the LED be if a 1 corresponds to 5 V and a 0 to less than 0.2 V?

25.15 A two-input 'multiplexer' has a single output which assumes the same logic state as a chosen input. The choice of input is determined by the logic state of the 'select' terminal, as shown below. If 'select' = 1, then input A is chosen; if 'select' = 0, then input B is chosen. Give the design of the multiplexer, using several logic gates.

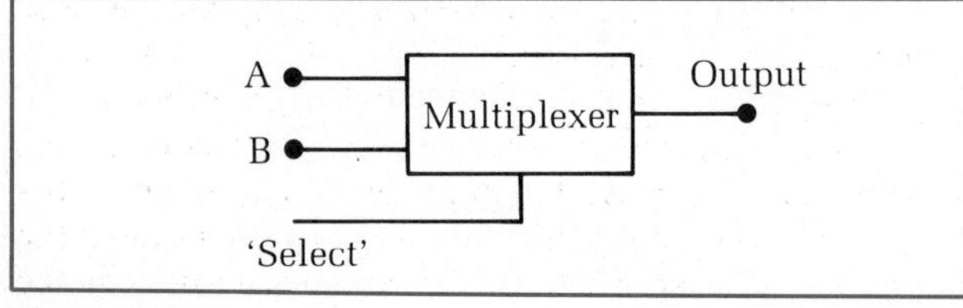

Figure 25.64

26

Electronics in control

26.1 Semiconductors

The huge advances in electronics over the last twenty years are due to the discovery and use of semiconducting materials. Semiconductors are a special class of materials with the common feature that they all conduct electricity better with increasing temperature. In other words, the resistance of a semiconductor falls as its temperature rises. At or near absolute zero, semiconductors become electrical insulators. Silicon is one of the most commonly used semiconductors, and is the basic material from which integrated circuit 'chips' are made. Chips used in **digital** electronics, where voltage levels are either 'high' or 'low', are described in the form of black boxes in Chapter 25. In this chapter, you can learn a little more about how these types of chip are made. Also, transistors and chips used for linear electronics are discussed. **Linear** electronics is the term used where voltages can take any value. Also, we shall see that semiconductors are used to make lots of other devices used in electronics. Examples include diodes, thermistors and LDRs, described in Chapter 23.

Understanding semiconductors

Intrinsic semiconductors are elements or compounds which are semiconductors. Pure silicon or germanium are examples of elements which are intrinsic semiconductors. A sample of pure silicon contains silicon atoms. An example of a compound semiconductor is cadmium sulphide. It contains cadmium and sulphur atoms. The atoms of any intrinsic semiconductor bond to one another to give the material its characteristic electrical property (i.e. resistance which falls with rise of temperature).

Consider the simple picture of an isolated silicon atom, as shown in Figure 26.1. The atom has four electrons in its outermost shell; the outermost shell is therefore half-full since it can accept up to eight electrons. In the solid state, silicon atoms bond to one another by forming covalent bonds (see p. 68). Each atom forms bonds with four neighbouring atoms, as in Figure 26.2. Each atom contributes one of its outermost shell electrons to form a covalent bond with a neighbouring atom which also contributes an electron to the bond. So the outer shell of each atom is filled with eight electrons; by sharing all the outer shell electrons in this way, the atoms are held together in a lattice.

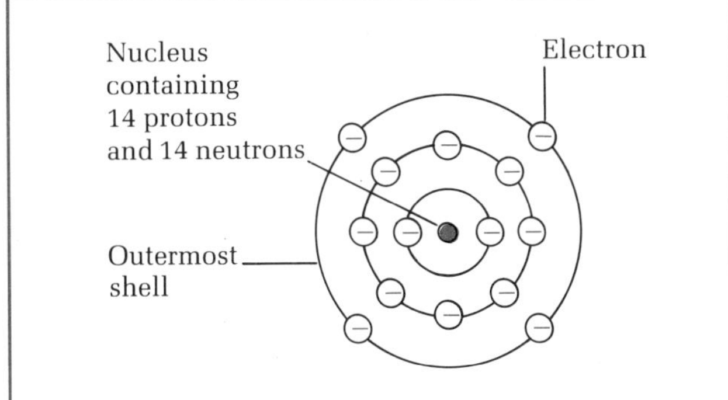

Figure 26.1 *An isolated silicon atom*

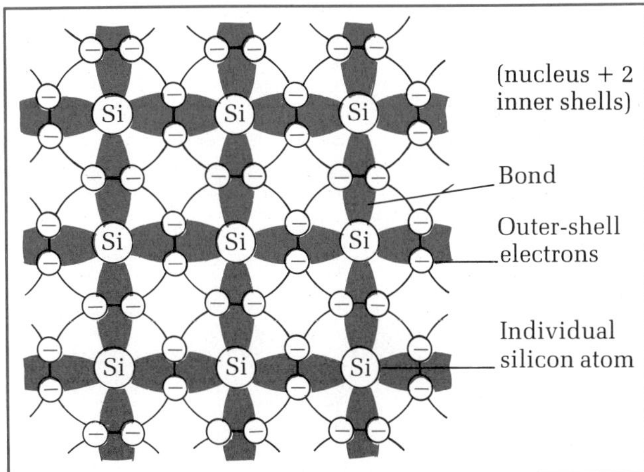

Figure 26.2 *Silicon atoms in the solid state*

At absolute zero, all the electrons are locked up in bonds so there can be no passage of electrons through the material. Hence the material is an electrical insulator at absolute zero. However, if heated, some of the electrons gain enough energy to break free from the bonds. These free electrons can carry charge through the material when a p.d. is applied across the material. The higher the temperature of the material, the greater the number of electrons that become free. Hence the resistance falls as the temperature rises.

Extrinsic semiconductors are made by adding carefully controlled amounts of chosen material to an intrinsic semiconductor. The chosen material is added in tiny amounts of the order of parts per million. In this way, the 'added' atoms are well-spaced from one another inside the 'host' material, so the basic structure of the intrinsic semiconductor is retained; but its electrical properties are greatly altered. The added atoms are sometimes called **impurity atoms**, but this does not mean their presence is accidental. Strictly controlled conditions are vital otherwise the material will produce unreliable devices.

There are two types of extrinsic semiconductor:

a) **n-type** in which the added atoms each have five electrons in the outermost shell. Phosphorus and antimony atoms each have five outer-shell electrons, so either of these two elements could be used to 'dope' the silicon. Each atom in the lattice needs four electrons only for bonding. Where an added atom 'replaces' a silicon atom in the lattice, a free electron is supplied to the material. The added atom needs four electrons only for bonding, so one of its outer-shell electrons is not held in a covalent bond. It therefore becomes free.

The added atoms are called **donor atoms** because they donate free electrons to the material, and are illustrated in Figure 26.3 on the next page. When a voltage is applied across n-type material, the free electrons in the material move towards the positive end of the material. So n-type semiconductors conduct due to the presence of free electrons as charge carriers in the material.

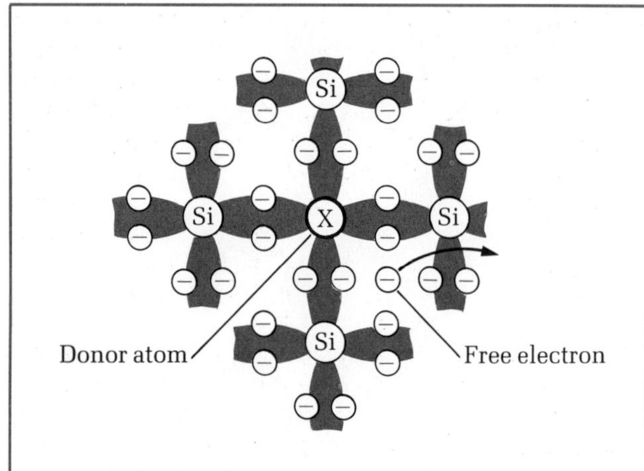

Figure 26.3 *n-type semiconductor*

b) p-type in which the added atoms each have three electrons in the outermost shell. Such an atom, in place of a silicon atom in the host lattice, is short of one electron for bonding purposes. Only three of the four bonds formed by the atom are complete, the other bond has only one electron instead of two. The vacancy in the bond is called a **hole**. The added atoms are referred to as **acceptors** because they accept any free electrons to fill the holes in the incomplete bonds. Aluminium or indium atoms each have three outer-shell electrons so either of these two elements may be used to supply acceptor atoms.

Conduction in p-type material takes place because incomplete bonds can be filled by electrons jumping across from other bonds. Figure 26.4 shows the idea. When an electron jumps from one bond to fill an incomplete bond, it creates a hole in the bond it left, so an electron moving in one direction is equivalent to a hole moving in the opposite direction.

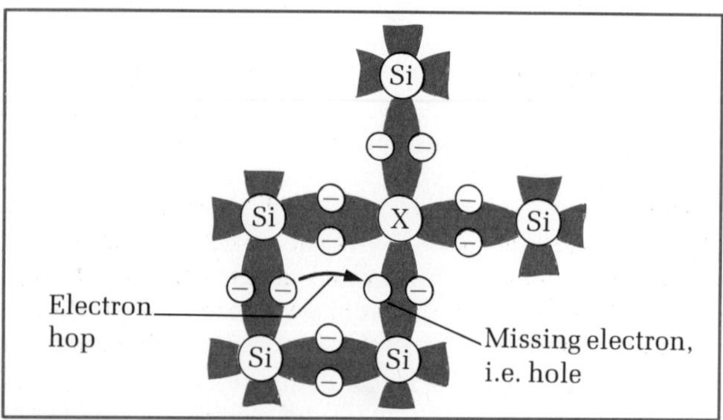

Figure 26.4 *p-type semi-conductor*

Energy bands can be used to explain conduction by intrinsic or extrinsic semiconductors. The electrons in the shell of an isolated atom have well-defined energies, depending on which shell each electron is in. An **energy ladder** gives a helpful picture of the possible energy values which such an electron can have. Imagine a ladder down a shaft below ground level. A person in the shaft can only have certain values of energy, depending on the rung he is on. Electrons in an isolated atom are like that. Suppose the ladder is wobbly, so each rung moves up or down a little. Hence the energy values spread out a little.

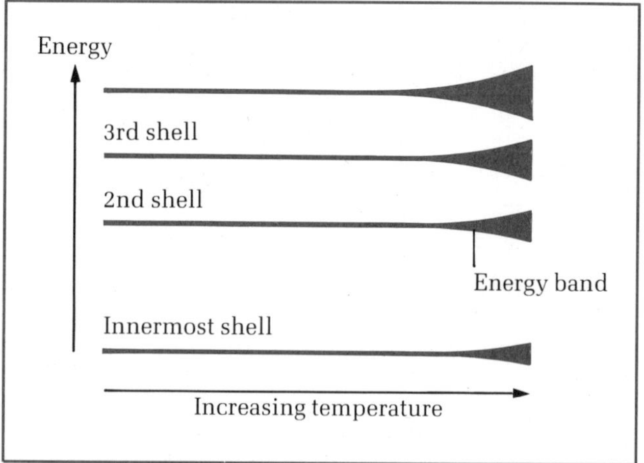

Figure 26.5 *Energy bands for a solid*

This happens when atoms join together. They vibrate and wobble because they have thermal energy. The energy levels spread out into bands, as shown in Figure 26.5. Electrons fill the bands from the lowest levels, up. We can use the band model to gain some understanding of why certain materials conduct whereas others do not.

Metals The highest band containing electrons is only partly-filled. So electrons in this band can easily move to unfilled levels further up the band. These electrons have broken free from individual atoms, and they are able to move through the material. So this band is called the **conduction band**. The **valence band** is immediately below the conduction band. Electrons in the valence band are firmly attached to individual atoms.

Insulators Each band containing electrons is filled completely. The highest filled band contains the outer shell electrons which are firmly held to atoms. This band, the valence band, is separated by a gap from the next band up, the conduction band, which is empty. The gap is too large for electrons from the valence band to jump across. Hence there can be no conduction.

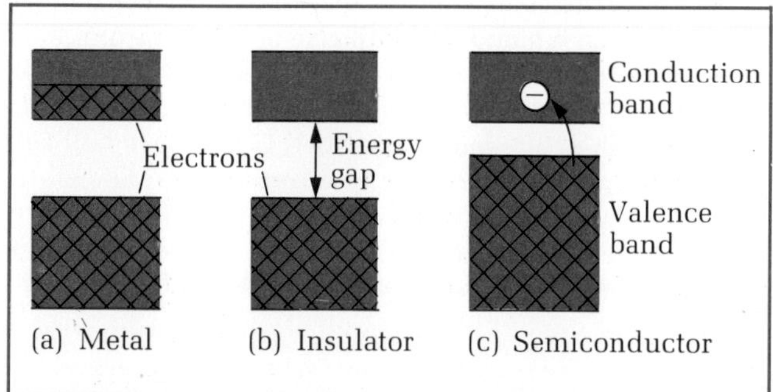

Figure 26.6 *Different types of solid*

Intrinsic semiconductors At absolute zero, the band picture is like the insulator. However, the energy gap between the valence and the conduction band is not so large as for an insulator. So at temperatures above absolute zero, valence electrons gain enough energy to jump across the gap to become conduction electrons. The higher the temperature, the greater the number of electrons which can jump across the gap. An electron jumping across the gap corresponds to the electron breaking away from an individual atom to become a conduction electron.

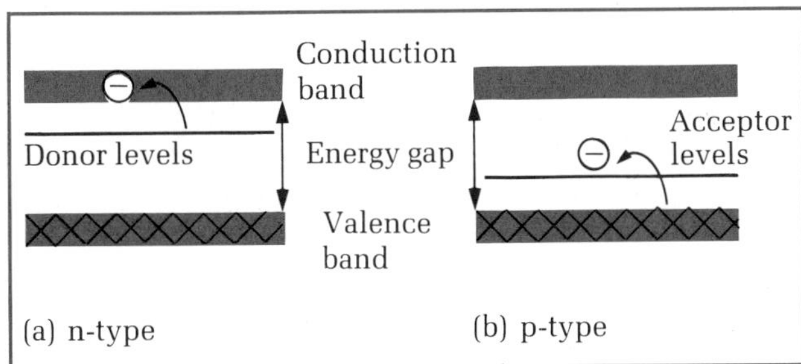

Figure 26.7 *Extrinsic semiconductors*

Extrinsic semiconductors There are two types to consider:

a) n-type semiconductors have added energy levels due to the donor atoms. The donor levels are in the energy gaps between the valence and conduction bands, below the conduction bands. Surplus electrons from the donor atoms occupy these levels, and can easily jump into a conduction band to become free electrons.

b) p-type semiconductors have added energy levels just above the valence band. So electrons from the valence band can jump into the acceptor levels, leaving holes in the valence band.

For both types of extrinsic semiconductor, conduction due to electrons jumping from the valence band to the conduction band is insignificant at room temperature. Only when the material is heated to much higher temperatures does 'jumping the gap' contribute to conduction more than the added levels do.

Semiconductor diodes

The p-n junction diode consists of a p-type semiconductor in contact with an n-type semiconductor. The device allows current through in one direction only, so it can be used to rectify alternating current (i.e. to convert a.c. to d.c.). Free electrons from the n-type material cross the junction to fill holes in the p-type material. So the n-type material loses electrons and becomes positively charged. The p-type material gains electrons to become negatively charged. The p.d. produced acts as a barrier to prevent any more electrons moving across the junction from the n-type to the p-type material, as in Figure 26.8(a).

Now consider the p-n junction connected to a battery.

1 If the battery makes the p-type material even more negative, then the barrier is made even stronger. No current can pass, and the diode is **reverse-biased**.

2 Now suppose the battery is reconnected to make the p-type material less negative, or even to make it positive. The barrier p.d. is reduced or even removed because electrons in the p-type material are removed from atomic bonds and attracted to the positive terminal of the battery. The battery supplies electrons to the n-type material at the same time so a continuous transfer of electrons across the junction from the n-type to the p-type material occurs. The positive terminal of the battery 'empties' the p-type material of electrons to create holes in the p-type material. The negative terminal of the battery 'fills' the n-type material with electrons to replace electrons that move across the junction into the p-type material. The diode is now said to be **forward-biased** since it allows current to pass through.

The way in which the current varies with applied p.d. for a silicon diode is shown in Figure 26.9. In the forward direction, the p.d. must be at least 0.6 V to make the diode conduct. The resistance of the diode in the forward direction decreases as the current increases. In the reverse direction, the resistance is very high, allowing only a small 'leakage' current to pass. If the p.d. in the reverse direction becomes greater than a certain value known as the **breakdown voltage** the barrier breaks and a large current is allowed through.

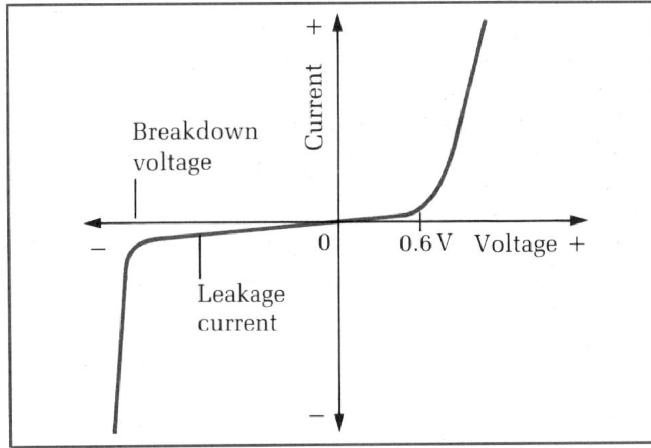

Figure 26.9 I v V *for a diode*

Zener diodes are designed with a specified breakdown voltage. Figure 26.10 shows the circuit symbols for a p-n junction diode and a zener diode. Zener diodes are used in 'stabilized' power supplies where constant p.d.s are required. Figure 26.11 shows a circuit where a zener

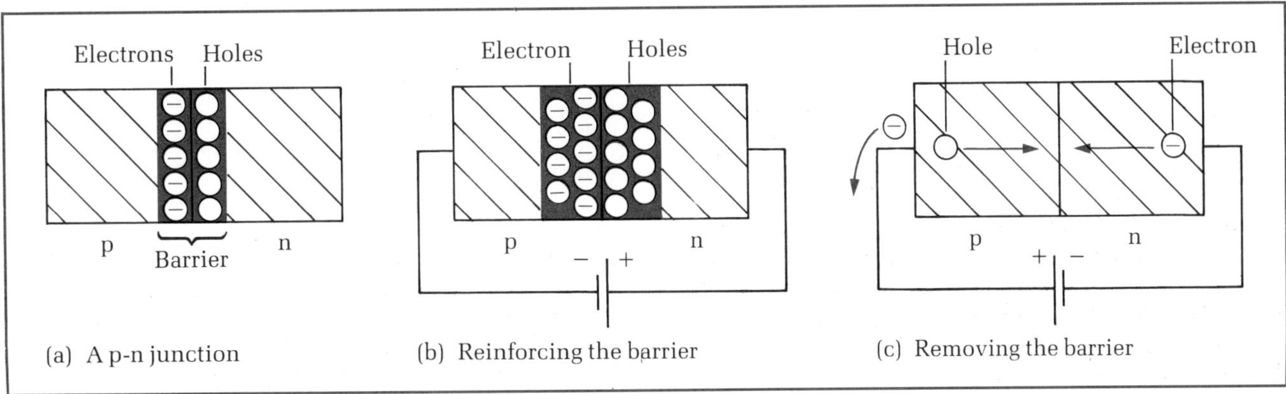

Figure 26.8

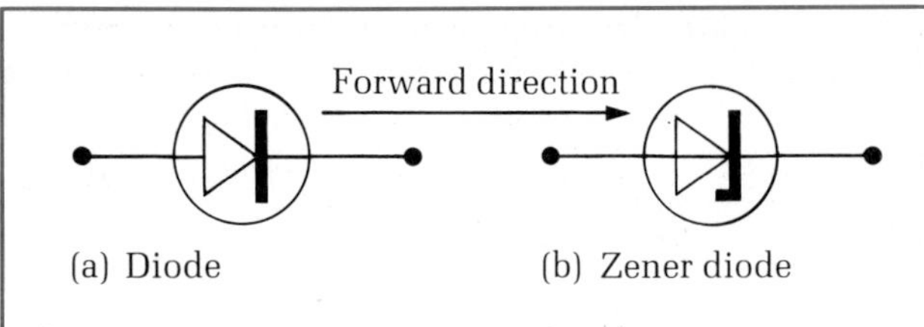

Figure 26.10 *Symbols*

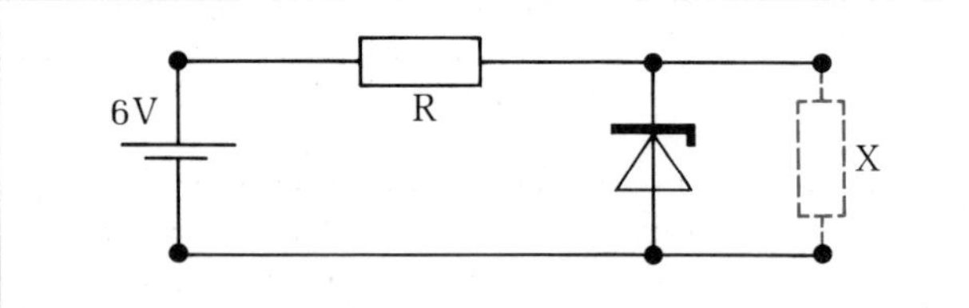

Figure 26.11 *A voltage stabilizer*

diode is used to give a constant output p.d. of 5.1 V. The zener diode has a breakdown voltage of 5.1 V. It is connected in series with a 6 V battery and a resistor. The diode is reverse-biased so it breaks down and passes a current. The resistor is used to limit the current so that the diode does not overheat and become damaged.

The p.d. across the zener diode is fixed at 5.1 V. The diode can therefore be used to supply a constant p.d. of 5.1 V to a device or circuit connected in parallel with the diode. For example, suppose the zener diode must not use more than 0.5 W of power. With a p.d. across its terminals of 5.1 V, the maximum current through it is 0.5/5.1 = 0.1 A (= power/p.d.). With a 6 V battery in the circuit, resistor R must take maximum current 0.1 A when its p.d. is 6 − 5.1 = 0.9 V. Hence the resistance of R = 0.9/0.1 = 9 Ω.

Now suppose a device X of constant resistance X is connected in parallel with the diode. The current taken by X is 5.1/X, since the p.d. across X is 5.1 V.

However, the current taken by X must be less than 0.1 A or the diode stops conducting. Hence 5.1/X must be less than 0.1 A. So X must be greater than 5.1/0.1 = 51 Ω for the circuit to work. Provided X is more than 51 Ω, the p.d. across it stays fixed at 5.1 V. So X could be a variable resistance but its resistance must be more than 51 Ω. If this is so, the p.d. across its terminals is always 5.1 V no matter how much current X draws from the battery.

Worked example In the circuit of Figure 26.11, a zener diode is used with a breakdown voltage of 2.7 volts. The diode is to be used with a 6 volt battery. The maximum power taken by the diode is 0.2 W. The diode is connected in series with a 'limiting' resistance *R* and the battery. Then a 100 ohm resistor is connected across the diode. Calculate

a) the value of *R*,

b) the current taken by the 100 ohm resistor.

Solution

a) Without the 100 ohm resistor, R must limit the current taken by the diode to 0.2/2.7 = 0.074 A (= power/p.d.).
The p.d. across R = 6 − 2.7 = 3.3 volts.
The current through R = 0.074 A.
Hence *R* = 3.3/0.074 = 44.5 ohms.

b) With the 100 ohm resistor in parallel with the diode, the p.d. across the diode is 2.7 volts.
Hence the p.d. across the 100 ohm resistor is 2.7 V.
So the current through the 100 ohm resistor = 2.7/100 = 0.027 A.

In this situation, the zener diode will take a current of 0.074 − 0.027 = 0.047 A. The power which it dissipates is therefore 0.047 × 2.7 = 0.127 W.

***Light-emitting diodes* (LEDs)** are made using compound semiconductors. When an LED is forward-biased, it conducts and emits light. The light is emitted as a result of electrons and holes recombining inside the material of the diode. Each recombination of an electron and a hole causes light energy to be released. The colour of the light emitted depends on the material used.

Photodiodes have a 'window' in the case which allows the p-n junction to be exposed to light. The diode is reverse-biased. When light falls onto the p-n junction, electrons are freed from the outer shells of the atoms so the material conducts. Each free electron gains enough energy from the light to jump from the valence to the conduction band.

26.2 Transistors

Transistors are key components in electronic circuits. They were invented in 1948, and transistor circuits soon replaced vacuum tube valve circuits which use much more power and need high voltages. Portable radios were among the early products of transistor circuits. The next major step in the development of transistors was their production in integrated form. Integrated circuits with several transistors on one chip were first produced in the 1960s. Since then, manufacturing processes have been developed to produce single chips with thousands of transistors on each one.

Junction transistors

A transistor is a 3-terminal device where a small current controls a much larger current. Junction transistors make use of the properties of p-n junctions to achieve such control. There are two forms of junction transistor.

The n-p-n transistor has a layer of p-type semiconductor sandwiched between two layers of n-type semiconductor, as in Figure 26.12. The middle layer is called the **base**, the smaller of the two outer layers is the **emitter** and the other outer layer is the **collector**. A small current into the base and out via the emitter controls a much greater current passing in at the collector and out at the emitter.

The p-n-p transistor has a layer of n-type semiconductor sandwiched between two layers of p-type semiconductor, as in Figure 26.13. The current directions are the reverse of the direction in the n-p-n transistor.

Consider the use of the most common type of transistor, the silicon n-p-n transistor. The small current which enters the base and leaves via the emitter is called the **base current**. The current which passes through from the collector to the emitter is called the **collector current**. In

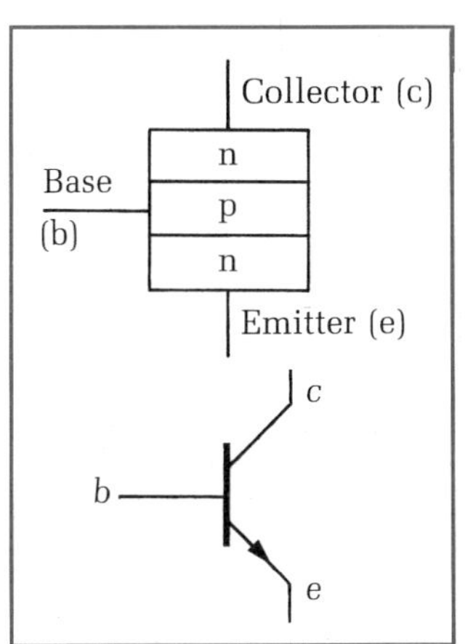

Figure 26.12 *n-p-n junction transistor*

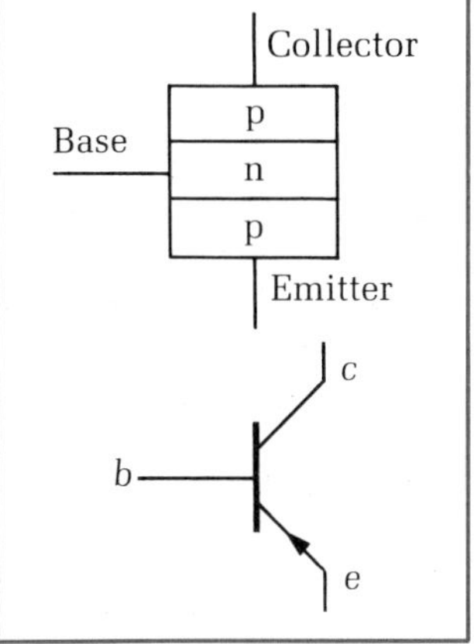

Figure 26.13 *p-n-p junction transistor*

use, the base current controls the collector current. If the base current is switched off, no collector current can pass through the transistor. Increasing the base current makes the collector current increase. Measurements show that the collector current is proportional to the base current. Figure 26.14 Shows typical results for a graph of collector current against base current.

The current gain h_{FE} is defined as

$$\frac{\text{the collector current } I_c}{\text{the base current } I_b}$$

The value of h_{FE} varies from one transistor to another, but is constant for any given transistor. The gradient of Figure 26.14 gives the current gain for the transistor used. So the collector current is equal to $h_{FE} \times$ the base current. For Figure 26.14, the gradient is 120; hence the collector current for the transistor used is equal to 120 × the base current.

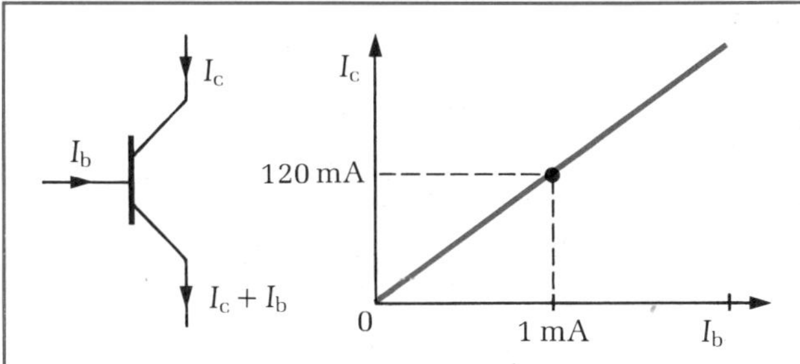

Figure 26.14 *Current gain*

Let's look at a simple alarm circuit to see how a single transistor can be used. Figure 26.15 shows a circuit for a temperature-sensitive alarm. A temperature-dependent resistor (i.e. a thermistor) is connected in series with the base of the transistor. The collector is in series with a relay coil. The relay switch is connected in series with the alarm bell and a battery. The alarm bell rings when current is passed through the relay coil because the coil's electromagnet closes the switch.

When the thermistor is cold, its resistance is high so the base current is small. Hence the collector current is too small to energize the relay. The alarm stays off.

Now consider what happens when the thermistor becomes warmer and warmer due to the temperature rise

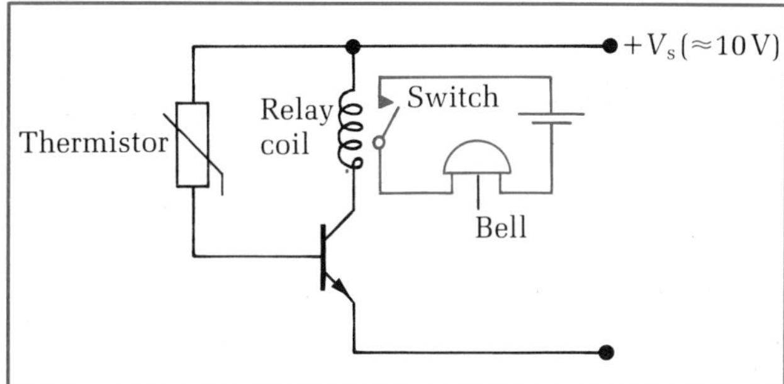

Figure 26.15 *A temperature alarm*

of its surroundings. As it becomes warmer, its resistance falls so the base current increases. Hence the collector current rises, the relay becomes energized and the alarm bell rings.

Integrated circuits contain transistors on a single chip. So-called TTL chips are made by forming junction transistors on a silicon wafer. The manufacturing process is in several stages, starting with a silicon 'wafer' and finishing with the wafer cut into separate chips. Each chip is a complete integrated circuit, and is packaged in a case and connected to the case pins. Each IC is tested since the failure rate can be as high as 70%, but since they are mass-produced their cost is kept down. The actual stages of the manufacturing process involve 'masking' the silicon wafer, then exposing the unmasked parts to vapour of doping material. In this way, n-type and p-type regions are formed on the wafer. The doped regions are then interconnected using different masks to deposit aluminium strips.

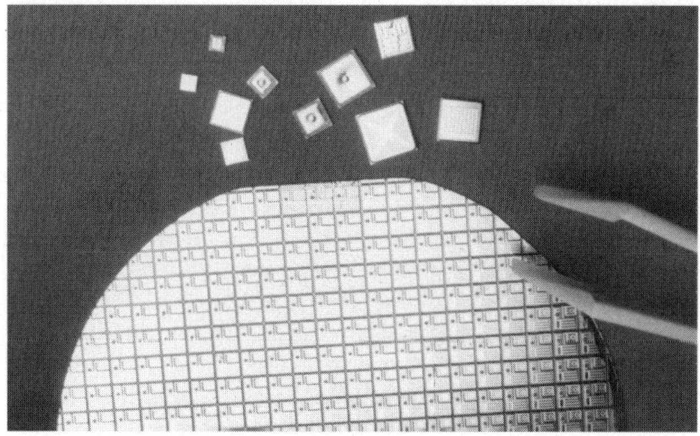

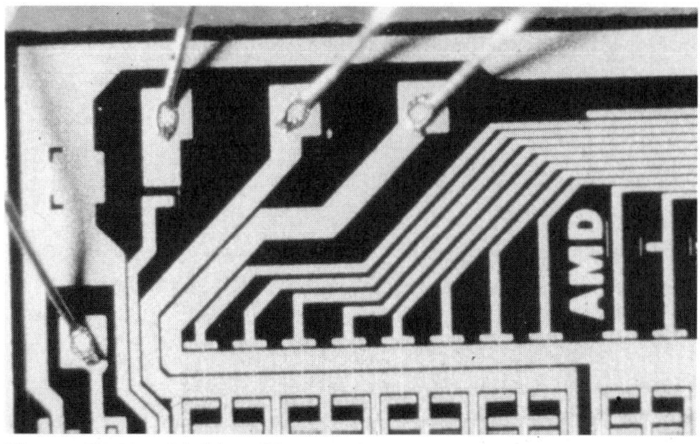

Figure 26.16 *Making ICs*

Measuring transistor characteristics

To use transistors, we need to know their characteristics. For example, in the circuit of Figure 26.15, it is important to know what the maximum value of the collector current is. The circuit is then designed to ensure that the collector current does not exceed the maximum value. Otherwise the transistor overheats and becomes damaged.

Input characteristic is defined as the base current I_b against the base-emitter voltage V_{be}. A potential divider is used to supply a variable p.d. between the base and the emitter. The p.d. is measured using a high-resistance voltmeter. The base current is measured using a microammeter (see Figure 26.17(a)). The p.d. is varied in steps, and at each step the readings of the meters are noted. Figure 26.17(b) shows typical results plotted as a graph of I_b against V_{be}.

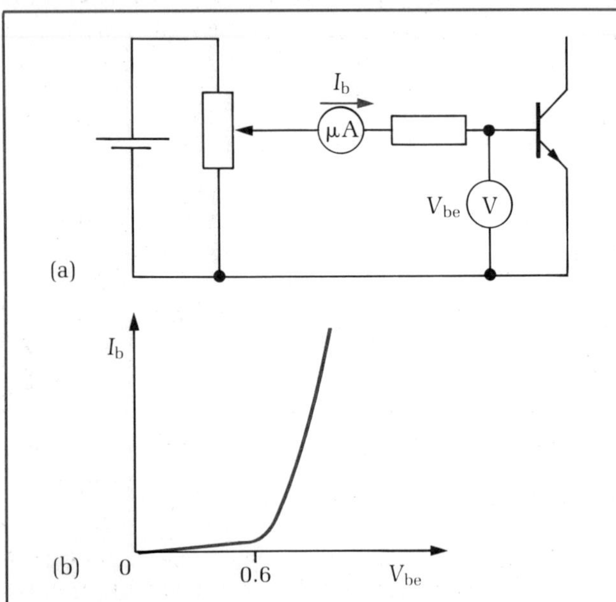

Figure 26.17 *Input characteristic*

The base-emitter is essentially a p-n junction diode so it gives a graph of current against p.d. which is the same as for a p-n junction diode. Two points are important.

a) To make the junction conduct, the p.d. must be at least 0.6 volts. If the p.d. is less than 0.6 volts, hardly any current enters the base. At p.d.s greater than 0.6 volts, the current changes very considerably with hardly any change of p.d. This is shown in Figure 26.17(b) where the curve becomes very steep.

b) Because the current is not proportional to the p.d. the base-emitter resistance is not constant.

Transfer characteristic is the collector current I_c against base current I_b. A potential divider R_2 is added to the circuit of Figure 26.17 to maintain a constant p.d. between the collector and the emitter (see Figure 26.18). A milliammeter in series with the collector enables the collector current to be measured. The potential divider R_1 is adjusted in steps to vary the base current. At each step, the base current and the collector current are measured. Typical results are shown in Figure 26.14. The graph shows that the collector current is proportional to the

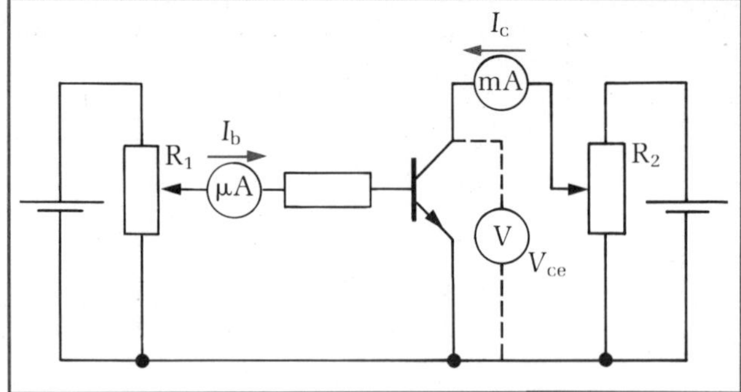

Figure 26.18 *Transfer and output characteristic*

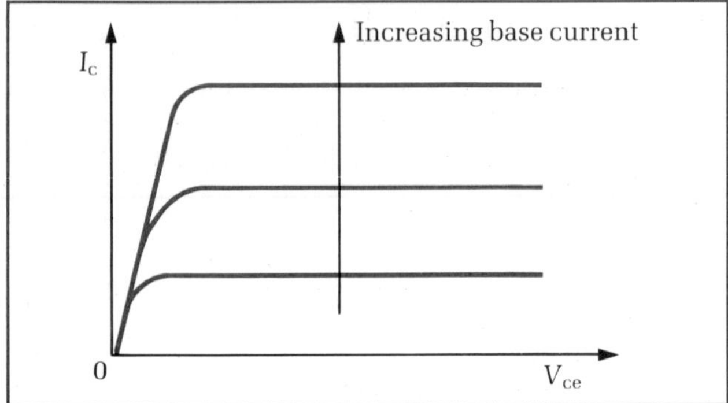

Figure 26.19 I_c *v* V_{ce}

base current. The gradient of the line gives the current gain h_{FE}.

Output characteristic is the collector current I_c against the collector-emitter voltage V_{ce}. The circuit of Figure 26.18 is used with a high resistance voltmeter connected between the collector and the emitter. R_1 is used to ensure the base current is unchanged while a set of readings are made. To obtain a set of readings, R_2 is adjusted in steps, and at each step the collector current and the collector-emitter voltage are measured. The procedure can be repeated for different values of base current.

The results can be plotted as a family of curves of I_c against V_{ce}. Each curve represents a constant value of base current, as shown in Figure 26.19.

Using transistors

A light-sensitive circuit Suppose we want to design a circuit to open a garage 'up-and-over' door automatically when the headlamps of the car shine on it. Figure 26.20 shows one possible circuit. A light-dependent resistor (LDR) is connected in series with the base of a transistor. The relay is connected in series with the collector.

a) In the dark, the LDR has a high resistance so the current supplied to the base is small. Hence the collector current is too small to energize the relay. The door stays closed.

b) When the LDR is illuminated, its resistance falls so the base current rises. Hence the collector current rises and the relay energizes. As a result, the electric motor operating the door is switched on and the door opens.

Consider the circuit in a little more detail. Suppose the relay current must be at least 0.2 A to energize it. Also, suppose the current gain of the transistor is 100. What

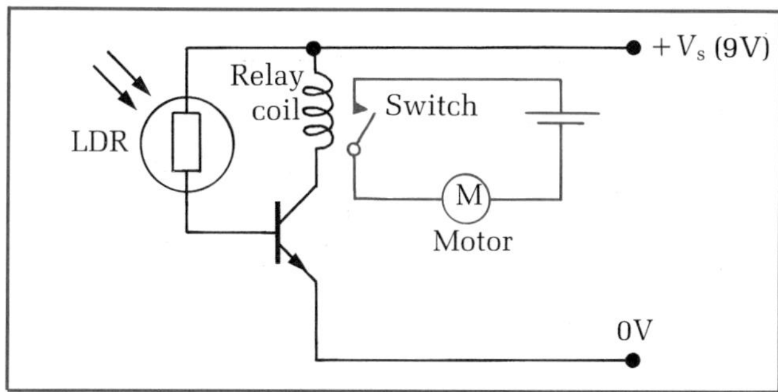

Figure 26.20 *A light-sensitive circuit*

must the resistance of the LDR be when it is illuminated? To operate the relay, the collector must take 0.2 A. Hence the base current $I_b = I_c/h_{FE} = 0.2/100 = 0.002\ A = 2.0\ mA$.

The base-emitter p.d. is 0.6 volts, assuming a silicon transistor is used. The p.d. across the LDR is therefore $V_s - 0.6 = 8.4$ volts. Hence the resistance of the LDR must fall to $\frac{8.4}{0.002} = 4200$ ohms.

An inverter The circuit shown in Figure 26.21 can be used as a logic 'inverter gate' provided the resistance R is small enough. Figure 26.22 shows how the output p.d. varies with the input p.d. for different values of R. For low resistance, the output p.d. switches abruptly from high to low when the input p.d. is increased above a certain value.

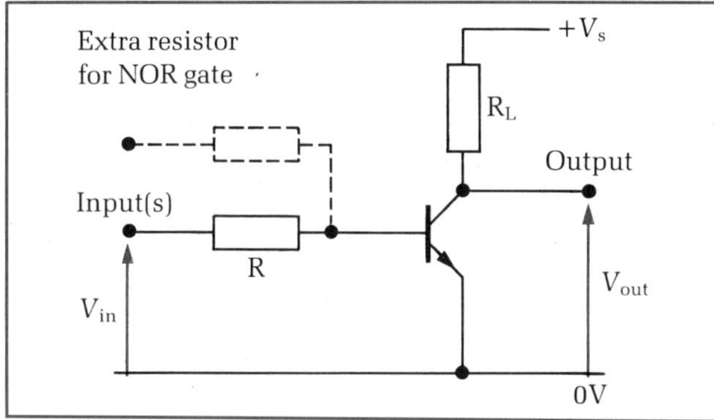

Figure 26.21 *A logic gate*

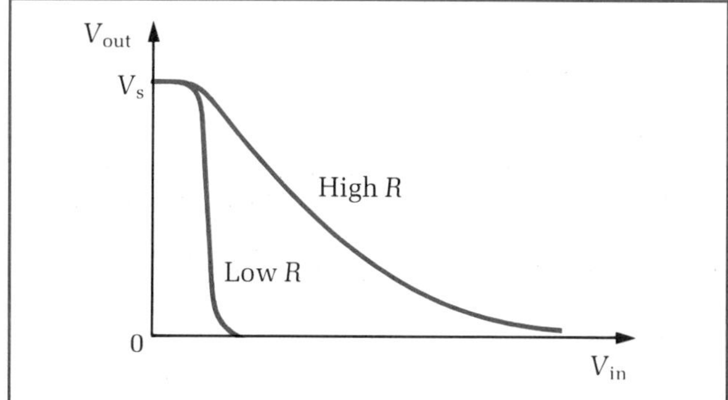

Figure 26.22 *Input v output*

If $V_{in} = 0$, then the base current is zero so no current passes through resistor R_L into the collector. Hence there is no p.d. across R_L. Since V_{out} + p.d. across R_L is always equal to the supply p.d. V_S, then $V_{out} = V_S$ when V_{in} is zero. So V_{out} is 'high' when V_{in} is 'low'.

If V_{in} is greater than 0.6 volts, then the transistor is supplied with current into the base so current enters the collector as well. Hence the p.d. across R_L is no longer zero so the output p.d. V_{out} is less than V_S. If the resistance R is small, the base current is large because the p.d. across R stays constant ($= V_S - 0.6$). Then the collector current is large, resulting in the p.d. across R_L equalling V_S. Hence the output p.d. V_{out} falls to zero.

The simple circuit shown in Figure 26.21 can be adapted to a NOR gate by using two base resistors in parallel. The dashed lines on Figure 26.21 show the idea. When either input is made 'high', the output becomes 'low' so the circuit acts as a NOR gate. See p. 365.

The transistor amplifier

Amplifiers are used to boost electrical signals. For example, a hearing aid contains an amplifier connected to an earphone. The amplifier is driven by electrical signals supplied by a microphone, as shown in Figure 26.23. The amplifier is designed to supply the earphone with an electrical signal which 'follows' the changes of the input signals from the microphone on a larger scale.

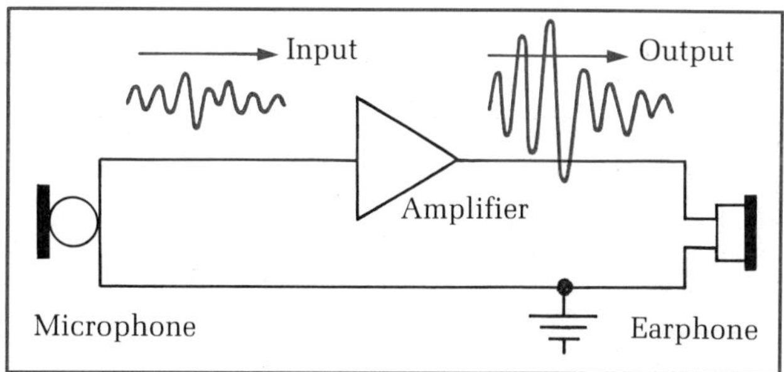

Figure 26.23 *An amplifier system*

A simple voltage amplifier is shown in Figure 26.24. A sine wave voltage applied to the input terminals produces a much larger sine wave voltage across the output terminals. The input voltage is said to have been **amplified** to produce the output wave form. The output voltage is also inverted compared with the input voltage; when the input voltage is at a peak, the output voltage is at a trough and *vice versa*.

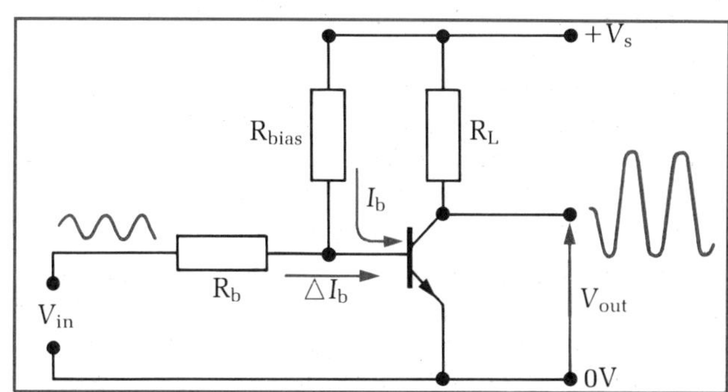

Figure 26.24 *A simple voltage amplifier*

The voltage gain is defined as

$$\frac{\text{the change of output p.d. } \Delta V_{out}}{\text{the change of input p.d. } \Delta V_{in}}$$

To understand how the amplifier shown in Figure 26.24 works, consider the following three input situations.

1 No input p.d. is supplied: the base bias resistor R_{bias} supplies a steady current to the base of the transistor, thus the transistor allows a steady current in at its collector via the load resistor. The resistance R_{bias} is chosen so that the p.d. across the load resistor R_L is exactly $0.5 \times$ the supply p.d. V_S. Hence the output p.d. is equal to $0.5\,V_S$ (since the output p.d. + the p.d. across R_L = supply p.d.).

For example, suppose $V_S = 10$ V and the current gain for the transistor used is 125. The value of R_L is governed by the maximum safe current which the transistor can pass without overheating and so becoming damaged. If the maximum safe current is 0.1 A, then the load resistance R_L must limit the current from the supply unit; the maximum current through R_L is when its p.d. is equal to the supply p.d. Hence R_L must be at least $10/0.1 = 100$ ohms. Choose $R_L = 200$ ohms to be well within the safety limits.

$V_S = 10$ V, $h_{FE} = 125$, $R_L = 200$ ohms.

What should the value of R_{bias} be? The output p.d. must be 5 V, so the p.d. across R_L must also be 5 V. Thus the current passing through $R_L = 5/200 = 0.025$ A. This is also the collector current I_c. Hence the base current $I_b = I_c/h_{FE} = 0.025/125 = 0.000\,20$ A $= 0.2$ mA. Now we need to determine the p.d. across R_{bias}. Since there is a base current, we can assume the base-emitter p.d. is 0.6 V (see p.386). The p.d. across R_{bias} and the base-emitter junction add up to the supply p.d. 10 V. Hence the p.d. across $R_{bias} = 10 - 0.6 = 9.4$ V. Hence $R_{bias} = 9.4/0.000\,2 = 47\,000\ \Omega$. Provided this value for R_{bias} is used, the output p.d. is 5 V when there is no input.

2 An input p.d. greater than 0.6 V is applied; the input p.d. supplies extra base current. So the base is supplied with current via R_{bias} as before and via the base resistor R_b from the input voltage V_{in}.

The extra base current causes the collector current to increase so the p.d. across the load resistor increases at the expense of the output p.d. Hence the output p.d. falls below 5 V.

3 An input p.d. less than 0.6 V is applied: the input p.d. 'taps off' some of the current through R_{bias} which would have entered the base. So the base current is reduced, hence the collector current and its p.d. fall. Therefore, the output p.d. rises above 5 V.

Changes of the input voltage cause changes of the base current. Hence changes of the collector current are produced, resulting in changes of the p.d. across the load resistor. Since the p.d. across the load resistor and the output voltage always add up to the supply p.d. V_S, then changes of the output voltage are caused. Consider the ideas step by step.

(i) Suppose the input voltage changes by a small amount ΔV_{in}. Assuming the base resistance R_b is much greater than the resistance of the base-emitter junction, then ΔV_{in} causes a change of base current $\Delta I_b = \Delta V_{in}/R_b$.

(ii) The change of the collector current $\Delta I_c = h_{FE}\Delta I_b$, hence $\Delta I_c = h_{FE}\Delta V_{in}/R_b$.

(iii) Now the p.d. across R_L + the output p.d. V_{out} = the supply p.d. V_S so the change of p.d. across $R_L = -$the change of the output p.d. ΔV_{out} (because the supply p.d. is fixed).

(iv) However, the change of the p.d. across R_L = change of current $\Delta I_c \times R_L$

$$= \frac{h_{FE}\Delta V_{in} R_L}{R_b}$$

Hence the change of the output p.d.

$$\Delta V_{out} = -\frac{h_{FE}\Delta V_{in} R_L}{R_b}$$

Therefore the voltage gain is given by

$$\frac{\Delta V_{out}}{\Delta V_{in}} = -\frac{h_{FE} R_L}{R_b}$$

where h_{FE} = current gain,
R_L = load resistance,
R_b = base resistance.

Investigating a transistor amplifier

Construct the circuit shown in Figure 26.24, if possible using a transistor with a known value for its current gain. As explained in the previous section, the value of R_L is chosen so that the transistor is limited to a safe current range. Most transistors are quite safe with a load resistance of 1000 ohms.

1 To select the value for R_{bias}, connect a high resistance voltmeter across the collector and emitter. Then, with no input p.d. connected, try different values of R_{bias}

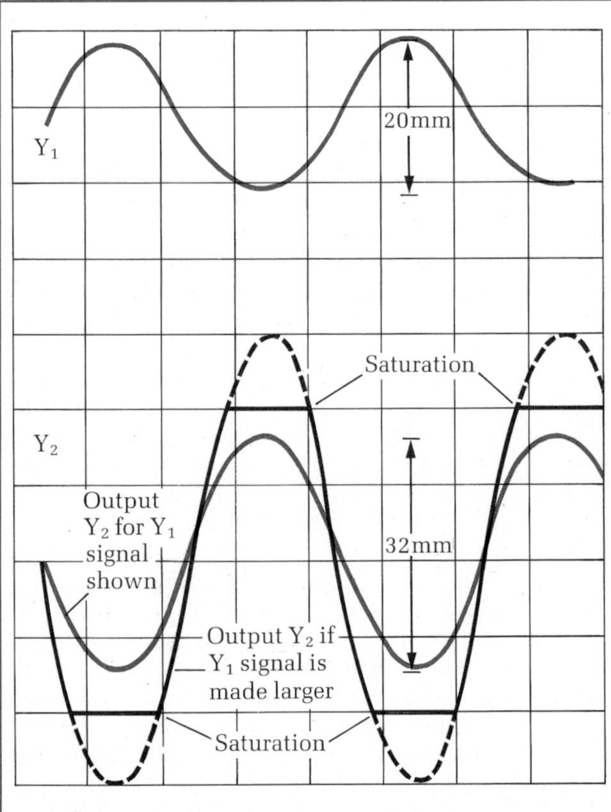

Figure 26.25 *Wave forms*

until the output voltage is 0.5 × the supply p.d. You could calculate the value of R_{bias} in the way explained in the previous section. See how the calculated value compares with the value determined by trial and error.

2 The value of R_b must be much greater than the base-emitter resistance when the transistor conducts. Try $R_b = 10\ k\Omega$ to start with.

3 Use a signal generator to apply a sine wave voltage to the input. Disconnect the voltmeter used in **1** and use a double-beam oscilloscope to display the input and output wave forms.

Measure the voltage gain by using the oscilloscope to measure the peak voltage of the input wave form and of the output wave form. Figure 26.25 shows an input wave form of peak height 10 mm. The peak height of the output wave form is 16 mm. The input wave form is applied to the Y_1 beam of the oscilloscope which has a voltage sensitivity of $0.5\ V\ cm^{-1}$. Hence the peak value of the input voltage is 0.5 V. The output wave form is displayed by the Y_2 beam which has a voltage sensitivity of $2.5\ V\ cm^{-1}$. So the peak value of the output voltage is $2.5 \times 1.6 = 4$ volts. Hence the voltage gain $= 4/0.5 = 8$.

Calculate the voltage gain using the equation $h_{FE}R_L/R_b$, assuming the value of h_{FE} is known. You ought to obtain agreement between the calculated and the measured value. For example, the transistor used in Figure 26.25 has a value of $h_{FE} = 160$, $R_L = 1\ k\Omega$, $R_b = 20\ k\Omega$. Hence its calculated voltage gain is $160 \times 1/20 = 8$.

Saturation of the output wave form results if the input voltage wave form is made larger and larger. The output voltage cannot rise above the supply p.d. and it cannot fall below 0 volts. If the input voltage is so large that it tries to drive the output voltage outside its limits, then the output voltage stays at the limit until the input voltage becomes smaller. Use the signal generator to increase the peak value of the input wave form. The output wave form increases its peak value too until its peaks and troughs are seen to flatten, as in Figure 26.25. Each flat section is where the output voltage has reached the limits set by the supply p.d. The output wave form is still inverted compared with the input wave form, but it is clipped at the peaks and troughs.

If the voltage gain and supply p.d. are known, you can calculate the maximum value of the input wave form which gives saturation. For example, suppose the supply p.d. is 10 V and the voltage gain is 8. The output voltage swings from 0 to 10 V at the point of saturation so the peak value of the output voltage at this point is 5 V. Hence the peak value of the input p.d. which just gives saturation is $5/8 = 0.625$ V. A larger input p.d. from the signal generator causes the output wave form to saturate. In practice, it may be necessary to alter the value of R_{bias} to clip the peaks and troughs equally. The reason is that the simple theory above assumes the sine wave input is in addition to a steady p.d. of 0.6 V at the input – necessary to give the mean output voltage at $0.5\ V_s$ unless R_{bias} is changed.

Try altering the values of R_{bias} and R_b. At saturation, changing the value of R_{bias} alters the mean output voltage which causes unequal saturation. In other words, the output wave form is clipped unequally at the peaks and troughs.

Distortion of the output wave form is produced if the value of R_b is too small. If so, the base-emitter junction's resistance is no longer insignificant. Because its resistance is variable, the voltage gain is no longer constant for different input voltages, so the output wave form is no longer a 'carbon copy' on a larger scale of the input wave form. Audio amplifiers which suffer from distortion spoil any cassette or disc recording.

Field effect transistors

Junction transistors are said to be **current-controlled** because a small current entering the base controls a much larger current entering the collector. The collector current depends on the base current.

***Field effect transistors* (FETs)** are designed with a very high input resistance. The input voltage controls the current passing through the transistor. The three terminals are called the **source**, **drain** and **gate**, as shown in Figure 26.26. The source and drain are at either end of a piece of n-type semiconductor. The gate is a piece of p-type semiconductor along the side of the n-type material between the source and the drain. At the boundary between the n-type and the p-type material, holes from the p-type and electrons from the n-type material recombine. So a **depletion layer** where there are no free electrons or holes is formed near the p-n boundary. The width of the depletion layer is controlled by a negative voltage applied to the gate, as in Figure 26.26.

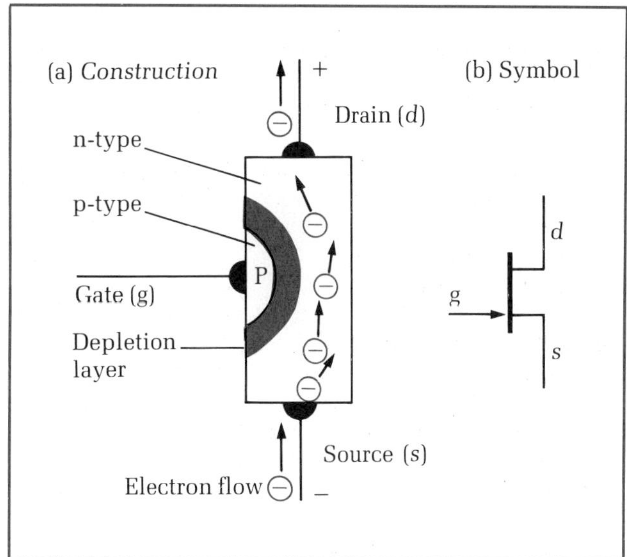

Figure 26.26 *A field effect transistor*

The drain is made positive with respect to the source, so the source injects electrons into the n-type material from the battery. The electrons pass through the n-type material to the drain. However, the presence of the depletion layer forces the electrons to pass along one side of the n-type material. The depletion layer constricts the flow of electrons from the source to the drain. Since the gate voltage controls the width of the depletion layer, the flow of electrons from the source to the drain is controlled by altering the gate voltage. In other words, the gate voltage controls the current from the drain to the source (i.e. the electron flow from source to drain).

The depletion layer is effectively an insulating barrier so very little current enters the gate. The device is therefore said to be **voltage-controlled** rather than **current-controlled**. The resistance between the gate and the source can be as high as 10^9 ohms.

The **drain current**, which is the current from the drain to the source, is determined by the voltage between the gate and the source. Figure 26.27 shows how the drain current varies with the gate-source voltage. When the gate is made increasingly negative, the depletion layer is made wider and wider so the drain current becomes smaller and smaller. When the gate is made positive, the depletion layer is made narrower so the drain current rises.

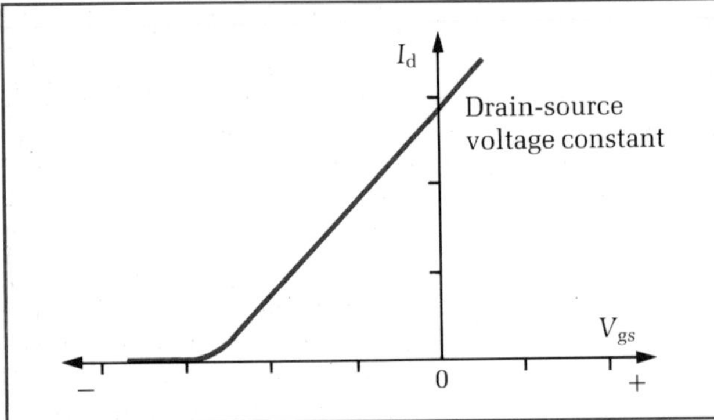

Figure 26.27 *FET characteristics*

The transconductance of an FET is defined as

$$\frac{\text{change of the drain current}}{\text{change of the gate-source voltage}}$$

$$g = \frac{\Delta I_d}{\Delta V_{gs}}$$

where g = transconductance,
ΔI_d = change of drain current,
ΔV_{gs} = change of gate-source voltage.

The unit of g is $A\,V^{-1}$ which equals Ω^{-1}. For the graph shown in Figure 26.27 the value of g is given by the gradient of the line.

An FET voltage amplifier has a much larger input resistance than a voltage amplifier which uses a junction transistor. A load resistor R_L is connected in series with the drain, as in Figure 26.28. The input p.d. to be amplified is applied between the gate and the source. However, the gate must be biased so that it is negative with respect to the source. Resistors R_g and R_s are used for this purpose.

Suppose the gate-source voltage changes by ΔV_{gs}. Hence the drain current changes by $\Delta I_d = g\Delta V_{gs}$.
The change of p.d. across $R_L = \Delta I_d R_L = g\Delta V_{gs} R_L$
Hence the change of the output p.d. is given by

$$\Delta V_{out} = -\text{change of p.d. across } R_L$$

So

$$\Delta V_{out} = -g\Delta V_{gs} R_L$$

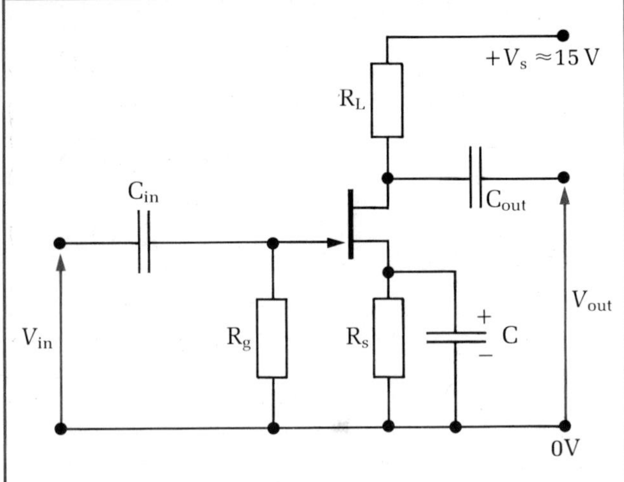

Figure 26.28 *A FET amplifier*

Therefore the voltage gain is given by

$$\frac{\Delta V_{out}}{\Delta V_{gs}} = -gR_L$$

The extra resistors R_g and R_s are essential if the gate is to be biased negatively with respect to the source. The drain current passes through R_s creating a p.d. across R_s, so the source is made positive with respect to the 0 volt line from the power supply. Resistor R_g 'pulls' the gate down to 0 volts because it draws negligible current from any input source. So the source is positive with respect to the gate, as required.

Capacitors are used in the circuit to allow a.c. signals to pass. The input and output capacitors are essential to prevent steady currents affecting the circuit operation. For example, if the input source is a sine wave signal varying from 1.0 to 1.2 V (with respect to the zero volt line), the input capacitor allows the varying component of ±0.1 V to pass. However, the steady component of 1.1 V is blocked. Also the output capacitor prevents the drain current from being diverted through any device connected across the output terminals.

The capacitor in parallel with the source resistance R_s is called a **decoupling capacitor** because it allows variations of the drain current (i.e. the a.c. component of the drain current) to bypass the source resistance. However, the steady component of the drain current cannot pass through the capacitor and must pass through R_s, so the biasing is not affected and the a.c. signal is not reduced.

CMOS integrated circuits contain field effect transistors rather than junction transistors. Hence CMOS chips take much smaller currents than TTL chips do, and so they use much less power (see p. 366).

26.3 Operational amplifiers

A simple voltage amplifier as described in the previous section gives a change of output voltage which is equal to a constant × the change of input voltage. The input signal is multiplied by a constant to give the output signal.

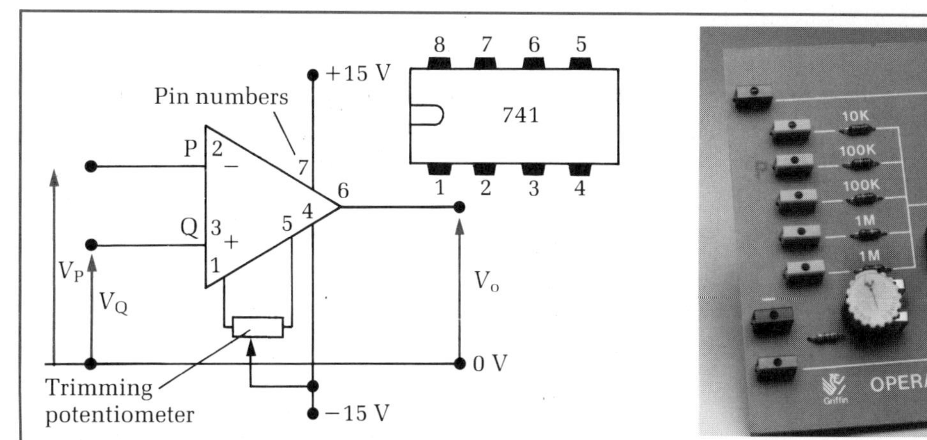

Figure 26.29 *The 741 op-amp*

Amplifiers which can be used for mathematical operations such as multiplication, addition or integration are called **operational amplifiers** (or 'op-amps' for short). Op-amps are now manufactured as integrated circuits; one of the most widely used is the 741 chip shown in Figure 26.29.

An op-amp has two input terminals and a single output terminal. The 741 op-amp needs a power supply of +15 V, 0, −15 V. All voltages are measured with reference to the earthed (i.e. 0 V) line from the power supply. So the output voltage is the p.d. between the output terminal and the 0 volt line. The output voltage is non-zero whenever a p.d. is applied between the two input terminals, so the voltage applied to one input terminal must be different to the voltage applied to the other input terminal if the output voltage is to be non-zero. The 741 op-amp is connected to a 'trimming' potentiometer, as in Figure 26.29, which is used to set the output voltage to zero when the inputs are at zero voltage. This extra feature is required because the internal circuitry of an op-amp is a complex network of transistors and resistors; the correct resistance values cannot be achieved exactly by the manufacturing process. The trimming potentiometer is added by the user to compensate for this.

The two inputs, labelled P and Q in Figure 26.29, are called the **inverting** and **non-inverting** inputs, respectively. They are usually labelled as '−' and '+', but don't confuse the labels with the voltage polarity at each input. The voltages at the inputs or at the output can range from +15 V through 0 to −15 V.

Suppose each input is used in turn, the other input being earthed. The polarity (i.e. sign) of the output voltage is

a) the same as the input voltage if the non-inverting input is used,

b) opposite to the input voltage if the inverting input is used.

For example, if the non-inverting input is made positive with the other input at 0 volts then the output voltage is positive as well. If the non-inverting input is then made negative, the output 'goes negative' too.

If the inverting input is used and the non-inverting input is earthed, then a positive input voltage gives a negative output voltage. A negative input voltage gives a positive output voltage.

Using an op-amp on open-loop

The op-amp is used without additional resistors, apart from the trimming potentiometer. Provided the output voltage does not saturate (i.e. reach the limits set by the power supply), then the output voltage is always proportional to the p.d. between the inputs.

$$V_0 = A_0(V_Q - V_P)$$

where V_0 = output voltage,
V_Q = voltage at the non-inverting input,
V_P = voltage at the inverting input,
A_0 = the constant of proportionality called the 'open-loop gain'.

The open-loop gain $\mathbf{A_0}$ is defined as

$$\frac{\text{the output voltage } V_0}{\text{the p.d. between the inputs } (V_Q - V_P)}$$

assuming no saturation occurs. For a 741 op-amp, the open-loop gain is about 100 000. You might well think that the op-amp on open-loop is therefore a very powerful amplifier. However, the very high gain means that the range of input p.d. is quite small if saturation is to be avoided. For example, with the power supply limiting the output voltage to the range ±15 V, an op-amp with $A_0 = 10^5$ saturates if the input p.d. is greater than $\pm 15/10^5$ V = ±150 μV. Since most practical situations involve input p.d.s greater than 150 μV, saturation will result frequently. We shall see in the next section how we can get round this problem by means of 'feedback'.

The variation of the output voltage with the input p.d. is shown for a typical op-amp in Figure 26.30. The open-loop circuit is shown in Figure 26.29 where the trimming potentiometer could have been left out to simplify the diagram. The graph shows that the output is saturated when the input p.d. is outside the range ±150 μV. Inside the ±150 μV range, the output voltage is proportional to the p.d. between the inputs. Over this section, the gradient of the line gives the open-loop gain A.

In most practical situations, the op-amp on open-loop gives saturation at the output if the input p.d. is non-zero (since 150 μV is very small in practical terms). Nevertheless, the open-loop op-amp can be useful in 'switching' between two states. Consider the following applications.

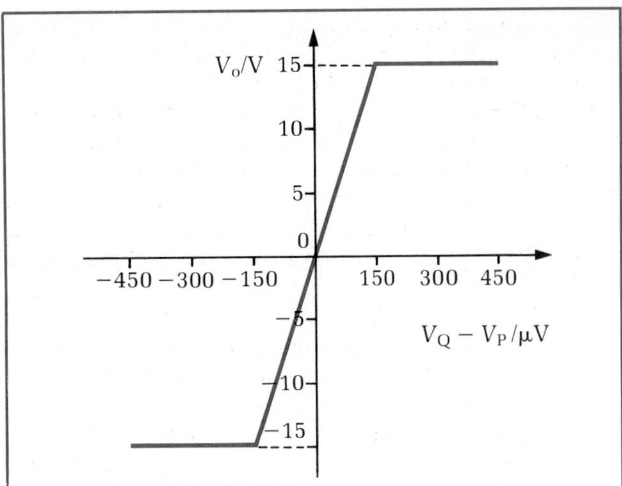

Figure 26.30 *Open-loop characteristics*

A voltage comparator is used to compare two voltages. The two voltages to be compared are applied to the inputs of the op-amp, as in Figure 26.29. The sign of the output voltage (i.e. its polarity) depends on which input voltage is more positive, as is shown above.

If $V_Q > V_P$, then V_0 is positive.
If $V_Q < V_P$, then V_0 is negative.

By reading the output voltage for its polarity, we can determine which of the two input voltages is the more positive. The circuit is used in a digital voltmeter where the voltage to be measured is compared with a 'ramp' voltage generated inside the meter. The voltage to be measured is applied to one of the input terminals of the op-amp comparator. The ramp voltage is applied to the other input terminal, as in Figure 26.31. When the ramp voltage becomes greater than the 'test' voltage, the output voltage from the op-amp switches from negative saturation to positive saturation. The change of state of the output is then used to display the pulse count from the pulse-producing circuit driving the ramp generator.

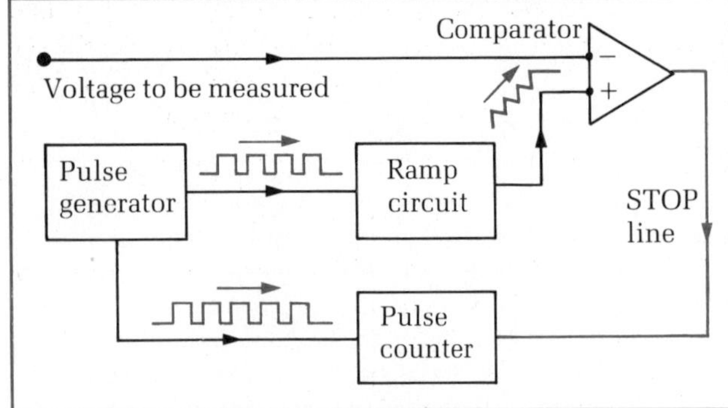

Figure 26.31 *Using an open-loop comparator*

Another application of the voltage comparator is to switch on an electric lamp when darkness falls. One of the two inputs of the op-amp is supplied from a potentiometer, as in Figure 26.32. The other input is supplied from another potential divider formed by using a light-dependent resistor (LDR) and a fixed resistor. The output from the op-amp is used to operate a relay via a transistor.

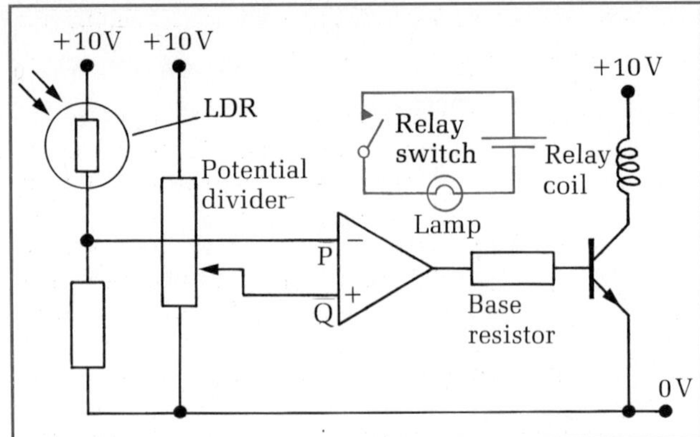

Figure 26.32 *A light-operated lamp*

When the LDR is fully illuminated, the voltage at P is greater than that at Q so the output voltage is negative; hence the transistor is off and the relay does not operate. However, in darkness, the LDR resistance is much greater than before and the voltage at P is now below the voltage at Q. Hence the output voltage is at positive saturation; therefore the transistor is switched on. This means that the relay operates and the lamp is switched on.

A sine/square wave converter used to convert a sine wave input to a square wave output. Figure 26.33(a) shows the circuit. The non-inverting input is set at 0 volts. The inverting input is connected to the output of a sine wave generator, as shown in the diagram.

a) When the sine wave input voltage is positive, the output voltage is negative. The input voltage only needs to exceed about 150 μV to saturate the output. So, in practice, the output is at negative saturation when the sine wave goes positive.

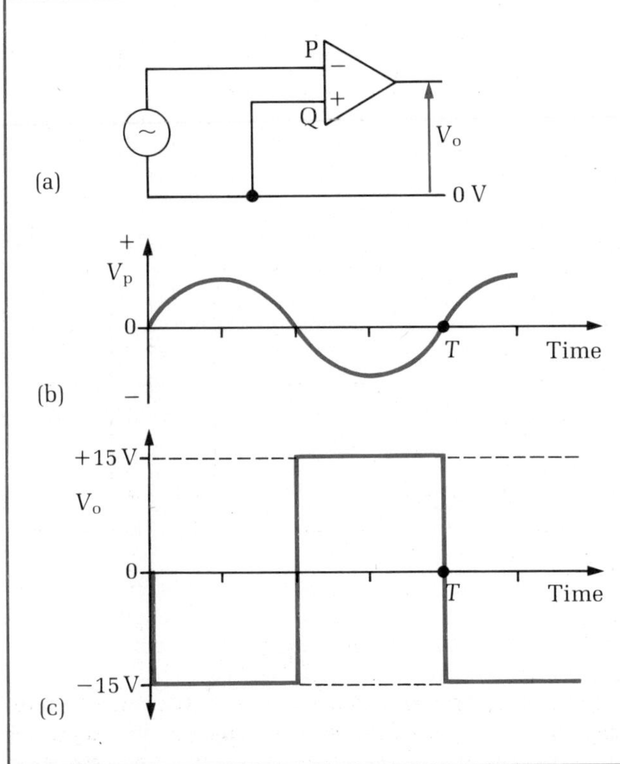

Figure 26.33 *A sine wave converter*

b) When the sine wave input voltage is negative, the output voltage is positive. Again, in practice, saturation is produced.

Hence the output voltage is at positive saturation for one-half cycle of the input sine wave. Then the output swings over to negative saturation for the next half-cycle. The output wave form is a square wave, as in Figure 26.33(b). The output frequency is the same as the frequency of the input wave form, and the output wave form is an 'even' square wave. In other words, the interval for positive saturation is the same as for negative saturation.

To make the output wave form uneven, a non-zero voltage must be applied to the non-inverting input. Figure 26.34(a) shows a potentiometer used to supply a positive voltage to the non-inverting input. The sine wave input is still applied to the other input. Whenever the sine wave input voltage is greater than the voltage from the potentiometer, V_P is greater than V_Q, so the output is at negative saturation. When the sine wave input is below the potentiometer input, the output is at positive saturation (because $V_Q > V_P$). Figure 26.34(b) shows the two input voltages on the same graph. The sine wave is above the potentiometer voltage for less than half of each cycle, so the output is at negative saturation for a shorter time than it is at positive saturation during each cycle. Hence the output is an uneven square wave.

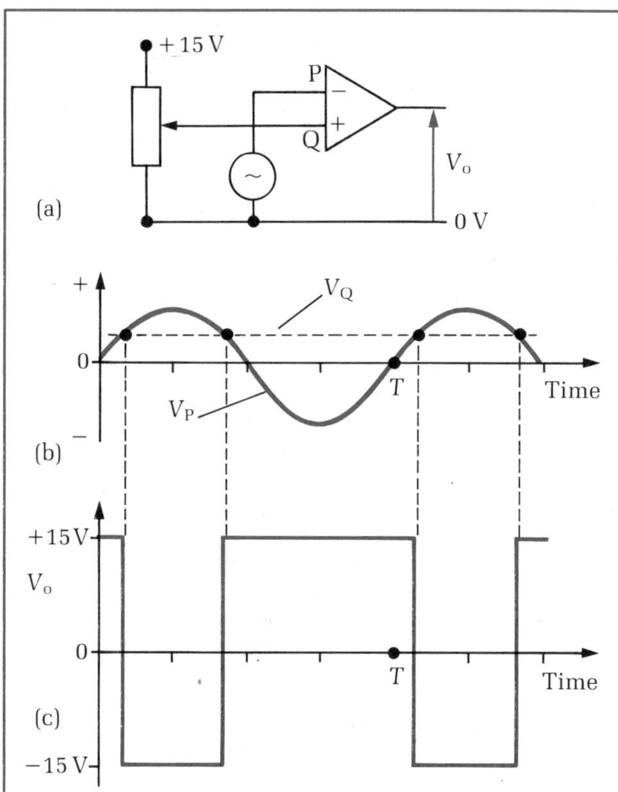

Figure 26.34 *Uneven pulses*

The inverting amplifier

Negative feedback is used to reduce the gain of an op-amp to a more practical value. Used on open-loop, an input p.d. as small as ±150 μV produces saturation at the output. Most practical situations involve amplifying input voltages which are greater than 150 μV. To avoid saturation in practical situations, the gain must therefore be reduced from the open-loop value.

Reducing the gain is achieved by **feeding back** part of the output voltage to the inverting input. The inverting input is used because it tries to make the output voltage the opposite polarity to its own. So by feeding back a proportion of the output voltage to the inverting input, the gain is reduced. If or when you have a bank account you may become familiar with negative feedback. If your account goes into the 'red', negative feedback from your bank will soon reach you to encourage you to move into the 'black'. Without that feedback, you might go deeper and deeper into debt!

The op-amp is supplied with negative feedback by using two resistors to form a **closed loop** from the output to the inverting input. The inverting amplifier uses the two resistors as a potential divider, as shown in Figure 26.35. In this way, a fraction of the output voltage is returned to the inverting input of the op-amp. The result is that the gain is reduced from the open-loop value to a more practical value. One of the two resistors, the feedback resistor R_f, is connected between the output and the inverting input. The other resistance, the input resistor R_i, is connected to the inverting input of the op-amp. The voltage to be amplified, V_i, is applied to the other terminal of R_i. The non-inverting input of the op-amp is set at 0 V.

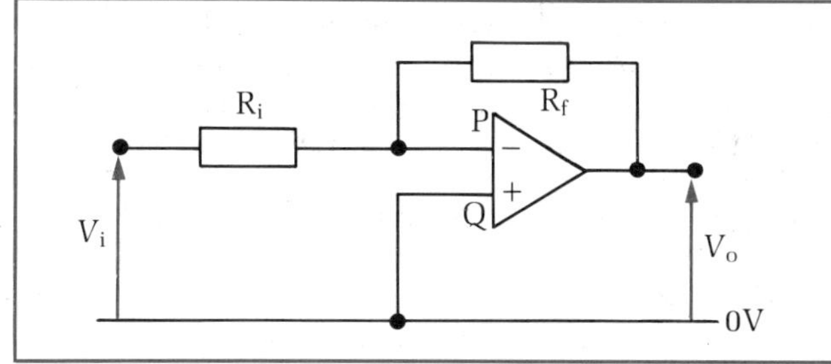

Figure 26.35 *The inverting amplifier*

To understand how the circuit works, suppose a negative voltage V_i is applied to the input. The output therefore becomes positive. However, the feedback network supplies a fraction of the (positive) output to the op-amp's inverting input, so the output is made less positive by the feedback network, as is shown below.

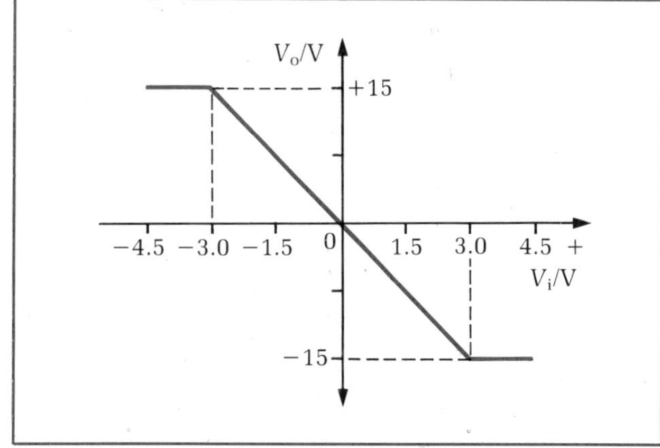

Figure 26.36 *Output v input characteristics for $R_f = 1$ MΩ, $R_i = 0.2$ MΩ*

The voltage gain is given by

$$\frac{V_0}{V_i} = -\frac{R_f}{R_i}$$

where V_0 = output voltage,
V_i = input voltage,
R_f = feedback resistance,
R_i = input resistance.

To prove the above equation, we need to make use of the following two features of the op-amp.

1 Its input resistance between the inverting and non-inverting inputs is very large. Hence current into the op-amp through either of its input terminals is negligible. Effectively, no current enters the op-amp through its input terminals.

2 Virtual earth provided the output does not saturate, the p.d. between the inverting and non-inverting inputs is very small (i.e. less than about 150 μV). In the inverting amplifier, the non-inverting input is set at 0 volts. So the inverting input P must be within ±150 μV if there is no saturation at the output. The voltage at P is said to be at **virtual earth** which means, practically at 0 volts – provided V_0 is not saturated.

Consider the operation of the circuit shown in Figure 26.35 when a constant voltage is applied to its input, and suppose V_i is negative. Hence current is drawn through the input resistance through the input voltage source, as shown in Figure 26.37. No current enters or leaves the op-amp via its inverting terminal P. Hence the same current passes through R_f and R_i. Hence $V_p - V_i = IR_i$, and $V_0 - V_P = IR_f$, where I is the current.

However, P is at virtual earth potential, so $V_p \approx 0$. Therefore $V_i = -IR_i$ and $V_0 = IR_f$.

Combining the two equations gives

$$V_0 = IR_f = -\frac{R_f V_i}{R_i}$$

Therefore, we can write

$$\frac{V_0}{V_i} = -\frac{R_f}{R_i}$$

If the input voltage V_i is positive, the current flow is reversed. The same theory applies, so the equation for the voltage gain is the same.

The equation shows that:

a) The output voltage has the opposite sign (i.e. polarity) to the input voltage.

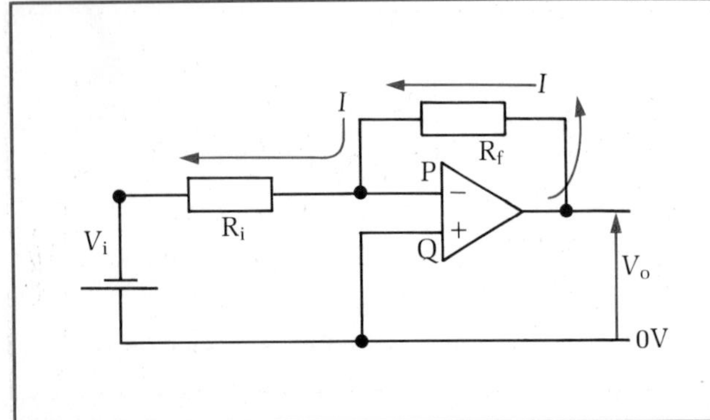

Figure 26.37 *Current directions*

b) The output voltage is equal to a constant $(-R_f/R_i) \times$ the input voltage.

c) The voltage gain does not depend on the open-loop gain. The values of R_i and R_f determine the voltage gain, so the gain can be set by the user by choosing suitable resistances.

d) The output saturates if the input voltage exceeds certain limits. Figure 26.36 shows the variation of the output voltage with the input voltage for $R_f = 1\,\text{M}\Omega$ and $R_i = 0.2\,\text{M}\Omega$. The graph is a straight line passing through the origin, provided the output voltage is not saturated. The gradient of the line passing through the origin is equal to the voltage gain, which equals −5 in this situation $(= -R_f/R_i)$. The output voltage is limited to the range ±15 V so saturation occurs if the input voltage is outside the range ±3 V in this example.

The non-inverting amplifier

Suppose we want an amplifier to give an output voltage with the same polarity as the input voltage. The op-amp with negative feedback is still used, but the resistors R_f and R_i are connected differently than in Figure 26.35.

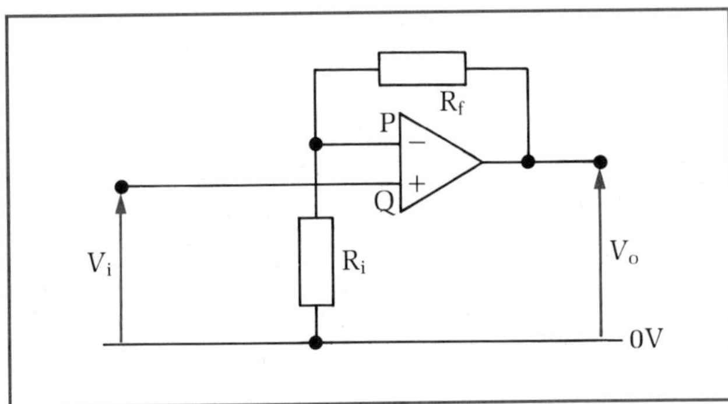

Figure 26.38 *Non-inverting amplifier*

The feedback resistance is still connected between the output and the inverting terminal of the op-amp. However, the resistor R_i is used to connect the inverting terminal to the earth line. The voltage to be amplified is connected directly to the non-inverting terminal of the op-amp.

Suppose the input voltage V_i is made positive; the output voltage therefore goes positive too. Hence current passes through R_f and R_i to earth. So a p.d. is created across R_i which makes input P of the op-amp go positive. The result of input P going positive is to make the output less positive. So the feedback network reduces the voltage gain to a practical value.

If a steady negative voltage V_i is applied, the output voltage goes negative, so current passes from the earth line through R_f and R_i. Therefore input P is made negative; so the output is made less negative. The voltage gain is given by

$$\frac{V_0}{V_i} = 1 + \frac{R_f}{R_i}$$

Worked example A sine wave voltage of peak value 5.0 V is applied to the input of a non-inverting amplifier with output against input characteristics as shown by Figure 26.39. Sketch the output wave form.

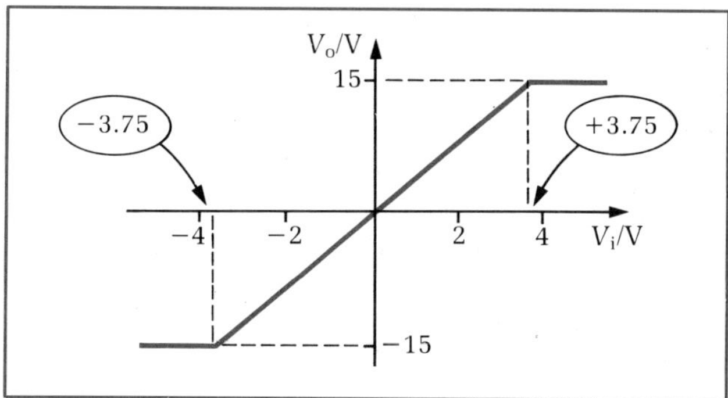

Figure 26.39

Solution From Figure 26.39, the voltage gain = gradient of line = +4.

If there were no saturation, the output wave form would be a sine wave of peak value 20 V (= 4 × 5 V) in phase with the input wave form.

However, the output cannot exceed the ±15 V limits, so it is clipped at +15 V and at −15 V. Figure 26.40 shows the actual output wave form.

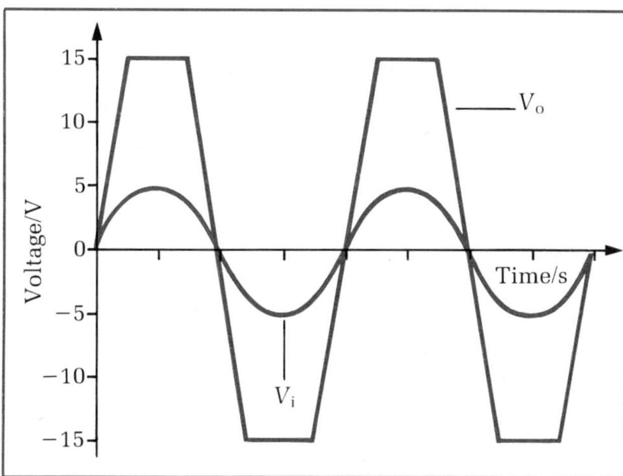

Figure 26.40 *Input and output wave forms*

Using op-amp amplifiers

The output against input characteristics of either the inverting amplifier or the non-inverting amplifier can be determined using the circuit of Figure 26.41. Potentiometer A is used to apply a variable input voltage V_i. Voltmeter V_A is used to measure the input voltage. Voltmeter V_B is used to measure the output voltage V_0.

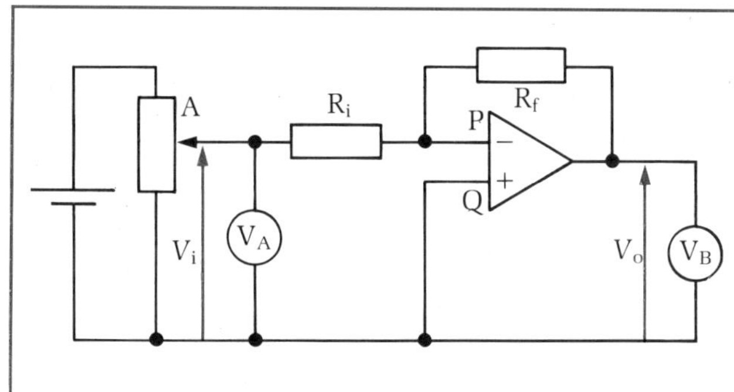

Figure 26.41 *Investigating op-amps*

The input voltage is changed in steps, at each step the voltmeter readings are measured. The results can be presented as a graph of V_0 against V_i, as shown in Figure 26.36 or Figure 26.39. The trimming potentiometer must be used to set V_0 at 0 volts when V_i is zero, before any measurements are made.

An electronic thermometer can be designed using an inverting amplifier, as in Figure 26.42. The inverting terminal of the op-amp is supplied with the voltage from a potential divider. The potential divider is formed by resistors VR_1 and thermistor T. VR_1 is a variable resistor. The voltage gain is adjusted by altering the setting of another variable resistor VR_2. The voltmeter across the output terminals is used to measure the temperature of the thermistor.

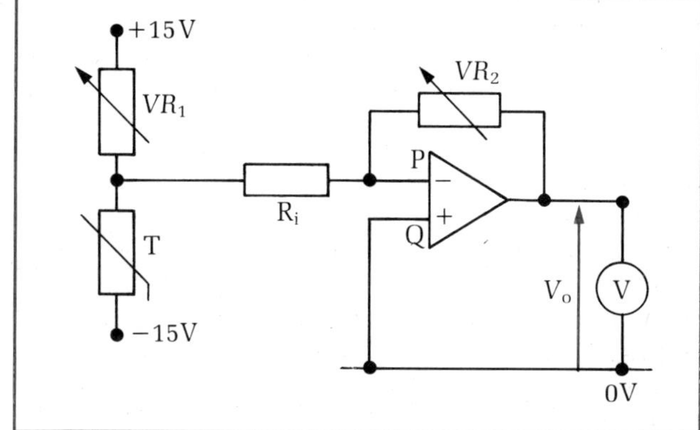

Figure 26.42 *An electronic thermometer*

With the thermistor in melting ice, VR_1 is adjusted so that the voltage at P is zero. This sets the output voltage at zero.

Then the thermistor is placed in boiling water. VR_2 is adjusted to alter the gain until the voltmeter reads exactly 10 V. The thermistor can then be used as a temperature probe, where the centigrade temperature is equal to 10 × the voltmeter reading.

A simple radio receiver uses an op-amp amplifier with a gain of about 1000. Suitable values of R_f and R_i would therefore be 1 MΩ and 1 kΩ respectively.

A tuning circuit can be made using a variable capacitor (see p. 170) and a low-resistance coil, as shown in Figure 26.43. An aerial is used to pick up radio signals, and the signals are supplied to the tuning circuit. A germanium 'point-contact' diode is used to rectify the radio signal

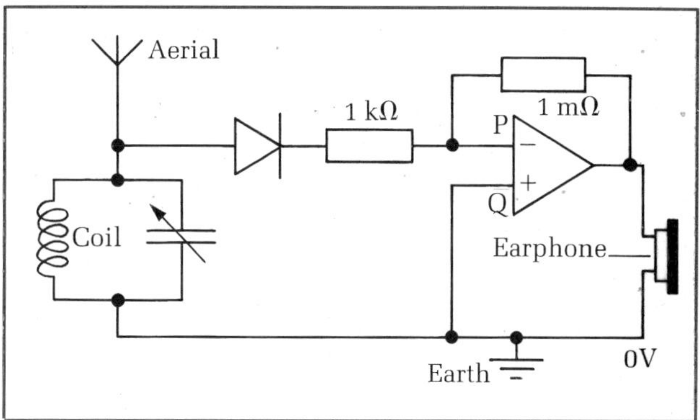

Figure 26.43 *A simple radio receiver*

selected by the tuning circuit. The rectified signal is therefore a variable d.c. signal which can then be amplified using the op-amp. Using an earpiece connected between the output terminal and earth, it ought to be possible to tune in to radio programmes.

Integrators

Op-amps are very versatile devices. Used with a feedback capacitor instead of a resistor, an op-amp can be used to integrate an input signal.

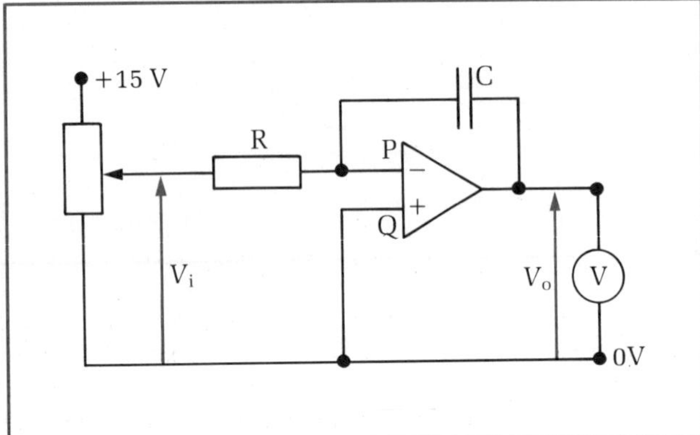

Figure 26.44 *An integrator*

Suppose the input voltage V_i is kept at a fixed value. If the inverting terminal P is set initially at 0 volts using a flying lead, then consider what happens when the lead is removed from P. The output voltage is initially zero.

1. Current passes through the input resistance to charge the capacitor up.
2. Assuming the output is not saturated, P is at virtual earth. So the p.d. across the input resistance remains steady. Hence the charging current is constant.
3. The capacitor charges at a steady rate, so its p.d. alters at a steady rate. Because P is at virtual earth, that side (side P) of the capacitor is at fixed voltage (≈ 0 V). So the other side of the capacitor, the output, must be at a voltage which changes at a steady rate.

In mathematical terms, assuming the input voltage is constant, then the current through R is given by $I = (V_i - V_P)/R$.
Since the voltage at P, V_P, is virtually 0, then $I = V_i/R$.
Hence the charging current to the capacitor, $I = V_i/R$.
The charge Q on the capacitor plates after time t is given by $Q = It$ so $Q = V_i t/R$.
Therefore the p.d. across the capacitor C is given by Q/C so $V_P - V_o = Q/C = V_i t/RC$. Since $V_P \approx 0$, then

$$V_o = -\frac{V_i t}{RC}$$

Suppose $R = 0.5\,\mathrm{M\Omega}$ and $C = 1\,\mu\mathrm{F}$. Then $RC = 0.5 \times 10^6 \times 1 \times 10^{-6} = 0.5$ s. Hence $V_o = -V_i t/0.5 = -2V_i t$. So every second, the output voltage changes by an amount equal to $-2V_i$.

Figure 26.45 shows the way the output voltage changes with time for a constant input voltage of -0.4 volts. In this case, $-2V_i = -2 \times -0.4 = 0.8$ volts. So the output voltage rises steadily at a rate of 0.8 volts per second, until saturation is reached.

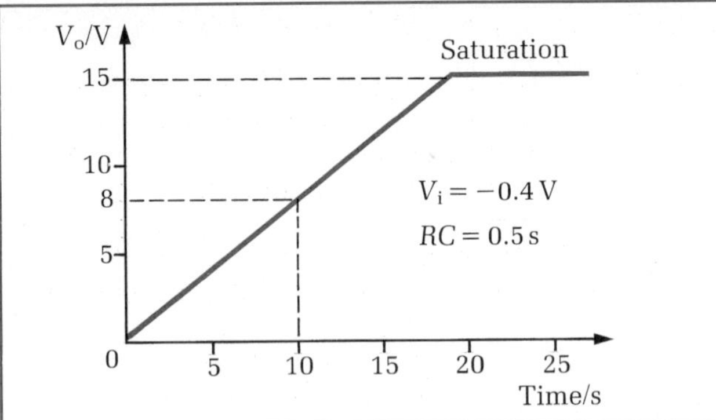

Figure 26.45 *Output of an integrator*

The circuit is called an **integrator** because the input signal is integrated mathematically to give the output signal. Integration in graphical terms represents the 'area under the curve'. Some examples from other branches of physics may be helpful here.

Vertical axis	Horizontal axis	Area under the curve
speed	time	distance travelled
acceleration	time	change of velocity
power	time	energy used
current	time	charge passed

The integrator can be used in any of the above examples. For example, if the input voltage represents the speed of an object, the output voltage represents the distance travelled. The link between input and output voltages still applies if the input voltage varies. If so, it is necessary to consider the changes of the output voltage in small intervals of time Δt. The above theory can be used provided t is replaced by Δt for a short interval of time.

Hence in time Δt the change of the output voltage is given by

$$\Delta V_o = -\frac{V_i}{RC}\Delta t$$

So the total change from $t = 0$ is given by adding the changes in each time interval. This is the mathematical process we call **integration**, written as

$$V_o = \int \Delta V_o = -\int \frac{V_i \Delta t}{RC} = -\frac{1}{RC}\int V_i \Delta t$$

Now we consider very small time intervals, so we write Δt as dt to give

$$V_o = -\frac{1}{RC}\int V_i\,dt$$

The term $\int V_i\,dt$ is the area under the V_i against t curve. So, whatever the variation of V_i with time t, the output voltage is given by

$$V_o = -\frac{1}{RC} \times \text{the area under the curve}$$

assuming $V_o = 0$ at $t = 0$.

Two examples of the use of integrators follow.

A ramp voltage generator can be achieved by supplying the integrator with a constant input voltage. As has been explained, the output voltage changes at a steady rate.

A 'staircase' ramp can be generated by supplying the integrator with a continuous train of pulses at its input. Assuming the pulses are of constant height, as in Figure 26.46, each pulse represents a constant input voltage for a fixed interval of time. So each pulse changes the output voltage by the same amount. Hence a continuous train of pulses makes the output voltage change in steps – like a staircase.

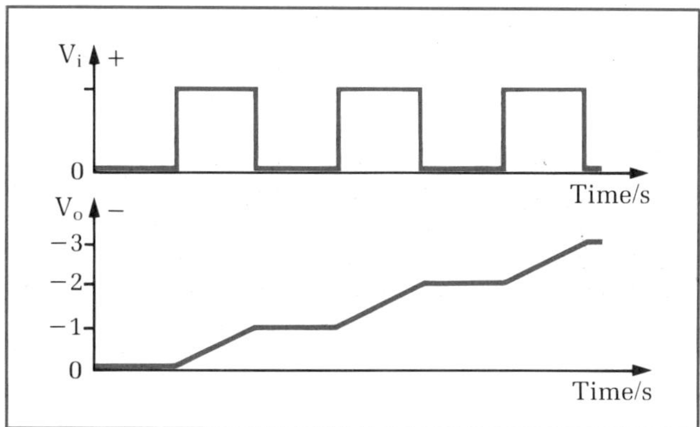

Figure 26.46 *A 'staircase' ramp*

A magnetic flux-meter uses a search coil connected to the input of the integrator. The coil is removed rapidly from the magnetic field being measured. The change of flux through the coil causes an induced voltage to be applied at the integrator input.

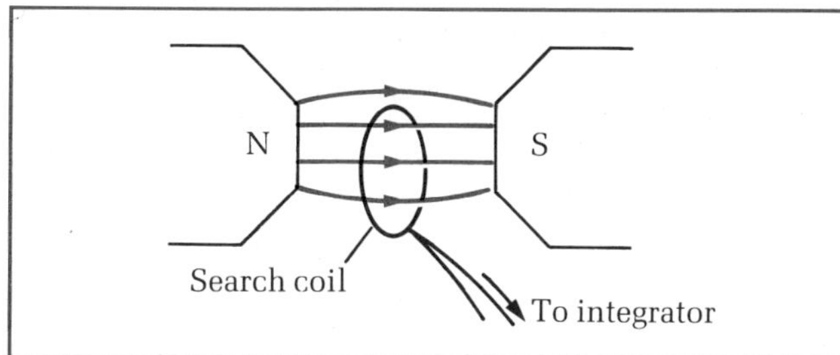

Figure 26.47 *Using an integrator*

Suppose the change of flux linkage due to removing the coil $= \Delta\Phi$. Hence the induced voltage is given by

$$V_i = \frac{\Delta\Phi}{\Delta t}$$

where Δt is the time taken.

So the area under the V_i against t curve $= \Delta\Phi$, the change of flux linkage. Therefore the output voltage is given by

$$V_0 = -\frac{1}{RC} \times \text{change of flux linkage}$$

assuming the output voltage is zero at the start.

The change of flux linkage $= -RCV_0$

Worked example A search coil of cross-sectional area 5.0 cm² and with 100 turns is placed between the poles of a strong horseshoe magnet, with the coil plane parallel to the pole faces. The coil is connected to an integrator with $R = 0.1\,\text{M}\Omega$ and $C = 0.1\,\mu\text{F}$. A voltmeter is connected across the output of the integrator. When the coil is rapidly removed from the field, the output voltage changes by 0.5 V. Calculate

a) the change of magnetic flux linkage through the coil when it is removed,

b) the magnetic field strength between the poles.

Solution

a) $RC = 0.1 \times 10^6 \times 0.1 \times 10^{-6} = 0.01$

Change of flux linkage $= RCV_0 = 0.01 \times 0.5 = 5 \times 10^{-3}$ Wb.

b) Magnetic flux linkage $= BAn$ where B = magnetic field strength, n = number of turns, A = area of cross-section of coil.

Since the final flux linkage through the coil = 0, initial flux linkage $= 5 \times 10^{-3}$ Wb.

Hence $B = \dfrac{5 \times 10^{-3}}{An} = \dfrac{5 \times 10^{-3}}{5 \times 10^{-4} \times 100} = 0.1\,\text{T}$.

The astable multivibrator

An op-amp can be used to generate square wave pulses continuously. The output switches repeatedly from positive saturation to negative saturation and *vice versa*. The frequency of the pulses depends on the values of the resistances and the capacitance in the circuit.

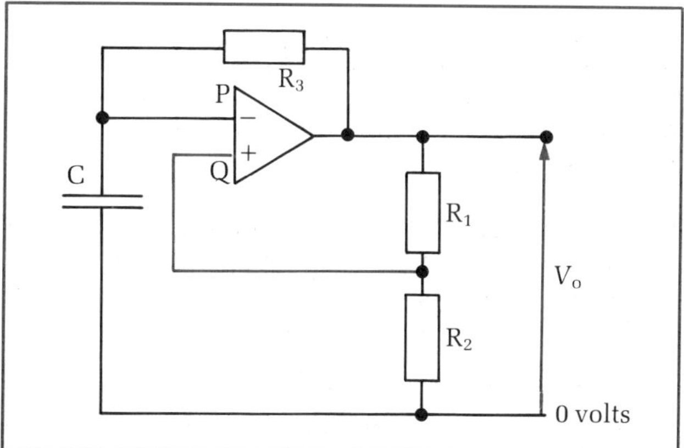

Figure 26.48 *The astable multivibrator*

To understand how the circuit (shown above) works, consider the sequence after the output voltage has just switched to positive saturation. If you prefer, suppose the output is held at the negative supply line using a flying lead; then the lead is used to connect the output briefly to the positive supply line.

1 At input Q, the voltage is a fixed fraction of the output voltage. Resistors R_1 and R_2 act as a potential divider. The 'top' of the potential divider is connected to the output; the 'bottom' is connected to the 0 volt line. Hence

$$V_Q = \frac{+V_sR_2}{(R_1 + R_2)}$$

where V_s is the saturation voltage.

So a positive voltage at input Q keeps the output at positive saturation, provided input P is less positive than Q.

2 At input P, the voltage rises as the capacitor charges up through R_3. With positive saturation at the output,

there is a constant p.d. across C and R_3 in series. So C charges up, and its p.d. gradually rises towards positive saturation.

However, when the voltage at input P rises above the voltage at input Q, the output switches to negative saturation. The op-amp is being used as a voltage comparator to compare the voltage at P with the voltage at Q. The output is at positive saturation for as long as it takes for the voltage at P to reach the fixed level at Q. Then the output voltage switches abruptly to negative saturation.

Now the potential divider supplies a fixed fraction of the negative voltage at the output to input Q. The voltage at P is still positive so the output is held at negative saturation. But now the capacitor discharges through R_3 and charges up with the opposite polarity. The flow of charge to C is now in the opposite direction. The voltage at P falls, becoming less positive then increasingly negative. When the voltage at P falls below the voltage at Q, the output voltage switches back to positive saturation, due to the comparator action of the op-amp. So we're back to the start for another cycle.

The output wave form is shown in Figure 26.49, which also shows the variation of the voltages at P and Q. The time period depends on the time taken for C to charge through resistor R_3 to the same voltage as supplied by the potential divider. Making R_3 or C larger increases the time period because the time constant of the CR_3 combination is increased. Also, reducing R_2 relative to R_1 reduces the time period because C does not have to charge up as much.

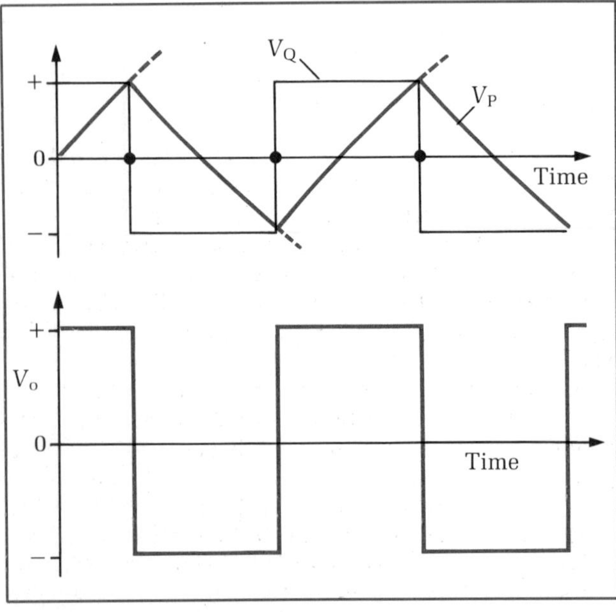

Figure 26.49 *Astable wave form*

26.4 Feedback and control

Feedback

Imagine you are at a live concert where an audio system is used to amplify the performance. The sounds from the performers are picked up by microphones, amplified by the audio system, then supplied to loudspeakers in the audience arena. The loudspeakers must be carefully sited, otherwise their sound is picked up by the microphones, fed back to the amplifier and the result is unpleasant screeching. The feedback is amplified and supplied to the loudspeakers so that the sound reaching the microphones is increased even more. Eventually the feedback increases until the amplifier becomes uncontrollable. The remedy is to position the loudspeakers so that their output does not reach the microphone. This link between the loudspeaker and the microphone is an example of feedback, where part of the output of a system is fed back to the input. Some other examples of feedback are discussed in the previous section.

Positive feedback takes place where the output is in phase with the input. The feedback reinforces the input, making the output even larger. The example discussed above is positive feedback. You can demonstrate positive feedback using the arrangement shown in Figure 26.50.

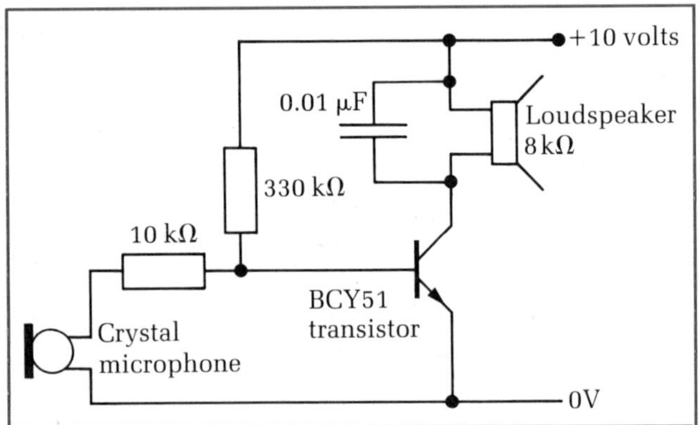

Figure 26.50 *Demonstrating feedback*

A moving coil loudspeaker is used as the load resistor of a transistor amplifier; the input is supplied using a microphone. Whistle into the microphone and the noise should be amplified by the circuit. A capacitor should now be connected in parallel with the loudspeaker. When the microphone is brought close to the loudspeaker, the loudspeaker emits a whistling sound without any help from you. Move the microphone too close and the sound becomes a screech. The loudspeaker sound is picked up by the microphone, amplified and returned to the microphone again via the loudspeaker. The loudspeaker and capacitor form a coil and capacitance in parallel. Charge oscillates from one plate of the capacitor, through the coil, to the other plate. So an alternating current passes round the parallel combination. The a.c. through the loudspeaker coil causes sound to be emitted from the loudspeaker; the microphone picks up the sound and reinforces the alternating current. So the sound emitted builds up to an audible whistle. The microphone must be connected so that its electrical signal, when reinforced, is in phase with the coil current.

Negative feedback takes place where the output from a system is out of phase by 180° with the input to the system, so the feedback cancels part of the input, making the output smaller. The microphone connected the 'wrong way' in Figure 26.50 gives negative feedback. We have already seen the use of negative feedback with op-amps in the previous section; the fraction of the output which is fed back reduces the output.

Control of any system requires feedback. A car driver approaching red traffic lights applies the car brakes to stop at the lights. The red traffic light is an input signal to the driver/car system; the driver 'feeds' a signal to the car to slow down by applying the brakes. If the car is approaching the lights too fast, the brakes must be applied harder. To control the output of the car (i.e. its speed), feedback must be applied from the driver.

Another example of the link between feedback and control is automatic speed control of an electric motor. Suppose we wish to maintain constant speed. This would be possible manually at low speed; if the motor speed alters from the required value, the control rheostat is readjusted so that it regains the required speed. The feedback is through observation (i.e. sensing) and adjusting the rheostat. If the motor is to run for hours and hours, it is obviously much more convenient to devise a means of automatic control.

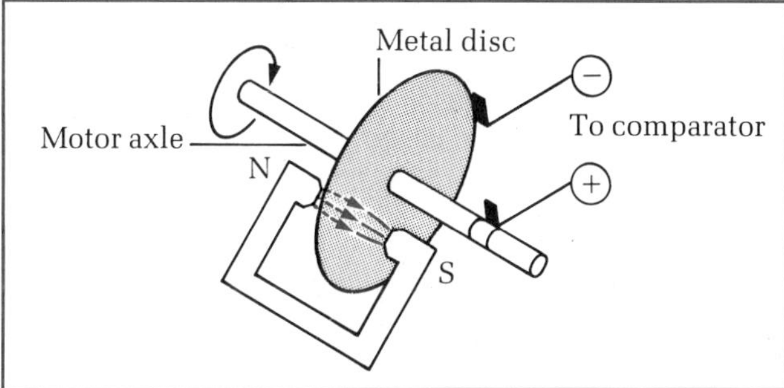

Figure 26.51 *Speed control*

One way to achieve automatic control is to 'sense' the motor speed by using the motor to drive a disc generator. A metal disc attached to the motor axis is arranged so that it cuts across the field lines of a horseshoe magnet. A voltage is induced between the disc centre and its rim (see p. 196 if necessary). The size of the induced voltage is proportional to the disc speed. The induced voltage is compared with a fixed voltage using a voltage comparator, as in Figure 26.51. A signal from the comparator is supplied to the motor control circuit to alter the motor speed if necessary. The comparator provides a feedback link to the input to keep the speed at the required level.

Transducers

A transducer is a device that converts energy from one form into another. Electrical transducers are designed to link circuits with non-electrical forms of energy.

1 **Input transducers** convert a non-electrical signal such as sound level or motor speed into an electrical signal.
2 **Output transducers** convert electrical energy into energy of some other form. For example, a lamp bulb is an electrical-to-light transducer.

If we want to measure or control a physical variable, such as temperature or light level, an input transducer must be used to give an electrical signal. Then the electrical signal can be compared with a control signal to give an output signal. The output signal can then be used to alter the physical variable using an output transducer.

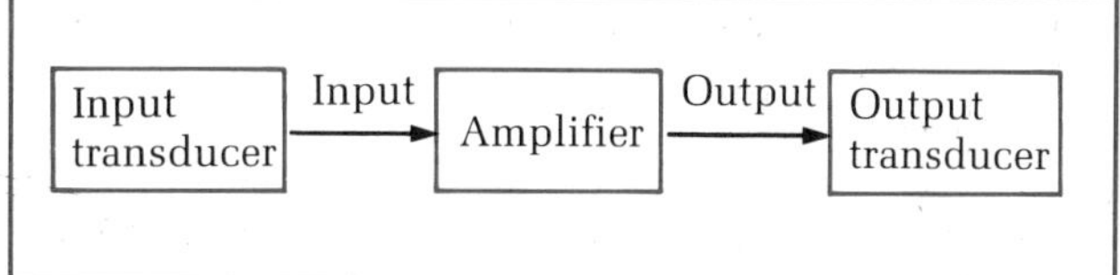

Figure 26.52 *Using transducers*

Resistance-to-voltage transducers are devices with resistance dependent on a physical variable. For example, a thermistor's resistance depends on its temperature. A light-dependent resistance (LDR) has resistance determined by the level of illumination falling on it. A strain gauge is another transducer; when it is flexed, its resistance changes because the network of wires in the strain gauge is stretched.

A Wheatstone bridge can be used to convert resistance changes into changes of voltage. The theory of the balanced bridge is treated in Chapter 23. The balance condition for the network shown in Figure 26.53 is $P/Q = R/S$. So the bridge is balanced if $R = P \times S/Q$.

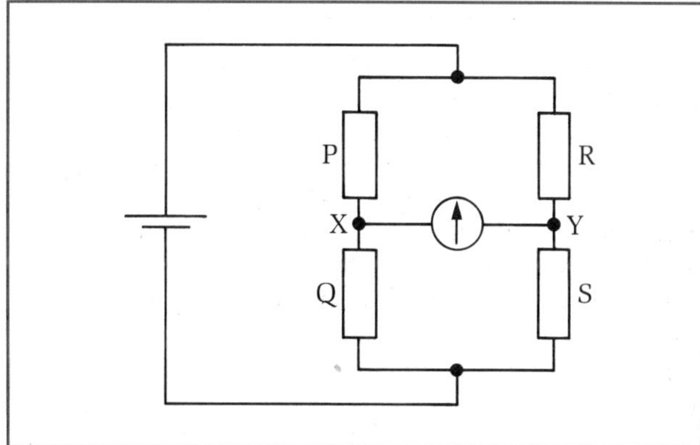

Figure 26.53 *A balanced bridge*

Suppose the value of R changes so that the balance condition no longer holds. For example, in Figure 26.53, increase of resistance R makes the voltage at X bigger than the voltage at Y. A centre-reading meter connected between X and Y could be used to monitor change of resistance R. If the current flows from X to Y, the resistance R must have increased from its balance value PS/Q.

An inverting amplifier as a comparator could be used in place of the centre-reading meter. The two inputs of the amplifier are connected to points X and Y of the bridge circuit. A voltmeter connected at the output of the amplifier could be used as an indicator of change of resistance, as is shown in Figure 26.54 on the next page.

1 Temperature can be measured by using a thermistor in place of resistor R in Figure 26.53. Suppose we want to use the measurement to control an electric heater in the same room as the thermistor. The output of the op-amp can be used to operate a relay, as shown in Figure 26.54. When current passes through the relay coil the relay switch closes and the heater is turned on.

   If the thermistor temperature falls below a certain value, the voltage at Y falls below the voltage at X, so the output of the op-amp goes high, which therefore causes current to flow into the base of the transistor. The transistor is switched on and current passes into

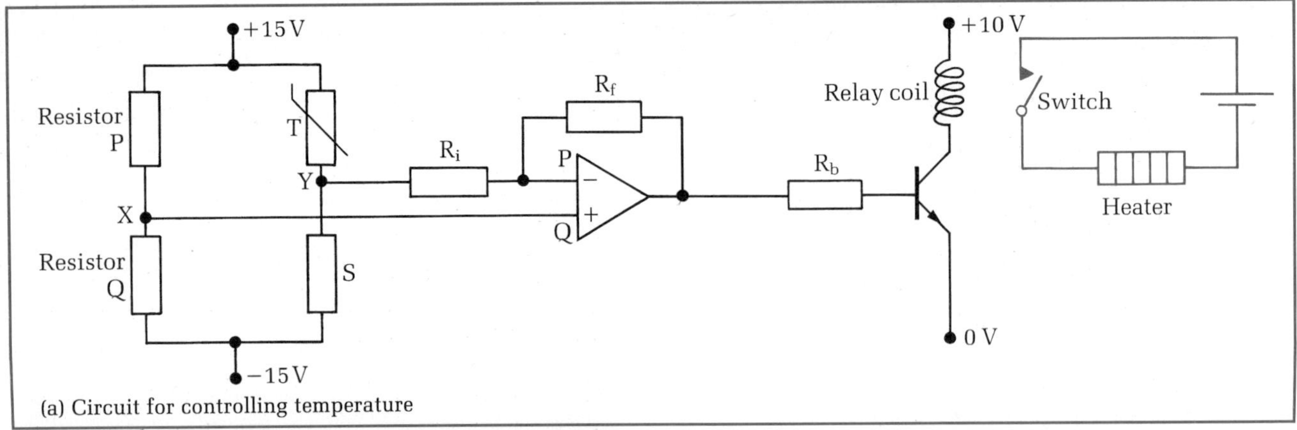

(a) Circuit for controlling temperature

(b) Heating and ventilation control panel

Figure 26.54 *Temperature control*

its collector through the relay coil. Hence the relay is energized and the heater is switched on.

2 Strain can be monitored by using a strain gauge in place of resistor R in the Wheatstone bridge of Figure 26.53. The strain gauge is bonded firmly to the surface of the component under test. When the component stretches or contracts under strain, the wires in the gauge alter in length, so their resistances alter. The voltmeter connected at the output of the op-amp is now a strain indicator.

3 Light levels can be controlled using an LDR in place of resistor R in the bridge network.

A microphone is an input transducer which converts sound energy into an electrical signal. There are several types of microphone.

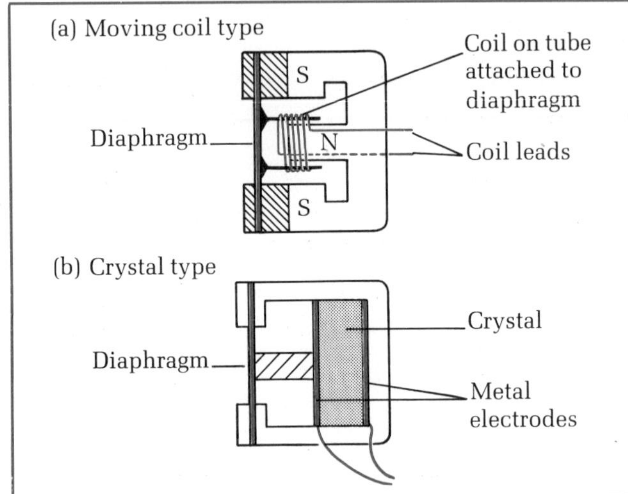

Figure 26.55 *Microphones*

The moving coil microphone consists of a coil attached to a diaphragm. Sound waves make the diaphragm vibrate, causing the coil to move in and out between the poles of a magnet. Hence an alternating voltage is induced in the coil. The impedance of a moving coil microphone is of the order of several hundred ohms.

The crystal microphone uses the **piezo electric effect** produced by certain types of crystals, such as quartz. When a piezo electric crystal is squeezed, a voltage is produced across opposite sides. The microphone diaphragm is designed to vibrate against the crystal. So an alternating voltage is produced by the crystal when sound waves fall on the diaphragm. The impedance of a crystal microphone is of the order of megohms.

A loudspeaker converts electrical energy into sound energy. Figure 26.56 shows the construction of a moving coil loudspeaker. When alternating current passes through the coil, the coil is forced to move in and out of the magnetic field, so the diaphragm attached to the coil is made to vibrate in and out, thus creating sound waves. The impedance of a loudspeaker increases with frequency but it is usually not more than about 100 ohms. Headphones used for personal stereo sets use the moving coil principle.

Matching microphones or loudspeakers to amplifiers is essential for effective performance. Impedance is the key property here; the impedance of a device is a measure

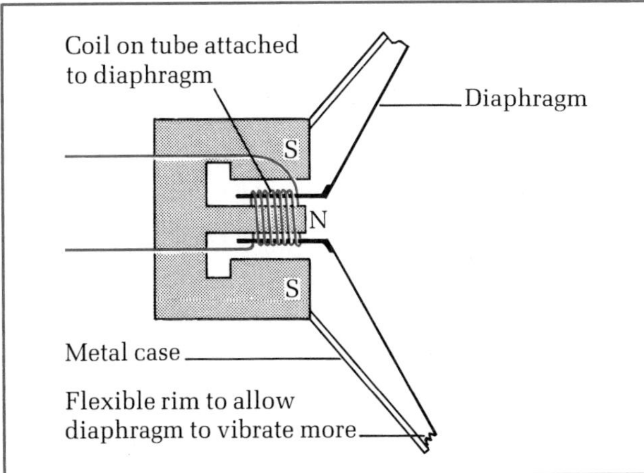

Figure 26.56 *Loudspeaker*

of the opposition of the device to passing alternating current. See p. 354 for the definition of impedance.

For microphones the input impedance of the amplifier must be much greater than that of the microphone. If not, a significant fraction of the microphone voltage is lost; the idea is similar to a cell with internal resistance. When the cell is connected to a load resistor, as in Figure 26.57, current passes through the cell and load resistor, so there is a drop of p.d. (i.e. a 'lost' p.d.) across the internal resistance of the cell. (This represents electrical energy converted into heat energy inside the cell.) Hence the p.d. across the terminals of the load resistor is less than the cell e.m.f.

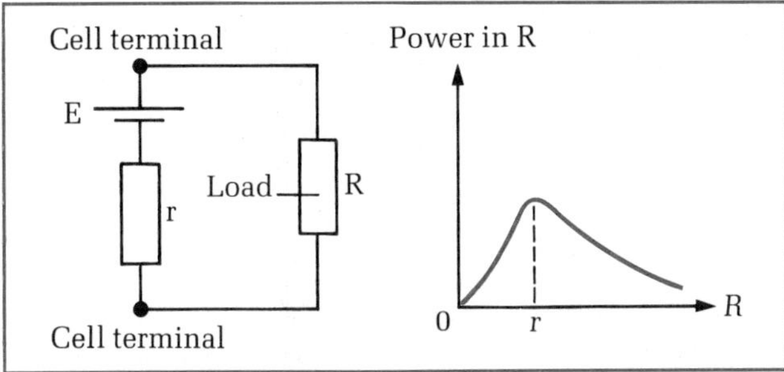

Figure 26.57 *Power*

The same idea applies to the microphone connected to an amplifier. The microphone is a source of e.m.f. and it has internal resistance. The amplifier is like a load resistor. The difference is that alternating current is involved here so the term 'impedance' is used instead of resistance. To keep the lost p.d. as small as possible, the microphone's impedance (i.e. its internal resistance) must be as low as possible. In practice, that means its impedance must be much less than the input impedance of the amplifier. In this way, maximum transfer of the microphone voltage to the amplifier is achieved.

For loudspeakers the maximum transfer of **power** from the amplifier to the loudspeaker is required. In practice, this is achieved by using a loudspeaker with impedance the same as the output impedance of the amplifier. So the loudspeaker must be **matched** to the amplifier. Figure 26.58 shows the idea.

To understand why the impedances must be matched, consider the equivalent d.c. circuit where a cell with inter-

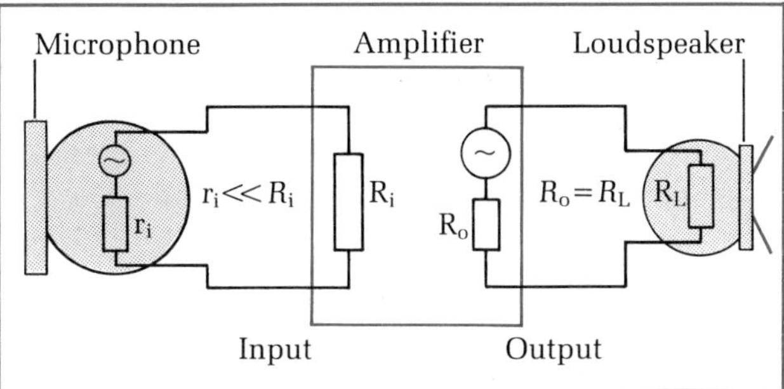

Figure 26.58 *Matching*

nal resistance delivers power to a load resistor. The d.c. circuit of Figure 26.57 will do. The current is given by

$$I = \frac{E}{R + r}$$

where E is the cell e.m.f.,
r is its internal resistance,
R is the load resistance.

Hence the power delivered to the load resistor is given by

$$P = \text{current} \times \text{p.d.} = I^2R = \frac{E^2R}{(R+r)^2}$$

Suppose the resistance of resistor R is varied. Figure 26.58 shows how the power delivered to R varies with R. The power reaches a peak when $R = r$. In other words, the cell delivers maximum power to the load resistor when the load resistor is matched (i.e. equal) to the cell's internal resistance.

The loudspeaker is the load 'resistor', and the amplifier is the power source like the cell. So the loudspeaker's impedance must equal the output impedance of the amplifier for maximum power transfer.

Relays are used in control circuits to switch devices on or off. A relay is a set of current-operated switches. When current passes through the relay coil, its iron core is magnetized. As a result, the soft-iron armature is pulled onto the core; the movement of the armature makes the switches operate. Figure 26.59 shows a relay in which one set of switches closes and a different set opens when the relay is energized. When the current through the coil is switched off, the iron core loses its magnetism so the relay switches revert to their previous positions. The relay can be used to switch any electrical device on or off, provided the switches can pass the current taken by the device without overheating.

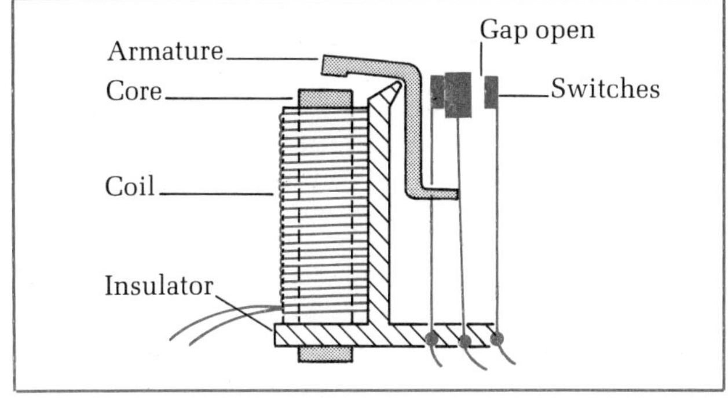

Figure 26.59 *A relay*

Using transducers

A light-operated buzzer The Wheatstone bridge contains an LDR, and is connected to an inverting amplifier, similar to the one in Figure 26.61. The output of the op-amp operates a transistor with the buzzer as its load resistor. Instead of the buzzer, a relay could be used to switch an electric bell on or off. When light falls on the LDR, the output from the op-amp goes high. Hence the transistor is switched on, making the buzzer operate. Another way of using transducers is shown in Figure 26.60.

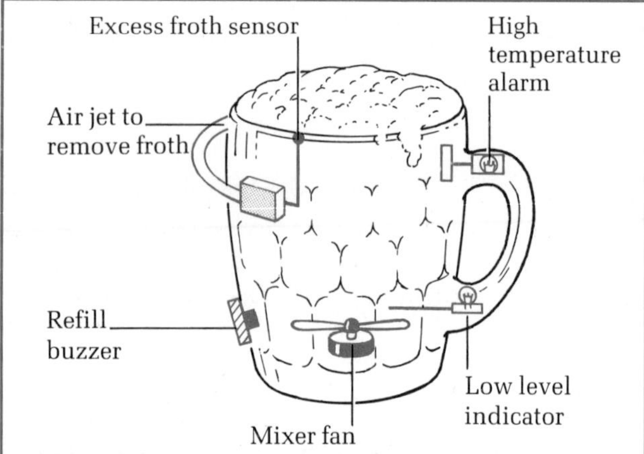

Figure 26.60

A temperature-operated cooling fan An automatic cooling fan fitted to a car helps to save fuel because the fan operates only when the engine temperature exceeds a safe value. A circuit for this purpose is shown in Figure 26.61. The op-amp is used as a comparator. The thermistor is attached to the engine block. When the engine gets too hot, the thermistor resistance falls so the voltage at X drops below the voltage at Y. Hence the output voltage from the op-amp goes high. The result is that the transistor switches on, switching the relay on to operate the cooling fan.

When the engine has cooled down, the voltage at X becomes greater than the voltage at Y so the transistor and relay switch off. Hence the fan switches off.

One problem here is that the temperature for switching on is the same as for switching off. It would be useful if the fan kept on working until the engine temperature was significantly lower than the 'switch on' temperature. Can you think of a modification to achieve this? One way would be to feed a fraction of the output back to the non-inverting input. To switch the fan off, the voltage at X must rise above the 'switch-on' voltage at X by a definite amount.

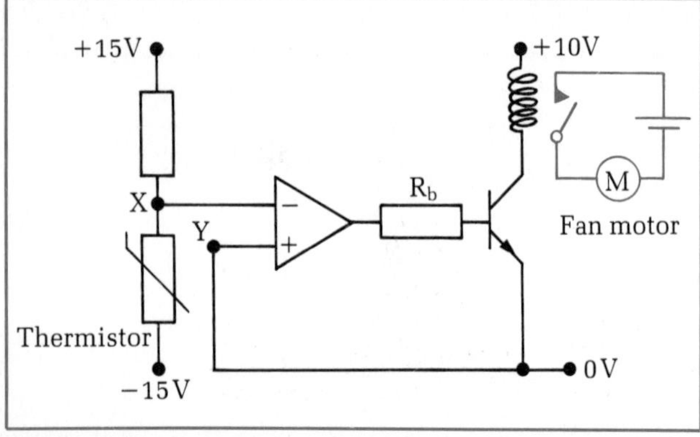

Figure 26.61 *A temperature-operated motor*

26.5 Microcomputers in control

So far we have looked at examples where the system in each case is designed for one purpose only. Each system is dedicated for one particular purpose only. Another example of a dedicated system is a temperature-operated alarm, useful for fire prevention. The temperature-operated alarm is a 'watchman', but unlike a real watchman, the alarm circuit has one task only. A real watchman could do other tasks as well, such as checking doors or windows or making a cup of tea. A microcomputer can be used like a real watchman since it can be programmed to check repeatedly on more than one input signal. More important, it can be programmed to operate one or more devices according to the state of the inputs.

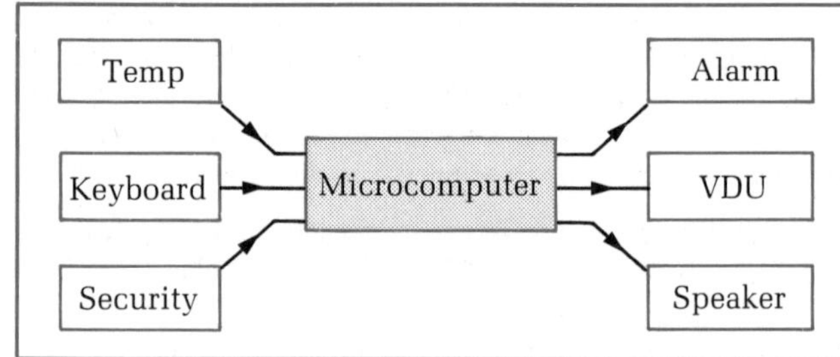

Figure 26.62 *Microcomputers in control*

To use a microcomputer for measuring, monitoring or controlling, it isn't essential to know what goes on inside. A microcomputer is a machine programmed by the user to follow a set of instructions. Such instructions might involve reading input signals or displaying an input value on its screen. The microcomputer can be instructed to respond to a given set of input signals by outputting a signal to operate a particular device.

Each type of microcomputer has its own particular code or language which must be used when keying in instructions. BASIC is one of the commonest languages but there are subtle variations from one type of micro to another. Run a program for a BBC micro on a PET or an APPLE and it won't get very far. The emphasis here is on the general ideas for using micros in physics and engineering; you will need to consult the user manual for your particular micro to put these ideas into practice. One or two examples are given, but only to illustrate the general ideas which are always needed, no matter how powerful your micro may be.

Interfacing for measurements

Information supplied to a micro must be in binary form. Instructions and information are processed by digital electronic circuits in the micro so signals supplied to a micro must be in bits (i.e. '0's and '1's). Micros can be used to measure physical variables such as temperature or pressure, but several stages are needed to convert the information into binary form.

A transducer must be used to convert the physical variable into a voltage signal. The voltage signal produced by a transducer can take any value between the limits set by the power supply to the transducer. The terms **analogue** or **linear** are used to describe circuits where voltages can take any value.

An analogue-to-digital (A/D) converter is then used to convert the voltage from the transducer into a digital signal in the form of a set of 0s and 1s. A simple A/D converter has one input line and several output lines. The size of the input voltage determines the state of the output lines. Figure 26.63 shows an 8-bit A/D converter; if the input voltage is 0, all eight output lines are at logic state 0. At maximum input voltage, all eight output lines are at logic state 1. Since 1111 1111 is the binary number for 255, the input range is divided into 256 intervals. Whichever interval the input voltage is in at a given time determines the output state. For example, if the input voltage is $0.2 \times$ the maximum, the output state is the binary number corresponding to 0.2×255 (i.e. 0011 001).

The way in which an A/D converter works is not important here. If you have understood how binary counters operate and what op-amps do, then you will be able to follow the description and circuit for an A/D converter given in reference manuals.

An interface is necessary between the A/D converter and the microcomputer. Then the digital signal from the converter can be fed into the microcomputer circuits when required. The interface terminals of the microcomputer are called **ports**, and can be used to transfer information to or from the micro computer as required.

The BBC microcomputer has a ready to use 4-channel A/D converter connected to one of its ports. The terminals of the converter are on the back of the micro, labelled 'analogue in', as shown in Figure 26.64. There are four separate input channels, selected by the ADVAL command, so four different physical variables can be measured using this facility. For example, the microcomputer could be used to measure temperature, light level, magnetic field strength (using a Hall device as a transducer) and pressure (using a suitable transducer). The ADVAL command is used to select one of the channels and the input from that channel is then measured by the microcomputer.

In fact, this particular converter gives an output on 12 lines; so work out for yourself the number of intervals

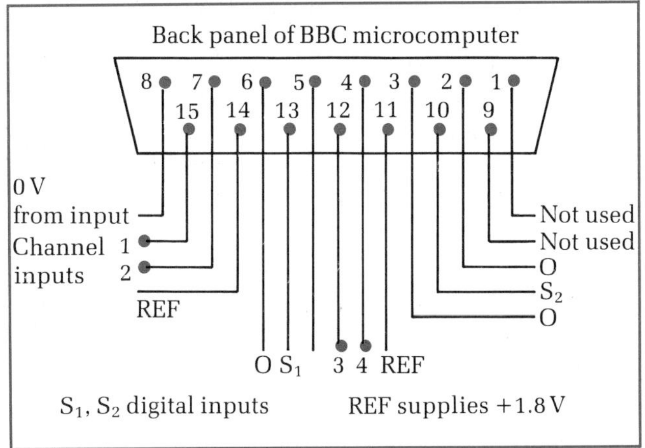

Figure 26.64 *BBC interface*

which the input range is divided into. The maximum input voltage on any channel is 1.8 V to send 1111 1111 1111 to the microcomputer circuits. If the microcomputer is instructed to read and then display an input voltage from one of the channels, it will do so to an accuracy of 1 part in 4095, because 4095 is equal to binary 1111 1111 1111.

A program of instructions must be supplied to the microcomputer before it can be used to read input data. The program must instruct the micro which port to use, and then whether to use the port as an input (as above) or as an output. Then it must instruct the microcomputer to read the data. The ADVAL command for the BBC micro supplies all these instructions in one word, but other commands are available for using other ports if required.

The program could also carry the instruction to display the data on the VDU (visual display unit). Or it could carry instructions for calculations, or for operating devices connected to other ports. Here then is the key advantage of using a microcomputer rather than a set of dedicated devices. The transducers, converters and the micro itself are parts of the **hardware** of the system; but its flexibility for use lies in the **software**, which is the program of instructions written by the user.

Consider, as an example, using the BBC micro to determine the changes of resistance of a lamp filament after it is switched on. Figure 26.65 (at the top of the next page) shows a suitable circuit using a 1.5 V, 0.3 A lamp bulb. The lamp resistance when in use is therefore $1.5/0.3 = 5\,\Omega$. The lamp is connected in series with a $1\,\Omega$ resistor and a switch. The circuit is connected to the micro's A/D converter as shown. The voltage at Y is monitored using channel 2, so channel 2 reads the voltage across the $1\,\Omega$ resistor. This is equal to the current because $V = IR$ and

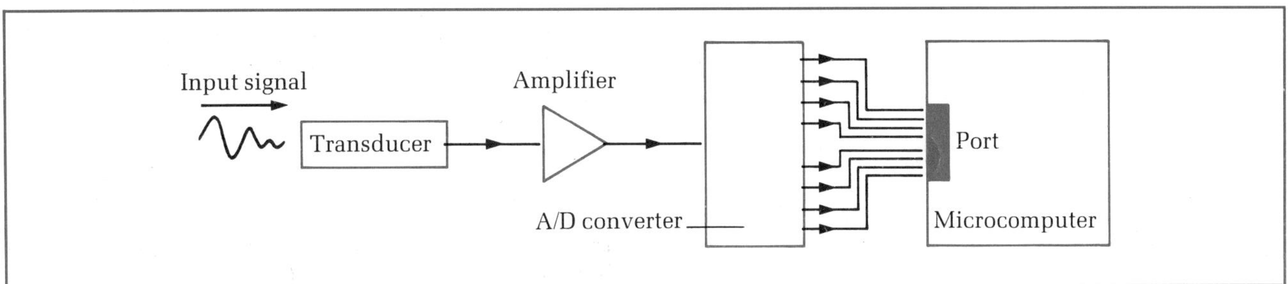

Figure 26.63 *Interfacing input transducers*

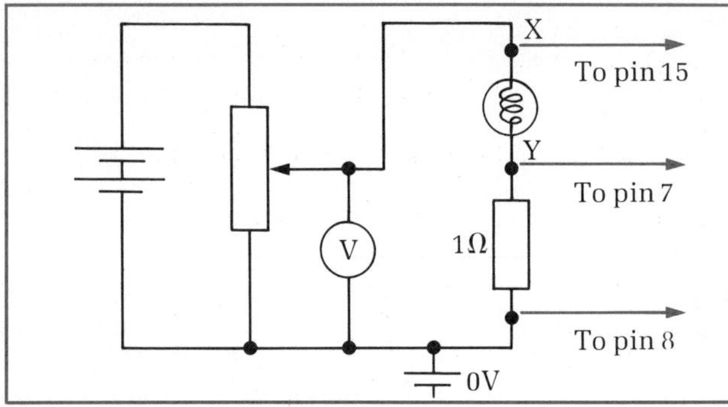

Figure 26.65 *Using a micro*

$R = 1\ \Omega$. Channel 1 monitors the voltage at X which is across the lamp and the resistor.

With the switch open, the potential divider is adjusted until the voltmeter reading is exactly 1.8 V. When the switch is closed we want the voltage levels at X and Y to be measured at rapid intervals, and the program below is designed to do this. One technical point before looking at a suitable program – readings using ADVAL are in 4096 intervals from 0 to 65 520 in steps of 16. So an input of 1.8 V gives a reading on the VDU of 65 520.

Computer fans could no doubt improve on the program given at the bottom of the page and include a graphical display of resistance against time.

Interfacing for control

Microcomputers can be programmed to produce signals as and when required. Used for control, a microcomputer would gather information from input sensors. The microcomputer could then produce output signals to operate devices to control the situation. Imagine a microcomputer used to control the heating system of a space-age house. Input sensors could monitor solar heating panels and room temperatures. If too cold, the microcomputer could switch on back-up electric heating or draw the curtains closed at night. If too hot, the heating panels could be shut off and air-conditioning switched on.

Output signals from a micro are in digital form, supplied by the micro to one of its ports. An output port with eight lines can be supplied with any binary number

Figure 26.66 *Microcomputer control*

up to 1111 1111 (=decimal 255). The logic level of any line can be used to operate circuits that will switch indicator lamps or relays on or off. Alternatively, the signal represented by the binary number at the port can be converted to an analogue signal using a digital-to-analogue (D/A) converter.

1 Indicator lamps or relays should not be connected directly to the output lines. Each line is connected to a non-inverter buffer which has a high input resistance. If the line supplies too much current, it will damage the port so it must be connected only to high resistance circuits. The buffer can then be used to operate a lamp or a relay, as in Figure 26.67.

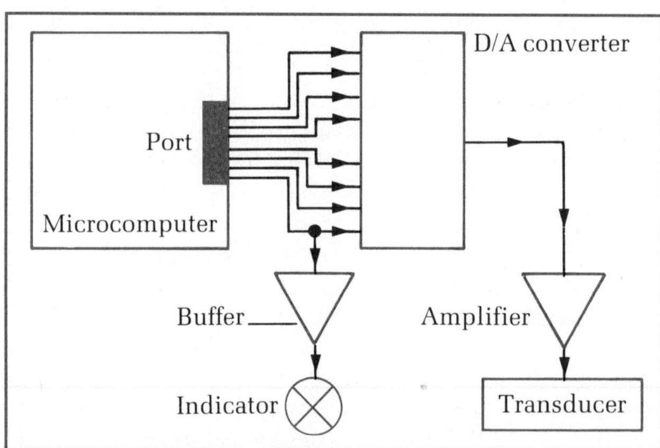

Figure 26.67 *Interfacing output transducers*

Program	**Comment**
`10 S = ADVAL(1)`	Read Channel 1 for the voltage at X
`20 IF S > 128 THEN 40`	If $V_x > 3.5$ mV ($= 128 \times 1.8/65\,520$) then jump to line 40
`30 GOTO 10`	Return to line 10
`40 T0 = TIME`	Read the micro's internal clock (in centiseconds)
`50 Y = ADVAL(2) : CURRENT = 1.8 * Y/65520`	Read the voltage at Y and calculate the current
`60 X = ADVAL(1) : VOLTAGE = 1.8*(X-Y)/65520`	Read the voltage at X, subtract the voltage at Y to give the lamp voltage
`70 T1 = TIME` `80 T = (T1 - T0) * 10`	Read the clock again and calculate the time from the start, (in milliseconds)
`90 R = VOLTAGE/CURRENT`	Calculate resistance R of the lamp
`100 PRINT"TIME in ms ="; T;`	Print the time from the start
`110 PRINT"RESISTANCE in ohms="; R`	Print the resistance of the lamp
`120 IF T > 1000 THEN 999`	If after 1 s from start, then end
`130 GOTO 50`	Repeat the readings, etc.
`999 END`	

2 D/A converters produce an analogue output from a digital input. For example, suppose a particular converter gives an output of 5.0 V for an input of 1111 1111 then the output is in steps of 5.0/255 = 19.5 mV. If the microcomputer presents the binary number 0110 1100 at the D/A input, the output will be equal to 108 × 19.5 = 2106 mV = 2.106 V.

The use of microcomputers for measurement and control is likely to increase as microcomputers become more and more sophisticated. But remember that the microcomputer is only as good as its user. If it is used with defective hardware or if the software is poor, then the microcomputer will not perform as it should. If, however, the user is fully aware of how it can be interfaced with hardware and can use software imaginatively the sky is the limit!

26.6 Summary

Definitions

1 *Current gain* $h_{FE} = \dfrac{\Delta I_c}{\Delta I_b}$ for a junction transistor

2 *Transconductance* $g = \dfrac{\Delta I}{\Delta V_{gs}}$ for an FET

3 *Matching* Maximum power is transferred from an amplifier to a loudspeaker when the impedance of the loudspeaker is equal to the output impedance of the amplifier.

Equations

Voltage gain for

a) a transistor amplifier $\dfrac{\Delta V_0}{\Delta V_i} = -\dfrac{h_{FE} R_L}{R_b}$

b) a FET amplifier $\dfrac{\Delta V_0}{\Delta V_i} = -gR_L$

Output voltage for an op-amp

a) on open-loop, $V_0 = A(V_Q - V_P)$

b) in an inverting amplifier, $V_0 = -\dfrac{R_F}{R_i} V_i$

c) in a non-inverting amplifier, $V_0 = \left(1 + \dfrac{R_F}{R_i}\right) V_i$

d) in an integrator, $V_0 = -\dfrac{1}{RC} \int V_i dt$

Short questions

26.1

a) What is the nature of the charge carriers in (i) n-type semiconductors, (ii) p-type semiconductors?

b) Explain the rectifying action of a p-n junction diode.

26.2 Copy the transistor circuit shown below, and on your copy mark the direction of the current passing through each terminal of the transistor. If the output voltage is at 5.0 V, calculate

a) the p.d. across the load resistor and hence the collector current,

b) the base-emitter current, assuming $V_{be} = 0.5$ V,

c) the current gain of the transistor.

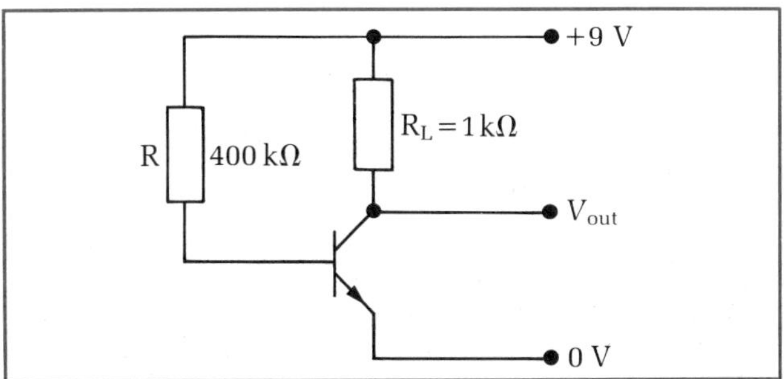

Figure 26.68

26.3 In a circuit like the one in question **26.2**, the resistor R is replaced by an LDR which has a 'dark' resistance of 800 kΩ. In daylight, the LDR's resistance falls to 100 kΩ. The transistor used has a current gain of 150. Calculate the collector current and the output voltage when the LDR is in

a) darkness,

b) daylight.

26.4 Consider the simple transistor amplifier shown below.

a) Explain why the bias resistor R_{bias} is essential.

b) The bias resistance is chosen so that the output voltage is 4.5 V for zero input voltage. Given the current gain of the transistor is 200 and the base-emitter p.d. is 0.5 V, calculate the resistance of R_{bias}.

c) Calculate the voltage gain if the base resistor is a 10 kΩ resistor.

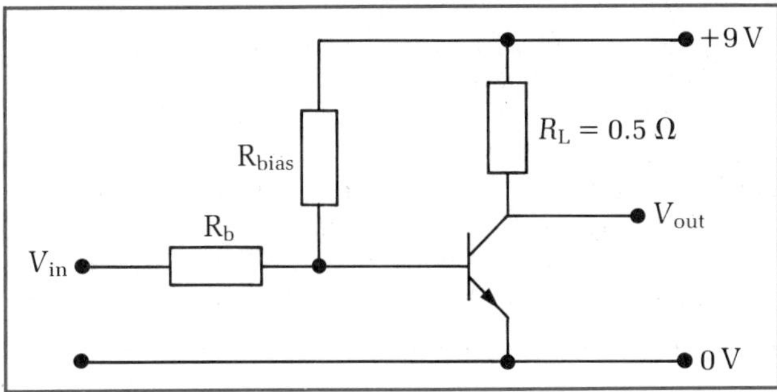

Figure 26.69

26.5 A transistor amplifier like the one in question **26.4** is designed with a voltage gain of −8 and its output is biased at 4.5 V (= ½ × supply p.d.).

a) Sketch a graph to show the variation of the output voltage with input voltage for the output voltage range from 0 to 9 V.

b) A 50 Hz sine wave voltage is applied across the input terminals of the amplifier. Sketch the input and output waveforms if the input voltage has a peak value of (i) 0.5 V, (ii) 1.0 V.

26.6 Give a circuit diagram to show the design of a temperature-operated relay using a transistor, a thermistor and any other components that you consider necessary. Assume the thermistor's resistance falls with increasing temperature, and that the relay is to operate an electric bell if the temperature of the thermistor becomes too high.

26.7 The circuit below shows an inverting amplifier with a feedback resistance of 2 MΩ and an input resistance of 0.5 MΩ. The op-amp is adjusted so that $V_{out} = 0$ when $V_{in} = 0$. Calculate the output voltage when
a) $V_{in} = +1$ V,
b) $V_{in} = -3$ V.

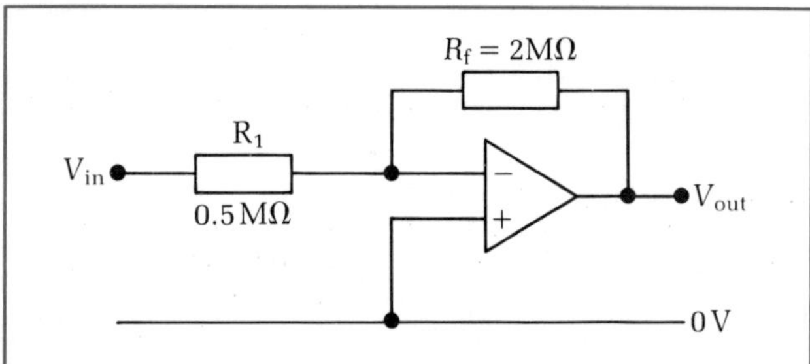

Figure 26.70

26.8 In the circuit of question **26.7**, a second 2 MΩ resistor is connected in parallel with the 2 MΩ feedback resistor. What would the output voltage be for V_{in} equal to
a) +1 V,
b) −3 V?

26.9 Suppose the feedback resistor of the circuit in question **26.7** is removed altogether. What would the output voltage now be for V_{in} equal to
a) +1 V,
b) −3 V?

26.10 If the resistor R_f of question **26.7** is replaced by a 2 μF capacitor, the circuit is made into an integrator. Suppose the output voltage is set at 0 V by a direct connection to the 0 V line; the input voltage is then set at +0.5 V.
a) Explain why the output voltage changes steadily when the output terminal is disconnected from the 0 V line.
b) Sketch a graph to show how the output voltage changes with time after the direct connection is removed.

26.11 A search coil is used with an integrator like the one in question **26.10** to measure the magnetic field strength of a horseshoe magnet. The integrator uses a capacitance of 0.1 μF and its input resistance is 100 kΩ. The search coil has 100 turns and its mean diameter is 34 mm. The coil terminals are connected to the integrator's input terminals, and the coil is then placed between the magnet poles with its plane parallel to the pole faces. When the coil is rapidly removed from the field, the integrator's output voltage changes by 1.50 V. Calculate the magnetic field strength between the poles of the horseshoe magnet.

26.12 Show how an op-amp can be used to make an astable multivibrator, and explain how the circuit works.

26.13 In the circuit below, an op-amp is used as a voltage comparator to operate a relay via a transistor.

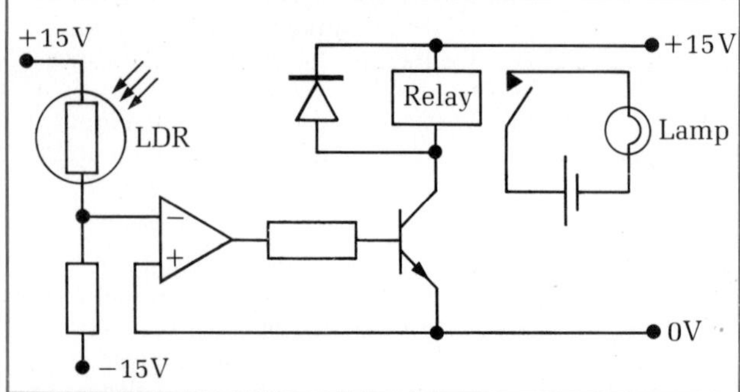

Figure 26.71

a) Explain why the transistor is necessary.
b) When the LDR is in darkness, the lamp operated by the relay is switched on; when the LDR is in daylight, the lamp is off. Explain how this works.
c) What is the diode for?

26.14 How would you modify the circuit of question **26.13** so that the relay switch is open or closed according to whether the voltage on a line from a microcomputer is at +5 V or 0 V respectively?

26.15 An analogue-to-digital converter has a single input line and eight output lines which are connected to a microcomputer as shown. The logic levels on the eight output lines represent the binary number corresponding to 100 × the input voltage.
a) If all the outputs are 0, the input voltage must be 0 V. If all the outputs are 1, what must the input voltage be?
b) The microcomputer is programmed to read the eight output lines and to calculate and display the input voltage. To what accuracy can the micro read the input voltage?
c) How would you use the system to measure
i) voltages up to 10 V,
ii) currents up to 5 A?

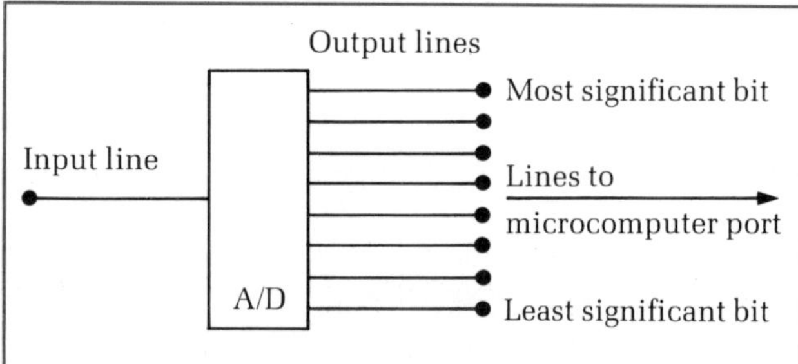

Figure 26.72

E

Multiple choice questions

EM.1 Which of the following statements explains why the resistance of a metal such as copper increases with temperature?
A Electrons begin to move more randomly at higher temperatures.
B Electrons are more likely to be captured by positive ions at higher temperatures.
C More positive ions are released at higher temperatures, thus counteracting the flow of electrons.
D Electrons move farther between collisions at higher temperatures.
E The positive ions impede the electrons more effectively at higher temperatures.
(London)

EM.2 If a metallic conductor, area of cross-section A, has n 'free' electrons per unit volume which each carry a charge e and move with drift velocity v
1 the current density is nev
2 the current flowing is evA
3 drift velocity increases with temperature
A 1, 2, 3 all correct **B** 1, 2 only correct **C** 2, 3 only correct
D 1 only correct **E** 3 only correct
(AEB June 82)

EM.3 A metal strip of uniform thickness has the shape shown in the diagram. The width of the narrow section is half the width of the wider section.
1 The current in the narrow section is less than the current in the wide section when current is passed along the strip.
2 When steady current is passed along the strip, the drift speed of the charge carriers in the narrow section is twice that in the wide section.
3 The resistance per mm length of the narrow section is twice that of the wide section.

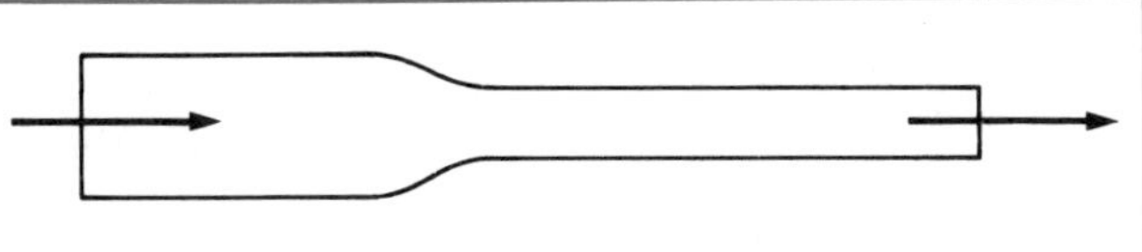

Figure EM1

A 1, 2, 3 all correct **B** 1, 2 only correct **C** 2, 3 only correct
D 1 only correct **E** 3 only correct

EM.4 A 3 μF capacitor is connected in series with a 6 μF capacitor, and a 6 V battery is connected across the combination. Which of the following statements is *not* correct?
A The combined capacitance is 2 μF.
B The total energy stored is 36 μJ.
C The p.d. across the 6 μF capacitor is 2 V.
D The p.d. across the 3 μF capacitor is twice the p.d. across the 6 μF capacitor.
E The 6 μF capacitor stores twice the charge that the 3 μF capacitor stores.

EM.5 A signal generator is connected across a series combination of a resistor and a capacitor as shown in the circuit diagram. The signal generator produces a square wave output as below. Which graph at the bottom of the page correctly shows the way the resistor p.d. varies with time?

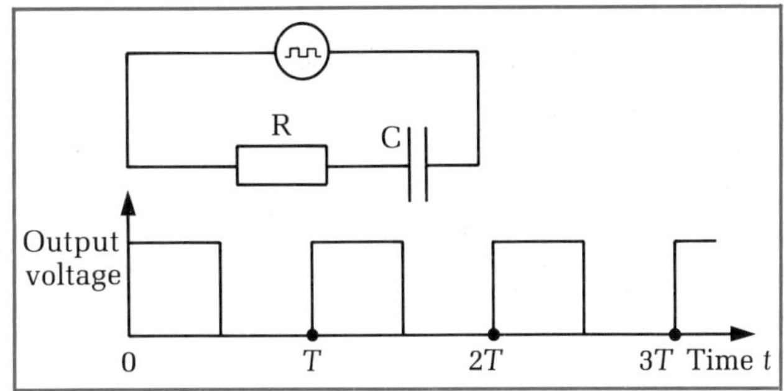

Figure EM2

EM.6 Which graph of question **EM.5** best shows how the power dissipated in the circuit resistor varies with time?

EM.7 A four-terminal box is connected as shown to a battery and two milliammeters. The currents in the two meters are identical.

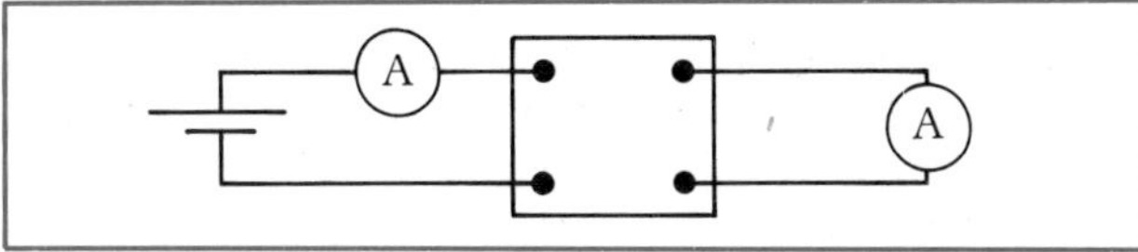

Figure EM3

Which of the circuits below, within the box, is the only one which will give this result?

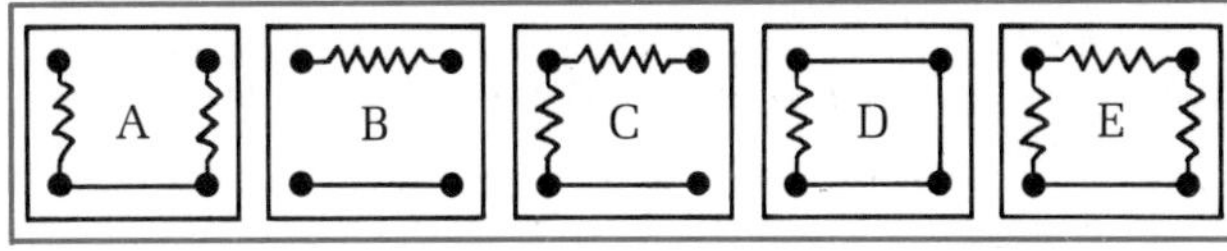

Figure EM4 *(Scottish H.)*

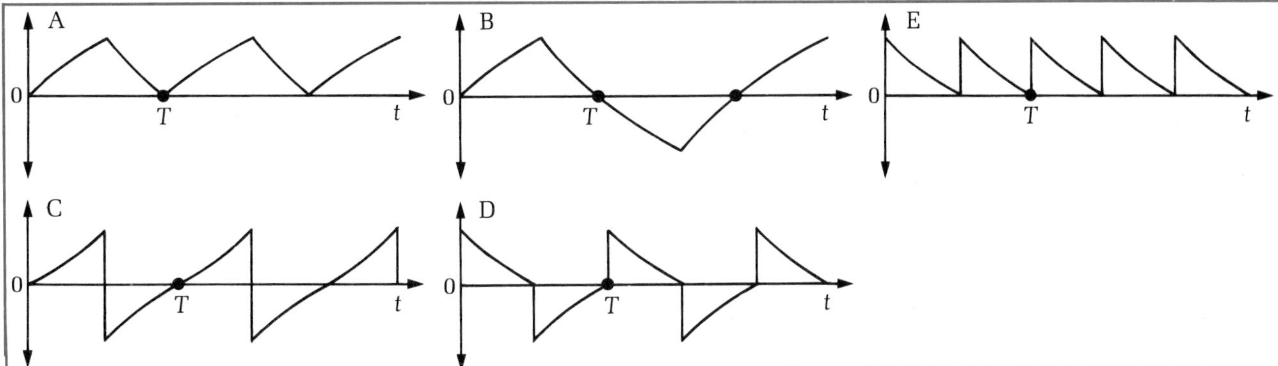

Figure EM5

EM.8 A digital voltmeter connected across the terminals of a certain battery gives a reading of 6.05 V. When a 10 ohm resistor is also connected across the battery terminals, the DVM reading falls to 5.85 V. The internal resistance of the battery, in ohms, must be

A 0.34 **B** 0.60 **C** 5.34 **D** 9.67 **E** 10.34

EM.9 The diagram shows a Wheatstone bridge circuit used to compare two resistances X and S, R_1 and R_2 being known resistances.

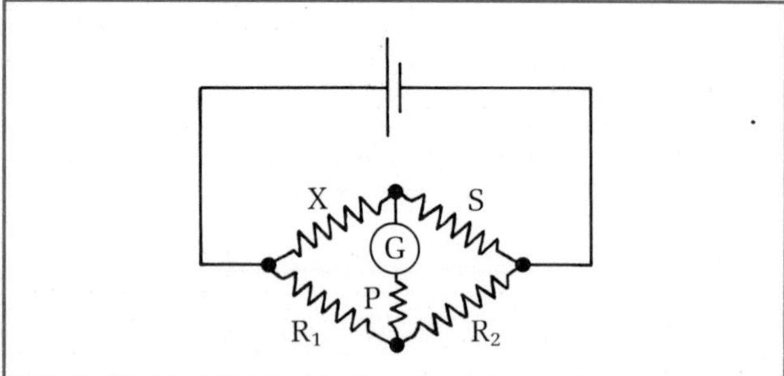

Figure EM6

When the bridge is balanced, which of the following statements is/are true?

1 The bridge becomes unbalanced if the e.m.f. of the cell varies.
2 The product of X and R_2 equals the product of S and R_1.
3 The resistor P is included to increase the sensitivity of the bridge.

A 2 only **B** 1 and 2 only **C** 1 only **D** 1 and 2 only
E 2 and 3 only

(Scottish H.)

EM.10 In the potentiometer circuit shown below, balance has been obtained.

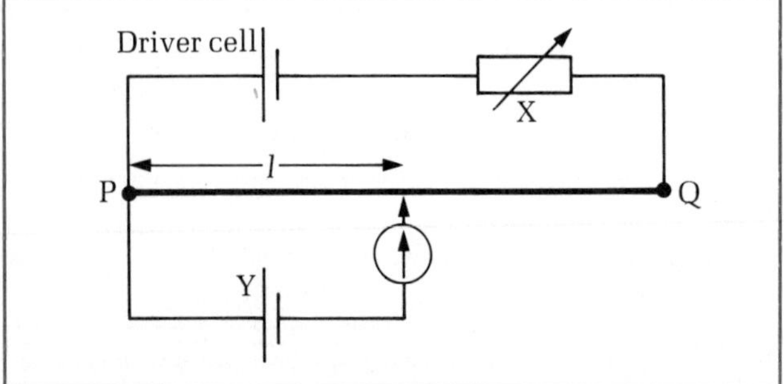

Figure EM7

1 The currents through the slide wire PQ and the cell Y are equal.
2 An increase in the internal resistance of cell Y would require an increase in l to restore balance.
3 An increase in the resistance of X would require an increase in l to restore balance.

A 1, 2, 3 all correct **B** 1, 2 only correct **C** 2, 3 only correct
D 1 only correct **E** 3 only correct

(London)

EM.11 With reference to the bridge rectifier circuit shown in the diagram

1 the current through the load resistor R flows from M to L only
2 at an instant at which P is positive with respect to Q the current flows along the path PABMLDAP
3 the output voltage across R may be smoothed by connecting a capacitor in series with R

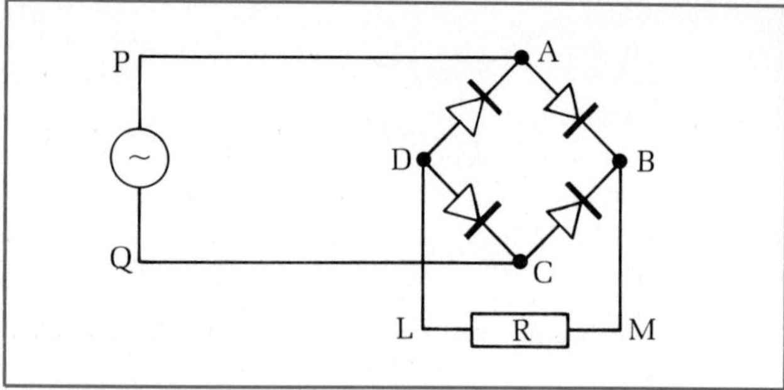

Figure EM8

A 1, 2, 3 all correct **B** 1, 2 only correct **C** 2, 3 only correct
D 1 only correct **E** 3 only correct

(A.E.B. June 82)

EM.12 The power dissipated in a resistor of resistance R supplied with alternating current of peak value I_0 is P. For a resistor 2R to dissipate power 4P when supplied with alternating current, the r.m.s. value must be

A $I_0/\sqrt{2}$ **B** I_0 **C** $I_0\sqrt{2}$ **D** $2I_0$ **E** $2\sqrt{2}I_0$

EM.13 The impedance of an inductor connected to an a.c. source

1 is equal to the r.m.s. potential difference across it divided by the r.m.s. current flowing through it
2 is greater than its d.c. resistance
3 decreases as the frequency of the source increases

A 1, 2, 3 all correct **B** 1, 2 only correct **C** 2, 3 only correct
D 1 only correct **E** 3 only correct

(AEB June 83)

EM.14 The output signal from an oscillator is set at amplitude 3.0 V and frequency 50 Hz. Which of the following pairs of oscilloscope settings gives the best opportunity to examine the shape of the oscillator signal using an ordinary bench oscilloscope with tube face of diameter 10 cm?

	A	B	C	D	E
Y gain/(V cm^{-1})	5	2	2	1	1
Time base/(ms cm^{-1})	1	1	10	1	10

(London)

EM.15 For the logic combination shown below, which alternative gives the correct logic output state Z for the given logic states at the inputs X and Y?

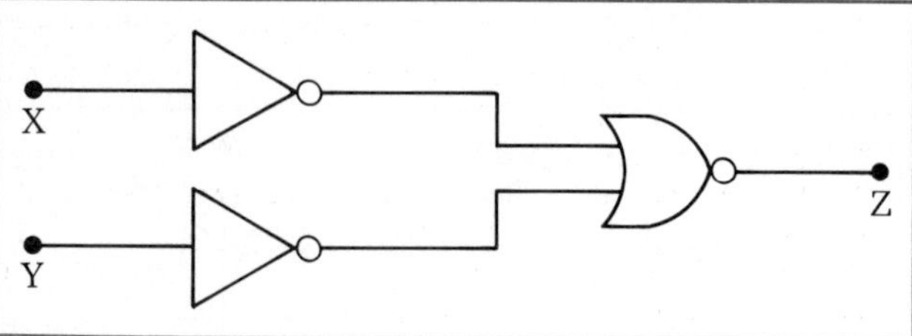

Figure EM9

	A	B	C	D	E
X	1	0	1	0	0
Y	0	1	1	1	0
Z	1	0	0	1	1

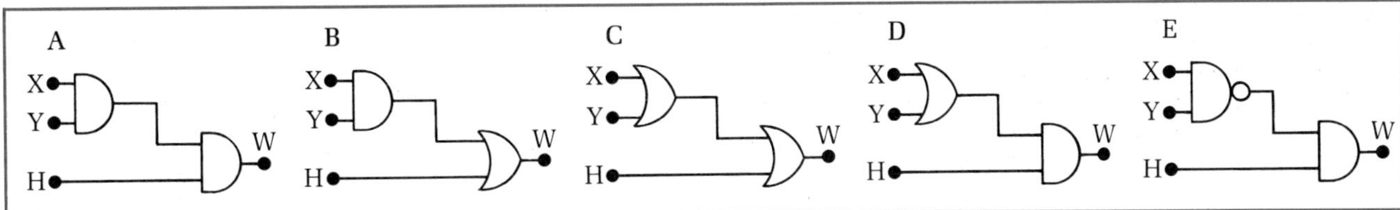

Figure EM10

EM.16 A warning light in a car is designed to be on if either of its two doors are open and the handbrake is off. The system is shown in block diagram form below.

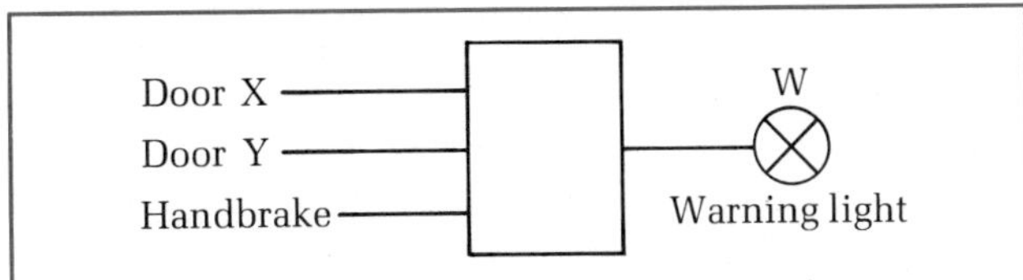

Figure EM11

Which logic combination from the top of the page is the only one suitable?
(Door open = 1;H/brake off = 1.)

EM.17 For the transistor circuit shown below, the output voltage is 5 V when resistor R is set at 200 kΩ. Neglecting the base-emitter voltage, which of the statements 1, 2 and 3 is (are) correct?

1 The collector current is 10 mA
2 The base current is 50 μA
3 The current gain h_{FE} is 200

A 1, 2, 3 all correct **B** 1, 2 only correct **C** 2, 3 only correct
D 1 only correct **E** 3 only correct

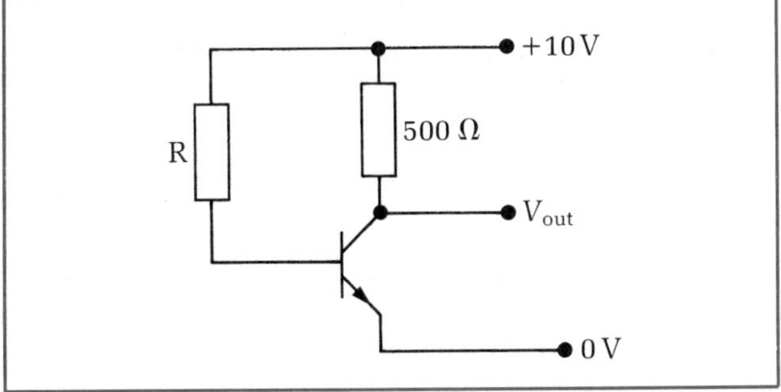

Figure EM12

EM.18 Resistor R in the circuit of question **EM.17** is replaced by a light dependent resistor with a dark resistance of 200 kΩ. When the LDR is illuminated, which of the following statements is (are) correct?

1 The collector current falls.
2 The output voltage falls.
3 The base current increases because the LDR resistance falls.

A 1, 2, 3 all correct **B** 1, 2 only correct **C** 2, 3 only correct
D 1 only correct **E** 3 only correct

EM.19 For the inverting amplifier shown in the circuit diagram in Figure EM.13, when the input voltage is +2.0 V, the output voltage is −8 V. When the input voltage is changed to +1.0 V, the output voltage changes to −4 V. Which one of the following statements is *not* correct?

A The voltage gain is −4.
B The voltage at P is 2 V when the input voltage is 2 V.
C The current through R_1 is 2 μA when the input voltage is 2 V.

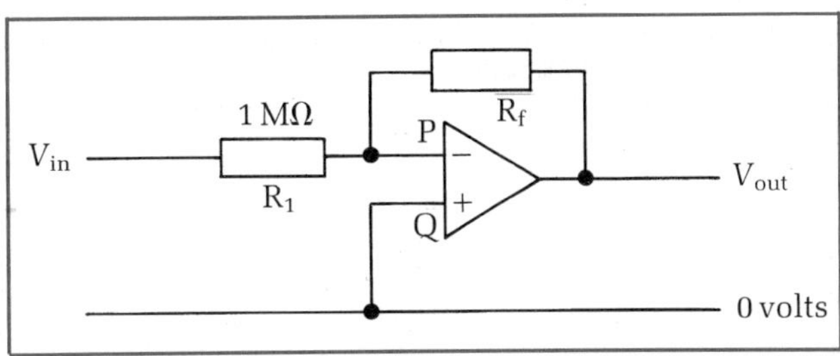

Figure EM13

D The resistance R_F is 4 MΩ.
E When the input voltage is −1.0 V the output voltage is +4.0 V.

EM.20 If the feedback resistance of question **EM.19** is changed to 2 MΩ, which alternative gives the correct values for the current I through the feedback resistor and the output voltage V_{out} when the input voltage is +2.0 V?

	A	B	C	D	E
I/μA	0.5	1	1	2	2
V_{out}/V	+4	−4	+4	−4	+4

E

Long questions

EL.1

a) i) Explain the origin of holes in intrinsic semiconducting materials. What is the process by which holes participate in current flow?

ii) Explain how the presence of the donor impurities in an *n*-type semiconducting material raises the number of free electrons per unit volume without increasing the number of mobile holes per unit volume.

b)

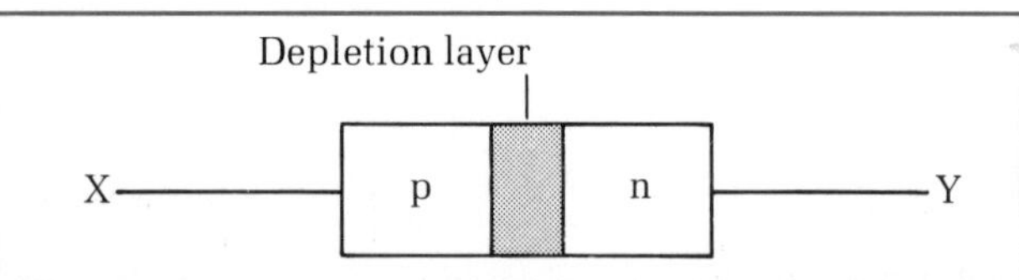

Figure EL1

The diagram shows two kinds of semiconducting material, p-type and n-type, in contact. What is the important characteristic which distinguishes the depletion layer from the rest of the assembly?
Explain the effect on the depletion layer of applying a *small* potential difference (about 0.1 V) across XY

i) if X becomes negative with respect to Y,
ii) if X becomes positive with respect to Y.

Hence explain the rectifying action of a p-n junction.

c)

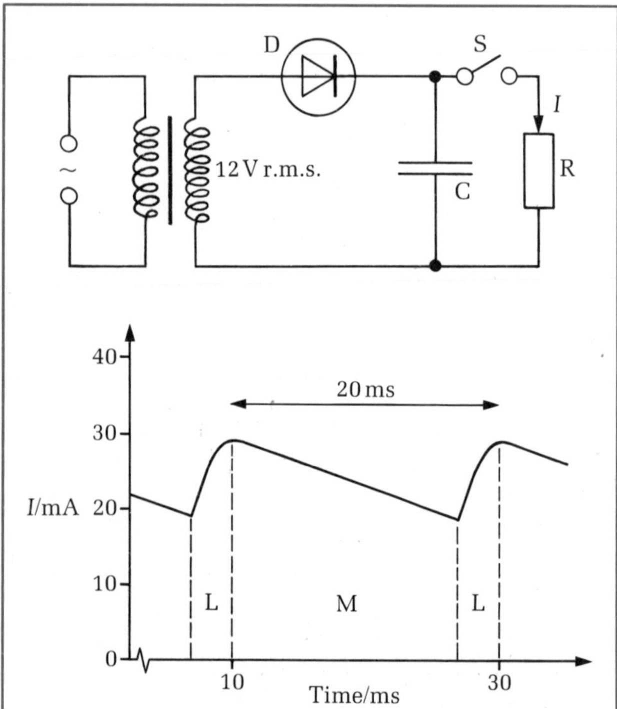

Figure EL2

The diagram above shows a half-wave rectifying circuit connected to a resistor R through a switch S. The graph shows how the current I in resistor R varies with time. Write down the source of the current in R

i) during the periods L,
ii) during the period M.

Calculate the maximum reverse bias potential difference the diode, D, must be capable of withstanding when the switch, S, is open.

(London)

EL.2 Deduce from first principles an expression for the rate at which heat is generated in a circuit of resistance R in which a current I passes. State clearly the basic definitions you employ.

A voltmeter is connected across a variable resistance R, which itself is in series with an ammeter and a battery. For one value of R, the readings are 1.00 V and 0.25 A. For another value of R, they are 0.90 V and 0.30 A. Calculate the e.m.f. of the battery, its internal resistance, and the two values of R. State any assumptions you make concerning the meters.

The value of R is now adjusted so that it equals the internal resistance of the battery. Calculate the rate at which heat is generated

a) in the battery,
b) in the whole system.

(Cambridge)

EL.3 State *Kirchhoff's laws*. Explain how each is based on a fundamental physical principle.

Use the laws to deduce values of the currents I_a, I_b, I_c and I_d as shown in the circuits below.

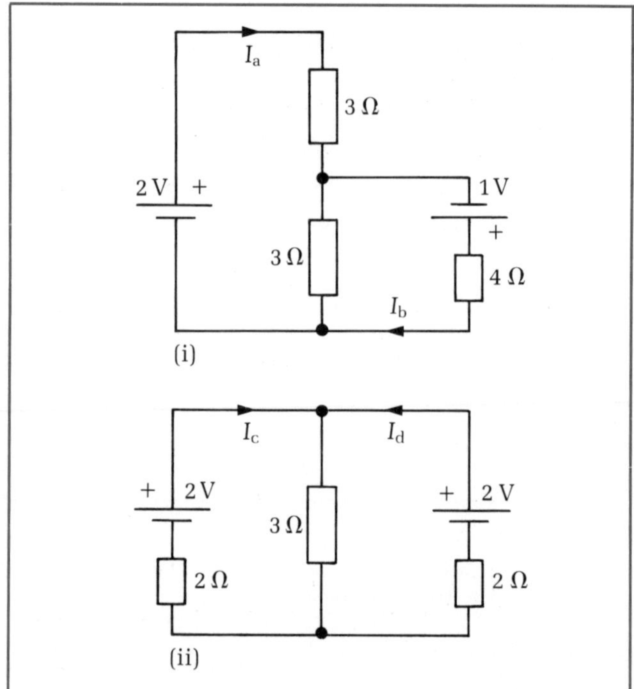

Figure EL3

A battery, switch and uniform wire of variable length l are connected in series. A bar magnet is suspended vertically above the wire by a fibre fixed to a torsion head fitted with a scale to measure angular rotation. The magnet is initially aligned along the wire. When the switch is closed, the magnet twists but it may be restored to its original position by rotation of the torsion head through a known angle φ. Values of φ for total lengths l of the wire are given.

$\varphi/°$	33.2	20.1	14.3	11.1
l/m	1.00	2.00	3.00	4.00

a) Explain why the magnet twists when current flows.

b) Explain why φ increases with increasing current.

c) Plot a graph of $1/\varphi$ against I and find the intercept on the I-axis.

d) Discuss the significance of this intercept with reference to the battery.

(Cambridge)

EL.4 Explain clearly the difference between e.m.f. and potential difference.

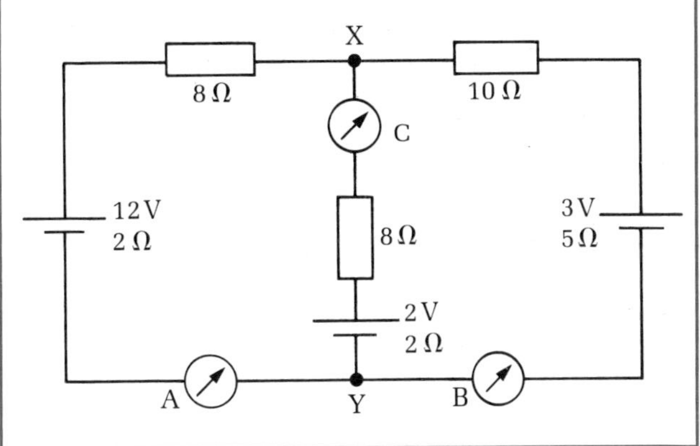

Figure EL4

Write down the Kirchhoff network laws and point out that each is essentially a statement of a conservation law.
Find, for the above circuit

i) the readings on the ammeters A, B and C (assumed to have effectively zero resistances),

ii) the potential difference between X and Y,

iii) the power dissipated as heat in the circuit,

iv) the power delivered by the 12 V cell.

Account carefully for the difference between (iii) and (iv).

(W.J.E.C.)

EL.5

a) Describe a method for measuring the relative permittivity of a material. Your account should include a labelled circuit diagram, brief details of the procedure and the method used to calculate the result.

b)

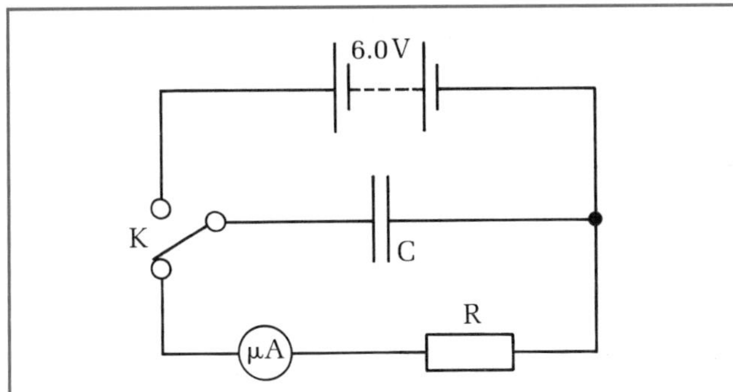

Figure EL5

In the circuit above, the capacitor C is first fully charged by using the two-way switch K. The capacitor C is then discharged through the resistor R. The graph at the top of the next column shows how the current in the resistor R changes with time. Use the graph to help you answer the following questions.

Calculate the resistance R. (The resistance of the microammeter can be neglected.) Find an approximate value for the charge on the capacitor plates at the beginning of the discharging process and hence calculate

i) the energy stored by the capacitor at the beginning of the discharging process, and

ii) the capacitance C.

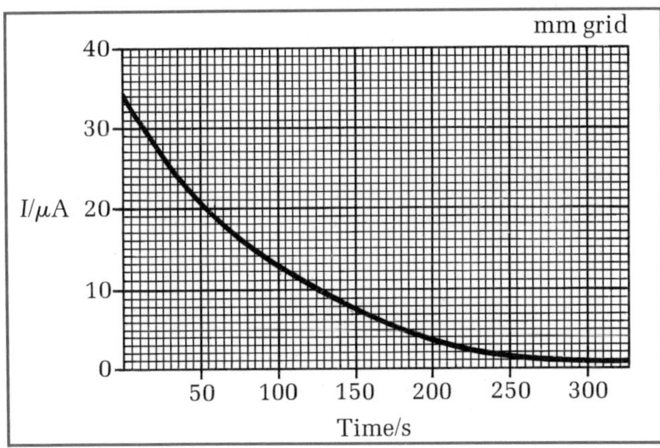

Figure EL6

(London)

EL.6 Explain what is meant by capacitance.

The plates of a parallel plate capacitor each have an area of 25 cm^2 and are separated by an air gap of 5 mm. The electric field intensity between the plates is 7×10^4 V m^{-1}. If one plate is at zero potential relative to earth, find the potential of the other plate and indicate on a sketch some equipotential surfaces in the gap. What is the potential half way between the plates? (Ignore end effects.)

What is the capacitance of the capacitor, and how much electrical energy is stored in it?

A slab of dielectric of relative permittivity 15 is introduced into the isolated capacitor so as to exactly fill the gap. What are the new values of

a) the potential difference,

b) the capacitance, and

c) the energy?

Give a labelled sketch of a practical form of variable capacitor and indicate *one* use.
[$\varepsilon_0 = 8.85 \times 10^{-12}$ F m^{-1}.]

(WJEC)

EL.7

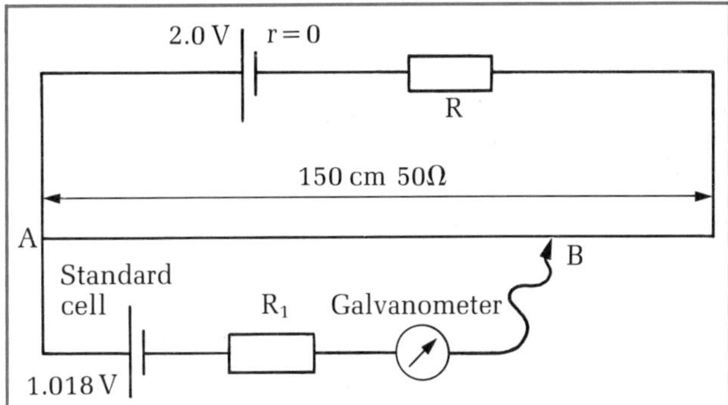

Figure EL7

A student wants to set up the above potentiometer so that at balance the potential drop along each 1 cm of wire is 0.01 V. Explain carefully how this can be achieved, and find the value of R.

Find the power provided by the driver cell, and the power

dissipation in the potentiometer wire. Point out any undesirable consequences of the latter and indicate how these can be minimised in practice.

Does the power provided by the driver cell depend on whether the potentiometer is balanced? Give reasons in support of your answer.

What is the purpose of the resistor R_1? Discuss whether its magnitude affects

a) the balance point, and

b) the precision with which the balance point can be located.

(WJEC)

EL.8

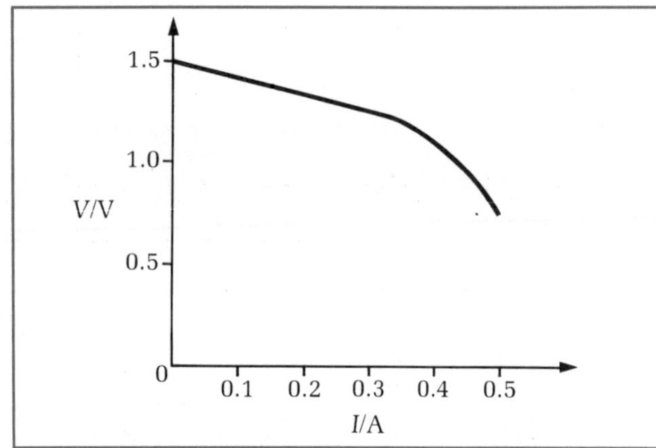

Figure EL8

a) The graph shows how the potential difference *V* between the terminals of a dry cell varies as the current *I* taken from the cell increases. Explain why the value of *V* decreases as *I* increases. To what does the value of *V* correspond when the current is zero? The first section of the graph is substantially straight and then the graph curves markedly. What does this imply?

b) Draw a diagram of a circuit, incorporating a slide-wire potentiometer to measure potential difference, which might have been used to take readings from which the graph in **a)** was plotted. Explain how the measurements would have been taken.

c) Explain why no current flows through the galvanometer when the balance point on a potentiometer wire has been found. What is the disadvantage of the high internal resistance associated with some types of standard cell used in potentiometer work?

d) A metre bridge has a resistor of resistance 7.30 Ω in the left-hand gap and the balance point is found to be 43.5 cm from the left-hand end of the wire. Due to an increase in temperature the resistance rises to 7.90 Ω. What is the position of the new balance point if the resistance of the resistor in the right-hand gap is unchanged?

(AEB Nov 83)

EL.9

a) Figure EL.9, at the top of the next column, shows two circuits intended for use in the determination of the resistance of R. Even if the values of current and voltage obtained are correct neither circuit gives an accurate value for the resistance. In each case state and explain whether the calculated resistance will be too large or too small.

b) Figure EL.10 shows the circuit diagram of the metre bridge form of the Wheatstone network.

i) Given that S is a standard resistor describe in detail how you would find the resistance of resistor R.

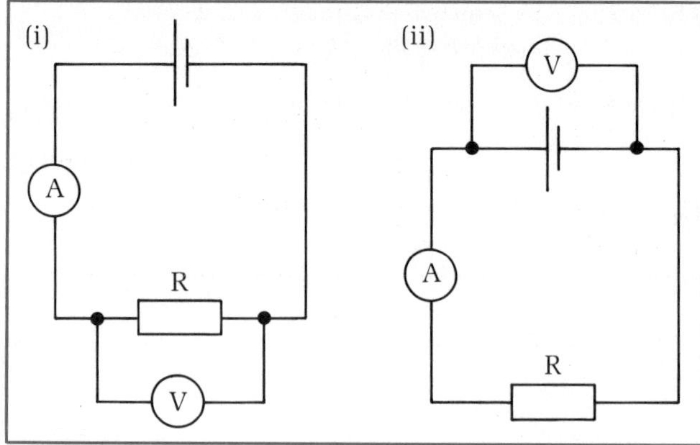

Figure EL9

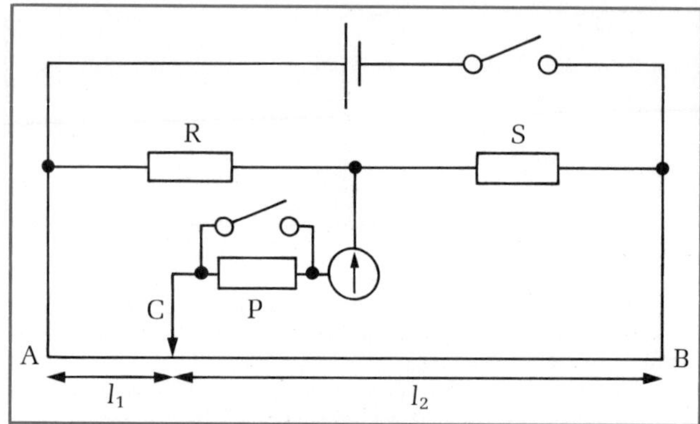

Figure EL10

ii) the value of the standard resistor is usually chosen so that the balance point C is near to the centre of the resistance wire. State and explain *two* advantages of such a choice.

iii) Explain why this method is unsuitable for determining values of resistance which are either very low or very high.

c) Shown below is the circuit diagram of another form of the Wheatstone network.
Suggest how it may be used for determing the resistance of the centre-zero galvanometer G, assuming that the values of *R*, *S* and *V* are known. Show how you would calculate the result explaining carefully why the circuit gives a valid result.

(AEB June 85)

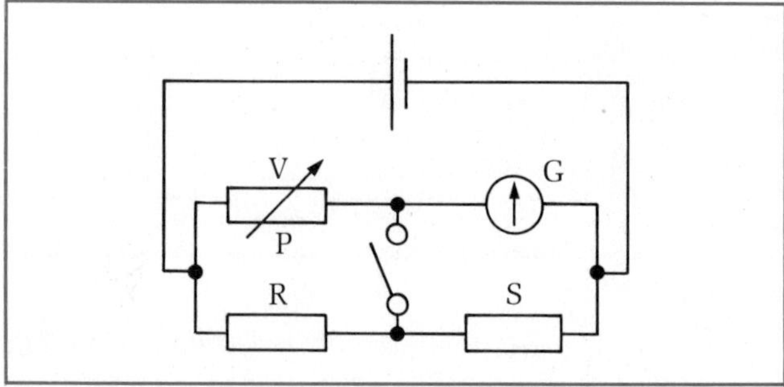

Figure EL11

EL.10

a) Explain what is meant by *electrical resistivity*, and show that its unit is Ωm.

b) 3.00 m of iron wire of uniform diameter 0.80 mm has a potential difference of 1.50 V across its ends.

i) Calculate the current in the wire.

ii) If E is the uniform electric field strength along the length of the wire, and J is the uniform current per unit cross-sectional area of the wire, calculate the magnitude of E and J. Calculate the ratio $\frac{E}{J}$ and comment on the result.
(Resistivity of iron $= 10.2 \times 10^{-8}\,\Omega$m.)

c)

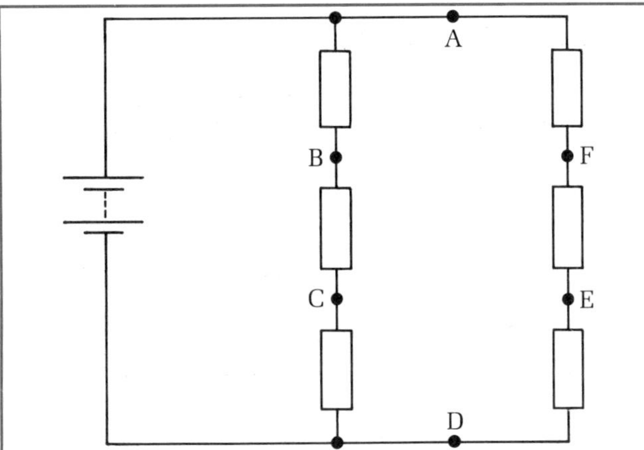

Figure EL12

In the circuit shown above, each resistor has a resistance of 10.0 Ω; the battery has an e.m.f. of 12.0 V and is of negligible internal resistance.

i) Calculate the potential difference between C and F.
ii) When a certain resistor is connected between D and F no current flows in a galvanometer connected between C and F. Calculate the resistance of this resistor and the total power delivered by the battery in this case.

(JMB)

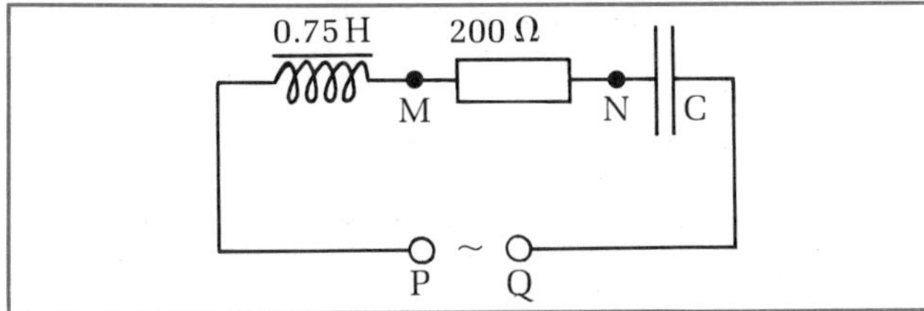

Figure EL13

EL.11
The circuit elements shown in the diagram above are in resonance when supplied with a signal of voltage 250 V r.m.s. at a frequency 600 Hz. Determine

a) the reactance of the inductor at resonance,
b) the reactance of the capacitor at resonance,
c) the capacitance of the capacitor,
d) the current in the circuit at resonance.

What are the phase relationships between the potential differences across each of the three components in the above circuit and the supply voltage?

What is the value of the voltage existing between points P and N? Describe the phase of this voltage relative to the current.

Explain why, in a practical tuned circuit, the series resistance is made as small as possible.

If the resistance of the circuit above were reduced to 20 Ω, what would be the percentage increase in the voltage between points P and N?

(London)

EL.12

a) The alternating voltage across a mains supply has an r.m.s. value of 240 V and a frequency of 50 Hz.
What is meant by *an r.m.s. value of 240 V* and *a frequency of 50 Hz*?

b) In alternating current circuits what is meant by
i) the *reactance* of a capacitor;
ii) the *impedance* of a circuit?

c) The diagram below shows a resistor, a capacitor and a pure inductor connected in series to a sinusoidal alternating current source.
A voltmeter is connected across each of the components in turn and the r.m.s. voltages are as shown on the diagram.

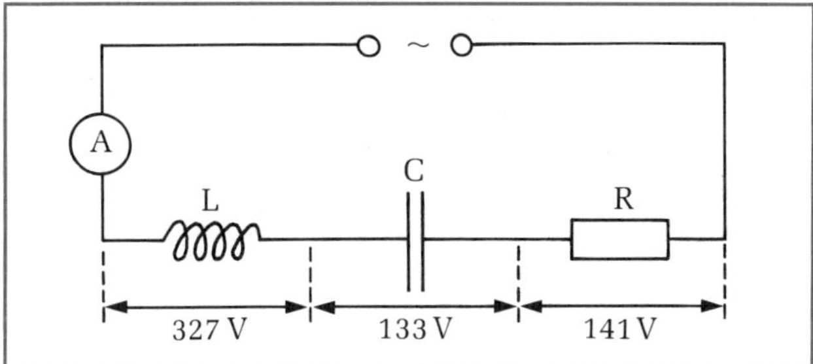

Figure EL14

i) By using a phasor diagram, or otherwise, calculate the r.m.s. voltage across the supply terminals.
ii) If the resistor has a resistance of 68 Ω calculate the reading you would expect to obtain on the ammeter.
iii) Calculate the values of the inductor and the capacitor assuming that the frequency of the supply is 50 Hz.

d) In the diagram below loudspeaker 1 is designed to operate at low frequencies and loudspeaker 2 operates best at high frequencies.

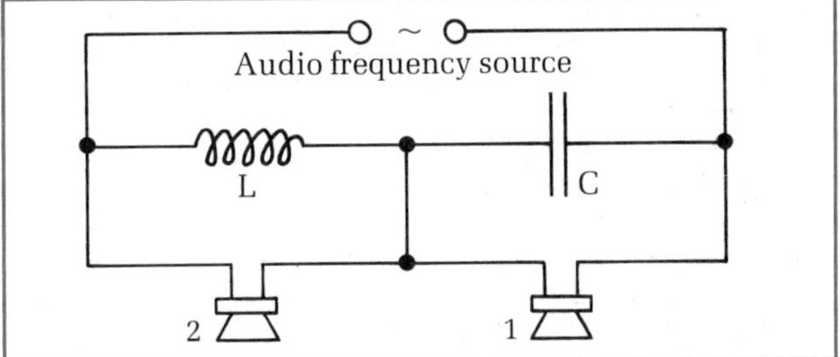

Figure EL15

Explain how such an arrangement can lead to good sound reproduction over a wide range of frequencies.

(AEB June 84)

EL.13 A certain sinusoidally-alternating e.m.f. of frequency 50 Hz has an r.m.s. value of 240 V. Write down the values of (i) the peak e.m.f., (ii) the period, and (iii) the amplitude of the e.m.f.

Explain what you understand by *the r.m.s. value of an alternating current*. Why are r.m.s. values used for public electricity supplies?

A certain special laboratory power-supply can provide varying e.m.f.s with wave forms as illustrated in Figure EL16. What are the r.m.s. values of these supplies?

A lamp is needed for use on 240 $V_{r.m.s.}$ 50 Hz supply but the only ones available are (*a*) an old 60 W lamp left over from when the supply was 120 V d.c. and (*b*) one from another country, where the frequency is 60 Hz, rated 120 $V_{r.m.s.}$ 60 W. Someone suggests putting a resistor in series to make

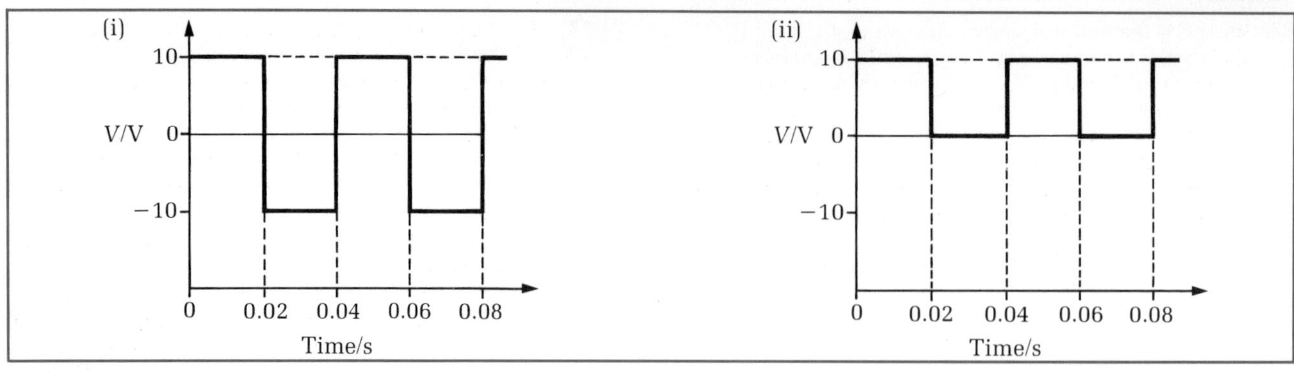

Figure EL16

one of the lamps work. Someone else suggests using a capacitor in series. For each suggestion explain why it will or will not work and which lamp or lamps it would work for, giving the value of resistor or capacitor needed for any arrangement which will work. Is one arrangement to be preferred to another? If so, why?

Calculate the impedance of each lamp (*a*) and (*b*) when working on its original intended supply.

(SUJB)

EL.14 Shown below is a logic diagram with 3 inputs.

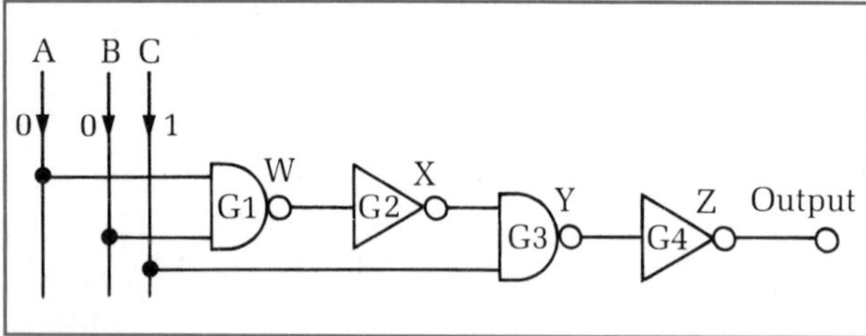

Figure EL17

a) What do the symbols G1 and G2 represent?
b) What will be the logical state at each of W, X, Y and Z for the input 001 shown?
c) Explain why the output at Z will always be 0 when the input C is 0 no matter what the inputs are to A and B.
d) Construct a truth table for the logic diagram drawn above showing the state of the output Z for all possible inputs to A, B and C.
e) State the name of the single gate which could replace the whole system and draw its symbol.
f) If a fourth input D is added, draw a diagram to show how an additional two gates of types G1 and G2 could be used to modify the system in such a way that the output Z will always be 0 when input D is 0, no matter what the inputs are to A, B and C.
g) State how a logic element of the type represented by the symbol G1 can be used to perform the same function as element G2.

(Scottish H.)

EL.15

a) A light emitting diode (LED) which is connected to the output of a logic gate is required to light when the output is logic 1 (+5 V). For this to be achieved a forward current of 10 mA together with a forward p.d. of 2.0 V is required. Draw a circuit diagram to show how the LED is connected and calculate the value of any additional component required.

b)

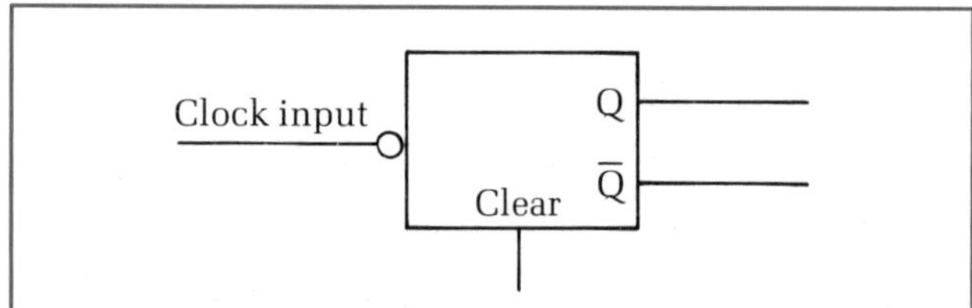

Figure EL18

The symbol shown in Figure EL.18 represents a T or toggle flip-flop circuit. The small circle on the clock input indicates that toggling occurs when an input pulse goes from logic 1 to logic 0. The output Q can be set to logic 0, i.e. cleared, by momentarily connecting the clear input to a zero line. With the aid of a truth table for input and outputs, explain what is meant by the toggling action of this circuit.

The clock pulses shown in Figure EL.19 are applied to the flip-flop circuit. Using the same time scale as Figure EL.19, sketch the outputs from Q and $\bar{Q}$.

c) Draw a diagram to show how four of the circuits shown in Figure EL.18 can connect together to make a four-bit binary counter to count upwards from zero. Show on your diagram where you would connect four LEDs to indicate the binary count, labelling them A to D with A being the least significant bit.

The clock pulses shown in Figure EL.19 are fed into the input of the four-bit counter after its outputs have been cleared. Using a common time scale, show by means of graphs how the output signal from each of the four flip-flops varies between logic 0 and logic 1 for nine input pulses.

(JMB)

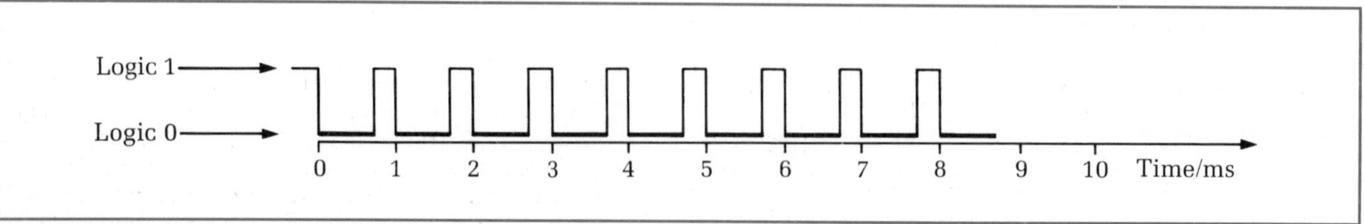

Figure EL19

EL.16

a) Explain what is meant by the terms *half-adder* and *full-adder*. Under what circumstances is it necessary to use a full-adder?

b) The circuit shown below contains four NAND gates. Construct a truth table for the circuit with *A* and *B* as input variables, *C*, *D* and *E* as intermediate variables and *F* as output variable. Hence show how the output *F* depends on the inputs *A* and *B*.

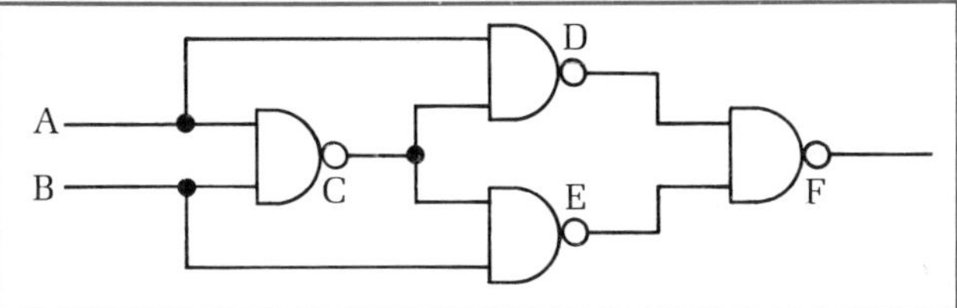

Figure EL20

c) Show, with the aid of a diagram, how two circuits, each identical to that of the circuit shown, can be combined with an extra two-input NAND gate to make a full-adder. Indicate on the diagram all the inputs and the sum and carry outputs. *(JMB)*

EL.17

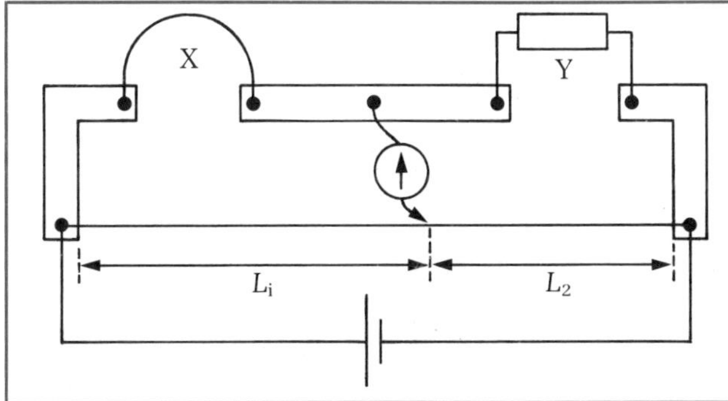

Figure EL21

a) In the Wheatstone bridge arrangement shown above, X is the resistance of a length of constantan wire and Y is the resistance of a standard resistor. Derive from first principles the equation relating X, Y, L_1 and L_2 when no current flows through the galvanometer, stating any assumptions you make.

b) A student plans to use the apparatus described in **a)** to determine the resistances of several different lengths of constantan wire of diameter 0.50 mm. If a 2 Ω standard resistor is available, suggest the range of lengths he might use given that the resistivity of constantan wire is approximately 5×10^{-7} Ωm. Give reasons for your answers.

c)

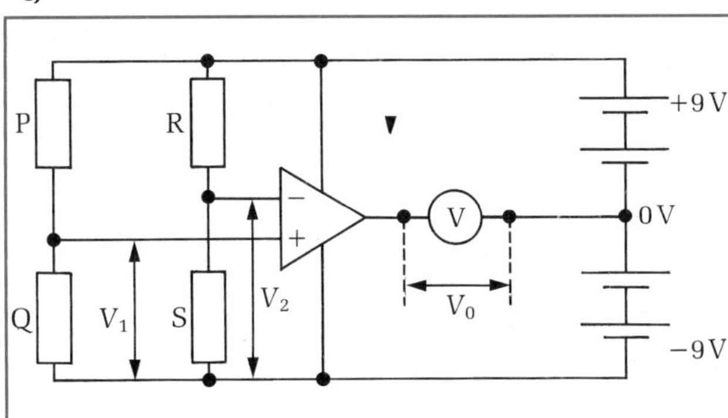

Figure EL22

In the circuit at the bottom of the adjacent column P, Q, R and S form part of a Wheatstone bridge network in which an operational amplifier circuit is used to detect the difference between the voltages V_1 and V_2.

i) The output voltage in this circuit, V_0, is given by $V_0 = A(V_1 - V_2)$ where *A* is the open-loop gain of the amplifier and V_0 must lie between 9 V and −9 V. Show that, if $A = 90\,000$, then V_0 should be either 9 V or −9 V if the difference between V_1 and V_2 exceeds 100 μV.

ii) If the resistances P and Q are each 10 kΩ and R is the resistance of a variable standard resistor, outline how you would use the arrangement to determine the resistance S, which is of the order of a few kΩ.

iii) A typical centre zero galvanometer has a resistance of 40 Ω and is graduated in divisions of 0.1 mA. If you were to choose between such a meter and the above operational amplifier as a null detector, which would you choose, and why? *(JMB)*

EL.18

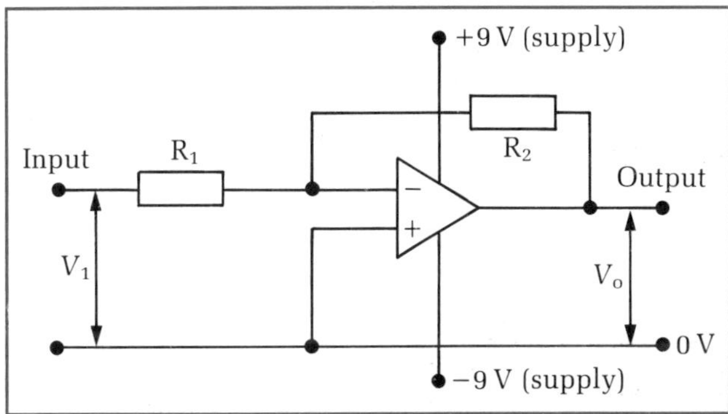

Figure EL23

a) The operational amplifier shown in the diagram above can be used as a voltage amplifier. Describe how you would determine experimentally its d.c. input/output characteristic for positive and negative input voltages. Include a labelled circuit diagram showing how the amplifier is connected to a suitable power supply and explain how you obtain different input voltages using a potential divider. Suggest suitable ranges of voltmeters you would use to measure V_1 and V_0 given that the resistance of $R_1 = 10$ kΩ and the resistance of $R_2 = 100$ kΩ.

b)

Input voltage V_1/V	Output voltage V_0/V
+2.0	−8.0
+1.5	−8.0
+1.0	−8.0
+0.5	−4.0
0	0
−0.5	+4.0
−1.0	+8.0
−1.5	+8.0
−2.0	+8.0

The table opposite shows typical results for a voltage amplifier similar to that shown above. Draw the input/output characteristic and, by reference to it, explain what is meant by (i) *voltage gain*, (ii) *saturation* and (iii) *inversion*. State the range of input voltages for which the amplifier has a linear response and calculate the voltage gain within this range.

c) A sinusoidal voltage of frequency 50 Hz is applied to the input terminals of the amplifier described in **b)**. Sketch graphs on one set of axes showing how the output voltage varies with time when the input voltage is (i) 0.5 V r.m.s., (ii) 1.0 V r.m.s.. In each case indicate the peak value of the output voltage and comment on the wave form. *(JMB)*

EL.19 Explain what is meant by *voltage gain* and *negative feedback* in relation to electronic circuits.

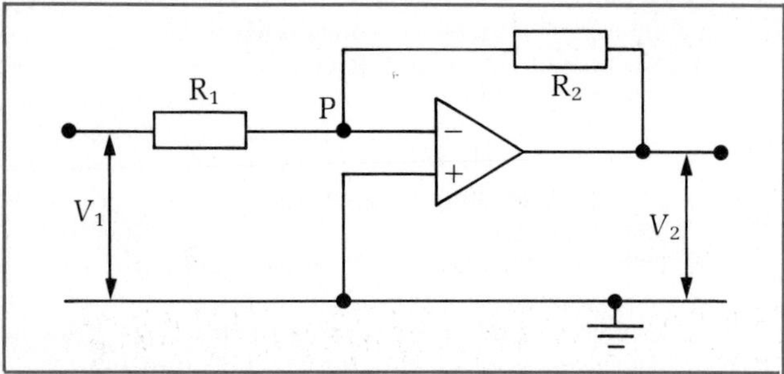

Figure EL24

The diagram above shows a circuit containing an ideal operational amplifier where the point P is usually referred to as a *virtual earth*. Explain what you understand by *virtual earth* in this context and hence derive an expression for V_2 in terms of V_1 and the values of the circuit components.

The current, *I*, through a certain device varies with applied potential difference, *V*, according to the relation

$$I = I_0 e^{kV}$$

where I_0 and k are constants. If R_2 is replaced by this device, write down an expression for the feedback current in terms of V_2. Hence show that V_2 is given by the expression

$$V_2 = 1/k \ln(V_1/R_1 I_0).$$

What is the possible advantage of this type of amplifier over a linear amplifier when a wide range of input signal amplitudes must be displayed?

(Cambridge)

EL.20

a) i) Explain with the aid of an energy band diagram what is meant by *intrinsic conduction* in a semiconductor.

ii) Explain the effect of a small temperature rise on the conductivity of an intrinsic semiconductor.

iii) Using further energy band diagrams, explain the two types of *extrinsic* conduction in semiconductors.

b) i) Sketch a labelled set of output characteristics (I_c against V_c) for a transistor, with typical values on the axes and curves. State whether the characteristics you have drawn are for a p-n-p or n-p-n transistor.

ii) Use the characteristic curves to indicate how the transistor may be used as a current amplifier.

c) Draw a circuit using *either* a p-n-p or an n-p-n transistor which can be used to amplify a small alternating voltage input. Indicate on the circuit diagram

i) the type of transistor used,

ii) the input and output terminals,

iii) the polarity of the d.c. supply.

(Component values are *not* required.)

d) i) When the transistor is used as a voltage amplifier in **c)** above, state the ideal relationship between the supply voltage V_S and the voltage V_T across the transistor when no input signal is applied. Explain why this voltage relationship is desirable.

ii) Explain why the transistor has only a limited range of input voltages which can be amplified with little distortion. Making reference to the characteristics in **b)**(i) above, indicate what this range is.

(N. Ireland)

SECTION F

Inside the atom

The properties of the electron · Radioactivity · Electrons inside the atom · Nuclear energy · Particles and patterns

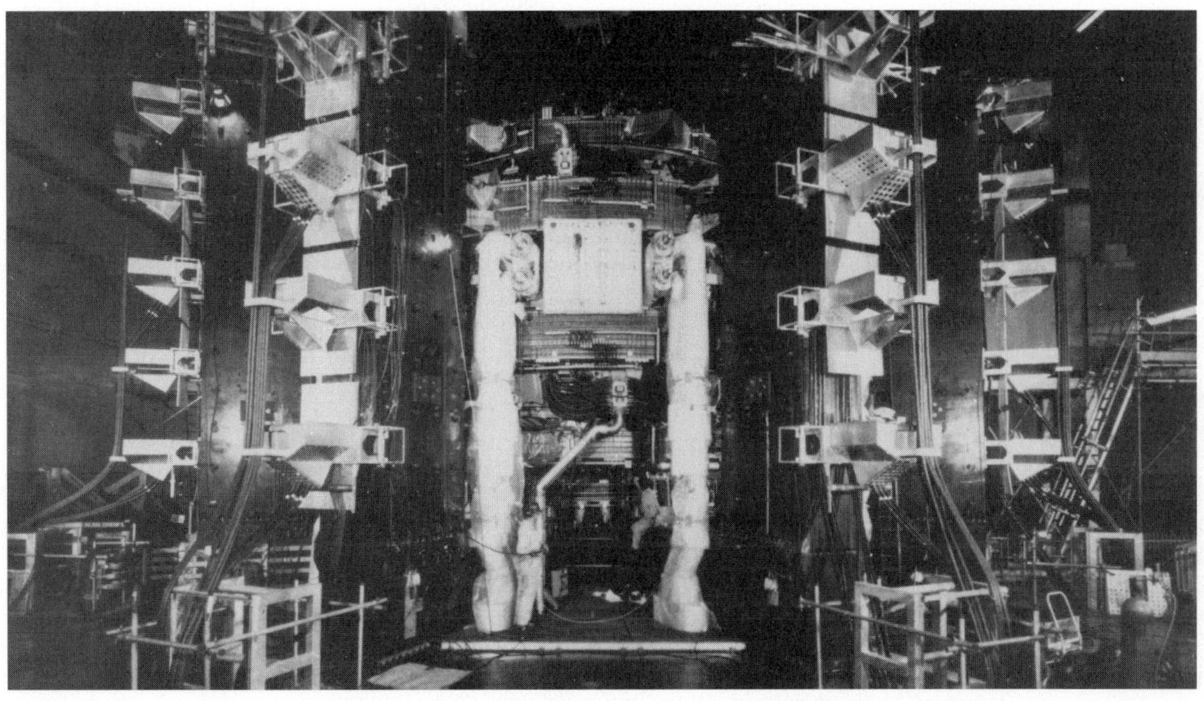

Figure F1 *The JET fusion reactor, shown during construction. This experimental reactor is designed to achieve temperatures of about 5×10^7K for fusion to start. A fusion power station would only need a few tonnes of fuel each year to produce the same output as the largest coal-fired station.*

Understanding the atom is one of the central aims of twentieth century physics. Ideas and techniques from other branches of physics have been put to use by scientists attempting to discover the nature of the atom. Often, the discoveries have led to a complete re-examination of other branches of physics as existing ideas are found to be inadequate. For example, the idea that energy is 'lumpy' (i.e. exists in basic amounts) is not apparent in everyday life. But when we look at events on an atomic scale, we are forced to re-examine our notion of energy and to admit that energy is quantized.

The applications of atomic physics are widespread, from X-rays to nuclear power, from archaeology to satellites. Ideas in atomic physics draw heavily on other branches of physics, especially mechanics and fields. The force that locks electrons into atoms is electrical in nature; the force that holds the atomic nucleus together, the nuclear force, is very different to either gravitational or electrical force. Its characteristics are discussed in the chapter on Radioactivity, and the consequences are developed in the chapter on nuclear power.

27

The properties of the electron

27.1 The discovery of the electron

Cathode rays

Figure 27.1 *A neon display*

Discharge tubes used for advertising displays add a colourful touch to any city centre. Imagine the excitement and curiosity of people in the early years of the 20th century when such displays first appeared. These tubes were developed first by William Crookes in the 1870s. He mastered the tricky problem of producing a high vacuum, i.e. reducing the gas pressure in the tube by removing gas from it. Crookes put his skill to good use by showing that gases at low pressure conduct electricity. One of the simplest tubes he devised is shown in Figure 27.2; the tube has a metal electrode at either end and is sealed after reducing the gas pressure in it. When a high voltage is set up between the electrodes, the remaining gas inside the tube glows with light. Crookes showed that the colour of light emitted depends on the type of gas used. Neon gas gives a bright red-orange colour, oxygen and nitrogen give a purple glow.

What causes the glow from a discharge tube? Crookes and other scientists took up the challenge to find out.

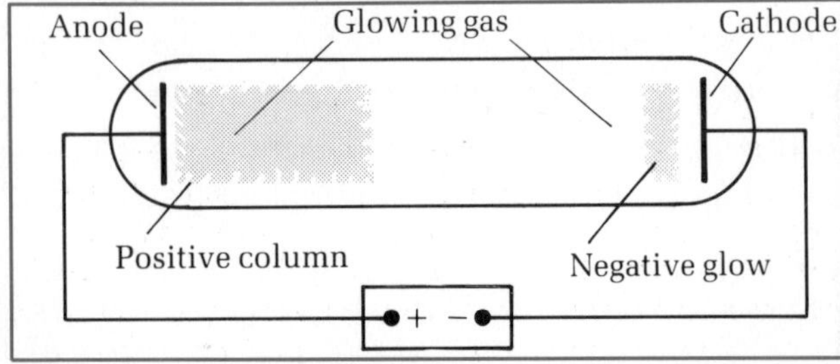

Figure 27.2 *A discharge tube*

Some ingenious experiments were devised, including a tube with a 'paddle wheel' in it. The paddle wheel in the tube was forced to rotate, bombarded by radiation from the cathode. Placing a magnet near the tube deflected these invisible cathode rays so the paddle wheel stopped turning. Further experiments showed that they caused objects to be charged negatively.

Exactly what cathode rays consist of was to puzzle scientists for 20 years or more, until Joseph J. Thomson turned his attention to the problem. Thomson carried out a series of experiments which proved that cathode rays were composed of tiny negatively charged particles which he called **electrons**. He showed that electrons are identical, whatever type of gas was used in the discharge tube. He measured the specific charge (i.e. its charge/mass value) of the electron, and proved beyond doubt that electrons were constituents of all atoms. No longer could the atom be regarded as indestructible; Thomson showed they were made up of smaller particles. His experiments provided the prototype for the modern TV tube and the oscilloscope.

Thomson's experiments

If cathode rays consist of fast-moving electrons, then what happens when they hit the anode or any other obstacle? Thomson made a slit in the anode so that a beam of electrons emerged through the anode. He devised means of collecting the electrons by deflecting the beam with a magnet, onto a metal electrode. The electrode gathers negative charge when the beam hits it, and it becomes hot. Thomson reasoned that electrons are negatively charged particles which are accelerated to high speeds in the discharge tube. When they hit the electrode, they lose their kinetic energy which is transformed into thermal energy of the electrode.

To determine the specific charge of the electron (i.e. its charge/mass value), Thomson devised a 'deflection tube' in which the electron beam was passed between two parallel plates (Figure 27.3). The impact of the beam on the inside of the glass bulb caused a spot of light to be emitted. When a p.d. is applied between the deflecting plates, the electrons in the beam are deflected towards the positive plate so the spot is displaced. Thomson measured the deflection of the spot for different plate voltages. He showed in theory how the deflection is related to the plate voltage and the speed of the electrons. He was able to work out the speed by deflecting the beam using a magnetic field. Then, by combining his theory and his

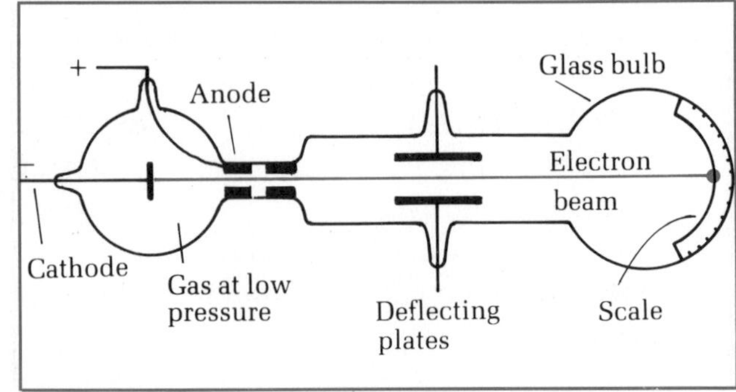

Figure 27.3 *Thomson's method for finding e/m*

measurements, he was able to calculate the specific charge of the electron. You can follow in Thomson's footsteps by trying the determination of specific charge using apparatus described later.

Thomson obtained a value of 1.76×10^{11} C kg^{-1} for the specific charge of the electron. How does this compare with the value for ions? Thomson knew that the hydrogen ion had the biggest value (of specific charge) of all the known ions. Careful measurements on electrolysis experiments had shown that 96 500 C of electric charge were needed to liberate 1 g of hydrogen gas in the electrolysis of acidified water. So the specific charge Q/M of a hydrogen ion is 9.65×10^{7} C kg^{-1}, the biggest known value until Thomson measured the specific charge of the electron. Thomson thought that the charge of an electron was equal and opposite to the charge of a hydrogen ion; this would mean that the electron is much lighter than the hydrogen ion, giving a much greater charge/mass value for the electron. But firm evidence for such an assumption was not to come for more than ten years. Nevertheless, Thomson's determination of the specific charge of the electron was a major step forward towards understanding the atom.

The electron gun

Thermionic emission is a much simpler way of producing an electron beam than using a discharge tube. When a metal is heated, its free electrons gain kinetic energy. If heated sufficiently, some of these electrons leave the metal. If the metal is isolated in a vacuum, the departing electrons gather above the surface like a cloud. But if there is a nearby positive electrode (i.e. anode), the electrons from the metal are pulled onto the anode. When a metal emits electrons in this way, the process is called 'thermionic emission'.

In practice, the metal is a filament of wire heated by an electric current through it, just like an ordinary lamp filament. An anode plate near the filament is made positive with respect to the filament, so electrons are drawn towards and onto the anode from the filament. By using an anode plate with a hole drilled through, as in Figure 27.4, some of the electrons pass through the anode to produce a beam of electrons. This is like releasing ball bearings at the top of a ramp with a wall along the lower edge of the ramp. The balls accelerate down the ramp and hit the wall, but if there is a gap in the wall at some position, then the balls heading for the gap pass straight through.

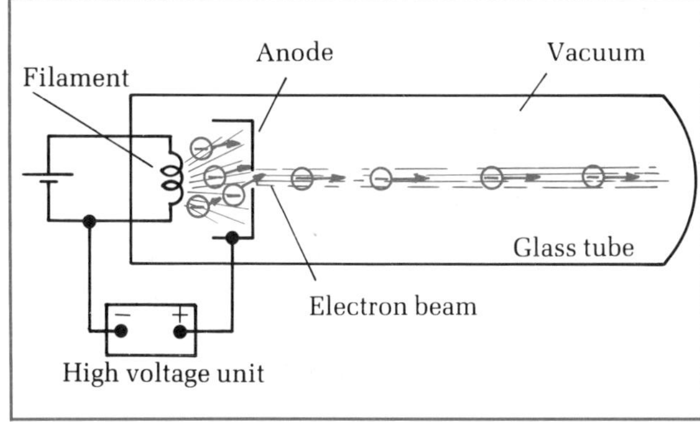

Figure 27.4 *An electron gun*

The work done on each electron when it is accelerated from the filament to the anode depends on the p.d. between the filament and the anode.

Let V_A = the p.d. between the filament and the anode; this is the work done per unit charge on a tiny charge moved from the filament to the anode. Each electron carries an identical charge, denoted by e. So the work done on each electron moved from the filament to the anode is equal to $e \times V_A$.

$$\text{Work done} = eV_A$$

where e = charge of an electron,
V_A = anode potential.

The electron-volt is an amount of energy equal to the work done on an electron moved through a p.d. of 1 volt. Since $e = 1.6 \times 10^{-19}$ C, then 1 electron-volt is equal to 1.6×10^{-19} J.

What happens to the work done on each electron? In a resistor, work done is converted to internal energy of the atoms, but in a vacuum tube there are no atoms in the path of the beam, so each electron is continuously accelerated from the filament to the anode. The electrons which emerge through the gap therefore have kinetic energy equal to the work done eV_A. The initial kinetic energy of each electron at the filament is negligible, compared with the gain of kinetic energy from the filament to the anode. So the speed at which each electron emerges through the anode gap is given by

$$\text{K.E. of each electron in the beam} = \tfrac{1}{2}mv^2 = eV_A$$

where m = electron's mass,
v = electron's speed.

Compare the electron gun with a ball rolling down the ramp and passing through the gap in the wall at the bottom onto a flat surface. Each ball has work done on it by gravity equal to mgH; so each ball passing through the gap has gained kinetic energy equal to mgH.

27.2 Electron beams in fields

Deflection in a uniform electric field

Consider an electron beam directed between two oppositely charged parallel plates, as in Figure 27.5, at the top of the next page. The beam is deflected towards the positive plate. The path of the beam can be displayed by using a fluorescent screen between the plates so the beam skims across the screen. The electrons in the beam cause atoms in the screen to emit light, so a visible trace is seen on the screen corresponding to the path of the beam. Provided all the electrons enter the space between the plates with the same speed, they all follow the same path, so a thin trace is seen on the screen.

With a constant p.d. between the two deflecting plates, the trace is a curve towards the positive plate. In Figure 27.5 the lower plate is positive so the trace is a curve like the flight path of a projecile (see p. 8), i.e. a parabola. The reason is that each electron experiences a constant force between the plates, just as a projectile above the Earth's surface is acted on by a constant force (i.e. the

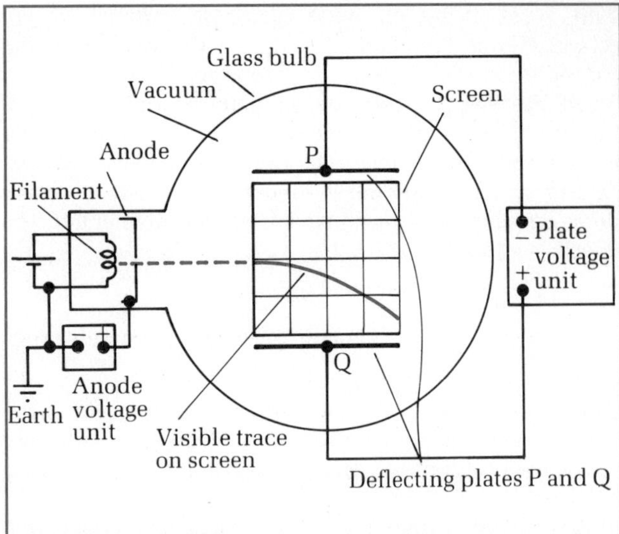

Figure 27.5 *Deflecting electrons in an electric field*

force of gravity). The difference is that the force on an electron in the beam is due to the electric field between the plates. The force due to gravity on the electron is negligible by comparison.

The force F on each electron in the field is given by

$$F = eE = \frac{eV_p}{d}$$

where E = electric field strength,
V_p = p.d. between plates,
d = plate spacing.

The equation is explained in detail on p. 167; in energy terms, an electron moved from the positive to the negative plate is acted on by a constant force F. So the work done $Fd = eV_p$.

Each electron in the beam experiences a constant force F, so its acceleration towards the positive plate is given by

$$\text{acceleration } a = \frac{F}{m} = \frac{eV_p}{md}$$

The direction of the acceleration is downwards towards the positive plate. Therefore the horizontal motion is unchanged by the electric field. So the time taken for each electron to travel a horizontal distance x from the point of entry O is given by distance x/initial (horizontal) speed v.

$$\text{Time taken } t = x/v$$

Now consider the vertical motion of each electron after entering the field.

The vertical displacement y is given by the dynamics equation $y = \frac{1}{2}at^2$. The initial vertical component of velocity is zero because the beam enters the field horizontally. So the full equation $y = ut + \frac{1}{2}at^2$ becomes $y = \frac{1}{2}at^2$ since u is zero for the vertical motion.

$$\text{Hence } y = \tfrac{1}{2}at^2 = \tfrac{1}{2}\left(\frac{eV_p}{md}\right)t^2 \qquad \text{since } a = \frac{eV_p}{md}$$

$$\text{Giving } y = \tfrac{1}{2}\left(\frac{eV_p}{md}\right)\frac{x^2}{v^2} \qquad \text{since } t = \frac{x}{v}$$

$$\text{Thus } y = \left(\frac{eV_p}{2mdv^2}\right)x^2$$

This is the equation for a parabola: $y = kx^2$ where k is a constant. For a projectile with initial velocity u in the horizontal direction, $x = ut$ and $y = \frac{1}{2}gt^2 = (g/2u^2)x^2$. The equations look alike because both situations involve constant (though different) force.

The vertical speed v_y is given by the dynamics equation $v_y = at$. The full equation $v_y = u + at$ becomes $v_y = at$ because the initial vertical speed u is zero.

$$\text{Hence } v_y = at = \left(\frac{eV_p}{md}\right)t = \frac{eV_p x}{mdv}$$

1 The direction of the beam at any point is given by

$$\tan\theta = v_y/v,$$

where θ is its direction to the horizontal. Note that v_y/v equals $2y/x (= eV_px/mdv^2)$; so where the beam emerges from the field, its direction is as if it had come straight from the centre of the field. Figure 27.6 shows this.

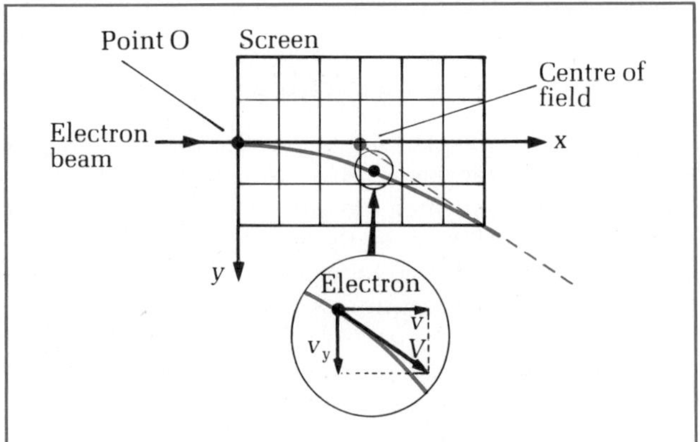

Figure 27.6 *Parabolic path*

2 The speed V of the beam at any point is given by

$$V = \sqrt{v^2 + v_y^2}$$

Worked example An electron gun operating at 3000 V is used to shoot electrons into the space between two oppositely charged parallel plates, as in Figure 27.5. The plate spacing is 50 mm and its length is 100 mm. Calculate the deflection of the electrons at the point where they emerge from the field when the plate p.d. is 1000 V. Assume the specific charge e/m for the electron is $1.76 \times 10^{11}\,\text{C kg}^{-1}$.

Solution First, we need to calculate the speed v of the electrons from the electron gun.

$$\tfrac{1}{2}mv^2 = eV_A$$

$$\text{Hence} \quad v^2 = 2\frac{e}{m}V_A = 2 \times 1.76 \times 10^{11} \times 3000$$
$$= 10.56 \times 10^{14}\,\text{m}^2\,\text{s}^2$$
$$v = 3.25 \times 10^7\,\text{m s}^{-1}$$

Now the deflection can be calculated using

$$y = \frac{(eV_p)}{2mdv^2}x^2$$

with $x = 100$ mm and $d = 50$ mm

Thus

$$y=\frac{1.76\times10^{11}\times1000\times(100\times10^{-3})^2}{2\times50\times10^{-3}\times10.56\times10^{14}}=1.67\times10^{-2}\,\text{m}$$

Deflection in a uniform magnetic field

Electrons passing across a magnetic field are deflected by the field. To demonstrate this, all you need do is to place a magnet near a TV tube – but **don't do it to a colour TV set** or you could spoil the colour pattern.

The fine beam tube shown in figure 27.7, can be used to demonstrate and measure the force on an electron beam due to a magnetic field. The tube contains hydrogen gas at low pressure. When a beam of electrons from the electron gun in the tube passes through the gas, the gas atoms along the path emit light due to collisions with electrons. So the path of the beam is seen as a fine trace of light through the gas. A pair of coils is placed either side of the tube and supplied with direct current to produce a uniform magnetic field through the tube.

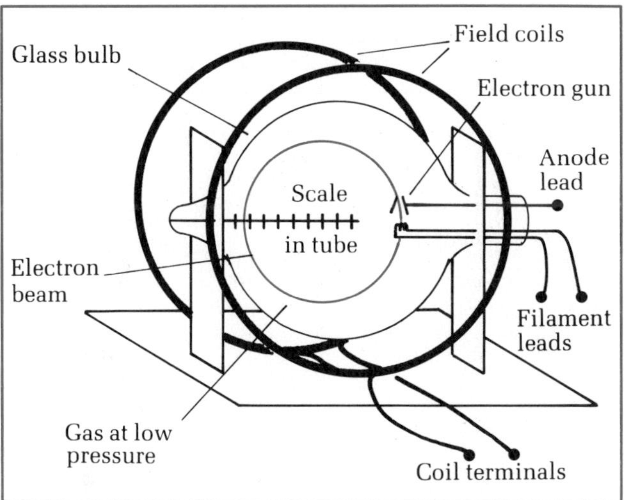

Figure 27.7 *Deflecting electrons in a magnetic field*

Provided the initial direction of the beam is at right angles to the magnetic field lines, the beam curves round into a circle. The stronger the field strength, the smaller the circle. Each electron in the beam is acted upon by a force due to the magnetic field. The force is always at right angles to the beam direction (i.e. its velocity); so the beam forms a circular path since the force provides the centripetal acceleration necessary for circular motion.

As explained on p. 183, the force F on an electron in a magnetic field is given by

$$F=Bev$$

where B = magnetic field strength,
e = electron's charge,
v = electron speed.

Since the magnetic force F is at right angles to the velocity direction, no work is done on the electron so its speed stays constant. It moves round a circular path of radius r, so its centripetal acceleration is equal to v^2/r.

Using $F=ma$, the centripetal acceleration can therefore be written as at the top of the next column.

Centripetal acceleration $a=\frac{F}{m}=\frac{Bev}{m}$

Hence $a=\frac{v^2}{r}=\frac{Bev}{m}$, which gives

$$r=\frac{mv}{eB}$$

where r = radius of electron path,
m = electron mass.

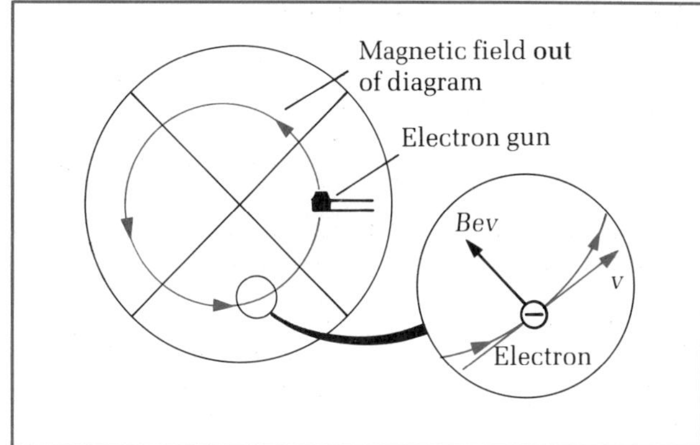

Figure 27.8 *Force due to the magnetic field*

To measure the specific charge e/m of the electron, using the fine beam tube, the magnetic field strength is adjusted in steps, using a rheostat to control the current in the field coils. At each step the beam diameter is measured from the scale inside the tube. In practice, the diameter is adjusted to easily-measured values. The coil current is measured at each step. The anode voltage must also be measured.

The electron gun equation gives the speed of the electrons.

$$\tfrac{1}{2}mv^2=eV_A \qquad (1)$$

The radius of the circle is given by,

$$\frac{mv^2}{r}=Bev$$

Hence $v=Ber/m$ (2)

Combining the two equations gives

$$\tfrac{1}{2}m\left(\frac{Ber}{m}\right)^2=eV_A$$

Rearranging this equation gives

$$\text{specific charge } \frac{e}{m}=\frac{2V_A}{B^2r^2}$$

where V_A = anode voltage,
B = magnetic field strength,
r = radius of electron path.

1 The anode voltage is measured directly using a voltmeter.
2 The radius of the electron path is calculated from the diameter which is measured directly.
3 The magnetic field strength B is calculated from the coil current I. If the coils are spaced at a distance equal to

their radius R, then the value of B is given by the formula $B=0.716\mu_0 NI/R$, where N is the number of turns of each coil. This arrangement is known as a pair of Helmholtz coils.

Alternatively, the magnetic field strength can be measured directly using a calibrated Hall probe (see p. 185). The tube must be removed from between the coils and the probe is placed at the centre of the pair of coils. The coil current is then adjusted to each of the values previously recorded, and at each step B is measured.

4 A graph of B against $\frac{1}{r}$ ought to give a straight line through the origin.

Since $B=\sqrt{\frac{2V_A}{e/m}}\times\frac{1}{r}$,

then the gradient of the graph is equal to $\sqrt{\frac{2V_A}{e/m}}$.

Hence the value of e/m can be determined from the graph.

Worked example An electron gun operating at 2000 V is used to shoot electrons into a magnetic field of strength 1.9 mT. The initial direction of the beam is at right angles to the field direction.

a) Calculate the specific charge on an electron if the beam forms a circle of diameter 160 mm.

b) Describe the beam path if the initial direction of the beam is changed to 30° to the field direction.

Solution

a) $V_A=2000$ V, $B=1.9\times10^{-3}$ T, $r=0.160/2$ m.

Hence

$$\frac{e}{m}=\frac{2V_A}{B^2r^2}=\frac{2\times2000}{(0.0019)^2\times(0.080)^2}=1.73\times10^{11}\text{ C kg}^{-1}.$$

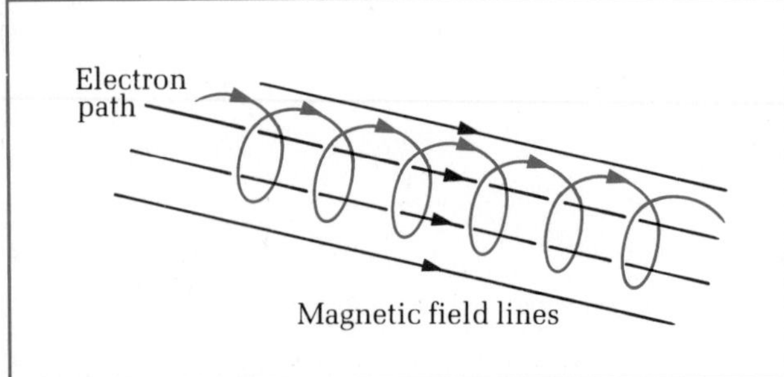

Figure 27.9 *Electron spirals*

b) The electrons spiral round the field lines as shown in Figure 27.9. The component of velocity parallel to the field, $v\cos30°$, does not change because the magnetic force is always at right angles to the field direction. The perpendicular component, $v\sin30°$, is the speed at which the electrons 'circle' in the field. This is exactly half the value of the speed when the direction was at right angles. So the 'circle' radius is half the previous value, giving 40 mm. In other words,

i) at right angles, $v=Ber_1/m$ where $r_1=80$ mm,

ii) at 30°, $v\sin30°=Ber_2/m$ where r_2 is the spiral radius.

Hence $r_2=r_1\sin30°=40$ mm.

The velocity selector

Suppose a beam of electrons is fired into a uniform electric field as in Figure 27.5. The beam curves towards the positive plate because each electron experiences a constant force towards that plate. The beam can also be deflected by a magnetic field. In Figure 27.10, the direction of the magnetic field is out of the plane of the diagram so the electron beam is pushed up – assuming no electric field is applied. But what happens if the magnetic field is applied at the same time as the electric field? Each electron is acted on by magnetic force (Bev) upwards and electric field force (eV_p/d) downwards. If the magnetic force is stronger than the electric force, the beam curves up. If the magnetic force is weaker than the electric force, the beam curves down. What if the two forces are equal in size? Then, the resultant force on each electron is exactly zero; so the beam passes through undeflected.

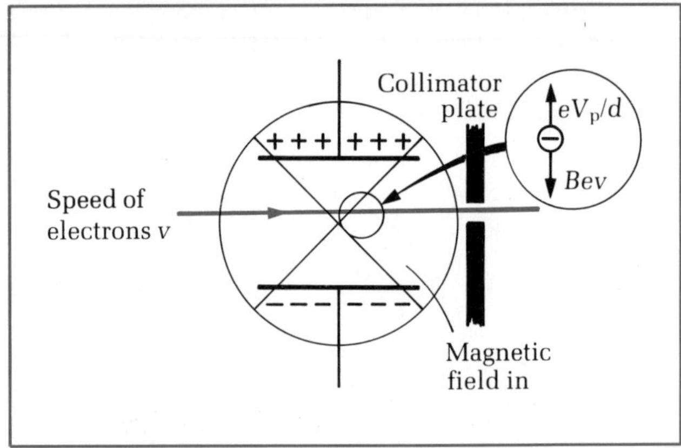

Figure 27.10 *The velocity selector*

For zero deflection

magnetic force Bev = electric force eV_p/d

Hence the speed at which the beam is undeflected is given by

$$v=\frac{V_p}{Bd}$$

where v = speed for zero deflection,
B = magnetic field strength,
V_p = plate voltage.

If the speed is changed by altering the anode voltage, all other factors being the same, then the forces are no longer in balance. Reducing the speed makes the magnetic force weaker than the electric force so the beam curves up. Increasing the speed from the value given by the above equation makes the beam curve down – the electric field force wins now!

The arrangement could be used as a velocity selector for a beam of charged particles with a range of speeds. With a 'collimator' plate at the edge of the field where the beam emerges, as in Figure 27.10, undeflected charged particles pass through the collimator gap. These particles are ones with just the right speed, given by the above equation. Particles with any other speed are deflected either up or down so they hit the collimator plate instead of passing through the gap.

Worked example A beam of electrons is fired into a uniform electric field as in Figure 27.10. The length of the plates is 100 mm and the plate spacing is 50 mm. When a constant p.d. of 720 V is applied across the deflecting plates, the beam is deflected vertically at the edge of the field by a distance of 25 mm. The beam is straightened out by applying a uniform magnetic field of strength 0.65 mT perpendicular to the beam with the same plate p.d. Calculate

a) the speed of the beam entering the fields,

b) the specific charge of the electron.

Solution

a) Use the velocity selector equation for the speed v.

$$v = \frac{V_p}{Bd} = \frac{720}{0.65 \times 10^{-3} \times 50 \times 10^{-3}} = 2.22 \times 10^7 \text{ m s}^{-1}$$

b) Use the equation for vertical displacement y in an electric field.

$$y = \frac{eV_p x^2}{2mdv^2}$$

$$\text{Hence } \frac{e}{m} = \frac{2ydv^2}{V_p x^2}$$

$$\text{So } e/m = \frac{2 \times 25 \times 10^{-3} \times 50 \times 10^{-3} \times (2.22 \times 10^7)^2}{720 \times (100 \times 10^{-3})^2}$$

$$e/m = 1.71 \times 10^{11} \text{ C kg}^{-1}$$

27.3 Electron tubes

The diode valve

One of the early applications of Thomson's research was the invention of the diode valve, used to rectify alternating current (i.e. convert a.c. to d.c.). Nowadays, solid state diodes are used in place of valves, except in a few specialist applications. A diode valve is an evacuated glass tube containing a filament and an anode. Figure 27.11 shows the construction. The filament is heated by passing an electric current through it. Electrons are attracted onto the anode whenever the anode is made positive with respect to the filament. When the anode is made negative, no electrons can pass from the filament to the anode. Figure 27.12 shows the variation of current (due to electrons reaching the anode) with anode voltage. The anode current increases to a maximum value as the anode

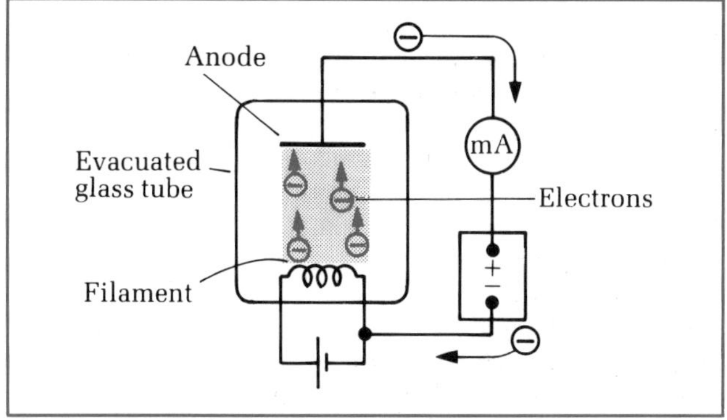

Figure 27.11 *A diode valve*

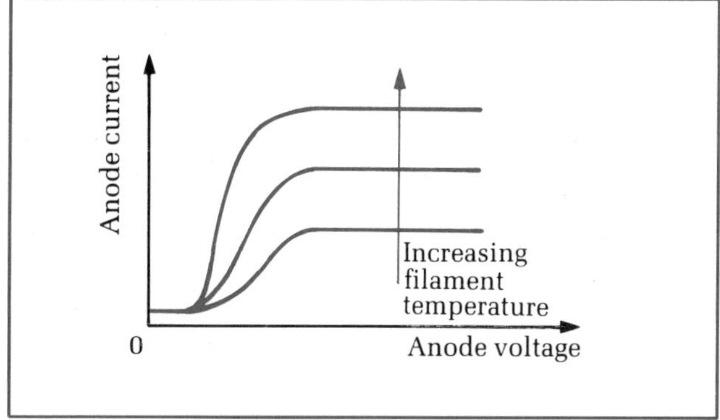

Figure 27.12 *Vacuum diode characteristics*

voltage is increased. The maximum value of the anode current is called the **saturation** current; at saturation, electrons leaving the filament are pulled straight across to the anode. The saturation current depends on the filament current. If the filament is made hotter by increasing the filament current, then more electrons leave the filament each second so the saturation current is increased.

To rectify an alternating voltage, the anode is connected in series with a resistor to one terminal of the a.c. supply, as in Figure 27.13. The other terminal of the a.c. supply is connected to one of the filament leads. When the anode goes positive each half-cycle, electrons are pulled onto it from the filament. These electrons pass through the resistor back to the supply unit. When the anode goes negative, no electrons are attracted to the anode so there is no current through the resistor.

You might like to make a diode yourself and ask your teacher to test it. Figure 27.14 shows the construction. Use a pyrex test tube with a rubber bung to fit. Ask your

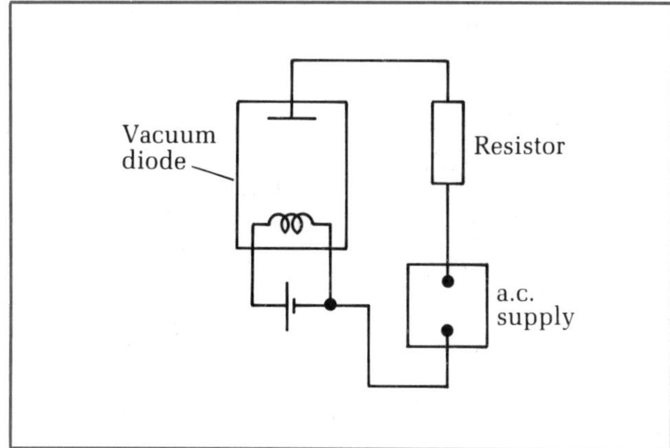

Figure 27.13 *Rectifying a.c.*

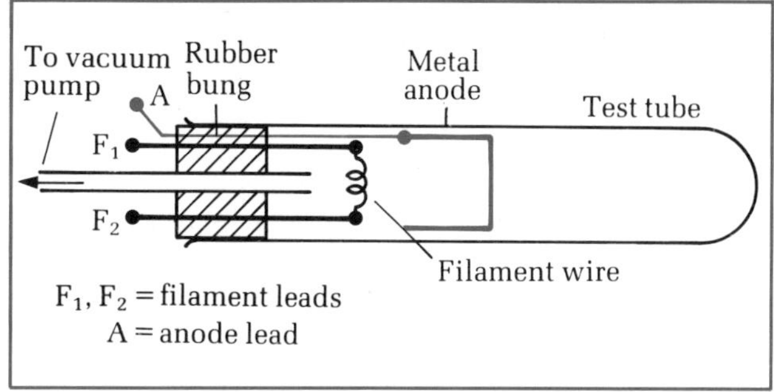

Figure 27.14 *Making a diode valve*

teacher to bore a hole through the bung to take a glass tube, and to bore pinholes for the leads. Then solder the wire filament and anode plate to the leads. Assemble the fittings in the tube, and connect up to a vacuum pump to test for leaks. Ask your teacher to test your device. It may not work too well, and you might well see a bluish glow above the filament. That tells you there is still too much air in the tube, because it is produced by electrons hitting air molecules.

The TV tube

Black and white TV tubes are easier to start with. An electron gun shoots electrons at the inside of the TV screen. The electrons hit a coating of fluorescent paint on the inside of the screen, causing light to be emitted from the point of impact. Magnetic deflecting coils are used to sweep the beam rapidly from left to right across the screen. At the same time, another pair of magnetic deflecting coils pushes the beam downwards. When it reaches the bottom of the screen, these coils flick the beam back to the top to start the downward sequence again. The TV signal alters the intensity (i.e. no. of electrons per second) of the beam as it sweeps across the screen. Line by line, the TV picture is built up from dots of different brightness on the screen. European TV sets are designed with 625 lines to build up the complete picture in 1/25 of a second.

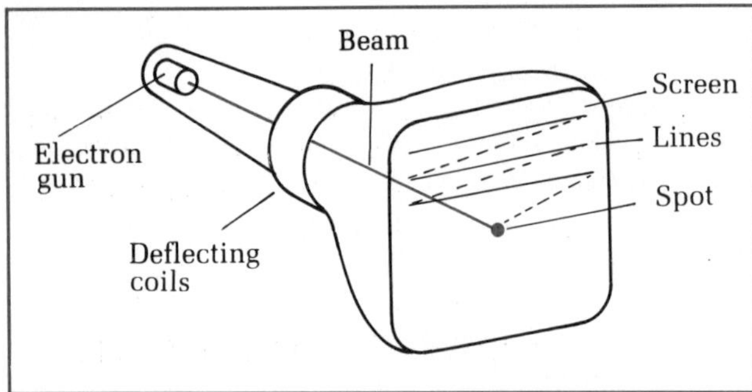

Figure 27.15 *A black and white TV tube*

Colour TV tubes use three electron guns. The screen is coated with alternate tiny dots of phosphor. There are three types of dot, each made of a different material selected to give red, green or blue light when the dot is struck by electrons. The dots are arranged in triangles, as shown in Figure 27.16. Between the screen and the electron guns is a 'shadow mask' of tiny holes. The holes are positioned so that each beam passing through always hits the same type of dot. So one beam gives the red picture, one beam the green and the other the blue picture. Provided all three beams are operating correctly, the viewer sees the correct colours on the screen; but if one or two beams stop working, false colours like blue trees and leaves may be seen!

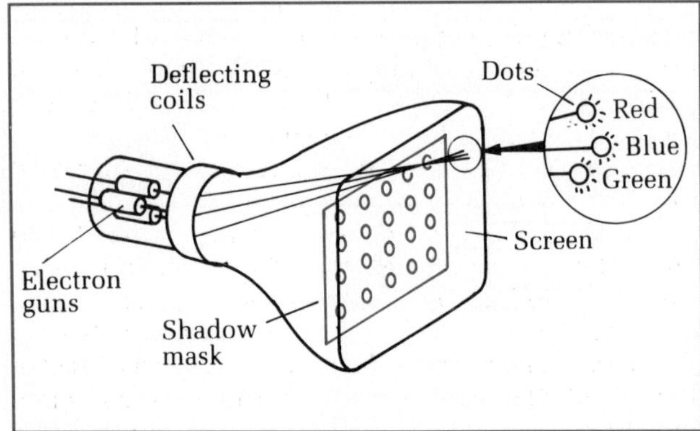

Figure 27.16 *A colour TV tube*

The oscilloscope tube

Two sets of deflecting plates are used to deflect the beam horizontally and vertically, as shown in Figure 27.17. The beam is produced from an electron gun, with a series of anodes used for focusing the beam and controlling its brightness. In this way, the trace on the screen is made narrow at the required brightness.

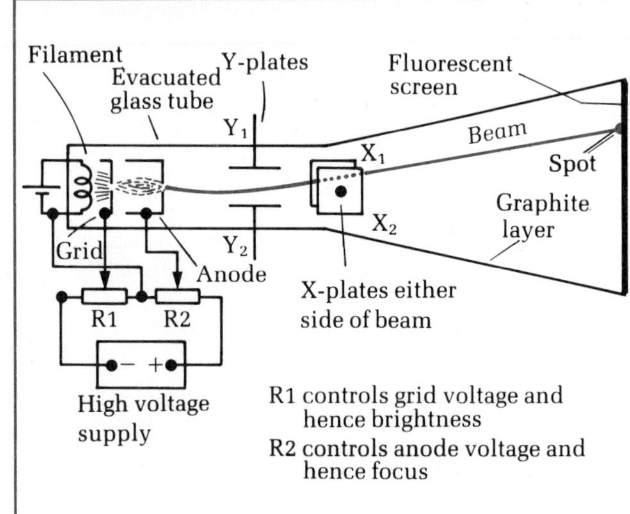

Figure 27.17 *The oscilloscope tube*

The signal under test is applied to the Y-plates to deflect the beam vertically. As shown later, the deflection is proportional to the p.d. between the two Y-plates. So the Y-scale can be calibrated using a 'known' signal to give the p.d. per cm deflection. Then the deflection produced by the unknown signal is measured, and the unknown p.d. can then be calculated. For example, suppose a 1.5 V cell is connected to the Y-input terminals to give a deflection of 30 mm. Then the p.d. per cm deflection is $0.5\ \text{V cm}^{-1}$. If an unknown p.d. is then applied in place of the 1.5 V cell, to give a deflection of 20 mm, the unknown p.d. must equal $0.5 \times 2.0 = 1.0$ V.

A time-base voltage is usually applied to the X-plates. Such a voltage sweeps the spot on the screen at constant speed from left to right then flicks the spot back to the left to start a new sweep. See p. 348 for details of using the oscilloscope for a.c. measurements.

The operation of the oscilloscope requires the deflection of the spot to be proportional to the p.d. across the deflecting plates. This applies to vertical and to horizontal deflection of the spot. Let's consider why the deflection is proportional to the p.d.

When the beam leaves the electric field between the plates, its direction is as if it had come straight from the centre of the field. This is explained in detail on p. 420.

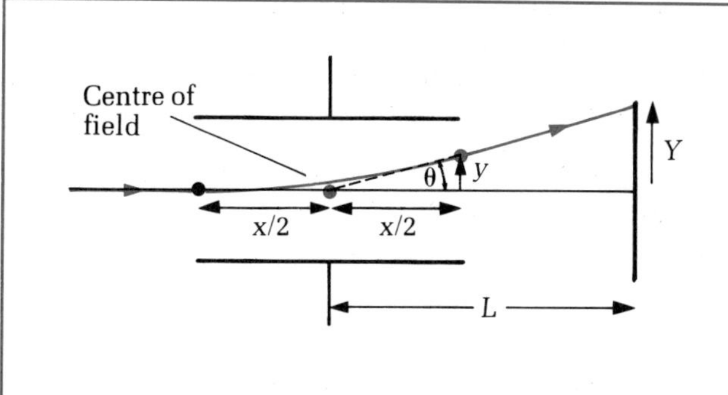

Figure 27.18 *Y-sensitivity*

Figure 27.18 shows the idea. The y-deflection at the edge of the field is given by

$$y = \frac{eV_p}{2mdv^2}x^2$$

where x = horizontal distance to edge,
V_p = plate p.d.,
d = plate spacing,
v = electron speed.

Hence the angle θ at which the beam emerges from the field, as if from the centre, is given by

$$\tan\theta = \frac{y}{(x/2)} = \frac{eV_p x}{2mdv^2}$$

The total deflection Y of the spot on the screen is given by

$$Y = L\tan\theta$$

where L is the distance from the centre of the field to the screen.

$$\text{So } Y = \left(\frac{Lex}{2mdv^2}\right)V_p$$

This complicated equation tells us that the total deflection Y is proportional to the plate p.d. V_p. In other words, double the plate p.d. and the deflection doubles, etc.

Lissajous' figures can be displayed using an oscilloscope tube. The time-base control is switched to *off*, and an a.c. signal of constant frequency is applied to the X-plates instead. The peak value of the signal is adjusted to make the spot trace out a line of the screen. Then, a signal generator is applied to the Y-plates, as in Figure 27.19(a). As the frequency of the signal generator is altered, different patterns are seen on the screen.

1 When the Y-frequency equals the X-frequency, the peak voltage of the signal generator can be adjusted to give a circle which seems to turn about its diameter. In fact, Figure 27.19(b) shows the sequence seen on the screen; when the two signals are in phase, a straight line at 45° is produced. If the phase between the two signals gradually changes at a steady rate, the line changes to a circle, then a line at −45°, then back to a circle, etc. The circle is produced when the two signals are 90° or 270° out of phase.

2 When the ratio of the two frequencies is a whole number, simple patterns called Lissajous' figures are seen on the screen. For example, if the Y-frequency is 2 × the X-frequency, then the spot goes up and down twice in the same time it takes to sweep left to right and back, so the result is a figure 8 on its side, as in Figure 27.19(c). Some other patterns are also shown.

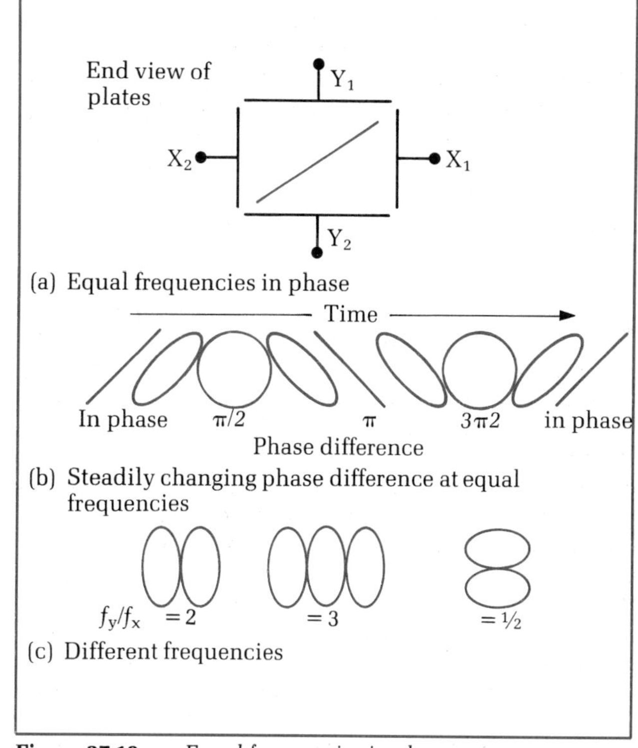

Figure 27.19 **a** *Equal frequencies in phase*
b *Steadily changing phase difference at equal frequencies*
c *Different frequencies*

27.4 Measurement of the charge of the electron

Balancing charged oil droplets

All electrons carry the same charge. If they did not, they would be deflected by differing amounts on passing through electric or magnetic fields. Precisely how much charge is carried by a single electron was determined first by Robert Millikan. Millikan was a brilliant experimenter who devised means for controlling and measuring the motion of tiny charged droplets of oil. Droplets from an oil spray were used. Such droplets are too small to be seen unless viewed in a beam of light using a microscope. Then they appear as pinpoints of light. Figure 27.20, at the top of the next page, shows an arrangement in simplified form like the one used by Millikan.

Two horizontal metal plates are arranged parallel to one another, one above the other. The top plate, P, has a tiny hole drilled through its centre. Oil droplets from the spray are allowed to fall through the tiny hole in the top plate into the space between the plates. A beam of light is directed into the space between the plates, and the microscope is used to observe the droplets as shown in Figure 27.20.

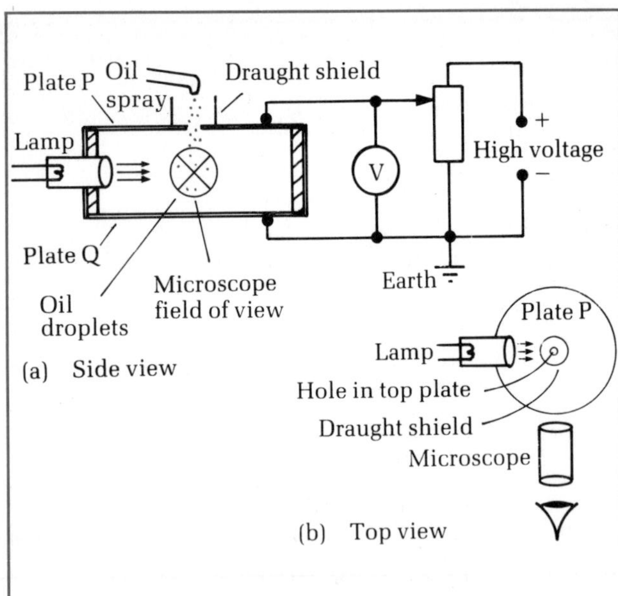

Figure 27.20 *Millikan's method for finding* e

1 Without any p.d. between the plates, the droplets fall at steady speed towards plate Q. Because the microscope inverts the field of view, plate Q is seen at the top of the field of view.

2 At the instant a p.d. is applied between plates P and Q, the motion of some of the droplets suddenly changes. These droplets carry charge; some droplets are negatively charged overall, and some are positively charged overall. The droplets which are not affected by switching the p.d. on are uncharged. Charging occurs when the droplets are forced through the nozzle of the spray. With plate P positive and plate Q negative, the negatively charged droplets experience a force towards plate P. So these droplets fall more slowly, or they stop moving, or they change direction to move towards P. What determines the change of motion? Each negatively charged drop is acted on by its weight (downwards) and by an electric field force (up towards P).

The drop moves

a) up towards P if the electric field force > its weight,
b) down away from P if the electric field force < its weight,
c) neither up nor down if the electric field force = its weight.

If the plate voltage can be increased gradually from zero, then it ought to be possible to observe directly the effect of increasing the electric force. As the plate voltage is increased, the drop falls more and more slowly until it is stopped; further increasing the plate voltage then makes the drop rise. When it was stationary, the plate voltage was such that the electric force balanced the weight exactly, as is shown in Figure 20.21.

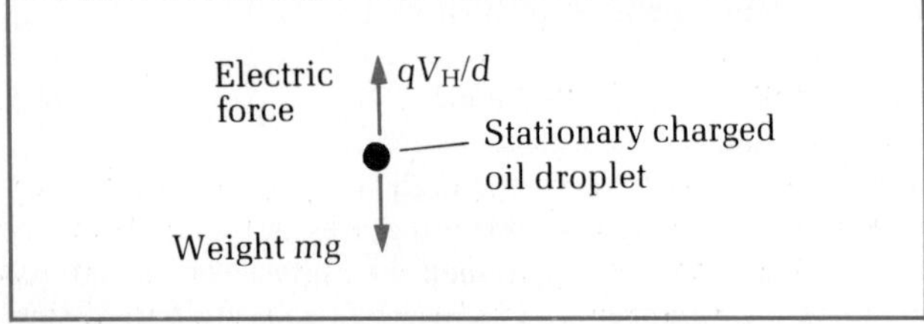

Figure 27.21 *A balanced droplet*

At balance, the droplet's weight (mg) is exactly equal and opposite to the electric field force. The electric field force is given by

Droplet charge $q \times$ the electric field strength V/d

$$\frac{qV_H}{d} = mg$$

where q = droplet charge,
d = plate spacing,
m = droplet mass,
g = acceleration due to gravity,
V_H = holding voltage (i.e. plate p.d. required to keep the drop stationary).

To determine the droplet's charge, the holding voltage V_H must be measured. In practice, the plate voltage is made large enough to make some droplets suddenly reverse direction when the voltage is applied. One of these droplets is then selected, and the plate voltage is reduced until that droplet is held stationary. The plate voltage is then measured, to give the holding voltage for that droplet. If the mass of the droplet is known, then the droplet charge can be calculated using the above equation.

Worked example An oil droplet of mass 4.9×10^{-15} kg is balanced and held stationary by the electric field between two parallel plates P and Q, as in Figure 27.20. The lower plate Q is earthed, and plate P is at a potential of +750 V. The spacing of the plates is 5.0 mm. Calculate the charge on the droplet, and state its sign. Assume $g = 9.8\ \text{m s}^{-2}$.

Solution Use the equation $qV_H/d = mg$ for the droplet charge q.
Its sign is negative because it is attracted by plate P which is positive.

$$\text{Hence } q = \frac{mgd}{V_H} = \frac{4.9 \times 10^{-15} \times 9.8 \times 5.0 \times 10^{-3}}{750}$$

$$= 3.2 \times 10^{-19}\ \text{C}.$$

Measuring the mass of a tiny oil droplet

How can the mass of a tiny droplet be determined accurately? Millikan showed that the speed of a falling droplet, with the plate voltage off, depends on the droplet's mass. So by timing the droplet as it falls with the plate voltage off, its mass can be calculated.

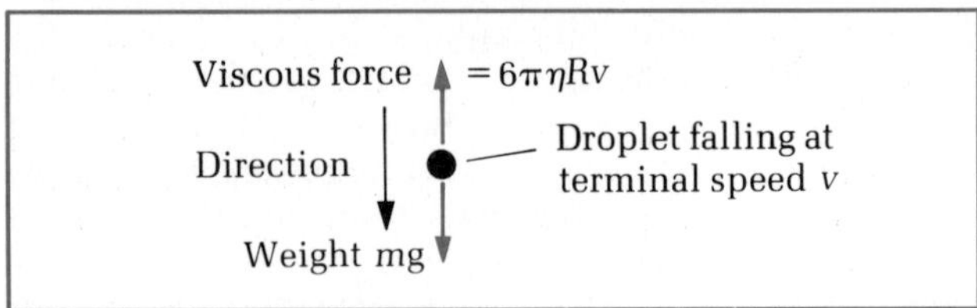

Figure 27.22 *Terminal speed*

Viscous forces act on any body moving through a fluid. The viscous force always opposes the motion. A simple example is when a steel ball bearing is released from rest at the surface of a long tube of water. The ball falls at constant speed because the viscous force increases as it moves faster. So the ball's speed builds up to a constant value when the viscous force is exactly equal (and opposite) to its weight. Thus no resultant force acts on the ball when it reaches this speed, called the **terminal speed**.

A tiny oil droplet falling through air is affected in the same way because air is viscous, although much less so than water. Stokes' law (see p. 100) states that the viscous force F on a sphere moving at speed v through a fluid is given by

$$F = 6\pi\eta Rv$$

where F = viscous force,
η = viscosity of fluid,
R = sphere radius,
v = sphere speed.

Consider what happens when the plate voltage holding a droplet stationary is switched off. Initially, the droplet is stationary so the viscous force is zero. The droplet therefore accelerates at g at first. However, as its speed builds up, the viscous force builds up too. So the resultant force (= weight − viscous force) becomes smaller as the drop speeds up. Its speed increases at a lower and lower rate until the viscous force balances the weight. The droplet is then moving at its terminal speed, given by

$$6\pi\eta Rv = mg$$

where m = droplet mass.

For a tiny oil droplet in air, the build-up to terminal speed is almost instant so the droplet may be assumed to move at steady speed from the moment the plate p.d. is switched off. The terminal speed is determined by timing the fall of the droplet over a measured distance.

To calculate the droplet's mass from the terminal speed, assume the droplet is a uniform sphere of density ρ. Its volume is given by $4\pi R^3/3$, so its mass is equal to volume × density = $\rho \times 4\pi R^3/3$.

Hence $6\pi\eta Rv = \rho \times 4\pi R^3/3$

So $$R^2 = \frac{9\eta v}{2\rho g}.$$

Hence the droplet radius can be calculated from its terminal speed, assuming values of the density of the oil (ρ) and viscosity of air (η) are known. Then the mass can be calculated from $\rho \times 4\pi R^3/3$.

Millikan's results obtained from measurements on many droplets showed that the droplet charge was always a whole number × 1.6×10^{-19} C. So he concluded that the charge carried by a single electron is -1.6×10^{-19} C. Thus a droplet carrying charge equal to $1.6 \times 10^{-19} \times N$ coulombs has either N electrons too many (if negative) or N electrons too few (if positive). Millikan thus proved that charge is **quantized**; in other words, charge is always in 'lumps' of 1.6×10^{-19} C.

Worked example Oil droplets are introduced into the space between two flat horizontal plates, as in Figure 27.20, set 5.0 mm apart. The plate voltage is then adjusted to exactly 780 V so that one of the droplets is held stationary. Then the plate voltage is switched off and the selected droplet is observed to fall a measured distance of 1.50 mm in 11.2 s. Given the density of the oil used is 900 kg m^{-3}, and the viscosity of air (at 20°C) is 1.8×10^{-5} N s m^{-2}, calculate
a) the mass and
b) charge of the droplet. Assume $g = 9.8$ m s^{-2}.

Solution

a) For the droplet mass m, first calculate its radius from

$$R^2 = \frac{9\eta v}{2\rho g}$$

The droplet's terminal speed $v = 1.50/11.2$ $= 1.34 \times 10^{-4}$ m s^{-1}.

Hence droplet radius is given by

$$R = \sqrt{\frac{9 \times 1.8 \times 10^{-5} \times 1.34 \times 10^{-4}}{2 \times 900 \times 9.8}}\ \text{m}$$

$$R = 1.11 \times 10^{-6}\ \text{m}$$

So droplet mass $m = \frac{4}{3}\pi R^3 \rho$

$$m = \tfrac{4}{3}\pi(1.11 \times 10^{-6})^3 \times 900 = 5.14 \times 10^{-15}\ \text{kg}$$

b) For the droplet charge q,

$$q = \frac{mgd}{V_H} = \frac{5.14 \times 10^{-15} \times 9.8 \times 5.0 \times 10^{-3}}{780}\ \text{C}$$

Hence

$$q = 3.2 \times 10^{-19}\ \text{C}.$$

Motion of charged droplets in an electric field

So far we have used the plate voltage to hold the selected droplet. Consider the same droplet when the plate voltage V is (a) less than V_H, (b) more than V_H.

a) Plate voltage < holding voltage V_H
The droplet falls at steady speed which is less than the speed when $V = 0$. The resultant force on the droplet is zero because its weight is balanced by the electric force *and* the viscous force. Figure 27.23 shows the idea.

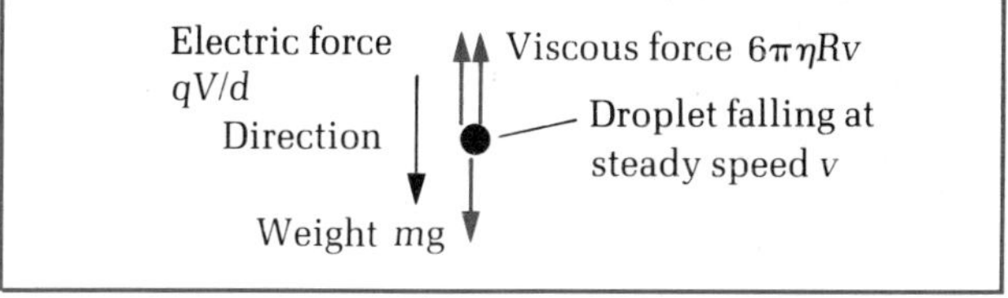

Figure 27.23 $V < V_H$

Weight = electric force + viscous force

$$mg = qV/d + 6\pi\eta Rv$$

$$\text{Hence } 6\pi\eta Rv = mg - \frac{qV}{d}$$

The droplet's speed depends on its weight and on the plate voltage. As the plate voltage is increased from 0 to V_H, the speed of fall is decreased to zero. So at any fixed value of voltage between 0 and V_H, the droplet falls at steady speed given by the above equation.

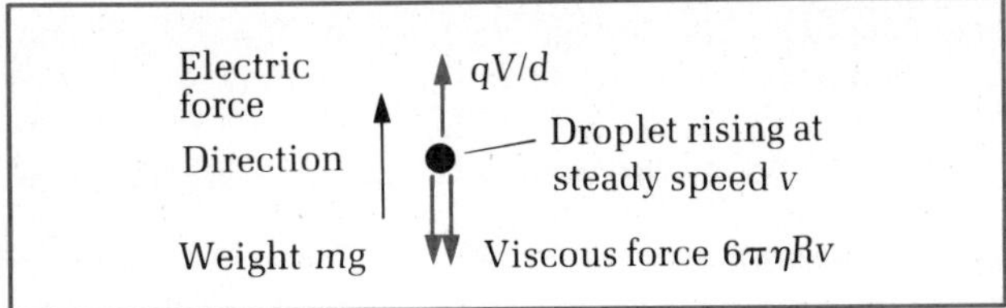

Figure 27.24 $V > V_H$

b) Plate voltage > holding voltage V_H
This time the droplet rises at steady speed. The viscous force acts downwards so the electric force is balanced by the weight + the viscous force.

Electric force = weight + viscous force

$$qV/d = mg + 6\pi\eta Rv$$

$$\text{Hence } 6\pi\eta Rv = \frac{qV}{d} - mg$$

27.5 Photoelectricity

The discovery of photoelectricity

Metals emit electrons when supplied with sufficient energy. Thermionic emission involves supplying the required energy by heating the metal. Another way of supplying the energy is to illuminate the metal with ultraviolet radiation. This was first discovered by Heinrich Hertz when investigating radio waves using a 'spark gap' detector (see p. 292). He found that the sparks were much stronger and thicker when ultra-violet radiation was directed at the spark gap contacts. Hertz was more interested in studying the radio waves so he left other scientists to follow up his discovery.

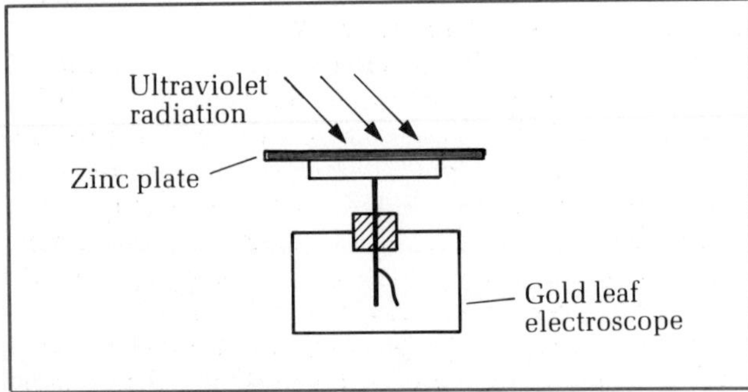

Figure 27.25 *Demonstrating photoelectricity*

A simple demonstration of photoelectric emission is shown in Figure 27.25. Ultra-violet radiation from a mercury lamp is to be directed onto a clean zinc plate placed on the cap of a gold-leaf electroscope.

a) Suppose the electroscope is given an initial negative charge so the leaf rises. With the lamp *off*, the leaf stays in the same position. However, when the lamp is switched on to allow ultra-violet radiation to fall on the zinc plate, the leaf gradually falls. The fall is stopped if a glass plate is placed between the lamp and the zinc plate. Ultra-violet radiation does not pass through glass so using the glass plate in this way is equivalent to switching the lamp off.
The leaf falls because the electroscope loses charge. This only happens when ultra-violet radiation falls on the zinc plate. Free electrons in the zinc plate gain sufficient energy from the ultra-violet radiation to leave the plate. These electrons are called **photoelectrons**.

b) If the plate and electroscope are made positive to start with, then no loss of charge takes place. The free electrons in the zinc plate need much more energy to leave the zinc plate because it is charged positively. The radiation cannot supply enough energy so no emission of electrons occurs.

Further investigations following Hertz's discovery showed that other metals can emit photoelectrons. If the metal is in a vacuum, a beam of light falling on it can cause photoelectric emission. Scientists showed that photoelectrons have the same specific charge as electrons emitted thermionically or produced in a discharge tube. So a photoelectron is identical to any other electron; we use the term 'photoelectron' to indicate that the electron has been emitted when light falls on the surface. Measurements showed that:

1. The number of photoelectrons emitted per second from any given metal is proportional to the intensity of the incident radiation (i.e. the light energy per second falling on the surface). The more intense the radiation beam is, the greater the number of photoelectrons leaving the metal each second.
2. Photoelectrons emitted from any metal have a range of kinetic energies up to a maximum value. The maximum kinetic energy was found to depend on the type of metal and on the frequency of the incident radiation.
3. For a given metal, no photoelectrons are emitted if the light frequency is below a certain value called the **threshold frequency**. If the light frequency is below the threshold frequency for that metal, no photoelectrons can be emitted, no matter how bright the light is. In fact, for zinc, the threshold frequency is in the ultraviolet range so light will not cause photoelectric emission from zinc.

The threshold frequency for a given metal is defined as the minimum frequency of electromagnetic radiation for which photoelectric emission occurs.

Understanding photoelectric emission

The results from investigations on photoelectricity presented a great puzzle to scientists for a time. Why does a threshold frequency exist at all? The wave theory of light tells us that emission of photoelectrons ought to take place whatever the light frequency. According to wave theory, each free electron should take some energy from each incoming light wave. So gradually each free electron ought to gain enough energy to leave the metal. With low frequency light, the emission should take longer – but it should, nevertheless, take place! It doesn't! Scientists could not explain photoelectric emission using wave theory. They entered the 20th century with photoelectricity unexplained, a weak point in the triumphs of 19th century science. Within a few years, these unexplained findings led scientists into a new era based on 'quantum physics'. It was as if photoelectricity provided a ladder to view the world from a new perspective, based on the no-

Figure 27.26 *Albert Einstein 1879–1955*

In 1905 Einstein published three papers, each of which changed existing ideas. His paper on Brownian motion showed that atoms must exist, and he established the photon theory of light in the second paper. Perhaps he is best remembered for his theories of relativity which he stated in the third paper.

tion that certain physical quantities such as charge and energy are 'lumpy'. These quantities can only have certain values; they are said to be 'quantized'.

The ladder to the new world was built in 1905 by a humble technical officer in a Swiss patent office. Albert Einstein was just 26 at the time; in 1905, he published three famous scientific papers, one of which explained photoelectric emission for the first time. The other two papers were rather important too! One was the theory of special relativity (see p. 465) which revolutionized our ideas of motion and energy. The other was the explanation of Brownian motion (see p. 64) which established once and for all that atoms do exist, even though they are too small to be seen.

Einstein's theory of photoelectric emission was developed from ideas first introduced by Max Planck to explain the laws of black body radiation (see p. 115). Planck showed that the laws could be explained by assuming light, and all other forms of electromagnetic radiation, is emitted in 'wavepackets'. Each wavepacket is a short burst of light energy from an atom. More important, Planck showed that when an atom emits light, its energy changes by certain allowed amounts only. But he didn't like the idea of 'lumpy' energy, even though it explained the laws of radiation so he kept trying to find an explanation where any value of energy was possible.

Einstein realized that photoelectricity could be explained using the wavepacket idea. He assumed that electromagnetic radiation consists of lumps of energy, each lump being the energy carried by a wavepacket. So when an atom emits light, it loses a lump of energy carried away by a wavepacket, which we call a photon.

A photon is a tiny packet of electromagnetic energy emitted by an atom. The term 'packet' is used because it gives a very different picture to waves on a water surface. The idea is that the wavepacket is a concentration of energy which travels in one direction only. So a point source of light emits photons in all possible directions, each photon travelling in one direction only away from the source.

The energy of a single photon is proportional to its frequency. By making this assumption, Einstein explained the existence of the threshold frequency of a metal.

Photon energy $E = hf$

where f = frequency,
h = Planck constant.

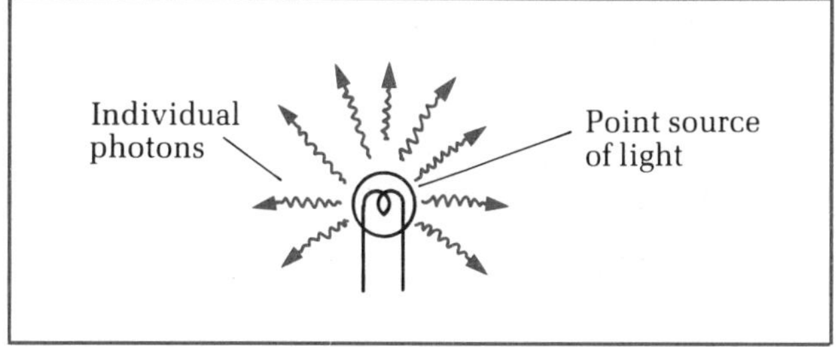

Figure 27.27 *Photons*

Consider what happens when light is directed at a metal surface. A stream of photons bombard the surface. Any free electron near the surface could be struck by a photon; such an electron would absorb the photon's energy so the electron would therefore gain kinetic energy. If the gain of energy is sufficient, the electron can leave the plate, as is shown in Figure 27.28. Each photoelectron from a metal plate has gained the whole amount of energy of a single photon. So the maximum energy gained by a photoelectron is equal to hf.

To escape from a metal, an electron must do a certain amount of work to remove itself from the surface to 'infinity'. The idea is not unlike the escape velocity from a planet, although the type of force is very different. An electron in the metal can only escape if it gains enough energy from a single photon to enable it to do the necessary work so individual photons must each supply more than a certain amount of energy. Since the light frequency determines the energy of each photon, the light frequency must therefore be greater than a minimum value.

Imagine you could see the sequence of events after an individual photon of energy hf is absorbed by a free electron in the metal. The electron's kinetic energy before absorbing the photon is much smaller than the photon's

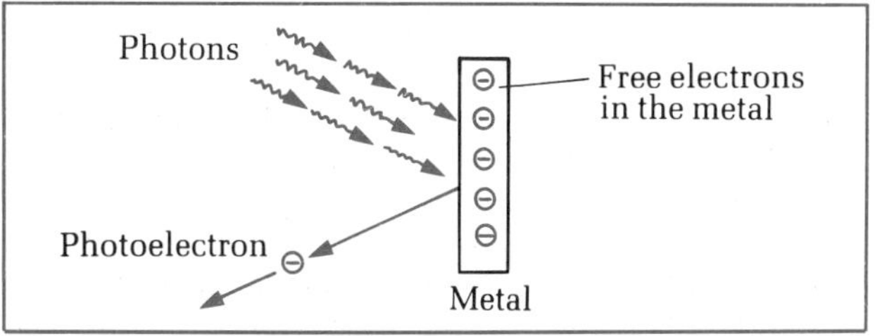

Figure 27.28 *Explaining photoelectricity*

energy so the energy gained by the electron increases its kinetic energy by a huge factor. If the electron's direction is out of the metal, the gain of energy enables it to shoot out through the surface. However, the electron must use some of its energy to do work to leave the metal. If its gain of kinetic energy is less than the work to be done to leave, then it is pulled back to the metal.

The work function W of a material is defined as the work necessary to remove an electron from the surface of the material.

So the maximum kinetic energy of a photoelectron is equal to the energy gained (hf) by absorbing a photon less the work done (W) to escape from the material.

$$\text{Maximum K.E.} = hf - W$$

where h = Planck constant,
f = light frequency,
W = work function.

The equation is known as **Einstein's equation for the photoelectric effect**. It shows that the maximum kinetic energy depends only on the light frequency. Making the light more intense increases the number of photons hitting the surface each second. But it doesn't make the light frequency – and hence the photon energy – change. So the maximum kinetic energy of the emitted photoelectrons is unaffected by making the light brighter.

The threshold frequency can now be explained. Electrons can only escape if the maximum kinetic energy is greater than zero.

Hence $hf - W > 0$

So $$f > \frac{W}{h}$$

In other words, the light frequency f must be greater than a value given by W/h if photoelectric emission is to occur.

Threshold frequency $f_0 = W/h$

So far we have considered photoelectric emission from earthed metals. What happens if the metal is at a fixed potential with respect to Earth? If the potential is negative, the photoelectrons escape more easily because they are repelled by the surface. But if the potential is positive, extra work must be done by each photoelectron to escape. For a metal at positive potential V, the extra work done is equal to charge (e) × potential (V). So the maximum kinetic energy which a photoelectron can have is given by

$$\text{Maximum K.E.} = hf - W - eV$$

What happens if the positive potential is increased gradually from zero? The maximum K.E. is reduced from $hf - W$ (when $V = 0$) to zero (when $hf - W = eV$). No electrons can escape when $eV = hf - W$ since even the most energetic photoelectron is pulled back onto the surface. Photoelectric emission is therefore stopped when $eV = hf - W$.

The stopping potential V_s of a surface is the positive potential which must be applied to the surface to stop photo electric emission at a particular light frequency.

$$eV_s = hf - W$$

where V_s = stopping potential,
W = work function.

To use Einstein's words, 'If the formula is correct, then V_s must be a straight-line function of the frequency of the incident light'. Any straight-line function is given by the equation $y = mx + c$ (see p. 502) where m is the gradient and c is the y-intercept. For the equation

$$V_s = \frac{hf}{e} - \frac{W}{e}$$

if V_s is represented by y, and f is represented by x, then the gradient is h/e. Einstein predicted that the gradient did not depend on the material since h and e are fundamental constants, the same for any material. However, the y-intercept does depend on the material used since its value is W/e.

Investigating photoelectricity

Experimental confirmation of Einstein's theory was obtained in 1916 by Robert Millikan. Millikan devised a vacuum tube in which clean metal surfaces could be cut. Figure 27.29(a) shows the arrangement. In fact, he used a drum with three different metals attached. The drum could be rotated so that each metal in turn could be given a freshly prepared surface using the rotating knife. A spectrometer was used to direct light of selected frequency onto the metal surface. Opposite the surface, a metal cylinder was positioned to collect the photoelectrons emitted from the metal surface. The potential of the clean metal was increased until the electroscope showed no further emission of photoelectrons. The potential, equal to V_s, was then measured using a voltmeter. Millikan

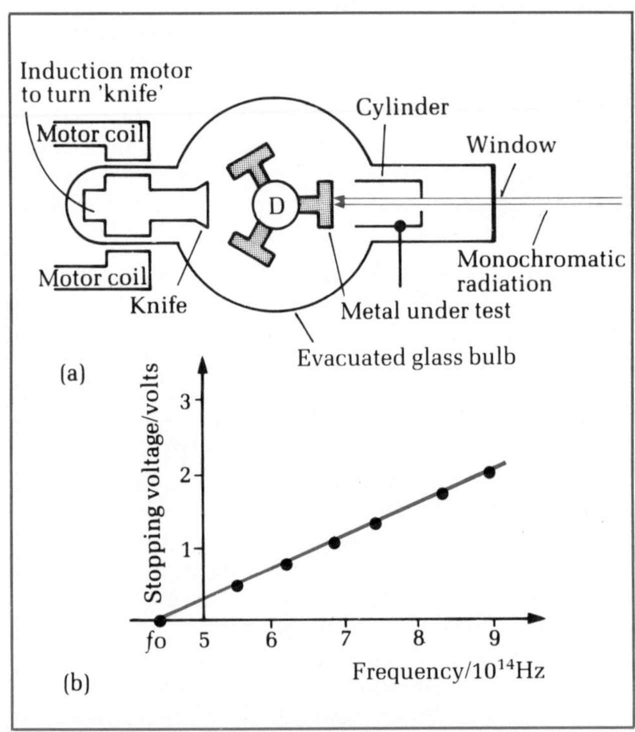

Figure 27.29 **a** *Millikan's photoelectricity investigation* **b** *Typical results*

made measurements using different frequencies for each metal in turn. Typical results for one of the metals are shown in Figure 27.29(b) as a graph of V_s (the stopping potential) against f (the frequency). The results obtained by Millikan fitted Einstein's predictions exactly, and gave a value for h equal to 6.6×10^{-34} J s.

The chief features of the graph are

1 The gradient for any metal is always equal to h/e.
2 The intercept on the f-axis gives the threshold frequency f_0. Hence the work function may be calculated from $W = hf_0$. This is because the equation for the stopping potential V_s is

$$V_s = \frac{hf}{e} - \frac{W}{e}$$

At the intercept on the f-axis, $V_s = 0$ so

$$\frac{hf_0}{e} - \frac{W}{e} = 0$$

since the frequency is equal to the threshold value f_0.

Hence $hf_0 = W$.

The wavelength λ of light is given by c/f, where c is the speed and f is the frequency of light. Hence the light wavelength λ corresponding to the threshold frequency f_0 is equal to c/f_0. Because of the inverse link between frequency and wavelength, it follows that for photoelectric emission to occur, the light wavelength must be *less* than c/f_0.

Consider two different metals A and B. Suppose the work function for A is less than that for B. Then

Threshold frequency for A $<$ threshold frequency for B.

For example, caesium has a much smaller work function than zinc. So caesium emits with visible light whereas zinc's threshold is in the ultra-violet range.

Worked example Potassium is illuminated with light of wavelength 500 nm and the stopping potential is measured at 0.226 V. The light wavelength is then altered to 380 nm and the new stopping potential is measured at 1.00 V. Calculate

a) the value of the Planck constant,

b) the work function in electron-volts of the metal.

Assume $e = 1.6 \times 10^{-19}$ C and $c = 3 \times 10^8$ m s^{-1}.

Solution

a) Use the equations $eV_s = hf - W$, and $f = c/\lambda$.

For $\lambda = 500$ nm, $f = 3 \times 10^8/(500 \times 10^{-9}) = 6.0 \times 10^{14}$ Hz, hence $e \times 0.226 = (h \times 6 \times 10^{14}) - W$.

For $\lambda = 380$ nm, $f = 3 \times 10^8/(380 \times 10^{-9}) = 7.89 \times 10^{14}$ Hz, hence $e \times 1.00 = (h \times 7.89 \times 10^{14}) - W$.

By subtraction,

$$e \times 1.00 - e \times 0.226 = h \times 7.89 \times 10^{14} - h \times 6 \times 10^{14}$$

$$\text{Hence } h = \frac{e \times 0.774}{1.89 \times 10^{14}} = \frac{1.6 \times 10^{-19} \times 0.774}{1.89 \times 10^{14}}$$

$$h = 6.55 \times 10^{-34} \text{ J s}$$

b) Either of the equations in **a)** may be used to calculate W, using the calculated value of h.

$e \times 1.00 = (6.55 \times 10^{-34} \times 7.89 \times 10^{14}) - W$

hence $W = 5.17 \times 10^{-19} - 1.6 \times 10^{-19} = 4.57 \times 10^{-19}$ J

Since 1 electron volt $= 1.6 \times 10^{-19}$ J, then

$$W = \frac{4.57 \times 10^{-19}}{1.6 \times 10^{-19}} = 2.86 \text{ eV}.$$

To estimate h a vacuum photocell may be used. The cell is part of a circuit, as in Figure 27.30. A variable positive potential is applied to the photoemitting plate P. The photoelectrons are collected by terminal Q, so a current of electrons passes through the microammeter.

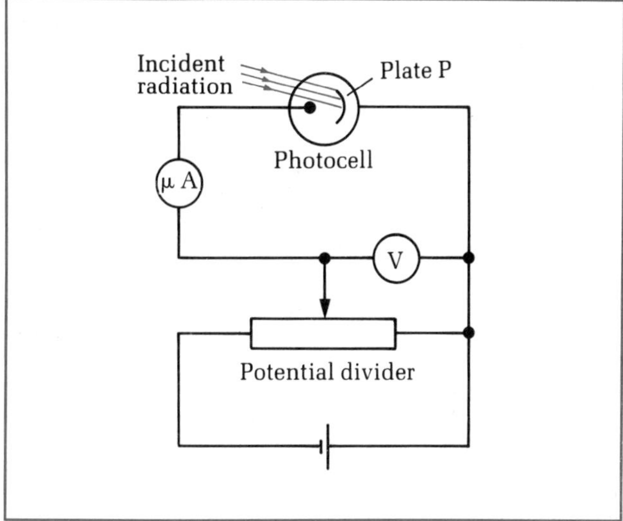

Figure 27.30 *Investigating photoelectricity*

Light of a selected known frequency from a spectrometer is directed at plate P. The potential of plate P is increased from zero using the potential divider. As the potential is increased, the microammeter reading falls to zero. The reason is that plate P is made increasingly positive, so photoelectrons must do more and more work to leave. Hence fewer and fewer photoelectrons reach terminal Q. The stopping potential is the potential of P when the microammeter reading is just zero. So the voltmeter reading is noted when the microammeter reading has become zero. The procedure is then repeated for other values of frequency. A graph of V_s against f ought to give results as in Figure 27.29(b). The value of h is given by the gradient $\times$ e.

27.6 Summary

Definitions

1 *Specific charge* (e/m) of an electron is its charge/mass value.
2 *The electron-volt* is a unit of energy defined as the work done when an electron is moved through a potential difference of 1 volt.
3 *The work function* W of a metal is the work which must be done by an electron to escape from the surface of the metal.
4 *The threshold frequency* of a metal is the minimum frequency of incident radiation that will cause photoelectric emission from that surface.

Equations

1 Electron gun equation $eV_A = \frac{1}{2}mv^2$
2 Electric field force on an electron $= eV_p/d$
3 Magnetic field force on a moving electron $= Bev$
4 Electron beam path in:
a) a uniform electric field is parabolic, given by the equation

$$y = kx^2$$

b) a uniform magnetic field is circular, given by the equation

$$Bev = mv^2/r$$

5 The velocity selector equation $Bev = eV_p/d$
6 For a stationary oil droplet in a uniform electric field

$$\frac{qV_P}{d} = mg$$

7 For a droplet falling at terminal speed in zero electric field, viscous force $6\pi\eta Rv = mg$
8 The energy of a photon $= hf$
9 The maximum kinetic energy of a photoelectron $= hf - W$
10 The stopping potential V_s is given by $eV_s = hf - W$

Short questions

Electron charge $e = 1.60 + 10^{-19}$C
Planck constant $h = 6.63 \times 10^{-34}$ J s
Speed of light $c = 3 \times 10^8$ m s^{-1}
Electron rest mass $= 9.1 \times 10^{-31}$ kg

27.1 Electrons from a hot-wire filament are accelerated onto a nearby positively-charged metal plate at a potential of 2000 V with respect to the filament.
a) Calculate the gain of K.E. of each accelerated electron just before impact at the plate.
b) Calculate the speed of each electron just before impact.
c) If the plate potential is increased to 4000 V, what is the speed of each electron now just before impact?

27.2 The specific charge (i.e. charge/mass) of the proton is twice that of the α-particle. A beam of protons is accelerated from rest through a p.d. of 1000 V to reach speed v. What p.d. must the α-particles be accelerated through to reach the same speed?

27.3 A beam of electrons is directed into uniform electric and magnetic fields that are at right angles to one another and to the beam direction. The strength of the magnetic field is 55 mT; the electric field strength is adjusted to 600 kV m^{-1} when the beam passes through the fields undeflected.
a) Give a diagram showing the directions of the fields and of the beam.
b) Calculate the speed of the electrons.

27.4 Electrons in a narrow beam travelling at a steady speed of 5×10^7 m s^{-1} are directed into the space between two oppositely-charged parallel plates 50 mm apart. The p.d. between the plates is 2500 V. The initial direction of the beam is parallel to the plates. Calculate
a) the force on each electron between the plates,
b) the time each electron spends between the plates if the plates are each 100 mm in length,
c) the speed and direction of the electrons on leaving the plates.

27.5 In question **27.4**, a fluorescent screen is positioned 300 mm from the end of the plates at right angles to the plates so that the beam electrons hit the screen after passing through the plates. Calculate the displacement of the spot on the screen produced by the electrons when the plate p.d. is switched from off to on.

27.6 A uniform magnetic field of strength 4 mT is used to deflect a beam of electrons into a circular path of diameter 65 mm.
a) Calculate the speed of the electrons.
b) If the field strength is halved, what will be the new diameter?
c) By what factor must the electron gun p.d. be increased at 2 mT field strength to bring the diameter in **b)** back to 65 mm?

27.7 An electron gun operating at a p.d. of 3000 V is used to direct a beam of electrons into a uniform magnetic field of strength 2.85 mT which forces the beam into a circular path of radius of curvature 65 mm. Use this information to calculate the specific charge of the electron.

27.8 An oscilloscope is used to display Lissajous figures by connecting a 50 Hz p.d. across the X-plates and a variable-frequency signal generator across the Y-plates. Sketch and explain the observed pattern when the signal generator frequency is
a) 25 Hz,
b) 50 Hz,
c) 100 Hz,
d) 150 Hz.

27.9 A charged oil droplet of mass 8.22×10^{-16} kg is held stationary by the electric field between two oppositely-charged horizontal parallel plates one above the other at a spacing of 50 mm. The top plate is at a potential of +840 V with respect to the lower plate.
a) Is the droplet charged positively or negatively?
b) Calculate the droplet's charge.
c) If the plate voltage is suddenly reversed, what will be the initial acceleration of the droplet?
Assume $g = 9.8$ m s^{-2}.

27.10
a) In question **27.9**, if the stationary droplet is suddenly allowed to fall by switching off the plate voltage, the droplet's speed quickly builds up to a constant value.
i) Sketch a graph to show how the speed builds up.
ii) Explain why the speed builds up to a constant value.
b) If the viscous force on a droplet falling through air is given by $3.34 \times 10^{-4} Rv$ where R is the droplet's radius and v is its speed, calculate the expected terminal speed for the droplet in question **27.9**. The density of the oil is 900 kg m^{-3}.

27.11 In question **27.9** with the droplet initially stationary, what would happen to the droplet if
a) the plates were moved further apart at the same p.d.,
b) an uncharged droplet coalesced with the charged droplet,
c) the charged droplet suddenly gained an extra electron with no significant increase of its mass?

27.12 Two vertical parallel plates are spaced 20 mm apart. A charged oil droplet is observed falling between the two plates at a constant speed of 0.08 mm s^{-1} vertically downwards.

Calculate
a) the droplet's radius, given the viscous force and oil density as in question **27.10**,
b) the droplet's mass,
c) the charge on the droplet if a p.d. of 480 V across the plates makes it move downwards at 45° to the vertical at steady speed.

27.13 Monochromatic light of wavelength 550 nm is incident on a metal surface in a vacuum photocell to produce photoelectric emission. The emission can be stopped by applying a positive potential of 1.40 V to the metal plate. Calculate
a) the work function of the metal,
b) the maximum K.E. of emitted photoelectrons when the plate potential is zero.

27.14
a) i) Explain why photoelectric emission from a given metal surface only occurs if the frequency of the incident light exceeds a certain minimum value.
ii) Explain why the emitted electrons have a range of energies up to a maximum value.
b) The work function of a certain metal is 2.1 eV. When monochromatic light of wavelength 430 nm is incident on the metal, calculate the maximum K.E. of the emitted photoelectrons.

27.15 The circuit shown below is used to investigate how the stopping potential for a certain metal affects the photoelectric current when the metal is illuminated by monochromatic light of constant intensity. The graph shows how the photoelectric current varies with the p.d. across the photocell.

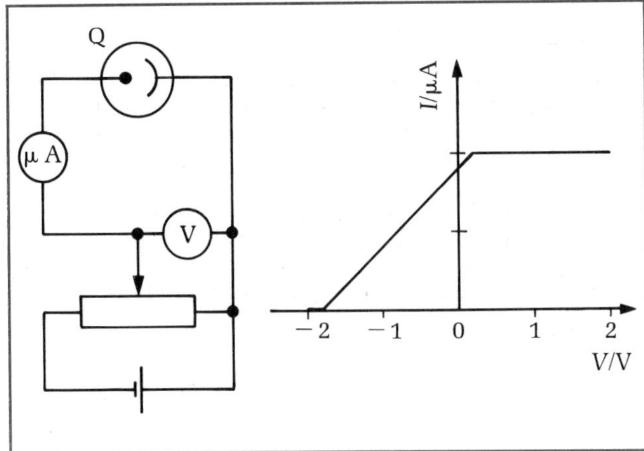

Figure 27.31

(a) Explain why the current stays constant if the battery in Figure 27.31 is reversed and Q is made more and more positive.
b) Explain why the photoelectric current falls when the p.d. V is made more and more negative.
c) If the light frequency is 6.67×10^{14} Hz, calculate the work function of the metal.
d) Copy the graph and sketch on it the curves you would expect for
i) increased light intensity,
ii) increased light frequency,
iii) increased light intensity and frequency.

28

Radioactivity

28.1 The discovery of radioactivity

The story of the discovery of radioactivity is yet another example of where natural curiosity leads. Henri Becquerel in 1896 was investigating materials which glow when placed in an X-ray beam. Curious to find out if strong sunlight caused uranium salts to glow, he prepared a sample ready for a sunny day. He put the sample into a drawer next to a wrapped photographic plate. A few days later, he decided to develop the film to see if it had been affected by his X-ray machine. To his amazement, the film had become blackened and showed the image of a key. He had put the key on the plate then put the uranium salts on top of the key when he originally put the salts in the drawer. It was clear that uranium salts emit penetrating radiation which can blacken photographic film. More interested in X-rays, Becquerel passed the challenge of further investigations on to one of his students, Marie Curie.

Within a few years, Madame Curie discovered other elements such as thorium which are radioactive. One of these elements, radium, was found to be over 1 million times more radioactive than uranium; to isolate 1 gram of pure radium, Madame Curie started with a tonne of pitchblende from which the radium was eventually isolated.

Figure 28.1 *Marie Curie*

Marie Curie established the nature of radioactive materials. She showed how radioactive compounds could be separated and identified. She and her husband Pierre won the 1903 Nobel Prize for their discovery of two new elements, ***polonium*** *and* ***radium****. After Pierre's death in 1906 she continued her painstaking research and was awarded a second Nobel Prize in 1911 – an unprecedented honour.*

The nature of the radiation was successfully investigated by Ernest Rutherford. Like Marie Curie, he was a skilful and imaginative scientist, and in 1899 he showed that there were two distinct types of radiation. One type, which he called **alpha** (α) radiation, was easily absorbed. The other type was more penetrating, and this type was called **beta** (β) radiation. A third type, called **gamma** (γ) radiation, even more penetrating than β-radiation, was discovered a year later.

Another way to identify each type of radiation is to use a magnetic field to deflect the radiations. Measurements showed that the alpha radiation consists of positively charged particles since a beam of α-particles crossing a magnetic field is deflected as shown in Figure 28.2. In the same way it was discovered that β-radiation consists of negative particles whereas gamma radiation is uncharged.

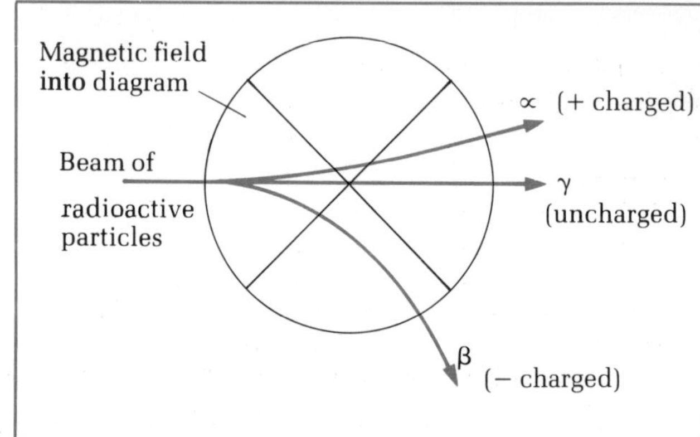

Figure 28.2 *Deflection by a magnetic field*

Rutherford and other scientists wanted to know exactly what each type of radiation was. What caused them? Thomson's discovery of the electron had shown that atoms were not indivisible. The puzzle of radioactivity added to the 'mystery' of the atom.

The nature of radioactive radiations

α-radiation was shown to consist of fast-moving ticles, each with mass about the same as a helium atom. Rutherford designed equipment to collect the α-particles emitted from a radioactive gas called thoron. The gas was held in a thin-walled glass tube in the middle of a special glass chamber. The α-particles emitted by the thoron atoms could pass through the wall of the tube. Eventually the outer chamber became filled with α-particles. The chamber was designed with two electrodes as shown in Figure 28.3, so that a high voltage could be applied between the electrodes. Light was emitted from the chamber when the voltage was applied. Using a spectrometer, Rutherford showed that the light emitted from the 'α-gas' was the same as from helium gas so he concluded that the α-particles were identical to helium ions. Once trapped in the chamber, the particles became neutralized by gaining electrons. So the particles became helium atoms.

β-radiation was shown to consist of fast-moving electrons. They can be deflected into a circular path by using a uniform magnetic field. Measurement of the specific charge of β-particles gave the same value as for

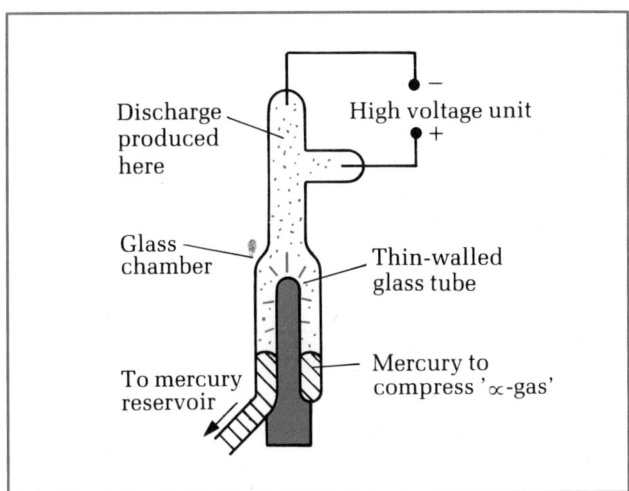

Figure 28.3 *Identifying α-particles*

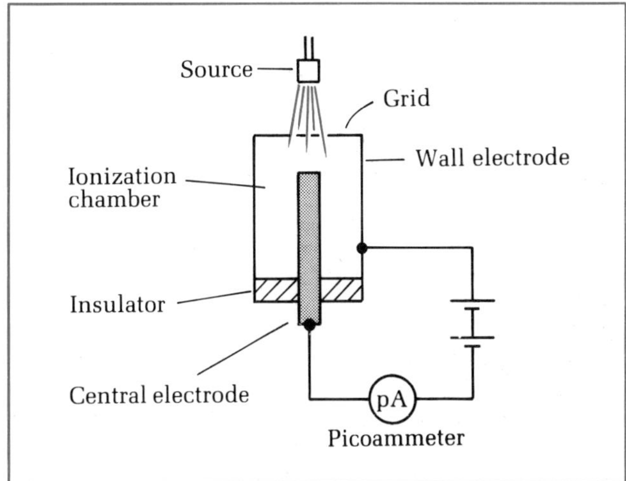

Figure 28.4 *Investigating ionization*

electrons so scientists concluded that β-particles are electrons, produced in a way which was yet to become clear.

γ-radiation was shown to be electromagnetic radiation with wavelength of the order of 10^{-10} m or less. Just as light can be analysed using a diffraction grating, so γ-radiation can be analysed using a 'crystal grating'. A crystal is a regular arrangement of layers of atoms at very close spacing so it acts like a diffraction grating, provided the wavelength is of the same order as the atom spacing (see p. 80).

The nature of each type of radiation was uncovered within ten years of Becquerel's initial discovery, but the information gained since then made the mystery of the atom even deeper. Why should big atoms like uranium emit alpha particles anyway? What was the mechanism which governed the emission of β-particles? The answers seemed to give rise to more and more questions.

28.2 The properties of radioactive radiations

Ionization

An ion is a charged atom. Negative ions are formed by adding electrons to uncharged atoms, positive ions by removing electrons. Radioactive radiations cause ionization of atoms in air by knocking electrons off atoms. To create a single pair of ions from an uncharged atom, work must be done to separate positive and negative charges. For air molecules, it takes about 30 electron-volts of energy to produce an ion pair. Radioactive particles are emitted with energies up to several Mev (i.e. millions of electron-volts). So a single radioactive particle could produce of the order of 10^5 ions in air.

The ionizing effect of each type of radiation can be investigated using an ionization chamber and a picoammeter, as shown in Figure 28.4. Ions created in the ionization chamber are attracted to the oppositely charged electrode. The negative ions reaching the positive electrode are neutralized by releasing electrons to the electrode. The positive ions at the other electrode take electrons off the electrode to become neutral. So a current of electrons passes through the picoammeter as a result of ionization in the chamber. The current is proportional to the number of ions per second created in the chamber.

1 Using an α-emitting source, the source is moved away from the grid of the ionization chamber in steps. At each step, the current and distance of the source from the chamber is measured. The current falls abruptly to zero if the source is moved beyond a certain distance from the grid. This proves that α-particles have a well-defined range. The range is determined by the number of ion pairs an alpha particle can produce; each ion pair requires the α-particle to do a fixed amount of work. So the initial kinetic energy of each α-particle determines how many ions it can produce and how far it will travel. Since the α-particles from a given source travel equal distances, they must have equal initial K.E.

   Different α-sources give different ranges and hence different energies. A method for measuring the energy of the α-particles from a given source is described later.

2 Using a β-emitting source, the ionizing effect is much weaker than with alpha particles. Also, there is no well-defined cut-off point as the source is moved away from the grid, and the range in air can vary by up to a metre or more. In general, α- and β-particles are emitted with initial kinetic energies of the same order so they both produce similar numbers of ions; however, the beta particles on average travel much further than alpha particles in air. So β-particles produce fewer ions per mm along the path.

3 Using a gamma-emitting source, the ionizing effect is much weaker than either α- or β-radiations. γ-radiation consists of electromagnetic waves which of course do not carry charge. So the ionizing effect is much less than α- or β-particles which exert electrical force on electrons in atoms.

Energy

The energy of alpha particles can be determined using a sealed ionization chamber with the α-emitting

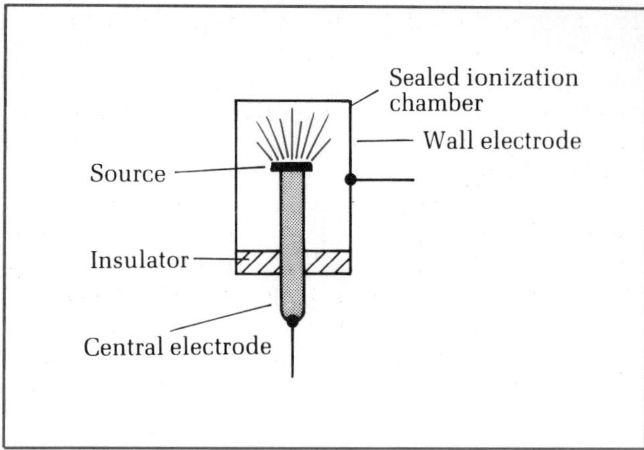

Figure 28.5 *Measuring α-energies*

source inside the chamber, as shown in Figure 28.5. The number of alpha particles emitted each second from the source must be counted separately. This can be done by using a microscope to observe and count the flashes of light produced when individual alpha particles hit a screen coated with zinc sulphide. Figure 28.6 shows the idea. With the source inside the sealed chamber, as shown in Figure 28.5, the ionization current is measured from the picoammeter. Let's consider some typical measurements to see how the energy of each emitted alpha particle is measured.

Ionization current $= 1 \times 10^{-11}$ A
Source activity $= 1.2 \times 10^{3}$ particles emitted per second

Assume that creating an ion pair involves removing an electron (charge 1.6×10^{-19} C) from an uncharged atom to leave a positive ion. Also, assume that 30 eV of work must be done to create a single ion pair.

In one second, 1.2×10^{3} α-particles leave the source, creating ions in sufficient number to give a current of 10^{-11} A. Hence the total charge created in one second equals 10^{-11} C (= current × time). So the total number of ion pairs produced is $10^{-11}/1.6 \times 10^{-19} = 6.25 \times 10^{7}$.

Thus a single alpha particle on average creates

$$\frac{6.25 \times 10^{7}}{1.2 \times 10^{3}} = 5.21 \times 10^{4} \text{ ion pairs.}$$

Since 30 eV of energy is needed for each ion pair, the total energy used by a single α-particle is

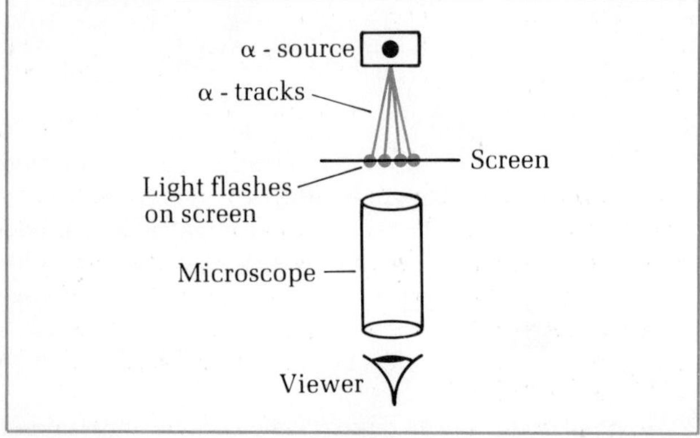

Figure 28.6 *Counting α-particles*

$5.21 \times 10^{4} \times 30 = 1.56 \times 10^{6}$ eV. Therefore, each α-particle is emitted with 1.56 MeV of kinetic energy which it uses to create ions. When it has used up its kinetic energy, it takes a couple of electrons from any air atom nearby so it becomes a helium atom.

For β-particles deflecting the particles using a magnetic field shows that there is a range of speeds from a given source. Accurate measurements show that there is a maximum speed from any given source, different for each type of source. Hence each source emits β-particles with a range of energies up to a maximum value. Using a uniform magnetic field, it is possible to measure the maximum speed (and hence K.E.) for any given source.

For γ-radiation the energy depends on the frequency. By measuring the wavelength of the radiation using a crystal grating, the frequency and hence the energy can be calculated (see p. 80).

Absorption

To investigate absorption by different materials, the arrangement shown in Figure 28.7 can be used. The radioactive radiation is detected using a geiger tube (see p. 439). attached to a scaler counter. Each particle that enters the tube is registered by the counter. The count rate, the number of particles per second that enter the tube, is determined by measuring the number of counts in a set time; hence count rate = number of counts/time taken. The arrangement could also be used to measure the range in air; to do so, the material is removed from between the source and tube. Then the count rate is determined for increasing distances between the source and tube.

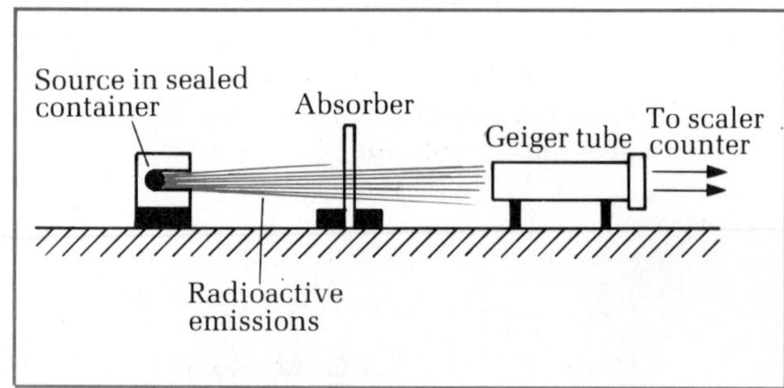

Figure 28.7 *Investigating absorption*

Background radiation must be taken into account using a geiger tube. Remove the source and see if the counter still operates. You should discover that it counts infrequently at random. The counting without the source present is due to background radiation. Most background radiation occurs naturally due either to cosmic radiation or, in certain areas, to rocks. To take account of background radiation, the count rate without the source present should be measured, timing over ten minutes or more, to give an average value for the number of counts per minute due to background radiation. The background count rate must then be subtracted from the count rate measured with the source present. This gives the true count rate due to the source only.

For α-particles Different materials are placed between the source and detector at fixed distance. You should find that metal foil or paper is capable of stopping α-particles. If possible, use sheets of thin paper to measure the count rate with different thicknesses of paper.

For **air**, remove the material used before. Place the end of the tube close to the source and then slowly move the tube away from the source. You ought to find that the counting abruptly stops when the tube is out of range of the source. Hence you can measure the range of the α-particles. Figure 28.8 shows how the range varies with the energy; the curve is drawn using energy and range values for different sources, the energy measured as explained earlier. As you might expect, the greater the energy, the longer the range. Use your measurement of range to determine the energy of the α-particles from the source used.

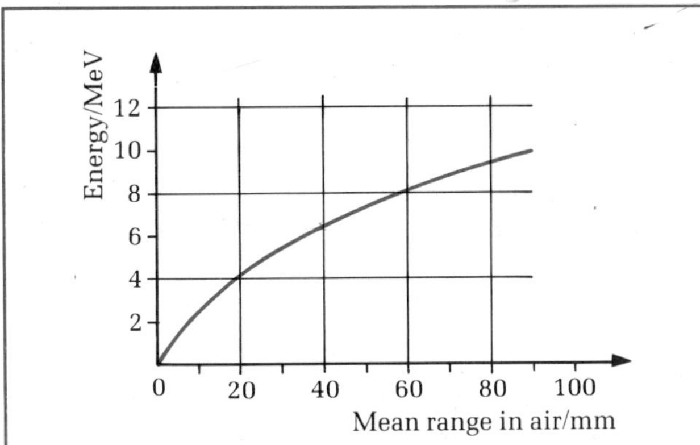

Figure 28.8 *Energy v range in air for α-particles*

For β-particles Use different thicknesses of aluminium foil with the source and detector at fixed separation. The count rate falls as the thickness increases. Figure 28.9 shows a graph of typical results of count rate against thickness using semi-log paper. The graph shows that the count rate (assumed corrected for background radiation) falls logarithmically with increasing thickness up to a certain value of thickness. Then the count rate evens out to a constant value. The reason for the change is that the logarithmic section represents β-particles being absorbed; the curve evens out where the most energetic β-particles are absorbed. After that, the count rate is due to γ-particles produced when the more energetic β-particles

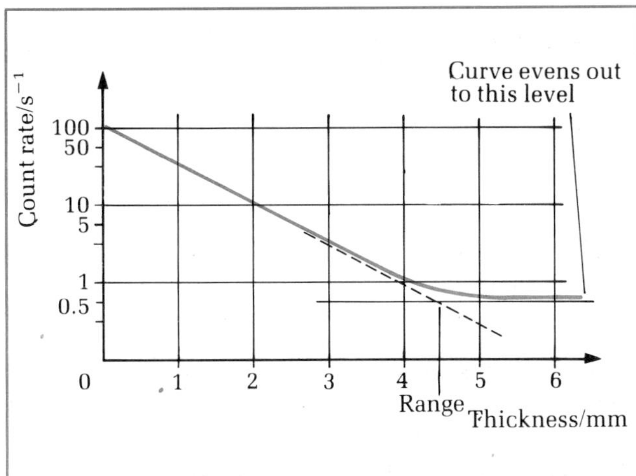

Figure 28.9 *Absorption curve for β-particles in aluminium*

are stopped. From the graph, **the range in aluminium** of the β-particles from that source can be estimated. Use your own results to determine the range for the source you have used. Then, use the energy against range curve of Figure 28.10 to determine the maximum energy of the β-particles from your source.

Other metals give the same pattern of results as aluminium, although the gradient of the sloped section of the graph in Figure 28.9 depends on the metal used. So too does the range. Lead is one of the best absorbers because it has a high density; its atoms are big with lots of electrons each so a β-particle passing through 1 mm of lead loses more energy than when passing through 1 mm of aluminium.

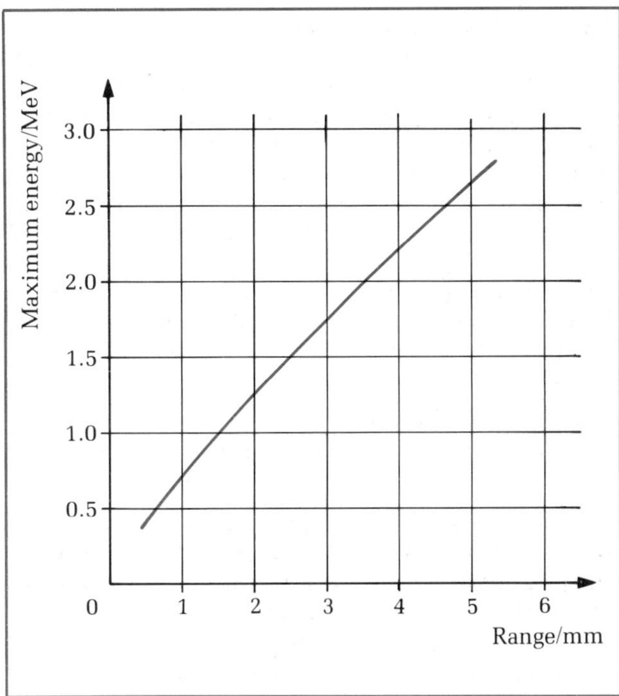

Figure 28.10 *Energy v range in aluminium for β-particles*

For γ-particles Use different thicknesses of lead rather than aluminium. γ-radiation is much more penetrating than β-radiation so lead is necessary – unless you have lots of aluminium available! Once again, determine the corrected count rate for different thicknesses of lead placed between the source and detector at fixed distance. Figure 28.11, at the top of the next page, shows typical results of count rate against thickness, this time using normal graph paper. The count rate decreases exponentially with increasing thickness. Suppose the count rate is C_0 for zero thickness, as shown by Figure 28.11. The count rate falls by a constant factor every time the thickness is increased by a constant amount.

The half-value thickness $X_{\frac{1}{2}}$ is the thickness of material which cuts the count rate by exactly 50%. For example, if the half-value thickness for a certain source is 5 mm, then thickness of material equal to $2X_{\frac{1}{2}}$ or 10 mm cuts the count rate to 25%. Figure 28.11 shows the idea. Use your own results to plot a graph like this, and find its half-value thickness.

Gamma radiation, like any form of electromagnetic radiation, consists of photons, each photon being a

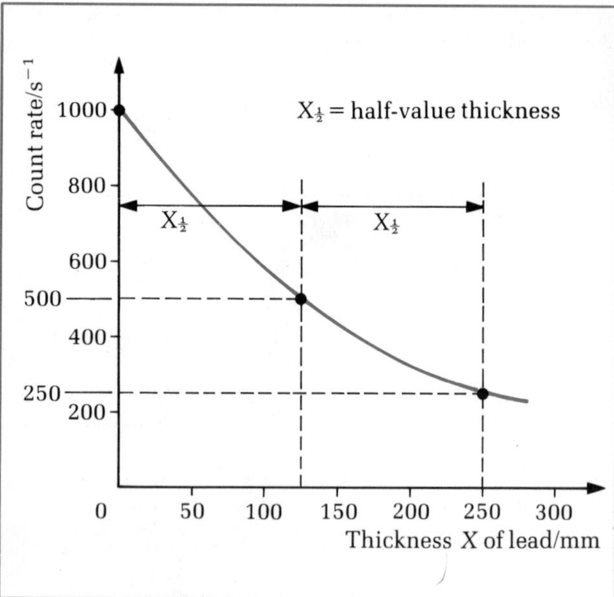

Figure 28.11 *Absorption of γ-radiation by lead*

wavepacket of electromagnetic energy (see p. 429). The photons from a given source have equal energies which differ from one source to another. Measurements of the half-value thickness and energy for different gamma sources shows that the half-value thickness increases with γ-photon energy. Figure 28.12 shows the link. Use the graph to determine the γ-photon energy for the source used in the half-value measurement above.

The inverse-square law for γ-radiation in air

Photons from a γ-emitting source spread out in all possible directions. If the source is in a box lined with thick lead sheet, then the γ-photons will be prevented from reaching the outside, provided the lead is very thick! But if the source is not in such a box, then the γ-photons spread out through the surrounding air with no absorption.

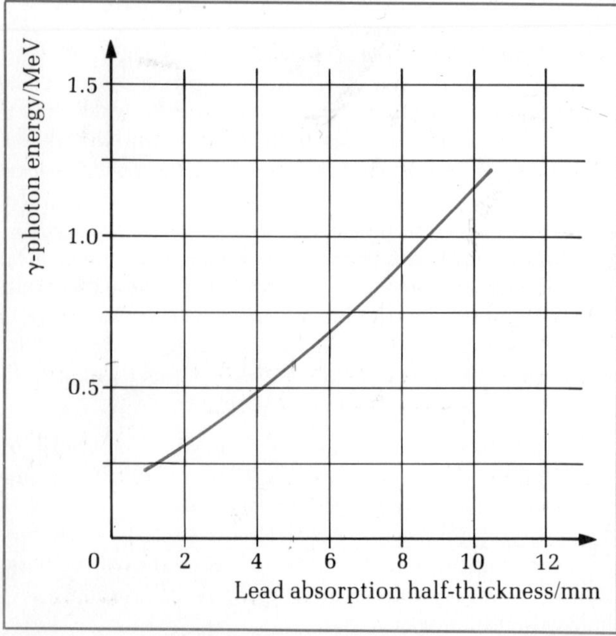

Figure 28.12 *Energy v half-value thickness for γ-radiation in lead*

Imagine holding a geiger tube close to a γ-source; move the tube away from the source and the count rate falls. The reason is that fewer photons enter the tube each second because the photons spread out from the source. If N γ-protons per second leave the source, then the count rate is proportional to N, but it doesn't equal N because most of the emitted photons miss the tube altogether. At distance r from the source, the gamma photons pass through the surface of a sphere of radius r centred on the source. So the number of γ-photons per second passing through unit area of the sphere is $N/4\pi r^2$ since the sphere area is $4\pi r^2$. Provided the tube always presents the same area A to the source, then the number of γ-photons per second entering the tube is $A \times N/4\pi r^2$. So the count rate is proportional to $1/r^2$.

In other words, the count rate follows the 'inverse-square' law; double the distance r and the count rate falls to 1/4 of its previous value, treble the distance and the count rate falls to 1/9, etc.

$$C = \frac{k}{r^2}$$

where C = (corrected) count rate,
r = distance from tube to point source,
k = constant.

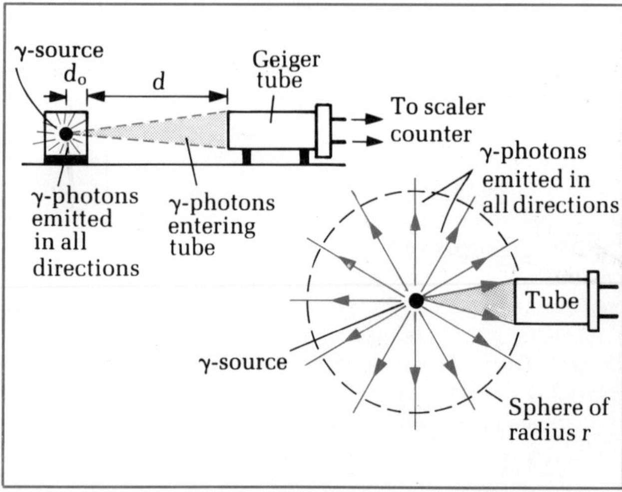

Figure 28.13 *Investigating the inverse square law for γ-radiation*

To verify the inverse-square law for γ-radiation, use the arrangement shown in Figure 28.13. The distance d between the end of the tube and the source holder is increased in steps. At each step the count rate is measured, taking three readings to give an average value at each distance.

The γ-source is inside a protective holder to prevent any radioactive material from the source contaminating the laboratory. Since its exact position is not known, assume the source is at distance d_0 inside the holder, as in Figure 28.13. Hence the distance from the source to the tube, r is equal to $(d + d_0)$. So the count rate $C = k/r^2 = k/(d + d_0)^2$. We do not know the value of d_0 as

Summary of the properties of radioactive radiations

Property	α	β	γ
Charge	+	–	Zero
Deflection by a magnetic field	Yes	Yes	No
Energy	Constant for a given source	Varies up to a maximum which depends on source	Constant for a given source
Range in air	Fixed range of the order of cm for a given source	Varies up to a metre or so in air	Follows the inverse square law
Ionization in air	Strong – produces about 10^3–10^4 ions per mm of path (at atmos. pressure)	Less strong than α-particles, about 100 ions per mm	Weak
Absorbed by	Paper or thin metal foil	Approx. 5 mm aluminium	Thick lead

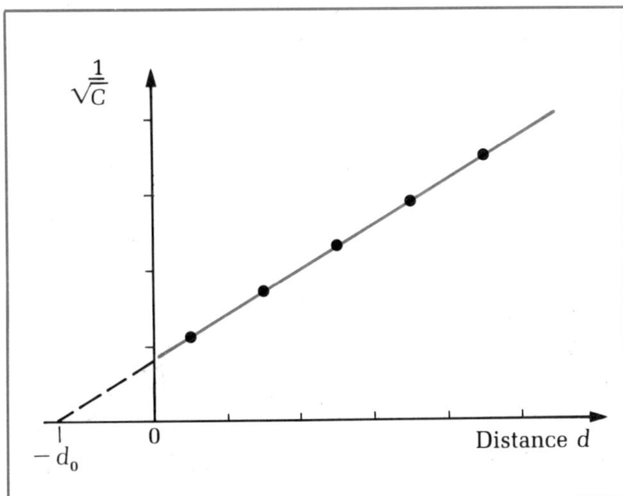

Figure 28.14 *Graph of $1/\sqrt{C}$ v d*

yet so we cannot calculate r. However, the count rate equation can be rearranged as

$$r^2 = \frac{k}{C}$$

Giving $$r = \frac{\sqrt{k}}{\sqrt{C}}$$

Or $$d = \frac{\sqrt{k}}{\sqrt{C}} - d_0$$

Use your results to plot a graph of d against $1/\sqrt{C}$. From the above equation, it should give a straight line with gradient $\sqrt{k}$. The equation predicts that the d-intercept is at $-d_0$, so from your graph you ought to be able to calculate how far the source is inside the holder. Assuming your results plotted as above give a straight line, then the inverse square law is valid.

The intensity of the radiation is the energy per second passing through unit area normal to the area. Since the energy of each γ-photon from a given source is always the same, then the energy per second is proportional to the number of photons per second. Hence the intensity is proportional to the count rate. The intensity therefore follows the inverse-square law.

$$I = \frac{\text{constant}}{r^2}$$

where I = intensity at a certain point,
r = distance from point to source.

The properties of α-, and β- and γ-radiations are tabulated at the top of this page.

28.3 Radiation detectors

The geiger tube

Although invented many years ago, Hans Geiger's device is still a commonly-used detector of ionizing radiations. Figure 28.15 shows a cross-section through a geiger tube.

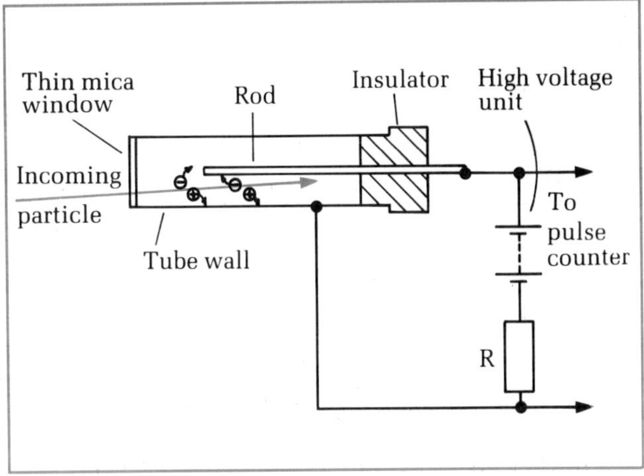

Figure 28.15 *A geiger tube*

The geiger tube is sealed and contains argon at low pressure. The thin mica window at the end of the tube allows alpha and beta particles to enter the argon gas from outside. Gamma radiation can enter via the tube walls as well. A metal rod down the middle is given a positive potential, as shown in Figure 28.15. The tube wall is connected to the negative terminal of the power supply unit, and the negative terminal is earthed.

When a radioactive particle enters the tube, the particle ionizes gas atoms along its track. The negative ions created are attracted to the rod, and the positive ions to the wall. The ions accelerate (because of the electric field between the tube and wall) and they collide with other atoms, producing more ions. These secondary ions are accelerated too and they then collide with other atoms to create more ions again. In a short time, lots of ions are created and discharged at the electrodes. A pulse of electrons passes round the circuit through the resistor R. Hence the p.d. between the tube wall and rod changes briefly to supply a voltage pulse to the electronic counter. The counter reading goes up by 1 when a single radioactive particle enters the tube.

The dead time of a tube is the time taken to regain its non-conducting state after a radioactive particle enters. Typically, the dead time is about 0.2 ms. If another particle enters during this time, it will not be registered because the tube is already conducting (i.e. ionized). So the count rate should not be greater than about 5000 s^{-1} ($=1/0.2$ ms).

To keep the dead time as short as possible, a 'quenching agent' such as bromine is added to the gas in the tube. Positive ions hitting the negative terminal (i.e. the wall of the tube) cause electrons to be emitted from the cathode. These secondary electrons would cause further ionization if the quenching agent was not present. Bromine atoms 'mop up' these unwanted electrons to reduce the dead time.

The working voltage of a geiger tube is an important characteristic. With a radioactive source at a fixed distance from the tube, the count rate can be measured for different values of voltage. Figure 28.16 shows a graph of count rate against tube voltage; the graph shows that the count rate is zero unless the tube voltage is greater than a certain value called the 'threshold value'. The working voltage should be on the flat part of the curve above the threshold value. The count rate is then independent of the tube voltage, so if the tube voltage drops slightly from the working voltage, the count rate does not change. Beyond the flat part of the curve, the count rate increases with increasing voltage; the tube becomes damaged if operated in this range.

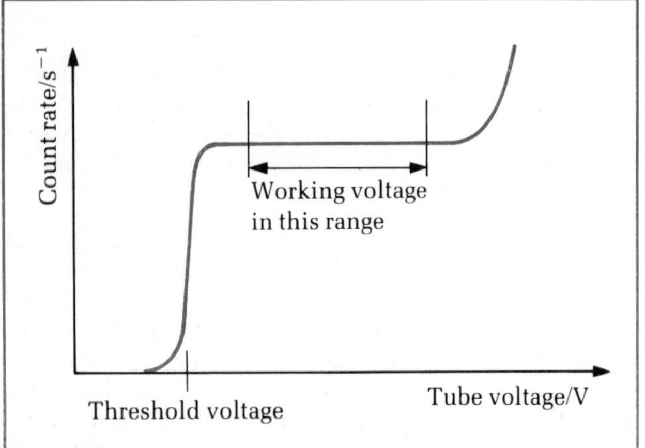

Figure 28.16 *Geiger tube characteristic*

The accuracy with which the count rate can be measured is greater for a large number of counts than for a small number. Radioactive particles are emitted at random from any source because the disintegration is random. Suppose you wish to determine the count rate from a given source; by counting up to a large number, the randomness is evened out, compared with a much smaller count. So operating the counter for 1 minute does not give as accurate a value as counting for 5 minutes then dividing by 5.

A ratemeter can be used as an alternative to a scaler counter. The reading of a ratemeter is a direct measure of the number of counts per second entering the tube. The ratemeter measures the average current round the circuit by using a capacitor across the tube terminals to smooth the pulses out.

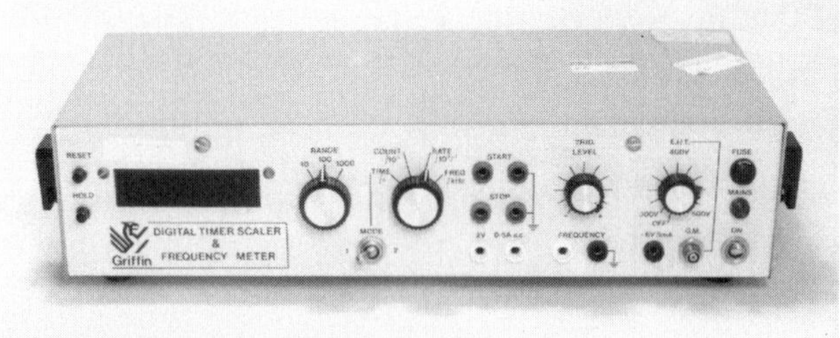

Figure 28.17 *A ratemeter*

Other types of detector

The solid state detector is useful as an α-particle detector. It consists of a p-n junction diode (see p. 383) which is reverse-biased slightly. When an ionizing particle hits the diode, pairs of electrons and holes are created in the junction region so a pulse of charge passes round the circuit. The pulse is then amplified and counted.

The scintillation detector shown in Figure 28.18, is useful as a γ-detector. The detector uses material such as sodium iodide which flashes with light when hit by a γ-photon. The γ-photon knocks electrons out of the atoms,

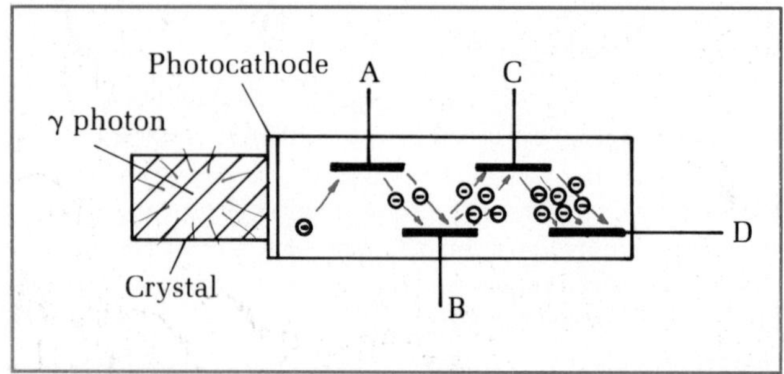

Figure 28.18 *A scintillation detector*

and when the electron vacancies are refilled, light is emitted (see p. 456). The flash of light causes a pulse of electricity to be produced by the photomultiplier tube behind the material. The reason is that photoelectrons are ejected from the photocathode by the light flash. These photoelectrons are attracted onto the anode A of the photomultiplier, causing more electrons to be emitted to hit the anodes B, C, D, etc., which are more positive. Therefore a measurable pulse is produced from the entry of a single γ-photon.

The cloud chamber

The diffusion-type cloud chamber is designed to create the same sort of conditions in the chamber as in the upper atmosphere, so a radioactive particle shooting across the chamber leaves a 'jet trail' behind, just like a plane high in the sky. Figure 28.19 shows the construction of a diffusion-type cloud chamber. The chamber is prepared by soaking the felt strip round the top of the chamber with alcohol. This makes the air in the chamber saturated. Then the air temperature is reduced greatly by using solid CO_2 'snow' under the chamber floor. The result is that the air space becomes supersaturated; when a radioactive particle shoots through, it leaves a trail of ions along its path. The ions act as condensation 'nuclei' to allow tiny droplets of alcohol to form.

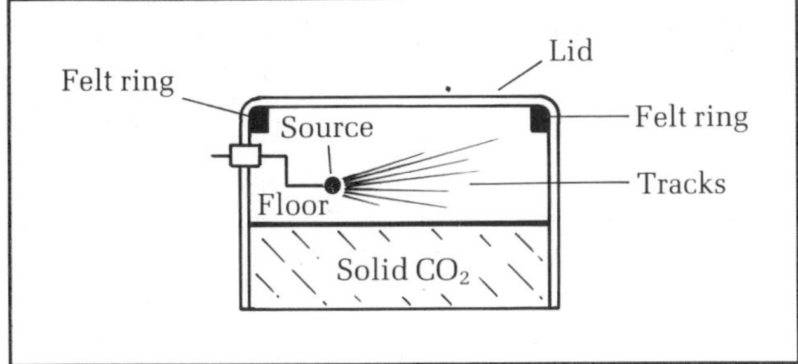

Figure 28.19 *The diffusion cloud chamber*

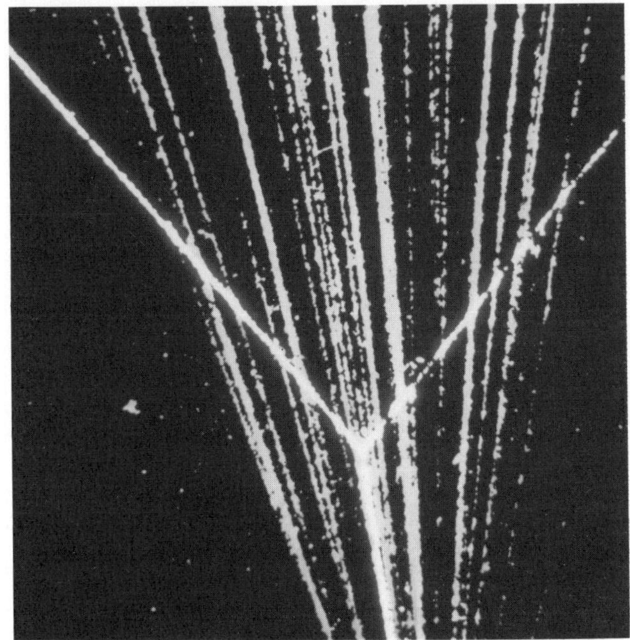

Figure 28.20 a *α-tracks in a cloud chamber*

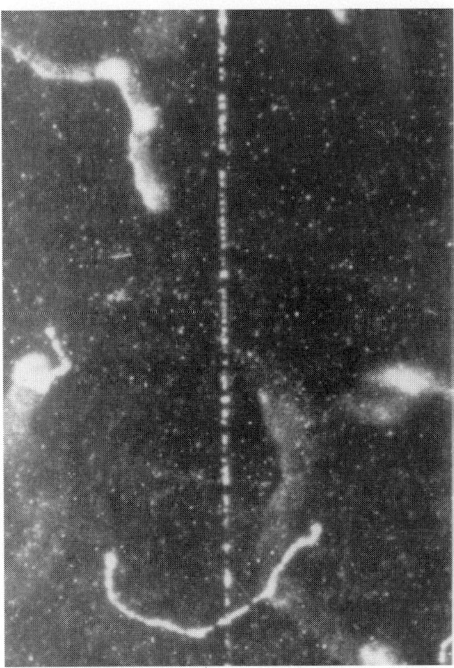

Figure 28.20 b *β-tracks in a cloud chamber*

For an α-source the tracks are straight and of well-defined length from a given source. Some sources produce several sets of α-tracks, each set of a different length. So each set has a different range and energy. Occasionally, a cosmic ray shoots across the chamber leaving a track at an odd angle to the other tracks.

For a β-source the tracks are much fainter because β-particles are not as strongly ionizing as α-particles. Also, β-particles are easily deflected by air molecules so the tracks are not as straight as the α-tracks.

28.4 The nucleus

Probing the atom

The discoveries in physics at the turn of the last century showed that atoms are made up of even smaller particles. Every atom contains electrons, each carrying a tiny negative charge. An uncharged atom must contain positive charge in just sufficient quantity to balance its negative charge. How is the atom structured? J. J. Thomson suggested that the atom might be like a 'currant bun' – with electrons dotted in the atom like currants in a bun. The positive charge was supposedly spread throughout the volume like the dough of the bun.

Alpha-scattering experiments were suggested by Rutherford in 1909 to test Thomson's theory. The idea was to shoot a beam of alpha-particles at thin metal foils. Scattered α-particles were detected using a small zinc sulphide screen viewed through a microscope. Each α-particle hitting the screen briefly produced a pinpoint of light. The foil and screen were in a vacuum chamber, designed so that the microscope and screen could be set at different angles to the incident beam, as in Figure 28.21.

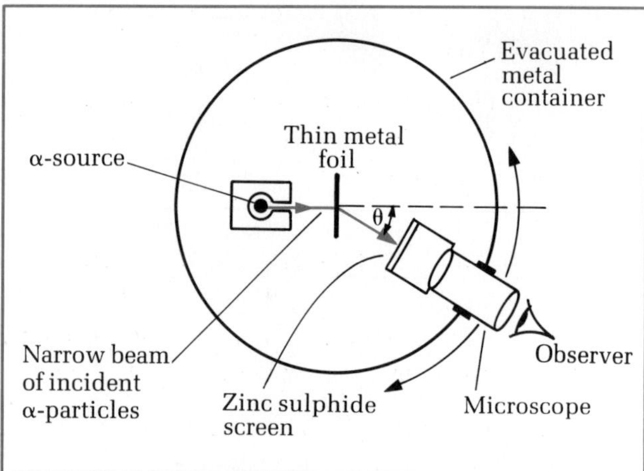

Figure 28.21 *Top view of α-scattering apparatus*

The foil scattered α-particles in directions at all possible angles to the incident beam. The screen was set at different angles to the undeflected beam at constant distance from the foil. At each angle, the number of α-particles per second hitting the screen was measured.

In fact, initial investigations by Geiger, working in Rutherford's laboratory, showed that most α-particles passed almost straight through the foil. This seemed reasonable to Geiger because alpha particles are much heavier than electrons, so electrons in the 'current bun atom' would just be pushed aside by the α-particles – just like shooting bullets at a real currant bun. Rutherford suggested that Geiger ought to make some measurements at much larger angles. Imagine the excitement when it was discovered that some α-particles did indeed deflect through very large angles. It was as if bullets fired at a real currant bun rebounded from inside the bun. If that happened, you would conclude that there must be something pretty hard inside the bun. That's what Rutherford and Geiger concluded about the atom! There is something 'hard' inside the atom and it must be very small because most α-particles miss it. Measurements were made for angles at intervals in the full range from 0° (i.e. undeflected) round to almost 180°. Different foils were used too. Figure 28.22 shows a set of the original results.

Angle of deflection θ	Number of scintillations per minute N	$N \times \sin^4 \frac{\theta}{2}$
150°	6.95	6.0
135°	8.35	6.1
120°	9.5	5.3
105°	14.6	5.8
75°	41.9	5.8
60°	101	5.3

Figure 28.22 *Typical results*

The nuclear model of the atom was devised by Rutherford to explain the α-scattering measurements. Rutherford assumed:

1 All the atom's positive charge is concentrated in a relatively small volume, called the 'nucleus' of the atom.
2 Most of the atom's mass is concentrated in its nucleus.
3 The electrons surround the nucleus at relatively large distances.

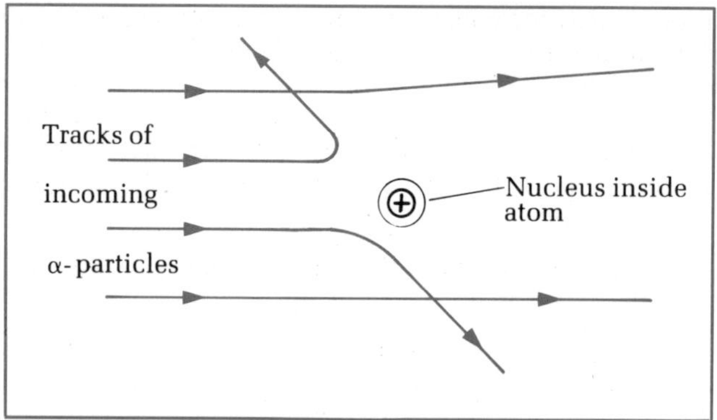

Figure 28.23 *α-scattering paths*

When α-particles are directed at the atoms of a thin metal foil, most α-particles are hardly deflected at all. Each atom's electrons are easily pushed aside by an incoming α-particle since the α-particles move so fast. Only if an α-particle passes close to the nucleus is its path affected significantly.

Rutherford knew that α-particles are positively charged, so an α-particle which rebounds from the foil must have been moving towards a nucleus. Because the nucleus carries positive charge too, then it repels the incoming α-particle. So the α-particle is deflected from its original path. Figure 28.23 shows the idea. Most α-particles are hardly deflected because the nucleus is so tiny inside the atom; few α-particles pass near the nucleus. Using Coulomb's law of force for point charges (see p. 171) and Newton's laws of motion, Rutherford calculated the expected paths of α-particles directed at the nucleus. He assumed that α-particles have the same initial kinetic energy. As an α-particle approaches a nucleus, the electrostatic force of repulsion from the nucleus acts on the α-particle to change its velocity and hence its path.

The potential hill is a simple analogy of α-scattering. Figure 28.24 shows the idea. A 'hill' is shaped so the height of a point on its surface varies inversely as the distance from the centre. In other words, the height varies as $1/r$ where r is the distance from the centre. Double the distance and the height drops to half; treble the distance and it drops to a third, etc.

The variation of height with distance is just like the variation of electrostatic potential near a nucleus; both follow the $1/r$ rule. Try rolling a ball bearing at the hill and study its path. Try it a few times. Most often, the ball is

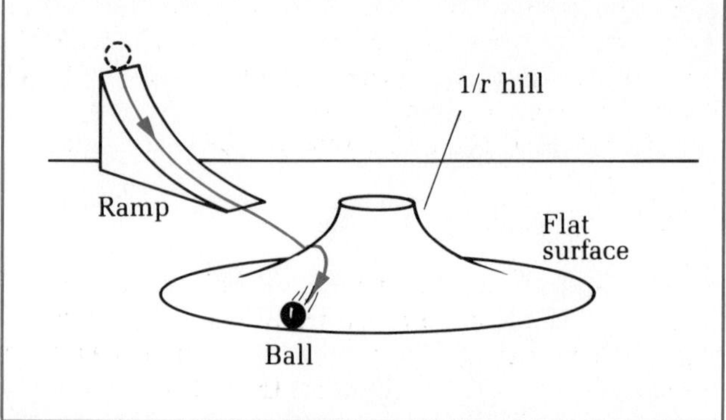

Figure 28.24 *Gravitational comparison*

hardly deflected unless it is directed carefully at the hill. The probability of a complete rebound from the hill is very small indeed.

Rutherford used the laws of motion and Coulomb's law to determine the α-particle paths. He showed that the number N of α-particles per second deflected by angle θ from the initial direction was given by

$N \sin^4(\theta/2) = $ constant (the same for all angles for each foil).

This prediction by Rutherford is in agreement with the experimental measurements, as shown in Figure 28.22. So Rutherford proved that every atom has a positively charged point-like nucleus where most of its mass is located. The charge of each nucleus was calculated by applying the measurements to the theory. In terms of the electron's charge e, the nuclear charge was found to equal $Z \times e$ where Z is the atomic number of the metal used as the foil. It's worth recalling that the Periodic Table arranges the elements in order of increasing atomic mass; the order number of each element is its atomic number Z. So Rutherford discovered that the positive charge on the nucleus is $Z \times e$.

An estimate of the size of the nucleus can be made in two ways.

1 Using α-scattering data approximately 1 in 8000 α-particles are deflected by more than 90° from a gold foil 0.6 μm thick.

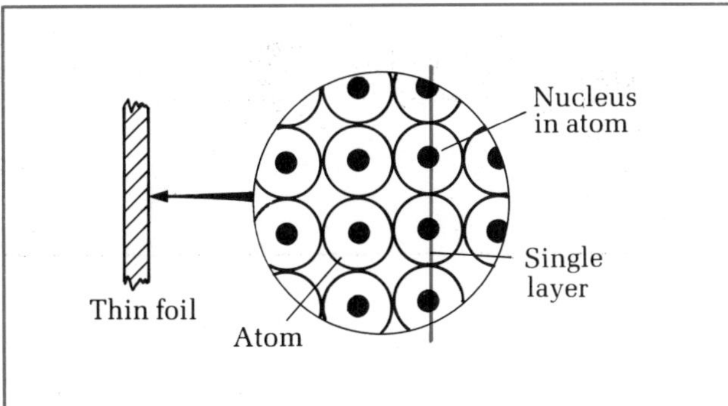

Figure 28.25 *Estimating nuclear size*

Suppose the foil has n layers so each atom's size is about $0.6/n$ μm. Also, assume each α-particle can be deflected once only. So each layer deflects 1 in $8000 \times n$ α-particles by more than 90°.

Each atom presents a cross-sectional area A_0 to the incoming α-particles but each particle only 'sees' the tiny nucleus with its area of cross-section A. So the fraction A/A_0 is the amount of each atom's area which 'reflects' the incoming α-particles. Hence A/A_0 is the fraction of each layer's area which acts as a reflector. This must be equal to the fraction of α-particles reflected by each layer.

i.e. $\dfrac{A}{A_0} = \dfrac{1}{8000n}$

What is the value of n, the number of layers in the foil? Each layer is one atom thick; what is a reasonable estimate for the size of a gold atom? Let's take a value of 0.3 nm as obtained on p. 67. Since the foil is 0.6μm thick, then the number of layers n is $0.6 \times 10^{-6}/0.3 \times 10^{-9} = 2000$ layers.

Hence $\dfrac{A}{A_0} = \dfrac{1}{8000 \times 2000} = \dfrac{1}{1.6 \times 10^7} = 6.25 \times 10^{-8}$.

Finally, make use of the fact that area is proportional to (radius)^2,

Hence $\dfrac{\text{nuclear radius}^2}{\text{atomic radius}^2} = \dfrac{A}{A_0} = 6.25 \times 10^{-8}$

So nuclear radius $= \sqrt{(6.25 \times 10^{-8})} \times$ atomic radius
$= 2.5 \times 10^{-4} \times$ atomic radius

In other words, the size of the nucleus is about 1/4000 ($= 2.5 \times 10^{-4}$) of the size of an atom. The ratio of sizes is about the same as a walnut in Wembley Stadium. With atoms about 0.3 nm in size, the nucleus is no more than about 10^{-14} m in size.

2 From the initial K.E. of an α-particle how closely can an α-particle approach a nucleus before it rebounds? The potential hill model is useful here. Suppose you roll a ball towards the precise centre; it climbs the hill then rolls back in the opposite direction. As it climbs up, it uses its K.E. to increase its P.E. At the highest point reached, it stops for an instant before rolling back. At this point, all its initial K.E. has been converted to P.E. This is the closest the α-particle gets to the nucleus; the point where all its initial K.E. is converted to P.E.

Suppose the closest distance of approach $= d$. The P.E. between the α-particle and the nucleus is given by

$$\text{P.E.} = \frac{Q_\alpha Q_{nuc}}{4\pi\varepsilon_0 d}$$

where Q_α is the α-particle's charge ($+2e$), Q_{nuc} is the nuclear charge ($+Ze$).

So the initial K.E. = P.E. at closest approach

Hence initial K.E. $= \dfrac{2e \times Ze}{4\pi\varepsilon_0 d}$

Giving $d = \dfrac{2e \times Ze}{4\pi\varepsilon_0} \Big/$ initial K.E.

Now we need to put some numbers in to estimate d. For gold, $Z = 79$. Also, the typical initial K.E. is about 5 MeV ($= 5 \times 10^6 \times e$ joules).

$$\text{So } d = \frac{2e \times 79e}{4\pi e_0} / 5 \times 10^6 e = \frac{2 \times 79 \times e}{4\pi\varepsilon_0 \times 5 \times 10^6}$$
$$= 4.5 \times 10^{-14}\,\text{m}$$

Assuming $\varepsilon_0 = 8.85 \times 10^{-12}$ F m^{-1},
$e = 1.6 \times 10^{-19}$ C.

The two methods give approximately the same value. More accurate methods can be used to determine nuclear size, but values are generally in the range 1×10^{-15} m to 10×10^{-15} m.

Inside the nucleus

Most of the mass of each atom is in its nucleus. Hydrogen is the lightest of all the elements. The next lightest is helium which is an inert gas. Then comes lithium, a reactive metal. The hydrogen atom is the simplest type of atom with one electron and its nuclear charge equal to $+e$. The helium atom has two electrons and nuclear charge $+2e$. Yet the atomic mass of helium is four times that of hydrogen. Rutherford reasoned that the nucleus of the hydrogen atom was a building block for all other species of nuclei. He introduced the term **proton** for the hydrogen nucleus, each proton therefore carrying charge $+e$. But he saw that another problem lay ahead; the mass of other nuclei could not be accounted for in terms of protons only. For example, a helium nucleus needs just two protons to make its charge $+2e$; yet its mass is approximately four times that of the proton. Every nucleus shows a discrepancy between its 'number' of charges and its 'number' of masses. Another example is carbon which has a nuclear charge of $+6e$. Yet the mass of a carbon atom is about twelve times the mass of a hydrogen atom, even though a carbon nucleus contains just six protons. Rutherford suspected that nuclei contain uncharged particles, each with mass about the same as a proton. He called them **neutrons**. So a helium nucleus (which is identical to an α-particle) is composed of two protons and two neutrons to give it an atomic mass of 4 and a nuclear charge $+2e$. A carbon (C-12) nucleus is composed of six protons and six neutrons, giving an atomic mass of 12 and nuclear charge $+6e$. Another example is the uranium-238 nucleus which has 92 protons and 146 neutrons; hence its nuclear charge is $+92e$ and its mass is approximately 238 atomic units.

	Proton	Neutron	Electron
Charge/e	1	0	−1
Mass/u	1	1	1/1850

Figure 28.26 *Constituents of the atom*

The atomic unit of mass is not based on the hydrogen atom because some hydrogen atoms have a neutron (or even two) in the nucleus. These 'heavy hydrogen' atoms make measuring atomic mass difficult in practice. So the scale of atomic mass is based on a more practical standard which is the **carbon-12 scale**. A carbon-12 atom has six protons and six neutrons in its nucleus; samples of pure carbon-12 are much easier to obtain than pure hydrogen without 'heavy' atoms. So the unit of atomic mass, 1 u, is defined as 1/12 of the mass of a carbon-12 atom. Hence proton and neutron mass are each about 1 u.

To sum up, consider a nucleus with charge $+Ze$ and atomic mass A. Such a nucleus is composed of Z protons and $(A-Z)$ neutrons. In an uncharged atom, the nucleus is surrounded by Z electrons at relatively large distances. Just one problem remained for Rutherford before the neutron-proton model could be accepted: there was no direct experimental evidence for neutrons at that time in 1920!

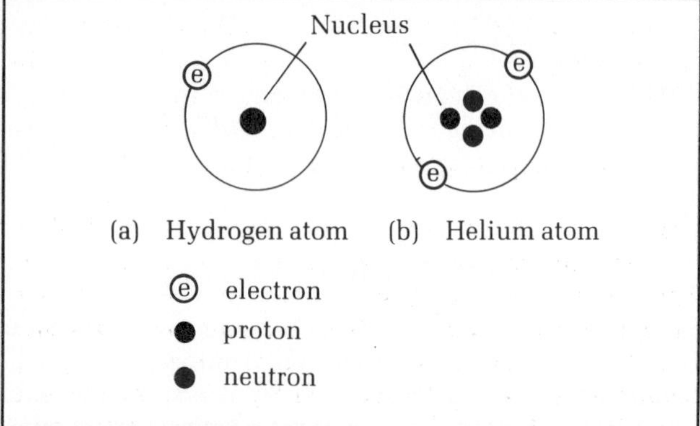

Figure 28.27 *Simple atoms*

The discovery of the neutron was made in 1932 by James Chadwick, yet another of Rutherford's protéges. Chadwick knew that when α-particles from polonium hit beryllium foil, very penetrating radiation was emitted. Some scientists thought this radiation was like X- or γ radiation (i.e. electromagnetic). But Chadwick proved that the radiation consisted of uncharged particles, each of mass about the same as a proton. In other words, Chadwick showed that the radiation was composed of high-speed neutrons.

Chadwick's proof involved several steps. The first step was to position a geiger tube in the path of the radiation from the beryllium foil. When a wax block was placed between the foil and the tube, as in Figure 28.28, the count rate *increased*. Chadwick showed that protons were knocked out of the wax block by the 'mystery' radiation; the count rate increases because protons carry charge, so they are readily detected by a geiger tube – unlike uncharged radiation. Chadwick measured the energy of the ejected protons and calculated the maximum speed as $3.3 \times 10^7\,m\,s^{-1}$.

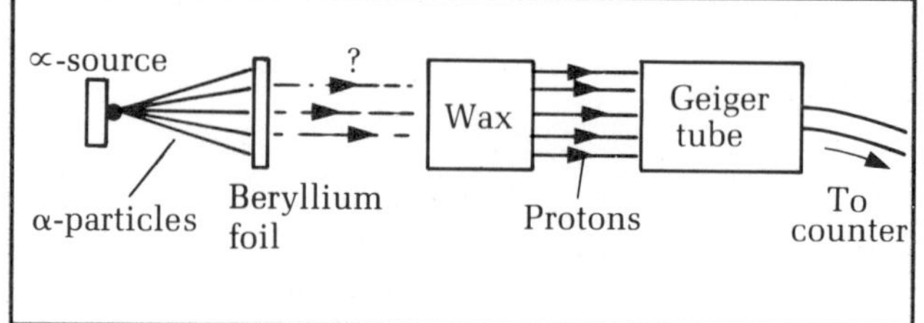

Figure 28.28 *Outline of Chadwick's experiment*

Next, he reasoned that if the mystery radiation was composed of neutrons then the maximum neutron speed must be the same as for the protons they knocked out. You can prove for yourself that when a moving billiard ball hits an identical ball at rest, the first ball stops and the target ball moves away with the same speed that the first ball had. Try it; otherwise, prove it theoretically (see p. 17).

Then Chadwick made a study of cloud chamber photographs taken when the mystery radiation from the beryllium foil passed through the chamber. The photographs showed short 'recoil' tracks, as in Figure 28.29.

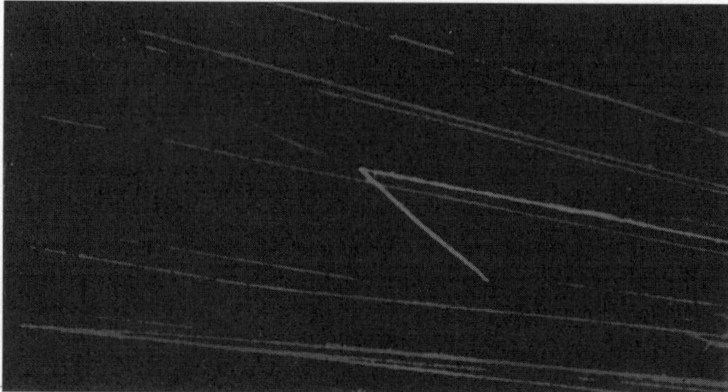

Figure 28.29 *Neutrons hitting nitrogen nuclei*

Chadwick measured the track lengths at 3.5 mm maximum; he assumed each track was due to a nitrogen nucleus recoiling after being hit by a neutron. From the track length, he worked out that the nitrogen atom recoiled with a maximum speed of $4.7 \times 10^6\ \mathrm{m\,s^{-1}}$. So a neutron moving at $3.3 \times 10^7\ \mathrm{m\,s^{-1}}$ collides with a nitrogen atom making it recoil at a speed of $4.7 \times 10^6\ \mathrm{m\,s^{-1}}$. The two speeds are in the ratio 7.5 to 1. Chadwick went back to billiard ball calculations to compare with the calculated ratio, worked out from collision theory.

A billiard ball of mass m moving at speed u makes a head-on impact with a target billiard ball of mass $14m$. What is the recoil speed of the target? Because the two balls have different masses, the theory is a bit more involved than where they have identical mass.

1 Momentum conservation gives

total initial momentum = total final momentum

$$mu = mv_1 + 14mv_2$$

where v_1 and v_2 are the speeds of the two balls after impact.

Hence $v_1 = u - 14v_2$ (1)

2 Conservation of kinetic energy is assumed; in other words, the impact is an elastic collision.

Total initial K.E. = total final K.E.

$$\tfrac{1}{2}mu^2 = \tfrac{1}{2}mv_1^2 + \tfrac{1}{2} \times 14mv_2^2$$

Hence $u^2 = v_1^2 + 14v_2^2$ (2)

Combining the two equations gives

$$u^2 = (u - 14v_2)^2 + 14v_2^2$$

So $u^2 = u^2 - 28uv_2 + 196v_2^2 + 14v_2^2$

To give $28uv_2 = 196v_2^2 + 14v_2^2 = 210v_2^2$

Dividing each term by $14v_2$ gives

$$2u = 15v_2$$

Hence $\dfrac{u}{v_2} = 7.5$ which is in agreement with the measurements.

Chadwick's measurements gave 7.5 for the speed ratio, the same as predicted by collision theory, assuming the mystery radiation consists of Rutherford's neutrons. So 12 years after Rutherford's suggestion of neutrons, Chadwick supplied convincing proof that neutrons really do exist.

A nucleus with Z protons and N neutrons is represented by the symbol $^{A}_{Z}X$, where X is the symbol for the element and $A = N + Z$.

The mass number A is the total number of neutrons and protons in the nucleus. The term 'nucleon' is often used for a neutron *or* a proton in the nucleus. So A is sometimes called the nucleon number.

The atomic number Z is the number of protons in the nucleus. An uncharged atom has Z protons and N neutrons in its nucleus; the nucleus is surrounded by Z electrons to balance the nuclear charge. The mass of each electron is 0.000 55 atomic units, compared with about 1 u for either the proton or the neutron mass. So only a tiny fraction of the mass of an atom is outside the nucleus.

The sub-atomic world seemed to have been sorted out by Rutherford and his colleagues. But in 1937, investigations of cosmic rays led to the discovery of particles with mass about 0.3 of the proton mass. These were called **mesons**, and they were the first of a bewildering variety of particles to be discovered. Just as the neat picture of neutrons, protons and electrons seemed complete, these new discoveries set scientists off once more, and even today, the picture is not yet complete.

28.5 Radioactive decay

Isotopes

Isotopic forms of an element have different atomic masses. The atoms of each isotope of an element have the same number of protons but a different number of neutrons for each form. For example, $^{238}_{92}U$ and $^{235}_{92}U$ are two of the isotopes of uranium. Both have 92 protons but one has three more neutrons per nucleus than the other.

Because isotopes have the same number of protons, they have the same number of electrons in each uncharged atom. Chemical reactions depend on the pattern of electrons round the atom so the isotopes of an element all have the same chemical properties; they cannot be separated from one another chemically. To separate them, their mass difference must be used. For example, liquid hydrogen contains a small percentage of 'heavy hydrogen' $^{2}_{1}H$ (i.e. deuterium). An ordinary hydrogen atom $^{1}_{1}H$ has half the mass of a heavy hydrogen atom, so when liquid hydrogen evaporates, the heavy hydrogen atoms get left behind.

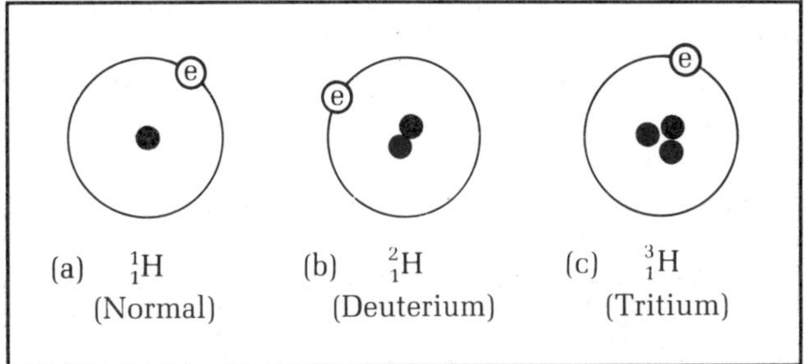

Figure 28.30 *Isotopes of hydrogen*

The abundance of an isotope in a given element is the percentage of atoms of the element which belong to that isotope. For example, chlorine gas contains isotopes $^{37}_{17}Cl$ and $^{35}_{17}Cl$, which are 25% and 75% abundant, respectively. So three out of every four chlorine atoms are $^{35}_{17}Cl$. The average atomic mass for chlorine gas atoms is therefore 35.5 ($= (3 \times 35 + 1 \times 37)/4$).

Equations for radioactive changes

α-emission An α-particle is identical to a helium nucleus with charge $+2e$ and mass of 4 atomic units. Hence it is composed of two protons and two neutrons. The symbol for an alpha particle is strictly $^{4}_{2}\alpha$, used rather than $^{4}_{2}He$ to signify the origin of the particle.

Suppose a nucleus $^{A}_{Z}X$ emits an α-particle; the nucleus loses two protons and two neutrons which carry away energy. After the emission, the nucleus has two units of charge less and four units of mass less. In other words, its atomic number changes to $Z-2$ and its mass number changes to $A-4$. So the new nucleus belongs to a different element Y.

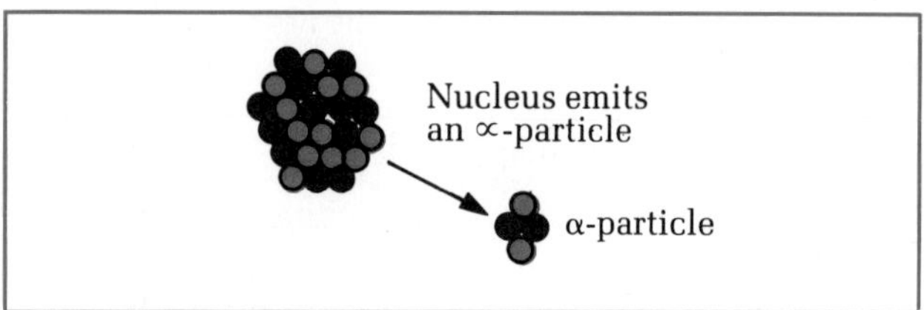

Figure 28.31 *α-decay*

^A_ZX	→	$^4_2\alpha$	+	$^{A-4}_{Z-2}Y$
(Z protons, A − Z neutrons)		(2 protons, 2 neutrons)		(Z − 2 protons A − Z − 2 neutrons)

β-emission A β-particle is identical to an electron. It is emitted with such high energy (of the order of MeV) that it can only have been produced by the nucleus. A β-particle is given the symbol $_{-1}^{0}\beta$, indicating that it has no protons or neutrons (so $A=0$) and its charge is negative (so $Z=-1$).

Consider the changes which occur when a nucleus ^A_ZX disintegrates by β-emission. Since the total charge after the emission must be equal to the total charge before, then the nuclear charge goes up by 1; this is because the β-particle takes away charge −1. So the nucleus formed after the emission has Z+1 protons; it belongs to a new element. However, the β-particle carries away no neutrons or protons since it has very little mass (compared with a neutron or a proton). So the new nucleus must have lost a neutron and gained a proton; in other words, β-emission occurs when a neutron in the nucleus suddenly switches to a proton; at the instant of change, a β-particle is created to conserve the balance of charge. The β-particle is instantly ejected from the nucleus.

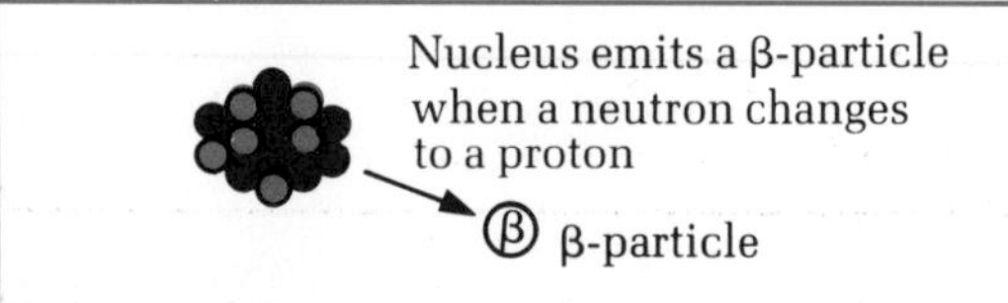

Figure 28.32 *β-decay*

^A_ZX	→	$_{-1}^{0}\beta$	+	$^{A}_{Z+1}Y$
(Z protons, A − Z neutrons)		(β-particle)		(Z + 1 protons, A − Z − 1 neutrons)

γ-emission A γ-photon carries neither charge nor mass. It takes energy away from the nucleus without altering the composition in terms of neutrons or protons. So γ-emission does not produce a new isotope.

Worked example Write down the equation of radioactive change to represent

a) the disintegration by α-emission of a $^{228}_{90}$Th nucleus,
b) β-emission from a $^{60}_{27}$Co nucleus.

Possible elements formed could be Ra (Z=88), Ac (Z=89), Pa (Z=91), U (Z=92) for (a). For (b), the product could be Mn (Z=25), Fe (Z=26), Ni (Z=28), Cu (Z=29).

Solution

a) $^{228}_{90}Th \rightarrow {}^4_2\alpha + {}^{224}_{88}Ra$

b) $^{60}_{27}Co \rightarrow {}_{-1}^{0}\beta + {}^{60}_{28}Ni$

Note that the numbers balance along each line on either side of each equation.

Half-life

The random nature of radioactive change can best be appreciated by studying the production of α-particle tracks in a cloud chamber. Figure 28.20 shows a photograph but the picture cannot show the unpredictable nature of the changes. You do need to see the cloud chamber in action; the individual tracks shoot out from the source at random. You cannot predict when a track will appear in a given direction. Each track is produced when a single nucleus in the source emits a radioactive particle. Hence radioactive disintegration of an individual nucleus is entirely unpredictable. If we could see the nuclei, it would be impossible to predict when any one particular nucleus will disintegrate (i.e. emit an alpha particle).

A radioactive nucleus is an unstable assembly of neutrons and protons. It becomes more stable by emitting a radioactive particle or a γ-photon, and this is the process we call 'radioactive disintegration'. Nuclei that emit α or β particles change their identity and become different isotopes. The newly-formed nucleus, the 'daughter' product, might be radioactive too, so it too in turn could disintegrate to form yet another isotope. This too could be radioactive. The sequence is called a **radioactive series**, and it ends only when a stable (i.e. non-radioactive) product is formed. An example of a radioactive series is shown below.

$$^{238}_{92}U \xrightarrow{\alpha} {}^{234}_{90}Th \xrightarrow{\beta^-} {}^{234}_{91}Pa \xrightarrow{\beta^-} {}^{234}_{92}U \xrightarrow{\alpha} {}^{230}_{90}Th \xrightarrow{\alpha} {}^{226}_{88}Ra \rightarrow$$

Suppose we start with a pure sample of a radioactive isotope ^A_ZX, containing N_0 atoms. Gradually, the number of atoms of X will decrease as individual nuclei disintegrate to form the daughter isotope. So the mass of X decreases. Measurements show that the number of atoms of a radioactive isotope (and hence its mass) always decreases *exponentially*. For example, suppose we start with 100 μg of X. If after time t_1, the mass of X has decreased to 80 μg, then it will decrease to 80% of 80 μg (=64 μg) in a further time t_1 (i.e. $2t_1$ from the start) etc.

Time from start	0	t_1	$2t_1$	$3t_1$	$4t_1$	...
Mass of X/μg	100.0	80.0	64.0	51.2	41.0	

The exponential decrease curve representing the variation of mass with time is shown in Figure 28.33. An exponential decrease takes place where the quantity (i.e. mass) decreases by a constant factor (e.g. × 0.8) in equal intervals of time (e.g. t_1). A convenient measure for the rate of decrease is the time for a decrease by half. This is the 'half-life' of the process.

The half-life $t_{\frac{1}{2}}$ of a radioactive isotope is the time taken for the number of atoms of that isotope to decrease to half the initial number.

Values of half-life vary enormously; $^{212}_{84}$Po has a half-life of 0.3 μs compared with 4.5×10^9 years for $^{238}_{92}$U.

Suppose a radioactive isotope X has N_0 atoms initially present. After one half-life, $0.5N_0$ atoms of X have disintegrated and $0.5N_0$ atoms of X remain. After two half-lives from the start, $0.25N_0$ atoms of X remain with $0.75N_0$ atoms having disintegrated. After three half-lives, $0.125N_0$ atoms remain etc.

Time from start	0	$t_{\frac{1}{2}}$	$2t_{\frac{1}{2}}$	$3t_{\frac{1}{2}}$
Number of atoms of X remaining	N_0	$0.5N_0$	$0.25N_0$	$0.125N_0$
Number of atoms of X disintegrated	0	$0.5N_0$	$0.75N_0$	$0.875N_0$

The theory of radioactive decay is based on the assumption that the disintegrations are entirely at random. Why should this assumption explain exponential decrease curves like Figure 28.33? To see why, consider a gigantic game of dice.

Start with 1000 dice, each one representing an atom of radioactive isotope X. Now throw the dice onto a table. Suppose all the dice with '1' uppermost represent atoms which disintegrate in the throw. How many have disintegrated? Pick out and count all the dice which show '1' uppermost; these are all disintegrated atoms. Each dice had a 1 in 6 change of coming up with '1'. Since they were thrown unpredictably, then 1000/6 should be picked out as having disintegrated after the first throw. So 5/6 of 1000 (=833) remain in the game for round 2.

Now use the 833 remaining dice for another throw. This time, you ought to find that 139 are picked out, showing '1' after being rolled. So 694 dice enter the third round. Does it work like this in practice? Yes, provided you have lots of dice; then, the law of chance predicts quite accurately *how many* disintegrate – but *not* which ones! In theory, the number of dice still in the game reduces by a factor of 5/6 each throw.

Number of throws	0	1	2	3	4	5	...
Number of dice remaining	1000	833	694	578	482	401	...

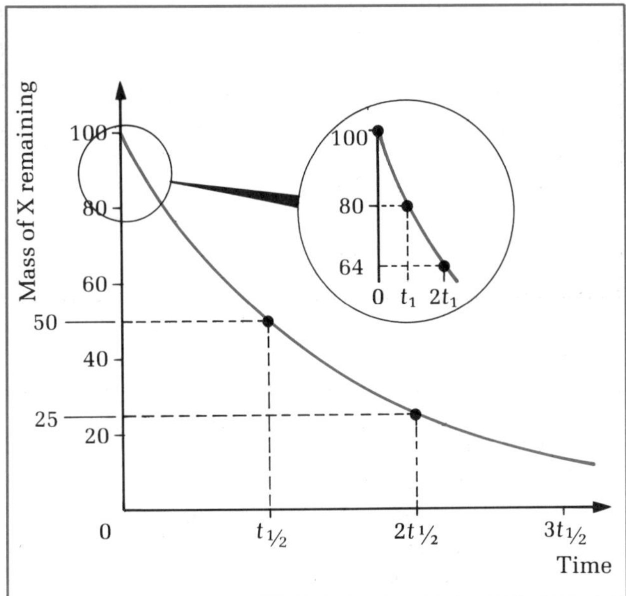

Figure 28.33 *Exponential decrease*

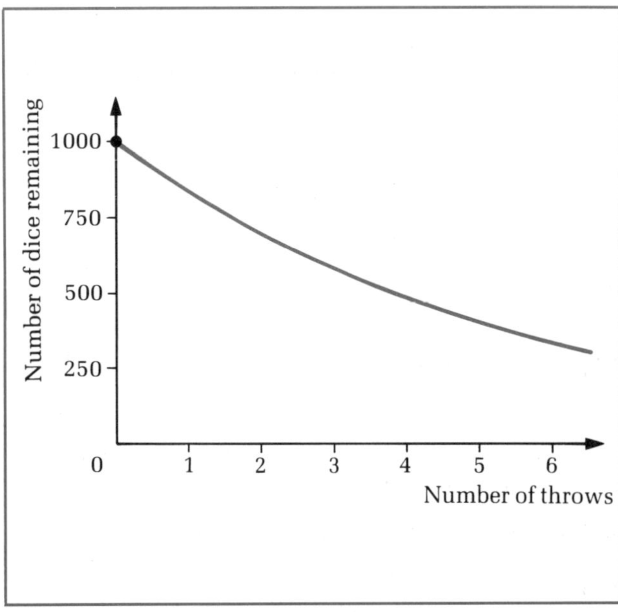

Figure 28.34 *Exponential dice*

Figure 28.34 shows the decrease of the number of dice with the number of throws. The curve is an exponential decrease with a half-life of about 3.8 throws.

The same ideas apply to radioactive atoms. Since the number of atoms is much greater than 1000 in most samples, the numbers game is even more predictable, but you cannot predict when an individual nucleus will disintegrate. Let's consider the ideas more generally.

Suppose we start with N_0 atoms of radioactive isotope X. Let the number of atoms of X remaining at time t be N.

In a short interval of time Δt, suppose ΔN atoms disintegrate. Since the process is entirely random, then ΔN is

a) proportional to N, the number still to disintegrate,

b) proportional to Δt, the short interval of time.

So we can write ΔN in terms of a constant of proportionality, λ, which is called the **decay constant**.

$$\Delta N = -\lambda N \Delta t$$

where the $-$ sign indicates a decrease.
The unit for λ is s^{-1}.

The activity of a radioactive isotope is defined as the number of disintegrations per second. With ΔN disintegrations in time Δt, then the activity is equal to $\frac{\Delta N}{\Delta t}$ provided Δt is very small.

In other words, the activity is $\left(\frac{\Delta N}{\Delta t}\right)_{\Delta t \to 0}$, which is usually written as $\frac{dN}{dt}$.

$$\text{Hence } \frac{dN}{dt} = -\lambda N$$

So the rate of disintegration $\frac{dN}{dt}$ is proportional to the number of atoms N still radioactive. Exponential processes always occur when the rate of change of a quantity is proportional to the quantity itself (see p. 507).

$$\frac{dN}{dt} = -\lambda N \quad \text{where } \frac{dN}{dt} = \text{rate of disintegration,}$$
N = number of atoms still radioactive.

The equation for radioactive decay shown above can be solved for N using the technique of integration.

$$\frac{dN}{N} = -\lambda dt$$

Hence $$\int_{N_0}^{N} \frac{dN}{N} = -\int_0^t \lambda dt$$

Giving $[\ln N]_{N_0}^{N} = -\lambda t$

So $\ln(N/N_0) = -\lambda t$

(where ln is the natural logarithm)

Therefore $N/N_0 = e^{-\lambda t}$

$$N = N_0 e^{-\lambda t}$$

where N_0 = initial number of atoms, λ = decay constant.

This is the exponential decrease equation for radioactive decay. Now it is possible to see why N decreases by a constant factor in equal intervals of time.

After time t_1, the number of atoms remaining $N_2 = N_0 e^{-\lambda t_1}$.

After time $2t_1$, the number remaining $N_2 = N_0 e^{-\lambda \times 2t_1}$ which equals $N_1 e^{-\lambda t_1}$.

After time $3t_1$, the number remaining, $N_3 = N_0 e^{-\lambda \times 3t_1}$ which equals $N_2 e^{-\lambda t_1}$.

So in each interval of time t_1, the number of atoms drops by a constant factor $e^{-\lambda t_1}$.

The decay constant λ is linked to the half-life $t_{\frac{1}{2}}$. After time equals to one half-life, the number of radioactive atoms falls to $0.5N_0$.

Using $N = N_0 e^{-\lambda t}$

$\frac{1}{2} N_0 = N_0 e^{-t_{\frac{1}{2}}}$

Hence $\frac{1}{2} = e^{-\lambda t_{\frac{1}{2}}}$

Thus $2 = e^{\lambda t_{\frac{1}{2}}}$ so $\ln 2 = \lambda t_{\frac{1}{2}}$

Half-life $t_{\frac{1}{2}} = \frac{\ln 2}{\lambda}$

where λ = decay constant.

The equation shows that a short half-life corresponds to a large decay constant. In other words, the rate of disintegration is high if the half-life is small. Take equal numbers of atoms of two radioactive isotopes, one with a short half-life and the other with a much longer half-life. The rate of disintegration of the short half-life material is much greater because its decay constant is larger.

Worked example A freshly prepared sample of a radioactive isotope X contains 10^{20} atoms. The half-life of the isotope is 12 hours. Calculate

a) the initial activity,

b) the number of radioactive atoms of X remaining after (i) 1 hour, (ii) 24 hours.

Solution

a) First, calculate the decay constant from the half-life.

$$\lambda = \frac{\ln 2}{t_{\frac{1}{2}}} = \frac{\ln 2}{12 \times 60 \times 60} = 1.60 \times 10^{-5}\,s^{-1}.$$

The activity $\frac{dN}{dt} = -\lambda N$

$= -1.6 \times 10^{-5} \times 10^{20}$

$= -1.6 \times 10^{15}\,s^{-1}$.

So the atoms of X disintegrate at the start at a rate of $1.6 \times 10^{15}\,s^{-1}$.

b) $N_0 = 10^{20}$ atoms, hence $N = 10^{20} e^{-\lambda t}$ gives the number of atoms which are still to disintegrate after time t.

i) For $t = 1$ hour $= 3600$ s, $N = 10^{20} \times e^{-1.6 \times 10^{-5} \times 3600}$

$= 10^{20} \times 0.944$.

$N = 0.944 \times 10^{20}$ atoms

ii) For $t = 24$ hours, the number of atoms decreases to $0.25N_0$.

This is because 24 hours in this case is exactly 2 half-lives.

Hence $N = 0.25 \times 10^{20} = 2.5 \times 10^{19}$.

Measuring the half-life of thoron gas

Thoron gas is an isotope of radon, $^{220}_{86}Rn$, produced in the radioactive series that starts with the long half-life isotope $^{232}_{90}Th$. All the other isotopes in the series have half-lives either much longer or much shorter than thoron gas, so they do not contribute to the activity of a sample of the gas. The long half-life isotope $^{232}_{90}Th$ is in powdered form in a sealed plastic bottle. So the thoron gas is produced in the air space in the bottle.

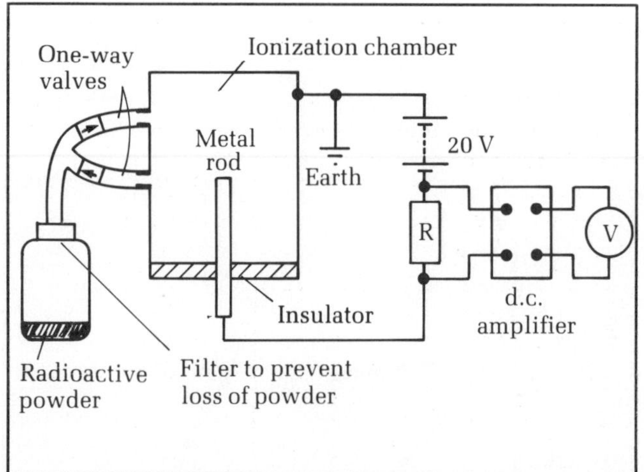

Figure 28.35 *Measuring half-life*

Using two tubes with one-way valves, thoron gas can be transferred safely into an ionization chamber, also sealed, by squeezing the bottle a few times. The arrangement is shown in Figure 28.35. The ionization chamber is connected in series with a large resistance and a 20 V power unit as shown. A direct current (d.c.) amplifier is connected across the resistor R to measure the p.d. The output voltage of the amplifier is measured using a voltmeter, as shown.

When a fixed quantity of thoron gas is transferred to the chamber, ionization of air molecules occurs in the chamber. The ionization is due to the radioactive particles emitted by the disintegrating thoron nuclei. The ions are attracted to the oppositely charged electrodes where they are discharged. Hence electrons pass round the circuit giving a p.d. across resistor R. The p.d. is then amplified to give a reading on the voltmeter in proportion to the activity in the chamber. Since the activity is proportional to the number of atoms of thoron, N, then the voltmeter reading is proportional to N.

The voltmeter reading is measured at 5 s intervals, and the results plotted as voltmeter readings against time. In this way, a curve representing the decrease of the number of atoms of thoron with time is obtained. Figure 28.36 shows a typical curve. From the curve the half-life may be determined as shown.

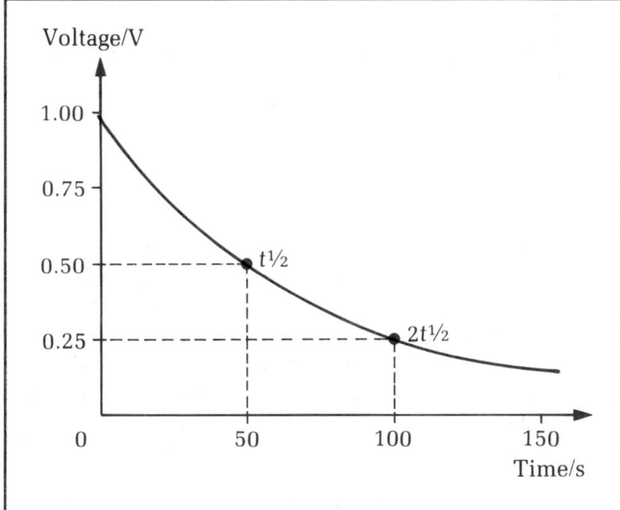

Figure 28.36 *Activity v time for thoron gas*

A chart recorder may be used instead of a voltmeter. From the chart produced, the half-life may be calculated if the chart speed is known. Figure 28.37 shows a chart recording. The curve shows random variations superimposed on the exponential decrease. This is to be expected because radioactive disintegration is a random process.

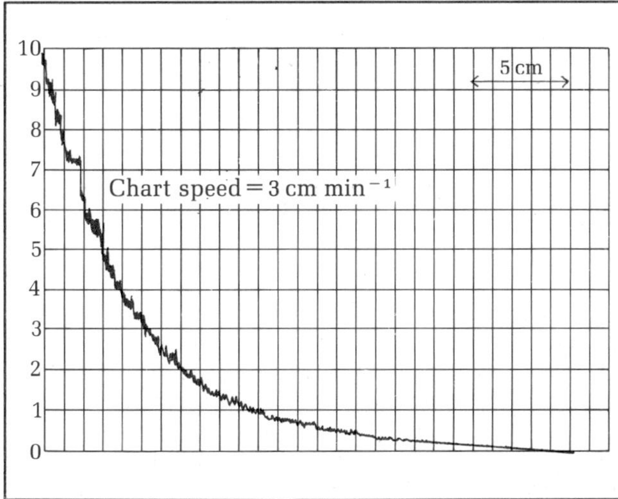

Figure 28.37 *Using a chart recorder*

Nuclear stability

Artificial radioactivity can be produced by bombarding stable elements with high-energy particles. An example of this is when metal tubes containing nuclear fuel are inserted into a nuclear reactor. The nuclei of the metal tubes are bombarded by neutrons from the fuel so the tubes become radioactive. Another example is when copper is placed in the core of a nuclear reactor. The copper nuclei absorb neutrons, to become unstable with a half-life of 13 hours, so the copper remains radioactive many hours after removal from the core. Each unstable nucleus disintegrates by β-emission to produce a zinc nucleus.

$$^{63}_{29}\mathrm{Cu} + ^{1}_{0}\mathrm{n} \rightarrow ^{64}_{29}\mathrm{Cu}\ \text{(unstable)}$$

$$^{64}_{29}\mathrm{Cu} \rightarrow ^{0}_{-1}\beta + ^{64}_{30}\mathrm{Zn}$$

The N-Z plot of all the known isotopes, shown in Figure 28.38, is a useful way to survey nuclear stability. The plot shows that stable nuclei lie along a belt from the origin curving up towards increasing neutron/proton ratio.

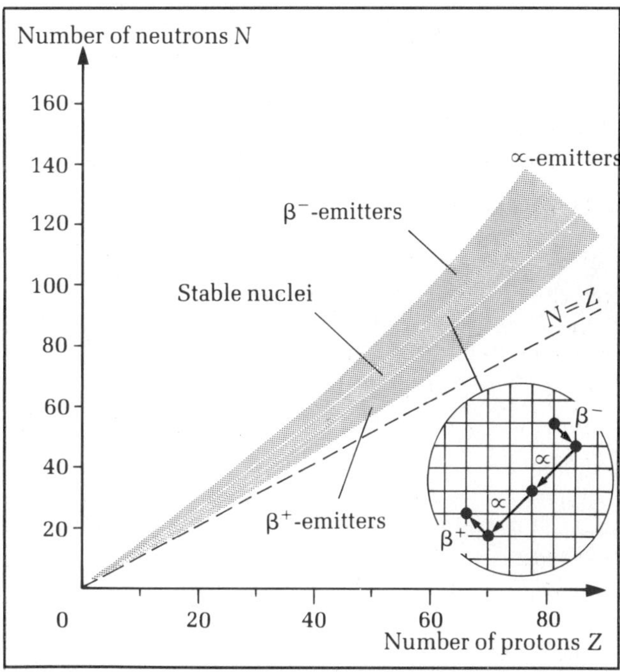

Figure 28.38 *The N-Z plot*

1. For light isotopes (Z up to about 20), the stable nuclei follow a straight line $N = Z$. Light nuclei which are stable have equal numbers of neutrons and protons.
2. For increasing atomic mass, the stable nuclei need more neutrons than protons. The neutron/proton ratio increases. Extra neutrons are essential for stability because they help to bind the nucleons together, without introducing repulsive forces as protons do.
3. α-emitters lie above the stability belt where the nuclei are very large.
4. β^--emitters lie to the left of the stability belt. Here the isotopes have too many neutrons compared with the stable nuclei, so they become stable by 'converting' a neutron into a proton and emitting a β^--particle at the same time.

5 Nuclei with too many protons compared with neutrons (i.e. to the right of the stability belt) become stable by converting a proton into a neutron. When this happens, a positively-charged β-particle is emitted from the nucleus. Such a particle is sometimes called a **positron**.

6 Radioactive series can be plotted on the graph. An α-emitter moves two places down and two places to the left, corresponding to losing two neutrons and two protons. A β^--emitter loses a neutron and gains a proton; so it moves one place down and one place to the right. A β^+-emitter changes a proton into a neutron, so it moves in the opposite direction to a β^--emitter. These changes are shown in Figure 28.38 as part of the N-Z plot.

28.6 Radioactivity in use

Tracers

A radioactive tracer is used to follow the path of a compound through a system. For example, leaks in underground pipes that carry water or oil can be detected by injecting radioactive tracer into the flow. Geiger tubes on the surface above the pipeline can then be used to detect leakage. What type of isotope should be chosen? The isotope should be short-lived with a stable daughter product. A γ-emitter would not be much use since γ-particles pass through pipes.

Another example is to improve recovery from oil reservoirs. Water injected into an oil reservoir at high pressure forces oil out. A suitable radioactive tracer can be added to the injected water; detectors at the production wells monitor breakthrough of the injected water. The results are used to build up a model of the reservoir to control and improve recovery of oil. Since the time from injection to breakthrough can be many months, the tracer must have a suitably long half-life. A suitable tracer is 'tritiated' water $^{3}H_{2}0$ – a β-emitter with a half-life of 12 years. It is diluted by a factor of 10^{10} when injected into the reservoir.

Tracers are often used in medicine. For example, a blocked kidney can be diagnosed using an isotope of iodine $^{131}_{53}I$ which has a half-life of eight days. The patient is given plenty of water to drink first, then half-an-hour later is injected with a solution containing a small quantity of iodine-131. The injection is into the vein and the iodine is extracted almost totally from the blood on passing through the kidneys. Geiger counters pointing at the kidneys ought to show an increase then a decrease in the count rate if the kidney function is normal, but a blocked kidney gives an increase with no decrease since it is unable to pass the iodine on to the bladder.

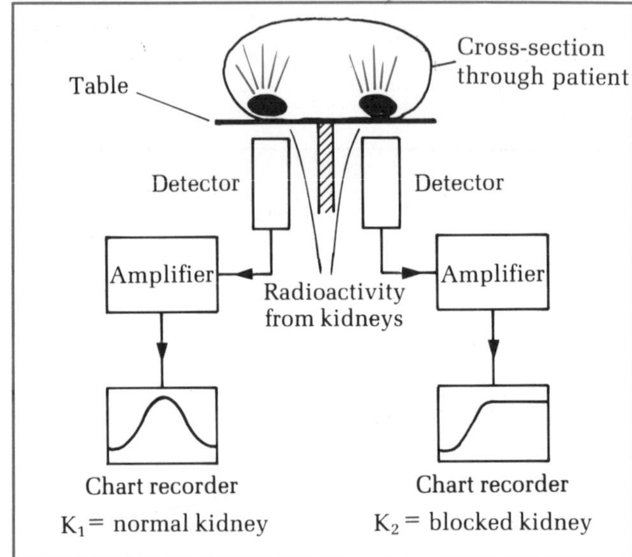

Figure 28.40 *Kidney testing*

Carbon dating

Living plants and trees contain a small percentage of radioactive carbon in the form of the isotope $^{14}_{6}C$. The isotope is formed in the atmosphere due to cosmic rays knocking out neutrons from nuclei. These neutrons then collide with nitrogen nuclei to form carbon-14 nuclei.

$$^{1}_{0}n + ^{14}_{7}N \rightarrow ^{14}_{6}C + ^{1}_{1}p$$

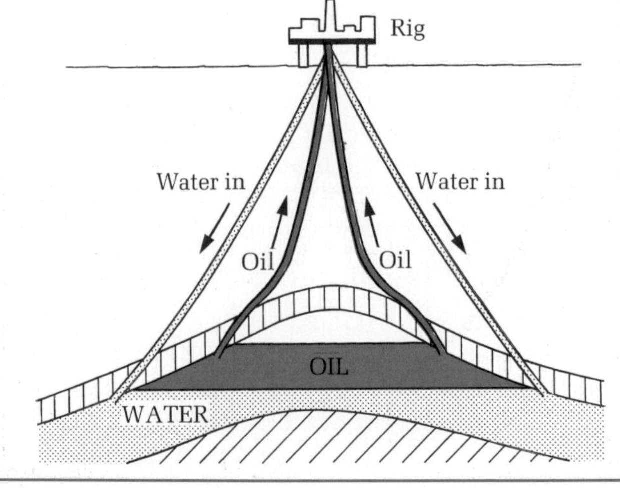

Figure 28.39 *Using tracers*

Carbon dioxide from the atmosphere is taken up by living plants as a result of photosynthesis, so a small percentage of the carbon contents of the plant material is carbon-14.

The half-life of carbon-14 is 5730 years so there is negligible disintegration over the lifetime of most plants. However, once the plant has died no further carbon-14 is taken in. Thus the proportion of carbon-14 in a dead plant or tree falls as the carbon-14 atoms disintegrate. After one half-life of 5730 years, the proportion is down to 50% of its initial value. Since the activity is proportional to the number of atoms still to disintegrate, then measuring the activity enables the age of the dead sample to be calculated. To do this, the measured activity is compared with the activity of the same mass of living wood. Then, using the known value of half-life of carbon-14, the age (i.e. time since death) can be determined.

Worked example Living wood has an activity of 15.3 counts per minute per gram of carbon. A certain sample of dead wood is found to have an activity of 17.0 counts per minute for 5.0 grams. Calculate the age of the sample of dead wood. Assume the half-life of carbon-14 is 5730 years.

Solution Initial activity = activity of live wood $= 15.3\ \text{min}^{-1}\ \text{g}^{-1}$.

Activity of dead sample $= 17/5 = 3.4\ \text{min}^{-1}\ \text{g}^{-1}$.

$$\text{Decay constant} = \frac{\ln 2}{t_{\frac{1}{2}}}$$

$$= \frac{\ln 2}{5730 \times 365 \times 24 \times 60 \times 60}$$

$$= 3.83 \times 10^{-12}\ \text{s}^{-1}.$$

To calculate the age of the sample, use $N = N_0 e^{-\lambda t}$ which gives $A = A_0 e^{-\lambda t}$ since activity A is proportional to N.

Hence $3.4 = 15.3\ e^{-\lambda t}$, thus $e^{-\lambda t} = 3.4/15.3 = 0.222$

So $e^{\lambda t} = 1/0.222 = 4.5$

there $\lambda t = \ln 4.5 = 1.504$

$$\text{So age of sample, } t = \frac{1.504}{\lambda} = \frac{1.504}{3.83 \times 10^{-12}} = 3.93 \times 10^{11}\ \text{s}$$

$$= 12\,444 \text{ years.}$$

Health hazards

All forms of ionizing radiations (e.g. α, β, γ, X, n) can damage body cells. Radioactive materials must be used with the utmost care to ensure they never enter the body accidentally. α-particles from material which is lodged inside the body can be very damaging to tissues, etc., all around the material. γ and β-radiations are much more penetrating so they can damage body cells from sources outside the body.

The absorbed dose delivered by a beam of ionizing radiation is defined as the energy absorbed from the incident radiation per kilogram of body matter. The unit is the **gray** (Gy), equal to $1\ \text{J kg}^{-1}$.

The biological effect of ionizing radiation is determined by the absorbed dose and the type of radiation. For example, α-radiation is ten times more damaging than X-radiation for the same absorbed dose.

The relative biological effectiveness (r.b.e.) of a certain type of radiation is defined as

$$\frac{\text{the absorbed dose with 250 kV X-rays for a certain effect}}{\text{the absorbed dose for the same effect with the radiation.}}$$

So the r.b.e. for α-radiation is 10. Hence 1 Gy of α-radiation does the same damage as 10 Gy of X-radiation.

The dose equivalent is defined as the absorbed dose × the r.b.e. The unit of dose equivalent is the **sievert** (Sv), also equal to $1\ \text{J kg}^{-1}$.

If a person is subjected to known doses of different types of radiation, the dose equivalent can be calculated for each type. Then the total dose is given by adding up dose equivalent values. People working in X-ray units in hospitals for example would need to be monitored this way. Each person in such an environment must wear a **film badge** which is replaced and tested periodically. For example, suppose a radiation worker in one year is subjected to 25 mGy of X-radiation and 0.2 mGy of α-radiation. The total dose equivalent is therefore $(25 \times 1) + (0.2 \times 10) = 27$ mSv.

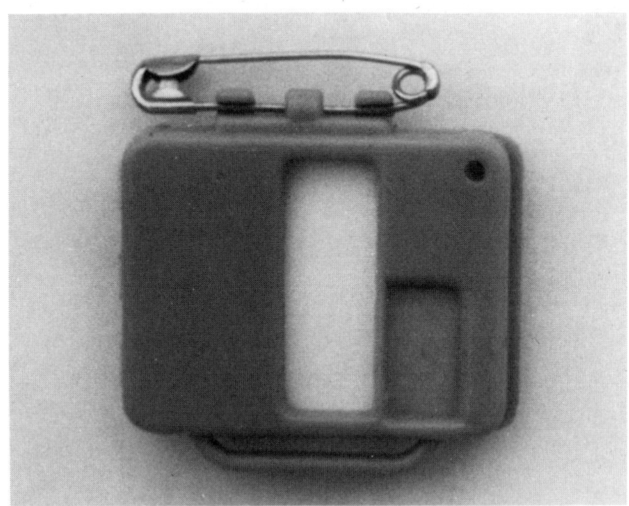

Figure 28.41 *A film badge*

28.7 Summary

Definitions

1 *The mass number A* of a nucleus is the number of neutrons and protons in the nucleus.
2 *The atomic number Z* of an atom is the number of protons in its nucleus.
3 *Isotopes* are forms of an element and have the same number of protons but different numbers of neutrons in each type of nucleus.
4 *The half-life* $t_{\frac{1}{2}}$ of a radioactive isotope is the time taken for half the initial number of atoms of that isotope to disintegrate.
5 *The activity* of a radioactive isotope is the number of disintegrations per second.

Equations

1 The inverse-square law for γ-radiation

Intensity = constant/r^2

2 Radioactive decay $\dfrac{dN}{dt} = -\lambda N$

where λ is the decay constant

3 Solution of equation for radioactive decay

$N = N_0 e^{-\lambda t}$

and activity $A = A_0 e^{-\lambda t}$

4 The half-life $t_{\frac{1}{2}} = \dfrac{\ln 2}{\lambda}$

5 Nuclear change caused by:

a) α-decay ${}^{A}_{Z}X \rightarrow {}^{4}_{2}\alpha + {}^{A-4}_{Z-2}Y$

b) β-decay ${}^{A}_{Z}X \rightarrow {}^{0}_{-1}\beta + {}^{A}_{Z+1}Y$

Short questions

Electronic charge $e = 1.6 \times 10^{-19}$ C,
The Avogadro constant $L = 6.02 \times 10^{23}$ mol^{-1}
Specific charge of electron $= 1.76 \times 10^{11}$ C kg^{-1}
1 u = 931 MeV.

28.1

a) A geiger tube is pointed at a small radioactive source with its window 30 mm from the source. The count rate is then measured. When paper is placed between the tube and the source, the count rate drops significantly although it is still several times the background rate. When the paper is replaced by lead plate of thickness 10 mm, the count rate falls to a level indistinguishable from background. What types of radiation are emitted by the source?

b) With no absorbers present in **a)**, the count rate was 73.1 s^{-1} after correcting for the background. The tube window area is 2.0 mm^2. Estimate the activity of the source. Give reasons why the actual activity is greater than your estimate.

28.2 An α-emitting source of activity 3×10^4 s^{-1} in an ionization chamber creates an ionization current of 0.8 nA. Assuming the energy required to create an ion pair in air is 30 eV, calculate the energy of each emitted α-particle.

28.3 α-particles emitted from a particular source S have a range of 40 mm in air at standard pressure. Use Figure 28.8 to determine

a) the initial K.E. in MeV of each emitted α-particle,

b) the number of ion pairs per mm along the α-particle's path, given that each ion pair uses 30 eV of the α-particle's energy.

c) Given the specific charge of an alpha particle is 4.79×10^7 C kg^{-1}, calculate the initial speed of the alpha particles.

28.4 In an experiment to determine the range of β-particles in aluminium, different thicknesses of aluminium sheets were interposed between a small β-source and the window of a geiger tube placed 20 mm apart.

Thickness/mm	0	0.45	0.90	1.35	1.80	5.40	7.20
Count rate/s^{-1}	85.0	59.5	41.6	29.2	20.4	1.5	1.5

a) Plot a graph of ln (count rate) against thickness.

b) Use your graph to determine the range of the β-particles in aluminium.

c) From Figure 28.10, estimate the energy of each emitted β-particle.

28.5 A small γ source placed 400 mm from a geiger tube gives a count rate of 115.3 s^{-1}. When a lead plate of thickness 12 mm is placed between the source and the tube, the count rate falls to 22.1 s^{-1}. When the source was removed, the count rate was 0.3 s^{-1}.

a) Calculate the half-value thickness of the lead for the γ-radiation used.

b) Use Figure 28.12 to estimate the energy of the emitted γ-photons.

28.6 In question **28.5**, what would the count rate be without lead present if the source is moved to a distance of

a) 1200 mm,

b) 1500 mm from the tube?

What should be the minimum distance between the tube and source if the count rate is to be less than 3 s^{-1}?

28.7

a) Explain why a geiger tube is a much more efficient detector of α-radiation than it is of γ-radiation.

b) To detect α or β radiation using a geiger tube, the tube window must face the source. This is not the case for γ-radiation. Explain why.

28.8

a) Describe with the aid of a sketch the appearance of α-particle tracks in a cloud chamber. Account for your observations.

b) β-particle tracks in a cloud chamber are much more difficult to see than α-particle tracks. Explain why.

28.9 A beam of α-particles is directed normally at a thin metal foil in an α-scattering experiment. Explain

a) why most α-particles pass straight through the foil,

b) why some α-particles are deflected through angles of more than 90°,

c) why multiple scattering of an individual α-particle is unlikely.

28.10 Calculate the least distance of approach of a 3.5 MeV α-particle to the nucleus of a gold (Z = 79) atom.

28.11 5 MeV α-particles are directed at a sheet of polythene, causing protons to be knocked out from the nuclei of the sheet. Calculate

a) the initial speed of the α-particles, given the specific charge as in question **28.3**,

b) the maximum speed of ejection of the protons from the sheet.

Assume a head-on elastic collision between an α-particle (mass = 4 u) and a stationary proton (mass = 1 u).

28.12

a) Copy and complete each of the following equations for radioactive disintegration.

i) ${}^{214}_{84}Po \rightarrow {}^{?}_{?}\alpha + {}^{?}_{?}Pb$

ii) ${}^{210}_{83}Bi \rightarrow {}^{?}_{?}\beta + {}^{?}_{?}Po$

b) Use the two equations in **a)** to describe the changes in the nucleus that occur as a result of (i) α-emission, (ii) β-emission.

28.13 A sample of copper (Z = 29, A = 63) is placed in the core of a nuclear reaction where the copper atoms are subjected to bombardment by neutrons. As a result, the sample becomes radioactive due to its nuclei absorbing neutrons.

a) Write down an equation to represent a copper nucleus absorbing a neutron.

b) The nucleus formed is unstable. What type of radioactive emission would you expect from these nuclei? Give an equation.

28.14

a) Radioactive disintegration is a random process yet it is possible to calculate reasonably accurately the number of atoms in a radioactive source of known activity and half-life. Explain why.

b) A radioactive isotope X with a half-life of 1.6×10^5 s disintegrates to form a stable product. A pure sample of X is prepared with an initial activity of 25 000 Bq. Calculate the activity of the sample after (i) 24 hours, (ii) 1 week.

28.15 The radioactive isotope $^{32}_{15}P$ has a half-life of 14.3 days and it disintegrates to form a stable product. A sample of the isotope is prepared with an initial activity of 2×10^6 Bq. Calculate

a) the number of P-32 atoms initially present,

b) the activity and number of P-32 atoms present after 30 days.

28.16

a) Living wood has an activity of 16.0 counts $min^{-1} g^{-1}$ which is due to the disintegration of carbon-14 atoms in the wood. The half-life of carbon-14 is 5568 years. Calculate the number of carbon-14 atoms in 1 g of living wood.

b) A sample of wood of mass 0.5 g from an ancient ship is found to have an activity of 6.5 counts min^{-1}. Calculate the age of the ship.

28.17 A sample of the radioactive isotope $^{131}_{53}I$ is to be used for medical diagnosis of the kidneys. The isotope has a half-life of 8.0 days, and the sample is required to have an activity of $8 \times 10^5\ s^{-1}$ at the time it is given to the patient. Calculate the mass of the I-131 present in the sample

a) at the time it is given,

b) when it is prepared 24 hours earlier,

c) 24 hours after being given.

28.18

a) Calculate the activity of 1 mg of pure $^{137}_{55}Cs$, given its half-life is 35 years.

b) Cs-137 decays by emitting 0.66 MeV γ-photons. Calculate the maximum power released by 1 mg of Cs-137.

28.19 The back-up power of a certain satellite is provided by the heat generated by a radioactive source. Calculate the mass of $^{226}_{88}Ra$ that will generate 50 W of power. The half-life of Ra-226 is 1620 years and it emits 4.78 MeV α-particles.

28.20

a) A small γ-source with activity $2 \times 10^6\ s^{-1}$ emits 3.5 MeV γ-photons. Calculate the energy per second falling normally on a surface of area 1 cm^2 placed 10.0 m from the source.

b) If a cube of material of thickness 30 mm is placed 10.0 m from the source in **a)**, estimate the energy absorbed from the beam if the half-value thickness for the material is 15 mm.

Electrons inside the atom

29.1 Energy levels

Ionization and excitation

Electrons are point-particles which carry negative charge. Every atom contains electrons; in an uncharged atom there are just sufficient electrons to balance the total positive charge carried by the nucleus. When an electron is removed from an atom, the atom's positive charge pulls on the electron to try to prevent it from leaving. To escape from an atom, an electron must be given energy to enable it to overcome the attraction of the atom.

Ionization is the process of creating charged atoms. A positive ion is produced when electrons are removed from an uncharged atom. One way to create ions is to allow radioactive particles to pass through air; radioactive particles are emitted with enough energy to ionize hundreds of thousands of gas atoms per particle. Another way to ionize air is to strike a match! Do so with the match near a charged electroscope and you ought to discover that the electroscope discharges quickly.

A more controlled way to ionize gas atoms is to direct a beam of electrons into the gas. An electron gun can be used to produce the electron beam. By adjusting the anode voltage on the gun, the speed of the electrons can be changed. Figure 29.1 shows a thyratron valve used to demonstrate ionization by collision with electrons. The valve contains an inert gas, xenon. As the anode voltage is increased from zero, little current is registered on the ammeter until the voltage reaches 12.1 V. At this value, the electrons from the filament are pulled to the grid anode and given 12.1 eV of kinetic energy. Some of these electrons pass through the holes in the grid; however, with 12.1 eV, each electron has just the right amount of energy to ionize a gas atom. The ions are attracted to the grid plate, so giving extra current to make the meter reading measurable, as is shown in Figure 29.2.

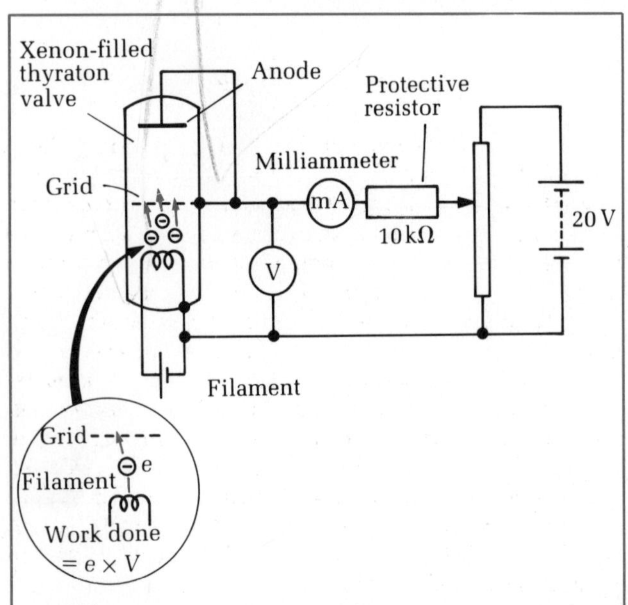

Figure 29.1 *Ionization by collision*

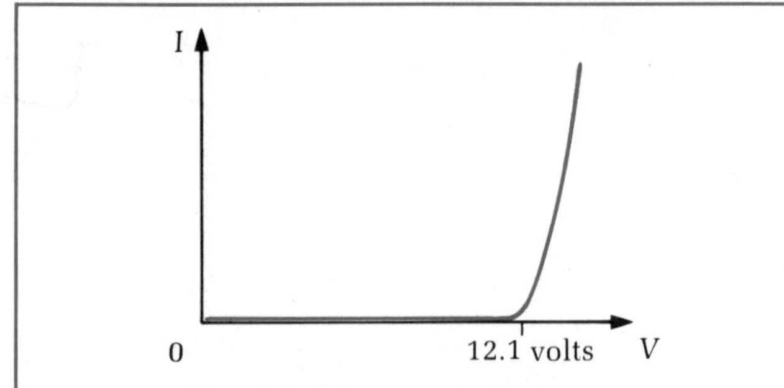

Figure 29.2 *Ionization results*

Excitation is the process whereby atoms absorb energy without ionization. The circuit of Figure 29.1 may be modified as in Figure 29.3 to show that gas atoms can absorb energy without being ionized. The idea is to measure the current due to electrons reaching the collector plate. As the grid voltage is increased, more and more electrons from the filament shoot through the grid to reach the collector plate. However, after a certain voltage, the current drops sharply as shown by Figure 29.4, to reach a minimum value. At this value of voltage, gas atoms will absorb energy from electrons shooting through the grid. For xenon, this happens at 8.4 V, less than the ionization potential of 12.1 V. The atoms are not ionized at 8.4 V but they absorb energy from the electrons

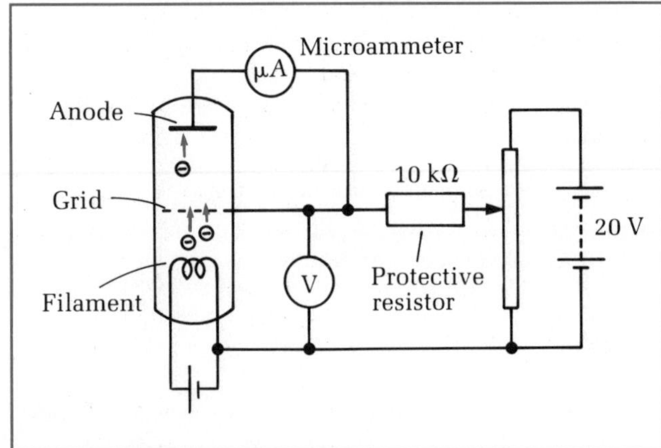

Figure 29.3 *Excitation by collision*

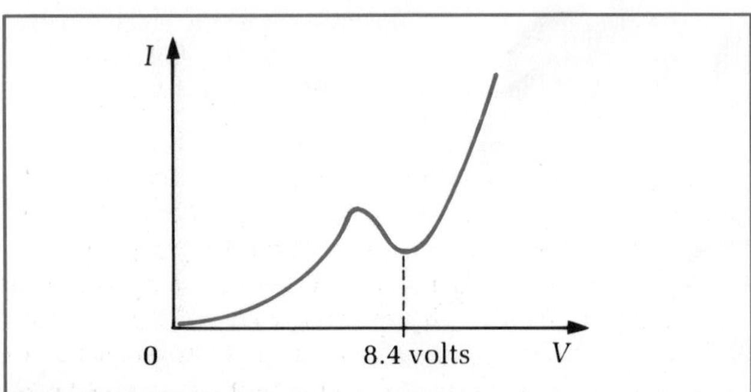

Figure 29.4 *Excitation results*

from the filament. The energy absorbed is not enough to ionize the atom.

This type of experiment was first performed in 1914. Different gases were tested in turn; all showed that atoms can absorb energy at certain values which depend on the type of atom. For example, mercury atoms absorb energy at 4.9 eV, 7.6 eV and 8.9 eV, and they ionize at 10.4 eV.

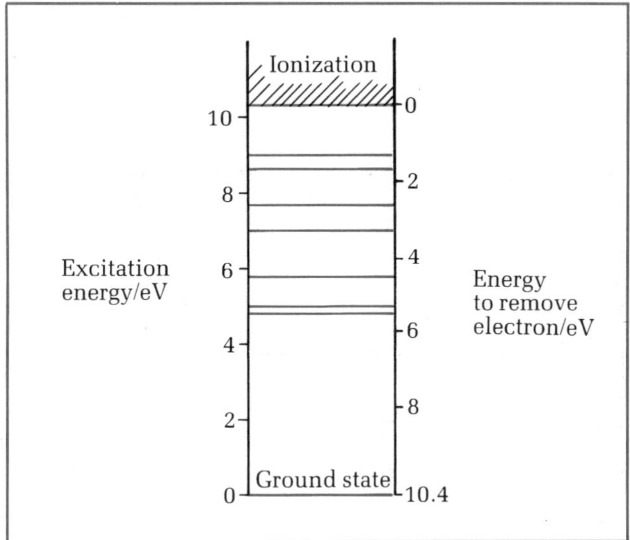

Figure 29.5 *The energy levels of the mercury atom*

An energy level ladder can be drawn up for each type of atom using such measurements. The ladder for mercury is shown in Figure 29.5. When a mercury atom absorbs 4.9 eV, it increases its potential energy by 4.9 eV. So its P.E. increases from the lowest possible level to a level 4.9 eV higher. What happens to this energy stored in each atom? It was discovered that ultra-violet radiation was emitted by the mercury vapour in the tube when the voltage was 4.9 V. Measuring the wavelength of the emitted radiation using a concave reflection grating (see p. 270) gave a value of 250 nm. Using Einstein's photon equation, the energy of each emitted photon was calculated.

$\lambda = 250\,\text{nm}$

So frequency $f = c/\lambda = \dfrac{3 \times 10^{8}}{2.5 \times 10^{-7}} = 1.2 \times 10^{15}\,\text{Hz}$

where c is the speed of light.

Given the Planck constant $h = 6.6 \times 10^{-34}\,\text{J s}$,

Then photon energy $= hf = 6.6 \times 10^{-34} \times 1.2 \times 10^{15}$
$= 7.92 \times 10^{-19}\,\text{J}$

$= \dfrac{7.92 \times 10^{-19}}{1.6 \times 10^{-19}} = 4.9\,\text{eV}.$

Other wavelengths emitted by mercury atoms can be measured using a mercury lamp. The measured values give photon energies corresponding to jumps on the energy level ladder for mercury built up from excitation experiments. So in the excitation experiment, mercury atoms absorb energy in fixed amounts from electrons shooting through the grid; each excited atom releases its extra energy by emitting photons.

So atoms absorb and emit energy in fixed amounts. Also, to pull electrons off atoms (i.e. to ionize), fixed amounts of energy are needed. It seems reasonable to conclude that electrons inside atoms have definite values of energy; when an atom absorbs a fixed quantity of energy, one of its electrons changes its potential energy.

Energy levels and spectra

The hydrogen spectrum can be observed using a spectrometer to view light from a hydrogen-filled discharge tube. The spectrum is a typical line spectrum as described on p. 272. Precise measurements of the lines give frequencies which can be fitted into a formula (first discovered in 1885 by Johann Balmer) for certain lines.

$$f = 3.29 \times 10^{5} \times \left\{\frac{1}{2^2} - \frac{1}{n^2}\right\}$$

where $n = 3, 4, 5,$ for the lines in the series.

Other lines could be fitted into formulae like Balmer's.

Twenty years after Balmer's discovery, the meaning in energy terms was uncovered using Einstein's photon theory. The light photons emitted from glowing hydrogen gas possess certain energy values only. From Einstein's photon equation,

Photon energy $E = hf$

where h is the Planck constant.

The light frequencies emitted from hydrogen gas fit the formula

$$f = 3.29 \times 10^{15} \times \left\{\frac{1}{m^2} - \frac{1}{n^2}\right\}$$

where $m = 2$ for the Balmer series, and $m = 3, 4, 5,$ etc., for other series.

So the photon energies are given by

$$E = hf = 6.6 \times 10^{-34} \times 3.29 \times 10^{15} \times \left\{\frac{1}{m^2} - \frac{1}{n^2}\right\}$$

$$= 21.7 \times 10^{-19} \times \left\{\frac{1}{m^2} - \frac{1}{n^2}\right\} \text{ (joules)}$$

Using the equation for photon energy, the energy level picture of hydrogen can be made up, as shown by Figure 29.6, at the top of the next page. Hydrogen atoms have just one electron each, so the energy level picture shows the energy levels of the single electron of the hydrogen atom. Normally, the electron is in the lowest level, called the **ground state**, but by making hydrogen atoms crash into each other at high speeds, their electrons can be excited to higher levels. If the electron from an atom escapes altogether, the atom is ionized; the electron then reaches the **ionization level** which is, for hydrogen, 13.6 eV above the ground state level. Why 13.6 eV? An electron which drops from the ionization level ($n = \infty$) to the innermost level ($m = 1$) loses $21.7 \times 10^{-19}\,\text{J}$ which equals 13.6 eV. So the ionization level is 13.6 eV above the ground state level.

When an electron returns to a lower level, it loses an exact amount of energy by emitting a photon. The photon

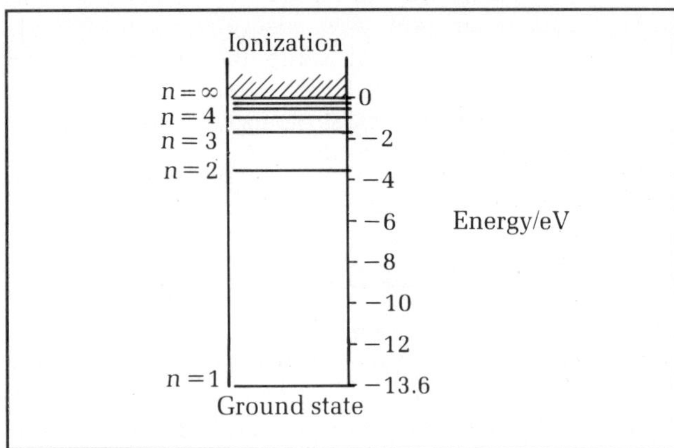

Figure 29.6 *Energy levels of the hydrogen atom*

carries away energy equal to the difference between the initial and final energy levels. The Balmer series of lines corresponds to electron **transitions** down to the level immediately above the ground state (i.e. down to the $n=2$ level). Other series of lines are produced by transitions down to $n=1$ (the Lyman series), $n=3$ (the Paschen series) etc., each series named after its discoverer.

The line emission spectrum of any element can be used to give the energy level picture for each element. However, no simple formula is possible for elements other than hydrogen. Atoms other than hydrogen have more than one electron each, and the electrons in these atoms interact with each other. So the energy level picture does not follow a simple pattern as for hydrogen. Figure 29.8 shows a graph of intensity of light emitted against wavelength for light from a certain element. The intensity is proportional to the number of photons per second emitted from the light source, and it can be measured using a light detector attached to a spectrometer. So the intensity against wavelength graph shows the number of photons per second emitted at each wavelength. An intense line corresponds to many more photons/second emitted than a weak line, so an intense line is due to electron transitions occurring frequently between two particular levels. The intensity against wavelength graph can be used to give the energies of the emitted photons; then, the energy level ladder can be worked out, as shown in Figure 29.9. The intense line at X is due to transitions from level P down to level Q.

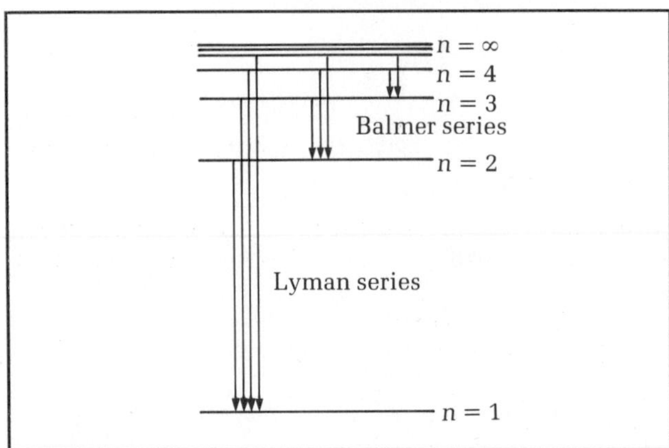

Figure 29.7 *Transitions*

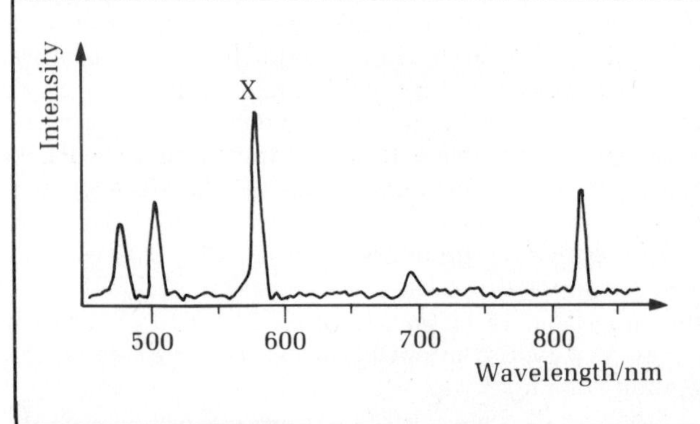

Figure 29.8 *Spectrum of a certain element*

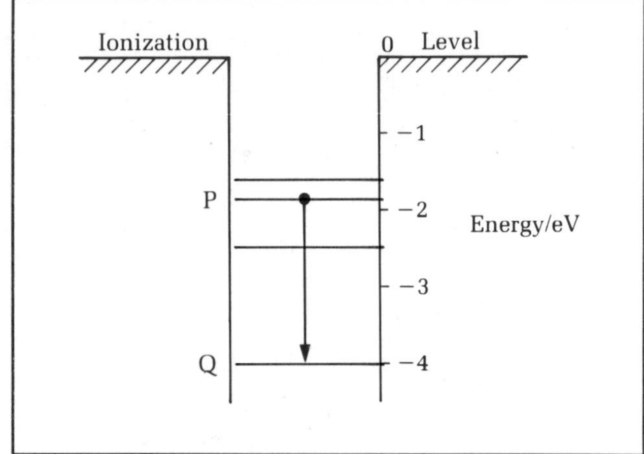

Figure 29.9 *Energy levels for figure 29.8*

Line absorption spectra can be explained using the energy level picture. Such spectra are produced when white light (i.e. continuous range of frequencies) is passed through suitable material. For example, the solar spectrum shows absorption lines due to material in the form of hot gas round the Sun. The Sun emits a continuous range of frequencies across the visible spectrum and beyond; when sunlight is analysed using a spectrometer, dark lines are visible against the continuous background of colours (see p. 273). These lines correspond to missing frequencies and hence missing photon energies. These missing photons, emitted from the Sun, have been absorbed by atoms in the hot envelope of gas around the Sun. They are absorbed because they have just the right amount of energy to make electrons in the gas atoms move to higher energy levels.

29.2 Models of the atom

The Bohr atom

To understand energy levels, it makes sense to start with the simplest type of atom, the hydrogen atom. Niels Bohr, in 1913, developed the model of the atom as a miniature solar system; he imagined the electron of the hydrogen atom whirling round the nucleus, like a planet revolves round the Sun. He said that each electron in its orbit had constant energy, corresponding to one of the energy levels of the atom. Other scientists objected to this statement because they knew that electrons being accelerated, such as in an electron gun, should change energy, but Bohr took the bold step of saying that the known ideas about electrons might not hold inside the atom. Bohr's whirling electron is accelerated because it moves along a circular path (see p. 42) but Bohr assumed that its energy was constant.

Using Newton's laws of motion and Coulomb's law of force Bohr derived an equation for the speed of the electron in terms of its radius of orbit.

For circular motion, the centripetal force F is given by

$$F = -mv^2/r$$

where m is the mass,
v is the speed,
r is the radius of orbit.

The − sign indicates an inward force.

The centripetal force F is provided by the electrostatic attraction between the electron and the nucleus, as shown in Figure 29.10.

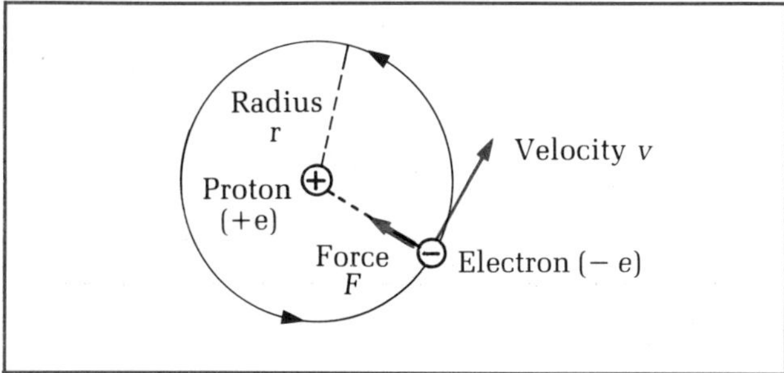

Figure 29.10 *The Bohr model of the hydrogen atom*

$$F = \frac{-e \times e}{4\pi\varepsilon_0 r^2}$$

$$\text{Hence } \frac{-mv^2}{r} = \frac{-e^2}{4\pi\varepsilon_0 r^2}$$

$$\text{So } v^2 = \frac{e^2}{4\pi\varepsilon_0 rm} \qquad (1)$$

The next step by Bohr involved making an assumption that seems strange at first. However, it led on to a full explanation of the Balmer formula for spectral lines so Bohr went ahead. In fact, he led scientist beyond a mere explanation of the Balmer formula into a new understanding of the 'atomic world'. What Bohr did was to show that the atom is part of quantum physics, discovered by Einstein and Planck a decade earlier. Einstein had shown that light is composed of photons, each carrying well-defined energy (=hf). Each photon carries a fixed amount, or quantum of energy. Electric charge too is quantized in units equal to the charge carried by an electron.

Bohr's assumption was that the angular momentum of the electron is quantized, always in lumps of $h/2\pi$. For an electron of mass m moving at steady speed v in a circular orbit of radius r, its angular momentum is equal to mvr (see p. 50). So Bohr wrote an equation for his assumption, given at the top of the next column.

$$mvr = \frac{nh}{2\pi} \qquad (2)$$

where $n = 1, 2, 3, \ldots$

This equation can be combined with the earlier equation (1) to give expressions for v and r separately.

$$\text{From (2), } m^2v^2r^2 = \frac{n^2h^2}{4\pi^2}$$

$$\text{So } m^2\frac{(e^2)}{4\pi\varepsilon_0 rm}r^2 = \frac{n^2h^2}{4\pi^2} \qquad \text{using (1)}$$

$$\text{Hence } r = \left(\frac{\varepsilon_0 h^2}{\pi me^2}\right) \times n^2$$

The equation tells us that the electron is allowed to orbit the nucleus at certain distances only. The innermost orbit is when $n = 1$.

$$\text{Hence radius of innermost of orbit } r_1 = \frac{\varepsilon_0 h^2}{\pi me^2}$$

Let's put some numbers in to calculate r_1.

$\varepsilon_0 = 8.85 \times 10^{-12}\ \text{F m}^{-1}$
(from electric field theory; see p. 168)
$h = 6.63 \times 10^{-34}\ \text{J s}$
(from photon energy; see p. 429)
$m = 9.11 \times 10^{-31}\ \text{kg}$
(from electron experiments; see p. 425)
$e = 1.60 \times 10^{-19}\ \text{C}$
(from Millikan's experiment; see p. 421)

Hence

$$r_1 = \frac{8.85 \times 10^{-12} \times (6.63 \times 10^{-34})^2}{\pi \times 9.11 \times 10^{-31} \times (1.6 \times 10^{-19})^2} = 5.29 \times 10^{-11}\ \text{m}$$

Do you think this value is of the right order? Bigger atoms such as copper atoms are about 3×10^{-10} m in size, so the value seems sensible (see p. 67).

The energy of the electron is partly kinetic energy and partly potential energy due to the electrostatic attraction.

1 Kinetic energy $= \frac{1}{2}mv^2$
Using equation (1) for v^2 gives

$$\text{K.E.} = \tfrac{1}{2}m\frac{e^2}{4\pi\varepsilon_0 rm} = \frac{e^2}{8\pi\varepsilon_0 r}$$

2 Potential energy due to electron near nucleus is given by

$$\text{P.E.} = \frac{-e \times e}{4\pi\varepsilon_0 r} = -\frac{e^2}{4\pi\varepsilon_0 r}$$

Hence the total energy E of the electron = K.E. + P.E.

$$= \frac{e^2}{8\pi\varepsilon_0 r} - \frac{e^2}{4\pi\varepsilon_0 r}$$

$$= -\frac{e^2}{8\pi\varepsilon_0 r}$$

$$\text{Since radius } r = \left(\frac{\varepsilon_0 h^2}{\pi me^2}\right)n^2$$

Then $E = -\frac{e^2}{8\pi\varepsilon_0} \times \frac{\pi m e^2}{\varepsilon_0 h^2 n^2} = -\frac{me^4}{8\varepsilon_0{}^2 h^2} \times \frac{1}{n^2}$

The equation tells us that the electron has fixed values of energy, depending on which orbit it is in. In the innermost orbit, its energy is given by

$$E_1 = -\frac{me^4}{8\varepsilon_0{}^2 h^2} \qquad \text{(with } n=1\text{).}$$

Use the values of m, e, ε_0 and h listed before to calculate E_1. The equation above gives E_1 in joules; convert to eV using the fact that $1.6 \times 10^{-19}\,\text{J} = 1\,\text{eV}$. You ought to obtain a value of -13.6 eV for E_1. So to remove the electron from the innermost orbit to infinity, its energy must be increased by 13.6 eV.

$$E = -13.6 \times \frac{1}{n^2}$$

where E = energy of electron in eV, $n = 1, 2, 3$, etc.

The above equation gives the energy levels of the hydrogen atom, and the values agree with the picture worked out by studying the line emission spectrum of hydrogen. In other words, the above equation gives an exact explanation of the Balmer and the other formulae. Using the energy equation above gives energy levels as follows:

n	1	2	3	4	5	...	∞
E/eV	−13.6	−3.4	−1.5	−0.85	0.54	...	0

The lowest energy level is $n=1$; this is called the **ground state** level because the electron normally occupies this level unless given sufficient energy to move up to a higher level.

The ionization level is the level corresponding to the electron just escaping from the nucleus. Its energy is then zero so it is the n = infinity (∞) level. Between the ground state level and the ionization level, there are an infinite number of other levels, but not equally spaced. Figure 29.11 shows the picture.

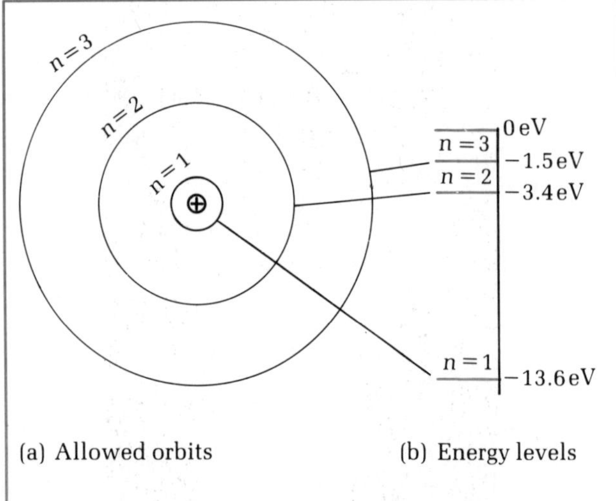

Figure 29.11 *Allowed orbitals of the hydrogen atom*

Excitation by collision occurs when atoms bump into each other violently. Some of the kinetic energy is used to allow an electron to move up to a higher level in one or both of the atoms that collide. An atom with an electron at a higher energy level than it normally occupies is said to be 'excited'. But its excited state is temporary, and the electron returns to the lower levels, eventually dropping into the ground state level.

When an electron drops from one level to a lower level, it loses energy, just like somebody climbing down a ladder from one rung to a lower rung. When an electron drops, it loses its surplus energy by emitting a photon. Suppose an electron makes a transition from level n_1 to lower level n_2.

Its initial energy $E_{n_1} = -E_0 \times \frac{1}{n_1{}^2}$ where $E_1 =$ 13.6 eV.

Its final energy $E_{n_2} = -E_1 \times \frac{1}{n_2{}^2}$

Hence the energy of the photon released as a result of the transition is given by

Photon energy hf = initial energy − final energy

$$= \left(-E_1 \times \frac{1}{n_1{}^2}\right) - \left(-E_1 + \frac{1}{n_2{}^2}\right)$$

$$= E_1 \times \left\{\frac{1}{n_2{}^2} - \frac{1}{n_1{}^2}\right\}$$

Which gives $\quad f = \frac{E_1}{h} \times \left\{\frac{1}{n_2{}^2} - \frac{1}{n_1{}^2}\right\}$

This is in agreement with Balmer's formula.

Beyond Bohr

Bohr made the atom understandable, just as Newton had done for the Solar System over two centuries earlier. Bohr's theory gave values for energy levels in agreement with the measurements – a remarkable achievement in itself. But Bohr's simple model could not be applied with equal success to atoms other than hydrogen. Even helium with just two electrons per atom is too complicated to be explained by the Bohr model exactly. The two electrons interact with one another so the energy levels for helium are not the same as for hydrogen. Nevertheless, the general ideas of the Bohr model can be used.

The ground state of an atom occurs when the electrons are in the lowest possible energy levels. For example, with helium, the two electrons of each atom are then in the innermost orbit. Supply the helium atom with enough energy and its electrons can move up to higher energy levels (i.e. outer orbits). As explained earlier, the move is temporary and photons are emitted when the atoms 'de-excite'.

Quantum mechanics was developed from the quantum theory created by Bohr and Einstein. Instead of picturing the atom as a miniature solar system, we now think in terms of **probability** to describe electrons inside atoms. Why probability? Suppose you devised a supermicroscope capable of pinpointing the electrons in an atom. To 'see' the electrons, you need to direct light at them and then observe the scattered light. But the scattering of light by an electron changes its position and energy, because the electron recoils. So the act of 'observing' changes the situation you are trying to observe. The ideas are discussed in a bit more detail on p. 481.

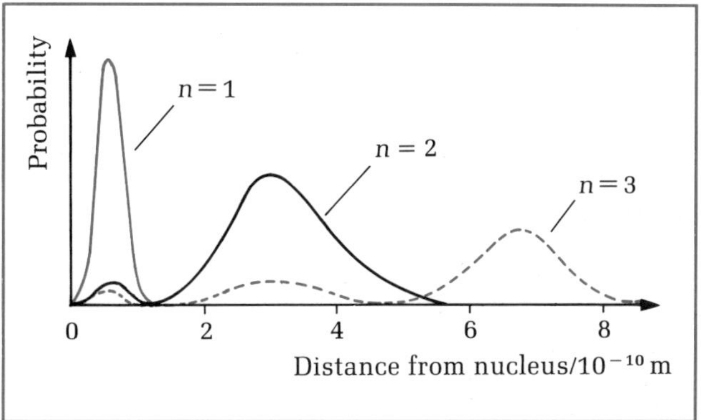

Figure 29.12 *Probability of an electron being at a given distance from the nucleus*

Since we cannot be sure where a moving electron is at any time, then we use the ideas of probability. Electrons in the atom are therefore best described in terms of the probability of finding one in any given position. The table below shows that the most probable locations, called **shells**, correspond to the Bohr 'orbitals'. Quantum mechanics gives the rules for working out the probability distribution for each allowed value of energy. The ideas can be applied to an 'ideal' atom where electrons are considered not to affect one another. The results enable us to understand many aspects of the behaviour of atoms, including the Periodic Table.

Type of atom	Number of electrons	Distribution of electrons in the ground state		
		1st shell ($n=1$)	2nd shell ($n=2$)	3rd shell ($n=3$)
hydrogen	1	1	—	—
helium	2	2	—	—
lithium	3	2	1	—
.				
.				
.				
neon	10	2	8	—
sodium	11	2	8	1

1. The innermost shell ($n=1$) is filled by two electrons only. So lithium atoms, with three electrons, have a single electron in the $n=2$ shell. Lithium is very reactive because this outer electron is easily 'picked off' by other atoms.
2. The second shell ($n=2$) is filled by eight electrons. So xenon atoms which have ten electrons each, have full shells. Xenon is an inert gas; it doesn't do much because its electrons are in 'full' shells.
3. The third and other shells each can take a fixed number of electrons. When each shell is filled, the next shell takes up electrons. The picture becomes more complicated now because the outer shells overlap.

Sodium atoms each have 11 electrons. So in the ground state, a sodium atom has two electrons in the $n=1$ shell, eight electrons in the $n=2$ shell and a single electron in the $n=3$ shell. Like lithium, sodium is very reactive because its atoms each have a single electron in an outer shell.

The energy level diagrams for an atom with more than one electron differ from those for the single-electron atom. The levels do not fit the simple Bohr pattern and they can be much deeper than in the hydrogen atom. If we could switch off the interaction between the electrons of a multi-electron pattern, the levels would fit the Bohr-type pattern. But they would be much deeper for big nuclei (i.e. large Z) because the electric field created by the nucleus would be much stronger (since there are many more protons in big nuclei). For example, a lead atom has a nuclear charge of $+82e$ because there are 82 protons in its nucleus. The Bohr model shows that the energy of an electron is proportional to (nuclear charge)^2; for lead, the levels are therefore 82×82 times deeper than for hydrogen, assuming no electron-electron interaction. Take account of electrons affecting one another and the picture is even more complicated.

29.3 X-rays

Producing X-rays

The discovery of X-rays was yet another consequence of research into discharge tubes. In 1895, Wilhelm Rontgen was working on a discharge tube in a darkened room when he noticed glowing material the other side of the room. He discovered that the glow was produced only when the discharge tube was operated and he found that covering the tube made no difference. He placed thin metal plates between the tube and material but found the glow persisted. The tube was emitting invisible penetrating radiation which became known as 'X-radiation'. Rontgen showed that X-rays blacken photographic film, and he X-rayed the bones of his hand.

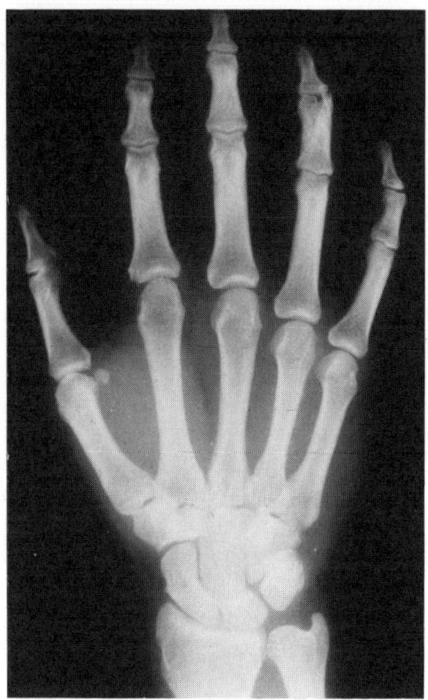

Figure 29.13 *An X-ray picture of a hand*

Rontgen's discoveries were greeted with amazement and soon caught the imagination of the world. Further investigations showed that X-rays are electromagnetic waves with wavelengths of the order of 10^{-10} m or less.

Modern X-ray tubes, as in Figure 29.14 are designed so that electrons from a hot-wire filament are accelerated onto a target anode. A high-voltage unit is used to set the anode at a large positive potential with reference to the filament. The tube is evacuated so that the beam of electrons from the filament is pulled onto the anode at very high speeds. These beam electrons are suddenly decelerated on impact, and some of the kinetic energy is converted into electromagnetic energy, as X-rays. So the target anode emits X-rays as a result of being bombarded with high-energy electrons. The X-rays spread out from the target and pass easily through the glass tube walls. Less than 1% of the energy supplied to an X-ray tube is converted to X-radiation; the rest is converted to internal energy of the target. So the tube is designed with a cooling system to prevent the target from melting.

The **efficiency** of an X-ray tube is given by

$$\frac{\text{The X-ray energy produced/second}}{\text{The electrical power supplied}} \times 100\%$$

For example, if a tube is 1% efficient and it is supplied with 3000 W of electrical power, then the X-ray energy produced each second is 1% of 3000 = 30 W. So the heat produced per second is 2970 W which must be removed by the cooling system.

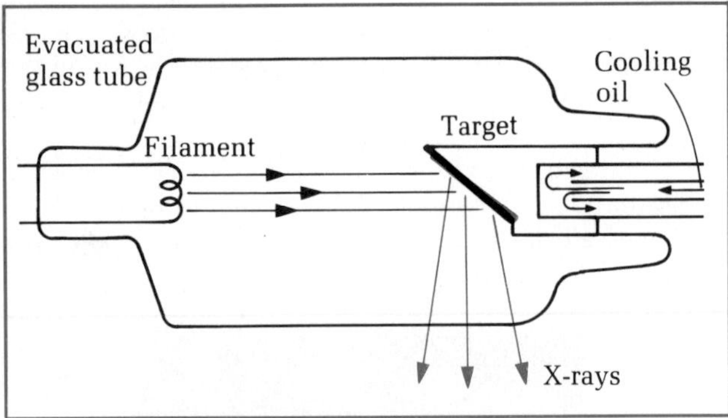

Figure 29.14 *An X-ray tube*

Properties of X-rays

Detectors of X-rays use one of the following properties of X-rays.

1 Photographic film is blackened by X-rays. The more energy per second an X-ray beam carries, the blacker the film is when it is developed.

2 Fluorescent materials glow when X-rays are directed at them. The X-ray photons cause excitation of the atoms of the material; electrons in the atoms jump to higher energy levels when each electron absorbs an X-ray photon. When the electrons return to lower energy levels, they do so in 'small steps', so light is emitted rather than X-radiation because light photons have less energy. The process is shown in Figure 29.15.

3 Photoelectric emission is produced by X-rays (see p. 428).

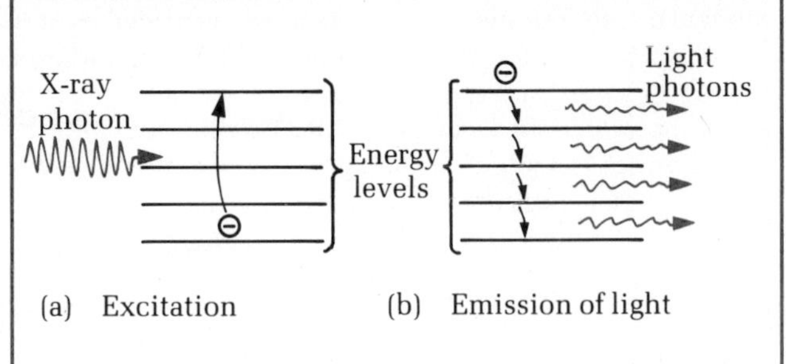

Figure 29.15 *Fluorescence*

4 Ionization of a gas results when an X-ray beam is passed through it. So X-rays may be detected by geiger tubes.

Diffraction of X-rays was discovered by Max Von Laue, in 1913, who showed that a crystal can diffract an X-ray beam into a pattern. If the spacing between the atoms is known, the X-ray wavelength can be calculated. A simpler technique was developed by William Bragg who showed how to use a crystal as a reflection grating. As shown in Figure 29.16, each plane of atoms acts as a weak reflector of X-rays; however, the weak reflections from parallel planes reinforce if the spacing between the planes is suitable. In Figure 29.16, the X-rays that reflect from plane B travel an extra distance of $2d \sin\theta$ compared with reflections from plane A. If the extra distance is a whole number of wavelengths, then the reflections from B reinforce those from A and all other parallel planes at that spacing (see p. 79).

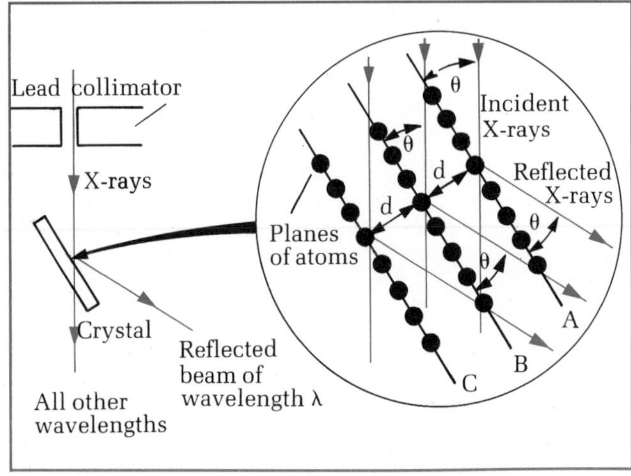

Figure 29.16 *Selecting X-rays*

$$2d \sin\theta = m\lambda$$

where d = crystal spacing,
θ = angle of reflected beam to the incident beam,
λ = X-ray wavelength,
$m = 1, 2, 3, \ldots$

The beam from an X-ray tube is composed of a continuous range of wavelengths. Using a crystal as a reflection grating, any part of the wavelength range can be selected. This is done by turning the crystal so that its angle to the beam fits the Bragg equation above. In other

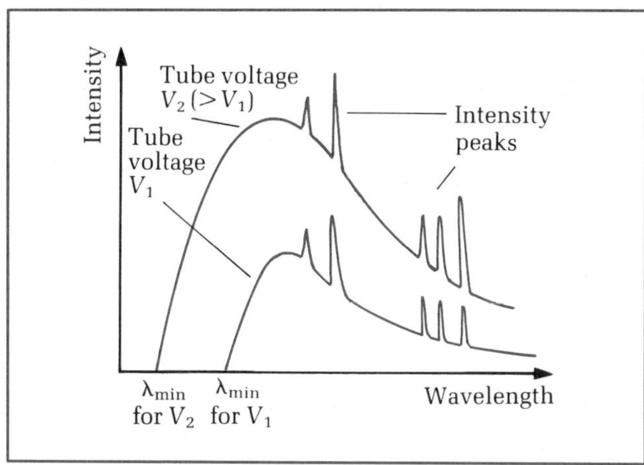

Figure 29.17 *Intensity v wavelength for an X-ray tube*

words, at any angle θ, the reflected X-rays have wavelength given by $2d \sin \theta = m\lambda$. All the other wavelengths in the incident beam pass through the crystal without reflection.

The intensity of an X-ray beam from an X-ray tube varies with wavelength. Intensity is a measure of the energy per second carried by the beam. Using a crystal as a wavelength selector, Bragg measured the intensity of different wavelengths. The results can be plotted as a graph of intensity against wavelength, shown in Figure 29.17. The graph shows the following features.

1 A continuous background of X-radiation in which the intensity varies smoothly with wavelength. The background intensity reaches a maximum value as the wavelength increases, then the intensity falls at greater wavelengths.

2 Minimum wavelength λ_{min} which depends on the tube voltage. The higher the voltage, the smaller the value of λ_{min}.

To explain the minimum wavelength, assume all the energy of a single beam electron is used to produce a single photon. The work done on each beam electron (when it is accelerated onto the anode) is eV. Then its kinetic energy just before impact is eV. (V is the tube voltage.)

Assuming all this kinetic energy is used to create a single photon then the maximum energy of an X-ray photon is given by

$$hf = eV$$

Hence the maximum frequency $= eV/h$.
Since the wavelength of an emitted photon is given by $\frac{\text{speed of light } c}{\text{frequency}}$, then maximum frequency corresponds to minimum wavelength.

So the minimum wavelength $\lambda_{min} = \frac{c}{(eV/h)} = \frac{hc}{eV}$

$$\lambda_{min} = \frac{hc}{eV}$$

where h = Planck constant,
c = speed of light,
e = electron charge,
V = tube voltage.

3 Sharp peaks of intensity occur at wavelengths unaffected by change of tube voltage. Different target materials give different wavelengths for these peaks. The peaks are due to beam electrons knocking out inner-shell electrons from target atoms. When these inner-shell vacancies are refilled by free electrons (see p. 460), X-ray photons are emitted. The inner-shell energy levels are so deep that, when refilled, the free electron dropping in emits a high-energy photon. Work out for yourself the energy carried away by an X-ray photon of wavelength 0.1 nm, given $c = 3 \times 10^8\,\text{m s}^{-1}$, $h = 6.6 \times 10^{-34}\,\text{J s}$. You ought to obtain a value of 12 000 eV (1 eV $= 1.6 \times 10^{-19}$ J).

The peaks for any target element define its **characteristic X-ray spectrum**. As explained in the next section, an element can be identified from the peaks.

Worked example An X-ray tube is operated at 120 kV with a beam current of 0.5 mA. Assuming its efficiency is 1%, calculate:

a) the number of electrons hitting the target each second,
b) the X-ray energy emitted each second,
c) the heat energy dissipated each second,
d) the minimum wavelength of the emitted X-radiation.

Solution

a) In 1 s, 0.5 mC of charge hits the target.
Each electron carries charge 1.6×10^{-19} C.
So the number of electrons hitting the target each second is $0.0005/(1.6 \times 10^{-19}) = 3.12 \times 10^{15}$.

b) The power supplied to the tube = current × voltage
$= 0.5\,\text{mA} \times 120\,\text{kV}$
$= 60\,\text{W}$.
Hence the X-ray energy emitted/second = 1% of $60\,\text{W} = 0.6\,\text{J s}^{-1}$.

c) The heat energy/second = power supplied − X-ray energy/s emitted
$= 60 - 0.6 = 59.4\,\text{J s}^{-1}$.

d) The minimum wavelength λ_{min} is given by $\lambda_{min} = \frac{hc}{eV}$

$$\lambda_{min} = \frac{6.6 \times 10^{-34} \times 3 \times 10^8}{1.6 \times 10^{-19} \times 100 \times 10^3} = 1.23 \times 10^{-11}\,\text{m}.$$

X-ray spectra and energy levels

The pattern of intensity peaks, such as Figure 29.17, depends on the target element. Henry Moseley, in 1913, measured these patterns for a number of elements, and he was able to explain them by modifying the Bohr model. For each element, he measured the wavelength of each peak by using a crystal as a reflection grating. By modifying the Bohr model, he showed that the peaks were due to electrons filling vacancies in the inner shells of the atoms. Beam electrons hitting the target knock out inner-shell electrons from the atoms, creating inner-shell vacancies. When these are refilled by free electrons, X-ray photons are emitted, as shown in Figure 29.18, at the top of the next page.

The K-series of peaks is due to the innermost shell, called the 'K-shell', being refilled by electrons from shells higher up in energy terms. In other words, an electron

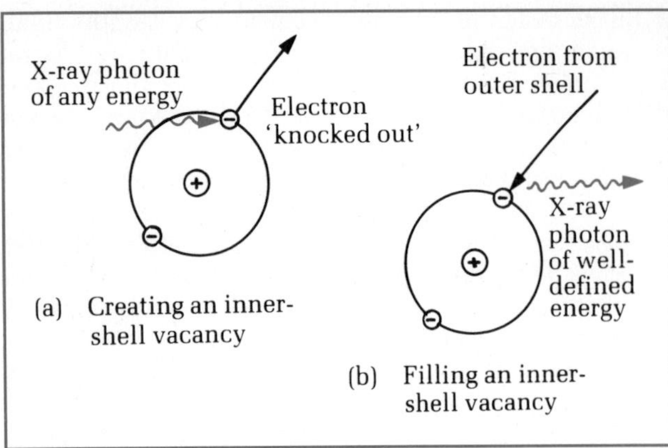

Figure 29.18 *Characteristic X-rays*

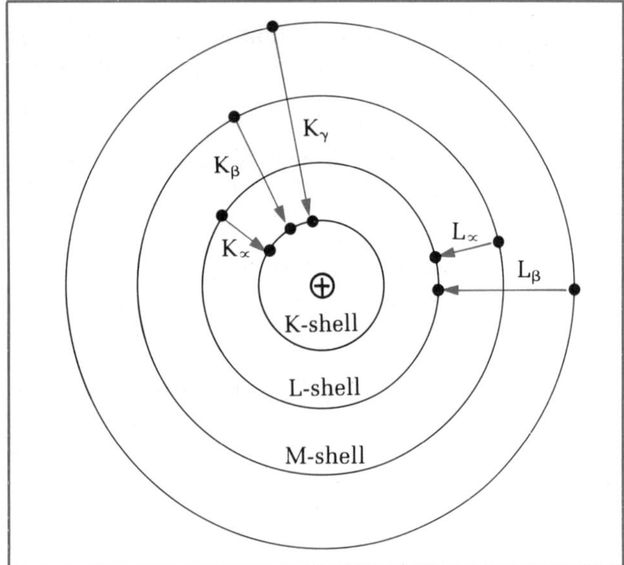

Figure 29.19 *Explaining characteristic X-rays*

making a transition from any one of the outer shells to the innermost shells emits an X-ray photon in this series. Figure 29.19 shows that the K_α and K_β lines in the K-series are due to transitions to the K-shell from the second and third shells respectively.

The L-series is due to the second shell being refilled by electrons from shells further out. As with the K-series, there are several lines in the series, each line caused by a particular transition to the second shell.

For the heavier elements,a third series is observed, referred to as the M-series. This is due to transitions down to the third shell.

Each set of lines is explained by considering an electron dropping from one particular level to another. Moseley showed that each set could be described by a Balmer-type formula, modified to take account of the atomic number being more than 1. For example, the K_α series of lines fits the formula

$$\text{Frequency } f=\frac{E_0}{h}(Z-1)^2\times\left\{\frac{1}{1^2}-\frac{1}{2^2}\right\}$$

where $Z=$ atomic number, $E_0/h=13.6$ eV.

The formula is like the Balmer formula discussed on p. 458, except for the presence of the term $(Z-1)^2$. This suggests that in an atom with nuclear charge $+Ze$, an electron dropping into the innermost shell is affected by the electron already in that shell. Effectively, the electron in the innermost shell shields some of the nuclear charge to make the effective nuclear charge equal to $Z-1$ not Z. So the energy is proportional to $(Z-1)^2$ not Z^2.

Using X-rays

In medicine X-rays are used to diagnose illnesses and for treatment. For example, damaged limbs are usually X-rayed to show any broken bones. X-ray imaging works by allowing X-rays to cast a 'shadow' onto a photographic film. The technique is relatively easy for bones because they absorb X-rays better than tissue does. The shadow of a bone inside a limb shows up well when such a film is developed. Other internal organs can be X-rayed provided they absorb X-rays better than surrounding tissue. Sometimes it is necessary to inject a 'contrast' medium into the organ to improve its absorbing ability.

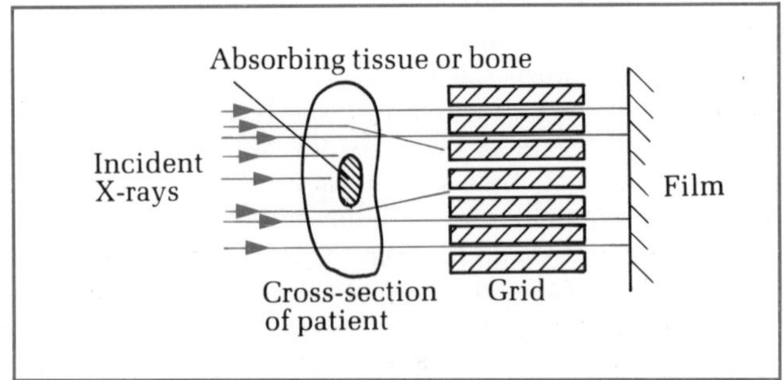

Figure 29.20 *Using a grid*

The X-ray beam is **filtered** before reaching the patient, to remove low-energy photons. This is achieved by passing the beam through a metal plate. If this was not done, low-energy photons would be absorbed by the patient's tissues, causing unnecessary heating and possible burns. By removing these photons, absorption by the tissues is reduced to a minimum. Also, after the beam has passed through the patient, the beam is passed through a **grid** of holes drilled in a thick lead plate. In this way scattered X-rays from the patient are prevented from reaching the film. Otherwise, these scattered photons might reach the 'shadow' areas of the film to reduce the contrast.

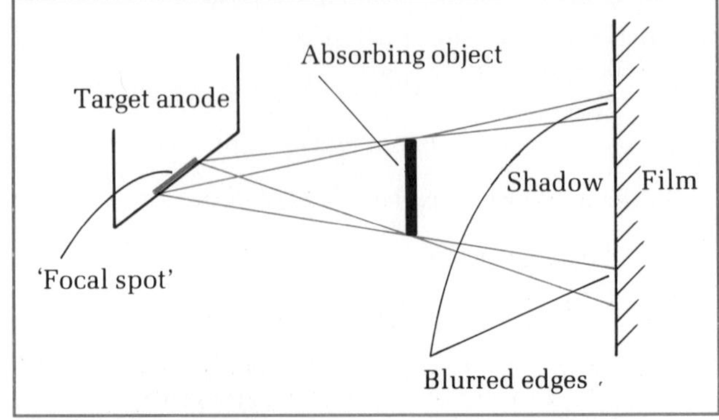

Figure 29.21 *The focal spot*

The target area of the anode must be made as small as possible to prevent the image from becoming blurred. This area, known as the **focal spot**, becomes hot so it cannot be made too small or it would melt, but if it is too large, the edges of the images on the film are not sharp. This is shown in Figure 29.21.

In engineering X-rays are used to locate cracks in metals. The cracks scatter X-rays so they show up on photographic film placed behind the metal. Figure 29.22 shows the idea.

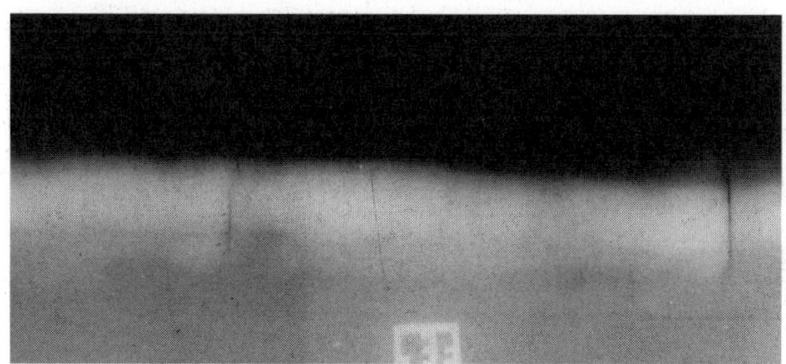

Figure 29.22 *Detection of cracks*

29.4 Summary

Definitions

1 *The energy level pattern* of a given type of atom or molecule is defined by the allowed values of energy which the atom or molecule can have.
2 *The characteristic X-ray spectrum* of an element is defined by the pattern of intensity peaks when that element is used as the target anode in an X-ray tube.

Equations

1 The energy levels of the Bohr atom are given by

$$-E_1/n^2$$

2 Photon frequency due to an electron dropping from energy level E_{n_1} to level E_{n_2} is given by

$$f = \frac{E_{n_1} - E_{n_2}}{h}$$

3 The minimum wavelength from an X-ray tube is given by

$$\lambda_{min} = \frac{hc}{eV}$$

Short questions

Electronic charge $e = 1.6 \times 10^{-19}$
Planck constant $h = 6.63 \times 10^{-34}$ J s
Speed of light in a vacuum $c = 3 \times 10^{-8}$ m s^{-1}

29.1 The current-voltage relationship for a xenon-filled diode valve is shown at the top of the next column.
a) Explain why the current is constant over the range from about 1 to 12 V.

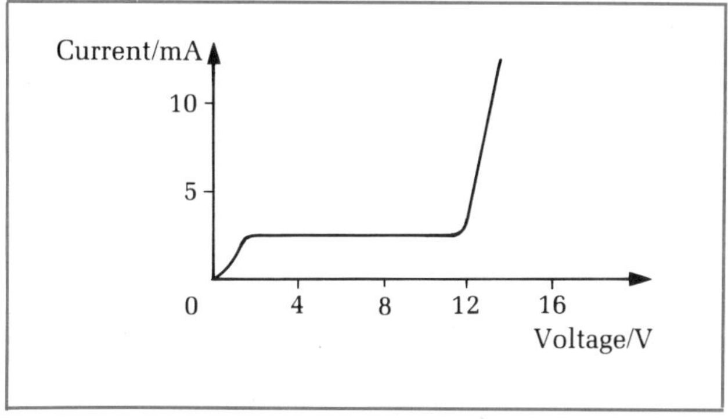

Figure 29.23

b) Why does the current increase rapidly for voltages greater than about 12 V?
c) Calculate the minimum energy, in J, needed to ionize a xenon atom.
d) Use the graph to estimate the ionization current at 14 V, and hence calculate the number of ions per second being produced at 14 V.

29.2 The energy level pattern for a particular type of atom is shown in the diagram below. Calculate
a) the p.d. which electrons, initially at rest, must be accelerated through in order to ionize such an atom,
b) the minimum wavelength for electromagnetic radiation emitted by this type of atom.

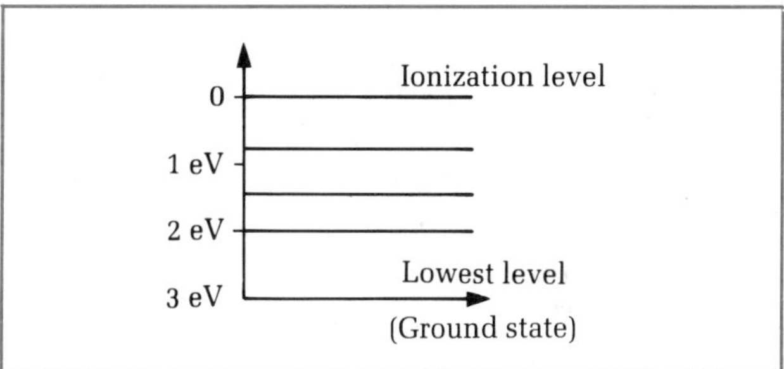

Figure 29.24

29.3 Use the energy pattern of question **29.2** to explain why atoms at high temperatures emit light of characteristic wavelengths.

29.4
a) Describe the appearance of
i) a line emission spectrum,
ii) a line absorption spectrum,
iii) a continuous spectrum.
b) A spectrometer is used to view light from a sodium flame produced by heating common salt in a bunsen flame. The spectrum shows two prominent yellow lines, closely spaced, at wavelengths of 589.0 and 589.6 nm. These lines are due to electron transitions from two closely spaced energy levels down to the same lower energy level. Calculate the energy difference between the closely spaced levels.

29.5 For the hydrogen atom, its energy levels are given by the formula $E_n/eV = -13.6 \div n^2$ where $n = 1, 2, 3$ etc. Identify the energy levels responsible for the 656 nm and 1026 nm wavelengths of the hydrogen spectrum.

29.6 Give two reasons why the spectral lines from atoms such as copper do not fit the formulae for the hydrogen spectrum.

29.7

a) Give a labelled diagram of an X-ray tube and explain the operation of such a tube.

b) For an X-ray tube operating at constant voltage and current, the beam electrons hit the target with the same kinetic energy. Yet the emitted X-rays have a range of energies up to a maximum value. Explain why the emitted X-rays do not have identical energies.

29.8

a) Sketch graphs to show how the intensity of the X-rays from an X-ray tube varies with wavelength when the tube voltage is

i) 10 kV,

ii) 40 kV.

In each case, calculate the minimum wavelength of the X-rays.

b) A 40 kV X-ray tube has a beam current of 5 mA. Calculate

i) the number of electrons hitting the anode each second,

ii) the electrical power supplied to the tube,

iii) the X-ray energy produced each second if the tube's efficiency is 1%,

(iv) the heat energy per second dissipated at the anode.

Why is the target anode usually a piece of tungsten set in a copper block?

29.9 A beam of X-rays from an X-ray tube is directed at a crystal grating so as to reflect X-rays of a particular wavelength off the crystal. The crystal is arranged so that a strongly reflected beam is produced at an angle of 20° to the direction of the incident beam, as below. If the layers of atoms responsible for the strong reflections are at a spacing of 0.31 nm, calculate the wavelength and energy of the X-ray photons in the reflected beam.

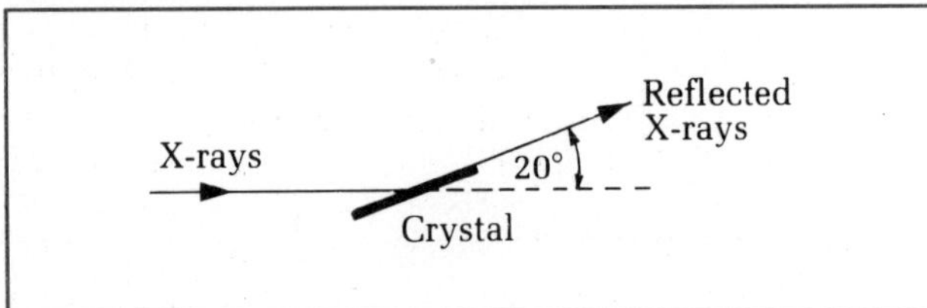

Figure 29.25

29.10 The spectrum of X-rays from an X-ray tube shows sharp intensity peaks at wavelengths that are characteristic of the anode material. Explain in atomic terms how these peaks are caused as a result of electron bombardment of the target anode atoms.

30

Nuclear energy

30.1 Mass and energy

$$E = mc^2$$

Einstein in 1905 was the first person to show that energy and mass are equivalent. After successfully explaining photoelectricity, he turned his creative mind to the laws of motion. Two centuries earlier, Newton had established these laws and they had been used very successfully since that time. But Einstein was unhappy about the basic assumptions behind these laws – assumptions so basic that they were never called into question until Einstein did so. For example, suppose you make a brief visit to a friend in another town; if you set off at noon and your watch shows 5.00 pm when you return, you would expect the clocks at your home to show 5.00 pm too when you return. But Einstein showed that travel at speeds approaching *c*, the speed of light, makes a difference. Let's leap ahead to the 21st century and suppose you make a brief visit to a friend on another planet; you set off at noon on the *Inter-planet Rocketbus* which gets you there and back in 5 hours – but your clocks at home show a much later time!

Newton's laws were first called into question towards the end of the 19th century. The laws predicted that the Earth's motion round the Sun ought to be detectable by measuring the speed of light; light travelling 1 m in the direction of the Earth's motion ought to take longer than light travelling 1 m at right angles to the Earth's motion, according to Newton's laws. The prediction was put to the test by Albert Michelson and Edward Morley in 1887. They designed an experiment using interference of light to detect the difference in speed. Despite checking and rechecking the measurements, no such difference was observed. The speed of light parallel to the Earth's motion is the same as the speed perpendicular to the Earth's motion. Michelson and Morley used Newton's laws to work out the expected difference. Was the experiment sensitive

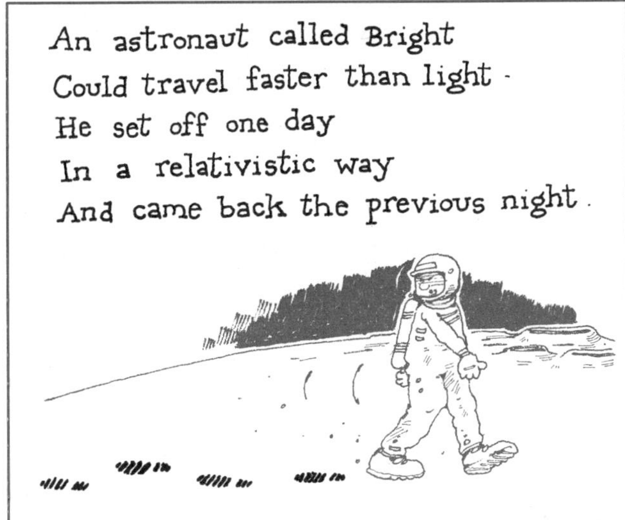

An astronaut called Bright
Could travel faster than light -
He set off one day
In a relativistic way
And came back the previous night.

Figure 30.1

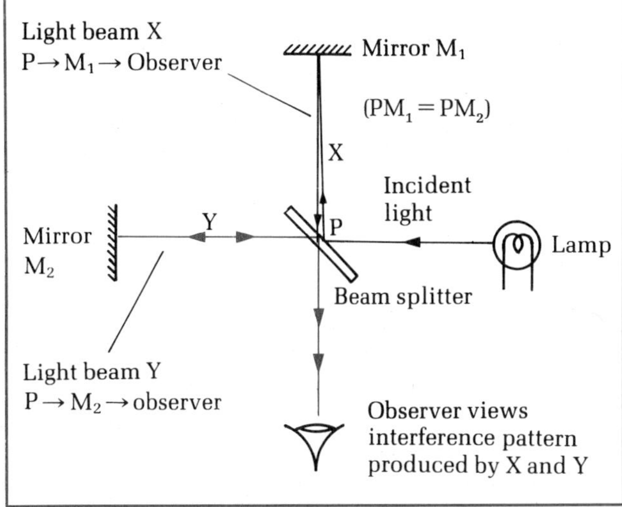

Figure 30.2 *The Michelson-Morley experiment*

enough? Their calculations showed that the experiment ought to have measured the difference easily. But no such difference could be detected, in spite of continual checking and rechecking.

The test of any scientific law or theory is whether or not it agrees with experimental results. Newton's laws didn't agree with the 'null' result from the Michelson-Morley experiment (Figure 30.2). Here was yet another unsolved problem carried into the 20th century. Einstein questioned the basic assumptions behind Newton's laws; imagine a cloud-covered Earth on which the inhabitants have never seen the stars, etc. It would be impossible for them to discover whether or not the Earth is in motion because a 'Michelson-Morley' type experiment would give a null result. So Einstein started by assuming that absolute motion is impossible to detect; only relative motion can be detected. By assuming that the speed of light does not depend on the motion of the light source or of the observer (of the light), Einstein rewrote the laws of motion. To do full justice to the *special theory of relativity*, as Einstein called his work, would require much more space than is available here. Let's concentrate on the 'energy' consequences of Einstein's ideas.

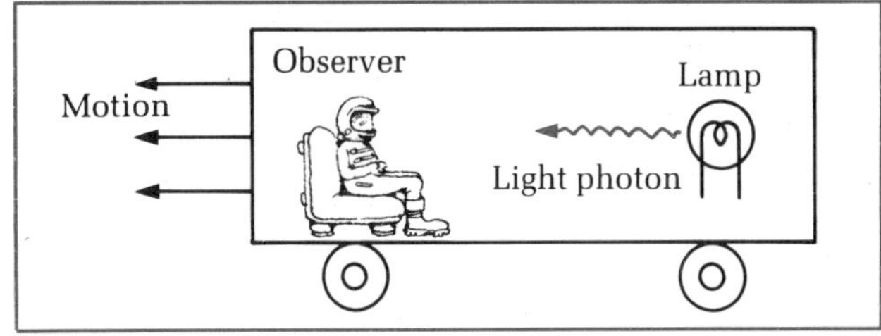

Figure 30.3 *The invariance of* c. *The speed of light is always the same*

Einstein showed that no material object can ever be made to travel faster than the speed of light. Suppose you're in a 'cosmic racing car' and you hold the gas pedal down to boost your speed. In fact the fuel is used to make you go faster **and** to increase the moving mass. As your speed approaches the cosmic speed limit, *c*, the mass becomes greater and greater. In theory, the mass becomes infinite when the speed equals *c* so in practice the speed never reaches c. Increasing the kinetic energy of an object increases its mass as its speed approaches the speed of light c. Not content with such a revolutionary idea, Einstein went further to show that supplying an object with *any* form of energy increases its mass. He showed that energy and mass are equivalent (i.e. interchangeable) on a scale given by his now-famous equation

$$E = mc^2$$

where E = energy,
m = mass,
c = speed of light.

Since the value of c is $3 \times 10^8\ \mathrm{m\,s^{-1}}$, then 1 kg of mass is equivalent to 9×10^{16} J $(= 1 \times (3 \times 10^8)^2)$. So an object supplied with 9×10^{16} J would increase its mass by 1 kg. Or if 9×10^{16} J of energy is removed from an object, its mass decreases by 1 kg. So mass is a bit like 'locked-up' energy which can be converted to or from other forms of energy. Consider some examples.

1 Heating an insulated metal block electrically, as in Figure 30.4. Suppose the total mass is measured when it is cold, then the block is heated electrically. Assume the heater power is 50 W and the heater is switched on for 10 minutes. The electrical energy supplied is equal to 30 000 J $(= 50 \times 10 \times 60)$. What increase of mass occurs as a result of supplying 30 000 J? From Einstein's mass-energy equation

$$m = \frac{E}{c^2} = \frac{30\,000}{(3 \times 10^8)^2} = 3.3 \times 10^{-13}\ \mathrm{kg}.$$

The increase is far too small to detect but nevertheless there is, in theory, an increase of mass.

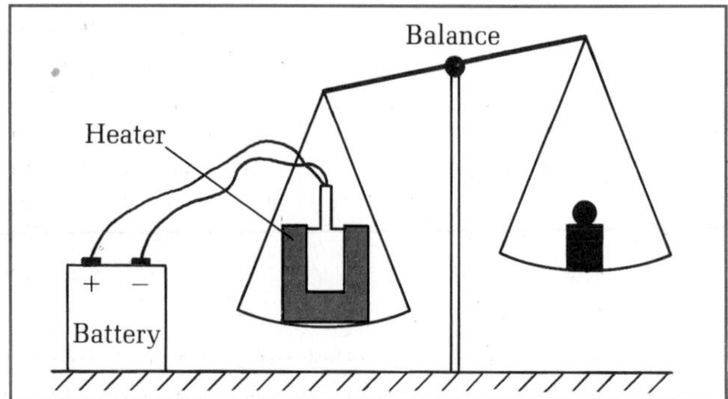

Figure 30.4 *Gaining mass*

2 Pulling two magnets apart increases their potential energy since you do work on them. So their mass is increased by pulling them apart. How much energy is used to pull two strong magnets apart? Suppose you need to pull on them with an average force of 10 N over a distance of about 0.1 m. The work done is 1 J (= force × distance). From $E = mc^2$, the mass increase is far too small to measure. But we shall see in the next section that big forces exist holding protons and neutrons together in the nucleus; these forces can and do cause significant changes of mass when they do work.

The principle of conservation of energy must be restated because of Einstein's ideas. Since mass and energy are equivalent, then the mass of any system must be included in the 'energy balance' in any situation.

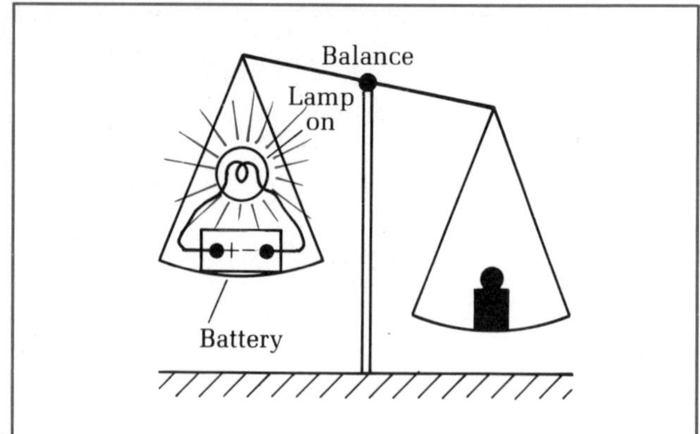

Figure 30.5 *Losing mass*

Energy can never be created or destroyed. By using $E = mc^2$ to calculate the equivalent energy of each mass in any situation, the principle of conservation of energy remains valid. A battery-operated torch lamp emits energy in the form of light. Since the battery is losing energy, its mass decreases although on a scale too small to detect in this situation. A star is a continuous emitter of light too; nuclear reactions deep inside its core cause massive amounts of energy to be emitted at a steady rate. Its mass decreases as its hydrogen atoms are fused to form heavier atoms (see p. 474).

Nuclear forces

The nucleus of any atom is composed of protons and neutrons, the only exception being the hydrogen nucleus $^1_1\mathrm{H}$ which is a single proton. In heavier nuclei, the protons try to repel each other because they have the same charge as each other. Most nuclei are stable and do not burst apart. What holds them together? There must be an attractive force between any two protons or neutrons in the nucleus, and this force is responsible for holding the nucleus together. For example, the stable isotope of lead $^{208}_{82}\mathrm{Pb}$ has 82 protons and 126 neutrons in each nucleus. The nuclear force, the force holding them all together, must be very strong to withstand the force of 82 protons repelling one another electrostatically.

To estimate the nuclear force, consider:

a) Its size must be big enough to overcome the repulsion between two protons 10^{-14} m apart. Coulomb's law of force gives the force of repulsion between two point charges q_1 and q_2 at distance r apart as

$$F = \frac{q_1 q_2}{4\pi\varepsilon_0 r^2}$$ (see p. 171)

Hence for $q_1 = q_2 = +1.6 \times 10^{-19}$ C and $r = 10^{-14}$ m,

$$F = \frac{1.6 \times 10^{-19} \times 1.6 \times 10^{-19}}{4\pi \times 8.85 \times 10^{-12} \times (10^{-14})^2} = 2.3\ \mathrm{N}.$$

So the size of the nuclear force is of the order of 2 N or more.

b) The range of the nuclear force can be estimated from α-scattering experiments. Low-energy α-particles are repelled and scattered by the electrostatic repulsion of the nucleus. However, if high-energy α-particles are used, the α-particles can penetrate this repulsive force

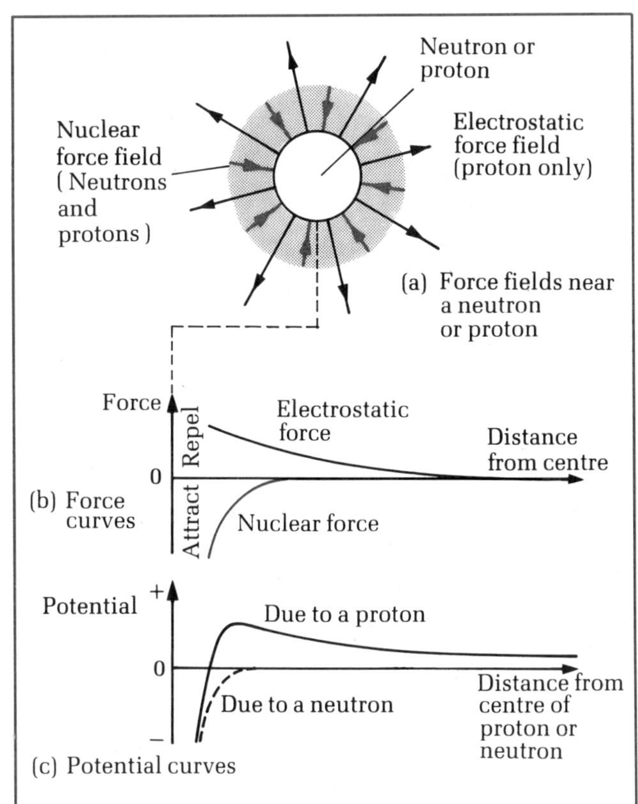

Figure 30.6 **a)** *Force fields near a neutron or proton* **b)** *Force curves* **c)** *Potential curves*

field surrounding the nucleus. Once inside, the α-particle is pulled into the nucleus and may cause a rearrangement leading to a neutron or proton being ejected. An example of such a rearrangement is when high-energy α-particles bombard beryllium foil.

$${}^{4}_{2}\alpha + {}^{9}_{4}\text{Be} \rightarrow {}^{12}_{6}\text{C} + {}^{1}_{0}\text{n}$$

Low-energy α-particles are unable to penetrate the repulsive field caused by the positive charge of the nucleus. So the least-distance of approach of a low-energy α-particle gives an approximate value for the range of the nuclear force. As explained on p. 443, the least distance of approach is about 10^{-14} m so the range of the nuclear force is no more than about 10^{-14} m.

To sum up, the force field surrounding the nucleus is made up of:

1 the nuclear force-field which is attractive and short-range,
2 the electrostatic force-field which repels positively charged particles and has a long range. Its strength decreases with increasing distance according to the inverse-square law.

The potential of a particle near the nucleus depends on whether or not the particle is charged. For example, the P.E. of a proton aimed at a nucleus increases as the proton uses its K.E. to move nearer the nucleus. But if the proton manages to penetrate the electrostatic shield, then the nuclear force grabs the proton and pulls it in. So the P.E. of the proton drops. Figure 30.6(c) shows how the P.E. of a proton varies with distance from a nucleus. It's a little like the P.E. of a ball rolled towards a well in the ground, when the well is on a mound or hill. When the ball is rolled towards the well, its P.E. rises as it moves uphill but when it falls into the well, its P.E. abruptly falls.

However, an uncharged particle such as a neutron does not feel the electrostatic shield. So neutrons penetrate the nucleus much more easily than protons. Figure 30.6 shows the variation of P.E. for a neutron near a nucleus. Once within range of the nuclear force, the neutron is pulled into the nucleus and trapped, just like a ball falling into a well.

Inside the nucleus, the individual protons and neutrons are trapped by the strong nuclear force which holds them together. Provided the nuclear force is stronger than the electrostatic force of repulsion between the protons, then the nucleus is stable. But if the nucleus is too large or if the balance of neutrons to protons is too large or small, then the nucleus is unstable.

Binding energy

The nuclear force can't tell the difference between neutrons and protons. The nuclear force between two protons is the same as between a neutron and a proton or between two neutrons. So we use the term **nucleon** to describe a neutron or a proton in the nucleus.

To remove a nucleon from the nucleus, work must be done on it to pull it away from the attractive force of the nucleus. This force, the **nuclear force**, is of the order of newtons and it acts over distances of the order of 10^{-14} m. So the work done is of the order of 10^{-14} J, equal to several MeV of energy (1 MeV $= 1.6 \times 10^{-13}$ J).

Suppose all the nucleons were separated from one another, removing each one from the nucleus in turn. As each one is lifted out of the nuclear potential well, its potential energy is increased by a few MeV. Increasing the potential energy increases the mass. So the mass of the separated nucleons is greater than the combined mass in the nucleus before separation. Effectively, the energy supplied to separate the nucleons increases the mass. The mass of the separated nucleons is always greater than the mass of the combined nucleons in the nucleus.

The mass defect of a nucleus is defined as the difference between the mass of the separated nucleons and the combined mass of the nucleus.

Since the change of P.E. of a nucleon is of the order of 10^{-14} J, does this give a significant change of mass of a nucleon when removed? Use $E = mc^2$.

For $E = 10^{-14}$ J, $m = E/c^2 = 10^{-14}/(3 \times 10^8)^2 = 10^{-31}$ kg.

But the mass of a nucleon is about 10^{-27} kg. So the change caused by removing the nucleon from its nucleus is significant.

Mass at an atomic level is usually expressed in atomic mass units (see p. 66) where 1 atomic mass unit, 1 u, is equal to 1.66×10^{-27} kg. Accurate measurement of atomic masses is achieved in practice using mass spectrometers. In a mass spectrometer, ions in a beam are deflected by known electric and magnetic fields. From the deflections, the mass of each type of ion present can be calculated. In this way, the atomic mass of each and every type of atom has been determined very precisely.

Consider an example to show that the measured mass of an atom is always less than the mass of its constituent particles, if separated.

The helium atom $^4_2$He has a mass equal to 4.002 604 u. The helium atom has two electrons, each of mass 0.000 55 u, and a nucleus composed of two protons and two neutrons. Each proton has a mass of 1.007 28 u and each neutron has a mass of 1.008 66 u.
Mass of $^4_2$He atom = 4.002 604 u,
hence mass of nucleus = 4.001 504 u
Mass of constituent nucleons if separated

$$\begin{aligned} &= 2 \times 1.007\,28 \quad \text{(2 protons)} \\ &+ 2 \times 1.008\,66 \quad \text{(2 neutrons)} \\ &= 4.031\,88 \quad \text{u} \end{aligned}$$

Hence the mass defect $m = 4.03188 - 4.001\,504 = 0.0303$ u. So when the nucleus is pulled apart, the mass increases from 4.001 504 to 4.031 88 u. The increase of mass is due to work done to separate the nucleons from one another. We can use $E = mc^2$ to calculate the work done.
For $m = 1\,\text{u} = 1.66 \times 10^{-27}$ kg,
$E = 1.66 \times 10^{-27} \times (3 \times 10^8)^2 = 1.494 \times 10^{-10}\,\text{J} = 931$ MeV,
since 1 MeV $= 1.6 \times 10^{-13}$ J
so **1 u is equivalent to 931 MeV.**

Therefore the work done to increase the mass by 0.0303 u is equal to $0.0303 \times 931 = 28.2$ MeV.

The binding energy of a nucleus is defined as the work done on the nucleus to separate it into its constituent neutrons and protons.

For example, the binding energy of the $^4_2$He nucleus is 28.2 MeV. To separate its nucleons from one another, 28.2 MeV of work must be done to overcome the (nuclear) force of attraction between the nucleons. The energy supplied for this purpose increases the mass from 4.001 504 to 4.031 88 u. So the binding energy equals the energy equivalent of the mass defect.

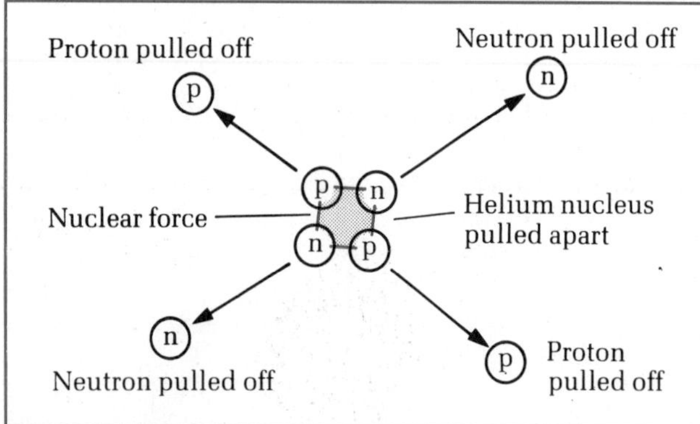

Figure 30.7 *Gaining mass*

Now consider the reverse process, i.e. the formation of a nucleus from separated protons and neutrons. The protons and neutrons form and fall into the nuclear potential well as the nuclear force of attraction pulls them together. So their potential energy falls. Assuming zero potential energy when separated, their P.E. falls to a negative value. In the formation of the $^4_2$He nucleus, the P.E. falls from zero to -28.2 MeV. To pull the nucleus apart again, 28.2 MeV of work must be done. So when a nucleus forms from separated neutrons and protons, it loses P.E. The P.E. lost is converted into K.E. so in effect the mass defect is used to give an equivalent amount of K.E.

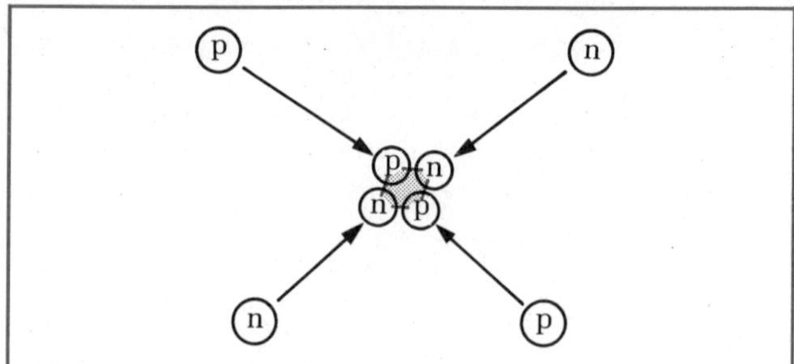

Figure 30.8 *Losing mass*

The binding energy curve of all the known isotopes is a plot of binding energy per nucleon against mass number (Figure 30.9). For each isotope, the binding energy (B.E.) per nucleon can be calculated by dividing the energy equivalent of the mass defect by the number of nucleons. In other words, the B.E. divided by the number of nucleons gives the B.E. per nucleon. For example, the $^4_2$He nucleus has B.E. of 28.2 MeV. So the B.E. per nucleon of the $^4_2$He nucleus is 8.05 MeV ($= 28.2/4$). The curve is shown in Figure 30.9.

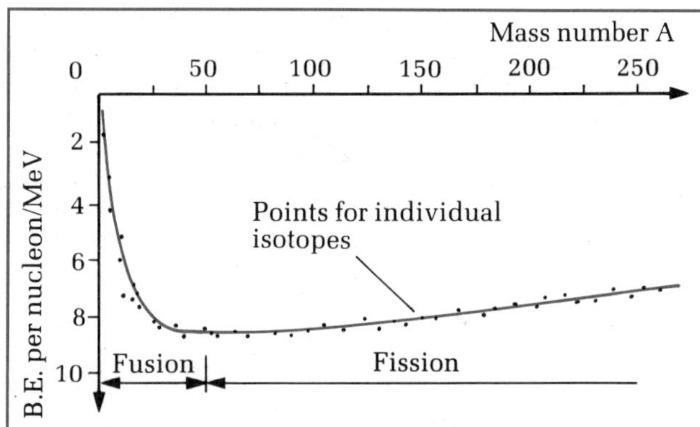

Figure 30.9 *The binding energy curve*

The important features of the curve are:

1 Maximum B.E. per nucleon is at about mass number $A = 50$. The greater the B.E. per nucleon, the greater the work that must be done to remove a nucleon. So maximum B.E. per nucleon corresponds to the most stable nuclei. At about $A = 50$, the potential well is at its deepest so the B.E. per nucleon (i.e. the energy needed to lift a nucleon out of the well) is at its greatest.
2 Either side of maximum B.E. per nucleon the nuclei are less stable. The less the B.E. per nucleon, the easier it is to separate a nucleus into its constituent nucleons. In other words, the nucleons are not bound together quite so tightly as at the maximum.
3 When a big nucleus disintegrates, each nucleon becomes a member of a smaller nucleus, so the B.E. per nucleon increases. The nucleons in the product nuclei are even more tightly bound together so energy is released, equal to the increase of B.E. The disintegration therefore creates a deeper potential well; the nucleons drop down even further so their binding energy increases. The loss of P.E. (i.e. increase of B.E.) is released as K.E. of the product nuclei. **Fission** occurs where the

original nucleus splits into two approximately equal halves to form two product nuclei of about equal size. **Radioactive disintegration** occurs where the original nucleus loses an α, β or a γ particle. Where an α or β particle is emitted, the product nucleus is lighter than the original nucleus so the B.E. per nucleon increases. So fission or radioactive decay both lead to an increase of B.E. per nucleon and hence to release of energy as K.E. of the products.

4 **Fusion** occurs where light nuclei are joined together. Where the product nucleus has a mass number smaller than 50, the B.E. per nucleon is increased by the fusion process. So energy is released when light nuclei are fused together. See p. 474.

Energy from nuclear reactions

When a nucleus becomes rearranged, the total energy (including mass) before the change must equal the total energy (including mass) after the change. If the change causes K.E. to be produced, then

K.E. produced = total mass before − total mass after (incl. K.E. equivalent)

Spontaneous reactions are what happen where a nucleus changes without any 'help' from outside. Radioactive disintegration is spontaneous, and fission can be too. Given the masses of the initial and final nuclei, the energy released can be calculated.

a) α-decay: consider the disintegration of a $^{212}_{83}\mathrm{Bi}$ nucleus by α-emission.

$$^{212}_{83}\mathrm{Bi} \rightarrow {}^{4}_{2}\alpha + {}^{208}_{81}\mathrm{Tl} + \text{energy released } Q$$

Atomic masses for each type of nucleus are as follows.

$^{212}_{83}\mathrm{Bi} = 211.991\,27\ \mathrm{u}$, $^{4}_{2}\mathrm{He} = 4.002\,604\ \mathrm{u}$,

$^{208}_{81}\mathrm{Tl} = 207.982\,01\ \mathrm{u}$.

Before The total mass of the Bi-212 nucelus is equal to $211.991\,27 - 83\,m_e$, where m_e is the mass of an electron.

After The total mass is equal to the mass of the α-particle $(4.002\,604 - 2m_e) +$ the mass of the Tl-208 nucleus $(207.982\,01 - 81\,m_e)$.

Hence the loss of mass

= total mass before − total mass after

$= (211.991\,27 - 83m_e)$
$\quad - (4.002\,604 - 2m_e + 207.982\,01 - 81m_e)$
$= 0.006\,656\ \mathrm{u}$.

Therefore the energy released Q

$= 0.006\,656 \times 931 = 6.2\ \mathrm{MeV}$.

b) β-decay: consider the decay of $^{90}_{38}\mathrm{Sr}$.

$$^{90}_{38}\mathrm{Sr} \rightarrow {}_{-1}^{0}\beta + {}^{90}_{39}\mathrm{Y} + Q$$

Atomic masses:

$^{40}_{38}\mathrm{Sr} = 89.9073\ \mathrm{u}$ $\quad {}_{-1}^{0}\beta = 0.000\,55\ \mathrm{u}$,

$^{90}_{39}\mathrm{Y} = 89.9067\ \mathrm{u}$.

Before The total mass = mass of Sr-90 nucleus $= 89.9073 - 38m_e$.

After The total mass = the mass of Y-90 nucelus + the mass of the β-particle

$= (89.9067 - 39m_e$.

Hence the loss of mass

= total mass before − total mass after

$= (89.9073 - 38m_e) - (89.9067 - 39m_e + m_e)$
$= 0.0006\ \mathrm{u}$.

Therefore the energy released

$= 0.0006 \times 931 = 0.56\ \mathrm{MeV}$.

Induced reactions occur where a nucleus changes as a result of being struck by a particle. If the products have greater mass than the nucleus and particle before the reaction, then the particle must supply enough K.E. to make up the mass difference.

30.2 Nuclear fission

Principles of nuclear fission

Fission of a nucleus occurs when the nucleus splits into two approximately equal fragments. This happens when uranium in the form of the isotope $^{235}_{92}\mathrm{U}$ is bombarded with neutrons, a discovery made by Otto Hahn and Frederic Strassmann in 1938. They knew that bombarding different elements with neutrons produced radioactive isotopes. Uranium is the heaviest of all the naturally-occurring elements; scientists thought that neutron bombardment could turn uranium nuclei into even heavier nuclei. Hahn and Strassmann undertook the difficult work of analysing chemically the products of uranium after neutron bombardment to try to discover any new elements even heavier than uranium. Instead, they discovered many lighter elements such as barium were present after bombardment, even though the uranium was pure before. The conclusion could only be that uranium nuclei were split into approximately equal fragments as a result of neutron bombardment. Such a fission reaction is shown below. (See Figure 30.10.)

$$^{235}_{92}\mathrm{U} + {}^{1}_{0}\mathrm{n} \rightarrow {}^{144}_{56}\mathrm{Ba} + {}^{90}_{36}\mathrm{Kr} + 2\,{}^{1}_{0}\mathrm{n} + \text{energy } Q \text{ released}$$

Further investigations showed that:

a) Several neutrons are released with the fission fragments.

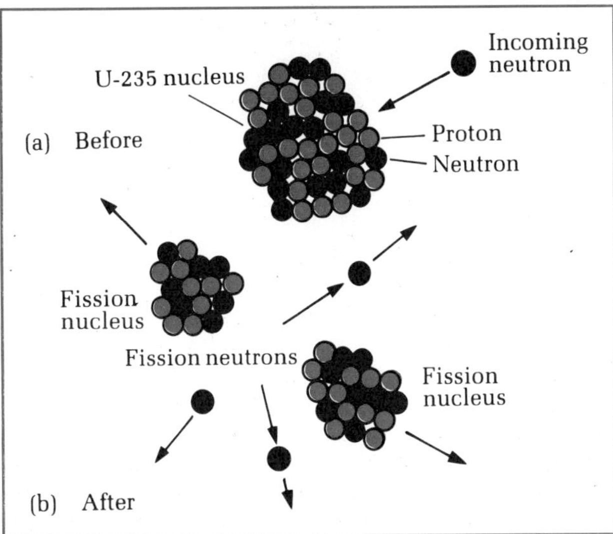

Figure 30.10 *Induced fission*

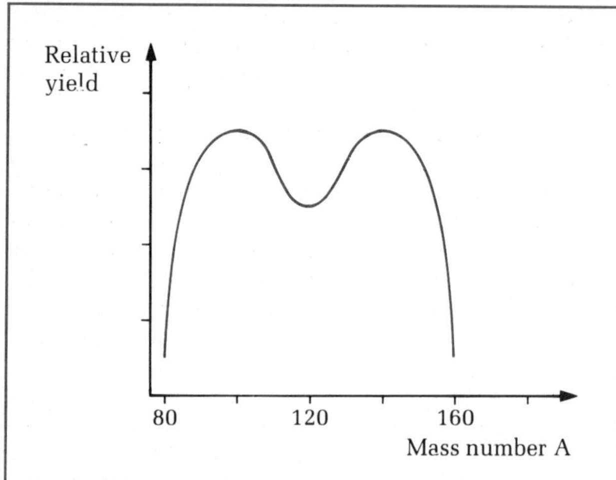

Figure 30.11 *Spread of fission fragments*

b) Many fission products are possible when U-235 is bombarded with neutrons. Figure 30.11 shows the spread of possible fission products. In all cases, the products, including fission neutrons, must 'add up' to give the same total of neutrons and protons as before the reaction (i.e. 92 protons and 144 neutrons).

c) The products themselves are radioactive; usually they have a higher neutron/proton ratio than stable isotopes of the same elements. Figure 30.12 shows that the product nuclei tend to be formed above the N-Z stability belt. So the product nuclei tend to be β^--emitters, see p. 449.

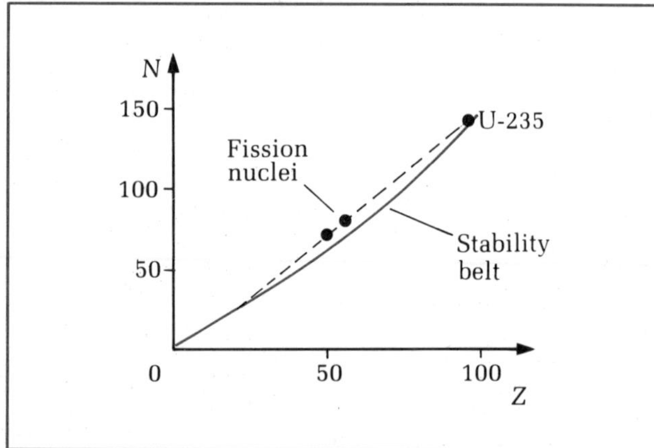

Figure 30.12 *Radioactivity of fission nuclei*

d) Slow neutrons are more effective in fissioning U-235 than fast neutrons.

e) Energy is released on a much greater scale than is released from chemical reactions.

To estimate the energy released consider the B.E. curve of Figure 30.9. When a U-235 nucleus fissions, each of its nucleons becomes part of a nucleus approximately half the size of the U-235 nucleus. So the B.E. per nucleon increases from about 7.5 MeV to about 8.5 MeV, as shown. So the energy released is about 1 MeV per nucleon (since the energy released = the increase of B.E.). With approximately 200 or so nucleons to start with, about 200 MeV is released as K.E. of the fragments when a single U-235 nucleus is fissioned.

Hence if one nucleus releases about 200 MeV when fissioned, then 6×10^{23} nuclei release about 12×10^{25} MeV $= 2 \times 10^{13}$ J. But 6×10^{23} nuclei of U-235 have a total mass of 235 g (since 6×10^{23} is the Avogadro constant; see p. 66).

Hence 235 g of U-235 releases 2×10^{13} J if fissioned completely. So 1 kg of U-235 can produce about 8×10^{13} J, about a million times as much energy as 1 kg of oil or coal. Work out for yourself how much U-235 would be needed to keep a 100 MW power station going for one year (which is about 3×10^7 s), assuming 100% efficiency. In practice, efficiency values are about 30% so about three times the amount worked out above would be needed.

Nuclear chain reactions involve fission neutrons producing further fission, giving, in turn, more neutrons, and so on. Suppose a single neutron fissions a U-235 nucleus to produce two more neutrons (and fission fragments) which then go on to fission other nuclei, etc. After 10 'generations' of fission, there will be 2^{10} neutrons moving about; after 100 generations, 2^{100} ($= 1.27 \times 10^{30}$) neutrons would be present. So a huge amount of energy would be released. Since each generation takes only a fraction of a second to be produced, then a huge amount of energy would be released in a very short time in an uncontrolled chain reaction.

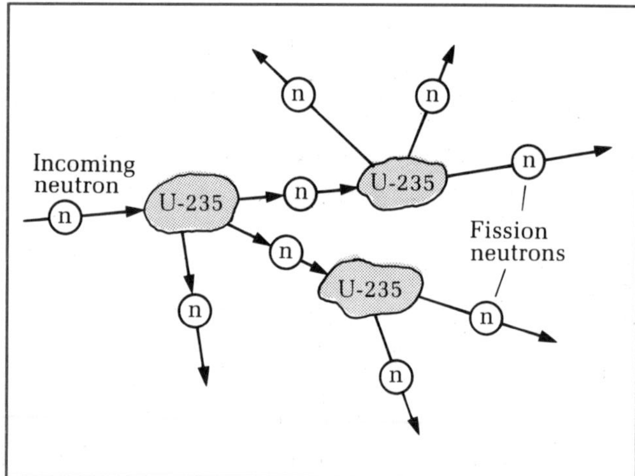

Figure 30.13 *A chain reaction*

Fortunately, the chain reaction can be controlled by inserting **absorber rods** into the uranium; the absorber rods are made of elements such as boron or cadmium which absorb neutrons without fissioning. The rods are used to ensure that an average of one further fission results from each fission. So if on average 2.5 neutrons are released per fission, then only one of these neutrons is to give further fission. Of the other 1.5 neutrons per fission, there will be some loss from the uranium anyway so the absorber rods must increase the total loss to 1.5 neutrons per fission. In this way, a steady rate of fissioning is achieved, so giving a steady power output.

The liquid drop model of the nucleus gives a helpful picture of how fission takes place. Imagine a uranium nucleus as being like a wobbly liquid drop. The nucleus vibrates due to thermal energy so its shape continually changes. Each time it becomes too distorted, the nuclear forces between its nucleons pull it back together again.

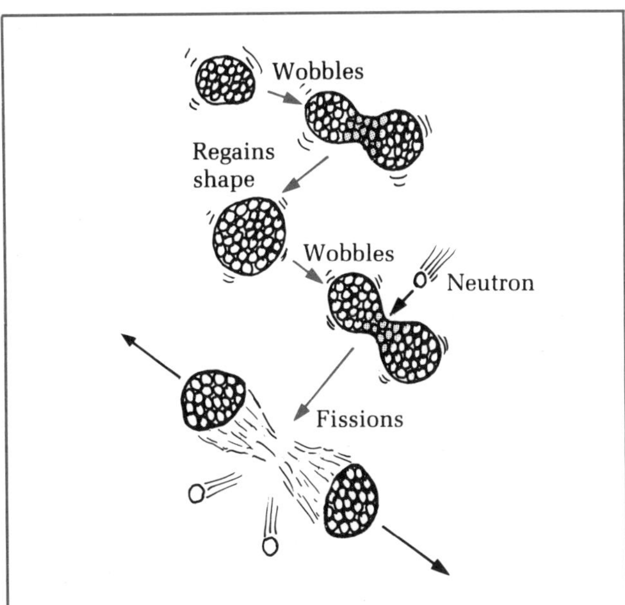

Figure 30.14 *The liquid drop model*

Figure 30.14 shows the idea; it's as if a liquid drop almost breaks into two but the forces between its molecules pull it back together again. However, if the drop is given a little extra energy just as it is at its most distorted, then it splits into two. This is what is thought to happen when a neutron hits a U-235 nucleus; if the neutron hits the nucleus at its most distorted state, then fission results. The two halves are charged positively so they repel each other once their nuclear forces are out of range of each other. The two halves gain K.E. as they repel each other so energy is released in this way. The energy is released because the total of the binding energies of the fragments is greater than of the original nucleus. So part of the mass of the initial nucleus is converted into K.E. of the fission nuclei and fission neutrons.

The nuclear reactor

The first nuclear reactor was designed and built in 1942 under the direction of Enrico Fermi. The fuel used was natural uranium which contains less than 1% U-235. The other 99% is U-238 which does not fission easily. Fermi had discovered that fissioning is increased if the neutrons are slowed down, so he designed the reactor using graphite blocks to slow the neutrons down. The fuel was in the form of chunks of uranium oxide regularly spaced in the graphite pile. Cadmium absorber rods inserted into channels through the pile prevented a chain reaction from building up until the assembly was ready. Fermi ordered the control rods all to be withdrawn except for one – sufficient to stop the build-up. Then that rod was pulled out step by step; at each step the calculated fission rate was checked; finally, one last step and as expected, the rate of release of energy built up to a steady level. Nuclear fission had been controlled.

Figure 30.15 *The world's first nuclear reactor*

Modern nuclear reactors use either natural uranium or enriched uranium as fuel. Enriched uranium contains a greater percentage of U-235 than natural uranium does. The fuel is in the form of rods enclosed in metal containers. These 'fuel pins' are spaced regularly in a **moderator**, such as graphite or water, chosen to slow fission neutrons down. Figure 30.16 shows a cross-section of a water-cooled reactor. **Control rods**, to maintain a steady rate of fissioning, are inserted into the moderator core, their position adjusted as required. A **coolant** is pumped through the channels in the moderator to remove heat energy to a heat exchanger. The moderator is enclosed in a **steel vessel** designed to withstand the high pressures and temperatures inside the reactor. A **concrete shield** round the vessel prevents escaping radiation from reaching the operators.

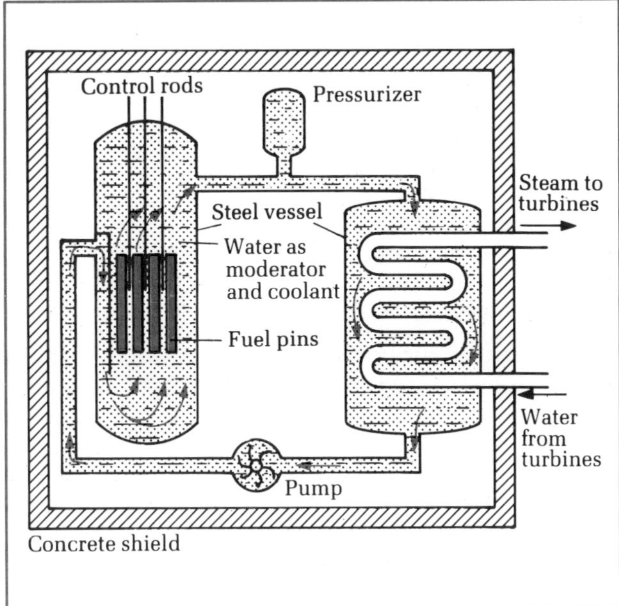

Figure 30.16 *The PWR reactor*

Consider the processes inside the reactor when it produces a steady output of power.

1 Each fission of a U-235 nucleus produces fission fragments including neutrons. About 200 MeV of kinetic energy is carried away by the fragments. The fission fragments recoil and carry away most of the K.E. released, transferring the K.E. to other atoms that they collide with. So the fuel pins get very hot.
2 The fission neutrons leave the fuel pins with K.E. of the order of MeVs. They enter the moderator and collide

with moderator atoms, transferring K.E. to these atoms. So the moderator gains energy and the neutrons slow down until the average K.E. of a neutron is about the same as that of a moderator atom vibrating in its lattice. The neutrons are then said to be **thermal** neutrons because they do not, on average, lose any more energy to the moderator atoms after that.

3 Slow neutrons re-enter the fuel pins and cause further fission of U-235 nuclei. The chance of fission occurring is measured in terms of the **collision cross-section** σ; Figure 30.17 shows how the value of σ varies with neutron energy for the two isotopes U-235 and U-238. High-energy neutrons are more likely to lose energy to U-238 nuclei without fission than to cause fission of U-235. By slowing the neutrons down, the chance of fission of U-235 is increased by a factor of $\times$ 100 or more.

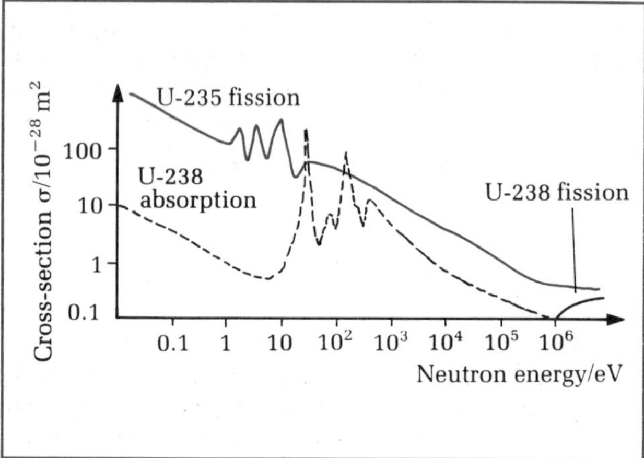

Figure 30.17 *Chance of fission v neutron energy*

Important features in the design of a nuclear reactor include:

a) the critical mass of fuel required,
b) the choice of the moderator,
c) the choice of control rods,
d) the type of coolant,
e) treatment of waste (i.e. spent fuel).

The critical mass of fuel is the minimum mass capable of producing a self-sustaining chain reaction. Imagine an isolated block of uranium in which U-235 nuclei are fissioning. The fission neutrons produced could be absorbed by the U-238 nuclei without producing further fission, or the fission neutrons could escape from the block without causing further fission. If too many neutrons escape, the chain reaction will die out.

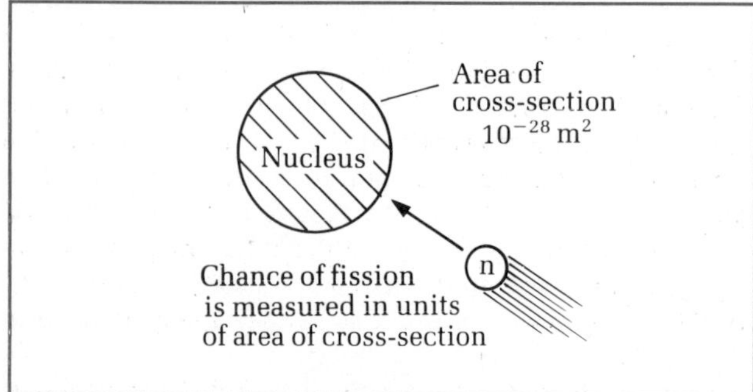

Figure 30.18 *Collision cross-section*

The chance of a neutron escaping depends on the surface area of the block in relation to its volume. Double the dimensions of a block and its surface area increases by a factor of 4 ($=2^2$) whereas its volume increases by a factor of 8 ($=2^3$). So the number of neutrons produced per second goes up by $\times$ 8 whereas the loss only goes up by $\times$ 4. So there is less chance of loss from a big block than from a small block. Hence the block must have a minimum size if the loss is to be reduced to a level where the chain reaction does not die out.

The shape of the block affects the critical mass; a flat disc of fuel has a much bigger surface area than a sphere of the same mass, so the loss from the disc is much greater. If the sphere is just self-sustaining, then the disc would not be. For a disc shape, the mass needs to be much greater.

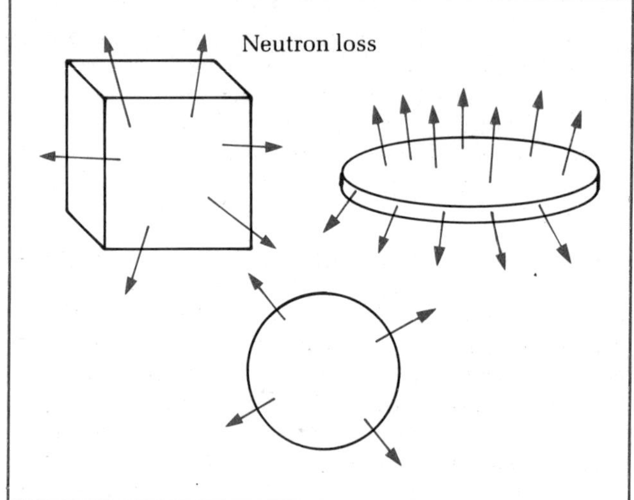

Figure 30.19 *Critical mass*

The choice of moderator is determined by the point that the atoms of an ideal moderator should have the same mass as a neutron. If so, a neutron colliding elastically with a moderator atom would lose almost all its K.E. to the moderator atom, just as a marble hitting a stationary identical marble does. In practice, graphite or water is chosen as a compromise between availability and effectiveness. A neutron hitting a stationary carbon nucleus and rebounding elastically can lose up to 25% of its K.E. So after two successive impacts with stationary carbon nuclei, the K.E. can be reduced to 75% of 75% of its initial value. Work out for yourself how many impacts are necessary to reduce its K.E. from 1 MeV to 10 eV.

The moderator atoms should not absorb neutrons but should scatter them instead. So the scattering cross-section (i.e. the chance of scattering) must be much greater than the absorption cross-section (i.e. the chance of absorption). Values for several light elements are given at the top of the next column. Clearly carbon and water are effective moderators.

The control rods are there to absorb rather than scatter neutrons. The values of σ above show that boron and cadmium are very suitable elements for control rods. Control rods are operated automatically, driven by motors controlled by sensors which measure the neutron 'flux'. If the number of neutrons per second hitting the

	$\sigma_{scatt}/10^{-28}\,m^2$	$\sigma_{abs}/10^{-28}\,m^2$
Hydrogen	20	0.3
Boron	4	760
Carbon	5	0.003
Oxygen	4	0.0003
Cadmium	6	2500

sensors rises suddenly, then the control rods are automatically pushed into the core, so reducing the neutron flux to the required level. To shut the reactor down, the control rods must be fully inserted into the core; then the rate of absorption (of neutrons) becomes much greater than the rate of production, so the rate of fissioning drops.

The coolant must flow easily and it must not be corrosive. Coolant atoms become radioactive when they pass through the core. So the coolant must be in a 'sealed circuit', pumped from the core of the reactor to the heat exchanger, as shown in Figure 30.16. Steam is then produced from the heat exchanger and is used to drive turbines to generate electricity.

The waste in the form of spent fuel and fuel cans is highly radioactive for a long time after removal from the core. The products of fission are neutron-rich, as explained on p. 449, so they are β-emitters. Also, U-238 nuclei absorb neutrons without fissioning; instead, the U-239 nuclei formed disintegrate by radioactive emission. The fuel can atoms also become radioactive because they are continually bombarded by neutrons in the core. So the spent fuel rods must be removed by remote control with great care from the reactor core. They are then treated in several stages.

1 Safe storage and cooling of the fuel rods is essential since the intense radioactivity makes them very hot. So they are stored in containers in cooling ponds until their activity has decreased and they are cooler.
2 The spent fuel is removed from the cans by remote control. The fuel is then reprocessed to recover unused fuel.
3 The unwanted material is then stored in sealed containers for many years until the activity has fallen to an insignificant level.

Types of nuclear reactor

Thermal reactors were the first generation of nuclear reactors. Many are still in operation after 30 years or more of continuous operation. They are called 'thermal' because the neutrons must be slowed down to 'thermal energies' to produce further fission. See p. 472. If no moderator were present, most of the fission neutrons would be absorbed by U-238 nuclei instead of fissioning U-235 nuclei. So they are designed to make use of either natural uranium or enriched uranium.

Different types of thermal reactor are in operation throughout the world. Features of some are shown at the top of the next column.

Type	Pressurized water reactor	Magnox	Advanced gas cooled reactor
Fuel	uranium oxide in zirconium alloy cans	uranium in magnesium (magnox) cans	uranium oxide in stainless steel cans
Moderator	water	graphite	graphite
Coolant	water	Co_2 gas	Co_2 gas
Coolant temp/K	600	670	900
Typical power output	700	600	1300

Fast reactors are designed to make use of the U-238 content of natural uranium. In a thermal reactor, the U-235 content (less than 1% of the total) is fissioned; most of the U-238 content (the other 99%) does not contribute to the power output. The spent fuel from a thermal reactor is mostly unused U-238 with fission products and also plutonium-239 (^{239}Pu). Each plutonium nucleus is produced as a result of a U-238 nucleus absorbing a neutron. The unstable isotope U-239 decays by β-emission to form Pu-239.

$$^{238}_{92}U + ^{1}_{0}n \rightarrow ^{239}_{92}U \rightarrow ^{0}_{-1}\beta + ^{239}_{93}Np$$

$$^{239}_{93}Np \rightarrow ^{0}_{-1}\beta + ^{239}_{94}Pu$$

Plutonium does not occur naturally. Pu-239 can be fissioned by fast neutrons, unlike U-235. So plutonium from thermal reactors is used as the fuel for fast reactors where no moderator is needed. Figure 30.20 shows the arrangement of a fast reactor.

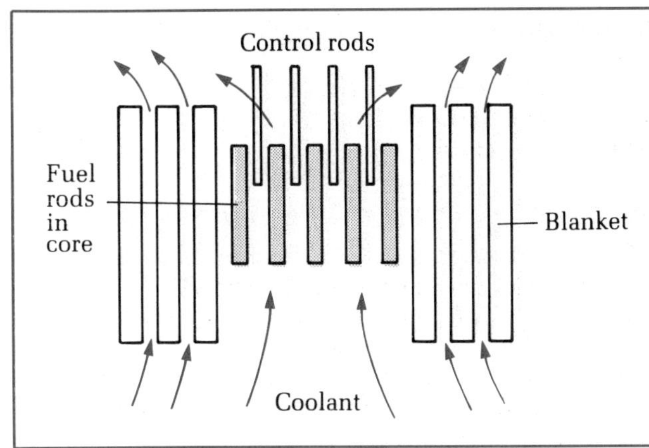

Figure 30.20 *Insidethe fast breeder*

Control rods are used to keep a steady rate of fissioning. Fission neutrons produce further fission of Pu-239 nuclei without needing to be slowed down. The reactor is designed so a blanket of U-238 surrounds the plutonium core. Neutrons that escape from the core bombard U-238 nuclei in the blanket. The result is that more Pu-239 is created; in other words, the plutonium fuel is created from unused U-238 in the blanket. **Breeding** the plutonium fuel in this way means that up to 60% or more of the U-238 is used. Hence natural uranium is used much more effectively.

In practice, fuel rods from the blanket are removed at intervals of about a year to be replaced by fresh fuel rods.

The removed rods are reprocessed to yield Pu-239. The plutonium is then used in the core.

Using thermal reactors only, the world supply of uranium will probably be exhausted in 50 to 100 years time. However, fast reactors should enable the uranium to be used more effectively, so extending the lifetime of the reserves by many centuries.

30.3 Nuclear fusion

Principles of fusion

Fusion takes place where two nuclei combine to form a bigger nucleus. The binding energy curve of Figure 30.9 shows that if two light nuclei are combined, the individual nucleons become more tightly bound. The binding energy of the product nucleus is greater than of the initial nuclei; in other words, the nucleons become trapped in an even deeper well when fusion occurs. So energy is released, equal to the increase of binding energy.

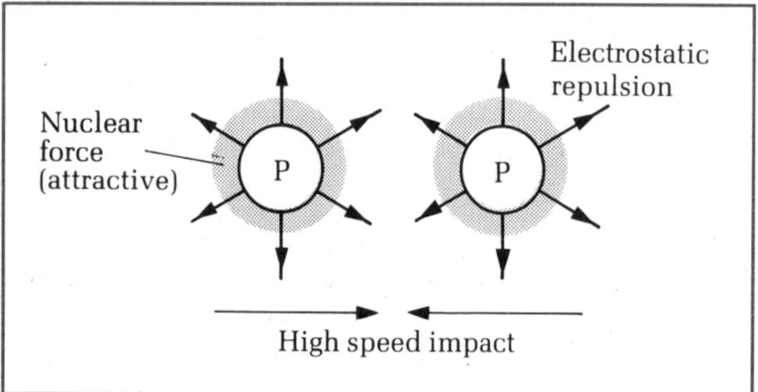

Figure 30.21 *Fusion of two protons*

Consider the simple example involving two protons being forced to fuse together to form a deuterium nucleus $^{2}_{1}\mathrm{H}$.

$$^{1}_{1}\mathrm{p} + {}^{1}_{1}\mathrm{p} \rightarrow {}^{2}_{1}\mathrm{H} + {}^{0}_{+1}\beta + \text{energy released } Q$$

(1.007 28 u) (1.007 28 u) (2.013 55 u) (0.000 55 u)

The mass of each particle involved is shown below the equation. The reaction produces a positively charged particle called a 'positron' which has mass equal to that of an electron (see p. 450).

The total mass before fusion
$= 1.007\,28 + 1.007\,28 = 2.014\,56$ u
The total mass after fusion
$= 2.013\,55 + 0.000\,55 = 2.014\,10$ u.
Hence the loss of mass $= 2.014\,56 - 2.014\,10 = 0.000\,46$ u.
Since 1 u is equivalent to 931 MeV, then the energy released Q is equal to $0.000\,46 \times 931 = 0.4$ MeV.

Solar energy is produced as a result of fusion reactions inside the Sun. The temperature at the centre of the Sun is thought to be 10^8 K or more, so high that atoms are stripped completely of electrons. Matter in this state is called 'plasma'. The nuclei of the plasma move at very high speeds because of the enormous temperature. When two nuclei collide, they fuse together because they approach each other at such high speeds, so they overcome the electrostatic repulsion. In other words, the speeds are so high that the nuclei penetrate each other's electrostatic force field. Once inside the electrostatic shield, the nuclear force makes the two nuclei fuse together.

Inside the Sun, protons (i.e. hydrogen nuclei) are fused together one by one to form nuclei, such as helium and carbon nuclei. For every helium nucleus formed from protons, about 28 MeV of energy is released (= the B.E. of the He-4 nucleus; see p. 468). Stars start life with loss of hydrogen with is gradually converted into helium and other heavier nuclei. When a star's store of hydrogen is used up, it swells up enormously to become a 'red giant'; at this stage it starts fusing helium nuclei into even bigger nuclei. When the Sun reaches the red giant stage, it will expand and swallow up the inner planets. After that it will shrink as it ceases to produce energy, although big stars (>10 solar masses) die in spectacular explosions called 'supernovas'.

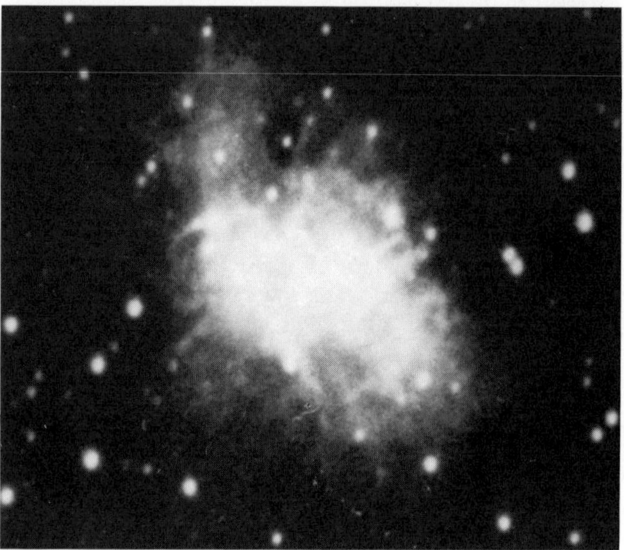

Figure 30.22 *The Crab Nebula – a supernova*

How long will the Sun continue to radiate energy? We can estimate its lifetime from its power output and its total mass. We know that each m² of surface at the Earth receives 1.4 kW of Solar power. The distance R from the Earth to the Sun is 1.5×10^{11} m. The Earth moves round the Sun on an orbit which is part of a sphere of radius R, as in Figure 30.23. Each m² of the surface area of the

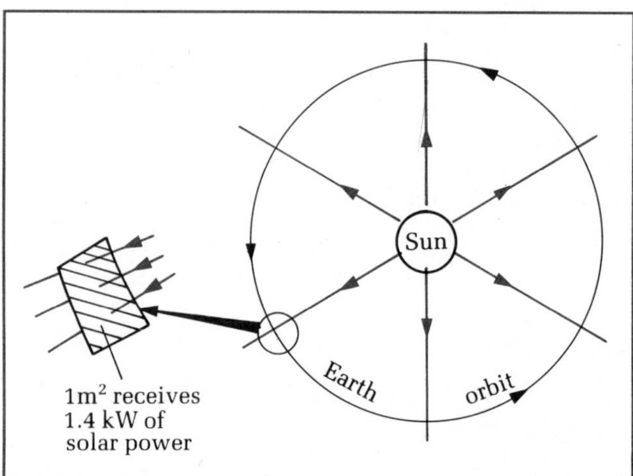

Figure 30.23 *Solar power*

sphere receives 1.4 kW so the total power P passing through the sphere surface is given by

$$P = \text{surface area of sphere} \times 1.4\,\text{kW}$$
$$= 4\pi R^2 \times 1.4 = 4\pi \times (1.5 \times 10^{11})^2 \times 1.4 = 4 \times 10^{23}\,\text{kW}.$$

Since there is just empty space between the Sun and Earth, then this value for P must equal the total power output from the Sun.

The Sun's mass is 2×10^{30} kg; when a helium nucleus is formed from four protons, the mass loss is about 1% $\left(= \frac{0.031}{4} \times 100\%\right.$; see p. 468$\left.\right)$. So the mass which is available for producing energy is about 2×10^{28} kg. Hence the total energy available can be calculated using $E = mc^2$.

$$E = mc^2 = 2 \times 10^{28} \times (3 \times 10^8)^2 = 18 \times 10^{44}\,\text{J}.$$

Since power = energy/time, then the time taken to lose all the energy is given by

$$\text{Time taken} = \frac{\text{energy } E}{\text{power } P} = \frac{18 \times 10^{44}}{4 \times 10^{26}}$$
$$\approx 5 \times 10^{18}\,\text{s} \approx 10^{11}\text{ years.}$$

So there is no need to worry yet about the Sun running out of fuel. More accurate calculations give a value of about 10^{10} years for the Sun's lifetime. The present age of the Sun is thought to be about 5×10^9 years so the Sun is approximately half-way through its lifespan. You could say that the Sun is a middle-aged star!

Worked example Make an estimate of the minimum speed required for protons to make them fuse on impact. Use the kinetic theory of gases to estimate the temperature to achieve such speeds by heating.

Solution Nuclear changes involve particle energies of the order of MeV or more.

1 MeV $= 1.6 \times 10^{-13}$ J and the mass of a proton is about 10^{-27} kg.

So the speed of a proton with K.E. about 1 MeV is given by

$$\text{K.E.} = \tfrac{1}{2}mv^2 = 1.6 \times 10^{-13}\,\text{J}$$

Hence $\frac{1}{2} \times 10^{-27} \times v^2 = 1.6 \times 10^{-13}$

$$\text{Giving } v^2 = \frac{1.6 \times 10^{-13}}{0.5 \times 10^{-27}}\,\text{m}^2\,\text{s}^{-2}.$$

So speed $v = 1.8 \times 10^7\,\text{m s}^{-1}$.

Hence at speeds of the order of $10^7\,\text{m s}^{-1}$, protons have enough K.E. to overcome electrostatic repulsion and so fuse when they collide.

From the kinetic theory of gases, $\frac{Nmv^2}{3} = RT$

(see p. 127.)

Hence the temperature required to produce speeds of the order of $10^7\,\text{m s}^{-1}$ is given by

$$T = \frac{Nmv^2}{3R} = \frac{Mv^2}{3R}$$

where M is the mass of 1 mole of protons (approximately 0.001 kg).

$$T = \frac{0.001 \times 10^{14}}{3 \times 8.3} \approx 5 \times 10^9\,\text{K}.$$

Fusion reactors

Practical fusion reactors are under development in several countries. The difficulties are considerable because of the enormous temperatures required.

The D-T reaction is the most likely way of achieving fusion on a practical scale. The idea is to fuse nuclei of deuterium ^2_1H and tritium ^3_1H as below.

$$^2_1\text{H} + {}^3_1\text{H} \rightarrow {}^4_2\text{He} + {}^1_0\text{n} + 17.6\,\text{MeV}$$

Deuterium is readily obtained from water since it forms 0.01% of naturally-occurring hydrogen. Tritium does not occur naturally so it is manufactured by allowing the neutrons from the D-T reaction to bombard and react with lithium nuclei ^6_3Li in a blanket round the fusion zone.

$$^6_3\text{Li} + {}^1_0\text{n} \rightarrow {}^4_2\text{He} + {}^3_1\text{H} + 4.8\,\text{MeV}$$

The raw materials are therefore deuterium and lithium, both of which are available in plentiful supply. The only waste product is the inert gas helium formed when α-particles are stopped. The tritium produced from the lithium blanket must be removed from the blanket and fed into the fusion zone. The reactor is surrounded with thick concrete to prevent escaping neutrons from reaching the operators.

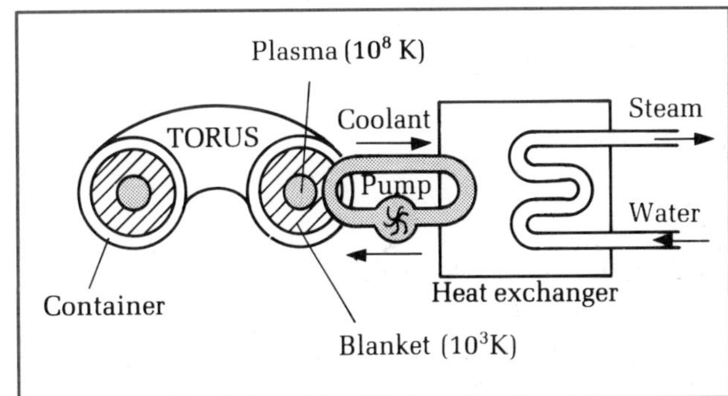

Figure 30.24 *Outline of JET*

In theory, the energy per second released is more than enough to maintain the high temperature of the plasma, so continuous supplies of deuterium and tritium (from the blanket) ought to give continuous power output. In practice, there are two major problems which must be solved before fusion power stations can be built.

1 Plasma heating This is achieved by passing a massive electric current of the order of 10^6 A through the plasma. The plasma, composed of charged particles moving about at high speeds, easily conducts electricity. The massive 'start-up' current heats the plasma up to the temperature range where fusion occurs. But massive amounts of electrical power are needed to supply the start-up current.

2 Plasma confinement The problem here is to contain a plasma at a temperature of the order of 10^8 K. The favoured method is to use magnetic fields to keep the charged particles of the plasma moving along a closed loop. Prototype reactors such as the Joint European Torus (JET) have a complicated field arrangement giving a helix-type field as shown in Figure 30.25. The plasma particles spiral round the field lines so travelling round the torus (a torus is a 'doughnut' shape; a Joint European Doughnut

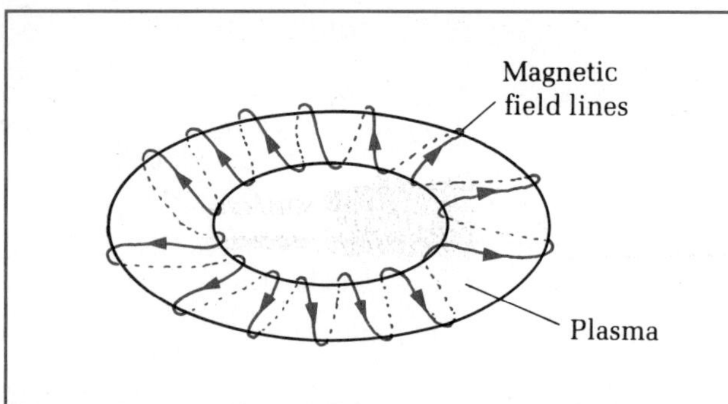

Figure 30.25 *The magnetic bottle*

doesn't sound as good!) to collide with one another. The plasma must be prevented from touching the sides of the steel container otherwise it just fizzles out and loses its energy. The idea is to make the plasma stable, using the magnetic field, so that it generates more power than it uses to keep it hot. However, the chief problem is to keep the plasma in position since its particles are very energetic and easily wriggle out of place.

Comparing fusion and fission reactors shows the importance of developing practical fusion power. Fusion offers the goal of energy from almost unlimited fuels, with the key advantage of non-radioactive waste. Also, fusion is more efficient than fission.

a) The D-T reaction and the lithium blanket reaction gives about 22 MeV from ^2_1H and ^6_3Li (i.e. for eight nucleons rearranged). So the energy released per nucleon used is about 3 MeV.

b) Fission gives approximately 200 MeV from the fission of each U-235 nucleus. So the energy released per nucleon used is about 0.8 MeV (=200/235 MeV).

Worked example Calculate the energy released in the D-T reaction, given the masses of deuterium, tritium and helium atoms are as follows.

$^2_1H = 2.014\,10$ u, $^3_1H = 3.016\,05$ u, $^4_2He = 4.002\,60$ u, $^1_0n = 1.008\,67$ u.

Assume 1 u = 931 MeV.

Solution $^2_1H + ^3_1H \rightarrow ^4_2He + ^1_0n + Q$ (energy released)

Initially Mass of ^2_1H nucleus $= 2.014\,10 - m_e$, where m_e is the mass of an electron.

It's necessary to subtract m_e because 2.014 10 u is the atomic mass of ^2_1H and we only want the nucleus mass.

Mass of ^3_1H nucleus $= 3.016\,05 - m_e$.

Finally Mass of ^4_2He nucleus $= 4.002\,60 - 2m_e$.

Mass of neutron = 1.008 67 u.

The loss of mass M = initial mass − final mass

$= (2.014\,10 - m_e + 3.016\,05 = m_e)$

$- (4.002\,60 - 2m_e + 1.0\,08\,67)$

$= 5.030\,15 - 5.011\,27 = 0.018\,88$ u.

Since 1 u = 931 MeV then the energy released Q is given by

$$Q = 0.018\,88 \times 931 = 17.6 \text{ MeV}.$$

30.4 Summary

Definitions

1 *Binding energy* of a nucleus is defined as the work done on the nucleus to separate it into its constituent neutrons and protons.

2 *Fission* happens when a heavy nucleus splits into two approximately equal fragments with the release of several neutrons as well.

3 *Critical mass* is the least amount of fuel mass to sustain a chain reaction.

4 *Fusion* happens when light nuclei are fused together to form a heavier nucleus.

Equations

1 Einstein's mass-energy equation $E = mc^2$

2 Energy conversions $1\text{ MeV} = 1.6 \times 10^{-13}$ J

$1\text{ u} = 931$ MeV

Short questions

Electronic charge $e = 1.5 \times 10^{-19}$C

1 atomic mass unit ≡ 931 MeV

The Avogadro constant $L = 6.02 \times 10^{23}$ mol^{-1}

30.1

a) Estimate the electrostatic force of repulsion between two protons at a separation of 10^{-14} m. ($\varepsilon_0 = 8.85 \times 10^{-12}$ F m^{-1}.)

b) The size of the nuclear force must be at least equal to your estimate in **a)**; the range of the nuclear force is of the order of 10^{-14} m. Estimate the work done by the nuclear force when two nucleons are separated.

30.2

a) Show that 1 u of mass converted to energy gives 931 MeV.

b) Calculate the binding energy of an α-particle given the masses of the proton, neutron and α-particle are respectively 1.007 28 u, 1.008 67 u and 4.001 50 u.

c) Calculate the binding energy per nucleon of a $^{12}_6C$ nucleus.

30.3

a) Sketch a graph to show how the binding energy per nucleon of the known nuclides varies with mass number.

b) Use the curve in **a)** to explain why energy is released when a $^{235}_{92}U$ nucleus fissions.

30.4 Calculate the energy released when a $^{210}_{84}Po$ nucleus disintegrates by α-emission. The *atomic* masses of the Po-210 and the He-4 atoms are 209.982 87 u and 4.002 60 u respectively, and the atomic mass of $^{206}_{82}Pb$ is 205.974 46 u. What fraction of the energy released is carried away by the α-particle?

30.5 Natural uranium contains 0.7% U-235 (i.e. $^{235}_{92}U$). When a U-235 nucleus is fissioned, approximately 200 MeV of energy is released.
Calculate the total energy released, in joules, when the U-235 content of 1 kg of natual uranium is completely fissioned.

30.6

a) Discuss how energy is released when the U-235 content of the fuel rods of a thermal reactor fissions at a steady rate.

b) What is the function of (i) the moderator, (ii) the absorber rods in a thermal reactor? Give an example of a suitable material in each case, stating the physical property that makes each material suitable for its function.

30.7

a) The advanced gas cooled reactor uses graphite as a moderator and CO_2 gas as its coolant. The pressurized water reactor uses water as both moderator and coolant. Discuss the relative merits of the two moderators and the two coolants.

b) A PWR reactor produces 700 MW of power from its core to be removed by the coolant. If the inlet temperature and the outlet temperature of the water coolant (under pressure) are 200°C and 650°C respectively, calculate the rate at which water must pass through the core to keep its temperature constant.
The specific heat capacity of water is $4200\,\mathrm{J\,kg^{-1}\,K^{-1}}$.

30.8

a) Describe the chief features of a thermal reactor that ensure safety when in use.

b) The mean lifetime of a fission neutron in a nuclear reactor is 10 ms. If the neutron multiplication factor in the core of a certain reactor increases from 1 to 1.01, calculate the time taken for the power generated to double.

30.9

a) One of the fusion processes by which energy is generated inside the Sun is called the proton-proton cycle. This involves fusing four protons step by step to form a helium-4 nucleus.

$${}^1_1\mathrm{p} + {}^1_1\mathrm{p} \rightarrow {}^2_1\mathrm{H} + {}^{\;0}_{+1}\beta\ \text{(positron)} + Q_1$$

$${}^2_1\mathrm{H} + {}^1_1\mathrm{p} \rightarrow {}^3_2\mathrm{He} + Q_2$$

$${}^3_2\mathrm{He} + {}^3_2\mathrm{He} \rightarrow {}^4_2\mathrm{He} + 2{}^1_1\mathrm{p} + Q_3$$

Calculate the energy released in each reaction, and hence calculate the total energy released when four protons are fused to form a ${}^4_2\mathrm{He}$ nucleus.

Masses/u: proton 1.007 28, positron 0.000 55,
${}^2_1\mathrm{H}$ atom 2.014 10,
${}^3_2\mathrm{He}$ atom 3.016 03,
${}^4_2\mathrm{He}$ atom 4.002 60

b) The Sun radiates energy into space at a rate of 4×10^{26} W. Its mass is 2×10^{30} kg. Estimate its lifetime, assuming that 0.7% of its mass can be converted into energy.

30.10

a) Calculate the energy released in the following fusion reaction.

$${}^3_1\mathrm{H} + {}^2_1\mathrm{H} \rightarrow {}^4_2\alpha + {}^1_0\mathrm{n} + Q$$

Masses/u: neutron 1.008 67,
α-particle 4.001 50,
${}^3_1\mathrm{H}$ nucleus 3.015 5,
${}^2_1\mathrm{H}$ nucleus 2.013 55

b) What are the chief difficulties in controlling fusion?

c) Describe briefly one of the methods currently being developed to produce fusion power. Discuss how the method, etc., attempts to overcome the difficulties you mention in **b)**.

31

Particles and patterns

31.1 Waves and particles

The nature of light

Isaac Newton put forward the 'corpuscular theory' of light in the 17th century. He imagined a light ray as a stream of tiny particles (Figure 31.1). He developed his ideas to explain reflection and refraction. Newton's theory of light was accepted for 150 years or more because there was little evidence to suggest otherwise.

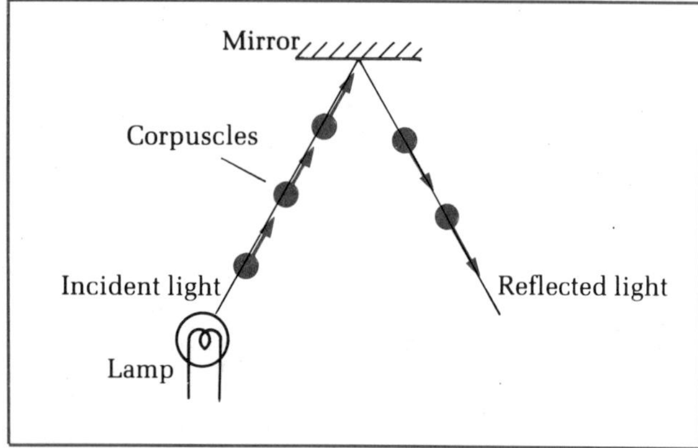

Figure 31.1 *Newton's corpuscular theory*

Thomas Young was the first person to demonstrate conclusively the wave nature of light. Young showed that light passed through double slits (Figure 31.2) produces an interference pattern. Since interference is a wave property, Young's experiment disproved Newton's theory of light established 150 years earlier.

Wave theory can be used to explain many properties of light (see p. 233) and of other forms of electromagnetic radiation. But at the end of the 19th century, investigations of photoelectricity gave measurements which could not be explained by wave theory (see p. 428).

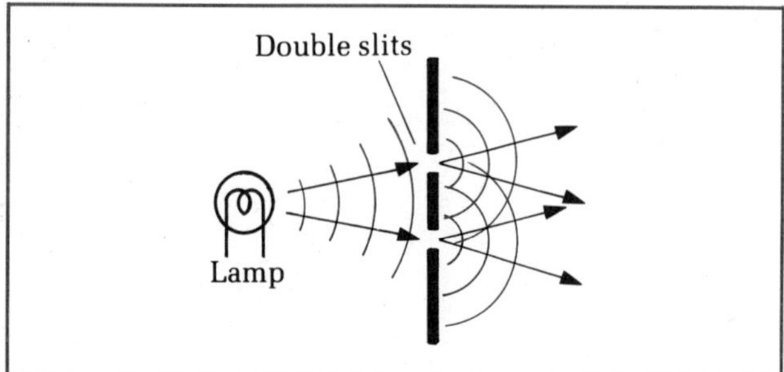

Figure 31.2 *Interference of light*

Albert Einstein provided the new ideas necessary to explain photoelectricity in 1905. According to Einstein, light is composed of photons, each photon being a wavepacket carrying a fixed amount of energy. By assuming that the energy is proportional to the frequency, Einstein was able to explain photoelectricity.

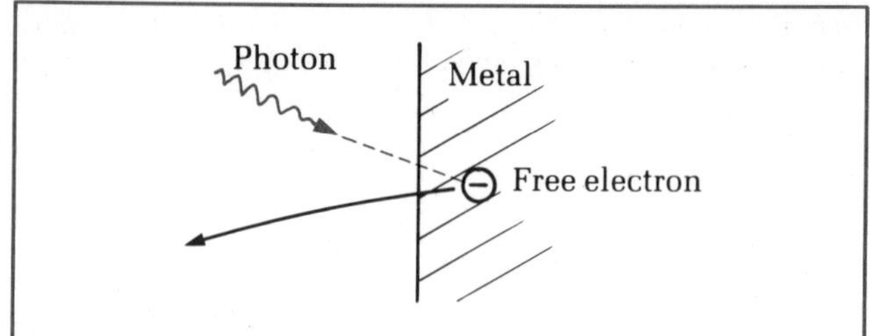

Figure 31.3 *Photons at work*

When an atom emits light, it does so in a short burst, lasting of the order of a nanosecond (10^{-9} s). So the emitted photon is a wavepacket of the order of half a metre in length (≈speed of light × time). The atoms of a light-emitting material emit photons in different directions so the light spreads out from the source.

How do photons produce interference patterns like the double slits pattern? When light passes through each of the slits, it is diffracted as shown in Figure 31.4. The two sets of diffracted waves overlap to give a pattern of reinforcement and cancellation. So an interference pattern is seen on a suitably placed screen. The intensity pattern on the screen shows maxima and minima, corresponding to reinforcements and cancellations respectively. Maximum intensity on the screen is where photons are most likely to arrive. But we don't know which route each photon takes to arrive at the screen from the lamp. Its route could be via slit S_1 or via slit S_2. The uncertainty is the reason for the interference pattern; if we could be sure which route the photons took for example, by blocking off one of the slits, we would destroy the interference pattern.

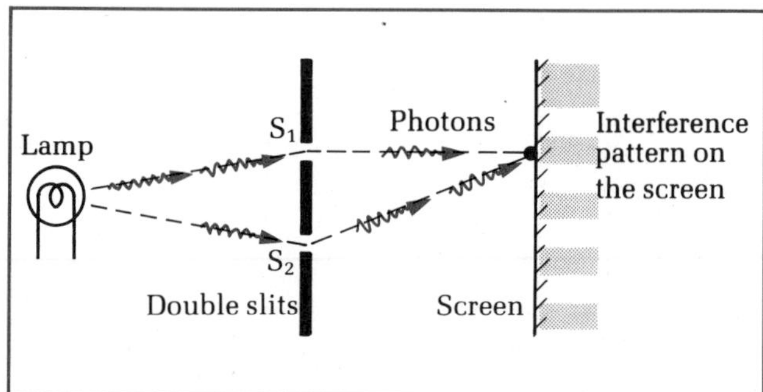

Figure 31.4 *Uncertain routes*

Wave-particle duality

If light waves have a 'particle-like' nature, then do particles have a wave-like nature? Louis de Broglie in 1925 suggested that they do. He put forward the theory that *all* particles have a wave-like nature. He said that the particle momentum was linked to the wavelength by the equation

Momentum × wavelength = h

where h = the Planck constant.

Why? de Broglie used the ideas developed by Einstein about light photons. Each photon has energy hf which is equivalent to mass m on a scale $mc^2 = hf$. So the momentum of a photon is equivalent to mc which equals $hf/c = h/\lambda$. The approach is not at all rigorous but it proved successful. Work out for yourself the 'de Broglie wavelength' λ of an electron moving at a speed of $10^7\,\mathrm{m\,s^{-1}}$. Given its mass is $10^{-30}\,\mathrm{kg}$ and $h = 6.6 \times 10^{-34}\,\mathrm{J\,s}$, then you ought to obtain a value of about 0.06 nm. The value is the same order as the wavelength of X-rays; X-rays can be diffracted by crystals (see p. 460). Can electrons be diffracted by crystals too?

Diffraction of electrons was discovered two years later when it was shown that a beam of electrons directed at a single crystal produced a diffraction pattern like an X-ray diffraction pattern. About the same time, George Thomson (son of J.J.) tested de Broglie's theory by directing electrons at a thin metal foil. Thin metal foils contain lots of tiny crystals called grains. Thomson showed that the electrons were diffracted to form a pattern of rings on a screen placed as in Figure 31.5. The same pattern is obtained when X-rays are directed at the foil. So de Broglie was proved right. Particles do have a wave-like nature; in other words, matter has a dual nature which is wavelike or particle-like.

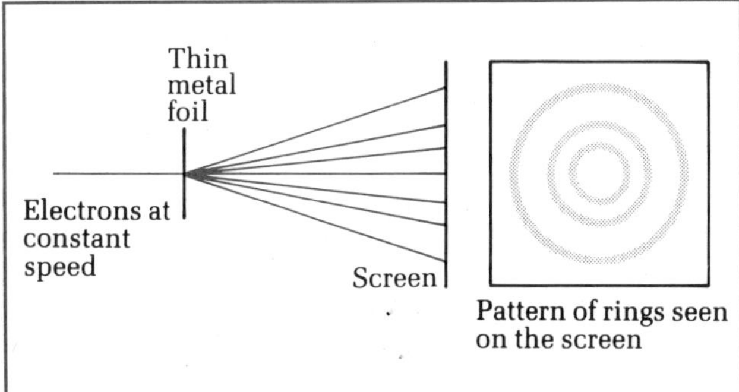

Figure 31.5 *Diffraction of electrons*

The de Broglie equation can be used to work out the angle of diffraction θ of each diffracted beam. From de Broglie's equation

$$\text{Wavelength } \lambda = \frac{h}{m_e v}$$

where m_e = mass of a particle, v = speed of the particles.

From the electron gun equation

$$\tfrac{1}{2} m_e v^2 = eV_A$$

where V_A is the anode voltage used to accelerate the electrons to speed v.

$$\text{Hence } \lambda^2 = \frac{h^2}{m_e^2 v^2} = \frac{h^2}{2m_e \times \frac{1}{2} m_e v^2} = \frac{h^2}{2m_e \times eV_A}$$

$$\text{Thus } \lambda = \frac{h}{\sqrt{2m_e eV_A}}$$

So the wavelength can be calculated from the measured value of anode voltage V_A. Given the spacing d between the layers of atoms in the crystal, then the Bragg equation (see p. 461) can be used to calculate the angle of diffraction θ.

$$2d \sin\theta = m\lambda$$

where m = order number = 1, 2, . . . etc.

The diffraction pattern for a thin metal foil shows a set of concentric rings. Each ring is produced by diffraction from one set of layers in the grains in the foil. The same set of layers may produce several orders of rings, depending on the wavelength and the spacing between the layers.

Worked example In an electron diffraction experiment, a beam of electrons from an electron gun operating at 3000 V was directed at a thin metal foil. Diffraction rings were observed on a suitably-placed fluorescent screen, giving measured angles of diffraction of 13.8° and 28.7°. Calculate

a) the de Broglie wavelength of the electrons,
b) the layer spacing for each diffracted ring.

Assume $h = 6.6 \times 10^{-34}\,\mathrm{J\,s}$, $m_e = 9.1 \times 10^{-31}\,\mathrm{kg}$, $e = 1.6 \times 10^{-19}\,\mathrm{C}$.

Solution

a) Use $\lambda = \dfrac{h}{meV_A}$

$$\lambda = \frac{6.6 \times 10^{-34}}{9.1 \times 10^{-31} \times 1.6 \times 10^{-19} \times 3000} = 3 \times 10^{-11}\,\mathrm{m}.$$

b) Use $2d \sin\theta = m\lambda$.

For $\theta = 13.8°$,

$$2d_1 = \frac{m_1 \times 3 \times 10^{-11}}{\sin 13.8°} = m_1 \times 1.26 \times 10^{-10}\,\mathrm{m}.$$

For $\theta = 28.7°$,

$$2d_2 = \frac{m_2 \times 3 \times 10^{-11}}{\sin 28.7°} = m_2 \times 0.63 \times 10^{-10}\,\mathrm{m}.$$

Since m_1 and m_2 are whole numbers, then $m_1 = 1$ (first order) and $m_2 = 2$ (second order) gives layer spacing equal to $0.63 \times 10^{-10}\,\mathrm{m}$.

The dual nature of matter has been confirmed by many other experiments. For example, in 1930 it was shown that beams of ions give an interference pattern when directed through double slits. Figure 31.6 shows an outline of such an experiment. The detector D gives a reading in proportion to the number of ions per second

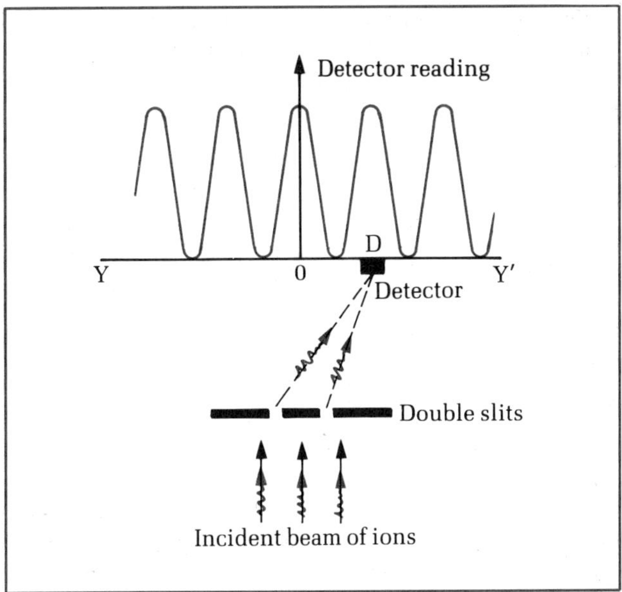

Figure 31.6 *Diffraction of ions*

arriving at the detector. The reading varies with the position of the detector along the line YOY′, giving a graph like the intensity distribution for the double slits experiment.

Models are used to explain the behaviour of matter and radiation in different experiments. We imagine electrons as being like tiny balls when considering their motion in electric or magnetic fields. But we cannot see the electrons; instead, we use a 'particle' model based on what we can see to imagine what is happening. When a beam of electrons is directed at a crystal, the electrons display wave properties; again, we cannot see what is happening in the crystal so we use a large-scale model to imagine the situation – the wave model.

We cannot say what an electron looks like. All we can do is to interpret its behaviour. The same is true for all matter at an atomic level, and for radiation too. Matter has a dual nature which can be wave-like or particle-like. Which side of its nature we see depends on the experiment we use to detect it. The situation is a bit like viewing a coloured object; what we see depends on the type of light we use to view it. If we fire electrons through a crystal, we interpret the behaviour of the electrons as wave-like; if we aim a beam of electrons at a target in an X-ray tube, we make the electrons knock out other electrons from atoms. So we make the beam electrons display particle-like behaviour.

Energy levels

Electrons inside the atom occupy well-defined energy levels. The wave nature of the electron can be used to link stationary waves (see p. 237) with energy levels.

For a square-well potential as in Figure 31.7, consider an electron trapped in an electric field which forms a potential well. Of course, the field round a nucleus has a $1/r$ potential, but the square well is much simpler.

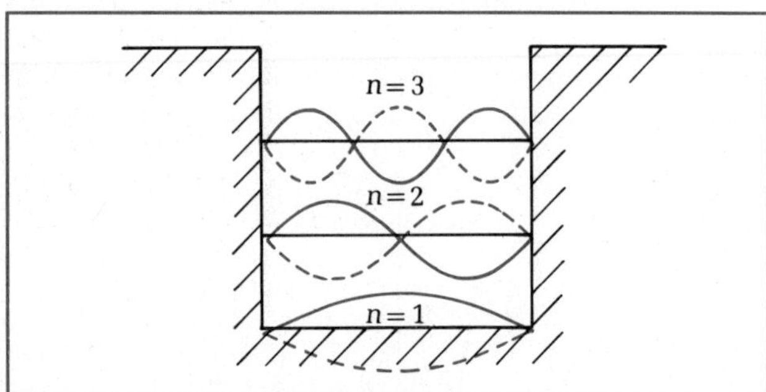

Figure 31.7 *A square-well potential*

A stationary wave has constant energy so imagine the de Broglie waves of an electron across the well, like waves on a string. Stationary waves are possible when the well width L is equal to a whole number of half-wavelengths.

$$L=\frac{n\lambda}{2}$$

where n is a whole number,

So the allowed wavelengths are given by $\lambda=\frac{2L}{n}$.

Using the de Broglie equation therefore gives the electron momentum as

$$\text{Momentum } mv=\frac{h}{\lambda}=\frac{h}{(2L/n)}=\frac{nh}{2L}$$

Therefore the kinetic energy of the electron is given by

$$\text{K.E.}=\tfrac{1}{2}mv^2=\frac{(mv)^2}{2m}=\frac{h^2}{8mL^2}\times n^2$$

This very simple model shows that the electron must have certain values of energy inside the well. Even though the energy level pattern for the square well is unlike the actual pattern for an atom, the key point is that it does give fixed energy values.

For a 1/r potential the Bohr theory assumes that the angular momentum mvr is 'quantized' in multiples of $nh/2\pi$.

$$\text{Angular momentum } mvr=\frac{nh}{2\pi}$$

$$\text{So the momentum } mv=\frac{nh}{2\pi r}$$

From the de Broglie equation, the wavelength of the electron waves is given by

$$\lambda=\frac{h}{mv}=\frac{h}{(nh/2\pi r)}=\frac{2\pi r}{n}$$

Hence $n\lambda=2\pi r$

So the allowed Bohr orbits correspond to stationary waves round the orbit. The circumference of the orbit is a whole number of electron wavelengths for each allowed orbit.

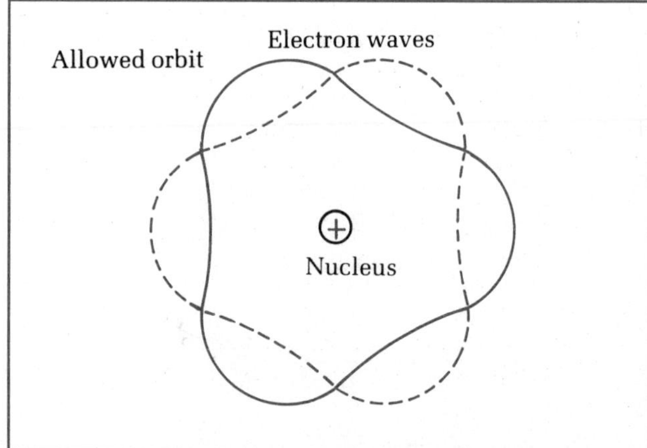

Figure 31.8 *de Broglie waves round the nucleus*

Wave mechanics

The ideas that de Broglie established were developed by Erwin Schrodinger in 1927. He reasoned that the amplitude ψ of a wave was given by a sinusoidal expression like $\sin\left(\frac{2\pi x}{\lambda}\right)$ where x is distance.

We can get the wavelength from this equation by differentiating twice.

$$\psi = \text{constant} \times \sin\left(\frac{2\pi x}{\lambda}\right)$$

$$\frac{d\psi}{dx} = \frac{2\pi}{\lambda} \times \text{constant} \times \cos\left(\frac{2\pi x}{\lambda}\right)$$

So $$\frac{d^2\psi}{dx^2} = \left(\frac{2\pi}{\lambda}\right)^2 \times \text{constant} \times -\sin\left(\frac{2\pi x}{\lambda}\right) = \left(\frac{2\pi}{\lambda}\right)^2 \times -\psi.$$

Writing $\frac{d^2\psi}{dx^2}$ as ψ'' our differentiation exercise gives

$$\left(\frac{2\pi}{\lambda}\right)^2 = -\frac{\psi''}{\psi}$$

Using the de Broglie equation, we can write the momentum mv in terms of ψ.

$$(mv)^2 = \frac{h^2}{\lambda^2} = \left(\frac{h}{2\pi}\right)^2 \times -\frac{\psi''}{\psi}$$

So the K.E. $= \frac{1}{2}mv^2 = \frac{(mv)^2}{2m} = \frac{\hbar^2}{2m} \times -\frac{\psi''}{\psi}$

where $\hbar = h/2\pi$.

The total energy of a 'particle' of mass m is therefore given by

$$\text{Total energy } E = \text{K.E.} + \text{P.E.} = \hbar^2/2m \times \left(-\frac{\psi''}{\psi}\right) + V$$

where V is the P.E. of the particle.

If this odd-looking equation is rearranged, it becomes

$$\psi'' + \frac{2m}{\hbar^2}(E - V)\psi = 0$$

This equation is known as the **Schrodinger equation**, and it is the guiding equation for the physics of the atom. It can be used to predict the properties of sub-atomic particles in any known potential V. From the equation, the energy values can be calculated. For example, it can be used to prove mathematically that the electrons in the shells round an atom fit a 2, 8, 8 ... pattern in accordance with the Periodic Table. In other words, the innermost shell can take two electrons maximum, the next one a further eight electrons etc. (see p. 65). The proof involves mathematics beyond the scope of this particular book but you can now perhaps appreciate the power of Schrodinger's ideas. His equation is the $F = ma$ of atomic physics.

Probability is used to interpret wave mechanics. We imagine a 'particle' as a wavepacket. The amplitude ψ of the wave at any point in the packet is a measure of where the particle is most probably located. So where the amplitude is greatest, that is where the particle is most likely to be located. For an electron round the atom, these locations correspond to the 'shells' round the nucleus. If we could knock electrons out of atoms by aiming photons precisely, the best position to aim for would be at the shells. But remember the shells are 'probability shells'; an electron knocked out of a shell must have been struck by a photon, but lots of photons might have passed through the shell before an impact took place.

31.2 Probing the nucleus

Uncertainty

Suppose we try to pinpoint a particle. The particle might be an electron in an atom, a nucleon in a nucleus or an ion in an ion beam. To pinpoint it, we can use a beam of light photons so that a photon hits the particle and rebounds into a detector. But the rebound alters the particle's speed and hence its energy. We cannot pinpoint its position *and* measure its speed at the same time. Making a measurement alters the situation being measured! Can you think of other situations where this happens? For example, when a thermometer is used to measure the temperature of a beaker of water, some energy is transferred from the beaker to the thermometer, so the water temperature is altered, even if only by a small amount.

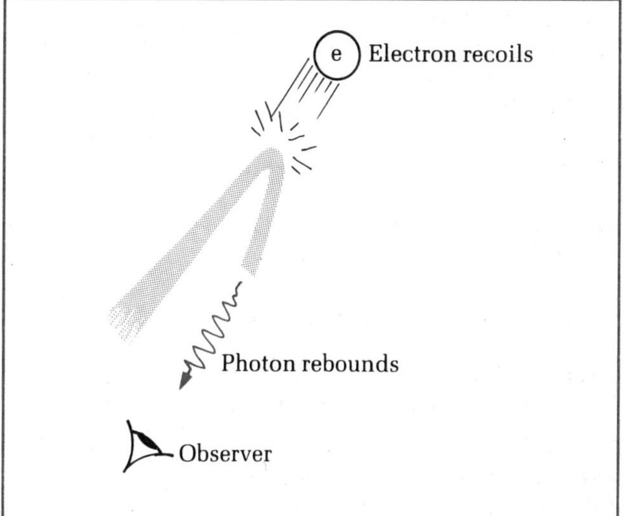

Figure 31.9 *Uncetainty*

Imagine we trap a particle in a box of length L. We don't know where the particle is in the box because if we look we would alter its energy. But we can say that its de Broglie wavelength is less than L. So its momentum is greater than h/L according to the de Broglie equation. Its speed is at least h/Lm since speed = momentum value/mass. In other words, the uncertainty in its speed Δv is given by

$$\Delta v \times L = \frac{h}{m}$$

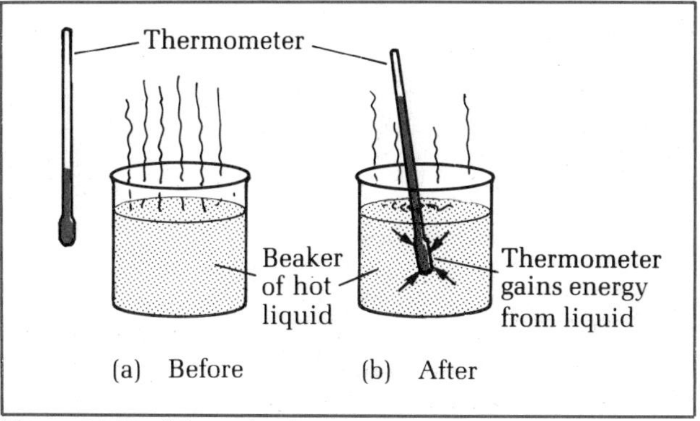

Figure 31.10 *Measuring temperature*

The smaller the value of L, the larger the uncertainty in its speed. These ideas were developed by **Werner Heisenberg** in 1928. He showed that the energy of a particle in a given time interval Δt could not be measured exactly. And since the energy can't be measured exactly, it could have any value over the range ΔE in that time. He proved that the uncertainty of the energy ΔE is given by

$$\Delta E \times \Delta t \approx h.$$

Why do we not notice these effects in everyday life? The reason is that our energy usage is on a large scale, so large that uncertainties are insignificant. You couldn't use the uncertainty principle above to claim your electricity meter reading was incorrect; for a time interval of three months ($=10^7$ s), the uncertainty of energy ΔE is about 10^{-41} J. But for a time interval of 10^{-22} s, ΔE is of the order of 10^{-12} J (i.e. order of MeV). So in the time taken for a photon to shoot across a nucleus, we can't be sure what the energy is.

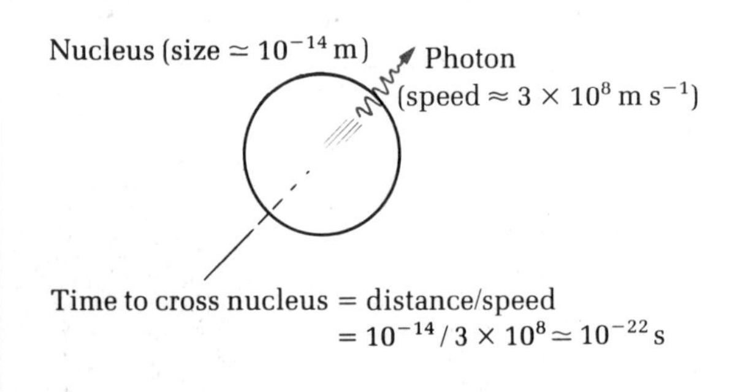

Figure 31.11 *Crossing the nucleus*

Borrowing energy from mass is possible because of the uncertainty principle. Energy and mass are equivalent; an α-particle formed briefly in a heavy nucleus needs energy to escape from the pull of the nuclear force field holding the nucleons in. By borrowing energy ΔE from its mass, the alpha particle can escape from the nucleus. But it must repay within a time $\Delta t = h/\Delta E$. If too much energy is borrowed, it isn't given enough time to escape, so it is pulled back (see Figure 31.12).

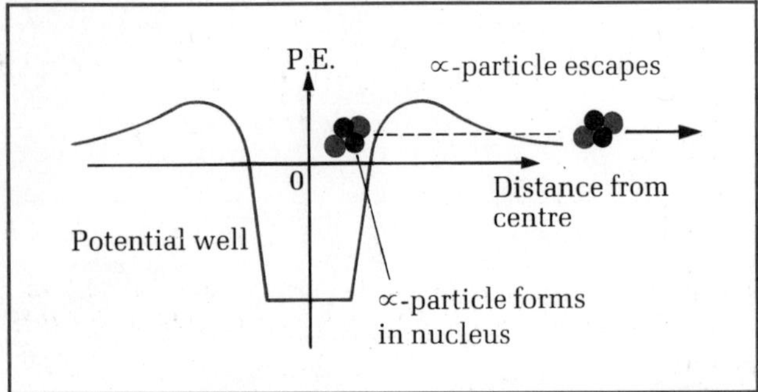

Figure 31.12 *Borrowing energy*

Investigating the nucleus

In the early decades of the 20th century, scientists had unlocked the structure of the atom. Attention turned to the nucleus, known to contain protons and neutrons. The general method used was similar to the approach used by Rutherford earlier. High energy particles were directed at matter and the bits that emerged were measured. Accelerators and cosmic rays give particles with high energies which can penetrate the nucleus. Over the past few decades, bigger and bigger accelerators have been built to achieve higher and higher energies.

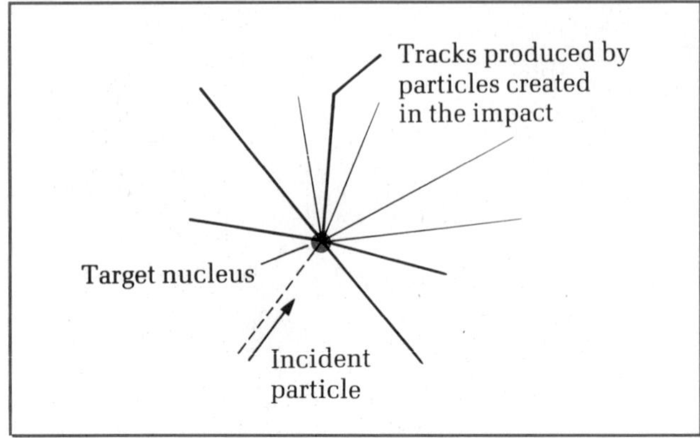

Figure 31.13 *Smashing a nucleus*

Cosmic rays are natural sources of high-energy particles produced by the Sun and other stars. Particles such as protons enter the Earth's atmosphere from space with huge energies. These are called 'primary' cosmic rays and the energies of individual particles can be up to 10^{12} MeV. Typical energies are of the order of 10^4 MeV so a primary particle entering the atmosphere creates a shower of 'secondary' cosmic rays by smashing into nuclei of atoms of the atmosphere. Thus a cascade of cosmic rays hits the surface from each primary particle. Background counting by a geiger tube is due mostly to cosmic rays.

By sending photographic plates high above the Earth's surface in rockets or balloons, cosmic ray collisions have been recorded and studied in detail. In 1937, Carl Anderson discovered the **positron** (the positive electron β^+) from cosmic ray plates. Particles called **mesons** with mass about 300 times the mass of the electron were also discovered. Scientists took up the challenge to find out more about the sub-atomic world.

Accelerators are used to boost particles to very high speeds. The particles must be charged because electric and magnetic fields are used. At speeds approaching the speed of light c the energy supplied by an accelerator goes into increasing the particle's mass (see p. 465). High-energy particles from an accelerator can be directed at nuclei in a target as required.

The Van der Graaf accelerator gives particle energies up to 20 MeV. A series of electrodes in a tube are supplied with high voltage in steps from the dome of the machine, as in Figure 31.14. Charged particles such as positive ions enter the accelerator near the dome. With the dome positive, the ions are accelerated down the tube. A positive potential of the order of 10 MV gives positive ions enough kinetic energy to smash into target nuclei.

The linear accelerator has a series of electrodes supplied by a high-frequency alternating voltage. Alternate electrodes are connected to the same terminal of the voltage supply (Figure 31.15). Positive ions from the

Figure 31.14 *The van der Graaf accelerator*

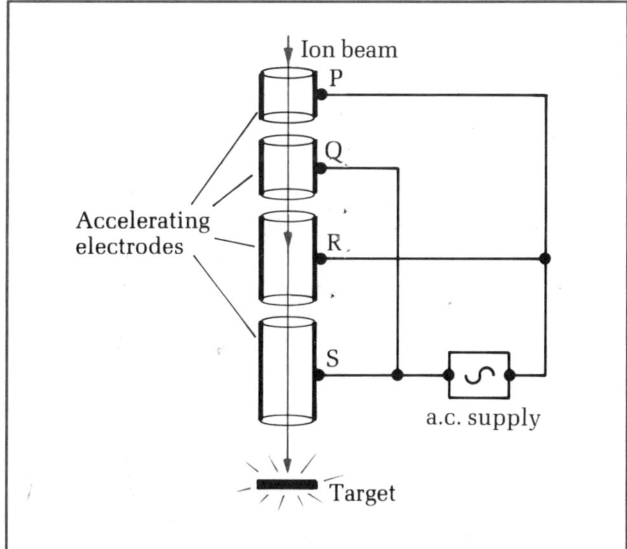

Figure 31.15 *The linear accelerator*

source are accelerated to electrode P when P is negative. As ions pass through P, the voltage supply reverses so P repels the ions and the next electrode Q attracts them. As the ions pass through Q, the voltage supply reverses again so Q now repels the ions and electrode R attracts them. The result is that a 10-stage accelerator with a 200 kV supply will boost protons to kinetic energies of the order of $10 \times 200 = 2000$ kV.

The electrodes must be designed with lengths determined by the speed at each stage. For example, at stage 2, a proton has twice the K.E. as at stage 1. So its speed is $\sqrt{2} \times$ its speed at stage 1. Since the time from one stage to the next is constant (= half a cycle of a.c.), the length of electrode 2 must be $\sqrt{2} \times$ the length of stage 1. The speed increases from one stage to the next so the electrode lengths must be longer from one stage to the next.

The **Stanford linear accelerator** can accelerate electrons to energies of the order of 20 GeV (1 GeV = 10^9 eV). The accelerator is about two miles (or 3 km) long and is like a giant TV tube, but instead of a screen, different targets are placed in the path of the beam.

The Cyclotron was invented in 1932 and it uses a magnetic field to keep charged particles circling. Two D-shaped electrodes, called 'dees', enclose an evacuated chamber containing the circling particles. A high-frequency alternating p.d. is applied between the dees. Each time the voltage reverses, charged particles crossing from one dee to the other are boosted by the change of voltage. For example, suppose a beam of protons is 'injected' into one of the dees, as in Figure 31.16, when it is negative. The protons are forced into a circular path by the magnetic field; if the voltage reverses as the protons cross to the other dee, the protons are accelerated as they cross. Then they travel round the second dee, and once again as they cross the voltage reverses, so they are boosted to even higher speeds. The frequency of the alternating p.d. must be at the correct value to accelerate the particles each time they cross over.

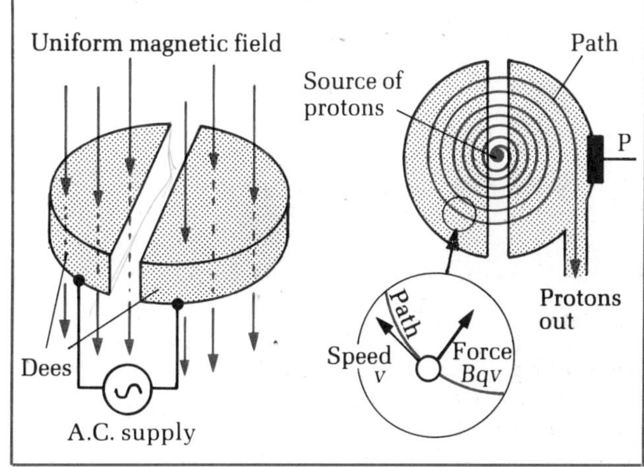

Figure 31.16 *The cyclotron*

Suppose a particle with charge q moving at speed v swaps sides just as the voltage reverses. The magnetic field, into the plane of the dees, makes the particle move round on a circular path of radius r. The force Bqv due to the field is given by

$$Bqv = \text{mass} \times \text{centripetal acceleration}$$

$$Bqv = m \times \frac{v^2}{r}$$

Hence $v = Bqr/m$.

Since the particle travels distance πr (= half the circumference) round each dee, then the time taken in each dee is given by

$$\text{Time} = \frac{\text{distance}}{\text{speed}} = \frac{\pi r}{(Bqr/m)} = \frac{m\pi}{Bq}.$$

So the time for one full cycle, $T = \dfrac{2m\pi}{Bq}$

Hence the frequency $f = \dfrac{1}{T} = \dfrac{Bq}{2m\pi}$

Each time the particle crosses the gap between the dees, its speed is increased so its radius increases too. So it is started off near the centre and its radius of orbit increases as it spirals out to near the edge. Then a suitable voltage applied to the deflecting plate P pulls the particles out from the magnetic field on to a target.

The limit for a cyclotron is about 1 MeV. This is because energy supplied increases the mass, as given by Einstein's equation. As the speed of the particles approaches c, the speed of light, the particles become more massive so they take longer to go round. Hence the reversal of polarity is no longer in phase with the particle's motion, so it can no longer be accelerated effectively.

Synchrotrons were developed after 1945 to boost particles to much higher energies than those produced by cyclotrons. The **super proton synchrotron** (SPS) at Geneva (Figure 31.17) was built in 1978 and is capable of accelerating protons to energies of the order of 500 GeV. The circumference of the SPS is about four miles (or 6 km). A ring of electromagnets is used to keep the protons on a circular path. The field strength of the magnets is increased to compensate for the gain of mass as the particles are accelerated. Each time the protons pass through the accelerating electrodes, they are boosted to even higher energies as they race round the ring.

Figure 31.17 *The SPS at Geneva*

The bubble chamber is used to detect the paths of charged particles. The chamber contains liquid hydrogen just below its boiling point. Sudden reduction of pressure on the liquid causes bubbles of hydrogen to form along the path of any charged particle passing through at that instant. The charged particle ionizes the hydrogen atoms along its path, and bubbles form more easily where there are ions. So the charged particle leaves a visible track of bubbles which is photographed at that instant. Modern bubble chambers are house-size constructions; charged particles from accelerators are directed into the chamber

Figure 31.18 *The bubble chamber*

Figure 31.19 *Tracks in a bubble chamber*

liquid and collide with the nuclei of the bubble chamber atoms. So the particles and the 'debris' created can be photographed and studied.

By applying a magnetic field into the chamber, as shown in Figure 31.18, the charged particles are forced to make curved tracks. From the tracks, the properties of the particles, such as charge, mass and energy, can be calculated. Figure 31.19 shows a typical bubble chamber photograph.

Quarks

Using accelerators and cosmic rays, scientists have discovered a multitude of short-lived sub-atomic particles. These particles are created when protons or electrons, etc, crash into nuclei. The kinetic energy of the 'bullets' is converted into mass to produce these short-lived particles. They disintegrate into more stable particles such as protons. Using bubble chambers and other techniques, the properties of these particles have been measured. From the measurements, the particles have been classified and scientists have found patterns which link their properties. These patterns have been used to predict and discover yet more particles. The explanation of the patterns has presented scientists with the exciting challenge of unravelling the sub-atomic world as they search out and test theories.

Out of the confusion of discoveries and the mysteries of the patterns, a simpler picture has emerged. Scientists now think that all matter belongs either to the 'quark family' or the 'electron family'. Each family has just a few members but they make up the multitude of particles discovered over the past 50 years.

The story of particle physics is like a re-run of 19th century science which led to the formation of the Periodic Table of the elements and its explanation. Scientists measured the properties of the elements and used the measurements to classify the elements in a pattern which we call the Periodic Table. New elements were predicted from spaces in the table, and in due course, these elements were discovered. Eventually the pattern of the Periodic Table was explained in terms of electrons in shells, etc. Now scientists work in teams using enormous equipment. As the search inside the atom has reached down to smaller and smaller scales, so the energies needed have become greater and greater. Teams in several continents have contributed to the search and it is impossible to describe their work in a few pages. So let's follow just one thread of this great challenge to appreciate their work.

The strong nuclear force is responsible for holding nucleons together in the nucleus. In 1935, Hideki Yukawa put forward a theory to explain the force; he said that the force is due to particles which he called 'pions' which are exchanged between the nucleons as in Figure 31.20. He used the uncertainty principle to estimate the mass of the pion. His calculation gave a predicted mass of about 0.15 u (= 140 MeV). Cosmic ray tracks on photographic plates were used to 'hunt' for pions.

In 1947, Cecil Powell took some photographic plates up to a mountain top in the Pyrenees. He suspected that pions from cosmic rays were prevented from reaching sea-level by the atmosphere. When the plates were developed, tracks produced by pions were discovered; from the track range, the particle mass was determined at about 140 MeV. Pions are now called π-mesons (or pi-mesons); the term 'meson' is used for any particle with mass between that of an electron and a proton. Further investigations showed that π-mesons are of three types, π^+, π^- and π^0, according to charge. The π^+ and the π^- carry the same amount of charge as the electron, although the charges are opposite. The π^0 is uncharged.

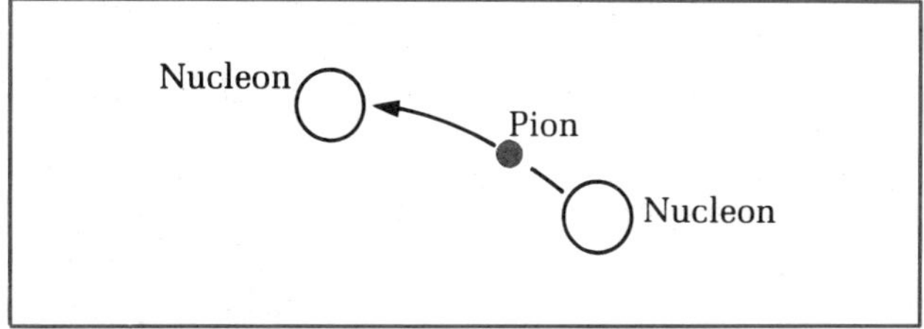

Figure 31.20 *Explaining the nuclear force*

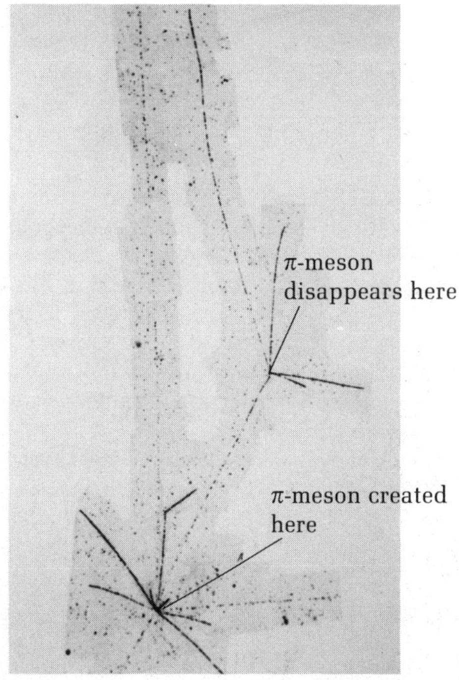

Figure 31.21 *The birth and death of a π-meson*

Strange particles were discovered in 1947 when π-meson tracks were studied in more detail. Scientists could tell a proton track from a π-meson track, and they discovered that pions often formed V-shaped tracks. Figure 31.22 shows an example.

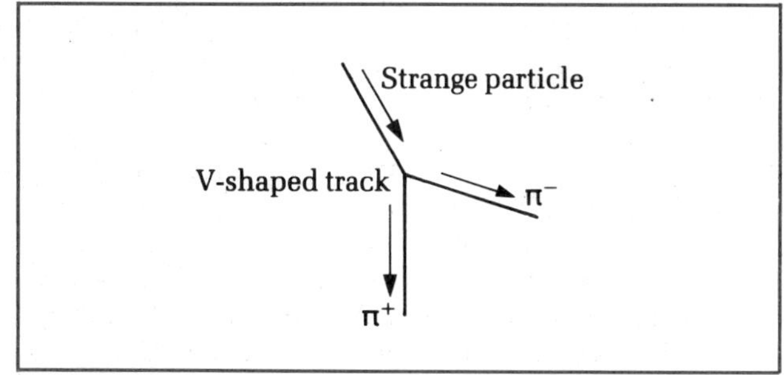

Figure 31.22 *Strange particles*

π-mesons interact strongly with protons or neutrons, and a π^--meson hitting a proton produces a pair of strange particles which decay to give the tell-tale V-shaped tracks. The particles produced were called 'strange' because they were always produced in pairs; and they always decayed into π-mesons, protons or neutrons. Sometimes the strange particles were uncharged so they did not leave tracks, but they disintegrated into charged particles, so giving away their existence as in Figure 31.23, by their decay into V-shaped tracks.

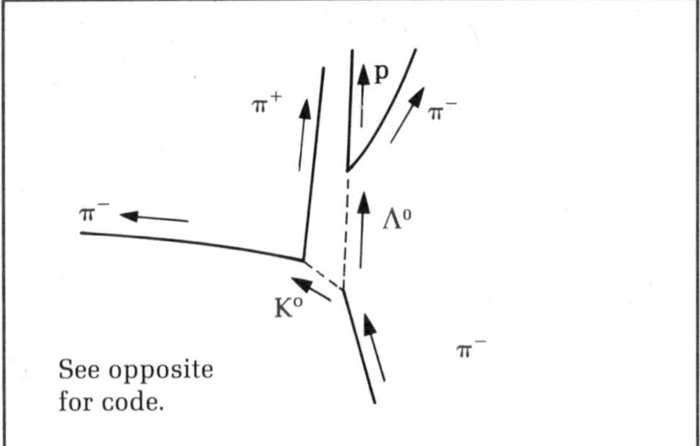

Figure 31.23 *'Hidden' strange particles*

Measurements were made on these strange particles and they were given names such as K-mesons and Λ particles (lambda particles). Using accelerators to smash nuclei with high-energy protons, scientists discovered yet more strange particles in addition to the ones discovered from cosmic ray photographs.

Because they were always produced in pairs, scientists invented a new property called **strangeness** to describe them. Starting with strangeness equal to $+1$ for the K^+-meson, the strangeness of other particles was worked out from the observed reactions. For example, consider this reaction

π^0	+	n	$\rightarrow$	K^+	+	Σ^-
neutral π-meson		neutron		positive K-meson		negative sigma particle

The π-meson and the neutron are not strange particles so they have zero strangeness. So by assuming total strangeness is conserved in this type of reaction, the sigma particle must have strangeness equal to -1. Then the total initial strangeness (0) = the total final strangeness ($+1+-1$). The idea of 'strangeness' is useful to make patterns out of the observations. But what is 'strangeness'? It's a property that helps to see patterns in the same sort of way that electric charge at an atomic or nucleon level does. 19th century scientists didn't know what the significance of atomic number was, apart from its use to make a pattern of the elements. Now we know that the atomic number is the number of protons in the nucleus.

Particles that experience the strong nuclear force are called **hadrons** and they fall into two groups according to whether each particle is heavier (called 'baryons') than a proton or lighter than a proton. By 1960, the list of hadrons was over 20 in number.

Name	**Symbol**	**Charge/e**	**Strangeness**	**Mass/u**	**Lifetime**
proton	p	1	0	1	∞ (stable)
neutron	n	0	0	1	15 mins
π-meson	$\pi^\pm$	±1	0	1/7	10^{-8} s
	π^0	0	0	1/7	10^{-16} s
K-meson	$K^\pm$	±1	±1	1/2	10^{-8} s
	$\overline{K^0}$	0	$+1$	1/2	10^{-8} s
	$\overline{K^0}$	0	-1	1/2	10^{-8} s
Sigma	$\Sigma^\pm$	±1	-1	1.2	10^{-10} s
	Σ^0	0	-1	1.2	10^{-20} s
Lambda	Λ^0	0	-1	1.1	10^{-10} s
Xi	Ξ^0	0	-2	1.3	10^{-10} s
	Ξ^-	-1	-2	1.3	10^{-10} s
Delta	Δ^{++}	$+2$	0	1.33	10^{-23} s
	Δ^+	$+1$	0	1.33	10^{-23} s
	Δ^0	0	0	1.33	10^{-23} s
	Δ^-	-1	0	1.33	10^{-23} s
Sigma-star	$\Sigma^{\pm*}$	±1	-1	1.49	10^{-23} s
	Σ^{0*}	0	-1	1.49	10^{-23} s
Xi-star	Ξ^{0*}	0	-2	1.64	10^{-23} s
	Ξ^{-*}	-1	-2	1.64	10^{-23} s

Can you work out any patterns here? The baryons (in black) are in two groups according to their lifetime. There are 8 in the long-lived group, and they can be plotted in an 'octet' as in Figure 31.24. The position of each member of the octet is plotted according to its charge and strangeness.

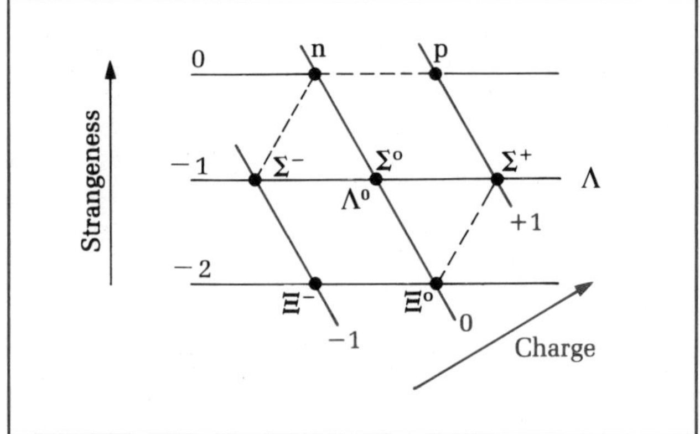

Figure 31.24 *The long-lived baryons*

The same can be done for the short-lived group, as in Figure 31.25. The pattern this time forms a large triangle, but in 1960 when the pattern was first drawn up, it was realized it was incomplete. To complete it, a particle with charge -1 and strangeness -3 should exist. Scientists called it the **omega minus Ω^-**. The race was on to hunt the Ω^-, and confirm its existence. Just as 19th century scientists predicted new elements from the Periodic Table when it was first drawn up, so here a new particle was predicted from the patterns. Its mass was predicted at 1.79 u, equivalent to about 1680 MeV. This was on the basis of the sequence of mass values 1.33, 1.49, 1.64, ... Sure enough, the Ω^- was duly discovered in 1963 from photographs of K^- particles smashing into protons.

The quark model was devised to explain the patterns. Murray Gell-Mann and George Zweig working independently realized that the patterns could be built up from combinations of three fundamental particles which they called 'quarks'. The three quarks were assumed to have different properties, as listed opposite (on page 487), and they were named as 'up', 'down' and 'strange'.

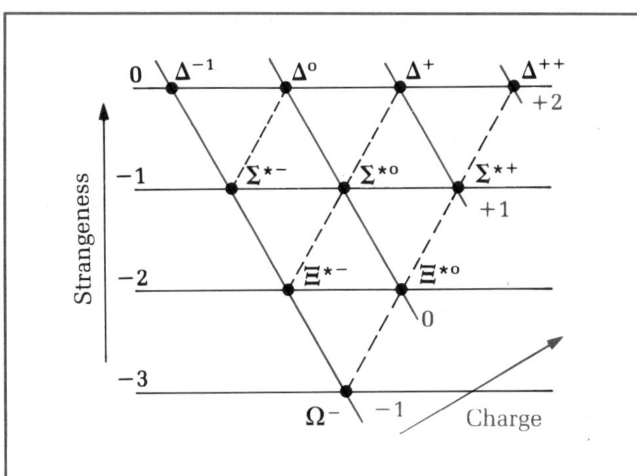

Figure 31.25 *The short-lived baryons*

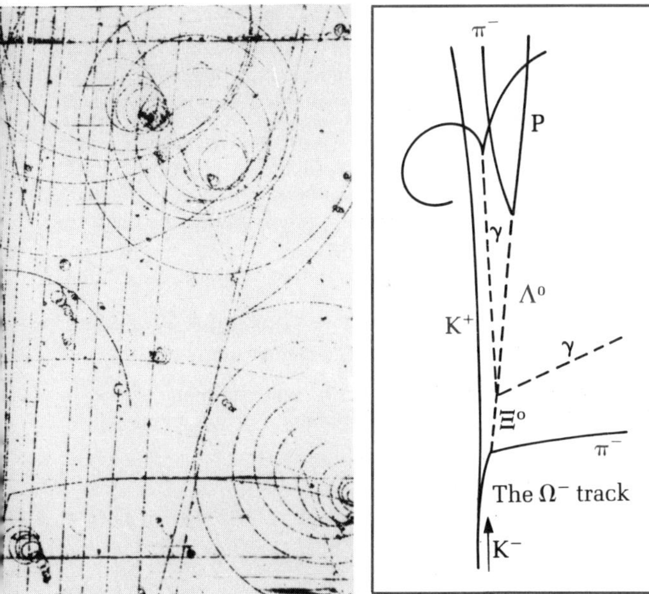

Figure 31.26 *The discovery of the* Ω^-

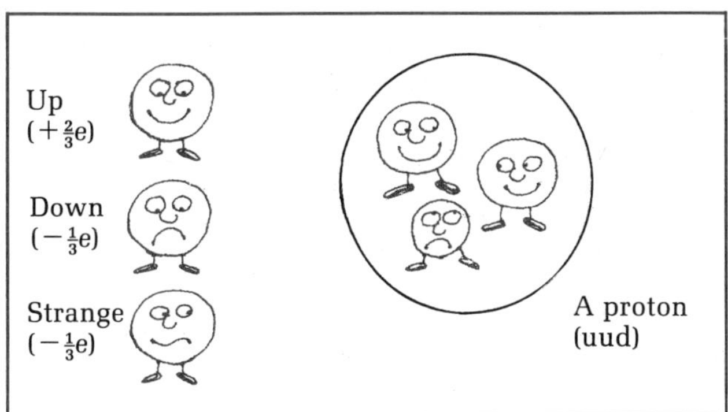

Figure 31.27 *Quark family*

Quark	Charge	Strangeness	Mass
up (u)	$+\frac{2}{3}e$	0	equal mass
down (d)	$-\frac{1}{3}e$	0	
strange (s)	$-\frac{1}{3}e$	−1	0.16 u more than an up or down quark

The following notes explain the patterns of Figures 31.24 and 31.25 and other similar patterns for mesons.

1 Each baryon is composed of three quarks. For example, a proton is made up of $u + u + d$ giving charge equal to $(2/3 + 2/3 - 1/3)e$ which is $+1e$. A neutron is made up of $u + d + d$ giving total charge equal to 0. The patterns of Figures 31.24 and 31.25 are explained by the various combinations of three quarks, as shown in Figures 31.28 and 31.29.

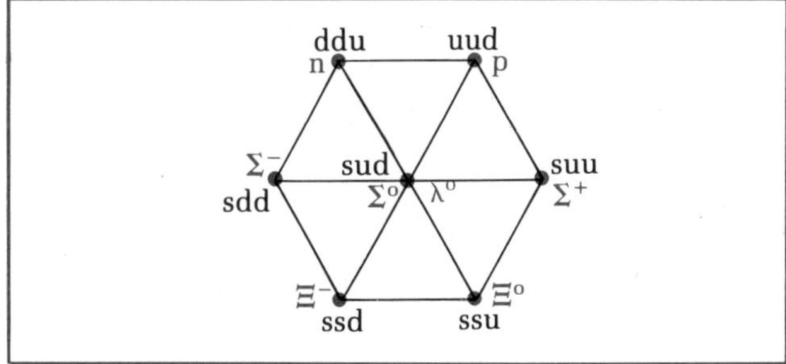

Figure 31.28 *The long-life pattern*

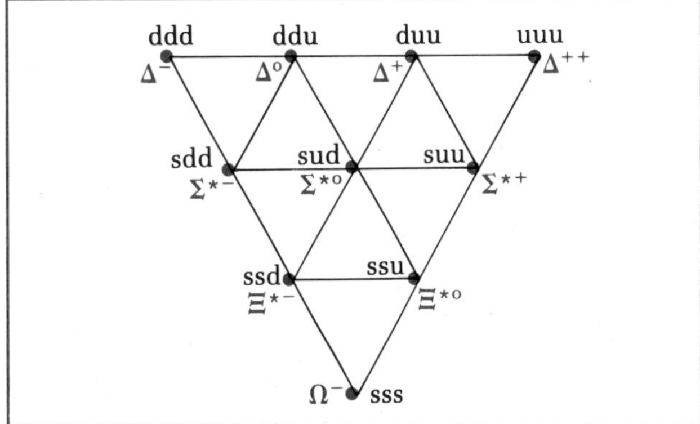

Figure 31.29 *The short-life pattern*

We have not considered why short-lived and long-lived combinations are possible. For example, the (*sdd*) combination can either be short or long-lived. The explanation involves another lengthy thread in the story so we won't follow it up here, except to say that it can be explained by the quark model.

2 Each meson is composed of two quarks. Again there are more threads to be followed up here to do with antiparticles (i.e. particles with equal mass but opposite charges). Each meson is thought to be made up of a quark and an antiquark. For example, an 'up' and an 'antidown' quark make up a π^+-meson.

Evidence for quarks in experiments was provided indirectly from accelerator experiments. High-energy electrons (≈ 20 GeV) fired at protons were found to be scattered occasionally through very large angles, not unlike Rutherford's α-scattering experiments years before, although on a much more energetic scale. So scientists deduced that protons are made up of point-like particles. Measurements gave results in agreement with calculations from the quark model. The success of the quark model is one of the triumphs of 20th century physics, but free isolated quarks have never been detected. The story has more threads to be followed, and it presents a continuing challenge. Who knows where it can lead?

31.3 Summary

Equations

1 De Broglie's equation momentum × wavelength $= h$
2 Crystal diffraction equation $2d \sin\theta = m\lambda$
3 Uncertainty principle $\Delta E \times \Delta t \approx h$
4 Cyclotron frequency $f = \dfrac{Bq}{2\pi m}$

Short questions

Electronic charge $e = 1.6 \times 10^{-19}$
Planck constant $h = 6.63 \times 10^{-34}$ J s
Speed of light in a vacuum $= 3 \times 10^{8}$ m s^{-1}

31.1 Radio waves and γ particles are forms of electromagnetic radiation with the same basic nature. Explain why we use the term 'wave' for one and 'particle' for the other.

31.2 A beam of electrons produced by an electron gun operating at 3.0 kV is directed at a thin crystal. A fluorescent screen on the other side of the crystal to the electron gun shows a pattern of rings due to diffraction of the electrons by the crystal. Measurement shows that the rings are produced by electrons diffracted through 11° and 5°, as in the diagram. Calculate
a) the speed of the electrons from the gun,
b) the de Broglie wavelength of the electrons from the gun,
c) the spacing between the layers of atoms in the crystal responsible for the rings.
(The rest mass of the electron $= 9.1 \times 10^{-31}$ kg.)

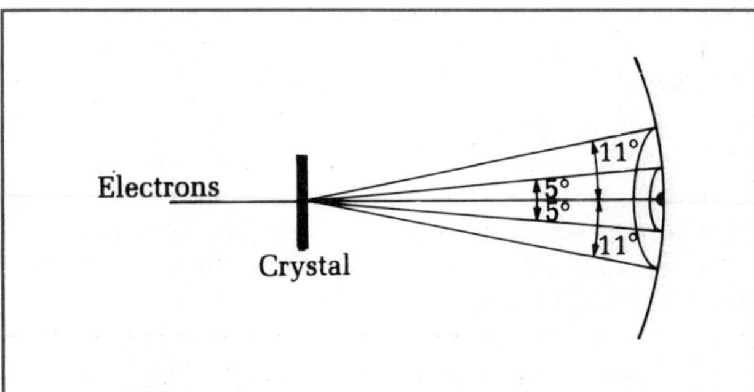

Figure 31.30

31.3 Consider a lattice of regularly arranged atoms which has an atom missing at one position. Assume the vacancy creates a potential well in the lattice that can trap an electron. Suppose the well is a 'square well' with width 0.1 nm and depth 12×10^{-18} J. Calculate
a) the maximum wavelength for an electron trapped in the well,
b) the minimum K.E. of a trapped electron,
c) the energy required to remove a trapped electron from the lowest energy level in the well.

31.4 A narrow beam of 5 MeV α-particles is directed at a slit of width 0.2 mm in a metal plate. Calculate
a) the momentum and wavelength of the α-particles before the slit,
b) the angular width of the central diffraction peak formed by α-particles being diffracted at the slit.

31.5
a) No particle can travel faster than the speed of light. Estimate the shortest time that a particle takes to cross the nucleus.
b) Use your answer to **a)** to estimate the maximum energy that a particle in the nucleus can 'borrow' to escape.

31.6 When a proton inside the nucleus suddenly turns into a neutron, a β-particle is created and emitted instantly from the nucleus.
a) Estimate the K.E. of a β-particle confined to a nucleus of size 10^{-14} m.
b) Why does the β-particle leave the nucleus when a β^- has opposite charge to the nucleus?

31.7 Calculate the de Broglie wavelength of 2 MeV protons, given the mass of a proton is 1.672×10^{-27} kg. Neglect relativistic effects.

31.8
a) Explain the operation of a cyclotron used to accelerate protons.
b) A cyclotron is designed with a magnetic field strength of 500 mT. Calculate the frequency of the alternating p.d. that should be applied to accelerate protons. The specific charge of the proton is 9.58×10^{7} C kg^{-1}.

31.9 Part of Figure 31.26 shows the tracks of a proton and a π^--meson produced by the disintegration of a Λ^0-meson. A uniform magnetic field applied to the plane of the diagram made the tracks curve as shown.
a) Was the magnetic field direction in or out of the plane of the diagram?
b) From the diagram, estimate the ratio of the proton's momentum to the π^--meson's momentum.

31.10 In terms of quarks, describe what happens when
a) a proton in the nucleus turns into a neutron and emits a β-particle,
b) a neutron collides with a neutral meson to produce a K^+-meson and a Σ^--meson.

F

Multiple choice questions

FM.1 A beam of electrons travelling in a horizontal direction is deflected downwards by an electrostatic field. The beam could be returned to its original direction by applying
A an electrostatic field at right angles to the directions of both the path of the electrons and the original electrostatic field
B a magnetic field directed vertically upwards
C a magnetic field in the same direction as the electrostatic field
D a magnetic field at right angles to the directions of both the electrostatic field and the path of the electrons
E a magnetic field in the same direction as that of the original beam

(AEB June 84)

FM.2 A thermionic diode operates at 100 V and the current through the diode is 20 mA. If the electron charge is -1.6×10^{-19} C the number of electrons striking the anode each second is
A 1.25×10^{17} **B** 1.60×10^{17} **C** 6.25×10^{18}
D 1.25×10^{20} **E** 1.60×10^{21}

(N. Ireland)

FM.3 In a Millikan's experiment an oil drop is held stationary in the space between the two metal plates. Which change would cause the drop to start moving downwards?

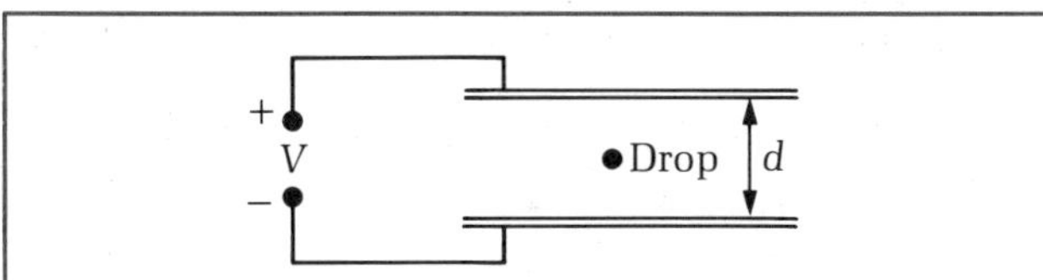

Figure FM1

A loss of oil from the drop by evaporation
B loss of negative charge from the drop
C a rise in the temperature of the drop causing it to expand
D an increase in the magnitude of the voltage, V
E a decrease in the separation, d, of the plates

(Scottish H.)

FM.4 An electron of mass m travelling with a speed u collides with an atom and its speed is reduced to v. The speed of the atom is unaltered, but one of its electrons is excited to a higher energy level and then returns to its original state, emitting a photon of radiation. If h is the Planck constant, the frequency of the radiation is

A $\frac{m(u^2 - v^2)}{2h}$ **B** $\frac{m(v^2 - u^2)}{2h}$ **C** $\frac{m(v^2 + u^2)}{2h}$

D $\frac{mu^2}{2h}$ **E** $\frac{mv^2}{2h}$

(London)

FM.5 In an experiment to investigate the maximum kinetic energy of photoelectrons emitted from a surface illuminated with monochromatic light, the stopping potential (the potential required to prevent any photoelectrons reaching the collecting electrode) was plotted against the frequency of the light.

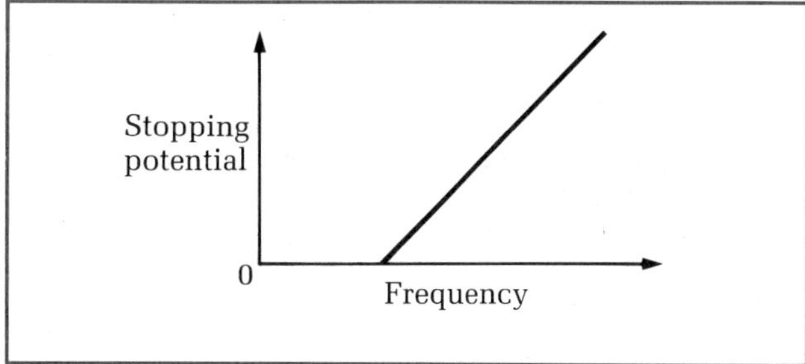

Figure FM2

The graph was a straight line, as shown in the figure above. The gradient of this graph is equal to
A the work function of the surface
B the work function of the surface multiplied by the electron charge
C the Planck constant multiplied by the electron charge
D the Planck constant
E the Planck constant divided by the electron charge

(N. Ireland)

FM.6 A radioactive nuclide Q, which has N neutrons and P protons in its nucleus, decays by a series of emissions to a stable element R. If two α-particles and one β-particle are emitted during the transition from Q to R the element R may be represented by
A ${}^{N-4}_{P-4}\text{R}$ **B** ${}^{N-4}_{P-3}\text{R}$ **C** ${}^{P+N-4}_{P-3}\text{R}$
D ${}^{P+N-8}_{P-3}\text{R}$ **E** ${}^{P+N-8}_{P+1}\text{R}$

(AEB June 84)

FM.7 A radioactive decay series, starting with thorium ${}^{232}_{90}\text{Th}$, involves the emission, in turn, of the following: alpha, beta, beta, gamma, alpha.
What is the final product of this series?
A ${}^{224}_{88}\text{Ra}$ **B** ${}^{230}_{82}\text{Pb}$ **C** ${}^{226}_{86}\text{Rn}$ **D** ${}^{227}_{85}\text{At}$ **E** ${}^{225}_{87}\text{Fr}$

(Scottish H.)

FM.8 An isotope has a half-life of 8 years. After a period of 24 years, the fraction of the isotope remaining is
A 2/3 **B** 1/3 **C** 1/4 **D** 1/8 **E** zero

(AEB Nov. 84)

FM.9 The diagram shows a G.M. tube which is connected to a counter. A radioactive source S, which can be assumed to radiate equally in all directions, is located on the axis of the tube at a distance d in front of the end window of the tube.

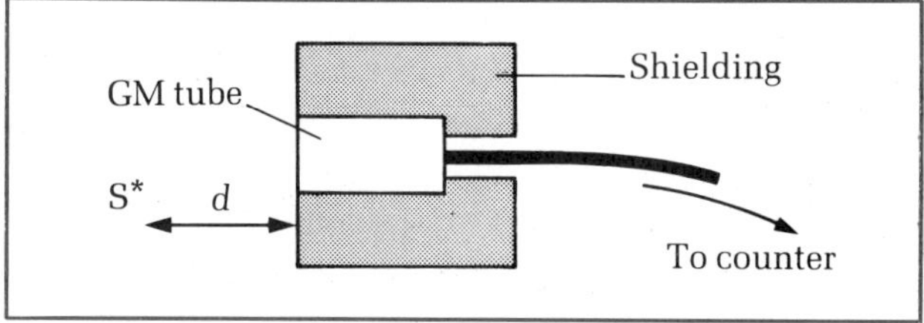

Figure FM3

The source S contains two radioactive isotopes, with half lives 1 h and 2 h respectively, both of which have stable decay products. If initially the count rate is 320 s^{-1}, to which the two isotopes contribute equally, 4 h later it will be
A 40 s^{-1} **B** 50 s^{-1} **C** 80 s^{-1} **D** 100 s^{-1} **E** 120 s^{-1}

(London)

FM.10 A stationary thoron nucleus ($Z = 90$, $A = 220$) emits an α-particle when it disintegrates. The disintegration energy E is shared between the α-particle and the recoiling

nucleus. What fraction of the disintegration energy is taken by the α-particle?
A 4/216 **B** 4/220 **C** 4/110 **D** 216/220 **E** 108/220

FM.11 Which type of electromagnetic radiation is emitted when an electron in an atom makes a transition from an energy level at -1.5 eV to an energy level at -3.5 eV? ($1\,\text{eV} = 1.6 \times 10^{-19}$ J, $hc = 2 \times 10^{-25}$ J m.)
A microwave **B** infra-red **C** visible **D** ultra-violet
E γ-radiation

FM.12 As a result of Geiger and Marsden's alpha-particle scattering experiments, Rutherford proposed that an atom had
A a plum-pudding type structure
B its mass evenly distributed
C a nucleus of small diameter and high density
D a nucleus of large diameter and low density
E a nucleus containing all the charged particles

(Scottish H.)

FM.13 When a tungsten target is bombarded with high energy electrons it becomes hot because
A X-rays are emitted
B a metal target is used
C the electrons are very penetrating
D kinetic energy of the electrons is absorbed on impact
E the moving electrons constitute an electric current

(AEB Nov. 84)

FM.14
The graph shows the spectrum of X-rays emitted from an X-ray tube. Which of the following statements is (are) correct?

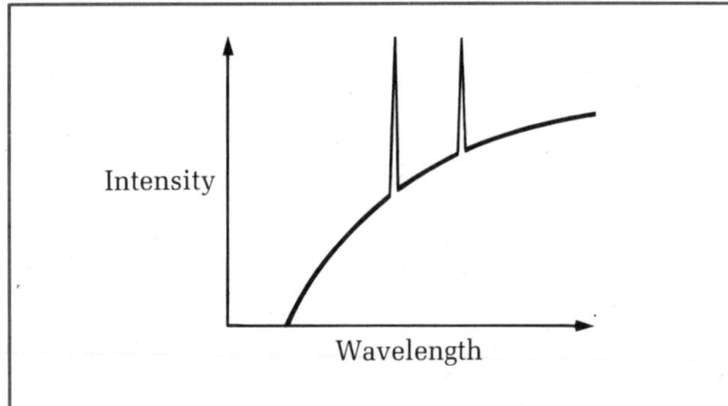

Figure FM4

1 The wavelengths at which the peaks appear are independent of the voltage across the tube.
2 The minimum (cut-off) wavelength is independent of the atomic number of the target in the X-ray tube.
3 The continuous part of the spectrum is due to the very high temperature attained by the target in the X-ray tube.
A 1 only **B** 2 only **C** 3 only **D** 1 and 2 only
E 2 and 3 only

(N. Ireland)

FM.15 Which alternative gives the correct expression for the minimum wavelength of the X-rays from an X-ray tube operating at potential difference V? (h = the Planck constant, c = the speed of light, e = electron charge.)
A eh/cV **B** hc/eV **C** eV/hc **D** ec/hV **E** cV/eh

FM.16 One possible fission reaction of $^{235}_{92}U$ following the capture of a slow neutron is shown by the equation

$$^{235}_{92}U + ^{1}_{0}n \rightarrow ^{236}_{92}U \rightarrow ^{98}_{40}Zr + ^{136}_{52}Te + 2^{1}_{0}n + 206\,\text{MeV}$$

In this equation
1 206 MeV is the binding energy of the $^{236}_{92}U$ nucleus
2 the combined rest mass of the fission products equals the rest mass of the $^{236}_{92}U$ atom.
3 the nucleon number is unchanged by the fission process
A 1, 2, 3 correct **B** 1, 2 only **C** 2, 3 only
D 1 only **E** 3 only

(London)

FM.17 When a $^{2}_{1}H$ nucleus is fused with a $^{3}_{1}H$ nucleus in a fusion reactor, a neutron is ejected. The product nucleus formed after ejecting the neutron has mass number A and proton number Z. Which alternative gives the correct values of A and Z?

	A	B	C	D	E
A	5	5	4	4	3
Z	1	2	1	2	2

FM.18 A particle of charge q and mass m is accelerated from rest through a potential difference V. If h is the Planck constant, the de Broglie wavelength associated with the particle is

A h/mV **B** mV/h **C** $\dfrac{h}{\sqrt{2qmV}}$

D $\dfrac{\sqrt{2qmV}}{h}$ **E** $\dfrac{h^2}{2qmV}$

(N. Ireland)

FM.19 Which of the following are explained by wave-like ideas?
1 Using an electron beam in an electron microscope to form images of very small objects
2 Using γ-radiation to knock electrons out of atoms
3 Deflecting an electron beam by using a magnetic field
A 1, 2, 3 correct **B** 1, 2 only **C** 2, 3 only
D 1 only **E** 3 only

FM.20 Ions of several isotopes are accelerated through the same potential difference V in an 'ion accelerator' then they are directed in a beam into a uniform magnetic field. The initial direction of the beam is at right angles to the field so the ions move on circular paths in the magnetic field. Which ion has the least curved path?
A $^{16}_{8}O^{+}$ **B** $^{16}_{8}O^{2+}$ **C** $^{14}_{7}N^{2+}$ **D** $^{6}_{3}Li^{+}$ **E** $^{12}_{6}C^{+}$

F

Long questions

FL.1

a) Electrons can be obtained from a cathode by thermionic emission.

i) Outline the process of thermionic emission from a suitable cathode material.

ii) Explain the difference between a *directly-heated* and an *indirectly-heated* cathode, and state the main advantage of an indirectly-heated cathode in a thermionic tube.

b) An electron of specific charge 1.76×10^{11} C kg^{-1} is initially at rest. Calculate the final velocity of the electron if it is subjected to a uniform electric field of strength 5000 V m^{-1} over a distance of 10 cm in vacuum.

c) Millikan carried out an experiment to measure the charge of a single electron. Describe the apparatus used in either the original experiment or in an equivalent school laboratory version, and draw a labelled diagram to illustrate your answer. Explain the principle of the experiment. Describe the procedures by which results were taken, including a qualitative account of how the radius of the oil-drop was determined.

d) A small oil droplet is observed between two horizontal metal plates which are 10 mm apart in air. This air is exposed to an ionizing radiation. The droplet is stationary when the voltage across the plates is 600 V. Suddenly, the drop moves; to restore equilibrium the voltage must be changed to 1200 V, and later again to 400 V and 300 V. Over a period of time, the maximum voltage which needs to be applied to hold the drop stationary is found to be 1200 V.

i) Explain what was happening to require the various changes in voltage mentioned.

ii) If the charge on an electron is -1.6×10^{-19} C, use the information above to obtain an estimate for the mass of the droplet.
(Take the acceleration of free fall as 10 m s^{-2}, and assume that the upthrust of the air is negligible.)

iii) If the experiment continues for a considerable time it is found that the voltages required to hold the droplet stationary are very slightly less than the values originally recorded. What is the most likely explanation for this? *(N. Ireland)*

FL.2 If a clean zinc sheet stands on the plate of a negatively-charged gold-leaf electroscope, the leaf will fall if ultra-violet radiation is directed on to the zinc. Name and explain the phenomenon involved. Why does it not happen with a beam of light from an ordinary electric light bulb, however powerful the beam?

Discuss whether or not it can happen with ultra-violet radiation falling on a positively-charged clean zinc plate.

Calculate a minimum value for the power of a beam of ultra-violet radiation of wavelength 200 nm if it takes 3.0 s to discharge the electroscope initially carrying a charge of -3.2 μC. Why does your calculation give only a minimum value for the power?

In a repeat of Millikan's famous experiment to measure the electronic charge, an oil-drop carrying a charge of 11.2×10^{-19} C was held at rest by the vertical electric field between two metal plates 4.0 cm apart when the potential difference between them was 1200 V. What was the weight of this drop? What would be its *initial* acceleration if the electric field were reversed in direction?

A student repeating this experiment, but using an incorrectly-calibrated voltmeter, obtained the following values for charges on drops:

2.8, 5.6, 4.2, 2.8, 5.6, 7.0 and 8.4, each $\times 10^{-19}$ C.

Using these results what value ought he to take for the electronic charge? What is the percentage error of his voltmeter readings, assuming it to be constant at all points on the scale? *(SUJB)*

FL.3

a) Explain how a beam of electrons may be produced in a vacuum tube, and describe an arrangement by which the beam may be deflected by a magnetic field. Draw a diagram showing clearly the directions of the beam, the field and the deflection.

b) An electron moving at velocity v passes simultaneously through a magnetic field of uniform flux density B and an electric field of uniform intensity E. It emerges with direction and speed unaltered.

i) Explain how the fields are arranged to achieve this result.

ii) Derive the relationship between v, B and E.

c) An electron is travelling at 2.0×10^6 m s^{-1} at right angles to a magnetic field of flux density 1.2×10^{-5} T; its path is a circle. Uniform circular motion of an electron is accompanied by the emission of electromagnetic radiation of the same frequency as that of the circular motion.
[Take the specific charge of the electron e/m_e to be 1.8×10^{11} C kg^{-1}, and the speed of electromagnetic waves in air c to be 3.0×10^8 m s^{-1}.]

i) Explain why the path of the electron is a circle.

ii) Calculate the radius of the circle.

iii) Calculate the frequency of the circular motion of the electron.

iv) Calculate the wavelength of the electromagnetic radiation emitted and identify in which part of the electromagnetic spectrum this radiation lies.

v) How would this wavelength be affected by a decrease in the speed of the electron? *(Oxford)*

FL.4

a) Write down the Einstein photoelectric equation and explain how it accounts for the emission of electrons from metal surfaces illuminated by light.

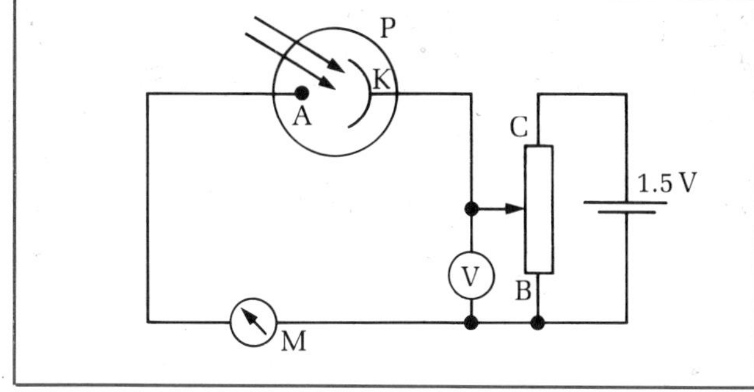

Figure FL1

b) P is a vacuum photocell with anode, A, and cathode, K, made from the same metal of work function 2.0 eV. The cathode is illuminated by monochromatic light of constant intensity and of wavelength 4.4×10^{-7} m. Describe and explain in general terms how the current shown by the microammeter, M, will vary as the slider of the potential divider is moved from B to C. What will be the reading of the high-resistance voltmeter, V, when photoelectric emission just ceases? Assume that no secondary emission of electrons occurs from the anode.

c) With the slider set midway between B and C, describe and explain how the reading of M would change if, in turn

i) the intensity of the light was increased,

ii) the wavelength of the light was changed to 5.5×10^{-7} m.

d) Outline *briefly* how you could use the arrangement described in **b)**, together with an additional light source of a different wavelength, to determine a value for the Planck constant assuming that the value of the work function was unknown.

(Charge of the electron $= 1.6 \times 10^{-19}$ C
Planck constant $= 6.6 \times 10^{-34}$ J s
speed of light in vacuo $= 3.0 \times 10^{8}$ m s^{-1}.) *(JMB)*

FL.5 Describe how you could detect and distinguish between beta-radiation and gamma-radiation from a radioactive source emitting both radiations.

A source of radioactive potassium is known to contain two isotopes, $^{42}_{19}$K and $^{44}_{19}$K, both of which decay by emission of beta-radiation to stable isotopes of calcium. Write a nuclear transformation equation for one of these decays.

The source is placed in front of a beta-radiation counter and the following count rates, corrected for background, are recorded.

Time/hours	0	0.5	1.0	1.5	2.0	2.5	3.0
Count rate/min^{-1}	10 000	3980	2125	1260	955	890	832

Time/hours	4.0	5.0	6.0	7.0	8.0	9.0	10.0
Count rate/min^{-1}	790	750	710	670	630	600	575

Plot the data on a graph of lg (count rate/min^{-1}) against time/hours.

From your graph estimate values for

a) the half life of $^{42}_{19}$K which is the longer lived isotope,

b) the half life of $^{44}_{19}$K which is the shorter lived isotope,

c) the initial count rates due to $^{42}_{19}$K and $^{44}_{19}$K,

d) the radio of the amounts of $^{42}_{19}$K and $^{44}_{19}$K present in the source at the start of the measurements.

(Oxford and Cambridge)

FL.6

a) i) Explain the nature of radioactivity and how it accounts for the observed exponential decay of the activity of a radioactive sample.

ii) Describe an experiment to measure the half-life of an element for which the value is known to be about two minutes.

b) In order to determine the volume of a patient's blood, 2 ml of a solution containing a radioactive sodium isotope of half-life 15 hours is injected into his bloodstream. The activity of the solution injected is 3000 counts per minute, and the activity of a blood sample of volume 12 ml taken 3 hours later is found to be 6 counts per minute.

i) Calculate the volume of the patient's blood given by these results.

ii) Discuss any assumptions made in the determination and suggest how they might be checked.

(Oxford)

FL.7

a) With the aid of a labelled sketch, describe a mass spectrometer. Explain how a mass spectrometer is used to separate isotopes of the same element.

b) Part of the radioactive series which starts with $^{238}_{92}$U is given below.

Isotope half-life:

$^{238}_{92}$U ⟶ $^{234}_{92}$U ⟶ $^{230}_{90}$Th ⟶
4.5 × 10^{9} years 2.5 × 10^{5} years 7.5 × 10^{4} years

$^{226}_{88}$Ra ⟶ $^{222}_{86}$Rn ⟶ $^{218}_{84}$Po
1600 years 3.8 days 3.0 min

By writing relevant equations, explain the transformation from $^{238}_{92}$U to $^{234}_{92}$U.

c) A sample of ore contains all the nuclides in the series in radioactive equilibrium, such that in a given time as many nuclei of a product nuclide are produced as are lost by decay, and the rates of decay at all stages are equal. The sample contains 5.0 kg of $^{238}_{92}$U.
[Take the Avogadro constant N_A to be 6.0×10^{23} mol^{-1} and the molar mass of $^{238}_{92}$U to be 0.238 kg. The radioactive decay constant $\lambda = \ln 2$/(half-life). 1 year $= 3.2 \times 10^{7}$ s.]
Calculate

i) the number of $^{238}_{92}$U atoms in the sample,

ii) the rate of decay of the $^{238}_{92}$U atoms,

iii) the rate of decay of the $^{226}_{88}$Ra atoms,

iv) the number of atoms of $^{222}_{86}$Rn in the sample.

d) A pure sample of $^{222}_{86}$Rn containing N_0 atoms is placed in a sealed container.

i) Show by means of a graph how the number of atoms of $^{222}_{86}$Rn will vary during the next 20 days.

ii) Calculate the number N of atoms of $^{222}_{86}$Rn remaining at the end of that time, given that $N_0 = 10^{8}$ atoms.

(Oxford)

FL.8 This question is about explaining ideas in physics.

Choose *one* of the three subjects **a)**, **b)** or **c)** below. For the one that you choose, you should give a careful explanation of each of the three topics listed. Your explanation should be suitable for a friend studying A-level physics who missed the teaching of this particular subject.
Show also how your explanations could help your friend to understand *one* everyday application of the subject.

Subjects:

a) *Electronics*	Topics: block diagrams, logic, feedback.
b) *Thermodynamics*	Topics: number of ways, temperature, entropy.
c) *Radioactivity*	Topics: decay, radiation, isotopes.

(O and C Nuffield)

FL.9

a) A G-M tube is exposed to a constant flux of alpha particles. The graph at the top of the next page shows how the recorded count rate depends on the potential difference across the tube.
Draw and label a diagram of a G-M tube. Outline its working principle with reference to what happens when an alpha particle enters the tube. Explain why there is an upper limit to the rate at which a G-M tube can detect α-particles.
How do you account for

i) the sharp rise in the recorded count rate at A,

ii) the 'plateau' at B, and

iii) the uncontrolled rise in the recorded count rate at C?

State what potential difference you would choose for the Geiger counter whose response is shown in the graph. Give one good reason for your choice.

b) A small amount of ^{24}Na is smeared on to a card and its activity falls by 87.5% in 45 h. What is the half-life of ^{24}Na? Describe how you would use a G-M tube in conjunction with a suitable counter to measure the half-life of ^{24}Na.

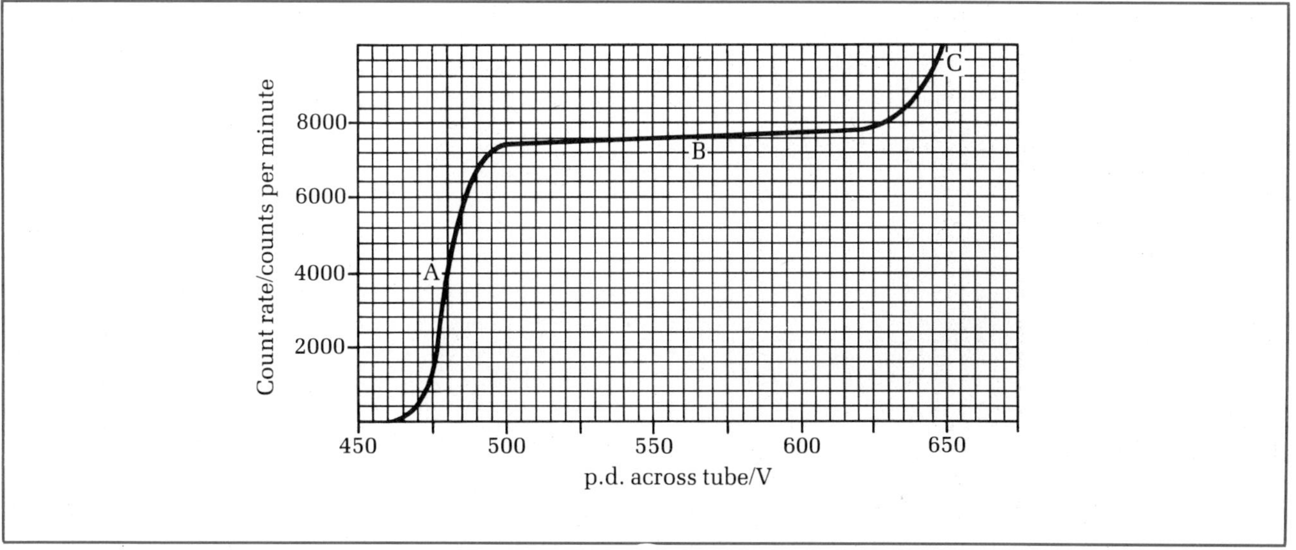

Figure FL2

Explain carefully how the result is found from the measurements.

(Use the fact that decay constant = 0.693/half-life.)

(London)

FL.10

a) The stability of a nucleus of an atom depends on the relative number of neutrons (N) and protons (Z) in the nucleus. Draw a clearly-labelled diagram involving a plot of N against Z (or Z against N) showing
   i) the region in which nuclei are stable,
   ii) the region in which the unstable nuclei decay by β^-emission.

b) If a radioactive source contains n_0 unstable nuclei of one type of radioisotope at a given instant ($t = 0$) then the number n remaining after time t is given by

$$n = n_0 \exp(-\lambda t)$$

where λ is the decay constant.

What is meant by the *half-life* $T_{\frac{1}{2}}$ of a radioisotope? Show that the half-life and the decay constant are related by

$$\lambda T_{\frac{1}{2}} = 0.693.$$

c) A thin radioactive foil which is emitting β-rays has an activity which is decreasing with a half-life of approximately one hour.
   i) Describe how you would make an accurate measurement of the half-life, giving an account of the apparatus, experimental procedure and analysis of data. State *briefly* what factor(s) limit the accuracy of the measurement.
   ii) The half-life of the foil activity in such an experiment is found to be 3240 s (0.90 hour). When the experiment is complete, the foil is found to have a residual activity of 6.4×10^5 Bq (1 Bq = 1 disintegration per second), but regulations forbid the disposal of the source with domestic waste until the activity has decreased to 3.7×10^4 Bq or less. Find the minimum time which must elapse before disposal of the foil.

d) The activity of a radioisotope in a source is equal to the product of the decay constant and the number of nuclei of the radioisotope in the source.

A certain source has an activity of 5.00×10^6 Bq due to the decay of the radioisotope ^{14}C, which has a half-life of 1.81×10^{11} s.

   i) How many atoms of ^{14}C are contained in the source?
   ii) Calculate the mass of ^{14}C in the source.
   iii) What is the specific activity (the activity per kilogram) of ^{14}C?

(Avogadro constant $N_A = 6.02 \times 10^{23}$ mol^{-1}; relative atomic mass of $^{14}C = 14.00$.)

(N. Ireland)

FL.11 The diagram shows the results of a Franck-Hertz experiment. Draw a circuit diagram of the apparatus which gives these results.

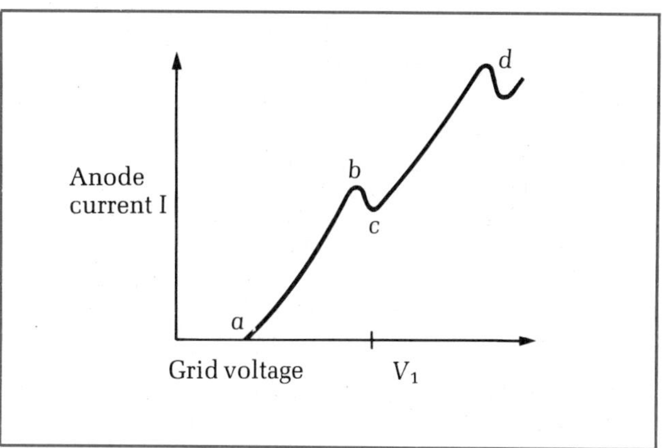

Figure FL3

The current I varies with V_1 as shown in the diagram, where I is the current collected at the anode and V_1 is the potential difference between grid and cathode.
Account for the shape of the curve, paying particular regard to
   i) the value of V_1 at a,
   ii) the increase ab in I,
   iii) the peak at b leading to the decrease bc,
   iv) the second peak at d.

When the experiment is done on sodium vapour the difference in V_1 between b and d is 2.10 V. Calculate one wavelength which will occur in the spectrum of hot sodium vapour.
(Planck constant $h = 6.6 \times 10^{-34}$ J s,
speed of light $= 3.00 \times 10^8$ m s^{-1},
charge on the electron $= -1.60 \times 10^{-19}$ C.)

(WJEC)

FL.12 Describe briefly an experiment to observe the most intense lines of the *absorption* spectrum of sodium. Why are some lines which are seen in the *emission* spectrum not seen in the absorption spectrum?

Three of the principal spectral lines in the sodium spectrum are at wavelengths 589 nm, 330 nm and 285 nm. Each of these lines in emission is caused when an excited atom returns to its ground state. Draw a simple energy level diagram of the four necessary energy levels for the transitions, showing the energy values. Give the lowest energy level of the value zero.

The *ionization energy* of a sodium atom is 5.12 eV. The *work function* of a clean sodium metal surface is 2.28 eV. Give the meaning of the two terms in italics. Add another line to your sodium energy level diagram to indicate the ionization energy.

An evacuated photocell has electrodes coated with sodium metal. Light from a sodium lamp is incident on one of the electrodes. Which of the three spectral lines given above will cause the emission of photoelectrons? Calculate the minimum voltage required between the electrodes to ensure that no electrons, caused by light at these wavelengths, can pass between them.

(1.00 eV $=1.60\times10^{-19}$ J $c=3.00\times10^{8}$ m s^{-1}
$h=6.63\times10^{-34}$ J s $e=1.60\times10^{-19}$ C.)

(Oxford and Cambridge)

FL.13

a) When atoms absorb energy by colliding with moving electrons, light or X-radiation may subsequently be emitted. For each type of radiation, state typical values of the energy per atom which must be absorbed and explain in atomic terms how each type of radiation is emitted.

b) State *one* similarity and *two* differences between optical atomic emission spectra and X-ray emission spectra produced in this way.

c) Electrons are accelerated from rest through a potential difference of 10 000 V in an X-ray tube. Calculate
- i) the resultant energy of the electrons in eV;
- ii) the wavelength of the associated electron waves;
- iii) the maximum energy and the minimum wavelength of the X-radiation generated.

Charge of electron $=1.60\times10^{-19}$ C
mass of electron $=9.11\times10^{-31}$ kg
Planck's constant $=6.62\times10^{-34}$ J s
speed of electromagnetic radiation in vacuo $=3.00\times10^{8}$ m s^{-1}

(JMB)

FL.14

a) A sodium vapour lamp is used in conjunction with a spectrometer and a diffraction grating so that an emission spectrum is seen.
- i) Explain, in atomic terms, how light is produced in the sodium lamp, when a current passes through the vapour.
- ii) Describe the appearance of this spectrum.

b) Light from a bright white source is now focused on the slit of the spectrometer. A transparent vessel containing sodium vapour is placed between the light source and the slit so that an absorption spectrum is seen. Describe the appearance of this spectrum and explain, in atomic terms, how it is formed.

c) The three lowest energy levels A, B, and C, of a certain atom have energies of -13.58 eV, -3.39 eV and -1.51 eV respectively.
- i) Identify the transitions between these levels corresponding to the shortest and the longest wavelengths of emitted radiation.
- ii) Calculate the wavelengths of each of these two radiations and state in which region of the electromagnetic spectrum each would be found.
- iii) An atom with an electron in level A absorbs 2.40×10^{-18} J of energy. Would this be sufficient to ionize the atom? Justify your answer.

(The Planck constant $=6.63\times10^{-34}$ J s;
speed of light in vacuo $=3.00\times10^{8}$ m s^{-1};
1 electronvolt $=1.60\times10^{-19}$ J.)

(JMB)

FL.15 What is meant by fusion and fission? Give one specific example of each kind of reaction explaining how each is induced.
Describe how the fission process is made use of in the production of energy, and produce a schematic diagram showing the main essentials of a nuclear fission reactor. What factors make fusion, in principle, a more desirable means for producing energy than fission?

(WJEC)

FL.16 This question is about the production of energy by the fusion process.
Stars, of which the Sun is a typical example, generate energy for most of their lives by combining protons (hydrogen nuclei) in a complex process out of which nuclei of helium are formed. The energy released keeps the star hot enough for the fusion process to work, but ultimately it escapes from the surface, causing the star to shine. If we know the size of the star and the temperature of its surface, we can work out roughly how much power it is *generating* by calculating the power *radiated* from its surface.

a) The power P radiated from a star of surface area A with a surface temperature T is given by

$$P=\sigma AT^4$$

where σ is called the Stefan-Boltzmann constant. Use the data given at the end of the question to calculate the power *radiated* from the Sun.

b) What assumptions are made in treating the answer to (a) as an estimate of the power *generated* by the Sun? Show that the power generated per unit mass of the Sun is approximately 2×10^{-4} W kg^{-1}. Given that the Sun is not cooling down, explain whether this value is an upper or a lower limit.

c) When four protons combine to form one helium nucleus, 4×10^{-12} J are liberated. How many protons are combining each second in unit mass of the Sun?

d) Assuming the Sun now consists of 75% hydrogen and 25% helium by mass, how many hydrogen nuclei per unit mass are there in the Sun at present?
If power generated has been the same throughout the Sun's life, and if it began life as a star with 100% hydrogen, how old is the Sun?

e)
- i) If the proton is regarded as a charged sphere of radius 1.5×10^{-15} m, and two protons cannot fuse until they 'touch', how much energy is needed to bring them together?
- ii) Assuming this energy is equal to the average thermal energy of motion of hydrogen atoms in a gas, and using the relationship

  Average thermal energy of an atom $\sim kT$

  where k is the Boltzmann constant, and T is the absolute temperature of the gas, show that T is approximately 5×10^{9} K.
- iii) The internal temperature of the Sun is, in fact, believed to be about 10^{7} K. Explain why, in spite of this, a few of the atoms have the energy required for the nuclei to make contact on collision.

f) Imagine that the energy needed to bring protons together were to be lowered suddenly by a factor of 10. Discuss, in as much detail as possible, the effects of this change on the subsequent behaviour of the Sun.

Data:

Stefan-Boltzmann constant σ	$\approx 6 \times 10^{-8}\,\mathrm{W\,m^{-2}\,K^{-4}}$
Radius of Sun	$\approx 7 \times 10^{8}\,\mathrm{m}$
Mass of Sun	$\approx 2 \times 10^{30}\,\mathrm{kg}$
Temperature of Sun's surface	$\approx 6 \times 10^{3}\,\mathrm{K}$
Mass of proton	$\approx 1.7 \times 10^{-27}\,\mathrm{kg}$
Charge of proton	$\approx 1.6 \times 10^{-19}\,\mathrm{C}$
One year	$\approx 3 \times 10^{7}\,\mathrm{s}$
Boltzmann constant k	$\approx 1.4 \times 10^{-23}\,\mathrm{J\,K^{-1}}$
$\dfrac{1}{4\pi\varepsilon_0}$	$\approx 9 \times 10^{9}\,\mathrm{N\,m^2\,C^{-2}}$

(O and C Nuffield)

FL.17 Given

$${}^{235}_{92}\mathrm{U} + {}^{1}_{0}\mathrm{n} \rightarrow {}^{x}_{45}\mathrm{Rh} + {}^{113}_{y}\mathrm{Ag} + 2{}^{1}_{0}\mathrm{n}$$

and

$${}^{2}_{1}\mathrm{H} + {}^{3}_{1}\mathrm{H} \rightarrow {}^{4}_{2}\mathrm{He} + A$$

i) explain what is meant by the 235 and 92 in ${}^{235}_{92}\mathrm{U}$;
ii) determine x, y, and A;
iii) describe the importance of the reactions;
iv) write down a similar equation for the fusion of two atoms of deuterium to form helium of atomic mass number 3.

Given the mass of the deuterium nucleus is 2.015 u, that of one of the isotopes of helium is 3.017 u and that of the neutron is 1.009 u, calculate the energy released by the fusion of 1 kg of deuterium. If 50% of this energy were used to produce 1 MW of electricity continuously, for how many days would the station be able to function?
(Speed of light $c = 3.00 \times 10^{8}\,\mathrm{m\,s^{-1}}$.)

(WJEC)

FL.18

a) In a nuclear reactor the following processes occur: nuclear fission; controlled chain reactions.
Explain what is meant by each process.
It may be helpful to draw diagrams to illustrate your answers.

b) Describe how the energy released in the core of a nuclear reactor is transferred and converted to mechanical energy.

c) Each fuel element in an Advanced Gas-cooled Reactor (AGR) contains 36 fuel pins, each made up from uranium dioxide-fuel pellets stacked inside stainless steel cans.
i) State *two* purposes of these fuel cans.
ii) Suggest *two* important reasons why stainless steel has been chosen as a suitable material for the fuel cans in an AGR.
iii) Describe briefly the sequence of operations to recover unused uranium from a fuel element which has exhausted its useful life in the reactor core.

d) When one $^{235}\mathrm{U}$ nucleus undergoes fission, $3.2 \times 10^{-11}\,\mathrm{J}$ of energy are released. Assuming the uranium dioxide pellets used in the fuel of an AGR contain 5.9×10^{22} nuclei of $^{235}\mathrm{U}$ in each kilogram, calculate the rate at which the fuel is 'used up' to generate 1200 MW of heat.

(Scottish H.)

FL.19 Describe an experiment to measure the maximum energy of photoelectrons emitted from a metal. With the aid of a sketch graph, summarize the results of such experiments with light of different frequencies and with different metals. Show how a value of the Planck constant may be deduced from the results, and explain how simple wave theory fails to account for them.

Outline an experiment to demonstrate the phenomenon of electron diffraction. Summarize the results of such experiments with electrons of various speeds. State the de Broglie relationship, and explain how simple particle theory fails to account for this relationship.

A certain electron stream and an X-ray beam produce identical diffraction patterns when they interact with the same object. Deduce an expression for the potential difference V required to accelerate the electrons from rest in terms of the wavelength of the X-ray beam, the charge and mass of the electron, and the Planck constant.

(Cambridge)

FL.20 This question is about clarifying the idea of wave-particle duality.
Physicists claim *both* that things commonly regarded as particles can behave like waves *and* that things commonly regarded as waves can behave like particles.

a) Outline experimental evidence which supports the claim that particles can behave like waves.

b) Outline experimental evidence which supports the claim that waves can behave like particles.

c) Explain why wave properties of particles may be important for electrons but not for tennis balls, and why particle properties of waves may be important for light waves but not for radio waves.

d) Answer a critic who objects that these ideas are absurd because something cannot be both a wave and a particle at the same time.

(O and C Nuffield)

SECTION G

Skills of physics

Essential mathematics · Data analysis ·
In the laboratory · Communication skills in physics

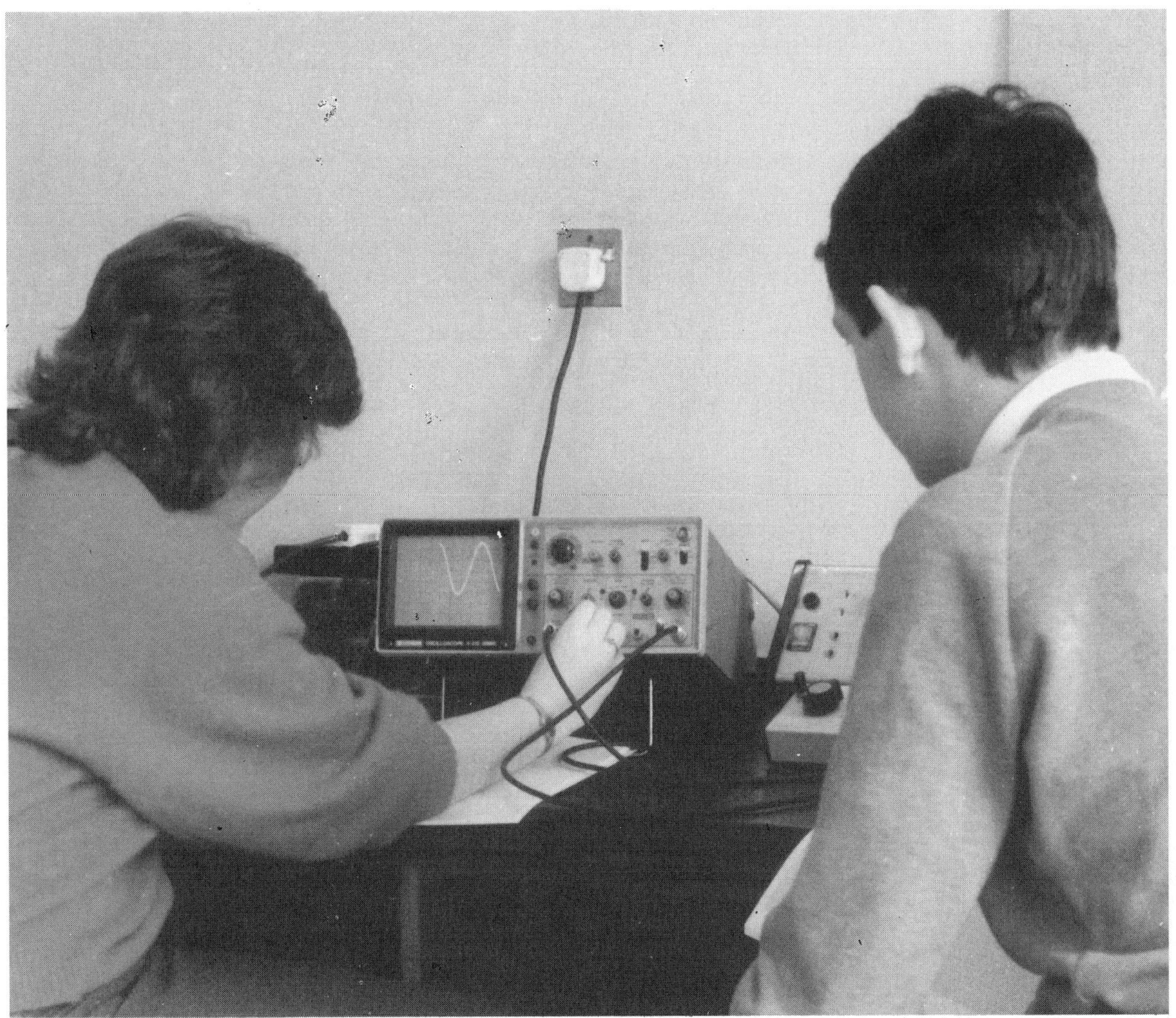

Figure G1 *Students working in a physics laboratory are expected to use mathematical and communications skills as well as practical skills.*

Physics is a demanding subject in terms of its depth of understanding and its wide range of applications. Studying physics involves developing a range of skills, notably laboratory skills, mathematical skills and the ability to communicate. Few courses teach such skills as separate topics but all examination boards assess these skills. The skills are usually taught as an integral part of the course. In this section, you can develop your capabilities in these key skills to assist your studies in the other sections.

32

Essential mathematics

32.1 Basic mathematics

Indices

Raising to a power is a common operation in calculations. For example $5 \times 5 \times 5$ is stated as '5 to the power 3' and written 5^3. In this example, 3 is the **index** and 5 is the **base.** Powers of ten are used to simplify very large or very small numbers. The **standard form** used for expressing any number, is as a number between 1 and 10 multiplied by the appropriate power of ten. It saves writing long strings of zeroes. For example, the speed of light is equal to $3.0 \times 10^8\ \mathrm{m\,s^{-1}}$; written as $300\,000\,000\ \mathrm{m\,s^{-1}}$, it is less than convenient. Try writing the value of the electronic charge ($= 1.6 \times 10^{-19}$C) in non-standard form! The power of ten indicates how many decimal places the decimal point is to be moved to convert the number to standard form. The sign of the index is important too, since it gives the direction the point is to be moved.

$\frac{1}{10000}$	$\frac{1}{1000}$	$\frac{1}{100}$	$\frac{1}{10}$	1	10	100	1000	10000
10^{-4}	10^{-3}	10^{-2}	10^{-1}	10^0	10^1	10^2	10^3	10^4

Finding the roots of a number is the inverse of 'raising to a power'. For example, in $3^2 = 9$ is where 3 is raised to the power 2. So the inverse operation is $9^{\frac{1}{2}} = \sqrt{9} = 3$. In other words, the square root of 9 is 3. The $\sqrt{\ }$ sign must be used carefully because it is usually taken to mean the square root (i.e. power $= \frac{1}{2}$) unless the sign is accompanied by a small number placed in the 'hook'. The cube root of 8 is written $8^{1/3}$ or $\sqrt[3]{8}$ which is equal to 2.

Suppose a number in standard form is to be raised to a given power. For example, suppose we wish to calculate

$$(3.4 \times 10^6)^3 = (3.4)^3 \times (10^6)^3$$
$$= (3.4 \times 3.4 \times 3.4) \times (10^6 \times 10^6 \times 10^6)$$
$$= 39.304 \times 10^{18} = 3.9304 \times 10^{19}.$$

The cube of 10^6 is $(10^6)^3 = 10^{6 \times 3} = 10^{18}$; the two powers are multiplied together.

The same rule applies to roots of numbers in standard form. For example, the fourth root of (2.64×10^8) is written as $(2.64 \times 10^8)^{1/4}$ which equals $2.64^{1/4} \times (10^8)^{1/4}$. Since $2.64^{1/4} = 1.2747$ and $(10^8)^{1/4} = 10^2$, then the final answer is 1.2747×10^2.

Multiplying or dividing numbers in standard form involves separating the 'power of ten' part from the rest of the number. Consider the following example.

$$\text{Calculate the value of } \frac{(1.6 \times 10^{-19}) \times (2.0 \times 10^3)}{9.1 \times 10^{-31}}$$

Separating the powers of ten gives

$$\left(\frac{1.6 \times 2.0}{9.1}\right) \times \left(\frac{10^{-19} \times 10^3}{10^{-31}}\right) = 0.3516 \times 10^{-19+3-(-31)}$$

Handling powers of ten is an important part of calculation skills. The longer method for the above example is not just tedious but liable to error.

$$\frac{10^{-19} \times 10^3}{10^{-31}} = \frac{0.000\,000\,000\,000\,000\,000\,1 \times 1000}{0.000\,000\,000\,000\,000\,000\,000\,000\,000\,000\,1}$$

$$= \frac{1000}{0.000\,000\,000\,000\,001} = 10^{15}$$

The example shows the advantages of being able to handle powers of ten; the rules are based on moving the decimal point according to the indices.

$$10^a \times 10^b = 10^{a+b}$$

$$\frac{10^a}{10} = 10^a \times 10^{-b} = 10^{a-b}$$

Significant figures reflect the accuracy of a number. The number of significant figures of a numerical value is the number of digits except for zeroes at the beginning. For example, the number 0.034 05 has four significant figures; 21 540 has 5. The number of significant figures which any numerical value is allowed must be consistent with the accuracy of the value. For example, suppose you wish to determine the average speed of a falling object by timing it as it falls through a measured vertical distance.

Distance fallen = 5.00 m
Time taken = 1.05 s

$$\text{Average speed} = \frac{\text{distance fallen}}{\text{time taken}}$$

$$= \frac{5.00}{1.05} = 4.761\,905\ \mathrm{m\,s^{-1}}$$

using a calculator. But the measurements cannot justify all the figures in the answer from the calculator. So we must round-off the value to the number of significant figures that can be justified. For example, if the measurements justify two significant figures only, then the average speed is written as $4.8\ \mathrm{m\,s^{-1}}$. Note that the next significant figure is rounded down if it is less than 5, and rounded up if it is 5 or more.

6.325 641 becomes 6.33 to 3 sig. fig.
0.001 527 becomes 0.001 53 to 3 sig. fig.
100.95 becomes 101 to 3 sig. fig.

Calculations

Using a calculator gives the wrong answer very easily unless you know how to use the calculator correctly. Different types of calculator have different rules for use so you must learn to use your own calculator effectively. Raising to a power can present problems using an unfamiliar calculator. Some guidelines for using calculators are listed below.

1 Don't hurry because it is all too easy to bounce a key twice by mistake or to press the wrong key.
2 For a lengthy calculation, work through in stages and write the values down at the end of each stage. Using

the memory facility of a calculator can cause errors if you forget to clear the memory first.

3 Write the answer down in standard form. Take care not to write any power of ten on the display as a power of the number. For example, if the display shows '5.6 13', then you write down 5.6×10^{13} *not* 5.6^{13} which is very different.

Worked example Calculate the density of a spherical ball bearing from the following measurements.
mass $m = 4.85 \times 10^{-3}$ kg, diameter $d = 1.05 \times 10^{-2}$ m.
Solution

Stage 1 The volume V of the ball bearing $= \frac{4}{3}\pi(d/2)^3$
Hence
$V = \frac{4}{3}\pi\,(0.525 \times 10^{-2})^3 = \frac{4}{3}\pi \times (0.525)^3 \times 10^{-2\times 3}$
So $V = 0.606 \times 10^{-6}\,\text{m}^3$

Stage 2 Write down the value for V from your calculator.

Stage 3 Calculate the density $\rho = \text{mass/volume}$.

$$\rho = \frac{4.85 \times 10^{-2}}{0.606 \times 10^{-6}} = 8000\,\text{kg m}^{-3}$$

Using tables for calculations is perhaps a skill not much needed now. But if your calculator develops a fault, then you may be forced to call on your skill in this area. Tables are available for **reciprocals** (i.e. $1/a$ is the reciprocal of a), squares, square roots and much more. The most useful are **logarithms** because they can be used in place of the other tables if you know the basic rules. The two important types of log tables are

a) base 10 logs, written as $\log_{10}$ or simply lg,
b) natural logs, written as $\log_e$ or just ln.

The logarithm in a given base of any number is the power the base must be raised to in order to equal the number.
For example $\log_{10}1000$ equals 3 because $10^3 = 1000$.

Multiplication using logs

If $a = 10^x$ where a is any positive number, then $x = \log_{10} a$ and if $b = 10^3$ where b is any positive number, then $y = \log_{10} b$.
Therefore $a \times b = 10^x \times 10^y = 10^{x+y}$, so $x + y = \log_{10}(ab)$
Hence $\log_{10}(ab) = \log_{10}a + \log_{10}b$

Worked example $a = 6.512$, $b = 0.3649$.
Calculate $c = ab$.
Solution

$\log_{10}a = 0.8137$
$\log_{10}b = -1 + 0.5621 = -0.4378$
(since $\log_{10} 0.3649 = \log_{10} 3.649/10$
$= \log_{10} 3.649 - \log_{10} 10 = 0.5621 - 1$)
Hence $\log_{10}ab = \log_{10}a + \log_{10}b$
$= 0.8137 - 0.4378 = 0.3759$

Now look up the antilog of 0.3759.
In other words, what number has its $\log_{10}$ equal to 0.3759? A look at Figure 32.1, at the bottom of the page, shows that the answer is 2.376.
So $ab = 2.376$.

Division by logs

If $a = 10^x$, then $x = \log_{10} a$. If $b = 10^y$, then $y = \log_{10}b$, so $a/b = 10^x/10^y = 10^{x-y}$.
Hence $x - y = \log_{10}(a/b) = \log_{10}a - \log_{10}b$

$\log(a/b) = \log a - \log b$

Worked example $a = 2.658$, $b = 34.23$
Calculate a/b.
Solution

$\log_{10}a = 0.4245$
$\log_{10}b = 1.5344$
(since $\log_{10} 34.23 = \log_{10}(10 \times 3.423)$
$= \log_{10} 10 + \log_{10} 3.423 = 1 + 0.5344$)
Hence $\log_{10}(a/b) = \log_{10}a - \log_{10}b = -1.1098$

To determine the antilog of this number using tables, it must be written as $-2 + 0.8901$ (or $\bar{2}.8901$). The antilog of 0.8901 is 7.764, and the '-2' (or $\bar{2}$) tells us to move the decimal point two places to the left. So the answer is 0.077 64.

$a/b = 0.077\,64$

A quick way to set out the calculation above is shown in Figure 32.2.

a = 2.658
b = 34.23

Calculate a/b using log tables.

Number	log
a = 2.658	0.4245
b = 34.23	1.5344
a/b	$\bar{2}.8901$

$a/b = 7.764 \times 10^{-2} = 0.07764$
Note: $\bar{2}.8901$ means $-2 + 0.8901$
The $\bar{2}$ tells you to move the decimal point of 7.764 two places to the left.

Figure 32.2 *Setting logs out*

Using logs to raise a number to a given power

Suppose we wish to raise n to a given power p (i.e. calculate n^p).

Let $x = n^p$, then $\log x = \log n^p = p \log n$.
So x is equal to the number whose log is $p \log n$.
The reason why $\log n^p$ equals $p \log n$ is that
$\log n^p = \log (n \times n \times \overset{p\ \text{times}}{\dots} \times n)$
$= \log n + \log n + \ldots + \log n$
$= p \log n$.

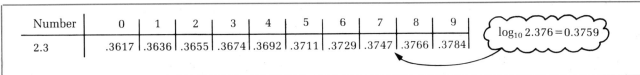

Number	0	1	2	3	4	5	6	7	8	9
2.3	.3617	.3636	.3655	.3674	.3692	.3711	.3729	.3747	.3766	.3784

$\log_{10} 2.376 = 0.3759$

Figure 32.1 *Using log tables*

Worked example Find $6^{1.4}$.
Solution If $x=6^{1.4}$, then $\log_{10} x=1.4 \log_{10} 6$
$=1.4\times 0.7781=1.08941$.

The antilog of 1.0894 is 12.29; this is because the antilog of 0.0894 is 1.229 and the '1' in front of the decimal point tells us to move the point one place to the right. Hence $x=12.29$.

Using logs to find a given root of a number
The root is written as a fraction. So the fifth root of 6 is written $6^{1/5}$. Then the same procedure as in the previous example is used.

Worked example Find $6^{1/5}$.
Solution Let $x=6^{1/5}$, so $\log_{10}x=(1/5)\times\log_{10} 6$
$=0.7781/5=0.1556$.

The antilog of this number is 1.4309.
Hence $x=1.4309$.

In the above examples, base 10 logs have been used. Natural logs could have been used although you can easily make an error in 'moving the decimal point'. Try some calculations using base 10 logs, and check your answers with a calculator. Then, you have a fall back if your calculator lets you down in an examination.

Trigonometry

Angles are measured in degrees or radians. The symbol for the radian is rad. The scale for conversion is 360 degrees $=2\pi$ radians, so 1 rad $=360/2\pi$ degrees.

The circumference of a circle of radius $r=2\pi r$, so the circumference can be written as $r\times$ the angle in radians round the circle.

For a segment of a circle, the length of the arc of the segment is in proportion to the angle θ subtended by the arc to the centre. Since the angle 2π radians corresponds to an arc length equal to the circumference ($2\pi r$), then

$$\frac{\theta}{2\pi}=\frac{\text{arc length } s}{\text{circumference } 2\pi r}$$

$$\text{So } \theta=\frac{s}{r}$$ (See Figure 32.5.)

Giving $s=r\theta$

where $s=$ arc length,
$r=$ radius,
$\theta=$ angle in radians.

The right-angled triangle is used to define sines, cosines and tangents. Figure 32.3 shows a right-angled triangle XYZ in which side XY is the hypotenuse (i.e. the side opposite the right angle), YZ is opposite to angle θ ($=\hat{YXZ}$) and XZ is adjacent to angle θ.

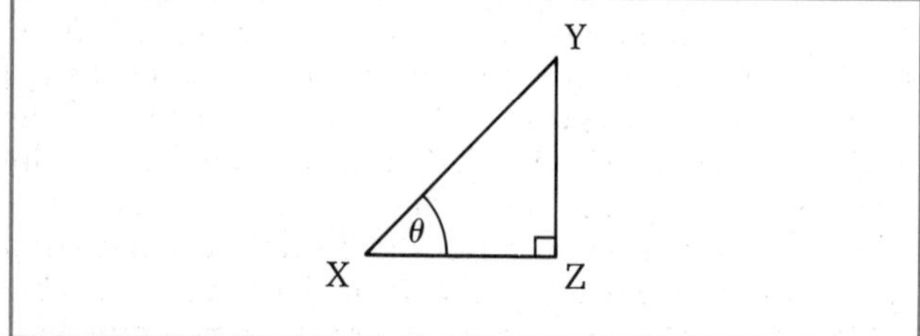

Figure 32.3 *The right-angled triangle*

$$\sin\theta=\frac{YZ}{XY}=\frac{o}{h}$$

$$\cos\theta=\frac{XZ}{XY}=\frac{a}{h}$$

$$\tan\theta=\frac{YZ}{XZ}=\frac{o}{a}$$

where $o=$ side opposite to θ (i.e. YZ)
$h=$ hypotenuse,
$a=$ side adjacent to θ (i.e. XZ).

Graphs of $\sin\theta$ and $\cos\theta$ against θ are shown in Figure 32.4. The maximum value of $\sin\theta$ is 1 which is when $\theta=90°=\pi/2$ radians. At this angle YZ = XY and XZ = 0 so $\sin\theta=1$.

The maximum value of $\cos\theta$ is when $\theta=0$; here YZ = 0 so XZ = XY hence $\cos\theta=1$ at $\theta=0$.

The value of $\tan\theta$ is 0 when $\theta=0$ because YZ is zero at $\theta=0$. When $\theta=45°$, YZ = XZ so $\tan 45°=1$. As θ approaches 90°, XZ tends to zero as YZ tends to 1; so $\tan\theta$ tends to infinity at 90°.

$$\tan\theta=\frac{o}{a}=\frac{o}{h}\div\frac{a}{h}=\frac{\sin\theta}{\cos\theta}$$

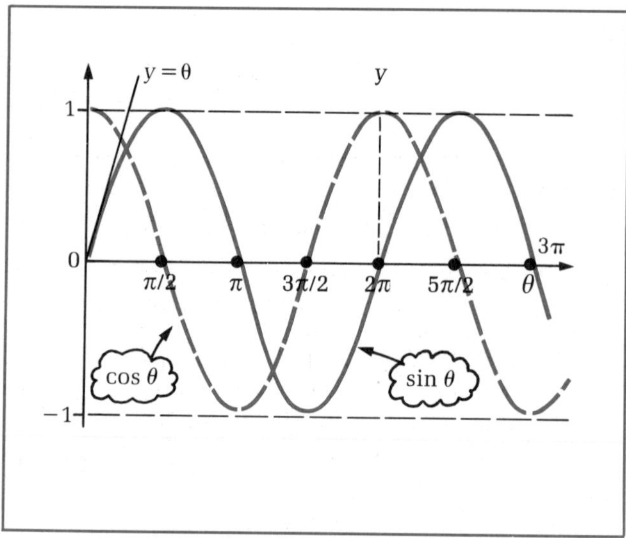

Figure 32.4 *Sine and cosine curves*

Pythagoras' theorem gives a useful link between $\sin\theta$ and $\cos\theta$. From Figure 32.3, Pythagoras' theorem can be written

$$YZ^2+XZ^2=XY^2$$

$$o^2+a^2=h^2$$

$$\text{So } \left(\frac{o}{h}\right)^2+\left(\frac{a}{h}\right)^2=1$$

Hence $\sin^2\theta+\cos^2\theta=1$

The small angle approximation for $\sin\theta$ is used in lots of topics in physics. The approximation is sometimes called the 'skinny triangle' rule. Figure 32.5 shows a triangle OAB which is part of segment OAC of a circle. The length of the arc from A to C is greater than the distance AB along the straight line.

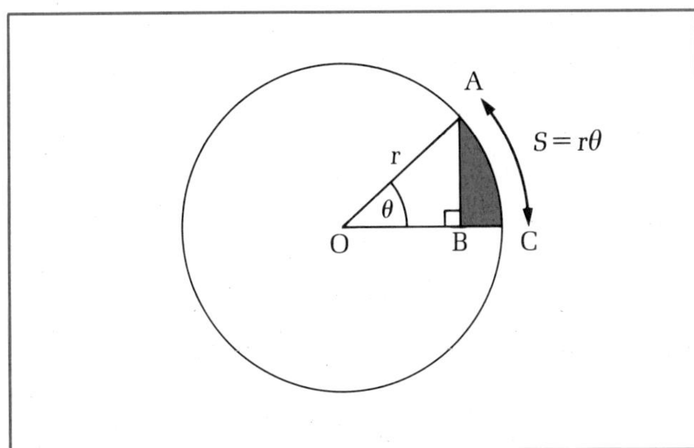

Figure 32.5 *Segments and triangles*

Arc length $s = r\theta$
Straight line distance $x = r \sin \theta$

Now suppose we make angle θ much smaller so the situation is like Figure 32.6. So the triangle OAB becomes 'skinny'; now the arc length AC is virtually the same as the straight line distance AB. So $r \sin \theta$ is approximately equal to $r\theta$ for small angles.

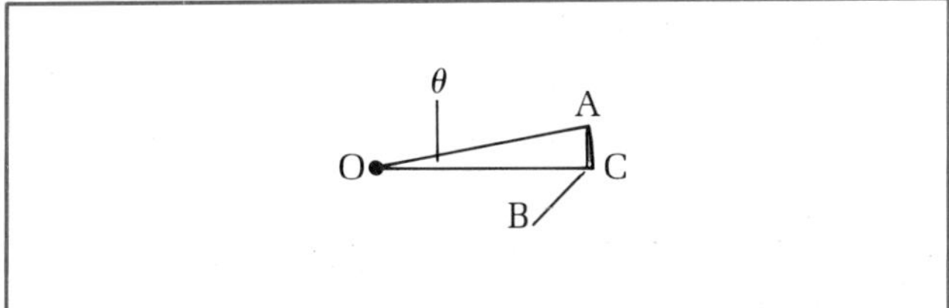

Figure 32.6 *sin θ≈θ*

$\sin \theta = \theta$ (in radians) for angles less than about 10°.

Figure. 32.4 shows the line $7 = \theta$ on the graph of the sine wave; up to about 10°, the sine wave is virtually straight and follows the line $\sin \theta = \theta$.

The cosine wave shows that $\cos \theta$ is approximately 1 for small angles; Figure 32.6 shows that OA is approximately equal to OB for small angles AOB.
So $\cos \theta = OB/OA$ is approximately equal to 1 here.

32.2 Mathematical links

Signs and symbols

Symbols used in equations and formulae represent physical variables. Any physical variable has a magnitude and a unit. So a symbol represents a number *and* a unit. The number is meaningless without the unit. If you received a bill that stated 'Amount owing = 77' you would be puzzled because no unit is given. So the unit, (£ or p in this case), is essential. The same applies to any physical variable; an instruction to move a lens holder by a distance of 30 is meaningless because you do not know whether 30 mm or 30 cm is intended. The same applies when you give answers to numerical problems; give the unit or else your answer is meaningless.

Inequality signs are often used in physics. You need to be able to recognize the meaning of the signs in the table at the top of the next column.

Sign	Meaning	Example of use	Meaning of example
$>$	greater than	$T > 300$ K	T is greater than 300 K
$<$	less than	$T < 300$ K	T is less than 300 K
$\geqslant$	greater than or equal	$T \geqslant 300$ K	T is greater than or equal to 300 K
$\leqslant$	less than or equal to	$T \leqslant 300$ K	T is less than or equal to 300 K
$\gg$	much greater than	$T \gg 300$ K	T is much greater than 300 K
$\ll$	much less than	$T \ll 300$ K	T is much less than 300 K
$\approx$	approximately equals	$T \approx 300$ K	T is about 300 K

The approximation sign is used where an estimate or an order-of-magnitude calculation is made rather than an accurate calculation. For an order-of-magnitude calculation, the final value is written with one significant figure only or even rounded up or down to the nearest power of ten. For example, the order of magnitude value of the Earth's mass is 10^{25} kg although a more exact value is 6×10^{24} kg. Order-of-magnitude calculations are useful as a quick check after using a calculator. For example, if you are asked to calculate the density of a 1.0 kg metal cylinder of height 0.100 m and diameter 0.071 m, you ought to obtain a value of 2530 kg m^{-3} using a calculator. Now let's check the value.

Volume $= \pi$ (radius)$^2 \times$ height
$\approx 3 \times (0.04)^2 \times 0.1 \approx 48 \times 10^{-5}$ m^3
Density = mass/volume
$\approx 1.0/50 \times 10^{-5} \approx 2000$ kg m^{-3}

(This 'confirms' our accurate calculation.)

Proportionality is represented by the $\propto$ sign. A simple example of its use in physics is for Hooke's law; the tension in a spring is proportional to its extension.

tension $T \propto$ extension x

By introducing a constant of proportionality k the link above can be made into an equation.

$T = kx$

where k is defined as the 'spring constant'.

With any proportionality relationship, if one of the variables is increased by a given factor (e.g. × 3), the other variable is increased by the same factor. So in the above example, if T is trebled then extension x is also trebled.

Straight line graphs

Links between two physical quantities can be established most easily by graph plotting. One of the physical quantities is represented by the vertical scale (the 'ordinate', often called the y-axis) and the other quantity by the horizontal scale (the 'abscissa' often called the x-axis). The simplest link is where the plotted points define a straight line. For example, Figure 32.7, at the top of the next page, shows the link between the tension and the extension of a spring; the gradient of the line is constant and the line passes through the origin. Any situation where the y-variable is proportional to the x-variable gives a straight line through the origin. For Figure 32.7, the gradient of the line is equal to the spring constant k.

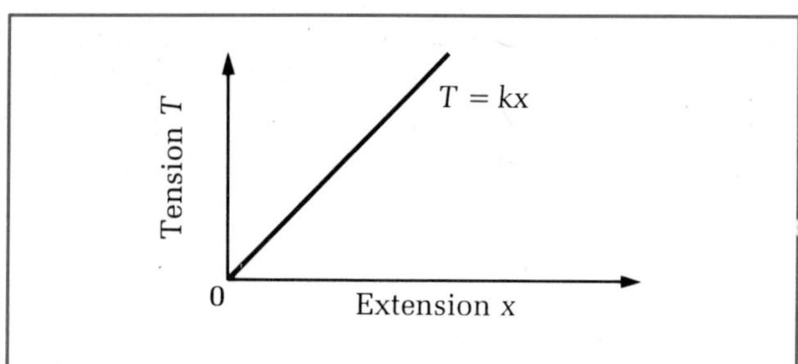

Figure 32.7 *Graph links*

The general equation for a straight line graph is usually written in the form

$y = mx + c$

where m = gradient,
c = y-intercept (i.e. where $x = 0$ and $y = c$).

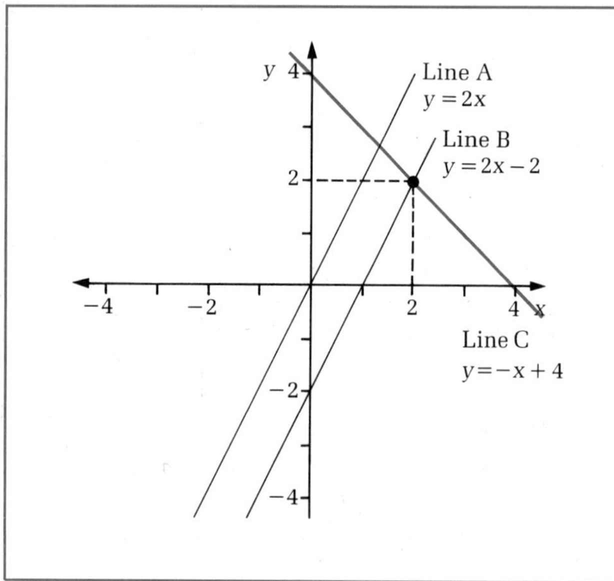

Figure 32.8 *Straight line graphs*

Figure 32.8 shows several straight lines.

Line A: $c = 0$ so the line passes through the origin; its equation is $y = 2x$.

Line B: $m > 0$ so the line has a positive gradient; its equation is $y = 2x - 2$.

Line C: $m < 0$ so the line has a negative gradient; its equation is $y = -x + 4$.

Simultaneous equations can be solved graphically by plotting the line for each equation. The solution of the equations is given by the coordinates of the point(s) where the lines meet. For example, Figure 32.8 shows that lines B and C meet at a single point, ($x = 2$, $y = 2$). These coordinates are the only ones to fit both equations.

Solving simultaneous equations doesn't require graph plotting if the equations can be arranged to eliminate one of the variables. Let's use the same example as above.

Line B: $y = 2x - 2$
Line C: $y = -x + 4$

At the point where they meet, their y-coordinates are equal so

$2x - 2 = -x + 4$.

Hence the x-coordinate at the meeting point is given by

$3x = 6$ so $x = 2$.

Since $y = 2x - 2$, then $y = 2 \times 2 - 2$ so $y = 2$.

In physics, simultaneous equations are used to solve circuit equations for currents (see p. 325).

Using graphs

To test a known link between two physical quantities, use the theory to give a straight line. For example, suppose you are asked to test Boyle's law, that pressure × volume is constant at constant temperature for a particular gas at a certain temperature. The measurements give a set of values of pressure and volume. A graph of pressure against volume does little to show if the pressure × volume is indeed constant for this particular gas. The equation linking pressure p and volume V for a gas that does obey Boyle's law is

$pV = \text{constant}$

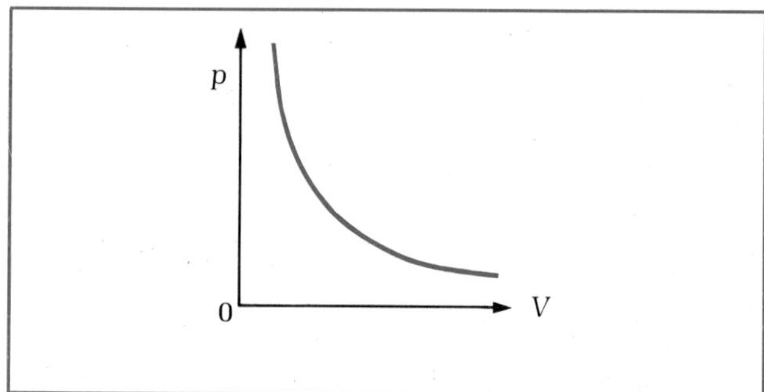

Figure 32.9 p v V *for an ideal gas*

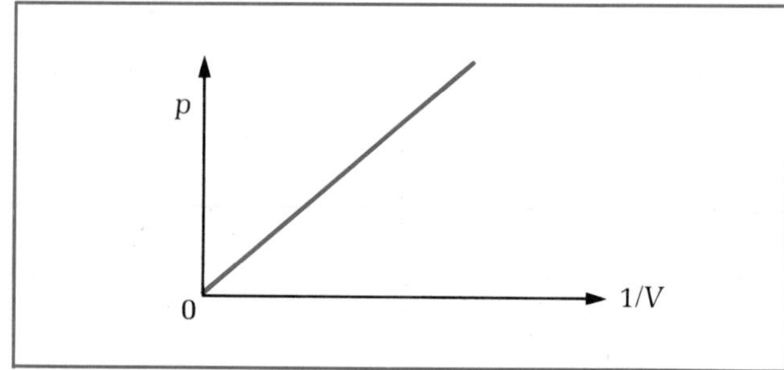

Figure 32.10 p v 1/V *for an ideal gas*

For a straight line graph, rearrange the equation to make p the subject.

$p = \text{constant}/V$

So by plotting a graph of pressure against 1/volume, if the points define a straight line, then the gas does obey Boyle's law.

The same idea applies in many other situations for which we wish to test the validity of an equation; make one of the variables the subject of the equation and plot accordingly. Consider the example of the simple pendulum, which uses the time period (T) and the length (L) to measure the acceleration due to gravity (g).

Time period $T = 2\pi\sqrt{L/g}$

where L is the length of the pendulum.

Measurements would give a set of values for T and L, but a graph of T against L would be of little help. Far more useful would be either a graph of T against $\sqrt{L}$ or T^2 against L, which ought to give a straight line. For T against $\sqrt{L}$, the gradient is $2\pi/\sqrt{g}$; for T against L^2 the gradient is $4\pi^2/g$.

To establish a link between two physical quantities, a log-log graph can be helpful. Suppose two quantities, Y and X, are linked by an equation of the form

$$Y = kX^n$$

where k is a constant and n is a power to be determined.

$$\log Y = \log k + \log X^n$$

so $\log Y = \log k + n \log X$

since $\log X^n = n \log X$.

Therefore a graph of log Y against log X gives a straight line with gradient equal to n and a log Y intercept equal to $\log k$. By measuring the gradient, assuming a straight line results from the log-log plot, then the power of X can be determined. If the log-log plot does not give a straight line, then the link is not of the form $Y = kX^n$.

Some common graph shapes

Parabolic curves describe the flight paths of projectiles or any other objects acted on by a constant force. The path of an electron beam between oppositely charged parallel plates is parabolic; each electron is acted on by a constant force (see p. 420).

The general equation for a parabola is $y = kx^2$. Figure 32.11 shows the shape of the parabola $y = 3x^2$. Equations of the form $x = ky^2$ are parabolic, but they are symmetrical about the x-axis not the y-axis. Figure 32.12 shows the curve for $x = y^2$.

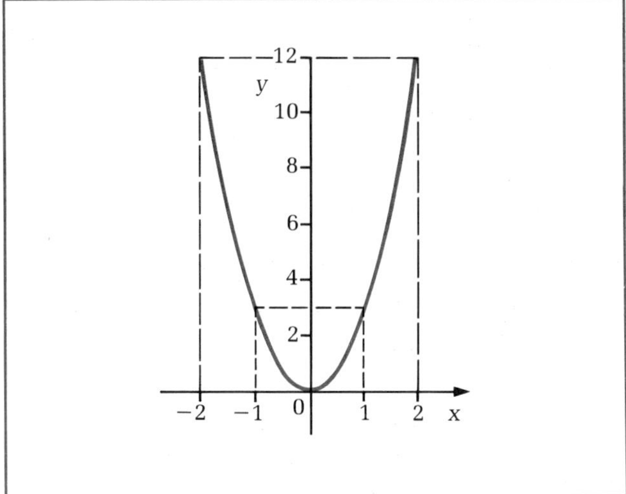

Figure 32.11 $y = 3x^2$

Compare the equations for projectiles and electron beams given in the table at the top of the next column. In both cases, suppose the initial velocity is u at right angles to the field lines.

Both equations are in the form $y = kx^2$, although k differs in each case. Figure 32.13 shows the shape of the trajectories.

	Projectile	**Electron beam**
Force	mg (g = gravitational field strength)	eE (E = electric field strength)
Acceleration (a)	g	$\frac{eE}{m}$
Time to travel distance x across field (t)	x/u	x/u
Displacement at right angles to field (y)	$\frac{1}{2}\left(\frac{g}{u^2}\right)x^2$	$\frac{1}{2}\left(\frac{eE}{mu^2}\right)x^2$ $(=\frac{1}{2}at^2)$

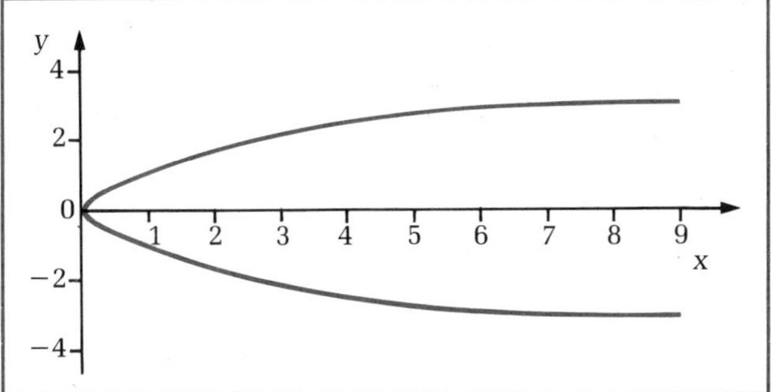

Figure 32.12 $x = y^2$

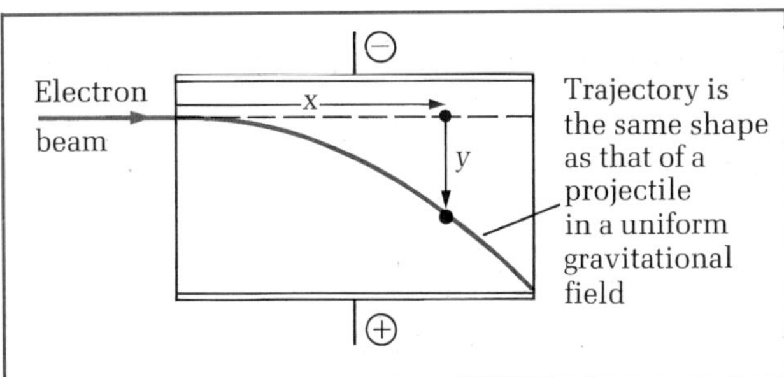

Figure 32.13 *Parabolic paths*

Hyperbolic curves are curves like p against V for Boyle's law (see Figure 32.9). The general equation for a hyperbola is $y = k/x$, where k is a constant. As Figure 32.9 shows, the curve tends towards either axis but never actually meets the axes; the correct mathematical word for 'tending towards but never meeting' is **asymptotic**.

Inverse-square law curves occur in gravitation and electric field theory and radioactivity (intensity of γ-radiation; see p. 438). The general equation is $y = k/x^2$, where k is a constant. So if x is doubled, y changes by a factor of 1/4. The shape of the curve is shown in Figure 32.14. Such curves show asymptotic behaviour at both axes.

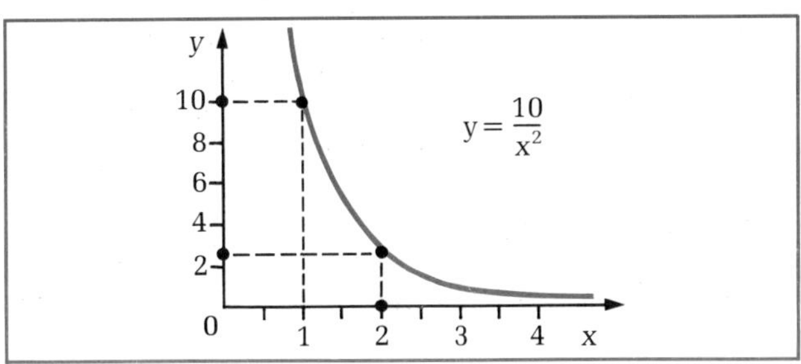

Figure 32.14 *An inverse square curve*

Exponential curves are described in more detail on p. 507. These curves arise whenever the rate of change of any quantity is proportional to the quantity itself. Figure 32.15 shows an exponential decrease curve. The key features of such curves are

a) At x = 0, the y-value is non-zero,

b) The y-value drops by 50% in equal intervals of the x-variable,

c) The curve is asymptotic at the x-axis; in other words, it never actually touches the x-axis.

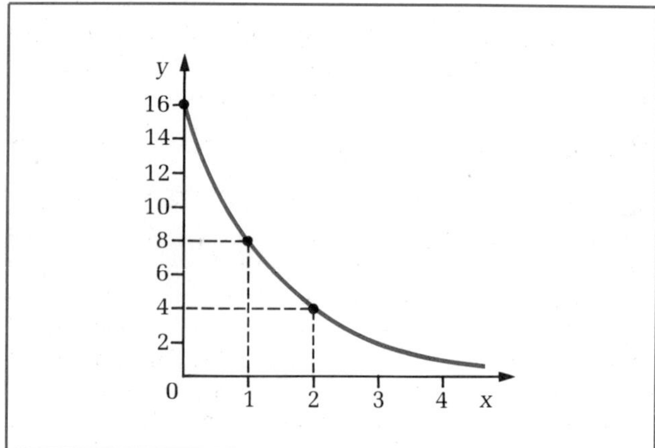

Figure 32.15 *An exponential decrease*

In fact, the y-value drops by a constant factor for *any* equal set of intervals along the x-axis; let the y-value at x = 0 be y_0. Then, the interval from y_0 to $0.7y_0$ is the same as from $0.7y_0$ to $0.49y_0$ $(= 0.7 \times 0.7y_0)$.

Labelling graphs

Scales along the axes of a graph are numerical. The axes must be labelled with the appropriate physical quantity named. The unit of the quantity and any prefix must be stated. Only if the axes are labelled can the physical quantities be read off the graph. The convention for labelling an axis is

Name of physical quantity, symbol/prefix + unit.

Some examples are

a) pressure, p/kPa

b) volume, V/m^3

c) 1/volume, $\frac{1}{V}/\text{m}^{-3}$

Powers of ten are included with the unit (and prefix, if any).

d) pressure, $p/10^5$ Pa

e) capacitance, $C/10^{-6}$ F

Although in this last case it would be more common to use the prefix μ to represent 10^{-6} and so it would be written as

f) capacitance, C/μF

Reading off the graph scale gives a number equal to the quantity/unit. Suppose the graph is pressure against volume and a point on the graph has coordinates (110, 0.0012) with the axes labelled as **a)** and **b)** above. Then the pressure/kPa = 110 hence pressure = 110 kPa. Also, volume/m³ = 0.0012 so volume = 0.0012 m³.

32.3 Rates of change

Gradients and graphs

For a straight line the gradient is constant. Any section of a straight line can be used to form a right-angled triangle as in Figure 32.16. The gradient is defined as the change of the y-value/the change of the x-value. To measure the gradient of a straight line, the triangle should be as large as possible, to increase accuracy. Examples of constant gradients in physics include

a) the fall of temperature along a uniform heat-conducting lagged bar (see p. 117).

b) the change of potential between two oppositely charged parallel plates along a line at right angles to the plates from one to the other (see p. 167).

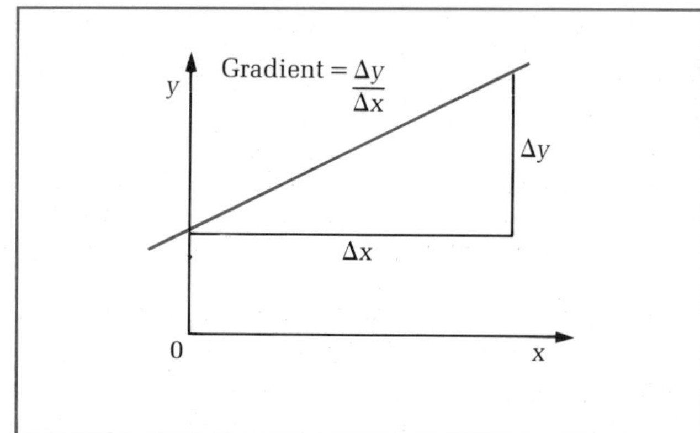

Figure 32.16 *Constant gradient*

For a curve the gradient changes along the curve. The gradient at any point on the curve is equal to the gradient of the tangent at that point. To see why, mark any two points on a curve and join them by a straight line. Now repeat with one of the points moved closer to the other; the straight line is now closer in direction to the curve. If the two points are very close, the straight line between them is almost along the curve; the gradient of the line is then virtually the same as the gradient of the curve at that position.

If the difference between the x-coordinates is Δx, and the difference in the y-coordinates is Δy for the two points, then the gradient of the line is Δy/Δx. As the two points

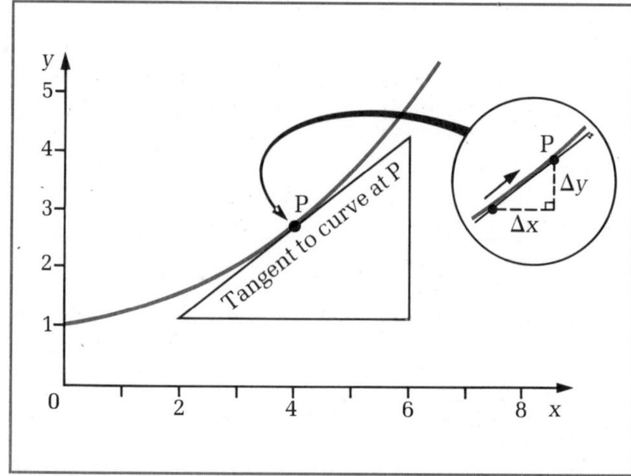

Figure 32.17 *Tangents and curves*

are moved closer and closer, $\Delta x \to 0$ and $\Delta y/\Delta x$ becomes equal to the gradient of the curve there. The curve gradient is written $\frac{dy}{dx}$ where $\frac{d}{dx}$ means 'rate of change'. So the gradient of the curve $\frac{dy}{dx}$ equals $\frac{\Delta y}{\Delta x}$ as $\Delta x \to 0$.

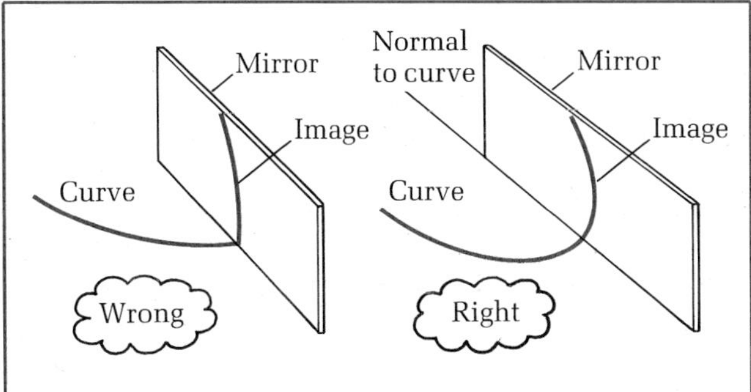

Figure 32.18 *Measuring gradients*

To measure the gradient of a curve, use a plane mirror as in Figure 32.18. The mirror is placed at right angles to the curve where the curve and its image are smoothly continuous at the mirror; then the normal to the curve is drawn. The mirror can be used to draw a line at right angles to the normal, the new line touching the curve to give the gradient of the tangent at that point.

Differentiation can be used to determine the gradient of a 'known' curve. For example, consider the curve $y = x^2$ shown in Figure 32.19.

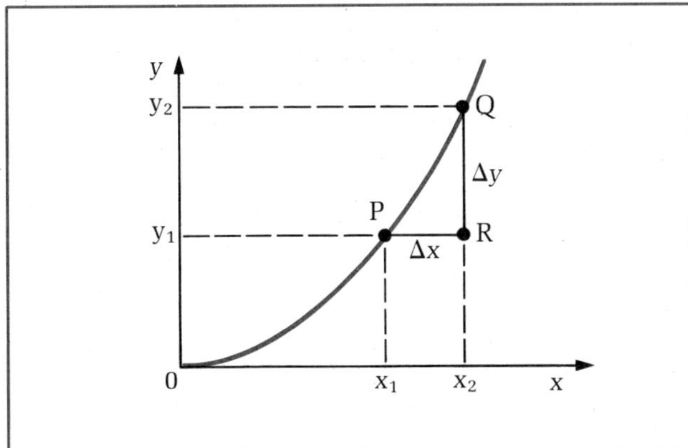

Figure 32.19 *Differentiation*

At point P, the coordinates are (x_1, y_1) where $y_1 = x_1^2$.
At point Q, the coordinates are (x_2, y_2) where $y_2 = x_2^2$.
The triangle PQR has height $\Delta y = y_2 - y_1$, and base $\Delta x = x_2 - x_1$.
The gradient of the triangle $= \Delta y/\Delta x$
Now $\Delta y = y_2 - y_1 = x_2^2 - x_1^2 = (x_1 + \Delta x)^2 - x_1^2$
so $\Delta y = 2x_1\Delta x + \Delta x^2$

Hence the gradient of the curve at x_1, $\left(\frac{dy}{dx}\right)_{x_1}$ is given by

$$\left(\frac{dy}{dx}\right)_{x_1} = \left(\frac{\Delta y}{\Delta x}\right)_{\Delta x \to 0} = 2x_1$$

So $\frac{dy}{dx} = 2x$ for $y = x^2$

The general rule for differentiation of the function $y = ax^n$ can be worked out in a similar way. The result is $\frac{dy}{dx} = nax^{n-1}$. Likewise, the rules of trigonometry can be used for the differentiation of sines and cosines.

y	$\frac{dy}{dx}$
ax^n	nax^{n-1}
$\sin x$	$\cos x$
$\cos x$	$-\sin x$

Figure 32.4 shows that the gradient of the sine curve follows a cosine curve and it shows that the gradient of the cosine curve follows a (−sine) curve.

Turning points on curves are where the gradient is zero. On one side of a turning point, the gradient is positive and on the other side it is negative. For example, consider an object projected directly upwards at initial speed *u*. Figure 32.20(a) shows how the vertical height gain y changes with time; at maximum height, the gradient of the curve is zero so the vertical component of velocity is momentarily zero at this point. The gradient at any point is equal to the rate of change of vertical displacement i.e. the vertical component of velocity. Figure 32.20(b) shows how the gradient varies with time; the gradient of *this* curve is constant, equal to the acceleration due to gravity.

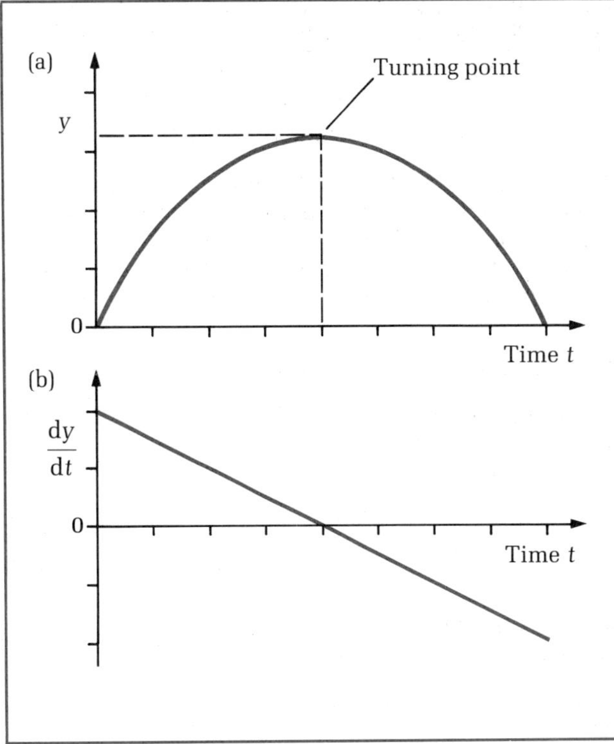

Figure 32.20 *Turning points*

Areas and integration

Areas under curves or under lines can give useful information if the product of the y-variable and the x-variable represent another physical variable. For example, consider Figure 32.7 which is a graph of tension against extension for a spring. Since tension × extension is 'force × distance' which equals work done, then the area under the line represents work done. Figure 32.21 shows a typical tension against extension curve for a rubber band; unlike Figure 32.7, the area under the curve is not a triangle so it is not so easy to determine, but it still represents work done. Each 'block' of area corresponds to a force of 0.1 N moved by a distance of 0.01 m. So each block represents 0.001 J of work (=0.1 N × 0.01 m). The total work done to extend the rubber band by 0.07 m is therefore 0.022 J since the area under the curve (shown shaded) up to 0.07 m is about 22 blocks.

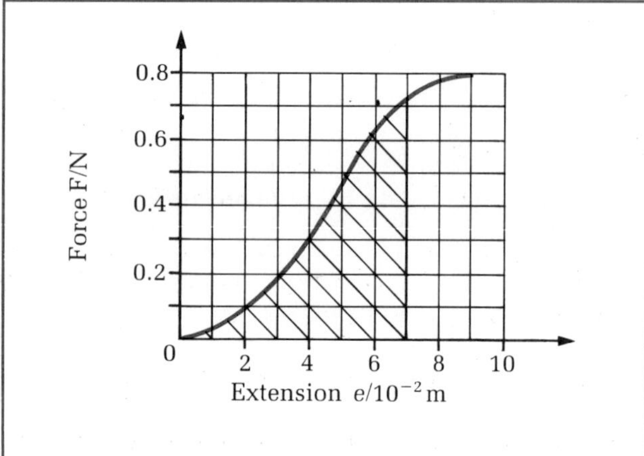

Figure 32.21 *Using areas*

The product of the y-variable and the x-variable *must* represent a physical variable if the area is to have any meaning or use. A graph of mass against volume for different sizes of the same material gives a straight line through the origin. The mass is proportional to the volume, and the gradient of the line gives the density. But the area under the line has no physical meaning since mass × volume does not represent a physical variable.

Other examples of curves where area is useful include

a) force against time where the area between the curve and the time-axis represents change of momentum,
b) power against time where the area between the curve and the time-axis represents energy.
c) potential difference against charge, where the area between the curve and the charge-axis represents energy.

Let's see why the area in each case represents a physical variable. And why is the area usually between the curve and the x-axis? A small increase of the x-variable, δx (δ for small) gives little or no change of the y-variable.

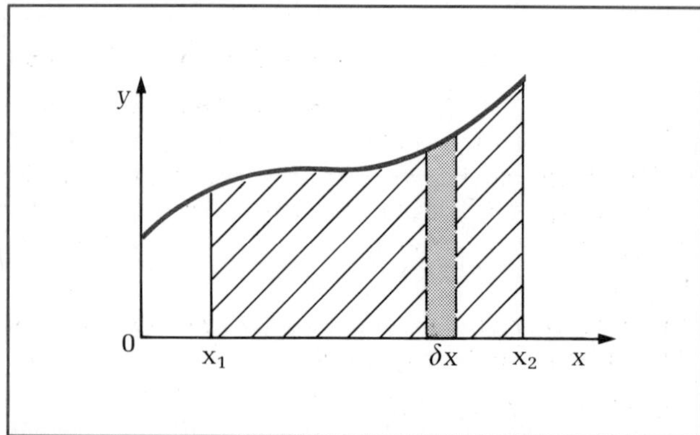

Figure 32.22 *Integration*

So the area under that section of the curve is a strip of width δx and height y which equals $y\delta x$. Hence the total area from x_1 to x_2, as in Figure 32.22, is equal to the area of all the small strips, each of width δx, from x_1 to x_2. Adding the individual strip areas together to give the total area is called **integrating** the curve.

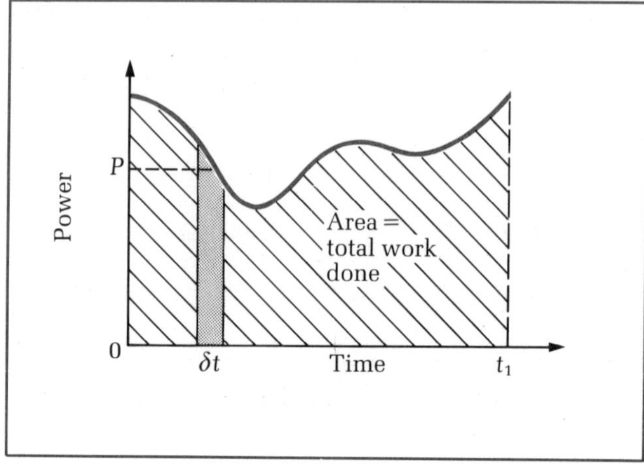

Figure 32.23 *Power curves*

For example Figure 32.23 shows a curve of power P against time t. The work done in a short interval δt is $P\delta t$, where P is the power at that point. So we write

$$\delta W = P\delta t$$

for the small amount of work done.

The total work done from $t=0$ to $t=t_1$ is represented by the total area under the curve from 0 to t_1.

$$\text{Work done } W = \int_{t=0}^{t=t_1} P\mathrm{d}t$$

where $\int$ is the integration sign.

Integration is the reverse process of differentiation. As Figure 32.23 shows, the work done in a short interval Δt is given by $\Delta W = P\Delta t$. Hence, P is $\Delta W/\Delta t$. So for $\Delta t \to 0$, $P = \frac{\mathrm{d}W}{\mathrm{d}t}$.

The y-variable must be equal to the rate of change of area with x. In general terms,

$$\text{Area } A = \int y\mathrm{d}x$$

$$\text{and } \quad y = \frac{\mathrm{d}A}{\mathrm{d}x}$$

Force-field curves representing the inverse square law of force give areas representing potential energy. Consider two point charges q_1 and q_2 at distance apart r. The force between the charges is given by

$$\text{Force } F = \frac{q_1 q_2}{4\pi\varepsilon_0} \times \frac{1}{r^2}$$

The work done by this force when the charges move apart from r to $r+\Delta r$ is $F\Delta r$. This amount of work is done by the charges so their P.E. (E_p) is lowered. Hence the change of P.E., $\Delta E_p = -F\Delta r$. So we can write

$$F = -\frac{\mathrm{d}E_p}{\mathrm{d}r}$$

Therefore, since the force is given by the inverse square law,

$$\frac{dE_p}{dr} = -\frac{q_1 q_2}{4\pi\varepsilon_0} \times \frac{1}{r^2} = \frac{k}{r^2}$$

where k is the constant $\frac{-q_1 q_2}{4\pi\varepsilon_0}$.

Because integration is the reverse of differentiation, the formula for E_p can be worked out accordingly. The result is

$$E_p = -\frac{k}{r}$$

Check that differentiating $-k/r$ gives K/r^2 (see p. 505).

Hence $E_p = \frac{q_1 q_2}{4\pi\varepsilon_0} \times \frac{1}{r}$

The inverse square law of force also applies to gravitation; the constant k is written as $-GMm$ for gravitational formulae.

32.4 Exponential processes

Exponential changes occur when the rate of change of a quantity is proportional to the quantity. Consider a physical quantity Q that changes with time. Its rate of change is $\frac{dQ}{dt}$, so an exponential process is where $\frac{dQ}{dt}$ is proportional to Q.

$$\frac{dQ}{dt} = kQ$$

where k is the constant of proportionality.

Suppose $k=1$; the solution of the equation is the function

$$Q = Q_0\left(1 + t + \frac{t^2}{2!} + \frac{t^3}{3!} + \ldots\right)$$

where Q_0 is a constant and $n!$ means $n(n-1)(n-2) \ldots 3 \times 2 \times 1$. The ! symbol is called 'factorial' so 3 factorial $= 3! = 3 \times 2 \times 1 = 6$. This complicated-looking function is called the *exponential* function; its special feature is evident when it is differentiated term by term.

$$\frac{dQ}{dt} = Q_0\left(0 + 1 + t + \frac{t^2}{2!} + \ldots\right)$$

So $\frac{dQ}{dt} = Q$

The exponential function shown above is written in short form as

$Q = Q_0 e^t$ where $e^t = 1 + t + t^2/2! + t^3/3! + \ldots$

Suppose $Q_0 = 1$; after a short interval Δt, $Q + \Delta Q = e^{t+\Delta t}$ where ΔQ is the change of Q in that time interval.

Hence $\Delta Q = e^{t+\Delta t} - Q = Qe^{\Delta t} - Q$

So $\Delta Q = Q(e^{\Delta t} - 1) = Q\left(1 + \Delta t + \frac{\Delta t^2}{2!} + \ldots - 1\right)$

$$= Q\left(\Delta t + \frac{\Delta t^2}{2!} + \frac{\Delta t^3}{3!} + \ldots\right)$$

Therefore $\frac{\Delta Q}{\Delta t} = Q\left(1 + \frac{\Delta t}{2!} + \frac{\Delta t^2}{3!} + \ldots\right)$

Now $\frac{dQ}{dt} = \left(\frac{\Delta Q}{\Delta t}\right)_{\Delta t \to 0}$

So $\frac{dQ}{dt} = Q$.

since all the terms but the first one in the bracket for $\frac{\Delta Q}{\Delta t}$ become zero as $\Delta t \to 0$.

So the equation $\frac{dQ}{dt} = Q$ has the solution $Q = Q_0 e^t$ where e^t is the exponential function equal to $1 + t + t^2/2! + t^3/3! + \ldots$ The exponential number, e, is equal to 2.718 which is the value of the above function when $t = 1$.

For the equation $\frac{dQ}{dt} = kQ$ the solution is $Q = Q_0 e^{kt}$.

All exponential processes involve functions of the form e^{kt} since these functions describe situations where the rate of change is proportional to the quantity.

Natural logarithms written ln or $\log_e$, are based on the exponential number e. The natural log of a number is the power that e must be raised to in order to equal the number.

From $Q = e^t$, then $t = \ln Q$ by definition of the natural log.

Exponential decrease

This is where a quantity **decreases** at a rate that is proportional to the quantity.

$$\frac{dQ}{dt} = -\lambda Q$$

where λ is called the **decay constant.**

The $-$ sign in the equation indicates that the rate of change is negative, so Q is a decreasing quantity.

The general solution of the equation is written

$$Q = Q_0 e^{-\lambda t}$$

where Q_0 is the initial value.

The graph of $Q = Q_0 e^{-\lambda t}$ is shown in Figure 32.24, at the top of the next page. As t becomes greater, the value of $e^{-\lambda t}$ becomes smaller so the curve tends towards the t-axis asymptotically. Another key feature of the curve is that Q drops by a constant factor in equal intervals of time.

The time constant for the process is the time taken for Q to fall to 1/e of its initial value.

$Q = Q_0 e^{-1}$ when $\lambda t = 1$.

Hence the time constant $= 1/\lambda$.

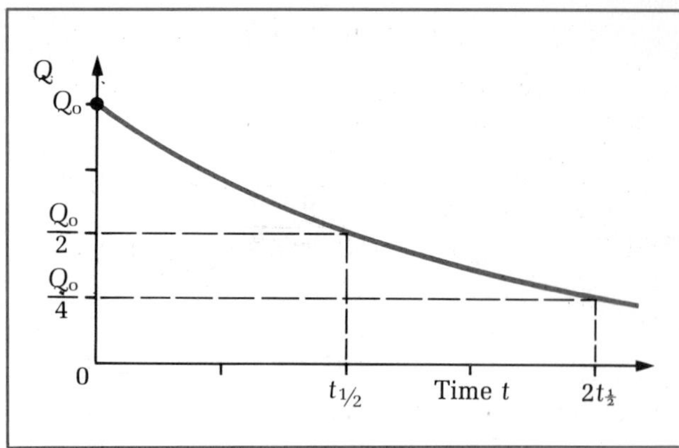

Figure 32.24 $Q = Q_0 e^{-\lambda t}$

Time constants occur in the theory of capacitor discharge circuits. The decay constant is $1/RC$ so the time constant is RC. (R = resistance, C = capacitance).

The half-life $t_{\frac{1}{2}}$ for the process is the time for Q to fall to half its initial value.

Hence $0.5Q_0 = Q_0 e^{-\lambda t}$
So $e^{\lambda t} = 2$
Giving $\lambda t = \ln 2$.

The bigger the decay constant, the shorter the half-life, so the faster the rate of decay (see p. 446).

A numerical approach to exponential decay is helpful for understanding the equations better. Rewrite the equation $\dfrac{dQ}{dt} = -\lambda Q$ as

$$\Delta Q = -\lambda Q \Delta t$$

Let $\lambda = 0.1$ and $Q_0 = 100$. Consider intervals $\Delta t = 1$.
So $\Delta Q = -0.1Q$ gives the fall of Q over each interval.
Now consider the sequence from $t = 0$.

t	0	1	2	3	4	5	6	7	8
Q	100	90	81	72.9	65.6	59.0	53.1	47.8	43.0
ΔQ	−10	−9	−8.1	−7.3	−6.6	−5.9	−5.3	−4.8	−4.3

Figure 32.25 shows how Q changes step-by-step with time. The half-life of the 'step' curve is 6.5. The theoretical value, $(\ln 2)/\lambda$ is 6.9. The step-by-step approach becomes more accurate the smaller Δt is made.

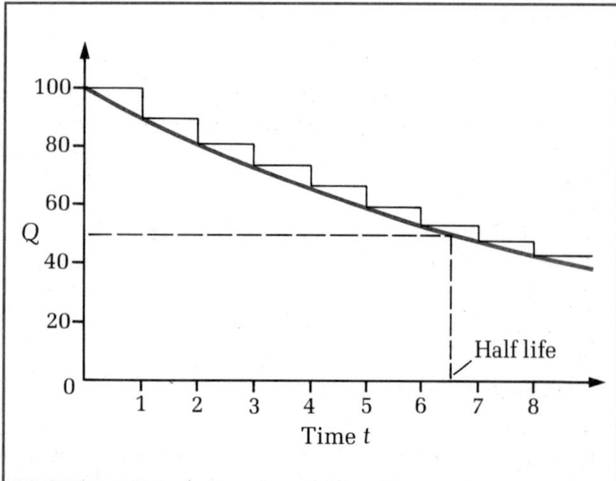

Figure 32.25 *Step-by-step approach*

Exponential increase

This is where the rate of **increase** of a quantity is proportional to the quantity.

$$\frac{dQ}{dt} = \alpha Q$$

where the constant $\alpha > 0$.

In this situation Q increases by a constant factor in equal intervals of time. Suppose $Q_0 = 100$, $\alpha = 0.1$ and $\Delta t = 1$. Then Q changes as below.

t	0	1	2	3	4	etc.
Q	100	110	121	133.1	146.4	
ΔQ	10	11	12.1	13.3	etc.	

Exponential increases are 'runaway' processes. Figure 32. 26 shows how Q increases with t for the numerical example given.

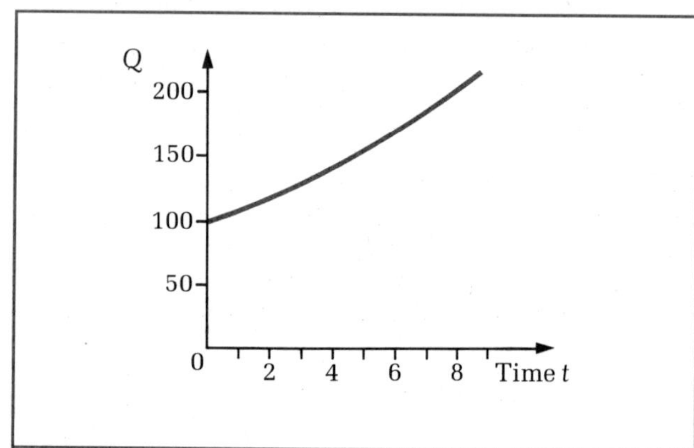

Figure 32.26 *Exponential increase*

Build-up exponentials

These occur where the rate of change **decreases from an initial non-zero value**. This type of situation is described by the equation

$$\frac{dQ}{dt} = A - \lambda Q$$

where A is a positive constant.

The rate of change, $\dfrac{dQ}{dt}$ is initially positive if we assume the initial value of Q is zero. So Q increases from zero, but increased Q makes $\dfrac{dQ}{dt}$ fall. So Q rises less and less rapidly, to reach a constant level as $t \to \infty$ when $\dfrac{dQ}{dt}$ becomes zero. The situation is shown by Figure 32.27 where the curve 'builds up' to a final level.

Physics situations where these type of curves occur include the following.

a) Terminal velocity reached when an object falls through a fluid. The drag force on the object due to the fluid is proportional to the object's speed v. So the resultant force F on the object is given by

$$F = \text{weight} - \text{drag force} = mg - kv$$

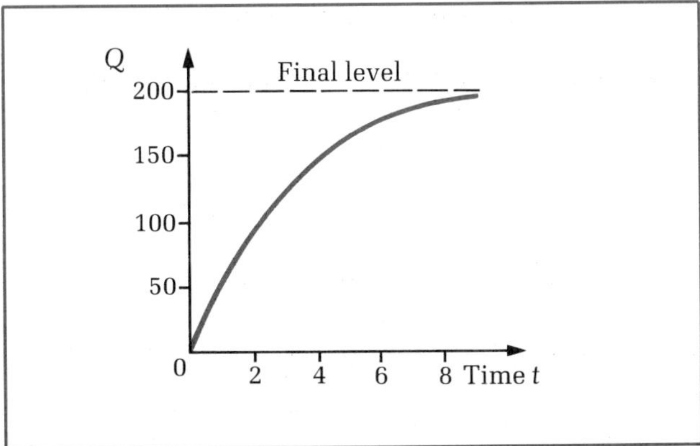

Figure 32.27 *A build-up exponential*

Also, acceleration $\frac{dv}{dt} = \frac{force}{mass} = g - \frac{kv}{m}$

where k is a constant.

So Figure 32.27 also shows how the speed of such an object released from rest increases with time. Eventually, $\frac{dv}{dt}$ becomes zero so the terminal speed is given by mg/k.

b) Growth of direct current in a circuit with an inductor and a resistor in series. Figure 32.28 shows a simple circuit where the current builds up to a limit after the switch is closed. At any time after the switch is closed, the battery e.m.f. E is dropped across the resistor R and the inductor L. Hence,

$$E = \text{p.d. across R} + \text{p.d. across L}$$

So $E = IR + L\frac{dI}{dt}$

Therefore $\frac{dI}{dt} = \frac{E}{L} - \frac{R}{L}I$

The final value of current is when $\frac{dI}{dt}$ is zero, which means that E/R is the final current.

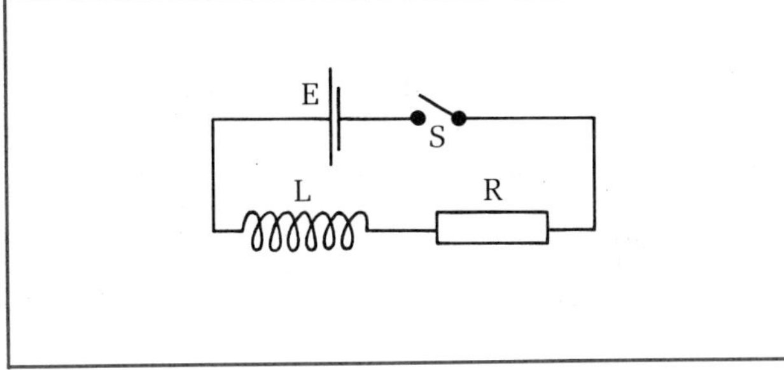

Figure 32.28 *Current growth*

32.5 Summary

Powers of ten

1 $10^a \times 10^b = 10^{a+b}$
2 $10^a/10^b = 10^{a-b}$
3 If $x = 10^a$, then $a = \log_{10}x$

Trigonometry

1 $\sin\theta$ = opposite/hypotenuse
2 $\cos\theta$ = adjacent/hypotenuse
3 $\tan\theta$ = opposite/adjacent
4 $\sin^2\theta + \cos^2\theta = 1$
5 arc length $s = r\theta$ for circle segment
6 small angle approximation: $\sin\theta \approx \theta$ for angles less than about 10°

Graphs

1 Equation for a straight line

$$y = mx + c$$

where m is the gradient and c is the y-intercept.
2 For $y = kx^n$, $\log y = n \log x + \log k$
so a graph of log y against log x gives a straight line.

Exponential decrease

1 General equation $\frac{dQ}{dt} = -\lambda Q$

where λ is the decay constant
2 Solution $Q = Q_0 e^{-\lambda t}$
$\ln Q = \ln Q_0 - \lambda t$

Short questions

32.1 Calculate
a) $(3.2 \times 10^4)^2 \times 2 \times 10^{-4}$
b) $(4.12 \times 10^{-3})^2 \times 1.4 \times 10^3$
c) $\frac{(2.8 \times 10^5)^3}{1.5 \times 10^{-3}}$
d) $\frac{1.4 \times 10^6}{(3.1 \times 10^5)^2}$
e) $\frac{6.4 \times 10^8}{(4.1 \times 10^{-7})^2}$
f) $(4.8 \times 10^2)^{\frac{1}{2}}$
g) $(5.6 \times 10^8)^{3/2}$
h) $(9.7 \times 10^5)^{-1/3}$.
i) $\frac{(6.3 \times 10^8)}{(2.4 \times 10^{-5})^{\frac{1}{2}}}$
j) $(7.6 \times 10^4)^{-\frac{1}{2}}$

32.2 Use log tables to make each of the following calculations and check your answers using a calculator.
a) 3.62 × 1.58
b) 4.91 × 21.4
c) 38.4/5.6
d) 61.4/0.159
e) 3.16 × 22.1/0.86
f) 0.156/0.35
g) 31.6 × 10^5/0.65
h) $2.15^{1/5}$
i) $36.9^{1.4}$
j) $(58.6/0.91)^{\frac{1}{2}}$

32.3
a) Measure the diameter of a 1p piece to the nearest mm. Calculate the angle subtended at your eye by a 1p piece held at a distance of 1 m from your eye.

b) Estimate the angular width of the Moon by holding a 1p piece at the distance from your eye at which it blocks out the lunar disc.

32.4

a) Plot the equations $y=x+5$ and $2y=7-x$ over the range from $x=-10$ to $+10$. Write down the coordinates of the point (P) at which the two lines intercept.

b) Write down the equation for the line OP, where O is the origin of the graph.

32.5

a) What is the equation of the straight line that passes through the two points ($x=5$, $y=0$) and ($x=-2$, $y=7$)? What is the value of the y-intercept of the line?

b) Plot the line in **a)** and determine the least distance from the line to the origin.

32.5 Solve each of the following pairs of simultaneous equations.

a) $3x+y=6 \quad y=2x+1$

b) $3a-2b=8 \quad a+b=2$

c) $5p+2q=18 \quad q=2p$

d) $7-y=2x \quad x/3=y$

e) $u=5/v \quad u+v=6$

32.7 Each of the following relationships is between two variable quantities. Given the variable to be plotted along one particular axis, what function of the other variable would you plot along the other axis to give a straight line?

Variables	Relationship	Horizontal axis	Vertical axis
a) f and T	$f=\frac{1}{2L}\left(\frac{T}{\mu}\right)^{\frac{1}{2}}$	T	?
b) I and z	$I=I_0e^{-\mu z}$	z	?
c) E and r	$E=k/r^2$	?	E
d) p and V	$p(V-b)=RT$	?	pV
e) C and t	$C-C_0=Ae^{-\lambda t}$	t	?

32.8 Which graph A to E fits each of the equations below?

a) $xy=$ constant

b) $x+y=$ constant

c) $y=Ae^{-kx}$ where A and k are constants

d) $y=$ (constant) x^2

e) $y-a=bx$ where a and b are constants.

32.9 Differentiate each of the following functions with respect to x.

a) $y=3x^2$

b) $y=2x-x^2$

c) $y=3/x^2$

d) $y=Ae^{-3x}$

e) $y=\sin 2x$

32.10 What physical variable is represented by the area under the curve for each of the following graphs?

a) induced e.m.f. against time

b) current against time

c) pressure against volume

d) stress against strain

e) acceleration against time

32.11 For the equation $C=Ae^{-\mu x}$, when $x=0$, $C=100$. Also, when $x=5$, $C=36$. Calculate:

a) the values of A and μ,

b) the value of C for (i) $x=1$, (ii) $x=10$,

c) the value of x when (i) $C=50$, (ii) $C=10$.

32.12

a) Plot a graph of the function $N=1000\,e^{-5t}$ from $t=0$ to $t=1$. What is the 'half-life' of the curve?

b) What is the ratio of the initial gradient to the gradient at one half-life?

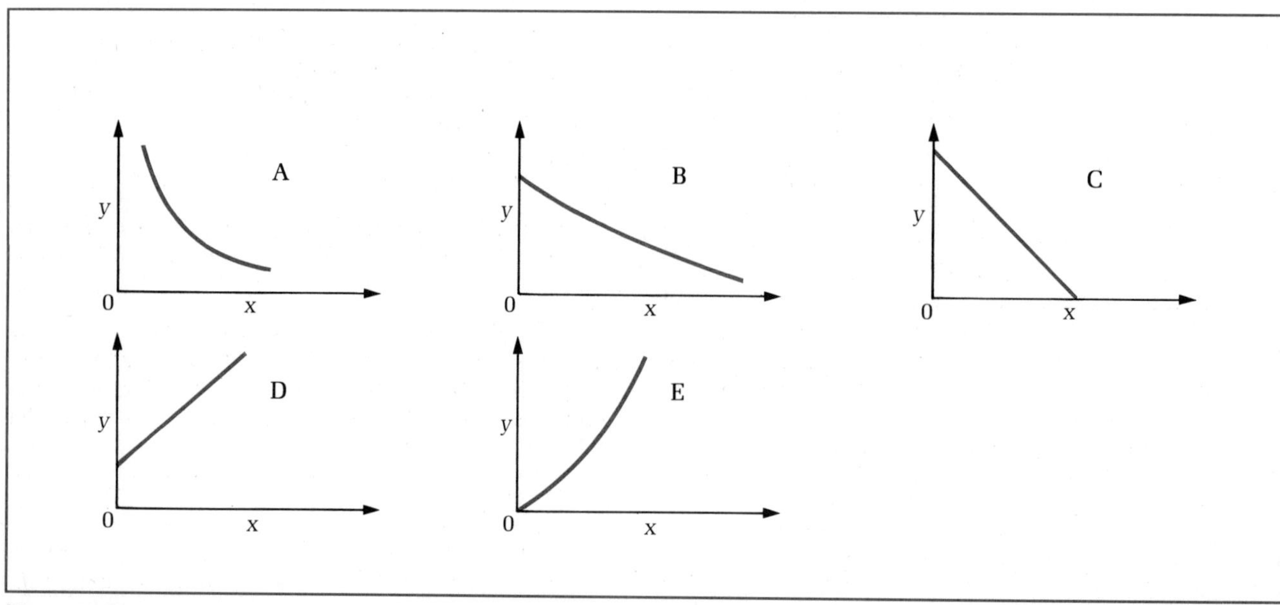

Figure 32.29

33

Data analysis

33.1 Units

Base units

Units for measurement have always played an important part in the life of any community. In the Middle Ages, the unit of length was the yard, based at first on the circumference of a person's body, then later the length of Henry I's arm! Small wonder that units in those times varied from country to country, sometimes even within the same country. Clearly, a nationally agreed 'yardstick' was an important development. Fortunately for us, scientists use an internationally agreed system of units called SI (*Systeme Internationale*) units. Communications between scientists in different countries or in different disciplines is much easier because of the use of a common system of units throughout the World.

The (SI) system is founded on seven **base units**, all of which are defined by international agreement. All other scientific units are derived from these seven base units.

The metre (m) is the unit of length, defined in terms of the wavelength of light from the krypton-86 atom.

The kilogram (kg) is the unit of mass, defined by an international prototype at Sevres, France.

The second (s) is the unit of time, defined in terms of the frequency of light from the caesium-133 atom.

The ampere (A) is the unit of current, defined in terms of the force between two parallel conductors (carrying equal current) 1 m apart (see p. 188).

The kelvin (K) is the unit of thermodynamic temperature (see p. 107).

The mole (mol) is the unit of amount of a substance (see p. 67).

The candela (cd) is the unit of luminous intensity. This unit does not feature in A-level courses.

The exact definitions of these base units are given in most physics reference books. Apart from the ampere, knowledge of the exact definitions is not required at A-level. Special laboratories in many countries use these definitions to calibrate instruments accurately for use by scientists outside these 'standards' laboratories. For example, the manufacturer of the 'humble' metre rule has a standard metre rule that is used to check the metre rules he makes; the standard metre rule would have been made in a 'standards' laboratory.

Each unit has a recognized symbol. The symbol is a small letter (or letters) unless it is the symbol of a unit named after a scientist (e.g. the ampere: A) where it is given a capital letter.

Prefixes used with SI units are as follows.

Derived units

All SI units that are not base units are defined in terms of base units. The derived units, as they are called, are built up 'step-by-step' from the base units using the known links between physical quantities. For example, the unit of speed is m s^{-1}; the negative index (e.g. '-1') is used for 'per'. So m s^{-1} means 'metres per second'. Figure 33.1 shows how some of the more common derived units are linked to the four base units, the metre, the kilogram, the second and the ampere.

Consider the following examples to see how they are linked to the base units.

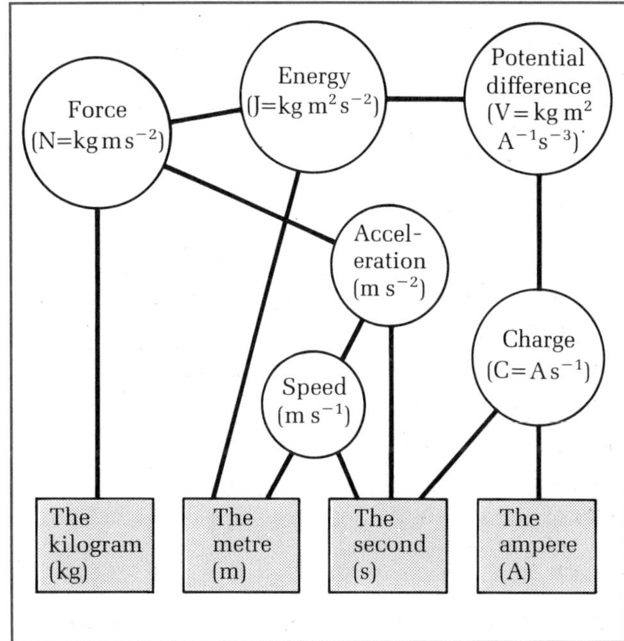

Figure 33.1 *Using base units*

The pascal (Pa) is the unit of pressure.
Pressure = force/area so 1 Pa = 1 N m^{-2}.
Force = mass × acceleration so 1 N = 1 kg m s^{-2}
Hence 1 Pa = 1 kg m s^{-2} m^{-2} = 1 kg m^{-1} s^{-2}.

The volt (V) is the unit of potential difference.
Potential difference = work done/charge so 1 V = 1 J C^{-1}.
Work = force × distance so 1 J = 1 N m = 1 kg m^2 s^{-2} since 1 N = 1 kg m s^{-2}.
Charge = current × time so 1 C = 1 A s.
Hence 1 V = 1 kg m^2 s^{-2} A^{-1} s^{-1} = 1 kg m^2 s^{-3} A^{-1}.

Now you can appreciate why we have *named* units. Imagine asking in your local shop for a 1.5 kg m^2 s^{-3} A^{-1} cell!

A list of the more common derived units is given at the top of the next page. The abbreviation for each unit and the symbol for each quantity is also given. Take care not to confuse symbols for quantities (e.g. *F* for force) with the abbreviations for units (e.g. F for farad).

Prefix	atto	femto	pico	nano	micro	milli	—	kilo	mega	giga	tera
Factor	10^{-18}	10^{-15}	10^{-12}	10^{-9}	10^{-6}	10^{-3}	1	10^3	10^6	10^9	10^{12}
Symbol	a	f	p	n	μ	m	—	k	M	G	T

For example, 6 μA = 6 microamperes = 6 × 10^{-6} A.

Note: MΩ is pronounced megohm

Quantity	Symbol	Unit	Abbreviation	Base units
Speed or velocity	v	metre per second	$\mathrm{m\,s^{-1}}$	$\mathrm{m\,s^{-1}}$
Acceleration	a	metre per second per second	$\mathrm{m\,s^{-2}}$	$\mathrm{m\,s^{-2}}$
Force	F	newton	N	$\mathrm{kg\,m\,s^{-2}}$
Energy or work	E or W	joule	J	$\mathrm{kg\,m^2\,s^{-2}}$
Power	P	watt	W	$\mathrm{kg\,m^2\,s^{-3}}$
Pressure	p	pascal	Pa	$\mathrm{kg\,m^{-1}\,s^{-2}}$
Electric charge	Q	coulomb	C	A s
Potential difference	V	volt	V	$\mathrm{kg\,m^2\,s^{-3}\,A^{-1}}$
Resistance	R	ohm	Ω	$\mathrm{kg\,m^2\,s^{-3}\,A^{-2}}$
Capacitance	C	farad	F	$\mathrm{A^2\,s^4\,kg^{-1}\,m^{-2}}$
Magnetic field strength	B	tesla	T	$\mathrm{kg\,s^{-2}\,A^{-1}}$
Electric field strength	E	volt per metre	$\mathrm{V\,m^{-1}}$	$\mathrm{kg\,m\,s^{-3}\,A^{-1}}$
Magnetic flux	Φ	weber	Wb	$\mathrm{kg\,m^2\,s^{-2}\,A^{-1}}$
Temperature gradient	—	kelvin per metre	$\mathrm{K\,m^{-1}}$	$\mathrm{K\,m^{-1}}$

Symbols and formulae

It is the normal convention that the symbol for a physical quantity represents the magnitude *and* the unit of the quantity. For example, if the speed of an object is $3\ \mathrm{m\,s^{-1}}$, then we write $v = 3\ \mathrm{m\,s^{-1}}$. Under the accepted convention, it is meaningless to write $v = 3$. Since each symbol represents magnitude and unit, then phrases such as 'Let the speed $= v\ \mathrm{m\,s^{-1}}$' are not correct; the statement would be the same as writing '$v = 3\ \mathrm{m\,s^{-1}\,m\,s^{-1}}$' if v was $3\ \mathrm{m\,s^{-1}}$. Symbols must be used carefully in formulae and calculations.

For graphs the unit and any prefix of a physical quantity is written near the appropriate axis in the form 'symbol/(prefix + unit)'. For example, if you wish to show that the vertical-axis of a graph represents speed in $\mathrm{mm\,s^{-1}}$, then the correct format to label the axis is

speed $v/\mathrm{mm\,s^{-1}}$.

Reading off the graph axis gives a number equal to $v/\mathrm{mm\,s^{-1}}$. So, for example, if the reading gives $v/\mathrm{mm\,s^{-1}} = 5$, then $v = 5\ \mathrm{mm\,s^{-1}}$. The prefix is sometimes replaced by powers of ten. So the label '$v/\mathrm{mm\,s^{-1}}$' could be written '$v/10^{-3}\ \mathrm{m\,s^{-1}}$'.

For calculations involving formulae, a helpful procedure is as follows.

1 List the value and unit of each 'known' quantity.
2 Convert all magnitudes to standard form, and all prefixes to powers of ten.
3 State the formula to be used.
4 Use the formula for the calculation.
5 Use the formula to work out the unit of the quantity to be calculated if its unit is not known.

For example, suppose you were given the heat radiation formula $P = \sigma A T^4$ for the power P radiated away from a hot wire of surface area A at surface temperature T, with the following instructions.

a) Show that the unit of σ, $\mathrm{W\,m^{-2}\,K^{-4}}$, is equivalent to $\mathrm{kg\,s^{-3}\,K^{-4}}$.

b) Calculate the surface temperature of a wire of length 326 mm and diameter 0.15 mm when the wire radiates power at 20 W, given $\sigma = 5.7 \times 10^{-8}\ \mathrm{W\,m^{-2}\,K^{-4}}$.

The equation can be rearranged to give $\sigma = P/AT^4$. So the base units of σ are the base units of P/ the base units of AT^4. Hence $1\ \mathrm{W\,m^{-2}\,K^{-4}}$ is equal to $1\ \mathrm{kg\,s^{-3}\,K^{-4}}$.

To calculate the surface temperature
$\sigma = 5.7 \times 10^{-8}\ \mathrm{W\,m^{-2}\,K^{-4}}$
$P = 20\ \mathrm{W}$,
A = surface area $= \pi \times$ diameter $\times$ length
$= \pi \times 0.15 \times 10^{-3} \times 0.326\ \mathrm{m^2}$
Hence

$$T^4 = \frac{P}{\sigma A} = \frac{20}{5.7 \times 10^{-8} \times \pi \times 0.326 \times 0.15 \times 10^{-3}}$$

So $T^4 = 2.29 \times 10^{12}$
Hence $T = 1230\ \mathrm{K}$

In the above example, the unit of the quantity to be calculated is known. Some formulae do not give such obvious units. Consider the equation for the drag force F on a sphere of radius r moving at speed v through a fluid of viscosity η.

$F = 6\pi\eta r v$

What are the base units of η? Rearrange the equation to make η its subject. So $\eta = F/(6\pi r v)$.

The unit of η is given by

$$\frac{\text{The unit of force}}{\text{Unit of } r \times \text{unit of } v} = \frac{\mathrm{kg\,m\,s^{-2}}}{\mathrm{m \times m\,s^{-1}}}$$

$= \mathrm{kg\,m^{-1}\,s^{-1}}$

33.2 Dimensional analysis

The method of expressing any derived unit in terms of the base units can be taken one stage further to make it more formal. Each of the base quantities is called a 'dimension', and is given a special symbol.

Dimension	Dimension symbol
Mass	M
Length	L
Time	T
Current	A
Temperature	θ

Every other quantity can be expressed in these basic dimensions without reference to the actual base units. The dimensions of a quantity are indicated by the use of square brackets. For example, [speed] means 'the dimensions of speed' which are written $\mathrm{LT^{-1}}$ since speed is distance per unit time. The dimensions of each of the quantities listed below can be worked out in the same way as its base units; the links between each quantity and the base units are used.

Speed	$[v] = L\,T^{-1}$
Acceleration	$[a] = L\,T^{-2}$
Force	$[F] = M\,L\,T^{-2}$
Energy	$[E] = M\,L^2\,T^{-2}$
Power	$[P] = M\,L^2\,T^{-3}$
Pressure	$[p] = M\,L^{-1}\,T^{-2}$
Charge	$[Q] = A\,T$
P.D.	$[V] = M\,L^2\,T^{-3}\,A^{-1}$
Resistance	$[R] = M\,L^2\,T^{-3}\,A^{-2}$
Capacitance	$[C] = A^2\,T^4\,M^{-1}L^{-2}$

Using dimensions

To determine the dimensions or units of an expression in a formula. For example, suppose you wish to determine the base units of thermal conductivity k. The equation for thermal conductivity is

$$Q/t = kA(T_1 - T_2)/L$$

where Q/t is the heat flow/second,
A is the area of cross-section,
$(T_1 - T_2)/L$ is the temperature gradient.

Hence $[Q/t] = [k] \times [A] \times [(T_1 - T_2)/L]$
$[Q/t] = [Q]/[t] = [\text{energy}]/[t] = [M\,L^2\,T^{-2}]/T = M\,L^2\,T^{-3}$.
$[A] = L^2$.
$[(T_1 - T_2)/L] = \theta\,L^{-1}$.
So $M\,L^2\,T^{-3} = [k] \times L^2 \times \theta L^{-1} = [k]\theta L$

$$\text{Giving}\quad [k] = \frac{ML^2T^{-3}}{L\theta} = M\,L\,T^{-3}\,\theta^{-1}$$

The base units for k are therefore $\text{kg m s}^{-3}\text{K}^{-1}$.

To check equations Suppose you want to check the link between the variables in the equation for the time constant of an inductor and resistor in series (i.e. growth of current in a series LR circuit; see p. 204). Perhaps you can't remember whether the time constant is RL or R/L or L/R.

$$[\text{Resistance}] = \frac{[\text{p.d.}]}{[\text{current}]} = \frac{M\,L^2\,T^{-3}\,A^{-1}}{A} = M\,L^2\,T^{-3}\,A^{-2}$$

$$[\text{Inductance}] = \frac{[\text{induced e.m.f.}]}{[\text{rate of change of current}]} = \frac{[V]}{[I]/[t]}$$

$$= \frac{M\,L^2\,T^{-3}\,A^{-1}}{A\,T^{-1}} = \frac{M\,L^2T^{-3}\,A^{-2}}{T^{-1}}$$

$$\text{Hence [inductance]} = \frac{[\text{resistance}]}{T^{-1}}$$

$$\text{So}\qquad T^{-1} = \frac{[\text{resistance}]}{[\text{inductance}]}$$

Therefore the time constant is L/R.

To establish links between quantities in a given situation. For example, suppose we are given that the drag force F on a sphere moving through a fluid at steady speed depends on

a) the fluid viscosity η,
b) the speed v,
c) the sphere's radius r.

Write the link as an equation $F = k\eta^a v^b r^c$ where k is a numerical constant and a, b and c are powers to be determined. Since k is a numerical constant, it has no dimensions.

$[F] = [\eta^a][v^b][r^c]$
but $[F] = M\,L\,T^{-2}$,
$[\eta] = M\,L^{-1}\,T^{-1}$
$= [\text{stress}]/[\text{velocity gradient}]$
$= [\text{force}]/[\text{area} \times \text{velocity gradient}]$
$= M\,L\,T^{-2}/(L^2\,T^{-1})$
$[v] = L\,T^{-1}$
$[r] = L$

Hence $M\,L\,T^{-2} = (M\,L^{-1}\,T^{-1})^a \times (L\,T^{-1})^b \times (L)^c$
so $M\,L\,T^{-2} = M^aL^{-a}T^{-a} \times L^bT^{-b} \times L^c$
i.e. $M\,L\,T^{-2} = M^aL^{-a+b+c}T^{-a-b}$

Dimensions of M: Left hand side $= 1$
Right hand side $= a$ hence $a = 1$
Dimensions of L: Left hand side $= 1$
Right hand side $= -a+b+c$
Hence $-a+b+c = 1$
Dimensions of T: Left hand side $= -2$
Right hand side $= -a-b$
Hence $-a-b = -2$

Since $a = 1$, the last equation $(-a-b = -2)$ gives $b = 1$.
So the second equation $(-a+b+c = 1)$ gives $c = 1$.
So $a = 1$, $b = 1$, $c = 1$.
Hence $F = k\eta rv$

The method cannot be used to determine the value of the numerical constant k but it does establish the link between the variables.

33.3 Treatment of errors

Types of error

An error in a measurement makes the measured value differ from the correct value. Measurement gives a value that may or may not be the correct value because errors may be present. So how can the correct value be determined? The best that can be done is to state **the probable error** when a measurement is made. For example, if the diameter of a wire is measured at 0.36 mm with a probable error of 0.02 mm, then the correct value is in the range 0.34 to 0.38 mm. The diameter is written as 0.36 ± 0.02 mm where ± 0.02 mm is the probable error and 0.36 mm is the value of the measurement.

The percentage probable error or just the percentage error, is the probable error converted to a percentage of the measurement. So the percentage error in the diameter measurement above is 5.5%. Resistors and capacitors are usually coded so that the percentage error or **tolerance** can be seen by inspection; for example, a resistor with a gold tolerance band has a tolerance of 5%. If the resistor's value is 100 ohms, then its resistance is in the range from 95 ohms to 105 ohms.

Errors in measurements are caused in many ways. If

the sensitivity of an instrument decreases after its initial calibration, then the instrument will consistently read 'low'; its reading will always be less than the correct value because the instrument has become less sensitive. This is an example of a **systematic error**. Such errors can be difficult to detect. Sometimes they only become obvious when the results are plotted on a graph. Recalibrating meters, checking zero errors, using a plane mirror to read a scale, as in Figure 33.2 are all ways of avoiding systematic errors.

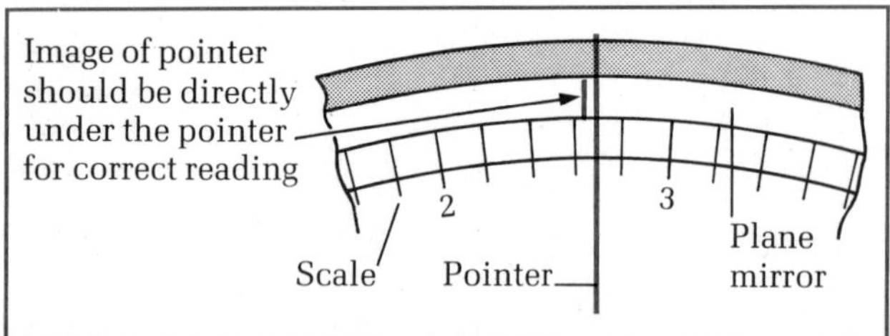

Figure 33.2 *Using a plane mirror*

Random errors are evident when repeated measurements of the same quantity in the same situation give different readings. Try timing an oscillating body for ten cycles; repeat your timing a few times to give several values of the time for ten cycles, and you ought to find there is a spread of values. The probable error gives the range which most of the readings lie in. The mean value of the readings is calculated by adding the readings together then dividing the sum by the total number of readings. By making as many timings as possible, the probable error is reduced and so the mean value is more accurate; in other words, the mean value is closer to the true value. Suppose the timings were:

20.2 s, 19.6 s, 20.0 s, 20.1 s, 19.8 s, 19.6 s, 19.9 s, 20.2 s, 20.0 s, and 20.1 s.

The mean value = sum of all ten timings/10 = 19.95 s.
The probable error = ± 0.2 s.

Hence the timing = 20.0 ± 0.2 s, since the value 19.95 is rounded up to 20.0 so that it is stated with the same number of decimal places as the probable error. Otherwise, 19.95 implies precision to within ±0.05 s which is not true.

If all the readings are the same, then the probable error is given by the precision with which the reading is made.

Plotting errors

When a variable quantity is measured at different intervals on a scale, the usual practice is to estimate the probable error from the precision with which the scale can be read. For example, Figure 33.2 shows the scale of a 0–5 A ammeter; the scale can be read to a precision of 0.2 A, so the probable error of a single reading is ± 0.2 A. Random errors that occur when a set of readings is made off a scale can be dealt with by graphs.

A straight line graph can be plotted if the relationship between the variables is known. For example, consider the following measurements made in a simple pendulum experiment to determine g (the acceleration due to gravity). See p. 224 for details of the theory of the simple pendulum.

Length of pendulum ***L*/mm**	200	400	600	800	1000 ± 4 mm
Time for 20 oscillations/s	18.2	25.4	31.4	35.5	40.0
	18.1	25.7	31.4	35.7	40.4
	17.8	25.3	30.9	35.8	40.4
Average time for 20 oscillations/s	18.0	25.5	31.2	35.7	40.3 ± 0.2 s
Time period, *T*/s	0.90	1.28	1.56	1.79	2.02 ± 0.01 s.

The time period is given by the equation

$$T = 2\pi\sqrt{\frac{L}{g}}$$

$$\text{Hence } T^2 = \frac{4\pi^2}{g}L$$

A graph of T^2 against L should give a straight line through the origin. The gradient of the line is equal to $4\pi^2/g$, so by measuring the gradient, the value of g can be calculated.

The probable errors in the measurements can be shown on the graph by using the 'box' method shown in Figure 33.3. Each box has width corresponding to the probable error of ±4 mm in the length measurement. The height of each box isn't so easy to deal with because the vertical axis represents T^2 not T. The most straightforward approach is to start with the range of each value.

For L = 200 mm, T is in the range 0.89 to 0.91 s. So T^2 is in the range from 0.79 to 0.83 s²; the value of T^2 is 0.81 ± 0.02 s². The height of the error box in this case is therefore 0.04 s². The other box heights can be calculated in the same way.

***L*/m**	0.200	0.400	0.600	0.800	1.000	±4 mm
***T*²/s²**	0.81	1.63	2.43	3.19	4.06	
Probable error in *T*²/s²	±0.02	±0.03	±0.03	±0.04	±0.04	

Figure 33.3 shows two straight lines drawn through the error boxes. Line A is as steep as possible, passing through each box. Line B is the line with the least possible gradient that passes through the boxes. The two lines therefore give the maximum and the minimum possible gradient, according to the errors. So an average value *and* a probable error can be determined for the gradient.

Line A gradient = 4.138 s² m⁻¹,
Line B gradient = 3.913 s² m⁻¹.

Hence the average value of the gradient = 4.03 s² m⁻¹, and the probable error is ±0.11 s² m⁻¹.

$$\frac{4\pi^2}{g} = 4.03 \pm 0.11\ \text{s}^2\,\text{m}^{-1}$$

The maximum value of $4\pi^2/g = 4.14$, so the minimum value of g is $4\pi^2/4.14 = 9.54$ m s⁻².

The minimum value of $4\pi^2/g = 3.91$, so the maximum value of g is $4\pi^2/3.91 = 10.12$ m s⁻².

Hence g = 9.83 ± 0.29 m s⁻².

Curve fitting is necessary if a straight line graph is not possible. The curve should be drawn so that it passes smoothly through all the error boxes. If the gradient changes sharply, then more measurements need to be made to define the curve more closely in that area. For example, suppose you are investigating the current against voltage characteristic for a silicon diode; typical results are shown in Figure 33.4 (opposite). The gradient

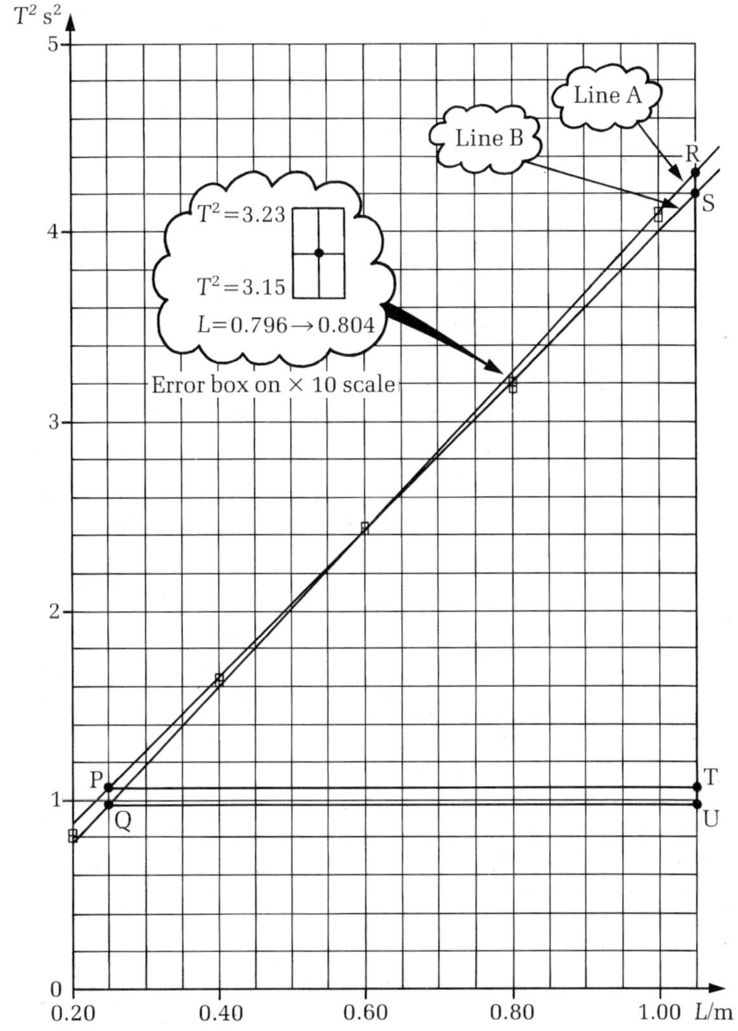

$$\text{Line A: gradient} = \frac{RU}{QU} = \frac{4.29 - 0.98}{1.02 - 0.25}\frac{(s^2)}{(m^2)} = 4.138\ s^2\ m^{-1}$$

$$\text{Line B: gradient} = \frac{ST}{TP} = \frac{4.19 - 1.06}{1.05 - 0.25}\frac{(s^2)}{(m^2)} = 3.913\ s^2\ m^{-1}$$

Figure 33.3 *Showing errors on graphs*

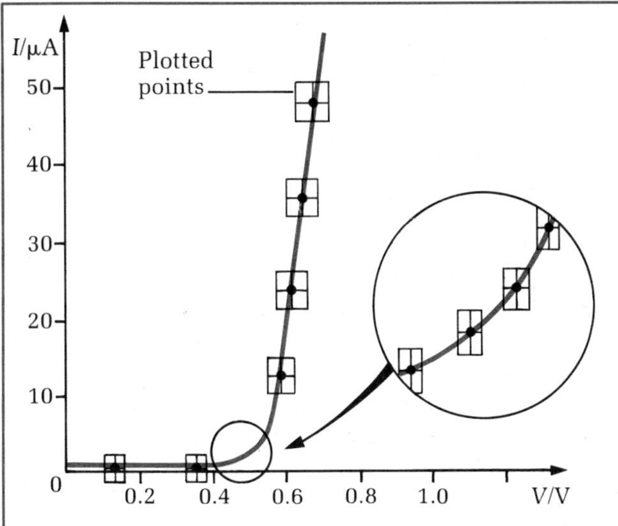

Figure 33.4 *Curve fitting*

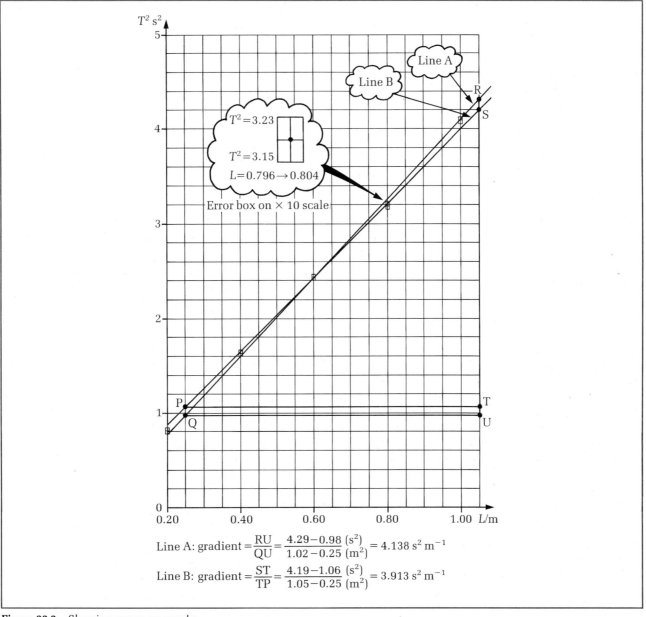

changes sharply at about 0.5 V. So as many measurements as possible in this part need to be made.

Combining errors

Add the probable errors where the change of a quantity is to be calculated. For example, suppose a travelling microscope is used to measure the distance across ten fringe widths in an interference experiment (see p. 258). The cross-wires of the microscope are centred on one of the dark fringes, as in Figure 33.5, at the top of the next page. The reading of the microscope's vernier scale, x_1, is recorded. Then the cross-wires are moved along a line at right angles to the fringes across ten fringe spacings. The reading at this new position, x_2, is then recorded. The precision which each reading is made with is determined by the contrast of the fringes. So a probable error e for each reading is estimated.

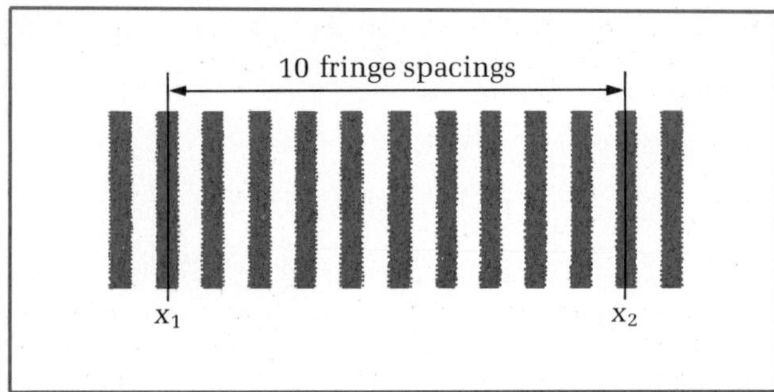

Figure 33.5 *Adding errors*

Suppose $x_1 = 3.5 \pm 0.1$ mm i.e. x_1 is from 3.4 to 3.6 mm.
$x_2 = 6.6 \pm 0.1$ mm i.e. x_2 is from 6.5 to 6.7 mm.
So the difference $x_2 - x_1$ is in the range from 2.9 (=6.5 − 3.6) to 3.3 mm (=6.7 − 3.4).
Hence $x_2 - x_1 = 3.1 \pm 0.2$ mm.

The probable error in the difference of the two readings is obtained by adding the individual errors.

Percentage probable errors are added where quantities are to be multiplied or divided by one another. For example, suppose we wish to calculate the average speed of a ball bearing falling through a vertical tube of water.

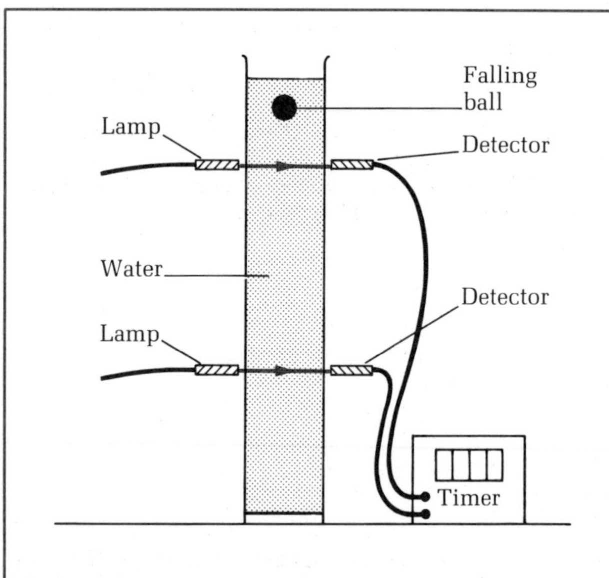

Figure 33.6 *Adding percentage errors*

Light gates could be used to time the ball as it falls through a measured distance. Several timings would need to be made, releasing the ball from rest at the same point each time. Suppose the readings are as follows.

Position of first light gate = 21 ± 2 mm
Position of second light gate = 224 ± 2 mm
Timings/ms 531 548 564 542 535
Hence distance fallen $D = 203 \pm 4$ mm
and the average time $T = 544 \pm 15$ ms
The average speed $= D/T = 0.373$ mm ms^{-1}
$= 0.373$ m s^{-1}.

The percentage error for the average speed
= percentage error for D + percentage error for T.
Percentage error for $D = 4 \times 100/203 = 2\%$.
Percentage error for $T = 15 \times 100/544 = 2.8\%$.
So the percentage error for the speed = 4.8%.

Therefore the probable error for the speed is $4.8 \times 0.373/100 = 0.018$ m s^{-1}.
Average speed $= 0.373 \pm 0.018$ m s^{-1}.

To understand why the percentage errors are added, consider the example again.

$$\text{Average speed} = \frac{D \pm \Delta D}{T \pm \Delta T} = \frac{(D \pm \Delta D)}{(T \pm \Delta T)} \times \frac{(T \mp \Delta T)}{(T \mp \Delta T)}$$

However $(T \pm \Delta T)(T \pm \Delta T) = T^2$, neglecting the second order term ΔT^2 since it is very small compared with the others.
Hence the average speed

$$= \frac{(D \pm \Delta D) \times (T \mp \Delta T)}{T^2}$$

$$= \frac{D}{T}(1 \pm \Delta D/D) \times (1 \mp \Delta T/T)$$

$$= \frac{D}{T}(1 \pm \Delta D/D \pm \Delta T/T)$$

again neglecting second order terms.

So if the average speed is $V \pm \Delta V$, which equals $V(1 \pm \Delta V/V)$,
then $\Delta V/V = \Delta D/D + \Delta T/T$.
Hence the percentage error in V is the sum of the percentage errors in D and T.

When quantities are multiplied or divided, the percentage error in the final value is the product of the individual percentage errors. If a quantity in the formula is to the power n, then that quantity contributes n times its individual percentage error to the total error. For example, the formula for the flow rate of a viscous liquid through a pipe is

$$Q/t = \frac{pR^4}{8\eta L}$$

where Q/t = volume per second,
p = pressure across ends,
L = pipe length,
R = pipe radius,
η = fluid viscosity.

To calculate η from measurements of all the other quantities, the formula must be rearranged to give

$$\eta = \frac{pR^4}{8LQ/t}$$

The total percentage error = percentage error in p + percentage error in R^4 + percentage error in L + percentage error in Q/t.

However, the percentage error in R^4 is four times the percentage error in R. So the contribution from the percentage error in R is much more significant than the other contributions. If R, L, Q/t, and p are each measured accurate to 2%, the total percentage error is 14%; R contributes 8% of the 14% total. To reduce the contribution from R^4 down to 2%, the error in R must be cut down to 0.5%. So R must be measured much more accurately than the other quantities.

33.4 Summary

Units

1 The most commonly-used base units are the metre (m); the kilogram (kg); the second (s); the ampere (A).
2 Each derived unit can be written as a combination of base units.
3 The base dimensions are mass M, length L, time T, current A and temperature. The dimensions of any physical quantity can be written in terms of the base dimensions.

Errors

1 The probable error of a measurement gives the range which the correct value will nearly always lie in.
2 The percentage error of a measurement is 100 × the probable error/the measurement.
3 Where quantities are added or subtracted from one another, the total probable error is the sum of the individual errors.
4 Where quantities are multiplied or divided, the total percentage error is the product of the individual percentage errors.

Short questions

33.1 Express each of the following quantities in terms of the base units of the SI system.
a) momentum
b) resistivity
c) thermal conductivity
d) density
e) electric field constant, ε_0.

33.2 Show that each of the following equations is dimensionally correct.
a) Time constant $= RC$,
where R = resistance,
C = capacitance.

b) Speed of light $c = \left(\dfrac{1}{\mu_0 \varepsilon_0}\right)^{\frac{1}{2}}$
where μ_0 = magnetic field constant,
ε_0 = electric field constant.

c) Energy stored $= \frac{1}{2}CV^2$,
where C = capacitance,
V = p.d.

d) Time period of a simple pendulum $T = 2\pi\sqrt{\dfrac{L}{g}}$
where L = length.

e) Speed of waves on a stretched wire $c = \left(\dfrac{T}{\mu}\right)^{\frac{1}{2}}$
where T = tension,
μ = mass per unit length.

33.3 In an experiment to measure the density of a ball bearing, the following readings were obtained
mass $= 7.06 \pm 0.02$ g
diameter $= 12.06 \pm 0.04$ mm
Calculate
a) the percentage error of each measurement,
b) the value of the density,
c) the percentage error in the density value.

33.4 In an experiment to determine g, the acceleration of free fall, a steel ball bearing was timed falling from rest through a distance D. The readings obtained were as follows.
$D = 1.215 \pm 0.004$ m.
Time taken = 495, 498, 503, 496, 501 ms.
Calculate
a) the percentage error in D and t, the time taken,
b) the value of g,
c) the percentage error in g.

33.5 Discuss the accuracy of the falling ball method for g compared with the simple pendulum method described on p. 514.

Longer questions

33.6 (45 min) This question is about drag, or resistance to motion acting on a high speed train. The drag is thought to be given by the equation

$$F = A + Bv + Cv^2$$

where F is the drag in kN, v is the velocity of the train in m s^{-1}, and A, B and C are constants.

The table below gives the observations made in a test in which the drag was measured at various train velocities.

Velocity in m s^{-1}	Drag in kN
5	3.3
10	4.4
15	6.0
20	8.2
25	10.9
30	14.0
35	17.8
40	22.0
45	26.8

a) Plot a graph of drag (y-axis) against velocity (x-axis).
b) Use your graph to estimate the value of A.
c) If the drag equation is correct, the slope S of the graph will be given by the equation

$$S = B + 2Cv$$

i) Measure the slope of the graph for train velocities of 10 m s^{-1}, 20 m s^{-1}, 30 m s^{-1} and 40 m s^{-1} showing how you obtained your values. Tabulate your results.
ii) Plot a suitable graph to test the relationship and use your graph to determine values for B and C. In each case explain how you obtained the result.

(AEB)

33.7 (Approx. 30 min.) A Geiger-Muller tube was fixed with its axis horizontal at a place on a bench well-removed from all known radioactive sources. Background count-rate measurements were made by recording counts over several ten-minute periods. The following counts were obtained for five such periods:
290, 277, 273, 263 and 247.
Find the mean background count-rate in counts per minute.

A weak radioactive source was then mounted on the axis of the tube with its protective grille facing the end-window of the tube, so that it could be moved along the axis to give various distances s between the grille and the end-window, as in the Figure 33.7.

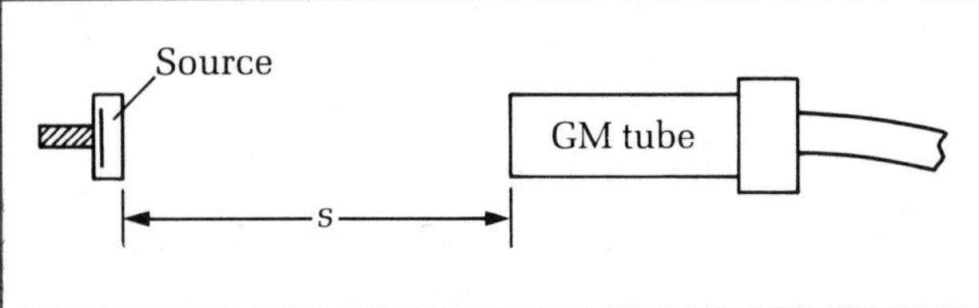

Figure 33.7

Counts, N, were taken over four-minute periods for the various values of s with results as follows:

s/mm	10	15	30	45	60	75	90	105
N	too rapid for the counter	7820	7980	4536	2942	2076	1554	1215

Copy this table adding further lines for n and $1/\sqrt{n}$, where n is the corrected count over four minutes (i.e. counts recorded minus background counts).

It is thought that the relationship between n and s is likely to be of the form $1/\sqrt{n} = k(s + x)$, where k and x are constants. Plot a graph of $1/\sqrt{n}$ against s/mm, and use it to obtain values for k and x.

What practical significance can be attached to x?

Plot a further graph of lg n against lg((s + x)/mm) and find its gradient.

(SUJB)

33.8 ($1\frac{1}{2}$ hours) Read the following account of an experimental investigation and then answer the questions at the end.

A thin metal wire is heated in a vacuum to a high temperature, T, by passing a steady current, I, through it. The potential difference, V, between the ends of the hot wire is also recorded. Varying I allows the wire to be heated to different temperatures.

At each temperature all the electrons which are emitted from the wire by thermionic emission are collected by a cylindrical metal anode surrounding the wire. The electrons constitute an emission current, I_e, which is also recorded for each temperature of the wire.

The data obtained are tabulated below.

I/A	V/V	T/K	I_e/mA
2.01	5.00	1575	30.4
1.88	4.53	1545	16.9
1.74	3.95	1490	6.06
1.67	3.59	1445	2.95
1.52	3.06	1385	0.913
1.40	2.61	1320	0.234
1.34	2.34	1270	0.087

Questions

1 Assuming the variation of resistance with temperature for the material of the wire is given by

$$R_T = R_0(1 + \alpha T + \beta T^2)$$

where R_T is the resistance at temperature T and R_0, α and β are constants of the wire, show that a graph of $\frac{(R_T - R_0)}{T}$ as ordinate against T as abscissa should yield a linear plot.

2 Assuming $R_0 = 0.20\,\Omega$, construct a table of values of T, R_T and $(R_T - R_0)/T$. By plotting a graph of $(R_T - R_0)/T$ as ordinate against T as abscissa, determine the value of α and β.

3 It is suspected that over the range of temperatures used, the emission current, I_e, is related to the temperature, T, of the wire by the relation

$$I_e = cT^n$$

where c and n are constants.

By plotting a graph of $\log_{10}(I_e/\text{mA})$ as ordinate against $\log_{10}(T/\text{K})$ as abscissa test this relation, and hence deduce a value for n.

4 It was hoped to stabilize the temperature of the wire at 1520 K, but this could only be achieved to ±1%.

a) What are the maximum and minimum temperatures of the wire for this 1% variation?

b) Find, using your graph in question 3 above or otherwise, the corresponding maximum and minimum emission currents.

(London)

34

In the laboratory

34.1 The role of measurements

Physics is an experimental science. Our present understanding of the natural world is based on theories and principles that have been tested by experiments. Theories which do not fit the facts obtained by experiments must be discarded or adapted. New discoveries by experimenters test existing theories; if the theory fails the test, its basic assumptions must be questioned. Science is littered with discarded ideas once held high as being correct; the 'caloric theory' is one such example. Heat was thought of as a fluid that flowed from hot to cold objects. Benjamin Thompson put paid to 'caloric' in the early 19th century when he showed that doing work by drilling on a metal caused it to heat up.

What do we mean when we say a theory is 'correct'? Essentially, the meaning is that no one has managed to disprove it. All the experimental tests on that theory have confirmed its correctness; if any one test had disagreed with the theory, then the theory must be adapted or replaced. We can never say that a given theory is 'true' because someone, somewhere might manage to disprove it as a result of a new experiment. Science works by disproving ideas and theories; our understanding of the natural world is in terms of ideas and theories that have not been disproved.

A famous example of the link between theory and experiment is the so-called **ultra-violet catastrophe** of the late 19th century. Up to that time, light was believed to travel in the form of continuous waves emitted by light sources. Wave theory successfully explained interference and diffraction effects (see p. 235). However, in the late 19th century, two experimenters, O. Lummer and E. Pringsheim, carried out a detailed investigation of the thermal radiation from 'black body radiators'. They measured the distribution of radiation energy with wavelength for the black body at different temperatures. Their results in graph form are known as black body radiation curves (see Figure 34.1); these curves presented the best scientists of that time with more than a headache! Wave theory could be used to explain the curve either side of the peak, but it predicted an infinitely high peak. The problem became known as the 'ultra-violet catastrophe' because the unwanted infinity was towards the UV region of the electromagnetic spectrum. Eventually, the problem was solved by Planck and Einstein, who rethought the nature of light and established the photon theory of light (see p. 428). Black body radiation curves are explained with complete success using the photon theory. And the photon theory predicts and explains much more too.

Experimental work in physics involves testing theories by investigating relationships between physical quantities. An equally important role for the experimenter is to investigate the behaviour of materials and devices. Once known, the material or device can be used correctly within its known limitations. If the behaviour is not fully known, the device or material could fail, with disastrous results! So experimenters have a dual role. Theories must be tested and materials and devices must be investigated. The first aspect, testing theories, is essentially pure science; testing materials and devices involves applications of science although there is often no well-defined border between applied and pure science. Both aspects involve identifying, controlling and measuring physical quantities, then looking for links between the quantities.

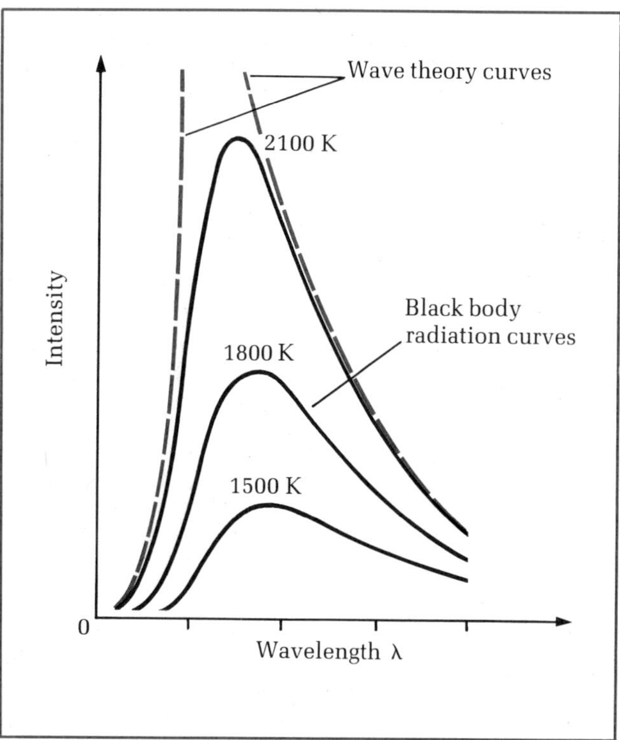

Figure 34.1 *The downfall of wave theory*

Measurements play a key role in science, so they must be reliable. Reliability means that a consistent value should be obtained each time the same measurement is repeated. An unreliable weighing machine in a shop would make the customers go elsewhere. In science, you can't go elsewhere so the measurements must be reliable. Each time a given measurement is repeated, it should give the same value within acceptable limits. What do we mean by acceptable limits? Consider the example of measuring the diameter of a uniform wire using a micrometer. Suppose the following readings are taken for different positions along the wire from one end to the other, diameter d: 0.34 mm 0.33 mm 0.36 mm 0.33 mm 0.35 mm.

The mean value, $\bar{d}$ (sometimes written $<d>$) is 0.34 mm, calculated by adding the readings together and dividing by the number of readings.

If the difference between each reading and $\bar{d}$ changed regularly from one end of the wire to the other, then it would be reasonable to conclude that the wire was non-uniform. The differences would be called 'systematic'. No such differences can be seen in the above set of readings. In other words, there is no pattern obvious, so the differences are random. Random differences may be due to the observer's judgement, the instrument or the non-uniformity of the wire. The spread of readings is from 0.33 to 0.36 mm, and most of them lie within 0.1 mm of the mean value. So the diameter is 0.34 ± 0.1 mm. The ± 0.1 mm is called the **probable error** of the value. Statistics formulae can be used to calculate probable errors. See p. 513 for more details about errors.

34.2 Measuring instruments

Achieving accuracy

Instruments used in the physics laboratory range from very basic (e.g. a mm scale) to the highly-sophisticated (e.g. multi-channel data recorders). Whatever type of instruments you use, you need to be aware of the following key points.

Zero error Does the instrument read zero when it is supposed to? For example, voltmeters and ammeters need to be checked before use to ensure they read zero when disconnected. Another example is using a micrometer; when the measuring gap is closed, the reading ought to be zero. If not, the zero reading must be taken into account when measuring the gap width.

Sensitivity In general terms, sensitivity is the reading per unit 'input'. For example, if the sensitivity of a microammeter is given as 75 mm μA^{-1}, then the meaning is that the reading in mm is 75 × the current in μA. An instrument that becomes less sensitive with age will give readings that are consistently low, causing systematic errors. Multi-range meters can be set for different ranges according to the situation. Each range has a different sensitivty. For example, when a multi-range voltmeter is used, the correct range to use is the one that gives the biggest reading on the meter scale without overloading the scale. Another example is when an oscilloscope is used to display a wave form; the Y-gain is adjusted to make the wave form cover as much as possible of the screen without any part of it disappearing off the screen.

Linearity is a design feature of most instruments. An instrument is 'linear' if its reading is proportional to the quantity being measured. So the sensitivity (= reading per unit quantity) is constant if the instrument is linear. For example, suppose a voltmeter which is linear is checked using a standard 1.08 V cell.

a) The zero reading is checked first.

b) The standard cell is connected to the meter, and the reading should be 1.08 V.

c) For measuring other voltages, the reading is proportional to the voltage since the meter is linear. So if the voltmeter gives a reading that is 0.5 × the reading for 1.08 V, then the voltage is 0.54 V. To check linearity, the instrument must be recalibrated over the whole of its range. See p. 339.

Accuracy is only possible if there are no systematic errors when a measurement is made. Precise readings are not necessarily accurate readings, since systematic errors could make precise readings lower than they ought to be. For example, suppose the hairsprings of a certain moving coil meter have become weaker with age; precise readings can still be made from the scale but they would be consistently greater than the correct readings since the coil becomes easier to deflect. So the readings would not be accurate.

The accuracy of a measurement is given by the probable error or the percentage error. By making several readings for a given measurement, the probable error can be estimated from the spread of readings; if all the readings are the same, the probable error is given by the precision of reading the scale. The same applies if just one reading is made. Figure 34.2 shows the scale of an ammeter where a reading is to be made. What do you think the reading is? How precise is the reading? You could use a lens as a magnifying glass to make it even more precise.

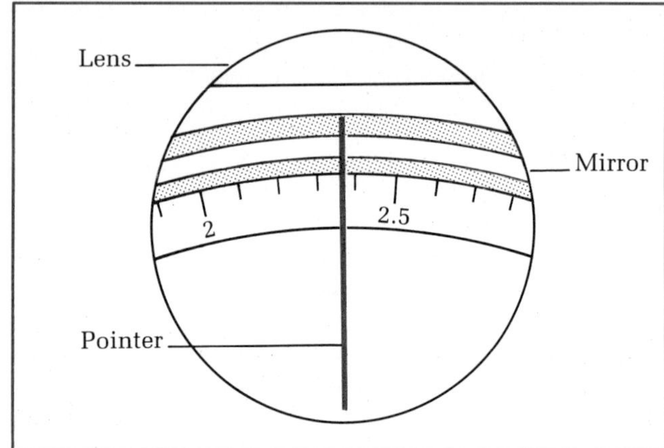

Figure 34.2 *Magnifying a scale*

Accuracy is easy to lose, even with simple equipment. For example, when investigating free fall (as on page 8), the metre rule must be vertical. To check that this is so, a plumb line is used, made of a string supporting a mass at rest. If the metre rule appears parallel to the plumb line from the front **and** the side, the rule is then vertical.

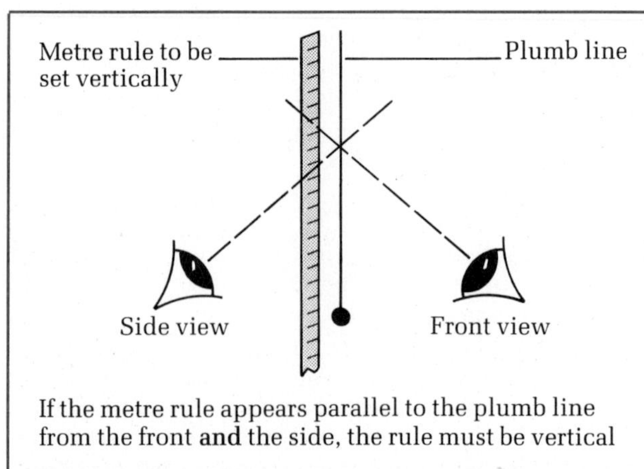

Figure 34.3 *Finding the vertical*

Precision can be lost through lack of care in reading the position of a pointer against a scale. The line of sight from the observer to the pointer must be at right angles to the scale. A plane mirror behind the scale is helpful. The observer reads the scale when the pointer's image is directly behind the pointer, so the scale is observed correctly.

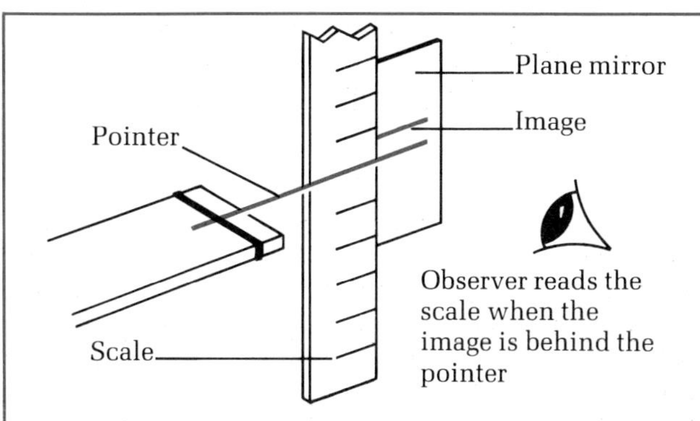

Figure 34.4 *Reading a scale*

Micrometers and verniers

Micrometers give readings with a precision of 0.01 mm. The barrel of a micrometer is a screw with a pitch of 0.5 mm. The edge of the barrel is marked in 50 equal intervals so each interval corresponds to changing the gap of the micrometer by 0.5/50 = 0.01 mm. The stem of the micrometer is marked with a linear scale graduated in $\frac{1}{2}$ mm intervals.

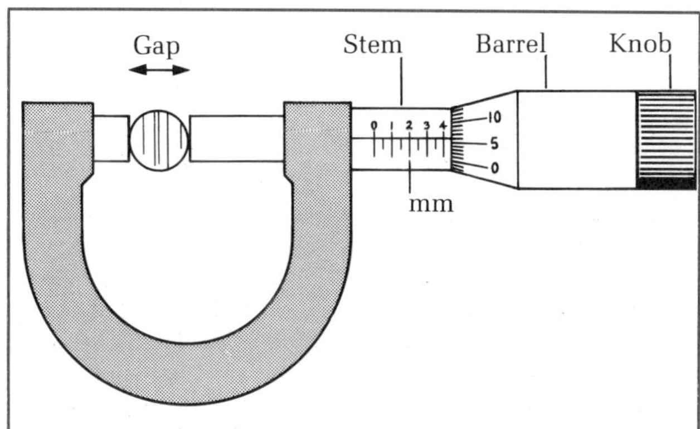

Figure 34.5 *Using a micrometer*

With the gap closed, the zero mark on the barrel scale should be exactly against the zero mark on the linear scale. When the gap is opened, its width is given by where the readings on the scales intersect. Figure 34.5 shows a reading of 4.56 mm. To use a micrometer correctly

a) check its zero reading,

b) open the gap by turning the barrel, then close the gap on the object to be measured. Turn the knob until it slips or clicks, when the gap has closed. Don't over-tighten the barrel. Take the reading, and calculate the gap width from the two readings. The accuracy of the measurement is ±0.01 mm because each reading is accurate to ±0.005 mm, and the difference between the readings is accurate to the sum of the individual errors.

Vernier calipers can be used for measurements of distances up to 100 mm, or more. The precision is not as great as with a micrometer since vernier readings are accurate to ±0.05 mm. However, vernier calipers are more versatile as they can be used for internal and external measurements.

The sliding scale of any vernier has ten equal intervals covering an exact distance of 9 mm, so each interval on the sliding scale is 0.1 mm less than a 1 mm interval. To make a reading, the zero mark on the sliding scale is used to read the main scale to the nearest millimetre. This reading is rounded down to the nearest mm. Then the sliding scale is inspected, using a lens as a magnifying glass if necessary. The number of the mark on the sliding scale closest to a mark on the main scale is noted. This number × 0.1 mm is the amount the zero reading was rounded down by. Figure 34.6 shows the idea; the 5th mark after the zero on the sliding scale is in line with a mm mark on the main scale. So the distance from the 39 mm mark on the main scale and the sliding scale zero is 0.5 mm. Hence the vernier reading is 39.5 mm. As with a micrometer, the zero reading should be checked and taken into account if necessary.

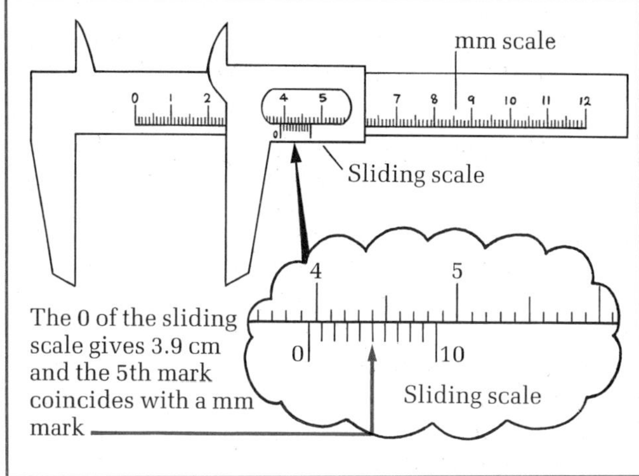

Figure 34.6 *Using a vernier*

Micrometers or verniers rarely need recalibrating. Thermal expansion is negligible over the range of temperatures that they are likely to be subjected to, although the instruments can become worn if used carelessly.

Timing

Stopwatches used for interval timing are subject to human error because reaction time, about 0.2 s, is variable for any individual. Starting and stopping a stopwatch involves similar delays at the start and finish, but even so, precision of less than 0.1 s is not likely because of random error. Digital stopwatches can given read-outs precise to within ±0.01 s. But unless they are operated from electronic gates, human error makes read-outs to ±0.01 s meaningless, and the reading must be considered accurate to ±0.1 s at most.

Timing oscillations requires timing for as many cycles as possible. The timing should be repeated several times to give an average value. Any timing that is significantly different to the other values is probably due to miscounting, so that timing is rejected. For example, suppose the timings for a given simple pendulum are as follows:

Time for 20 complete cycles/s: 20.1 19.8 19.7 20.2 20.0 18.3 19.8 20.4

The timing of 18.3 s is rejected because it was probably due to miscounting. The mean value is then 20.0 s with a spread from 19.7 to 20.4 s. Most of the readings lie in the range from 19.8 to 20.2 s so the timing is 20.0 ± 0.2 s.

For accurate timing of oscillations, a **fiducial mark** is essential. The mark acts as reference to count the number of cycles as the object swings past each cycle.

Electronic timers use gates to start and stop an electronic counter. The counter is supplied with 1.0 kHz pulses when the 'condition' of the gates allows. So the counter read-out is precise to 1 ms (i.e. 1 pulse). Figure 34.7 shows a single light gate to time a card attached to a trolley as the trolley passes under the gate. When the card interrupts the light beam, the counter operates, so the counter read-out gives the time in ms for the card to pass through the gate. Suppose the timing was repeated several times, each time the trolley being released from rest at the same point. The probable error is estimated from the spread of the readings; if the readings are all the same (unlikely in this situation), then 10 kHz pulses should be used so that the read-out is more precise.

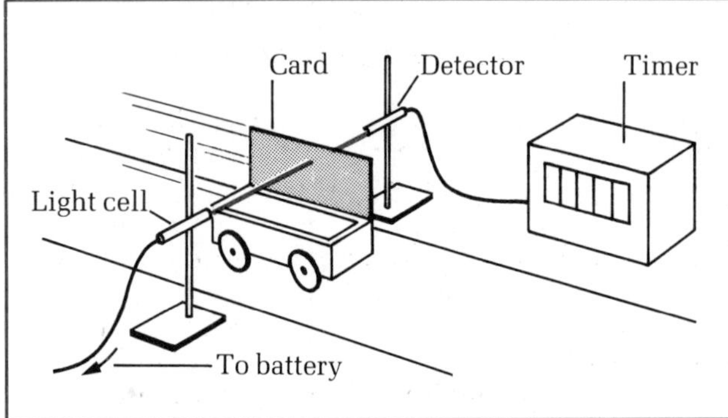

Figure 34.7 *Electronic timing*

Another method of using electronic timers is to allow one gate to switch the counter on and then a different gate to switch it off. For example, a falling object could be timed over a measured distance in this way (see p. 516).

Light gates can be interfaced to a microcomputer to use the micro's internal clock. The microcomputer must be programmed to start timing when it receives a signal from the light gate, and then to stop when a second signal is received. The program must include the instruction to print the timing on the VDU screen, and usually a calculation in the program is necessary to give the timing in seconds.

Checking the accuracy of a timer can be done using a time signal from the radio or 'phone to start the timer. Then a second time signal is used to stop the timer. For example, suppose a stopwatch is allowed to run for 30 minutes exactly, and when it is stopped it shows a reading of 30 min 10 s. The stopwatch has 'gained' 10 s in 30 minutes so each stopwatch second is $\frac{1800}{1810}$ s. Timing using this stopwatch needs to be corrected for systematic error.

Top-pan balances

Electronic balances for measuring mass usually give digital read-outs. Before using a top-pan balance, its zero reading must be checked and taken into account if necessary. Top-pan balances need to be checked for accuracy at regular intervals; a set of standard masses is essential for this purpose. Use tweezers to lift standard masses to prevent any deposit or corrosion that might be caused if the masses are handled.

Voltmeters and ammeters

Whether digital or analogue (i.e. pointer-type), electrical meters should be checked for accuracy at regular intervals and recalibrated if necessary.

Voltmeters can be calibrated using a potentiometer and a standard cell. Figure 34.8 shows a suitable circuit for this purpose. The potentiometer wire is calibrated using a standard cell (see p. 339). Then the voltmeter reading is adjusted in steps to different values. At each value, the balance point on the wire is located and the balance length l is measured. Then the correct p.d. across the voltmeter is calculated from

$$V = \frac{\text{standard e.m.f.}}{\text{standard balance length}} \times l.$$

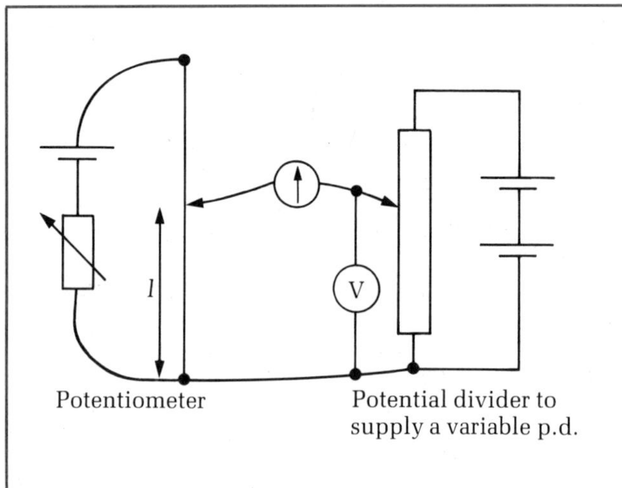

Figure 34.8 *Calibrating a voltmeter*

A calibration curve of calculated p.d. against voltmeter readings is a convenient way to display and use the results.

If the range of the voltmeter is beyond the range of p.d.s from the potentiometer, then the circuit is modified. Figure 34.9 shows the modification; two accurate resistance boxes in series are connected across the voltmeter terminals. By setting the resistance boxes at known values, a known fraction of the voltmeter p.d. is dropped

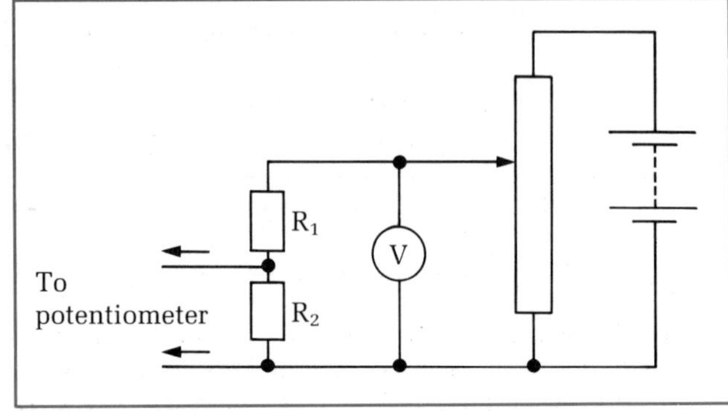

Figure 34.9 *Extending the range*

across each box. The potentiometer p.d. is balanced against the p.d. across one of the boxes. The potentiometer p.d. is equal to $VR_2/(R_1+R_2)$ where V is the voltmeter reading, and R_1 and R_2 are the box resistances as in Figure 34.9.

Ammeters can be calibrated or checked by connecting the meter in series with a standard resistor. A potentiometer is then used to measure the p.d. across the standard resistor. Figure 34.10 shows the arrangement. The current is then calculated from the measured p.d./resistance of the standard resistor. The calculated value is compared with the measured value of current. By making measurements at steps over the ammeter's range, a calibration curve can be drawn, as shown in Figure 34.11. The curve can then be used to determine the correct current for a given ammeter reading if there is a difference.

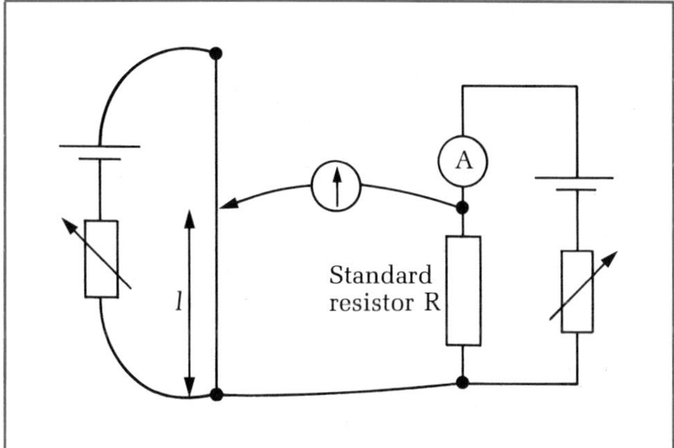

Figure 34.10 *Calibrating an ammeter*

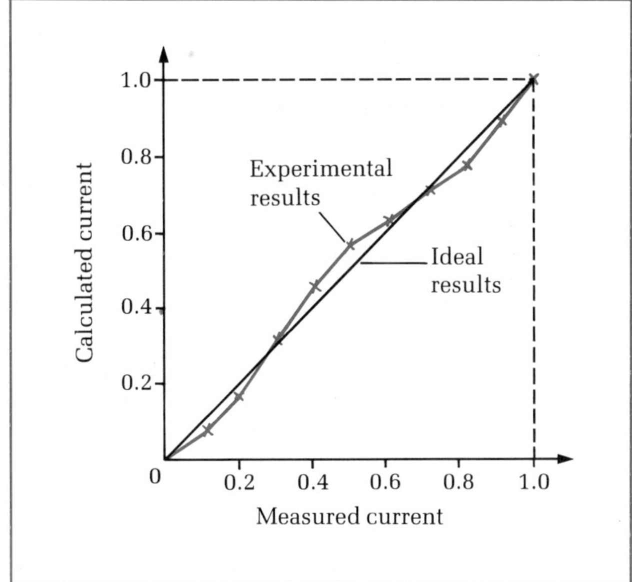

Figure 34.11 *A calibration curve*

Oscilloscopes

Oscilloscopes are used to measure p.d.s and time intervals, as explained on p. 348. To check its accuracy for p.d.s, the same method as for a voltmeter can be used; in theory, the vertical deflection of the trace on the screen is proportional to the input p.d. To check the Y-gain sensitivity, the deflection of the trace is measured for a given input p.d. from a standard cell or potentiometer. The sensitivity is equal to the deflection/input p.d.

To check the accuracy for timing intervals, a signal of known frequency (e.g. 50 Hz) is applied to the Y-input; with the time-base on, a stable wave form is displayed on the screen. The time period of the signal (= 1/frequency of the signal) is then used to check the time-base setting; this is done by measuring the distance across the screen for one complete cycle. This distance × the time-base setting (in ms cm^{-1} for example) gives the time period measurement.

34.3 Projects and investigations

Short experiments

You have probably done lots of short experiments in your physics classes by now. Some short experiments might require more than the usual 1 to 2 hour laboratory session but in general, short experiments are usually done in a single session. Short experiments generally involve one or more of the following tasks.

a) Practising a particular skill (e.g. using an oscilloscope for measuring).
b) Measuring some property of a material (e.g. the density of a solid).
c) Determining the characteristics of a device (e.g. a transistor).
d) Testing a suspected link between two physical quantities.

Usually a short experiment involves carrying out specific tasks by following a set of instructions. The instructions might tell you which variables to control and which to measure. You may need to select suitable instruments for the measurements from a range of instruments available. You might need to select the most suitable range for an instrument.

Consider the example of using the ammeter-voltmeter method to measure the resistance of a wire-wound resistor. Suppose you are given a suitable voltmeter, a multimeter, a variable resistor and a 6 V battery and switch. Your instructions might be as follows.

> *Use the equipment given to measure the current through resistor X for several different values of p.d. across X up to 6 V. Plot a graph of your results, and use the graph to calculate the resistance of X.*

Step 1 Sketch the circuit diagram you propose to use.
Step 2 Make out a table for your results.
Step 3 Connect the circuit together but keep the switch open. Check the zero readings on the meters. Set the multimeter range switch for maximum d.c. current.
Step 4 Close the switch and check that the current can be varied by altering the variable resistor. For maximum current, set the multimeter range switch to give the maximum on-scale reading.
Step 5 Alter the p.d. in regular steps. At each step, measure the p.d. using the voltmeter and measure the

current using the multimeter. Note the probable error for each meter. Open the switch but don't disconnect the circuit. More readings may need to be made.

Step 6 Plot p.d. against current, including the zero readings. For each point, outline its error box as in Figure 33.3. Then determine the resistance from the maximum and minimum gradient, as explained on p. 505. If the steepest line through the error boxes gives 6.3 ohms, and the least steep line gives 5.9 ohms, then the resistance of X is 6.1 ± 0.2 ohms.

Whatever the experiment, the procedure is along the lines indicated. If the instructions are brief, as in the example above, then you need to plan the procedure step-by-step. You may find it is helpful to write your account of the experiment as you proceed with the experiment.

Title e.g. 'Measurement of the resistance of a wire wound resistor by the ammeter-voltmeter method'.

Diagram e.g. circuit diagram, labelled with the range of the meters shown.

Method e.g. write up details as in Steps 4, 5 and 6.

Results e.g. the table of measurements.

Treatment of results e.g. a brief explanation of relevant theory, leading to the chosen graph.

Graph e.g. as in Step 6.

Calculations e.g. as in Step 6.

Conclusions e.g. give the value of resistance, and comment on accuracy.

Short investigations

Investigations involve decisions being made by the experimenter. What are the variable quantities involved? Which one should be controlled to measure its effect on other quantities? What method should be used? These are all questions you need to ask yourself before you start using any apparatus. For example, suppose you are asked to investigate the factors that determine the fundamental frequency of vibration of a given wire. You might suspect the factors include the vibrating length, the tension and the amplitude of vibration. You could vary each factor in turn, keeping the other ones constant, to discover the effect of each factor individually on the frequency of vibration. You could choose the a.c. sonometer method (see p. 251) for this experiment. Then you need to decide on the procedure step by step.

Step 1 List the factors that you think affect the frequency. Consider and note how each factor can be varied.

Tension: use different load weights.

Length: use a movable bridge to alter the vibrating length.

Amplitude: change the amplitude of the alternating current in the wire.

Step 2 How can you measure the frequency and the other factors? Note briefly the instrument and procedure for each factor.

Frequency: use a frequency meter connected across the output terminals of the signal generator.

Tension: use a top pan balance to measure the mass of each of the load weights if necessary.

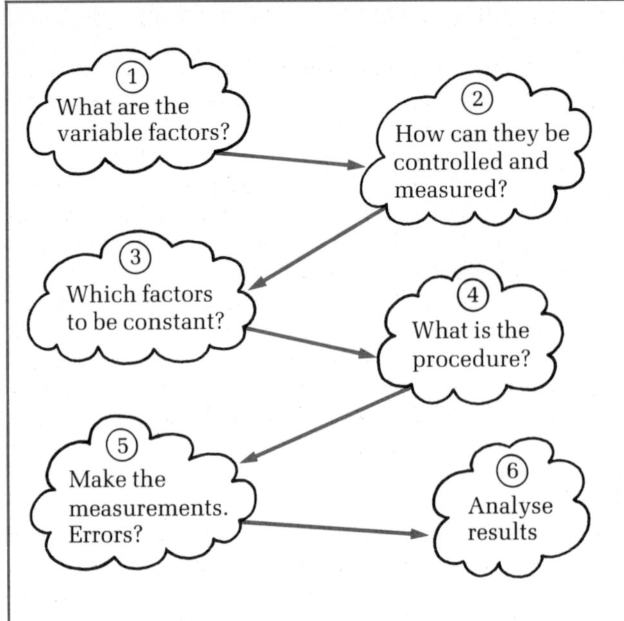

Figure 34.12 *Short investigations*

Length: use a mm scale.

Amplitude: use an a.c. ammeter in series with the wire.

Step 3 Choose one of the factors to vary and keep the other two factors constant. Alter the chosen variable in steps, and at each step adjust the frequency until the wire resonates in its fundamental mode. Check that the other factors (tension and current if the length is the chosen variable) remain unchanged. Then measure the frequency and the chosen variable. Repeat for different values of the chosen variable. Note the measurements in table form, and note the probable errors.

Step 4 If you suspect a link between the frequency and the chosen variable, plot a graph to give a straight line. For example, if you suspect that the frequency is proportional to 1/length, then plot frequency against 1/length. A straight line confirms your suspected link.

A better approach to establish a link is to plot a log-log graph. The idea of log-log graphs is explained on p. 503. Suppose the link is in the form

$$f = kx^n$$

where x is the variable factor,
f is the frequency
k is a constant.

The power n is to be determined to establish the link.

$$\log f = n \log x + \log k$$

Hence a graph of $\log f$ against $\log x$ should give a straight line with gradient n if the link is of the form $f = kx^n$. If the graph does give a straight line, then the value of n can be determined.

Step 5 Summarize any links you have established. In this example, the frequency is

a) proportional to 1/length,
b) proportional to $\sqrt{\text{tension}}$,
c) independent of amplitude.

Longer investigations and projects

The skills and methods used in short investigations are important in any long investigation or project. But other techniques and skills are important too. Short investigations are usually carried out with a limited aim in mind (e.g. measure the density, etc., investigate the variation of, etc.) so you start with that aim in mind. Longer investigations are often open-ended so you cannot predict where the investigation will lead when you start. If you start with a vague, general aim in mind, you may find difficulty in deciding the direction to take; there might be too many variables to sort out. So if you are choosing a project, you should not be too ambitious to involve too many factors.

With a suitable investigation chosen, now you need to organize yourself. No set instructions can cover a project so start by deciding on the general procedure to follow. If necessary, buy a note book with alternate graph and lined pages. The note book is useful as a diary in which to plan your project. You can use it to record your thoughts, decisions, ideas on design etc., as you proceed with the project. Measurements, errors, theory, graphs and calculations should all be included in your note book. Here are some hints to help you get organized.

1 **Research** the topic area by selective reading from your text book and any other suitable books. Note down any promising 'lines' of investigation and any potential difficulties.
2 **Narrow** the area of interest down to an investigation with just a few variables involved. Make sure that you have the facilities to control and measure the variables. Perhaps a little more reading-up is necessary here. Check that your teacher thinks your project is feasible in the time available.
3 **Note** the variables you intend to measure or control. Plan to operate on a 'cause and effect' basis with one variable being changed causing another variable to alter; the other variables need to be monitored to make sure they do not change.
4 **List** the equipment necessary for your measurements, etc. Check it is available when you wish to use it. List all the materials and components required. If necessary, obtain materials from home or a local supplier.
5 **Design**, construct and test any special pieces of equipment you need for your measurements.
6 **Modify** your plans or designs as a result of preliminary tests.

Figure 34.13 *Ask permission to borrow!*

Let's see how this procedure might work in practice. Suppose you wish to investigate the friction and slipperiness of floor coverings used in kitchens.

Researching should enable you to discover that friction between solid surfaces is measured by the coefficient of friction, μ. So your investigation narrows down to measuring μ for different surfaces. The surfaces need to be clean and dry to test for maximum friction.

Measurements to determine the coefficient of friction need to be considered next. μ is defined as the frictional force at the point of slipping/the normal reaction at the surface. On a level surface, the normal reaction is equal to the force pressing down. So the frictional force at the point of slipping must be measured for different loadings on the surfaces (assumed level).

Equipment and design is the next step. A top-pan balance is needed to measure the load weights. A spring balance or an uncalibrated spring as a forcemeter is needed to measure the frictional force at the point of slipping. To apply a gradually increasing frictional force, a 'test rig' along the lines of Figure 34.14 must be designed. One of the surfaces to be tested is fixed on the bed of the test rig. The other surface is stuck onto the lower surface of the test box. The box is to be loaded with different weights. For each load weight, the forcemeter is used to apply a gradually increasing force until the box slips. The reading of the forcemeter, just on the point of slipping, is equal to the frictional force. Modify the design if necessary to detect the point of slipping more accurately.

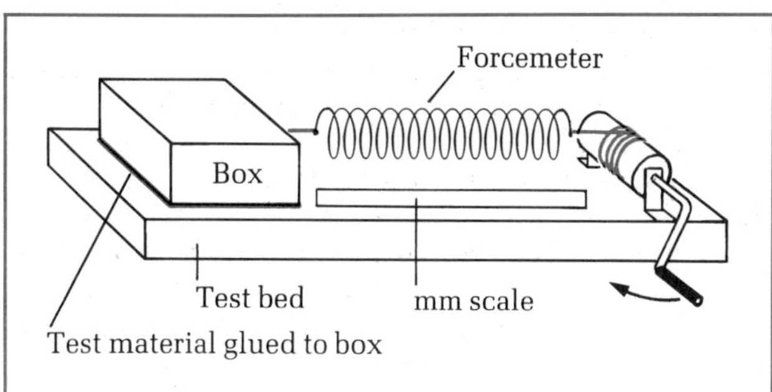

Figure 34.14 *A test rig*

Calibrate the forcemeter by suspending known weights from it when it is held vertical. Plot a calibration graph of tension against extension. Now follow the same general steps as for any short investigation.

i) Obtain sets of readings of the frictional force F and load weight W, using fresh samples of the same material. Note the errors involved in your measurements.
ii) Plot graphs of F against W to determine μ for each sample from the gradient of the line. Hence obtain an average value of μ for the pair of materials under test.

iii) Repeat the procedure for other materials used as the surface of the test bed but keep the same material on the box throughout. In this way, the values of μ measure the friction between a given material as a 'shoe' and different floor surfaces.

Conclusions from your results must be justified by the results. For example, measuring μ for two different floor coverings might give the result that μ for floorcovering X is greater than μ for floorcovering Y. But you need to consider the probable error of each value. If $\mu_X = 0.55 \pm 0.10$ and $\mu_Y = 0.48 \pm 0.08$, then you would not be justified in claiming μ_X is greater than μ_Y. You would need to test for more samples of X and Y and you would need to improve the less accurate features of the experiment. If time does not permit further investigations, then you should outline in your conclusions what further investigations you would have done.

Writing a report on your investigation is an important part of the work in itself. The note book should contain all the day-to-day details of the project; planning, organization, decision making, design details, measurements, errors, graphs, calculations should all be in the note book. The book might cover 50 or more pages, and its details are vital for your report. But the report itself should be based on **selective** use of your note book. Details of abandoned ideas would not be an essential part of your report, so only a brief reason need be given in your report to explain why you changed your ideas or approach. Your note book might contain lots of tables of measurements; these can be given as an appendix to your report. Report writing is discussed in more detail on p. 528. Don't neglect this important aspect of any project; your experimental skills might be 'honed to perfection' but if your report is ill-prepared, the reader will probably fail to appreciate your skills. It's a bit like a superb footballer who forgets which end to attack – and scores an own goal!

35

Communication skills in physics

35.1 The written report

Explaining yourself

Scientists need to communicate their ideas to one another and to the general public. Sometimes the writers of scientific articles in popular magazines forget the intended reader and write as if to another scientist. Then the unfortunate reader is left to struggle with the article. Sometimes the writer 'waters down' the scientific content to a trivial level, and the reader gains little from the article. It is not just scientists who make this sort of mistake though; Inland Revenue returns can be taxing in more ways than one! Here are some simple rules.

1 Use short words if possible.
2 Use familiar words if possible.
3 Avoid 'padding' with unnecessary words.
4 Avoid long sentences if possible.

Perhaps you have met someone who deliberately breaks the rules. Don't be impressed! Pompous expressions might make the user feel terribly clever and superior, but anyone who resorts to such expressions is probably hiding his or her lack of knowledge. Communicating ideas is too important to allow barriers to be erected. Physics is a subject with lots of difficult ideas; no one gains if the subject is expressed in ways that make it seem more difficult. Your studies in science would be incomplete if you were unable to express your ideas effectively.

Technical terms are terms that have a special meaning in the context in which they are used. For example, 'coherence' has a special meaning when it is used in physics. Its general meaning is 'holding together'; its use in physics extends that meaning to waves with a constant phase difference. Scientists often use technical terms without explanation when communicating with each other; sometimes, an unfamiliar technical term needs explaining. Usually though, the user assumes the reader or listener is familiar with the term. The writer must judge which technical terms need explaining. Too many explanations would interrupt the train of thought of the reader; too few explanations would make the reader stumble along, unable to develop a train of thought.

Where a technical term is introduced and explained, it can then be used subsequently without explanation. The more often the term is used after being introduced, the more familiar the reader becomes with that term. But if several terms are introduced and explained close together, then the reader can become confused.

When writing a scientific article, the writer must keep the scientific background of the typical reader in mind.

a) Specialists in the same subject would need newly-discovered ideas and methods to be explained. A survey of knowledge up to the point of discovery is helpful. But explanations of unnecessary terms would hide the impact of the discovery.

b) If intended for non-specialists with a scientific background, then the specialist terms would need to be explained. For example, suppose you were to read an article about Astronomy, and you met the unit 'the parsec' without an explanation. Your progress would be halted whilst you thought about its meaning, and you might need to consult a reference book. So a brief explanation, perhaps in a separate diagram, would be more than helpful. However, with a scientific background, you ought to be familiar with SI units, prefixes, use of formulae, graphs and general scientific terms. In other words, 'scientific literacy' is assumed.

c) For intended readers who are not scientists, only a limited scientific background can be assumed. For example, an article about 'World Fuel Supplies' might perhaps include a paragraph or two about energy transformations. The article might discuss chemical energy, heat energy, nuclear energy and electrical energy; but the writer would probably assume that the reader is aware that there are other forms of energy.

Essays and essay-style answers written in physics examinations need careful planning. The intended reader is either your teacher (for internal exams) or a distant examiner; so your reader is a highly-qualified scientist whose task is to assess your knowledge and understanding of the topic in question. The purpose of your answer is not to enlighten the reader, but to demonstrate your knowledge. Suppose you don't explain your ideas fully because you think the examiner is bound to know what you mean anyway. The examiner can only judge what you have written so will assume that you cannot understand the ideas. 'But I meant . . .' may keep your teacher happy, but your examiner judges only what is on paper.

So what should be your approach? As a useful guide, write your answer as if for a classmate who missed the topic. You can then assume a scientific background, but you must explain technical terms relevant to that topic.

Many essay questions provide little information and expect you to write your answer with little guidance from the question.

> *This question is about explaining ideas. Give a careful explanation of the topics listed, as if for a classmate who missed the subject.*
> **Subject** *Digital electronics*
> **Topics in the subject** *Logic levels, logic gates, flip-flops, indicators.*

An answer plan is essential; start by listing relevant points against each topic heading. Then link the subtopics together. Then decide the order you think best for the points in your answer. When you write out your answer, don't forget that labelled diagrams save lots of unnecessary writing; you need only refer to key features of the diagrams in your written answer. A possible answer plan is shown in Figure 35.1, on the next page.

Project reports

Planning your report Before you start writing a report about a project or investigation, your first step should be to make up a 'plan of attack'. Your report will be based on the information in your project note book; all the day-by-

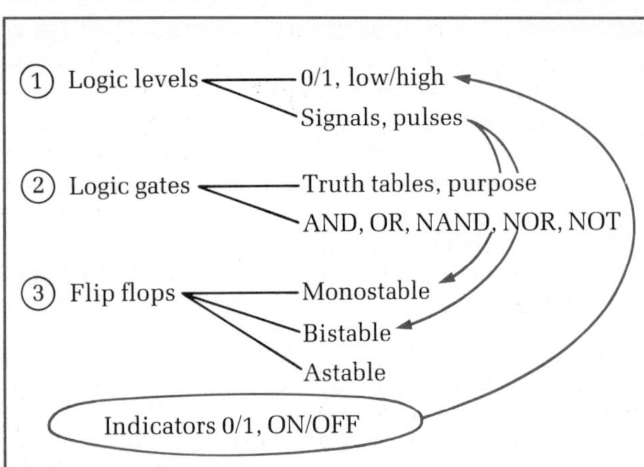

Figure 35.1 *Answer plan*

day details should be in the note book. Your plan could take the form of the main headings you intend to use in your report, with important points jotted down under each heading. If you restrict your plan to a single side of paper, then you can spot anything left out of the plan. The plan then gives you an overall framework that will guide you through the actual writing of the report.

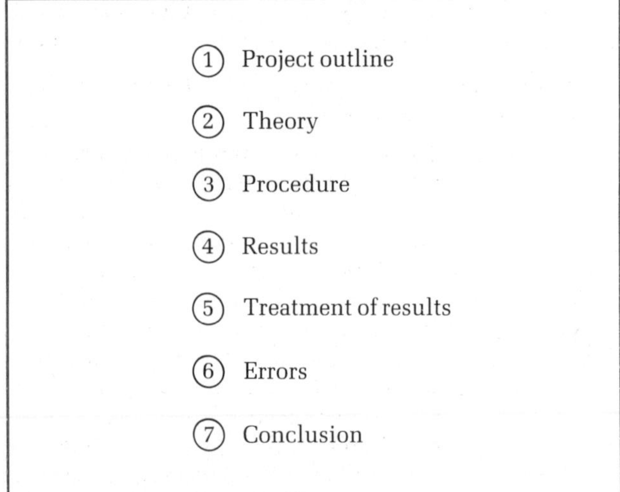

① Project outline
② Theory
③ Procedure
④ Results
⑤ Treatment of results
⑥ Errors
⑦ Conclusion

Figure 35.2 *Report plan*

The main headings, in order, would probably be as follows:

1 Project outline or 'abstract' At the start of your report, after the title, give an outline of the project in no more than half a page or so. The outline should state the aim of the project, and the general method used to achieve your aim. If the aim is in general terms (e.g. to investigate methods of preventing 'thermal runaway' of transistors), then more precise details should be given. In addition, a brief explanation of specialist terms (like 'thermal runaway' in the above example) should be given.

2 Theory Explain what quantities you intend to control and measure. Give the theory that shows how the measurements relate to your aim.

3 Apparatus Describe, with the aid of labelled diagrams, the apparatus you used. Number your diagrams (e.g. *Figure 1*) so you can refer to them in your writing. Include details of any equipment you designed, tested and used. If you changed the apparatus during the project, state why the change was necessary.

4 Procedure The next step is to describe how you used the apparatus. How did you control and measure the variable quantities? List the sequence in which the measurements were made. List the instruments used for each measurement. What steps did you take to check the consistency of your readings? Refer to diagrams and tables of measurements by number. How did you estimate the probable errors in your measurements? List the precautions taken to ensure consistent and accurate readings.

5 Results Give your measurements in the order you took them in. Tabulate sets of measurements and number each table. If your measurements take up more than a page or so, plan them as an 'appendix' to the report, and refer to the tables in the results section. List units and probable errors.

6 Treatment of results

a) Calculations from the measurements and related theory are to be given here. For calculations on a set of measurements, give the results in a table.

b) Graphs to be plotted are listed here, with a brief explanation of why you have chosen each graph. Each graph can be given in an appendix with a reference number. Calculation of gradients and intercepts of straight lines can be given on the graph sheets, but state the results of your calculations in the main report. Explain how you measured the gradient of any curves. Show probable errors and gradient triangles on the graphs, but comment on the features of the graph in the main report.

c) Final calculations should include any theory necessary to explain your use of previous calculations or data from measurements or graphs. In other words, if your final calculation draws data from different parts of your report, explain the theory behind the calculations.

7 Accuracy and errors The measurements and graphs ought to show probable errors. Describe the steps you took to eliminate systematic errors (e.g. calibrating meters) and random errors (e.g. repeating readings). Then show how the errors are treated in the final calculations to give the total error.

8 Conclusions State the results of your final calculations, including your estimate of the total error. Comment on any unusually large errors in the measurements, and suggest ways to reduce any such errors.

State any mathematical links between quantities established or confirmed by your project. Discuss how well you have achieved your initial aim; be critical here, and don't claim without justification. Describe further tests you could have done or improvements you would have made if more time had been available.

The style and presentation of your report are important. Aim to make your report easy to follow. The layout should include clear, numbered headings. Tables of measurements and diagrams should be clearly set out, numbered and given a caption (e.g. *Figure 1 The test rig*). If there are lots of measurements to report, keep them in a separate appendix after the main report; in this way, the flow of your report is not broken by pages of measurements. You can then refer to any table by its reference number. Avoid the need for the reader to jump forward or refer back to key sections of the report; a brief quote from the key section is preferable.

A plain style of writing allows the scientific content to stand out. Think about the words you choose so you can avoid lapses like '... the thermometer rose by ...'. Be precise when you explain how a measurement was made: '... the time taken for 20 oscillations was measured.' What was used for the timing? Was a fiducial mark used? Why were 20 oscillations timed? The report should tell the reader how you made the measurements, not just what the measurements were. 'The wave form was measured ...' does not convey what aspect of the wave form was measured, so more information is needed.

A useful idea is to prepare a draft report and ask a friend to read it through; your friend may pinpoint phrases where the meaning is not clear or where there is insufficient explanation.

Figure 35.3 *Misunderstandings*

35.2 Comprehension

Reading skills need to be developed, like any other form of skills. Background reading is one way to further your reading skills; choose articles that interest you from popular scientific magazines or from 'Science Extra' features in newspapers. If an article is particularly relevant to your studies, write a summary of it to include with your notes. Reading articles in this way should help you to develop your writing skills too!

What sort of skills are involved in effective reading? The aim of reading a scientific article is to absorb information or to develop understanding. If an article is entertaining too, then so much the better. It's useful to mark long articles into sections, to be read through section by section.

1 Read each section at your normal reading speed without 'backtracking'. As you read through, mark any unfamiliar terms and underline any points you consider important.
2 Now consider the unfamiliar terms one by one; make sure that you know the meaning of each. If necessary, refer to your text book although a well-written article ought to explain unfamiliar terms.
3 Re-read the section critically, looking for links between key points.

Reading is more effective if you think for yourself as you read through. A scientific article is meant to convey knowledge; you should ask yourself what you have gained in knowledge through reading the article. Has the article widened your scientific background or developed your understanding? If not, you might as well have read a short story! Sharpen your reading skills and your studies in science will become much more effective.

Examination questions can easily be misread under pressure in an examination room. Missing the point of an article or question is all too easy when under strain. So you need to practise your reading skills alongside laboratory skills, mathematical skills, etc. All too often, candidates answer the question they had hoped for rather than the one actually asked. If the topic in a question is seen as its key feature, then the instructions may be misread. Consider the following example.

Describe the motion of smoke particles in air when viewed using a microscope.

If you write all about air molecules moving at random, you have missed the point of the question. The question asks you to **describe** the motion of the smoke particles, not to say what you think is the cause of their motion. If the question had read '*Describe and explain* ...'

Comprehension papers and longer examination questions are scientific articles so the skills mentioned above are important. A typical comprehension paper develops a topic in the syllabus by introducing and using unfamiliar ideas. The passage explains such ideas so you have to take the meaning in. What's the best way to do this? Use the same sort of approach as for any scientific article – as described earlier. Here's a summary.

1 Read the passage through and mark unfamiliar terms.
2 Look for the meaning of these terms in the passage.
3 Then re-read the passage before tackling the questions.

Usually, the questions follow the paragraphs in sequence. As you come to each question, identify which part of the passage the question is drawn from, and re-read that section before you attempt your answer. If necessary, jot the key points of your answer down on rough paper and link them together in an answer plan; base your answer on the plan.

The questions in a typical comprehension paper test a wide range of skills. Your reading skills are obviously under test. But so too are your skills of writing, graph work, calculation and so on. You may be asked to explain an unfamiliar term in your own words; or you may be asked to use an unfamiliar formula given in the passage. You could be asked to plot a graph based on the formula. You will undoubtedly be asked to show your knowledge of physics, and perhaps to apply that knowledge to an unfamiliar situation. Comprehension papers and longer questions are included in this book to help you develop your skills of physics. Even if you do not pursue your studies in physics beyond your present course, the skills you develop will help you in any career. That's why physics students are sought after by such a wide range of employers!

Figure 35.4 *Good luck!*

G

Long questions

Comprehension paper 1 (1 hour)

The following passage is based on an extract from the book *Energy: A Guidebook. Read the passage carefully and then answer the questions at the end.**

The winds that blow

The energy of the winds is stored solar energy. A few per cent of the radiation reaching the Earth is absorbed in the atmosphere, and the resulting uneven heating of the air leads to large-scale circulation patterns. Not all this is available for use, however. We must ask how, when and where the energy is available, how much of it we can extract, and what it will cost in financial and environmental terms.

Calculating the energy delivered by the wind is easy. It is kinetic energy. Given that the mass of a cubic metre of air is about 1.3 kg, you find that each cubic metre of air in a 9.0 m s^{-1} wind carries slightly more than 50 J of energy. If we put up a windmill having an area of about one square metre facing this wind, then the wind power input is about 500 W. But Figure GL.1 brings out the extremely important fact that the power delivered by the wind rises very steeply indeed as the wind speed increases. It's not difficult to see why. The kinetic energy of a given mass of air depends on the square of its speed, and then the mass arriving each second depends on the speed again. So doubling the speed multiplies the wind power input by eight.

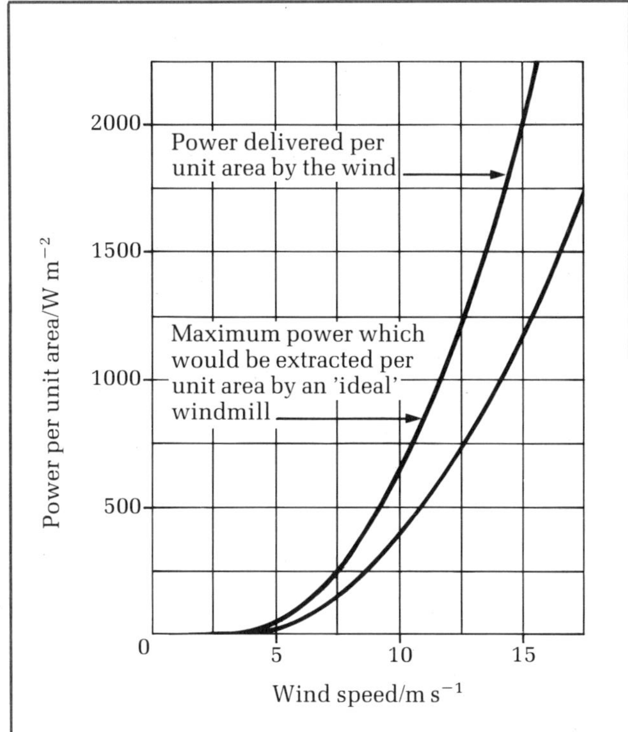

Figure GL1

*Energy: *A Guidebook*, Janet Ramage, *OUP Opus*, 1983. Reproduced by permission.

The next question is how much of this power can be extracted. For maximum power extraction it can be shown that the optimum condition is that the wind speed be reduced to one third of its initial value. The lower curve in Figure GL.1 corresponds to this situation, i.e. an 'ideal' machine, and it is valid for a wide range of wind machines and speeds.

Wind turbines

Windmills come in many shapes and sizes but the general idea is always the same. Forces induced by the air flow keep the blades or sails moving, causing an axle to rotate, and this transmits a twisting force, or torque, to machinery driven by the mill. Our subject here is: what is the origin of the force which keeps the blade moving?

The first answer is an obvious one. It is the force of the wind hitting the blade which drives it. However, with this simple picture it is difficult to explain why windmills with broad sails actually extract less power than those with thin airfoils.

The Savonius rotor

The Savonius rotor, or S-rotor, illustrated in Figure GL.2 is a very simple vertical axis machine (sometimes made from an oil drum sawn in half down the centre). The impact of the air on the concave side of the vane A will push it backwards. The other side of A pushes against the air, which naturally pushes back. And we mustn't forget that the vane is moving. If the wind speed is 9.0 m s^{-1} but the outer tip of A is moving at 7.0 m s^{-1}, the relative wind speed at the tip is, at minimum, only 2.0 m s^{-1}.

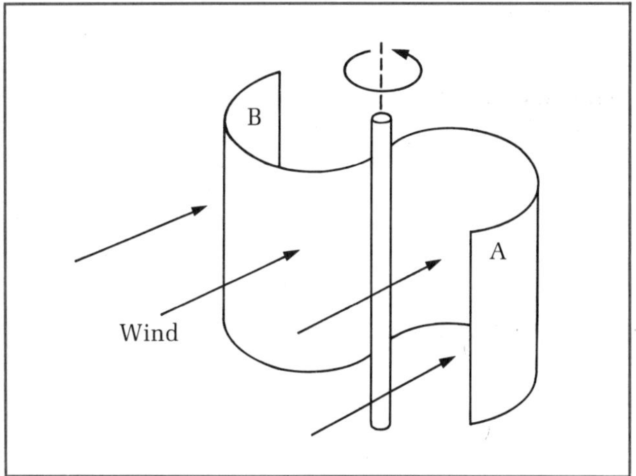

Figure GL2

If the machine is to produce useful power, the drag force of the 2.0 m s^{-1} wind flowing past A must be greater than the reverse drag force of the 16.0 m s^{-1} wind against B. Hence the need for curved vanes. A moving car, for example, is also subject to a drag force and information used to calculate the energy dissipated by a vehicle moving *against* the drag could equally well be used to find the energy extracted by a vane moving *with* the drag.

Airfoil turbines

By contrast, airfoil windmills or turbines have two or three thin blades shaped like the wings of aircraft, and are really flying machines.

Figure GL.3, at the top of the next page, shows a three-bladed, horizontal axis turbine. (The largest operating wind

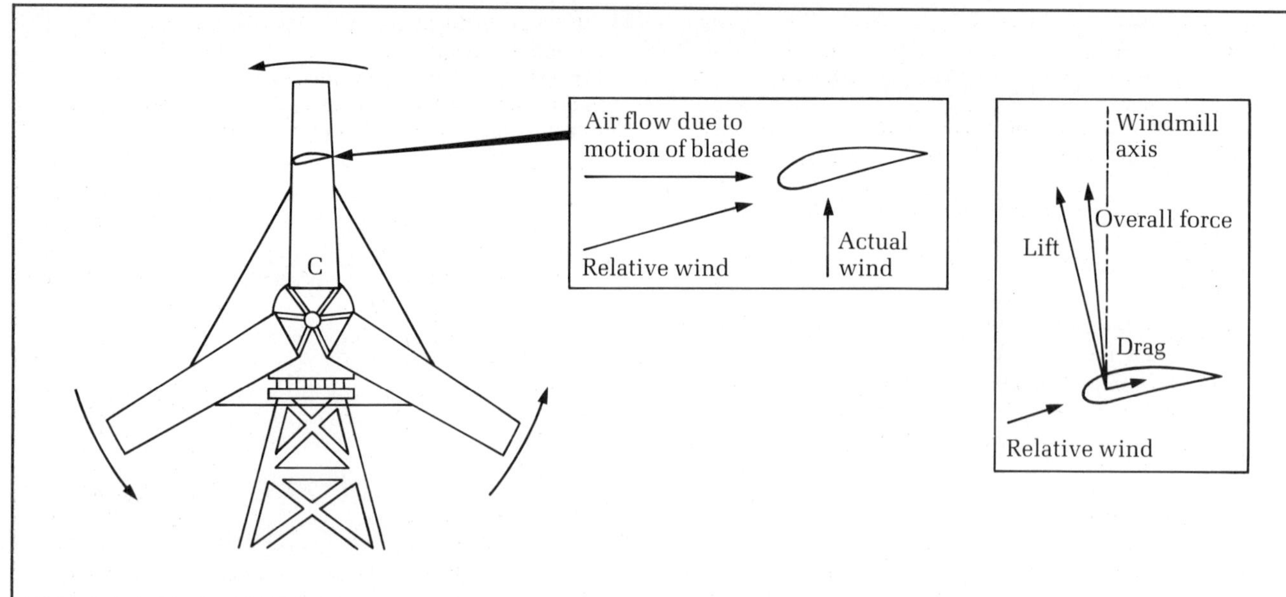

Figure GL3

generator in Britain is a 100-kW machine of this type, with blades about 7.5 m long.) Let us suppose that the axis of the turbine is pointing straight into the wind and the blades are rotating at high speed. If the blade speed is several times the wind speed, the *relative* wind will be coming almost directly at the leading edge of the airfoil.

Under these conditions the air moves faster over the rear surface of the airfoil, with a consequent reduction in air pressure in this region. The airfoil is subject to a lift force (Figure GL.3) as is the wing of an aircraft. The moving airfoil is also subject to a drag force, but with good design the lift-drag ratio can be high, so that the overall force is almost in the direction of the lift force and slightly forward of the axis line of the turbne. This overall force pulls the blade round in the direction of motion. (Unfortunately it is also trying to bend the blade back along the windmill axis.)

The ratio of lift to drag also depends on the angle of attack, i.e. the angle that the relative wind makes with the airfoil section (almost zero in Figure GL.3). If this angle becomes large there is an abrupt loss of lift and the airfoil can stall, just like an aircraft. As the blade speed is very different at the tip and the inner end, the angle of attack decreases along the blade. Introducing a twist in the blade – making it more like an aircraft propeller than a wing – can improve efficiency.

Questions

1 Justify the claims that
 a) '... each cubic metre of air in a 9.0 m s^{-1} wind carries slightly more than 50 J of energy.'
 b) '... a windmill having an area of about one square metre facing this wind [has a] wind power input of about 500 W.'

2 A horizontal wind turbine has an effective area of 180 m^2 and is operating in a wind speed of 11.0 m s^{-1}. The actual power output is 40% of the total wind power delivered by the wind. Use Figure GL.1 to determine:
 a) the total power per unit area delivered by the wind,
 b) the actual power output of the turbine in kW,
 c) the maximum power output from the equivalent-area 'ideal' windmill.

3 Sketch a plan view of the S-rotor of Figure GL.2 when it has the plane containing the tips of the vanes and the axis of rotation perpendicular to the wind direction.
 Indicate on and alongside your sketch, using the data of the passage,
 a) the magnitude and direction of the wind,
 b) the direction of rotation,
 c) the relative wind velocities of the vane tips,
 d) the directions of the drag forces at the vane tips.

4 **a)** What is the essential difference between the mode of action of the S-rotor and that of the horizontal-axis wind turbine?
 b) Why must the blades of the turbine be made of a high strength material?

5 A wind turbine of the type shown in Figure GL.3, with blades of length 7.5 m, is operating such that the speed of the outer tip of the blade is 67.0 m s^{-1}. If the end of the blade closest to the axis of rotation is 0.50 m from the axis, calculate the speed of this part (C in Figure GL.3).
 Why will the angle of attack increase from the outer tip of the blade towards the centre of rotation? What step is taken to keep the angle of attack as small as possible along the length of the blade?

(London)

Comprehension paper 2 (1 hour)

First read the passage below carefully and afterwards answer the questions which follow.

Superconductivity

Adapted from *Superconductivity* by E. A. Lynton, Methuen Physical Monographs, 1962.

Basic phenomenon

The behaviour of electrical resistivity at low temperatures was among the first problems investigated by Kamerlingh Onnes after he succeeded in liquefying helium for which the boiling point is 4.2 K. In 1911, while measuring the way the resistivity of mercury diminishes as the temperature diminishes, he found that at about 4 K the resistivity falls abruptly to a value which his best efforts could not distinguish from zero, and remains at this value at any lower temperature. He called this phenomenon *superconductivity* and the highest temperature at which it occurred the *critical temperature*, T_c. He and other workers found a number of other metals which behaved in the same way. They were called *superconductors*, and each was characterized by its own critical temperature (see Table).

Detection of superconductivity

When a metallic ring is exposed to a changing magnetic field, a current, I_0, is induced in the ring. If the magnetic field is subsequently kept constant then, at any later time, t, the current, I_t, in the ring is less than I_0. The current decays according to

$$I_t = I_0 \exp\left[-\frac{R}{L}t\right] \qquad \textbf{(1)}$$

where R is the resistance and L the inductance of the ring. If the ring is mounted vertically, I_t can be measured with great precision by observing the deflection of a square coil which is suspended concentrically with the ring and with its plane initially at right angles to that of the ring. The measurement of I_t enables the resistance of the ring and hence the resistivity of the material of the ring to be determined. This allows the detection of a much smaller resistance than any potentiometric method. A long series of such measurements on superconducting rings culminated in an experiment by Collins (1956) in which a superconducting ring, kept below T_c, carried a current without any detectable decay for about three years. This allowed Collins to place an upper limit on the resistivity of the superconductor of about $10^{-23}\,\Omega\,\text{m}$. This can be compared with a value of $10^{-12}\,\Omega\,\text{m}$ for the low temperature resistivity of copper, which is not a superconductor. There is therefore little doubt that a superconductor has negligible resistivity and that a current induced in a superconducting ring will persist almost indefinitely without decay.

The effect of a magnetic field

At any temperature T, below T_c, superconductivity will cease and normal conductivity will be restored by placing the specimen in a magnetic field of sufficiently large flux density. The minimum value of the external flux density for this to occur is called the *critical flux density*, B_c. B_c varies with T according to

$$B_c = B_0\left[1 - \left(\frac{T}{T_c}\right)^2\right] \qquad \textbf{(2)}$$

where B_0 is the value of B_c when T is 0 K. This relation holds for all metals which exhibit superconductivity. Values for B_0 for some of these are shown in the Table.

Table: Values of T_c and B_0 for some elements

Element	T_c/K	B_0/tesla
Aluminium	1.19	0.010
Lead	7.18	0.080
Mercury	4.15	0.041
Niobium	9.46	0.19
Tin	3.72	0.030
Zinc	0.88	0.005

The critical current

If the current in a superconducting wire which is maintained at a temperature below T_c is gradually increased, a value is reached at which superconductivity ceases. This value is the *critical current*.

Superconducting magnets

An important application of superconductivity is in the production of high magnetic flux densities. Because no power is dissipated in a superconductor it is attractive to wind a solenoid with superconducting wire and generate a very high magnetic field. A limitation is that the field produced must not be so large that superconduction in the material of the wire ceases.

It has been found that certain metallic alloys have significantly higher critical flux densities than pure metals and in many cases higher critical temperatures also. The alloy Nb_3Sn, for which $T_c = 18.5$ K, remains superconducting in flux densities up to about 20 tesla and has been wound into a solenoid producing a flux density of about 10 tesla. To produce the same flux density using a conventional electromagnet having copper windings would require about 1 MW of electrical power and about 350 litres of cooling water per minute.

Questions

1 Explain what is meant by *superconductivity* and by *critical temperature*.

2 a) Being provided with a coil, a rheostat and a battery, describe how you would induce a current in a metal ring.

b) Explain why the square coil behaves like the coil of a moving-coil galvanometer, in the arrangement described in lines 25 to 28. Hence explain why the arrangement can detect the current in the ring.

3 a) State **one** advantage and **one** disadvantage in using a solenoid wound with superconducting wire rather than one wound with copper wire for producing a magnetic flux density.

b) State **two** advantages of using the alloy Nb_3Sn rather than the metal niobium for the wire of a solenoid to produce a magnetic flux density.

4 a) Sketch a graph of equation (2) with temperature T/K as abscissa and critical flux density B_c/tesla as ordinate. What is the significance of each of the two regions of the graph separated by the line you draw?

b) Calculate the critical flux for lead maintained at a temperature of 5.0 K.

5 a) The current in a superconducting ring does not diminish over a long period of time. Explain why this means that the ring has negligible resistance.

b) A ring of diameter 0.050 m is made from wire of cross-sectional area $1.00 \times 10^{-6}\,\text{m}^2$. A current of 2.00 A induced in the ring decays to 1.00 A in 0.31 s. If the inductance of the ring is 1.32×10^{-7} H, calculate the resistivity of the material of the ring.

6 Explain why superconductivity ceases when the current in a superconducting wire is increased to the value of the *critical current*.

(J.M.B.)

Comprehension paper 3 (1 hour)

Optical fibre communication

Adapted from 'Devices for Optical Communication Systems' by M. J. Robertson, published in *Physics Education*, January 1982.

Optical communication systems are at present undergoing very rapid development. Optical communication takes places by the transmission of pulses of light along transparent glass or other suitable fibres of circular cross-section. In optical communication the radiation used is not only that in the visible spectrum but may be ultraviolet or infrared radiation just outside the visible spectrum. In this passage 'light' is used in this sense.

The main advantages over conventional wire-based transmission, in which electric currents pass through a cable, are the small diameter of the fibre cable required, its small mass and the low loss achievable. At present, in the UK inter-city telephone network, repeaters (amplifiers) have to be buried every 2 km along the cable route. With optical fibres it is hoped to increase repeater spacing to 30 km, because losses are so low.

In most optical systems the information to be transmitted is converted into a series of pulses, pulses of '1' and '0' ('on' and 'off') forming a binary code. Obviously the rate of information transfer will depend on the time duration of the pulse (or pulse length). Each pulse consists of a narrow range of wavelength and the shorter the pulse length the greater is the range of wavelengths needed to transmit the pulse.

Operation of the fibre

Light which strikes the curved boundary of a circular fibre can undergo total internal reflection so that such light is not lost. Total internal reflection occurs when light travelling in one material reaches the surface of an optically less dense material at an angle of incidence too great for refraction to occur. All the light is reflected back into the optically denser material. Hence light can be made to travel along fibres which are not straight. There are three types of optical fibre, known as step-index, graded-index and monomode. In all three, a central core of modified silica glass is surrounded by a concentric glass cladding of slightly lower, constant refractive index. In a step-index fibre the refractive index of the core is the same throughout the core for a given wavelength, but in the graded-index fibre the refractive index of the core varies from a maximum at the centre to a minimum at the cladding. In the monomode fibre the core has a constant refractive index, but the diameter of core is particularly small.

Length of fibre

In spite of the guiding properties of optical fibres there will be some reduction in transmitted light due to scattering and absorption. Minimum absorption occurs when the light which enters the fibres has wavelengths of 0.85, 1.30 and 1.55 μm. In early work on fibres the loss at 0.85 μm was found to be least. Systems designed at this wavelength showed the intensity of the light transmitted down the fibre to be reduced by 40% per kilometre of fibre. As the silica has been made more pure, the losses at 1.30 and 1.55 μm have been reduced until now at 1.55 μm intensity is reduced by only 4% per kilometre.

The maximum length of fibre which can be used is limited by dispersion, since dispersion smooths out pulses that are initially square and runs adjacent pulses into each other. There are basically two types of dispersion, modal and material. The former results from the different optical pathlength that light entering along the axis of the fibre will travel compared with light entering at an angle and undergoing multiple reflections. These different paths are known as modes. Clearly a pulse of light in an axial mode will arrive at the detector before that in an angular mode. Material dispersion occurs in a fibre because light of different wavelengths in a pulse of light travel at different speeds and arrive at the detector at different times.

Modal dispersion is reduced in graded-index fibres. As the speed of light is inversely proportional to the absolute refractive index, the off-axis rays travel faster, compensating for their longer paths, and different modes can be made to arrive at the detector together. Non-uniformities in the structure of the fibre limit the performance with this technique but dispersion is between 20 and 100 times less than for a step-index fibre. In the monomode fibre the diameter of the core is so small (5 μm–10 μm) that only one (the axial) mode can propagate and this immediately eliminates modal dispersion.

Questions

1 a) An advantage of an optical transmission system compared with a wire-based transmission system is the small diameter of the fibre required (line 11). Why is it not practical to use small diameter wire in the wire-based transmission system?

b) State **two** other advantages of an optical transmission system compared with a wire-based transmission system.

2 a) What determines the choice of wavelengths used in optical communication?

b) The article refers to systems operating at 0.85 μm and 1.55 μm (lines 50–51). In what part of the electromagnetic spectrum will these radiations be found?

3 a) Sketch graphs showing the variation of refractive index with distance measured radially outwards from the centre of the core to the outside of the cladding for (i) a step-index fibre, (ii) a graded-index fibre, in which the rate of change of refractive index with distance is least at the centre and greatest at the edge of the fibre.

b) Draw diagrams, one in each instance, showing the path of a non-axial ray in (i) a step-index fibre, (ii) a graded-index fibre. Account for the difference between the two diagrams.

c) Justify the statement that 'the speed of light is inversely proportional to the absolute refractive index' (lines 69–72).

4 a) What is meant in the passage by 'material dispersion'?

b) How can information be transmitted along a glass fibre with minimum material dispersion?

5 a) A series of pulses, as shown in the diagram, enters a length of fibre in which modal dispersion is significant. Sketch the pulses as they are likely to emerge from the fibre.

b) Why will modal dispersion limit the frequency with which pulses can be transmitted?

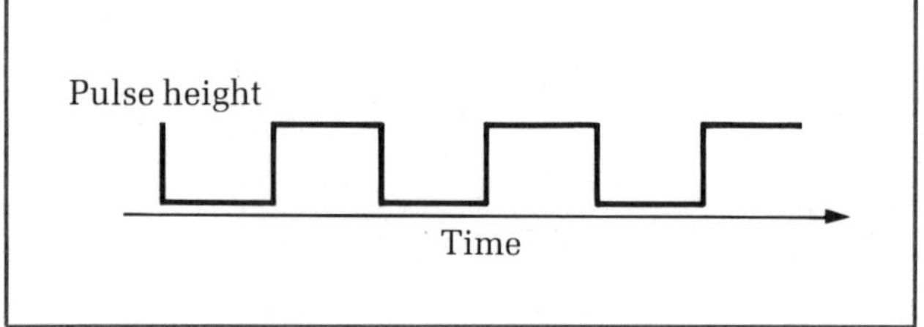

Figure GL4

6 **a)** State an advantage of using pulses of short length in the transmission of information.

b) What considerations determine the choice of pulse length?

7 For light of wavelength 1.55 μm the intensity is reduced by 4% per kilometre (line 56). What is the intensity of the signal after 30 km, expressed as a percentage of the initial intensity? Sketch a graph showing how the intensity varies between 0 and 30 km.

(J.M.B.)

Table of physical constants

The Avogadro constant L	$= 6.022 \times 10^{23}\ \mathrm{mol^{-1}}$
Molar gas constant R	$= 8.314\ \mathrm{J\,K^{-1}\,mol^{-1}}$
Boltzmann's constant k	$= 1.381 \times 10^{-23}\ \mathrm{J\,K^{-1}}$
Stefan's constant σ	$= 5.670 \times 10^{-8}\ \mathrm{W\,m^{-2}\,K^{-4}}$
Gravitational constant G	$= 6.672 \times 10^{-11}\ \mathrm{N\,m^{-2}\,kg^{-2}}$
Electric field constant ε_0 (permittitivy of free space)	$= 8.854 \times 10^{-12}\ \mathrm{F\,m^{-1}}$
Magnetic field constant μ_0 (permeability of free space)	$= 4\pi \times 10^{-7}\ \mathrm{H\,m^{-1}}$
Speed of light in a vacuum c	$= 2.998 \times 10^{8}\ \mathrm{m\,s^{-1}}$
Planck constant h	$= 6.626 \times 10^{-34}\ \mathrm{J\,s}$
Electronic charge e	$= 1.602 \times 10^{-19}\ \mathrm{C}$
Specific charge of the electron e/m_e	$= 1.759 \times 10^{11}\ \mathrm{C\,kg^{-1}}$
Mass of the electron m_e	$= 9.11 \times 10^{-31}\ \mathrm{kg}$
Mass of the proton m_p	$= 1.673 \times 10^{-27}\ \mathrm{kg}$
Mass of the neutron m_n	$= 1.675 \times 10^{-27}\ \mathrm{kg}$
Atomic mass unit 1 u	$= 1.661 \times 10^{-27}\ \mathrm{kg}$

Appendix 1
More about waves

A1.1 Waveforms

If you were to use a camera to take a 'snapshot' of waves travelling along a string, the photograph would look like Figure A1.1. This shows the position of the string at a given instant when the snapshot was taken. It shows the waveform which is the shape of the waves.

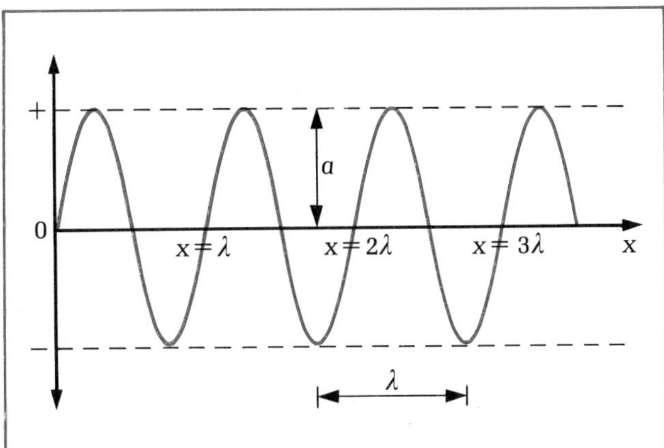

Figure A1.1 *A snapshot of a wave*

The waveform is **sinusoidal**. This means that the displacement, y, of each particle of the string is given by an equation of the form

$$y = a \sin \frac{(2\pi x)}{\lambda}$$

where a = the amplitude of the wave
x = distance along the wave from the fixed point, 0
λ = the wavelength of the waves

The displacement is zero at distances $x = \lambda/2$, λ, $3\lambda/2$, etc., as well as at $x = 0$ (i.e. at 0). At each of these distances, $2\pi x/\lambda$ is a whole number $\times\ \pi$. These values of $2\pi x/\lambda$ always give zero for y.

The displacement is maximum at distances $x = \lambda/4$, $3\lambda/4$, $5\lambda/4$, etc. At each of these distances, $2\pi x/\lambda$ is an odd number $\times\ \pi/2$. These values of $2\pi x/\lambda$ always give either $+a$ (a 'crest') or $-a$ (a 'trough').

x	0	$\frac{\lambda}{4}$	$\frac{\lambda}{2}$	$\frac{3\lambda}{4}$	λ	$\frac{5\lambda}{4}$	$\frac{3\lambda}{2}$	$\frac{7\lambda}{4}$	2λ
$\frac{2\pi x}{\lambda}$	0	$\frac{\pi}{2}$	π	$\frac{3\pi}{2}$	2π	$\frac{5\pi}{2}$	3π	$\frac{7\pi}{2}$	4π
y	0	$+a$	0	$-a$	0	$+a$	0	$-a$	0

Figure A1.2 $y = a \sin \left(\frac{2\pi x}{\lambda}\right)$

A1.2 Progressive waves

The snapshot shown in Figure A1.1 does not show which direction the waves are moving in. Successive snapshots as in Figure A1.3 are needed to show which way the waves travel.

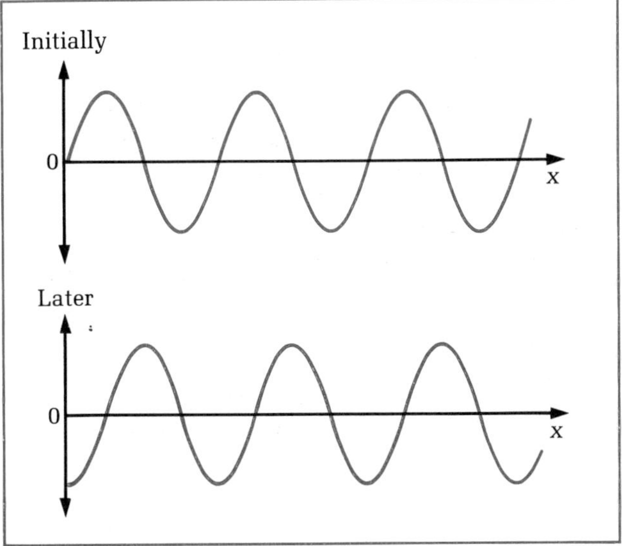

Figure A1.3 *Successive snapshots of a progressive wave*

Waves that travel through a medium carry energy through the medium. The waves are said to be **progressive**. In Figure A1.4, the two snapshots are superimposed to show that the second snapshot is to the right of the first.

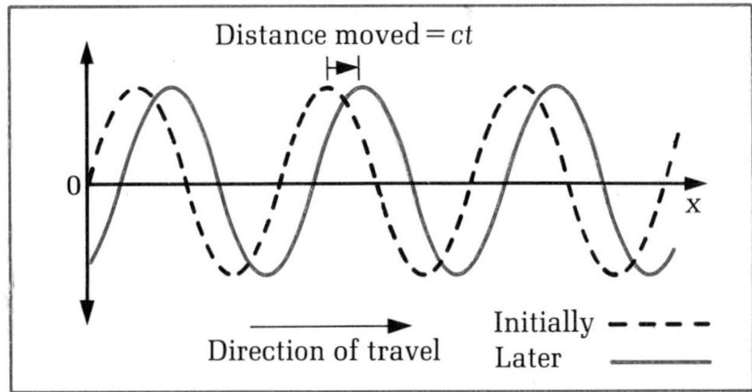

Figure A1.4 *Superimposed snapshots*

Suppose the speed of the waves is c. If the time between the snapshots is t, each wave crest and wave trough must have moved through the same distance $c \times t$ along the x-axis.

The equation for displacement for the wave in the first 'snapshot' is $y = a \sin (2\pi x/\lambda)$. For the second 'snapshot', distance ct must be subracted from x to give the following equation for displacement y:

$$y = a \sin \frac{2\pi}{\lambda} (x - ct)$$

This is the equation for waves of amplitude a and wavelength λ travelling at speed c **in the $+x$ direction**. The equation gives the displacement at distance x from 0 and at time t.

If the waves had been moving from right to left, the snapshots would be as in Figure A1.5. Distance ct is now added onto x to give the following equation for displacement y:

$$y = a \sin\frac{2\pi}{\lambda}(x + ct)$$

This is the equation for waves travelling **in the $-x$ direction**. These equations give the displacement of the particles in any medium through which sinusoidal waves travel.

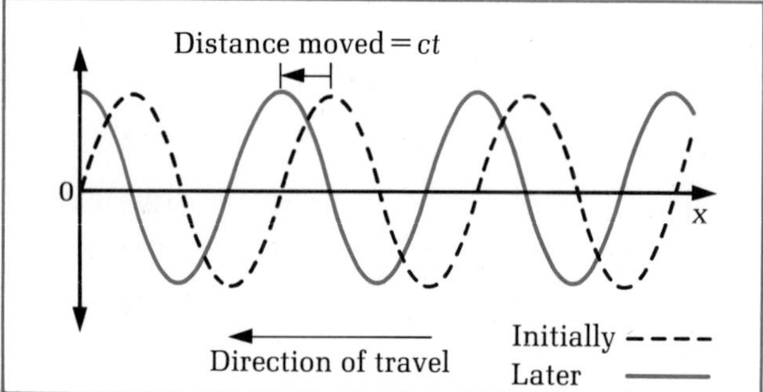

Figure A1.5 *A progressive wave travelling right to left*

When a progressive wave travels through a medium, each particle of the medium oscillates. **Each particle** on the string oscillates in simple harmonic motion as the waves travel along the string.

Figure A1.6 shows how the displacement of the particle at 0 varies with time when a wave travels **right to left** along the string. The displacement at $x = 0$ (ie. at 0) may also be described by the following equation:

$$y = a \sin\frac{2\pi\, ct}{\lambda} = a \sin 2\pi ft$$

where $f = c/\lambda$

This equation is obtained by making $x = 0$ in the equation for a progressive wave travelling from right to left.

The amplitude of oscillation is the maximum displacement. This is equal to a since the maximum value of sin $2\pi ct/\lambda$ is 1.

The frequency of oscillation, f, equals c/λ. This is because the time T for one complete cycle of oscillation is such that $2\pi\, cT/\lambda$ equals 2π for one complete cycle. Hence $cT = \lambda$, and thus frequency $f = 1/T = c/\lambda$.

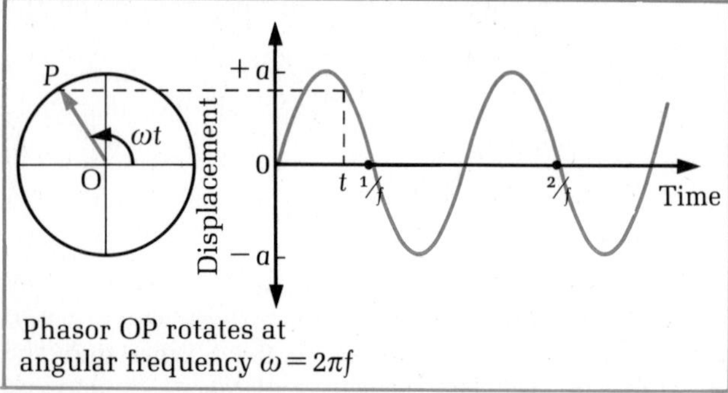

Figure A1.6 *Displacement v. time*

The wavespeed, c is therefore equal to $f\lambda$. A simpler proof of this equation is given on p. 232. Here, the analysis shows how $c = f\lambda$ is 'buried' within the progressive wave equation.

When a continuous train of progressive waves passes through a medium, all the particles of the medium oscillate with the same amplitude and at the same frequency. However, any two particles oscillate in phase only if their separation along the x-axis is a complete number of wavelengths.

In general, for two particles at separation Δx, maximum displacement of one particle occurs at time $\Delta x/c$ later than for the other particle. The second particle 'lags' behind the first by a fraction of a cycle equal to $\Delta x/cT$ or $\Delta x/\lambda$ (since $cT = \lambda$). The phase difference between the oscillations is therefore $2\pi\, \Delta x/\lambda$.

The progressive wave equation

$$y = a \sin\frac{2\pi}{\lambda}(x + ct)$$

may be rewritten as

$$y = a \sin\left(2\pi ft + \frac{2\pi x}{\lambda}\right) = a \sin(2\pi ft + \phi)$$

where the phase angle

$$\phi = 2\pi x/\lambda$$

The phase angle formula $\phi = 2\pi x/\lambda$ gives the phase difference between a particle at distance x and the particle at the origin. Thus, for two particles at separation Δx, the phase difference is $2\pi\, \Delta x/\lambda$ as worked out in the previous paragraph.

The variation of displacement with time may also be represented using a **phasor diagram**, as in Figure A1.6. For two particles at separation Δx, their phase difference $2\pi\Delta x/\lambda$ is the angle between the phasor of one particle and the phasor of the other. Both phasors are the same length and rotate at frequency f.

A1.3 The principle of superposition

Where waves pass through each other, the resultant displacement is equal to the sum of the individual displacements.

Figure A1.7 shows a sequence of snapshots of two waves as they pass through each other. At any given instant, the resultant displacement at each position is the sum of the displacements each wave alone would give at that position.

Where two sine waves of equal amplitude pass through each other, cancellation occurs where crests from one set of waves arrive at the same time as troughs from the other set. The resultant displacement is zero where this happens. Reinforcement occurs where the crests from one set of waves arrive at the same time as the crests from the other set.

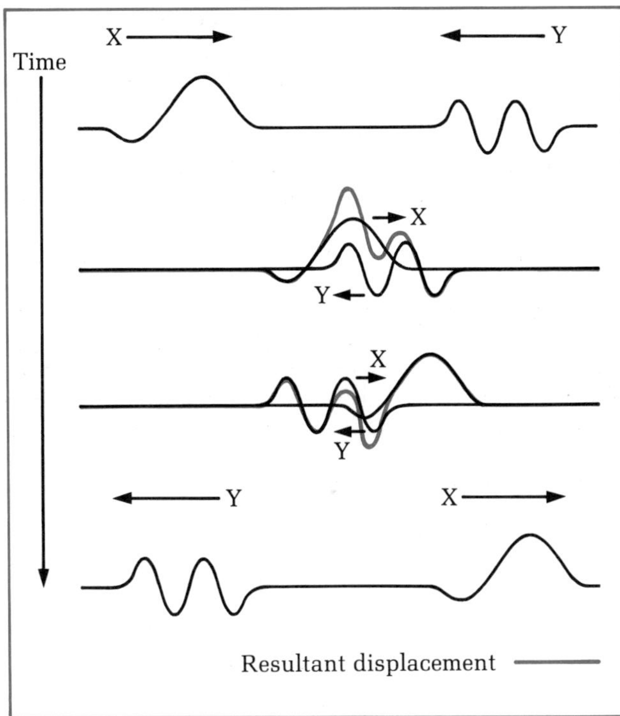

Figure A1.7 *Superposition*

For two adjacent point sources emitting waves at the same frequency, the waves from one source pass through the waves from the other source. If the sources are coherent (i.e. emit waves with a constant phase difference), the points of cancellation and reinforcement are fixed, so an interference pattern is formed. Figure A1.8 shows the idea.

The displacement y_1 at distance x_1 from source S_1 may be written

$$y_1 = a \sin 2\pi \left(\frac{x_1}{\lambda} - ft \right)$$

The displacement y_2 at distance x_2 from source S_2 may be written

$$y_2 = a \sin 2\pi \left(\frac{x_2}{\lambda} - ft + \phi \right)$$

where ϕ is the phase difference between waves emited by S_1 and S_2.

The resultant displacement is

$$y = y_1 + y_2$$

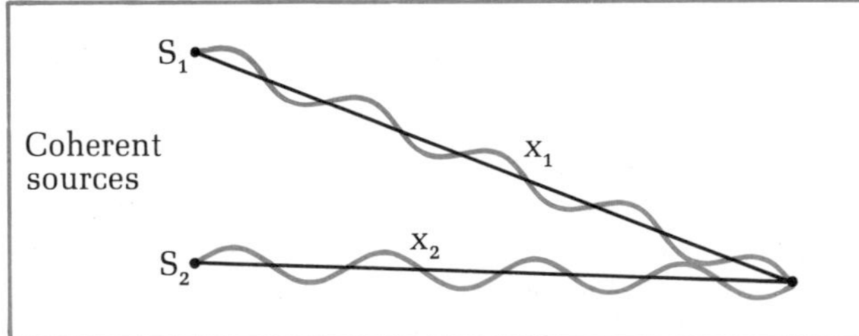

Figure A1.8 *Interference*

1 Suppose the sources emit in phase. Hence $\phi = 0$. The condition for reinforcement is then $x_1 = x_2 + m\lambda$ where m is an integer. This condition ensures that at any point of reinforcement, waves from S_1 arrive either in phase (for $m = 0$) or m cycles earlier or later than from S_2. Thus the central fringe is a bright fringe corresponding to $m = 0$ and hence $x_1 = x_2$.

2 Suppose the sources emit out-of-phase exactly. Hence $\phi = \pi$. The condition for reinforcement is now $x_1 = x_2 + m\lambda + \frac{1}{2}\lambda$. Because the waves are emitted exactly out-of-phase, this condition gives the points where they arrive exactly in phase. Each bright fringe is now where a dark fringe would have been had ϕ been zero.

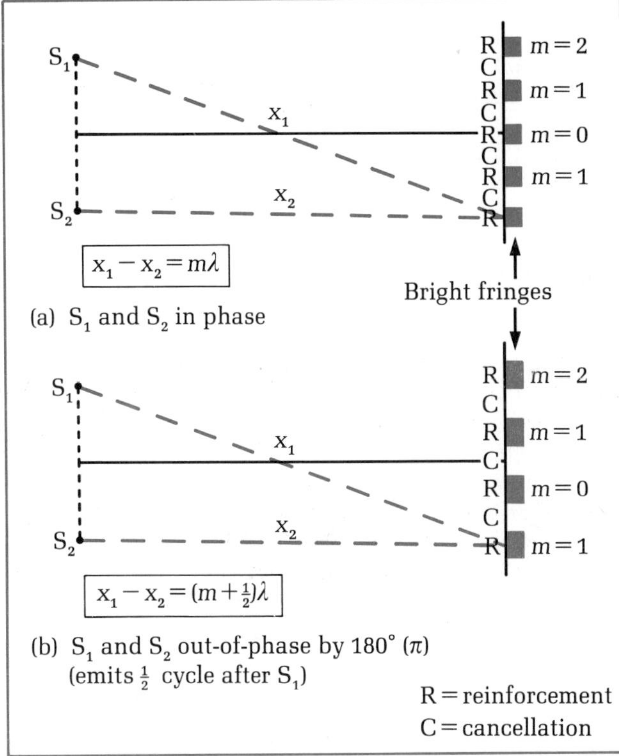

Figure A1.9 *Phase difference*

3 Suppose it is possible to change ϕ gradually from 0 to 2π. The effect would be to shift the fringes through one fringe spacing.

4 If ϕ changes at random, the points of cancellation and reinforcement change positions at random. No interference pattern would then be observable. This is why no interference pattern is seen when light from two separate lamps overlaps. Each lamp emits photons at random so the phase difference changes at random.

5 Interference of light is achieved by illuminating a pair of double slits with light from a single slit. This ensures the double slits emit light waves with a constant phase difference.

If slit S_2 is covered with a thin piece of transparent material, the phase difference ϕ is increased and the fringes are displaced. The reason is that light travels more slowly through the material than through air, thus delaying waves from S_2 compared with those from S_1.

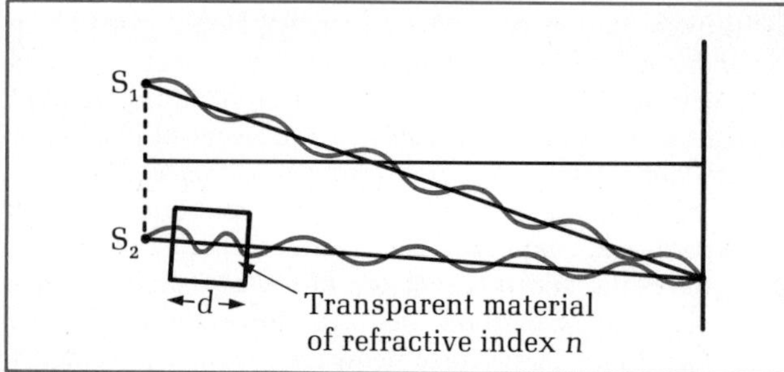

Figure A1.10 *Optical path difference* $=(n-1)d$

For material of thickness d, the delay can be worked out from the wavespeed in air, c_0, and the refractive index n of the material.

Time taken to travel through distance d in air $= d/c_0$

Time taken to travel through distance d in the material $= d/(c_0/n)$ since the wavespeed in the medium is c_0/n.

Hence the delay $= nd/c_0 - d/c_0 = (n-1)d/c_0$

This is equivalent to the waves from S_2 travelling an extra distance through air equal to $c_0 \times$ delay $=(n-1)d$. This is sometimes called the **optical path difference** of the piece of material.

A1.4 *Michelson's interferometer*

This may be used to show the effect of change of phase on an interference pattern more easily than in a double slits arrangement. It may also be used to measure the refractive index of a gas by measuring the change of phase. Figure A1.11 shows the essential arrangement for the Michelson interferometer.

Two sets of waves from the same light source produce an interference pattern consisting of alternate bright and dark circular fringes. The two mirrors M_1 and M_2 are exactly perpendicular so the image of M_1, M_1', is parallel to M_2. Hence M_1' and M_2 are effectively a parallel-sided film. With an extended monochromatic light source, circular fringes are seen.

Mirror M_2
M'_1
R_2
Beam of monochromatic light
X
Y
R_1
Mirror M_1
Semi-reflecting mirror C
Observer

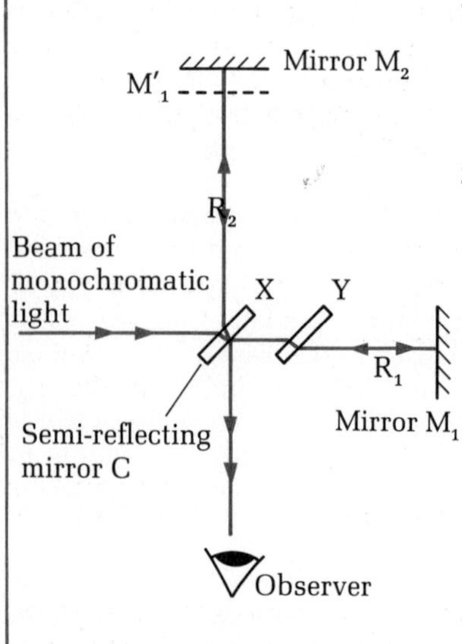

1. The semi-reflecting mirror C on glass plate X splits the beam into R_1 and R_2.
2. R_1 reflects at M_1 then at C to enter the observer's eye. R_2 reflects at M_2 then passes through C to enter the observer's eye.
3. The observer sees an interference pattern due to R_1 and R_2 interfering. Block Y is necessary to make the path length of R_1 the same as R_2.

Figure A1.11 *Michelson's interferometer*

The central spot of the fringe pattern is dark if the path difference between the two beams of light is $(m+\frac{1}{2})\lambda$ where m is an integer. This means that the distance M_2M_1' must be $\frac{1}{2}(m+\frac{1}{2})\lambda$.

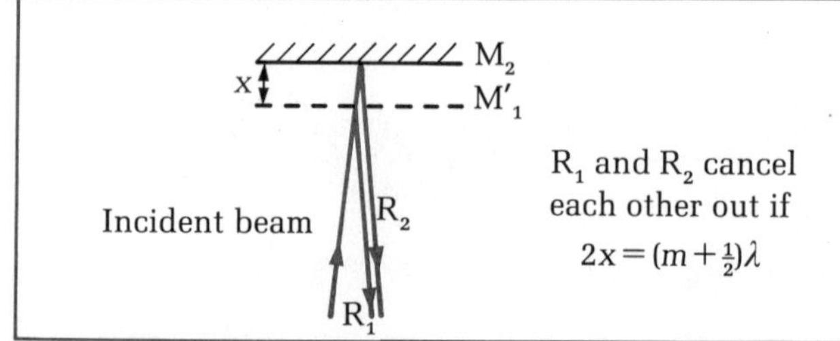

Figure A1.12 *Cancellation for a thin film*

1 If distance CM_2 is gradually decreased, the fringes move inwards towards the centre, and the centre alternates between dark and bright. Each change from dark to bright to dark corresponds to a change of path difference of one wavelength. If the wavelength is known, movement of M_2 by a fraction of a wavelength can be detected.

2 If a thin transparent plate is inserted into the path of beam R_2, as in Figure A1.13, the beam is further delayed compared with R_1. This is equivalent to moving M_2 away from C. The change of optical path difference $(n-1)d$ causes the fringes to move outwards.

If the centre spot alternates in brightness through m cycles (dark→bright→dark), the change of the optical path difference must be $m\lambda$. This gives an equation from which n may be calculated if λ and d are known:

$m\lambda = (n-1)\,d$ — λ = wavelength in air; d = thickness of material; n = refractive index

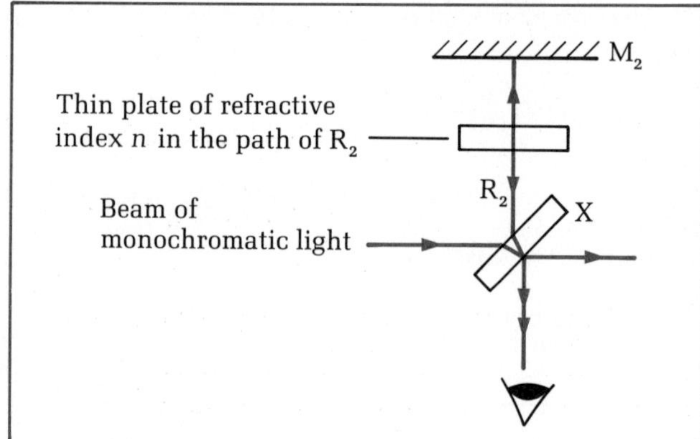

Figure A1.13 *Delaying* R_2

3 Suppose the thin plate is replaced by a tube of gas, as in Figure A1.14. The refractive index of the gas may be determined by measuring the phase shift (i.e. the number of cycles the brightness of the centre spot passes through) when the gas is pumped from the tube.

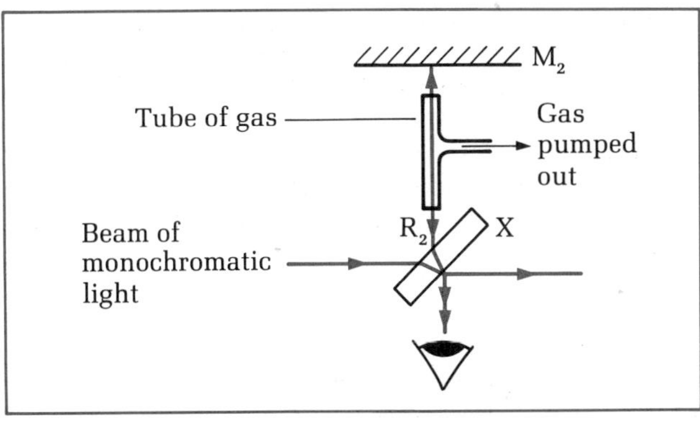

Figure A1.14 *Measuring the refractive index of a gas*

A1.5 Stationary waves

When two sets of waves of the same wavelength and amplitude pass through each other, a stationary wave pattern is set up. Fixed points where the displacement is always zero are characteristic of a stationary wave pattern. These points are called **nodes**. Points of maximum displacement are called **antinodes**. The distance from a node to the nearest antinode is $\frac{1}{4}\lambda$.

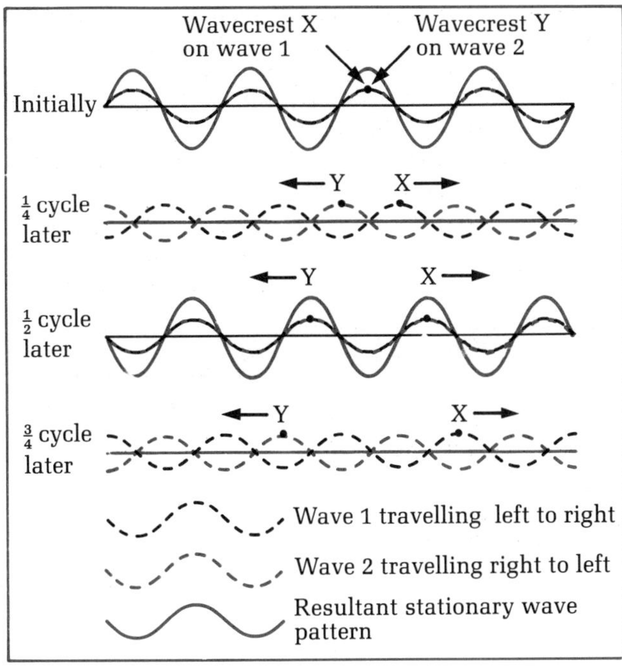

Figure A1.15 *Formation of a stationary wave pattern*

To understand more fully how stationary waves are formed, consider the displacement at distance x for each wave alone.

$$y_1 = a \sin 2\pi \frac{(x - ct)}{\lambda}$$

for the progressive waves travelling from left to right,

$$y_2 = a \sin 2\pi \frac{(x + ct)}{\lambda}$$

for the progressive waves travelling from right to left.

The total displacement, y, caused by the two waves passing through each other is $y_1 + y_2$, hence

$$y = y_1 + y_2$$
$$= a \sin 2\pi \frac{(x - ct)}{\lambda} + a \sin 2\pi \frac{(x + ct)}{\lambda}$$

To add together two sine functions, it is necessary to use the following mathematical formula. The proof will be found in an A-level maths textbook.

$$\sin A + \sin B = 2 \sin \tfrac{1}{2}(A + B) \cos \tfrac{1}{2}(A - B)$$

Let $A = 2\pi \frac{(x - ct)}{\lambda}$ and $B = 2\pi \frac{(x + ct)}{\lambda}$

Hence $\frac{1}{2}(A + B) = 2\pi \frac{x}{\lambda}$ and $\frac{1}{2}(A - B) = 2\pi \frac{ct}{\lambda}$

Therefore, the total displacement is

$$y = 2a \sin 2\pi \frac{x}{\lambda} \cos 2\pi \frac{ct}{\lambda}$$

The equation for y has a time-dependent part ($\cos 2\pi ct/\lambda$) multiplied by a position-dependent part ($2a \sin 2\pi x/\lambda$). This shows that the particles of the medium vibrate at frequency $f = c/\lambda$ with an amplitude $2a \sin 2\pi x/\lambda$.

1. The frequency of vibration is the same as the frequency of the progressive waves that form the pattern.
2. The amplitude of vibration varies with distance x. For $x = 0, \frac{1}{2}\lambda, \lambda, 1\frac{1}{2}\lambda, 2\lambda$, etc., the amplitude is zero because $2\pi x/\lambda$ is always a whole number $\times \pi$ for these values. Hence $\sin 2\pi x/\lambda$ is zero for any of these values of x. These points of no displacement are the nodes. The distance from one node to the next is $\frac{1}{2}\lambda$.
3. Between two adjacent nodes, all the particles vibrate in phase with each other. The amplitude of vibration is equal to $2a \sin 2\pi x/\lambda$. This is a maximum, equal to $2a$, whenever x equals an odd number $\times \frac{1}{4}\lambda$. These values of x make $2\pi x/\lambda$ equal to $\pi/2$ or $3\pi/2$ or $5\pi/2$, etc., and hence $\sin 2\pi x/\lambda$ equal to $+1$ or -1.

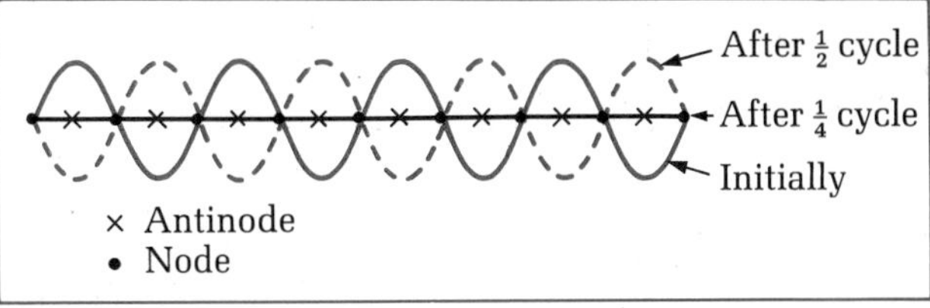

Figure A1.16 *Time sequence of the stationary wave pattern*

The pattern is referred to as *stationary* because the nodes and antinodes are at fixed positions. Energy does not pass through the medium even though the particles are vibrating. All the particles between two adjacent nodes vibrate in phase so energy is not transferred from one particle to another as happens with a progressive wave.

Type of wave	Amplitude	Phase difference between two points
Progressive	Same at all points	Proportional to distance i.e. $\phi = \frac{2\pi\Delta x}{\lambda}$
Stationary	Zero at nodes Max. at antiodes	$\phi = 0$ if separated by an even no. of nodes $\phi = \pi$ if separated by an odd no. of nodes

Figure A1.17 *Comparison between progressive and stationary waves*

You will find a descriptive treatment of stationary waves on pp. 237–38. Measurement of the wavelength of microwaves using a stationary wave pattern is explained on p. 239. Stationary waves of sound in tubes are explained on pp. 249–50 and stationary wave patterns on a vibrating wire on pp. 251–52. In all these examples, waves from the wave source are reflected so that reflected waves pass through waves moving towards the reflector. Thus two sets of waves travel through each other to set up a stationary wave pattern

A1.6 Beats

Where two sets of waves of different frequencies overlap, the vibrations of the particles of the medium vary in amplitude. In other words, the amplitude varies regularly between a minimum and a maximum.

For sound waves, the amplitude variations are detected as **beats** of sound, each beat being heard when the amplitude reaches a maximum. The beat frequency is equal to the difference in frequency between the two sets of waves producing the effect. See p. 247 for experimental details.

The reason why the resultant amplitude at any point varies periodically is because the two sets of progressive waves differ in frequency. At any point, the two sets of waves move steadily in and out of phase because they have different frequencies. A non-mathematical proof of why the beat frequency is equal to the difference in frequency of the two sets of waves is given on p. 248.

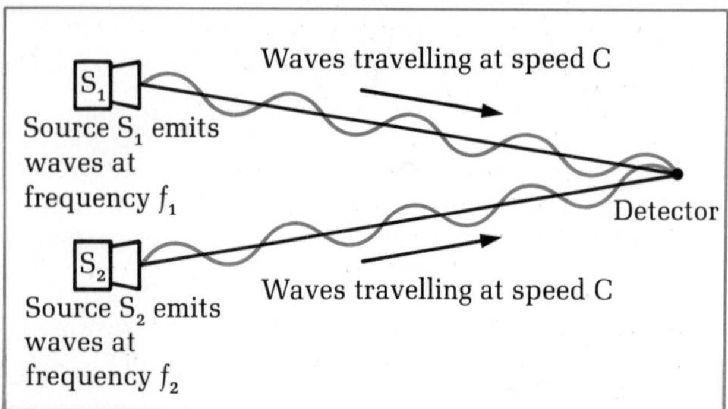

Figure A1.18 *Producing beats*

When progressive waves pass a fixed point, the displacement at that point varies sinusoidally according to the equation $y = a \sin 2\pi ct/\lambda$. Substituting c/f for λ into this equation gives $y = a \sin 2\pi ft$.

Hence for two sets of waves travelling in the same direction at speed c, the displacement at distance x for each set of waves is given by the following equations:

displacement $y_1 = a \sin 2\pi f_1 t$ for waves of frequency f_1
displacement $y_2 = a \sin 2\pi f_2 t$ for waves of frequency f_2

The wavelength of the first set of waves is $\lambda_1 = c/f_1$, and $\lambda_2 = c/f_2$ for the second set of waves.

The resultant displacement $y = y_1 + y_2$

$$= a \sin 2\pi f_1 t + a \sin 2\pi f_2 t.$$

Using the mathematical formula on p. 541 to add together two sine functions,

$$y = 2a \sin 2\pi \frac{(f_1 - f_2)}{2} t \cos 2\pi \frac{(f_1 + f_2)}{2} t.$$

This may be written in the form

$$y = A \cos 2\pi f_{ave} t$$

where $f_{ave} = \frac{1}{2}(f_1 + f_2)$ and $A = 2a \sin 2\pi \frac{(f_1 - f_2)}{2} t.$

1 The **carrier wave** term, $\cos 2\pi f_{ave} t$ represents waves of frequency $\frac{1}{2}(f_1 + f_2)$ travelling at speed c.
2 The **amplitude** of the carrier wave, A, varies sinusoidally with frequency $\frac{1}{2}(f_1 - f_2)$. The amplitude waveform travels at the same speed as the carrier waves.

At any fixed point, the amplitude increases from zero to a maximum and back to zero each half-cycle. Thus the time between successive maxima is $\frac{1}{2}T$ where T is the time period of the amplitude waveform. The frequency

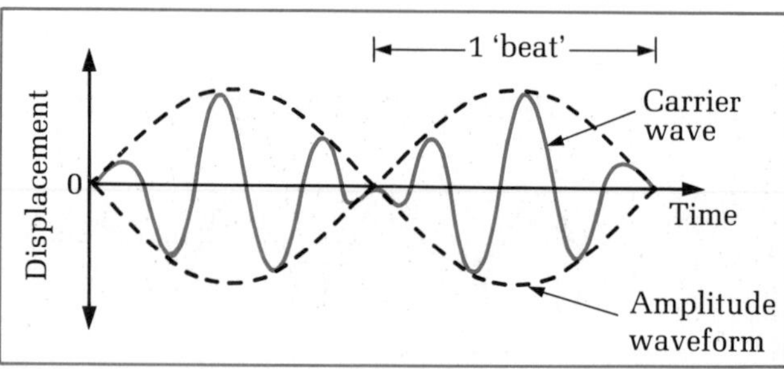

Figure A1.19 *Variation of detector signal with time*

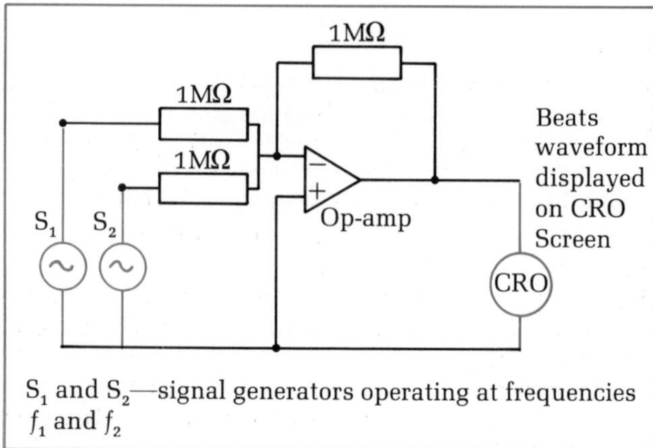

Figure A1.20 *Electrical beats*

of successive maxima (i.e. the beat frequency) is $1/(\frac{1}{2}T)$ which equals $2/T$. Since $1/T = \frac{1}{2}(f_1 - f_2)$, then the beat frequency is equal to $(f_1 - f_2)$.

A1.7 Waveform synthesis

A beats waveform like Figure A1.19 can be made by adding two separate sine waves together. Any waveform can be made by adding a suitable combination of sine or cosine waves together. Figure A1.21 shows how a square waveform can be made by adding sine waves together. The square wave is said to be *synthesised* by adding suitable sine waves together.

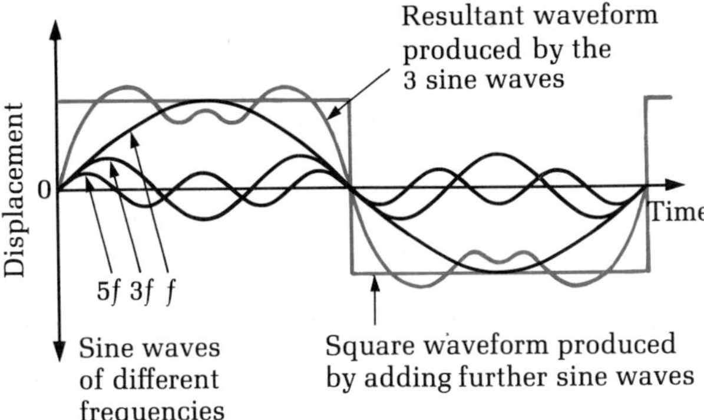

Figure A1.21 *Wavefront synthesis*

Different sound waveforms may be produced using an electronic synthesiser to add sine and cosine waves together. Each waveform is made up of a unique combination of sine and cosine waves.

Figure A1.22 *An electronic synthesiser*

Analysis of waveforms is possible because each different waveform can be made by adding a unique combination of sine and cosine waves together. Analysing a waveform involves finding out what the combination is.

Knowing the frequency and amplitude of each sine or cosine wave that makes up a given waveform is useful if the waveform is to be amplified or filtered. If the amplifier has an uneven frequency response (i.e. amplifies some frequencies more than others), the waveform will be distorted by the amplifier.

For example, an electrocardiograph (ECG) machine is used to amplify and display 'heart beat' signals. These signals contain frequencies up to about 20 Hz. The amplifier must therefore have an even frequency response up to about 20 Hz. However, if its frequency response extends beyond about 20 Hz, then unwanted signals from muscle activity (EMG) will be amplified too.

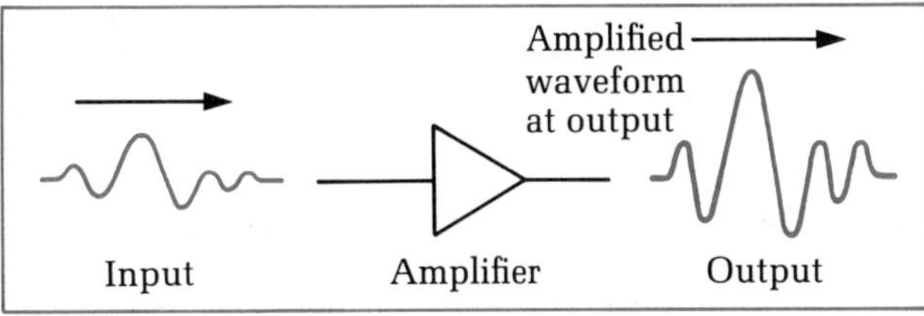

Figure A1.23 *Using an amplifier*

Questions

1 A train of waves travelling on a string is represented by the equation

$$y = 0.25 \sin 2\pi (0.25x - 2.0t)$$

where y is the displacement in metres at t seconds at distance x metres from the origin.

a) What is the amplitude and wavelength of the waves?
b) What is the speed and frequency of the waves?
c) What direction are the waves moving in?
d) What is the phase difference between x = 0 and (i) x = 1.0 m, (ii) x = 3.0 m, (iii) x = 6.0 m?

2 A cine camera operating at 25 Hz is used to film a vibrating string. Two frames X and Y of the developed film are shown in Figure A1.24.

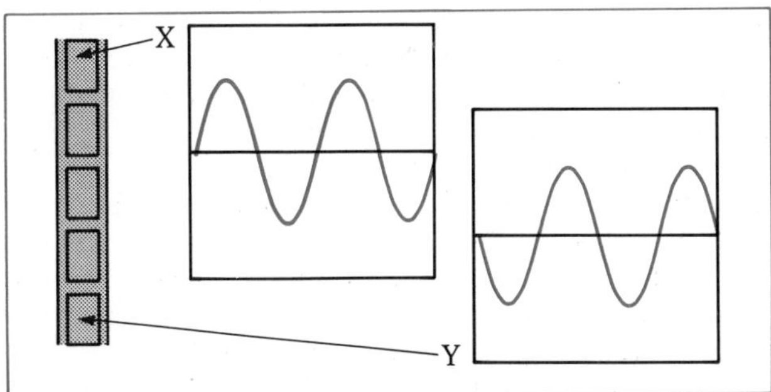

Figure A1.24

a) There are three other frames between X and Y. What was the time interval from X to Y?
b) What is the least possible frequency of vibration of the string?
c) How would the intervening three frames appear if the string had been vibrating at the frequency worked out in (b) if the waves had been (i) progressive, (ii) stationary waves?

3 Two sets of waves moving in opposite directions pass through each other. The amplitude of each set of waves is 0.04 m and they have the same wavelength of 0.040 m and the same frequency of 15 Hz.

a) Describe the resultant waveform as they pass through each other.
b) What is the distance between adjacent nodes?
c) Write down an equation for the displacement at distance x from a node.
d) What is the phase difference between a particle at an antinode and another particle 0.10 m away?

4 In a microwave analogue experiment of the Michelson experiment, reflector R_1 is moved towards the semi-reflecting plate S until the detector signal is a minimum. When R_1 is moved a further distance of 15 mm towards S, the detector signal rises then falls back to a minimum.

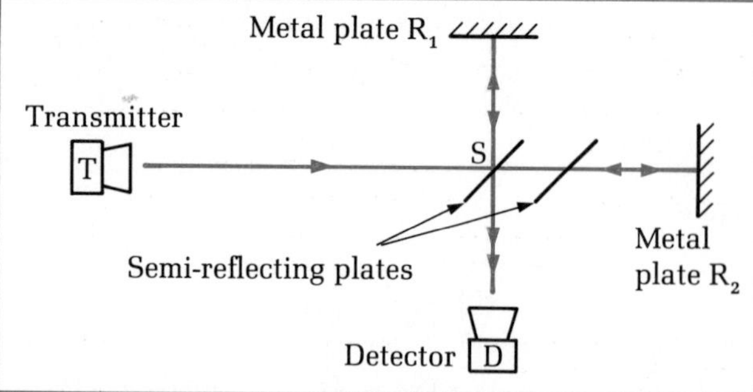

Figure A1.25

a) Explain why the signal rises then falls as R_1 is moved.
b) Work out the wavelength of the microwaves.
c) A glass block of thickness 25 mm is inserted into the path of the beam between R_1 and S. To regain minimum signal, R_1 must be moved towards S by a distance of 12 mm. Work out the wavelength of the microwaves in the glass.

5 a) In the Michelson experiment, the movable mirror is displaced towards the semi-reflecting mirror by a distance of 0.12 mm. How many fringes pass through the centre of the fringe pattern if light of wavelength 589 nm is used?
b) A shift of 50 fringes is observed when air is removed from a tube of length 0.10 m placed in the path of one of the beams. Work out the refractive index of the air that was in the tube.

6 a) What type of wave is represented by the equation

$$y = 0.50 \cos 4\pi t \sin 0.5\pi x$$

b) Draw a graph of y against x for x from 0 to 4.0 m when $t = 0$.
c) Sketch, on the same axes, the graph 0.10 s later.

7 The equation $p = A \sin 2\pi f_1 t$ represents the variation of pressure at a fixed point where sound waves of frequency f_1 pass that point. Sound waves of frequency f_2 and maximum pressure A also pass that point. Derive an expression for the resultant variation of pressure with time at the fixed point.

8 A dipper vibrates on the surface of the water in a ripple tank, sending out circular waves at a frequency of 10 Hz. Sketch graphs to show how the displacement of the water surface varies
a) with time at a fixed distance from the dipper,
b) with distance from the dipper at a fixed moment in time.

9 A second dipper is placed 100 mm from the first and is set into vibration on the water surface. Sketch graphs to show how the displacement midway between the two dippers varies with time if the second dipper vibrates
a) exactly in phase with the first one,
b) at a steady frequency of 11 Hz.

10 Square wave pulses at a frequency of 1.0 kHz are simplified by an amplifier which has a frequency response as in Figure A1.26. Sketch the approximate shape of the output waveform.

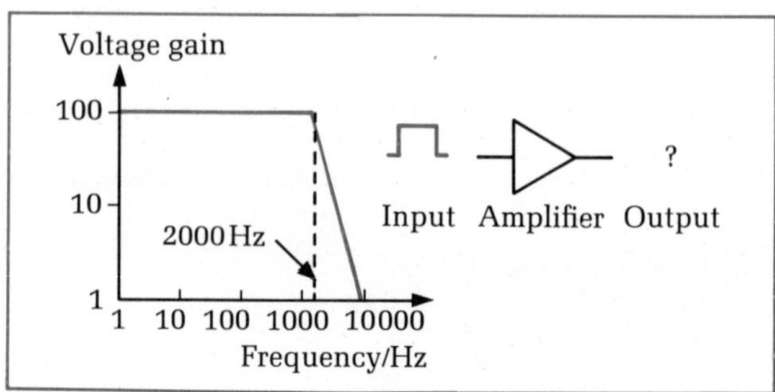

Figure A1.26

Appendix 2 Telecommunications

A2.1 Information transfer

Every four years, the Olympic Games dominate radio and TV channels for several weeks. Events happening in distant countries can be watched on TV as they happen. Electromagnetic waves are used to carry the TV pictures at a speed of 300 000 km s^{-1}.

Figure A2.1 *Media coverage of the Olympics*

Information can be carried by any type of wave. When you speak to someone, sound waves created by your vocal cords carry information to the listener. Sound waves are of little use for sending information beyond a few hundred metres.

Messages carried by light signals can be sent up to several kilometres in clear conditions at night. Using a shutter, the light beam from a high intensity lamp can be switched on and off repeatedly to send out a message in **Morse code**.

Figure A2.2 *Morse code*

Telecommunications means sending information from one place to another. Radio and TV broadcasts send information out from the transmitting stations to the receivers. Transfer of information is in one direction only.

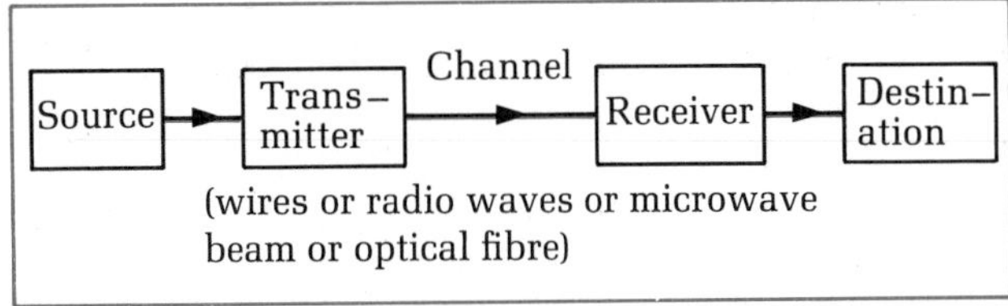

Figure A2.3 *Sending information*

The telephone enables people to speak to each other over long distances. The telephone microphone converts sound waves into electrical signals which are guided along wires to the receiving telephone. The electrical signals are referred to as **audio** waves. Figure A2.4 shows how a simple telephone link works. The development of the telephone system is described in more detail on p. 546.

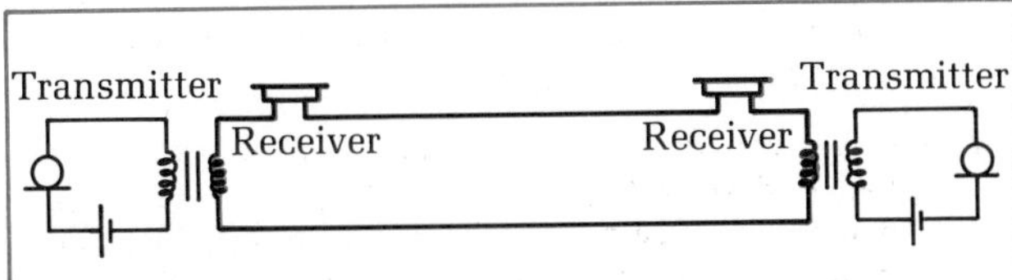

Figure A2.4 *A simple phone link*

To make a telephone call to a distant country is as easy as making a local call. Such a long distance call may involve using a **cable** link and a **microwave** link between your local telephone exchange and the nearest international exchange. The international exchange then uses a microwave beam via a satellite to link up to an international exchange in the country being called. A cable link

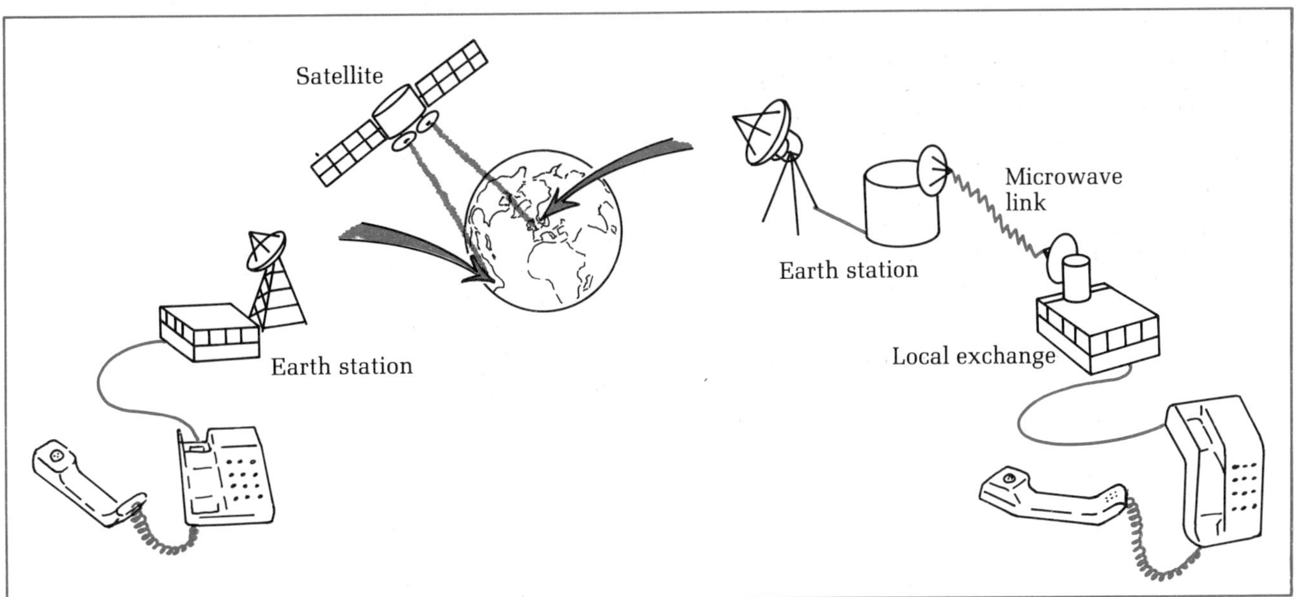

Figure A2.5 *International calls*

completes the route for the call. Increasing use is being made of **optical fibre** cables in the telephone network. These are described in more detail on pp. 550–2.

Satellite links are also used to transmit TV programmes from one country to annother. TV signals from a transmitting aerial have a range of 100 km or less. When you watch a distant event on TV as it happens, the TV signal is being carried by a microwave beam via a satellite to a receiving station in this country. The receiving station then sends the TV signals out from its local transmitting aerial.

A2.2 Carrier waves

Electromagnetic waves travel much faster and much further than sound waves. When you listen to a radio programme, your radio is detecting electromagnetic waves transmitted by the radio station. In the studio at the station, sound waves are converted into audio waves using microphones. These waves are then used to **modulate** the amplitude of the electromagnetic waves sent out by the station. See p. 293 for further explanation of the general idea.

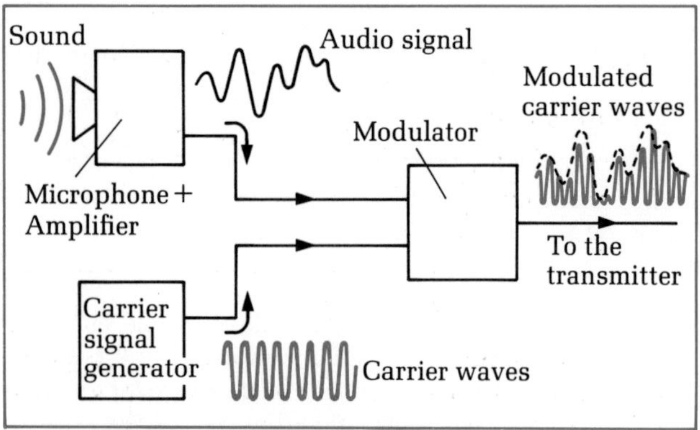

Figure A2.6 *Amplitude modulation*

Each radio station broadcasts its programmes using its own carrier wave frequency. It is important that nearby radio stations do not broadcast on the same carrier frequency. If this happened, you would hear two different programmes at the same time.

Telephone conversations can be carried by electromagnetic waves. When the phone network was first set up, an operator in each local exchange had to connect each caller manually. Each call required one pair of wires. As more people wanted their own phones, more and more wires were needed and costs rose. Engineers realised that it was possible to transmit several calls down the same pair of wires at the same time by using carrier waves. Each call was carried by a high frequency alternating signal; a different carrier frequency, called a **channel**, was used for each call. The first multi-channel cable was put into successful operation between Bristol and Plymouth in 1936. It was capable of carrying 12 simultaneous phone conversations. As explained on p. 551, modern optical fibre cables are capable of carrying more than 10 000 simultaneous calls on each cable.

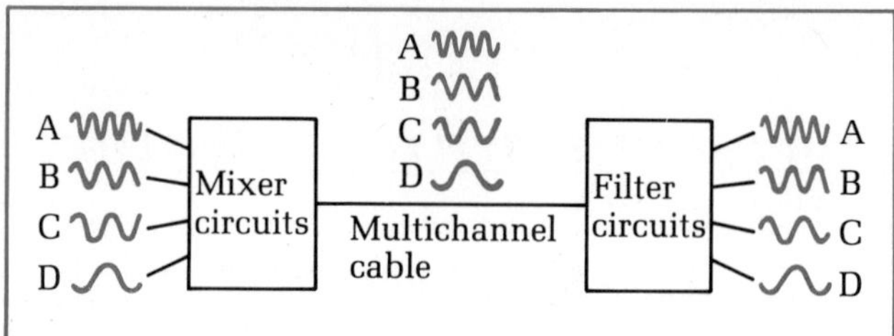

Figure A2.7 *Simultaneous calls*

How many channels can be carried by a single pair of wires? How close can the frequency of a radio station be to that of another station without causing interference? Consider a carrier wave of frequency f_c which is modulated by a sinusoidal waveform of frequency f_m. The resultant waveform is shown in Figure A2.8.

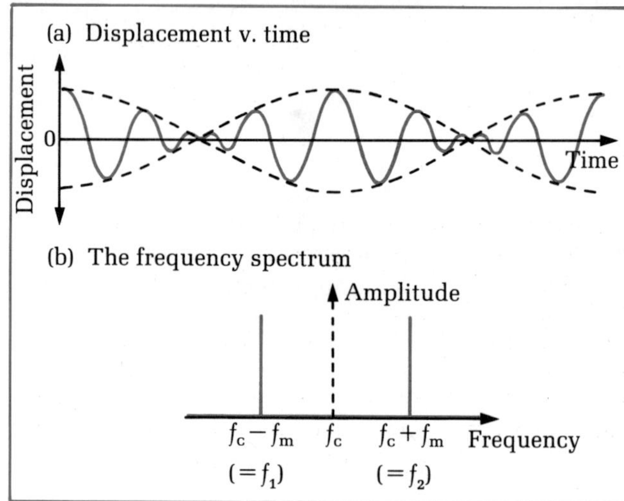

Figure A2.8 *A modulated sine wave*

This is the same as the 'beats' waveform on p. 542 and so it can be produced by adding sine waves of frequency f_1 $(= f_c - f_m)$ to sine waves of frequency $f_2 (= f_c + f_m)$. As explained on p. 542, the carrier frequency f_c equals $\frac{1}{2}(f_1 + f_2)$ and the modulating signal frequency f_m equals $\frac{1}{2}(f_1 + f_2)$. Thus the resultant waveform is made up of sine waves of two frequencies, f_1 and f_2, which are either side of the carrier frequency f_c.

Audio waveforms contain a spread of frequencies up to more than 10 kHz. When audio waveforms are used to modulate a carrier wave, the resultant waveform is made up of sine waves in two frequency bands either side of the carrier frequency. These are called **sidebands**.

In practice, the width of each sideband is limited to 4 kHz. This is achieved by filtering out speech frequencies above 4 kHz. Thus, a radio station broadcasting using a carrier frequency f_c transmits a spread of frequencies from 4 kHz above f_c to 4 kHz below f_c. The frequency spread is thus 8 kHz. This is called the **bandwidth**. Each radio station is allocated its own frequency channel, 8 kHz wide, which no other nearby radio station is permitted to use.

Coaxial cables used in the telephone network are capable of transmitting alternating voltages at frequencies up to 4 MHz. This means a coaxial cable can carry about 1000

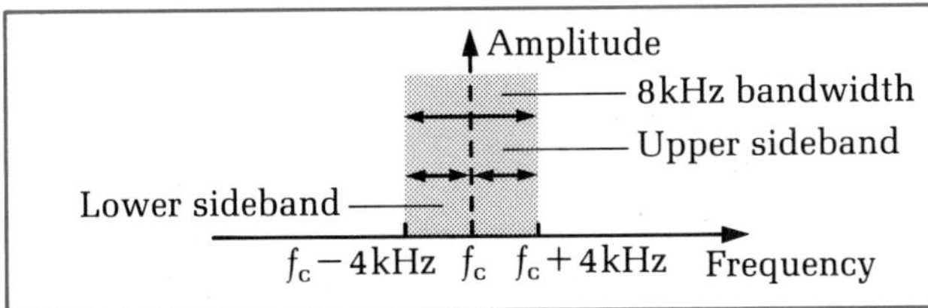

Figure A2.9 *Side bands*

simultaneous phone conversations, each one occupying a different frequency channel 4 kHz wide. The carrier frequencies are at 4 kHz intervals, each frequency carrying audio waveforms with a 4 kHz bandwidth from one sideband only.

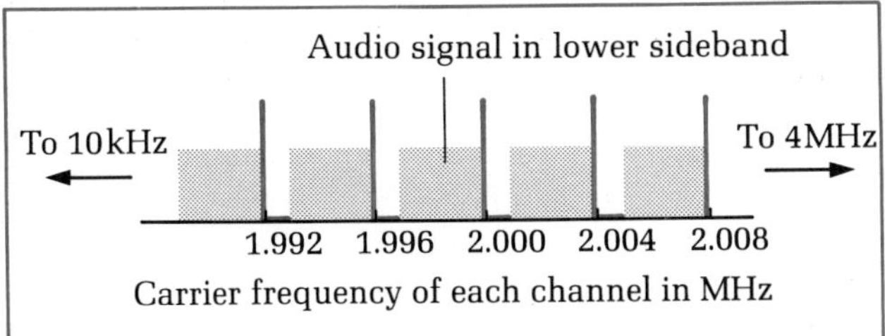

Figure A2.10 *Frequency channels*

Pulse amplitude modulation (PAM) is a technique that has been developed to increase the capacity of telephone links further. PAM involves sampling the audio signal once every 125 μs. Each sample is transmitted as a very brief pulse of carrier waves lasting no more than 3 μs. At the receiver end, the pulses are used to reconstruct the audio signal. In the 125 μs between samples from a given audio signal, other audio signal samples can be sampled. Up to 30 audio signals can be transmitted in this way in one single frequency channel.

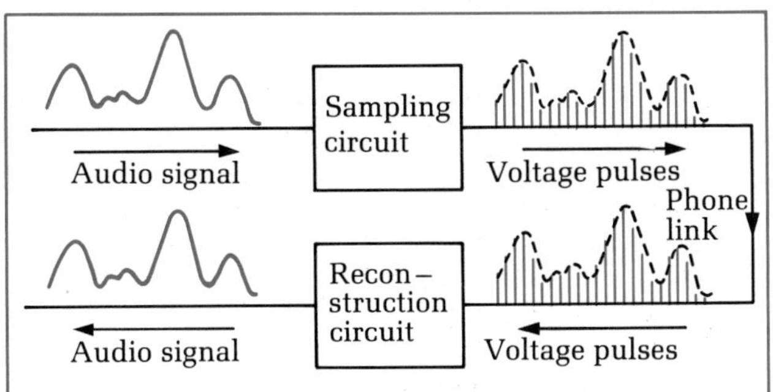

Figure A2.11 *Pulse amplitude modulation (PAM)*

The technique of interleaving different signals is called **multiplexing**. A multiplexer is used to take samples in sequence from each audio signal. A demultiplexer is needed to separate the samples to reconstruct each audio signal.

Telephone links need *amplifiers* at intervals along each line. This is because the signals diminish in strength the further they travel. This is called **attenuation**. The cause of attenuation along telephone wires is the resistance of the wires. This causes the voltage at the far end of the wire to be less than it would be if the wires had zero resistance.

Electrical noise sometimes makes telephone conversation very difficult. This causes a hissing noise which can be loud enough to 'drown' conversation. Such noise is produced in the amplifiers or in the wires. Noise due to the wires is called **resistance noise**, and is caused by the random motion of conduction electrons in the wires. In the amplifiers, noise is generated by electrons being accelerated and decelerated in components such as transistors. This is called **shot noise**.

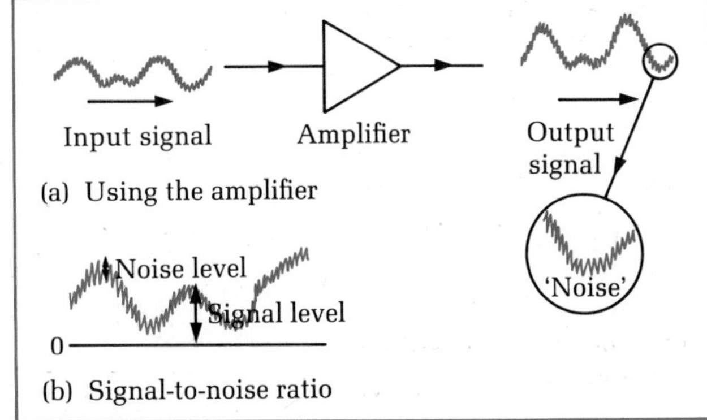

Figure A2.13 *Signal-to-noise ratio*

Signal strengths need to be much stronger than noise levels if people are to be able to hold a conversation without difficulty. In practice, the signal voltage must be at least 60 times stronger than the noise voltage. This means that the signal power must be 3600 times greater than the noise power (since power ∝ (voltage)²).

The signal-to-noise ratio may be expressed in decibels (dB) according to the equation:

$$\text{Signal-to-noise ratio in decibels} = 10 \log_{10} (\text{signal power/noise power})$$

Thus the minimum signal-to-noise ratio that is acceptable is 35 dB ($= 10 \log_{10} 3600$).

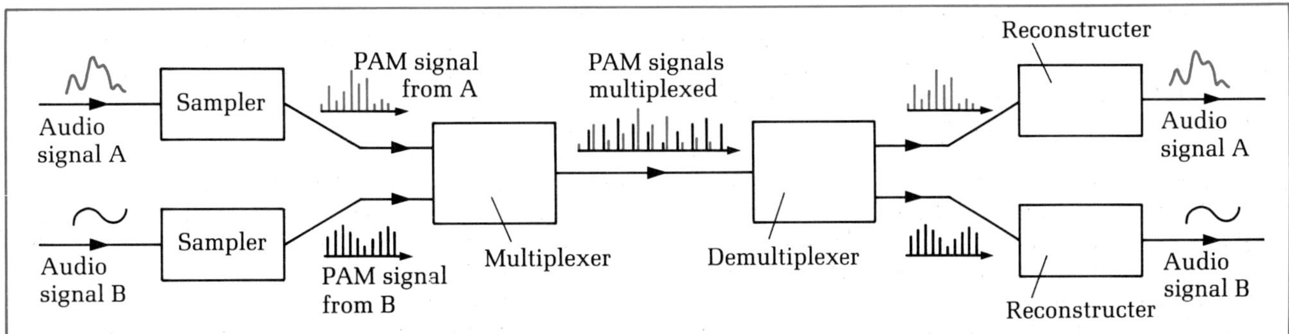

Figure A2.12 *Multiplexing*

Radio and TV signals also become attenuated with distance from the transmitter. If you travel away from your home area in a car listening to your local radio station, you will discover that the signal becomes weaker and weaker the further you go. The signal becomes masked by the noise and you only hear a hissing sound.

A2.3 Radio and TV broadcasting

Broadcasting is essentially one-way communication from the broadcasting station to receivers in all directions. Radio programmes are carried on frequencies up to about 150 MHz. Higher frequencies are used for TV broadcasting. Carrier wave frequencies for radio and TV programmes are divided into **wavebands**. Most radios are designed to receive programmes on at least the LW (long wave) and MW (medium wave) bands.

Figure A2.14 *Wavebands*

As mentioned on p. 293, radio waves are affected by the Earth's atmosphere and surface. The effect varies with frequency as summarised below.

1 **Ground waves** are low frequency radio waves in the LW/MW wavebands. These waves stay near the ground as they spread out from the transmitter and they travel further over the sea than over land. They are used for long-distance broadcasts and for maritime communications.
2 **Sky waves** reflect from the ionosphere in the upper atmosphere back to the ground. The ionosphere is capable of reflecting radio waves of frequencies up to 30 MHz. Above this frequency, radio waves pass through the ionosphere into space. Hence sky waves cover the MW and part of the HF (high frequency) wavebands.
3 **Space waves** are radio waves of frequencies greater than 30 MHz. They can only be used for direct 'line-of-sight' communications or for satellite links. They cover the VHF, UHF and microwave bands and have a limited range for broadcasting. For example, TV programmes are carried in the UHF band and have a range of about 100 km. To pick up space waves, the receiving aerial must be in line of sight to the transmitter.

Radio channels in the LW and MW wavebands have a bandwidth of 4 kHz, sufficient to carry audio waveforms without undue distortion. Higher quality transmission for music programmes does not use amplitude modulation. Instead, the audio signal is used to modulate the *frequency* of the carrier waves. This is known as **frequency modulation** (FM).

The frequency of the carrier waves is changed in proportion to the amplitude of the audio signal. Compared with AM receivers, FM receivers suffer much less from noise. This is because noise has a random spread of frequencies so has little effect on changes of the carrier frequency caused by the audio signal. However, the bandwidth for FM transmission is 250 kHz which is much wider than 4 kHz for AM transmission. Hence FM transmission must use the VHF waveband.

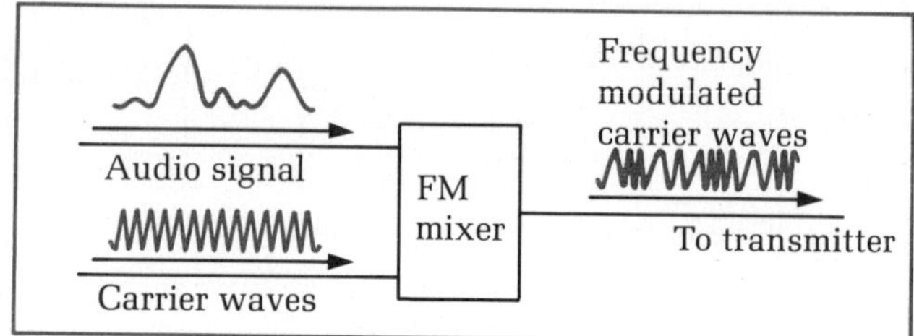

Figure A2.15 *FM transmission*

A colour TV channel has a bandwidth of 8 MHz. Carrier frequencies of TV channels must be in the UHF waveband to accommodate 8 MHz channels. Figure A2.16 shows some TV channel frequencies in use at present.

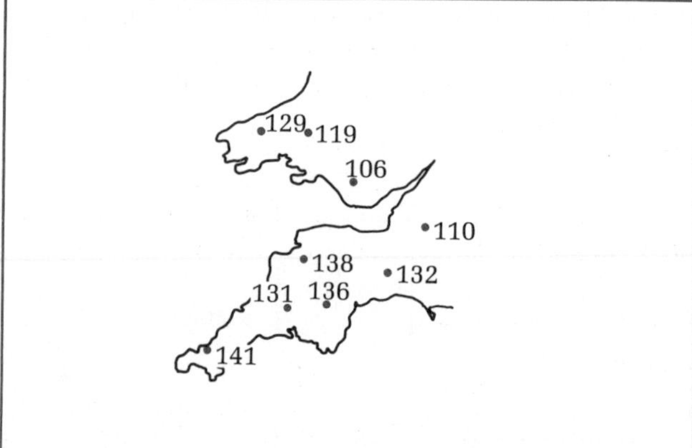

Channels		BBC1	BBC2	IBA1	IBA2
106	Wenvoe	44	51	41	47
110	Mendip	58	64	61	54
119	Carmel	57	63	60	53
129	Presely	46	40	43	50
131	Caradon Hill	22	28	25	32
136	Beacon Hill	57	63	60	53
138	Huntshaw Cross	55	62	59	65
141	Redruth	51	44	41	47

Channel 21... 22... 31... 41... 51...
Vision (MHz) 471.25...479.25...551.25...631.25...711.25...

Figure A2.16 *The main UHF transmitters in SW England and Wales*

Aerials and dishes

A **transmitter aerial** produces radio waves as a result of a high-frequency alternating voltage being applied to the aerial. The waves produced have the same frequency as the alternating voltage applied. These waves are the carrier waves. Most broadcasting transmitters use a *half-wave* or a *quarter wave* aerial.

The **half-wave aerial**, also called a **dipole** aerial, has a length equal to one-half wavelength of the carrier waves. A standing wave pattern for the voltage is set up on the aerial, giving maximum efficiency for transmission of radio waves. Figure A2.17(b) is a **polar** diagram that shows how the radiation varies with direction from the aerial.

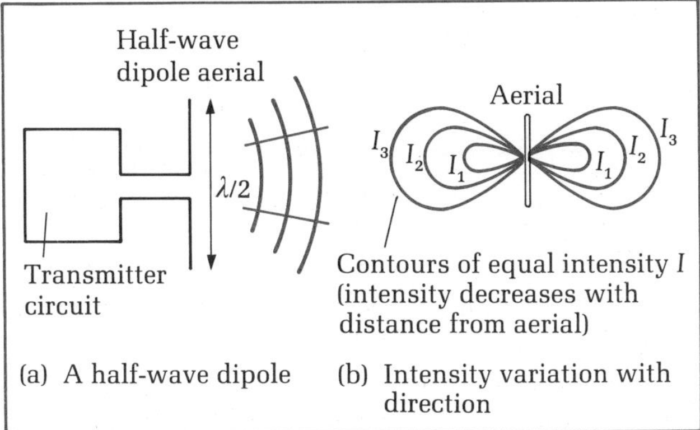

Figure A2.17 *The dipole*

By including reflector and director rods in parallel with the dipole, the radiation can be made strongest in a certain direction. The polar diagram shows contours of equal intensity. In Figure A2.18, most of the radiation is in the direction shown by the main lobe. At constant distance from the aerial, the intensity decreases with angle from the direction of greatest intensity. The **beamwidth** is defined as 2 × the angle from maximum intensity to half of maximum intensity at constant distance (Figure A2.18(c)).

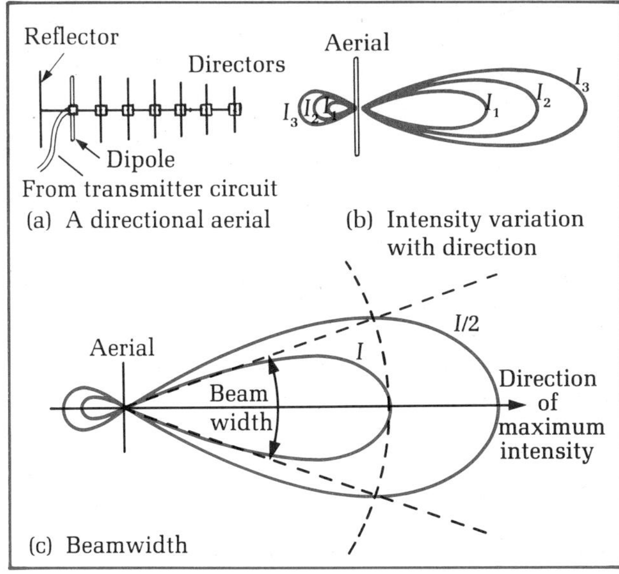

Figure A2.18 *Directing radiation*

The dipole can also be used as a receiving aerial. By including reflector and director rods in parallel with the aerial, the aerial is made much more sensitive. The sensitivity varies with direction in the same way as intensity would if the aerial was used to transmit instead. Thus, the polar diagram of Figure A2.18(b) shows that a receiver aerial needs to be pointed directly towards the broadcasting transmitter aerial. If it is out of line by 10° or more, the sensitivity is greatly reduced from its maximum possible value.

Direct broadcasting satellites (DBS) beam programmes to Earth using microwave beams at frequencies of 10 GHz or more. A single microwave beam can carry several TV channels. For 10 GHz microwaves, the wavelength of the carrier waves is 3 cm (using $\lambda = c/f$). To detect this frequency, a dipole aerial of length 15 mm is needed. However, the aerial needs to be at the focal point of a parabolic dish which thus reflects microwaves from the satellite onto the aerial.

The dish must be aimed directly at the satellite. Due to diffraction by the dish, its sensitivity varies with direction. As explained on p. 295, for a dish of width W, the alignment must be less than a certain angle θ where $\sin\theta = 1.22\,\lambda/W$. Thus for a 1 m dish detecting 3 cm waves, $\theta = 2°$. A narrower dish would give a larger value of θ but less power would be reflected onto the aerial.

DB satellites are in geostationary orbits above the equator. The signal strength from such a satellite decreases with increasing latitude so users in Northern Europe need bigger dishes than those in Southern Europe and hence must aim them more precisely.

Communications satellites transmit and receive information carried by microwave beams. Three of these satellites are needed to link round the world. They are positioned at the same height at equal intervals over the equator in geostationary orbits. To prevent interference between the transmitted and received beams, the satellite changes the frequency of the microwave beam so that it transmits on a different carrier frequency to the frequency received.

Communication satellites are powered by **solar panels**. Rechargeable batteries are needed to supply power when the satellite passes into the Earth's shadow. The batteries are recharged by the satellite when it is in sunlight.

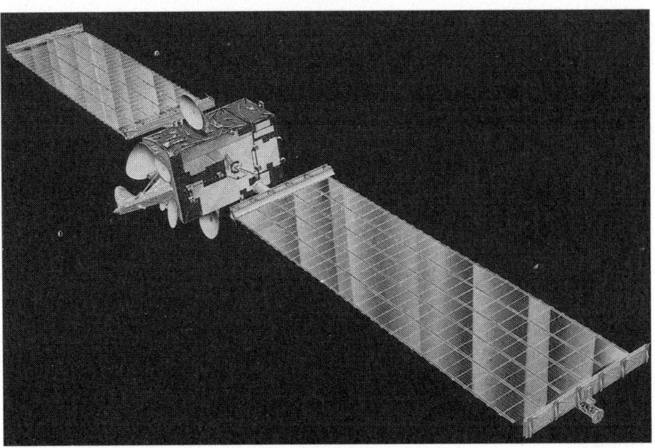

Figure A2.19 *Solar panels in space*

A2.4 Digital transmission

Push button phones, computers and fax machines are just a few of the devices that transfer information in digital form. Digital transmission effectively eliminates noise as a problem. Each pulse does become weaker and does gain noise as it travels from the transmitter to the receiver. However, by using **regenerators**, the noise accompanying each pulse can be cleaned off and the pulse restored to its original strength.

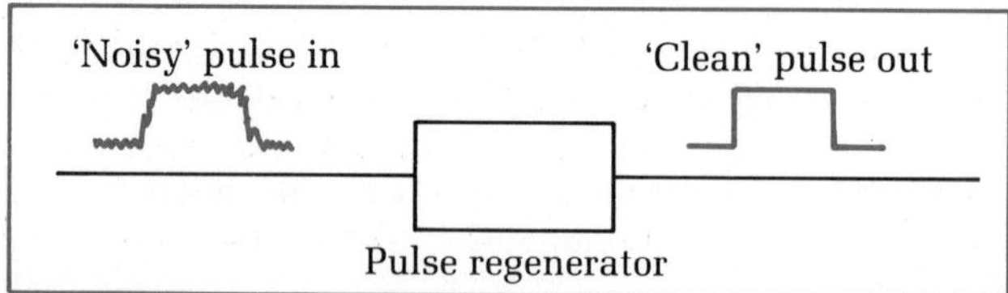

Figure A2.20 *A pulse regenerator*

As explained on pp. 239 and 362–3, audio signals can be converted into digital signals using the technique of **pulse code modulation** (PCM). Telephone networks using PCM are now widespread. The audio signal is sampled every 125 μs by an **analogue-to-digital** (A/D) converter. The A/D converter produces an 8-bit binary number at its output to represent the voltage at its input. See p. 405. An **encoder** is then used to send the 8-bit number as a sequence of pulses to the receiver. Figure A2.21 shows the idea.

At the receiver, each 8-bit binary number is *decoded* and then reconverted by a **digital-to-analogue** (D/A) converter to reconstruct the audio signal. Figure A2.22 shows how a D/A converter works.

At a sampling rate of once every 125 μs with eight bits representing each sample, the time between successive bits is 16 μs ($\simeq$125/8 μs). In practice each bit lasts 0.5 μs so other audio signals can be converted and carried between these bits. In other words, at a rate of 1 bit every 16 μs with each bit lasting just 0.5 μs, 32 separate phone channels can be carried. A multiplexer is used to interleave the 32 channels. At the receiver end, a demultiplexer is needed.

In practice, two of the channels are used to synchronise the signals to ensure the signals are decoded correctly so 30 channels are available, each able to carry a separate phone conversation. The overall bit rate at one bit every 0.5 μs is 2 million bits per second. A coaxial cable is capable of carrying 140 million bits per second.

With more multiplexing, a single cable can be made to carry about 2000 phone channels. Figure A2.24 gives a comparison between different types of links. An international phone call would reach its destination via several of these links in 'series' with each other.

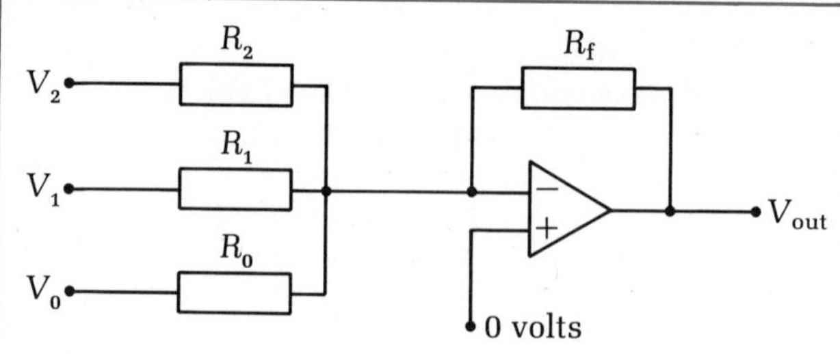

Op-amp theory gives

$$V_{out} = -\frac{R_f}{R_2}V_2 - \frac{R_f}{R_1}V_1 - \frac{R_f}{R_0}V_0$$

To operate the circuit as an A/D converter, values of R_0, R_1 and R_2 are selected so $R_0 = R_f$, $R_1 = R_f/2$, $R_2 = R_f/4$

Hence $\quad V_{out} = -(4V_2 + 2V_1 + V_0)$.

To convert a binary number, a '0' or a '1' is applied to inputs V_0, V_1, V_2 as follows:

V_2	V_1	V_0	$-V_{out}$
0	0	1	1
0	1	0	2
0	1	1	3
1	0	0	4
1	0	1	5
1	1	0	6
1	1	1	7

Figure A2.22 *A 3-bit D/A converter*

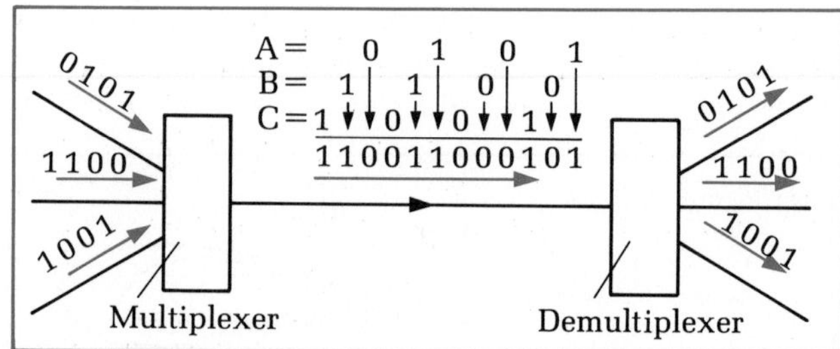

Figure A2.23 *A PCM multiplexer*

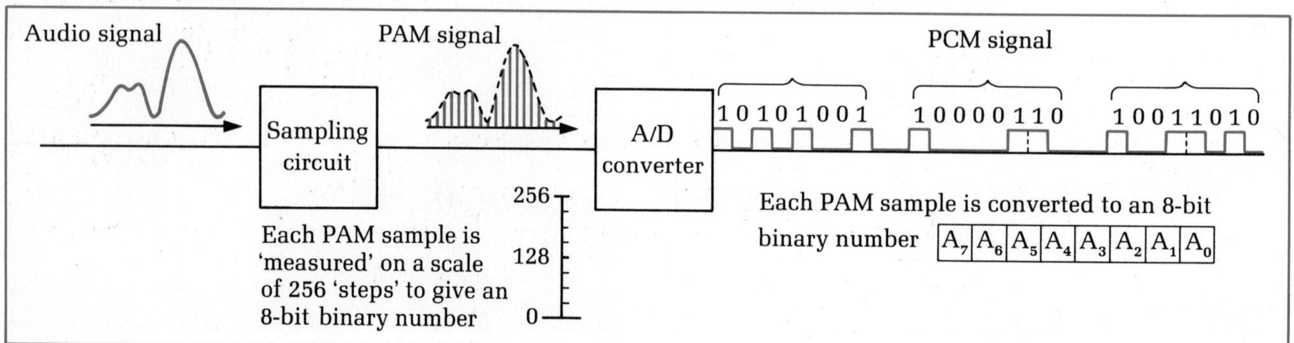

Figure A2.21 *Pulse code modulation*

Type of link	Capacity in M bit/s	Number of channels phone or TV	
Pair of wires	0.01	—	
Coaxial cable	140	1920	2
Microwave beam	70	960	1
Optical fibre	565 (2400 planned)	7680	8

Figure A2.24 *Comparing links*

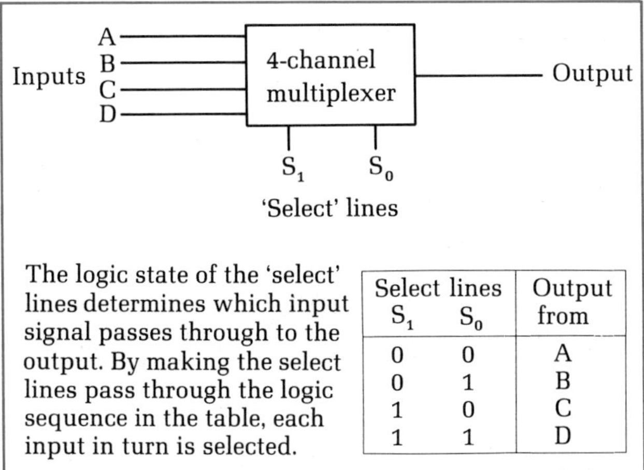

The logic state of the 'select' lines determines which input signal passes through to the output. By making the select lines pass through the logic sequence in the table, each input in turn is selected.

Select lines S_1	S_0	Output from
0	0	A
0	1	B
1	0	C
1	1	D

Figure A2.25 *Using a multiplexer*

Amplitude Shift Keying (ASK)

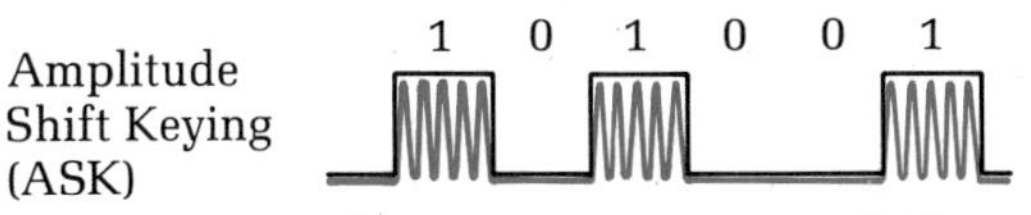

The carrier wave is switched on and off to send pulses

Used for optical fibre links and short cable links but unsuitable for microwaves and long cable links due to noise and fading

Frequency Shift Keying (FSK)

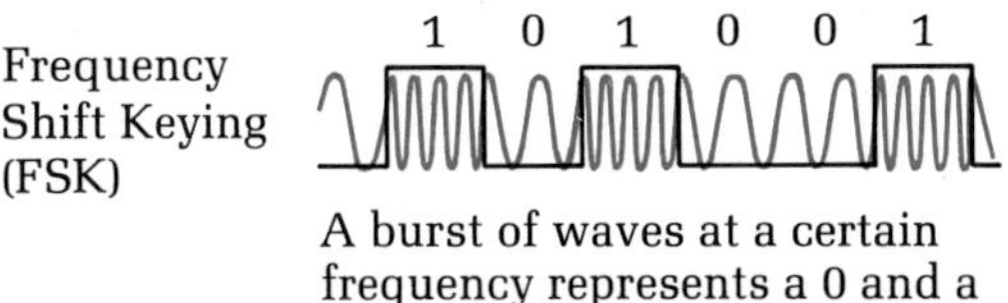

A burst of waves at a certain frequency represents a 0 and a burst of equal duration at a different frequency represents a 1

Used for computer links via the phone network (modems). Also used to link microcomputers to cassette recorders (the BBC micro uses 1200/2400 Hz).

Phase Shift Keying (PSK)

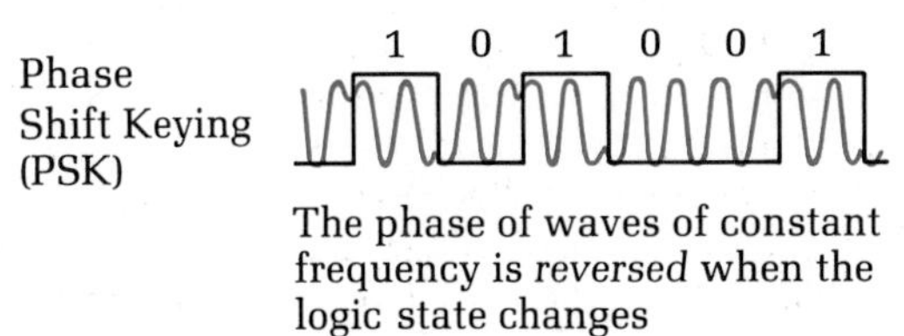

The phase of waves of constant frequency is *reversed* when the logic state changes

Used for microwave links where the frequency is constant. Atmospheric effects would cause distortion of ASK pulses.

Figure A2.26 *Methods of pulse transmission using carrier waves*

Computers and electronic terminals transmit data in the form of binary numbers, each number representing a character. Each number is transmitted as a sequence of pulses representing a 7-bit binary number. This allows 128 ($=2^7$) characters to be transmitted.

Each bit is accompanied by a 'start' bit and two 'stop' bits and a **parity** bit. The start and stop bits are necessary so that the decoder circuit in the receiver knows where the number starts and ends. The parity bit enables the decoder to check for errors in transmission. Such errors may occur due to noise or electrical interference. The encoder circuit sets the parity bit at 1 if the 7-bit byte has an odd number of 1's. For an even number of 1's, the parity bit is set at 0. Thus the 7-bit number and the parity bit are always transmitted with an even number of 1's. The decoder checks this and ignores any such 8-bit sequence with an odd number of 1's since an error must have occurred between the transmitter and the receiver. In this way, errors can be checked and monitored.

PCM links use a more complicated system to detect errors but the basic principle is the same: bits are added so the decoder circuit can check if the sequence has been transmitted correctly.

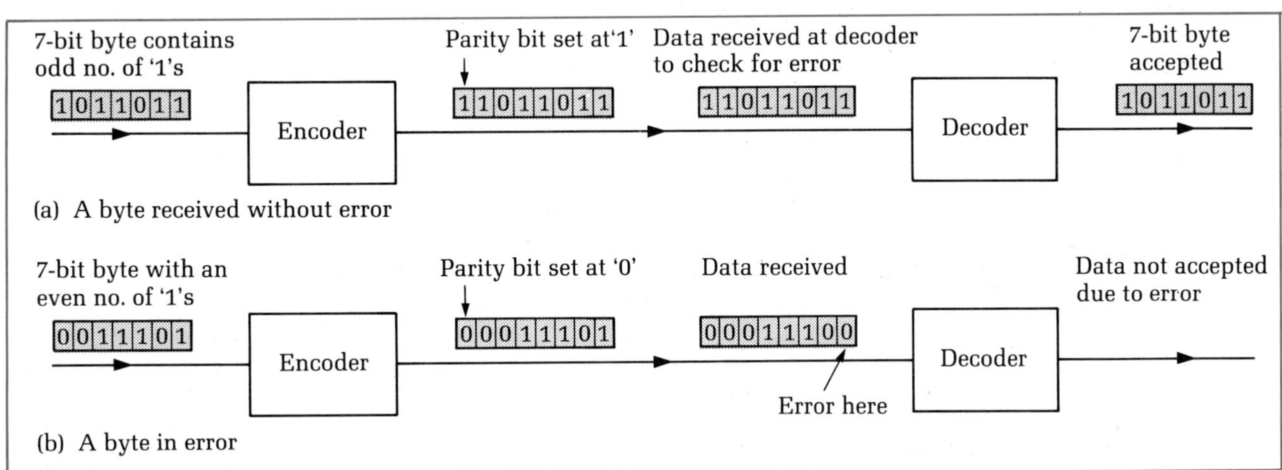

Figure A2.27 *Parity check*

A2.5 Optical fibre systems

A coaxial cable is capable of carrying 2000 simultaneous phone calls. Optical fibres under development have a potential capacity of more than 30 000 simultaneous calls per fibre. Coaxial cables are made from copper which is expensive to refine. Optical fibres are made from material that is much less expensive. It is not surprising that optical fibres are likely to dominate telecommunications in future.

Why is the capacity of an optical fibre so large? Light waves are used to carry digital signals in optical fibre. Even if a fibre twists and bends, the light rays reflect internally, as in Figure A2.28, and pass along the fibre. Using pulse regenerators at regular intervals, pulses of light can be transmitted over long distances. Noise is eliminated as a problem because the regenerators reshape each pulse and remove noise from it before transmitting it to the next regenerator.

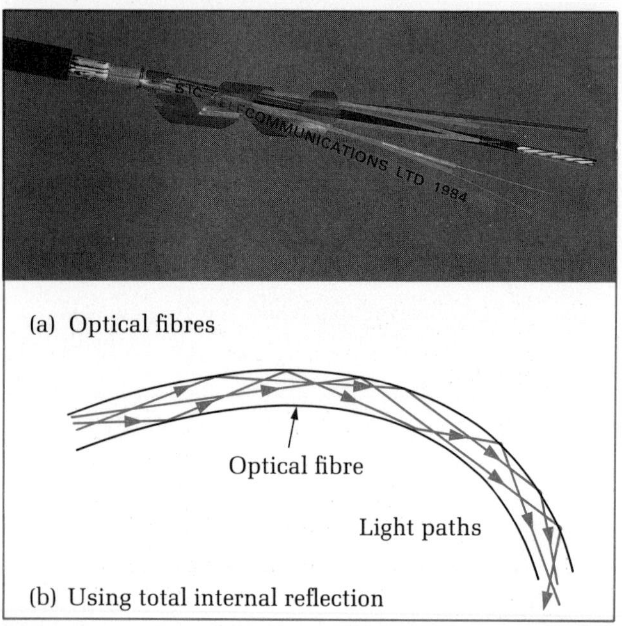

Figure A2.28 *Optical fibres and cables*

Considerable research is underway to develop glass for optical fibres. Such glass must be extremely clear. Attenuation of light travelling along a fibre decreases with wavelength. At a wavelength of 1.3 μm, the clearest glass available absorbs 50% of the light entering it after a distance of 50 km! Thus, wavelengths used for optical fibre transmission are in the near infra-red part of the spectrum. Nevertheless, they are still referred to as light waves even though not part of the visible spectrum.

Figure A2.29 shows how the light intensity varies with distance for a fixed wavelength. The intensity decreases exponentially with distance. The distance needed to reduce the intensity from 100% to 50% is the same as the distance to reduce the intensity from 50% to 25% and from 25% to 12.5%. The intensity I at distance x is given by the following equation:

$$I = I_0 e^{-\alpha x}$$

where I_0 is the initial intensity and α is the **attenuation coefficient**.

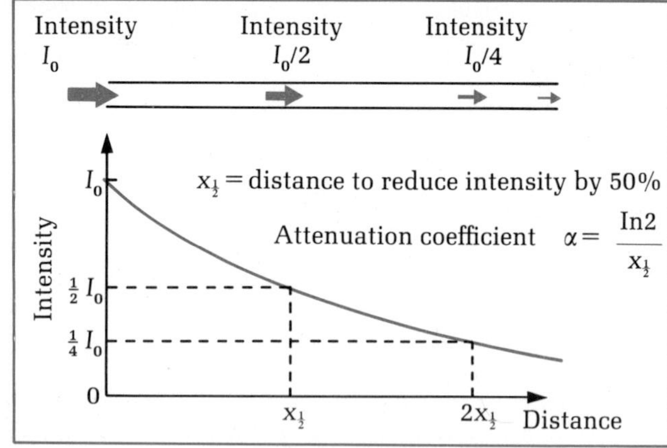

Figure A2.29 *Variation of intensity with distance*

To prevent leakage from the sides of a glass fibre, the fibre is coated in a **cladding**. This material has a lower refractive index than the core so the core/cladding boundary is where total internal reflection occurs. The cladding therefore prevents light passing from one fibre to another. This type of fibre is called a **step index** fibre.

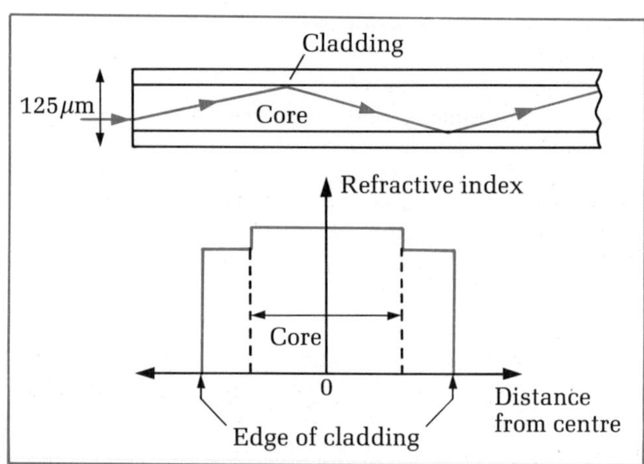

Figure A2.30 *A step index fibre*

Step index fibres with core diameters as small as 50 μm create problems due to rays travelling along the axis taking less time than rays undergoing multiple reflections. Thus a pulse of light entering at one end spreads out as it travels along the fibre. This is called **modal dispersion**. Over 1 km, the time difference can be as large as 30 ns, which creates problems if pulses are being transmitted at frequencies of more than 30 MHz. Therefore, this type of fibre can only be used for short-distance links.

The **graded index** fibre reduces modal dispersion to less than 1 ns per km. The refractive index of the fibre decreases with distance from its centre outwards, as in Figure A2.31.

A light-emitting diode may be used to send pulses of light down either a step index or a graded index fibre. LEDs which emit at a wavelength of 850 nm are used. Because the light intensity from an LED is low, the width of the fibres cannot be made too small or not enough light will be transmitted in each pulse.

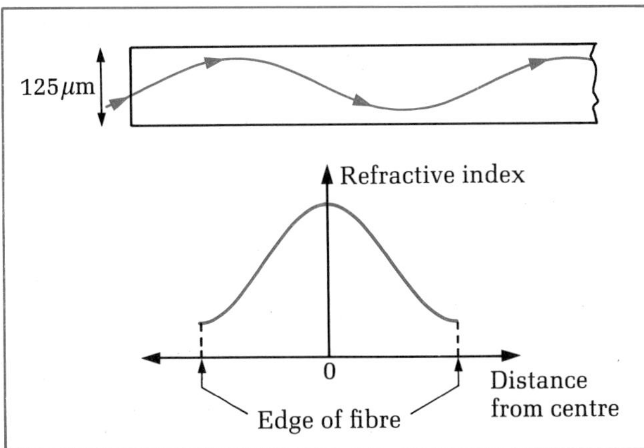

Figure A2.31 *A graded index fibre*

LEDs emit a spread of wavelengths of about 40 nm. Since the speed of light in glass varies with wavelength, the longer wavelengths in a pulse of light from an LED reach the end of the fibre before the shorter wavelengths (blue light travels more slowly in glass than red light does). The pulse thefore spreads out and broadens as it travels due to this effect. This is known as **material dispersion** and can be as much as 0.5 ns per km.

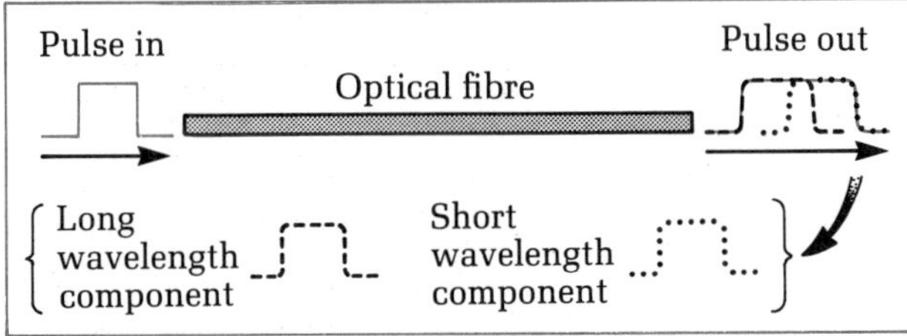

Figure A2.32 *Material dispersion*

The **monomode** fibre is a step index fibre with a very thin core of about 5 μm. A laser source has to be used to send light down such a thin fibre. Since laser light has a much smaller spread of wavelengths (about 1 nm) than light from an LED, material dispersion is much reduced. Also, modal dispersion is eliminated because the core is so thin. This means that pulse rates can be more than 10^9 per second (= 1 G bit/s). An optical fibre transmitting pulses at this rate will be able to carry more than 14 TV channels or 14 000 phone calls. Laser sources as small as a grain of sand have been developed for use in optical systems.

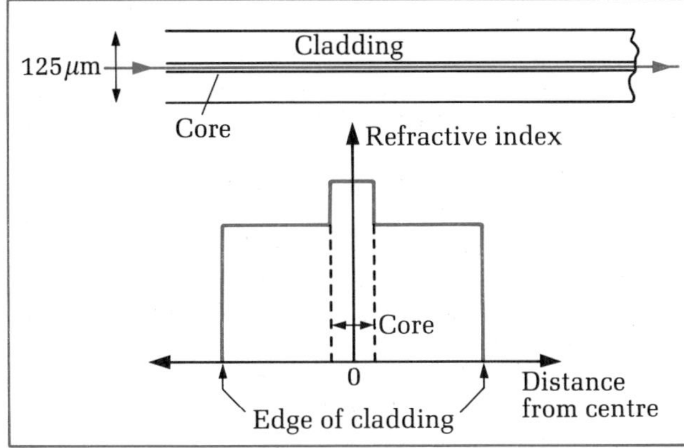

Figure A2.33 *A monomode fibre*

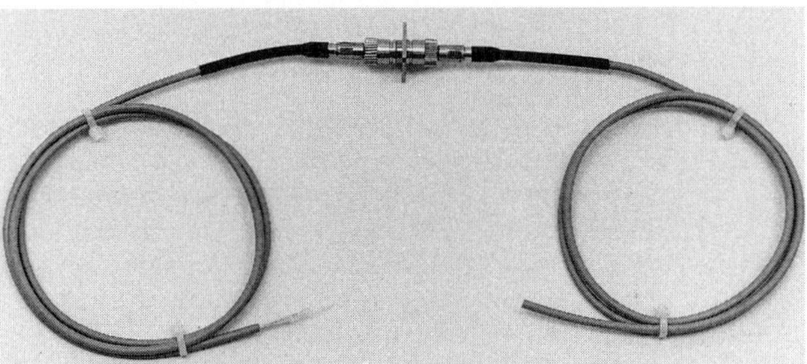

Figure A2.34 *Two single fibre cables mated through a bulkhead coupler*

To detect the pulses at the 'receiver' end of the fibre, a **PIN diode** is used. This is a p–n junction diode with a thin layer of intrinsic semiconductor between the p-type and the n-type, material. See pp. 381–3. The device is reverse-biased so that when light is absorbed by the intrinsic region, charge carriers break free from the atoms and a tiny current passes through the diode. The PIN diode has an extremely fast response which makes it suitable for receiving pulses at a fast rate.

Questions

1 a) A radio broadcast is transmitted using *amplitude modulation* at a *carrier frequency* of 680 kHz using the *lower side band*. Explain the meaning of each of the italicised terms.
b) Sketch the frequency spectrum for 1500 m radio waves modulated by a 4 kHz audio signal. The speed of electromagnetic waves through air is 3×10^8 m s^{-1}.

2 Explain how a coaxial cable can carry many phone calls at the same time, each call carried by a different frequency.

3 a) What is the purpose of amplifiers in a phone link?
b) Why is a high signal-to-noise ratio essential for phone conversations?
c) Work out the ratio of the signal voltage/noise voltage for a signal-to-noise ratio of (i) 10 dB, (ii) 40 dB.

4 National radio programmes such as Radio 2 are available on the VHF waveband yet VHF broadcasts have limited range. VHF transmitters in different parts of the country broadcast such programmes to the surrounding area.
a) Why should nearby transmitters broadcasting the same programme use different carrier frequencies?
b) How does the VHF signal reach each transmitter from the radio station?

5 a) A DBS system is designed to transmit 11 GHz microwaves. A 1.5 m dish is needed to pick up the signals.
i) What length should the aerial at the centre of the dish be?
ii) How accurately does the dish need to be aligned?
b) i) How much more power would a 3 m dish collect compared with a 1.5 m dish?
ii) Why would a 3 m dish need to be aligned more accurately?

6 a) What is meant by pulse code modulation?
b) What are the advantages of PCM over amplitude modulation?

7 a) For a 4-bit D/A converter similar to Figure A2.22, work out the output voltage for each of the following input voltage combinations:
(i) 0001, (ii) 0010, (iii) 0100, (iv) 1000, (v) 1001, (vi) 1101.

b) The following bytes were received at the end of a data link. The parity bit of each byte is at the end of the byte. Which bytes contain a single error each?
(i) 11010100, (ii) 11000110, (iii) 01101011, (iv) 11001111

8 a) Explain why an optical fibre transmits light round bends.

b) What are the advantages of monomode fibres over graded index and step index fibres?

c) Work out the critical angle for light at the boundary between the core and the cladding of a step index fibre if the refractive index of the core is 1.48 and of the cladding is 1.45.

9 a) What is the frequency of light of wavelength 1.3 μm, given the speed of light is $3 \times 10^5 \, \text{m s}^{-1}$?

b) How many cycles are in a pulse of duration 0.1 ns?

c) Estimate the fastest bit rate capable of being carried by light of wavelength 1.3 μm.

d) How many phone calls could be carried at this bit rate?

10 a) The intensity of a pulse travelling along a fibre decreases exponentially with distance according to the equation $I = I_0 e^{-\alpha x}$ where I_0 is the intensity at $x = 0$ and α is the attenuation coefficient. Show that the intensity reduces by 50% over a distance equal to $\ln 2/\alpha$.

b) Attenuation can be expressed in decibels (dB) according to the equation $10 \log_{10} (I_0/I)$. Work out attenuation in dB per km for a fibre in which the intensity falls by 50% over a distance of 50 km.

Table of physical constants

The Avogadro constant L	$= 6.022 \times 10^{23}\ \mathrm{mol^{-1}}$
Molar gas constant R	$= 8.314\ \mathrm{J\,K^{-1}\,mol^{-1}}$
Boltzmann's constant k	$= 1.381 \times 10^{-23}\ \mathrm{J\,K^{-1}}$
Stefan's constant σ	$= 5.670 \times 10^{-8}\ \mathrm{W\,m^{-2}\,K^{-4}}$
Gravitational constant G	$= 6.672 \times 10^{-11}\ \mathrm{N\,m^2\,kg^{-2}}$
Electric field constant ε_0 (permittitivy of free space)	$= 8.854 \times 10^{-12}\ \mathrm{F\,m^{-1}}$
Magnetic field constant μ_0 (permeability of free space)	$= 4\pi \times 10^{-7}\ \mathrm{H\,m^{-1}}$
Speed of light in a vacuum c	$= 2.998 \times 10^{8}\ \mathrm{m\,s^{-1}}$
Planck constant h	$= 6.626 \times 10^{-34}\ \mathrm{J\,s}$
Electronic charge e	$= 1.602 \times 10^{-19}\ \mathrm{C}$
Specific charge of the electron e/m_e	$= 1.759 \times 10^{11}\ \mathrm{C\,kg^{-1}}$
Mass of the electron m_e	$= 9.11 \times 10^{-31}\ \mathrm{kg}$
Mass of the proton m_p	$= 1.673 \times 10^{-27}\ \mathrm{kg}$
Mass of the neutron m_n	$= 1.675 \times 10^{-27}\ \mathrm{kg}$
Atomic mass unit 1 u	$= 1.661 \times 10^{-27}\ \mathrm{kg}$

Answers to numerical questions

Short questions

Chapter 1 **1.1** (a) 300 m (b) 9 km **1.2** 13.33 km, S 40°54′ E, 13.22 km h^{-1} **1.3** (a) 7.1 m s^{-1} (b) 46.7 min **1.4** 14.40 GMT **1.5** (a) 1.75 m s^{-1} (b) 300 m (c) 333 s **1.6** 37°, 417 s **1.7** (b) 0.4 m s^{-2}, 0, −0.267 m s^{-2}, 80 m, 800 m, 120 m (c) 6.67 m s^{-1} **1.8** 26.25 m **1.9** (a) 1.18 s (b) 6.32 m s^{-1} (c) 5.48 m s^{-1} **1.10** 0.2 s **1.11** (a) 210 m s^{-1}, 3.15 km (b) 5.36 km (c) 327 m s^{-1} (d) 83.7 s **1.12** 1.73 m **1.14** 80 m, 96 m **1.15** (a) 11.25 m (b) 77.9 m (c) 3.00 s **1.16** (a) 7.75 s (b) 92.2 m s^{-1} (c) 388 m **1.17** 92.0 m along the ground

Chapter 2 **2.1** (a) 120 N (b) 46.7 s, 17.1 N **2.2** (a) 37.5 N (b) 56.25 W **2.3** 20 m s^{-2} up, 10 m s^{-2} down **2.4** (a) and (b) 13.9 kN (c) 8.0 kN **2.5** 1.25 m s^{-2} **2.6** (a) 0.625 ms (b) 0.08 kg m s^{-1} (c) 128 N **2.7** (a) 40 N s, 8 m s^{-1} (b) 24 s (c) 96 m **2.8** (a) 4.08 m s^{-1}, N 11° E (b) 0.8 J (c) 8 m **2.9** 20 m s^{-1} (a) 8 kgs^{-1} (b) 160 Ns (c) 160 N **2.10** (a) 13.3 m s^{-1}, 44.4 kJ (b) 171 kJ (c) 126.6 kJ (d) 1266 N **2.11** (a) 1.41 m s^{-1} (b) 1.73 m s^{-1} (c) 0.718 m s^{-1} (d) 26 mm (e) 0.03 J **2.12** (a) 5 m (b) 450 J (c) 1.2 kW **2.13** (a) 8.94 m s^{-1} (b) 0.730 m s^{-1} (c) 653 N **2.14** (a) 14.5 kW (b) 4 kW (c) 18.5 kW **2.15** 5.17 m s^{-2} **2.16** (a) 0.5 m s^{-1} (b) 2.25 kJ **2.17** 0.25 m s^{-1}, 5.06 kJ **2.18** (a) 54 N (b) 48.6 W (c) 97.2 W **2.19** (a) 1.33 kg (b) 0.0256 J (c) 0.253 m s^{-1} **2.20** 25%

Chapter 3 **3.3** 640 tonnes **3.4** 18 GJ, 0.043 **3.5** 2.67 × 10^{6} m^{3}s^{-1} **3.6** 714 m^{2} **3.7** 1.76 kW **3.8** 112 kW **3.9** 6.51 GW **3.10** 300 W

Chapter 4 **4.1** (a) 5 N (b) 3.61 N **4.2** 6 N **4.3** (a) 110 N (b) 110 N **4.4** 1.15 N **4.5** (a) 0.3 N (b) 2500 kg m^{-3} **4.6** 200 N **4.7** (a) 844 N (b) 422 N **4.8** (a) 32.4 kN (b) No **4.9** (a) 0.4 (b) 60 N **4.10** 0.364 **4.11** (b) 39.6 N **4.12** 10.75 kN, 12.25 kN **4.13** (a) 369 N (c) (i) XY: 422 N (tension) (ii) YZ: 259 N (compression) **4.14** 0.375, 0.625 **4.15** 21.8°

Chapter 5 **5.1** (a) 33.5 m s^{-1} (b) 2807 m s^{-2} **5.2** (a) 20 m s^{-1} (b) 20 m s^{-2} (c) 2 g **5.3** (a) 6.37 m s^{-1} (b) 22.2 m s^{-2} (c) 3.70 N **5.4** (a) 1520 m s^{-2} (b) 602 N, 610 N **5.5** 21° **5.6** 0.034 m s^{-2} **5.7** 10.0 m s^{-1} **5.10** (a) 0.031 rad s^{-2} (b) 0.31 rad s^{-1} **5.11** (a) 5.21 m, 2.08 kJ (b) 8.33 m s^{-1} **5.12** (a) 14.4 MJ (b) 170 Hz **5.13** (a) 2.51 rad s^{-2} (b) 320 **5.14** (a) 6.98 × 10^{-2} rad s^{-2} (b) 2.23 × 10^{-2} N m (c) 1.12 W **5.15** 0.219 kg m^{2}

Chapter 6 **6.1** 0.3 nm **6.2** 11; 1 **6.3** (a) 9.45 × 10^{24} (b) 1.18 × 10^{-29} m^{3} (c) 0.23 nm **6.7** (a) 6.02 × 10^{28} (b) 0.26 nm (c) 1.5 × 10^{19} **6.9** 6.0 × 10^{23}

Chapter 7 **7.1** 50 MN m^{-2} **7.2** (a) 325 mm (b) 312.5 mm **7.3** (a) 2.81 mm (b) 84 mJ **7.4** (a) 2.98 mm (b) 14.9 J **7.5** 35 N **7.6** (a) 424 MN m^{-2} (b) 2.12 × 10^{-3}, 3.26 × 10^{-3} (c) 7.43 mm **7.7** (a) 650 kN (b) 81.25 J **7.8** 32.6 MN **7.9** (a) 6.5 mJ (b) 1.5 mJ **7.10** 25 MJ m^{-3} **7.11** 0.17 nm.

Chapter 8 **8.1** 8.33 km **8.2** (a) 21.2 × 10^{-6} m^{3}, 0.021 kg (b) 0.021 kg (c) 938 kg m^{-3} **8.3** (a) 10 cm^{3} (b) 2600 kg m^{-3} **8.4** (a) 5.25 MPa (b) 2625 N (c) 2125 N **8.5** (a) 3.65 m s^{-1} (b) 4.24 × 10^{-3} m^{2}, 0.127 mJ **8.6** 0.075 N m^{-1} **8.7** (a) 10^{19} (b) 2.33 × 10^{-21} J **8.8** 693 Pa **8.9** laminar **8.10** (a) 0.0209 N (b) 2.41 × 10^{-3} N (c) 0.69 N s m^{-2} **8.12** (a) 3.46 m s^{-1} (b) 1.732 × 10^{-6} m^{3} s^{-1} **8.13** 1.89 m s^{-1} **8.14** 2.30 kPa, 64.3 kN **8.15** (a) 3.02 m s^{-1}, 1.73 m s^{-1} (b) 5.94 × 10^{-3} m^{3}

Chapter 9 **9.1** (a) 62.2°C (b) 63.0° **9.2** 80° **9.4** 47.3° **9.5** 15.6 K **9.6** 301.6 kJ **9.7** 179.2 s **9.8** (a) 108 kJ kg^{-1} (b) 1800 J kg^{-1}K^{-1} **9.9** 554 J s^{-1} **9.10** 19.3 g **9.11** 750 kJ **9.12** (a) 4.4 × 10^{-3} m^{2} (b) 795 K **9.13** 317 W **9.14** 9.60 × 10^{5} J **9.15** 2.22 × 10^{5} J **9.16** 0.0375, 0.125 K W^{-1}, 2.3 K **9.17** 6365 K W^{-1}, 1061 K W^{-1}, 14.3 K **9.18** (a) 2.3 kW m^{-2} (b) 7.1 μm s^{-1} **9.19** (a) 1.33 × 10^{-3} K W^{-1}, 4 × 10^{-3} m^{2} K W^{-1} (b) 5.15 W m^{2} K^{-1} **9.20** 2.96 W m^{-2} K^{-1}

Chapter 10 **10.1** (a) 129 kPa (b) 0.225 **10.2** (a) 1.43 kg m^{-3} (b) 1.04 kg m^{-3} **10.3** 1.37 p_0, 1.15 p_0 **10.4** (a) 1127 kPa (b) 0.48 kg **10.6** 2.44 × 10^{22} m^{-3} **10.7** 1.66 × 10^{23} **10.10** (a) (i) 461 m s^{-1} (ii) 539 m s^{-1} (b) (i) 3.40 kJ (ii) 4.64 kJ **10.11** (a) 510 m s^{-1} **10.12** 10^{5} **10.13** (a) 3.65 kJ (b) (i) 6.06 × 10^{-21} J (ii) 3.10 × 10^{-21} J **10.14** (a) 0.377 nm (b) 3.43 nm **10.15** 2.69 × 10^{25} m^{-3}, 3.9 × 10^{-7} m **10.18** 102 kPa

Chapter 11 **11.1** (a) 2.26 MJ (b) 0.18 MJ (c) 2.08 MJ **11.2** (a) 180 K (b) 400 J (c) 0.4 (d) 600 J (e) 1000 J **11.3** 16.7 kPa, 510 J **11.4** 40%, 7.5 kW **11.8** (a) 3500 J s^{-1} (b) 30% (c) 4.5 J K^{-1} (d) 57% **11.9** (a) 1.2 kW (b) 0.2 kW

Chapter 12 **12.1** (a) 2.8 × 10^{-3} N kg^{-1} (b) 3.39 × 10^{8} m **12.2** (a) −250 J (b) −200 J (c) −150 J (d) 50 J (e) 0 J **12.3** (a) 16 MJ kg^{-1} (b) −400 MJ (c) 3.2 N kg^{-1} **12.5** (a) 1.7 × 10^{20} N (b) 27 days **12.6** (a) 5 × 10^{6} m s^{-1} (b) 2.1 × 10^{45} kg (c) 10^{15} **12.7** (a) 7.74 × 10^{11} m (b) −178 MJ kg^{-1} **12.8** 1690 km **12.9** 36 000 km **12.10** −473 MJ

Chapter 13 **13.2** (a) −200 nJ (b) 600 nJ (c) 25.0 kV m^{-1} **13.3** 24 kV m^{-1}, 1.6 × 10^{-18} C **13.4** (a) (i) 1.7 × 10^{-11} C (ii) 2.12 × 10^{-7} C m^{-2} (b) (i) same (ii) 480 V **13.5** (a) 0.235 J (b) 0.012 J **13.6** (a) 84.4 μN (b) (i) 180 kV m^{-1}, 900 V (ii) 3.00 kV m^{-1}, 0 **13.7** (a) 225 kV m^{-1} (b) 0 **13.8** (a) 5.02 × 10^{-4} N (b) 4.37 × 10^{-5} N (c) 5.83 nC **13.9** 10 kV m^{-1} **13.10** (a) (i) 66.4 nC (ii) 1.11 nF (b) ε_r = 3.07, 6.76 μA

Chapter 14 **14.1** (a) 57.9 μT, 72° to vertical (b) (i) 36 μN due E (ii) 116 μN, 18° above due N **14.2** (a) 1.275 mN (b) 2.83 mT (c) 1000 m^{-1} **14.3** (a) 1.403 × 10^{-3} Nm (b) 7.01 × 10^{-4} Nm **14.4** (a) 3 mT (b) 0.075 mT **14.5** 0.6 mN **14.6** (a) 0.22 mT (b) 173 μT at 71° to the line joining the wires **14.7** (a) 12.7 A (b) 17.4 mT **14.8** (a) 51 μN, 102 μN (b) 51 μN **14.9** (a) (ii) 7.0 × 10^{21} m^{-3}

Chapter 15 **15.1** 144 μV **15.2** (b) 5 mWb (c) 55 mT **15.3** (a) 91.7 mm μC^{-1} (b) 61 mT **15.4** (a) 28.9 Hz (b) 0–75 mV **15.5** (a) 6.41 V **15.7** (a) 7 mH (b) 0.42 s (c) smaller **15.8** (b) (i) 15 A s^{-1} (ii) 1 A (iii) 0.05 J **15.9** (a) 0.3 H (b) 0.033 μWb (c) 83.4 μH **15.10** (b) (i) 4.5 V (ii) 26.88 W (iii) 11.25 W (c) 2.8 A

Chapter 16 **16.1** (a) 19.3 rad s^{-1} (b) 0.677 m s^{-1} (c) 13.0 m s^{-2} **16.2** (a) 0.649 m s^{-1} (b) 0.556 m s^{-1} **16.3** (a) 15 mm, 0.2 s (b) 4.64 mm, 448 mm s^{-1} **16.4** (a) 34.6 m s^{-1} (b) 24 200 m s^{-2} **16.5** (a) 27.3 N m^{-1} (b) 0.466 s (c) 405 mm s^{-1}, 12.3 mJ (d) 0.68, 2.32 N **16.6** (a) 1.67 × 10^{4} N m^{-1} (b) 1.68 Hz (c) 25.1 m s^{-1} **16.7** (a) 50 N m^{-1} (b) 0.421 s, 0.373 m s^{-1} (c) 0.421 s, 0.186 m s^{-1} **16.8** 2.7 kg **16.9** (a) 2.43 s (b) 0.632 m s^{-1} (c) 0.246 N, 0 **16.10** (a) 0, 4.8 mJ (b) 4.8 mJ, 0 (c) 2.4 mJ, 2.4 mJ **16.11** 253 mm **16.13** (a) 200 N m^{-1} (b) 8 Hz **16.14** (a) 1.88 mm s^{-1} (b) 36 m s^{-2}, 3.6 Pa **16.15** 40 mm

Chapter 17 **17.2** 66.7 m, 25 m s^{-1}, 0.375 Hz **17.5** 48.6° **17.6** (b) 16 m s^{-1} **17.7** (c) 1 m, 85 m s^{-1} **17.8** (a) 0.8 m (b) 375 MHz **17.9** (a) 1.79 s (b) 5.09 km s^{-1}

Chapter 18 **18.1** (a) 2 × 10^{-17} J s^{-1} (b) 113 mm **18.2** (a) 10^{-5} W m^{-2} (b) 60.5 dB **18.4** (a) 272 Hz (b) 16 Hz **18.5** 520 Hz **18.6** (a) 344 m s^{-1} (b) 645 Hz **18.7** 21 to 2833 Hz **18.8** (a) 341 m s^{-1} (b) 4.5 mm **18.9** (a) 7.94 × 10^{-4} kg m (b) 100 Hz **18.10** (a) 1.52 m (b) 22.8 m s^{-1} **18.11** (a) 4.57 km s^{-1} **18.12** 212 Hz **18.13** (a) to X (b) 514 Hz (c) 493 Hz **18.14** 810 Hz **18.15** 17.9 m s^{-1}, 448 m

Chapter 19 **19.2** (b) (i) 0.3 m (ii) 6×10^5 **19.3** (a) 0.49 mm **19.4** (a) 550 nm **19.5** (b) 101 nm **19.7** 0.065° 0.051 mm **19.8** 0.143 mm **19.10** (b) 13.9 mm **19.11** (a) 3 km (b) 0.1 m **19.12** (a) 10.8 mm **19.13** (a) 1090 (b) 69.9° **19.14** (a) 2 (b) 35′ **19.15** (a) 599 mm^{-1} (b) 3

Chapter 20 **20.1** (a) 600 mm (b) 240 mm **20.2** (a) −0.91 m (b) −0.83 m **20.3** (a) 300 mm, ×1 (b) −300 mm, ×−3 **20.4** 7 mm **20.5** (a) 38° 28′ (b) 27°56′ **20.6** (a) 0.33 m (b) 50 mm **20.7** (a) 0.248 m (b) (i) 0.50 m (ii) 5.90 **20.8** (a) −0.25 D (b) +2 D **20.9** (a) 225 mm (b) 1575 mm **20.11** (b) (i) 81.1 mm (ii) 35.3 mm **20.13** (a) 40, 41.67 mm (b) 81.67 mm (c) −24 **20.14** (a) 200 mm from eyepiece (b) −25 **20.16** (a) 700 mm (b) 6.5° **20.17** 692 mm **20.18** (a) 704 mm (b) 79.4 mm

Chapter 21 **21.1** (b) 1.33 $nW\,m^{-2}$ **21.2** (a) 200 kHz (b) 0.2 m **21.4** (b) 112.5 $m\,s^{-1}$ **21.5** (a) 30 mm, 10 GHz (b) 10.8 $m\,s^{-1}$ **21.6** 400 m **21.8** (b) 56° **21.9** (b) 250 Hz (c) $3 \times 10^8\,m\,s^{-1}$ **21.10** $3 \times 10^8\,m\,s^{-1}$

Chapter 22 **22.1** (a) 90 C (b) 6.25×10^{15} **22.2** 2.49 $mm\,s^{-1}$ **22.3** 0.864 MJ **22.4** (a) 480 C (b) 1.5×10^{21} (c) 0.165 g **22.6** *A* (a) 6 Ω (b) 1 A (c) 3 Ω, 3 V, 1 A, 3 W; 4 Ω, 3 V, 0.75 A, 2.25 W; 12 Ω, 3 V, 0.25 A, 0.75 W (d) 6 W *B* (a) 3.6 Ω (b) 3.33 A (c) 6 Ω, 2 A, 12 V, 24 W; 4 Ω, 1.33 A, 5.33 V, 7.1 W; 5 Ω, 1.33 A, 6.67 V, 8.9 W (d) 40 W *C* (a) 3 Ω (b) 1 A (c) 0.5 Ω, 1 A, 0.5 V, 0.5 W; 1Ω, 1 A, 1 V, 1 W; 2 Ω, 0.75 A, 1.5 V, 1.125 W; 6 Ω, 0.25 A, 1.5 V, 0.375 W (d) 3 W **22.7** (a) 0.83 Ω (b) 0.9 W (c) 0.75 W **22.9** 1.37 V, 1.14 Ω **22.10** 0.133 A **22.11** (a) 7.2 kC, 108 kJ (b) 1.5 Ω (c) 86.4 kJ **22.12** (b) (i) 1.5 Ω (ii) 1.7 Ω (iii) 2.95 Ω **22.13** *C* **22.14** 4.99 $\Omega\,m^{-1}$, 0.20 m **22.15** (a) 1.99 mΩ, 40 μΩ (b) 1.01 mΩ, 38.5 μΩ **22.16** (a) 0.571 A, 0.143 A (b) 0.428 A (c) 4.28 V **22.17** 2 A **22.18** 15.5 **22.19** (a) 15 mC (b) 3 V **22.20** *A* (a) 2 μC, 4 μC, 2 V (b) 6 μC, 6 μJ *B* (a) 3 μF: 4 μC, 1.33 V; 6 μF: 4 μC, 0.67 V (b) 4 μC, 4 μJ *C* (a) 1 μF; 2 μC, 2 V; 3 μF; 4 μC, 1.33 V; 6 μF; 4 μC, 0.67 V (b) 6 μC, 6 μJ **22.21** (a) 180 mJ (b) 0.3 ms **22.22** 1.8 mC, 10.8 mJ **22.23** (a) 36 μJ (b) 12 μA (c) 2.21 V **22.25** (a) 72 μC, 432 μJ (b) 7.2 V (c) 104 μJ, 156 μJ

Chapter 23 **23.1** (a) 75.1 mΩ in parallel (b) (i) 4925 Ω (ii) 99 925 Ω **23.2** (a) P = 2, Q = 3 (b) P = 1, Q = 1 **23.2** 6.11 V, 111 Ω **23.4** 0.4 V, 0.8 V **23.5** 25 MΩ **23.6** (a) 20 Ω (b) 0.4 Ω **23.7** (a) 1.53 V (b) 0.338 Ω **23.8** 1.60 V **23.9** 4.4 Ω **23.10** 1.08 V **23.11** (a) 2 mA (b) 0.8 mA (c) 0 mA **23.12** (a) 2.12 mA (b) 6.78 mV (c) 5.11 mV **23.13** (a) 22.5 Ω (b) 25.7 Ω **23.14** (a) 4950 Ω (b) 860 Ω

Chapter 24 **24.1** (a) 20 ms, 50 Hz (b) 1.50, 0.53 V (c) 1.06 mA (d) 0.563 mW **24.2** (a) 1.41 A (b) 2.12 A (c) 0.1 A (d) 1.73 V **24.4** (a) 4.76 A (b) −2.94 A (c) 2.94 A **24.6** (a) 565 Ω (b) 2.82 μF **24.11** (a) 9 Ω (b) 127 mH **24.12** 100 Ω, 19.9 μF **24.13** (a) 0.1 A (b) 13 V (c) 0.477 H, 15.9 μF **24.14** (a) 333 μF (b) 2.4 A (c) 25.7 V

Chapter 25 **25.4** output = 1 only when A = B **25.5** (a) output = 1 only when A = B = C = 0 (b) As question 25.4 **25.11** 1905 > X > 335 Ω **25.14** R > 125 Ω

Chapter 26 **26.2** (a) 4 V, 4 mA (b) 21.25 μA (c) 188 **26.3** (a) 1.6 mA, 7.4 V (b) 13 mA, 0 V **26.4** (b) 189 kΩ (c) −10 **26.7** (a) −4 V (b) +12 V **26.8** (a) −2 V (b) +6 V **26.9** (a) −saturation (b) +saturation **26.10** (b) output decreases at 0.5 $V\,s^{-1}$ **26.11** 165 mT **26.15** (a) 2.55 V (b) 0.01 V

Chapter 27 **27.1** (a) 3.2×10^{-16} J (b) $2.65 \times 10^7\,m\,s^{-1}$ (c) $3.75 \times 10^7\,m\,s^{-1}$ **27.2** 2000 V **27.3** $1.09 \times 10^7\,m\,s^{-1}$ **27.4** (a) 8×10^{-15} N (b) 2.0 ns (c) $5.30 \times 10^7\,m\,s^{-1}$ at 19.4° to the initial direction **27.5** 123 mm **27.6** (a) $2.29 \times 10^7\,m\,s^{-1}$ (b) 130 mm (c) × 1/4 **27.7** $1.75 \times 10^{11}\,C\,kg^{-1}$ **27.9** (a) negative (b) 4.70×10^{-19} C (c) 19.6 $m\,s^{-2}$ **27.10** (b) $4.09 \times 10^{-5}\,m\,s^{-1}$ **27.12** (a) 8.50×10^{-7} m (b) 2.32×10^{-15} kg (c) 9.47×10^{-19} C **27.13** (a) 1.38×10^{-19} J (b) 2.24×10^{-19} J **27.14** (b) 1.27×10^{-19} J **27.15** (c) 1.54×10^{-19} J

Chapter 28 **28.1** (b) $4.13 \times 10^5\,s^{-1}$ **28.2** 5.0 MeV **28.3** (a) 6 MeV (b) 5000 mm^{-1} (c) $1.7 \times 10^7\,m\,s^{-1}$ **28.4** (b) 5.1 mm (c) 2.7 MeV **28.5** (a) 5.0 mm (b) 0.6 MeV **28.6** (a) 13.0 s^{-1} (b) 8.5 s^{-1}; 2,61 m **28.10** 6.5×10^{-14} m **28.11** (a) $1.55 \times 10^7\,m\,s^{-1}$ (b) $2.48 \times 10^7\,m\,s^{-1}$ **28.14** (b) (i) $1.72 \times 10^4\,m\,s^{-1}$ (ii) 1820 s^{-1} **28.15** (a) 3.56×10^{12} (b) 4.67×10^5 Bq, 8.32×10^{11} **28.16** (a) 6.76×10^{10} (b) 1668 yrs **28.17** (a) 1.76×10^{-10} g (b) 1.92×10^{-10} g (c) 1.61×10^{-10} g **28.18** (a) 2.76×10^9 Bq (b) 0.291 mW **28.19** 1.82 kg **28.20** (a) $8.91 \times 10^{-14}\,J\,s^{-1}$ (b) $6 \times 10^{-13}\,J\,s^{-1}$

Chapter 29 **29.1** (c) 1.92×10^{-18} J (d) 10 mA, 6×10^{16} **29.2** (a) 3 V (b) 414 nm **29.4** (b) 3.4×10^{-22} J **29.5** 3 to 2, 7 to 3 **29.8** (a) (i) 1.24×10^{-10} m (ii) 0.31×10^{-11} m (b) (i) 3.125×10^{16} (ii) 200 W (iii) 2 W (iv) 198 W **29.9** 0.107 nm, 1.84×10^{-15} J

Chapter 30 **30.1** (a) 2.30 N (b) 2×10^{-14} J **30.2** (b) 28.3 MeV (c) 7.4 MeV **30.4** 5.41 MeV, 98% **30.5** 5.66×10^{11} J **30.7** (b) 370 kg s^{-1} **30.8** (b) 0.70 s **30.9** (a) 0.43 MeV, 5.49 MeV, 12.86 MeV, 24.7 MeV (b) 3.15×10^{18} s **30.10** (a) 17.6 MeV

Chapter 31 **31.2** (a) $3.25 \times 10^7\,m\,s^{-1}$ (b) 2.24×10^{-11} m (c) 0.117, 0.257 nm **31.3** (a) 0.2 nm (b) 6.03×10^{-18} J (c) 5.97×10^{-18} J **31.4** (a) $1.03 \times 10^{-19}\,kg\,m\,s^{-1}$, 6.42×10^{-15} m (b) 3×10^{-11} rad **31.5** (a) 10^{-23} s (b) 10^{-10} J **31.6** (a) 10^{-9} J **31.7** 2.03×10^{-14} m **31.8** (b) 7.6 MHz **31.9** (a) in (b) 2

Chapter 32 **32.1** (a) 2.045×10^8 (b) 2.376×10^{-2} (c) 1.463×10^{19} (d) 1.457×10^{-5} (e) 3.807×10^{21} (f) 21.91 (g) 1.325×10^{13} (h) 1.01×10^{-2} (i) 1.286×10^{11} (j) 3.627×10^{-3} **32.2** (a) 5.720 (b) 1.051×10^2 (c) 6.857 (d) 386.2 (e) 81.20 (f) 0.446 (g) 4.862×10^6 (h) 1.165 (i) 156.3 (j) 8.025 **32.4** (a) −1, 4 (b) $y = -4x$ **32.5** (a) $y = -x + 5$, 5 (b) 3.54 **32.6** (a) $x = 1$, $y = 3$ (b) $a = 2.4$, $b = -0.4$ (c) $p = 2$, $q = 4$ (d) $x = 3$, $y = 1$ (e) $u = 1$, $v = 5$ and $u = 5$, $v = 1$ **32.7** (a) f^2 (b) ln *I* (c) $1/r^2$ (d) p (e) ln $(C - C_0)$ **32.8** (a) A (b) C (c) B (d) E (e) D **32.9** (a) $6x$ (b) $2 - 2x$ (c) $-6/x^3$ (d) $-3Ae^{-3x}$ (e) $2\cos 2x$ **32.10** (a) flux linkage (b) charge (c) work (d) energy per unit volume (e) velocity **32.11** (a) $A = 100$, $\mu = 0.204$ (b) (i) 81.5 (ii) 13.0 (c) (i) 3.40 (ii) 11.29 **32.12** (a) 0.139 (b) 2

Chapter 33 **33.1** (a) $kg\,m\,s^{-1}$ (b) $kg\,m^3\,s^{-3}\,A^{-2}$ (c) $kg\,m\,s^{-3}\,K^{-1}$ (d) $kg\,m^{-3}$ (e) $A^2\,s^4\,m^{-3}\,kg^{-1}$ **33.3** (a) 0.28%, 0.33% (b) 7690 $kg\,m^{-3}$ (c) 1.3% **33.4** (a) 0.3%, 0.6% (b) 9.76 $m\,s^{-2}$ (c) 0.15 $m\,s^{-2}$

Longer questions

Chapter 33 (Data analysis questions) **33.6** (b) 2.5 (c) (i) 0.27, 0.48, 0.69, 0.91 $kN\,s\,m^{-1}$ (ii) 0.053 $kN\,s\,m^{-1}$, 10.7 $N\,s^2m^{-2}$ **33.7** 27.2 min; $k = 2.36 \times 10^{-4}\,mm^{-1}$, $x = 18.5$ mm; gradient = −2 **33.8** (2) $\alpha = 1.34 \times 10^{-3}\,K^{-1}$, $\beta = 3.75 \times 10^{-6}\,K^{-2}$ (3) 27 **(4)** (a) 1505, 1535 K (b) 8.6, 14.3 mA

Multiple choice questions

Section A **AM1** A **AM2** D **AM3** D **AM4** B **AM5** C **AM6** B **AM7** E **AM8** B **AM9** E **AM10** D **AM11** C **AM12** D **AM13** E **AM14** E **AM15** D **AM16** D **AM17** E **AM18** E **AM19** A **AM20** E

Section B **BM1** B **BM2** A **BM3** E **BM4** E **BM5** A **BM6** B **BM7** D **BM8** B **BM9** D **BM10** E **BM11** A **BM12** D **BM13** A **BM14** B **BM15** A **BM16** A **BM17** B **BM18** D **BM19** C **BM20** B

Section C **CM1** D **CM2** C **CM3** C **CM4** C **CM5** C **CM6** D **CM7** C **CM8** B **CM9** A **CM10** D **CM11** D **CM12** D **CM13** B **CM14** B **CM15** B **CM16** A **CM17** D **CM18** C **CM19** E **CM20** E

Section D **DM1** A **DM2** A **DM3** D **DM4** A **DM5** A **DM6** C **DM7** E **DM8** C **DM9** C **DM10** E **DM11** D **DM12** A **DM13** A **DM14** C **DM15** D **DM16** C **DM17** C **DM18** B **DM19** D **DM20** D

Section E **EM1** E **EM2** D **EM3** C **EM4** E **EM5** D **EM6** E **EM7** B **EM8** A **EM9** A **EM10** E **EM11** D **EM12** B **EM13** B **EM14** E **EM15** B **EM16** D **EM17** A **EM18** C **EM19** B **EM20** D

Section F **FM1** D **FM2** A **FM3** B **FM4** A **FM5** E **FM6** D **FM7** A **FM8** D **FM9** B **FM10** D **FM11** C **FM12** C **FM13** D **FM14** D **FM15** B **FM16** E **FM17** D **FM18** C **FM19** D **FM20** D

Long questions

Section A **AL1** (a) (i) 250 m (ii) 4 kN (iii) 2×10^4 kg m s^{-1}, 2 kN (b) 8 m s^{-1}, 200 kJ, 80 kJ **AL3** (a) 25.5 kN, 6.375 MJ s^{-1} (b) 21.5 kN, 1.72 MJ s^{-1} **AL5** (d) 2000 m^3 **AL7** (c) 1.85 kN, 3.33 m s^{-2} **AL8** (i) 44.6° (ii) 1.40 m (iii) 5 J **AL9** (i) 1.2 m s^{-1} (ii) 0.57 m from one end **AL10** (i) −40 kJ (ii) 30 kJ (iii) 10 kJ **AL11** 360 J, 36 kg m s^{-1}, 2.4 N **AL12** 0.09 m s^{-1}, 1.08 mJ **AL13** (i) 18 kg s^{-1} (ii) 540 N (iii) 54 MJ (iv) 16.2 kW, 9 litres **AL14** (i) 0.45 s, 0.41 s (ii) 0.41 s (iii) A 4.47 m s^{-1}; B 4.07 m s^{-1} (iv) 4.07 m s^{-1} **AL15** (a) 0.707 s, 5 **AL16** (c) 0.6 ohms, 0.338 W (d) 17% (e) (iii) 3.38 W (v) 0.42 W (vi) 12 m^2 **AL18** (a) (i) 2.4 J (ii) 0.2 m (iii) 0.38 kg m^2 rad s^{-2} (b) 2.4 rad s^{-2} **AL19** (b) 31.4 rad s^{-2} (c) 0.463 J **AL20** (i) 2.4 N m (ii) 120 J (iii) 240 J

Section B **BL2** (b) (i) 0.32 nm (ii) 3.4×10^{-19} J **BL4** (a) (i) 2×10^{11} Pa (iii) 48.8 N, 9.75 J **BL5** (b) 3.5 m (c) 7.5 kJ (d) 10.7 m **BL6** 4.4 mK **BL8** (b) (iii) $1.2 \times 10^{-3} \pm 0.36 \times 10^{-3}$ N s m^{-2} **BL9** (c) 15.8 m s^{-1} **BL10** (c) (i) 2.83 m s^{-1} (ii) 5.66×10^{-3} m^3 s^{-1} **BL11** (c) 0.17 MJ, 2.13 MJ **BL12** 4120 J kg^{-1} K^{-1} **BL13** 4 **BL14** (a) (i) 507 m s^{-1} (b)(ii) 450 m s^{-1} **BL15** (b) 1830 m s^{-1} **BL16** 14 kPa **BL17** 152 K **BL18** 50°C, 55°C **BL19** (i) 4×10^{26} W (ii) 5700 K (iii) 4.4×10^9 kg s^{-1}

Section C **CL1** (a) B (b) C (c) D, E **CL2** (iv) 358 MJ (v) 8.50 N **CL4** (i) 6.98 km s^{-1} (ii) 1.46×10^{10} J (iii) -2.93×10^{10} J **CL5** (ii) 4.17 km s^{-1}, 1.74 m s^{-1} **CL6** (i) 1406 V m^{-1}, 3.375 kV (ii) 0.1 m from Q, 0.3 m from −3Q **CL7** (c) (i) 5.25×10^6 m s^{-1} (iii) 143 kV **CL8** (d) 4 **CL9** (c) 8 MJ, 1.78 MJ **CL10** (ii) 1.33 nC **CL11** (b) (i) 39 mT (ii) 7.85 μN m **CL12** 0.4 **CL13** 2.76×10^{-8} N m, 6.89×10^{-10} N m **CL15** 0.5 T **CL16** (c) (i) 0.17 mm μC^{-1} (ii) 50.1 mT **CL17** (b) 15 A, 60 μT **CL18** (b) (ii) 67.5 Hz **CL19** (b) 1.4 A **CL20** (b) (i) 1.35 mN (ii) 0.18 V (iii) 8.1 mW

Section D **DL1** 5.0 g, 0.44 s **DL2** 1.58 **DL3** 87.3 mm, 3.16 rad s^{-1}, $\pi/2$ rad; 276 mm s^{-1}, 873 mm s^{-2}; 1.99 N, 2.02 N; 54 μJ **DL4** (a) 1.25 s (c) 1.02 m, 1.14 ms^{-1} **DL5** (c) (i) all (ii) 7.85 m s^{-1} (iii) 0.62 m s^{-2} (d) 1.25, 31 m s^{-1} **DL6** (c) (i) 2000 Hz **DL7** (b) (ii) 291 K **DL8** (c) 66.7 Hz **DL9** 1.40 M km **DL11** (a) 3, 55°47′ (b) 10′ (c) 11.2 mm **DL12** (c) 1.43′ (d) 0.46 mm **DL13** 2.55′, 0.10 mm **DL14** 200 m **DL15** (a) (ii) 5.9×10^{-3} rad **DL16** (ii) 442 mm (iii) 8.6×10^{-2} rad, × 9.6 (iv) 32.4 mm **DL17** (iii) 850 mm (iv) 94 mm **DL18** (b) 120, 62.5 mm, 4 mm, 12.5 mm

Section E **EL1** (c) (ii) 33.9 V **EL2** 2 Ω, 1.5 V, 4 Ω, 3 Ω (a) 0.281 W (b) 0.562 W **EL3** (i) 0.515, 0.364 A (ii) 0.25, 0.25 A **EL4** (i) A 0.6, B −0.2, C 0.4 A (ii) 6 V (iii) 5.8 W (iv) 7.2 W **EL5** 171 KΩ, 3.0 mC (use 'area'), 9.3 mJ, 510 μF **EL6** 350 V; 175.5 V, 4.43 pF, 271 nJ; 23.3 V, 66.5 pF, 18 nJ **EL7** 17 Ω 60 mW, 15 mW **EL8** (d) 45.5 cm from left-land end **EL10** (b) (i) 2.46A (ii) 0.5 Vm^{-1}, 4.89×10^6 Am^{-2} (c) (i) 4 V (ii) 6.66 Ω, 14.4 W **EL11** (a) and (b) 2.83 MΩ (c) 93.8 nF (d) 1.25A; V_{PN} = 3.45 kV, 85°57′, 1000% **EL12** (c) (i) 240 V (ii) 2.07 A (iii) 0.502 H, 49.5 μF **EL13** (i) 339 V (ii) 20 ms (iii) 339 V; (Figure (i)) 10 V (Figure (iii)) 7.07 V; 7.65 μF or 240 ohms in series; 240 ohms **EL14** (b) 1, 0, 1, 0 **EL15** (a) 300 ohms in series **EL17** (b) 0.4 to 1.6 m **EL18** (b) −8 (c) (i) 5.66 V (ii) 8 V

Section F **FL1** (b) 1.3×10^7 m s^{-1} (d) (ii) 1.92×10^{-15} kg **FL2** 6.63 μW, 3.36×10^{-14} kg, 20 m s^{-2}, 1.4×10^{-19} C, $12\frac{1}{2}$% **FL3** (c) (ii) 0.926 m (iii) 344 kHz (iv) 872 m (radio) **FL4** (b) 0.825 V **FL5** (a) 12.04 h (b) 20.63 min (c) 966, 9034 (d) 3.74 **FL6** (b) (i) 5045 ml **FL7** (c) (i) 1.26×10^{25} (ii) and (iii) 60.7 MBq (iv) 2.92×10^{13} (d) (ii) 2.60×10^6 **FL9** (b) 15 h **FL10** (c) (ii) 3.70 h (d) (i) 1.31×10^{18} (ii) 30.4 μg (iii) 1.65×10^{14} kg^{-1} **FL11** 589 nm **FL12** 285, 330 nm; 2.08, 1.49 V **FL13** (c) (i) 10^4 eV (ii) 1.23×10^{-11} m (iii) 1.6×10^{-15} J, 1.24×10^{-10} m **FL14** (c) (i) shortest C to A; longest C to B (ii) 103 nm, 661 nm **FL16** (a) 5×10^{26} W (c) 5×10^{38}(d) 10^{57}, 2×10^{10} years (e) (i) 1.54×10^{-13} J **FL17** (ii) 121, 47, ${}^1_1$p; 8.93×10^{13} J, 517 days **FL18** 0.636 g s^{-1}

Comprehension papers (following Chapter 35)

Paper 1 **2** (a) 850 W m^{-2} (b) 61.2 kW (c) 90 kW **5** 4.5 m s^{-1}
Paper 2 **4** (b) 41.2 mT **5** (b) 9.4×10^{-13} ohm metres
Paper 3 **7** 29.4%

Appendix 1 **1** (a) 0.25 m, 4.0 m (b) 8.0 m s^{-1}, 2.0 Hz (c) +x direction (d) (i) $\pi/2$ (ii) $3\pi/2$ (iii) π **2** (a) 0.16 s (b) 3.125 Hz **3** (b) 0.02 m (d) 0 **4** (b) 30 mm (c) 20.3 mm **5** (a) 41 (b) 1.000 15
Appendix 2 **3** (c) (i) 3.2 (ii) 100 **5** (a) (i) 1.36 cm, (ii) <1.27° (b) ×4 **7** (a) (i) 1 (ii) 2 (iii) 4 (iv) 8 (v) 9 (vi) 13 (b) 3 **8** (c) 78.4° **9** (a) 2.31×10^{14} Hz (b) 2.31×10^4 (c) $\simeq 10^{13}$ (d) 137 million **10** (b) 0.060 dB km^{-1}

Index